Fourth Edition

Engineering Properties of Foods

Engineering Properties of Foods

Fourth Edition

EDITED BY

**M. A. Rao • Syed S. H. Rizvi
Ashim K. Datta • Jasim Ahmed**

CRC Press
Taylor & Francis Group
Boca Raton London New York

CRC Press is an imprint of the
Taylor & Francis Group, an **informa** business

CRC Press
Taylor & Francis Group
6000 Broken Sound Parkway NW, Suite 300
Boca Raton, FL 33487-2742

© 2014 by Taylor & Francis Group, LLC
CRC Press is an imprint of Taylor & Francis Group, an Informa business

No claim to original U.S. Government works

Printed on acid-free paper
Version Date: 20140117

International Standard Book Number-13: 978-1-4665-5642-3 (Hardback)

This book contains information obtained from authentic and highly regarded sources. Reasonable efforts have been made to publish reliable data and information, but the author and publisher cannot assume responsibility for the validity of all materials or the consequences of their use. The authors and publishers have attempted to trace the copyright holders of all material reproduced in this publication and apologize to copyright holders if permission to publish in this form has not been obtained. If any copyright material has not been acknowledged please write and let us know so we may rectify in any future reprint.

Except as permitted under U.S. Copyright Law, no part of this book may be reprinted, reproduced, transmitted, or utilized in any form by any electronic, mechanical, or other means, now known or hereafter invented, including photocopying, microfilming, and recording, or in any information storage or retrieval system, without written permission from the publishers.

For permission to photocopy or use material electronically from this work, please access www.copyright.com (http://www.copyright.com/) or contact the Copyright Clearance Center, Inc. (CCC), 222 Rosewood Drive, Danvers, MA 01923, 978-750-8400. CCC is a not-for-profit organization that provides licenses and registration for a variety of users. For organizations that have been granted a photocopy license by the CCC, a separate system of payment has been arranged.

Trademark Notice: Product or corporate names may be trademarks or registered trademarks, and are used only for identification and explanation without intent to infringe.

Library of Congress Cataloging-in-Publication Data

Engineering properties of foods. -- Fourth edition / [edited by] M.A. Rao, Syed S.H. Rizvi, Ashim K. Datta, Jasim Ahmed..
    pages cm
Includes bibliographical references and index.
ISBN 978-1-4665-5642-3 (hardback)
1. Food--Analysis. 2. Food industry and trade. I. Rao, M. A., 1937-

TP372.5.E54 2014
664'.07--dc23
2014001340

Visit the Taylor & Francis Web site at
http://www.taylorandfrancis.com

and the CRC Press Web site at
http://www.crcpress.com

CONTENTS

Preface — vii
Editors — ix
Contributors — xi

1 Mass–Volume–Area-Related Properties of Foods — 1
  M. Shafiur Rahman

2 Surface Properties — 37
  Karl F. Schilke and Joseph McGuire

3 Food Microstructure Analysis — 63
  Bryony James

4 Glass Transition in Foods — 93
  Jasim Ahmed and M. Shafiur Rahman

5 Rheological Properties of Fluid Foods — 121
  M. A. Rao

6 Rheological Properties of Solid Foods — 179
  V. N. Mohan Rao and Ximena Quintero

7 Thermal Properties of Unfrozen Foods — 223
  Paul Nesvadba

8 Thermal Properties of Frozen Foods — 247
  Gail Bornhorst, Arnab Sarkar, and R. Paul Singh

9 Properties Relevant to Infrared Heating of Food — 281
  Ashim K. Datta and Marialuci Almeida

CONTENTS

10	Mass Transfer Properties of Foods *George D. Saravacos and Magda Krokida*	311
11	Thermodynamic Properties of Foods in Dehydration *Syed S. H. Rizvi*	359
12	Physicochemical and Engineering Properties of Food in Membrane Separation Processes *Dipak Rana, Takeshi Matsuura, and Srinivasa Sourirajan*	437
13	Electrical Conductivity of Foods *Sudhir K. Sastry and Pitiya Kamonpatana*	527
14	Dielectric Properties of Foods *Ashim K. Datta, G. Sumnu, and G. S. V. Raghavan*	571
15	Ultrasound Properties of Foods *Donghong Liu and Hao Feng*	637
16	Kinetic Data for Biochemical and Microbiological Processes during Thermal Processing *Ann Van Loey, Stefanie Christiaens, Ines Colle, Tara Grauwet, Lien Lemmens, Chantal Smout, Sandy Van Buggenhout, Liesbeth Vervoort, and Marc Hendrickx*	677
17	Kinetics and Process Design for High-Pressure Processing *Tara Grauwet, Stijn Palmers, Liesbeth Vervoort, Ines Colle, Marc Hendrickx, and Ann Van Loey*	709
18	Gas Exchange Properties of Fruits and Vegetables *Bart M. Nicolaï, Jeroen Lammertyn, Wendy Schotsmans, and Bert E. Verlinden*	739
Index		771

PREFACE

We are pleased to present the fourth edition of *Engineering Properties of Foods*. During the last few years, food structure/microstructure has remained a subject of research interest. Furthermore, significant developments have taken place in the area of high-pressure processing (HPP), and the process has been approved by the Food and Drug Administration (FDA) for pasteurization of food. Kinetic data related to HPP play a crucial role for validating pressure-assisted pasteurization. On the basis of these developments, three new chapters: "Food Microstructure Analysis," "Glass Transition in Foods," and "Kinetics and Process Design for High-Pressure Processing" have been added in the fourth edition. Most of the existing chapters were revised to include recent developments in each subject. The chapter on colorimetric properties of food was removed from the earlier edition. Data on physical, chemical, and biological properties have been presented in the book to illustrate their relevance and practical importance.

We included Dr. Jasim Ahmed as a coeditor to help with this rather large undertaking. In looking for experts on topics, we also made an effort to expand the international participation of authors. We made a special effort to follow a consistent format for the chapters so that readers can follow each chapter easily. Thus, each chapter includes an introduction, property definition, measurement procedure, modeling, representative data compilation, and applications.

We concentrated on the clear physical understanding of the properties and their variations, supplemented by representative data. We avoided extensive data collection, for which electronic formats are likely to be more suitable. By providing a succinct presentation of each property in a consistent format, we hope to make it useful to students as well as professionals. As computer-aided engineering (modeling) is becoming more commonplace, the primary use of food properties data is expected to be in computer modeling of food processes. Data correlations with composition of food and temperature are particularly useful in this context since they allow easy inclusion of variation of properties in computer models. We have included such correlations as much as possible.

Our sincere thanks to all the authors and reviewers whose participation and comments have certainly helped improve the chapters. We thank the editors at Taylor & Francis for their patience and help.

M. A. Rao, Syed S. H. Rizvi, Ashim K. Datta, and Jasim Ahmed

EDITORS

M. A. Rao, PhD, is a professor emeritus of food engineering at Cornell University, Ithaca, New York. He focuses his research on the measurement and interpretation of rheological properties of foods. Dr. Rao and his coworkers have contributed many original articles, reviews, and book chapters, including the book: *Rheology of Fluid and Semisolid Foods: Principles and Applications,* 2nd edition, Springer Publishers Inc., 2007. Professor Rao serves as the scientific editor of the food engineering and physical properties section of the *Journal of Food Science.* He was elected a fellow of the Institute of Food Technologists in 1997 and the Association of Food Scientists and Technologists (India) in 2000. He received the Scott Blair Award for Excellence in Rheology from the American Association of Cereal Chemists in 2000, the Distinguished Food Engineer award from IAFIS/FPEI-ASAE in 2003, and the Lifetime Achievement award from the International Association of Engineering and Food in 2011.

Syed S. H. Rizvi, PhD, is a professor of food process engineering and international professor of food science at Cornell University, Ithaca, New York. His teaching and research focus on engineering aspects of food processing and manufacturing operations. He has authored or coauthored over 100 research articles, coedited five books, and holds seven patents. He previously taught at Clemson University and The Ohio State University and worked at Nestle and Glaxo. A fellow of the Institute of Food Technologists, Professor Rizvi has received many awards in recognition of his teaching and research efforts, including the Chancellor Medal for Excellence in Teaching, State University of New York; International Dairy Foods Association Research Award in Food Processing; and Marcel Loncin Research Prize, Institute of Food Technologists. He earned his BS and MS from Punjab University, MEng (chemical engineering) from the University of Toronto, and PhD from The Ohio State University.

Ashim K. Datta, PhD, is a professor of biological engineering at Cornell University, Ithaca, New York. He earned a PhD (1985) in agricultural (food) engineering from the University of Florida, Gainesville. His research interests are in the fundamental studies of heat and mass transport relevant to food processing—developing mechanistic, physics-based understanding and optimization of process, quality, and safety. Prediction of food properties in the context of modeling is of particular interest to him. Dr. Datta is a fellow of the Institute of Food Technologists and member of the American Society of Agricultural and Biological Engineers, and the American Institute of Chemical Engineers. He has authored and coauthored two textbooks and two monographs on heat and mass transfer and its modeling, with application to food and biomedical processes.

Jasim Ahmed, PhD, is a research scientist of food process engineering at Kuwait Institute for Scientific Research, Kuwait. He earned his BTech and MTech (food and biochemical engineering) from Jadavpur University, India, and a PhD (2000) in food technology from

EDITORS

Guru Nanak Dev University, India. He previously taught at Guru Nanak Dev University, India and United Arab Emirates University, UAE. Dr. Ahmed has served as a visiting professor at McGill University, Canada, and research director, Biopolymer Division at Polymer Source Inc., Montreal, Canada. His research focus is the structure/microstructure of food. Dr. Ahmed is a professional member of the Institute of Food Technologists and a life member of the Association of Food Scientists & Technologists (AFSTI), India. He has authored or coauthored more than 100 peer-reviewed research papers, coedited six books including *Handbook of Food Process Design, Starch-Based Polymeric Materials and Nanocomposites: Starch Chemistry, Processing and Applications*. He is one of the editors of the *International Journal of Food Properties,* and he is on the editorial boards of several international journals.

CONTRIBUTORS

Jasim Ahmed
Food and Nutrition Program
Kuwait Institute for Scientific Research
Kuwait

Marialuci Almeida
Cornell University
Ithaca, New York

Gail Bornhorst
University of California
Davis, California

Sandy Van Buggenhout
Laboratory of Food Technology
Leuven, Belgium

Stefanie Christiaens
Laboratory of Food Technology
Leuven, Belgium

Ines Colle
Laboratory of Food Technology
Leuven, Belgium

Ashim K. Datta
Cornell University
Ithaca, New York

Hao Feng
University of Illinois
Urbana, Illinois

Tara Grauwet
Laboratory of Food Technology
Leuven, Belgium

Marc Hendrickx
Laboratory of Food Technology
Leuven, Belgium

Bryony James
University of Auckland
Auckland, New Zealand

Pitiya Kamonpatana
Ohio and Kasetsart University
Bangkok, Thailand

Magda Krokida
National Technical University
Athens, Greece

Jeroen Lammertyn
Katholieke University Leuven
Heverlee, Belgium

Lien Lemmens
Laboratory of Food Technology
Leuven, Belgium

Donghong Liu
Zhejiang University
Hangzhou, China

Ann Van Loey
Laboratory of Food Technology
Leuven, Belgium

Takeshi Matsuura
University of Ottawa
Ottawa, Canada

Joseph McGuire
Oregon State University
Corallis, Oregon

Paul Nesvadba
Rubislaw Consulting Limited
Aberdeen, United Kingdom

Bart M. Nicolaï
Katholieke University Leuven
Heverlee, Belgium

Stijn Palmers
Laboratory of Food Technology
Leuven, Belgium

CONTRIBUTORS

Ximena Quintero
PepsiCo American Foods
Plano, Texas

G. S. V. Raghavan
McGill University
Quebec, Canada

M. Shafiur Rahman
Sultan Qaboos University
Muscat, Sultanate of Oman

Dipak Rana
University of Ottawa
Ottawa, Canada

M. A. Rao
Cornell University
Ithaca, New York

V. N. Mohan Rao
PepsiCo American Foods
Plano, Texas

Syed S. H. Rizvi
Cornell University
Ithaca, New York

George D. Saravacos
National Technical University
Athens, Greece

Arnab Sarkar
PepsiCo R&D
Dubai, United Arab Emirates

Sudhir K. Sastry
The Ohio State University
Columbus, Ohio

Karl F. Schilke
Oregon State University
Corallis, Oregon

Wendy Schotsmans
Katholieke University Leuven
Heverlee, Belgium

R. Paul Singh
University of California
Davis, California

Chantal Smout
Laboratory of Food Technology
Leuven, Belgium

Srinivasa Sourirajan
University of Ottawa
Ottawa, Ontario, Canada

G. Sumnu
Middle East Technical University
Ankara, Turkey

Bert E. Verlinden
Katholieke University Leuven
Heverlee, Belgium

Liesbeth Vervoort
Laboratory of Food Technology
Leuven, Belgium

1

Mass–Volume–Area-Related Properties of Foods

M. Shafiur Rahman

Contents

1.1	Introduction	2
1.2	Fundamental Considerations	3
	1.2.1 Volume	3
	1.2.1.1 Boundary Volume	3
	1.2.1.2 Pore Volume	3
	1.2.2 Density	3
	1.2.2.1 True Density	3
	1.2.2.2 Material Density	3
	1.2.2.3 Particle Density	4
	1.2.2.4 Apparent Density	4
	1.2.2.5 Bulk Density	4
	1.2.3 Porosity	5
	1.2.3.1 Open Pore Porosity	5
	1.2.3.2 Closed Pore Porosity	5
	1.2.3.3 Apparent Porosity	5
	1.2.3.4 Bulk Porosity	6
	1.2.3.5 Bulk-Particle Porosity	6
	1.2.3.6 Total Porosity	6
	1.2.4 Surface Area	6
	1.2.5 Pore Size Distribution	6
1.3	Measurement Techniques	6
	1.3.1 Density Measurement	6
	1.3.1.1 Apparent Density	6
	1.3.1.2 Material Density	11

		1.3.1.3	Particle Density	13
		1.3.1.4	Bulk Density	14
	1.3.2	Measurement Techniques of Porosity		14
		1.3.2.1	Direct Method	14
		1.3.2.2	Optical Microscopic Method	14
		1.3.2.3	Density Method	14
	1.3.3	Surface Area		14
		1.3.3.1	Boundary Surface Area	14
		1.3.3.2	Pore Surface Area	15
		1.3.3.3	Cross-Sectional Area	16
1.4	Specific Data			16
	1.4.1	Predictions of Density		16
		1.4.1.1	Gases and Vapors	16
		1.4.1.2	Liquid Foods	17
		1.4.1.3	Density of Solid Foods	18
	1.4.2	Predictions of Porosity		21
	1.4.3	Prediction of Surface Area		24
		1.4.3.1	Euclidian Geometry	24
		1.4.3.2	Non-Euclidian or Irregular Geometry	24
		1.4.3.3	Theoretical Prediction	25
		1.4.3.4	Size Distribution	28
1.5	Summary			31
Acknowledgments				31
List of Symbols				31
	Greek Symbols			32
	Subscripts			33
	Superscripts			33
References				33

1.1 INTRODUCTION

Mass–volume–area-related properties are one of five groups (acoustic, mass–volume–area-related, morphological, rheological, and surface) of mechanical properties (Rahman and McCarthy, 1999). These properties are needed in process design, for estimating other properties, and for product characterization or quality determination. The geometric characteristics of size, shape, volume, surface area, density, and porosity are important in many food materials handling and processing operations. Fruits and vegetables are usually graded depending on size, shape, and density. Impurities in food materials are separated by density differences between impurities and foods. Knowledge of the bulk density of food materials is necessary to estimate floor space during storage and transportation (Mohsenin, 1986; Rahman, 1995). When mixing, transportation, storing, and packaging particulate matter, it is important to know the properties of bulk material (Lewis, 1987). The surface areas of fruits and vegetables are important in investigations related to spray coverage, removal of residues, respiration rate, light reflectance, and color evaluation, as well as in heat-transfer studies in heating and cooling processes (Mohsenin, 1986). In many physical and chemical processes, the rate of reaction is proportional to the surface area; thus, it is often desirable to maximize the surface area. Density and porosity have a direct effect on the other physical properties. Volume change and porosity are important parameters in estimating the diffusion coefficient of shrinking systems. Porosity and tortuosity are used to calculate effective diffusivity during mass transfer processes. Mechanical properties of agricultural materials also vary with porosity. This chapter provides terminology, measurement techniques, and prediction models of selected mass–volume–area-related properties.

1.2 FUNDAMENTAL CONSIDERATIONS

1.2.1 Volume

1.2.1.1 Boundary Volume
Boundary volume is the volume of a material considering the geometric boundary. A material's volume can be measured by buoyancy force; liquid, gas, or solid displacement; or gas adsorption; it can also be estimated from the material's geometric dimensions. Estimation equations of the boundary volume of shapes of regular geometry are given in Table 1.1.

1.2.1.2 Pore Volume
Pore volume is the volume of the voids or air inside a material.

1.2.2 Density

Density is one of the most important mechanical properties and so is widely used in process calculations. It is defined as mass per unit volume:

$$\text{Density} = \frac{\text{Mass}}{\text{Volume}} = \frac{m}{V} \tag{1.1}$$

The SI unit of density is kg/m^3. In many cases foods contain multicomponent phases, such as solid, liquid, and gaseous or air. In this case, a simple definition such as that given above cannot be sufficient to relate the mass and volume. In this case, different terminology should be defined. Rahman (1995) clearly explained different forms of density used in process calculations and characterizing food products. The definitions are given as follows.

1.2.2.1 True Density
True density (ρ_T) is the density of a pure substance or a composite material calculated from its components' densities considering conservation of mass and volume.

1.2.2.2 Material Density
Material density (ρ_m) is the density measured when a material has been thoroughly broken into pieces small enough to guarantee that no closed pores remain.

Table 1.1 Volume and Surface Area of Some Common Shapes[a]

Sphere

$$V = \frac{4}{3}\pi r^3 \quad \text{and} \quad A = 4\pi r^2$$

Cylinder

$V = \pi r^2 L \quad \text{and} \quad A = 2\pi r^2 + 2\pi r L$

Cube

$V = a^3 \quad \text{and} \quad A = 6a^2$

Brick

$V = abc \quad \text{and} \quad A = 2(ab + bc + ca)$

Prolate spheroid

$$V = \frac{4}{3}(\pi a b^2) \quad \text{and} \quad A = 2\pi b^2 + \frac{2\pi ab}{e} \sin^{-1} e$$

Oblate spheroid

$$V = \frac{4}{3}(\pi a^2 b) \quad \text{and} \quad A = 2\pi b^2 + \frac{\pi b^2}{e} \ln\left(\frac{1+e}{1-e}\right)$$

Frustam right cone

$$V = \frac{\pi}{3} L(r_1^2 + r_1 r_2 + r_2^2) \quad \text{and} \quad A = \pi(r_1 + r_2)\sqrt{L^2 + (r_1 - r_2)^2}$$

[a] Where a and b, respectively, are major and minor semi-axes of the ellipse of rotation, e is the eccentricity given by $e = \sqrt{1 - (b/a)^2}$, r_1 and r_2, respectively, are the radii of base and top, and L is the altitude.

1.2.2.3 Particle Density

Particle density (ρ_p) is the density of a particle, which includes the volume of all closed pores but not the externally connected pores. In this case, the particle is not modified structurally, as in the case of material density.

1.2.2.4 Apparent Density

Apparent density (ρ_a) is the density of a substance including all pores remaining in the material.

1.2.2.5 Bulk Density

Bulk density (ρ_B) is the density of a material when packed or stacked in bulk. The bulk density of packed materials depends on the geometry, size, and surface properties of individual particles (Lewis, 1987).

1.2.3 Porosity

Porosity indicates the volume fraction of void space or air in a material and is defined as

$$\text{Porosity} = \frac{\text{Air or void volume}}{\text{Total volume}} \quad (1.2)$$

Different forms of porosity are used in food process calculations and food products characterization (Rahman, 1995). These are defined below.

1.2.3.1 Open Pore Porosity

Open pore porosity is the volume fraction of pores connected to the exterior boundary of a material and is given by (ε_{op}):

$$\text{Open pore porosity} = \frac{\text{Volume of open pore}}{\text{Total volume of material}}$$

$$\varepsilon_{op} = 1 - \frac{\rho_a}{\rho_p} \quad (1.3)$$

There may be two types of open pores: one type is connected to the exterior boundary only, and the other type is connected to the other open pores as well as to the exterior geometric boundary. The level of open and closed pores depends on what component (helium, nitrogen, toluene, or mercury) is used in the measurement.

1.2.3.2 Closed Pore Porosity

Closed pore porosity (ε_{cp}) is the volume fraction of pores closed inside the material and not connected to the exterior boundary of the material. It can be defined as

$$\text{Closed pore porosity} = \frac{\text{Volume of closed pores}}{\text{Total volume of material}}$$

$$\varepsilon_{cp} = 1 - \frac{\rho_p}{\rho_m} \quad (1.4)$$

1.2.3.3 Apparent Porosity

Apparent porosity is the volume fraction of total air or void space in the material boundary and is defined as ($\varepsilon_a = \varepsilon_{op} + \varepsilon_{cp}$):

$$\text{Apparent porosity} = \frac{\text{Volume of all pores}}{\text{Total volume of material}}$$

$$\varepsilon_a = 1 - \frac{\rho_a}{\rho_m} \quad (1.5)$$

1.2.3.4 Bulk Porosity

Bulk porosity (ε_B) is the volume fraction of voids outside the boundary of individual materials when packed or stacked as bulk:

$$\text{Bulk porosity} = \frac{\text{Volume of voids outside materials' boundary}}{\text{Total bulk volume of stacked materials}}$$

$$\varepsilon_B = 1 - \frac{\rho_b}{\rho_a} \tag{1.6}$$

1.2.3.5 Bulk-Particle Porosity

Bulk-particle porosity is the volume fraction of the voids outside the individual particle and open pore to the bulk volume when packed or stacked as bulk.

$$\varepsilon_{BP} = \varepsilon_B + \varepsilon_{op} \tag{1.7}$$

1.2.3.6 Total Porosity

Total porosity is the total volume fraction of air or void space (i.e., inside and outside of the materials) when material is packed or stacked in bulk.

$$\varepsilon_T = \varepsilon_a + \varepsilon_B = \varepsilon_{op} + \varepsilon_{cp} + \varepsilon_B \tag{1.8}$$

1.2.4 Surface Area

Two types of surface area are used in process calculations: the outer boundary surface of a particle or object, and the pore surface area for a porous material. An object can be characterized using Euclidian or non-Euclidian geometries. Euclidian geometric shapes always have characteristic dimensions and have an important common peculiarity of smoothness of surface; examples include spheres, cubes, and ellipsoids.

1.2.5 Pore Size Distribution

In addition to size and shape, particle or pore size population needs to be determined. Pore size distribution is most commonly used to characterize populations. Similar techniques are also used for particle size distribution.

1.3 MEASUREMENT TECHNIQUES

1.3.1 Density Measurement

1.3.1.1 Apparent Density
1.3.1.1.1 Geometric Dimension Method
The apparent density of a shape of regular geometry can be determined from the volume calculated from the characteristic dimensions and mass. This method is not suitable for

soft and irregularly shaped materials, where it is not easy to measure the characteristic dimensions (Rahman, 1995). Little information exists in the literature about the density measurement of frozen foods. Keppeler and Boose (1970) used a thick-walled cylindrical metal container to measure the density of a frozen sugar solution. The density determination method consists of finding the mass of a frozen sample with a known volume. The unfrozen sample is placed in the cylindrical container and then frozen at the desired temperature. The excess frozen sample can be removed with a sharp knife. Then the cylinder and frozen sample should be weighed immediately. From the mass of the sample and the volume of the cylinder, the density can be calculated. Rahman and Driscoll (1994) wrapped the metal container with electrical tape to reduce the heat gain during weighing. This method is only suitable for liquid and soft materials, where no void exists in the packing.

1.3.1.1.2 Buoyant Force Method

In this procedure, buoyant force can be determined from sample weight in air and liquid. The apparent density can be calculated from the equation:

$$\rho_a = \rho_w \times \frac{m}{G} \tag{1.9}$$

where m and G are the mass (in kilograms) of the sample in air and liquid (i.e., water), respectively, and ρ_w is the density of the liquid. The methods of weighing the samples are shown in Figure 1.1 for a top-loading balance and an analytical balance. Two errors may often occur with this method, and hence precautions should be taken during measurement. The first may be due to mass transfer from the sample to liquid, that is, the exchange of solid, liquid, or gas from the sample to liquid. This can be avoided by enclosing the sample in cellophane or polythene or coating with a thin layer of polyurethane varnish or wax. The sample could be tied with a thin string and dipped in wax a couple of times

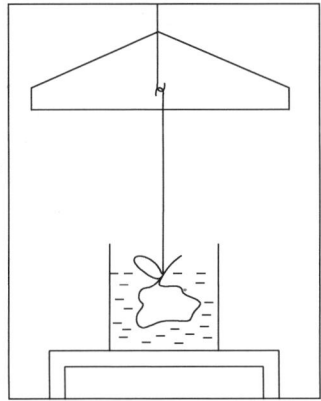

Figure 1.1 Top-loading balance for measurement of buoyant force for a sample lighter than liquid. (From Mohsenin, N.N. 1986. *Physical Properties of Plant and Animal Materials.* Gordon and Breach Science Publishers, New York.)

ENGINEERING PROPERTIES OF FOODS

and solidified before measurement. Samples can also be covered with silicon grease to make them impervious to water (Loch-Bonazzi et al., 1992). Lozano et al. (1980) measured plastic film-coated samples of fresh and dried fruits and vegetables and samples without coating and found that no significant moisture was taken up when the uncoated samples were used. The above authors noted that this was due to the very brief time required for measurement. However, coating is the best possible option for accuracy, and care must be taken to prepare it.

The second error may be due to partial floating of the sample. In this case, a liquid with a lower density than that of the sample can be used. Mohsenin (1986) described a simple technique with a top-loading balance that applies to large objects such as fruits and vegetables (Figure 1.2). The sample is first weighed on the scale in air and then forced into water by means of a sinker rod. Second readings are then taken with the sample submerged. The density can then be calculated as

$$\rho_a = \left[\frac{m_s}{G_1/\rho_w - G_2/\rho_w} \right] \tag{1.10}$$

where G_1 refers to the sinker plus the sample, and G_2 refers to the sinker only in water or liquid. Again, if the solid is lighter than the liquid, another solid can be attached, heavier than the liquid, to the object as a sinker. In the case of coated sample with a sinker, the following equation can be used:

$$\rho_a = \left[\frac{m_s}{G_1/\rho_w - G_2/\rho_w - G_3} \right] \tag{1.11}$$

where G_1 refers to the sinker plus the wax-coated sample, G_2 the sinker only, and G_3 the wax only. G_3 can be calculated as

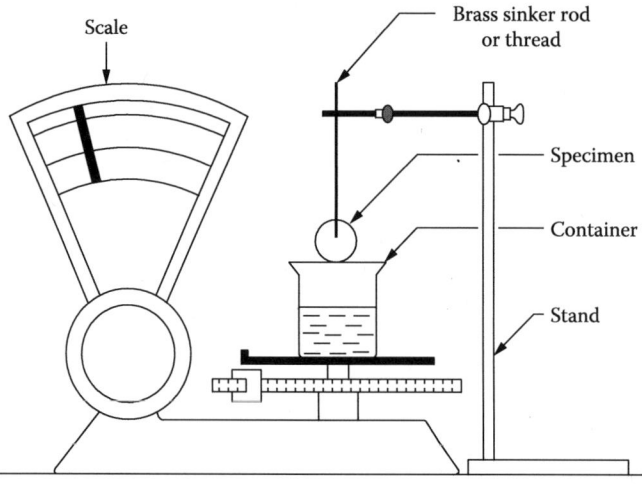

Figure 1.2 Analytical balance for measurement of buoyant force.

$$G_3 = \frac{m_{wax}}{\rho_{wax}} \qquad (1.12)$$

The mass of wax can be calculated from the difference between the masses of coated and uncoated samples. The density of wax is usually around 912 kg/m^3; however, it could vary based on the source. Thus, the density of wax should be measured separately. Mohsenin (1986) also suggested that a solution of 3 cm^3 wetting agent in 500 cm^3 distilled water can reduce errors due to surface tension and submergence in water. Ramaswamy and Tung (1981) used a buoyant force determination technique to measure the apparent density of frozen apple. They used water at 2–3°C, and the apple was frozen at –20 to –35°C with a sinker. There is a real need to develop accurate and easy measurement techniques for frozen samples at subzero temperature.

1.3.1.1.3 Volume Displacement Method
1.3.1.1.3.1 Liquid Displacement Method

The volume of a sample can be measured by direct measurement of the volume of liquid displaced. The difference between the initial volume of the liquid in a measuring cylinder and the volume of the liquid plus immersed material (coated) is the volume of the material. Coating is necessary so that liquid does not penetrate in the pores. A nonwetting fluid such as mercury is better to use for displacement since in this case samples do not need coating.

The use of a specific gravity bottle and toluene has been practiced for many years (Bailey, 1912). A small-neck specific gravity bottle is not suitable for large objects; thus, a special design is required. The volume of a specific gravity bottle can be measured using distilled water. Toluene has many advantages when used as reference liquid (Mohsenin, 1986):

- Little tendency to soak on the sample
- Smooth flow over the surface due to surface tension
- Low solvent action on constituents, especially fats and oils
- Fairly high boiling point
- Stable specific gravity and viscosity when exposed to the atmosphere
- Low specific gravity

Toluene is carcinogenic; thus, adequate precautions need to be taken in using it. The experiment should be performed inside a fume chamber (Rahman et al., 2002).

Rahman and Driscoll (1994) reported the use of a method for irregular and small frozen food particles such as grain or cereals. The procedures were as follows: eight cylindrical glass bottles of diameter 2 cm with small necks filled three-fourths full (20 g) with the sample and the rest of the way with toluene were frozen at –40°C. After freezing, the bottles were immediately placed inside glass wool insulation columns of inner and outer diameter of 2 and 7 cm, respectively. The temperature was then recorded from one bottle by a thermocouple placed inside the center of the bottle. At different temperatures, the bottles were taken out, one at a time, from the glass wool insulation, and toluene was added to completely fill the bottle. The bottle was closed immediately, and the weight was determined. From the mass and volume of the sample, which was estimated by subtracting the volume of toluene from the volume of the bottle, the density was calculated. The volume of

ENGINEERING PROPERTIES OF FOODS

toluene was estimated from the mass and density at the respective temperatures. Rahman and Driscoll (1994) used this method to measure the density of frozen seafood at different temperatures and found reproducibility within 1%.

Commercial mercury porosimeters are available to measure the volume of porous and nonporous solids. The principle of mercury intrusion porosimetry is based on the fact that mercury ordinarily behaves as a nonwetting liquid (i.e., the contact angle of mercury is larger than 90°). Because it is nonwetting, mercury will not flow into the openings of porous solid bodies unless it is forced to do so by a pressure gradient (Clayton and Huang, 1984). The mercury injection method of measuring effective porosity is based on the fact that, due to the surface tension and nonwetting properties of mercury, a porous sample can be immersed in mercury without entry of mercury into the sample at atmospheric pressure. Thus, the apparent volume of the sample can be determined by displacement of mercury from a sample chamber of known volume.

1.3.1.1.3.2 Gas Pycnometer Method

Different commercial gas pycnometers for volume measurement are available. The gases air, nitrogen, and helium can be used. Mohsenin (1986) described a method to measure volume using high-pressure air (Figure 1.3). The test material is placed in tank 2, and air is supplied to tank 1 when valve 2 is closed. When suitable manometer displacement is achieved, valve 1 is closed and equilibrium pressure P_1 is read. Now valve 3 is closed and valve 2 is opened, and pressure P_3 is read. Under this condition with valves 1 and 3 closed, the volume of sample in tank 2 is measured as V_s. Then the volume of the sample in tank 2 is estimated based on ideal gas law as

$$V_s = V_1 + V_1 \times \left(\frac{P_3 - P_1}{P_3} \right) \tag{1.13}$$

where V_1 is the empty volume of tanks 1 or 2.

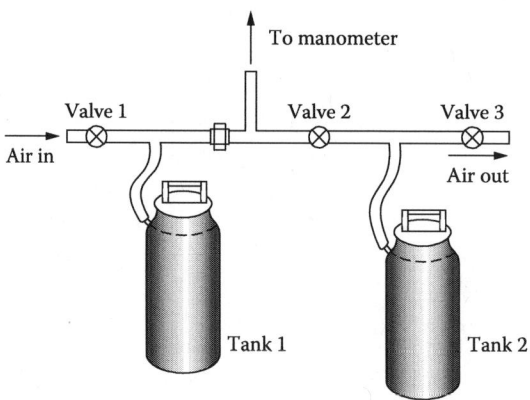

Figure 1.3 Air comparison pycnometer. (From Mohsenin, N.N. 1986. *Physical Properties of Plant and Animal Materials.* Gordon and Breach Science Publishers, New York.)

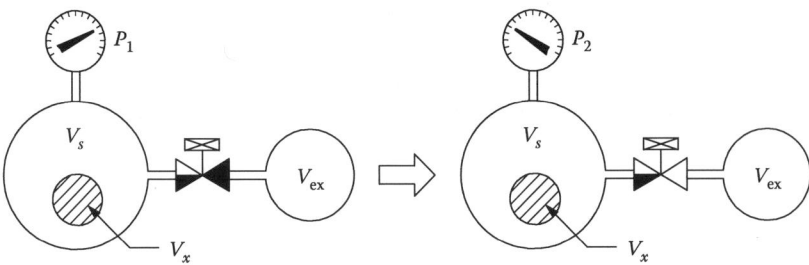

Figure 1.4 Operating principle of Horiba helium pycnometer VM-100. (From Horiba Bulletin: HRE-8815A. Horiba, Koyoto, Japan.)

Commercial automatic helium gas pycnometers are available to measure volumes of samples. Figure 1.4 shows the operating principle of the Horiba helium pycnometer VM100. If a sample of volume V_s is placed in a sample cell with volume V_{sc}, pressure P_1 is applied to the sample cell, and the valve is then opened and gas passed through an expansion cell with volume V_{ec}, the pressure will decrease to P_2 due to the expansion of the gas. The volume V_s of the sample may be obtained from the known volumes V_{sc} and V_{ec} and the ratio of pressures P_1 and P_2 using the following formula (Horiba Bulletin: HRE-8815A):

$$V_x = V_{xc} + \left[\frac{1}{1 - (P_1/P_2)}\right] V_{ec} \qquad (1.14)$$

The above equation is derived based on the ideal gas law. In order to measure the apparent density, the sample needs to be coated with wax before being placed inside the pressure chamber.

1.3.1.1.3.3 Solid Displacement Method
The apparent volume of an irregular solid can be measured by a solid displacement or glass bead displacement method. Glass beads have an advantage over sand due to their uniform size and shape, thus producing reproducible results.

1.3.1.2 Material Density
1.3.1.2.1 Pycnometer Method
Material density can be measured when a material is ground enough to guarantee that no closed pores remain. Both liquid and gas displacement methods (pycnometer) can be used to measure the volume of ground material. When a liquid is used, care must be taken to use sufficient liquid to cover the solid's surface or pores. This difficulty can be overcome by

- Gradually exhausting the air from the bottle by use of a vacuum pump to promote the escape of the air trapped under the surface
- When air bubbles escape after several cycles of vacuuming and releasing the vacuum, filling the bottle with toluene and allowing the temperature to reach 20°C

1.3.1.2.2 Mercury Porosimetry

Pore volumes can be measured by gas adsorption techniques or mercury porosimetry. In addition, pore characteristics and size distribution can also be determined by these methods. Both techniques are very well defined and are generally accepted methods for characterization of pores.

Modern mercury porosimeters can achieve pressures in excess of 414 MPa, which translates into a pore size of 0.003 nm. The upper end of this technique can measure pores up to 360 nm. Gas adsorption, on the other hand, can measure pores as small as 4 Å and up to 5000 Å (Particulars, 1998). In comparison to gas adsorption, mercury porosimetery takes less time, with a typical analysis taking less than an hour. Gas adsorption can take from a couple of hours to 60 hours to complete (Particulars, 1998).

Both techniques involve constructing an isotherm, either of pressure versus intrusion volume in the case of mercury porosimetry or volume adsorbed versus relative pressure in the case of gas adsorption. Total pores or pore size distribution can be derived from these isotherms. As pressure is applied, the pressure at which the mercury intrudes into the pores is measured. The smaller the pore, the higher is the pressure required to force the mercury into the pores. The pressure is increased and the amount of mercury that intrudes is monitored at a series of pressure points that represent corresponding pore sizes. If decreasing pressures are included in the analysis, then a series of volumes and pore sizes can be obtained as the mercury extrudes out of the pores. Very often, a difference exists between the intrusion and extrusion curves because the pores may form bottlenecks or restrictions. In this case, some of the mercury may be left in the sample. This evidence can provide valuable data regarding the shape of pores in sample (Particulars, 1998). Mercury porosimetry uses the Washburn equation, which is based on the capillary law governing the penetration of a liquid into small pores (Figure 1.5):

$$D_v = \frac{4\tau\cos\theta}{P} \tag{1.15}$$

Mercury porosimetry is suitable for the measurement of smaller open pores since it uses high pressure. In the case of mercury porosimetry for material density, the sample need not be ground since high pressure forces mercury to penetrate into the pores. However, it does not guarantee that the mercury has intruded into all pores even at very high pressure (Rahman et al., 2002). When high pressure is used, the compressibility of solids should be considered for accuracy. If the sample chamber is closed and the hydrostatic pressure of mercury in the chamber is increased to a very great value, the mercury will enter the pores, compressing the trapped air in the pores to negligible volume. The volume of mercury injected is therefore equal to the pore volume.

An advantage of mercury porosimetry is that both apparent volume and pore volume are directly determined without coating the sample. This method may not be very precise if the volume occupied by compressed air is not determined. The sample cannot be used for further tests, even after extensive cleaning procedures, due to contamination with mercury.

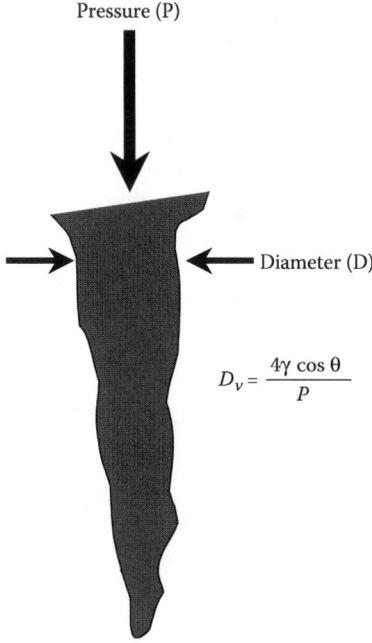

Figure 1.5 The relation of intrusion pressure to the pore diameter. (From Rahman, M.S. 2000. Mechanisms of pore formation in foods during drying: present status. Papers presented at the *Eighth International Congress on Engineering and Food (ICEF8)*, Puebla, Mexico, April 9–13, 2000.)

1.3.1.2.3 Gas Adsorption Method
In the gas adsorption method, the sample is cooled, usually to cryogenic temperatures, and then is exposed to the inert adsorptive gas—typically nitrogen—at a series of precisely controlled pressures. As the pressure increases, a number of the gas molecules are adsorbed onto the surface of the sample. This volume of adsorbed gas is measured using sensitive pressure transducers. As the pressure increases, more and more gas is adsorbed. First the very small micropores, less than 20 Å, are filled, followed by the free surface and finally the rest of the pores. Each pressure corresponds to a pore size, and from the volume of gas adsorbed, the volume of these pores can be obtained. By collecting data as the pressure reduces, a desorption isotherm can be derived as well and observed for any hysteresis and, hence, again for information on pore shape. Obviously, using this technique the BET surface area can also be derived (Particulars, 1998).

1.3.1.3 Particle Density
Particle density can be measured by the volume displacement method used in apparent density measurement without coating the particle or object. Care should be taken to avoid any structural changes during the measurement. Large particles cannot be placed in a side narrow-neck specific gravity bottle or pycnometer; thus, a special design is required for foods containing such particles.

1.3.1.4 Bulk Density

Bulk density can be determined by stacking or placing a known mass of particles into a container of known volume, such as a measuring cylinder. Material can be filled into a specific geometric container of known volume, and the excess amount on the top of the cylinder can be removed by sliding a string or ruler along the top edge of the cylinder. Gentle tapping of the cylinder vertically down to a table may also be done. This method considers all pores inside as well as outside the individual particles. After the excess has been removed, the mass of the sample can be measured and the bulk density can be estimated as

$$\rho_B = \frac{m}{V_B} \tag{1.16}$$

1.3.2 Measurement Techniques of Porosity

1.3.2.1 Direct Method

In this method, the bulk volume of a piece of porous material is measured first, and then the volume is measured after compacting the material to destroy all its voids. Porosity can be determined from the difference of the two measured volumes. This method can be applied if the material is very soft and no repulsive or attractive force is present between the surfaces of solid particles.

1.3.2.2 Optical Microscopic Method

In this method the porosity can be determined from the microscopic view of a random section of the porous medium. This method is reliable if the sectional (two-dimensional) porosity is same as the three-dimensional porosity. Image analysis is necessary to estimate the surface area of pores.

1.3.2.3 Density Method

Porosity can also be estimated from the densities of the materials from Equations 1.3 through 1.6. Alternatively, pore volume can be measured directly by liquid or gas displacement methods, described earlier in the discussion of density measurements.

1.3.3 Surface Area

1.3.3.1 Boundary Surface Area

The boundary surface area is mainly estimated from geometric dimensions or measured by image analysis or contour analysis. Leaf and stalk surface area are measured by contact-printing the surface on a light-sensitive paper and then estimating the area with a planimeter, or by tracing the area on graph paper and counting the squares or determining the mass of the paper. In this method, the mass–area relationship of the paper should be developed first. Another method is the use of an air-flow planimeter, which measures the area as a function of the surface obstructing the flow of air (Mohsenin, 1986).

The surface area of fruits and vegetables can be estimated from the peeled or the skin area. In this method, fruit is peeled in narrow strips and the planimeter sum of the areas of tracings of the strips is taken as the surface area. Similarly, strips of narrow masking tape can be used to cover the surface; from the length and width of the tape, the surface area can be estimated. A transient heat-transfer study can also be used to estimate the surface area (Mohsenin, 1986).

The simplest method of obtaining the surface area of a symmetrical convex body such as an egg is the projection method using a shadow-graph or photographic enlarger. Using the profile of the egg, equally spaced parallel and perpendicular lines can be drawn from the axis of symmetry to the intersection with the profile. Then using manual computation, by integration, the surface area can be obtained by summing up the surfaces of revolution for all of the divided segments (Mohsenin, 1986).

1.3.3.2 Pore Surface Area

The pore surface area can be defined as the surface of the pores in a porous medium exposed to fluid flow either per unit volume or per unit mass of the solid particles.

1.3.3.2.1 Methods Based on Adsorption

The quantity of an inert vapor that can be adsorbed on the pore surface is dependent on the area of the surface. The quantity of a gas or vapor adsorbed is proportional to a surface area that inclines the tiny molecular interstices of the porous material, whereas the surface area pertinent to fluid flow does not include this portion of surface area.

1.3.3.2.2 Methods Based on Fluid Flow

Mohsenin (1986) noted that the Carman–Kozeny equation can be employed to measure the specific surface of the nonuniform pore space. Kozeny showed that permeability can be written for steady state and stream line flow ($Re < 2.0$) through porous media:

$$k = \frac{f \, \varepsilon_{op}^3}{A^2} \quad (1.17)$$

$$Re = \frac{\rho_a \, u}{\mu \, (1 - \varepsilon_{op}) A} \quad (1.18)$$

where f is a dimensionless constant, ε_{op} is the open pore porosity, A is the specific surface area (m^2/m^3), and u is the approach or velocity of the fluid in the empty column (m/s). Carman modified the above equation considering $f = 1/5$, the result is known as the Carman–Kozeny equation given by

$$k = \frac{\varepsilon_{op}^3}{5A^2(1 - \varepsilon_{op})^2} \quad (1.19)$$

The value of f depends on the particle shape, porosity, and particle size range and lies between 3.5 and 5.5 with a common value of 5 (Holland, 1973). The values of A can be calculated from the above equation by knowing the porosity and permeability. The permeability can be calculated from the well-known Darcy's equation:

$$Q = k \left[\frac{A \, (\Delta P)}{\mu \, L} \right] \quad (1.20)$$

An inert fluid needs to be used. This method is more commonly used for rocks and is not suitable for fragile solids, such as foods. Also, the surface of dead-end pores cannot be included in the fluid flow method.

1.3.3.2.3 Mercury Intrusion
Mercury intrusion measures the characteristics of pores. Surface area is determined from the intruded volume using geometric dimensions of a preassumed geometric shape of the pores.

1.3.3.3 Cross-Sectional Area
The cross-sectional area is the area of a surface after longitudinal or transverse section of a material. It is necessary when fluid is flowing over an object. The methods of boundary surface area can also be used.

1.4 SPECIFIC DATA

1.4.1 Predictions of Density

In most engineering design, solids and liquids are assumed to be incompressible, that is, density moderately changes with temperature and pressure. Actually, the density of solids and liquids changes with temperature and pressure. Gases and vapors are compressible and thus are affected by temperature and pressure.

1.4.1.1 Gases and Vapors
The ideal gas equation is commonly used to estimate the density of gases and vapors. The ideal gas equation is based on unhindered movement of gas molecules within a confined space, thus at constant temperature when molecular energy is constant, the product of pressure and volume is constant. At low pressure, most gases obey the ideal gas equation, which can be written as

$$PV = nRT \tag{1.21}$$

One kg mole ideal gas occupies 22.4 m³ at 273 K and 1 atm. As pressure is increased, the molecules are drawn closer together, and attractive and repulsive forces between the molecules affect molecular motion. At low pressure when molecules are far apart, only an attractive force exists. At high pressures, the pressure–volume–temperature relationship deviates from ideality (Toledo, 1993). Gases that deviate from ideal behavior are considered real gases. One of the most commonly used equations of state for real gases is Van der Waal's equation. For n moles of gas, Van der Waal's equation of state is

$$\left(P + \frac{n^2 a}{V^2}\right) \times (V - nb) = nRT \tag{1.22}$$

Values of a and b for different gases are given in Table 1.2 (Toledo, 1993).

Table 1.2 Values of Van der Waal's Constants for Different Gases

Gas	a (Pa [m^3/kg mol]^2)	b (m^3/kg mol)
Air	1.348×10^5	0.0366
Ammonia	4.246×10^5	0.0373
Carbon dioxide	3.648×10^5	0.0428
Hydrogen	0.248×10^5	0.0266
Methane	2.279×10^5	0.0428
Nitrogen	1.365×10^5	0.0386
Oxygen	1.378×10^5	0.0319
Water vapor	5.553×10^5	0.0306

Source: Toledo, R. T. 1993. *Fundamentals of Food Process Engineering.* 2nd ed. Chapman & Hall, New York.

1.4.1.2 Liquid Foods
1.4.1.2.1 Milk and Dairy Products

The density of whole and skim milk as a function of temperature can be calculated from the equations developed by Short (1955):

Whole milk:
$$\rho_a = 1035.0 - 0.358t + 0.0049t^2 - 0.00010t^3 \quad (1.23)$$

Skim milk:
$$\rho_a = 1036.6 - 0.146t + 0.0023t^2 - 0.00015t^3 \quad (1.24)$$

Phipps (1969) reported an equation for the estimation of the density of cream as a function of temperature and fat content with an accuracy of ±0.45%:

$$\rho_a = 1038.2 - 0.17t - 0.003t^2 - \left(133.7 - \frac{475.5}{t}\right)X_f \quad (1.25)$$

where X_f is the mass fraction of fat. Roy et al. (1971) reported equations for estimating the density of fat from buffalo's and cow's milk:

Buffalo milk: $\rho_a = 923.84 - 0.44t$ (1.26)

Cow's milk: $\rho_a = 923.51 - 0.43t$ (1.27)

1.4.1.2.2 Fruit Juices and Purees

The apparent density of a sucrose solution can be estimated as a function of sucrose concentration (X_w: 0 to 1.0) at 20°C (Chen, 1989):

$$\rho_a = \sum_{j=0}^{5} 100 C_j (X_w)^j \quad (1.28)$$

where $C_o = 997.2$, $C_1 = 3.858$, $C_2 = 1.279 \times 10^{-2}$, $C_3 = 6.192 \times 10^{-5}$, $C_4 = -1.777 \times 10^{-7}$, and $C_5 = -4.1997 \times 10^{-10}$, respectively. For fruit juices, the density versus the refractive index of sugar solution can be estimated as (Riedel, 1949):

$$\rho_a = \frac{\vartheta^2 - 1}{\vartheta^2 + 2} \times \frac{62.4}{0.206} \times 16.0185 \tag{1.29}$$

where ϑ is the refractive index. For the density of tomato juice, Choi and Okos (1983) developed a predictive equation based on the water (X_w) and solids (X_s) fractions:

$$\rho_a = \rho_w X_w + \rho_s X_s \tag{1.30}$$

$$\rho_w = 9.9989 \times 10^2 - 6.0334 \times 10^{-2}\, t - 3.6710 \times 10^{-3}\, t^2 \tag{1.31}$$

$$\rho_s = 1.4693 \times 10^3 + 5.4667 \times 10^{-1} t - 6.9643 \times 10^{-3} t^2 \tag{1.32}$$

Bayindirli (1992) proposed a correlation to estimate the apparent density of apple juice as a function of concentration (B: 14–39°Brix) and temperature (20–80°C) as

$$\rho_a = 830 + 350\left[\exp(0.01B)\right] - 0.564t \tag{1.33}$$

Ramos and Ibarz (1998) correlated the apparent density of peach (Equation 1.34) and orange (Equation 1.35) juices as a function of concentration (B: 10–60°Brix) and temperature (0–80°C) as

$$\rho_a = 1006.6 - 0.5155t + 4.1951B + 0.0135B^2 \tag{1.34}$$

$$\rho_a = 1025.4.6 - 0.3289t + 3.2819B + 0.0178B^2 \tag{1.35}$$

Ibarz and Miguelsanz (1989) correlated the apparent density of depectinized and clarified pear juice as (B: 10–71°Brix and t: 5–70°C):

$$\rho_a = 988.8 + 5.13B - 0.546t \tag{1.36}$$

Telis-Romero et al. (1998) correlated the apparent density of Brazilian orange juice as affected by temperature (0.5–62°C) and water content (X_w: 0.34–0.73) as

$$\rho_a = 1428.5 - 454.9X_w - 0.231t \tag{1.37}$$

1.4.1.3 Density of Solid Foods

The density of food materials depends on temperature and composition. Choi and Okos (1985) presented correlations for the densities of the major food components at a temperature range of –40–150°C (Table 1.3). Density of food materials varies nonlinearly with moisture content. Lozano et al. (1983) developed a general form of correlation to predict the density of fruits and vegetables during air drying. They found a wide variation in values and shapes of the curves when plotting the apparent density against M_w/M_w^o of a carrot, a potato, a sweet potato, and whole and sliced garlic. They found that the following form of the equation can predict the density of all fruits and vegetables considered.

Table 1.3 Density of Major Food Components at the Temperature[a] Range of −40–150°C

Material	Equation
Air	$\rho_T = 1.2847 \times 10^1 - 3.2358 \times 10^{-3}t$
Protein	$\rho_T = 1.3300 \times 10^3 - 0.5184t$
Carbohydrate	$\rho_T = 1.5991 \times 10^3 - 0.31046t$
Fat	$\rho_T = 9.2559 \times 10^2 - 0.41757t$
Fiber	$\rho_T = 1.3115 \times 10^3 - 0.36589t$
Ash	$\rho_T = 2.4238 \times 10^3 - 0.28063t$
Water	$\rho_T = 9.9718 \times 10^2 + 3.1439 \times 10^{-3}t - 3.7574 \times 10^{-3}t^2$
Ice	$\rho_T = 9.1689 \times 10^2 - 0.1307t$

Source: Choi, Y. and Okos, M. R. 1985. In *Food Engineering and Process Applications, Vol. 1, Transport Phenomena.* Le Maguer, M. and Jelen, P., Eds. Elsevier Applied Science, London.

[a] t in degrees C.

$$\rho = g + hy + q[\exp(-ry)] \quad (1.38)$$

The parameters of the models are provided by Lozano et al. (1983) for different food materials.

1.4.1.3.1 Fruits and Vegetables

The particle density of granular and gelatinized corn starches was measured in the range $0 < M_w < 1.0$ and correlated as (Maroulis and Saravacos, 1990):

$$\rho_p = 1442 + 837M_w - 3646M_w^2 + 448M_w^3 - 1850M_w^4 \quad (1.39)$$

Lozano et al. (1979) correlated the apparent density of apple above freezing as

$$\rho_a = 636 + 102[\ln M_w] \quad (1.40)$$

Singh and Lund (1984) noted that the above correlation is not valid up to zero moisture content and they correlated the apparent density of apple up to zero moisture content above frozen as

$$\rho_a = 852 - 462[\exp(-0.66M_w)] \quad (1.41)$$

Again Lozano et al. (1980) correlated the apparent and particle density of apple during air drying by an exponential form of equation for whole range of moisture content above frozen as

$$\rho_a = 684 + 68.1[\ln(M_w + 0.0054)] \quad (1.42)$$

$$\rho_p = 1540[\exp(-0.051M_w)] - 1150[\exp(-2.40M_w)] \quad (1.43)$$

Lozano et al. (1980) noted that the apparent density changed in an almost linear manner between full turgor and $M_w = 1.5$. Beyond that as moisture content decreased, the change in apparent density became steeper and showed that the apparent volume change became slower. The material density increased as moisture content decreased from full turgor to $M_w = 1.5$. Then it showed a sharp decrease, converging to an intercept similar to that of apparent density. Thus, one exponent form of the equation is not suitable for the entire range of moisture content. Maroulis and Saravacos (1990) measured the particle density of starch granules from $M_w = 0$ to $M_w = 1.0$ and found a similar peak at $M_w = 0.15$.

Madamba et al. (1994) measured the bulk density (outside void) of garlic slices and correlated it as a function of water content (X_w: 0.03 to 0.65) and slice thickness [l: (2–5) × 10^{-3} m] at room temperature as

$$\varepsilon_B = 0.865 - 30X_w - 0.8 \times 10^3 l + 2.0 \times 10^3 X_w l \tag{1.44}$$

$$\rho_B = 200.6 + 280X_w + 1.24 \times 10^4 l + 2.0 \times 10^4 X_w l \tag{1.45}$$

Bulk and particle densities of grapes as a function of water content (M_w: 0.176 to 4.0) (Ghiaus et al., 1997) are given as

$$\rho_B = 775.99 - 228.1M_w + 133.6M_w^2 - 22.19M_w^3 \tag{1.46}$$

$$\rho_p = 1480.4 - 382.2M_w + 131.8M_w^2 - 15.48M_w^3 \tag{1.47}$$

Table 1.4 presents coefficients of the quadratic form of density versus moisture content data. Most of the above density data were measured at room temperature; thus, a temperature term was not included in developing the above correlations.

Table 1.4 Parameters of the Quadratic Equations for Density ($\rho = a + bM_w + cM_w^2$)

Material	Density	Range	a	b	c	Ref.
Gorgon nut[a]	Bulk	M_w:0.15–0.60	369.2	1290	–1020	1
Gorgon nut[b]	Bulk	M_w:0.15–0.60	396.7	1330	–1080	1
Gorgon nut[c]	Bulk	M_w:0.15–0.60	434.0	1660	–8420	1
Gorgon nut[a]	True	M_w:0.15–0.60	995.9	830	–640	1
Gorgon nut[b]	True	M_w:0.15–0.60	1039.7	690	–420	1
Gorgon nut[c]	True	M_w:0.15–0.60	1025.7	930	–800	1
Macadamia nut[d]	Bulk	M_w:0.02–0.24	605.2	–92	800	2
Macadamia nut[d]	Apparent	M_w:0.02–0.24	1018.6	–44	1300	2

Sources: Jha, S. N. and Prasad, S. 1993. Physical and thermal properties of gorgon nut. *Journal of Food Process Engineering* 16(3): 237–245; Palipane, K. B., Driscoll, R. H. and Sizednicki, G. 1992. Density, porosity, and composition of macadamia in shell nuts. *Food Australia*. 44(6): 276–280.

[a] Large size.
[b] Medium size.
[c] Small size.
[d] In-shell.

1.4.1.3.2 Meat and Fish

Rahman and Driscoll (1994) correlated the density of fresh seafood by a quadratic equation (X_w: 0.739–0.856 and t: 20°C):

$$\rho_a = 2684 - 3693 X_w^o + 2085(X_w^o)^2 \tag{1.48}$$

The above equation was developed using the density data of different types of fresh seafood. Similarly, the density of frozen seafood at −30°C can be estimated as

$$\rho_a = 1390 - 520 X_w^o + 31.56(X_w^o)^2 \tag{1.49}$$

The density of frozen squid mantle (X_w^o: 0.814) below its freezing point up to −40°C can be estimated as (Rahman and Driscoll, 1994):

$$\rho_a = 1047 + 3.603\, t + 0.057\, t^2 \tag{1.50}$$

Sanz et al. (1989) proposed the density of fresh meat products above and below freezing as

$$\rho_a = 1053 \quad (t \geq t_F) \tag{1.51}$$

$$\rho_a = \frac{1053}{0.9822 + 0.1131 X_w^o + [0.2575(1 - X_w^o)]/t} \quad (t < t_F) \tag{1.52}$$

1.4.2 Predictions of Porosity

Negligible theoretical methods to predict porosity exist. Kunii and Smith (1960) derived from theoretical concepts $\varepsilon_a = 0.476$ (for loose packing) and $\varepsilon_a = 0.260$ (for close packing) when spheres are packed (Figure 1.6). Hence, it was not possible for the void fraction to be less than 0.260. Therefore, if the observed void fraction is less than 0.260, it might be considered as a result of clogging by small particles in void spaces. Similarly, observed void fractions larger than 0.476 would be caused by the presence of exceptionally large hollow spaces compared with the average void space.

Lozano et al. (1980) proposed the geometric model of Rotstein and Cornish (1978) to predict the porosity of fruits and vegetables. The above authors noted that change in

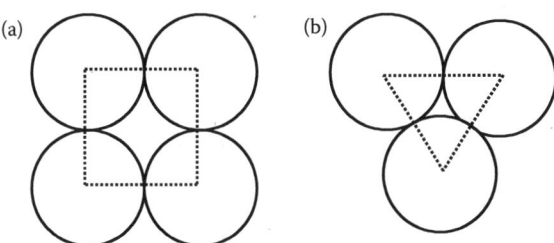

Figure 1.6 (a) Cubic packing (loosest); (b) hexagonal packing (tightest). (From Rahman, M.S. 1995. *Food Properties Handbook*. CRC Press, Boca Raton, FL.)

porosity is the result of two processes: the shrinkage of the overall dimensions and the shrinkage of the cells themselves. The model is based on cubically truncated spheres of radius r, where the cube side is 2α. On the basis of geometric considerations, the porosity at full turgor is

$$\varepsilon_a^o = 1 - \frac{\pi \zeta^o}{6(\alpha^o)^3} \qquad (1.53)$$

where ζ^o is a geometric value corresponding to the full turgor situation and can be calculated as

$$\varsigma^o = -2 + 4.5\alpha^o - 1.5(\alpha^o)^3 \qquad (1.54)$$

The formation of pores in foods during drying can be grouped into two generic types: one with an inversion point and another without an inversion point (Figures 1.7 and 1.8). Figure 1.7a shows that during drying pores are initially collapsed, causing decrease in porosity, and reached at a critical value, and further decrease of moisture causes the formation of pores again until the food is completely dried. The opposite condition exists in Figure 1.7b. Figure 1.8 shows that the level of pores is increased or decreased as a function of moisture content (Rahman, 2000).

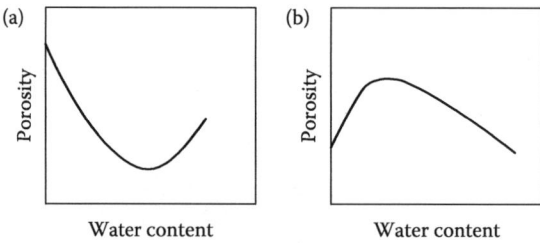

Figure 1.7 Change of porosity as a function of water content (with inversion point). (From Rahman, M.S. 2000. Mechanisms of pore formation in foods during drying: present status. Papers presented at the *Eighth International Congress on Engineering and Food (ICEF8)*, Puebla, Mexico, April 9–13, 2000.)

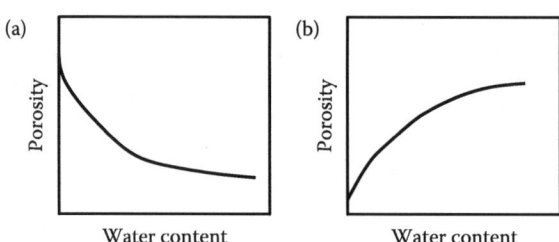

Figure 1.8 Change of porosity as a function of water content (no inversion point). (From Rahman, M.S. 2000. Mechanisms of pore formation in foods during drying: present status. Papers presented at the *Eighth International Congress on Engineering and Food (ICEF8)*, Puebla, Mexico, April 9–13, 2000.)

Most of the porosity is predicted from the density data or from empirical correlations of porosity and moisture content. Rahman et al. (1996) developed the following correlations for open and closed pores in calamari during air drying up to zero moisture content as

$$\varepsilon_{op} = 0.079 - 0.164\lambda + 0.099\lambda^2 \tag{1.55}$$

$$\varepsilon_{cp} = 0.068 - 0.216\lambda + 0.138\lambda^2 \tag{1.56}$$

Rahman (1991) developed an equation for the apparent porosity of squid mantle during air drying up to zero moisture content as

$$\varepsilon_a = 0.109 - 0.219\lambda + 0.099\lambda^2 \tag{1.57}$$

Lozano et al. (1980) developed a correlation for open-pore porosity of an apple during air drying as (X_w: 0.89–0.0):

$$\varepsilon_{op} = 1 - \frac{852.0 - 462.0[\exp(-0.66M_w)]}{1540.0[\exp(-0.051M_w)] + -1150.0[\exp(-2.4M_w)]} \tag{1.58}$$

The above authors found a peak at low moisture content. Ali et al. (1996) studied the expansion characteristics of extruded yellow corn grits in a single-screw extruder with various combinations of barrel temperature (100–200°C) and screw speed (80–200 r/min). They observed that open and total pore volume increased with the increase of temperature and screw speed when moisture content during extrusion was 0.64 (wet basis). Correlations for total and open pore volume are

$$(\phi_v)_T = -4.8 \times 10^{-3} + 6.7 \times 10^{-5}t + 1.97 \times 10^{-7}t^2 + 6.7 \times 10^{-5}\psi$$
$$- 2.0 \times 10^{-7}\psi^2 - 7.98 \times 10^{-7}t\psi + 2.43 \times 10^{-9}t^2\psi \tag{1.59}$$
$$- 7.55 \times 10^{-12}t\psi^2$$

$$(\phi_v)_{op} = -5.91 \times 10^{-3} + 8.3 \times 10^{-5}t - 2.53 \times 10^{-7}t^2 + 8.5 \times 10^{-5}\psi$$
$$- 2.68 \times 10^{-7}\psi^2 - 1.06 \times 10^{-7}t\psi + 3.43 \times 10^{-9}t^2\psi \tag{1.60}$$
$$- 1.08 \times 10^{-11}t\psi^2$$

where volume is in m³/kg. Rahman et al. (1996) recommended that further detailed studies beyond the empirical correlations are necessary to understand the physicochemical nature of the interactions of component phases and the collapse and formation of pores in food materials during processing. Hussain et al. (2002) developed a general porosity prediction model of food during air drying as

$$\varepsilon_a = 0.5X_w^2 - 0.8X_w - 0.002\,t^2 + 0.02\,t - 0.05(1 - \varepsilon_a^o)F \tag{1.61}$$

where the values of F are the numeric values assigned to the types of product—1 for sugar-based products, 2 for starch-based products, and 3 for other products.

1.4.3 Prediction of Surface Area

1.4.3.1 Euclidian Geometry

The boundary surface area and volume of some common shapes are given in Table 1.1. Surface area can also be predicted from the empirical correlations between the surface area and mass of the food materials. Mohsenin (1986) compiled linear correlations of surface area of apple, pear, and plum. Besch et al. (1968) proposed an empirical equation of

$$A_s = \beta \, m^{2/3} \tag{1.62}$$

where β is a constant with reported values varying from 0.0456 to 0.0507 for fresh eggs and a value of 0.0566 for apple (Frechette and Zahradnik, 1965). Avena-Bustillos et al. (1994) correlated the surface area of noncylindrical- and cylindrical-shaped zucchini as

$$A = (139.89 + 0.62m) \times 10^{-4} \tag{1.63}$$

$$A = (56.68 + 0.83m) \times 10^{-4} \tag{1.64}$$

Equation 1.62 is for noncylindrical (m: 0.23–0.40 kg) and Equation 1.63 is for cylindrical (0.08–0.12 kg) zucchini. These authors also correlated actual surface area with the surface area calculated from geometric dimension considering cylindrical shape as

$$A = (77.65 + 0.70 \, A_c) \times 10^{-4} \tag{1.65}$$

$$A = (37.58 + 0.61 \, A_c) \times 10^{-4} \tag{1.66}$$

where A_c [$= \pi D (H + D/2)$] is the cylindrical shape for zucchini. Equation 1.64 is for noncylindrical (A_c: 290–480 m²) and Equation 1.65 is for cylindrical (A_c: 130–1900 m²) zucchini. These authors concluded that a correlation of surface area with zucchini mass generally produced a better fit than using the assumption of a right cylinder formula for area estimation for noncylindrical zucchini. However, for cylindrical-shaped zucchini, the right cylinder assumption fitted better than using the initial fruit mass. The surface areas of apples, plums, and pears are $(43.7–64.0) \times 10^{-4}$, $(13.7–17.8) \times 10^{-4}$, $(56.4–58.4) \times 10^{-4}$ m², respectively (Mohsenin, 1986).

1.4.3.2 Non-Euclidian or Irregular Geometry

Fractal analysis can be used in characterizing non-Euclidian geometry (Mandelbrot, 1977; Takayasu, 1990; Rahman, 1995). The fractal dimension can be estimated by structured walk (Richardson's plot), bulk density–particle diameter relation, sorption behavior of gases, and pore size distribution. Richardson's method was used by Graf (1991) for fine soil particles and by Peleg and Normand (1985) for instant coffee particles to estimate fractal dimension. Yano and Nagai (1989), and Nagai and Yano (1990) used a gas adsorption method for native and deformed starch particles, and Ehrburger-Dolle et al. (1994) used a porosity method for activated carbon particles formed by different techniques or treatments. However, none of the above authors used all methods to estimate the fractal dimensions for the same material particles so that a clear physical meaning of fractal dimensions could be drawn. Nagai and Yano (1990) used two gas adsorption methods and found different fractal dimensions for the same starch particles. Thus, fractal dimension interpretation without physical understanding can be misleading or incorrectly applied. Rahman (1997) tried to obtain a better understanding of the fractal dimension by reviewing different methods available in the literature. An example is presented for starch and modified starch particles considering experimental data from the literature. This analysis can minimize confusion and avoid misinterpretation of fractal dimensions and can provide the theoretical limitations of the fractal dimensions. More details on fractal analysis are provided by Rahman (1995).

In general, for any nonfractal (Euclidian) object, the following relation holds between its length (L), area (A), and volume (V):

$$L \infty A^{1/2} \infty V^{1/3} \tag{1.67}$$

The relation for the fractal-shaped objects can be written as

$$L \infty A^{\delta/2} \infty V^{\delta/3} \tag{1.68}$$

or

$$L \infty A^{\delta/d} \infty V^{\delta/d} \tag{1.69}$$

where δ is the fractal dimension and d is the Euclidian dimension; δ is the perimeter fractal dimension, $\delta + 1$ is the area fractal dimension, and $\delta + 2$ is the volume fractal dimension. However, perimeter, area, and volume fractal dimension may not be the same in all cases. In the above relationships it is assumed to be the same.

1.4.3.3 Theoretical Prediction
1.4.3.3.1 Based on Conservation of Mass and Volume
Most of the density prediction models in the literature are empirical in nature. Fundamental models exist based on the conversion of both mass and volume, and thus a number of authors have proposed models of this kind. Food materials can be considered as multiphase systems (i.e., gas–liquid–solid systems). When the mixing process conserves both mass and volume, then the density of the multiphase system can be written as

$$\frac{1}{\rho_T} = \sum_{i=1}^{N} \frac{X_i}{(\rho_T)_i} = \xi \tag{1.70}$$

where $(\rho_T)_i$ and ρ_T are the true density of component i and the composite mixture, respectively, X_i is the mass fraction of component i, and N is the total number of components present in the mixtures. Miles et al. (1983) and Choi and Okos (1985) proposed the above equation for predicting the density of food materials. However, this equation has limited use in cases where no air phase is present and no interaction between the phases occurs. Rahman (1991) has extended the theoretical model, introducing the pore volume and an interaction term into the above equation.

The apparent density of a composite mixture can be divided into three parts based on the conservation law for mass and volume:

Apparent volume =
Actual volume of the pure component phases +
Volume of pores or air phase +
Excess volume due to the interaction of phases

The excess volume can be positive or negative depending on the physicochemical nature of the process of concern, whereas porosity is always positive. The above equation can be written as

$$\frac{1}{\rho_a} = \sum_{i=1}^{N} \frac{X_i}{(\rho_T)_i} + V_v + V_{ex} \tag{1.71}$$

where V_v is the specific volume of the void or air phase (m³/kg) and V_{ex} is the specific excess volume (m³/kg). The total porosity is defined as

$$\varepsilon_a = \frac{V_v}{V_a} \tag{1.72}$$

Based on density, apparent volume is defined as

$$V_a = \frac{1}{\rho_a} \tag{1.73}$$

The excess volume fraction can be defined as

$$\varepsilon_a = \frac{V_v}{V_a} \tag{1.74}$$

Equation 1.71 can be transformed to the following:

$$\frac{1}{\rho_a} = \frac{\xi}{(1 - \varepsilon_{ex} - \varepsilon_a)} \tag{1.75}$$

When total porosity and excess volume fraction are negligible, Equation 1.75 is reduced to

$$\frac{1}{\rho_a} = \frac{1}{\rho_T} = \xi \tag{1.76}$$

Total porosity (or total air volume fraction) and excess volume fraction can be calculated from the experimental density data as

$$\varepsilon_a = 1 - \frac{\rho_a}{\rho_m} \quad \text{and} \quad \varepsilon_{ex} = 1 - \frac{\rho_m}{\rho_T}$$

The shrinkage can be written as

$$S_a = \frac{V_a}{V_a^o} = \frac{\xi}{\xi^o} \times \frac{(1 - \varepsilon_{ex} - \varepsilon_a)}{(1 - \varepsilon_{ex}^o - \varepsilon_a^o)} \tag{1.77}$$

The excess properties and porosity are usually not easy to correlate in the case of complex systems such as foods (Rahman et al., 1996).

1.4.3.3.2 Mechanisms of Collapse

Genskow (1990) and Achanta and Okos (1995) mentioned several mechanisms that affect the degree of collapse or shrinkage and formation of pores. Understanding these mechanisms would aid in achieving desired shrinkage or collapse in the products. The following physical mechanisms play an important role in the control of shrinkage or collapse (Rahman and Perera, 1999):

- Surface tension (considers collapse in terms of the capillary suction created by a receding liquid meniscus)
- Plasticization (considers collapse in terms of the plasticizing effect on various polymer solutes)
- Electrical charge effects (considers collapse in terms of van der Waals electrostatic forces)
- The mechanism of moisture movement in the process
- Gravitational effects

1.4.3.3.3 Glass Transition Concept

Slade and Levine (1991) first applied the concept of glass transition to identify or explain the physicochemical changes in foods during processing and storage. The glass transition theory is one of the concepts that have been proposed to explain the process of shrinkage, collapse, fissuring, and cracking during drying (Cnossen and Siebenmorgen, 2000; Karathanos et al., 1993; Krokida et al., 1998; Rahman, 2001). The hypothesis indicates that a significant shrinkage can be noticed during drying only if the temperature of the drying is higher than the glass transition of the material at that particular moisture content (Achanta and Okos, 1996). The methods of freeze-drying and hot air drying can be compared based on this theory. In freeze-drying, with the drying temperature below or close to t'_g (maximally freeze-concentrated glass transition temperature; it is independent of solids content) or t_g (glass transition as a function of solids content), the material is in the glassy state. Hence shrinkage is negligible. As a result, the final product is very porous. With hot air drying, on the other hand, with the drying temperature above t'_g or t_g, the material is in the rubbery state, and substantial shrinkage occurs causing a lower level of pores. During the initial stage of freeze-drying, the composition of the freeze-concentrated phase surrounding the ice dictates the t'_g. In initial or early stage of drying, t'_g is very relevant, and the vacuum must be sufficient to ensure that sublimation is occurring. At the end of initial stage of drying, the pore size and the porosity are dictated by ice crystal size, if collapse of the wall of the matrix that surrounded the ice crystal does not occur. The secondary stage of drying, on the other hand, refers to removal of water from the unfrozen phase. After sublimation is completed, the sample begins to warm up to the shelf temperature. At this stage, t_g of the matrix is related to the collapse and no longer to t'_g because $t_g > t'_g$ (t_g increases from t'_g as the concentration of solids increases during the process of drying). Recently, it was found that the concept of glass transition is not valid for freeze-drying of all types of biological materials indicating the need to incorporate other concepts (Sablani and Rahman, 2002); thus, a unified approach needs to be used.

In many cases during convection air drying, the observations related to collapse are just the opposite of the glass transition concept (Del Valle et al., 1998; Ratti, 1994; Wang and Brennan, 1995). The mechanism proposed for this was the concept of case hardening (Achanta and Okos, 1996; Ratti, 1994). These authors indicated that at a low drying rate (low temperature), the moisture gradient within the product is small and internal stresses are low; hence, the material shrinks down fully onto a solid core, and shrinkage is uniform. At a high drying rate (higher temperature), the surface moisture decreases very fast so that the surface becomes stiff (i.e., case hardening phenomenon), limiting subsequent shrinkage, thus increasing pore formation.

1.4.3.3.4 Rahman's Hypothesis

After analyzing experimental results from the literature, Rahman (2001) stated that the glass transition theory does not hold true for all products or processes. Other concepts, such as surface tension, pore pressure, structure, environment pressure, and mechanisms of moisture transport also play important roles in explaining the formation of pores. Rahman (2001) hypothesized that as capillary force is the main force responsible for collapse, so counterbalancing this force causes formation of pores and lower shrinkage. The counterbalancing forces are due to generation of internal pressure due to vaporization of water or other solvents, variation in moisture transport mechanism, and pressure outside the material. Other factors could be the strength of solid matrix (i.e., ice formation; case hardening; permeability of water through crust; change in tertiary and quaternary structure of polymers; presence or absence of crystalline, amorphous, and viscoelastic nature of solids; matrix reinforcement; residence time). Figure 1.9 shows how the different factors act on a single pore.

1.4.3.4 Size Distribution

In addition to size, the size distribution of particles or pores is also an important characteristic. The distribution can be plotted in terms of cumulative percent greater than or less than the size versus the size. It can also be plotted as a distribution of the amounts that fall within a given diameter range. The normal procedure is to express the amount of each range on a mass basis, but in some cases the particle frequency is used.

The mean, median, standard deviation, and distribution of size are most commonly used to characterize the particle size. In a symmetric distribution, the mean and median coincide and their ratio is 1. In a distribution skewed to the right this ratio is greater than 1, and in a distribution skewed to the left it is less than 1. The deviation from unity is a measure of the degree of skewness but is not formally used as a statistical criterion (Barrett and Peleg, 1992). The standard statistical test for skewness is based on the coefficient of skewness as

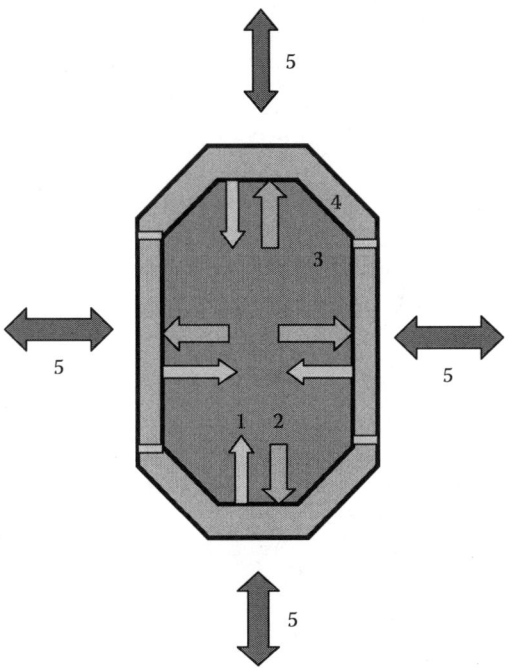

Figure 1.9 Rahman's (2001) hypothesis on the mechanism of pore formation. (1) Pore pressure, (2) vapor pressure inside pore, (3) permeability of crust or pore wall, (4) strength of pore wall, (5) pressure outside pore. (From Rahman, M.S. 2000. Mechanisms of pore formation in foods during drying: present status. Papers presented at the *Eighth International Congress on Engineering and Food (ICEF8)*, Puebla, Mexico, April 9–13, 2000.)

$$\text{Skewness} = \frac{\sqrt{N}\left[\sum_{i=1}^{N}(x-\bar{x})^3\right]}{\left[\sum_{i=1}^{N}(x-\bar{x})^2\right]^{3/2}} \quad (1.78)$$

where x is the size and N is the number of observations. The cell size distribution of polyurethane synthetic foam had only minor skewness to the left with a mean to median ratio of 0.97 as compared with 1.5–2.17 for extruded product. Skewness should be treated as a characteristic property and not a general feature of cellular solids. Many different types of size distribution functions exist. The application of these functions depends on their mathematical simplicity, adequate statistical properties, and usefulness in describing the distribution (Ma et al., 1998).

Gates–Gaudin–Schuhmann function (GGS)—The GGS function is expressed as

$$Y = \left(\frac{x}{k}\right)^m \quad (1.79)$$

where Y is the cumulative fraction with a size less than x, k is the characteristic size of the distribution, and m is a measure of the distribution spread, which is also called the Schuhmann slope.

Rosin–Rammler—Many distributions skewed to the right can also be described by Rosin–Rammler distribution. Its most familiar form is

$$f(x) = n^b x^{n-1} \left[\exp(-b\, x^n) \right] \qquad (1.80)$$

where n and b are constants. Testing for this distribution by standard nonlinear regression procedures can sometimes be cumbersome. In cumulative form, it is a two-parameter function given

$$Y = 1 - \exp\left[-\left(\frac{x}{x_r} \right)^n \right] \qquad (1.81)$$

where Y is the mass fraction of material finer than size x, x_r is a constant that characterizes the particle size range, and n is another constant, which is a measure of the uniformity of particle sizes. Lower values of n are associated with a more scattered distribution, while higher values of n will imply a more uniform particle size.

Modified Gaudin–Meloy—It is expressed as

$$Y = \left[1 - \left(1 - \frac{x}{x_o} \right)^r \right]^m \qquad (1.82)$$

where x_o is the measure of the maximum particle size, m is the Schuhmann slope, and r is the ratio of x_o to the size modulus.

Log-Normal (LN) function is

$$f(z) = \frac{1}{\sigma_z \sqrt{2\pi}} \exp\left[-\frac{(z - \bar{z})^2}{2\sigma_z^2} \right] \qquad (1.83)$$

where $z = \ln x$ (x is the size parameter), $\bar{z}$ is the mean of $\ln x$, and σ_z is the standard deviation of $\ln x$. The LN function is perhaps the most frequently used function among the different types (Ma et al., 1998). One of its characteristics is that the frequency–log size plot shows a symmetric bell shape (Gaussian distribution). The lack of any appreciable skewness is also evident in the magnitude of the ratio of mean log (size) to median log (size). Another test for a log-normal distribution is the linearity of the plot between cumulative number, or fraction, and the log (size) when constructed on probability paper. For many powders for which the population mode and median spread vary independently and the size distributions have a finite range, the Modified Beta function is more appropriate than the LN function.

The **Modified Beta** function is defined as

$$f(y) = \frac{y^{am}(1 - y)^m}{\int_0^1 y^{am}(1 - y)^m \, dy} \qquad (1.84)$$

where a and m are constants. y is the normalized length given by

$$y = \frac{x - x_{min}}{x_{max} - x_{min}} \qquad (1.85)$$

where x_{min} and x_{max} are the smallest and largest particle sizes, respectively. Therefore,

$$x_{min} < x < y_{max}, \quad 0 < y < 1$$

An advantage of the beta or log beta distribution is that it can be written in a form that makes the mode independent of the spread while providing the same fit as the log normal or Rosin–Rammler distribution.

1.5 SUMMARY

Mass–volume–area properties are needed for process design, estimation of other properties, and product characterization. This chapter presents the terminology, measurement techniques, and prediction models of selected mass–volume–area-related properties, such as volume, density, porosity, surface area, and size distribution. First, clear definitions of the terms used are presented, followed by a summary of measurement techniques and prediction models. In the case of compiling models, both empirical correlations and a theoretical approach are considered.

ACKNOWLEDGMENTS

I would like to acknowledge the researchers who have developed new concepts, compiled data, and developed generic rules for mass–volume–area-related properties, which made my task easy.

LIST OF SYMBOLS

A	Surface area (m²)
a	Parameter in Equation 1.22 or Equation 1.83 or major axis
B	Degrees Brix
b	Parameter in Equation 1.22 or parameter for size-distribution or minor axis
C	Parameter in Equation 1.28
D	Diameter (m)
d	Dimension of Euclidian geometry
e	Eccentricity
F	Product type
f	Friction factor
G	Buoyant force (kg)
g	Parameter in Equation 1.38

h Parameter in Equation 1.38
i ith component
k Permeability (m²/s) or size characteristics
L Length (m)
l Thickness (m)
M Water content (dry basis, kg/kg solids)
m Mass (kg) or distribution spread
N Number of components of observation
n Moles of gas or parameter for size distribution
P Pressure (Pa)
Q Volumetric flow rate (m³/s)
q Parameter in Equation 1.38
R Ideal gas constant
r Parameter in Equation 1.38 or ratio of size modulus or radius (m)
Re Reynolds number
S Shrinkage (V_a/V_a^o)
T Temperature (K)
t Temperature (°C)
u Velocity (m/s)
V Volume (m³)
X Moisture content (wet basis, kg/kg sample)
x Size
Y Cumulative fraction
y $(x - x_{min})/(x_{max} - x_{min})$
z $\ln x$

Greek Symbols

ρ Density (kg/m³)
ψ Screw speed (r/min)
σ Standard deviation
λ X_w/X_w^o
ϕ Volume (m³/kg)
α Parameter in Equation 1.54
δ Fractal dimension
β Parameter in Equation 1.61
τ Surface tension (N/m)
ς Geometric value
ε Porosity (dimensionless)
μ Viscosity (Pa s)
ξ Specific volume based on composition (Equation 1.69)
Δ Difference
ϑ Refractive index
θ Contact angle

Subscripts

a	Apparent
B	Bulk
BP	Bulk particle
c	Cylindrical shape
cp	Closed pore
ec	Expansion cell
ex	Excess
F	Freezing point
f	Fat
g	Glass transition
j	Number of coefficient
m	Material
min	Minimum
max	Maximum
o	Size modulus
op	Open pore
p	Particle
r	Characteristics
s	Sample or solids in X
sc	Sample cell
T	True or total
v	Void
w	Water

Superscripts

o	Initial
$\prime$	Maximal freeze-concentration conditions

REFERENCES

Achanta, S. and Okos, M.R. 1995. Impact of drying on the biological product quality. In *Food Preservation by Moisture Control: Fundamentals and Applications*. Barbosa-Canovas, G.V. and Welti-Chanes, J., Eds. Technomic Publishing, Lancaster, PA. p. 637.

Achanta, S. and Okos, M.R. 1996. Predicting the quality of dehydrated foods and biopolymers: Research needs and opportunities. *Drying Technology*. 14(6): 1329–1368.

Ali, Y., Hanna, M.A., and Chinnaswamy, R. 1996. Expansion characteristics of extruded corn grits. *Food Science and Technology*. 29: 702–707.

Avena-Bustillos, R.D.J., Krochta, J.M., Saltveit, M.E., Rojas-Villegas, R.D.J., and Sauceda-Perez, J.A. 1994. Optimization of edible coating formulations on zucchini to reduce water loss. *Journal of Food Engineering*. 21(2): 197–214.

Bailey, C.H. 1912. A method for the determination of the specific gravity of wheat and other cereals. USDA Bureau of Plant Industry, Circular N. 99.

Barrett, A.M. and Peleg, M. 1992. Cell size distributions of puffed corn extrudates. *Journal of Food Science.* 57: 146–154.

Bayindirli, L. 1992. Mathematical analysis of variation of density and viscosity of apple juice with temperature and concentration. *Journal of Food Processing and Preservation.* 16: 23–28.

Besch, E.L., Sluka, S.J. and Smith, A.H. 1968. Determination of surface area using profile recordings. *Poultry Science.* 47(1): 82–85.

Chen, C.S. 1989. Mathematical correlations for calculation of Brix-apparent density of sucrose solutions. *Food Science and Technology.* 22: 154–156.

Choi, Y. and Okos, M.R. 1983. The thermal properties of tomato juice. *Transactions of the ASAE.* 26: 305–311.

Choi, Y. and Okos, M.R. 1985. Effects of temperature and composition on the thermal properties of foods. In *Food Engineering and Process Applications, Vol. 1, Transport Phenomena.* Le Maguer, M. and Jelen, P., Eds. Elsevier Applied Science, London.

Clayton, J.T. and Huang, C.T. 1984. Porosity of extruded foods. In *Engineering and Food. Volume 2. Processing and Applications.* McKenna, B., Ed. Elsevier Applied Science Publishers, Essex. pp. 611–620.

Cnossen, A.G. and Siebenmorgen, T.J. 2000. The glass transition temperature concept in rice drying and tempering: Effect on milling quality. *Transactions of the ASAE.* 43: 1661–1667.

Del Valle, J.M., Cuadros, T.R.M. and Aguilera, J.M. 1998. Glass transitions and shrinkage during drying and storage of osmosed apple pieces. *Food Research International.* 31: 191–204.

Ehrburger-Dolle, F., Lavanchy, A. and Stoeckli, F. 1994. Determination of the surface fractal dimension of active carbons by mercury porosimeby. *Journal of Colloid and Interface Science.* 166(2): 451–461.

Frechette, R.J. and Zahradnik, J.W. 1965. Surface area–weight relationships for McIntosh apples. *Transactions of the ASAE.* 9: 526.

Genskow, L.R. 1990. Consideration in drying consumer products. In *Drying '89*, Mujumdar, A.S. and Roques, M., Eds. Hemisphere Publishing, New York.

Ghiaus, A.G., Margaris, D.P. and Papanikas, D.G. 1997. Mathematical modeling of the convective drying of fruits and vegetables. *Journal of Food Science.* 62: 1154–1157.

Graf, J.C. 1991. The importance of resolution limits to the interpretation of fractal descriptions of fine particles. *Powder Technology.* 67: 83–85.

Holland, F.A. 1973. *Fluid Flow for Chemical Engineers.* Edward Arnold, London.

Horiba Bulletin: HRE-8815A. Horiba, Koyoto, Japan.

Hussain, M.A., Rahman, M.S., and Ng, C.W. 2002. Prediction of pores formation (porosity) in foods during drying: Generic models by the use of hybrid neural network. *Journal of Food Engineering.* 51: 239–248.

Ibarz, A. and Miguelsanz, R. 1989. Variation with temperature and soluble solids concentration of the density of a depectinised and clarified pear juice. *Journal of Food Engineering.* 10: 319–323.

Jha, S.N. and Prasad, S. 1993. Physical and thermal properties of gorgon nut. *Journal of Food Process Engineering.* 16(3): 237–245.

Karathanos, V., Anglea, S., and Karel, M. 1993. Collapse of structure during drying of celery. *Drying Technology.* 11(5): 1005–1023.

Keppeler, R.A. and Boose, J.R. 1970. Thermal properties of frozen sucrose solutions. *Transactions of the ASAE.* 13(3): 335–339.

Krokida, M.K., Karathanos, V.T., and Maroulis, Z.B. 1998. Effect of freeze-drying conditions on shrinkage and porosity of dehydrated agricultural products. *Journal of Food Engineering.* 35: 369–380.

Kunii, D. and Smith, J.M. 1960. Heat transfer characteristics of porous rocks. *AIChE Journal.* 6(1): 71.

Lewis, M.J. 1987. *Physical Properties of Foods and Food Processing Systems.* Ellis Horwood, England and VCH Verlagsgesellschaft, FRG.

Loch-Bonazzi, C.L., Wolf, E., and Gilbert, H. 1992. Quality of dehydrated cultivated mushrooms (*Agaricus bisporus*): A comparison between different drying and freeze-drying processes. *Food Science and Technology*. 25: 334–339.

Lozano, J.E., Urbicain, M.J., and Rotstein, E. 1979. Thermal conductivity of apples as a function of moisture content. *Journal of Food Science*. 44(1): 198.

Lozano, J.E., Rotstein, E. and Urbicain, M.J. 1980. Total porosity and open-pore porosity in the drying of fruits. *Journal of Food Science*. 45: 1403–1407.

Lozano, J.E., Rotstein, E. and Urbicain, M.J. 1983. Shrinkage, porosity and bulk density of foodstuffs at changing moisture contents. *Journal of Food Science*. 48: 1497.

Ma, L., Davis, D.C., Obaldo, L.G., and Barbosa-Canovas, G.V. 1998. *Engineering Properties of Foods and Other Biological Materials*. American Society of Agricultural Engineers. St. Joseph, MI.

Madamba, P.S., Driscoll, R.H., and Buckle, K.A. 1994. Bulk density, porosity, and resistance to airflow of garlic slices. *Drying Technology*. 12: 937–954.

Mandelbrot, B.B. 1977. *Fractals: Form, Chance, and Dimension*. W.H. Freeman and Company, San Francisco.

Maroulis, S.N. and Saravacos, G.D. 1990. Density and porosity in drying starch materials. *Journal of Food Science*. 55: 1367–1372.

Miles, C.A., Beek, G.V., and Veerkamp, C.H. 1983. Calculation of thermophysical properties of foods. In *Thermophysical Properties of Foods*, Jowitt, R., Escher, F., Hallstrom, B., Meffert, H.F.T., Spiess, W.E.L., and Vos, G., Eds., Applied Science Publishers, London, pp. 269–312.

Mohsenin, N.N. 1986. *Physical Properties of Plant and Animal Materials*. Gordon and Breach Science Publishers, New York.

Nagai, T. and Yano, T. 1990. Fractal structure of deformed potato starch and its sorption characteristics. *Journal of Food Science*. 55(5): 1336–1337.

Palipane, K.B., Driscoll, R.H. and Sizednicki, G. 1992. Density, porosity, and composition of macadamia in shell nuts. *Food Australia*. 44(6): 276–280.

Particulars. 1998. Newsletter of Particle and Surface Sciences, issue 3, Gosford, New South Wales.

Peleg, M. and Normand, M.D. 1985. Characterization of the ruggedness of instant coffee particle shape by natural fractals. *Journal of Food Science*. 50: 829–831.

Phipps, L.W. 1969. The interrelationship of viscosity, fat content, and temperature of cream between 40°C and 80°C. *Journal of Dairy Research*. 36: 417–426.

Rahman, M.S. 1991. *Thermophysical Properties of Seafoods*. PhD thesis, University of New South Wales, Sydney.

Rahman, M.S. 1995. *Food Properties Handbook*. CRC Press, Boca Raton, FL.

Rahman, M.S. 1997. Physical meaning and interpretation of fractal dimensions of fine particles measured by different methods. *Journal of Food Engineering*. 32: 447–456.

Rahman, M.S. 2000. Mechanisms of pore formation in foods during drying: present status. Papers presented at the *Eighth International Congress on Engineering and Food (ICEF8)*, Puebla, Mexico, April 9–13, 2000.

Rahman, M.S. 2001. Toward prediction of porosity in foods during drying: A brief review. *Drying Technology*. 19(1): 1–13.

Rahman, M.S. and Driscoll, R.H. 1994. Density of fresh and frozen seafood. *Journal of Food Process Engineering*. 17: 121–140.

Rahman, M.S. and McCarthy, O.J. 1999. A classification of food properties. *International Journal of Food Properties*. 2(2): 93–99.

Rahman, M.S. and Perera, C.O. 1999. Drying and food preservation. In *Handbook of Food Preservation*. Rahman, M.S., Ed. Marcel Dekker, New York. pp. 173–216.

Rahman, M.S., Al-Amri, D., and Al-Bulushi, I.M. 2002. Pores and physicochemical characteristics of dried tuna produced by different methods of drying. *Journal of Food Engineering*. 53: 301–313.

Rahman, M.S., Perera, C.O., Chen, X.D., Driscoll, R.H., and Potluri, P.L. 1996. Density, shrinkage and porosity of calamari mantle meat during air drying in a cabinet dryer as a function of water content. *Journal of Food Engineering*. 30: 135–145.

Ramaswamy, H.S. and Tung, M.A. 1981. Thermophysical properties of apples in relation to freezing. *Journal of Food Science*. 46: 724.

Ramos, A.M. and Ibarz, A. 1998. Density of juice and fruit puree as a function of soluble solids content and temperature. *Journal of Food Engineering*. 35: 57–63.

Ratti, C. 1994. Shrinkage during drying of foodstuffs. *Journal of Food Engineering*. 23: 91–105.

Riedel, L. 1949. Thermal conductivity measurement on sugar solutions, fruit juices, and milk. *Chemical Engineering and Technology*. 21(17): 340–341.

Rotstein, A. and Cornish, A.R.H. 1978. Prediction of the sorption equilibrium relationship for the drying of foodstuffs. *AIChE Journal*. 24: 966.

Roy, N.K., Yadav, P.L., and Dixit, R.N. 1971. Density of buffalo milk fat. II. Centrifuged fat. *Milchwissenschaft*. 26: 735–738.

Sablani, S.S. and Rahman, M.S. 2002. Pore formation in selected foods as a function of shelf temperature during freeze-drying. *Drying Technology*. 20: 1379–1391.

Sanz, P.D., Domonguez, M. and Mascheroni, R.H. 1989. Equations for the prediction of thermophysical properties of meat products. *Latin American Applied Research*. 19: 155.

Short, A.L. 1955. The temperature coefficient of expansion of raw milk. *Journal of Dairy Research*. 22: 69.

Singh, R.K. and Lund, D.B. 1984. Mathematical modeling of heat- and moisture transfer-related properties of intermediate moisture apples. *Journal of Food Processing and Preservation*. 8: 191.

Slade, L. and Levine, H. 1991. A food polymer science approach to structure property relationships in aqueous food systems: Non-equilibrium behavior of carbohydrate–water systems. In *Water Relationships in Food*, Levine, H. and Slade, L. Eds. Plenum Press, New York. pp. 29–101.

Takayasu, H. 1990. *Fractals in the Physical Sciences*, Manchester University Press, Manchester, UK.

Telis-Romero, J., Telis, V.R.N., Gabas, A.L., and Yamashita, F. 1998. Thermophysical properties of Brazilian orange juice as affected by temperature and water content. *Journal of Food Engineering* 38: 27–40.

Toledo, R.T. 1993. *Fundamentals of Food Process Engineering*. 2nd ed. Chapman & Hall, New York.

Wang, N. and Brennan, J.G. 1995. Changes in structure, density and porosity of potato during dehydration. *Journal of Food Engineering*. 24: 61–76.

Yano, T. and Nagai, T. 1989. Fractal surface of starchy materials transformed with hydrophilic alcohols. *Journal of Food Engineering*. 10: 123–133.

2
Surface Properties

Karl F. Schilke and Joseph McGuire

Contents

2.1	Introduction	37
2.2	Fundamental Considerations	38
	2.2.1 Definitions	38
	2.2.2 Gibbs Adsorption Equation	40
	2.2.3 Contact Angle	43
	2.2.3.1 Critical Surface Tension	44
	2.2.3.2 Polar and Dispersive Contributions to Surface Energy	44
	2.2.3.3 An Equation of State Relationship among Interfacial Energies	47
	2.2.4 Effects of Adsorbed Layer Composition and Structure on Interfacial Energy	48
2.3	Measurement Techniques	51
	2.3.1 Evaluation of the Contact Angle	51
	2.3.2 Evaluation of Liquid Surface Tension	51
	2.3.3 Evaluation of γ_L^d and γ_S^d	53
	2.3.4 Polymer Surfaces	54
2.4	Surface Property Data	54
2.5	Modification of Surface Properties	56
	2.5.1 Superhydrophobic Surfaces	57
	2.5.2 Chemical and Physical Modifications	58
2.6	Summary	59
References		60

2.1 INTRODUCTION

Surface and colloid science is receiving increasing recognition as an important aspect of food science and technology. Issues surrounding fouling and cleaning in food processing, physical and chemical interactions between food and packaging materials, and the stability and function of food foams and emulsions, among many other issues in food technology, have origins appropriately described by interfacial science.

Thermodynamics dictates that in order for a boundary between phases to be stable, it must possess an interfacial free energy; that is, work must be done to extend or enlarge the boundary. If that were not true, then any random force would distort and convolute the interface until the phases became mixed. Thermodynamics also dictates that systems tend to minimize their total free energy. Concerning interfacial free energy, one way to do this is by minimizing interfacial area. Another is by adsorption, the preferential location of a substance at an interface. Properties and predictive relationships relevant to these two natural tendencies in food systems constitute the subject of this chapter. The focus is on the fundamental surface property termed the "excess interfacial free energy," also referred to as (interfacial or) surface energy, or surface tension. Measurement of this property and others related to it, their interpretation and application in food technology, and their variation with processing conditions, food composition, and so on, are summarized. In addition, some commercially relevant methods for the manipulation of surface properties of materials used in food handling and packaging are briefly discussed.

2.2 FUNDAMENTAL CONSIDERATIONS

2.2.1 Definitions

Interfaces, as well as the interactions that take place in interfacial regions, can be complex. The properties of atoms or atomic groups at a material surface are different from those of the bulk material. The first layer of surface atoms or atomic groups in contact with another phase is particularly unique, as the group spacing and orientation at the interface can differ substantially from either bulk material (Figure 2.1).

The free energy change associated with the isothermal, reversible formation of a surface is termed the excess interfacial free energy, γ (J/m²). The explicit thermodynamic definition for surface energy is

$$\gamma = (dA/da)_{T,V} \tag{2.1}$$

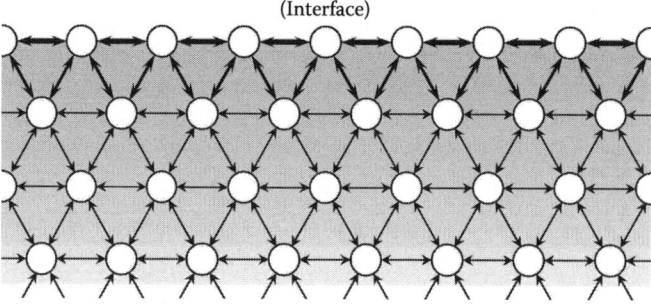

Figure 2.1 Differences in molecular bonding at the interface (heavier arrows) produce a higher free energy at the surface than in the bulk. Unfulfilled, strained, or other differences in the molecular bonding at the surface may also cause changes in its chemistry.

where A is the Helmholtz free energy of the system and a is area [1]. Consider the process of "cleaving" a bulk phase, such as liquid water, along a plane (Figure 2.2). Initially, water molecules in the bulk phase experience a reasonably uniform force field due to nearest-neighbor interactions. When the bulk phase is cleaved, this force field is no longer uniform for molecules along the plane of separation. As the separation increases, molecular interactions across the plane decrease until the new "surfaces" are sufficiently far apart that they do not interact at all, and a new equilibrium is achieved where molecules along the plane of separation experience fewer nearest-neighbor interactions. The removal of such interactions by the formation of a new surface thus results in an increase in the free energy of molecules along the plane of separation. For the system as a whole, that increase is proportional to the area of the new surface formed [2,3]. If the area of each new surface formed is a, the total additional energy (J) in (reversibly) proceeding from the initial to final states is then $\gamma \times 2a$.

This leads to the definition of two other important terms [2,3]. The work of cohesion, W_c (J/m^2), is the reversible work per unit area required to separate (i.e., "create") two surfaces of a bulk material. With reference to the illustration below (Figure 2.2), and in general, $W_c = 2\gamma$. Similarly, the work of adhesion, W_a (J/m^2), which appears in several contexts elsewhere in this chapter, is the reversible work required to separate a unit area of interface between two dissimilar phases 1 and 2, or

$$W_a = \gamma_1 + \gamma_2 - \gamma_{12} \tag{2.2}$$

These concepts can also be defined schematically, according to the "reactions" shown in Figure 2.2. Note that like W_a, the value of W_c can also be determined using the right side of Equation 2.2 as well; in this case phases 1 and 2 are identical, and thus $\gamma_{12} = 0$.

The higher energy state of surface molecules subjects them to an inward attraction normal to the surface. The surface is thus in a state of lateral tension, and in the context of liquids this gave rise to the concept of "surface tension" (N/m, dimensionally equivalent

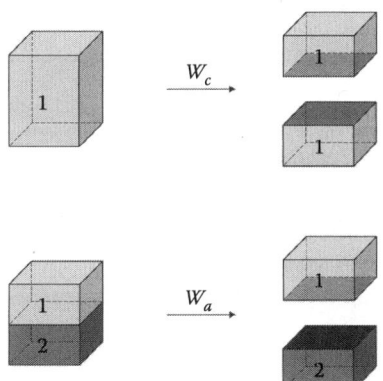

Figure 2.2 Schematic illustration of the work of cohesion, W_c, and the work of adhesion, W_a, associated with formation of a new surface (darker) by cleavage of a single, homogeneous phase (top), or the separation of two dissimilar phases (bottom).

to J/m^2). For a pure liquid in equilibrium with its vapor, surface tension and excess interfacial free energy are numerically equal. For interfaces involving solid materials, however, surface tension is often defined with reference to the nonequilibrium structure of such surfaces, and the term is thus not interchangeable with surface energy. The details of this are provided in several texts on surface chemistry and physics [1,2,4]. In any event, the treatment here will adhere to properties and predictive relationships relevant to the surface energy of a material as defined in Equation 2.1, and in this way, findings and conclusions will remain applicable to any interface.

Evaluation of surface energetics in a system is important. For synthetic processing and packaging materials, as will be shown in Section 2.2.3, such evaluation makes determination of surface hydrophobicity possible and allows prediction of the extent of adsorption and adhesion of macromolecules and cells to expect with use of such materials. Also, food foams and emulsions are stabilized by adsorption of components such as small-molecule surfactants and proteins at the relevant phase boundaries (e.g., air–water in the case of some foams, oil–water in the case of some emulsions). It is the minimization of surface energy that drives adsorption in such systems. The ability to understand and control interfacial energies and/or adsorption processes constitutes the foundation of modern technology related to colloidal stability, emulsification, foaming, adhesion, and many other areas. In the following section we describe the quantitative relationship between adsorption and surface energy.

2.2.2 Gibbs Adsorption Equation

Several models have been developed to predict the equilibrium-adsorbed amount of some substance at some interface as a function of the amount of nonadsorbed substance present. The Gibbs adsorption equation is the thermodynamic expression which relates the adsorbed amount (or surface excess concentration) of a species to γ and the bulk activity or fugacity of that species. It is widely applied to study of adsorption phenomena, in and outside of food science and technology, especially at the air–water interface. The Gibbs adsorption equation is written as

$$-d\gamma = \sum_i \Gamma_i d\mu_i \qquad (2.3)$$

where Γ_i (mol/m²) is the excess surface concentration of component i, and μ_i (J/mol) is its chemical potential [1]. A more useful form of Equation 2.3 can be developed as follows. Consider adsorption of a component i, dissolved in a liquid, α, at the liquid–vapor interface. For this system, Equation 2.3 becomes $-d\gamma = \Gamma_\alpha d\mu_\alpha + \Gamma_i d\mu_i$. If the three-dimensional interfacial region is defined such that the excess concentration of liquid α contained within it is zero [1–4], then $-d\gamma = \Gamma_i d\mu_i$. The chemical potential, μ_i, is equal to $\mu_i = \mu_i^0 + RT\ln(a_i)$, where μ_i^0 is the standard chemical potential of component i in solution, R is the gas constant, T is temperature, and a_i is the activity of component i in solution. Thus, $d\mu_i = RTd\ln(a_i)$, and Equation 2.3 becomes

$$\Gamma_i = -\left(\frac{1}{RT}\right)\frac{d\gamma}{d\ln(a_i)} \qquad (2.4)$$

For the majority of systems characterized by strongly adsorbed monomolecular films, the surface excess concentration and the total surface concentration of the adsorbate are more or less identical. Adsorption involves a profound reduction in γ, even for dilute solutions, and in that case a_i can be approximated by c_i, the concentration of species i in solution. Equation 2.4 can then be written in its simplest and most used form,

$$\Gamma_i = -\left(\frac{1}{RT}\right)\frac{d\gamma}{d\ln(c_i)} = -\left(\frac{c_i}{RT}\right)\frac{d\gamma}{dc_i} \quad (2.5)$$

Thus adsorption reduces γ, and a plot of this surface tension reduction vs. the equilibrium solution concentration of a given adsorbate allows for the determination of the adsorption isotherm: an expression for Γ as a function of solution concentration.

However, proper application of the Gibbs adsorption equation requires an understanding of its origin. For example, equilibrium "spreading pressure" data have been recorded as a function of solution concentration for a wide variety of proteins, surfactants, and other molecules used as stabilizers in food foams and emulsions. The spreading pressure, denoted by Π (J/m^2), is simply a positive measure of surface energy reduction at a given solution concentration; that is, Π is the difference between γ evaluated when $\Gamma = 0$, and γ evaluated at a selected solution concentration (i.e., $\Pi = \gamma|_{c_i} - \gamma|_{c_i=0}$). The concentration dependence of Π is shown in Table 2.1 for each of two milk proteins at the air–water interface: α-lactalbumin and bovine serum albumin. These data were recorded at 25°C in each case, from 0.010 M sodium phosphate buffer, pH 7.0 [5,6]. One could use these data to derive an expression for the isotherm, $\Gamma = f(c_i)$, according to Equation 2.5, where $-d\gamma = d\Pi$, and compare the isotherm to independent experimental determinations of adsorbed amounts for each of these proteins. Table 2.2 lists the results of such an exercise, providing a comparison of adsorbed amounts of α-lactalbumin and BSA recorded by direct

Table 2.1 Concentration Dependence of Equilibrium Spreading Pressure for α-Lactalbumin and Bovine Serum Albumin at the Air–Water Interface

Protein Concentration, C (mg/mL)	Equilibrium Spreading Pressure, Π (mJ/m^2)	
	α-Lac	BSA
0.05	23.9	11.8
0.10	26.0	13.0
0.30	27.6	14.0
0.60	28.4	14.9
1.00	29.4	15.4
1.50	30.1	15.5
2.00	30.5	15.9
3.00	31.0	16.1
10.0	32.6	16.4
20.0	34.1	17.1

Table 2.2 Adsorbed Amounts of α-Lactalbumin and BSA, Determined Experimentally at a Model Hydrophobic Interface and Calculated with Application of the Gibbs Adsorption Equation

Protein Concentration, C (mg/mL)	Adsorbed Amount, Γ (nmol/m^2)			
	Experimental Determination		Gibbs Adsorption Equation	
	α-Lac	BSA	α-Lac	BSA
0.10	114	35	569	305
0.30	115	60	604	325
0.60	221	81	628	338
1.00	205	75	646	348
1.50	236	78	660	357
2.00	218	74	671	363

surface spectroscopy at a hydrophobic, silanized silica–water interface (fundamentally similar to the model hydrophobic air–water interface), and calculated with Table 2.1 and Equation 2.5 as a function of equilibrium concentration. The comparison reveals important issues surrounding the utility of the Gibbs adsorption equation in this context. In particular, although it did provide estimates within an order of magnitude of actual values, the Gibbs adsorption equation clearly overpredicted the actual adsorbed amounts in each case. Also, the plateau value of adsorbed mass for α-lactalbumin is about twice that of BSA, consistent with the respective plateau values of Π, but not consistent with experimental values of adsorbed mass, where $\Gamma_{plateau,\alpha-lac}$ is several times greater than $\Gamma_{plateau,BSA}$. The Gibbs adsorption equation treats every incremental change in Π as an increase in Γ. This feature serves to limit its application in analysis of protein adsorption, where one protein molecule can bind to the surface by multiple noncovalent contacts [7,8]. In this way, through unfolding and formation of new contacts with the apolar interface, γ may continue to decrease even after adsorption has ended and Γ is constant. However, the Gibbs adsorption equation can be a useful starting point for analysis of small-molecule surfactant adsorption.

The study of adsorption from solution onto solids is important and wide-ranging, and in many cases the most relevant to practical applications. The solid–liquid interface is certainly the least understood of the four major interfaces, however, due to its complexity, and this contributes to it being the most actively studied interface today. Solid surfaces have different electrical and optical properties than found in the bulk, and can be characterized by atomic- or molecular-level textures and roughnesses. They are generally energetically heterogeneous. For example, although a solid surface may be assigned a particular "wettability," it would most likely be the result of a distribution of heterogeneous surface regions of varying wettability [9].

Measurement of γ at liquid–liquid and liquid–gas interfaces is straightforward, and is briefly summarized in Section 2.3. Evaluation of solid surface energetics is considerably less straightforward. In the following section, we discuss evaluation of properties relevant to γ for solid surfaces, through analysis of a very widely used macroscopic thermodynamic approach.

2.2.3 Contact Angle

The importance of hydrophobic–hydrophilic balance at interfaces has prompted numerous investigators to develop techniques to measure this property at solid surfaces. Contact angle methods have been prominent in this regard [10]. Contact angle analysis is inexpensive, rapid, and fairly sensitive. However, contact angle data can be difficult to interpret, and the technique is subject to artifacts due to macroscopic, energetic heterogeneities in the surface; hysteresis; and drop-volume effects among others. Still, useful conclusions regarding biological interactions with surfaces have been based on the results of contact angle analysis in numerous areas relevant to food technology [11].

When a liquid phase contacts both a second fluid phase and a solid surface there occurs a net, characteristic orientation of the liquid–fluid interface with respect to the solid surface. This orientation is reflected in the so-called contact angle [1,2,4]. Under controlled equilibrium conditions, the contact angle can be considered to be an intensive property, dependent only on the natures of the three component phases and independent of the geometry and quantities present. Placing a drop of liquid on a solid surface is a convenient way to create this kind of three-phase system. The contact angle, θ, is identified in Figure 2.3 for this solid–liquid case. Theoretically, the contact angle identified in this way would be the same characteristic angle defining a meeting of the same three phases in any geometry.

All of the interfacial "tensions," γ, shown in Figure 2.3, are acting to minimize the overall interfacial energy (i.e., reduce interfacial area), such that

$$\gamma_S = \gamma_{SL} + \gamma_L \cos\theta \qquad (2.6)$$

Equation 2.6 is Young's equation, essentially a force balance on the drop at rest. An energy balance defines the work of adhesion between the solid and liquid (also see Figure 2.2 and Equation 2.2), such that

$$W_a = \gamma_S + \gamma_L - \gamma_{SL} \qquad (2.7)$$

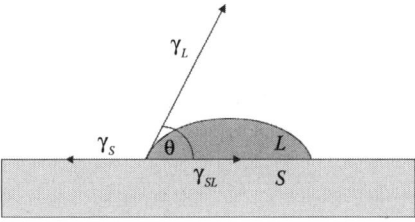

Figure 2.3 The contact angle, θ, between a liquid (L) and a solid (S). Contributions of the tension forces, γ, between the phases are indicated by arrows.

Equation 2.7 is sometimes called the Dupré equation. Young and Dupré equations are not very useful by themselves, as γ_S and γ_{SL} are difficult, if not practically impossible, to measure. They can be combined, however, to yield a very useful equation for calculating the work of adhesion,

$$W_a = \gamma_L(1 + \cos\theta) \tag{2.8}$$

Equations 2.6 through 2.8 are valid only at equilibrium: the liquid should be saturated with the solid, the vapor and solid must be at adsorption equilibrium, and the solid surface must be energetically homogeneous and smooth (effects of surface roughness are discussed below). Although that is not always the case, and θ for a given three-phase system, in practice, can depend somewhat on judgment, Equations 2.6 through 2.8 are fully functional, and form the basis for evaluation of surface energetics. We summarize three different but widely used approaches to this end in the next section.

2.2.3.1 Critical Surface Tension

The most common interpretation of contact angle data to gain a measure of surface energy is completely empirical. Given an uncharacterized solid surface one determines θ for each of a series of homologous liquids contacted with the surface. The cosine of each angle is plotted against the surface tension of the corresponding liquid. The result, a (typically) rectangular band of data, is called a Zisman plot [11]. The intercept of a line drawn through the data at the $\cos\theta = 1$ axis is termed the critical surface tension of the solid, γ_c.

Figure 2.4 shows a Zisman plot constructed for an acetal surface, and one constructed for a polyethylene surface [11]. The data indicate that $\gamma_{c,acetal} = 23.7$ and $\gamma_{c,polyethylene} = 21.8$ mN/m. That implies that a liquid with $\gamma_L = 23.7$ mN/m would completely spread on acetal, whereas a liquid of $\gamma_L > 23.7$ mN/m would yield a nonzero value of θ. In general, for any combination of solid and liquid, the higher the value of θ, the higher the interfacial energy between them (γ_{SL}). Put another way, the lower the value of γ_c, the higher the interfacial energy between that surface and water.

Although data derived in this manner are widely considered to be a function of the solid surface alone, and therefore related to the "true" surface energy of the solid, a Zisman plot can yield highly misleading results [12]. In particular, it can be difficult to determine the best value of the slope and therefore, $\cos\theta = 1$ intercept, which represent the material surface properties. A major advantage of the method is, however, that it is very simple.

2.2.3.2 Polar and Dispersive Contributions to Surface Energy

Approaches to interpretation of contact angle data were much improved by Fowkes [13]. With a focus on liquid–liquid and liquid–vapor interfaces, he proposed resolving liquid surface tension into two components:

$$\gamma_L = \gamma_L^d + \gamma_L^p \tag{2.9}$$

where γ_L^d refers to surface tension contributions arising only from the London–van der Waals dispersion forces, and γ_L^p refers to those contributions arising from electrostatic forces, dipole–dipole, dipole–induced dipole, so-called donor–acceptor interactions, and

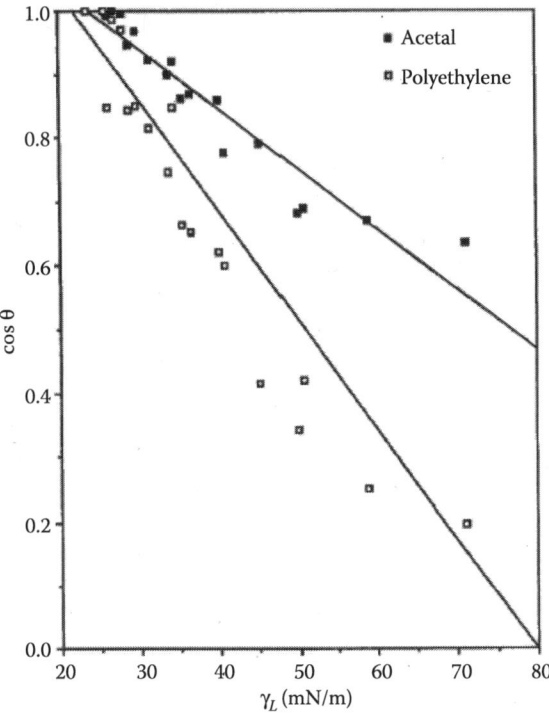

Figure 2.4 A Zisman plot constructed with contact angle data recorded for an acetal and a polyethylene surface. (Reproduced with permission from J McGuire, V Krisdhasima. In H Schwartzberg, R Hartel, Eds. *Physical Chemistry of Foods*. New York: Marcel Dekker, 1992, pp. 223–262.)

so on. Superscripts "*d*" and "*p*" stand for dispersive and polar contributions, respectively. The following development is based on two truisms:

- At the interface between any two liquids where the intermolecular attraction of any one of them is entirely due to London dispersion forces, the only appreciable interfacial interactions present will be due to London dispersion forces.
- If we think of the interface as comprising two interfacial regions, the interfacial energy is the sum of the energies in each region.

When the intermolecular attraction of any one of two phases is entirely due to dispersion forces, it can be shown [13] that the geometric mean of the dispersion force attractions is an adequate representation of the magnitude of the interaction between the two phases. That is, when liquid 1 contacts liquid 2, the energy in "interfacial region 1" is $\gamma_1 - \sqrt{\gamma_1^2 \gamma_2^2}$ and the energy in "interfacial region 2" is $\gamma_2 - \sqrt{\gamma_1^2 \gamma_2^2}$. Thus, the interfacial energy where the two phases 1 and 2 meet is

$$\gamma_{12} = \gamma_1 + \gamma_2 - 2\sqrt{\gamma_1^2 \gamma_2^2} \tag{2.10}$$

Note that with reference to Equation 2.2, $W_a = 2\sqrt{\gamma_1^2 \gamma_2^2}$. Since only dispersive interaction is possible in the present case, we can write

$$W_a^d = 2\sqrt{\gamma_1^2 \gamma_2^2} \tag{2.11}$$

where W_a^d is the dispersive component of the total work of adhesion, W_a. Of course, if only dispersive forces interact at the interface, then $W_a = W_a^d$. In any event, the same rationale for splitting γ into γ^d and γ^p components allows us to write $W_a = W_a^d + W_a^p$.

Kaelble [14] applied these concepts to solid–liquid interfaces, stating that $\gamma_S = \gamma_S^p + \gamma_S^d$. That is appropriate, but the analogy was made complete by stating, as above, that the geometric mean of the polar force attractions is an adequate representation of the magnitude of the total polar interaction between two phases, leading to $W_a^p = 2\sqrt{\gamma_1^p \gamma_2^p}$. While mathematically convenient, there is no theoretical justification for this and in fact, although two surfaces may be characterized as "polar," there may well be no polar attraction between them [15]. Fowkes [15] argued that all "polar" interactions are of the "acid–base" type. "Acids," here, are electron acceptors, while "bases" are electron donors (i.e., Lewis acids and bases). A pure liquid exhibiting only acidic or basic character, while certainly polar, for example, will be miscible with an organic solvent. In summary, two polar bodies interact only when one is acidic and the other is basic. It is worth noting that researchers never try to estimate W_a^p for two liquids in contact according to $2\sqrt{\gamma_{L_1}^p \gamma_{L_2}^p}$ but literally hundreds of researchers have estimated $W_a^p = 2\sqrt{\gamma_S^p \gamma_L^p}$ for liquid–solid contact (15).

In this regard, Equation 2.9 is best written as $\gamma_L = \gamma_L^{ab} + \gamma_L^d$, and it is also fair to write $\gamma_S = \gamma_S^{ab} + \gamma_S^d$. For a solid–liquid contact system, Equation 2.3 can be expanded with Equation 2.11, such that

$$W_a = \gamma_L (1 + \cos\theta) = W_a^d + W_a^{ab} = 2\sqrt{\gamma_L^d \gamma_S^d} + W_a^{ab} \tag{2.12}$$

Just as the dispersive contribution to W_a can be written as a function of γ_L^d and γ_S^d, the acid–base contribution to W_a (i.e., W_a^{ab}) is a function of γ_L^{ab} and γ_S^{ab}. But in the case of W_a^{ab}, the functional relationship is not known, and calculation of γ_S (and therefore γ_S^{ab}) is not possible using contact angle methods. Nevertheless, W_a^{ab} is a very useful property, in and of itself. We know that in addition to capacity to take part in dispersive attractions, solid surfaces can take part in acid–base attractions; they can be acidic, basic, or harbor both types of sites, and W_a^{ab} for a given solid surface necessarily depends on acid–base properties of the liquid contacting it. With reference to food science and technology, we would be most concerned with processes where solid surfaces are contacting fluids with some biological relevance, that is, aqueous solutions and suspensions. Water acts as an acid as well as a base, and often, surface hydrophobicity is of most interest. The acid–base component of the work required to remove water from a surface, $W_{a,water}^{ab}$, is clearly related to its hydrophobicity: a high work of adhesion would correspond to a hydrophilic surface, while a low value would correspond to a hydrophobic surface. As summarized in Section

2.3, measurement of θ, γ_L, γ_L^d, and γ_S^d is straightforward. Thus W_a^{ab} is readily determined, according to Equation 2.12 when rearranged as follows:

$$W_a^{ab} = \gamma_L(1 + \cos\theta) - 2\sqrt{\gamma_L^d \gamma_S^d} \tag{2.13}$$

Measurement of θ for a drop of water on the surface of a material of interest yields $W_{a,water}^{ab}$ directly. Alternatively, $W_{a,water}^{ab}$ can be evaluated using a series of test liquids. In this case, W_a^{ab} of the test liquid is calculated according to Equation 2.13 for each solid–liquid contact, and then plotted against $\gamma_L - \gamma_L^d$ (which equals γ_L^{ab}) of each corresponding liquid. A reasonably straight line is usually observed, with a better coefficient of determination than found in the Zisman plot [16]. From the straight line fit, one can calculate surface hydrophobicity according to

$$W_{a,water}^{ab} = k(\gamma_{L,water}^{ab}) + b \tag{2.14}$$

where k is the slope of the line, and b its ordinate intercept.

In other cases, it may be a material's capacity for acid–base attraction that is of interest, rather than its hydrophobicity. This capacity is evaluated by contacting the surface with liquids of only basic character, and then with liquids of only acidic character [15]. W_a^{ab} is calculated for each solid–liquid contact, to determine the dominant character of the surface. This information provides direction for selecting the kinds of polymers to place adjacent to another in layered packaging materials, the kinds of adhesives to use in a given situation, and so on. Unfortunately, only very few such "monopolar" diagnostic liquids exist. But for solid–fluid contact in food processing, Equation 2.13 can be quite useful, since proteins and other "solutes" in fluid foods can take part in acid–base interactions with acidic and basic sites, as can water. This analysis can be used to explain some observations, in spite of the fact that the relative numbers of acidic and basic sites on a solid are often unknown.

To this point, we have discussed ways to measure properties relevant to solid surface energy (e.g., γ_c, $W_{a,water}^{ab}$), without seriously exploring calculation of γ_S directly. Below we discuss the existence of an equation of state relationship among solid–liquid, solid–vapor, and liquid–vapor interfacial energies, allowing direct calculation of γ_{SL} and γ_S.

2.2.3.3 An Equation of State Relationship among Interfacial Energies

Ward and Neumann [17] stated that an equation of state relationship must exist among γ_{SL}, γ_S, and γ_L in a two-component, three-phase system, such as that illustrated in Figure 2.3. That is,

$$\gamma_{SL} = f(\gamma_S, \gamma_L) \tag{2.15}$$

If the functional relationship of Equation 2.15 were known, its combination with Young's equation (Equation 2.6) would yield two equations and the two unknowns γ_{SL} and γ_S (since θ and γ_L are readily measurable). Neumann et al. [18], using contact angle data recorded on low-energy surfaces, obtained an explicit empirical formulation of Equation 2.15:

$$\gamma_{SL} = \frac{(\sqrt{\gamma_S} - \sqrt{\gamma_L})^2}{1 - 0.015\sqrt{\gamma_S \gamma_L}} \tag{2.16}$$

They then combined Equation 2.16 with Young's Equation 2.6, to arrive at

$$\cos\theta = \frac{(0.015\gamma_S - 2)\sqrt{\gamma_S\gamma_L} + \gamma_L}{\gamma_L(0.015\sqrt{\gamma_S\gamma_L} - 1)} \tag{2.17}$$

With Equation 2.17, one can measure θ for any liquid of known γ_L, and thus determine γ_S directly. However, it is very important to note that Equation 15.17 was developed using very low-energy solids, for which $\gamma_S \ll \gamma_{water}$. Such solids would be most accurately characterized as "hydrophobic," with values of $W_{a,water}^{ab}$ close to zero.

2.2.4 Effects of Adsorbed Layer Composition and Structure on Interfacial Energy

Relative to small-molecule stabilizers and surfactants used in food systems, in a quantitative sense we know very little about how the molecular properties of a complex food polymer such as a globular protein influence its adsorption and eventual interfacial function. Proteins constitute one of two classes of surface-active agents used in formulating and stabilizing industrial and colloidal food systems. Phospholipids (e.g., lecithins) and other nonprotein macromolecules are used as well, but only a few polysaccharides, such as modified cellulose derivatives or acetylated pectin, are considered sufficiently surface active for practical purposes [19]. Interfacial behavior is a cumulative property of a protein, influenced by many factors. Among these are its size, shape, charge, and structural stability. Experimentally observed differences in interfacial behavior among different protein molecules have been particularly difficult to quantify in terms of these because proteins can vary substantially from one another in each category. Still, many experimental observations have been explained in terms of protein size, shape, charge, and tendency to unfold, as well as hydrophobicity of the interface itself [9]. Important findings in this regard are summarized in Table 2.3. The relevant interfacial dynamics contributing to changes in interfacial energy are strongly concentration dependent as well.

It is well accepted that a given protein can exist in multiple adsorbed conformational "states" at an interface [7–9]. These states can be distinguished by differences in occupied area, binding strength, propensity to undergo exchange events with other proteins, and catalytic activity, or function. All of these features of adsorbed protein are interrelated, and can be time dependent. For example, decreases in surfactant-mediated elution of proteins from an adsorbed layer (an indirect measure of binding strength) are observed as protein–surface contact time increases [20]. This time dependence is illustrated in Figure 2.5. As conformational change proceeds, the likelihood of desorption decreases. Another important observation is that protein adsorption is often a practically irreversible process, at least in the sense that it is often irreversible to dilution, or buffer elution. The adsorbed mass remains constant or decreases very little when the solution in contact with the interface is depleted of protein. However, although spontaneous desorption is not generally observed, adsorbed protein can undergo exchange reactions with similar or dissimilar protein molecules, surfactants, or other sufficiently surface-active species adsorbing from solution [7–9]. Such exchange reactions are shown schematically in Figure 2.6. Adsorbed protein exchange rates are likely state dependent, being slower for more conformationally altered proteins.

SURFACE PROPERTIES

Table 2.3 Factors that Affect Protein Adsorption

Factor	Mechanism of Action	Ref.
Size	Small molecules have higher diffusivities than large molecules, but tighter binding is consistent with multiple noncovalent contacts	[20]
Flexibility	An ability to bind to a relatively small unoccupied area is a requirement for adsorption at a crowded interface	[21]
Structural stability	Protein unfolding exposes hydrophobic residues which can associate with the surface	[7,8,22]
Hydrophobicity	Hydrophobic association is consistent with high binding strength in aqueous media	[20,23]
Electrostatic charge	When electrostatic interactions predominate, the attractive force between a protein and an oppositely charged surface can be strong	[24–26]
	Low ionic strength: protein and surface charge contrast determines the electrostatic interaction	
	High ionic strength: reduces electrostatic effects between protein and surface	

With reference to oil-in-water emulsions, qualitatively, the higher the "emulsifying activity" of a given protein or other stabilizer, the greater the amount of oil droplets "solubilized" during emulsion formation. After homogenization, dispersed oil droplets will rise and coalesce to form a floating layer. "Emulsion stability" is commonly measured in terms of the amount of oil separating from an emulsion during a certain period of time under certain conditions, reflecting the rate of this rise and coalescence. These properties can be measured by turbidimetric techniques or changes in suspension conductivity, for example [19]. In any event, higher emulsion activities and emulsion stabilities are achieved by molecules better able to reduce interfacial energy, as generally suggested by the guidelines provided in Table 2.3.

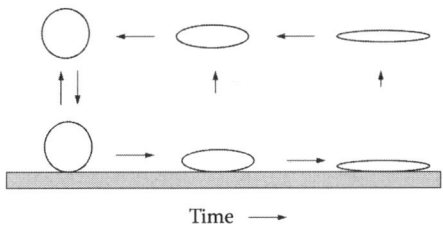

Figure 2.5 Surface-induced conformational changes undergone by adsorbed protein, resulting in multiple noncovalent bonds with the surface, and coverage of greater interfacial area/molecule. (Reproduced with permission from J McGuire, CK Bower, MK Bothwell. In A Hubbard, Ed. *Encyclopedia of Surface and Colloid Science*. New York: Marcel Dekker, 2002, pp. 4382–4395.)

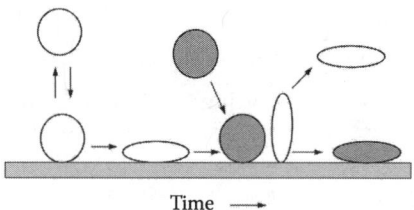

Figure 2.6 Exchange reaction between a conformationally altered, adsorbed protein (white) and a dissimilar protein (gray) adsorbing from solution. (Reproduced with permission from J McGuire, CK Bower, MK Bothwell. In A Hubbard, Ed. *Encyclopedia of Surface and Colloid Science*. New York: Marcel Dekker, 2002, pp. 4382–4395.)

The molecular-level dynamics contributing to changes in interfacial energy with mixed protein–surfactant systems are strongly concentration-dependent, and difficult to predict. These will be very briefly summarized by considering the change in interfacial energy at the air–water interface that might be produced by increasing surfactant concentration in a protein–surfactant mixture [27]. In general, at very low surfactant concentrations, the interfacial energy is the same as it would be for pure protein. Interfacial energy then decreases, due to surfactant occupation of "empty sites" at the interface, as well as formation of surface active, surfactant–protein complexes. At higher surfactant concentrations, interfacial energy is seen to "plateau," presumably because it is energetically favorable for surfactant to bind to protein at these concentrations. In this high concentration range, the critical micelle concentration (*cmc*) for pure surfactant preparation may be exceeded. Interfacial energy then decreases again with increasing surfactant concentration, a result of complete displacement of protein from the interface by surfactant. After this decrease, further increases in surfactant concentration have no effect on interfacial energy, indicating that the *cmc* of the protein–surfactant mixture has been met.

For pure liquids, surface energy decreases with increasing temperature and becomes zero at the critical point [28]. The surface tension of liquid water, for example, decreases from 75.6 to 58.9 mN/m as temperature is increased from 0 to 100°C, in a fairly linear manner [29]. Very little literature exists on the temperature dependence of properties related to solid surface energy. Dispersive forces exist in all types of matter and always give an attractive force between adjacent atoms or atomic groups, no matter how dissimilar their chemical natures may be. The forces depend on electrical properties of the volume elements involved and the distance between them, and are independent of temperature [13]. Thus, while γ_L^d and γ_S^d are expected to be temperature independent, $W_{a,water}^{ab}$ would be expected to decrease with increasing temperature. Temperature effects on surface energetics of some solid materials which are commonly encountered in food processing have been reported by McGuire et al. [30].

For liquid foods, protein solutions, mixed protein–surfactant systems, and so on, the effect of temperature on surface properties is not predictable in any quantitative sense, and must usually be measured experimentally. In such cases, the solution chemistry and time- and concentration-dependent denaturation and aggregation phenomena near the interface will affect the observed interfacial energetics.

2.3 MEASUREMENT TECHNIQUES

In this section, techniques to measure θ, γ_L, γ_L^d, and γ_S^d are briefly summarized. Properties relevant to solid surface energy, for example, $W_{a,water}^{ab}$, and γ_S (according to the equation of state approach), are readily determined from these properties, as described in Section 2.2.3.2.

2.3.1 Evaluation of the Contact Angle

The contact angle, θ, is routinely measured using a contact angle goniometer. This instrument consists of a light source which illuminates a stage on which the liquid drop/solid material three-phase system rests (see Figure 2.3). An image of the drop is acquired using a still or video camera, and computer software analyzes the drop shapes to give accurate contact angle data with minimal operator intervention (constituting a significant improvement over older manual instrumentation). In particular, prealigned, fixed optical systems are used, lighting is preset, and fluid is dispensed from an industry-standard syringe using a mechanical ratchet to provide predetermined drop volumes. These instruments can also accommodate relatively large sample sizes and provide precise sample positioning. Furthermore, the time-dependent behavior of the drop, advancing and receding contact angles, and other data can be automatically derived by some systems.

2.3.2 Evaluation of Liquid Surface Tension

Liquid surface energy (or surface tension), γ_L, can be readily measurable by various methods. Its measurement is simple, and several common methods have been thoroughly described in most surface chemistry texts [1,2,4]. Three of the most common methods are briefly summarized here.

The capillary rise method is generally considered to be the most accurate method available, but its scope is limited in the sense that it is best suited to the measurement of the surface tension of pure liquids [1]. Other methods are preferable for mixtures, even for dilute solutions. For the rise of a liquid in a capillary tube, it can be shown that

$$\gamma_L = r_c \frac{h \Delta \rho g}{2 \cos \theta} \quad (2.18)$$

where r_c is the radius of the capillary, h is the height of the liquid rise, $\Delta \rho$ is the difference in density between the liquid and the medium in contact with it (theoretically, its vapor), and g is the gravitational constant. θ is the measured contact angle between the liquid and the wall of the capillary tube. For practical reasons, if this method is to be accurate, θ must be close to zero. This is because nonzero contact angles within the capillary are generally very difficult to measure and reproducibility is poor.

The pendant drop method involves the determination of the volume (or more accurately, the weight) of a drop of liquid surrounded by its vapor, hanging from a wetted capillary of known radius. Ideally, for a drop of volume V_d which just becomes detached from a capillary of radius r_c, a force balance shows

$$\gamma_L = V_d \frac{\Delta \rho g}{2\pi r_c} \qquad (2.19)$$

Equation 2.19 represents a rather crude approximation, however, mainly because in practice an entire drop never falls completely from the tip. A correction factor, which takes the form of a proportionality constant, is often incorporated into Equation 2.19 to account for this effect. This correction factor is typically some function of $r_c/\sqrt[3]{V_d}$ (1).

Two methods are of particular utility in food science, where solutions containing surfactants, proteins, or any of a variety of other surface-active components relevant to formation and stabilization of foams and emulsions are of interest. These are the Wilhelmy plate method and Du Nouy ring tensiometry. The principle of the Wilhelmy plate method is illustrated schematically in Figure 2.7a. With this method, a thin plate of (typically) platinum or any solid, for which θ between its surface and the liquid in question is zero, is suspended from one arm of a balance or other force detection device and partially immersed in the liquid of interest. The liquid is lowered until the plate becomes detached from the surface and the maximum "pull," W_t, on the balance is recorded. The difference between this pull and the weight, W_p, of the plate in air is thus the weight of the liquid in the meniscus. If θ = 0, then

$$\gamma_L = \frac{(W_t - W_p)}{P} \qquad (2.20)$$

where P is the perimeter of the plate. This method is often applied to insoluble monolayers, in which case detachment of the plate from the interface is not desired. Instead, the force required to keep the plate at a constant depth of immersion is recorded, as surface tension is altered. Similar to the Wilhelmy plate method is Du Nouy ring tensiometry, which is widely accepted as giving satisfactory results for colloidal suspensions. In this case, the horizontally oriented, platinum–iridium ring is dipped into the liquid and then raised

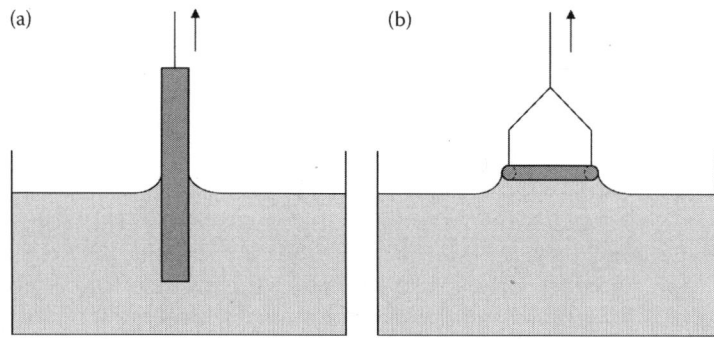

Figure 2.7 Schematic illustration of the principle underlying the (a) Wilhelmy plate and (b) Du Nouy ring techniques for measuring liquid surface tension.

until a liquid collar, which rises with the ring, collapses. The net force required to lift the ring is measured and is equal to the downward pull resulting from the liquid surface tension. Figure 2.7b shows a cross-sectional schematic of a submerged ring with the liquid collar. Correction factors are typically introduced to Wilhelmy plate and Du Nouy ring estimations of surface tension, to account for the hydrostatic weight of liquid underneath the plate or ring at the time of detachment. Several reliable, computer-controlled automatic tensiometers are available for the measurement of interfacial tensions by the corrected Du Nouy ring and Wilhelmy plate methods. With automated instruments, the time-dependent behavior of surface tension kinetics accompanying adsorption or desorption can also be measured [31,32].

2.3.3 Evaluation of γ_L^d and γ_S^d

In order to estimate γ_L^d for a given liquid, we recall that for any solid–liquid contact, if the surface energy of either the solid or the liquid has only a dispersive contribution, then the interaction between the two can be entirely attributed to only dispersive forces. Therefore, for a drop of liquid formed on a solid surface (for which only dispersive forces are at play at the interface), the general expression of Equation 2.12 becomes

$$W_a = \gamma_L(1 + \cos\theta) = W_a^d = 2\sqrt{\gamma_L^d \gamma_S^d} \tag{2.21}$$

Equation 2.21 can be rearranged to

$$\gamma_L^d = \gamma_L^2 \frac{(1 + \cos\theta)^2}{4\gamma_S^d} \tag{2.22}$$

Equation 2.22 indicates that γ_L^d could be found through a single contact angle measurement, if γ_S^d were known. Since water and aqueous or otherwise "polar" solutions are often used in contact angle analysis, Equation 2.22 would be best applied at a solid surface for which $\gamma_S = \gamma_S^d$. Paraffin wax is one example of such a solid (16). However, to calculate the value of γ_S^d for any solid, γ_L^d must be known. Fortunately, there are various liquids (e.g., hydrocarbons) for which $\gamma_L^d = \gamma_L$, and γ_L (and hence γ_L^d) can be readily measured as described previously.

A different rearrangement of Equation 2.21 yields

$$\cos\theta = \frac{2\sqrt{\gamma_L^d \gamma_S^d}}{\gamma_L} - 1 \tag{2.23}$$

Hence, by choosing any of a series of completely nonpolar liquids and any solid surface, a plot of $\cos\theta$ vs. $\sqrt{\gamma_L^d}/\gamma_L$ (or, equivalently, $1/\sqrt{\gamma_L}$) can be constructed. This plot should be a straight line, with slope $2\gamma_S^d$, and intercepting the ordinate at $\cos\theta = -1$. Using a diagnostic solid, with which only dispersive interactions are possible at the interface, one can analyze liquids of both polar and dispersive character, and calculate γ_L^d for each according to Equation 2.22.

A similar method can be used to measure the effective contact angle of solid particles, by measuring the wicking behavior of different probe liquids through a thin layer of the particles. This technique can also be used on porous particles (such as are often found in foods) by precontacting with solvent vapor and application of a correction factor [33].

A number of references and reviews [1–4,34] discuss surface forces and techniques for their measurement in detail.

2.3.4 Polymer Surfaces

In practice, water and alkanes and other organic liquids are used for the generation of Zisman plots. In the case of polymers, the solid surface structure will often be energetically and conformationally different when in contact with water than when it is in contact with an apolar organic solvent or other hydrophobic interface (e.g., air). Moreover, liquids composed of relatively small molecules, such as water or low molecular weight solvents, may penetrate into some polymer materials and cause swelling or changes of other properties (e.g., modulus or dielectric constant).

Rearrangements of polymer surface structure are often observed, especially in response to different handling or storage conditions, and this can lead to confounding results. These effects are particularly well known for polyurethane block copolymers, in which the "hard" and "soft" segments will segregate to form microdomains of hydrophilic or hydrophobic character [35,36]. The size and surface distribution of these domains depends on many factors, including the exact polymer blend used in the bulk material, and the environment to which the surface is exposed prior to and during analysis. Segregated domains strongly influence properties such as contact angle and adsorption behavior at the surface [37]. Rearrangements of surface polymer chains in response to adsorption of proteins have also been reported [38].

Polydimethylsiloxane (PDMS, silicone) chains have an extremely low glass transition temperatures (–127°C) [39], and are hence extremely mobile under ambient conditions. Oxidation or other chemical modification of the PDMS surface can be used to change surface properties such as wettability, but the effect is typically short lived and time dependent due to rapid subduction of the more hydrophilic oxidized polymers into the bulk, leading to a "hydrophobic recovery" of the native surface [40,41]. Great care must thus be exercised in measuring and reporting surface properties of polymers which may undergo molecular rearrangements.

2.4 SURFACE PROPERTY DATA

Values of surface tension for a variety of pure liquids and some simple liquid mixtures are available, in handbooks of chemistry and physics. However, properties relevant to solid surface energy are not generally tabulated. In any event, for solids and all but pure liquids, properties relevant to surface energy must be measured experimentally and not simply secured from a published table, unless only a crude approximation is desired.

Table 2.4 lists properties relevant to solid surface energy (γ_S^d and $W_{a,water}^{ab}$) for several materials, most of which are commonly encountered in food processing and packaging [16]. Contact angle data needed to generate values of γ_S^d for each surface were recorded

Table 2.4 Values of γ_S^d and $W_{a,water}^{ab}$ Estimated for Selected Materials

Material	γ_S^d (mJ/m²)	$W_{a,water}^{ab}$ (mJ/m²)
Paraffin	31.6	0
Viton	26.7	0
Polytetrafluoroethylene	22.6	0
Polyethylene	30.5–32.1	0–19.7
High-density polyethylene	31.8	25.6
Linear low-density polyethylene	32.4	27.0
Ultra-high molecular weight polyethylene	32.4	28.1
Glycol-modified polyethylene terephthalate	32.8	36.0
Ethylene vinyl alcohol	32.1	36.0
Polypropylene (2)	30.0–30.7	3.1–29.5
Nylon (3)	32.0–46.5	32.7–38.8
Germanium	32.4	43.0
Acetal	46.4	35.1
#304 Stainless steel	39.0	56.9
Glass (2)	27.5–31.6	66.4–69.3

Note: In some cases, materials were secured from multiple suppliers. Numbers shown within parentheses identify the number of suppliers in such cases, and the range of surface energy values recorded is provided.

Source: J McGuire. *J Food Eng* 12:239–247, 1990.

using a manual contact angle goniometer and a series of "nonpolar" liquids, as described with reference to Equation 2.23. The surface tensions of these "diagnostic" liquids, measured by Du Nouy ring tensiometry, are depicted in Table 2.5. Values of $W_{a,water}^{ab}$ were estimated according to Equation 2.14. In this case, W_a^{ab} was estimated for each of a series of ethanol–water solutions in contact with a given surface, using Equation 2.13. The surface

Table 2.5 Surface Tensions Recorded for Nonpolar "Diagnostic" Liquids

Liquid	$\gamma_L = \gamma_L^d$ (mJ/m²)
Nonane	22.78
Decane	23.86
Dodecane	25.20
Dicyclohexyl	25.62
Hexadecane	27.36
1-Bromonaphthalene	44.41
Diiodomethane	52.24

Table 2.6 Surface Tension Components of Aqueous Ethanol Solutions

Ethanol Concentration (%)	γ_L	γ_L^d	γ_L^{ab}
0 (Water)	71.17	34.03	37.14
10	49.86	22.63	27.23
20	40.72	24.08	16.64
30	35.40	26.70	8.70
40	34.22	28.17	6.05
50	29.64	24.80	4.84
60	28.00	22.20	5.80
70	26.96	22.49	4.47
80	25.79	23.26	2.53
90	24.40	21.53	2.87
100 (Ethanol)	22.40	19.45	2.95

Source: J McGuire. *J Food Eng* 12:239–247, 1990.

tension of each of these liquids is listed in Table 2.6, along with the associated values of γ_L^d (and γ_L^{ab}, or $\gamma_L - \gamma_L^d$). These values were estimated according to Equation 2.22, using paraffin as the diagnostic solid. In any case, the range typically observed in values of γ_L^d among liquids and solids is quite narrow, in comparison to those of γ_L and $W_{a,water}^{ab}$.

While a value of $W_{a,water}^{ab}$ for a given material describes a property of its surface that is more or less independent of the processing or packaging environment in which it is applied, as discussed in Section 2.2.4, the surface or interfacial tension of protein-containing solutions is strongly dependent on solution conditions such as pH, ionic strength, and temperature, as well as solute concentrations, denaturation or aggregation states, and time. Conformational changes occur at the interface, a function of time, molecular stability, and separation from neighboring molecules, and this adds appreciably to our current inability to predict reduction in interfacial energy in a quantitative way. Experimental results vary widely among different food proteins and stabilizers, and even among genetic variants of a single protein. For example, the adsorption behaviors of two β-lactoglobulin variants were observed to differ significantly from each other [42,43]. Thus, accurate accounting of interfacial energetics requires some experimentation with the system of interest.

2.5 MODIFICATION OF SURFACE PROPERTIES

While the properties of an intact surface are initially defined by the bulk material, a variety of chemical and physical processes (e.g., etching, chemical derivatization, oxidation, or adsorption) at the surface may substantially change the properties of the interface. Controlling these processes offers opportunities to minimize undesirable effects (e.g., fouling), or to improve material properties such as wettability, printability, or adhesion. Some important surface modification processes of interest in food processing and packaging are briefly described below.

2.5.1 Superhydrophobic Surfaces

Microscopic roughness or surface structures can drastically change the properties of a surface. The most common example is the so-called "lotus effect" or the Cassie state (Figure 2.8), which is explained by microscopic structures on the surface which entrap pockets of air under a liquid droplet. This trapped air produces a superhydrophobic surface, for which the apparent water contact angle can approach 180° (i.e., that of pure air). In addition, any debris or contaminants at the surface are mainly in contact with this trapped air, rather than associating with the underlying solid. This loosely held material is easily picked up and removed by droplets as they roll along the surface [44,45].

Such micro-textured surfaces may be of particular interest in food processing and food handling surfaces, heat exchanger plates, and so on, due to their inherent nonfouling and "self-cleaning" attributes. Under high-shear conditions, such as fast fluid flow within a pipe or standard spray-washing operations, proteins and other contaminants can be nearly completely desorbed from such surfaces [44]. Textured surfaces can be economically and easily applied to materials through a variety of stamping, etching, lithography, and deposition techniques. These approaches show great promise for improved sanitation and release characteristics of food-contacting surfaces [44,45].

In many situations, the Cassie state is undesirable (e.g., nonuniform coating of liquids on dry food surfaces, difficulties in dispersing hydrophobic powders in water, or poor printing or adhesion on textured surfaces). Pretreatment of a textured solid surface with a miscible, low-γ_L liquid (e.g., ethanol) "breaks" the surface tension, allowing for complete wetting of the surface (a Cassie–Wenzel state transition). This may be an expedient method to improve wettability [46]. Other approaches to induce wetting of the air-trapping cavities include application of pressure, electrical charge, vibration, and the like. In the absence of adsorbed contaminants, a wetted superhydrophobic surface may potentially be regenerated by briefly heating the substrate to vaporize the liquid film at the surface (Leidenfrost effect) [47]. Importantly, prolonged exposure to surfactant-like molecules such as proteins can lead to complete wetting and exclusion of the air from the surface cavities (the Wenzel state) and a loss of superhydrophobicity. This effect may limit the usefulness of such surfaces in contact with complex, surfactant-rich media such as foods, marine waters, or biological fluids.

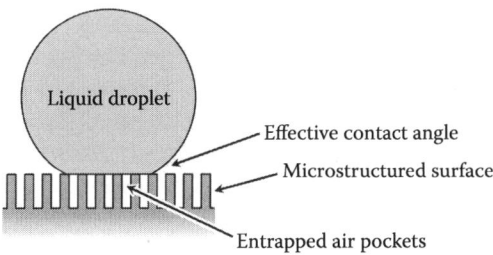

Figure 2.8 Schematic illustration of a superhydrophobic microtextured surface. The "Lotus effect" or Cassie state is caused by entrapped air underneath a liquid droplet which, for sufficiently small and widely spaced features, can produce an effective contact angle that approaches the air–liquid interface (180°).

2.5.2 Chemical and Physical Modifications

Polymer materials offer low cost and light weight, with good impact and abrasion resistance, and are easy to mold and seal. However, most native polymer surfaces have a very low surface energy (Table 2.4) and are chemically inert, which can cause difficulties with adhesion, sealing, and printing. Several physical methods have been developed to change the surface characteristics of polymers, while retaining the bulk properties which make them desirable in food processing and packaging applications [48–50]. These processes can be divided into two major categories: oxidation and grafting.

Oxidation of a polymer surface generally creates a variety of chemical species, including hydroxyl (–OH), carboxyl (–COOH), carbonyl (C=O), and other polar groups. Most of these species can participate in hydrogen-bonding interactions, thus improving adhesion, wettability, and printability of polar liquids, compared to the unmodified surface. They can also form covalent bonds with neighboring molecules, thus serving as chemical anchors for further modifications. While wet-chemical etching methods can be used to create oxidized chemical groups on the surface of a polymer, these processes tend to be difficult to control and create a liquid waste disposal problem. In contrast, physical methods produce little waste and are relatively efficient.

One simple and rapid method to oxidize a polymer is to simply pass it through an oxygen-rich flame. The process creates a variety of oxidized species by reaction of the surface polymer chains with the abundant free radicals found in the flame. By controlling the contact time, fuel-oxygen mixture, and temperature of the flame, the extent of surface modification can be controlled. However, precise control of modification is difficult, and the process can result in contamination by combustion products, mechanical weakening, or even burning of the polymers [48].

Corona discharge methods use a high-frequency, high-voltage AC electric field to ionize the air or other gas molecules near the polymer surface. These ionized species are highly reactive and form a variety of polar, oxidized groups on the polymer surface. The extent of modification must be carefully controlled by changing the voltage and frequency, residence time, gas composition, and other variables. Oxidized surfaces produced by flame or corona methods are highly wettable and exhibit good printing and sealing behavior at first, but these properties typically diminish over time as the oxidized polymer surface relaxes and the hydrophobic chains reorient toward the air to lower the overall surface energy, as previously described [48,50].

Plasma treatments offer various advantages over flame or corona methods, including rapid, uniform functionalization of the surface at low temperatures, and the ability to process arbitrarily shaped materials. Because surface penetration is very limited, plasma treatment produces very thin (tens of nm), dense layers of surface functional groups. Air or oxygen plasmas produce highly oxidized and hydrophilic surfaces, by the reaction of surface polymer chains with ionized and radical species in the plasma. The treated surfaces exhibit excellent adhesion and peel strength, and improved wettability and printing properties. In contrast, the use of fluorinated gases (e.g., CF_4 or SF_6) produces hydrophobic and oleophobic ("non-stick") surfaces with excellent barrier and release properties. Other gases (CO_2, N_2, Ar, NH_3, CH_4, etc.) are also widely used to impart various desirable characteristics to surfaces [50]. Plasma polymerization processes can also be used to deposit uniform, thin, and highly cross-linked polymer films with a variety of useful properties. One potential disadvantage of plasma processes is the requirement for the surface to be under vacuum during processing [49,50].

Ultraviolet (UV) radiation uses high-energy photons to initiate various radical reactions. Irradiation by shorter wavelengths (e.g., 185 nm from Hg lamps) in the presence of oxygen produces ozone, which promotes the rapid oxidation of polymers and renders their surface strongly hydrophilic. Similar UV methods are also commonly used to cross-link or graft acrylate-based polymers on surfaces, enhancing adhesion, printability, mechanical strength, and protein repulsion [44,51].

Regulatory considerations limit the application of grafted polymers on food-contacting surfaces, and the technique is most often used for enhanced printing, labeling, and adhesion on packaging. Various other techniques (ion-beam, electron beam, gamma irradiation, laser, etc.) can also be used to produce controlled oxidation, chemical functionalization, or polymerization on polymer surfaces [50,51].

Another exciting and rapidly growing area is that of active food packaging. Various antimicrobials, antioxidants, chelators, enzymes, and other bioactive molecules can be incorporated into or immobilized on the surface of food packaging. In particular, nonmigratory strategies such as covalent attachment of bioactive agents to a surface avoids the loss of activity caused by migration into the bulk, and can mitigate regulatory concerns [52]. Plasma and UV/ozone oxidation is routinely used to activate substrates for covalent immobilization of enzymes and proteins, antimicrobial peptides, and other bioactive molecules [51,53–55]. Active packaging shows promise for improved food safety, extended shelf-life and freshness, and even "in-package" processing by immobilized enzymes.

2.6 SUMMARY

Properties and predictive relationships relevant to the interfacial energy of solid–fluid and liquid–fluid interfaces have been discussed in this chapter, with reference to the relationship between adsorption and interfacial energy in a system. For solid materials, contact angle methods and their interpretation were highlighted in terms of the following:

1. An empirical "critical surface tension," γ_c
2. An indication of surface hydrophilic–hydrophobic balance, $W_{a,water}^{ab}$
3. An equation of state relationship for γ_S

Values of $W_{a,water}^{ab}$ were presented for a variety of "low energy" (e.g., hydrophobic polymers) and "high energy" (e.g., stainless steel, glass) materials used in food processing and packaging. Measurement of the interfacial energy of liquids and liquid mixtures was also described and some values presented for several pure organic compounds and aqueous ethanol solutions. However, the need for experimentation was presented as necessary for accurate accounting of interfacial energetics in systems containing soluble surface-active components.

While surface properties are initially defined by the bulk properties of the materials, they can be modified and controlled by careful choice of surface topography, and chemical or physical treatments. Processes of importance in food applications include oxidation (to increase adhesion, wettability, or printability), polymerization (to enhance mechanical strength or modify barrier properties), and functionalization to produce bioactive surfaces.

REFERENCES

1. R Aveyard, DA Haydon. *An Introduction to the Principles of Surface Chemistry*. Cambridge: University Press, 1973, pp. 1–30.
2. D Myers. *Surfaces, Interfaces, and Colloids: Principles and Applications*. 2nd ed. New York: Wiley VCH, 1999, pp. 8–20.
3. J Israelachvili. *Intermolecular and Surface Forces*. 3rd ed. New York: Academic Press, 2011, pp. 253–289, 415–467.
4. PC Hiemenz, R Rajagopalan. *Principles of Colloid and Surface Chemistry*. 3rd ed. Baton Rouge, LA: CRC Press, 1997, pp. 286–342.
5. P Suttiprasit, V Krisdhasima, J McGuire. The surface activity of α-lactalbumin, β-lactoglobulin, and bovine serum albumin: I. Surface tension measurements with single-component and mixed solutions. *J Colloid Interface Sci* 154:316–326, 1992.
6. P Suttiprasit, J McGuire. The surface activity of α-lactalbumin, β-lactoglobulin, and bovine serum albumin: II. Some molecular influences on adsorption to hydrophilic and hydrophobic silicon surfaces. *J Colloid Interface Sci* 154:327–336, 1992.
7. J Brash, T Horbett. Proteins at interfaces: An overview. In: T Horbett, J Brash (eds.), *Proteins at Interfaces II*. Washington, DC: American Chemical Society, 1995, pp. 1–23.
8. T Horbett, J Brash. *Proteins at Interfaces: Current Issues and Future Prospects*. Washington, DC: American Chemical Society, 1987, pp. 1–35.
9. J McGuire, C Bower, M Bothwell. Protein films. In AT Hubbard (ed.), *Encyclopedia of Surface and Colloid Science*, Marcel Dekker, New York, 2002, pp. 4382–4395.
10. RJ Good. Contact angle, wetting, and adhesion: a critical review. *J Adhes Sci Technol* 6:1269–1302, 1992.
11. J McGuire, V Krisdhasima. Surface thermodynamics, protein adsorption and biofilm development. In: HG Schwartzberg, RW Hartel (eds.), *Physical Chemistry of Foods*. Marcel Dekker, 1992, pp. 223–262.
12. J Andrade. Surface and interface analysis of polymers—polymer surface dynamics. In: D Lund, E Plett, C Sandu (eds.), *Fouling and Cleaning in Food Processing*. Madison, WI: University of Wisconsin, 1985, pp. 79–87.
13. FM Fowkes. Attractive forces at interfaces. *Ind Eng Chem* 56:40–52, 1964.
14. DH Kaelble. Dispersion-polar surface tension properties of organic solids. *J Adhes* 2:66–81, 1970.
15. FM Fowkes. Quantitative characterization of the acid-base properties of solvents, polymers, and inorganic surfaces. *J Adhes Sci Technol* 4:669–691, 1990.
16. J McGuire. On evaluation of the polar contribution to contact material surface energy. *J Food Eng* 12:239–247, 1990.
17. CA Ward, AW Neumann. On the surface thermodynamics of a two-component liquid-vapor-ideal solid system. *J Colloid Interface Sci* 49:286–290, 1974.
18. AW Neumann, RJ Good, CJ Hope, M Sejpal. An equation-of-state approach to determine surface tensions of low-energy solids from contact angles. *J Colloid Interface Sci* 49:291–304, 1974.
19. P Suttiprasit, K Al-Malah, J McGuire. On evaluating the emulsifying properties of protein using conductivity measurements. *Food Hydrocolloids* 7:241–253, 1993.
20. JL Bohnert, TA Horbett. Changes in adsorbed fibrinogen and albumin interactions with polymers indicated by decreases in detergent elutability. *J Colloid Interface Sci* 111:363–377, 1986.

21. J Andrade, V Hlady, L Feng, K Tingey. Proteins at interfaces: Principles, problems and potentials. In: JL Brash, PW Wojciechowski (eds.), *Interfacial Phenomena Bioproducts*. New York: Marcel Dekker, 1996.
22. A Kondo, F Murakami, K Higashitani. Circular dichroism studies on conformational changes in protein molecules upon adsorption on ultrafine polystyrene particles. *Biotechnol Bioeng* 40:889–894, 1992.
23. H Elwing, S Welin, A Askendahl, I Lundström. Adsorption of fibrinogen as a measure of the distribution of methyl groups on silicon surfaces. *J Colloid Interface Sci* 123:306–308, 1988.
24. SH Lee, E Ruckenstein. Adsorption of proteins onto polymeric surfaces of different hydrophilicities—A case study with bovine serum albumin. *J Colloid Interface Sci* 125:365–379, 1988.
25. J-K Luey, J McGuire, RD Sproull. The effect of pH and NaCl concentration on adsorption of β-lactoglobulin at hydrophilic and hydrophobic silicon surfaces. *J Colloid Interface Sci* 143:489–500, 1991.
26. W Norde, J Lyklema. The adsorption of human plasma albumin and bovine pancreas ribonuclease at negatively charged polystyrene surfaces: I. Adsorption isotherms. Effects of charge, ionic strength, and temperature. *J Colloid Interface Sci* 66:257–265, 1978.
27. G Narasimhan. Emulsions. In: HG Schwartzberg, RW Hartel (eds.), *Physical Chemistry of Foods*. New York: Marcel Dekker, 1992, pp. 307–386.
28. A Adamson. *Physical Chemistry of Surfaces*. New York: Interscience, 1967, pp.
29. R Weast. *Handbook of Chemistry and Physics*. 61st ed. Boca Raton, LA: CRC Press, 1980, pp. F45–F46.
30. J McGuire, E Lee, RD Sproull. Temperature influences on surface energetic parameters evaluated at solid–liquid interfaces. *Surf Interface Anal* 15:603–608, 1990.
31. N Wu, J Dai, FJ Micale. Dynamic surface tension measurement with a dynamic Wilhelmy plate technique. *J Colloid Interface Sci* 215:258–269, 1999.
32. R Wüstneck, J Krägel, R Miller, VB Fainerman, PJ Wilde, DK Sarker, DC Clark. Dynamic surface tension and adsorption properties of β-casein and β-lactoglobulin. *Food Hydrocolloids* 10:395–405, 1996.
33. Z-G Cui, BP Binks, JH Clint. Determination of contact angles on microporous particles using the thin-layer wicking technique. *Langmuir* 21:8319–8325, 2005.
34. JC Fröberg, OJ Rojas, PM Claesson. Surface forces and measuring techniques. *Int J Miner Process* 56:1–30, 1999.
35. J-H Chen, E Ruckenstein. Solvent-stimulated surface rearrangement of polyurethanes. *J Colloid Interface Sci* 135:496–507, 1990.
36. F Huijs, J van Zanten, T Vereijken, M Wouters. Surface rearrangement of tailored polyurethane-based coatings. *J Coat Technol Res* 2:435–443, 2005.
37. L-C Xu, CA Siedlecki. Microphase separation structure influences protein interactions with poly(urethane urea) surfaces. *Journal of Biomedical Materials Research Part A* 92A:126–136, 2010.
38. M Berglin, E Pinori, A Sellborn, M Andersson, M Hulander, H Elwing. Fibrinogen adsorption and conformational change on model polymers: Novel aspects of mutual molecular rearrangement. *Langmuir* 25:5602–5608, 2009.
39. L Sperling. *Introduction to Physical Polymer Science*. 3rd ed. Hoboken, NJ: Wiley, 2001, pp. 300–309.
40. E Rangel, G Gadioli, N Cruz. Investigations on the stability of plasma modified silicone surfaces. *Plasmas Polym* 9:35–48, 2004.
41. M Farrell, S Beaudoin. Surface forces and protein adsorption on dextran—and polyethylene glycol-modified polydimethylsiloxane. *Colloids Surf B Biointerfaces* 81:468–475, 2010.
42. U Elofsson. Protein adsorption in relation to bulk phase properties: β-lactoglobulins in solution and at the solid/liquid interface. Lund, Sweden: Lund University, 1996.
43. UM Elofsson, MA Paulsson, P Sellers, T Arnebrant. Adsorption during heat treatment related to the thermal unfolding/aggregation of β-lactoglobulins A and B. *J Colloid Interface Sci* 183:408–415, 1996.

44. T Mérian, JM Goddard. Advances in nonfouling materials: Perspectives for the food industry. *J Agric Food Chem* 60:2943–2957, 2012.
45. A Lafuma, D Quere. Superhydrophobic states. *Nat Mater* 2:457–460, 2003.
46. RM Blanco, P Terreros, N Muñoz, E Serra. Ethanol improves lipase immobilization on a hydrophobic support. *J Mol Catal B: Enzym* 47:13–20, 2007.
47. G Liu, L Fu, AV Rode, VSJ Craig. Water droplet motion control on superhydrophobic surfaces: Exploiting the Wenzel-to-Cassie transition. *Langmuir* 27:2595–2600, 2011.
48. S Farris, S Pozzoli, P Biagioni, L Duó, S Mancinelli, L Piergiovanni. The fundamentals of flame treatment for the surface activation of polyolefin polymers—a review. *Polymer* 51:3591–3605, 2010.
49. KT Lee, JM Goddard, JH Hotchkiss. Plasma modification of polyolefin surfaces. *Packag Technol Sci* 22:139–150, 2009.
50. M Ozdemir, CU Yurteri, H Sadikoglu. Physical polymer surface modification methods and applications in food packaging polymers. *Crit Rev Food Sci Nutr* 39:457–477, 1999.
51. JA Barish, JM Goddard. Polyethylene glycol grafted polyethylene: A versatile platform for nonmigratory active packaging applications. *J Food Sci* 76:E586–E591, 2011.
52. D Dainelli, N Gontard, D Spyropoulos, E Zondervan-van den Beuken, P Tobback. Active and intelligent food packaging: Legal aspects and safety concerns. *Trends Food Sci Technol* 19, Supplement 1:S103–S112, 2008.
53. P Appendini, JH Hotchkiss. Review of antimicrobial food packaging. *Innov Food Sci Emerg Technol* 3:113–126, 2002.
54. M Ozdemir, JD Floros. Active food packaging technologies. *Crit Rev Food Sci Nutr* 44:185–193, 2004.
55. J Vartiainen, M Rättö, S Paulussen. Antimicrobial activity of glucose oxidase-immobilized plasma-activated polypropylene films. *Packag Technol Sci* 18:243–251, 2005.

3
Food Microstructure Analysis

Bryony James

Contents

3.1	Introduction	64
3.2	Light Microscopies	65
	3.2.1 Sample Preparation, LM	66
	3.2.2 Confocal Laser Scanning Microscopy	66
	3.2.3 Sample Preparation, CLSM	68
	3.2.4 Autofluorescence	69
	3.2.5 Photobleaching	69
	3.2.6 Dynamic and *In Situ* Studies Using CLSM	70
3.3	Scanning Electron Microscopy	70
	3.3.1 Cryo-SEM	72
	3.3.2 Variable Pressure SEM	74
	3.3.3 Dynamic and *In Situ* Studies Using VP-SEM	76
3.4	Transmission Electron Microscopy	78
	3.4.1 Sample Preparation, TEM	80
	3.4.2 Cryo-TEM	81
3.5	Micro-CT and X-Ray Microscopy	81
3.6	Quantifying Food Microstructure	83
3.7	Other Methods of Structural Analysis	84
	3.7.1 Scattering Techniques, DLS, SAXS, SANS	85
	3.7.2 Atomic Force Microscopy	85
	3.7.3 Magnetic Resonance Techniques, MRI/NMR	86
	3.7.4 Optical Coherence Tomography	87
3.8	Summary	88
	References	88

3.1 INTRODUCTION

As long as there have been microscopes, there have been scientists examining the structure of foods for both interest and application. The earliest food-engineering application of structural analysis was to assess foods for possible contaminants and adulteration; light microscopy (LM) particularly coming into its own in the late 1800s and the early years of the twentieth century (Clayton and Hassel 1909; Greenish 1903; Hassel 1857). With no small modesty Arthur Hassel (1857, p. 43) described his application of the microscope to this purpose as being "certainly the most practical and important use which has ever been made of that instrument."

In more recent years, food structure has become a subject of research in its own right, as the manufacture, and consumption, of food can be considered as a set of de-structuring and re-structuring steps. Food materials consist of structural hierarchies, both natural and manufactured, that dictate the performance of the foodstuff in several important regards. One of the great challenges of food structure analysis is that these structural hierarchies cover eight orders of magnitude in length scale (Figure 3.1). The structural components often encompassed by the catch all term "microstructure" (though occasionally differentiated by "nanostructure" for submicron components) range from the nanometer scale to a few hundreds of microns. This subset of length scales influences properties including fracture mechanics, rheology, mass transport phenomena, and perception of texture (Aguilera 2005), what might be considered the "Engineering Properties of Foods." The impact that food structure at this range of length scales has on nutritional outcomes is also being increasingly recognized (Ferguson et al. 2010; Nehir and Simsek 2012; Turgeon and Rioux 2011).

Techniques for analyzing the microstructure of food are increasing in number and sophistication, although access to some techniques for a practicing food engineer might be less common than others. Table 3.1 summarizes examples of the structural components that might exist at certain length scales and the techniques available for analysis. Not all techniques can be covered in detail in a single chapter, hence this chapter will emphasize the more common techniques used for the length scales of interest. The aim of the chapter

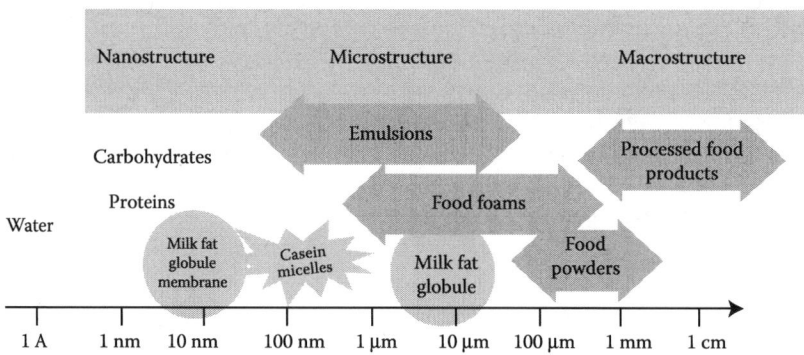

Figure 3.1 The wide range of relevant length scales in the structural hierarchy of food materials.

Table 3.1 Example Structural Features in Food Materials at Different Length Scales

Size Range	Example Structural Feature	Influence on Functional Property (Example Reference)	Possible Analysis Technique
μm–mm	Air cells in bread	Texture effects through apparent Young's modulus, "resilience," shelf-life, through moisture transport (Ramirez and Aguilera 2011)	Flatbed scanner, micro-CT, MRI
μm–mm	Parenchyma cells in fruit	Ripening processes, firmness (Verboven et al. 2008)	SEM, LM, micro-CT, OCT, MRI
μm	Ice crystals in ice cream	Pumpability in plant, shelf-life (ice crystal ripening), texture effects (Bollinger et al. 2000)	Cryo-SEM
μm	Fat globules in cheese	Baking performance on pizza (Ma et al. 2013)	CLSM, cryo-SEM, XRM
μm	Emulsion droplets in mayonnaise	Rheology, texture, nutrition (Stokes and Telford 2004)	CLSM, SEM, low-vacuum SEM
μm–nm	Surface fat structure of chocolate	Shelf stability, bloom performance, appearance, texture (Sonwai and Rousseau 2010)	SEM, AFM
μm–nm	Casein micelle network in yoghurt	Texture, processing, shelf-stability possibly nutrition (McCann et al. 2011)	SEM, cryo-SEM
nm	Casein micelle structure	Building block of most dairy-based systems (Marchin et al. 2007)	TEM, scattering techniques
nm	Plant cell wall structure	Digestibility bio-availability, ripening, texture (Sila et al. 2008)	TEM, scattering techniques, NMR

is not to be an exhaustive guide to operation of these techniques, rather to summarize the type of information that can be gleaned (and the artifacts that are risked) giving the reader a place to start in their own investigations. A brief introduction to alternative methods for structural characterization is included so that the reader has a starting point to explore these options.

3.2 LIGHT MICROSCOPIES

LM is the oldest of the microstructural analysis techniques applied to food materials and also the technique most likely to be found in a food quality, or new product development laboratory. Although the variety of configurations for specific contrast mechanisms is large, the basics of LM remain effectively unchanged from its inception. A light source produces visible light that passes through a condenser lens, through a thin sample (unless working in reflectance) and is focused by an objective lens to form an enlarged image. Resolution of the LM (the smallest distance between two features where the features are still visibly separate) is around 1000 times better than the naked eye.

ENGINEERING PROPERTIES OF FOODS

3.2.1 Sample Preparation, LM

Sample preparation for LM can be moderately simple as the samples can be imaged in atmospheric conditions, meaning hydrated samples can be easily examined. An excellent text for anyone starting LM for food structure analysis is *Food Microscopy* by Olga Flint (1994). Samples must be sectioned to thin slices for transmission LM, and various microtomes are available to do this without introducing substantial artifacts. Alternatively, liquid samples can be smeared or squashed between a slide and cover slip to create a thin sample. Frequently, samples are stained to improve contrast between components. These histochemical techniques rely on known reactions between the stain and the food components, for example iodine is a stain used to highlight the presence of starch and protein (Figure 3.2). The choice of stain depends on the sample and the component of interest and, with experience, very subtle details of food structure can be ascertained, as shown in Table 3.2.

Fluorescent stains for specific components are useful for differentiating, for example, cell wall components in plant-based foods. These flurophores are covered in the next section on confocal laser scanning microscopy (CLSM). Various contrast techniques can also enhance contrast without staining. For example, in differential interference contrast (also called Nomarski contrast), the phase of the light is altered to create contrast. Polarizing filters can be used to track specific changes in birefringent food components; native starch gives a distinctive "maltese cross" contrast that disappears as the starch gelatinizes (Aguilera et al. 2001).

3.2.2 Confocal Laser Scanning Microscopy

Since the 1990s CLSM has become increasingly relevant to food microstructure analysis and is now often the "go to" technique before all others. One of the disadvantages of

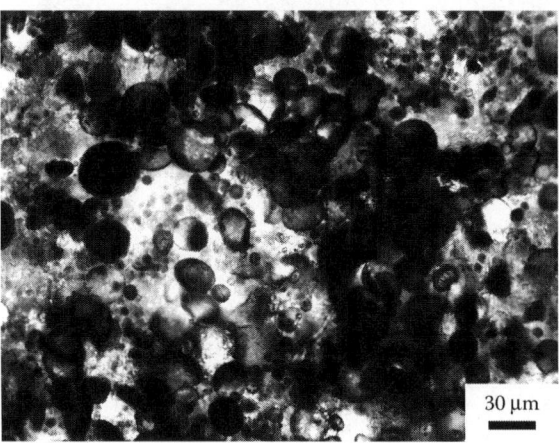

Figure 3.2 Iodine-stained chewed bolus of biscuit showing staining of starch granules. (Courtesy of Sophia Rodrigues, University of Auckland.)

Table 3.2 Stains for Contrast in Light Microscopy

Food Component	Stain	Result
Muscle fibres	Toluidine blue[a]	Pale blue (fresh meat)
Raw collagen	Toluidine blue	Pale pink
Cooked collagen	Toluidine blue	Pale lilac
Elastin fibres	Toluidine blue	Turquoise
Soya protein	Toluidine blue	Dark purple-blue
Wheat protein (gluten)	Toluidine blue	Pale blue-green
Lignified cellulose	Toluidine blue	Dark blue or blue-green
Lignified cellulose	Trypan blue	Dark blue
Cellulose	Trypan blue	Pale blue
Fat	Toluidine blue	Unstained (unless acidic)
Fat	Sudan red or Oil Red O	Red (solid fats can stay unstained but can be identified by birefringence)
Fatty acids	Toluidine blue	Pale blue
Raw amylose starches	Aqueous iodine	Blue (birefringent between crossed polars)
Raw amylopectin starches		Reddish (birefringent between crossed polars)

Source: Adapted from Flint O. 1994. *Food Microscopy.* Oxford, UK: BIOS Scientific Publishers Ltd.
[a] Toluidine blue is a very sensitive stain for gums differentiating gums based on stain color and intensity.

standard LM is the need for samples to be thin for good transmittance and high resolution (avoiding blurring from features out of the focal plane). Producing thin sections, or smears, of sample can introduce structural artifacts. CLSM instruments use a focused laser, scanned across the specimen, to image a subsurface layer. Information from this focal plane is projected onto a detector, with out-of-plane information blocked by a "pinhole" or confocal aperture. The resolution of any optical microscopy technique can be described by the Rayleigh criterion, and a large amount theory can be accessed elsewhere for readers particularly interested in optics (Mertz 2010). In terms of practical application of CLSM, the parameters of relevance to optimal resolution are the magnification, the numerical aperture (NA) of the objective lens, and the working distance. NA determines how much reemitted light is collected and works in combination with the confocal aperture to determine the thickness of the optical "slice" from which the image is built. The lateral and axial resolution can be estimated by (Lorén et al. 2007):

$$R_{lateral} = \frac{0.4_{emission}}{NA}$$

and

$$R_{axial} = \frac{1.4 n_{emission}}{NA^2}$$

where $\lambda_{emission}$ is the wavelength of the detected light, NA = n sin (opening angle), and n is the refractive index of the mount medium. Typical values give $R_{lateral}$ of around 150 nm and R_{axial} of around 550 nm. One of the great strengths of CLSM is the ability to collect a number of optical "slices" and recombine them to generate a three-dimensional (3D) image of the sample. This is frequently referred to as a z-stack.

There are numerous configurations of CLSM and different laser and filter combinations can be selected in almost infinite variety from single photon (point scanning) instruments with a single laser and a filter prior to the detector (to collect/exclude wavelengths) to very advanced multi-photon instruments with multiple lasers and spectral imaging detectors. Instruments are developing all the time and a recent summary can be found in the excellent introductory text *Basic Confocal Microscopy* (Price and Jerome 2011).

3.2.3 Sample Preparation, CLSM

Confocal microscopy is so frequently used in combination with fluorescent stains (fluorophores) that the terms are often conflated (it is important to remember that CLSM can also be used in other contrast modes such as brightfield and differential interference contrast). Staining for specific food components is common, with basic stains such as Nile Red (for lipids), Rhodamine B or Fast Green (for proteins), and so on. Figure 3.3 shows typical images for CLSM of various cheeses stained to separate fat and protein phases, the images are recombinations of the individual excitation signals.

Suitable stains must be selected based not only on the component that needs to be labeled but also giving consideration to the excitation and emission wavelengths of the fluorescence. These details can be sourced from literature (Auty et al. 2001; Brooker 1995) or suppliers (Invitrogen 2013). Diffusion of stain into solid samples can be time-consuming, although in some cases melting the sample and stirring the dye through prior to resolidification can be suitable. As with all preparation techniques for microstructural characterization, consideration must be given to the introduction of artifacts. For example, swelling

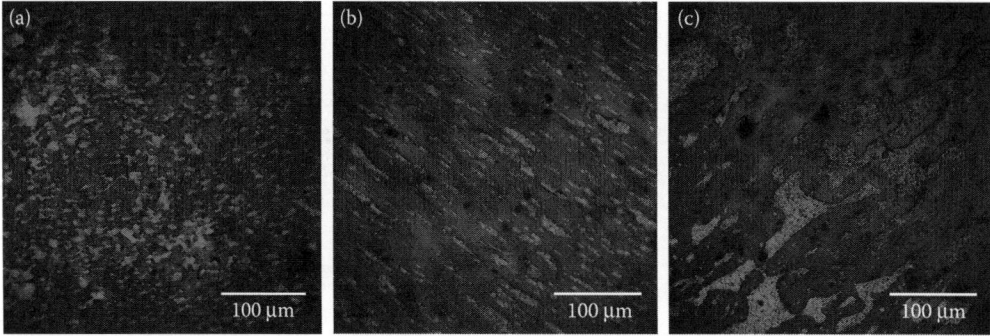

Figure 3.3 CLSM images of cheese. Samples were sectioned to 50-μm thick using a cryotome and stained with Nile Red and Fast Green. Contrast between fat and protein phases is clear. Samples are (a) cheddar, (b) mozzarella, and (c) provolone. (Image courtesy of Xixiu Ma, University of Auckland.)

when using liquid stains is possible and this needs to be acknowledged in any subsequent quantification of structural features.

Two particular issues that need to be considered when using CLSM and fluorescent stains are autofluorescence and photobleaching.

3.2.4 Autofluorescence

Fluorophore dyes are unlikely to be the only fluorescent compounds present in a sample of food. Many naturally occurring compounds fluoresce including chlorophyll, carotene, collagen, and elastin. The fixatives often used in sample preparation can also autofluoresce (including formaldehyde and glutaraldehyde). Generally, autofluorescence is an unavoidable nuisance that can be worked around by optimizing selection of excitation source and filters. Suppression of autofluorescence using heparin has been reported for CLSM of bread doughs (Dürrenberger 2001). Figure 3.4 shows autofluorescence in chocolate, arising from the plant material present in the cocoa nibs.

3.2.5 Photobleaching

Photobleaching occurs as the fluorophores are excited by the laser light. Chemical damage to the structure of the fluorophore itself can occur and reactions with surrounding molecules are more likely with the fluorophore in its excited state. This results in fading of the signal and can be a particular issue when conducting extended observations

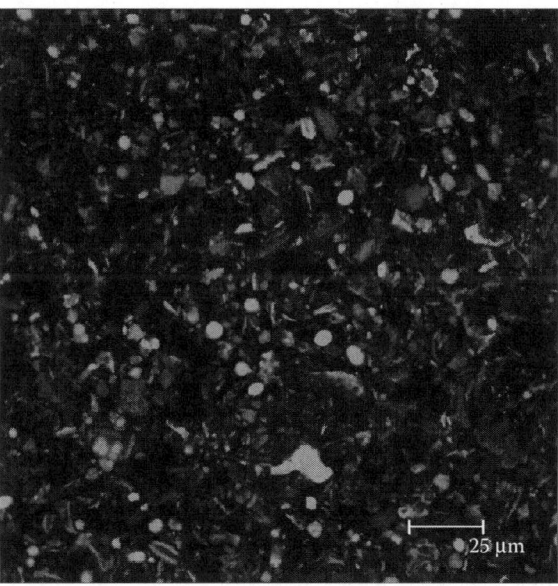

Figure 3.4 CLSM image of dark chocolate (72% cocoa solids) showing autofluorescence of the cocoa nibs. Sample was unstained. (Image courtesy of Ying Jing Tan, University of Auckland.)

(e.g., generating a z-stack sequence). Some fluorophores are more susceptible to this problem than others and therefore stain selection can address the issue. Chemicals that limit the phenomenon (antifade reagents such as *p*-phenylenediamine (PPD), *n*-propyl gallate (NPG), and 1,4-diazobicyclo(2.2.2)octane (DABCO)) are available, although not all are compatible with every fluorophore (Price and Jerome 2011). An alternative that is gaining increased interest is the use of quantum dots for high-resolution, fade-resistant labeling of cellular components (Weng et al. 2006).

Photobleaching is not all bad news. It has led to a dynamic imaging technique known as "fluorescence recovery after photobleaching." Using this technique the mobility of molecules within a sample can be monitored by deliberately photobleaching an area of the sample using high intensity excitation then watching the recovery of fluorescence in that region (using less intense excitation) as labeled molecules diffuse back into the area (Braga et al. 2004; Presley et al. 1997).

3.2.6 Dynamic and *In Situ* Studies Using CLSM

As the sample in CLSM can be, effectively, a bulk sample (rather than a thin specimen), it is feasible to conduct *in situ* experiments while observing dynamic changes in the specimen. These experiments require the use of specific sample stages, which are available (or can be made) for mechanical or, rheological testing and heating/cooling studies. The fracture behavior of cheese has been followed during testing using CLSM (Lorén et al. 2007), as have the mechanical and fracture properties of model biopolymer gels (Plucknett et al. 2001). Freezing and thawing can be followed moderately easily using a temperature-controlled stage, as the ice crystal formation excludes the fluorophores from the water phase.

3.3 SCANNING ELECTRON MICROSCOPY

Scanning electron microscopy (SEM) has been comprehensively applied to the analysis of food structure since the 1960s and earlier applications are well summarized in a number of books, book chapters (Aguilera and Stanley 1999; Holcomb and Kalab 1981), and review articles (Heertje 1993; Holcomb 1990; Sargant 1988). Unlike transmission electron microscopy (TEM), the sample for SEM does not have to be electron transparent, removing many of the sample preparation difficulties inherent in TEM. However, until relatively recently, SEM (like TEM) has been a technique that required the sample be exposed to high vacuum conditions.

The basic components of a conventional SEM are shown in Figure 3.5 with the enlarged area highlighting the aspects of the technique that provide both its main advantages (high resolution, high depth of field) and disadvantages (high vacuum required, conducting specimen best). SEM is a high-resolution imaging technique, the ultimate resolution being dictated by a number of parameters from the electron source to the sample itself. Image contrast in SEM is created by one of two mechanisms, the creation of back scattered electrons (BSE) or secondary electrons (SE). BSE contrast is sensitive to both local topography and atomic number (Z) variation, the BSE coefficient, η, rising with Z. SE are almost insensitive to atomic number but are sensitive to local topography, the SE coefficient, δ, rises as

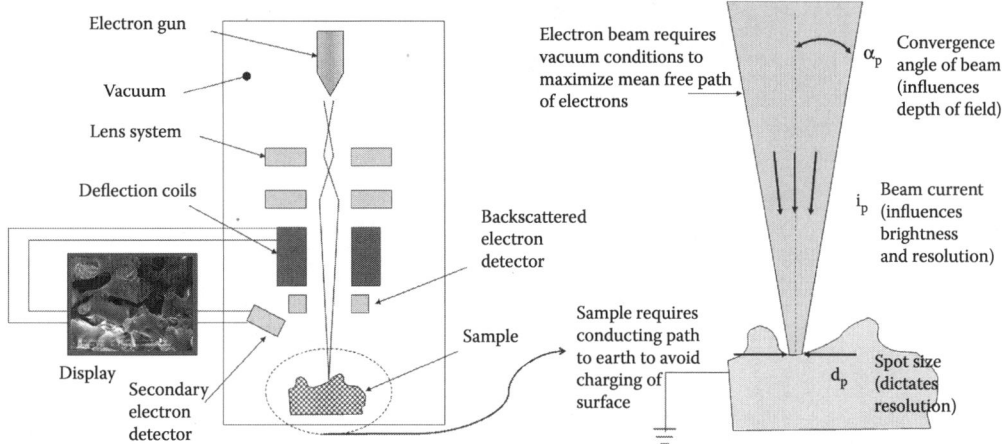

Figure 3.5 Basic components of a conventional scanning electron microscope.

the angle of the incident beam to the surface decreases. BSE have higher energy (up to the energy of the incident beam) and as such can emerge from deeper within the specimen, SE are low energy electrons and can only escape from within a few mean free path interactions of the surface, this means the resolution and surface sensitivity of an SE image is higher than that of a BSE image. Figure 3.6 shows the effect of these two contrast mechanisms.

As $\eta + \delta < 1$ more electrons are injected into the specimen than are emitted, as such a conductive path must be made from the sample surface to ground to avoid sample charging. Even with a field emission source, that can operate at low accelerating voltage hence

Figure 3.6 SEM images of "honey-roasted" cashew nut showing effect of contrast mechanism on image formation. Contrast in SE image (left) arises predominantly from topography, whereas contrast in BSE image (right) arises from a combination of topography and atomic number contrast. As such the NaCl present (example highlighted by arrow) in the flavoring appears brighter than the sugar crystals (example highlighted by "thumbtack").

ENGINEERING PROPERTIES OF FOODS

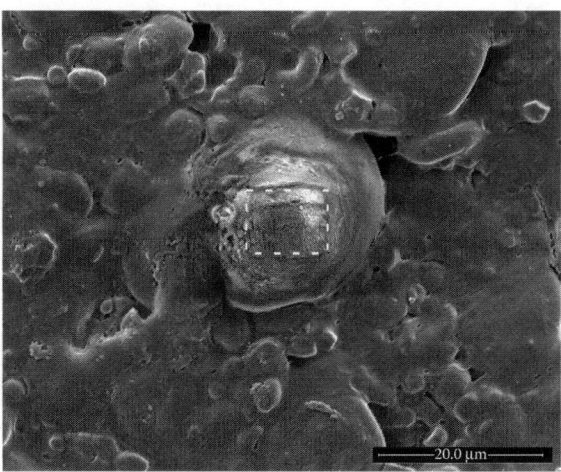

Figure 3.7 SEM image of dried uncooked pasta showing beam damage. Beam damage was caused by imaging at higher magnification so the beam only rastered across the indicated area, as such electron dose was high in this area and thermal damage caused the swelling and blistering shown.

lower electron "dose" to the sample, insulating materials are best imaged with a conductive coating. This is usually a sputter-coated layer of platinum, gold, or gold-palladium a few nanometers thick. The interaction of the beam with the sample does not only generate an electron signal, it can also damage the sample through thermal effects, called "beam damage." Beam damage occurs during imaging as the beam scans across an area of the sample and can be minimized by lowering the electron dose (by working at lower accelerating voltage or lower magnification). Beam damage can be dramatic and fast, and as such easily recognized as an artifact, or more subtle, in which case it is good practice to work at variety of magnifications to check the impact of the beam on the specimen (Figure 3.7).

The need for a high vacuum in the sample chamber has led to various techniques for imaging high vapor pressure food materials, mainly based on fixing and drying. Any dehydration of a specimen will introduce artifacts by the simple act of removing the water. Shrinkage is the main artifact but the surface tension created as water evaporates can mechanically damage delicate structures. Critical point drying and freeze drying offer alternate routes for sample preparation that introduce less shrinkage and can preserve more delicate features, but still change the structure of the food material considerably. Figure 3.8 illustrates this with an example of freeze-dried carrot compared to an LM image of the same material. Both CPD and FD usually require the sample to be "fixed" in some way, for example, using glutaraldehyde, to immobilize components.

3.3.1 Cryo-SEM

To avoid the need for drying and fixing, samples can be prepared by rapid freezing and examined in the SEM in the frozen state using a cold stage. This is cryo-SEM and preserves the structure of a sample in a state far closer to the spatial arrangement found in the native

FOOD MICROSTRUCTURE ANALYSIS

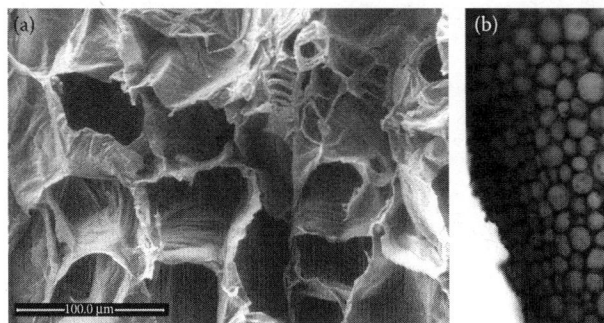

Figure 3.8 Comparison of micrographs from LM and conventional SEM. (a) SEM image of freeze-dried carrot; carrot cubes were freeze dried and coated with platinum prior to imaging, (b) LM image of fresh carrot; thin sections were hand-cut using a razor blade and stained with toluidine blue (0.05%w/v in 20 mM sodium benzoate buffer, pH 4.4). (Image courtesy of Bronwen Smith, University of Auckland).

state. Cryo-SEM requires specialized accessories to a conventional SEM, usually consisting of an *ex situ* freezing station, a cryo-transfer device to keep the sample under vacuum while moving it to the SEM, and a cold stage able to maintain cryogenic temperatures. Ice crystal formation during the freezing step is inevitable unless special steps are taken to produce glassy ice (high-pressure freezing or introducing cryo-protectants) (Echlin 1992). The freezing step needs to be as fast as possible to produce micro-crystalline ice and the most common method for this is plunging the sample into liquid nitrogen slush (never simply liquid nitrogen as the cooling rate would be far too slow). Cryo-SEM can preserve very fine features and, when coupled with a Field Emission Gun SEM, is a high-resolution technique (Figure 3.9).

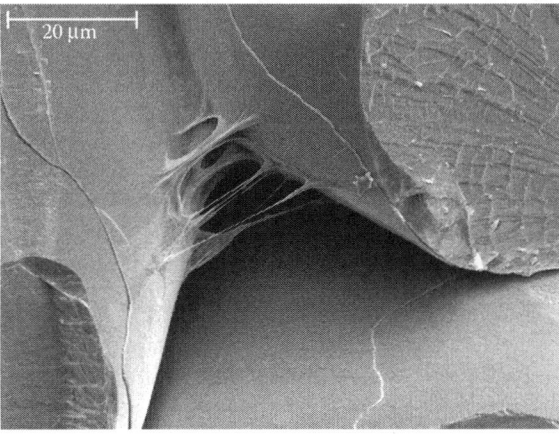

Figure 3.9 Cryo-SEM image of apple parenchyma cells. Sample was plunge frozen in liquid nitrogen slush, fractured under vacuum, sublimed at −95°C for 10 min then sputter coated with gold.

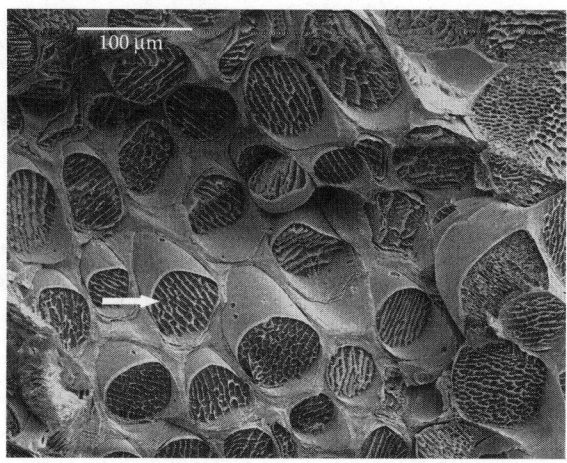

Figure 3.10 Cryo-SEM image of banana skin. Sample was plunge frozen in liquid nitrogen slush, fractured under vacuum, sublimed at −95°C for 30 min then sputter coated with gold. Network structure within cells (highlighted with arrow) is a result of ice crystal growth during freezing.

Many cryo-transfer systems include a freeze-fracture capability so that internal structures can be revealed. This is usually located in a chamber off the main sample chamber so that sample temperature can be controlled for limited sublimation of the ice. This is known as "etching" and is used to reveal structures within the water phase. It also reveals the most distinctive artifact of cryo-SEM: ice crystal ghosts. As the micro-crystalline ice grows during freezing, as it will to some extent unless glassy ice is produced, components of the food are excluded from the liquid phase during solidification. The network structure that is revealed on sublimation is very distinctive (Figure 3.10) and care should be taken when interpreting structural information as a result.

All the advantages of conventional SEM are preserved with cryo-SEM, including the high resolution and great depth of field. The disadvantages are also preserved, with beam damage being exacerbated by the presence of water (Walther et al. 1995). One of the main disadvantages of cryo-SEM and conventional SEM with a dried specimen is that the sample is literally or figuratively "frozen," meaning that dynamic processes need to be interrupted to follow progress (Harker et al. 2006; James and Fonseca 2006). The newer variable pressure SEM instruments offer a complementary answer to this last issue.

3.3.2 Variable Pressure SEM

Developments in the 1970s led to new SEMs that could operate with higher pressures in the sample chamber than previously (Danilatos and Robinson 1979; Moncrieff et al. 1979), the first commercial instrument (the Electroscan Environmental SEM or ESEM), becoming available in the 1980s. Due to the fact that this design was purchased and widely commercialized by Philips Electron Optics (later FEI) who continued to use the ESEM nomenclature these instruments are often collectively referred to as ESEMs. The more generic name is variable pressure SEM or, awkwardly, low vacuum SEM.

Variable pressure SEM (VP-SEM) currently refers to instruments that can operate with sample chamber pressures up to around 2500 Pa (20 Torr), whereas conventional SEMs operate at pressures of around 10^{-4} Pa (10^{-6} Torr). The development and details of these instruments have been thoroughly reviewed elsewhere (Danilatos 1993; Donald 2003; Stokes 2003) but a brief summary of these details is warranted here so operators in the food engineering space can be aware of any limitations.

The vacuum management that allows the sample chamber (and the sample chamber only) to be at variable pressure is summarized in Figure 3.11. By mounting the sample on a peltier cooled stage the relevant portion of the water phase diagram can be accessed, allowing water to be liquid in the sample chamber. While pressures of up to 2500 Pa (20 Torr) are easily achieved by bleeding some water vapor into the chamber realistically, at these pressures, imaging is challenging so frequently, to keep samples fully hydrated, it is useful to cool them to 4–5° and operate at slightly lower pressure.

The presence of gas in the sample chamber requires alternative detectors to those used in high vacuum. Each manufacturer has used a slightly different approach but in every case the gas is an important part of the imaging mechanism. This is most easily explained when considering the FEI design of "Environmental SE detector." The primary beam is scattered to some extent by the gas but the majority of the beam still lands in the original "spot" meaning resolution does not suffer. The SE that are generated interact with the gas, as indicated in Figure 3.12, create an amplification cascade of "environmental" electrons that are subsequently detected by an annular solid state detector. This means that the imaging gas selection is a critical part of imaging in these conditions as the ionization cross-section of the gas will alter the amplification of different parts of the signal (Fletcher et al. 1999). Fortunately, water vapor is an excellent imaging gas for most purposes; nitrous oxide has a similar ionization potential and can be usefully employed for imaging low

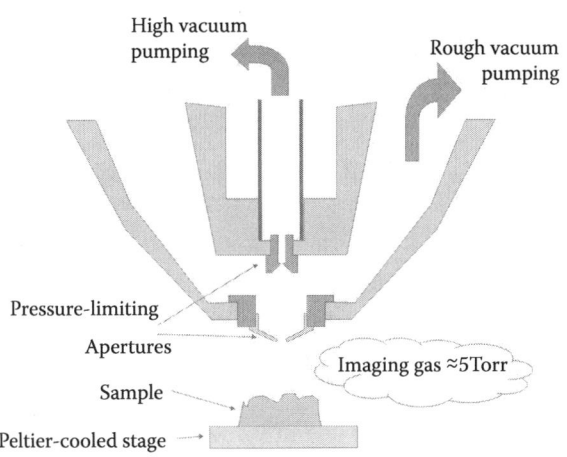

Figure 3.11 Differential pumping and a series of pressure-limiting apertures allows the sample in a VP-SEM to be at pressures up to 20 Torr, while maintaining the electron column at high vacuum (around 10^{-9} Torr).

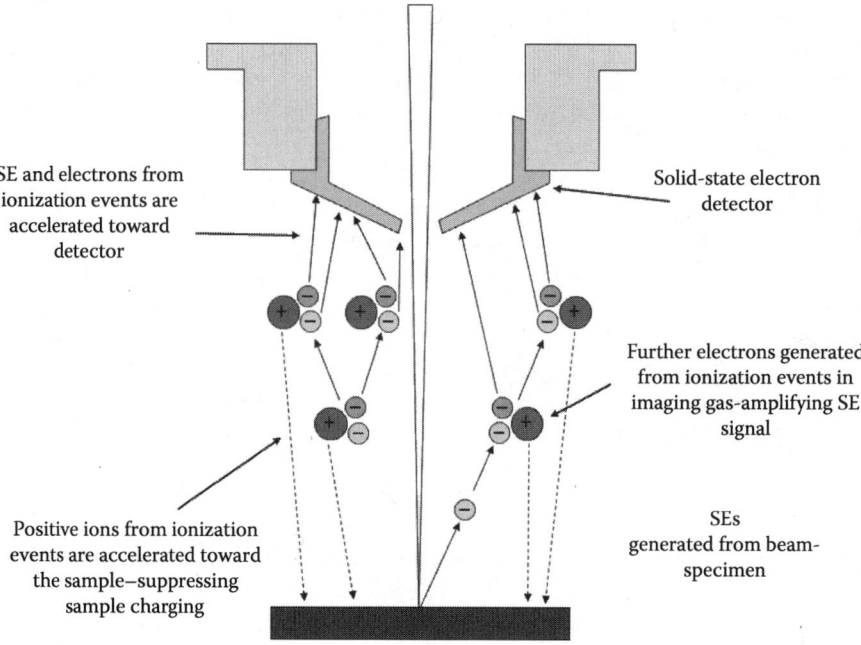

Figure 3.12 The gas in the sample chamber of a VP-SEM is part of the imaging system. Secondary electrons ionize the gas amplifying the electron signal and simultaneously suppressing sample charging.

water activity foods at slightly subzero temperatures, such as chocolate (James and Smith 2009).

Many of the applications of VP-SEM to food structure have been reviewed elsewhere (James 2009) though often details of sample preparation and imaging conditions make some reported studies difficult to interpret. In general, the sample preparation for VP-SEM for food materials is relatively straightforward. Samples are not fixed and not coated for VP-SEM as the positive ions generated during the ion cascade in the imaging gas suppress sample charging. To examine internal structures, a sample section can be prepared by fracturing (which will follow natural lines of weakness) or cutting (which can introduce knife marks and crushing artifacts). This is simple with brittle foods but nearly impossible with soft foods, which are better fractured in the cryo-stage and examined using cryo-SEM.

3.3.3 Dynamic and *In Situ* Studies Using VP-SEM

The sample in VP-SEM is far closer to its native state than in the other iterations of SEM. As such it can be manipulated to follow dynamic changes in structure. Dehydrating/rehydrating processes are an obvious area to exploit the technique's ability to move around the gas–liquid isotherm. Drying studies of apples (Chen et al. 2006) and studies of rehydration behavior of carrots (Smith et al. 2007) have both been followed in real time in the VP-SEM.

The scattering of the primary beam in the VP-SEM can introduce artifacts in studies that rely on close measurement of swelling during hydration, through two thorough studies (Tang et al. 2007a, b) give reasons why any introduced errors are negligible.

Heating studies are possible in the VP-SEM, the detectors are not "blinded" by infrared radiation as are those of conventional SEM and hot stages are available that go to 1500°C. However, the pressure limitation means that to maintain a wet sample temperatures are not yet achievable above around 20–30°C. As such the idea of watching the changes during, say, baking of bread is not yet possible, whereas the melting of chocolate is quite accessible.

Mechanical deformation is possible with *in situ* straining stages (Figure 3.13) (Donald et al. 2003; James and Yang 2012; Thiel and Donald 2000). The challenge for these experiments is the limited geometry of many SEM sample chambers and substages mean that an overly long working distance limits the pressure that is achievable with good image

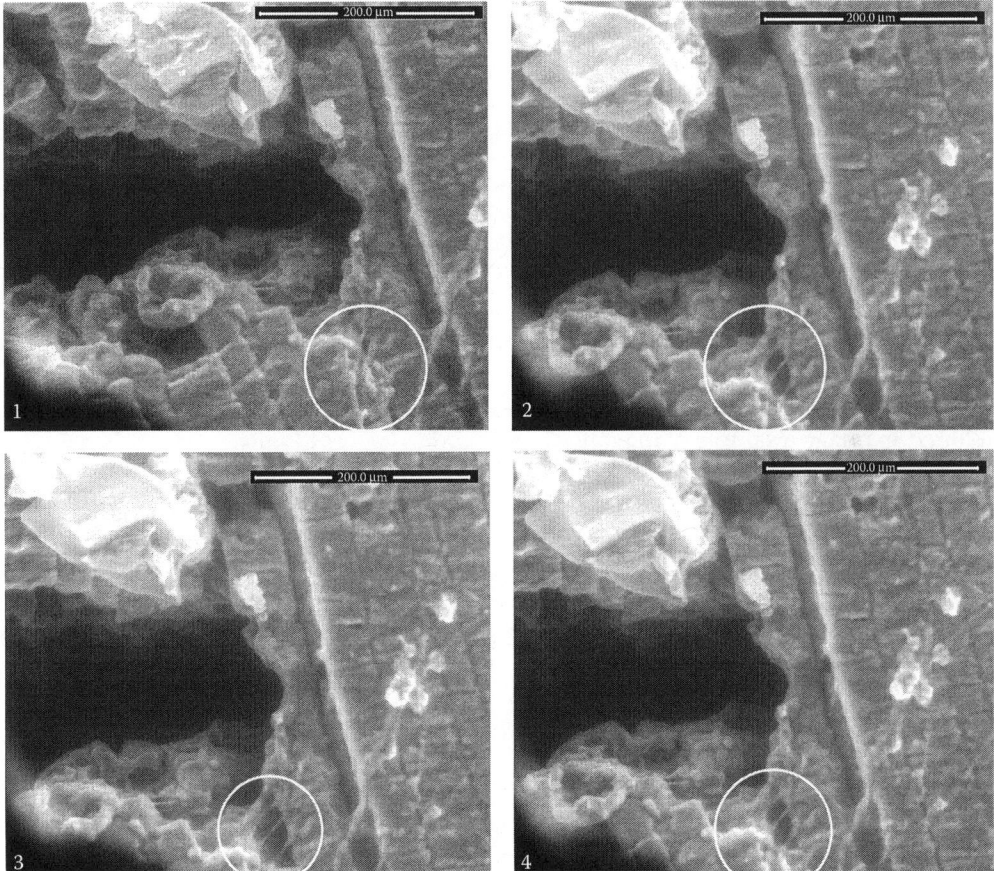

Figure 3.13 VP-SEM sequence of images from *in situ* tensile test of bovine *M. Semitendinosus*. (Image courtesy of Seo Won Yang, University of Auckland.)

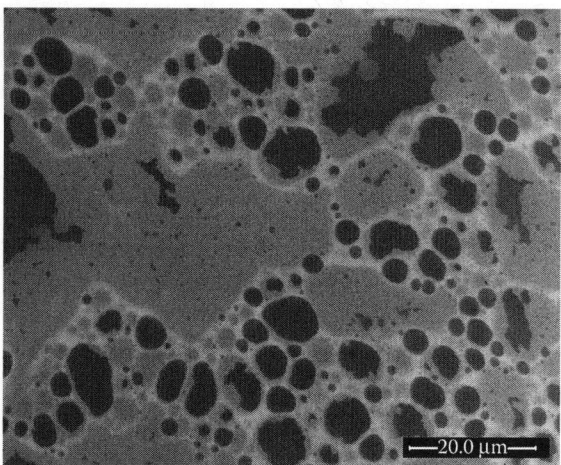

Figure 3.14 VP-SEM image of mayonnaise. Sample imaged at 5 °C using water vapor as imaging gas (pressure 5.0 Torr). Light gray continuous phase is water and mid-gray dispersed phase is oil. Dark areas are air bubbles.

quality. It is also worth remembering that the sample will still need to be in a controlled temperature environment, for example, by resting against a peltier stage, which may introduce nonuniform stress conditions.

Contrast variations in images can result from localized electronic properties of the sample, and these can be exploited in the VP-SEM for imaging emulsions. The source of the contrast is a combination of intrinsic SE emission characteristics and extrinsic charging artifacts (Stokes et al. 1998, 2000), which can be deliberately induced by operating with barely enough gas pressure to suppress all sample charging. Figure 3.14 shows this effect exploited to image a sample of mayonnaise.

VP-SEM is not free of artifacts, beam damage will still be an issue (though the mechanism is not simply thermal, but includes hydrolysis damage from hydroxyl free radicals (Royall et al. 2001)). In many cases, the major impediment to successful VP-SEM imaging is the one thing that makes VP-SEM an attractive technique in the first place. The ability to keep water in its liquid state is vital to maintaining a sample at close to native state, but water is just as likely to obscure the details of interest. This is simply illustrated by watching NaCl go into solution and recrystallize (Figure 3.15). The liquid phase is not transparent as the imaging energy is not visible light.

3.4 TRANSMISSION ELECTRON MICROSCOPY

For very high-resolution imaging of internal structures of food materials, TEM is still the gold standard. The resolution of current TEM instruments is extremely high, with commercially available ultra-high-resolution instruments offering sub-angstrom imaging. This resolution is not practically achievable when imaging food materials because

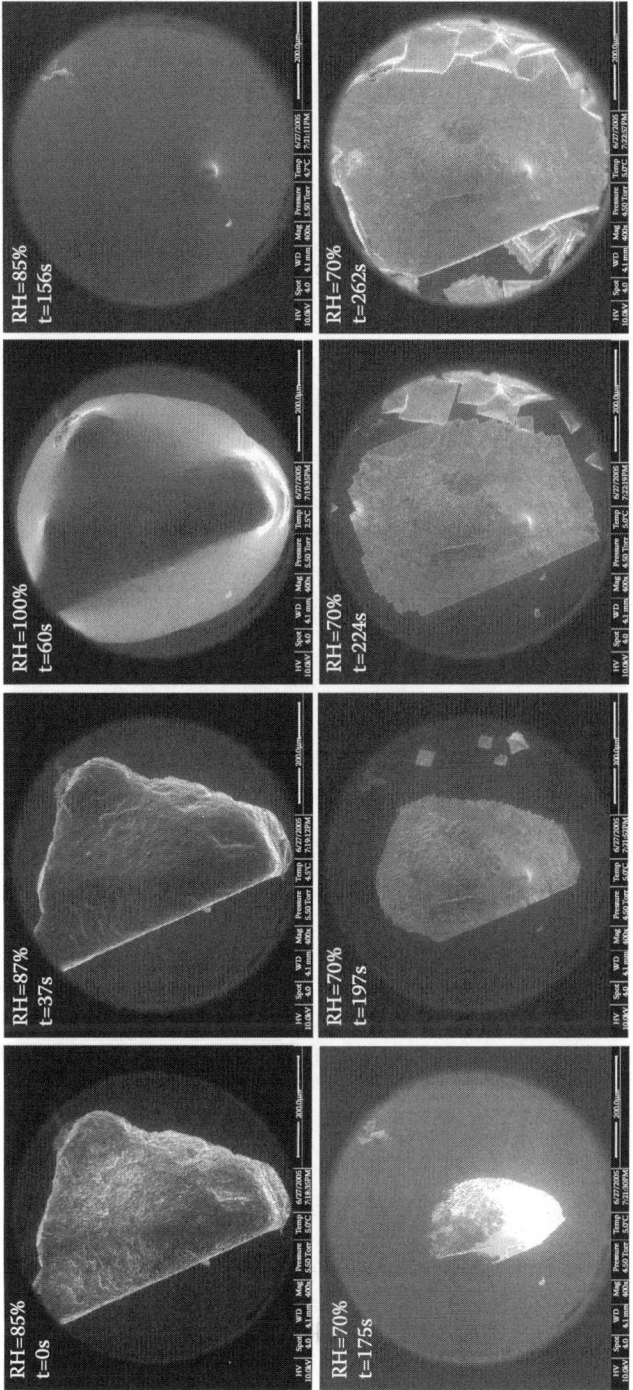

Figure 3.15 VP-SEM sequence of images illustrating the presence of water in the liquid phase obscuring details of the sample. Sample is a grain of NaCl, table salt, humidity was increased and the salt went into solution (left to right, top row). The humidity was then decreased and the salt recrystallized (left to right, bottom row).

of limitations introduced by the sample itself. As the name implies TEM requires electron transmission through a sample, thus the sample must be very thin (<100 nm). TEMs operate at ultra-high vacuum so the samples must be fixed and dehydrated, or a cryo-TEM might be used. For good contrast samples need to have some electron stopping power conveyed by density, food materials are frequently stained with heavy salts such as osmium tetroxide. As such the sample preparation for TEM introduces sample characteristics that limit the resolution. With all microscopy techniques sample preparation is the most important consideration and as such requires the most effort and thought. With TEM it might be speculated that sample preparation represents 90% of the scientific effort of acquiring a meaningful image.

3.4.1 Sample Preparation, TEM

Samples for TEM are supported on copper grids, often with carbon films present (these present a set of issues of their own regarding sample wetting that is outside the scope of this chapter but can be investigated further through (Dubochet et al. 1982)). Dilute liquid samples can be applied to the grid using a dropper with the excess blotted off. Samples are then stained frequently using the negative staining technique. This involves immersing or adding a droplet of a solution of heavy metal salts such as uranyl acetate, leaving for a moment then blotting away the excess. The sample is then left to dry and the salt forms a glassy film around the components of the sample (thus serving the secondary purpose of stabilizing the sample). When imaged in bright-field mode in the TEM the beam is scattered by the heavy atoms of the stain, creating an image of the sample components of interest by the unscattered portion of the beam (hence the term *negative* staining). This technique is frequently employed for imaging the structure of proteins and protein gels and resolution down to 2 nm is achievable (Heertje and Pâques 1995), though higher resolution (0.4 nm) has been reported under particular conditions (Kiselev et al. 1990). Figure 3.16 shows TEM images of β-lactoglobulin aggregates prepared by negative staining, coupled with particle size distributions determined by dynamic light scattering (DLS) as described below (Zúñiga et al. 2010).

For many solid and soft solid samples thin sectioning is used to prepare an electron transparent sample. In order to support the structure the sample is fixed, thereafter embedded in resin prior to sectioning using an ultramicrotome. The steps involved include fixing, dehydrating, infiltrating and embedding, sectioning, followed by staining; the detail of these steps has been thoroughly discussed elsewhere (Glauert and Reid 1974). To examine the structure of a fracture surface replication techniques can be used, where a carbon film is evaporated onto the surface of interest and peeled or floated off. Contrast comes from variations in the thickness of the replica and this is often enhanced by shadowing the replica with a directional evaporation of metal. This latter technique is frequently coupled with freeze fracturing for food structure studies and the basic steps for this are represented schematically in Figure 3.17. Very few systematic studies are published comparing sample preparation techniques but one good example is that of (Hermansson and Buchhein 1981). The authors of the article prepared protein gels using the freeze-etching technique (both with and without fixing the sample in glutaraldehyde, and with and without a cryoprotectant) and compared the resulting images with samples prepared via

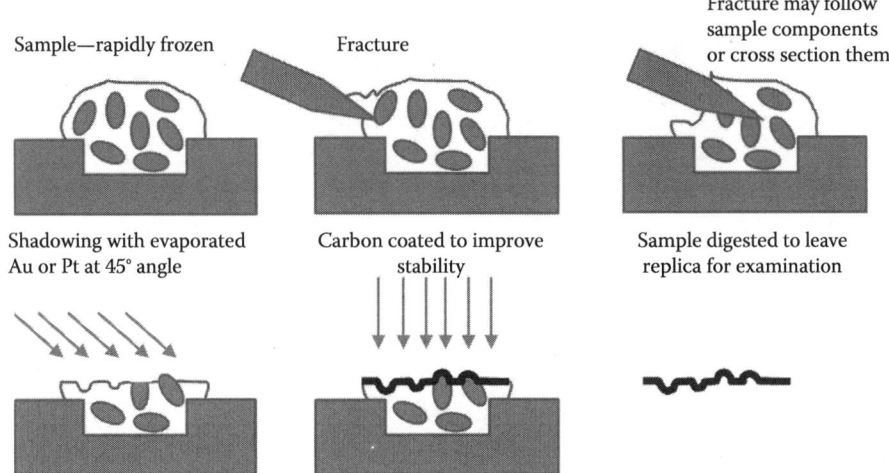

Figure 3.16 Steps in freeze-fracture replication for TEM sample preparation.

fixing, embedding and thin sectioning. The authors went on to recommend the use of the oil-emulsion technique to avoid the use of cryo-protectants and for the finest details to use the thin-sectioning technique.

3.4.2 Cryo-TEM

Cryo-TEM offers the opportunity to avoid many possible sources of image artifact that can arise from the fixing, dehydrating, and embedding steps of sample preparation. Other artifacts are possible from the freezing steps and several routes are available that attempt to produce vitreous ice. It is worth noting that it is less challenging to produce vitreous ice in the very thin samples required for TEM than it is for the larger samples used in cryo-SEM. High-pressure freezing is one route to ensure vitreous ice, but the equipment is expensive. The most common route is to plunge the thin sample, on the sample grid, into liquid ethane (giving a cooling rate of approximately 1,000,000 K/s which is needed to avoid ice crystal nucleation (Echlin 1992)). The sample is then transferred under vacuum conditions to the TEM and imaged using "low-dose" techniques. Numerous factors will impact on the resolution achievable, even after careful sample preparation. These include instrumental issues such as vibration (from the TEM cooling systems) movement of the sample (through temperature differentials) and contamination within the instrument. When everything is addressed very high-resolution imaging of food systems is possible, with, for example, the ability to image the milk fat globule membrane (Waninge et al. 2004).

3.5 MICRO-CT AND X-RAY MICROSCOPY

Examining internal structures of food with as little sample intervention as possible is the ideal method for quantifying microstructure. Micro-CT is an x-ray-based technique

ENGINEERING PROPERTIES OF FOODS

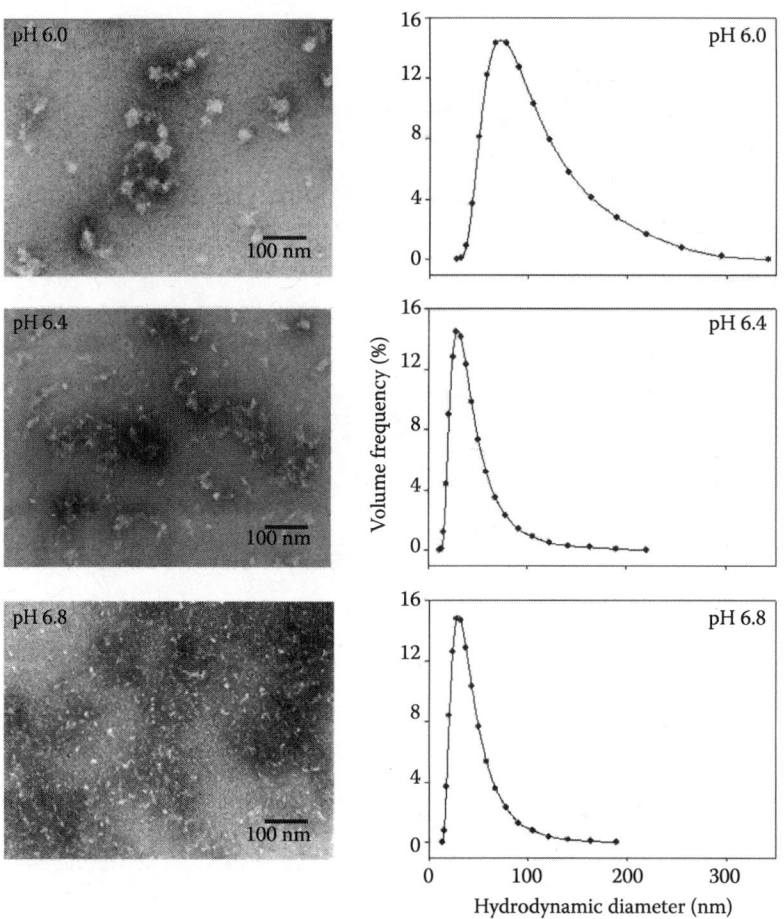

Figure 3.17 TEM images and DLS size distributions corresponding to 15 min of thermal treatment (80°C) of bovine β-lactoglobulin. (Image courtesy of Rommy Zúñiga, Universidad Tecnológica Metropolitana.)

related to medical CAT scanners but with greatly increased resolution. The basic concept of computer-aided tomography is to reconstruct a 3D image based on a number of two-dimensional (2D) slices through a solid. In micro-CT a micro-focused x-ray source is used to illuminate the sample and a set of planar x-ray detectors capture images as the sample rotates. Sample size for most bench-top micro-CT scanners is around a few to tens of millimeters, allowing "bulk" samples to be imaged in their native state.

Micro-CT is being increasingly applied to food structure with recent examples including the air spaces in biscuits and breadsticks (Frisullo et al. 2010), fat structure in yoghurt (Laverse et al. 2011), chocolate (Frisullo et al. 2010), and freezing behavior of food (Mousavi et al. 2007). The spatial resolution of bench-top micro-CT scanners is highly sample- and setting-dependent but is in the order of tens of micrometers. One of the challenges when

FOOD MICROSTRUCTURE ANALYSIS

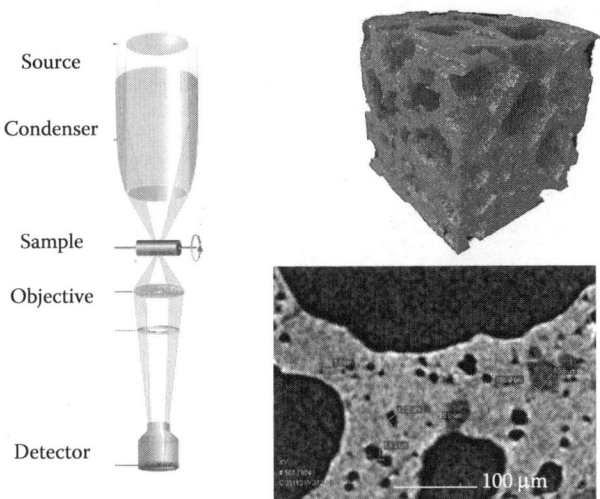

Figure 3.18 Configuration of x-ray microscope and example image of nougat. (Image courtesy of SH Lau, Xradia.)

examining food materials is the low contrast of many food components. Differentiating water from air or fat, for example, can be challenging. Samples can be stained for enhanced contrast, using x-ray dense materials such as osmium tetroxide. Also approaches such as that of Mousavi et al. (2007) in the freezing study might be used, where the samples were freeze dried prior to analysis and the air spaces assumed to correspond exactly to the ice crystal size.

Alternatively, enhanced contrast (and increased resolution) can be attained by using synchrotron-based micro-CT. Synchrotron radiation is high intensity and highly collimated, a divergent beam results in image blurring and loss of resolution. This technique has been applied to a study of the structure of pome fruit, imaging air spaces down to 1 μm in size (Verboven et al. 2008). Access to synchrotron sources can be limited but new developments in "x-ray microscopes" (XRM) bring some of their advantages to the laboratory. XRMs are micro-CT units with the addition of focusing systems. In conventional CT, magnification results from the geometric arrangement of the sample and detection plane, limiting resolution. In an XRM resolution and contrast are greatly enhanced (Figure 3.18).

3.6 QUANTIFYING FOOD MICROSTRUCTURE

Irrespective of the type of microscopy employed results are meaningless unless some effort is made to quantify the structural features revealed. Micrographs should never be published without scale bars (magnification numbers are mostly useless unless the absolute size of reproduction of the image is known). Extracting meaningful numerical data from an image is an area of research of its own and there are many books on image analysis. The most relevant of these texts is that of John Russ *Image Analysis of Food Microstructure*

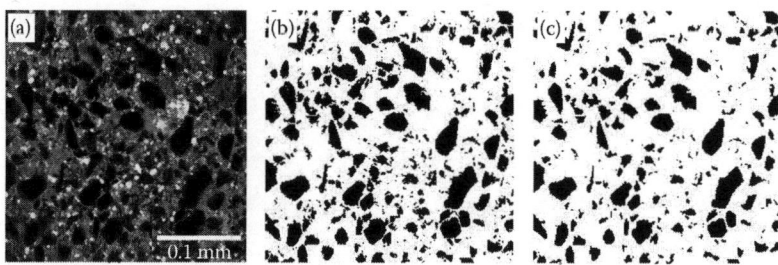

Figure 3.19 Steps in image processing for particle size distribution of sugar particles in chocolate. (a) CLSM image, the sugar is unstained (black), (b) binarized image, (c) thresholded image. (Image courtesy of Jennie Pottbäcker, University of Auckland and TU München.)

(Russ 2004) and the reader is directed to this book for a very practical course of instruction in this area. The images from LM, CLSM (excluding z-stacks), and TEM (excluding tomography) are 2D representations of 3D structures, as such any image analysis needs to take this into account. Extracting data from full 3D representations of structure (as from CLSM z(-stacks and micro-CT images) are not without its challenges. Image analysis software packages are available for quantifying almost all of the parameters one could imagine for a food material but understanding what these parameters mean is always up to the researcher. For example, particle size distribution is an important parameter for a large number of rheological, mechanical, and textural attributes of food. If particles are spherical then a measure of their diameter is sufficient to describe their size. However, particles are rarely purely spherical; as such consideration needs to be given to whether the *relevant* particle size is the longest diameter (maximum ferret diameter) or the diameter of a sphere with equivalent volume, or projected area. Analysis of micrographs is usually based on projections of a diameter and sampling might skew this toward the largest diameter (if samples settle on a flat surface say).

Image analysis usually begins by binarizing and thresholding an image to isolate features of interest (Figure 3.19). The common parameters then determined from the image include feature size (diameter, perimeter, or area) and shape factors (aspect ratio, roundness) (Aguilera and Germain 2007). Included in any consideration of image quantification needs to be the idea of representative sample size, meaning the number of features that need be analyzed to ensure a statistically significant result (Ramirez et al. 2009; Ramirez et al. 2009).

3.7 OTHER METHODS OF STRUCTURAL ANALYSIS

As stated in the introduction a single chapter cannot hope to do justice to the wide range of techniques available, and now being applied, for the study of food microstructure. The techniques described above were selected for their ubiquity in the literature and in the laboratory, being those techniques a food engineer is most likely to encounter. There are many other possibilities for examining the micro- (and nano-)structure of food and a few more techniques are briefly introduced here.

3.7.1 Scattering Techniques, DLS, SAXS, SANS

Scattering techniques, based on the scattering of some form of radiation (usually light, x-rays, or neutrons) are a natural extension to the imaging techniques for analyzing structures present in liquid foods and dispersions. Typically, the information acquired from these systems includes particle size distribution and interaction kinetics between the particles. While some of the microscopy-based techniques offer sufficient resolution for the length scale of interest the sample preparation required to access this length scale is intensive and intrusive.

The scattering techniques are based on the elastic interaction of radiation and particles. Radiation that is scattered from different particles within a dispersion will reach a detector with either constructive or destructive interference depending on the path length, giving information about the particle size distribution of the sample. Even if the scattering is entirely elastic the wavelength of the scattered beams can be altered by the Doppler effect, thus giving information about the dynamics of the particles (Nicolai 2007). DLS can give information on length scales down to around 5 nm (though it must be noted that this measurement relies on particle hydrodynamics, Dalgleish and Hallet 1995). While DLS can be used for concentrated systems multiple scattering effects and opacity make this far harder (Horne 1995). In these cases, and in cases where greater resolution is required, x-rays or neutrons can be used as the incident radiation: small-angle scattering of x-rays or small-Angle scattering of neutrons (Eastoe 1995).

One of the advantages of scattering techniques is the ability to follow reactions in real time, as the samples are not immobilized in any way. For example, protein gelation can be followed (Takat et al. 2000) and DLS is the basis of microrheological techniques (Amin et al. 2012).

3.7.2 Atomic Force Microscopy

Atomic force microscopy (AFM) is one of the microscopy techniques that is a subset of scanning probe microscopy (SPM). SPM encompasses numerous high-resolution imaging techniques including scanning tunnelling microscopy (STM) (only marginally applicable to food structure due to the need for a conducting specimen) and scanning near-field optical microscopy (SNOM). SNOM uses an illumination source very close to the sample surface to overcome the diffraction limitations on resolution of conventional LM (Morris et al. 1999). Applications to food samples are far from routine using SNOM (Morris 2004) and the SPM technique most commonly applied to food systems is AFM.

An AFM instrument scans a sharp tip over a sample (though often the tip is stationery and the sample is scanned using a piezoelectric stage) to track the morphology of the surface in a similar manner to a contact profilometer. Figure 3.20a shows an AFM tip (imaged using SEM), and Figure 3.20b shows how an image is formed by using a laser to locate the position of the tip (Figure 3.19c, d illustrate the image signal and possible source of artifacts from sharp surface features). Successful AFM imaging relies on the sample being relatively flat (usually with a z-axis variation of less than 5 μm) and its application to food systems have been quite specific as a result. Among other studies the technique has been used to determine the surface roughness changes as chocolate bloom forms (Hodge and

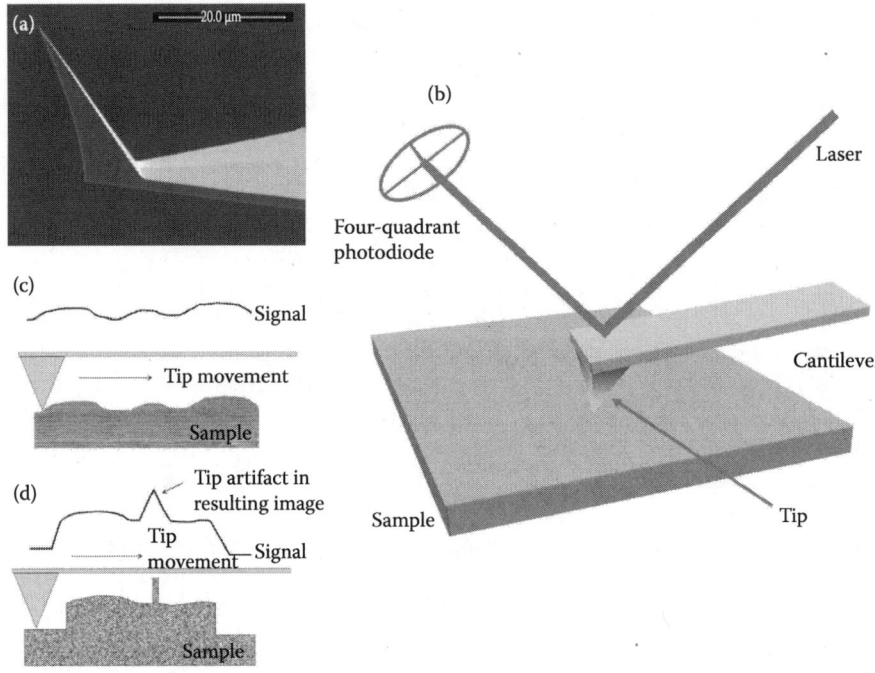

Figure 3.20 Components of an AFM. (a) SEM image of AFM tip, (b) arrangement of tip and cantilever with respect to the laser and photo-diode, (c) image signal as sample is scanned over smooth surface, (d) image signal as sample is scanned over surface with feature sharper than the tip.

Rousseau 2002) and how polysaccharides rearrange during gelation (Gunning et al. 1996). As AFM can offer resolution comparable to TEM it has been used to support proposed models for starch structure first suggested from TEM studies (Morris 2007).

In many ways, the imaging aspects of AFM are so sample specific that many researchers are dissuaded from using the technique. However, imaging is only one result from AFM; as the probe itself interacts with the surface of the sample force–distance curves (Figure 3.21) can be used to quantify the molecular scale adhesion and repulsion forces between the tip and the sample. If the tip is functionalized by attaching, for example, a protein then the interaction forces can be directly assessed (Tran and James 2012).

3.7.3 Magnetic Resonance Techniques, MRI/NMR

Magnetic resonance imaging is most frequently considered a medical imaging procedure in which a strong magnetic field, coupled with weaker fluctuating magnetic field, are used to image slices through a solid. Magnetic resonance imaging (MRI) is based on nuclear magnetic resonance; NMR is a spectroscopic technique that can give information on the molecular state of components within a food.

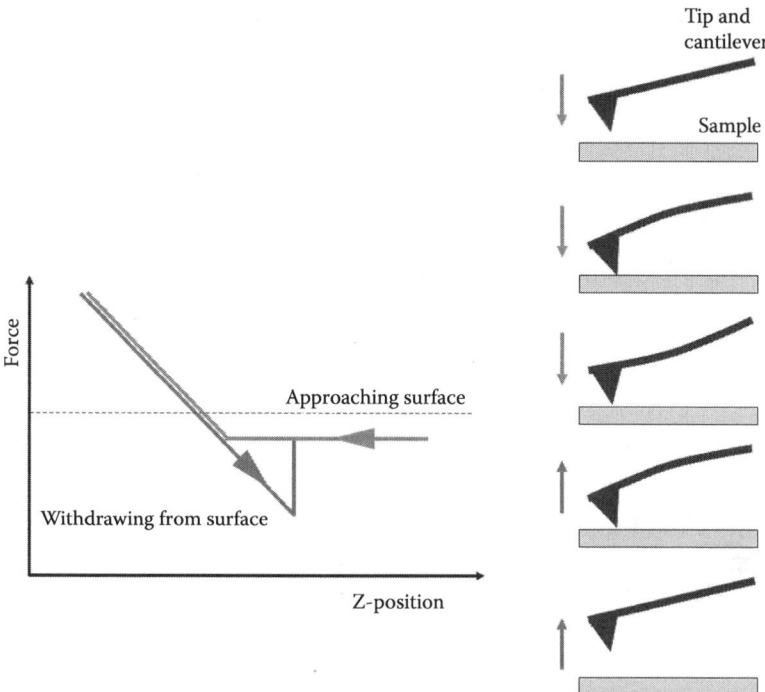

Figure 3.21 Generating a "force–distance" curve as the AFM tip approaches, and is then withdrawn from, the sample surface gives information about the magnitude of force interactions between the tip and the sample. Tips can be activated by attaching molecules to probe-specific interactions.

The advantage of MRI is that whole samples can be imaged without having to create sections. While the spatial resolution is not high (a few μm) the noninvasive nature of the technique means that processes can be followed in real time (Hills 2012).

3.7.4 Optical Coherence Tomography

The emerging technique of optical coherence tomography (OCT) is related to CLSM inasmuch as a light source (usually near infrared) penetrates into a sample to create "optical sections." In OCT the echo time of the reflected light is measured using an interferometer and as such the technique is somewhat analogous to ultrasound imaging. Spatial resolution of a few μm is achievable although a few tens of μm is more usual (Fujimoto aet al. 2000). Depth of penetration is greater than with CLSM (though not so great as ultrasound), typically for plant-based foods being in the order of 1.5 mm (Clements et al. 2004). As with MRI one of the main advantages envisaged for OCT is the ability to image rapidly and noninvasively, therefore with the possibility of in-line sorting and measuring (Meglinski et al. 2010).

3.8 SUMMARY

The influence of microstructure on properties and processes of food-based material has been increasingly recognized in the past few decades, in parallel with the development of tools and techniques that can be used to probe that microstructure. Equally, the materials engineering concept of using processing to manipulate structure, to give desirable functional properties is now the basis of a great deal of nutritional research (Morgenstern et al. 2012).

When deciding what techniques to use to quantify the microstructural features of food there are several golden rules, but no silver bullet. Simply put:

- A single technique is highly unlikely to give the complete answer, and in fact might mislead. All structural imaging techniques are complementary and progressing in techniques of increasing spatial resolution is the best approach.
- Sample preparation will alter the sample. The major effort of any microstructural imaging technique is the sample preparation. There are noninvasive techniques (e.g., MRI) that are often limited by resolution but even then image artifacts can be introduced.
- Quantification of microstructure is essential for meaningful interpretation. Computer-based solutions that allow automated counting and measuring of numerous features are the ideal approach, although outputs should always be checked against an inspection of the structure.

Techniques are developing all the time, at an accelerating rate. Frequently these techniques are not developed specifically with food structural analysis in mind, as such may need adapting for appropriate application within food engineering.

REFERENCES

Aguilera JM. 2005. Why food microstructure? *Journal of Food Engineering*, 67, 3–11.

Aguilera JM, Cadoche L, López C, Gutierrez G. 2001. Microstructural changes of potato cells and starch granules heated in oil. *Food Research International* 34 (10), 939–947.

Aguilera JM, Germain JC. 2007. Advances in image analysis for the study of food microstructure. In: D. Julian McClements (Ed.) *Understanding and Controlling the Microstructure of Complex Foods* (pp. 261–287). Boca Raton, USA: CRC Press.

Aguilera JM, Stanley DW. 1999. *Microstructural Principals of Food Engineering*. Gaithersburg, USA: Aspen Publications, Inc.

Amin S, Rega CA, Jankevics H. 2012. Detection of viscoelasticity in aggregating dilute protein solutions through dynamic light scattering-based optical microrheology. *Rheologica Acta* 51 (4), 329–342.

Auty MAE, Twomey M, Guinee TP, Mulvihull DM. 2001. Development and application of confocal scanning laser microscopy methods for studying the distribution of fat and protein in selected dairy products. *Journal of Dairy Research* 68 (3), 417–427.

Bolliger S, Kornbrust B, Goff HD, Tharp BW, Windhab EJ. 2000. Influence of emulsifiers on ice cream produced by conventional freezing and low-temperature extrusion processing. *International Dairy Journal* 10 (7), 497–504.

Braga J, Desterro JMP, Carmo-Fonseca M. 2004. Intracellular macromolecular mobility measured by fluorescence recovery after photobleaching with Confocal Laser Scanning Microscopes. *Molecular Biology of the Cell* 15, 4749–4760.

Brooker BE. 1995. Imaging food systems by Confocal Laser Scanning Microscopy. In: E. Dickinson (Ed.) *New Physico-Chemical Techniques for the Characterization of Complex Food Systems* (pp. 53–68). London: Blackie.

Chen XD, Chiu YL, Lin SX, James B. 2006. In situ ESEM examination of microstructural changes of an apple tissue sample undergoing low-pressure air-drying followed by wetting. *Drying Technology*, 24(8), 965–972.

Clayton EG, Hassel AH. 1909. *A Compendium of Food-Microscopy with Sections on Drugs, Water, and Tobacco*. London: Baillière, Tindall and Cox.

Clements JC, Zvyagin AV, Silva KKMBD, Wanner T, Sampson DD, Cowling WA. 2004. Optical coherence tomography as a novel tool for non-destructive measurement of the hull thickness of lupin seeds. *Plant Breeding* 123, 266–270.

Dalgleish DG, Hallet FR. 1995. Dynamic light scattering: Applications to food systems. *Food Research International*, Vol. 28, No. 3, pp. 181–193.

Danilatos GD. 1993. Introduction to the ESEM instrument. *Microscopy Research and Techniques* 25, 354–361.

Danilatos GD, Robinson VNE. 1979. Principles of scanning electron microscopy at high specimen pressures. *Scanning* 2, 72–82.

Donald AM. 2003. The use of environmental scanning electron microscopy for imaging wet and insulating materials. *Nature Materials* 2, 511–516.

Donald AM, Baker FS, Smith AC, Waldron KW. 2003. Fracture of plant tissues and walls as visualized by environmental scanning electron microscopy. *Annals of Botany* 92, 73–77.

Dubochet J, Groom M, Mueeker-Neuteboom S. 1982. Mounting of macromolecules for electron microscopy with particular reference to surface phenomena and treatment of support films by glow discharge. In: R. Barrer and V.E. Cosslett (eds.) *Advances in Optical and Electron Microscopy*. London: Academic Press.

Dürrenberger MB, Handschin S, Conde-Petit B, Escher F. 2001. Visualization of food structure by confocal laser scanning microscopy (CLSM). *LWT* 34, 11–17.

Eastoe J. 1995. Small-angle neutron scattering and neutron reflection. In: E. Dickinson (Ed.) *New Physico-Chemical Techniques for the Characterization of Complex Food Systems* (pp. 268–295). London: Blackie.

Echlin P. 1992. *Low Temperature Microscopy and Analysis*. New York: Plenum Press.

Ferguson LR, Smith BG, James BJ. 2010. Combining nutrition, food science and engineering in developing solutions to inflammatory bowel diseases—Omega-3 polyunsaturated fatty acids as an example. *Food and Function* 1 (1), 60–72.

Fletcher AL, Thiel BL, Donald AM. 1999. Signal components in the environmental scanning electron microscope. *Journal of Microscopy* 196(1), 26–34.

Flint O. 1994. *Food Microscopy*. Oxford, UK: BIOS Scientific Publishers Ltd.

Frisullo P, Conte A, Del Nobile MA. 2010. A novel approach to study biscuits and breadsticks using x-ray computed tomography. *Journal of Food Science* 75, 6, 353–358.

Frisullo P, Licciardello F, Muratore G, Del Nobile MA. 2010. Microstructural characterization of multiphase chocolate using x-ray microtomography. *Journal of Food Science* 75, 7, 469–476.

Fujimoto JG, Pitris C, Boppart SA, Brezinski ME. 2000. Optical coherence tomography: An emerging technology for biomedical imaging and optical biopsy. *Neoplasia*, 2(1–2), 9–25.

Glauert AM, Reid N. 1974. Fixation, dehydrating and embedding of biological specimens and ultramicrotomy. In: AM Glaurt (Ed) *Practical Methods in Electron Microscopy* (Vol. 3, pp. 1–207). Amsterdam: North-Holland.

Greenish HG. 1923. *The Microscopical Examination of Foods and Drugs: A Practical Introduction to the Methods Adopted in the Microscopical Examination of Foods and Drugs, in the Entire, Crushed and Powdered States*. London: J & A Churchill.

Gunning AP, Kirby AR, Ridout MJ, Brownsey GJ, Morris VJ. 1996. Investigation of gellan networks and gels by atomic force microscopy. *Macromolecules* 29 (21), 6791–6796.

Harker FR, White A, Gunson FA, Hallett IC, De Silva HN. 2006. Instrumental measurement of apple texture: A comparison of the single-edge notched bend test and the penetrometer. *Postharvest Biology and Technology* 39, 185–192.

Hassel AH. 1857. *Adulterations Detected; Or, Plain Instructions for the Discovery of Frauds in Food in Medicine*. London: Longman, Green, Longman, and Roberts.

Heertje I. 1993. Microstructural studies in fat research. *Food Structure* 12 (1), 77–94.

Heertje I, Pâques M. 1995. Advances in electron microscopy. In E. Dickinson (Ed) *New Physico-Chemical Techniques for the Characterization of Complex Food Systems* (1–52). London: Blackie.

Hermansson A-M, Buchheim W. 1981. Characterization of protein gels by scanning and transmission electron microscopy—A methodology study of soy protein gels. *Journal of Colloid and Interface Science* 81 (2), 519–530.

Hills B. 1995. Food processing: An MRI perspective. *Trends in Food Science and Technology* 6 (4), 111–117.

Hodge SM, Rousseau D. 2002. Fat bloom formation and characterization in milk chocolate observed by atomic force microscopy. *JAOCS, Journal of the American Oil Chemists' Society* 79 (11), 1115–1121.

Holcomb DN. 1990. Food microstructure—Cumulative index. *Food Structure* 9(3), 155–173.

Holcomb DN, Kalab M. 1981. *Studies of Food Microstructure*. O'Hare, USA: Scanning Electron Microscopy, Inc.

Horne DS. 1995. Light scattering of colloid stability and gelation. In E. Dickinson (Ed) *New Physico-Chemical Techniques for the Characterization of Complex Food Systems* (240–267). London: Blackie.

Invitrogen. 2013. Fluorescence Spectra Viewer. Retrieved 20/2/2013 from:http://www.invitrogen.com/site/us/en/home/Products-and-Services/Applications/Cell-Analysis/Labeling-Chemistry/Fluorescence-SpectraViewer.html

James BJ. 2009. Advances in "wet" electron microscopy techniques and their application to the study of food structure. *Trends in Food Science and Technology* 20, 114–124.

James BJ, Fonseca CA. 2006. Texture studies and compression behaviour of apple flesh. *International Journal of Modern Physics B* 20, 3993–3998.

James BJ, Yang SW. 2012. Effect of cooking method on the toughness of bovine *m. Semitendinosus*. *International Journal of Food Engineering* (8) 2, 19.

James BJ, Smith BG. 2009. Surface structure and composition of fresh and bloomed chocolate analysed using x-ray photoelectron spectroscopy, cryo-scanning electron microscopy and environmental scanning electron microscopy. *LWT Food Science and Technology*, 42 (5), 929–937.

Kiselev NA, Sherman MB, Tsuprun VL. 1990. Negative staining of proteins. *Electron Microscopy Reviews*, 3, 1, 43–72.

Laverse J, Mastromatteo M, Frisullo P, Albenzio M, Gammariello D, Del Nobile MA. 2011. Fat microstructure of yogurt as assessed by x-ray microtomography. *Journal of Dairy Science* 94, 668–675.

Lorén N, Langton M, Hermansson A-M. 2007. Confocal fluorescence microscopy (CLSM) for food structure characterisation. In D. Julian McClements (Ed.) *Understanding and Controlling the Microstructure of Complex Foods* (pp. 232–260). Boca Raton, USA: CRC Press.

Ma X, James B, Zhang L, Emanuelsson-Patterson EAC. 2013. Correlating mozzarella cheese properties to its production processes and microstructure quantification. *Journal of Food Engineering* 115 (2), 154–163.

Marchin S, Putaux J-L, Pignon F, Léonil J. 2007. Effects of the environmental factors on the casein micelle structure studied by cryo transmission electron microscopy and small-angle x-ray scattering/ultrasmall-angle x-ray scattering. *Journal of Chemical Physics* 126 (4), art. no. 045101.

McCann TH, Fabre F, Day L. 2011. Microstructure, rheology and storage stability of low-fat yoghurt structured by carrot cell wall particles. *Food Research International* 44 (4), 884–892.

Meglinski IV, Buranachai C, Terry LA. 2010. Plant photonics: Application of optical coherence tomography to monitor defects and rots in onion. *Laser Physics Letter* 7, No. 4, 307–310.

Mertz J. 2010. *Introduction to Optical Microscopy*. Greenwood Village, Colorado USA: Roberts & Company Publishers.

Moncrieff DA, Barker PR, Robinson VNE. 1979. Electron scattering by gas in the scanning electron microscope. *Journal of Physics* 12, 481–488.
Morgernstern M, Foster K, James B. 2012. Food structure breakdown in the mouth to enhance bioactivity. *Food Science and Technology*, 26, 1, 32–34.
Morris VJ. 2004. Probing molecular interactions in foods. *Trends in Food Science & Technology*, 15, 6, 291–29.
Morris VJ. 2007. Atomic Force Microscopy (AFM) techniques for characterising food structure. In D. Julian McClements (Ed.) *Understanding and Controlling the Microstructure of Complex Foods* (pp. 209–231). Boca Raton, USA: CRC Press.
Morris VJ, Kirby AJ, Gunning AP. 1999. *Atomic Force Microscopy for Biologists*. London: Imperial College Press.
Mousavi R, Miri T, Cox PW, Fryer PJ.. Imaging food freezing using x-ray microtomography. *International Journal of Food Science and Technology* 2007, 42, 714–727.
Nehir El S, Sinsek S. 2012. Food technological applications for optimal nutrition: An overview of opportunities for the food industry. *Comprehensive Reviews in Food Science and Food Safety* 11 (1), 2–12.
Nicolai T. 2007. Food characterisation using scattering methods. In D. Julian McClements (Ed.) *Understanding and Controlling the Microstructure of Complex Foods* (pp. 288–310). Boca Raton, USA: CRC Press.
Plucknett KP, Pomfret SJ, Normand V, Ferdinando D, Veerman C, Frith WJ, Norton IT. 2001. Dynamic experimentation on the confocal laser scanning microscope: Application to soft-solid, composite food materials. *Journal of Microscopy*, 201, 2, 279–290.
Presley JF, Cole NB, Schroer TA, Hirschberg K, Zaal KJM, Lippincott-Schwartz J. 1997. ER-to-Golgi transport visualized in living cells. *Nature* 389, 81–85.
Price RL, Jerome WG. 2011. *Basic Confocal Microscopy*. New York: Springer.
Ramírez C, Aguilera JM. 2011. Determination of a representative area element (RAE) based on non-parametric statistics in bread. *Journal of Food Engineering* 102 (2), 197–201.
Ramírez C, Germain JC, Aguilera JM. 2009. Image analysis of representative food structures: Application of the bootstrap method. *Journal of Food Science* 74, 6, 65–72.
Ramírez C, Young A, James B, Aguilera JM. 2010. Determination of a representative volume element based on the variability of mechanical properties with sample size in bread. *Journal of Food Science* 75 (8), 516–521.
Royall CP, Thiel BL, Donald AM. 2001. Radiation damage of water in environmental scanning electron microscopy. *Journal of Microscopy* 204(3), 185–195.
Russ JC. 2004. *Image Analysis of Food Microstructure*. Boca Raton, USA: CRC Press.
Sargant JA. 1988. The application of cold stage scanning electron microscopy to food research. *Food Microstructure* 7(2), 123–135.
Sila DN, Duvetter T, De Roeck A, Verlent I, Smout C, Moates GK, Hills BP, Waldron KK, Hendrickx M, Van Loey A. 2008. Texture changes of processed fruits and vegetables: Potential use of high-pressure processing. *Trends in Food Science and Technology* 19 (6), 309–319.
Smith BG, James BJ, Ho CAL. 2007. Microstructural characteristics of dried carrot pieces and real time observations during their exposure to moisture. *International Journal of Food Engineering* 3(4), Art7.
Sonwai S, Rousseau D. 2010. Controlling fat bloom formation in chocolate—Impact of milk fat on microstructure and fat phase crystallisation. *Food Chemistry* 119 (1), 286–297.
Stokes DJ. 2003. Recent advances in electron imaging, image interpretation and applications, environmental scanning electron microscopy. *Philosophical Transactions of the Royal Society A*, 361, 2771–2787.
Stokes DJ, Thiel BL, Donald AM. 1998. Direct observation of water-oil emulsion systems in the liquid state by environmental scanning electron microscopy. *Langmuir* 14, 4402–4408.
Stokes DJ, Thiel BL, Donald AM. 2000. Dynamic secondary electron contrast effects in liquid systems studied by environmental scanning electron microscopy. *Scanning* 22, 357–365.

Stokes JR, Telford JH. 2004. Measuring the yield behaviour of structured fluids. *Journal of Non-Newtonian Fluid Mechanics* 124 (1–3 SPEC. ISS.), 137–146.

Takata S-I, Norisuye T, Tanaka N, Shibayama M. 2000. Heat-induced gelation of β-lactoglobulin. 1. Time-resolved dynamic light scattering. *Macromolecules* 33 (15), 5470–5475.

Tang X, de Rooij MR, de Jong L. 2007. Volume change measurements of rice by environmental scanning electron microscopy and stereoscopy. *Scanning* 29, 197–205.

Tang X, de Rooij MR, van Duynhoven J. 2007. Dynamic volume change measurements of cereal materials by environmental scanning electron microscopy and videomicroscopy. *Journal of Microscopy* 230(1), 100–107.

Thiel BL, Donald AM. 2000. Microstructural failure mechanisms in cooked and aged carrots. *Journal of Texture Studies* 31, 437–455.

Tran ATT, James BJ. 2012. Study of the interaction forces between the bovine serum albumin protein and montmorillonite. *Surface Colloids And Surfaces A*, 414, 104–114.

Turgeon SL, Rioux L-E. 2011. Food matrix impact on macronutrients nutritional properties. *Food Hydrocolloids* 25 (8), 1915–1924.

Verboven P, Kerckhofs G, Mebatsion HK, Ho QT, Temst K, Wevers M, Cloertens P, Nicolaï BM. 2008. Three-dimensional gas exchange pathways in pome fruit characterized by synchrotron x-ray computed tomography. *Plant Physiology* 147, 518–527.

Verboven P, Kerckhofs G, Mebatsion HK, Ho QT, Temst K, Wevers M, Cloetens P, Nicolaï BM. 2008. Three-dimensional gas exchange pathways in pome fruit characterized by synchrotron x-ray computed tomography. *Plant Physiology* 147, 2, 518–527.

Walther P, Wehrli E, Hermann R, Muller M. 1995. Double-layer coating for high-resolution low-temperature scanning electron microscopy. *Journal of Microscopy* 179 (3), 229–237.

Waninge R, Kalda E, Paulsson M, Nylander T, Bergenståhl B. 2004. Cryo-TEM of isolated milk fat globule membrane structures in cream. *Physical Chemistry Chemical Physics*, 6, 1518–1523.

Weng J, Song X, Li L, Qian H, Chen K, Xu X, Cao C, Ren J. 2006. Highly luminescent CdTe quantum dots prepared in aqueous phase as an alternative fluorescent probe for cell imaging. *Talanta* 70 (2), 397–402

Zúñiga RN, Tolkach A, Kulozik U, Aguilera JM. 2010. Kinetics of formation and physicochemical characterization of thermally-induced β-lactoglobulin aggregates. *Journal of Food Science* 75 (5), E261–E268.

4

Glass Transition in Foods

Jasim Ahmed and M. Shafiur Rahman

Contents

4.1	Introduction	94
4.2	Fundamental Considerations	95
	4.2.1 Heat Flow and Heat Capacity	95
	4.2.2 Glass Transition and Molecular Mobility	96
	4.2.2.1 Molecular Mobility above T_g	96
	4.2.2.2 Molecular Mobility below T_g	98
4.3	Analysis of Glass Transition Temperature of Miscible Components	100
4.4	Factors Affecting the Glass Transition Temperature	102
	4.4.1 Molecular Weight	102
	4.4.2 Structure and Architecture	104
	4.4.3 Plasticizers	104
	4.4.4 Pressure	105
	4.4.5 Glass Transition of Water	105
4.5	State Diagram of Food	106
4.6	Characteristic Points of State Diagram	108
	4.6.1 Glass Transition (Samples Containing Freezable Water)	108
	4.6.2 Freezing Curve	109
	4.6.3 Maximal-Freeze-Concentration Condition	109
	4.6.4 Glass Transition (Samples Containing Unfreezable Water)	110
	4.6.5 Solids-Melting Characteristics	111
4.7	Glass Transition Measurement Techniques	111
	4.7.1 Differential Scanning Calorimetry	112
	4.7.2 Dynamic Mechanical Thermal Analysis	113
	4.7.3 Thermomechanical Analysis	114
	4.7.4 Dielectric Thermal Analysis	115
References		116

4.1 INTRODUCTION

Solids in food materials are either completely amorphous or partially crystalline. Amorphous materials can exist in a glassy or rubbery state. Glass formation is in fact just a failure of crystallization (Turnbull and Cohen, 1958; Angel and Donnell, 1977). The glass can be viewed as a supercooled liquid, which is not in internal equilibrium, with modulus >10^9 N.m^{-2}. The glassy state of matter and the glass transition itself have still remained unsolved problems in various disciplines of science and engineering. The existence of glassy or rubbery state is related to the molecular motion inside the material that is mostly governed by temperature, timescale of observation, plasticization, and other factors. A glassy material lies below the temperature at which molecular motions exist on the timescale of the experiment, and a rubbery material is above the temperature at similar conditions (Andrews and Grulke, 1999). A glassy material is formed when a melt or liquid is cooled below its crystalline melting temperature, T_m, at a faster rate to avoid crystallization. The change between rubbery liquid and glassy behavior is known as the glass transition, and the critical temperature, which separates glassy behavior from rubbery behavior is known as the glass transition temperature, T_g (Figure 4.1). This transition occurs with no change in order or structural reorganization of the liquid and is not a thermodynamic first-order process since there is no change in entropy, enthalpy, or volume (Haward, 1973). The transition is considered as a thermodynamic second-order phase transition, which implies a jump in the heat capacity or expansivity of the sample that occurs over a temperature range. The glass transition is the most important property of amorphous materials, both practically and theoretically, since it involves a dramatic slowing down in the motion of chain segments, which rarely one can observe in the static state. Glass transition leads to affect many physical properties including density, specific heat, heat flow, specific volume, mechanical modulus, viscosity, dielectric properties, and so on (Andrews and Grulke, 1999).

The technological importance of the glass transition in amorphous food products is enormous. The polymer science approach of glass transition has been proved useful in

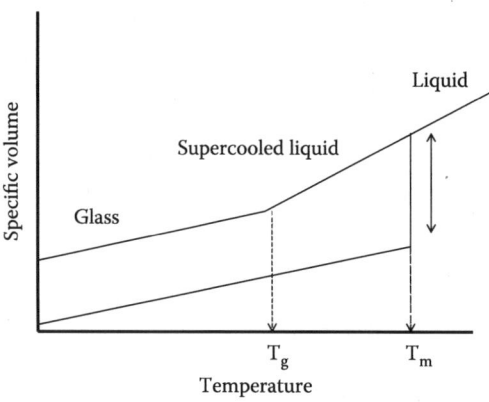

Figure 4.1 Specific volume versus temperature plot for pure amorphous and crystalline materials.

understanding the structure–function relationships of food materials, and can govern food processing, product properties, quality, safety, and stability (Slade and Levine, 1994). Glass transitions of food materials have gained importance in recent years, particularly in the area of dried and frozen products (Truong et al., 2005). In the rubbery state, crispy products undergo a loss of crunchiness and lose texture quality (Martínez-Navarrete et al., 2004) whereas food powders exhibit solute crystallization and/or experience collapse-related phenomena, such as stickiness and caking (Bhandari and Howes, 1999; Telis and Martínez-Navarrete, 2009). Applications of the concept have been currently explored in biomaterials such as starches and their hydrolyzates, and proteins. Processes such as baking, air- and freeze drying, extrusion, flaking, and in the ice cream manufacturing also operate through the glass transition range (Ahmed, 2010). The approach could be used to understand the relationships between moisture and reaction rates (Nelson and Labuza, 1994).

In this chapter, attention is focused on the fundamentals of the glass transition temperature including molecular mobility, compositional variations of glass transition temperature, glass transition in frozen foods, and state diagram of foods.

4.2 FUNDAMENTAL CONSIDERATIONS

4.2.1 Heat Flow and Heat Capacity

Heat flows spontaneously from an object having a higher temperature to an object with a lower temperature. The transfer of heat may occur from an object, to another object with an equal or higher temperature with the aid of an external source. Simply, heat flow provides how much of heat (q) flows per unit time (t).

$$Heat\ flow = \frac{q}{t} \tag{4.1}$$

There is another term "heating rate" commonly used in calorimetry. It provides information on how fast we can heat an object. It is represented by

$$Heating\ rate = \frac{Increase\ in\ temperature}{Time} = \frac{\Delta T}{t} \tag{4.2}$$

The heat capacity of a substance is a measure of how well the substance retains heat. While a material is heated, it obviously causes an increase in the material's temperature. The heat capacity is an extensive variable since a large quantity of matter will have a proportionally large heat capacity. Mathematically, heat capacity is the ratio of an amount of heat increase, to the corresponding increase in its temperature. Heat capacity is a ratio of heat flow to heating rate as shown below:

$$\frac{Heat\ flow}{Heating\ rate} = \frac{q/t}{\Delta T/t} = \frac{q}{\Delta T} = heat\ capacity \tag{4.3}$$

The heat capacity is not only of theoretical interest but also has considerable importance for practical application. Commonly, heat capacity is used to calculate enthalpy difference between two temperatures or as the basis for interpretation of differential scanning calorimeter (DSC) curves. Heat flow and heat capacity are mostly measured by DSC.

4.2.2 Glass Transition and Molecular Mobility

The glass transition temperature (T_g) is considered as a reference temperature: below T_g, the food is expected to be stable; above this temperature, the difference between the storage temperature, T and T_g, that is, ($T–T_g$) often controls the rate of physical, chemical, and biological activities (Champion et al., 2000; Silalai and Roos, 2010). Furthermore, this transition is one of the most important characteristics of the material concerning the mechanical properties.

Amorphous food materials are complex and different modes of molecular mobility have been reported related to the stability or instability of food products during processing or on storage. An asymmetric molecule with N atoms has three modes of translational, three modes of rotational, and 3N-6 modes of internal vibrational mobility (Castellan, 1983). Small proteins or carbohydrates (N > 1500) have over 4500 vibrational modes (Ludescher et al., 2001). Furthermore, different types of motions such as molecular displacement or deformation resulting from a solvent migration or mechanical strain; migration of solvent or solute molecules influenced by a chemical potential gradient, or electric field; molecular diffusion reflecting Brownian movements; and rotation of atom groups or polymeric segments around covalent bonds (Roudaut et al., 2004). Generally, molecular mobility is mainly driven by hydration and temperature. Further, moisture content and interactions between water and other food ingredients direct both thermodynamic and dynamic properties of all aqueous-phase elements.

It is recognized that almost all glass-forming polymers/liquids display both primary-relaxation processes and secondary-relaxation processes (Johari and Goldstein, 1970). The primary process is solely related to the glass transition whereas the secondary processes are more local in nature and remain active to temperatures much lower than T_g. The glass transition phenomenon is a kinetic and relaxation process associated with the so-called α-relaxation of the material (Le Meste et al., 2002). In contrast to α-relaxations, secondary relaxations follow Arrhenius kinetics, and are characterized by activation energy values that are much smaller than those of α-relaxations. A given polymer/biopolymer sample does not have a unique value of T_g because the glass phase is not at equilibrium. The measured value of T_g depends on the structure, molecular weight of the polymer, side chain of the polymer, crystalline or cross-linking, plasticization, thermal history and age, on the measurement method, and on the rate of heating or cooling.

4.2.2.1 Molecular Mobility above T_g

The specific volume of the material increases to accommodate the increased motion of the wiggling chains above glass transition temperature. A theoretical interpretation of the temperature dependence of polymeric materials above glass transition is described based on free volume and viscosity (Gibbs and DiMarzio, 1958).

The total volume per mole of a material measures volume occupied and the free volume. The space within the polymer domain that is available for rotation and translational movements is termed as a free volume that will favor the mobility of macromolecules (Abiad et al., 2010). It may also be considered as the excess volume that is occupied by voids. In polymeric melts, free volume occupies about 30% of the total volume, and the theory predicts that it collapses to about 3% at the T_g (Ferry, 1980). The fraction of free volume (f) can be written as

$$f = f_g + (T - T_g)\Delta\alpha \quad \text{when } T \geq T_g \tag{4.4}$$

$$f = f_g \quad \text{when } T < T_g \tag{4.5}$$

The value of f and f_g remains constant below the T_g. The volume expansion coefficient (α) measures an increase in amplitude of molecular vibration with an increase in temperature. An increase in free volume is expected above T_g and can be estimated by expansion coefficient.

Williams et al. (1955) proposed that the log viscosity is a function of $1/f$ and varies linearly above T_g as shown below:

$$\ln\left(\frac{\eta}{\eta_g}\right) = \frac{1}{f} - \frac{1}{f_g} \tag{4.6}$$

Substitution into Equation 4.1 results

$$\log\left(\frac{\eta}{\eta_g}\right) = -\frac{a(T - T_g)}{b + (T - T_g)} \tag{4.7}$$

The above equation is known as the Willams, Landel, and Ferry (WLF) equation. The equation was fitted with available literature data for polymeric materials to obtain numeric values of a and b. These coefficients were also found to be valid for food materials (Soesanto and Williams, 1981). Equation 4.7 can be rewritten as

$$\log a_T = \frac{-17.44\,(T - T_g)}{51.6 + (T - T_g)} \tag{4.8}$$

The shift factor (a_T) is the ratio of the viscosity at T relative to that of T_g. The latter one is about 10^{14} Pa.s for many materials.

The WLF equation fails to predict the material properties at temperature 100°C above T_g, and, therefore, the valid temperature range is T_g–T_g + 100 K.

The WLF equation describes the profound range of kinetics between T_g and T_m, with correspondingly profound implications for process control, product quality, safety, and the shelf life of food products. The WLF equation has been applied for various food materials (Sopade et al., 2002). It was observed that the relationship fitted well for amorphous sucrose and lactose with the "universal" constants (Roos and Karel, 1991). However, the validity of these universal constants has been challenged by Peleg (1992) and Yildiz and Kokini (2001) after experimenting with some polymers and amorphous sugars. These constants are dependent on various conditions such as moisture content, water activity, and so on. Some serious errors were observed in the magnitude of viscosity during fitting the model over 20–30 K above the T_g (Peleg, 1992). Further, Peleg (1996) advocated that the upward concavity of changes in a transitional region, which cannot be predicted by WLF or Arrhenius equation, can be described by another model with the structure of Fermi's function. Fermi's model can be written as

$$Y(T) = Y_S \cdot \frac{1}{1 + e^{(T-T_c)/a}} \qquad (4.9)$$

where $Y(T)$ is the magnitude of the stiffness or any other mechanical property, Y_s is the magnitude of that parameter at the glassy state, T_c is the characteristic temperature, and a is a constant. The above discussions inferred that though WLF equation is well fitted for various food systems, still, it needs a careful examination before considering as universal for a complex food system.

4.2.2.2 Molecular Mobility below T_g

Glass is rigid and brittle below its T_g since the molecular mobility is restricted below T_g. Although molecular mobility is reduced by orders of magnitude below T_g (relaxation time constant, τ goes from seconds to hours from above to below glass transition), there may still be sufficient mobility below T_g to enable degradation over a long-time frame that may be anywhere from months to years. Mostly, the mobility is local, restricted to vibrations of atoms or bonds, reorientation of small groups of atoms, and thus does not directly involve the surrounding atoms or molecules (Chan et al., 1986). Molecular mobility of freeze-dried formulations containing polymer excipients indicated that glass transition occurs partially even at temperatures below T_g, enhancing molecular rearrangement motions and leading to decreases in the physical and chemical stability (Yoshioka et al., 2004). In many situations, the time required for molecular motion at temperatures below T_g is of great interest due to the usage of stabilizers, bulking agents, and coatings, which usually exist as glasses at usual storage conditions (Shamblin et al., 1999). Furthermore, the importance of chemical reactions and protein denaturation in glassy food systems has created interest in the molecular motions below T_g.

To achieve an infinite stability of food products, the temperature is of particular interest where the timescales for molecular motion (i.e., relaxation time) exceed expected lifetimes, and where structural and configurational mobility ceases (Shamblin et al., 1999). As a first estimation of this temperature, the temperature where the thermodynamic properties of the crystal and the equilibrium supercooled liquid converge is known as the Kauzmann temperature, T_K (Kauzmann, 1948). A direct measurement of T_K is not possible since an equilibrium supercooled liquid is never accessible. T_K of a semicrystalline can be estimated from thermodynamic properties from either the enthalpy or entropy of melting along with C_p values (Alba et al., 1990; Hodge, 1994). Shamblin et al. (1999) described well the timescales of molecular motion below T_g in sorbitol, sucrose, and trehalose system using T_K. An illustration for determination of T_K is shown in Figure 4.2. T_K values estimated from heat capacity and enthalpy of fusion data ranged between 40 and 190 K below the calorimetric T_g (Table 4.1). For more details, readers can consult the original paper of Shamblin et al. (1999).

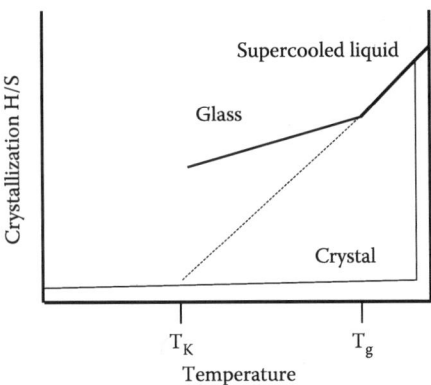

Figure 4.2 Determination of Kauzmann temperature.

Table 4.1 Melting, Glass Transition, and Kauzmann Temperatures for Selected Sugars

Sugar Type	T_m (K)	ΔH_m (J/g)	T_{KS} (K)	T_{KH} (K)	T_g (K)
Trehalose	—	—	—	—	378
Sorbitol	367	162	251	231	264
Sucrose	461	103	331	311	348/351

Source: Adapted from Shamblin, S. et al. 1999. *Journal of Physics Chemistry B*, 103: 4113–4121.

During aging at a temperature in the range 16–47 K below T_g, significant signs of relaxation were noticed for sucrose (Hancock et al., 1995). Relaxation below glass transition is driven by the nonequilibrium state of the glassy system that is trying to relax to the equilibrium state (Hodge, 1995). Relaxation processes can be observed in the glassy state with dynamical mechanical thermal analysis (DMTA) or dielectric spectroscopy (DES); they also give rise to endothermic features on DSC curves. Molecular relaxation time constant (τ) is used to compare molecular mobility among different materials that can be calculated from enthalpy relaxation data (Montserrat, 1994). The maximum enthalpy recovery can be calculated from the following expression:

$$\Delta H_\infty = (T_g - T)\Delta C_p \tag{4.10}$$

where H_∞ is the maximum enthalpy recovery at any given temperature T and ΔC_p is the change of heat capacity at T_g. The extent of a material relaxation (φ_t) can be obtained from the maximum enthalpy recovery, at a preset time–temperature condition as

$$\varphi_t = 1 - \frac{\Delta H_t}{\Delta H_\infty} \tag{4.11}$$

where H_t is the measured enthalpy recovered under the above conditions. Further, the molecular relaxation time, t, can be computed from the empirical Willams–Watts equation that is basically used for dielectric relaxation data (Williams and Watts, 1970).

$$\varphi_t = exp\left(-t/\tau\right)^\beta \tag{4.12}$$

where β is a relaxation time distribution parameter and the value ranges between 0 and 1. The material shows a single relaxation time when $\beta = 1$. Using all these equations and curve-fitting method, Hancock et al. (1995) obtained $\tau = 100$ and $\beta = 0.5$ for sucrose. The β value for glucose, maltose, and maltotriose is reported as 0.23, 0.33, and 0.24, respectively (Syamaladevi et al., 2012).

A special class of secondary relaxations is associated with the α-relaxation in many ways and may play an essential role as the precursor of the structural α-relaxation. These secondary processes are noncooperative motions involving essentially all parts of the molecule, and are termed as the Johari–Goldstein (JG) β-relaxations in current literature (Ngai, 1998). In addition to β-relaxations, several other secondary relaxations such as γ and δ have also been reported (Roudaut et al., 2004). For biopolymers and low-molecular-weight sugars, the secondary β-relaxation has been extensively studied, although its origin is still under investigation. β-relaxations have been reported for several starches (Roudaut et al., 1999; Borde et al., 2002), and polysaccharides such as dextran, pullulan, and so on (Montès et al., 1998). For polysaccharides, these relaxations are believed to be either rotational of lateral (γ relaxation at low temperature) or to local conformational changes of the main chain (β-relaxation closer to T_g) (Montès et al., 1998). A secondary relaxation is observed in both dry white bread and in gelatinized dry starch, but not in gluten in the temperature range of $-60°C$ and $-40°C$ (depending on the frequency) with DES (Roudaut et al., 1999). It is inferred that secondary relaxation is associated with the carbohydrate ingredients (sucrose and starch) of bread. Moreover, in starch–sucrose mixtures, the amplitude of the secondary relaxation exhibited sensitivity to sucrose concentration (Roudaut et al., 1999).

4.3 ANALYSIS OF GLASS TRANSITION TEMPERATURE OF MISCIBLE COMPONENTS

Blending of two components is a very common phenomenon in polymer and biopolymer system. The miscibility of binary polymer blends is commonly ascertained by measuring their glass transition temperature(s). Knowledge of T_g values is particularly needed as a function of composition for binary polymer blends: they provide information about miscibility, compatibility, or immiscibility. Full miscibility of the blend is characterized by a single T_g for all the blends whereas, two T_gs indicate a two-phase system.

The general equation for a binary mixture assuming Δc_{pi} independent of temperature is given by Couchman and Karasz (CK) (1978) as

$$\ln T_g = \frac{W_1 \Delta C_{p1} \ln T_{g1} + W_2 \Delta C_{p2} \ln T_{g2}}{W_1 \Delta C_{p1} + W_2 \Delta C_{p2}} \tag{4.13}$$

T_g is the glass transition temperature of the blend polymer, W_1 and W_2 are the weight fraction of two components, and ΔC_{pi} is the difference in specific heat between the liquid and glassy states as T_{gi}.

The above equation can be modified with some assumptions. Assuming equal heat capacity, Equation 4.13 can be converted as

$$\ln T_g = W_1 \ln T_{g1} + W_2 \ln T_{g2} \tag{4.14}$$

Again, if T_{g2}/T_{g1} is approximately one, then the logarithmic expansion of Equation 4.14 can be limited to the first term and the equation will be

$$T_g = W_1 T_{g1} + W_2 T_{g2} \tag{4.15}$$

Again, considering $\Delta C_{p1} = \Delta C_{p2}$, after rearrangement and expansion of the logarithmic term of Equation 4.14, Fox equation yields

$$\frac{1}{T_g} = \frac{W_1}{T_{g1}} + \frac{W_2}{T_{g2}} \tag{4.16}$$

Further, instead of considering $\Delta C_{p1} = \Delta C_{p2}$ that is not a good approximation, we can consider $\Delta C_{p1}/\Delta C_{p2} = k$, a constant results

$$\ln T_g = \frac{W_1 T_{g1} + k W_2 T_{g2}}{W_1 + k W_2} \tag{4.17}$$

Finally, with the first term of the expansion of the logarithmic term of Equation 4.13 and if $\Delta T_{g1}/\Delta T_{g2}$ approaches to unity, therefore

$$T_g = \frac{W_1 T_{g1} + k W_2 T_{g2}}{W_1 + k W_2} \tag{4.18}$$

where $k = \Delta C_{p1}/\Delta C_{p2}$; Equation 4.18 is known as Gordon–Taylor (GT) equation.

In the GT equation, the ratio of change in heat capacity is replaced by a constant (k) for a binary mixture (water and solute) to be similar in form as CK equation. However, it is difficult to obtain exact values of ΔC_p experimentally especially while small molecules such as water are involved. The ΔC_p reported values vary significantly depending on the method of determination (Roos, 1995a) and, therefore, authenticity of experimentally determined k value is doubtful. Roos (1995b) suggested the following equation to predict k values for carbohydrates:

$$k = 0.0293 T_g, °C + 3.61 \tag{4.19}$$

The k value represents the sensitivity to the water-plasticizing effect; the higher the k value, the greater the moisture content dependence of T_g. The k value of mono- and disaccharides increases linearly with an increase in T_g (Crowe et al., 1996; Roos, 1995a,b); however, a linear decrease in k values has been reported for inulin with increases in T_g (Kawai et al., 2011). The nature and number of sugar moiety plays an important role on dependency of k with T_g for various sugars. The T_g of common sugars such as glucose, fructose, and sucrose is considerably low and, therefore, these sugar moieties function as a plasticizer for multicomponent food systems. Table 4.2 provides a compilation of T_g values of common sugars.

4.4 FACTORS AFFECTING THE GLASS TRANSITION TEMPERATURE

There are many factors that affect the glass transition temperature of any material. These include the composition of the material, presence of plasticizers, molecular weight, crystallinity thermal history, heating/cooling rate, pressure, and structural features. There are a number of structural features that have bearing on the T_g. The polymeric chain backbone, nature of groups adjacent to the backbone, rotation pattern, and stereochemistry influence the T_g since these factors influence the chain mobility and flexibility. In the food system, water activity has a crucial role in controlling glass transition temperature.

4.4.1 Molecular Weight

In the polymeric system, the T_g of a homopolymer increases systematically as a function of molecular weight (more precisely number average molecular weight, M_n) up to a limiting value and leveled off, thereafter. Food is a very complex system with various compositions; so, it is not so simple to follow the polymeric system pattern. However, low-molecular-weight sugars such as fructose, glucose, maltose, maltotriose, and sucrose exhibited an increasing trend of T_g with an increase in molecular weight (Roos, 1995a,b; Syamaladevi et al., 2012). The T_g values of glucose, maltose, and maltotriose increased from 35°C to 129°C with increase in molecular weights from 180 to 504 g/mol (Syamaladevi et al., 2012). Fox and Flory equation (Equation 4.20) has been to describe linear increase in T_g (100–188°C) of maltodextrins with increase in molecular weight (500– 3600 g/mol) (Roos and Karel, 1991).

$$T_g = T_{g\infty} - \frac{A}{M} \quad (4.20)$$

where $T_{g(\infty)}$ and B are the maximum anhydrous T_g, M is the molecular weight and constant, respectively.

Another approach has been used to correlate dextrose equivalent (DE) and T_g. A linear relationship observed between T_g and DE over a wide DE ranges from 2 to 100 (Busin et al., 1996). These authors have advocated that the degree of hydrolysis of starch could be predicted from the measured T_g using the following equation:

$$T_g = -1.4DE + 449.5 \quad (4.21)$$

For higher molecular weight of food materials, it is difficult to measure T_g that needs heating for a wider range of temperatures. These food products commonly decompose at higher temperature and T_g can be predicted using the equation. Slade and Levine (1994) mentioned that although a general correlation between molecular weight and T_g of carbohydrates has been widely accepted, one should be aware that the T_g can vary substantially, even within a series of compounds of the same molecular weight due to their chemical

Table 4.2 Characteristic Parameters of the State Diagram

			Freezing Characteristics				Glass Transition (Unfreezable Water)				Melting	
Material	E	B	$(T_m')_u$ (°C)	$(T_g''')_u$ (°C)	T_g' (or T_g^{iv}) (°C)	X_w' (= $1 - X_s'$)	T_{gs} (°C)	T_c or T_{gw} (°C)	k_c	T_{ms} (°C)	$RV_u/\Delta H_u V_w$	χ
Gelatine[1]	0.026	0.050	−11.9	−14.9	34.0[a]	0.20	153.7	6.8[b]	17.3	235[c]	7.77 × 10⁻³	2.1500
Date[2,9]	0.129	0.053	−38.2	−41.4	−48.0	0.22	9.7	−135.0	2.6	—	—	—
PSE[3]	0.075	0.732	−32.2	−33.4	−76.0	0.48	1.2	−135.0	1.2	—	—	—
Spaghetti[4]	—	—	−10.3	−14.0	40.0[a]	0.19	150.6	7.4[b]	20.3	—	—	—
Date pits[5]	—	—	—	—	—	—	—	—	—	140[d]	2.30 × 10⁻³	0.0068
Starch[6]	—	—	—	—	—	—	—	—	—	—	1.90 × 10⁻³	0.5000
King fish[7]	0.028	0.303	−17.4	−21.4	—	0.31	—	—	—	—	—	—
Garlic[8]	0.080	−0.062	−26.0	−48.6	−38.6	0.18	40.1	−135.0	3.7	—	—	—

B: unfreezable water (kg/kg dry solids).
E: molecular weight ratio of water and solids.
[a] T_g^{iv}.
[b] $T_{c'}$.
[c] Onset melting.
[d] Peak melting.
[1] Rahman et al. (2010).
[2] Guizani et al. (2010).
[3] Al-Rawahi et al. (2013).
[4] Rahman et al. (2011).
[5] Suresh et al. (2013).
[6] Farhat and Blanshard (1997).
[7] Sablani et al. (2007).
[8] Rahman et al. (2005).
[9] Rahman (2004).

structure. The microstructure of polymers plays a significant role in the T_g value whereas very little information is available on food matrix and its architecture.

4.4.2 Structure and Architecture

The structure (e.g., isomerism) and architecture of a polymer affect the glass transition temperature. The kinetic flexibility of a macromolecule is directly related to the ease with which conformational changes between *cis* and *trans* isomers can take place (Rudin, 1999). Two isomers of polylactides (PLAs) (L and DL) with similar molecular weight (7000 Da) revealed that they differed significantly in their T_g (Ahmed et al., 2009). The amorphous sample (DL) showed a T_g of about 44°C whereas the value increased to 68°C for L-form of the polymer. It is further observed that the T_g of PLA differed with the position of side chain to the polymer backbone since the glass-to-rubber transition reflects the onset of movements of sizeable segments of the polymer backbone. Further, insertion of an aromatic ring in the main chain increases the T_g. Cross-linking increases the T_g while the average size of the segments between cross-links is the same or less than the length of the main chain that can initiate to move at a temperature near to T_g (Rudin, 1999).

4.4.3 Plasticizers

Plasticizers are amorphous materials that increase the flexibility of the polymer by decreasing the glass transition temperatures of the blend. Water plays an important role as a plasticizer in amorphous carbohydrate materials. The T_g of amorphous carbohydrate materials dropped significantly with increasing water content, and therefore, the transition from a glassy state to a rubbery state occurs even at constant temperature. The plasticization of carbohydrates by addition of water can be described by the CK and GT equation as described earlier. The glass transition temperature of miscible polymers containing the plasticizer can be calculated from the following equation:

$$\frac{1}{T_g} = \frac{w_A}{T_{gA}} + \frac{w_B}{T_{gB}} \tag{4.22}$$

where T_g and w_i are the T_g and weight fraction of component i of the compound.

A mathematical model suitable for a wider composition of plasticizers was given by Kelley and Bueche (1961). The model was based on the effects of polymer–diluent viscosity and free volume on the glass transition temperature.

$$T_g = \frac{\alpha_s . \bar{V}_s . T_g + \alpha_{p.(1-\bar{V}_s)} T_g^p}{\alpha_s . \bar{V}_s + \alpha_{p.(1-\bar{V}_s)}} \tag{4.23}$$

where $\bar{V}_s$ is the volume fraction of the plasticizer and α_s and α_p are the thermal coefficients of expansion of the plasticizer and the polymer, respectively.

The decrease of the T_g was most significant as moisture content increased from 0 to 5 g water per 100 g dry matter. At that moisture content, the glass transition of all models was close to or below room temperature. In amorphous dry foods, this leads to rapid collapse of structure, stickiness, and probably to increased rates of deteriorative reactions in the plasticized rubbery state (Simatos and Karel, 1988).

4.4.4 Pressure

The effect of pressure has been more extensively studied in recent years due to the availability of new experimental devices and interest from the industry. In general, an increase in pressure on an amorphous material results in a decrease in the total free volume and decreases its entropy. An increase in T_g is expected for postprocess biopolymers. The effect of pressure on the glass transition temperature is expressed as (O'Reilly, 1962)

$$\frac{dT_g}{dP} = \frac{\Delta \beta}{\Delta \alpha} \tag{4.24}$$

where Δ denotes the difference between the properties above and below the T_g, and β and α are the compressibility and expansion coefficients of the material, respectively. However, mostly, experimental results showed significant deviations from the above equation. While working on different isomers of PLAs under high-pressure treatment (up to 650 MPa), Ahmed et al. (2009) observed a drop in T_g of postprocess samples.

4.4.5 Glass Transition of Water

Glass transition of water plays a vital role in food processing, and storage of food products. In drying operation, water is removed to control the microorganisms, and increases the shelf life of the product. Water is being frozen in freezing operation where microbial activity is substantially reduced. A better understanding of glass transition of water provides more insight of water mobility inside food materials and phase diagrams of food during freezing.

Water is an unusual liquid whose structure and properties are not well understood yet although numerous research works have been carried out on the subject. Whether water exhibits a glass transition before crystallization has been much debated over five decades (Angell, 2002; Pryde and Jones, 1952; Velikov et al., 2001). Similar to other hyperquenched glasses, water exhibits a large relaxation exotherm on reheating at the normal heating rate of 10 K.min^{-1}. The release of heat indicates the transformation of a high-enthalpy state to a lower-enthalpy state found in slow-cooled glasses (Velikov et al., 2001). The existence of a glass transition at 136 K has been widely accepted (Johari et al., 1987). Water can exist in more than one distinct amorphous form (Jenniskens et al., 1997). The conversion between different glass structures, the different routes producing glass structures, and the relation between the liquid and the glass phases are under active debate (Giovambattista et al., 2004). The major concern is identification of the glass transition temperature at ambient pressure and the magnitude of the associated jump of the specific heat.

DSC scans reported conflicting values of T_g. The T_g has been either not detected at all (MacFarlane and Angell, 1984) or clearly observed (Hallbrucker and Mayer, 1987). An exothermal peak in the specific heat of properly annealed hyperquenched water supports the estimate of a T_g value at –137°C (Johari et al., 1987), with a specific heat jump from 1.6 to 1.9 J/mol/K. The reported T_g value has been challenged by several researchers (Hallbrucker et al., 1989; Ito et al., 1999). It has been argued (Yue and Angell, 2004) that the small peak measured in the work carried out by Johari et al. (1987) is a prepeak typical of annealed hyperquenched samples preceding the true glass transition located at $T_g = 165$ K. Assigning T_g value of 108°C would explain some of the puzzles related to the glass transition in water (Ito et al., 1999). The T_g at 108°C proposal cannot be experimentally verified due to the homogeneous nucleation of the crystal phase at T_x at 123°C. A value of T_g 136 K was finally assigned on the basis of both extrapolations of unambiguous glass transitions measured in binary aqueous solutions (MacFarlane and Angell, 1984) and of a weak and broad thermal effect observed after extended annealing of the initial deposit at 130 K (Hallbrucker et al., 1989).

4.5 STATE DIAGRAM OF FOOD

State diagram is a stability map of different phases or state of a food as a function of water or solids content and temperature (Levine and Slade, 1986; Rahman, 2004, 2006). The main advantages of drawing a map are to help in understanding the complex changes that occur when the water content and temperature of a food are changed. It also assists in identifying the food's stability during storage as well as selecting a suitable condition of temperature and moisture content for processing (Roos, 1995a,b; Rahman, 2006, 2010, 2012; Slade and Levine, 1988). Levine and Slade (1986) presented a state diagram of polyvinyl pyrrolidone (PVP) that contained the glass line, freezing curve, and the intersection of these lines as T_g''. The characteristic temperature T_g'' was defined as the intersection of the extension of freezing curve to the glass line by keeping similar curvature. This is most probably the first state diagram in the food science literature. Many of their publications later presented the state diagram of starch by quoting the reference of van den Berg (1986) (Slade and Levine, 1988). This state diagram provided four macroregions: region I (i.e., below glass transition temperature line), region II (below maximal-freeze-concentration temperature and above glass transition, i.e., completely frozen), region III (above maximal-freeze-concentration temperature and below freezing curve, i.e., partially frozen), and region IV (i.e., above glass transition and freezing curve) (Rahman, 2009).

A macro–micro region concept combining water activity and glass transition concepts in the state diagram was first presented in the 18th International Congress of Chemical and Process Engineering (CHISA, 2008), August 24–28, 2008, Prague, Czech Republic and was subsequently published in the *International Journal of Food Properties* (Rahman, 2009). On the basis of this, Rahman (2009) developed the state diagram as shown in Figure 4.3 and hypothesized 13 microregions having the highest to the lowest stability based on the location from glass, freezing, and Brunauer, Emmett and Teller (BET)-monolayer lines (Rahman, 2010, 2012). For example, region-1 (relatively nonreacting zone, below the BET-monolayer line and glass line) is the most stable and region-13 (highly reacting zone, far from BET-monolayer line and glass line) is the least stable. The stability decreased as the zone number increased. The most unstable microregion is region-13 since it is the most reactive mobile region. The macro–micro region concept could be the scientific, systematic,

GLASS TRANSITION IN FOODS

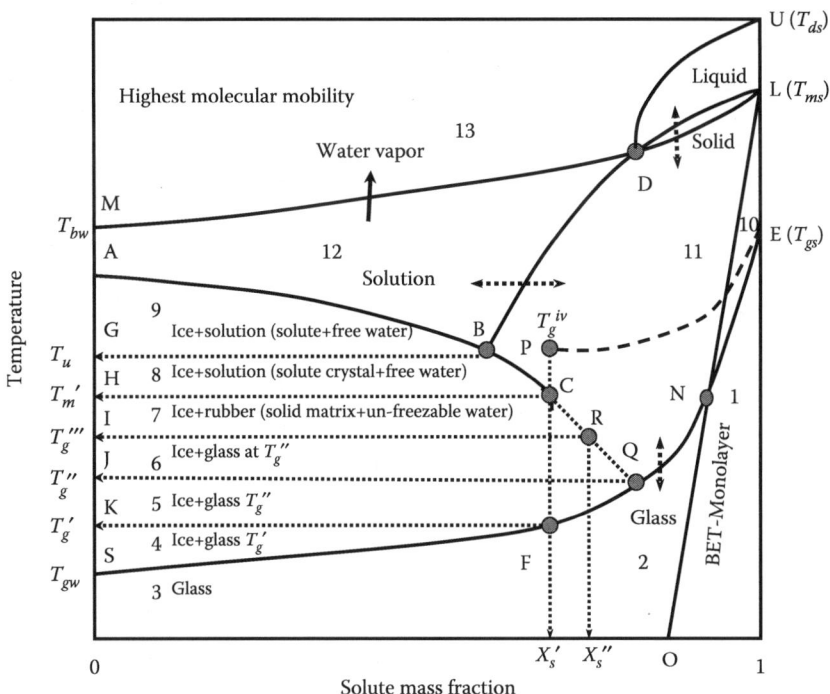

Figure 4.3 State diagram showing different regions and state of foods. (Updated from Rahman, M. S. 2006. *Trends in Food Science and Technology*, 17: 129–141; Rahman, M. S. 2009. *International Journal of Food Properties*, 12(4): 726–740.) T_{ds}: solids-decomposition temperature, T_{ms}: solids-melting temperature, T_{gs}: solids–glass transition temperature, T_g^{iv}: end of solids-plasticization temperature, T_{gw}: glass transition of water, T_u (solute crystallization temperature during freeze concentration), T_m' (maximal-freeze-concentration condition, i.e., end point of freezing), T_g''' (glass transition of the solids matrix in the frozen sample as determined by DSC), T_g'' (intersection of the freezing curve to the glass line by maintaining the similar curvature of the freezing curve), and T_g' (glass transition at maximal-freeze-concentration, i.e., glass transition from the samples without freezable water), T_{bw}: boiling temperature of water.

and rational approaches to determine the stability and has potential since it could incorporate other preservation hurdles within a narrow and/or wide range of water content (i.e., solids content) and temperature (Rahman, 2012). Each microregion could be studied for specific quality attributes separately or numbers of attributes considering characteristics of a medium in the microregion (Rahman, 2010). Numbers of applications are presented earlier to utilize the state diagram in determining suitable conditions for processing and to determine storage stability, for example, determining drying paths for air-, vacuum-, spray-, and freeze drying; determining tempering conditions to reduce rice kernel damages during drying; determining baking paths for bread; determining the conditions to avoid solutes crystallization; and determining the location of lowest sugar hydrolysis rate during storage (Rahman, 2012).

ENGINEERING PROPERTIES OF FOODS

To develop the state diagram (Figure 4.3), it is necessary to generate data for the freezing curve ABC (i.e., freezing points *versus* solids content), glass transition line ENS or EP (i.e., glass transition temperature *versus* solids content), solids-melting line LDB (i.e., solids-melting temperature versus solids content), vapor line LDM or UDM, maximal-freeze-concentration conditions (T_m', T_g', and X_s'), and other characteristic points (T_u, T_g'', T_g''', T_g^{iv} and X_s'', and X_s'''). These lines and characteristic points are mainly developed by thermal analysis technique (i.e., DSC). However, many instances more sensitive modulated DSC (MDSC) are being used. Considering this as a reference, it is possible to measure foods' thermal, thermodynamic, mechanical, structural, relaxation, electrical and electromagnetic properties by other techniques. Other commonly used techniques in foods are DMA or DMTA, DES, and nuclear magnetic resonance (NMR). The parameters measured by these techniques are used to relate the characteristic points of the state diagram and to explore different characteristics or attributes of the phases and states in the 13 microregions.

4.6 CHARACTERISTIC POINTS OF STATE DIAGRAM

The point C (T_m' and X_s') is the central point of the state diagram indicating the boundary of freezable and unfreezable water (Rahman, 2006). Line ABC is the freezing curve, BDL is the solids-melting line, ENQFS is the glass transition line, and LNO is the BET-monolayer line.

4.6.1 Glass Transition (Samples Containing Freezable Water)

Typical DSC heating thermogram of a food is shown in Figure 4.4 for samples containing freezable water (Rahman, 2004). A sample containing freezable water shows two shifts: one at low temperature (i.e., first glass transition, marked as (A) and another one (i.e., second glass transition, marked as (B) just before the melting endothermic peak of ice (marked as

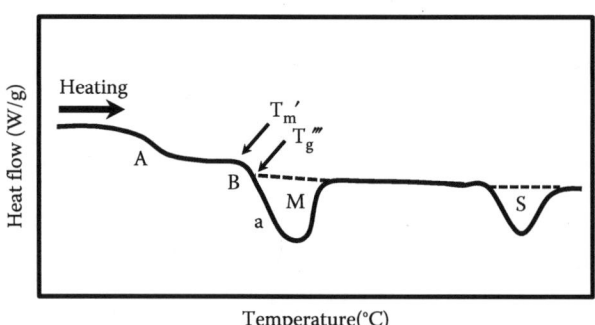

Figure 4.4 A typical DSC heating thermogram of a food containing freezable water. A: First glass transition shift, B: second glass transition shift, M: ice-melting endotherm, T_m' and T_g': apparent maximal-freeze-concentration condition, and S: solids-melting endotherm.

M). Solids-melting endotherm is also observed as identified S in Figure 4.4. The values of T_m' and T_g''' are determined from the second transition as shown in Figure 4.4. Moisture loss (i.e., leak from the DSC pan) in the case of high moisture content samples may cause difficulty in measuring this peak. The freezing point and melting enthalpy can be determined from the endothermic peak. The initial or equilibrium freezing point was considered as the maximum slope in the ice-melting endotherm (marked as a) as suggested by Rahman (2004). This freezing point data are plotted as a function of solids content and considered as freezing curve (i.e., line ABC in Figure 4.3).

4.6.2 Freezing Curve

The theoretical Clausius–Clapeyron equation is usually used to estimate the freezing point (i.e., line ABC) and the equation can be written as

$$\delta = -\frac{\beta}{\lambda_w} \ln\left[\frac{1 - X_s^o}{1 - X_s^o + EX_s^o}\right] \quad (4.25)$$

where δ is the freezing point depression $(T_w - T_F)$, T_F is the freezing point of food (°C), T_w is the freezing point of water (°C), β is the molar freezing point constant of water (1860 kg K/kg mol), λ_w is the molecular weight of water, X_s^o is the initial solids mass fraction, and E is the molecular weight ratio of water and solids (λ_w/λ_s). The Clausius–Clapeyron equation is limited to the ideal solution (i.e., for a very dilute solution). Theoretical models can be improved by introducing parameters for nonideal behavior when a fraction of total water is unavailable for forming ice. The unfreezable water content B can be defined as the ratio of unfrozen water even at very low temperature to total solids (kg unfrozen water/kg dry solids). Equation 4.25 can be modified based on this concept as (Chen, 1986)

$$\delta = -\frac{\beta}{\lambda_w} \ln\left[\frac{1 - X_s^o - BX_s^o}{1 - X_s^o - BX_s^o + EX_s^o}\right] \quad (4.26)$$

In the literature, the values of B were determined as negative or positive values. This indicates that the value of B shows less sensitive (as compared to E) to the model and nonlinear optimization could estimate within wide variations, although the overall good accuracy for prediction is obtained. This is one of the generic problems when a theoretical-based model is extended to fit the experimental data by nonlinear regression (Rahman, 2004).

4.6.3 Maximal-Freeze-Concentration Condition

Specific procedure is used to determine maximal-freeze-concentration condition as compared to the freezing. Samples with freezable water are first scanned from low temperature to determine freezing point and apparent maximal-freeze-concentration condition $[(T_m')_a$ and $(T_g''')_a]$. After knowing the apparent $(T_m')_a$ and $(T_g''')_a$, the sample was scanned similarly with 30 min annealing at $[(T_m')_a - 1]$ and then annealed maximal-freeze-concentration temperatures $(T_m')_n$ and $(T_g''')_n$ are determined. The use of annealing condition allowed to maximize the formation of ice before the second heating cycle and to avoid the appearance of exothermic or endothermic peak before the glass transition. The ultimate-freeze-concentration condition is then determined from the freezing curve and glass transition line. The ultimate maximal-freeze-concentration (X_s') is determined from the intersection point of the extended freezing curve (i.e., ABC in Figure 4.3) by maintaining the similar curvature and drawing a horizontal line passing through the ultimate $(T_m')_u$ [average value of the lowest possible $(T_m')_n$]. Finally, X_s' is read on the x-axis by drawing a vertical line passing through point C in Figure 4.1. The unfreezable water (i.e., key characteristic point in the state diagram, C) can be estimated from X_s' using $X_w' = 1 - X_s'$.

4.6.4 Glass Transition (Samples Containing Unfreezable Water)

Typical DSC thermogram of dates is shown in Figure 4.5 for the sample containing unfreezable water. It shows a shift for the glass transition (marked as A) followed by an endothermic peak for solids melting (marked as S). The glass transition temperature of foods and biological materials is commonly modeled by the GT equation as mentioned earlier (Equation 4.18) and the equation can be rearranged as

$$T_g = \frac{X_s T_{gs} + k_c X_w T_c}{X_s + k_c X_w} \qquad (4.27)$$

where T_g used here is the T_g, T_{gs} and T_c (usually considered as T_{gw}, glass transition of water) are the glass transition temperature of mixture, solids, and characteristic, respectively; k_c is the GT parameter; and X_w and X_s are the mass fraction of water and solids (wet basis). In the original GT (1952) equation, the parameters T_{gs} and k_c are usually determined by considering that T_g of pure water as $-135°C$ (Johari et al., 1987). However, in many cases, defining a critical point fitted or represented experimental data better; thus, Rahman (2009) defined critical point instead of T_{gw}. In the above Equation 4.3, T_c is considered as

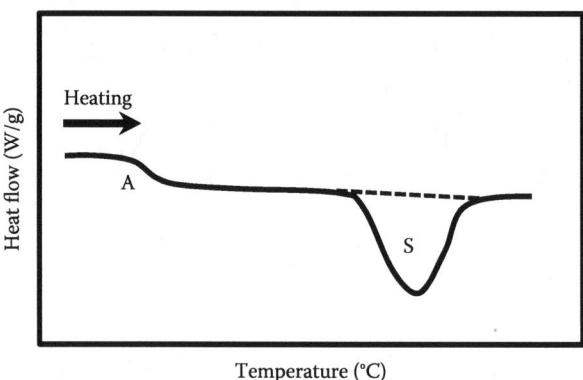

Figure 4.5 A typical DSC heating thermogram of a food containing unfreezable water. A: Glass transition shift and S: solids-melting endotherm.

critical temperature and it is related to T_g^w as defined by Rahman (2009). The values of T_g^w are determined from the intersection point of the glass transition line and a vertical line through X_s' (Figure 4.1). The values of T_c can be related as (Rahman et al., 2010)

$$T_c = T_g^{iv}(1 - X_s') \tag{4.28}$$

4.6.5 Solids-Melting Characteristics

The melting point of a polymer in a diluent can be modeled by Flory–Huggins as (Flory, 1953)

$$\frac{1}{T_{mp}} - \frac{1}{T_{ms}} = \left(\frac{R}{\Delta H_u}\right)\left(\frac{V_u}{V_w}\right)\left(\varepsilon_w - \chi\varepsilon_w^2\right) \tag{4.29}$$

where, T_{mp} and T_{ms} are the peak of melting temperature for the polymer with diluent, and pure polymer (i.e., only solids) (K), R is the gas constant (8.314 J/g mol K), ΔH_u is the heat of fusion for repeated polymer units in the diluent (J/g), V_w is the molar volume of the diluent (m³/g mol), V_u is the molar volume of polymer unit (m³/g mol), ε_w is the volume fraction of the diluent, and χ is the Flory–Huggins polymer–diluent interaction parameter, respectively. The volume fraction of water was calculated from the following equation:

$$\varepsilon_w = \frac{X_w/\rho_w}{X_w/\rho_w + X_s/\rho_s} \tag{4.30}$$

where, X_w is the mass fraction of water (wet basis, kg/100 kg sample), ρ_w is the density of water (kg/m³), X_s is the solids content (kg/100 kg sample), and ρ_s is the density of dry solids (kg/m³), respectively. The density of dry solids can be determined using helium gas pycnometer (Rahman et al., 2005). Table 4.2 shows the characteristic parameters of the state diagrams of selected foods.

4.7 GLASS TRANSITION MEASUREMENT TECHNIQUES

The glass transition temperatures of amorphous food materials can be measured by incessantly measuring various physical properties as a function of temperature. These measurements may include specific volume, conductivity, deformation, elasticity, and thermal properties (e.g., heat capacity). Table 4.3 summarizes T_g measurement methods and a description of each technique follows. The change of these properties can be identified at a particular temperature range. The T_g can vary widely depending on the property being measured and the sensitivity of the instrument. Mostly, three temperatures are assigned to describe the T_g: (i) the onset temperature where the first changes in the monitored properties are observed, or (ii) as the inflection point, midpoint of the steepest slope connecting the onset, and (iii) offset horizontals. These methods are briefly discussed below.

Table 4.3 Common Thermal Analysis Methods Used to Investigate the Glass Transition, the Technique or Concept of Each Method, and the Event That Signals the Glass Transition

Method	Technique	Event-Signaling Glass Transition
DSC	Thermodynamic	Step change in the heat flow or heat capacity
Differential thermal analysis (DTA)	Thermodynamic	Temperature of sample differs from reference temperature
TMA	Volumetric	Increased penetration or dimension change
DMTA	Mechanical	Decrease in storage modulus and maximum in loss modulus
DETA	Mechanical	Dielectric loss constant goes through a maxima

Source: Carter, B. P., Schmidt, S. J. 2012. *Food Chemistry*, 132: 1693–1698.

4.7.1 Differential Scanning Calorimetry

DSC is a thermoanalytical technique, which measures the differences in the amount of heat flow or the difference in temperatures between a sample and a reference when both are heated or cooled at a constant rate. The differential signal is the basic characteristic of DSC. The most important differentiation with other calorimeters is the dynamic mode of operation. On the mechanism of operation, DSCs can be classified into two types: (i) the heat flux DSC and (ii) the power compensation DSC. In the heat flux DSC, a defined exchange of the heat is measured with the environment that takes place through a well-defined heat conduction path with known thermal resistance. In this type of DSC, the sample material sealed in a pan, and an empty reference pan are positioned on a thermoelectric disk surrounded by a furnace. The furnace is heated at a linear heating rate, and the heat is transferred to the sample and reference pan through the thermoelectric disk. The temperature difference is associated to the enthalpy change in the sample and can be transformed into heat flow information through calibration runs and mathematical equations built into most of the programs supplied by the instrument manufacturers. For the power compensation DSC, the heat to be measured is compensated with the electrical energy, by changing Joule's heat. In this type of DSC, the sample and reference pans are placed in separate furnaces heated by separate heaters. The difference of power input between these two specimens is measured as a function of time and recorded as the sample temperature is increased at a constant rate. With modulated temperature DSC (MTDSC), a recent extension of conventional DSC using a modulated temperature input signal and information on the "amplitude of the complex heat capacity" is obtained, both in (quasi)isothermal and nonisothermal conditions. This complementary information, giving rise to a deconvolution of the (total) heat flow signal into "reversing" and "nonreversing" contributions, enables a more detailed study of complicated material systems (De Meuter et al., 1999).

During a thermal scan, thermally induced conformational changes and phase transitions are detected through peaks and inflection points. The direction of the peak indicates the nature of reaction, that is, exothermic or endothermic. A typical DSC thermogram is illustrated in Figure 4.6.

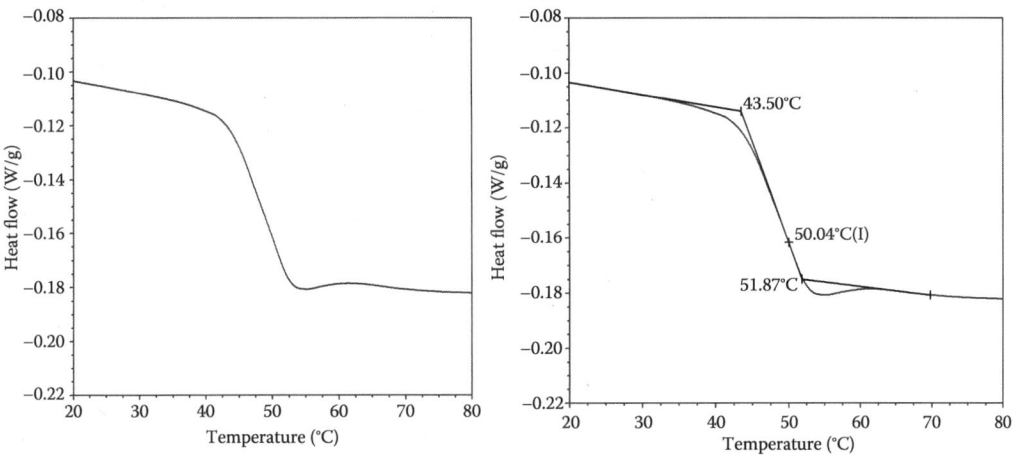

Figure 4.6 Typical glass transition temperature for a biopolymer at a heating rate of 5°C/min.

4.7.2 Dynamic Mechanical Thermal Analysis

DMTA/dynamic–mechanical analysis (DMA) is a nondestructive powerful mechanical technique for developing insight into structure, morphology, and viscoelastic behavior of polymeric materials with a high level of sensitivity in the detection of changes in mobility of the polymer chain segments within the linear viscoelastic region (LVR). This technique uses an oscillating stimulus (load or deformation) to a sample and analyzes the response of the material. DMTA test generates some important physical parameters: (i) storage modulus (E′)—a measure of the maximum energy, stored in the material during one cycle of oscillation, (ii) loss modulus (E″)—dissipated energy by the sample, and (iii) mechanical damping factor (tan δ)—the ratio of the loss modulus to storage modulus and is related to the degree of molecular mobility in the polymer material. A nonelastic (i.e., viscous) material is characterized by high tan δ, whereas a low value of tan δ is a feature of solid material. Generally, E′, E″, and tan δ are plotted against temperature in the DMTA test. Typically, the glass transition (T_g) is detected at the temperature(s) where either a maximum in the mechanical damping parameter (tan δ) or loss modulus (E″) observes. Sometimes, the T_g is obtained from the onset of the drop of storage modulus E′. It has been reported that DMTA is 100 times more sensitive than DSC in detection of the T_g (Chartoff, 1997); however, a larger sample size is required.

A typical DMTA curve is shown in Figure 4.7. A primary relaxation corresponding to the T_g is observed at a characteristic temperature, which is also known as an α-relaxation (T_α). The T_α for oat flour and rice flour as well as for their blend ranged between in the order of 65°C and 90°C, at about 12% (d.b.) moisture content (Sandoval et al., 2009). Furthermore, some other temperature-dependent relaxation transitions (β and γ) associated with largely amorphous systems that may occur due to certain structural features, for example, temperature-dependent mobility of side groups are also evident. The α transition in DMTA occurs as the most intense transition at the highest temperature, whereas the transition at

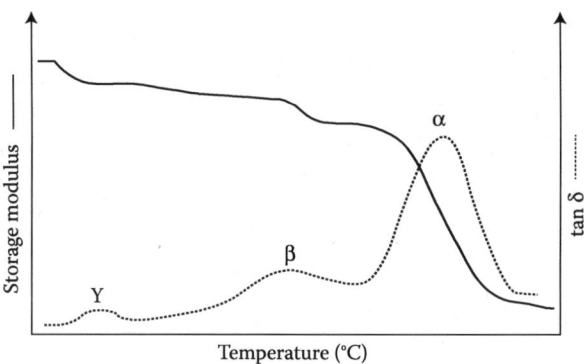

Figure 4.7 Typical DMTA curve showing different transitions at constant frequency. (From Jones, D. S. 2012. *Advanced Drug Delivery Reviews*, 64: 440–448.)

the second highest temperature is called the β transition and the third transition is termed the γ-transition and occurs due to rotation of the side chains, commonly methyl groups (Lafferty et al., 2002).

4.7.3 Thermomechanical Analysis

Thermomechanical analysis (TMA) is a useful technique for detecting T_g and it has been extensively applied in the measurement of thermal stability of polymers. TMA can be defined as the measurement of a specimen's dimensions (e.g., length) as a function of temperature when it is subjected to a constant mechanical stress. Thermal expansion coefficients (α) can be determined and changes in this property with temperature (and/or time) monitored (Equation 4.31). The unit of α is K^{-1}. Some authors still use the conventional unit as $\mu m \cdot m^{-1} \cdot °C^{-1}$. The TMA glass transition is considered as the change in the coefficient of thermal expansion when the polymer transforms from the glass to rubber state with the associated change in the free molecular volume. In a TMA experiment, a probe is dipped onto the surface of a test sample and the movement of the probe is measured as the sample is heated. With a load applied to the probe, a combination of modulus changes and expansion of the sample are observed. Depending on the probe/sample contact area and the load applied, the T_g can be detected by either an upward (expansion) or downward (penetration) movement of the probe. With large contact areas and low forces, expansion is primarily observed, whereas, for small contact areas and high forces, penetration is primarily observed.

$$\alpha(\Delta T) = \frac{1}{l_o} \cdot \frac{l_2 - l_1}{T_2 - T_1} = \frac{1}{l_o} \cdot \frac{\Delta l}{\Delta T} \qquad (4.31)$$

The linear expansion and the differential coefficient of linear expansion of the polymer at the glass transition are illustrated in Figure 4.8. The expansion above T_g depends on probe area, applied load, and the temperature-dependent modulus of elasticity.

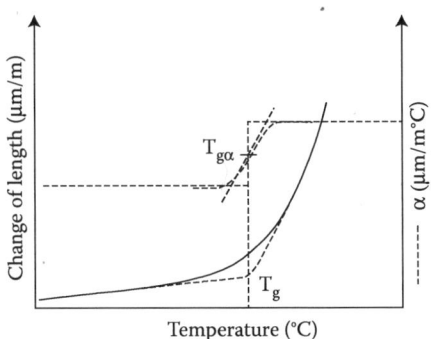

Figure 4.8 Typical TMA glass transition temperature of a material following expansion.

4.7.4 Dielectric Thermal Analysis

Dielectric thermal analysis (DETA) is a valuable tool extensively used to study molecular dynamics in dispersed systems, and to identify the glass transition region for polymers and biopolymers. Dielectric measurements are the electrical analog of dynamic mechanical measurements. A test sample can be subjected to a constant or oscillating electric field rather than a mechanical stress during measurements. The test specimen is generally placed as a thin film between two metal electrodes so as to form a parallel plate capacitor, and a sinusoidal alternating current is applied across the plates to establish an electric field in the specimen. As a result, a current will pass through the capacitor partially as a capacity current and partially as a resistive current. The resistive portion of the current that flows through the capacitor will be in phase with applied alternating voltage but the capacity current will be 90° out of phase with the applied alternating voltage. Measurement of the amplitude and individual phase differences of the voltage and current through the capacitor provide dielectric data for the polymer and the extent of ionic conductance. With increasing frequency, the dielectric constant (ε') decreases regularly whereas, the dielectric loss (ε'') exhibits a peak due to the delay in dipole moments; this phenomenon is known as dielectric relaxation (Maxwell, 1873). Both the dipole peak that occurs in the dielectric loss factor and a sharp transition in permittivity during a temperature scan have been correlated with the glass transition temperature. Dielectric relaxation has been reported for many systems, especially concentrated sugar–water, bovine serum albumin/myoglobin–water systems (Noel et al., 1996; Shinyashiki et al., 2009; Jansson et al., 2011). The variation of tan δ with temperature, at a frequency of 1 kHz, is shown in Figure 4.9 for D-mannose and its 10% dispersion. The peak in tan δ was recorded at a temperature of 20°C above the calorimetric T_g, which is the dielectric α-relaxation. The calorimetric and dielectric techniques probe exhibited similar α-relaxation process although shifted in temperature due to difference in relaxation time. At temperatures below the T_g, β-relaxation peaks were observed for D-mannose. The addition of 10% water (w/w) on the dielectric response has shifted the α-relaxation peak to lower temperatures and increase the strength of the β-relaxations.

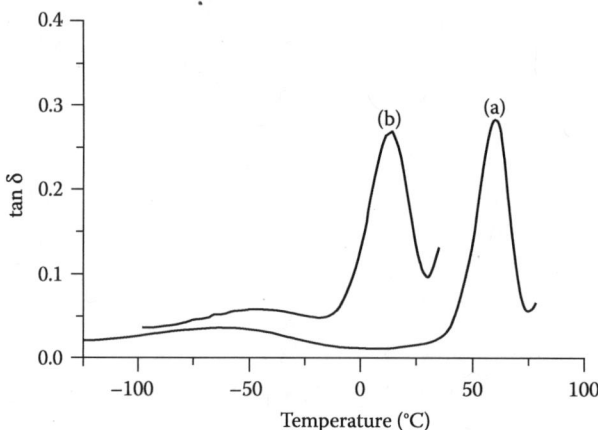

Figure 4.9 Variation of tan d with temperature for (a) amorphous D-mannose and (b) water mixture (10%) at 1 kHz. (Adapted from Noel, T. R., Parker, R., Ring, S. G. 1996. *Carbohydrate Research*, 282: 193–206.)

REFERENCES

Abiad, M. G., Campanella, O. H., Carvajal, M. T. 2010. Assessment of thermal transitions by dynamic mechanical analysis (DMA) using a novel disposable powder holder. *Pharmaceutics*, 2: 78–90.

Alba, C., Busse, L. E., List, D. J., Ange, C. A. 1990. Thermodynamic aspects of the vitrification of toluene, and xylene isomers, and the fragility of liquid hydrocarbons. *Journal of Chemistry Physics*, 92: 617–625.

Ahmed, J., Varshney, S. K., Zhang, J. X., Ramaswamy, H. S. 2009. Effect of high pressure treatment on thermal properties of polylactides. *Journal of Food Engineering*, 93: 308–312.

Ahmed, J. 2010. Thermal phase transitions in food. In: *Mathematical Modeling of Food Processing*. Mohammed, F., ed. CRC Publication, USA.

Al-Rawahi, A., Rahman, M. S., Waly, M., Guillemin, G. J. 2013. Thermal characteristics of a water soluble extract obtained from pomegranate skin: Developing a state diagram for determining stability. *Industrial Crops and Products*, 48: 198–204.

Andrews, R. J., Grulke, E. A. 1999. Glass transition temperatures of polymers. In: *Polymer Handbook*, 4th ed. Brandrup, J., Immergut, E. H., Grulke, E. A. Eds. John Wiley & Sons, NY. pp. 193–198.

Angell, C. A. 2002. Liquid fragility and the glass transition in water and aqueous solutions. *Chemistry Review*, 102: 2627–2649.

Bhandari, B. R., Howes, T. 1999. Implication of glass transition for the drying and stability of dried foods. *Journal of Food Engineering*, 40: 71–79.

Borde, B., Bizot, H., Vigier, G., Buleon, A. 2002. Calorimetric analysis of the structural relaxation in partially hydrated amorphous polysaccharides. I. Glass transition and fragility. *Carbohydrate Polymer*, 48:83–96.

Busin, L., Buisson, P., Bimbenet, J. J. 1996. Notion de transition vitreuse appliquee au sechage par pulverisation de solutions glucidiques [Concept applied to the glass transition by drying spraying of carbohydrate solutions]. *Sciences des Aliments*, 16: 443–459.

Carter, B. P., Schmidt, S. J. 2012. Developments in glass transition determination in foods using moisture sorption isotherms. *Food Chemistry*, 132: 1693–1698.

Castellan, G. W. 1983. *Physical Chemistry*. The Benjamin/Cummings Publishing, Menlo Park.

Champion, D., Le Meste, M., Simatos, D. 2000. Towards an improved understanding of glass transition and relaxations in foods: Molecular mobility in the glass transition range. *Trends in Food Science and Technology*, 11(2): 41–55.

Chan, R. K., Pathmanathan, K., and Johari, G. P. 1986. Dielectric relaxations in the liquid and glassy states of glucose and its water mixtures. *Journal of Physics Chemistry*, 90: 6358–6362.

Chartoff, R. P. 1997. Thermoplastic polymers. In *Thermal Characterization of Polymeric Materials*, 2nd ed., Vol. 1, E. A. Turi (ed.). pp. 483–743. Academic Press, New York.

Chen, C. S. 1986. Effective molecular weight of aqueous solutions and liquid foods calculated from the freezing point depression. *Journal of Food Science*, 51(6): 1537–1553.

Couchman, P. R., Karasz, F. E. 1978. A classical thermodynamic discussion of the effect of composition on glass-transition temperatures. *Macromolecules*, 11(1): 117–119.

Crowe, L. M., Reid, D. S., Crowe, J. H. 1996. Is trehalose special for preserving dry biomaterials? *Biophysical Journal*, 71(4): 2087–2093.

De Meuter, P., Rahier, H., Van Mele, B. 1999. The use of modulated temperature differential scanning calorimetry for the characterisation of food systems. *International Journal of Pharmaceutics*, 192: 77–84.

Farhat, I. A., Blanshard, J. M. V. 1997. On the extrapolation of the melting temperature of dry starch from starch–water data using the Flory–Huggins equation. *Carbohydrate Polymers*, 34(4): 263–265.

Ferry, J. D. 1980. *Viscoelastic Properties of Polymers*. John Wiley and Sons, New York.

Flory, D. J. 1953. *Principles of Polymer Chemistry*, Cornell University Press, Ithaca, New York.

Gibbs, J. H., DiMarzio, E. A. 1958. Nature of the glass transition and the glassy state. *Journal of Chemistry Physics*, 28: 373–383.

Giovambattista, N., Angell, C. A., Sciortino, F., Stanley, H. E. 2004. Glass-transition temperature of water: A simulation study, *Physics Review Letters* 93: 047801.

Gordon, M., Taylor, J. S. 1952. Ideal copolymers and second-order transitions in synthetic rubbers. I. Non-crystalline polymers. *Journal of Applied Chemistry*, 2: 493–500.

Guizani, N., Al-Saidi, G., Rahman, M. S., Bornaz, S., Al-Alawi, A. A. 2010. State diagram of dates: Glass transition, freezing curve and maximal-freeze-concentration condition. *Journal of Food Engineering*, 99: 92–97.

Hallbrucker, A., Mayer, E. 1987. Calorimetric study of the vitrified liquid water to cubic ice phase transition. *Journal of Physics Chemistry*, 91(3): 503–505.

Hallbrucker, A., Mayer, E., Johari G. P. 1989. Glass–liquid transition and the enthalpy of devitrification of annealed vapor-deposited amorphous solid water. A comparison with hyperquenched glassy water. *The Journal of Physical Chemistry*, 93: 4986–4990.

Hancock, B. C., Shamblin, S. L., Zografi, G. 1995. Molecular mobility of amorphous pharmaceutical solids below their glass transition temperatures. *Pharmaceutical Research*, 12(6):799–806.

Haward, R. N. 1973. *The Physics of Glass Polymers*. Applied Science Publishers Ltd., London.

Hodge, I. M. 1994. Enthalpy relaxation and recovery in amorphous materials. *Journal of Non-Crystalline Solids*, 169(3): 211–266.

Hodge, I. M. 1995. Science. *New Series*, 267: 1945–1947.

Ito, K., Moynihan, C. T., Angell, C. A. 1999. Thermodynamic determination of fragility in liquids and a fragile-to-strong liquid transition in water. *Nature*, 398: 492–495.

Jansson, H., Bergman, R., Swenson, J. 2011. Role of solvent for the dynamics and the glass transition of proteins. *Journal of Physics Chemistry B*, 115 (14): 4099–4109.

Jenniskens, P., Banham, S. F., Blake, D. F., McCoutra, M. R. S. Jenniskens, P. et al., 1997. Liquid water in the domain of cubic crystalline ice I_c. *Journal of Chemistry Physics*, 107: 1232.

Johari, G. P., Goldstein, M. 1970. Viscous liquids and the glass transition. II. Secondary relaxations in glasses of rigid molecules. *Journal of Chemical Physics*, 53: 2372–2388.

Kauzmann, W. 1948. The nature of the glassy state and the behavior of liquids at low temperatures. *Chemistry Review*, 43: 219–256.

Kawai, K., Fukami, K., Thanatuksorn, P. 2011. Effects of moisture content, molecular weight, and crystallinity on the glass transition temperature of inulin—*Carbohydrate Polymer*, 83(2): 934–939.

Kelley, F. N., Bueche, F. 1961. Viscosity and glass temperature relations for polymer–diluent systems. *Journal of Polymer Science*, 50: 549–556.

Lafferty, S. V., Newton, J. M., Podczeck, F. 2002. Dynamic mechanical thermal analysis studies of polymer films prepared from aqueous dispersion. *International Journal of Pharmaceutics*, 235(1–2):107–111.

Le Meste, M., Champion, D., Roudaut, G., Blond, G., Simatos, D. 2002. Glass transition and food technology: A critical appraisal. *Journal of Food Science*, 67: 2444–2458.

Levine, H., Slade, L. 1986. A polymer physico-chemical approach to the study of commercial starch hydrolysis products (SHPs). *Carbohydrate Polymer*, 6: 213–244.

Ludescher, R. D., Shah, N. K., McCaul, C. P., Simon, K. V. 2001. Beyond T_g: Optical luminescence measurements of molecular mobility in amorphous solid foods. *Food Hydrocolloids*, 15: 331–339.

MacFarlane, D. R., Angell, C. A. 1984. Nonexistent glass transition for amorphous solid water. *Journal of Physics Chemistry*, 88: 759–762 .

Martínez-Navarrete, N., Moraga, G., Talens, P., Chiralt, A. 2004. Water sorption and the plasticization effect in Wafers International. *Journal of Food Science and Technology*, 39(5):555–562.

Maxwell, J. C. 1873. *A Treatise on Electricity and Magnetism*. Clarendon Press, London, UK.

Montès, H., Mazeau, K., Cavaillé, J. Y. 1998. The mechanical β relaxation in amorphous cellulose. *Journal of Non-Crystalline Solids*, 235–237(2): 416–421.

Montserrat, S. 1994. Physical aging studies in epoxy resins. I. Kinetics of the enthalpy relaxation process in a fully cured epoxy resin. *Journal of Polymer Science Part B*, 32: 509–522.

Nelson, K. A., Labuza, T. P. 1994. Water activity and food polymer science: Implications of state on Arrhenius and WLF models in predicting shelf life. *Journal of Food Engineering*, 22: 271–289.

Ngai, K. L. 1998. Relation between some secondary relaxations and the α relaxations in glass-forming materials according to the coupling model. *The Journal of Chemical Physics*, 109: 6982–6995.

Noel, T. R., Parker, R., Ring, S. G. 1996. A comparative study of dielectric relaxation behaviour of glucose, maltose, and their mixtures with water in the liquid and glassy states. *Carbohydrate Research*, 282: 193–206.

O'Reilly, J. 1962. The effect of pressure on glass temperature and dielectric relaxation time of polyvinyl acetate. *Journal of Polymer Science*, 57: 429–444.

Peleg, M. 1996. On modeling changes in food and biosolids at and around their glass transition temperature range. *Critical Reviews in Food Science and Nutrition*, 36 (1/2): 49–67.

Peleg, M. 1992. On the use of the WLF model in polymers and foods. *Critical Reviews in Food Science and Nutrition*, 32 (1): 59–66.

Pryde, J. A., Jones, G. O. 1952. Properties of vitreous water. *Nature*, 170: 635–639.

Rahman, M. S. 2004. State diagram of date flesh using differential scanning calorimetry (DSC). *International Journal of Food Properties*, 7(3): 407–428.

Rahman, M. S. 2006. State diagram of foods: Its potential use in the food processing and product stability. *Trends in Food Science and Technology*, 17: 129–141.

Rahman, M. S. 2009. Food stability beyond water activity glass transition: Macro–micro region concept in the state diagram. *International Journal of Food Properties*, 12(4): 726–740.

Rahman, M. S. 2010. Food stability determination by macro–micro region concept in the state diagram and by defining a critical temperature. *Journal of Food Engineering*, 99: 402–416.

Rahman, M. S. 2012. Applications of macro–micro region concept in the state diagram and critical temperature concepts in determining the food stability. *Food Chemistry*, 132: 1679–1685.

Rahman, M. S., Al-Saidi, G., Guizani, N., Abdullah, A. 2010. Development of state diagram of bovine gelatin by measuring thermal characteristics using differential scanning calorimetry (DSC) and cooling curve method. *Thermochimica Acta*, 509: 111–119.

Rahman, M. S., Sablani, S. S., Al-Habsi, N., Al-Maskri, S., Al-Belushi, R. 2005. State diagram of freeze-dried garlic powder by differential scanning calorimetry and cooling curve methods. *Journal of Food Science*, 70(2): E135–E141.

Rahman, M. S., Senadeera, W., Al-Alawi, A., Truong, T., Bhandari, B., Al-Saidi, G. 2011. Thermal transition properties of spaghetti measured by Differential Scanning Calorimetry (DSC) and thermal mechanical compression test (TMCT). *Food and Bioprocess Technology*, 4(8): 1422–1431.

Roos, Y., Karel, M. 1991. Phase transitions of mixtures of amorphous polysaccharides and sugars. *Biotechnology Progress*, 7: 49–53.

Roos, Y. H. 1995a. Glass transition-related physicochemical changes in foods. *Food Technology*, 49: 97–102.

Roos, Y. 1995b. Characterization of food polymers using state diagrams. *Journal of Food Engineering*, 24: 339–360.

Roudaut, G., Maglione, M., van Duschotten, D., Le Meste, M. 1999. Molecular mobility in glassy bread: A multi spectroscopic approach. *Cereal Chemistry*, 1:70–77.

Roudaut, G., Simatos, D., Champion, D., Contreras-Lopez, E., Le Meste, M. 2004. Molecular mobility around the glass transition temperature: A mini review innovative. *Food Science and Emerging Technologies*, 5: 127–134.

Rudin, A. 1999. *Polymer Science and Engineering*. Academic Press, New York.

Sablani, S. S., Rahman, M. S., Al-Busaidi, S., Guizani, N., Al-Habsi, N., Al-Belushi, R., Soussi, B. 2007. Thermal transitions of king fish whole muscle, fat and fat-free muscle by differential scanning calorimetry. *Thermochimica Acta*, 462: 56–63.

Sandoval, A. J., Nuñez, M., Müller, A. J., Valle, G. D. 2009. Glass transition temperatures of a ready to eat breakfast cereal formulation and its main components determined by DSC and DMTA. *Carbohydrate Polymer*, 76(4): 528–534.

Shamblin, S., Tang, X., Chang, L., Hancock, B., Pikal, M. 1999. Characterization of the time scales of molecular motion in pharmaceutically important glasses. *Journal of Physics Chemistry B*, 103: 4113–4121.

Shinyashiki, N., Yamamoto, W., Yokoyama, A., Yoshinari, T., Yagihara, S., Kita, R., Ngai, K., Capaccioli, S. 2009. Glass transitions in aqueous solutions of protein (bovine serum albumin). *The Journal of Physical Chemistry B*, 113 (43): 14448–14456.

Syamaladevi, R. M., Barbosa-Cánovasa, G. V., Schmidt, S. J., Sablani, S. S. 2012. Influence of molecular weight on enthalpy relaxation and fragility of amorphous carbohydrates. *Carbohydrate Polymers*, 88(1): 223–231.

Silalai, N., Roos, Y. H. 2010. Roles of water and solids composition in the control of glass transition and stickiness of milk powders. *Journal of Food Science*, 75(5): E285–E296.

Simatos, D., Karel, M. 1988. Characterization of the condition of water in foods—Physico-chemical aspects. In: *Food Preservation by Water Activity*, p. 1, C. C. Scow (ed.). Elsevier, Amsterdam.

Slade, L., Levine, H. 1994. Water and the glass transition dependence of the glass transition on composition and chemical structure: Special implications for flour functionality in cookie baking. *Journal of Food Engineering*, 22: 143–188.

Slade, L., Levine, H. 1988. Non-equilibrium behavior of small carbohydrate–water systems. *Pure Applied Chemistry*, 60: 1841–1864.

Soesanto, T., Williams, M. C. 1981. Volumetric interpretation of viscosity for concentrated and dilute sugar solutions. *The Journal of Physical Chemistry*, 85: 3338–3341.

Sopade, P. A., Halley, P., Bhandari, B., D'Arcy, B., Doebler, C., Caffin, N. 2002. Application of the Williams–Landel–Ferry model to the viscosity–temperature relationship of Australian honeys. *Journal of Food Engineering*, 56: 67–75.

Suresh, S., Guizani, N., Al-Ruzeiki, M., Al-Hadhrami, A., Al-Dohani, H., Al-Kindi, I., Rahman, M. S. 2013. Thermal characteristics, chemical composition and polyphenol contents of date-pits powder. *Journal of Food Engineering*, 119: 668–679.

Telis, V. R. N., Martínez-Navarrete, N. 2009. Collapse and color changes in grapefruit juice powder as affected by water activity, glass transition, and addition of carbohydrate polymers. *Food Biophysics*, 4: 83–93.

Truong, V., Bhandari, B. R., Howes, T. 2005. Optimization of co-current spray drying process of sugar-rich foods. Part I—Moisture and glass transition temperature profile during drying. *Journal of Food Engineering*, 71(1): 55–65.

Turnbull, D., Cohen, M. H. 1958. Molecular transport in liquids and glasses. *Journal of Chemistry Physics*, 29: 1049.

Van den Berg, C. 1986. Water activity. In: *Concentration and Drying of Foods*. MacCarthy, D. ed. Elsevier Applied Science, London. pp. 11–36.

Velikov, V., Borick, S., Angell, C. A. 2001. The glass transition of water, based on hyperquenching experiments. *Science*, 294: 2335–2338.

Williams, G., Watts, D. C. 1970. *Transactions of Faraday Society*, 66: 80.

Williams, M. L., Landel, R. F., Ferry, J. D. 1955. The temperature dependence of relaxation mechanisms in amorphous polymers and other glass-forming liquids. *Journal of the American Chemical Society*, 77: 3701–3706.

Yildiz, M. E., Kokini, J. L. 2001. Determination of Williams–Landel–Ferry constants for a food polymer system: Effect of water activity and moisture content. *Journal of Rheology*, 45(4): 903–916.

Yoshioka, S., Aso, Y., Kojima, S. 2004. Temperature and glass transition temperature-dependence of bimolecular reaction rates in lyophilized formulations described by the Adam–Gibbs–Vogel equation. *Journal of Pharmaceutical Science*, 93: 1062–1069.

Yue, Y., Angell, C. A. 2004. Clarifying the glass-transition behaviour of water by comparison with hyperquenched inorganic glasses. *Nature*, 19(427): 717–720.

5
Rheological Properties of Fluid Foods

M. A. Rao

Contents

5.1	Introduction	122
5.2	Rheological Classification of Fluid Floods	123
	5.2.1 Rheological Models for Viscous Foods	127
	5.2.1.1 Models for Time-Independent Behavior	127
	5.2.1.2 Rheological Models for Thixotropic Foods	131
	5.2.1.3 Effect of Temperature on Viscosity	132
	5.2.1.4 Combined Effect of Temperature and Shear Rate	132
	5.2.1.5 Effect of Concentration on Viscosity	134
	5.2.2 Rheological Models for Viscoelastic Fluid Foods	135
	5.2.2.1 Normal Stress Data on Fluid Foods	135
	5.2.2.2 Creep Compliance Studies on Foods	137
5.3	Structure of Fluid Foods via Solution Viscosity and Physicochemical Approach	137
	5.3.1 Solution Viscosity	137
	5.3.2 Physicochemical Approach	139
5.4	Measurement of Flow Properties of Fluid Foods	140
	5.4.1 Fundamental Methods	140
	5.4.1.1 Capillary Flow	140
	5.4.1.2 Couette Flow Viscometers	141
	5.4.1.3 Plate-and-Cone Viscometers	143
	5.4.1.4 Parallel Plate Geometry	144
	5.4.1.5 Slit (Channel) Rheometers	145
	5.4.1.6 Extensional Flows	146
	5.4.2 Empirical Methods	149
	5.4.2.1 Adams Consistometer	149
	5.4.2.2 Bostwick Consistometer	149
	5.4.2.3 Efflux Tube Viscometer	150
	5.4.3 Imitative Methods	150

	5.4.3.1 Mixers for Determining Flow Properties	151
	5.4.3.2 In-Plant Measurement of Rheological Behavior of Fluid Foods	154
5.5	Flow of Fluid Foods in Tubes	158
	5.5.1 Isothermal Flow of Fluids in Tubes	158
	5.5.1.1 Velocity Profiles and Volumetric Flow Rate Relationships	158
	5.5.1.2 Friction Losses for Power Law Foods in Pipes	160
	5.5.1.3 Pressure Drop Across Valves and Fittings	162
	5.5.1.4 Friction Losses for Herschel–Bulkley Fluids	163
	5.5.1.5 Calculation of Kinetic Energy for Non-Newtonian Fluids	164
5.6	Dynamic Rheological Measurement of Viscoelastic Behavior of Fluid Foods	164
5.7	Conclusion	169
List of Symbols		170
Greek Symbols		171
Subscripts		171
Superscript		171
References		172

5.1 INTRODUCTION

Fluid foods are encountered widely in everyday life. In this chapter, foods that flow under gravity and do not retain their shape are considered to be fluid foods. Some of these foods, such as ice cream and shortenings, exist as solids at certain temperatures and liquids at other temperatures. Foods such as applesauce, tomato puree, baby foods, and some soups and dressings are suspensions of solid matter in fluid media. Following Sherman (1970), these will be called dispersions. When liquid droplets, instead of solid particles, are dispersed in fluid media, we have *emulsions*. Foods that are emulsions include milk and ice cream mix.

Because of the wide variation in their structure and composition, foods exhibit flow behavior ranging from simple Newtonian to time-dependent non-Newtonian and viscoelastic. Further, a given food may exhibit Newtonian or non-Newtonian behavior, depending on its origin, concentration, and previous history. For example, raw whole egg at 21°C was found to be a Newtonian fluid. However, thawed frozen whole egg was found to be a shear-thinning fluid (Cornford et al., 1969). Likewise, single-strength apple juice is a Newtonian liquid, but concentrated (undepectinized and filtered) apple juice is a shear-thinning fluid (Saravacos, 1970).

The complex nature of foods, their variability, and their diverse behavior are some of the reasons for cataloging separately the flow behavior of specific foods (Sherman, 1970). For example, the rheological behaviors of milk, butter, fruit juices, and so on, are described separately. However, attempts have been made to describe foods under specific rheological behavior (Rao, 1999).

Flow properties of foods are determined for a number of purposes, such as quality control (Kramer and Twigg, 1970), understanding the structure (Sherman, 1966), process engineering applications (Boger and Tiu, 1974; Rao and Anantheswaran, 1982), and correlations with sensory evaluation (Szczesniak and Farkas, 1962). Correlation with sensory evaluation is a unique area of rheology of foods. In particular, food rheologists have made unique contributions to the study of mouthfeel and its relation to basic rheological parameters.

One objective of this work is to identify the types of rheological behavior and the fluid foods that fall into each class of behavior. Detailed descriptions of the characteristics of specific foods are not provided here because they can be found elsewhere (Holdsworth, 1993; Rao, 1977a,b; 1999; Sherman, 1970). Rheological models employed for characterizing fluid food behavior are covered. The use of solution viscosity and physicochemical studies to understand the structure of fluid foods is discussed. Engineering aspects of flow of foods in pipes and heat transfer to foods in pipe flow are also discussed.

Another objective of the study is to describe the experimental methods for measuring flow properties of fluid foods. The methods are discussed under three categories: fundamental, empirical, and imitative. The problems associated with the proper measurement of flow properties are intimately connected with understanding the applicable flow equations, continuity and motion, and the boundary conditions in the test geometries as well as the structure of the food and the transitions it undergoes. However, these topics have been covered elsewhere (Rao, 1977a,b; 1999) and are therefore not discussed here. Likewise, because well-designed modern commercial viscometers employ the basic techniques described later and are automated, they are not described here. Descriptions of some of the early viscometers and the basic measurement techniques can be found in Van Wazer et al. (1963) and Whorlow (1980).

The problems associated with the proper measurement of flow properties are intimately connected with understanding the applicable flow equations: continuity and motion, and the boundary conditions in the test geometries as well as the structure of the food and the transitions it undergoes. However, these topics have been covered elsewhere (Rao, 1977a,b; 1999). Likewise, because well-designed modern commercial viscometers employ the basic techniques described later and are automated, they are not described here. Descriptions of some of the early viscometers and the basic measurement techniques can be found in Van Wazer et al. (1963) and Whorlow (1980).

5.2 RHEOLOGICAL CLASSIFICATION OF FLUID FLOODS

The different types of rheological behavior of fluid foods, and, to some extent, the techniques for determining such behavior can be explained by considering simple shearing flow. Simple shearing flow is that encountered in geometries such as the capillary, Couette, cone-and-plate, and parallel-plate. For these systems, the deformation tensor, $\tilde{\tilde{e}}$, and the stress tensor, $\tilde{\tilde{\tau}}$, are expressed in Equations 5.1 and 5.2, respectively (Brodkey, 1967; Han et al., 1975).

$$\tilde{\tilde{e}} = \begin{Vmatrix} 0 & \dot{\gamma} & 0 \\ \dot{\gamma} & 0 & 0 \\ 0 & 0 & 0 \end{Vmatrix} \tag{5.1}$$

$$\tilde{\tau} = \begin{Vmatrix} \sigma_{11} & \sigma_{21} & 0 \\ \sigma_{12} & \sigma_{22} & 0 \\ 0 & 0 & \sigma_{33} \end{Vmatrix} \tag{5.2}$$

In Equations 5.1 and 5.2, $\dot{\gamma}$ is the shear rate; σ_{11}, σ_{22}, and σ_{33} are normal stresses that are equal to zero in the case of Newtonian fluids; σ_{12} and σ_{21} are tangential components and are symmetrical. In dealing with the state of stress of an incompressible fluid under deformation, the total stress (τ_{ij}) can be divided into two parts:

$$\tau_{ij} = p\delta_{ij} + \sigma_{ij}$$

where δ_{ij} are components of the Kronecker delta, p is the isotropic pressure, and σ_{ij} are components of the deviatoric (or extra) stress. The isotropic pressure is defined as

$$-p = \frac{1}{3}(\tau_{11} + \tau_{22} + \tau_{33})$$

In an incompressible fluid, the state of stress is determined by the strain or strain history, and the absolute value of any normal component is not of rheological interest. Further, the values of the differences of normal stress components are not altered by the addition of isotropic pressure. Thus, three independent material functions, $\acute{\eta}(\dot{\gamma}) = \sigma/\dot{\gamma}$, $N_1(\dot{\gamma}) = \sigma_{11} - \sigma_{22}$, and $N_2(\dot{\gamma}) = \sigma_{22} - \sigma_{33}$, are satisfactory to describe the relationship between the shear rate on the one hand and the shear and normal stresses on the other.

To obtain information on the structure of a material in its undisturbed state, rheological tests must be conducted at small magnitudes of shear rates so that results from the tests can be extrapolated to zero shear rate. Such an extrapolation in the case of the viscosity function leads to the concept of zero-shear-rate viscosity, $\acute{\eta}_0$. In the case of the first and second normal stress differences, one needs to define the corresponding functions ψ_1 and ψ_2, respectively. The advantage of ψ_1 and ψ_2 is that they can be expected to attain finite magnitudes as the shear rate approaches zero. Thus, three material functions are sufficient to correlate the stress components to the shear rate. They are (Brodkey, 1967; Han et al., 1975)

$$\sigma = \eta(\dot{\gamma})\dot{\gamma} \tag{5.3}$$

$$\sigma_{11} - \sigma_{22} = \psi_1(\dot{\gamma})\dot{\gamma}^2 \tag{5.4}$$

$$\sigma_{22} - \sigma_{33} = \psi_2(\dot{\gamma})\dot{\gamma}^2 \tag{5.5}$$

Procedures for determining the viscosity function $\acute{\eta}$ and the first normal stress function, ψ_1, are well accepted. After considerable controversy regarding the existence of the second normal stress function, it is generally recognized that it is much smaller in magnitude than the first normal stress function and may also exhibit negative values.

Most of the rheological studies on foods have dealt with the viscosity function because it plays an important role in food engineering and processing applications. Early studies on viscoelastic behavior of foods interpreted the results with the aid of mechanical models, such as the Maxwell and Kelvin bodies (Sherman, 1966); later, normal stress functions were used (Dickie and Kokini, 1982; Genovese and Rao, 2003a; Kokini and Dickie, 1981), where the results were interpreted in terms of constitutive equations.

More recently, viscoelastic behavior of foods has been studied extensively in terms of dynamic rheological behavior mainly due to the availability of affordable, automated, controlled-stress rheometers, and the phenomenological nature of the parameters (Rao, 1999). Dynamic rheological tests are nondestructive. They are conducted by applying small strains or stresses at known frequency, and measuring the corresponding strain or stress response of the test materials, respectively. The important rheological parameters obtained are the storage modulus, G', that reflects the energy stored during a cycle, and loss modulus, G'', that reflects the viscous energy dissipation during that cycle. Because the time involved in the measurements is short relative to the characteristic times of the gelation and softening processes, dynamic rheological tests are used extensively to study phase transitions (Rao, 1999).

From G', G'', and ω, other useful viscoelastic parameters can be derived, such as the complex modulus, G^*, loss tangent, δ, complex viscosity, η^*, and dynamic viscosity, η:

$$G^* = \sqrt{(G')^2 + (G'')^2} \tag{5.6}$$

$$\tan \delta = (G''/G') \tag{5.7}$$

$$\eta^* = (G^*/\omega) = \eta' - i\frac{G'}{\omega} \tag{5.8}$$

where $i = \sqrt{-1}$.

The viscosity function η can be used to classify the flow behavior of several foods. The viscosity of Newtonian foods is influenced only by temperature and composition; it is independent of the shear rate and previous shear history. Foods known to be Newtonian are listed in Table 5.1.

Fluids that do not follow Newtonian behavior are called non-Newtonian fluids. The flow properties of non-Newtonian fluids are influenced by the shear rate. Instead of the Newtonian viscosity η, for non-Newtonian fluids the apparent viscosity, η_a, at a specified

Table 5.1 Examples of Newtonian Foods

Milk	Egg products
Total solids, 8.36–29.07%	Whole egg (unfrozen)
Clear fruit juices	Stabilized egg white
Depectinized apple juice (15–75°Brix)	Plain yolk
Filtered orange juice (10–18°Brix)	Sugared and salted yolk
Concord grape juice (15–50°Brix)	Sucrose solutions
	Most honeys
	Corn syrups

Source: From Rao (1977a).

shear rate can be used. In this chapter, η_a is defined as the ratio of the shear stress to the shear rate ($\eta_a = \sigma / \dot{\gamma}$).

In the SI system, the units of and are Pa s. This is based on expressing the shear stress in Pascals (Pa) and the shear rate in reciprocal seconds ( ). For low-viscosity fluids, the values are expressed in mPa s that are equal to values in centipoises in the metric system.

For several fluids, particularly polymer solutions, the complex and apparent viscosities are found to be equal at the same frequency and shear rate, respectively. This relationship is known as the Cox–Merz rule, named after the authors of that observation:

$$\eta^* = \eta_a \big|_{\omega = \dot{\gamma}} \tag{5.9}$$

However, many materials do not exhibit the Cox–Merz rule; often, the values of η^* versus and η_a versus $\dot{\gamma}$ are parallel to each other on double logarithmic plots and one can apply a modified rule (Rao, 1999):

$$\eta^* = A(\eta_a)^B \tag{5.10}$$

where A and B are empirical constants determined from experimental data for a specific material.

Non-Newtonian foods can be divided into two categories: *time-independent* and *time-dependent*. At a constant temperature, η_a for the former depends only on the shear rate; for the latter, η_a also depends on the duration of shear. *Time-independent* flow behavior can be divided into *shear-thinning (pseudoplastic)* and *shear-thickening (dilatant)* categories, depending on whether η_a decreases or increases, respectively, with an increase in shear rate. A large number of non-Newtonian fluid foods exhibit *pseudoplastic* behavior, and these foods are listed in Table 5.2. Figure 5.1 illustrates the flow curves of Newtonian and *time-independent* non-Newtonian fluids. It should be noted that often the shear rate is plotted on the abscissa and the shear stress on the ordinate.

Shear-thickening foods are rarely encountered. Pryce-Jones (1953) observed *shear-thickening* behavior for honeys from *Eucalyptus ficifolia, Eucalyptus eugeniodes, Eucalyptus corymbosa*, and *Opuntia engelmanni*. Bagley and Christianson (1982) observed shear-thickening behavior for cooked starch suspensions. Although these studies do indicate that shear-thickening behavior can be found among foods, very often instrument artifacts and a limited amount of data have been interpreted as indicators of shear-thickening behavior.

Table 5.2 Examples of Shear-Thinning Foods

Concentrated fruit juices	Dairy cream
Undepectinized apple juice (50–65°Brix)	Thawed frozen whole egg
Passion fruit juice (15.6–33.4°Brix)	Unmixed egg white
Orange juice (60–65°Brix)	Fruit and vegetable purees
Melted chocolate	Gum solutions—high concentrations
French mustard	Protein concentrates

Source: From Rao (1977a).

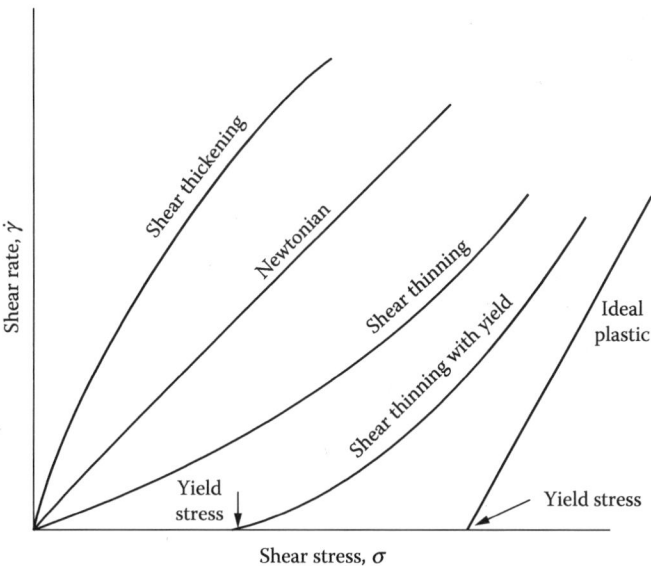

Figure 5.1 Newtonian and time-independent non-Newtonian fluids.

Non-Newtonian foods with *time-dependent* flow properties are subdivided into *thixotropic* and *rheopectic* fluids. In thixotropic fluids, at a fixed shear rate, the viscosity decreases with time, whereas the viscosity of a rheopectic fluid increases with time. *Thixotropic* behavior has been noted for condensed milk (Higgs and Norrington, 1971), mayonnaise (Figoni and Shoemaker, 1983; Tiu and Boger, 1974), and egg white (Tung et al., 1970). Rheopectic behavior is also referred to as anti-thixotropic behavior. Anti-thixotropic behavior was observed with gelatinized cross-linked waxy maize starch dispersions (Chamberlain et al., 1999; Tattiyakul and Rao, 2000).

5.2.1 Rheological Models for Viscous Foods

5.2.1.1 Models for Time-Independent Behavior

The power law model with or without a yield term (Equations 5.11 and 5.12) has been employed extensively to describe the flow behavior of viscous foods over wide ranges of shear rates (Vitali and Rao, 1984a, b):

$$\sigma = K\dot{\gamma}^n \tag{5.11}$$

$$\sigma - \sigma_0 = K_H \dot{\gamma}^{n_H} \tag{5.12}$$

where σ_0 is the yield stress, $K(K_H)$ is the consistency coefficient, and $n(n_H)$ is the flow behavior index. Equation 5.12 is also known as the Herschel–Bulkley model (Brodkey, 1967; Sherman, 1970).

Holdsworth (1971) and Steffe et al. (1986) compiled magnitudes of the power law parameters reported in the literature, and these references should be consulted for data on specific foods. Typical magnitudes of viscosities of Newtonian foods are 4.8 Pa s for honey at 25°C, 0.0064 Pa s for whole egg at 30°C, 0.0027 Pa s for stabilized egg white, and 0.40 Pa s for salted yolk. The flow behavior index can be found to vary from 1.0 for Newtonian foods to about 0.2 for highly shear-thinning pureed foods such as tomato concentrates. The units of the consistency coefficient, sometimes called the consistent index, are Pa s^n, while the flow behavior index is dimensionless. Very high magnitudes of the consistency coefficient were reported for tomato concentrates (Rao et al., 1981a). Values of the consistency coefficient and the flow behavior index of a few fruit products taken from an extensive list in Rao (1999) are given in Table 5.3. Because magnitudes of the power-law parameters are affected by the temperature of the sample and often dependent on the range of shear rates used, they are also given in the table. Values of

Table 5.3 Rheological Properties of Fruit Products

Product	Concentration (% Solids)	Method	Temperature (°C)	K (Pa s^n)	n	Yield Stress (Pa)	Reference
Tomato juice (pH 4.3)	5.80	Conc cylinder	32.2	0.22	0.59		Harper and El-Sahrigi (1965)
			48.9	0.27	0.54		
			65.6	0.37	0.47		
	12.80		32.3	2.10	0.43		
			48.9	1.18	0.43		
			65.6	2.28	0.34		
			82.2	2.12	0.35		
	16.00		32.3	3.16	0.45		
			48.9	2.27	0.45		
			65.6	3.18	0.40		
			82.2	3.27	0.38		
	25.00		32.3	12.9	0.41		
			48.9	10.5	0.42		
			65.6	8.0	0.43		
			82.2	6.1	0.44		
Tomato paste	29.7 °Brix	Haake mixer	32.0	208.0	0.27	206	Rao et al. (1993)
	29.7 °Brix		39.0	179.0	0.31	180	
	23.8 °Brix		33.0	48.0	0.47	40	
	23.8 °Brix		39.0	34.0	0.52	29	
	16.3 °Brix		25.0	24.0	0.23	38	
Applesauce		Haake mixer	32.0	200.0	0.42	240	Rao et al. (1993)
Apple juice	69.8 °Brix	Haake RV2/	25.0	0.24	1.00		Rao et al. (1993)
	65.3 Brix	Deer Rheom.	23.0	0.09	1.00		
	69.8 Brix		43.0	0.03	1.00		
	51.1 Brix		25.0	0.02	1.00		

Table 5.3 (continued) Rheological Properties of Fruit Products

Product	Concentration (% Solids)	Method	Temperature (°C)	K (Pa s^n)	n	Yield Stress (Pa)	Reference
Orange juice Pera orange, 3.4% pulp	64.9 Brix	Haake RV2	−18.0	18.3	0.80		Vitali and Rao (1984b)
		Conc cylinder	−14.0	10.6	0.81		
			−10.0	7.1	0.79		
			−5.0	4.9	0.78		
			0.0	3.2	0.79		
			10.0	1.6	0.79		
			20.0	0.7	0.83		
			30.0	0.4	0.82		
Valencia orange 21.2% pulp	65.3 Brix		−18.3	109.9	0.55		
			−14.2	59.7	0.61		
			−9.7	40.6	0.60		
			−5.1	24.5	0.63		
			−0.4	18.5	0.62		
			10.2	8.3	0.65		
			19.7	6.1	0.61		
			29.6	2.5	0.68		

the Casson model parameters for a few chocolate samples from Chevalley (1991) are provided in Table 5.4.

The Casson (1959) model (Equation 5.13) has been used for foods, particularly for estimating the yield stress:

$$\sigma^{0.5} = K_{0C} = K_C \dot{\gamma}^{0.5} \tag{5.13}$$

The magnitude of K_{0C}^2 has been used as the yield stress by several workers (Charm, 1963; Hermansson, 1975; Rao et al., 1981a; Tung et al., 1970). It seems that for foods, such as tomato paste and concentrated orange juice, the Casson yield stress is much higher than either that predicated by the Herschel–Bulkley model (Equation 5.12) (Rao and Cooley, 1983) or that determined experimentally (Vitali and Rao, 1984a). Casson's model has been adopted as the official method for the interpretation of chocolate flow data by the International Office of Cocoa and Chocolate. The two parameters frequently discussed are the Casson plastic viscosity, $\eta_{CA} = K_C^2$, and the Casson yield, $\sigma_{CA} = K_{0C}^2$.

Mizrahi and Berk (1972) derived a modified Casson equation and employed it to describe shear rate–shear stress data on concentrated orange juice. The development was

ENGINEERING PROPERTIES OF FOODS

Table 5.4 Casson Model Parameters of Commercial Chocolate Samples

Brand Name	Method	Temp. (°C)	Shear Rate (s^{-1})	Plastic Viscosity, $\acute{\eta}_\infty$ (Pa s)	Yield Stress (Pa)
ArniCoop	Conc cylinder	40	5–60	2.5	10.9
			60–5	2.5	9.1
Cailler (Nestle)			5–60	3.7	16.5
Cadbury Dairy			5–60	4.4	33.7
Milk GB			60–5	4.8	20.5
Galaxy GB			5–60	4.7	78.0
			60–5	4.9	49.2
Lindt F			5–60	1.8	10.3
			60–5	1.9	8.9
Nestle F			5–60	2.6	19.8
			60–5	2.6	15.9
Poulain F			5–60	2.7	19.4

Source: Chevalley, J. 1991. *J. Texture Studies* 22:219–229.

based on the model of a suspension of interacting particles in a pseudoplastic solvent. The equation developed was

$$\sigma^{1/2} - K_{0M} = K_M \dot{\gamma}^{n_M} \qquad (5.14)$$

It was found that K_{0M} was affected by the concentration of the suspended particles and the concentration of soluble pectin, and K_M and n_M were determined mainly by the properties of the solvent.

The Bingham relationship (Equation 5.15) has been employed to describe the flow behavior of apricot puree (Schaller and Knorr, 1973), minced fish paste (Nakayama et al., 1980), and cooked cassava starch paste (Odigboh and Mohsenin, 1975):

$$\sigma - \sigma_0 = \eta' \dot{\gamma} \qquad (5.15)$$

where $\acute{\eta}$ is the plastic viscosity.

Among the aforementioned models, the power law model (Equation 5.11) has been employed extensively for characterizing foods, including shear-thinning foods. However, shear-thinning foods exhibit a Newtonian viscosity at very low shear rates (zero-shear viscosity) that the power law model fails to predict. In addition, shear-thinning fluids may also exhibit a limiting viscosity at very high shear rates that may be too low to be measured with confidence. There are a number of models that predict Newtonian viscosities at low and high shear rates. The flow model of Cross (1965) (Equation 5.16) was employed by Doublier and Launay (1974) to describe the shear rate–shear stress data of guar gum solutions over a wide range of shear rates (0.16–17,600 s^{-1}), whereas the Powell–Eyring model (Equation 5.17) was employed to describe the data on serum from concentrated orange juice (Vitali and Rao, 1984a):

$$\eta = \eta_\infty + \frac{\eta_0 - \eta_\infty}{1 + \dot{\alpha}\gamma^{2/3}} \qquad (5.16)$$

$$\sigma = \eta_\infty \dot{\gamma} + \frac{\eta_0 - \eta_\infty}{\beta} \sinh^{-1}(\beta\dot{\gamma}) \qquad (5.17)$$

In Equations 5.16 and 5.17, $\dot{\eta}$ is the apparent viscosity at a specific shear rate, $\dot{\eta}_0$ is the limiting viscosity at zero rate of shear, $\dot{\eta}\infty$ is the limiting viscosity at infinite rate of shear, and α and β are constants.

The above equations have been used for describing time-independent flow behavior of fluid foods. The equations for characterizing thixotropic behavior are described in the next section.

5.2.1.2 Rheological Models for Thixotropic Foods

Higgs and Norrington (1971) studied the thixotropic behavior of sweetened condensed milk. They measured the coefficient of thixotropic breakdown with time, B, which indicates the rate of breakdown with time at a constant shear rate, and the coefficient of thixotropic breakdown due to the increasing shear rate, M, which indicates the loss in shear stress per unit increase in shear rate (Green, 1949; Wilkinson, 1960). The coefficients were estimated from the equations

$$B = \frac{\eta_1 - \eta_2}{\ln(t_2 - t_1)} \qquad (5.18)$$

and

$$M = \frac{\eta_1 - \eta_2}{\ln(N_2/N_1)} \qquad (5.19)$$

In Equation 5.18, $\dot{\eta}_1$ and $\dot{\eta}_2$ are viscosities measured after times t_1 and t_2, respectively; in Equation 5.19, they are measured at the angular speeds N_1 and N_2, respectively.

Tiu and Boger (1974) employed a kinetic rheological model to characterize the thixotropic behavior of a mayonnaise sample. It was based on the Herschel–Bulkley model (Equation 5.12) multiplied by a structural parameter, λ, which ranges between an initial value of unity for zero shear time to an equilibrium value, λ_e, which is less than unity:

$$\sigma = \lambda\left(\sigma_0 + K_H \dot{\gamma}^{n_H}\right) \qquad (5.20)$$

The decay of the structural parameter was assumed to obey the second-order rate equation

$$\frac{d\lambda}{dt} = -K_1(\lambda - \lambda_e)^2 \quad \text{for } \lambda > \lambda_e \qquad (5.21)$$

In Equation 5.21, K_1 is a parameter that is a function of the shear rate. If the parameters σ_0, K_H, n_H, K_1, and λ_e are determined from experimental data, Equations 5.20 and 5.21 can be used for the complete rheological characterization of a thixotropic food product.

Tiu and Boger (1974) provided additional equations to facilitate estimation of the various parameters. They also estimated the magnitudes of the parameters for a mayonnaise sample.

Tung et al. (1970) studied the thixotropic properties of fresh, aged, and gamma-irradiated egg white. The mathematical models of Weltman (1943) (Equation 5.22) and of Hahn et al. (1959) (Equation 5.23) were used to describe the thixotropic behavior:

$$\sigma = A_1 - B_1 \log t \qquad (5.22)$$

$$\log(\sigma - \sigma_e) = A_2 - B_2 t \qquad (5.23)$$

where σ_e is the equilibrium shear stress, t is time in seconds, and A_1, A_2, B_1, and B_2 are constants. The coefficients A_1 and A_2 indicate initial shear stresses, whereas the coefficients B_1 and B_2 indicate rates of structural breakdown. The magnitudes of the coefficients were lower for aged and gamma-irradiated samples than for fresh egg white.

5.2.1.3 Effect of Temperature on Viscosity

Fluid foods are subjected to different temperatures during processing, storage, transportation, marketing, and consumption. For this reason, the rheological properties are studied as a function of temperature. In general, the effect of temperature on the viscosity (η) or apparent viscosity (η_a) determined at a specific shear rate can be expressed by the Arrhenius relationship

$$\eta_a = \eta_\infty^{E_a/RT} \qquad (5.24)$$

where E_a is the activation energy in kJ g mol. For non-Newtonian fluids, in addition to the apparent viscosity at a specific shear rate, the consistency index of the power law model (Equation 5.11) can be employed for determining the effect of temperature (Harper and El-Sahrigi, 1965; Vitali and Rao, 1984b).

Magnitudes of activation energy have been tabulated (Holdsworth, 1971; Rao, 1977a) for a number of fluid foods: sucrose solutions, dilute fruit juices, egg products, pureed foods, and concentrated fruit juices. Table 5.5 contains magnitudes of the activation energy of flow for selected foodstuffs. It is important to note that the magnitude of E_a for most foods such as concentrated fruit juices depends on the range of temperatures considered. The magnitudes of E_a for concentrated fruit juices were higher than those in Table 5.5 when relatively low temperatures (–15–40°C) were employed (Rao et al., 1984). Holdsworth (1971) examined the published activation energy values of a number of pseudoplastic fruit purees and juices and concluded that the lower the value of the flow behavior index, the less the effect of temperature on viscosity.

5.2.1.4 Combined Effect of Temperature and Shear Rate

In applications such as sterilization, a fluid food is subjected to temperature and shear gradients, and models that describe their combined effect are necessary for solving the pertinent transport equations (Rao and Anantheswaran, 1982; Simpson and Williams, 1974). Two models based on the power law and the Arrhenius models have been proposed (Christiansen and Craig, 1962; Harper and El-Sahrigi, 1965):

Table 5.5 Energy of Activation for Flow of Selected Fluid Foods

Fluid food	°Brix or wt% water	Flow behavior index (n)	Activation energy [kJ/(g mol)]
Depectinized apple juice[a]	75	1.0	59.5
	50	1.0	35.2
	30	1.0	26.4
Cloudy apple juice[a]	40	1.0	24.3
	30	1.0	21.4
Concord grape juice[a]	50	1.0	28.9
	30	1.0	26.0
Cloudy apple juice[a]	65.5	0.65	38.1
	50.0	0.85	25.5
Applesauce[a]	11.0	0.30	5.0
Peach puree[a]	11.7	0.30	7.1
Pear puree[a]	16.0	0.30	8.0
Passion fruit concentrates[a]	15.6	0.74	18.8
	20.7	0.59	17.2
	25.3	0.52	16.7
	30.6	0.49	15.9
	33.4	0.45	13.4
Tomato (Nova) concentrates[a]	6.0–32.0	0.26	9.2
Whole egg[b]	75	1.0	24.7
Stabilized egg white[b]	88	1.0	16.3
Plain yolk[b]	55	1.0	26.8

Sources: From Saravacos (1970), Scalzo et al. (1970), Vitali et al. (1974), Rao et al. (1981a).
[a] Concentration in °Brix.
[b] Water content, wt %.

$$\sigma = K_{TC} \left(\dot{\gamma} \exp \frac{E_{aC}}{RT} \right)^{\bar{n}} \quad (5.25)$$

$$\sigma = K_{TH} \exp\left(\frac{E_{aH}}{RT}\right) \dot{\gamma}^{\bar{n}} \quad (5.26)$$

where $\bar{n}$ indicates an average value for data of all the temperatures. The two models are not identical in that (Vitali and Rao, 1984b)

$$E_{aH} = \bar{n}(E_{aC}) \quad (5.27)$$

Equations 5.24 and 5.25 are called thermorheological models and should be applicable to most fluid foods that do not undergo a phase transition. In the special case of dispersions of starch that undergo the phase transition, gelatinization during continuous heating to sterilization temperatures, the apparent viscosity increases during the initial stage, reaches a maximum, and then decreases to a value lower than the maximum (Liao et al., 1999; Yang and Rao, 1998a). Equations 5.24 and 5.25 cannot describe such data, and more complex models are needed in studies on heat transfer to starch dispersions (Liao et al., 2000; Tattiyakul et al., 2002; Yang and Rao, 1998b).

5.2.1.5 Effect of Concentration on Viscosity

For a given fluid food, the viscosity increases as the concentration of the fluid food is increased. Figure 5.2 illustrates the increase in the apparent viscosity (at a shear rate of 100 s^{-1}) with an increase in total solids of concentrates of Nova tomatoes (Rao et al., 1981a).

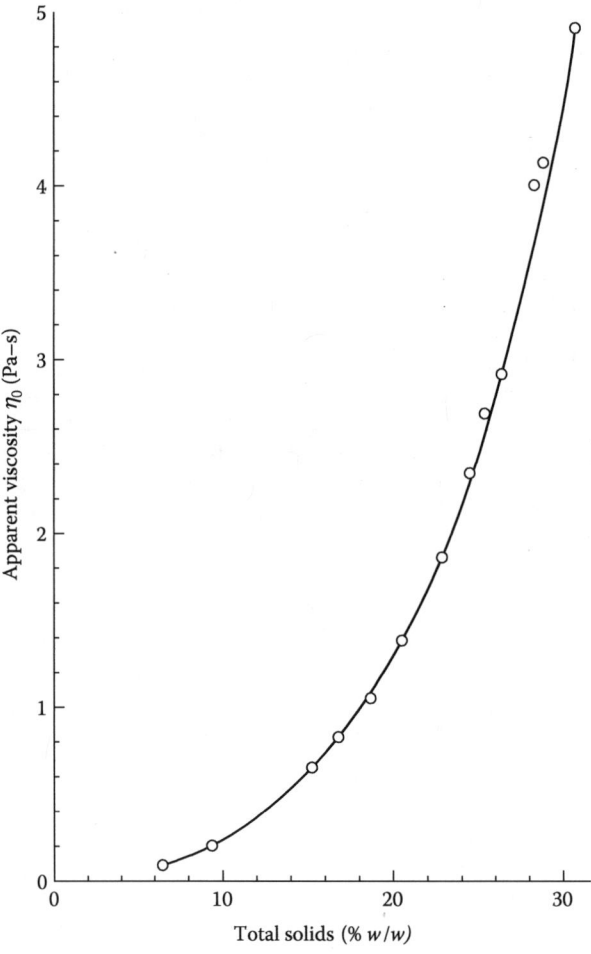

Figure 5.2 Apparent viscosity (at 100 s^{-1}) as a function of total solids content of concentrates of Nova tomatoes.

Over limited ranges of concentration, the effect of concentration on apparent viscosity can be described by either an exponential relationship (Cervone and Harper, 1978; Vitali and Rao, 1984b) or a power type of relationship (Harper and El-Sahrigi, 1965; Rao et al., 1981a). The former has been found to be suitable for doughs and concentrated fruit juices, whereas the latter is suitable for viscous pureed foods, such as tomato concentrates.

One can combine the effect of temperature (Equation 5.24) and the effect of concentration into a single equation. In the case of the power dependence on concentration, the combined equation has the form

$$\eta_a = \alpha \exp\left(\frac{E_a}{RT}\right) c^\beta \tag{5.28}$$

This equation is useful in the computation of apparent viscosity as a function of temperature and concentration. Harper and El-Sahrigi (1965) suggested that the magnitude of β would be nearly constant for pureed foods and that of α would depend on the cultivar of the fruit and processing methods. The data of Rao et al. (1981a) confirm, in part, these predictions; the magnitude of β was found to be 2.5 for the concentrates of four tomato varieties. However, this magnitude of β was higher than the value of 2.0 that one can calculate from the data of Harper and El-Sahrigi (1965) on VF6 tomatoes. Generally, equations describing the combined effect of temperature and concentration are valid over limited ranges of the variables (Vitali and Rao, 1984b), and caution must be exercised in using a single equation over wide ranges of temperatures and concentrations. For good predictive capabilities, one may employ several equations valid over narrow ranges of temperatures and concentrations.

5.2.2 Rheological Models for Viscoelastic Fluid Foods

Some fluids demonstrate both viscous and elastic properties; these fluids are called *viscoelastic*. Examples of viscoelastic foods are dairy cream (Prentice, 1968, 1972) and ice cream mix, frozen product, and melt (Shama and Sherman, 1966; Sherman, 1966), as well as the prepared foods such as peanut butter, whipped butter, and marshmallow cream (Dickie and Kokini, 1982). Viscoelastic behavior of foods can be studied by measuring the normal stresses and computing the appropriate differences ψ_1 and ψ_2. These data are then employed in conjunction with constitutive equations to obtain knowledge regarding the structure of the food. Viscoelastic behavior can also be studied by means of the creep compliance test, where the results are interpreted in terms of mechanical models (Sherman, 1970).

5.2.2.1 Normal Stress Data on Fluid Foods

Primary normal stress measurements have been reported for xanthan gum solutions (Whitcomb and Macosko, 1978), zein dispersions in ethanol (Menjivar and Rha, 1980), and prepared foods such as peanut butter and canned frosting (Dickie and Kokini, 1982; Kokini and Dickie, 1981). To illustrate typical magnitudes of the primary normal stress difference, data on the aforementioned foods at 25°C as a function of shear rate are shown in Figure 5.3.

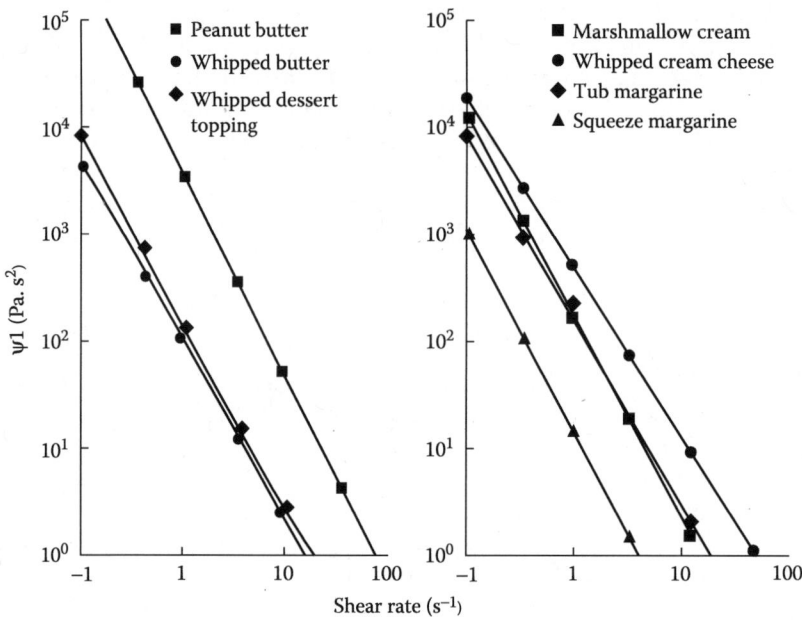

Figure 5.3 Primary normal stress difference data of selected foods. (From Dickie, A. M., and Kokini, J. L. 1982. *J. Food Process Eng.* 5: 157–184.)

Primary normal stress differences and the corresponding coefficients are dependent on the shear rate. Dickie and Kokini (1982) employed a power relationship to describe the effect of shear rate on the primary normal stress difference:

$$\sigma_{11} - \sigma_{22} = m' \gamma^{n'} \tag{5.29}$$

The magnitudes of m' and n' for selected foods are given in Table 5.6.

Table 5.6 Magnitudes of m' and n' (Equation 5.25) for Selected Foods

	m' (Pa·s$^{n'}$)	n'	R^2
Whipped butter · 1.1 ± 0.1 ± 0.1 × 10²	0.476	0.99	
Whipped cream cheese	3.6 ± 0.2 × 10²	0.418	0.99
Squeeze margarine	1.6 ± 0.4 × 10²	0.168	0.99
Tub margarine	1.8 ± 0.3 × 10²	0.358	0.99
Whipped dessert topping	1.4 ± 0.5 × 10²	0.309	0.99
Marshmallow cream	1.9 ± 0.3 × 10²	0.127	0.99
Peanut butter	3.8 ± 0.1 × 10³	0.175	0.99

Source: From Dickie, A. M., and Kokini, J. L. 1982. *J. Food Process Eng.* 5: 157–184.

In general, the normal stress dates are used to test different constitutive equations that have been developed by different research groups (Han et al., 1975). The first normal stress difference plays an important role in extrudate swell and the rod-climbing (Weissenberg) effect of viscoelastic foods. In this respect, when the normal stress coefficient, ψ_1, cannot be determined experimentally, the relationships derived by Bird et al. (1977) to predict it from apparent viscosity versus shear rate data should be useful (Genovese and Rao, 2003a).

5.2.2.2 Creep Compliance Studies on Foods

In the studies on cream and ice cream, creep measurements were performed at low rates of shear (Sherman, 1966). The experimental data were plotted in terms of creep compliance, $J(t)$, as a function of time, t, where J is the ratio of strain to stress.

For ice cream mix and the frozen product, the creep compliance versus time curve was defined by the equation

$$J(t) = J_0 + J_1\left(1 - e^{-t/\tau_1}\right) + J_2\left(1 - e^{-t/\tau_2}\right) + \frac{t}{\eta_N} \tag{5.30}$$

In Equation 2.30, $J_0 = 1/E_0$ is the instantaneous elastic compliance, E_0 is the instantaneous elastic modulus; $J_1 = 1/E_1$ and $J_2 = 1/E_2$ are the compliances associated with retarded elastic behavior; $\tau_1 = \acute{\eta}_1/E_1$ and $\tau_2 = \acute{\eta}_2/E_2$ are the retardation times associated with retarded elasticity; E_1 and E_2 are the elastic moduli associated with retarded elasticity; $\acute{\eta}_1$ and $\acute{\eta}_2$ are the viscosity components associated with retarded elasticity; and $\acute{\eta}_N$ is the viscosity associated with Newtonian flow.

For melted ice cream, the equation defining the creep compliance curve was Equation 5.30, with $J_2 = 0$.

Figure 5.4 depicts the models for ice cream mix and frozen product and melted ice cream. It is seen that for the mix and frozen product, two Kelvin–Voigt bodies connected in series were needed. In contrast, melted ice cream could be represented by one Kelvin–Voigt body.

5.3 STRUCTURE OF FLUID FOODS VIA SOLUTION VISCOSITY AND PHYSICOCHEMICAL APPROACH

One of the goals of food rheologists is to obtain a better understanding of the behavior, structure, and composition of foods through rheological measurements and equations. In this section, the use of the solution viscosity and the physicochemical approach are discussed in brief.

5.3.1 Solution Viscosity

Viscosities of dilute solutions of materials, particularly polymers, can be used to improve understanding of the molecular properties. The viscosities of the solution and the solvent are used to compute the reduced viscosity, $\acute{\eta}_{red}$, defined by the equations

$$\eta_{rel} = \frac{\eta_g}{\eta_s} \tag{5.31}$$

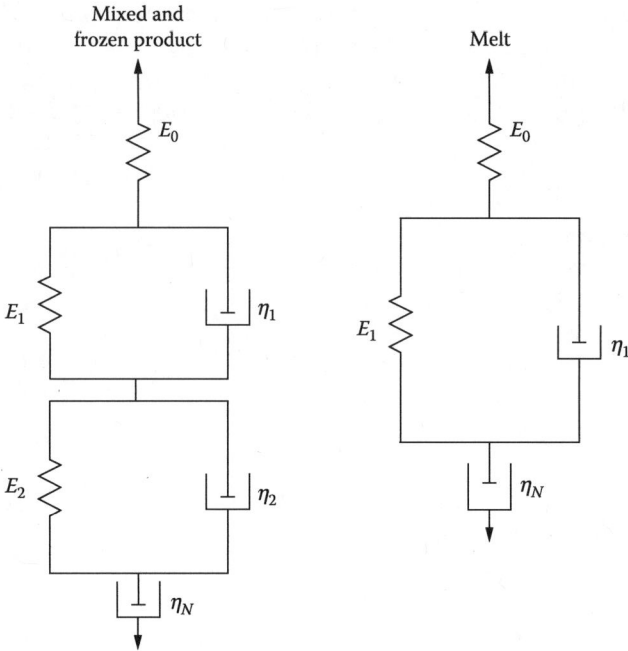

Figure 5.4 Mechanical models for the viscoelastic behavior of ice cream mix, frozen ice cream, and ice cream melt.

$$\eta_{sp} = \eta_{rel} - 1 \tag{5.32}$$

$$\eta_{red} = \frac{\eta_{sp}}{c} \tag{5.33}$$

In Equations 5.31 through 5.33, $\acute{\eta}_{rel}$ is the relative viscosity, $\acute{\eta}_g$ is the apparent viscosity of the solution, $\acute{\eta}_s$ is the apparent viscosity of the solvent, $\acute{\eta}_{sp}$ is the specific viscosity, $\acute{\eta}_{red}$ is the reduced viscosity, and c is the concentration of the solution. A plot of concentration versus reduced viscosity can be described by the series

$$\frac{\eta_{sp}}{c} = [\eta] + k_1 [\eta]^2 c + k_2 [\eta]^3 c^2 \tag{5.34}$$

where $[\acute{\eta}]$ is the intrinsic viscosity and k_1 and k_2 are the interaction coefficients. When only the first-order concentration terms are retained, Equation 5.34 is called the Huggins relation.

For polyelectrolytes, the variation of reduced viscosity with concentration follows the empirical equation (Elfak et al., 1978)

$$\frac{\eta_{sp}}{c} = \frac{A}{1 + Bc^{0.5}} + D \tag{5.35}$$

where A, B, and D are constants.

The intrinsic viscosity, $[\eta]$, is a measure of the hydrodynamic volume of the solute molecules. For many polymers, the intrinsic viscosity can be related empirically to the molecular weight by the Mark–Houwink relation.

$$[\eta] = a(\text{MW})^b \tag{5.36}$$

where a and b are constants and MW is the molecular weight. The interaction coefficient k_1 (Equation 5.34) reflects how much the viscosity of the system increases with an increase in solute concentration.

As one example of the use of solution viscosity in the study of foods, we consider the study of Elfak et al. (1977) with solutions of guar gum and locust bean gum. The magnitude of the intrinsic viscosity of the guar gum was higher than that of the locust bean gum; however, the interaction coefficient of the guar gum was lower. These results suggest that for guar gum the solute–solvent interactions are greater than for locust bean gum, but the solute–solute interactions are smaller.

Heat treatment (at 121.1°C, 10 min) of aqueous solutions of guar gum and carboxymethylcellulose (CMC) resulted in permanent loss of the viscosity of the solutions (Rao et al., 1981b). Solution viscosity data showed that heat treatment resulted in lower values of intrinsic viscosity and the interaction coefficient in the Huggins relationship. The lower magnitudes of intrinsic viscosity indicate reduction in the hydrodynamic volumes of the solute molecules due to partial hydrolysis during heat treatment. The intermolecular interaction was also less for the heat-treated solutions. An extended discussion of intrinsic viscosity of a polymer dispersion, its calculation, and its values for typical food polymers can be found in Rao (1999).

5.3.2 Physicochemical Approach

The physicochemical approach was developed by Mizrahi and coworkers (Mizrahi, 1979; Mizrahi and Berk, 1970, 1972; Mizrahi and Firstenberg, 1975). Its main goal is to understand the flow mechanism in fluid foods and to control as well as synthesize a desired consistency (Mizrahi, 1979). This approach has been applied with success to concentrated orange juice.

Three methods have been employed to correlate the rheological properties of a product with the physicochemical properties of the components (Mizrahi, 1979). In the first method, flow and physicochemical data are obtained for samples by varying a single factor at a time. From these experiments, an empirical correlation is derived that is then used in practice after proper verification.

The second method, put forward by Corey (1972), makes use of analogy between the flow behavior of an experimental model and that of a commercial product. In this approach, it is hoped that the experimental models will provide rheological and physicochemical data on various fluid foods.

In the third method, theoretical models are used. The success of this approach depends on how closely the model approximates the real system. In the case of concentrated orange juice, better insight into product structure was obtained from a modified Casson's model (Equation 5.13).

5.4 MEASUREMENT OF FLOW PROPERTIES OF FLUID FOODS

Following Scott Blair's classification (1958) of instruments for the study of texture, the instruments for measuring the flow properties of fluid foods can be classified into three categories: (1) fundamental, (2) empirical, and (3) imitative.

Fundamental tests measure well-defined properties, utilizing geometries that are amenable to analysis of fluid flow. Empirical tests measure parameters that are not clearly defined but that past experience has shown to be useful. Imitative tests measure properties under test conditions similar to those encountered in practice (White, 1970). Here, the emphasis is on fundamental methods.

5.4.1 Fundamental Methods

Several instruments are employed for measuring flow properties via fundamental methods; some of these are available commercially. Many commercial instruments that were designed for application to materials other than food can be used for studying many fluid foods. It appears that the word *viscometer* is used for an instrument designed solely to provide information on the viscosity function (Equation 5.1), whereas the word *rheometer* is used for an instrument that can also provide information on other rheological parameters such as those related to viscoelastic behavior.

The fundamental methods can be classified under the specific geometry employed: capillary, Couette (concentric cylinder), plate-and-cone, and parallel plate. Many of the commercially available instruments and their principles of operation have been described by Van Wazer et al. (1963) and Whorlow (1980), and hence have not been described here. The underlying equations and assumptions have been detailed by Walters (1975). Rao (1999) discussed the use of the fundamental methods for fluid foods, whereas Prentice (1984) outlined problems encountered with specific food products. Here, the fundamental methods are considered in brief. Where possible, studies on foods are cited for each geometry.

Three requirements are common to all the geometries listed above. These are (1) laminar flow of the fluid, (2) isothermal operation, and (3) no slip at solid–fluid interfaces. Additional requirements specific to each geometry are discussed later.

5.4.1.1 Capillary Flow

Capillary viscometers made of glass and operating under gravity are used mainly for Newtonian liquids. For non-Newtonian fluids, the design must allow operation over a wide range of flow rates, and the shear stress must be determined for fully developed flow conditions. Figure 5.5 illustrates capillary flow for use in a viscometric system. The requirements to be satisfied are (1) steady flow, (2) no end effects, and (3) absence of radial and tangential components of velocity; the axial velocity is a function of axial distance only.

The sheer stress and shear rate at the wall can be determined from Equations 5.37 and 5.38, respectively (Brodkey, 1967):

$$\sigma_w = \frac{D \Delta P}{4L} \quad (5.37)$$

RHEOLOGICAL PROPERTIES OF FLUID FOODS

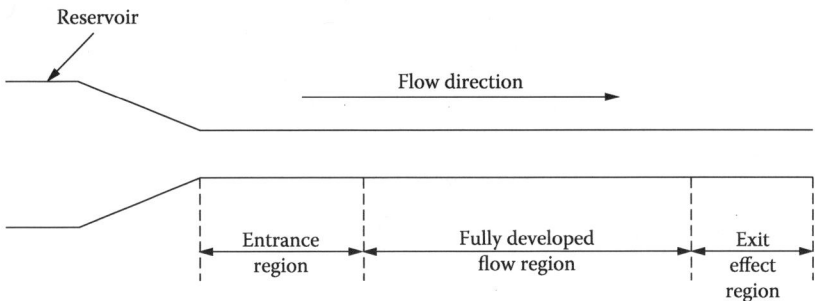

Figure 5.5 Capillary flow viscometer.

$$\left(\frac{dv}{dr}\right)_w = \frac{3}{4}\left(\frac{32Q}{\pi D^3}\right) + \frac{\sigma_w}{4}\frac{d(32Q/\pi D^3)}{d\tau_w} \tag{5.38}$$

For a capillary aligned vertically, the measured pressure drop must be corrected for the column of liquid of height L. When the shear stress, σ_w, and the apparent shear rate, $(32Q/\pi D^3)$, data follow a power relationship over a wide range of values, the simpler form of Equation 5.38 can be used:

$$\left(\frac{dv}{dr}\right)_w = \left(\frac{3n+1}{4n}\right)\left(\frac{32Q}{\pi D^3}\right) \tag{5.39}$$

Many foods such as applesauce, baby foods, and tomato puree are suspensions with relatively large particles. The flow behavior of a variety of food suspensions have been studied by using tubes having a diameter of 6–10 mm (Charm, 1960; Escardino et al., 1972; Rao et al., 1974; Saravacos, 1968; Scheve et al., 1974; Vitali and Rao, 1982). The pressure drop over a given length, required to compute the shear stress, has been measured by means of manometers and pressure transducers as well as the load cell of a universal testing machine (Blake and Moran, 1975).

In situations where the pressure drop cannot be determined for fully developed flow conditions, corrections for the undeveloped flow conditions (end effects) must be made by following a procedure such as that of Bagley (1975). The study of Jao et al. (1978) on soy flour dough is an example of such an effort.

5.4.1.2 Couette Flow Viscometers

A number of Couette (concentric cylinder) viscometers are available commercially (Van Wazer et al., 1963; Whorlow, 1980), and several have been employed for studies on fluid foods (Rao et al., 1981a; Saravacos, 1970; Tung et al., 1970; Vitali and Rao, 1984a,b). Corey and Creswick (1970) presented a design in which the revolutions per minute (rpm) of the rotating cylinder could be increased or decreased in a continuous manner. This option is now available with some commercial viscometers, such as Haake Rotovisco RV100, RV2, and RV3 models.

Griffith and Rao (1978) described the manner in which an inexpensive viscometer was modified so that rheological data could be obtained over a wide range of shear rates. The specific assumptions for this geometry (Figure 5.6) are (1) steady flow, (2) the absence of radial and axial components of velocity, and (3) the absence of distortion of the flow field by the ends of the cylinders.

Either the outer or the inner cylinder can be rotated. However, when the outer cylinder is rotated, the transition to turbulent flow occurs at a higher speed than when the inner cylinder is rotated (Schlichting, 1960). For the case of the outer cylinder rotating and the inner cylinder being stationary, it can be shown that (Brodkey, 1967)

$$2\sigma_i \frac{d\Omega}{d\sigma_i} = \dot{\gamma}_i - \dot{\gamma}_o \tag{5.40}$$

where Ω is the angular velocity of the outer cylinder (radians/second), γ is the shear rate, and the subscripts i and o denote the inner and outer cylinders, respectively.

Several methods have been published for determining the shear stress and the corresponding shear rate for the Couette geometry (Brodkey, 1967; Van Wazer et al., 1963). Here, we note that shear rate can be determined easily for fluids obeying a power-type

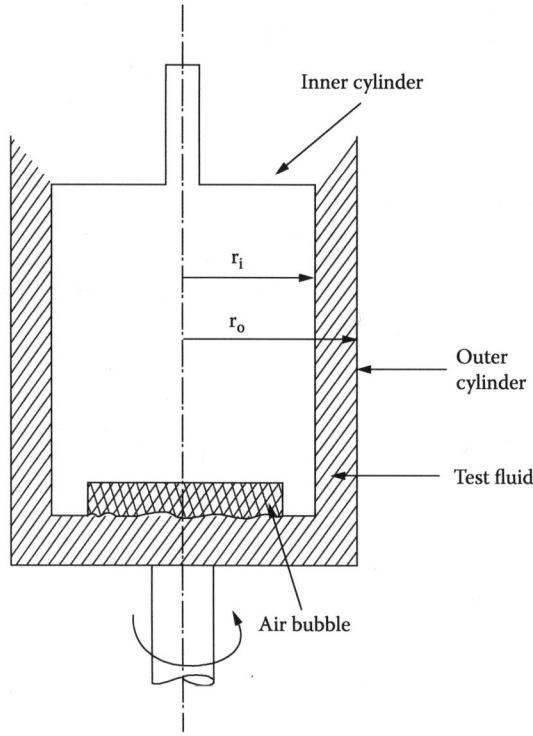

Figure 5.6 Concentric cylinder viscometer.

relationship between the angular velocity and shear stress, including Newtonian fluids. For a Newtonian fluid, the shear rate can be calculated from the expression:

$$\dot{\gamma}_{iN} = \frac{2\Omega}{\left[1 - (r_i / r_o)^2\right]} \tag{5.41}$$

Also of interest, here are the facts that (1) the gap between the cylinders must be narrow to determine the shear rate of non-Newtonian fluids accurately, and (2) for a cylinder rotating in an infinite fluid, Equation 5.40 reduces to

$$2\sigma_i \frac{d\Omega}{d\sigma_i} = \dot{\gamma}_i \tag{5.42}$$

Smith (1984) studied the effect of gap width in concentric cylinder viscometers and showed that significant errors occur in the case of systems with wide gaps, particularly in the case of highly shear-thinning fluids. For many concentric cylinder systems, the non-Newtonian shear rates at the rotating bob (γ_b) can be calculated from the magnitudes of bob angular velocity (Ω), shear stress at the bob (σ_B), and the quantity e = cup radius/bob radius, using the expression of Krieger and Elrod (Van Wazer et al., 1963):

$$\dot{\gamma}_B = \frac{\Omega}{\ln e}\left[1 + \ln e \frac{d(\ln \Omega)}{d(\ln \sigma_B)} + \frac{(\ln e)^2 d^2\Omega}{3\Omega d(\ln \sigma_B)^2}\right] \tag{5.43}$$

The requirement of a narrow gap between the cylinders precludes the study of foods containing large particles, such as the suspension indicated earlier, and fermentation broths (Bongenaar et al., 1973; Rao, 1975; Roels et al., 1974).

5.4.1.3 Plate-and-Cone Viscometers

As the name indicates, a plate-and-cone viscometer consists of a circular flat plate and a cone (Figure 5.7). The cone angle is about 3° or less. When the angle is larger than 3°, edge effects can distort the flow field.

For the case of a fixed plate and a rotating cone with a small angle, Brodkey (1967) showed that

$$\sigma_p = \frac{3T}{D} \tag{5.44}$$

$$\dot{\gamma}_p = \frac{\Omega}{\theta} \tag{5.45}$$

where σ_p is the shear stress at the plate, T is the torque per unit area, D is the plate diameter, Ω is the angular velocity, γ_p is the shear rate at the plate, and θ is the cone angle in radians. It is readily observed that the shear rate can be computed easily. Further, when the cone angle is small, the shear rate across the gap between the cone and plate is constant. Therefore, the plate–cone geometry is well suited for studying rheology of homogeneous

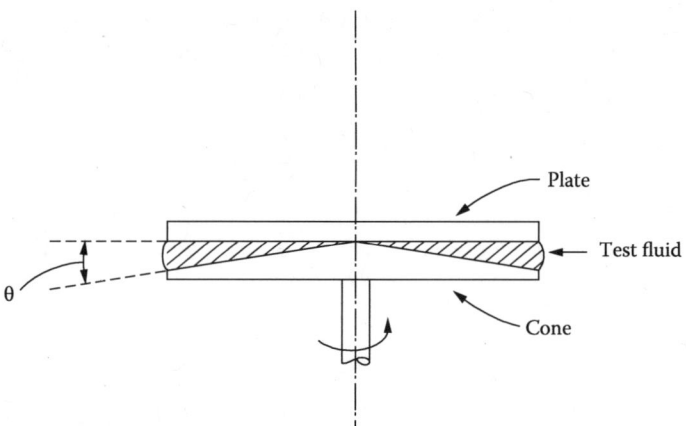

Figure 5.7 Plate-and-cone system.

non-Newtonian fluid foods and has been used in numerous studies. With foods, one concern with this geometry is the possibility of dehydration at the edge.

5.4.1.4 Parallel Plate Geometry

Instead of a plate-and-cone system, a parallel plate geometry can also be used. The pertinent equations for the shear rate ($\dot{\gamma}$) and the shear stress (σ) are (Walters, 1975)

$$\dot{\gamma} = \Omega \frac{r}{h} \qquad (5.46)$$

$$\sigma = \frac{3C}{2\pi r^3}\left[1 + \frac{1}{3}\frac{d\ln C}{d\ln \dot{\gamma}}\right] \qquad (5.47)$$

where Ω is the angular velocity, r is the radius, h is the gap width between the plates, and C is the torque exerted. The parallel plate geometry is well suited for studying foods containing particles of about 50 to 60 μm when gaps of the order of 500 to 600 μm can be used (Yang and Rao, 1998a). As with the plate-cone geometry, dehydration of the sample at the edge is a concern. In dynamic rheological tests, because the strains are small, mineral or silicone oil may be used to minimize dehydration of a sample.

Viscoelastic behavior of fluid and semisolid foods can be studied using small-amplitude oscillatory tests, popularly known as dynamic rheological tests. Dynamic rheological experiments are conducted with the cone–plate, parallel plate, and concentric cylinder geometries. With modern rheometers, it is not too difficult to conduct these tests accurately. Due to availability of powerful computers and developments in hardware and software, many steps have been automated. Those steps include setup (e.g., gap of a parallel plate geometry), temperature control (fixed values or ramps), conduct of the tests, and data collection, as well as calculation of the various rheological parameters (e.g., G', $G \leq$, tan δ, and $\dot{\eta}^*$), taking into consideration the inertia of measuring systems.

An experiment in which the strain or stress is varied over a range of values is essential to determine the linear viscoelastic range. The limit of linearity can be detected when dynamic rheological properties (e.g., G' and G'') change rapidly from their almost constant values in the linear viscoelastic region. Two tests that can be used to examine phase transitions in foods are the following (Rao, 1999):

1. Temperature sweep studies in which G' and G'' are determined as a function of temperature at fixed frequency (ω). This test is well suited for studying gel formation during cooling of a heated food polymer (e.g., pectin, gelatin) dispersion and gel formation by a protein dispersion and gelatinization of a starch dispersion during heating.
2. Time sweep studies in which G' and G'' are determined as a function of time at fixed ω and temperature. This type of test, often called a gel cure experiment, is well suited for studying structure development in physical gels.

Additionally, frequency sweep studies in which G' and G'' are determined as a function of frequency (ω) at a fixed temperature are useful for probing the structure of viscoelastic foods.

Once a food gel or product has been created, a frequency sweep, conducted over a wide range of oscillatory frequencies, can be used to examine its properties. The behavior of the moduli G' and G'' over a wide range of frequencies provides insights into the nature of the gel, such as whether it is a "weak" gel, a "strong" gel, or a "viscoelastic liquid."

Oscillatory shear experiments with a plate-and-cone geometry were employed by Elliott and Ganz (1977) to study the viscoelastic behavior of salad dressings. Davis (1973) employed a parallel plate system to study lard and shortening. With the availability of automated rheometers during the past few years, the dynamic rheological test has become an accessible and valuable tool for studying the rheological behavior of fluid and semisolid foods (Rao, 1999; Rao and Steffe, 1992).

5.4.1.5 Slit (Channel) Rheometers

Slit or channel geometries have been used for studying primarily the rheological behavior of products processed in extruders. The applicable equations have been derived by Dealy (1982), Steffe (1992), and others. For the fully developed flow of a Newtonian fluid in a slit of length L, width W, and height (thickness) h (Figure 5.8), the relationship between the shear stress (σ_{yx}) and the gradient ($\Delta P/L$) is given by

$$\sigma_{yx} = \frac{\Delta p \, y}{L} \tag{5.48}$$

where y is the coordinate in the direction of height. In order to neglect edge effects, the ratio of width to height should be greater than 10. The velocity profile for a Newtonian fluid is given by

$$v_x = \frac{3Q}{2hW}\left[1 - 4\left(\frac{y}{h}\right)^2\right] \tag{5.49}$$

ENGINEERING PROPERTIES OF FOODS

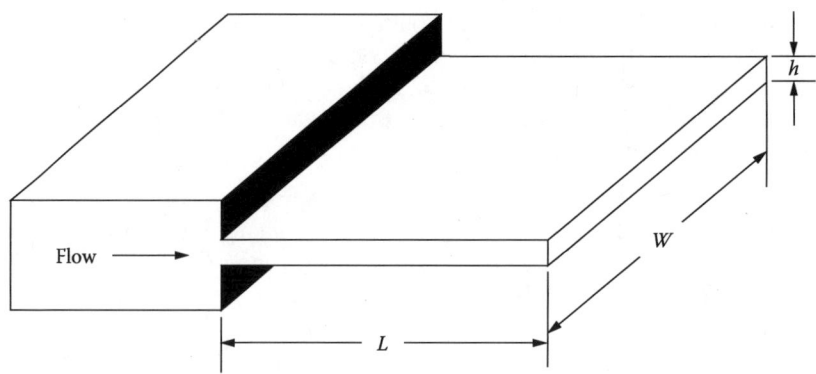

Figure 5.8 Slit rheometer geometry.

At the wall ($y = h/2$), the shear rate and shear stress of a Newtonian fluid are given by Equations 5.50 and 5.51, respectively.

$$\dot{\gamma}_a = \frac{6Q}{h^2 W} \tag{5.50}$$

$$\sigma_w = \left(\frac{\Delta p}{L}\right)\left(\frac{h}{2}\right) \tag{5.51}$$

The shear rate for a non-Newtonian fluid can be obtained from

$$\dot{\gamma}_w = \left(\frac{6Q}{h^2 W}\right)\left(\frac{2+\beta}{3}\right) \tag{5.52}$$

where β equals

$$\beta = \frac{d\ln(6Q/Wh^2)}{d\ln[(h/2)(\Delta p/L)]} = \frac{d\ln \dot{\gamma}_a}{d\ln \sigma_w} \tag{5.53}$$

Equation 5.52 is analogous to the Weissenberg–Rabinowitsch–Mooney equation for laminar flow in a tube or pipe. As in pipe flow, the shear stress and the Newtonian shear rate are calculated first using Equations 5.50 and 5.51, respectively. Using the slope (β) of a plot of $\ln \sigma_w$ and $\ln \dot{\gamma}_a$ in Equation 5.52, the non-Newtonian shear rates ($\dot{\gamma}_w$) can be calculated. It is of interest to note that the slit rheometer has been used also for determining normal stress differences. The study by Senouci and Smith (1988) is an example of the use of the slit rheometer for characterizing the rheological properties of foods.

5.4.1.6 Extensional Flows

Extensional flow is another type of deformation used to obtain information regarding the rheological behavior of foods. For uniaxial extension or simple extension of a liquid sample of length L, one meaningful measure of strain is (Dealy, 1982)

$$de = dL/L$$

The strain rate is given by

$$\dot{e} = \frac{de}{dt} = \frac{1}{L}\frac{dL}{dt} = \frac{d\ln L}{dt} \tag{5.54}$$

Because dL/dt is velocity, the above equation can be written as

$$\dot{e} = V/L \tag{5.55}$$

When the strain rate $\dot{e}$ is maintained constant, the deformation obtained is called steady simple extension or steady uniaxial extension. It can be shown (Dealy, 1982) that the extensional viscosity ($\acute{\eta}_E$) is related to the normal stress difference:

$$\eta_E(\dot{e}) = \frac{\sigma_{11} - \sigma_{22}}{\dot{e}} \tag{5.56}$$

From the above equation, it is clear that to obtain magnitudes of $\acute{\eta}_E$ one must have the means to measure both the normal stress difference and the strain rate $\dot{e}$. For the special case of a Newtonian fluid with viscosity $\acute{\eta}$,

$$\eta_E = 3\eta \tag{5.57}$$

For a shear thinning fluid with a zero-shear viscosity $\acute{\eta}_0$,

$$\lim_{\dot{e} \to 0}\left[\eta_E(\dot{e})\right] = 3\eta_0 \tag{5.58}$$

5.4.1.6.1 Biaxial Extension

Often, it is convenient to work with circular disks that can be subjected to biaxial extension (Figure 5.9) with a universal testing machine (Chatraei et al., 1981). This technique, called squeezing flow, was adapted to measurements on foods using lubrication between the food sample and the metal platens. Squeezing flow between two disks with lubricated surfaces can generate a homogeneous compression or equal biaxial extension in a high viscosity polymer and other materials, such as cheese and peanut butter (Campanella et al., 1987; Casiraghi et al., 1985; Chatraei et al., 1981). The extensional viscosity can be calculated assuming a homogeneous deformation, which implies that there is perfect slip at the wall. The requirement of perfect slip at the wall can be an asset in melted cheese and other foods that exhibit slip at walls in the conventional viscometric flows. The pertinent equations for obtaining biaxial extensional viscosity are obtained assuming an incompressible material undergoing homogeneous deformation; the latter implies perfect slip at the wall (Chatraei et al., 1981; Campanella et al., 1987). For the system illustrated in Figure 5.9, the velocity field is

$$V_z = \dot{e}_T H(t); \quad V_r = -\dot{e}_T \frac{r}{2}; \quad V_\theta = 0 \tag{5.59}$$

ENGINEERING PROPERTIES OF FOODS

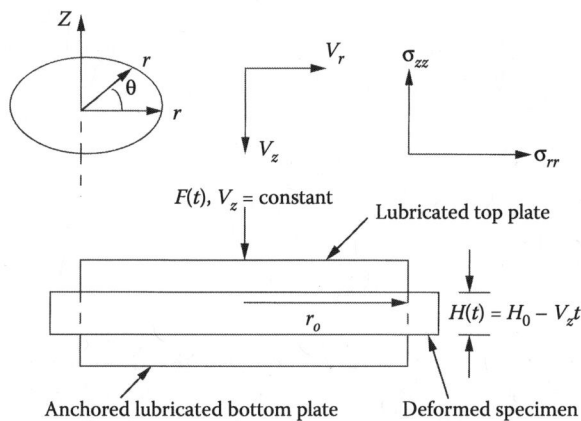

Figure 5.9 Biaxial extensional flow test setup. (From Campanella et al., 1987.)

where V_z, V_r, and V_θ are the vertical, radial, and the angular velocity components; $H(t)$ is the instantaneous height at time t, and r is the radial distance. In a universal testing machine operated at a constant crosshead speed, the strain rate is given by

$$\dot{e}_T = \frac{V_z}{H_0 - V_z t} \tag{5.60}$$

where H_0 is the initial height of the test specimen.

The experiment can be carried out at either constant or time-dependent force. With the area of the plates being constant and neglecting edge effects, the normal stress difference is related to the applied force:

$$\sigma_{zz} - \sigma_{rr} = \frac{F(t)}{\pi R^2} \tag{5.61}$$

The elongational viscosity ($\dot{\eta}_b$) was defined as the ratio of the normal stress difference and the radial extension rate assumed to be half the strain rate:

$$\eta_b = \frac{2F(t)(H_0 - V_z t)}{\pi R^2 V_z} \tag{5.62}$$

For an unvulcanized polydimethylsiloxane, the biaxial viscosity was approximately six times the shear viscosity over the biaxial extensional rates from 0.003 to 1.0 s^{-1} (Chatraei et al., 1981), a result expected for Newtonian fluids. That is, the relationship between the limiting value of biaxial extensional viscosity ($\dot{\eta}_b$) at zero strain rate and the steady zero-shear viscosity ($\dot{\eta}_a 0$) of a non-Newtonian food is

$$\lim_{\dot{e}_b \to 0}\left[\eta_b(\dot{e})\right] = 6\eta_0 \tag{5.63}$$

The lubricated squeezing flow technique was applied to melting American cheese where an Instron universal testing machine with lubricated platens was employed (Campanella et al., 1987) to obtain experimental data. This technique is particularly well suited for materials prone to slip, such as melted cheese.

5.4.2 Empirical Methods

Rotational viscometers with spindle geometries that are difficult to analyze mathematically have been employed in the empirical tests. The geometries include spindles with protruding pins and flags. These geometries are available from the manufacturers of rotational viscometers: Brookfield, Haake, and Epprecht Rheomat. Utilizing these geometries, one obtains for non-Newtonian fluids a magnitude of apparent viscosity in arbitrary units. The use of these complex geometries has been limited to quality control.

Several instruments, described below, have been developed for characterizing fruit and vegetable purees and baby foods. These instruments are used to measure the consistency of pureed foods. The word *consistency* is used with the understanding that it is a property related to the apparent viscosity of suspensions. However, few studies have pointed out the specific shear rate that must be employed to determine the apparent viscosity of a given product. These empirical methods are used for quality control of these products and in studies dealing with the effects of changing processing conditions on the consistency of purees.

5.4.2.1 Adams Consistometer

This instrument consists of a hollow truncated cone that can be filled with 200 mL of the product to be studied. The cone containing the sample is seated on a graduated disk. For determining the consistency of the sample, the cone is suddenly raised, and after 30 s the extent of flow at four equidistant points on the disk is recorded. The four values are averaged, and this value is considered to be the consistency of the product.

5.4.2.2 Bostwick Consistometer

The Bostwick consistometer consists of a rectangular trough whose floor is graduated in millimeters. One end of the trough has a holding compartment 100 mL in capacity, in which the test sample is held in place by a spring-loaded gate. The Bostwick consistometer is used in the food industry to evaluate the consistency of fruit or vegetable purees off-line. The food product, such as a pureed baby food, is allowed to flow through a channel; the Bostwick consistency is the length of flow recorded in a specified time, typically 30 s. McCarthy and Seymour (1993) evaluated the flow observed in a Bostwick consistometer as a balance of gravitational and viscous forces dependent on the height of the fluid, fluid density, and fluid viscosity. For Newtonian fluids, they derived an expression showing the Bostwick measurement, L, to be a function of trough geometry and fluid properties:

$$L = \xi_N \left(\frac{g q^3}{3 \upsilon} \right)^{0.2} t^{0.2} \tag{5.64}$$

where ξ_N is a similarity variable with a theoretical value of 1.411, g is the gravitational constant, q is fluid volume per unit width, υ is the kinematic viscosity, and t is time of flow. Experimental measurements with four Newtonian fluids (corn syrup and silicone oils of different viscosities) verified the theoretically predicted dependence of the extent of flow on kinematic viscosity (McCarthy and Seymour, 1993).

Additional work (McCarthy and Seymour, 1994) showed that the extent of flow at a given time was greater in a wider consistometer than in the standard consistometer. Another significant cause of deviation was that the flow analysis did not consider the inertial and gravitational flow regimes in the early stages of flow in the consistometer. After the experimental measurements for Newtonian and power law fluids were compared to theoretical predictions, it was suggested that the theory could be used to evaluate values for power law parameters. The Bostwick measurement length, L, after 30 s was found to be linearly related to $(\acute{\eta}_a/\rho)^{-0.2}$.

5.4.2.3 Efflux Tube Viscometer

The efflux tube viscometer is used to measure the time necessary for a fixed quantity of fluid to pass through a tube or capillary.

Davis et al. (1954) showed that the data obtained with the Adams and Bostwick consistometers were linearly related and that the efflux viscometer measured a parameter different from those of the consistometers. It must be emphasized that the empirical method employed must be identified when the investigator is dealing with consistency values of a product.

5.4.3 Imitative Methods

In the case of imitative methods, the properties are measured under test conditions that simulate those in practice. There are several instruments that perform imitative tests, and most of these are applicable to solid foods. Examples of instruments that perform imitative tests are butter spreaders, the General Foods Texturometer, and the Brabender Farinograph (White, 1970). Here, only the Brabender, Visco-Amylo-Graph, which is used for evaluating the thickening power of ingredients, is discussed.

The Visco-Amylo-Graph is used in the evaluation of the thickening power of pastes of flours, starches, and gums. With this apparatus, the sample to be studied is placed in a container, and the drag exerted on a rotating paddle is recorded on a chart paper. The temperature of the sample can be raised or lowered at 1.5°C/min (Van Wazer et al., 1963). The Rapid Visco Analyzer (RVA), in which a smaller amount of sample is used, has found widespread use in recent years. One feature of the RVA is that its agitator can be operated at different rotational speeds.

It appears that the sample temperature in the Visco-Amylo-Graph and the RVA can be made to imitate the conditions in practice. For realistic simulation of the shear rate conditions in practice, they should be determined ahead of time. Subsequently, an attempt should be made to reproduce those conditions in the test instruments.

A few refinements to empirical and imitative methods have been developed and they are considered next.

5.4.3.1 Mixers for Determining Flow Properties

Quantitative shear stress–shear rate data can be obtained with agitators having complex geometries if one assumes that the shear rate is directly proportional to the rotational speed of the agitator and if the flow behavior of the fluid can be described by the power law model. The procedure is based on the studies of Metzner and Otto (1957), Bongenaar et al. (1973), and Rieger and Novak (1973). It has been described in detail by Rao (1975) and is considered only in brief here.

It is assumed that the impeller (agitator) exerts an average shear rate that is directly proportional to the rotational speed:

$$\dot{\gamma} = cN \tag{5.65}$$

where c is the proportionally constant between the shear rate and the rotational speed of the impeller. For any impeller, the constant c can be determined from a plot of $1-n$ versus $\log [p/(kN^{n+1}d^3)]$; the slope of the line is equal to $-\log c$ (Rieger and Novak, 1973). It is clear that the tests for a given impeller must be conducted such that the following data are obtained; p, the power (Nm/s); N, the rotational speed (s^{-1}); d, the diameter of the impeller (m); and the power law parameters of several test fluids.

The method proposed by Rieger and Novak (1973) to determine the magnitude of the proportionally constant c is relatively simple. However, because the magnitude of the constant is equal to the antilogarithm of the slope of the plot indicated earlier, a small error in determining the magnitude of the plot indicated earlier, a small error in determining the magnitude of the slope will result in a large error in the magnitude of c (Rao and Cooley, 1984). Compared to the method of Rieger and Novak (1973), the method of Metzner and Otto (1957) involves more work. It is contingent on finding a Newtonian and a non-Newtonian fluid such that the power consumed by an impeller at a given speed is identical in the two fluids. Also, the experimental conditions, such as the dimensions of the mixing vessel, must be the same in the experiments with the Newtonian and non-Newtonian fluids.

5.4.3.1.1 Power Law Parameters Using a Mixer

Once the proportionality constant in Equation 5.64 is known for a specific impeller, one can determine the power law parameters of a test suspension (x). Because the shear stress and the shear rate are directly proportional to the torque (T) and the rotational speed (N), respectively, a plot of $\log (T)$ versus $\log (N)$ would yield the magnitude of the flow behavior index n_x. The consistency index of a test fluid (K_x) can be calculated using the torque values at known rotational speeds of the test fluid and another fluid (y) whose power law parameters (K_y, n_y) are known:

$$\frac{T_x}{T_y} = \frac{\sigma_x}{\sigma_y} = \frac{K_x N_x^{n_x} c^{n_x}}{K_y N_y^{n_y} c^{n_y}} \tag{5.66}$$

where the subscripts x and y refer to the test and standard fluids, respectively. Because of the assumption that the impeller exerts an average shear rate, the method is approximate. Nevertheless, it is the only one available for the quantitative study of food suspensions.

Because viscoelastic fluids climb up rotating shafts (the Weissenberg effect), mixer viscometers may not be suitable for studying these fluids.

5.4.3.1.2 Yield Stress with a Mixer

As stated earlier, yield stress is also an important property of many foods whose determination requires considerable care. In the relaxation method using concentric cylinder or cone-and-plate geometries, it can be determined (Van Wazer et al., 1963) by recording the shear stress level at a low rpm at which no stress relaxation occurs on reducing the rpm to zero (Vitali and Rao, 1984b). This method, however, is time-consuming and requires great care to obtain reliable results. For the most part, magnitudes of yield stress were determined by extrapolation of shear rate–shear stress data according to several flow methods such as those of Casson (Equation 5.9), Herschel and Bulkley (Equation 5.8), and Mizrahi and Berk (Equation 5.10) (Rao and Cooley, 1983; Rao et al., 1981a).

Dzuy and Boger (1983) employed a mixer viscometer for the measurement of yield stress of a concentrated non-food suspension and called it the "vane method." This method is relatively simple because the yield stress can be calculated from the maximum value of torque recorded at low rotational speeds with a controlled shear-rate viscometer. The maximum torque (T_m) value recorded with a controlled shear-rate viscometer and the diameter (D_V) and height (H) of the vane were used to calculate the yield stress (σ_V) using the equation

$$T_m = \frac{\pi D_V^3}{2}\left(\frac{H}{D_V} + \frac{1}{3}\right)\sigma_V \tag{5.67}$$

Equation 5.67 was derived by conducting a torque balance on the surface of the impeller (Dzuy and Boger, 1983).

Yield stress of applesauce samples was determined according to the vane method by Qiu and Rao (1988) using two vanes. With both impellers, the maximum torque value increased slightly with rotational speed over the range 0.1–2.0 rpm. In the derivation of Equation 5.66, the shear stress was assumed to be uniformly distributed everywhere on the cylinder and the test material yields at the impeller surface. In a comprehensive review of the vane method for yield stress, Dzuy and Boger (1985) concluded that these assumptions are satisfactory. In contrast, there is limited experimental evidence (Keentok, 1982) to suggest that some materials may yield along a diameter (D_S) that is larger than the actual diameter of the vane; the ratio D_S/D_V can be as large as 1.05 for some greases and is apparently dependent on the plastic, thixotropic, and elastic properties of the material. However, for inelastic and plastic substances, the data published by these workers suggest that D_S/D_V is very close to 1.0.

5.4.3.1.3 Role of Structure on Yield Stress

A sample with undisrupted structure has a higher value of yield stress, called static yield stress (σ_{0s}), whereas that whose structure has been disrupted by shear has a lower magnitude, called dynamic yield stress (σ_{0d}) (Rao, 1999; Yoo et al., 1995). In two different foods with equal magnitudes of yield stress (σ_{0d}), it is very likely that they are due to contributions of different magnitudes by differing forces. In turn, these forces are affected by the composition of the two foods and their manufacturing methods.

From an energy balance, the shear stress necessary to produce deformation at a constant shear rate of a dispersion of particle clusters or flocs is (Metz et al., 1979; Michaels and Bolger, 1962):

$$E_{sT} = E_{sb} + E_{sy} + E_{sn} \tag{5.68}$$

where E_{sT} is the total energy dissipation rate to produce deformation, E_{sb} is the energy dissipation rate required to break bonds, E_{sv} is the energy dissipation rate due to purely viscous drag, and E_{sn} is the energy dissipation rate required to break the aggregate network. Because $E = \sigma \dot{\gamma}$ yield stress (σ_0) is calculated from vane mixer data it can be written as

$$\sigma_0 = \sigma_b + \sigma_v + \sigma_n \tag{5.69}$$

where σ_b, σ_v, and σ_n are components of yield analogous to the energy components.

The failure stress of an undisturbed food dispersion is:

$$\sigma_0 = \sigma_{0s} \tag{5.70}$$

The total energy required for sample deformation at yield can be calculated:

$$E_{sT} = \sigma_{0s}\dot{\gamma} \tag{5.71}$$

The subscript s of the energy terms is used to denote shear-based quantity. The continuous phase of a typical food dispersion is an aqueous solution of solutes, such as sugars, and polymers, such as amylose or pectins. In the special case of a starch dispersion, it arises from association of amylose and a few amylopectin molecules into double-helical junctions, with further association of helices into aggregated assemblies (Genovese and Rao, 2003b). Bonding, sometimes called adhesivity, in a food dispersion can be associated with bridging between particles (e.g., irregularly shaped insoluble-in-water particles) and their interactions with the continuous phase. The stress required to break the bonds, σ_b, can be calculated as follows:

$$\sigma_b = \sigma_{0s} - \sigma_{0d} \tag{5.72}$$

Values of E_b can be calculated as

$$E_b = \sigma_b \dot{\gamma} \tag{5.73}$$

The viscous stress component σ_v is given by (Metz et al., 1979)

$$\sigma_v = \eta_\infty \dot{\gamma} \tag{5.74}$$

where η_∞ is the viscosity of the dispersion at infinite shear rate. Therefore, E_v, the energy dissipation rate due to purely viscous drag, is calculated as

$$E_v = \sigma_v \dot{\gamma} = \eta_\infty \dot{\gamma}^2 \tag{5.75}$$

It is known that at zero shear rate, the network yield stress σ_n equals σ_0; however, no relationship to estimate it at low finite values of shear rate exists (Metz et al., 1979), and it can only be estimated by difference

$$\sigma_n = \sigma_{0s} - (\sigma_b + \sigma_v) \tag{5.76}$$

Based on a study on three starch dispersions (Genovese and Rao, 2003b) and commercial samples of mayonnaise and tomato ketchup and concentrates (Genovese and Rao, 2004), it can be said that, in general, compared to the contributions of bonding (σ_b) and network (σ_n), the contribution of the viscous component (σ_v) would be small.

5.4.3.1.4 *Effective Shear Rate in a Brabender Viscograph*
Wood and Goff (1973) determined the effective shear rate to be 40 s^{-1} in a Brabender viscograph having the following characteristics: model, VSK 4; bowl speed, 75 rpm; bowl internal diameter, 8.8 cm; length of bowl pin, 7.0 cm; length of sensing pin, 9.75 cm; and depth of liquid with 450 g water, 7.5 cm. It is necessary to note these characteristics because the viscograph is not an absolutely standard instrument.

5.4.3.1.5 *Effective Shear Rate in a Rapid Visco Analyzer*
Using the principle of mixer viscometry, Lai et al. (2000) determined the average shear rate in the mixing system (impeller-cup combination) of the RVA. A relationship between the impeller Reynolds number and the power number was established with Newtonian standards. Using the matching viscosity technique and non-Newtonian fluids consisting of various aqueous solutions of guar gum and methylcellulose, the average value of the mixer viscometer constant (c) was 20.1 per revolution over speeds of 1.0 to 3.5 r/s (60 to 210 r/min). Hence, the average shear rate in the RVA can be estimated as 20.1 multiplied by the angular velocity given in r/s.

5.4.3.2 In-Plant Measurement of Rheological Behavior of Fluid Foods
An in-line measurement is performed in a process line; an online measurement is performed in a bypass loop from the main process line, and the food may be returned to the main process line after the measurement is performed. A near-line measurement is performed on a sample taken from a process line, which is often discarded after measurement. Because foods are complex materials (e.g., suspensions, emulsions, gels), structural changes may take place during sampling (e.g., flow through a valve) for online and near-line measurements (Roberts, 2003). Nevertheless, in principle, the previously described capillary flow, concentric cylinder, plate–cone, and mixer viscometers may be used for in-line, online, and near-line measurements. The empirical measurement methods described previously are used primarily in near-line measurements.

Roberts (2003) listed several requirements that both in-line and online measuring systems for foods should satisfy, including:

Free of hygiene risk. The system must be constructed with a food-grade material, permit standard clean-in-place procedures, and be free of dead flow zones.
Nondestructive. The system should not alter the quality of the product or perturb the production schedule and process.
Real-time operation. To minimize down time and waste or rework of a product, the response time should be short, typically seconds.
Physically robust and stable. In general, the system must require little maintenance and withstand the process operating conditions (e.g., temperature and pressure). The sensor signal must be unaffected by the typical environment in a processing

plant (e.g., mechanical vibration, electrical interference) and amenable to control operations.

Easy operation. It would be desirable that the sensor and system be easy to operate and provide an acceptable signal for process control, and that the results not be dependent on operator skills. However, determination of non-Newtonian rheological behavior also requires knowledge of the flow characteristics of the fluid food and its structure, as well as potential changes that can occur due to the shear rate and temperature prevalent in the measurement system. Thus, in addition to a skilled operator, it would be desirable to have supervisors with a thorough knowledge of the rheological and physicochemical behavior of the food product being manufactured.

5.4.3.2.1 Tomographic Techniques

Some, if not all, of the requirements of in-line measurement techniques are satisfied by tomographic techniques that provide spatially and temporally resolved data. The techniques utilize the inherent properties of the food material and include those based on magnetic, acoustic, optical, and electrical signals (Choi et al., 2002). These techniques have also been used in measurement of velocity profiles in tubes and rheology of stationary materials. Here, the emphasis is on determination of in-line measurement of rheological behavior of fluid foods using tube flow.

Magnetic resonance imaging (MRI) is a spectroscopic technique based on the interaction between nuclear magnetic moments and applied external magnetic fields. A sample is placed in a magnetic field within a radio frequency probe and the response of the test material in terms of attenuation, frequency, and phase to energy added in that frequency range is recorded. Two notable constraints of MRI are the need to include a nonmetallic and nonmagnetic test section in the flow system and the high cost of setting up MRI systems in processing plants.

Ultrasonic refers to sound waves with frequencies of 20,000 Hz or greater, which are beyond the range of human hearing. The sound waves are transmitted through the wall of a pipe, and the reflections are analyzed. In principle, there are two different kinds of ultrasonic flow meters: transit time and Doppler flow meters. Both kinds measure primarily velocity. The principal advantages of ultrasonic Doppler velocity (UDV) meters over other types, such as turbine and conductivity meters, are as follows:

- No moving parts involved
- Nonintrusive
- Low maintenance
- Hard to block
- Can be used with nonconductive media

The UDV and MRI methods offer similar potential for rheological measurements under fully developed, steady, pressure-driven tube flow. In addition, the data-processing techniques for MRI and UDV are somewhat similar.

Doppler meters measure the frequency shifts caused by liquid flow. The frequency shift is proportional to the liquid's velocity. Time-of-flight meters use the speed of the signal traveling between two transducers, which increases or decreases with the direction of transmission and the velocity of the liquid being measured. Important parameters to consider when specifying ultrasonic flow meters include flow rate range, operating pressure, fluid temperature, and accuracy. One-beam Doppler flow meters are widely used, but multibeam profiling Doppler flow meters have been reported.

A schematic diagram of an UDV system is shown in Figure 5.10. The relationships among fluid velocity, v, and UDV data are given by

$$v = \frac{cf_D}{2f_0 \cos\theta} \quad (5.77)$$

where v is the velocity component along the axis of the ultrasound transducer, f_D is the Doppler shift frequency, f_0 is the frequency of the transmitted pulse, c is the speed of sound, and θ is the angle between the transducer and the flow direction, typically 45° (Dogan et al., 2003).

The spatial location, d, of the velocity component in the above equation can be identified by a time-of-flight, Δt, measurement that relates the speed of the reflected wave to the distance traveled:

$$d = \frac{c\,\Delta t}{2} \quad (5.78)$$

In turn, the values of d can be converted to the radial location in the pipe so that the velocity profile in a pipe can be obtained. The velocity profile is used to calculate velocity gradients (shear rates), (dv/dr), at specific locations using an even-order polynomial curve fit to the velocity data.

$$v(r) = a + br^2 + cr^4 + dr^6 + er^8 \quad (5.79)$$

Resolution of the velocity data and removal of data points near the center of the tube, which are distorted by noise, aid robustness of the curve fit; the polynomial curve fit introduced a systematic error when plug-like flow existed at radial positions smaller than 4 mm in a tube of 22 mm diameter. The curve fit method correctly fit the velocity data of

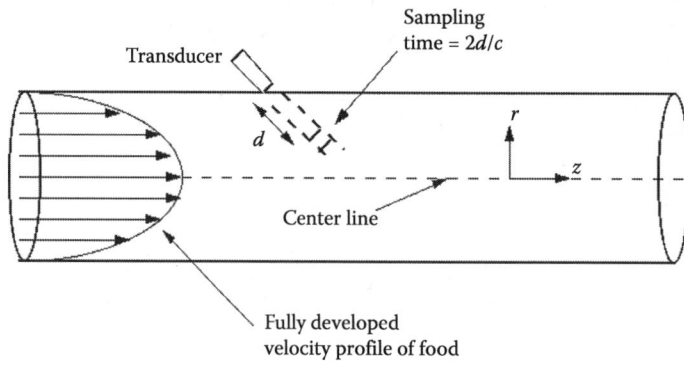

Figure 5.10 Schematic diagram of in-line measurement of flow properties using an ultrasonic Doppler velocity meter.

Newtonian and shear-thinning behaviors but was unable to produce accurate results for shear-thickening fluids (Arola et al., 1999).

With velocimeter-based or pointwise rheological characterization, in addition to calculation of shear rate profile, the corresponding shear stress distribution is obtained by combining pressure drop measurements and the linear relationship between the shear stress and the radial position in a pipe, to be discussed later in this chapter.

UDV and pressure drop (ΔP) measurements were carried out on tomato concentrates with total solids of 8.75%, 12.75%, and 17.10% (Dogan et al., 2003); the rheological parameters deduced from these data agreed well with those based on capillary flow data obtained at four different flow rates. Also, pointwise rheological characterization using MRI and UDV of 4.3 °Brix tomato juice and 65.7% corn syrup agreed well with off-line data obtained using a rheometer (Choi et al., 2002). The UDV technique was also used on 1- to 3-mm diced tomatoes suspended in tomato juice, and the yield stress of the suspension was characterized in terms of the Herschel–Bulkley model (Equation 5.12) and the apparent wall slip region modeled as a Bingham fluid (Dogan et al., 2002).

For reliable characterization of a specific food by the UDV technique, extensive studies would be necessary to establish the operating parameters for that food and flow system. A change in the type of food or the composition of a specific food may necessitate a thorough evaluation of all operating parameters. Nevertheless, this technique may find a place in in-line characterization of rheological properties in food processing plants.

5.4.3.2.2 Vibrational Viscometers

Vibrational viscometers are robust, easy to install for in-line sensing of viscosity, offer minimal disruption of flow in a process line, operate over a wide range of temperatures (e.g., –40 to 400°C), and provide real-time data. Vibrational viscometers actually measure kinematic viscosity (viscosity/density), and units are available capable of measuring kinematic viscosities ranging from 0.1 to 10^6 mPa s/g cm^{-3} (centistokes). Typically, a vibrational viscometer employs a high frequency (e.g., 650 Hz) torsional oscillation of a sphere- or rod-shaped probe that undergoes damping by the fluid whose viscosity is of interest. The amplitude of oscillation is small, of the order of a micrometer, and the power consumed is converted to viscosity. When the viscosity of a fluid changes, the power input to maintain constant oscillation amplitude is varied.

However, vibrational viscometers may not indicate the true bulk viscosity of a suspension that forms a thin layer of the continuous phase (e.g., serum of tomato juice) around the immersed probe or when the probe is covered by a higher viscosity gel due to fouling. Vibrational viscometers are suitable for measuring viscosities of Newtonian fluids, but not the shear-dependent rheological behavior of a non-Newtonian fluid (e.g., to calculate values of the power law parameters).

Vibrational viscometers may also be suitable for following gelation in near-line or laboratory experiments at a constant temperature. For example, a vibrational viscometer was used to determine the coagulation time of renneted milk at fixed temperatures (Sharma et al., 1989, 1992). However, in nonisothermal physical gelation, the elastic modulus depends on the temperature dependence of the resonant response, so the precise correction for the influence of temperature must be known.

5.5 FLOW OF FLUID FOODS IN TUBES

Rheological behavior of fluid foods plays an important role in unit operations of food processing. Treatment of Newtonian fluids can be found in standard texts on unit operations (McCabe and Smith, 1976). Applications related to chemical engineering have been discussed by Brodkey (1967), Skelland (1967), and Wohl (1968). In the area of food engineering, fluid flow and heat transfer have been covered in several texts, including those by Charm (1971), Rao (1999), and Singh and Heldman (2001). Datta (2002) covered heat and mass transfer in biological, including food, systems. Here, fluid flow of non-Newtonian fluid foods with respect to pressure drop due to friction is emphasized.

5.5.1 Isothermal Flow of Fluids in Tubes

Isothermal flow takes place in circular tubes (pipes) in the conveyance of foods between locations of a plant and in the holding tubes of pasteurizers. Because non-Newtonian foods are very viscous, laminar flow is encountered very often (Rao and Anantheswaran, 1982). For this reason, laminar flow applications are emphasized although some of the relationships to be discussed, such as the Reynolds number–friction factor chart, are applicable also to turbulent flow.

5.5.1.1 Velocity Profiles and Volumetric Flow Rate Relationships

Equations describing velocity profiles can be used to examine the influence of rheological models on the distribution of velocities and for determining the residence time distribution of the fluid particles. Pressure drop in pipes can be estimated from volumetric flow rate–pressure drop relationships.

For the system illustrated in Figure 5.11, the velocity profile for a fluid flowing in a pipe can be derived from the relationship

$$v = \int_r^R -\frac{dv}{dr} dr \qquad (5.80)$$

The volumetric flow rate, Q, in pipes can be obtained from the general equation (Brodkey, 1967; Skelland, 1967)

$$Q = \frac{\pi R^3}{\sigma_w^3} \int_0^{\sigma_w} \sigma_{rz}^2 \left(-\frac{dv}{dr}\right) d\sigma_{rz} \qquad (5.81)$$

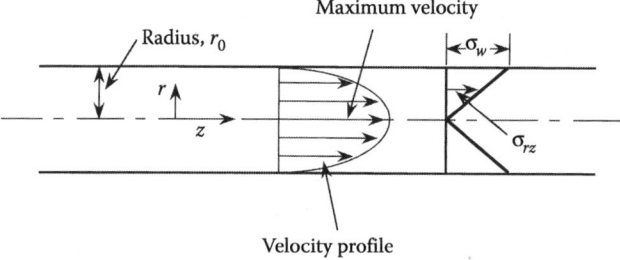

Figure 5.11 Coordinate system for velocity profile in pipe flow.

This general equation can be obtained from the relation for the volumetric flow rate,

$$Q = \int_0^R 2\sigma r v \, dr \tag{5.82}$$

and the relation between the shear stress at any radial position, σ_{rz}, and the magnitude at the wall, σ_w,

$$\sigma_{rz} = \sigma_w \frac{r}{R} = -\frac{r}{2}\frac{dP}{dz} \tag{5.83}$$

In the general equation for the volumetric flow rate, one can substitute appropriate expressions from different rheological models for the shear rate and derive equations relating Q and pressure drop ΔP. Table 5.7 contains the equations for velocity profiles

Table 5.7 Velocity Profiles and Volumetric Flow equations for Laminar Flow of Power Law, Herschel–Bulkley, and Bingham Plastic Fluid Foods in Circular Pipes[a]

Power Law Model (Equation 5.11)

Velocity profile[b]:
$$v = \left(\frac{n}{n+1}\right)\left(\frac{\Delta P}{2KL}\right)^{1/n}\left[R^{(n+1)/n} - r^{(n+1)/n}\right]$$

Volumetric flow rate:
$$\frac{Q}{\pi R^3} = \left(\frac{n}{3n+1}\right)\left(\frac{\sigma_w}{K}\right)^{1/n}$$

Herschel–Bulkley model (Equation 5.12)

Velocity profile[c]:
$$v = \frac{2L}{\Delta P(m+1)K_H^m}\left[(\sigma_w - \sigma_0)^{m+1} - \left(\frac{r\Delta P}{2L} - \sigma_0\right)^{m+1}\right],$$

when $r_0 \leq r \leq R$

Volumetric flow rate:
$$\frac{Q}{\pi R^3} = \frac{(\sigma_w - \sigma_0)^{m+1}}{\sigma_w^3 K_H^m}\left[\frac{(\sigma_w - \sigma_0)^2}{m+3} + \frac{2\sigma_0(\sigma_w - \sigma_0)}{m+2} + \frac{\sigma_0^2}{m+1}\right],$$

$$m = \frac{1}{n_H}$$

Bingham Plastic model (Equation 5.15)

Velocity profile[c]:
$$v = \frac{1}{\eta'}\left[\frac{\Delta P}{4L}(R^2 - r^2) - \sigma_0(R - r)\right], \text{ when } r_0 \leq r \leq R$$

Volumetric flow rate:
$$\frac{4Q}{\pi R^3} = \frac{\sigma_w}{\eta'}\left[1 - \frac{4}{3}\left(\frac{\sigma_0}{\sigma_w}\right) + \frac{1}{3}\left(\frac{\sigma_0}{\sigma_w}\right)^4\right]$$

[a] Average velocity, $\bar{v}$, can be obtained by dividing the equation for volumetric flow rate by the area of cross section of the pipe.
[b] Maximum velocity, v_m occurs at the center at $r = 0$.
[c] Velocity profiles are valid for $r_0 \leq r \leq R$, where the radius of the plug $r_0 = 2\sigma_0 L/\Delta P$. The maximum velocity occurs when $0 \leq r \leq r_0$ and is obtained by substituting r_0 for r.

and volumetric flow rates for the power law, Bingham plastic, and Herschel–Bulkley models.

The role of rheological behavior on pressure drop per unit length can be understood from its relationship to Q and R:

$$\frac{\Delta P}{L} \propto \frac{Q^n}{R^{3n+1}} \tag{5.84}$$

From this relation, it can be seen that for Newtonian foods ($n = 1$) the pressure gradient is proportional to R^{-4}. Therefore, a small increase in the radius of the tube will result in a major reduction in the magnitude of the pressure gradient. In contrast, for a highly pseudoplastic food (e.g., $n = 0.2$), increasing the pipe radius does not have such a profound effect on the pressure gradient.

In the case of fluids following the Herschel–Bulkley and Bingham plastic models, there will be a zone of fluid surrounding the centerline that moves as a plug; the reason is that the shear stress is zero at the centerline. The volumetric equations for these two models can also be expressed in terms of the radius of the plug r_o and the radius of the tube R instead of the yield stress (σ_0) and the wall shear stress (σ_w), respectively.

The volumetric flow equation for the Bingham plastic model in Table 5.7 is known as the Buckingham equation. The error in omitting the term $(1/3)(\sigma_0/\sigma_w)^4$ in this equation is 5.9% when $\sigma_0/\sigma_w = 0.5$ and 1.8% when $\sigma_0/\sigma_w = 0.4$ (Skelland, 1967).

5.5.1.2 Friction Losses for Power Law Foods in Pipes

The pressure drop in the pipe flow can be estimated from either the capillary diagram or the applicable volumetric flow rate equation in Table 5.7. In the case of the latter, because the shear rate is not uniform across the cross section of the tube, the power law parameters must be applicable over the range of shear rates prevailing in the tube (Skelland, 1967). In addition to these methods, one can use the Reynolds number–friction factor chart developed by Dodge and Metzner (1959).

The Fanning friction factor, f, is defined by the relation

$$f = \frac{D \Delta P / 4L}{\rho \bar{v}^2 / 2} \tag{5.85}$$

For a Newtonian fluid in laminar flow, the friction factor, f, and the Reynolds number, Re, are related by

$$f = \frac{16}{\text{Re}} \tag{5.86}$$

Metzner and Reed (1955) pointed out that for non-Newtonian fluids one can use the apparent viscosity, $\acute{\eta}_a$, from the capillary diagram ($32Q/\pi D^3$ vs. σ_w) in the Reynolds number:

$$\eta_a = K'\left(\frac{8\bar{v}}{D}\right)^{n'-1} \tag{5.87}$$

When the above expression for $\acute{\eta}_a$ is substituted for the Newtonian viscosity in the Reynolds number, an expression for the generalized Reynolds number, GRe, can be obtained:

$$GRe = \frac{D^{n'-1}\bar{v}^{2-n'}\rho}{K'8^{n'-1}} \tag{5.88}$$

It should be noted that if the power law parameters obtained using true shear rates are to be used, a different expression, given later, should be used to calculate GRe.

For a non-Newtonian fluid in laminar flow, the Fanning friction factor can be calculated by substituting GRe for Re in Equation 5.86:

$$f = \frac{16}{GRe} \tag{5.89}$$

Dodge and Metzner (1957) presented a chart of GRe versus f for laminar and turbulent flow in pipes. The correlation for laminar flow was based on the work of Metzner and Reed (1955).

It appears that with most non-Newtonian fluid foods, one rarely encounters true turbulent flow (Simpson and Williams, 1974). Therefore, either Equation 5.89 or the straight line representing it in Figure 5.12 should find frequent use.

Often, power law parameters based on shear rate–shear stress data obtained with a rotational viscometer will be available instead of K' and n' based on a capillary shear diagram. In this case, the parameters n and n' will be equal, and K' can be calculated from the Weissenberg–Rabinowitsch–Mooney equation:

$$K' = K\left(\frac{3n+1}{4n}\right)^n \tag{5.90}$$

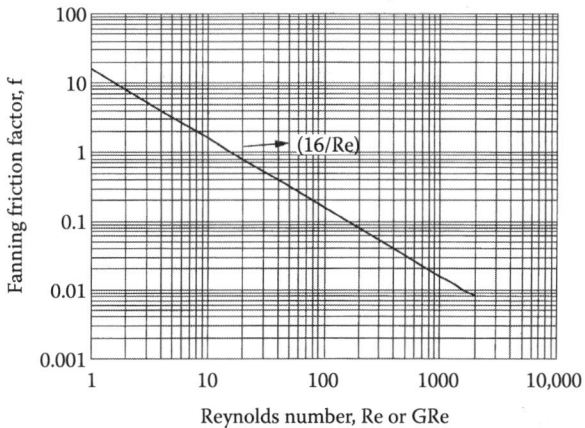

Figure 5.12 Chart of friction and generalized Reynolds number. (From Dodge, D. W., and Metzner, A. B. 1957. *AIChE J.* 5: 189–204. Errata. *AIChE J.* 8: 143.)

5.5.1.2.1 Friction Loss Data on Shear-Thinning Foods
With respect to the pressure drop data on fluid foods in pipe flow, Rozema and Beverloo (1974) measured the pressure drop in a pipe for laminar flow of several floods: starch gel (4%, 5%, and 6% by weight), mayonnaise, beet syrup concentrate, orange juice concentrates, apricot pulp, pineapple pulp, and mustard. In addition, solutions of CMC and paper pulp suspensions were also employed. Rheological data on the test fluids were obtained with a rotational viscometer, and they were used to determine parameters of the power law model (Equation 5.11) and the Powell–Eyring model (Equation 5.17).

The experimental volumetric flow rate–pressure drop data were found to be in better agreement with predictions based on the power law model than with those based on the Powell–Eyring model. The data were presented in the form of three plots of generalized Reynolds number versus the Fanning friction factor: (1) $0.1 < Re' < 10$; (2) $10 < Re' < 10^3$; and (3) $1 < Re' < 10^3$. In these plots, the constant in Equation 5.89, instead of being 16, was found to range between 13.8 and 16.1, between 11.7 and 15.0, and between 9.0 and 18.0, respectively. Time-dependent flow behavior and settling of solids in fruit pulps were cited as possible reasons for the deviations. Nevertheless, the pressure drop for all the fluids was estimated to within an error of 10%.

Steffe et al. (1984) conducted experiments with applesauce samples and found the constant in Equation 5.89 to be 14.1. Phase separation in the test samples was indicated as a possible reason for the lower value of the constant in Equation 5.89.

On the basis of the two studies cited above, it can be said that the generalized Reynolds number–friction factor chart (Figure 5.12) can be used for shear-thinning fluid foods that can be characterized by the power law model.

5.5.1.2.2 Friction Loss for Shear-Thickening Fluids
Relatively few shear-thickening fluids are encountered in practice. Therefore, few studies dealing with pressure losses for shear-thickening fluids can be found in the literature. Griskey and Green (1971) measured the pressure drop for laminar flow of shear-thickening fluids in pipes and found that the generalized Reynolds number–friction factor chart successfully predicted the pressure drop. The fluids were made by mixing cornstarch in ethylene glycol or ethylene glycol/glycerin with small amounts of water. The magnitude of the flow behavior index of the fluids was reported to be between 1.15 and 2.50. The study of Griskey and Green (1971) is an important contribution because it showed the applicability of the generalized Reynolds number–friction factor chart to the flow of shear-thickening fluids.

5.5.1.3 Pressure Drop Across Valves and Fittings
The pressure drop due to fittings can be expressed in terms of a friction loss coefficient, k_f, or in terms of the equivalent length of a straight section of a pipe. The friction loss coefficient, k_f, is defined as

$$\frac{\Delta P}{\rho} = \frac{k_f \bar{v}^2}{2} \qquad (5.91)$$

where ΔP is the pressure drop across a fitting, ρ is the density of the fluid, and $\bar{v}$ is the average velocity of the fluid.

Steffe et al. (1984) presented Equations 5.79 through 5.81 for k_f for the flow of shear-thinning fluids through a three-way plug valve, a tee line to branch, and a 90° elbow, respectively.

$$k_f = 30.3 \, Re' - 0.492 \tag{5.92}$$

$$k_f = 29.4 \, Re' - 0.504 \tag{5.93}$$

$$k_f = 191.0 \, Re' - 0.896 \tag{5.94}$$

For other fittings in laminar flow, the friction coefficients determined for Newtonian fluids and tabulated by Perry et al. (1984) and other handbooks can be used until data with shear-thinning fluids are reported.

For shear-thickening fluids, few data are available on friction losses for flow through pipe fittings. Griskey and Green (1971) reported that the equivalent length for a 90° elbow was dependent on both the magnitude of the generalized Reynolds number and the magnitude of the flow behavior index of the power law model.

5.5.1.4 Friction Losses for Herschel–Bulkley Fluids

For the laminar flow of a Herschel–Bulkley fluid (Equation 5.12), the friction factor can be written as

$$f = \frac{16}{\Psi GRe} \tag{5.95}$$

where the generalized Reynolds number, GRe, is defined in terms of the consistency coefficient and flow behavior index of the Herschel–Bulkley model:

$$GRe = \frac{D^n v^{2-n} \rho}{8^{n-1} K} \left(\frac{4n}{3n+1} \right)^n \tag{5.96}$$

Ψ is related to the yield stress (σ_0) and the flow behavior index (n) (Garcia and Steffe, 1987):

$$\Psi = (3n+1)^n (1-\xi_0)^{1+n} \left[\frac{(1-\xi_0)^2}{3n+1} + \frac{2\xi_0(1-\xi_0)}{2n+1} + \frac{\xi_0^2}{n+1} \right]^n \tag{5.97}$$

where

$$\xi_0 = \frac{\sigma_0}{\sigma_w} = \frac{\sigma_0}{(D\Delta P / 4L)} \tag{5.98}$$

ξ_0 can be calculated as an implicit function of GRe and the generalized Hedsrom number, GHe:

$$GRe = 2GHe \left(\frac{n}{3n+1} \right)^2 \left(\frac{\Psi}{\xi_0} \right)^{(2/n)-1} \tag{5.99}$$

where

$$GHe = \frac{D^2\rho}{K}\left(\frac{\sigma_0}{K}\right)^{(2/n)-1} \quad (5.100)$$

For power law and Newtonian fluids, the friction factor can be estimated directly from Equation 5.95 because $\xi_0 = 0$ and $\Psi = 1$ when $\sigma_0 = 0$. For fluids that can be described by the Bingham plastic and Herschel–Bulkley models, ξ_0 is calculated through iteration of Equation 5.99 using Equation 5.95 to Equation (5.97).

5.5.1.5 Calculation of Kinetic Energy for Non-Newtonian Fluids

The kinetic energy of a fluid must be known for solving the mechanical energy balance equation (Brodkey, 1967; Heldman and Singh, 1971). The kinetic energy (KE) for laminar flow can be calculated from the equation

$$KE = \frac{\bar{v}^2}{\alpha} \quad (5.101)$$

Osorio and Steffe (1984) presented analytical and graphical solutions for the kinetic energy correction factor, α, for a Herschel–Bulkley fluid. It was shown that the solutions satisfied the limiting form for a power law fluid:

$$\alpha = \frac{(4n+2)(5n+3)}{3(3n+2)^2} \quad (5.102)$$

5.6 DYNAMIC RHEOLOGICAL MEASUREMENT OF VISCOELASTIC BEHAVIOR OF FLUID FOODS

This section is drawn from an earlier work by the author (Rao, 2007). The viscoelastic behavior of many foods has been studied by means of mostly small-amplitude oscillatory shear (SAOS), also called dynamic rheological experiment, technique. The experiments can be conducted with the Couette, plate and cone, and parallel plate geometries. While uniform shear is achieved in the gap of a cone-plate geometry, however, the narrow gap may not be suitable for studying foods containing solids with dimensions greater than about 100 μm. In an SAOS experiment, a sinusoidal oscillating stress or strain with a frequency ω is applied to the material and the phase difference between the oscillating stress and strain as well as the amplitude ratio are measured:

$$\gamma(t) = \gamma_0 \sin(\omega t) \quad (5.103)$$

where, γ_0 is the strain amplitude and ω the angular frequency. The applied strain generates two stress components in the viscoelastic material: an elastic component in line with the strain and a 90° out-of-phase viscous component. Differentiation of Equation 5.103 yields Equation 2.55, which shows the strain rate $\dot{\gamma}(t)$ for evaluating the viscous component to be $\pi/2$ radians out of phase with the strain.

$$\dot{\gamma}(t) = \gamma_0 \omega \cos(\omega t) \quad (5.104)$$

For deformation within the linear viscoelastic range, Equation 5.105 expresses the stress generated (σ_0) in terms of an elastic or storage modulus G' and a viscous or loss modulus G''.

$$\sigma_0 = G' \gamma_0 \sin(\omega t) + G'' \gamma_0 \cos(\omega t) \tag{5.105}$$

For a viscoelastic material the resultant stress is also sinusoidal but shows a phase lag of δ radians when compared with the strain. The phase angle δ covers the range of 0 to $\pi/2$ as the viscous component increases. Equation 5.106 also expresses the sinusoidal variation of the resultant stress.

$$\sigma(t) = \sigma_0 \sin(\omega t + \delta) \tag{5.106}$$

The following expressions that define viscoelastic behavior can be derived from Equations 5.105 and 5.106:

$$G' = \left[\frac{\sigma_0}{\gamma_0}\right] \cos \delta \tag{5.107}$$

$$G'' = \left[\frac{\sigma_0}{\gamma_0}\right] \sin \delta \tag{5.108}$$

$$\tan \delta = \frac{G''}{G'} \tag{5.109}$$

where G' (Pa) is the storage modulus, G'' (Pa) is the loss modulus, and $\tan \delta$ is the loss tangent.

The storage modulus G' expresses the magnitude of the energy that is stored in the material or recoverable per cycle of deformation. G'' is a measure of the energy which is lost as viscous dissipation per cycle of deformation. Therefore, for a perfectly elastic solid, all the energy is stored, that is, G'' is zero and the stress and the strain will be in phase (Figure 5.13). On the contrary, for a liquid with no elastic properties all the energy is dissipated as heat, that is, G' is zero and the stress and the strain will be out of phase by 90° (Figure 5.13).

For a specific food, magnitudes of G' and G'' are influenced by frequency, temperature, and strain. For strain values within the linear range of deformation, G' and G'' are independent of strain. The loss tangent, is the ratio of the energy dissipated to that stored per cycle of deformation. These viscoelastic functions have been found to play important roles in the rheology of structured polysaccharides. One can also employ notation using complex variables and define a complex modulus $G^*(\omega)$:

$$|G^*| = \sqrt{(G')^2 + (G'')^2} \tag{5.110}$$

We note that the dynamic viscosity and the dynamic rigidity are components of the complex dynamic viscosity, η^*:

$$\eta^* = (G^*/\omega) = \eta' - i(G'/\omega) \tag{5.111}$$

ENGINEERING PROPERTIES OF FOODS

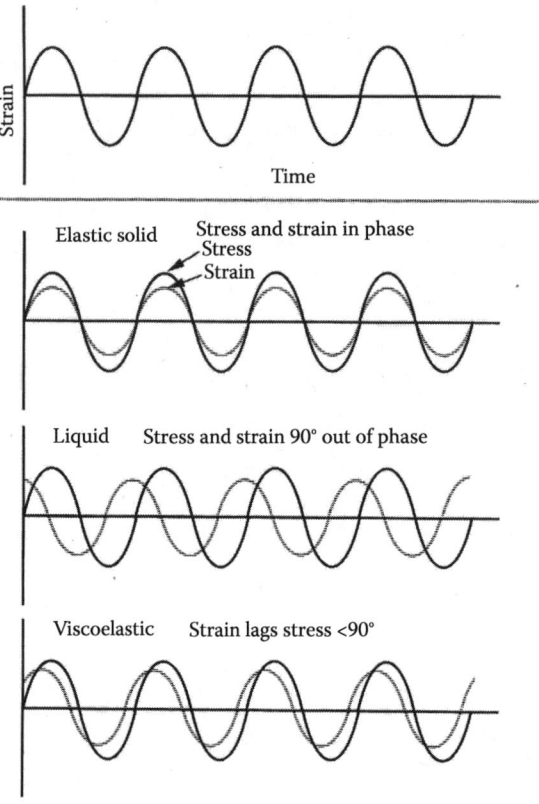

Figure 5.13 The responses of various liquids and a solid to applied dynamic strain: Newtonian, viscoelastic, and perfect elastic solid.

where ω is the frequency of oscillation; the real part of the complex viscosity η' is equal to (G''/ω) and the imaginary part η'' is equal to (G''/ω). One can also determine the loss modulus G'' from oscillatory shear data using the expression:

$$G'' = \omega \eta' \tag{5.112}$$

It should be noted that if G' is much greater than G'', the material will behave more like a solid, that is, the deformations will be essentially elastic or recoverable. However, if G'' is much greater than G' the energy used to deform the material is dissipated viscously and the materials behavior is liquid like. The viscoelastic moduli determined over a range of frequencies will indicate transition zones corresponding to relaxation processes dependent on the material's structure (Ferry, 1980).

Types of dynamic rheological tests: Dynamic rheological experiments provide suitable means for monitoring the gelation process of many food polymers and for obtaining insight into gel/food structure because they satisfy several conditions (Lopes da Silva et al., 1998): (1) they are nondestructive and do not interfere with either gel formation or

RHEOLOGICAL PROPERTIES OF FLUID FOODS

softening of a structure, (2) the time involved in the measurements is short relative to the characteristic times of the gelation and softening processes, and (3) the results are expressible in fundamental terms so that they be related to the structure of the network (e.g., Durand et al., 1990). For the same reasons, they were used to follow the changes induced on potato cells by cellulase (Shomer et al., 1993), α-amylase (Champenois et al., 1998) on wheat starch, and can be used in studying other phase transitions in foods, such as heat-induced starch gelatinization (Yang and Rao, 1998a).

Once the linear viscoelastic region is established, three types of dynamic rheological tests can be conducted to obtain useful properties of viscoelastic foods, such as gels, and of gelation and melting. The linear viscoelastic region can be verified with a given sample by increasing the amplitude of oscillation and noting the magnitude of phase lag and the amplitude ratio. The limit of linearity can be detected when dynamic rheological properties (e.g., G' and G'') change rapidly from their almost constant values.

The three types of dynamic rheological tests that can be conducted are

1. Frequency sweep studies in which G' and G'' are determined as a function of frequency (ω) at a fixed temperature. When properly conducted, frequency sweep tests provide data over a wide range of frequencies. However, if fundamental parameters are required, each test must be restricted to linear viscoelastic behavior. Figure 5.14 illustrates such data obtained on unmodified and acid-hydrolyzed amylopectin dissolved in 90% DMSO/10% water with a plate-cone geometry (Chamberlain, 1999). Frequently, one observes increase in G' and G'' with increasing frequency. On the basis of frequency sweep data, one can designate "true gels" when the molecular rearrangements within the network are much reduced over the time scales analyzed, such that G' is higher than G'' throughout the frequency range, and is almost independent of frequency (ω)

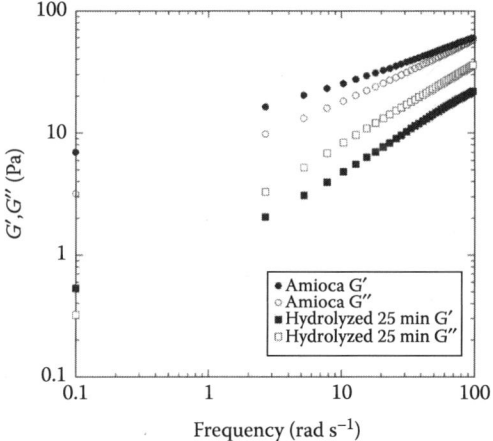

Figure 5.14 Values of G' and G'' from a frequency sweep on acid hydrolyzed 8% waxy maize starch dissolved in 90% DMSO/10% water (Chamberlain, 1999).

ENGINEERING PROPERTIES OF FOODS

(Clark and Ross-Murphy, 1987). In contrast for "weak gels," there is a higher dependence on frequency for the dynamic moduli, suggesting the existence of relaxation processes occurring even at short time scales, and lower difference between moduli values.

2. Temperature sweep studies in which G' and G'' are determined as a function of temperature at fixed ω. This test is well suited for studying gel formation during cooling of a heated dispersion (Rao and Cooley, 1993), and gelatinization of a starch dispersion during heating (Figure 5.15) (Tattiyakul, 1997) and gel formation of proteins (Owen et al., 1992); the gelatinization of starch was studied with a parallel plate (disc) geometry and considerable care was exercised to minimize moisture loss from the dispersions. The existence of the linear viscoelastic range can be verified in preliminary experiments conducted on a given product at several temperature steps (Owen et al., 1992).

3. Time sweep study in which G' and G'' are determined as a function of time at fixed ω and temperature. This type of test, often called a gel-cure experiment, is well suited for studying structure development in physical gels. Figure 5.16 is an example of a gel-cure experiment on low-methoxyl pectin + Ca^{2+} gels containing sucrose with a cone-plate geometry (Grosso and Rao, 1998).

Additional insight into gelation and melting phenomena can be obtained from the temperature and time sweep data by calculation of structure development rates (dG'/dt or $d\eta^*/dt$) or structure loss rates ($-dG'/dt$ or $-d\eta^*/dt$) (Rao and Cooley, 1993; Lopes da Silva et al., 1998). From the time-sweep data, since G' values of gels were recorded at the same time intervals, the influence of sugars and calcium concentrations on them could be examined quantitatively in terms of differences ($\Delta G'$) in G' values (Grosso and Rao, 1998).

Dynamic rheological studies can also be used to study the characteristics of solid foods. Often it is convenient to impose a sinusoidal compressive or elongational strain,

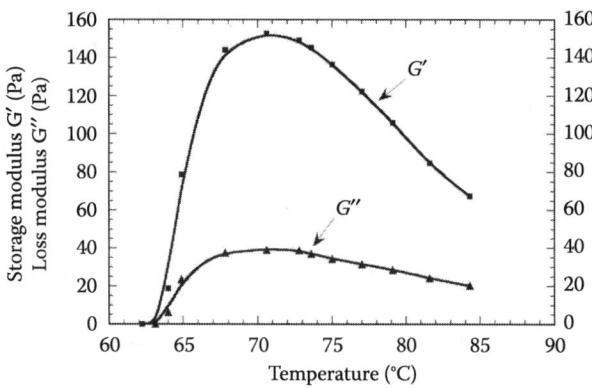

Figure 5.15 Values of G' and G'' at 6.28 rad s^{-1} 3% strain obtained from a temperature sweep, 60–85°C, on a 8% tapioca starch dispersion (Tattiyakul, 1997).

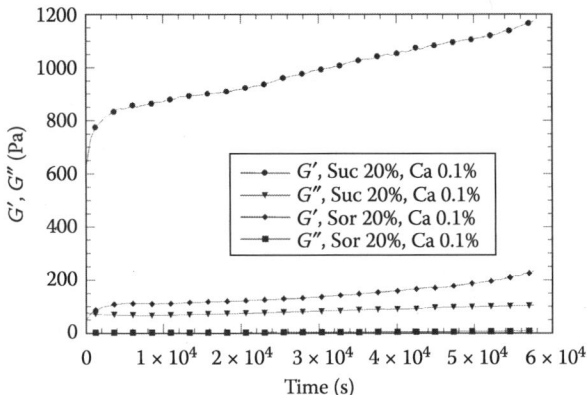

Figure 5.16 Values of G' and G'' recorded during a gel-cure experiment on low-methoxyl pectin + Ca^{2+} gels with 20% sucrose, 0.10% and 0.15% calcium content, and $\Delta G'$ due to changing the calcium content.

so that storage and loss moduli obtained are denoted as E' and E'', respectively (Rao et al., 1995).

5.7 CONCLUSION

Rheological properties of several fluid foods have been determined. Studies have been conducted for quality control, engineering applications, and better understanding of the structure. The effect of temperature on apparent viscosity has been described by the Arrhenius relationship. Many rheological equations have been used to describe the flow properties of fluid foods, but the power law and the Casson flow models have found the most extensive use.

Although most studies on fluid foods simply document the flow properties, those using the physicochemical approach and determination of intrinsic viscosity have provided useful insights into structure and molecular interactions.

Experimental methods for determining flow properties fall into three categories: fundamental, empirical, and imitative. Empirical methods, to a large extent, are confined to quality control applications.

Viscoelastic behavior of fluid foods has been studied in terms of normal stresses, creep compliance, and oscillatory shear flow experiments.

It is anticipated that the number of studies designed to achieve a better understanding of foods via fundamental rheological studies will increase and will utilize some of the developments in the study of high polymers and biopolymers. Nevertheless, one can expect the pragmatic approach of food rheologists to continue in the future.

Friction-loss studies in pipe flow and across fittings have been conducted for shear-thinning foods. The applicability of the generalized Reynolds number–friction chart has been demonstrated for fluid foods.

LIST OF SYMBOLS*

a	Constant; radius of plate, m
A_1, A_2	Constants
b	Constant
B	Constant; time coefficient of thixotropic breakdown
B_1, B_2	Constants
c	Constant; concentration, microbes/m^3, °Brix, kg/m^3; phase lag, deg
C	Total torque, N · m
D	Constant; diameter, m
E_0, E_1, E_2	Elastic moduli, N/m^2
E_a, E_{aC}, E_{aH}	Activation energy, kcal/mol, kJ/mol
f	Fanning friction factor, dimensionless
G'	Storage modulus, Pa
$G \leq$	Loss modulus, Pa
h	Width between plates, m
GHe	Generalized Hedstrom number, dimensionless
i	Square root of –1
J	Creep compliance, Pa^{-1}
J_0, J_1, J_2	Compliances, Pa^{-1}
k_f	Friction loss coefficient
k_1, k_2	Interaction coefficients in Huggins's equation
K	Consistency index in power law model, Pa · s^n; restoring constant of torsion wire
K'	Consistency index based on capillary diagram, Pa · s$^{n'}$
K_H	Consistency index in Herschel–Bulkley model, Pa · s^n
K_C, K_{0C}	Casson model constants
K_{TC}	Constant in Christiansen–Craig model
K_{TH}	Constant in Harper–El Sahrigi model
L	Length, m
m'	Constant relating shear rate and first normal stress difference, Pa · s$^{n'}$
M	Shear coefficient of thixotropic breakdown
n	Flow behavior index in power law model
n'	Constant relating shear rate and first normal stress difference; flow behavior index based on capillary shear diagram
n_H	Flow behavior index in Herschel–Bulkley model
n_M	Constant in modified Casson (Mizrahi–Berk) model
N_*	Rotational speed of impeller, rev/s
p	Power, N · m/s
Q	Volumetric flow rate, m^3/s
r	Radial coordinate
r_o	Radius of plug, m
R	Radius, m; gas constant, kJ/(kg · mol · K)
Re	Reynolds number, dimensionless
GRe	Generalized Reynolds number, dimensionless
Re, ή'	Reynolds number for Bingham plastic fluids, dimensionless
t	Time, s
T	Temperature, °C, K; torque per unit area, N · m/m^2

---

* Symbols not defined here have been defined in the text.

T_m	Maximum torque, N · m
v	Local velocity, m/s
$\bar{v}$	Average velocity, m/s
w	Mass flow rate, kg/s
z	Axial coordinate of pipe
z_L	Length of an HTST sterilizer

Greek Symbols

α	Constant; kinetic energy correction factor
β	Constant
$\dot{\gamma}$	Shear rate, s^{-1}
ΔP	Pressure drop, N/m^2
$\acute{\eta}$	Viscosity, Pa · s
$\acute{\eta}_1$	Dynamic viscosity, Pa · s
$\acute{\eta}_a$	Apparent viscosity, Pa · s
$\acute{\eta}'$	Plastic viscosity, Pa · s
$\acute{\eta}0$	Zero-shear viscosity, Pa · s
$\acute{\eta}\infty$	Constant in Arrhenius equation; infinite shear viscosity, Pa · s
λ	Structural parameter
ρ	Density, kg/m^3
σ	Shear (tangential) stress, Pa
τ_1, τ_2	Retardation times, s
$\sigma_{11}, \sigma_{22}, \sigma_{33}$	Normal stresses, Pa
σ_{12}, σ_{21}	Shear (tangential) stresses, Pa
σe	Equilibrium shear stress, Pa
σ_0	Yield stress, Pa
ψ_1	First normal stress function, Pa · s^2
ψ_2	Second normal stress function, Pa · s^2
ω	Frequency, rad/s
Ω	Angular velocity, rad/s

Subscripts

i	Inner cylinder
o	Outer cylinder
L	Outlet or total value
p	At surface of plate
rel	Relative
red	Reduced
sp	Specific

Superscript

⁻	Average value

REFERENCES

Arola, D. F., Powell, R. L., Barrall, G. A., and McCarthy, M. J. 1999. Pointwise observations for rheological characterization using nuclear magnetic resonance imaging. *J. Rheol. 43:* 9–30.

Bagley, E. B. 1975. End correction in the capillary flow of polyethylene. *J. Appl. Phys. 28:* 624–627.

Bagley, E. B., and Christianson, D. D. 1982. Swelling capacity of starch and its relationship to suspension viscosity—Effect of cooking time, temperature and concentration. *J. Texture Stud. 13:* 115–126.

Bird, R. B., Armstrong, R. C., and Hassager, O. 1977. *Dynamics of Polymeric Liquids—Fluid Mechanics*, John Wiley and Sons, New York.

Blake, J. A., and Moran, J. J. 1975. A new approach to capillary extrusion rheometry. *J. Texture Stud. 6:* 227–239.

Boger, D. V., and Tiu, C. 1974. Rheological properties of food products and their use in the design of flow systems. *Food Technol. Australia 26:* 325–335.

Bongenaar, J. J. T., Kossen, N. W. F., Metz, B., and Meijboom, F. W. 1973. A method for characterizing the rheological properties of viscous fermentation broths. *Biotech. Bioeng. 15:* 201–206.

Brodkey, R. S. 1967. *The Phenomena of Fluid Motions*, Chapter 15. Addison-Wesley, Reading, MA.

Campanella, O. H., Popplewell, L. M., Rosenau, J. R., and Peleg, M. 1987. Elongational viscosity measurements of melting American process cheese. *J. Food Sci. 52:* 1249–1251.

Casiraghi, E. M., Bagley, E. B., and Christianson, D. D. 1985. Behavior of mozzarella, cheddar and processed cheese spread in lubricated and bonded uniaxial compression. *J. Texture Stud. 16:* 281–301.

Casson, N. 1959. A flow equation for pigmented-oil suspensions of the printing ink type. In *Rheology of Disperse Systems*, C. C. Mill (Ed.), pp. 82–104. Pergamon, New York.

Cervone, N. W., and Harper, J. M. 1978. Viscosity of an intermediate moisture dough. *J. Food Process Eng. 2:* 83–95.

Chamberlain, E. K. 1999. Rheological properties of acid converted waxy maize starches: Effect of solvent, concentration and dissolution time. PhD thesis, Cornell University, Ithaca, NY.

Chamberlain, E.K., Rao, M.A., and Cohen, C. 1999. Shear thinning and antithixotropic behavior of a heated cross-linked waxy maize starch dispersion. *Int. J. Food Properties 2:* 63–77; errata, 2: 195–196.

Champenois, Y.C., Rao, M.A., and Walker, L.P. 1998. Influence of α-mylase on the viscoelastic properties of starch–gluten pastes and gels. *J. Sci. Food Agric.* 127–133.

Charm, S. E. 1960. Viscometry of non-Newtonian food materials. *Food Res. 25:* 351–362.

Charm, S. E. 1963. The direct determination of shear stress–shear rate behavior of foods in the presence of a yield stress. *J. Food Sci. 28:* 107–113.

Charm, S. E. 1971. *Fundamentals of Food Engineering*, 2nd ed. AVI, Westport, Conn.

Chatraei, S. H., Macosko, C. W., and Winter, H. H. 1981. A new biaxial extensional rheometer. *J. Rheol. 25:* 433–443.

Chevalley, J. 1991. An adaptation of the Casson equation for the rheology of chocolate. *J. Texture Stud. 22:* 219–229.

Choi, Y.J., McCarthy, K.L., and McCarthy, M.J. 2002. Tomographic techniques for measuring fluid flow properties. *J. Food Sci. 67:* 2718–2724.

Christiansen, E. B., and Craig, S. E. 1962. Heat transfer to pseudoplastic fluids in laminar flow. *AIChE J. 8:* 154–160.

Clark, A. H. and Ross-Murphy, S. B. 1987. Structural and mechanical properties of biopolymer gels. *Adv. Polymer Sci. 83:* 57–192.

Corey, H. 1972. The contribution of fiber–liquid interactions to the rheology of fibrous suspensions. Ph.D. Thesis, Rutgers University. University Microfilms, Ann Arbor, Mich.

Corey, H., and Creswick, N. 1970. A versatile recording Couette-type viscometer. *J. Texture Stud. 1:* 155–166.

Cornford, S. J., Parkinson, T. L., and Robb, J. 1969. Rheological characteristics of processed whole egg. *J. Food Technol. 4:* 353–361.

Cross, M. M. 1965. Rheology of non-Newtonian Flow: equation for pseudoplastic systems. *J. Colloid Sci.* 20: 417–437.

Datta, A.K. 2002. *Biological and Environmental Heat Transfer,* Marcel Dekker, New York.

Davis. S. S. 1973. Rheological properties of semi-solid foodstuffs. *J. Texture Stud.* 4: 15–50.

Davis, R. B., De Weese, D., and Gould, W. A. 1954. Consistency measurements of tomato puree. *Food Technol.* 8: 330–334.

Dealy, J. M. 1982. *Rheometers for Molton Polymers—A Practical Guide to Testing and Property Measurement.* Van Nostrand Reinhold, New York.

Dickie, A. M., and Kokini, J. L. 1982. Use of the Bird–Leider equation in food rheology. *J. Food Process Eng.* 5: 157–184.

Dodge, D. W., and Metzner, A. B. 1957. Turbulent flow of non-Newtonian systems. *AIChE J.* 5: 189–204. Errata. *AIChE J.* 8: 143.

Dodge, D. W., and Metzner, A. B. 1959. Turbulent flow of non-Newtonian systems. *AIChE J.* 5:189–204

Dogan, N., McCarthy, M.J., and Powell, R.L. 2002. In-line measurement of rheological parameters and modeling of apparent wall slip in diced tomato suspensions using ultrasonics. *J. Food Sci.* 67: 2235–2240.

Dogan, N., McCarthy, M.J., and Powell, R.L. 2003. Comparison of in-line consistency measurement of tomato concentrates using ultrasonics and capillary methods. *J. Food Process Eng.* 25(6): 571–587.

Doublier, J. L., and Launay, B. 1974. Proprietes rheologiques des solutions de gomme guar. In *Lebensmittel-Einfluss der Rheologie,* pp. 197–210. Dechema Monographien, Band 77, Dechema, Frankfurt.

Durand, D., Bertrand, C., Busnel, J. P., Emery, J., Axelos, M. A. V., Thibault, J. F., Lefebvre, J., Doublier, J. L., Clark, A. H., and Lips, A. 1990. Physical gelation induced by ion complexation: Pectin–calcium systems, in *Physical Networks,* eds. W. Burchard and S.B. Ross-Murphy, pp. 283–300, Elsevier Applied Science Publishers.

Dzuy, N. Q., and Boger, D. V. 1983. Yield stress measurement for concentrated suspensions. *J. Rheol.* 27: 321–349.

Dzuy, N. Q., and Boger, D. V. 1985. Direct yield stress measurement with the vane method. *J. Rheol.* 29: 335–347.

Elfak, A. M., Pass, G., Phillips, G. O., and Morley, R. G. 1977. The viscosity of dilute solutions of guar gum and locust bean gum with and without added sugars. *J. Sci. Food Agric.* 28: 895–899.

Elfak, A. M., Pass, G., and Phillips, G. O. 1978. The viscosity of dilute solutions of carrageenan and sodium caroboxymethylcellulose. *J. Sci. Food Agric.* 29: 557–562.

Elliott, J. H., and Ganz, A. J. 1977. Salad dressing—preliminary rheological characterization. *J. Texture Studies* 8: 359–337.

Escardino, A., Fito, P., and Molina, A. 1972. Determination de las propriedades reologicas de alimentos liquidos. I. Compartamiento de las soluciones acuosas de algunas espesantes. *Rev. Agroquimica Tec. Alimentos* 11: 418–426.

Ferry, J. D. 1980. *Viscoelastic Properties of Polymers,* John Wiley, New York.

Figoni, P. I., and Shoemaker, C. F. 1983. Characterization of time dependent flow properties of mayonnaise under steady shear. *J. Texture Stud.* 14: 431–442.

Ferry, J. D. 1980. *Viscoelastic Properties of Polymers,* John Wiley, New York.

Garcia, E.J. and Steffe, J.F. 1987. Comparison of friction factor equations for non-Newtonian fluids in pipe flow. *J. Food Process Eng.* 9: 93–120.

Genovese, D.B. and Rao, M.A. 2003a. Apparent viscosity and first normal stress of starch dispersions: role of continuous and dispersed phases, and prediction with the Goddard–Miller model. *Appl. Rheol.* 13(4): 183–190.

Genovese, D.B. and Rao, M.A. (2003b). Vane yield stress of starch dispersions. *J. Food Sci.* 68: 2295–2301.

Genovese, D.B. and Rao, M.A. 2004. Contributions to vane yield stress of structured food dispersions. *Paper to be presented at the IFT Annual Meeting,* Las Vegas, July 13–16, 2004.

Grosso, C. R. F. and Rao, M. A. 1998. Dynamic rheology of structure development in low-methoxyl pectin + Ca^{2+} + sugar gels. *Food Hydrocoll.* 12: 357–363.

Green, H. 1949. *Industrial Rheology and Rheological Structures.* Wiley, New York.

Griffith, D. L., and Rao, V. N. M. 1978. Flow characteristics of non-Newtonian foods utilizing a low-cost rotational viscometer. *J. Food Sci.* 43: 1876–1877.

Griskey, R. G., and Green, R. G. 1971. Flow of dilatant (shear-thickening) fluids. *AIChE J.* 17: 725–728.

Grosso, C. R. F. and Rao, M. A. 1998. Dynamic rheology of structure development in low-methoxyl pectin+Ca^{2+}+sugar gels. *Food Hydrocoll.* 12: 357–363.

Hahn, S. J., Ree, T., and Eyring, H. 1959. Flow mechanism of thixotropic substances. *Ind. Eng. Chem.* 51: 856–857.

Han, C. D., Kim, K. U., Siskovic, N., and Huang, C. R. 1975. An appraisal of rheological models as applied to polymer melt flow. *Rheol. Acta* 14: 533–549.

Harper, J. C., and El-Sahrigi, A. F. 1965. Viscometric behavior of tomato concentrates. *J. Food Sci.* 30: 470–476.

Hermansson, A. M. 1975. Functional properties of proteins for foods—Flow properties. *J. Texture Stud.* 5: 425–439.

Higgs, S. J., and Norrington, R. J. 1971. Rheological properties of selected foodstuffs. *Proc. Biochem.* 6(5): 52–54.

Holdsworth, S. D. 1971. Applicability of rheological models to the interpretation of flow and processing behavior of fluid food products. *J. Texture Stud.* 2: 393–418.

Holdsworth, S. D. 1993. Rheological models used for the prediction of the flow properties of food products: a literature review. *Trans. Inst. Chem. Eng.* 71(C): 139–179.

Jao, Y. C., Chen, A. H., Lewandowski, D., and Irwin, W. E. 1978. Engineering analysis of soy dough rheology in extrusion. *J. Food Process Eng.* 2: 97–112.

Keentok, M. 1982. The measurement of the yield stress of liquid. *Rheol. Acta* 21: 325–332.

Kokini, J. L., and Dickie, A. M. 1981. An attempt to identify and model transient viscoelastic flow in foods. *J. Texture Stud.* 12: 539–557.

Kramer, A., and Twigg, B. A. 1970. *Quality Control for the Food Industry*, Vol. 1. AVI, Westport, Conn.

McCabe, W. L., and Smith, J. C. 1976. *Unit Operations of Chemical Engineering*, 3rd ed. McGraw-Hill, New York.

Lai, K.P., Steffe, J.F. and Ng, P.K.W. 2000. Average shear rates in the Rapid Visco Analyser (RVA) mixing system. *Cereal Chem.* 77(6): 714–716.

Liao, H-J., Tattiyakul, J., and Rao, M.A. 1999. Superposition of complex viscosity curves during gelatinization of starch dispersion and dough. *J. Food Process Eng.* 22: 215–234.

Liao, H-J., Rao, M.A. and Datta, A.K. 2000. Role of thermorheological behavior in simulation of continuous sterilization of a starch dispersion. *IChemE Trans. Part C — Food Bioprod. Proc.* 78(C1): 48–56.

Lopes da Silva, J. A., Rao, M. A., and Fu, J.-T. 1998. Rheology of structure development and loss during gelation and melting, in *Phase/State Transitions in Foods: Chemical, Rheological and Structural Changes*, eds. M. A. Rao and R. W. Hartel, pp. 111–156, Marcel Dekker, Inc., NY.

McCarthy, K.L. and Seymour, J.D. 1993. A fundamental approach for the relationship between the Bostwick measurement and Newtonian fluid viscosity. *J. Texture Stud.* 24(1): 1–10.

McCarthy, K.L. and Seymour, J.D. 1994. Gravity current analysis of the Bostwick consistometer for power law foods. *J. Texture Stud.* 25(2): 207–220.

Menjivar, J. A., and Rha, C. K. 1980. Viscoelastic effects in concentrated protein dispersions. *Rheol. Acta* 19: 212–219.

Metz, B., Kossen, N.W.F. and Suijdam, J.C. 1979. The rheology of mould suspensions. In Ghose, T.K., Fiechter, A., and Blakebrough, N., Eds. *Advances in Biochemical Engineering*, Vol. 2. Springer Verlag, New York, pp. 103–156.

Metzner, A. B., and Otto, R. E. 1957. Agitation of non-Newtonian fluids. *AIChE J.* 3: 3–10.

Metzner, A. B., and Reed, J. C. 1955. Flow of non-Newtonian fluids—Correlation of the laminar, transition, and turbulent-flow regions. *AIChE J.* 1: 434–440.

Michaels, A.S. and Bolger, J.C. 1962. The plastic flow behavior of flocculated kaolin suspensions. *Indust. Eng. Chem. Fundam.* 1: 153–162.

Mizrahi, S. 1979. A review of the physicochemical approach to the analysis of the structural viscosity of fluid food products. *J. Texture Stud.* 10: 67–82.

Mizrahi, S., and Berk, Z. 1970. Flow behavior of concentrated orange juice. *J. Texture Stud.* 1: 342–355.

Mizrahi, S., and Berk, Z. 1972. Flow behaviour of concentrated orange juice: mathematical treatment. *J. Texture Stud.* 3: 69–79.

Mizrahi, S., and Firstenberg, R. 1975. Effect of orange juice composition on flow behaviour of six-fold concentrate. *J. Texture Stud.* 6: 523–532.

Nakayama, T., Niwa, E., and Hamada. I. 1980. Pipe transportation of minced fish paste. *J. Food Sci.* 45: 844–847.

Odigboh, E. V., and Mohsenin, N. N. 1975. Effects of concentration on the viscosity profile of cassava starch pastes during the cooking–cooling process. *J. Texture Stud.* 5: 441–447.

Osorio, F. A., and Steffe, J. F. 1984. Kinetic energy calculations for non-Newtonian fluids in circular tubes. *J. Food Sci.* 49: 1295–1296, 1315.

Owen, S. R., Tung, M. A., and Paulson, A. T. 1992. Thermorheological studies of food polymer dispersions. *J. Food Eng.* 16: 39–53.

Perry, R. H., Green, D. W., and Maloney, J. O. 1984. *Perry's Chemical Engineers' Handbook*, 6th ed. McGraw-Hill, New York.

Prentice, J. H. 1968. Measurement of some flow properties of market cream. In *Rheology and Texture of Foodstuffs*, pp. 265–279. SCI Monograph No. 27, Society of Chemical Industry, London.

Prentice, J. H. 1972. Rheology and texture of dairy products. *J. Texture Stud.* 3: 415–458.

Prentice, J. H. 1984. *Measurements in the Rheology of Foodstuffs*. Elsevier, New York.

Pryce-Jones, J. 1953. The rheology of honey. In *Foodstuffs: Their Plasticity, Fluidity and Consistency*, G. S. Scott Blair, Ed., pp. 148–176. North-Holland, Amsterdam.

Qiu, C.-G., and Rao, M. A. 1988. Role of pulp content and particle size in yield stress of applesauce. *J. Food Sci.* 53: 1165–1170.

Rao, M. A. 1975. Measurement of flow properties of food suspensions with a mixer. *J. Texture Stud.* 6: 533–439.

Rao, M. A. 1977a. Rheology of liquid foods—A review. *J. Texture Stud.* 8: 135–168.

Rao, M. A. 1977b. Measurement of flow properties of fluid foods—Developments, limitations, and interpretation of phenomena. *J. Texture Stud.* 8: 257–282.

Rao, M. A. 1999. *Rheology of Fluid and Semisolid Foods: Principles and Applications*, Kluwer Academic/Plenum Publishers, New York, p. 433.

Rao, M. A. 2007. *Rheology of Fluid and Semisolid Foods: Principles and Applications*, 2nd edition, p. 483, Springer, New York.

Rao, M. A., and Anantheswaran, R. C. 1982. Rheology of fluid foods in food processing. *Food Technol.* 36(2): 116–126.

Rao, M. A., and Cooley, H. J. 1983. Applicability of flow models with yield for tomato concentrates. *J. Food Process Eng.* 6: 159–173.

Rao, M. A., and Cooley, H. J. 1984. Determination of effective shear rates of complex geometries in rotational viscometers. *J. Texture Stud.* 15: 327–335.

Rao, M. A. and Cooley, H. J. 1993. Dynamic rheological measurement of structure development in high-methoxyl pectin/fructose gels. *J. Food Sci.* 58: 876–879.

Rao, M.A., Cooley, H.J., Ortloff, C., Chung, K., and Witjs, S.C. 1993. Influence of rheological properties of fluid and semisolid foods on the performance of a filler. *J. Food Proc. Eng.* 16: 289–304.

Rao, V.N.M., Delaney, R.A.M., and Skinner, G.E. 1995. Rheological properties of solid foods, in *Engineering Properties of Foods*, eds. M.A. Rao and S.S.H. Rizvi, 2nd edition, pp. 55–97, Marcel Dekker, Inc., New York.

Rao, M. A., and Steffe, J. F. (Eds.) 1992. *Viscoelastic Properties of Foods*. Elsevier, New York.

Rao, M. A., Palomino, L. N., and Bernhardt, L. N. 1974. Flow properties of tropical fruit purees. *J. Food Sci. 39:* 160–161.

Rao, M. A., Bourne, M. C., and Cooley, H. J. 1981a. Flow properties of tomato concentrates. *J. Texture Stud. 12:* 521–538.

Rao, M. A., Walter, R. H., and Cooley, H. J. 1981b. The effect of heat treatment on the flow properties of aqueous guar gum and sodium carboxymethylcellulose (CMC) solutions. *J. Food Sci. 46:* 896–899, 902.

Rao, M. A., Cooley, H. J., and Vitali, A. A. 1984. Flow properties of concentrated juices at low temperatures. *Food Technol. 38*(3): 113–119.

Rieger, F., and Novak, V. 1973. Power consumption of agitators in highly viscous non-Newtonian liquids. *Trans. Inst. Chem. Eng. 51:* 105–111.

Roberts, I. 2003. In-line and on-line rheology measurement of food. In *Texture in Food, Volume 1: Semi-Solid Foods*, B.M. McKenna (Ed.), pp. 161–182. Woodhead Publishing Ltd., Cambridge, U.K..

Roels, J. A., Van den Berg, J., and Voncken, R. M. 1974. The rheology of mycelial broths. *Biotech. Bioeng. 16:* 181–208.

Rozema, H., and Beverloo, W. A. 1974. Laminar isothermal flow of non-Newtonian fluids in a circular pipe. *Lebensm.-Wiss. Technol. 7:* 223–228.

Saravacos, G.D. 1968. Tube viscometry of fruit juices and purees. *J. Food Sci. 35:* 122–125.

Saravacos, G. D. 1970. Effect of temperature on viscosity of fruit juices and purees. *J. Food Sci. 35:* 122–125.

Scalzo, A. M., Dickerson, R. W., Peeler, J. T., and Read, R. B. 1970. The viscosity of egg and egg products. *Food Technol. 24:* 1301–1307.

Shomer, I., Rao, M. A., Bourne, M. C., and Levy, D. 1993. Rheological behaviour of potato tuber cell suspensions during temperature fluctuations and cellulase treatments. *J. Sci. Food. Agric. 63:* 245–250.

Schaller, A., and Knorr, D. 1973. Ergebnisse methodolischer untersuchungen zur schatzung der fliessgrenze und plastischen viskositat am beispiel von aprikosenpuree unter zugrundelegung lines idealplastischen fliessverhaltens. *Confructa 18:* 169–176.

Scheve, J. L., Abraham, W. H., and Lancaster, E. P. 1974. A simplified continuous viscometer for non-Newtonian fluids. *Ind. Eng. Chem. Fundam. 13:* 150–154.

Schlichting, H. 1960. *Boundary Layer Theory*, 4th ed. McGraw-Hill, New York.

Scott Blair, G. W. 1958. Rheology in food research. In *Advances in Food Research*, Vol. VIII, E. M. Mrak and G. F. Stewart (Eds.), pp. 1–61. Academic, New York.

Senouci, A., and Smith, A. C. 1988. An experimental study of food melt rheology. I. Shear viscosity using a slit die viscometer and a capillary rheometer. *Rheol. Acta 27:* 546–554.

Shama, F., and Sherman, P. 1966. The texture of ice cream. 2. Rheological properties of frozen ice cream. *J. Food Sci. 31:* 699–706.

Sharma, S.K., Hill, A.R., Goff, H.D., and Yada, R. 1989. Measurement of coagulation time and curd firmness of renneted milk using a Nametre viscometer. *Milchwissenschaft. 44*(11): 682–685.

Sharma, S.K., Hill, A.R., and Mittal, G.S. 1992. Evaluation of methods to measure coagulation time of ultrafiltered milk. *Milchwissenschaft. 47*(11): 701–704.

Sherman, P. 1966. The texture of ice cream. 3. Rheological properties of mix and melted ice cream. *J. Food Sci. 31:* 707–716.

Sherman, P. 1970. *Industrial Rheology*. Academic, New York.

Shomer, I., Rao, M. A., Bourne, M. C., and Levy, D. 1993. Rheological behaviour of potato tuber cell suspensions during temperature fluctuations and cellulase treatments. *J. Sci. Food. Agric. 63:* 245–250.

Simpson, S. G., and Williams, M. C. 1974. An analysis of high temperature/short time sterilization during laminar flow. *J. Food Sci. 39:* 1047–1054.

Singh, R. P. and Heldman, D. R. 2001. *Introduction to Food Engineering*. Elsevier, New York.

Skelland, A. H. P. 1967. *Non-Newtonian Flow and Heat Transfer.* Wiley, New York.

Smith, R. E. 1984. Effect of gap errors in rotational concentric cylinder viscometers. *J. Rheol.* 28: 155–160.

Steffe, J. F. 1992. *Rheological Methods in Food Process Engineering.* Freeman, East Lansing, Mich.

Steffe, J. F., Mohamed, I. O., and Ford, E. W. 1984. Pressure drop across valves and fittings for pseudoplastic fluids in laminar flow. *Trans. Am. Soc. Agric. Eng.* 27: 616–619.

Steffe, J. P., Mohamed, I. O., and Ford, E. W. 1986. Rheological properties of fluid foods. In *Physical and Chemical Properties of Food,* M. R. Okos (Ed.), pp. 1–13. *Am. Soc. Agric. Eng.,* St. Joseph, MI.

Szczesniak, A. S., and Farkas, E. 1962. Objective characterization of the mouth-feel of gum solutions. *J. Food Sci.* 27: 381–385.

Tattiyakul, J. 1997. Studies on granule growth kinetics and characteristics of tapioca starch dispersion during gelatinization using particle size analysis and rheological methods. *MS thesis*, Cornell University, Ithaca, NY.

Tattiyakul, J. and Rao, M.A. 2000. Rheological behavior of cross-linked waxy maize starch dispersions during and after heating. *Carbohyd. Polym.* 43: 215–222.

Tattiyakul, J., Rao, M.A., and Datta, A.K. 2002. Heat transfer to three canned fluids of different thermo-rheological behavior under intermittent agitation. *IChemE Trans. Part C—Food Bioprod. Proc.* 80: 20–27.

Tiu, C., and Boger, D. V. 1974. Complete rheological characterization of time-dependent products. *J. Texture Stud.* 5: 329–338.

Tung, M. A., Richards, J. F., Morrison, B. C., and Watson, E. L. 1970. Rheology of fresh, aged and gamma-irradiated egg white. *J. Food Sci.* 35: 872–874.

Van Wazer, J. R., Lyons, J. W., Kim, K. Y., and Colwell, R. E. 1963. *Viscosity and Flow Measurement: A Laboratory Handbook of Rheology.* Interscience, New York.

Vitali, A. A., and Rao, M. A. 1982. Flow behavior of guava puree as a function of temperature and concentration. *J. Texture Stud.* 13: 275–289.

Vitali, A. A., and Rao, M. A. 1984a. Flow properties of low-pulp concentrated orange juice: Serum viscosity and effect of pulp content. *J. Food Sci.* 49: 876–881.

Vitali, A. A., and Rao, M. A. 1984b. Flow properties of low-pulp concentrated orange juice: Effect of temperature and concentration. *J. Food Sci.* 49: 882–888.

Vitali, A. A., Roig, S. M., and Rao, M. A. 1974. Viscosity behavior of concentrated passion fruit juice. *Confructa* 19: 201–206.

Walters, K. 1975. *Rheometry.* Chapman and Hall, London.

Weltman, R. N. 1943. Breakdown of thixotropic structure as function of time. *J. Appl. Phys.* 14: 343–350.

Whitcomb, P. J., and Macosko, C. W. 1978. Rheology of xanthan gum. *J. Rheol.* 22: 493–505.

White, G. W. 1970. Rheology in food research. *J. Food Technol.,* 5: 1–32.

Whorlow, R. W. 1980. *Rheological Techniques.* Wiley, New York.

Wilkinson, W. L. 1960. *Non-Newtonian Liquids.* Pergamon, New York.

Wohl, M. A. 1968. Designing for non-Newtonian fluids. *Chem. Eng.* 75(2): 148; 75(4): 130; 75(7): 99; 75(8): 143; 75(10): 183; 75(12): 95; 75(14): 81; 75(15): 127; 75(18): 113.

Wood, F. W., and Goff, T. C. 1973. The determination of the effective shear rate in the Brabender viscograph and in other systems of complex geometry. *Die Stärke* 25: 89–91.

Yang, W.H. and Rao, M.A. 1998a. Complex viscosity–temperature master curve of cornstarch dispersion during gelatinization. *J. Food Proc. Eng.* 21: 191–207.

Yang, W.H. and Rao, M.A. 1998b. Transient natural convection heat transfer to starch dispersion in a cylindrical container: numerical solution and experiment. *J. Food Eng.* 36: 395–415.

Yoo, B., Rao, M.A. and Steffe, J.F. 1995. Yield stress of food suspensions with the vane method at controlled shear rate and shear stress. *J. Texture Stud.* 26: 1–10.

6

Rheological Properties of Solid Foods

V. N. Mohan Rao[*] and Ximena Quintero

Contents

6.1	Introduction	180
6.2	Quasistatic Tests for Solid Foods	181
	6.2.1 Introduction	181
	6.2.2 Some Simple Tests	182
	6.2.3 Rheological Modeling	185
	6.2.4 Creep	186
	6.2.5 Relaxation	188
6.3	Dynamic Testing of Solid Foods	190
	6.3.1 Introduction	190
	6.3.2 Theoretical Considerations	191
	6.3.2.1 Resonance	191
	6.3.2.2 Direct Stress–Strain Tests	193
	6.3.3 Application of Resonance	197
	6.3.4 Application of Direct Stress–Strain Tests	199
6.4	Failure and Glass Transition in Solid Foods	202
	6.4.1 Failure in Solid Foods	202
	6.4.2 Glass Transition of Solid Foods	204
	6.4.2.1 Factors that Affect Glass Transition	204
	6.4.2.2 Measurement of Glass Transition	204
	6.4.2.3 Importance of Glass Transition in Solid Foods	206
6.5	Empirical and Imitative Tests	207
	6.5.1 Introduction	207
	6.5.2 Texture Profile Analysis	208

---

[*] Drs. Mohan Rao and Ximena Quintero are employees of PepsiCo Inc. The views expressed in this chapter are those of the authors and do not necessarily reflect the position or policy of PepsiCo, Inc.

6.5.3	Texture (Shear) Press	210
6.5.4	Warner–Bratzler Shear	210
6.5.5	FMC Pea Tenderometer	211
6.5.6	Penetrometer	211
6.5.7	Bend testing	211
	6.5.7.1 Testing method	212
6.5.8	Other Empirical Methods	213
6.5.9	Structure and Porosity as related to Snack Texture	213
	6.5.9.1 Structure	213
	6.5.9.2 Porosity	215
6.6 Conclusions		215
References		217

6.1 INTRODUCTION

Rheological properties of solid foods are a subject of concern to researchers, food industries, and consumers alike. Muller (1973) enumerated four main reasons for the study of rheological properties:

1. To enable an insight into the structure of the material because the physical manifestation of a material is due to its chemical composition. The relationship between cross-linkage of polymeric materials and their elasticity is an example of this.
2. To improve quality control in the food industry.
3. To design machinery for handling solid foods.
4. To correlate consumer acceptance with some definite rheological property. Many food industries now have standard tests for correlating some rheological aspect of solid foods—for example, hardness of peanut brittle—with consumer acceptance.

The evaluation of rheological properties of solid foods can be divided into two broad classes. *Fundamental tests* measure properties that are inherent to the material and do not depend on the geometry of the test sample, the conditions of loading, or the apparatus. Examples of these properties are modulus of elasticity, Poisson's ratio, relaxation time, and shear modulus. *Empirical* or *imitative tests* are used to determine properties such as puncture force and extrusion energy, where the mass of the sample, the geometry, speed of the test, and so on, also determine the magnitude of the parameter estimated.

The fundamental tests as applied to solid foods may again be classified into two essentially different groups: those conducted under conditions of static or quasistatic loading, and those conducted under dynamic conditions. The use of a universal testing machine, such as an Instron, in determining the modulus of elasticity in compression, constitutes a *quasistatic* test. However, if in determining the modulus of elasticity a vibrating device is used at a frequency of 200 Hz, the testing method is termed *dynamic*. In general, loading rates are used to determine whether a test is dynamic or quasistatic. The term *quasistatic* is used instead of *static* because theoretically there can be no test that can be termed *static*, as imposition of any force, however small, will always induce relative motion of the particles.

6.2 QUASISTATIC TESTS FOR SOLID FOODS

6.2.1 Introduction

Before we discuss the fundamental rheological tests that have been used on solid foods, we present a short review of fundamental rheology. Two extremes of behavior may result (from a rheological viewpoint) when a force is applied to a material; the pure elastic deformation of a solid and the pure viscous flow of a liquid.

Pure elastic behavior is defined such that when a force is applied to a material, it will instantaneously and finitely deform; and when the force is released, the material will instantaneously return to its original form. Such a material is called a *Hookean solid*. The amount of deformation is proportional to the magnitude of the force. The rheological representation for this type of solid is a spring. A material of this nature can be given a rheological constant, termed the elastic modulus. The *elastic modulus* is the ratio of stress to strain in a material, where stress is equal to force per unit area and strain is the observed deformation due to the force, divided by the original length of the material. Three types of moduli may be calculated for a Hookean solid, depending on the method of applying the force. The modulus calculated by applying a force perpendicular to the area defined by the stress is called the *modulus of elasticity* (E). The modulus calculated by applying a force parallel to the area defined by the stress, or a shearing force, is called the *shear modulus* or *modulus of rigidity* (G). If the force is applied from all directions (isotropically) and the change in volume per original volume is obtained, then one can calculate the *bulk modulus* (K). Thus, these are material constants, because the deformation is proportional to the applied force, and unit area and length are considered in the calculations.

A *pure viscous flow of a liquid* means that the liquid begins to flow with the slightest force and that the rate of flow is proportional to the magnitude of force applied. This liquid flows infinitely until the force is removed, and upon removal of the force, it has no ability to regain its original state. Such a material is called a *Newtonian liquid*. The rheological representation for this type of liquid is a *dashpot*, which can be thought of as a piston inside a cylinder. When a force is applied to the piston, it moves in or out of the cylinder at constant velocity, the rate depending on the magnitude of the force. When the force is removed, the piston remains fixed and cannot return to its original position. A material of this nature has a rheological constant called the coefficient of viscosity (η). The *coefficient of viscosity* is defined as the shearing stress applied divided by the resulting rate of strain. In this way, it is very similar to the modulus for Hookean solids.

If foods were either Hookean solids or Newtonian liquids, determination of their rheological constants would be simple. However, foodstuffs possess rheological properties associated with both the elastic solid and the viscous fluid. Such materials are called *viscoelastic* (Mohsenin, 1978). The rheological representation of this type of material is a body incorporating at least one spring (representing the solid character) and at least one dashpot (representing the viscous character). The number of springs and dashpots in the body and the manner in which they are connected can be manipulated to represent different types of viscoelastic materials, and to demonstrate how they will behave under a stress or strain. Thus, a viscoelastic material has several rheological constants, depending on the number of springs and dashpots that represent its behavior. There is no simple constant for viscoelastic materials such as modulus, because the modulus will change over time. Thus, if one subjects a viscoelastic material to a constant stress, the manner in which the material is strained will change over time. The rheological constants for a viscoelastic material are represented by an equation to give modulus as a function of time. The theory of viscoelasticity is discussed in detail by Reiner (1960, 1971), Christensen (1971), and Flugge (1975).

Because foods are viscoelastic, both time-dependent and time-independent measurements are required. Alfrey (1957) lists three methods that use experimental curves to "map out" the viscoelastic character of a material:

1. The creep curve, showing strain as a function of time at constant stress,
2. The relaxation curve, showing stress as a function of time at constant strain, and
3. The dynamic modulus curve, consisting of the dynamic modulus as a function of the frequency of the sinusoidal strain.

For linear viscoelastic materials, these three types of experimental curves should yield consistent results; that is, the moduli and coefficients of viscosity from the relaxation, creep, and dynamic tests should be interconvertible mathematically and should be independent of the magnitude of the imposed stress or strain.

The following review includes publications that have employed one or more of the three methods listed above as well as other studies that lend insight into the fundamental rheological behavior of solid foods.

6.2.2 Some Simple Tests

The simplest of all the quasistatic tests is perhaps the uniaxial compression/tension test. In this test, a sample with a convenient geometry (e.g., cylinder or rectangular prism) is subjected either to a deformation or to a force, and the corresponding force or deformation is recorded. If the magnitudes of force and deformation are small, then the body may be assumed to be elastic. The resultant stress (σ) and strain (e) may be calculated as

$$\sigma = \frac{F}{A} \tag{6.1}$$

and

$$e = \frac{\Delta L}{L} \tag{6.2}$$

where F is the force, A is the cross-sectional area of the body, ΔL is the deformation, and L is the original length of the body (Figure 6.1). The modulus of elasticity (also called Young's modulus), usually denoted by E or Y, can be computed as σ/e. In a similar manner, it is also possible to determine the shear modulus (G) or bulk modulus (K). The bulk modulus (K) can be evaluated as the ratio of isotropic stress (σ in Figure 6.1) to the volumetric strain. The volumetric strain is defined as the change in volume divided by original volume. The essential point is that the deformations and forces have to be extremely small in order to assume elastic behavior.

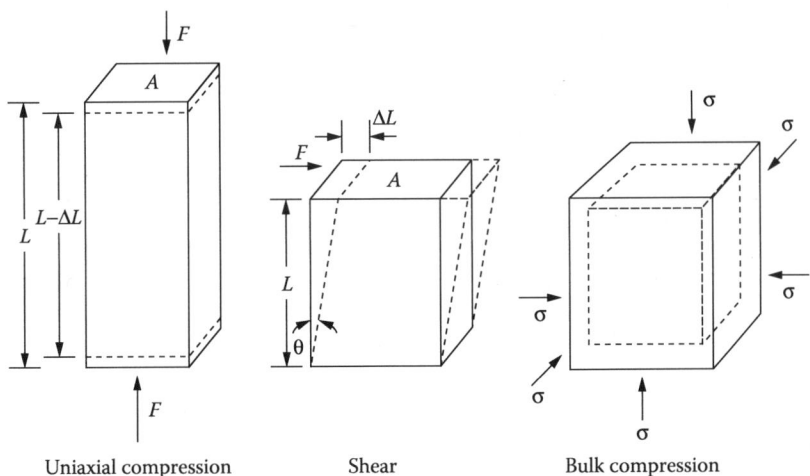

Figure 6.1 Uniaxial compression, shear, and isotropic (bulk) compression of an elastic solid.

Uniaxial compression and tension tests provide the researcher with an extremely simple test for the determination of material properties under conditions of both elastic and viscoelastic behavior. However, due to the effects of bonding and lubrication, such tests have certain serious problems with the computation of the elastic moduli. In most uniaxial compression tests, the effect of frictional forces between the loading plates and the material under test need to be considered in the computation of stress–strain relationships. Extensive work in this area has been conducted by Bagley and Christianson (1987). They have shown that bonding the material to the test platen by adhesives leads to improved and consistent reproducibility of the results. The other alternative is to lubricate the test platens to eliminate or minimize the effect of friction. This approach has been used by many researchers (Montejano et al., 1983). In an extensive and pioneering study, Casiraghi et al. (1985) examined the effect of lubrication of the test platen and bonding of the material to the test platen in uniaxial compression tests using cheese. In bonded compression they recommend that strain be calculated as

$$e = -\delta h/h \tag{6.3}$$

where δh and h are the change in height and the height after deformation. (Note that this definition of strain is different from Equation 6.2, where the original undeformed dimension was used.) The calculation of stress (σ_B) for bonded compression is as usual:

$$\sigma_B = F/\pi R_0^2$$

where R_0 is the radius. For comparing the bonded response with the lubricated response, the correction for stress (σ_{BC}) was

$$\sigma_{BC} = \frac{\sigma_B}{\sqrt{1 + R_0^2/2h^2}} \tag{6.4}$$

For lubricated materials they recommend calculating stress (σ_L) as

$$\sigma_L = \frac{F}{\sqrt{\pi R^2}}$$

where R is the radius of the deformed cylinder with an initial radius of R_0 and height h_0 (assuming that there is no change in volume). Using these equations, it appears that the agreement between results from lubricated and bonded compression of mozzarella cheese samples are valid at least until the strain levels correspond to 60% deformation. In conclusion, Casiraghi et al. (1985) recommend that all uniaxial compression of foods be carried out under all conditions (lubrication, bonding) before meaningful results are obtained. This appears to be a good rule, especially if the strain levels are considerable (usually more than a few percent). This procedure of correcting for the effects for bonding and lubrication is equally applicable to uniaxial tests for elastic and viscoelastic solid foods.

Another test that can be applied to foods that are fairly brittle is the bending test (Figure 6.2). The advantage of this method lies in the extremely small true deformations in the material in addition to measurable deflections. The calculations are as follows.

In a loading of a material with two symmetrical vertical supports, the bending moment (M) at any point x is

$$M = \frac{P}{2}x \qquad (6.5)$$

where P is the force.

If the effects of shearing force and shortening of the beam axis are neglected, the expression for the curvature of the axis of beam is

$$EI\frac{d^2y}{dx^2} = -\frac{P}{2}x \qquad (6.6)$$

where I is the moment of inertia. Integrating Equation 6.6 twice yields

$$y = \frac{Px^3}{12\,EI} + Ax + B$$

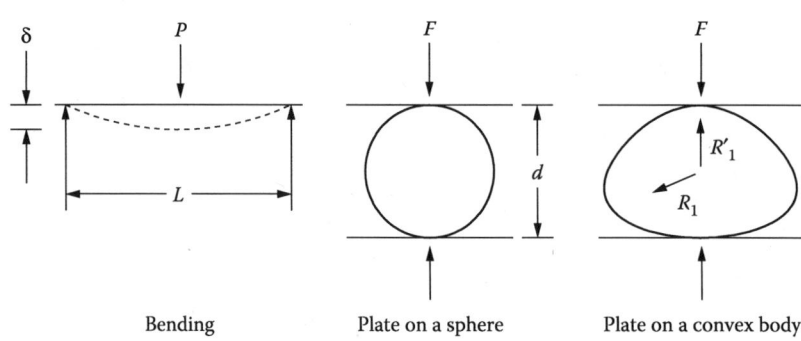

Figure 6.2 Application of bending and Hertz's equations to measure the modulus of elasticity (E).

where A and B are integration constants. Now, $y = 0$ at $x = 0$, and hence $B = 0$. Also, $dy/dx = 0$ at $x = L/2$. Therefore,

$$A = \frac{PL^2}{16 EI}$$

and

$$y = \frac{PL^2 x}{16 EI} - \frac{Px^3}{12 EI}$$

For this problem, the maximum deflection (δ, Figure 6.2) occurs in the center, that is, $x = L/2$.

$$y = \delta = \frac{PL^3}{48 EI}$$

Hence, the modulus of elasticity (E) is

$$E = \frac{PL^3}{48 I \delta} \tag{6.7}$$

For materials that cannot be modified to yield a sample possessing convenient geometry (e.g., some fruits, vegetables, grains), the application of Hertz's equations is appropriate (Mohsenin, 1978). The necessary equations for a spherical and a convex body (Figure 6.2) are as follows:

For axial loading of a spherical sample between flat plates,

$$E = \frac{0.531 F(1 - \mu^2)}{D^{1.5}} \left(\frac{4}{d}\right)^{0.5} \tag{6.8}$$

where F is the force corresponding to deformation D, d is the diameter of the sphere, and μ is Poisson's ratio.

For a plate on a convex body,

$$E = \frac{0.531 F(1 - \mu^2)}{D^{1.5}} \left(\frac{1}{R_1} + \frac{1}{R'_1}\right)^{0.5} \tag{6.9}$$

where R_1 and R'_1 are radii of curvature.

The only problem in using the Hertz equations is the prior knowledge of the Poisson ratio (μ). However, the error introduced by assuming an approximate value would be minimal (Mohsenin, 1978). The above technique has been widely used in the literature for a variety of convex-shaped foods.

6.2.3 Rheological Modeling

Food materials seem to behave as viscoelastic materials when they are exposed to various conditions of stress or strain (Chappell and Hamann, 1968; Chen and Fridley, 1972;

Clevenger and Hamann, 1968; Datta and Morrow, 1983; Hammerle and Mohsenin, 1970; Hundtoft and Buelow, 1970; Mohsenin, 1978; Morrow and Mohsenin, 1966; Peleg, 1976a; Skinner, 1983). Many researchers have designed experimental procedures that provide insight into the rheological modeling of these materials in order to characterize them and predict their behavior under specific physical conditions. These viscoelastic models contain various combinations of Hookean solid elements (springs) and Newtonian fluid elements (dashpots), and show complex behavior that can represent various food materials. If a material is found to be linearly viscoelastic, this property allows transformation of the individual element constants to fit different arrangements of these elements into other equivalent models. This linearity is guaranteed only at very small levels of strain. Thus, the moduli of viscoelastic elements is a function of time, not stress, at these small strains. The rheological constants for viscoelastic materials are represented by mathematical equations for different models where the modulus is expressed as a function of time.

If an accurate model is made to represent food material, it can be used to predict changes in the material that may occur during mechanical harvesting or handling, and perhaps further be used to reduce the risk of damage and other structural defects in raw agricultural commodities. Researchers such as Peleg (1976a) found rheological models to be extremely useful tools in predicting mechanical response of foods to specific stress–strain conditions.

Peleg (1976a) provided a list of conditions to satisfy in constructing a rheological model to represent a food material:

1. The model must enable the prediction of a real material behavior under any force–deformation history.
2. The model should be able to respond to both positive and negative forces and deformations (i.e., tension and compression). However, this is limited to the instance where the "physical structure" of the elements themselves would be under stress.
3. Changes and variations occurring in the behavior of the real material must be explained in terms of the model parameters.

Two of the most useful physical tests used in the determination of a rheological model and in the computation of the individual model constants incorporated into the chosen model are static creep and stress relaxation (Datta and Morrow, 1983; Gross, 1979; Skinner, 1983). It is therefore imperative to discuss these two tests in detail. The elementary models used to build more complex models are termed *Maxwell* (a spring and a dashpot in series) and *Kelvin* (a spring and a dashpot in parallel), after the inventors.

6.2.4 Creep

Static uniaxial normal creep is a condition in which the constant shear or dynamic forces involved are all parallel to the longitudinal axis of the specimen. This imposed stress must not be so great as to yield large sample deformations to the point where elastic limit of the material is exceeded and it no longer behaves as a linearly viscoelastic material. In such a case, representation of these materials by rheological models would no longer be valid.

Shama and Sherman (1973) state that for a material to be linearly viscoelastic, (1) the strain must be linearly related to the stress; (2) the stress–strain ratio must be a function of the time for which the stress is applied, not the magnitude of the stress itself; and (3) the Boltzmann superposition principle should be followed (i.e., the strain at any time t after a stress has been applied depends on the shear history of the sample).

In the creep experiment, when the load (force) is applied to the sample instantaneously, the sample is rapidly deformed, imposing a strain on the material, which continues to increase at a decreasing rate as a function of time. Creep recovery is another case of a creep test whereby at some point the constant stress is released, causing the strain to decrease and approach a zero value depending on the sample properties and its predicted mode.

Regardless of sample dimensions, when the sample is deformed in compression, the strain generated will decrease the height (or length) of the sample and result in an increase in sample diameter or width to a value dependent on the bulk modulus of the material or its Poisson ratio. In many cases, the transverse strain can be neglected because of the partly compressible nature of most agricultural materials, which causes the resultant lateral strain to be negligible compared to the uniaxial strain.

A plot of uniaxial strain or deformation as a function of time results in a plot known as a *creep curve*. Creep curves and other computational methods can be effectively used to study a material's physical properties (Bloksma, 1972; Chappell and Hamann, 1968; Datta and Morrow, 1983; Finney et al., 1964; Mohsenin et al., 1963; Skinner, 1983).

Creep experiments demonstrate the fact that strain exhibited by fruits and vegetables is not independent of time (Mohsenin, 1978); thus, a single Hookean element is not sufficient in representing the physical behavior of most agricultural materials. A typical creep behavior curve of cylindrical samples of flesh from apples, potatoes, and frankfurters, as observed by Skinner (1983), can be seen in Model III of Figure 6.3. The typical graph shape agrees with the work conducted by Mohsenin et al. (1963) and Finney et al. (1964).

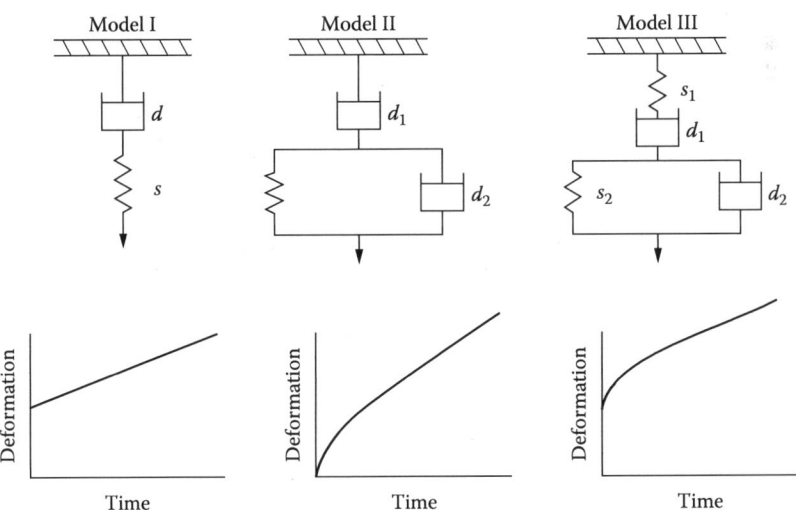

Figure 6.3 Typical creep curves for Maxwell (Model I), three-element (Model II), and Burgers (Model III) models.

Datta and Morrow (1983) showed that a generalized Kelvin model (a series of Kelvin bodies) in series with a Maxwell model best represents the creep data obtained from apples, potatoes, and cheese. In this same work, they offer a detailed solution to the graphical and computational applications of the method of successive residuals for the solutions to creep curves. Extensive early work on bread doughs (Halton and Scott Blair, 1937; Lerchenthal and Muller, 1967) used creep tests to obtain information about their physical properties. Bloksma (1972) found the model best representing dough to be a three-element model (Model II, Figure 6.3) consisting of a dashpot in series with a Kelvin body.

Numerous attempts to find a rheological model to represent the flesh of apple, potato, pear, and other fruits as well as low-methoxyl pectin gel preparations (Gross, 1979) under conditions of static creep have yielded the Burgers model (Mohsenin, 1978; Reiner, 1960), which can be seen in Figure 6.3 (Model III). The creep curves of apples (Skinner, 1983) exhibited behavior identical to that of the Burgers model. There was an immediate deformation upon application of a constant load, due to the spring s_1. The immediate deformation is followed by a curved portion of the creep curve characteristic of the Kelvin body behavior. A typical linear deformation at extended periods of time can be represented by a single viscous element.

The Burgers model can be mathematically represented (Gross, 1979) as

$$D(t) = D_0 + D_1 t + D_2 \left[1 - \exp\left(-\frac{t}{\tau}\right) \right] \quad (6.10)$$

where $D(t)$ is the compliance for the entire model at time t and is equal to the strain at time t divided by the constant stress applied to the model (reciprocal of E); D_0 is the compliance of spring element s_1 and is equal to the strain at time $t = 0$, divided by the constant stress value; D_1 represents the flow of dashpot element d_1 and is equal to $1/\acute{\eta}_1$ or the reciprocal of the coefficient of viscosity of that viscous element; D_2 is the compliance of spring element s_2 and is thus equal to the compliance of the Kelvin element at $t = \infty$ (infinity); τ represents the retardation time of the Kelvin element and is defined as the time required for this sample to deform to 64% of its total length. The retardation time is defined at $\acute{\eta}_2/E_2$, where $\acute{\eta}_2$ is the coefficient of viscosity for dashpot element d_2, and E_2 is the modulus of the spring element s_2.

Mohsenin (1978) gives a graphical solution method for generated creep curves. Creep curves not only provide information about the physical properties of agricultural materials but can also be used to predict a material's behavior under a dead load.

6.2.5 Relaxation

The other major test that is frequently used to determine the model constants of an agricultural material is stress relaxation (Peleg, 1979). *Stress relaxation* can be described as the ability of a material to alleviate an imposed stress under conditions of constant strain. Strains must again be kept very limited, less than 1.5–3.0% for vegetables (Mohsenin and Mittal, 1977) or even less than 1.5% for materials such as potatoes (Skinner, 1983). This low strain will ensure that the sample is within the elastic range of that material, and rheological modeling can be used.

A plot of the rate at which the stress is dissipated as a function of time is known as a relaxation curve. A typical Instron universal testing machine-generated relaxation curve for Red Delicious and Stayman winesap apples, Idaho potatoes, frankfurters (Skinner, 1983), and low-methoxyl pectin gels (Gross, 1979) can be seen for Model III in Figure 6.4. If the stress dissipates completely to a zero value, then no residual spring need be included in the representing model.

The model most often used to represent most agricultural products subjected to stress relaxation is a generalized Maxwell model composed of a finite number of Maxwell elements (a spring and a dashpot in series) in parallel with each other. Finney et al. (1964) and Skinner (1983) found that potato tissue is approximately linearly viscoelastic at small strains (~2%) and can effectively be characterized by a generalized Maxwell model. Chen and Fridley (1972) found pear tissue to be effectively represented by a generalized Maxwell model. Mohsenin (1978) and Chen and Fridley (1972) presented a method for determining the model constants for a Maxwell model. The model best representing cooked potato flesh was a generalized Maxwell model containing a residual spring in parallel with the other Maxwell elements (Davis et al., 1983). Duration of cooking affected the individual model constants in this modified generalized Maxwell model (Chen and Fridley, 1972).

Whether or not a residual spring need be included in the model is easily determined. If the stress relaxation test is run for an extended period of time and the stress imposed is totally alleviated, then no residual spring is required in the model representing the test material (Models I and III, Figure 6.4). If the stress generated is not totally alleviated after long periods of time, but in fact levels off at some finite stress value, then the residual spring must be present in the model and have an elastic modulus equal to that finite stress

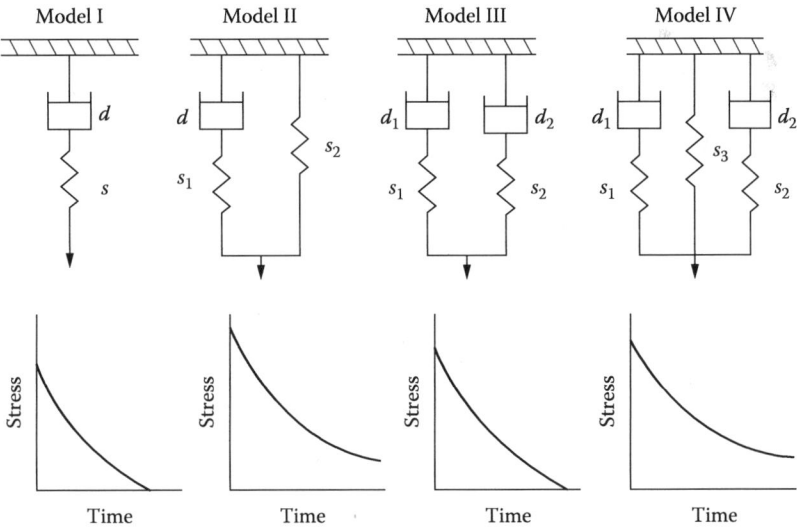

Figure 6.4 Typical relaxation curves for Maxwell (Model I), three-element (Model II), four-element (Model III), and five-element (Model IV) models.

value divided by the constant strain of the material in that particular test (Models II and IV, Figure 6.4).

The number of Maxwell elements required to represent the sample efficiently can be determined by the method of successive residuals (Mohsenin, 1978) or by comparing the R^2 (percent explained variation) values for relaxation data fit to individual model equations by a statistical analysis system nonlinear procedure (Barr et al., 1976). Using both methods for Red Delicious and Stayman winesap apples, Idaho potatoes, and frankfurters, Skinner (1983) found that a four-element model (Model III, Figure 6.4) was sufficient to represent the samples because R^2 values for the sample replications were all greater than 0.99. Gross (1979) found that this same four-element model was sufficient to represent low-methoxyl pectin gels. The mathematical representation of this model is as follows (Skinner, 1983; Gross, 1979):

$$E(t) = E_1 \exp\left(-\frac{t}{\tau_1}\right) + E_2 \exp\left(-\frac{t}{\tau_2}\right) \tag{6.11}$$

where $E(t)$ is the modulus of elasticity of the entire model at time t and is equal to the stress at time t divided by the constant strain; E_1 is the elastic modulus of spring s_1; E_2 is the elastic modulus of spring s_2; τ_1 is the relaxation time of the first Maxwell element (s_1, d_1) and is equal to the ratio of $\dot{\eta}_1/E_1$, where $\dot{\eta}_1$ is the coefficient of viscosity of dashpot element d_1 and E_1 is the elastic modulus of spring element s_1; τ_2 is the relaxation time of the second Maxwell element (s_2, d_2) and is equal to the ratio of $\dot{\eta}_2/E_2$, where $\dot{\eta}_2$ is the coefficient of viscosity of dashpot element d_2 and E_2 is the elastic modulus of spring element s_2.

Goodness of fit (or an increase in R^2 values) of the relaxation model to the relaxation data can, in fact, be increased by adding one or more Maxwell elements to the four-element model (or exponential terms to the mathematical equation). However, it was found that this addition of exponential terms (Skinner, 1983) did not increase the value of R^2 significantly enough to aid the model or help in distinguishing between two products or varieties of materials.

Pitt and Chen (1983) stated that higher strain rates result in apparent tissue stiffness, and Murase et al. (1980) showed that the elastic modulus of potato increases with increasing deformation rates. By defining the physical behavior of plant tissues such as apples, potatoes, and pears and realizing that the model constants can be correlated to parameters such as vegetable firmness, one may find these parameters and rheological models to be useful in evaluating the storage quality of these commodities.

6.3 DYNAMIC TESTING OF SOLID FOODS

6.3.1 Introduction

The measurement of dynamic mechanical properties of foods offers a very rapid test with minimal chemical and physical changes. In addition, mechanical properties such as Young's modulus may be determined at various frequencies and temperatures within a short time frame. Another important advantage of dynamic tests is the extremely small strains imposed on the foods (usually well within 1%), which ensures a linear stress–strain behavior. These small strains are essential to the use of linear elastic and viscoelastic theories in predicting material behavior.

Dynamic properties have been extensively studied by chemists, engineers, and physicists (Ferry, 1970) in analyzing the composition and behavior of a variety of nonfood materials. Dynamic properties have been particularly valuable for the study of the structure of high polymers (Nielsen, 1962) because they are very sensitive to glass transition, cross-linking, phase separation, and molecular aggregation of polymer chains. The measurement of the dynamic properties of rubberlike materials by Nolle (1948) and Marvin (1952) are considered to be pioneering works in this field. Nolle (1948) used frequencies from 0.1 Hz to 10,000 Hz in determining the Young's modulus in the temperature range of −60°C to 100°C. Although dynamic testing of foods is not widely used, it offers potential in the field of food texture evaluation. Voisey (1975, p. 256) states as follows: "Recent emphasis on the importance of texture test conditions points to the desirability of using much higher deformation rates." One method of attaining these rates is the use of dynamic testing.

Morrow and Mohsenin (1968) divided the methods of dynamic testing into four types: (1) direct measurement of stress and strain, (2) resonance methods, (3) wave propagation methods, and (4) transducer methods. The direct measurement of stress and strain and resonance methods has been used by a number of researchers to characterize the chemical composition, texture, and maturity of various foods. Therefore, this chapter deals primarily with these two methods of measuring dynamic properties of foods. The theory and instrumentation are reviewed, and a comparison of results of various researchers is presented.

6.3.2 Theoretical Considerations

6.3.2.1 Resonance

Resonance methods have been primarily used for determining resonant frequencies and thereby also determining Young's modulus, shear modulus, loss coefficient, and so on, at the resonant frequency. The theory of resonance has been practiced on a discrete system with a single degree of freedom. The system consists of a spring with a spring constant k, a damper (dashpot) with a coefficient of viscous damping c, and a mass m (Figure 6.5). The general equation for representing the system behavior in response to a sinusoidally varying force $P_0 \sin \omega t$, with an amplitude P_0 and circular frequency ω, as given by Meirovitch (1967), is

$$m\ddot{x} + c\dot{x} + kx = P_0 \sin \omega t \qquad (6.12)$$

where x is the displacement and $\dot{x}$ and $\ddot{x}$ are the first and second derivatives of x with respect to time. If the damping constant is negligible and external exciting force is absent (Finney, 1972), then

$$m\ddot{x} + kx = 0 \qquad (6.13)$$

and the undamped natural frequency of vibration, f_n, is

$$f_n = \frac{1}{2\pi}\left(\frac{k}{m}\right)^{0.5}$$

ENGINEERING PROPERTIES OF FOODS

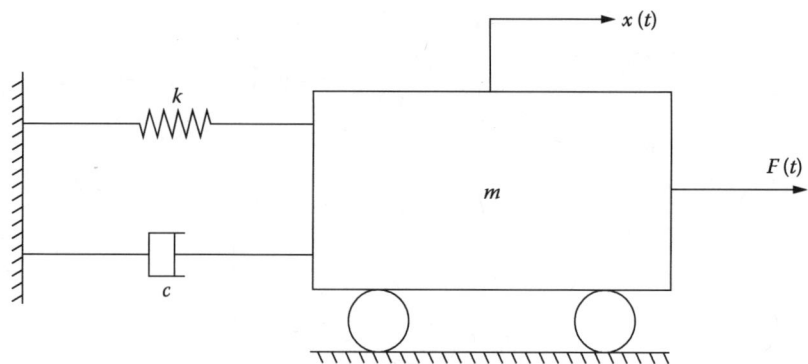

Figure 6.5 Schematic diagram of a spring (k), damper (dashpot, c), and mass (m) system with an external force, F(t), acting in the direction of motion, x(t).

Drake (1962) used basically the above system for determining the modulus of elasticity of foods in the form of a rectangular beam (6 × 12 × 50 mm). The formula used was

$$E = 48\pi^2 \rho l^2 \frac{f^2}{m^4 a^2} \tag{6.14}$$

where E is the modulus of elasticity; ρ, the density of the material; l, the length of the sample; f, the frequency; m, a factor with certain eigenvalues for the different modes of vibration; and a, the thickness of the beam.

The equation was further simplified by using $m^4 = 12.36$ for the fundamental frequency. Spinner and Tefft (1961) provided extensive literature on the computation of elastic moduli from mechanical resonance frequencies for variously shaped specimens. The equations for computing Young's modulus from the first three modes of flexural resonance of cylinders are

$$E = 1.261886 \rho l^4 f_1^2 \frac{T_1}{d^2} \tag{6.15a}$$

$$E = 0.1660703 \rho l^4 f_2^2 \frac{T_2}{d^2} \tag{6.15b}$$

$$E = 0.04321184 \rho l^4 f_3^2 \frac{T_3}{d^2} \tag{6.15c}$$

where f_n is the resonant frequency of the nth mode of vibration, d is the diameter of the cylinder, and T_n ($n = 1,2,3$) are correction factors depending on d, l, and Poisson's ratio. The authors provided values for T_1, T_2, and T_3 for Poisson's ratio ranging from 0.0 to 0.5 and d/l ratios from 0.00 to 0.60.

Spinner and Tefft (1961) also provided the necessary equations for determining Young's modulus of cylinders and bars of square cross section in longitudinal resonance. They recommended the use of longitudinal resonance as a more accurate method for determining

Young's modulus than flexural resonance. For cylindrical specimens, Young's modulus is given by

$$E = \frac{\rho}{K_n}\left(\frac{2lf_n}{n}\right)^2 \qquad (6.16)$$

where K_n is a correction factor for the nth mode of longitudinal vibration. The correction factor for a Poisson ratio of μ is given by

$$K_n = 1 - \frac{\pi^2 n^2 \mu^2 d^2}{8\,l} \qquad (6.17)$$

provided $d/\lambda ! 1$, where λ is the wavelength. If the correction factor K_n is not small compared to 1, they recommended the use of tables provided in the manuscript. For bars of square cross section, they used the same equation as that for a cylinder, except for the substitution for d as

$$d^2 = \frac{4}{3}b^2$$

where b is the width of the square. For bars of rectangular cross section,

$$d^2 = \frac{2}{3}(a^2 + b^2)$$

where a and b are the sides of the rectangle. The authors stated that the use of these equations, except in extreme cases, was estimated to be within an error of 1%.

The general instrumentation for resonance tests for foods used by most of the researchers is more or less the same as the one described by Finney (1972). It consists of a beat frequency generator and a power amplifier that drives the vibration exciter (Figure 6.6). The signal from the detector (accelerometer) is amplified and measured by an ac voltmeter. An oscilloscope and a recorder are optional units used to record the frequency response curves and to observe waveforms and phase relationships. More detailed descriptions of the instrumentation and characteristics are given by Finney and Norris (1968) and Finney (1972).

6.3.2.2 Direct Stress–Strain Tests

In direct stress–strain tests, a linear viscoelastic material (most foods show this behavior to some extent) is subjected to a sinusoidally varying strain, and the resulting stress as a function of time is observed. If a material is subjected to a strain (e) variation, such that

$$e = e_0 \sin \omega t \qquad (6.18)$$

where e_0 is the amplitude of strain and ω is the angular frequency, the stress will then vary with the same frequency as the strain but will lag behind the strain by an angle θ, which is usually referred to as the *phase angle*.

$$\sigma = \sigma_0 \sin(\omega t - \theta) \qquad (6.19)$$

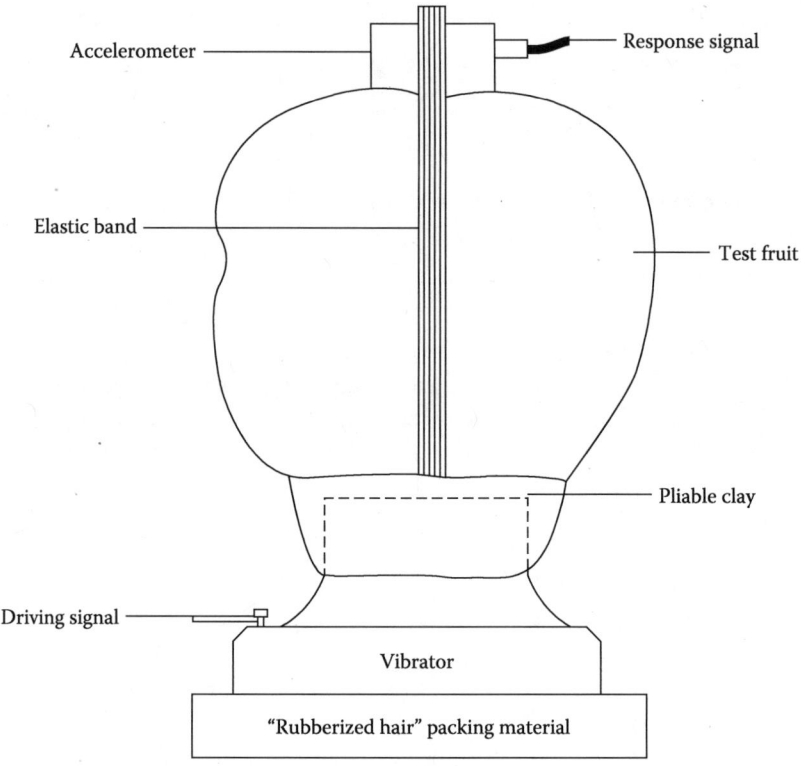

Figure 6.6 Schematic diagram of a resonance type of dynamic tester. (From Finney, E. E., Jr., and Norris, K. H. 1968. *Trans. ASAE* 11(1): 94–97.)

The dynamic viscoelastic behavior can be determined by measuring the complex dynamic modulus $E(i\omega)$ and the phase angle θ. The dynamic modulus can be written as a complex (real and imaginary) quantity.

$$E(i\omega) = E_1 + iE_2 \qquad (6.20)$$

where $E(i\omega)$ is the complex modulus, i is the imaginary unit, and E_1 and E_2 are the storage (real) and loss (imaginary) moduli, respectively. The ratio of amplitude of stress to the amplitude of strain is the absolute value of the complex modulus:

$$|E(i\omega)| = \frac{\sigma_0}{e_0} \qquad (6.21)$$

Therefore, the storage modulus, E_1, can be expressed as

$$E_1 = |E(i\omega)|\cos\theta$$

and the loss modulus, E_2, can be expressed as

$$E_2 = |E(i\omega)|\sin\theta$$

Substituting these relationships into the equation for stress, one obtains

$$\sigma = e_0[E_1 \sin\omega t - E_2 \cos\omega t] \tag{6.22}$$

The above equation shows the decomposition of stress into two components, one in phase with the strain and the other 90° out of phase. At very high frequencies the storage modulus has a constant maximum value and the material exhibits a perfectly elastic, solid-like behavior. At low frequencies, the storage modulus decreases, and eventually at very low frequencies the material exhibits an almost perfectly viscous type of behavior.

The tangent of the phase angle (loss tangent) is sometimes referred to as the *coefficient of internal friction*. It is also possible to calculate from the phase angle the energy lost (E_L) and the energy stored (E_S), because

$$K\tan\theta = \frac{E_L}{E_S}$$

where K is a proportionality constant. The value of K depends on how much of a cycle is being considered. As an example, for 1/2 cycle, $K = \pi$ if E_s is the maximum energy stored during the half-cycle. If K is set equal to 1, the percent energy loss is

$$\frac{E_L}{E_L + E_S} \times 100 = \frac{\tan\theta}{1 + \tan\theta} \times 100 \tag{6.23}$$

In evaluating the dynamic modulus and phase angle of a food, the calculation of the parameters is fairly simple if the geometry of the food is a cylinder or a rectangular beam. The stress amplitude is calculated as the ratio of the peak-to-peak force to the cross-sectional area of specimen. The strain amplitude is obtained as the peak-to-peak displacement over the length of the specimen. The only concern in this type of geometry is the length of the sample. If the length of the sample is sufficiently large, one has to consider the effects of inertia. However, if the length of the sample is small in comparison to the wavelength λ, then inertia effects can be neglected without appreciable error. The wavelength can be calculated as (Hamann, 1969)

$$\lambda = \frac{(|E|)^{0.5}}{\rho/f} \tag{6.24}$$

where ρ and f are the density of the material and the frequency of testing, respectively.

In testing a material that does not possess a convenient geometry (such problems often arise with foods when an intact product such as a fruit, grain, or egg is tested), the use of Hertz's equation (Mohsenin, 1978) is not valid. Although the use of Hertz's equation is justified for use on viscoelastic materials with convex surfaces, the test conditions require static or quasistatic loading. For dynamic tests, it has been shown (Hamann and Diehl, 1978) that the use of these equations yields erroneous results. However, Hamann and Diehl (1978), via dimensional analysis, arrived at an empirical equation, the accuracy of which has been verified for a variety of materials. The equation can be written as

$$|E| = \frac{1.51F}{Dd} \qquad (6.25)$$

where F is the peak-to-peak force, D is the peak-to-peak displacement, and d is the mean contact diameter between the sample and contacting plates.

The basic instrumentation for a dynamic, direct, stress–strain test requires a sinusoidal force provided by an electrodynamic vibrator. The vibrator is driven by a sine-wave generator signal after power amplification. An accelerometer is mounted on the vibrator shaft. The specimen is held between two disks, one attached to the vibrator and the other to a rigid vertical support via a dynamic force transducer. A static force transducer may be added between the dynamic force transducer and the rigid support to monitor static load. The accelerometer and force transducer signals are amplified and fed into an oscilloscope on the horizontal and vertical channels, producing a Lissajous ellipse. A typical apparatus used by Gross (1979) is shown in Figure 6.7. From the ellipse the peak-to-peak force (vertical height of ellipse) and the peak-to-peak acceleration (horizontal width of ellipse) can be obtained. The acceleration can be converted to displacement by dividing it by ω^2 (due to the sinusoidal input). The sine of the phase angle is obtained as the ratio of vertical height at the center of the ellipse to the total height.

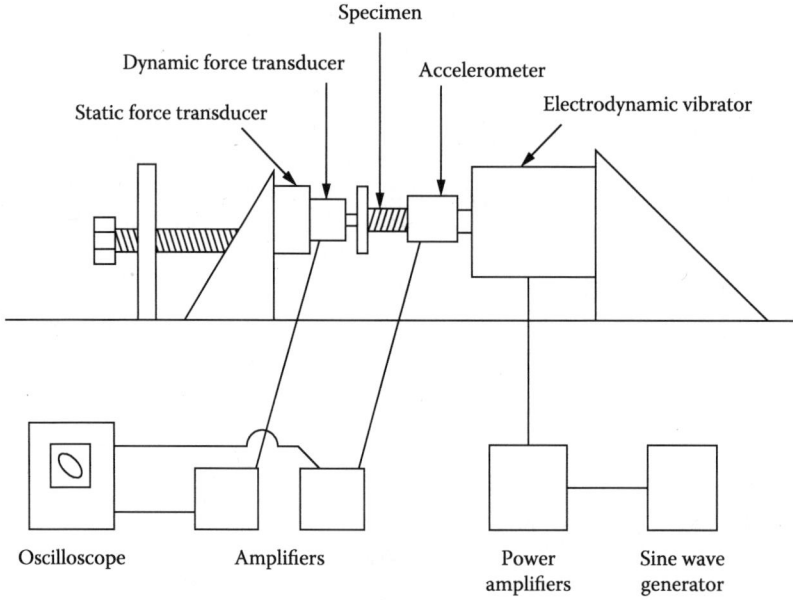

Figure 6.7 Schematic diagram of a direct stress–strain type of dynamic test apparatus in uniaxial compression. (From Gross, M. O. 1979. Chemical, sensory and rheological characterization of low-methoxyl pectin gels. PhD dissertation, University of Georgia, Athens, GA.)

6.3.3 Application of Resonance

Extensive work on the application of resonance tests for food has been carried out by Finney (1972). The application of vibration and resonance to sort fruits and vegetables according to maturity seems to be a promising approach. Hamann and Carroll (1971) used the vibration of a rigid platform with varying frequencies to sort muscadine grapes successfully ($R^2 > 0.88$). Holmes (1979) used a similar technique based on resonance to sort green tomatoes from ripe ones, with a rejection rate of 75–98% in various experimental runs. Montijano-Gaitan et al. (1982) found that the specification of Young's modulus and loss tangent of the vibrating surface in sorting small fruits by resonance was an effective way of improving sorting.

A mathematical analysis of resonance in intact fruits and vegetables was conducted by Cooke and Rand (1973). They used apples, peaches, and cantaloupes as approximate elastic spheres and developed the relationship between intact resonant frequencies and the shear modulus for the core, flesh, and skin.

Resonance-type tests were conducted by Finney et al. (1967) on bananas to monitor firmness during the ripening process. Cylindrical specimens of banana were vibrated longitudinally, and the resonant frequency was determined. They also measured the spectral reflectance on the peel and converted the data to the Commission Internationale de l'Eclairage system and Munsell color notation. The starch and reducing sugar contents of bananas were determined. The authors found that the modulus of elasticity and starch content declined during ripening, whereas the reducing sugar content increased during ripening. The modulus of elasticity was linearly correlated with starch content and inversely correlated to the logarithm of reducing sugar and luminous reflectance. The resonant frequencies for bananas (diameter 1.5 ± 0.02 cm; length 4.3–7.45 cm) ranged from 375 to 470 Hz. A summary of dynamic properties for several foods obtained via resonance tests is given in Table 6.1

The resonance methods were applied to Irish and sweet potatoes by Finney and Norris (1967) to monitor texture. They used cylindrical specimens of potato (diameter 1.52 ± 0.02 cm; length 2.54–7.62 cm) suspended by cotton threads. A horn-type speaker driven by an audio generator was used to vibrate the specimens. A stylus-type cartridge was used to measure the response. The modulus of elasticity and internal friction coefficient of potatoes were calculated. The results in this study showed that the coefficient of variation for the modulus of elasticity was greater *between* potatoes than *within* a given potato, and the resonant frequencies were in the range of 900–1800 Hz. In a similar study on whole apples and cylindrical specimens of apple flesh, Abbott et al. (1968b) determined resonant frequencies and stiffness coefficients. The frequency range investigated was 20–4000 Hz. The authors concluded that the second- and higher-order resonant frequencies were affected by the ripeness of the apples. They recommended the use of cylinders because the resonant frequency decreased with time.

Mechanical resonance within Red Delicious apples was investigated by Finney (1970, 1971b). The tests were conducted on whole apples of varying maturity and on cylindrical specimens of apple flesh. Results were compared with results of the Magness–Taylor pressure tester and taste panel evaluation. The author determined significant correlations between resonance parameters (Young's modulus, shear modulus, and the loss coefficients)

ENGINEERING PROPERTIES OF FOODS

Table 6.1 Dynamic Properties of Foods Determined by Resonance Tests

Food	Property[a]	Magnitude	Reference
Apples (whole)	Index of firmness	$1.7–2.2 \times 10^8$ Hz$^2 \cdot$ g	Finney (1971b)
		$1.0–1.5 \times 10^8$ Hz$^2 \cdot$ g	Abbott et al. (1968b)
	Elastic modulus	6.2–10.0 MPa	Finney (1971b)
	Internal friction	0.034–0.160	Abbott et al. (1968a)
Apples (cylinders)	Elastic modulus	5.8–17.0 MPa	Finney (1970)
		1.7–4.5 MPa	Abbott et al. (1968b)
	Internal friction	0.054–0.091	Finney (1970)
		0.09–0.125	Abbott et al. (1968b)
	Index of firmness	$1.2–2.0 \times 10^8$ Hz$^2 \cdot$ g	Finney (1970)
	Shear modulus	2.1–3.9 MPa	Finney (1970)
Bananas (cylinders)	Elastic modulus	0.85–2.72 MPa	Finney et al. (1967)
Irish potatoes (cylinders)	Absolute modulus	5.9–13.0 MPa	Finney and Norris (1967)
	Internal friction	0.082–0.118	Finney and Norris (1967)
	Elastic modulus	5.9 MPa	Finney and Norris (1968)
	Poisson ratio	0.58	Finney and Norris (1968)
Irish potatoes (cubes)	Elastic modulus	7.4–13.3 MPa	Jasper and Blanshard (1973)
	Vibration response	45.3–73.3 db at 2000 Hz	Finney (1971a)
Peaches (whole)	Index of firmness	$0.39–3.4 \times 10^8$ Hz$^2 \cdot$ g	Finney (1971a)
	Vibration response	33–76 db at 2000 Hz	Finney and Abbott (1972)
	Resonant frequency	426–737 Hz	Shackelford and Clark (1970)
	Stiffness coefficient	$0.54–1.6 \times 10^5$	Shackelford and Clark (1970)
Peaches (cylinders)	Elastic modulus	1.95–19.3 MPa	Finney (1967)
	Internal friction	0.090–0.143	Finney (1967)
	Poisson ratio	0.020–0.391	Finney (1967)
Pears (whole)	Stiffness coefficient	$9.0–10.7 \times 10^6$	Amen et al. (1972)
Pears (cylinders)	Elastic modulus	12–29 MPa	Finney (1967)
	Internal friction	0.072–0.099	Finney (1967)
	Shear modulus	4.6–5.8 MPa	Finney (1967)
	Poisson ratio	0.252–0.354	Finney (1967)
Sweet potatoes (cylinders)	Elastic modulus	9.4–18.4 MPa	Finney and Norris (1967)
Internal friction	0.095–0.198	Finney and Norris (1967)	

[a] Index of firmness = $f^2 - m$; stiffness coefficient = $f_{n-2}^2 - m^{2/3}$.

and sensory panel results. Cooke (1970), using classical results from linear elasticity, investigated the analysis of the resonance of intact apples. The calculated shear moduli were independent of μ and density and compared well with the results obtained by Finney (1970).

The use of resonance methods for peaches was investigated by Shackelford and Clark (1970), Finney and Abbott (1972), and Finney (1971a). In the study of Shackelford and Clark (1970), the relationship between resonant frequencies of whole Kembo peaches and peach maturity was investigated. They concluded that the square of the second resonant frequency was correlated to maturity as determined by the conventional Magness–Taylor tester. They did not attempt to determine mechanical properties from the resonance characteristics. Finney (1971a) used random vibration tests of 5 Hz to 20 kHz on Elberta peaches to evaluate peach firmness. One-third octave vibration signals at 2 kHz correlated with the firmness determined by the Magness–Taylor pressure tester. Finney suggested the use of any frequency in the range of 2–5 kHz to determine peach firmness.

Resonance techniques for measuring texture of three varieties of apples were investigated by Abbott et al. (1968b). These investigators used cylindrical sections of apple flesh to determine their natural frequencies. They then calculated Young's modulus and internal friction, utilizing the equations of Spinner and Tefft (1961). They also calculated a "stiffness coefficient" from the mass and second resonant frequency of whole fruit. Young's modulus calculated from flexural and longitudinal modes of vibration did not differ appreciably (20% or less), although the frequencies were at least 10 times larger for the longitudinal vibration. The results are perplexing, as the apple flesh in other studies has been shown to be frequency dependent.

In other work related to resonance methods, Virgin (1955) used the resonant frequency as a measure of the turgor of plant tissues. The author used wheat root, potato tuber parenchyma, and leaf of *Helodea densar* to determine the relationship between resonant frequency and osmotic value of the cell sap and cell permeability. Jasper and Blanshard (1973) developed a simple instrument for resonance tests and, using potato cubes and assuming Kelvin–Voight behavior, determined the Young's modulus of potato.

6.3.4 Application of Direct Stress–Strain Tests

The pioneering work in the use of direct stress–strain dynamic tests with applications to food was carried out by Hamann (1969). Other modifications of the method are also reported; however, most researchers set up their tests to determine the complex modulus in compression and shear of solid foods in a manner similar to that used by Hamann (1969). Baird (1981), in determining the dynamic moduli of soy isolate dough, employed a commercial mechanical spectrometer. Using the storage and loss modulus in shear, Baird (1981) determined that the dough does not form cross-links between protein molecules as a result of heating.

Hamann (1969) studied the dynamic properties of cylindrical apple flesh specimens to determine the modulus of elasticity and modulus or rigidity (shear modulus) at several frequencies. In a follow-up study, Morrow et al. (1971) used Red Delicious apples and determined the absolute modulus of apple flesh from 20 to 330 Hz. The numerical values of modulus and phase angle at selected frequencies for several foods are summarized in Table 6.2. The Red Delicious apples were tested by the above investigators to determine

ENGINEERING PROPERTIES OF FOODS

Table 6.2 Dynamic Properties of Selected Foods by Direct Stress–Strain Tests

Food	Property	Frequency (Hz)	Magnitude	Reference
Apples (red delicious; cylinders)	Absolute modulus	20 330	6.7–12.6 MPa 18.7–23 MPa	Morrow et al. (1971)
Apples (3 varieties, cylinders)	Storage modulus	50 230	11–16 MPa 17–21 MPa	Hamann (1969)
	Absolute modulus	2 80	9.14 MPa 11.6 MPa	Hamann and Diehl (1978)
Corn (slab)	Absolute modulus	1	213–675 MPa	Wen and Mohsenin (1970)
	Phase angle	1	9°	
Frankfurters (cylinders)	Absolute modulus	40 240	1.4–2.3 MPa 3.8–6.0 MPa	Webb et al. (1975)
	Energy loss	40 240	11–18% 12–18%	
Irish potatoes (cylinders)	Absolute modulus	50 500	6.96–9.18 MPa 7.6–11.2 MPa	Peterson and Hall (1974)
Peaches (cylinders)	Absolute modulus	70	4–10 MPa	Clark and Rao (1978)
Pears (Bartlett; cylinders)	Absolute modulus	50 210	2.4 MPa 6.7 MPa	Marinos (1983)
Pears (canned; D'Anjou cylinders)	Absolute modulus	50 210	160–516 kPa 420–1281 kPa	Marinos (1983)
Pectin gels (cylinders)	Absolute modulus	100 280	5.9–11.4 MPa 59.9–77.2 MPa	Gross (1979)
	Phase angle	100 280	6.65–34° 20.6–29.8°	
Rice (cylinders)	Absolute modulus	100 1000	840–4200 MPa 1890–6710 MPa	Chattopadhyay et al. (1978)
	Loss tangent	100 1000	0.085–0.263 0.038–0.088	
Sweet potatoes (cylinders)	Absolute modulus	40 240	15–19 MPa 19–23 MPa	Rao et al. (1976)
Turnips (cylinders)	Phase angle	60 240	4–6° 5–7°	Hamann and Diehl (1978)
	Absolute modulus	2 80	7.69 MPa 7.67 MPa	

the effect of static preload on the absolute modulus of elasticity. The results indicated a significant increase in modulus (from 7 to 13 MPa) with an increase in preload (from 10 to 40 N).

The direct stress–strain testing was investigated by Peterson and Hall (1974) to show thermorheologically simple behavior of Russet Burbank potatoes. The device used was similar to that employed by Hamann (1969), and cylindrical specimens of potato were uniaxially compressed with frequencies ranging from 50 to 500 Hz and temperatures ranging from 2 to 30°C. The complex modulus decreased with an increase in temperature and increased with an increase in test frequency. They determined that the stem end of the tuber was found to have a higher phase angle and lower modulus than the bud end and the center of the tuber. The modulus was also very dependent on preload, just as was the case with apples (Morrow et al., 1971), with higher modulus values associated with higher preloads. Rao et al. (1974a) determined the absolute modulus of sweet potato flesh, using cylindrical specimens at frequencies of 30–170 Hz. The results from this study were used to develop a uniaxial modulus master curve for sweet potatoes over an extended time scale. In a related study, Rao et al. (1976) determined the absolute modulus and phase angle of sweet potato flesh taken from the stem, middle, and root end at frequencies of 40–240 Hz. The results indicated that for some sweet potato cultivars there were no significant differences between absolute modulus in the three locations, whereas for others there were significant differences. They also found that the crude fiber in sweet potato was significantly correlated to the absolute modulus.

Measurement of dynamic viscoelastic properties of corn horny endosperm was investigated by Wen and Mohsenin (1970). They used a slightly different instrumentation, with an eccentric cam producing the vibration. The frequency of all the tests was 1 Hz, and the tests were conducted on corn with a moisture content of 14.50–24.75% dry basis. The absolute modulus at these moisture contents varied from 630 to 188 MPa, respectively. They found an appreciable drop in absolute modulus at 16–17% moisture levels. Chattopadhyay et al. (1978) determined the dynamic mechanical properties of brown rice at frequencies from 100 to 1000 Hz at four different moisture levels from 12% to –29% dry basis. The modulus was dependent on the moisture content, and at any particular moisture content the behavior of rice grain was equivalent to a Maxwell model. The modulus values were significantly higher (840–5100 MPa) for rice than for corn (200–600 MPa), but this could be attributed to the higher frequencies used for rice.

Dynamic uniaxial compression tests were conducted on low-methoxyl pectin gels by Gross (1979) at seven frequencies ranging from 100 to 280 Hz. The magnitudes of absolute modulus were highly dependent on test frequency, and a resonance dispersion effect was revealed at several frequencies. The frequency that exhibited the greatest resonance dispersion was also the most sensitive in distinguishing between the gels on the basis of absolute modulus. The absolute modulus was also significantly correlated to the chemical composition of the pectin and sensory characteristics of the gels. Clark and Rao (1978) used the direct stress–strain type of tests on fresh peaches to determine texture. They used frequencies from 70 to 300 Hz to determine absolute modulus and correlated it to hardness and elasticity of fresh peaches at a significance level of 1% (correlation coefficients were 0.96 or higher). They concluded that the green and ripe peaches had a much higher modulus than the overripe peaches for all varieties, and hence the modulus could be used to determine harvest dates. Marinos (1983) used direct stress–strain dynamic tests on canned and fresh pears to evaluate texture and chemical composition. He found significant correlations between absolute modulus and hardness and between elasticity and mouthfeel as evaluated by a sensory panel. Also, significant correlations were detected between absolute modulus, titratable acidity, and total solids.

Webb et al. (1975) used the direct stress–strain type of dynamic tests to evaluate the texture of frankfurters. A direct sinusoidal force was applied at six frequencies (40–240 Hz) to four brands of commercial frankfurter samples, and the magnitude of deformation was sensed to determine the complex dynamic modulus and energy loss. Instrumental values were individually correlated with 23 characteristics identified by a trained texture profile panel as the texture profile for frankfurters. The results indicated that the dynamic testing method provided a highly significant means of determining differences for eight of the frankfurter texture characteristics—initial hardness, cohesiveness, elasticity, hardness of the cross section, ease of swallowing, moisture release, coarseness, and skin texture. The results indicate that the dynamic test variables describe a substantial amount of the total texture profile, probably as complete as any single instrumental method reported to date.

Extensive research has been conducted in the area of dynamic mechanical properties of flour doughs. Hibberd and Wallace (1966) determined the complex shear modulus of wheat flour dough over a frequency range of 0.032–32 Hz. At lower strains, they found a linear stress–strain behavior. Departures from linear response at higher strain levels were explained by protein–starch interaction. In a related study, Hibberd (1970) determined the influence of starch granules on the dynamic behavior of wheat gluten. A similar dependence of complex modulus on strain levels was also shown by Smith et al. (1970) for wheat flour doughs. They found that the increased protein content and protein/water ratio increased the shear modulus. Similar findings were reported by Navickis et al. (1982) on wheat flour doughs.

6.4 FAILURE AND GLASS TRANSITION IN SOLID FOODS

6.4.1 Failure in Solid Foods

Failure or fracture mechanisms in solid foods play a very important role with consumers, as most solid foods are broken down in the mouth and subjected to several repeated breakdowns in structure before mastication. Fracture properties of solid foods affect the sensory perception of foods, their transportation during manufacture and distribution, and their ability to maintain integrity and quality during storage. The mode of failure in foods can be similar or very dissimilar to that of common engineering materials. Most engineering materials fail when the stress imposed on the material exceeds the failure strength of the material. Additionally, engineering materials that are isotropic and homogenous fail in the shear plane even when the subjected load is tension or compression, as the shear strength of the material is usually less than the tensile or compressive strength. Furthermore, the region of maximum shear stress will be in a plane 45° to the application of a load, and hence the failure occurs in that plane.

The general modes of failure for a solid body can be classified into four main groups:

- Excessive deformation of a deformable body
- Fracture of a brittle body, where the deformation may be quite small
- Fatigue, when stresses less than the yield strength of the material are applied repeatedly
- Chemical failure, where the basic solid body has reacted with water or other solvents to result in a weaker structure and hence failure

A common example of the last mode of failure is observed in the form of rust development in ferrous materials that can significantly reduce the strength of the material. Malcolm Bourne (2002) has described fracture as Types 1, 2, and 3. Type 1 is described as a simple fracture, where the imposed stress has exceeded the strength of the material and the body has separated into two or more pieces, or occasionally the fracture may be partial and the body may not separate into pieces. Type 2 is described as brittle fracture, where there is little or no deformation before fracture, and the original undeformed body may result in many pieces. Type 3 is described as ductile fracture, where there is substantial plastic deformation and low energy absorption prior to fracture.

In an effort to describe the fracture mechanisms in foods, Lilliford (2001) has provided an excellent insight into the fracture behavior of brittle foods (e.g., crackers), vegetables, meat, and cheese. He has postulated that in fracture in dense solid food materials, the failure is not directly related to the intrinsic properties (such as modulus of elasticity) but to the existence of microscopic voids or cracks that are inherent in the material. Therefore, the fracture strength of such materials must include bulk density of the foam structure in addition to the intrinsic properties such as modulus of elasticity. In fact, the relationship of modulus and bulk density is in the form of a quadratic equation as illustrated by Attenburrow et al. (1989):

$$E = K_1^2 \rho$$

where E is the modulus, K_1 is a constant varying with water activity, and ρ is the bulk density of the foamy solid material. It was further demonstrated by Attenburrow et al. (1989) that the critical fracture stress decreases tenfold with increasing incubation at higher equilibrium relative humidities of 75% versus 57% at the same bulk density. In addition, these authors showed that the stress decreases tenfold as the bulk density decreases from 0.8 to 0.3 g/cc for foamy foods such as bread and cakes.

The modes of fracture and failure in vegetables and fruits are similar to those of foams filled with a dilute solution of water as the continuous phase. However, in fruits and vegetables the fracture stress decreases with ripening, as the turgor is lower as a result of ripening. In raw fruits and vegetables, the mode of fracture is through cell walls while in cooked vegetables the mode of fracture is between cells, as there is no turgor and the adhesive forces between cell walls are significantly reduced in the cooking process. Diehl and Hamann (1979), in a comprehensive review of structural failure in solid foods, have provided detailed discussions of potato and apple as case studies in fracture. They have cited several works and concluded that failure in shear is the general mode of fracture even in uniaxial compression. In addition, the values for true shear stress in torsion always appear to be more than for shear stress in compression (e.g., for red-skinned potato, 530 versus 454 kPa). They also reported that for apples, the gas volumes are much higher than for potatoes, and hence the true shear stresses at fracture are about half of that for potatoes.

It has been shown that for meat products the failure of single fibers appears to be more like that of brittle bodies, wherein the strain at failure is less than 2%. However, bundles of fibers behave very differently, requiring almost 25% strain to initiate failure and approximately 75% strain for total fracture. In analyzing the failure modes in cooked meats, Lilliford (2001) concluded that failure occurs by delamination within and between connective tissue, almost resembling a flaking process. The failure modes in cheeses can vary depending on the variety and age of the cheese. Most cheese varieties at initial stages of storage behave like a rubbery material with a constant level of stress with increasing strain prior to failure. Furthermore, the fracture occurs around curd particles (Lilliford, 2001). Matured cheese, however, fractures through the particle as the bonds between particles become stronger because of the proteolysis of the casein.

6.4.2 Glass Transition of Solid Foods

Glass transition is a significant change in modulus that an amorphous (noncrystalline) material undergoes over a certain temperature range. Crystalline materials do not undergo glass transition. The components of solid foods such as proteins and polysaccharides are polymers that undergo glass transition. To understand the behavior of these polymers, one must be familiar with their glass transition temperature (Tg).

The phenomenon of glass transition can be explained based on free volume theory. Let us picture a solid food system composed of relatively high-molecular-weight polymers that are entangled like snakes. At room temperature, each polymer chain undergoes several motions that keep other polymer chains from invading its space. If the temperature of the food system is lowered, the motions of the chains are slowed, and other chains invade the polymer's space. This results in a significant reduction of the volume of the system. Also, as the polymer segments of the various polymers interact with each other, the viscosity of the system increases. The system is now glassy (hard). When the temperature increases, the opposite occurs. The volume of the food system increases, the viscosity decreases, and the food goes from glassy to leathery, to rubbery, and finally to a flowable material.

6.4.2.1 Factors that Affect Glass Transition

The Tg of a polymer is affected by the presence of solvents or plasticizers. Plasticizers are small molecules that get in between the polymer chains and increase the space between them (increasing the free volume). The plasticizer causes the polymer chains to slide past each other more easily at lower temperatures. In this way, the Tg of a polymer is lowered. Water is the most common plasticizer in foods. Sugars are also plasticizers but they are not as efficient as water.

Other factors that affect Tg are the number and rigidity of polymer side chains, pressure, polymer molecular weight, and cross-linking. More rigid side chains increase the Tg. High pressure decreases the Tg (decreases free volume). Cross-linking increases molecular weight, thus increasing Tg.

6.4.2.2 Measurement of Glass Transition

Differential scanning calorimetry (DSC) is a popular technique used to study the thermal transitions of polymers such as Tg. The polymer under study is heated in a device (Figure 6.8) that has two pans—the sample pan and the reference pan that is left empty. Each pan sits on top of a heater, which heats the pans at a specific rate, usually 10°C per minute. Since the sample pan has a polymer in it and the reference pan is empty, it takes more heat to keep the temperature of the sample pan increasing at the same rate as that of the reference pan.

RHEOLOGICAL PROPERTIES OF SOLID FOODS

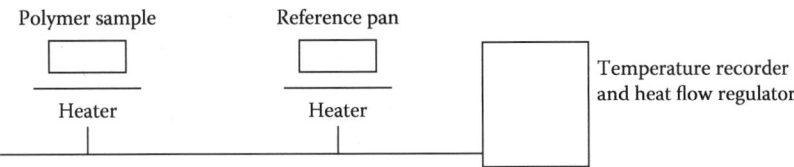

Figure 6.8 Schematic diagram of a DSC.

The heater underneath the sample pan puts out more heat than the heater under the reference pan does. How much more heat it has to put out is what is measured in a DSC experiment. As the temperature increases, a plot of temperature versus heat flow is made. When the pans are being heated, the DSC will plot the difference in heat output of the two heaters against temperature. In other words, the DSC is plotting the heat absorbed by the polymer against temperature.

The heat flow is shown in heat units, q supplied per time unit, t. The heating rate is temperature increase T per time unit, t.

- q/t = heat flow
- $\Delta T/t$ = heating rate

The heat supplied divided by the temperature increase is obtained by dividing the heat flow by the heating rate. The amount of heat it takes to get a certain temperature increase is called heat capacity, or Cp. The Cp can be obtained from a DSC plot.

$$(q/t)/(\Delta T/t) = Cp$$

Much more than just a polymer's Cp can be learned with DSC. When a polymer is heated a little more, after a certain temperature the DSC plot shifts upward suddenly. This means that there is more heat flow due to an increase in the heat capacity of the polymer. This happens because the polymer has just gone through the glass transition. This change does not occur suddenly, but takes place over a temperature range. This makes choosing a discrete Tg troublesome, but usually, the middle of the incline is taken as the Tg (Figure 6.9). Above the Tg the polymer's mobility increases.

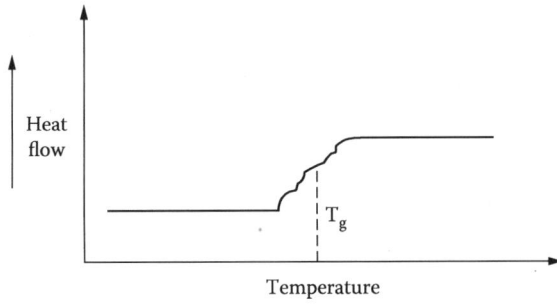

Figure 6.9 A typical DSC plot.

6.4.2.3 Importance of Glass Transition in Solid Foods

Understanding the glass transition of solid food systems allows us to design better processes, innovative products, and better packages to increase shelf life (Table 6.3).

For example, in breakfast cereal and extruded snack production, the understanding of Tg is key. These products are made with corn or other cereals in the glassy state (i.e., cornmeal), which are fed into an extruder, where the moisture and temperature are raised to cause the material to go through glass transition, becoming a flowable mass. As the flowable material comes out of the extruder, a significant amount of moisture is lost and the material cools rapidly, becoming glassy.

Many products, such as tortilla chips and popcorn, deteriorate in quality if they undergo glass transition due to moisture gain. Designing a package that can keep this from occurring will increase shelf life and consumer acceptability. The "state diagram" is very useful in understanding the properties of solid foods (Figure 6.10).

Table 6.3 Glass Transition Temperatures of Selected Foods

	Glass Transition Temperature	References
Fresh tortillas	51 to 89°C	Quintero-Fuentes (1999)
Bread	−12°C	LeMeste et al. (1992)
Ice cream	−34.5°C	Livney et al. (1995)
Honey	−42 to −51°C	Kantor et al. (1999)

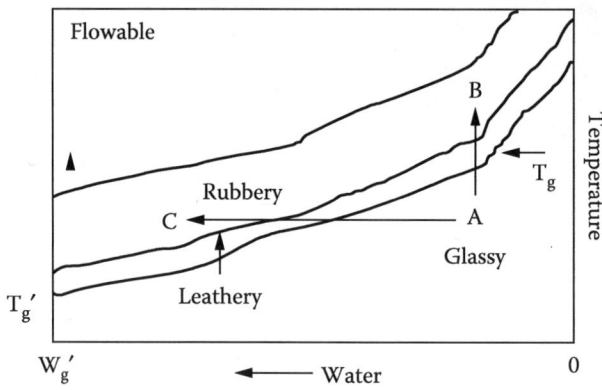

Figure 6.10 Idealized-state diagram as a function of moisture and temperature. Tg is the glass transition at zero moisture, and Tg′ is the glass transition at the moisture content at which free water starts to appear as a separate phase (Wg′). A, assumed starting point of the polymer; arrow to B, effect of increased temperature; arrow to C, effect of increased moisture. (From Hoseney, R.C. 1994. In *Principles of Cereal Science and Technology*. Hoseney, R.C., Ed. AACC, Inc: St. Paul, MN.)

6.5 EMPIRICAL AND IMITATIVE TESTS

6.5.1 Introduction

Texture, appearance, and flavor are the three major components involved in food acceptability (Bourne, 1978). Therefore, it can be seen that an accurate method for determining food texture is of vital importance to the food scientist. A main goal of many texture studies is to devise one or more mechanical tests with the capacity to replace human sensory evaluation as a tool to evaluate food texture (Peleg, 1983). Many attempts have been made to define precisely the term *texture* as well as to explain how it is actually perceived (Friedman et al., 1963; Szczesniak, 1975; Peleg, 1983). Szczesniak (1963) stated that "texture is composed of a number of different sensations/parameters," and that the "sensory evaluation of texture is a dynamic property." Peleg (1983) perceived texture as a "response to different kinds of physical and physiochemical stimuli." Many early instruments were used to aid in the texture evaluation of foods.

A review of the literature by Friedman et al. (1963) enumerated numerous instruments that have been used to study textural properties. These include a shear press (Decker et al., 1957; Kramer, 1961), gelometers (Schachat and Nacci, 1960; Billheimer and Parette, 1956; Dorner, 1955; Fellers and Griffiths, 1928; Pintauro and Lang, 1959), various types of viscometers (Cunningham et al., 1953; Becker and Clemens, 1956; Bauer et al., 1959), compressimeters (Combs, 1944; Crossland and Favor, 1950; Kattan, 1957), consistometers (Bloom, 1938; Birdsall, 1946; Clardy et al., 1952; Eolkin, 1957), and tenderometers (Cain, 1951; Clarke, 1951; Doesburg, 1954; Kramer, 1948; Proctor et al., 1955, 1956a,b; Lovegren et al., 1958; Davison et al., 1959), to mention a few.

The measurement of engineering parameters to describe brittle and crunchy foods has always been a challenge. The textural attributes such as hardness, crispness, and fracturability of fried corn chips, potato chips, pretzels, and so on, have been well described from a sensory point of view. However, mechanical stress–strain relationships for such foods have been restricted to empirical tests. In a novel and creative approach, Peleg and Normand (1992) attempted to develop a mathematical treatment of the stress–strain relationship of crunchy foods. They used symmetrized dot pattern displays to visualize textural differences between crunchy foods. The jaggedness of the stress–strain relationship of dry and moist pretzels was characterized and quantified by their apparent fractal dimension. They recommend using this method on a variety of crunchy foods and establishing a reference for visualizing certain aspects of texture.

In a separate study, Peleg (1993) examined the applicability of an apparent fractal dimension or parameters derived from the Fourier transform of the stress–strain plots to determine the degree of jaggedness of brittle foods. He used attractor plots to transform the stress–strain relationships of three different crunchy foods: dry breadsticks, pretzels, and zwiebacks. He showed that the morphology of the attractors of the three foods did not follow a strictly random pattern. However, he concluded that with the resolution available it was not possible to identify the characteristic morphology and relate it to structural features and failure mechanisms of crunchy foods. Despite drawbacks encountered using this approach, it is perhaps one of the keys to understanding textural characteristics of brittle and crunchy foods. This is especially true as the use and power of computational techniques continue to grow.

6.5.2 Texture Profile Analysis

One early texturometer, called the denture tenderometer, was built and used by the Food Technology Laboratory of the Massachusetts Institute of Technology (MIT) (Proctor et al., 1956a,b). This instrument employed strain gauges connected to the jaws of a dental articulator. One important breakthrough in food texture evaluation came with the development of the General Foods texturometer (Friedman et al., 1963), which was a modification of the original MIT denture tenderometer (Szczesniak, 1975). The General Foods texturometer was designed to simulate the masticating action of the human mouth. Szczesniak and Hall (1975) described the texturometer as a unit composed of a plate supported by a flexible arm that is attached to a strain gauge and a plunger. The plunger acts upon the food sample. The strain gauges detect the force generated, which is recorded on a strip recorder.

The curve generated by the General Foods texturometer is a plot of force as a function of time; it became known as a *texture profile*. This curve, in conjunction with specific terms defined by Szczesniak (1963), is known as the *texture profile analysis technique* and is still used for food samples.

A typical texture profile curve can be seen in Figure 6.11. Bourne (1978) has listed (and modified) the five measured and two calculated parameters, originally named by Szczesniak (1975) and Friedman et al. (1963):

1. *Fracturability* (once called *brittleness*) is defined as the force at the first significant break in the first positive bite area (PA1).
2. *Hardness* is defined as the peak force (PP1) during the first compression cycle.
3. *Cohesiveness* is defined as the ratio of the positive force area during the second compression cycle to the positive force area during the first compression cycle, or PA2/PA1.

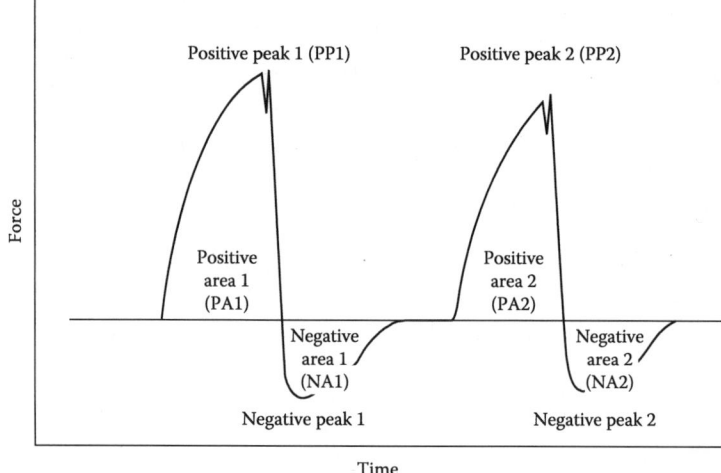

Figure 6.11 A typical texture profile curve with a two-bite compression cycle.

4. *Adhesiveness* is defined as the negative force area for the first bite (NA1), representing the work required to pull the plunger away from the food sample.
5. *Springiness* (originally called *elasticity*) is defined as the height to which the food recovers during the time that elapses between the end of the first bite and the start of the second bite.
6. *Gumminess* is defined as the product of hardness and cohesiveness.
7. *Chewiness* is defined as the product of gumminess and springiness.

Texturometer evaluation of mechanical parameters of a food correlated well with scores obtained by the use of a trained texture profile panel (Szczesniak et al., 1963). This correlation indicates that the General Foods texturometer has the capacity to measure certain characteristics with a type and intensity similar to those perceived by the human mouth. The original General Foods texturometer method has undergone many modifications and changes since its inception, as various researchers adapt their own interpretations of the definitions and force curves.

Sherman (1969), in a critical review of the texture profile definitions, concluded that hardness and cohesiveness are not independent of elasticity and therefore should not be treated separately. He also stated that the definitions of elasticity, gumminess, and chewiness have very little practical significance and that only adhesiveness was correctly defined among the original descriptor definitions. Drake (1966) disliked the use of the descriptive words and claimed that some of them were misleading (e.g., *elasticity*, which has a different meaning rheologically from the way it was originally used in the texture profile method). Some of these discrepancies led to the modified descriptors listed by Bourne (1978), who changed Szczesniak's original term *elasticity* to *springiness* and the term *brittleness* to *fracturability*.

Bourne (1978) was the first to apply the texture profile analysis technique to the Instron universal testing machine (Breene, 1975; Bourne, 1978). He concluded that the Instron was a better tool for determining texture profile analysis parameters. Bourne followed the individual descriptors of Friedman et al. (1963) quite closely, except for his definition of cohesiveness. He measured the positive areas of compression and excluded the negative areas of the decompression for each cycle. Drake (1966), Olkku and Rha (1975), and Peleg (1976b) changed the definition of cohesiveness further by actually subtracting the negative decompression areas (NA1 and NA2 in Figure 6.12) from the corresponding compression cycles (PA1 and PA2, respectively).

Bourne (1978) described the differences between the original General Foods texturometer curves and the Instron-generated curves and pointed out that the latter show sharp peaks at the end of each compression, whereas the General Foods texturometer yielded rounded peaks. He also stated that each method, if successfully used, has its own advantages and disadvantages. The General Foods texturometer more closely imitates the actual movement of the human jaw, however. Bourne (1978) noted also that the Instron-generated force curves can be calibrated into known units such as pounds, kilograms, or newtons. He stated finally that no conclusive evidence has been found that favors one method over the other and that they both are useful in evaluating the textural properties of foods.

ENGINEERING PROPERTIES OF FOODS

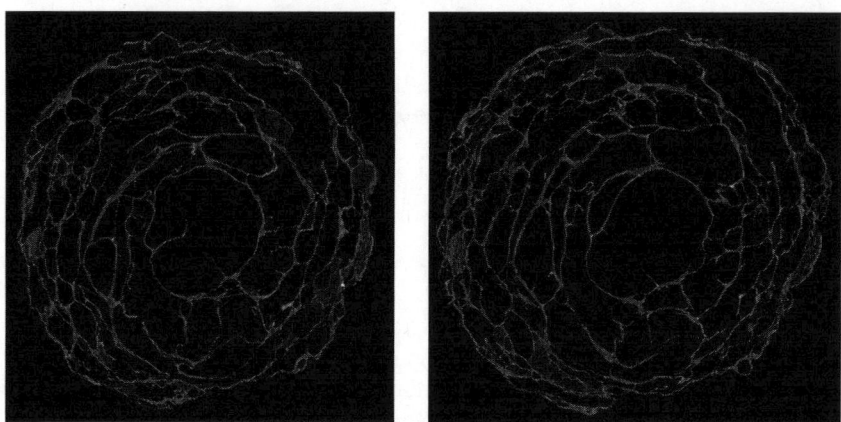

Figure 6.12 Fried extrudate CT x-ray scanning analysis.

6.5.3 Texture (Shear) Press

This is a versatile and well-known instrument that has undergone several name changes. Among the popular ones are Kramer shear press, FTC shear press, shear press, and Lee Kramer shear press. This particular test apparatus is manufactured by several companies and is also adaptable to the Instron universal testing machine. The standard test cell of the press is a metal box with internal dimensions of 6.6 × 7.3 × 6.4 cm. A set of 10 blades, each 3 mm thick and 42 mm wide and spaced 3 mm apart, are attached to the press driving head. The food is placed in the box, and the blades are moved into the box by means of slits in the lid of the box. As the blades traverse through the box, they compress the food and exit through the bottom slits in the box. During the duration of this operation, the force is continuously monitored.

The shear press was initially developed for quality evaluation of fresh vegetables (Kramer, 1961). Bailey et al. (1962) used the shear press to measure the tenderness of beefsteak. Szczesniak et al. (1970) tested 24 different foods with the shear press at various weight (quantity) levels. They inferred that the food is extruded in addition to being sheared and compressed and that the weight of the food has a power law effect on the maximum force recorded. Rao et al. (1974a, b) used the shear press to measure the maximum force and shear energy (the area under the force deformation curve) of baked sweet potatoes. They found that the moistness of the sweet potatoes, as evaluated by a sensory panel, was well correlated to both the shear press measurements (force and energy) and was significant at the 99% confidence level. There are numerous applications using the shear press that it is impossible to cite all of them here. A detailed list of products used with the shear press is given by Bourne (1982). The shear press is used almost as a standard instrument for measuring maturity and/or textural properties of fruits and vegetables today.

6.5.4 Warner–Bratzler Shear

The Warner–Bratzler shear is probably the most widely used instrument in the United States for measuring toughness of meat. The test cell consists of a thin stainless-steel blade with a hole (an equilateral triangular hole approximately 25 mm long on each side). Two metal anvils move down on each side of the blade and shear the sample material. The test cell can be mounted on any universal testing machine capable of recording force and motion of the crosshead.

In an extensive survey of the texture of meat, Szczesniak and Torgeson (1965) reported that the most popular technique for evaluating meat tenderness is the Warner–Bratzler shear. Furthermore, they concluded that of the 50 research studies, 41 established highly significant correlations between meat tenderness and Warner–Bratzler shear. The advantages of this shear device are its simplicity and low cost. In spite of some drawbacks, such as poor correlations with raw meat evaluation of shear force and cooked meat tenderness, this instrument will no doubt be used for measuring texture of meat and other nonisotropic foods.

6.5.5 FMC Pea Tenderometer

This instrument measures the quality and maturity of fresh green peas. It was developed by the Food Machinery Corporation and has been widely used by the pea processing industry for quality control purposes. The instrument has a motor-driven grid of 19 stainless-steel blades rotated through a second reaction grid of 18 blades. The peas are placed between these grids and extruded through the slits. Because of some problems with its operation and maintenance, as well as contradictory results in comparison with other tenderometers, the manufacture of this instrument has been discontinued.

6.5.6 Penetrometer

The penetrometer was developed for measuring the firmness or yield point of semisolid foods such as thick puddings and gels. The penetrometer consists of a cone, needle, or sphere attached to a short rod that can be mounted on the crosshead of a universal testing machine. As the penetrometer is allowed to come down either at a constant speed or by the force of gravity, it contacts the food specimen and registers a force. The force can be measured either by a suitable transducer or by a spring balance. If the penetrometer is allowed to fall freely under the influence of gravity, the depth of penetration is also recorded.

Tanaka and DeMan (1971) used the constant-speed cone penetrometer to evaluate margarine and butter. They found good correlation between hardness of butter and margarine with penetrometer force. DeMan (1969) studied the texture of processed cheese, butter, margarine, and peanut butter via penetration tests with two types of plungers. This instrument has been extensively used in evaluating the quality of food gels and jellies (Gross, 1979).

6.5.7 Bend testing

Bend testing (also flex or flexural testing) is commonly performed to measure the flexural strength and modulus of all types of materials and products. This test is performed on a universal testing machine (tensile testing machine or tensile tester) with a 3- or 4-point bend fixture. Most common for product testing is the 3-point test. Blahovec et al. (1999)

used deformation curves from a 3-point bending test on potato crisps to determine Young's modulus (EM), maximum relative deflection (MRD), maximum tensile stress (MTS), and surface energy. The MRD and MTS correspond to rupture point values of deflection and stress, respectively. They were able to use some of the parameters from the 3-point bending test to differentiate the crispness of the potato crisps.

The key analyzes when performing bending test are

1. Flexural modulus: This measures the slope of a stress/strain curve and is an indication of a material's stiffness.
2. Flexural strength: This measures the maximum force that a material withstand before it breaks or yields. Yield is where you have pushed a material past its recoverable deformation and it will no longer go back to the shape it once was.
3. Yield point: The yield point is the point where the material essentially "gives up" or the point where if you were to continue to bend the product, the force will not continue to increase and will then start to decrease or break.

The 3-point bending flexural test provides values for the modulus of elasticity in bending E_f, flexural stress σ_f, flexural strain ε_f, and the flexural stress–strain response of the material. The main advantage of a 3-point flexural test is the ease of the specimen preparation and testing. However, this method has also some disadvantages: the results of the testing method are sensitive to specimen and loading geometry and strain rate (Bower, 2009).

6.5.7.1 Testing method

Calculation of the flexural stress σ_f

$$\sigma_f = \frac{3PL}{2bd^2} \text{ for a rectangular cross section}$$

$$\sigma_f = \frac{PL}{\pi R^3} \text{ for a circular cross section}$$

Calculation of the flexural strain ε_f

$$\varepsilon_f = \frac{6Dd}{L^2}$$

Calculation of flexural modulus E_f

$$E_f = \frac{L^3 m}{4bd^3}$$

In these formulas, the following parameters are used:

- σ_f = stress in outer fibers at midpoint (MPa)
- ε_f = strain in the outer surface (mm/mm)
- E_f = flexural Modulus of elasticity (MPa)
- P = load at a given point on the load deflection curve (N)
- L = support span (mm)
- b = width of test beam (mm)
- d = depth of tested beam (mm)

- D = maximum deflection of the center of the beam (mm)
- m = the gradient (i.e., slope) of the initial straight-line portion of the load deflection curve (P/D) (N/mm)

6.5.8 Other Empirical Methods

There are numerous other instruments for empirical and imitative measurement of solid foods. Some of those commonly used are discussed briefly. The ridgelimeter is a very common, simple instrument for measuring the sag of gels. It was originally developed for measuring the grade of fruit pectins. The Institute of Food Technologists has adopted this instrument as a standard for measuring pectin gels. The Magness–Taylor pressure tester is another simple instrument that can be operated by hand, even in the field, for determining the maturity of fruits and vegetables. This instrument measures the force required for a plunger of some specified shape to penetrate the surface of the fruit or vegetable to a fixed distance. Because of the simplicity and low cost, this instrument has very wide acceptance among producers of fresh produce. The Cherry–Burrell meter has been used by the dairy industry for measuring the firmness of cottage cheese. A consistometer for measuring the spreadability of butter and margarine is also used by some industries. The Baker compressimeter is probably the most widely used instrument for measuring the firmness of bread. The force–deformation curve obtained is used to characterize the bread. A squeeze tester has also been evaluated for measuring the bread texture (Finney, 1969).

The development of empirical instruments to measure physical properties will continue to grow as the food industry introduces new foods. A promising area is extruded snack foods, which still lack a good objective test. However, most of the results obtained with the use of empirical and imitative methods are very difficult to compare and express in terms of known engineering dimensions.

6.5.9 Structure and Porosity as related to Snack Texture

6.5.9.1 Structure

Computational tomography (CT) scanning can be a useful tool when determining the role of structure as related to texture. Small cross-sectional slices of three different extruded samples were analyzed using a CT scanner (Bortone et al., 2012). Figure 6.1 is a cross-sectional CT scan of a fried product. Figure 6.1 illustrates expanded pores in the outer layer. These pores comprise many pockets of air. The outer portions of the pores comprise oil, cornmeal, and water. The shaded gray portions beneath the surface illustrate the presence of oil. This illustrates the depth which the oil has penetrated the product. The light gray masses illustrate pockets of oil.

Figure 6.2 is a cross-sectional CT scan of the hydrated collet baked in a convection oven in one embodiment. The hydrating step comprised adding water in an amount of about 15% by weight of the collet and oil in an amount of about 11% by weight of the collet. Figure 6.2 reveals that the baked product had compressed pores in the outer layers compared to the pores of the fried product in Figure 6.12.

ENGINEERING PROPERTIES OF FOODS

Figure 6.3 is a cross-sectional CT scan of a hydrated collet baked in an air impingement oven. This figure illustrates compressed and shrunken pores in the outer layers. This product had more compressed and shrunken pores than the fried product and the convection oven baked product. Also illustrated are some dense outer regions of cornmeal and oil. This results in a collapsing of the pores resulting in a denser outer layer. This denser outer layer attributes to the increased crunch associated with a hydrated then baked product.

Further, comparing Figures 6.13 and 6.14 it can be seen that the product baked in the impingement oven has more compressed and shrunken pores compared to the pores of the product baked in the convection oven. It is believed that the intensity of the thermal treatment of the impingement oven causes the pores to collapse at a quicker rate than the

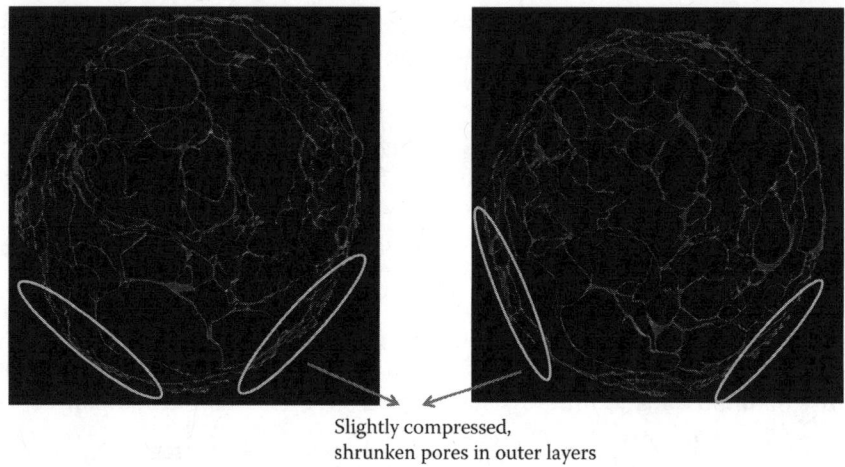

Slightly compressed, shrunken pores in outer layers

Figure 6.13 Baked extrudate CT x-ray scanning analysis.

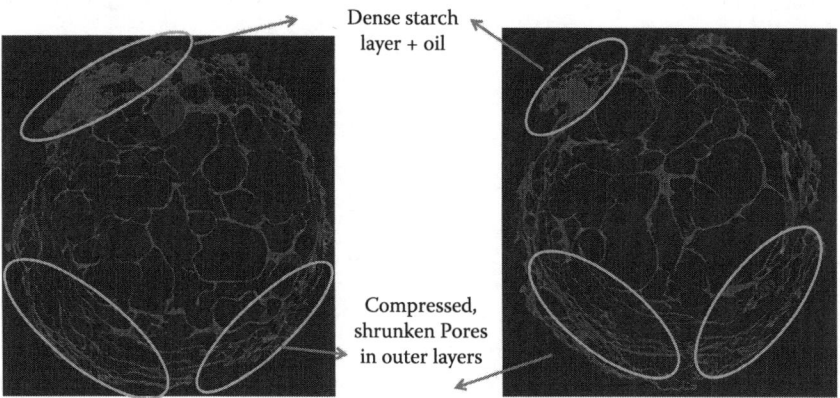

Dense starch layer + oil

Compressed, shrunken Pores in outer layers

Figure 6.14 Air impingement-baked extrudate CT x-ray scanning analysis.

Table 6.4 Pore Size Distribution of Baked Extrudates

Product	Average pore size, μm	Pore size range, μm	% Porosity
Fried	875	21–1729	71
Convection oven	709	20–1397	74
Impingement oven	648	20–1276	66

convection oven. Further, it is believed that the air bearing down upon the product in an impingement oven cause the pores to collapse. When the pores collapse, they have a shingle effect in that they collapse on adjacent pores causing additional pores to collapse. Accordingly, these figures illustrate the effect that the impingement oven has on the final texture.

6.5.9.2 Porosity

The samples were also analyzed to determine the pore size distribution. An x-ray CT scanning was utilized to analyze the pore size. As above, all three products comprised cornmeal and water, and were extruded at identical conditions. The first sample was fried. The second sample was baked in a convection oven. The third sample was baked in an air impingement oven. As seen, both baked products had smaller average pore sizes compared to the fried product (Table 6.4). Further, both baked products had a narrower range of pore sizes compared to the fried product. Additionally, the impingement oven had smaller average pore sizes and a narrower range of pore sizes than the convention oven. As stated above, it is believed that this is due to the intensity of the thermal treatment of the impingement oven as well as the force of the air in the impingement oven. A more collapsed pore results in increased crunch.

Table 6.4 also illustrates the percent porosity. This is calculated by dividing the total pore volume by the total piece volume. As can be seen, the impingement oven results in decreased porosity compared to the convection and fried products.

6.6 CONCLUSIONS

The rheological properties of solid foods play a vital role in the manufacture, quality control, and product development of foods. The knowledge of fundamental properties is important as well as that of empirical and imitative properties. In a discussion of the relative importance of each, Bourne (1975) correctly stated: "Food researchers should realize that fundamental rheological tests describe only a portion of the physical property sensed in the mouth during mastication." It is true that, when masticated, a food undergoes several tests at the same time, including, but not limited to, grinding, shearing, extrusion, compression, and tension. However, one should understand that fundamental properties cannot always be used as a measure of texture perception in the mouth but may be necessary for other purposes. For example, consider the process of drying grains. The grains are normally subjected to heat by means of forced circulation of hot air. In this process, the grains are subjected to thermal gradients and moisture gradients, resulting in thermal and hydro stresses. If the stresses in the material exceed the failure stress of the material, then the material will crack. In order to optimize this operation to ensure minimal mechanical damage, it is imperative to understand and evaluate the fundamental rheological properties of the grains.

The importance of both the fundamental and empirical/imitative rheological properties is well stated by Bourne (2002) and Voisey and Kloek (1975). The empirical can usually be carried out more quickly and are perhaps a better measure of the texture as perceived by a taste panel. They are used in a wide array of quality control measures in the food industry. The fundamental tests, however, are time-consuming and require detailed technical calculations. However, sometimes it is necessary to evaluate the fundamental properties, as they can provide a much better insight into the chemistry of the food and sometimes they are the only meaningful measurement one can make. In addition, they are the only means of predicting the behavior of the food to an unknown, untested physical force or deformation. They also provide a means of verifying the results obtained by different research investigations on a specific food.

Dynamic testing methods, as evidenced by the literature, offer a promising new technique in the evaluation of physical properties of foods. Resonance methods have been primarily used for determining the maturity or texture of fruits and vegetables. The dynamic parameters from resonance appear to be well related to texture, although the numerical values obtained by different investigators are difficult to compare. The calculation of modulus of elasticity from resonance (assuming ideal elastic behavior in the case of fruits and vegetables) may be introducing significant errors, as these foods have been shown to possess viscoelastic behavior. In many ways, the assumption that a material possesses no damping is a mathematical convenience rather than a reflection of physical evidence. Nevertheless, the concept of undamped systems not only serves a useful purpose in analysis but can also be justified in certain circumstances. For example, if the damping is small and one is interested in the free vibration of a system over a short interval of time, there may not be sufficient time for the effect of damping to become noticeable.

The direct stress–strain tests, however, assume, in fact, that the sample possesses damping. These tests have also been used successfully in predicting some of the sensory and chemical parameters of a wide variety of foods. They are extremely useful in determining the modulus of materials over an extended time scale. The use of dynamic stress–strain tests in shear has provided an insight into the mechanical behavior of wheat and other flour doughs and has illustrated the effect of protein, water, and starch interactions.

However, dynamic tests of the direct stress–strain type are difficult to analyze with intact fruits such as apples because of the contact stresses. Attempts are being made to determine complex modulus of spheroidal foods (Hamann and Diehl, 1978) via dimensional analysis. However, despite these attempts, foods are too complex, with their non-homogeneous structure, and any attempt will yield only approximate solutions. The resonance tests appear to be better suited for intact foods such as apples, whereas direct stress–strain methods are better suited for homogeneous food with convenient geometry such as frankfurters and pectin gels. The dynamic tests have so far provided excellent results in relating the mechanical properties to the texture and composition of a variety of foods. The dynamic tests will contribute even more significantly to the field of food rheology with further research, better instruction, and data analysis.

REFERENCES

Abbott, J. A., Bachman, G. S., Childers, R. F., Fitzgerald, J. V., and Matusik, F. J. 1968a. Sonic techniques for measuring texture of fruits and vegetables. *Food Technol. 22:* 101–112.

Abbott, J. A., Childers, N. F., Bachman, G. F., Fitzgerald, J. V., and Matusik, F. J. 1968b. Acoustic vibration for detecting textural quality of apples. *Am. Soc. Hort. Sci. 93:* 725–737.

Alfrey, T., Jr. 1957. *Mechanical Behavior of High Polymers.* Interscience, New York.

Amen, R. J., Ivannou, J., and Haard, N. F. 1972. Comparison of acoustic spectrometry, compression and shear force measurements in ripening pear fruit. *Can. Inst. Food Sci. Technol. 5*(2): 97–100.

Attenburrow, G. E., Goodband, R. M., Taylor, L. J., and Lilliford, P. J. 1989. Structure, mechanics and texture of a food sponge. *J. Cereal Sci. 9:* 61–70.

Bagley, E. B., and Christianson, D. D. 1987. Stress relaxation of chemically leavened dough data reduction using BKZ elastic fluid theory. *J. Rheol. 31:* 404–413.

Bailey, M. E., Hedrick, H. B., Parrish, F. C., and Naumann, H. D. 1962. Lee Kramer shear force as a tenderness measure of beef steak. *Food Technol. 16*(12): 99–101.

Baird, D. G. 1981. Dynamic viscoelastic properties of soy isolate doughs. *J. Texture Stud. 12:* 1–16.

Barr, A. J., Goodnight, J. H., Sall, J. P., and Helwig, J. T. 1976. *A User's Guide to SAS— 1976.* Sparks Press, Raleigh, NC.

Bauer, W. H., Finkelstein, A. P., Larom, C. A., and Wiberley, S. E. 1959. Modification of a cone–plate viscometer for direct recording of flow curves. *Rev. Sci. Instr. 30:* 167–169.

Becker, E., and Clemens, W. 1956. Determination of viscosity of processed cheese in the Brabender viscograph. *Dairy Sci. Abstr. 18*(1): 91.

Billheimer, G. S., and Parette, R. 1956. Apparatus for automatic determination of gel time. *Anal. Chem. 28:* 272–273.

Birdsall, E. L. 1946. Food consistency; a new device for its estimation. *Food 15:* 268.

Bloksma, A. H. 1972. Rheology of wheat flour doughs. *J. Texture Stud. 3:* 3–17.

Blahovec, J., Vacek, J., and Patocka, K. 1999, Texture of fried potato tissue as affected by pre-blanching in some salt solutions. *J. Texture Stud., 30:* 493–507.

Bloom, O. T. 1938. Apparatus for testing the consistency of food shortenings. U.S. Patent 2,119,669.

Bortone, E., Quintero-Fuentes, X., Rao, V. N. M., and Weller, W. C. 2012. Method for improving the texture of baked snack food. U.S. Patent Application US2012/0093993A1.

Bourne, M. C. 1975. Is rheology enough for food texture measurement? *J. Texture Stud. 6:* 259–262.

Bourne, M. C. 1978. Texture profile analysis. *Food Technol. 32*(7): 62–66, 72.

Bourne, M. C. 1982. *Food Texture and Viscosity: Concept and Measurement.* Academic Press, New York.

Bourne, M. C. 2002. Physics and texture. In *Food Texture and Viscosity. Concept and Measurement*, 2nd ed. Bourne, M. C., Ed. Academic Press, San Diego, CA, pp. 101–102.

Bower, A. F. 2009. *Applied Mechanics of Solids.* CRC Press. Boca Raton, FL.

Breene, W. M. 1975. Application of texture profile analysis to instrumental food texture evaluation. *J. Texture Stud. 6:* 53–82.

Cain, R. F. 1951. Influence of storage time and temperature on the tenderometer reading, drained weight, free starch, and color index of southern peas. *Canner 112*(14): 11–12.

Casiraghi, E. M., Bagley, E. B., and Christianson, D. D. 1985. Behavior of mozzarella, cheddar and processed cheese spread in lubricated and bonded uniaxial compression. *J. Texture Stud. 16:* 281–301.

Chappell, T. W., and Hamann, D. D. 1968. Poisson's ratio and Young's modulus apple flesh under compressive loading. *Trans. ASAE 11:* 608–610, 612.

Chattopadhyay, P. K., Hamann, D. L., and Hammerle, J. R. 1978. Dynamic stiffness of rice grain. *Trans. ASAE 21:* 786–789.

Chen, P., and Fridley, R. B. 1972. Analytical method for determining viscoelastic constants of agricultural materials. *Trans. ASAE 15*(6): 1103–1106.

Christensen, R. M. 1971. *Theory of Viscoelasticity: An Introduction.* Academic, New York.

Clardy, L., Pohle, W. D., and Mehlenbacker, V. C. 1952. A shortening consistometer. *JAOCS 29:* 591–593.

Clark, R. C., and Rao, V. N. M. 1978. Dynamic testing of fresh peach texture. *Trans. ASAE 21:* 777–781.

Clarke, B. W. 1951. Instruments for objective measurement of quality control factors in processed food products. *Food Technol. 5:* 414–416.

Clevenger, J. T., and Hamann, D. D. 1968. The behavior of apple skin under tensile loading. *Trans. ASAE 11*(1): 34–40.

Combs, Y. F. 1944. An instrument for determining the compressibility and resistance to shear of baked products. *Cereal Chem. 21:* 319–324.

Cooke, J. R. 1970. A theoretical analysis of the resonance of intact apples. Presented at ASAE Meeting, Minneapolis, Minn., Paper No. 70–345.

Cooke, J. R., and Rand, R. H. 1973. A mathematical study of resonance in intact fruits and vegetables using a 3-media elastic sphere model. *J. Agric. Eng. Res. 18:* 141–157.

Crossland, L. B., and Favor, H. H. 1950. The study of the effects of various techniques on the measurement and firmness of bread by the Baker compressimeter. *Cereal Chem. 27:* 15–25.

Cunningham, J. R., Hlynka, I. H., and Anderson, J. A. 1953. An improved relaxometer for viscoelastic substances applied to the study of wheat dough. *Can. J. Technol. 31:* 98–108.

Datta, A., and Morrow, C. T. 1983. Graphical and computational analysis of creep curves. *Trans. ASAE 26*(6): 1870–1874.

Davison, S., Brody, A. L., Proctor, B. E., and Felsenthal, P. 1959. A strain gauge pea tenderometer. I. Instrument description and evaluation. *Food Technol. 13:* 119–123.

Decker, R. W., Yeatman, J. N., Kramer, A., and Sidwell, A. P. 1957. Modifications of the shear-press for electrical indicating and recording. *Food Technol. 11:* 343–347.

Diehl, K. C. and Hamann, D. D. 1979. Structural failure in selected raw fruits and vegetables. *J. Texture Stud. 10:* 371–400.

DeMan, J. M. 1969. Effect of mechanical treatment on the hardness of butter and margarine. *J. Texture Stud. 1:* 109–113.

Doesburg, J. J. 1954. Instruments for the determination of harvesting maturity and quality of green peas. *Food Sci. Abstr.* 26, No. 872.

Dorner, H. 1955. The determination of the thickening capacity of foods with the gelometer. *Food Sci. Abstr.* 27, No. 381.

Drake, B. 1962. Automatic recording of vibrational properties of foodstuffs. *J. Food Sci. 27:* 182–188.

Drake, B. 1966. Advances in the determination of texture and consistency of foodstuffs. SIK Report No. 207.

Eolkin, D. 1957. The plastometer—A new development in continuous recording and controlling consistometers. *Food Technol. 11:* 253–257.

Fellers, C. R. and Griffiths, F. P. 1928. Jelly strength measurement of fruit jellies by the bloom gelometer. *J. Ind. Eng. Chem. 20:* 857–859.

Ferry, J. D. 1970. *Viscoelastic Properties of Polymers.* Wiley, New York.

Finney, E. E., Jr. 1967. Dynamic elastic properties of some fruits during growth and development. *J. Agric. Eng. Res. 12:* 249–256.

Finney, E. E., Jr. 1969. Objective measurements for texture in foods. *J. Texture Stud. 1:* 19–37.

Finney, E. E., Jr. 1970. Mechanical resonance within Red Delicious apples and its relation to fruit texture. *Trans. ASAE 13*(2): 177–180.

Finney, E. E., Jr. 1971a. Random vibration techniques for non-destructive evaluation of peach firmness. *J. Agric. Eng. Res.* 16(1): 81–87.

Finney, E. E., Jr. 1971b. Dynamic elastic properties and sensory quality of apple fruit. *J. Texture Stud.* 2: 62–74.

Finney, E. E., Jr. 1972. Vibration technique for testing fruit firmness. *J. Texture Stud.* 3: 263–283.

Finney, E. E., Jr., and Abbott, J. A. 1972. Sensory and objective measurements of peach firmness. *J. Texture Stud.* 3: 372–378.

Finney, E. E., Jr., and Norris, K. H. 1967. Sonic resonant methods for measuring properties associated with texture of Irish and sweet potatoes. *Am. Soc. Hort. Sci.* 90: 275–282.

Finney, E. E., Jr., and Norris, K. H. 1968. Instrumentation for investigating dynamic mechanical properties of fruits and vegetables. *Trans. ASAE* 11(1): 94–97.

Finney, E. E., Jr., Hall, C. W., and Mase, G. E. 1964. Theory of linear viscoelasticity applied to the potato. *J. Agric. Eng. Res.* 9(4): 307–312.

Finney, E. E., Jr., Ben-Gera, I., and Massie, D. R. 1967. An objective evaluation of changes in firmness of ripening bananas using a sonic technique. *J. Food Sci.* 32: 642–646.

Flugge, W. 1975. *Viscoelasticity.* Springer-Verlag, Berlin, West Germany.

Friedman, H. H., Whitney, J. E., and Szczesniak, A. S. 1963. The texturometer—a new instrument for objective texture measurement. *J. Food Sci.* 28: 390–396.

Gross, M. O. 1979. Chemical, sensory and rheological characterization of low-methoxyl pectin gels. PhD dissertation, University of Georgia, Athens, Ga.

Halton, P., and Scott Blair, G. W. 1937. A study of some physical properties of flour doughs in relation to bread making qualities. *Cereal Chem.* 14: 201–219.

Hamann, D. D. 1969. Dynamic mechanical properties of apple fruit flesh. *Trans. ASAE* 12(2): 170–174.

Hamann, D. D., and Carroll, D. E. 1971. Ripeness sorting of muscadine grapes by use of low frequency vibrational energy. *J. Food Sci.* 36: 1049–1051.

Hamann, D. D., and Diehl, K. C. 1978. Equation for the dynamic complex uniaxial compression modulus of spheroidal shaped foods. *Trans. ASAE* 21(5): 1009–1014.

Hammerle, J. R., and Mohsenin, M. N. 1970. Tensile relaxation modulus of corn horny endosperm as a function of time, temperature, and moisture content. *Trans. ASAE* 13(3): 372–375.

Hibberd, G. E. 1970. Dynamic viscoelastic behavior of wheat flour doughs. III. The influence of the starch granules. *Rheol. Acta* 9: 501–505.

Hibberd, G. E., and Wallace, W. J. 1966. Dynamic viscoelastic behavior of wheat flour doughs. I. Linear aspects. *Rheol. Acta* 5: 193–198.

Holmes, R. G. 1979. Vibratory sorting of process tomatoes. Presented at the annual meeting of the ASAE, Chicago, IL, Paper No. 79–6543.

Hoseney, R. C. 1994. Glass transition and its role in cereals. In *Principles of Cereal Science and Technology.* Hoseney, R. C., Ed. AACC, Inc., St. Paul, MN, p. 317.

Hundtoft, E. B., and Buelow, F. H. 1970. The development of a stress relaxation model for bulk alfalfa. ASAE Paper No. 70516, ASAE, St. Joseph, Mich.

Jasper, R. F., and Blanshard, J. M. V. 1973. A simple instrument for the measurement of the dynamic elastic properties of foodstuffs. *J. Texture Stud.* 4: 269–277.

Kantor, Z., Pitsi, G., and Theon, J. 1999. Glass transition temperature of honey as a function of water content as determined by differential scanning calorimetry. *J. Agric. Food Chem.* 47(6): 2327–2330.

Kattan, A. A. 1957. The firm-o-meter, an instrument for measuring firmness of tomatoes. *Arkansas Farm Res.* 6(1): 7.

Kramer, A. 1948. Make the most of your tenderometer in quality work, estimating yields. *Food Packer* 29: 34–38.

Kramer, A. 1961. The shear press, a basic tool for the food technologist. *Food Sci.* 5: 7–16.

LeMeste, M., Huang, V. T., Panama, J., Anderson, G., and Lentz, R. 1992. Glass transition of bread. *Cereal Foods World.* 37(3): 264–267.

Lerchenthal, C. H., and Muller, H. G. 1967. Research in dough rheology at the Israel Institute of Technology. *Cereal Sci. Today 12:* 185–192.

Lilliford, P. J. 2001. Mechanisms of fracture in foods. *J. Texture Stud.* 32(5): 397–417.

Livney, Y. D., Donhowe, D. P. and Hartel, R. W. 1995. Influence of temperature on crystallization of lactose in ice-cream. *Int. J. Food Sci. Technol.* 30: 311–320.

Lovegren, N. U., Guice, W. A., and Fengl, R. O. 1958. An instrument for measuring the hardness of fats and waxes. *JAOCS 35:* 327–331.

Marinos, G. 1983. Texture evaluation of fresh and canned pears by sensory, chemical and rheological characterization. MS thesis, University of Georgia, Athens, GA.

Marvin, S. 1952. Measurement of dynamic properties of rubber. *Ind. Eng. Chem. 44:* 696–702.

Meirovitch, L. 1967. *Analytic Methods and Vibrations.* Macmillan, New York.

Mohsenin, N. N. 1978. *Physical Properties of Plant and Animal Materials.* Gordon and Breach, New York.

Mohsenin, N. N., and Mittal, J. P. 1977. Use of rheological terms and correlation of compatible measurements in food texture research. *J. Texture Stud. 8:* 395–408.

Mohsenin, N. N., Cooper, H. E., and Tukey, L. D. 1963. Engineering approach to evaluating textural factors in fruits and vegetables. *Trans. ASAE* 6(2): 85–88, 92.

Montejano, J. G., Hamann, D. D., and Lanier, T. C. 1983. Final strengths and rheological changes during processing of thermally induced fish muscle gels. *J. Rheol. 27:* 557–579.

Montijano-Gaitan, J. G., Hamann, D. D., and Giesbrecht, F. G. 1982. Vibration sorting of simulated small fruit. *Trans. ASAE* 25(6): 1785–1791.

Morrow, C. T., and Mohsenin, N. N. 1966. Consideration of selected agricultural products as viscoelastic materials. *J. Food Sci.* 33(6): 686–698.

Morrow, C. T., and Mohsenin, N. N. 1968. Dynamic viscoelastic characterization of solid food materials. *J. Food Sci. 33:* 646–651.

Morrow, C. T., Hamann, D. D., Mohsenin, N. N., and Finney, E. E., Jr. 1971. Mechanical characterization of Red Delicious apples. Presented at the annual ASAE Meeting, Pullman, Wash., Paper No. 71-372.

Muller, H. G. 1973. *An Introduction to Food Rheology.* Crane, Russak, New York.

Murase, H., Merva, G. E., and Segerlind, L. J. 1980. Variation of Young's modulus of potato as a function of water potential. *Trans. ASAE* 23(3): 794–796, 800.

Navickis, L. L., Anderson, R. A., Bagley, E. B., and Jasberg, B. K. 1982. Viscoelastic properties of wheat flour doughs: Variation of dynamic moduli with water and protein content. *J. Texture Stud.* 13: 249–264.

Nielsen, L. D. 1962. *Mechanical Properties of Polymers.* Reinhold, New York.

Nolle, A. W. 1948. Methods for measuring dynamic mechanical properties of rubberlike materials. *J. Appl. Phys.* 19: 753–774.

Olkku, J., and Rha, C. K. 1975. Textural parameters of candy licorice. *J. Food Sci. 40:* 1050–1054.

Peleg, M. 1976a. Considerations of a general rheological model for the mechanical behavior of viscoelastic solid food materials. *J. Texture Stud. 7:* 243–255.

Peleg, M. 1976b. Texture profile analysis parameters obtained by an Instron universal testing machine. *J. Food Sci. 41:* 721–722.

Peleg, M. 1979. Characterization of the stress relaxation curves of solid foods. *J. Food Sci.* 44(1): 277–281.

Peleg, M. 1983. The semantics of rheology and texture. *Food Technol. 11:* 54–61.

Peleg, M. 1993. Do irregular stress–strain relationships of crunchy foods have regular periodicities? *J. Texture Stud. 24:* 215–227.

Peleg, M., and Normand, M. D. 1992. Symmetrized dot patterns (SDP) of irregular compressive stress–strain relationships. *J. Texture Stud. 23:* 427–438.

Peterson, C. L., and Hall, C. W. 1974. Thermorheological simple theory applied to the Russet Burbank potato. *Trans. ASAE 17:* 546–552, 556.

Pintauro, N. D., and Lang, R. E. 1959. Graphic measurement of unmolded gels. *Food Res. 24:* 310–318.

Pitt, R. E., and Chen, H. L. 1983. Time-dependent aspects of the strength and rheology of vegetative tissue. *Trans. ASAE 26:* 1275–1280.
Proctor, B. E., Davison, S., Malecki, G. J., and Welch, M. 1955. A recording strain gauge denture tenderometer for foods. I. Instrument evaluation and initial tests. *Food Technol. 9:* 471–477.
Proctor, B. E., Davison, S., and Brody, A. L. 1956a. A recording strain gauge denture tenderometer for foods. II. Studies on the mastication force and motion, and the force penetration relationship. *Food Technol. 10:* 327–331.
Proctor, B. E., Davison, S., and Brody, A. L. 1956b. A recording strain gauge denture tenderometer for foods. III. Correlation with subjective tests and the denture tenderometer. *Food Technol. 10:* 344–346.
Quintero-Fuentes, X. 1999. Characterization of corn and sorghum tortillas during storage. PhD dissertation. Texas A&M University, College Station, TX.
Rao, V. N. M., Hammerle, J. R., and Hamann, D. D. 1974a. Uniaxial modulus of sweet potato flesh using various types of loading. *Trans. ASAE 17:* 956–959.
Rao, V. N. M., Hamann, D. D., and Humphries, E. G. 1974b. Mechanical testing on a measure of kinesthetic quality of raw and baked sweet potatoes. *Trans. ASAE 17:* 1187–1190.
Rao, V. N. M., Hamann, D. D., and Purcell, A. E. 1976. Dynamic structural properties of sweet potato. *Trans. ASAE 29:* 771–774.
Reiner, M. 1960. *Deformation, Strain, and Flow.* Lewis, London.
Reiner, M. 1971. *Advanced Rheology.* Lewis, London.
Schachat, R. E., and Nacci, A. 1960. Transistorized bloom gelometer. *Food Technol. 14:* 117–118.
Shackelford, P. S., Jr., and Clark, Rex L. 1970. Evaluation of peach maturity by mechanical resonance. Presented at ASAE Meeting, Chicago, IL, Paper No. 70–552.
Shama, F., and Sherman, P. 1973. Stress relaxation during force–compression studies on foods with the Instron universal testing machine. *J. Texture Stud. 4:* 353–362.
Sherman, P. 1969. A texture profile of foodstuffs based upon well-defined rheological properties. *J. Food Sci. 34:* 458–462.
Skinner, G. E. 1983. Rheological modeling using linear viscoelastic assumptions in static creep and relaxation. Master's thesis, Food Science Dept., Univ. of Georgia, Athens, Ga.
Smith, J. R., Smith, T. L., and Tschoegl, N. W. 1970. Rheological properties of wheat flour doughs. III. Dynamic shear modulus and its dependence on amplitude, frequency and dough composition. *Rheol. Acta 9:* 239–252.
Spinner, S., and Tefft, W. E. 1961. A method for determining mechanical resonance frequencies for calculating elastic moduli from these frequencies. *Proc. Am. Soc. Test. Mater. 61:* 1221.
Szczesniak, A. S. 1963. Classification of textural characteristics. *J. Food Sci. 28:* 385–389.
Szczesniak, A. S. 1975. General Foods texture profile revisited—Ten years perspective. *J. Texture Stud. 6:* 5–17.
Szczesniak, A. S., and Hall, B. J. 1975. Application of the general foods texturometer to specific food products. *J. Texture Stud. 6:* 117–138.
Szczesniak, A. S., and Torgeson, K. W. 1965. Methods of meat texture measurement viewed from the background of factors affecting tenderness. *Adv. Food Res. 14:* 33–165.
Szczesniak, A. S., Brandt, M. A., and Friedman, H. H. 1963. Development of standard rating scales for mechanical parameters of texture and correlation between the objective and sensory methods for texture evaluation. *J. Food Sci. 28:* 397–403.
Szczesniak, A. S., Humbaugh, P. R., and Block, H. W. 1970. Behavior of different foods in the standard shear compression cell on the shear press and effect of sample weight on peak area and maximum force. *J. Texture Stud. 1:* 356–378.
Tanaka, M., and DeMan, J. M. 1971. Measurement of textural properties of foods with a constant speed cone penetrometer. *J. Texture Stud. 2:* 306–315.
Virgin, Hemmin I. 1955. A new method for the determination of the turgor of plant tissues. *Physiol. Plantarum 8:* 954–962.

Voisey, P. W. 1975. Selecting deformation rates in texture tests. *J. Texture Stud.* 6: 253–257.
Voisey, P. W., and Kloek, M. 1975. Control of deformation in texture tests. *J. Texture Stud.* 6: 489–506.
Webb, N. B., Rao, V. N. M., Civille, G. V., and Hamann, D. D. 1975. Texture evaluation of frankfurters by dynamic testing. *J. Texture Stud.* 6: 329–342.
Wen, P. R., and Mohsenin, N. N. 1970. Measurement of dynamic viscoelastic properties of corn horny endosperm. *J. Mater.* 5(4): 856–867.

7

Thermal Properties of Unfrozen Foods

Paul Nesvadba

Contents

7.1	Introduction	224
	7.1.1 Importance of Thermal Properties for the Quality and Safety of Foods	224
	7.1.2 Modeling and Optimization of Processes	224
7.2	Sources of Data on Thermal Properties	225
	7.2.1 Measurement	225
	7.2.2 Literature	225
	7.2.3 Computerized and Online Databases	226
	7.2.4 Software for Predicting Thermal Properties of Foods	226
7.3	Density	226
	7.3.1 Definition of Powder Bulk Density	228
7.4	Specific Heat Capacity	228
	7.4.1 Latent Heat of Melting	230
	7.4.2 Specific and Latent Heat of Fats	230
7.5	Thermal Conductivity	231
	7.5.1 Predictive Equations	231
	7.5.2 Influence of Structure of Food on Thermal Conductivity	234
7.6	Measurement Methods for Thermal Conductivity	234
	7.6.1 Basis of Operation of the Needle Probe	235
	7.6.2 Reference Materials	238
7.7	Other Properties Relevant to Thermal Processing of Foods	239
	7.7.1 Compressibility and Thermal Expansion	239
	7.7.2 Glass Transitions	239
	7.7.3 Sorption and Hydration Properties	240
7.8	Conclusions	240
Symbols, Names, and Dimensions		241
References		241

7.1 INTRODUCTION

7.1.1 Importance of Thermal Properties for the Quality and Safety of Foods

Food-processing operations such as blanching, cooking, pasteurization, and sterilization involve temperature-dependent biochemical or chemical changes. The safety and quality of foods critically depend on correct temperature regimes; for example, in canning, the classical problem is finding the optimum heating regime that inactivates any microorganism while still preserving nutritional quality (avoiding overprocessing and destruction of vitamins). In water-containing foods, heat transfer is often accompanied by a significant water transfer. Thus, the quality and safety of foods critically depend on the entire temperature history and the state and distribution of water in the food. Other physical properties and variables such as pressure, flow, electric fields, and water activity also greatly influence processing of foods.

Table 7.1 summarizes the thermal processes in unfrozen foods. The temperature range is wide, from the freezing point (around –1°C in most foods) in chilled foods up to 135°C in sterilization under pressure (canning retort). In baking, the temperature of the oven reaches 250°C, although the temperature of the food (e.g., bread) is lower than the oven temperature because of evaporative cooling by water loss. In scraped surface heat exchangers, the wall temperature can be as high as 150°C.

Understanding, predicting (modeling), and controlling these processes, further described in Ref. [1], and designing the processing equipment require knowledge of the thermal properties of foods. Calculations of the temperature profiles inside a solid food as a function of time require the values of the specific heat capacity, c_p (the subscript p denoting constant pressure); thermal conductivity, k; and density, ρ of the food. These properties are routinely used in sizing thermal processing equipment (freezers, coolers, ovens, etc.).

7.1.2 Modeling and Optimization of Processes

A full understanding and prediction of the thermal properties of foods and their dependence on composition, structure, and interaction with other variables influencing the

Table 7.1 Thermal Processes in Foods above Freezing

Process	External Medium	Temperature of the External Medium (°C)	Temperature of the Food (°C)
Chilling	Air	0 (<0 for superchilling)	–1 to 5
Blanching	Water	70–90	60–80
Pasteurization	Water/steam	80–100	70–90
Sterilization	Water/steam	100–140	100–135
UHT	Scraped heat exchanger	160	130–150
Frying	Oil	165	100–120
Roasting	Air/infrared transfer	250	120–150
Baking	Air/steam/infrared transfer	200–250	100–200

Note: UHT, ultrahigh temperature.

quality and safety of foods presents a formidable challenge. Researchers are making progress by gradually accumulating and pooling data, forming models, testing the models against new data, and making better models. Fortunately, the thermal properties of foods are relatively easy to model. The specific heat capacity is especially easy to model because it is an additive property. The specific heat of mixtures is the sum of the specific heat capacities of the components comprising the mixture. Modeling of the thermal conductivity is much more difficult because it involves the structure of the food, for example, the porosity. Useful models for thermal conductivity can be constructed using models from dielectric theory (the Maxwell–Eucken model), in which two component mixtures are modeled in terms of a major (continuous) phase and a minor (dispersed) phase. Such models are the basis of the predictive computer program COSTHERM [2,3].

Modeling of food processes requires various types of physical property data. Density and viscosity data are used in computational fluid mechanics, specific heat and thermal conductivity for heat-transfer calculations, and, if moisture movement is important (as in many baked goods), moisture diffusivity and sorption isotherms are also required. These properties are often functions of the temperature and composition, in particular, moisture or air content. Also, if the food undergoes a phase change during processing (ice formation, fat crystallization, etc.), data on the associated enthalpy change are required.

7.2 SOURCES OF DATA ON THERMAL PROPERTIES

7.2.1 Measurement

Measurements (as opposed to modeling) are the primary source of data. Much progress has been made in developing measurement techniques for thermal properties of foods [4,5]. Steady accumulation of data [6,7] has resulted in bibliographies [8,9], monographs critically evaluating data and equations [10–12], and databases of physical properties of agro-food materials [13].

The EU (European Union) project COST93 [8] addressed the lack of thermal properties data especially above 100°C, measuring in a "round-robin" exercise the following materials: Nylon 66 (used as a calibration material but found unsuitable because of infrared transmissivity and varying absorbed moisture content and crystallinity), sodium caseinate gel, chemically modified starch gel, meat paste, tomato paste, and apple puree. The advance made in these measurements over those of the COST90 action is the increased upper temperature limit of 135°C. A pressurized apparatus had to be built to avoid losses of moisture by evaporation [8].

7.2.2 Literature

A vast amount of literature on thermal properties of foods exists. However, the data tend to be scattered, and the composition and origin (variety, cultivar), processing conditions, and structure of the foods are often not well documented. This detracts from the value of the data.

Textbooks [14–17] give background information and limited general data. There are also books [17,18] devoted to physical (including thermal) properties and critical evaluations of data and equations, for example, Refs. [19,20]. Articles that provide comprehensive data sets of thermal conductivity values for groups of food products include Refs. [21–32].

Databases or databanks of physical properties of foods are another source of information. A database that has been in existence at the Food Research Institute, Prague, since 1960s in paper form is being computerized [33,34]. An Excel database [35] presents data on thermal conductivity of foods collected during a literature survey [36].

7.2.3 Computerized and Online Databases

The recommendation of the EU project COST90 in 1983 was to create a computerized database of the available data on physical properties of agro-food materials. Professor R.P. Singh assembled the first such database in the United States [37]. This was followed by an in-house database by Unilever (UK) in 1995, leading to a project funded by the European Commission [38].

The EU database of physical properties of agro-food materials is available online at http://www.nelfood.com. The database contains more than 11,000 bibliographic records. About one in five of these records have numerical tables and equations attached. The novel and unique feature of the database is that it specifies both the measurement methods (in terms of their principle, accuracy, and precision) and the foods (in terms of their composition and structure). Two four-point scales indicate the quality of this specification (1–4 for method, A–D for food definition). There are five main categories of data on physical properties of agro-food materials, including thermal properties, the second most populated after mechanical properties. The thermal properties data, especially on specific heat, are less sensitive to changes in physical microstructure than mechanical or diffusion-related properties are, and are therefore more robust.

7.2.4 Software for Predicting Thermal Properties of Foods

Several computer programs are available for estimating the thermal properties of foods from their proximate chemical composition and density. The most widely distributed is COSTHERM. Hans Pol of the Spelderholt Institute in the Netherlands wrote the first version of this program based on the work of Miles et al. [2]. COSTHERM was further developed by Miles and Morley [3]. They reexamined the models for thermal conductivity and the initial freezing point. The accuracy of the predictive equations is about ±10%, sufficient for most food engineering calculations. The predictive equations are valid over the temperature range from −40°C to about 90°C.

7.3 DENSITY

Density is important in most food process calculations and in determining the thermal diffusivity, a, from the heat capacity and thermal conductivity $a = k/\rho c_p$. Table 7.2 gives the values of density for selected food materials.

THERMAL PROPERTIES OF UNFROZEN FOODS

Table 7.2 Density of Selected Foods

Food or Food Component	Water Content (% Wet Basis)	Temperature (°C)	Density (kg m⁻³)	Reference
Air (1 atm pressure)	0	20	1.2	40
Water	100	0	999.84	40
		80	971.79	40
Ice	100	0	920	40
Fats and oils				
Olive oil	0	20	900	40
Meats	74–84			
Beef, fat content x_f		5–20	1076–1.37 x_f	41
Fish	79–82			
Anchovy, $x_f = 10\%$, $x_p = 14\%$	76	21	930	42
Lemon fish $x_f = 0.65\%$, $x_p = 24.6\%$		15	1055	
Fruits and vegetables	84–90	25		2
Apples	84–87	20	775–925ᵃ	43
Pears (Maxine)	83–85	20	1000	43
Dairy products				
Milk, fat content x_f 3–6%	19	0–10	1028.9 – 0.195 t + 1.432 x_f	44
Cheese $x_f = 24.6$	65	20	1230	45
Cheese $x_f = 22.0$		20	1080	45

ᵃ Very poor correlation with water content; none is expected due to the influence of porosity.

Choi and Okos [39] reviewed the densities of the main constituents and the equations for their temperature dependence.

	ρ (kg/m³)
Water	$997.18 + 3.1439 \cdot 10^{-3} t - 3.7574 \cdot 10^{-3} t^2$
Ice	$916.98 - 1.3071 \cdot 10^{-1} t$
Protein	$1329.9 - 5.1840 \cdot 10^{-1} t$
Fat	$925.59 - 4.1757 \cdot 10^{-1} t$
Carbohydrate	$1599.1 - 3.6589 \cdot 10^{-1} t$
Fiber	$1311.5 - 3.6589 \cdot 10^{-1} t$
Ash	$2423.8 - 2.8063 \cdot 10^{-1} t$

The high precision (to five significant digits) with which the coefficients in the above equations are given for proteins, fats, and carbohydrates is probably not warranted because the density depends on the type of these complex and not well-defined components (e.g., fats are blends of various types of triglycerides). Miles et al. [2] gave the following densities (in kg/m³): protein 1380 – 0.5 t, fat 930 – 0.6 t, and carbohydrate 1550 – 0.3 t. The decrease of density with increasing temperature is due to thermal expansion. However, in starch-containing foods, there is an additional contribution due to swelling and imbibing of water by starches during heating [46].

The law of addition of specific volumes allows good estimation of the density of a mixture:

$$1/\rho = \sum x_i/\rho_i$$

where ρ is the density of the mixture, the values of x_i are the mass fractions of the various components, and the values of ρ_i are the densities of the constituents.

In porous products, air makes an important contribution, lowering the density of the food with increasing porosity ε

$$\rho = (1-\varepsilon)\Big/\sum x_i/\rho_i$$

The perfect gas law provides an estimate of the density of air:

$$\rho_{air} = M_{air} p/(RT)$$

where ρ_{air} is in kilograms per cubic meter, p is the pressure in pascals, T is the absolute temperature in degrees kelvin, and $M_{air} = 0.02895$ kg/mol is the average molecular weight of air. At normal atmospheric pressure of 101.3 kPa and temperature 20°C, ρ_{air} is 1.2 kg/m³.

7.3.1 Definition of Powder Bulk Density

When measuring the thermal properties of powders and comparing values, it is important to establish a systematic and repeatable method for determining the powder density because of the different ways in which powders can pack down when inserted into a container.

The bulk density of a powder, defined as the mass divided by the total bulk volume occupied, depends on how it is packed into a container [47]. The four basic "packing density states" are

- Aerated bulk density
- Poured bulk density
- Tap density
- Compacted bulk density

These can vary by up to 30%. Different standards and techniques exist for defining and measuring these different forms of packing density. ASTM defines a measurement procedure for tap density that involves repeatedly dropping the container onto a hard surface from a given height [48].

7.4 SPECIFIC HEAT CAPACITY

Specific heat capacity at constant pressure, c_p, and its temperature integral, enthalpy, H, are the fundamental properties for heat balance calculations. Water has a much higher

Table 7.3 Representative Values of Specific Heat Capacity

Food or Food Component	Water Content (% Wet Basis)	Temperature (°C)	c_p (kJ kg^{-1} K^{-1})	Reference
Air (1 atm pressure)	0	20	1.012	50
Water	100	0	4.217	40
		15	4.186	
Ice	100	0	2.06	50
Fats and oils	0	25	2.1	2
	3.3	25	2.28	
Meats	74–84	10	3.4–3.6	2
Fish	79–82	25	3.65–3.75	2
Fruits and vegetables	84–90	25	3.8–3.95	2
Potato puree	73.3	5	3.70	46
		80	3.78	46
Dairy products Cheese	79.5	24	3.72	2

specific heat and thermal conductivity than the other major food constituents. Therefore, water greatly influences the thermal properties of foods, albeit not as dramatically in unfrozen food compared with frozen foods, where the water–ice phase change dominates [49].

Calorimetry [51–53] and more conveniently, differential scanning calorimetry (DSC) [54,55] are the usual techniques for measuring the specific heat capacity and phase transitions. Table 7.3 shows representative data for various foods.

The specific heat capacity of the various components of the food can be calculated from empirical equations [39].

	c_p (J/(kg°C))
Water	$4176.2 - 0.0909\,t + 5.4731 \times 10^{-3}\,t^2$
Ice	$2062.3 + 6.0769\,t$
Protein	$2008.2 + 1.2089\,t - 1.3129 \times 10^{-3}\,t^2$
Fats	$1984.2 + 1.4373\,t - 4.8008 \times 10^{-3}\,t^2$
Carbohydrates	$1548.8 + 1.9625\,t - 5.9399 \times 10^{-3}\,t^2$
Fibers	$1845.9 + 1.8306\,t - 4.6509 \times 10^{-3}\,t^2$
Ash	$1092.6 + 1.8896\,t - 3.6817 \times 10^{-3}\,t^2$

The high value of the specific heat capacity of water makes the main contribution to the specific heat capacity of water-containing foods. This dominance of water enables the construction of simple linear predictive equations in the form

$$c_p = a + bx_w$$

where x_w is the mass fraction of water in the food, and a and b are empirical constants specific to the food. Riedel [52] refined this by introducing temperature dependence and introducing a nonlinear term improving predictions at low water contents:

$$c_p \text{ (kJ/kg)} = c_w x_w + 4.19 \, (\alpha + 0.001 \, t)(1 - x_w) - \beta \exp(-43 \, x_w^{2.3})$$

Riedel [52] determined the constants α and β for eight different foods. However, if mean values of $\alpha = 0.37$ and $\beta = 0.9$ are taken, the maximum deviation of c_p from the individual eight equations does not exceed 10%, and food engineers may find this approximation useful for some calculations [2].

As the specific heat capacity is an additive property, one can use the summation formula to predict the specific heat capacity of a food with known composition:

$$c_p = \sum x_i \, c_{pi}$$

where c_p is the specific heat capacity of the mixture, the values of x_i are the mass fractions of the various constituents, and the values of c_{pi} are the specific heat capacities of the constituents.

7.4.1 Latent Heat of Melting

Water and fat are two components of foods that exhibit solidification/melting-phase transition. Water has a very high enthalpy of melting. That of fats is smaller but still very significant, up to 200 J/g over the melting range of up to 60°C wide. In foods, which are complex "impure" substances, the latent heat is released or absorbed not at one well-defined temperature (as in very pure substances) but over a range of temperatures. Fats are mixtures of different triglycerides, and this is the reason why the latent heat peak is broadened. The reason for broadening the latent heat peak during freezing or thawing of a water-containing food is different. Water in a matrix such as meat or fish muscle crystallizes as pure ice; however, this leaves behind a more concentrated "solution" with a depressed freezing point (Raoult's law). Therefore, further crystallization takes place at progressively lower temperatures down to about −40°C. This is covered in detail in Chapter 8.

7.4.2 Specific and Latent Heat of Fats

The predictive equations of the preceding section are only applicable if there is not an appreciable contribution of the latent heat of melting of fat, that is, in low-fat foods or when the temperature range of interest does not include the melting range of the fats in the food. In foods of high-fat content and in the temperature range where phase transitions in fats occur, it is necessary to take into account the latent heat of crystallization of fats. The latent heat of fusion of fats is in the range from 80 to 200 J/g. It increases with increasing length of the fatty acid chain and the degree of saturation.

Latyshev and Ozerova [56] proposed an empirical equation for the specific heat of pork and beef fat:

$$c_p = A + Bt + \sum A_i \Big/ \left\{ 1 + B_i (t - t_i)^2 / A_i \right\}$$

where A, B, A_i, B_i, and t_i are fat-specific constants and $i = 1, 2$. Miles et al. [2] remarked that this equation may fit other types of fats, for example, the data of Riedel [51]. The absence in the literature of equations relating the thermal properties of fats to their fatty acid composition indicates the need for research in this area.

Miles et al. [2] cautioned against uncritical use of the additive-type equation for the specific heat of fatty foods,

$$c_p = (1 - x_f)c_{ff} + x_f c_f$$

where the index *ff* refers to fat-free food and *f* refers to fat. Melting of lipids is influenced by impurities, and the simple additive equation does not account for interactions between the fat-free food matrix and the added fat. Moreover, the melting and solidification of fats depend on the heating and cooling rates and on hysteresis caused by time dependence of crystallization and polymorphism. Therefore, rigorous modeling of the thermal properties of fats is difficult. A simple model, employed in COSTHERM, assumes that fats, because they are mixtures of triglycerides, melt over a temperature range and that the latent heat of the phase change evolves uniformly over that range.

7.5 THERMAL CONDUCTIVITY

Thermal conductivity is an essential property determining the rate of transmission of heat through foods during thermal processing, through the Fourier law of heat conduction:

$$q = k\nabla T$$

where q is the heat flux and ∇T is the temperature gradient. Water (and ice) have thermal conductivities much higher than those of the other food components (protein, fat, and carbohydrates), and thus, the water content of foods has a great influence on the thermal conductivity of foods. On the other hand, air has a low value of thermal conductivity, and thus, porous foods are poor heat conductors. Table 7.4 gives representative values of thermal conductivity of the main components of foods and main types of foods.

7.5.1 Predictive Equations

Choi and Okos [39] gave the following equations for estimating the thermal conductivity of food components, valid in the range 0–90°C:

	k (W/(m°C))
Water	$0.57109 + 1.762 \cdot 10^{-3} t - 6.7036 \cdot 10^{-6} t^2$
Ice	$2.21960 - 6.2489 \cdot 10^{-3} t + 1.0154 \cdot 10^{-4} t^2$
Proteins	$0.17881 + 1.1958 \cdot 10^{-3} t - 2.7178 \cdot 10^{-6} t^2$
Fats	$0.18071 - 2.7604 \cdot 10^{-3} t - 1.7749 \cdot 10^{-7} t^2$
Carbohydrates	$0.20141 + 1.3874 \cdot 10^{-3} t - 4.3312 \cdot 10^{-6} t^2$
Fibers	$0.18331 + 1.2497 \cdot 10^{-3} t - 3.1683 \cdot 10^{-6} t^2$
Ash	$0.32961 + 1.4011 \cdot 10^{-3} t - 2.9069 \cdot 10^{-6} t^2$
Air (still)	0.025

ENGINEERING PROPERTIES OF FOODS

Table 7.4 Representative Values of Thermal Conductivity

Food or Food Component	Water Content (% Wet Basis)	Temperature (°C)	k (W m^{-1} K^{-1})	Reference
Air	0	0	0.023	50
Water	100	0	0.554	50
		80	0.686	
Ice	100	0	2.24	57
Fat	0	>0	0.175	55a
Meats Beef (4–9% fat)	74–84	10	0.40 ± 0.15	55b
	75–79	0	0.47	58
		20	0.50	58
Fruits and vegetables	80	20	0.40 ± 0.20	55b
Dairy products	80	20	0.5 ± 0.1	55b
Fats and oils	0	20	0.2 ± 0.05	55b
Fat/water dispersions	75	20	0.45 ± 0.15	55b
Water–alcohol mixtures	0	20	0.2 ± 0.05	55b
	80	20	0.55 ± 0.05	
Porous materials (grain and flour)	5	20	0.10 ± 0.01	55b

Again, water has the highest thermal conductivity of all the constituents (the value for ice is almost 4 times higher). This is why the thermal properties of foods above freezing do not depend as much on temperature as on water content [59]. Porosity of foods also has a significant effect on thermal conductivity because air, a constituent of porous foods, has a very small thermal conductivity.

Unfortunately, unlike for specific heat capacity, there is no single, straightforward model for predicting the thermal conductivity of mixtures of constituents. This is because the thermal resistance that food offers to the heat flow depends on the structure (geometrical arrangement) of the components of the food. Figure 7.1 shows this for a food composed of two components. Various models for multicomponent foods have been proposed, one of which is the geometric mean model:

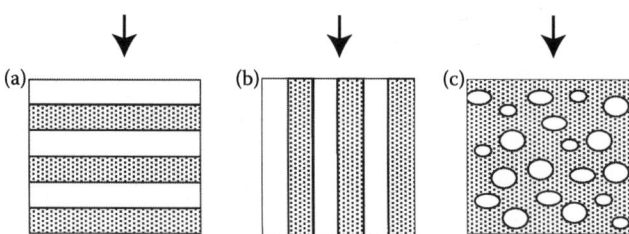

Figure 7.1 Models for thermal conductivity of a two-component mixture: (a) *perpendicular*—layered food with layers perpendicular to heat flow (arrows show the direction of heat flow), (b) *parallel*, and (c) *dispersed*.

$$k = \prod k_i^{\varepsilon_i}$$

where k is the thermal conductivity of the composite food, the values of ε_i are the volume fractions of the constituents (x_i/ρ_i), and the values of k_i are the thermal conductivities of the constituents.

The thermal conductivity of multicomponent foods lies between two limiting values. The lower limit is given by a perpendicular model that assumes that the constituents are disposed in layers perpendicular to the flow of heat. The law of addition of thermal resistances in series leads to the equation

$$k\perp = 1 \Big/ \sum \varepsilon_i/k_i$$

where $k\perp$ is the thermal conductivity of the food according to the series model.

The upper limit on the thermal conductivity comes from the parallel model, in which the constituents are arranged as layers parallel to the heat flow:

$$k\| = \sum \varepsilon_i k_i$$

A mixing model combining the two limiting values is often used as an estimate of the thermal conductivity,

$$k_{mix} = g\,k\perp + (1-g)\,k\|$$

where g is a number between 0 and 1. For $g = 0.5$, the conductivity is the arithmetic mean of $k\perp$ and $k\|$. Miles and Morley [3] have refined this model for foods that contain air, such as powdered milk or coffee, where the included air significantly lowers the thermal conductivity because the thermal conductivity of air is much lower than that of the other components of the food. Analysis of data from the COST90 project showed that the best fit was for

$$g = 0.04 + 0.8\,\varepsilon_a$$

where ε_a is the volume fraction of air in the food. The thermal conductivity of the porous product is then

$$k_{porous} = (0.04 + 0.8\,\varepsilon_a)\,k\perp_{porous} + (0.96 - 0.8\varepsilon_a)\,k\|_{porous}$$

where $1/k\perp_{porous} = \varepsilon_a/k_a + (1-\varepsilon_a)/k_p$, $k\|_{porous} = \varepsilon_a k_a + (1-\varepsilon_a) k_p$, and k_a and k_p are the thermal conductivities of air and air-free product, respectively.

For a two-component food consisting of a continuous and a dispersed phase (Figure 7.1c), a formula adopted from the dielectric theory developed by Maxwell and adapted by Eucken [60], provides a model for the thermal conductivity of the food.

$$k = k_c\big(2k_c + k_d - 2\varepsilon_d(k_c - k_d)\big) \Big/ \big(2k_c + k_d + \varepsilon_d(k_c - k_d)\big)$$

where k is the thermal conductivity of the mixture, k_c is the thermal conductivity of the continuous phase, k_d is the thermal conductivity of the dispersed phase, and ε_d is the volume fraction of the dispersed phase (x_d/ρ_d).

Miles et al. [2] tested the robustness of the predictive equations by comparing the measured and calculated thermal conductivities for 11 foods. The parallel model and the Maxwell–Eucken model gave the smallest deviations (on the order of 10%) between the measured and calculated values. Such accuracy is acceptable or useful for most food engineering heat-transfer calculations.

Sakiyama et al. [61,62] used a three-dimensional finite-element model to analyze the effective thermal conductivity of dispersed systems of spheres of various sizes. When the results were compared with several known models, it was found that the Maxwell–Eucken model gave a fairly good approximation for the effective thermal conductivity when the dispersed spheres were arrayed in a simple cubic lattice and a body-centered cubic lattice. The ratio of the thermal conductivity of the spheres to the continuous phase varied between 0.01 and 100. The applicability of the Maxwell–Eucken model suggested by the finite-element analysis was confirmed by experimental measurements for gels with dispersed solid or liquid paraffin particles or droplets.

In chilling and storage of fruits or vegetables stored in boxes on pallets, the produce has to be treated as an assembly of near-spherical particles with air between them. Such a system has an effective thermal conductivity [63,64].

7.5.2 Influence of Structure of Food on Thermal Conductivity

Apart from the influence of porosity, in nonporous materials such as meat and fish, the protein fiber orientation plays a role. In frozen meat, the thermal conductivity is usually up to 30% higher when the heat flows parallel to muscle fibers because the ice crystals are dendrites growing along the fibers. In unfrozen meat, the heat flows more easily in the direction across the fibers. The difference in conductivity is up to 10% [12].

7.6 MEASUREMENT METHODS FOR THERMAL CONDUCTIVITY

Thermal conductivity is difficult to measure because of the usual problems posed by foods—nonhomogeneous materials with complex structure (e.g., fibrous structures) on the macro-, meso-, and microscales, perishable and containing water that migrates under thermal gradients. Therefore, rapid transient methods using small temperature gradients are needed to avoid spoilage and changes of the food during the measurement [4].

Measurements of thermal conductivity are much more difficult than those of specific heat capacity because the heat flow pattern in the sample must be carefully defined. Moreover, in fluid foodstuffs, precautions (limiting temperature rises) have to be taken to suppress gravity-/buoyancy-driven convection currents. The *steady-state* methods involve constant heat flow generated by a heat source and absorbed by a heat sink. Thermal guards have to be employed to eliminate heat flow from the sides of the specimen not in contact with the heat source and heat sink. The time to achieve constant conditions is usually long (several hours) and not well suited for perishable foods that may change chemically or physically during that time. From this point of view, *transient* methods are much better suited and widely used to measure thermal conductivity of foods. A general feature of the transient methods is that the sample is subjected to time-varying heat flow, and temperature is measured at one or more points within the sample or on the surface of the sample [4]. There are now a number of contact transient methods having a common principle [65], some of them developed commercially and available globally. The heated disk is one relatively recent addition [66] to the battery of transient methods. One or more of the methods may be used for a broad range of conditions or properties. However, a particular method may be more suited to a certain type, size, or form of food or a range of thermal conductivity.

A widely used method for measuring thermal conductivity of foods is the heated probe or needle probe method. The proponents are Sweat [31,32], who developed one of the best-known probes, Murakami et al. [67], and others [68,69], who explored the optimal design of the probe and analyzed the uncertainties of measurements.

The heated probe method is a well-established and reliable method for determining thermal conductivity of materials. The method is easy to use and measurements can be made *in situ*. This makes the method suitable for use in industry, especially if the measurement is automated so that nonspecialists can use it. For small temperature rises of the heated probe, convection currents do not arise in fluid samples. This makes the method suitable for liquid foods.

7.6.1 Basis of Operation of the Needle Probe

Figure 7.2 shows a diagram of the heated needle probe. Closing the switch applies a heat pulse to the sample through the heater encased in the needle probe. The temperature rise of the probe depends on the thermal conductivity of the sample (increases with decreasing k). The temperature rise of the probe is measured as a function of time, and from this, the conductivity is deduced using mathematical analysis. To simplify the mathematical

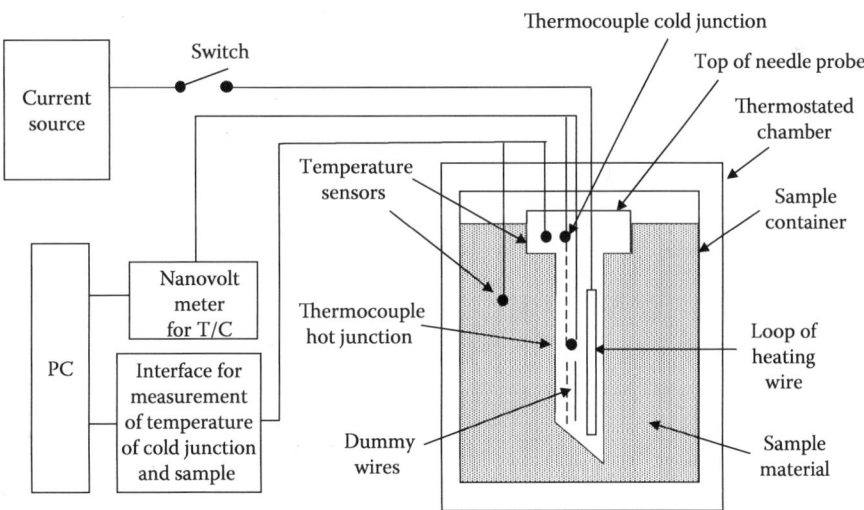

Figure 7.2 A block diagram of the heated needle probe for measuring thermal conductivity of foods.

model, the heat flow should be such that the isotherms are concentric cylinders with the needle probe as the axis. Therefore, the needle probe has to be straight, and the ratio of the length to diameter sufficiently large (over 150) and the heater wire dissipation uniform along its length. The time of measurement and the distance of propagation of the heat pulse increase with the measurement time. The measurement time must be smaller than the time taken by the heat pulse from the heater to reach the wall of the measurement cell. This imposes conditions on the various dimensions of the experimental arrangement, as discussed by Salmon et al. [70] (Figure 7.3).

The temperature rise of foods should be in the range 0.2–5 K, depending on the type of food and sample size. Low-power input is preferred to avoid excessive moisture migration in the sample. For short heating times, the temperature rise is

$$\Delta\theta = A \cdot \ln(\tau) + B + (D + E \cdot \ln(\tau))/\tau$$

where τ is the time from the moment the heater is energized and A, B, C, D are coefficients dependent on probe geometry, heater power per unit length, and probe/specimen thermophysical properties.

For sufficiently large times, this relation approximates to a simple straight-line equation. The linear portion of the graph of $\Delta\theta$ versus $\ln(\tau)$ (selected by visual inspection or by software according to defined criteria) yields the thermal conductivity using the equation:

$$k = q/4\pi S$$

where q is the heater power dissipated per unit length (W/m) and S is the slope of the linear portion of the graph. The needle probe method is an absolute method because q and S

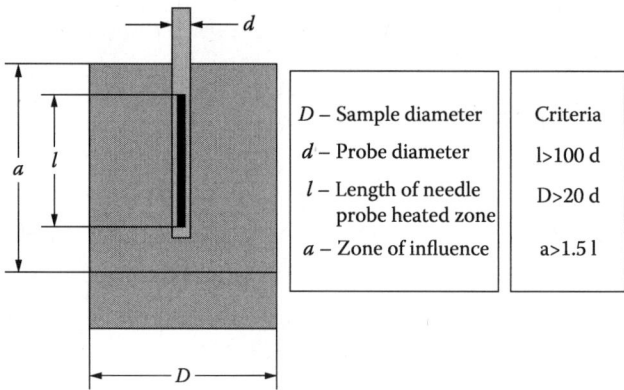

Figure 7.3 A schematic view of the needle probe method showing the critical dimensions and criteria for ideal operation. (Adapted from Salmon, D. et al. 2001. Thermal conductivity standards for powders, sludges, and high moisture materials. National Physical Laboratory Report CBTLM S49 for the DTI National Measurement Project 3.4, with permission of NPL—National Physical Laboratory, Teddington, UK.)

THERMAL PROPERTIES OF UNFROZEN FOODS

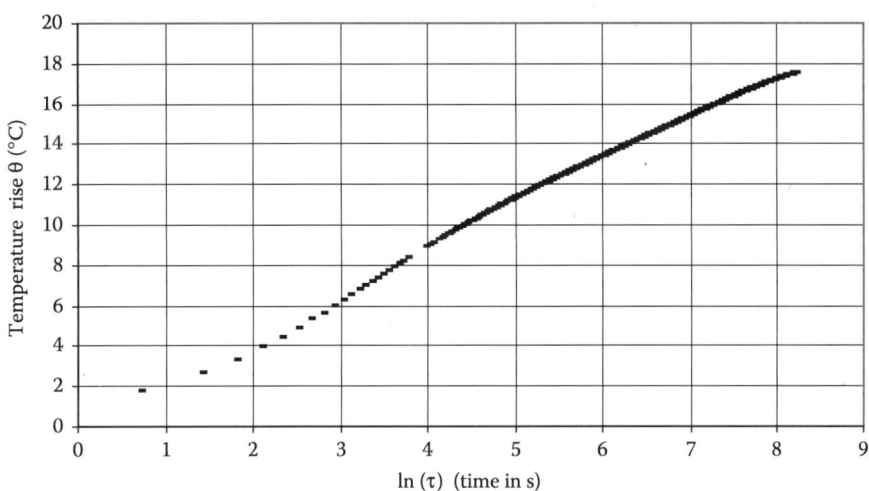

Figure 7.4 Temperature rise Δθ versus ln (τ), τ is the time in seconds. Alumina grinding grit, heater power 6.8 W/m. (From Salmon, D. et al. 2001. Thermal conductivity standards for powders, sludges, and high moisture materials. National Physical Laboratory Report CBTLM S49 for the DTI National Measurement Project 3.4 with permission of NPL.)

can be calculated. Alternatively, the probe may be calibrated by determining q by measuring a reference material with a known thermal conductivity.

Figure 7.4 shows the linear region of the graph Δθ versus ln(τ) between log time 5 and 7. The curve starts to deviate from linearity at about log time 7.5. Care has to be taken in choosing which part of the curve to use in the analysis. In general, the temperature rise versus log time curve converges toward a straight line as time increases. However, if the heat capacity of the sample is less than that of the probe material, it can be shown from the equation for Δθ that there are two turning points in the curve before linearity is reached, with a relatively straight region between them. This can be seen between log time 3 and 4. Care must be taken to ensure that this region is not used in the analysis, as it will give erroneous results.

Salmon et al. [65] discussed the design specification for the needle probe and its associated apparatus. The recommended features of the apparatus that provide high accuracy (uncertainties <2%) thermal conductivity data are as follows:

- Construction of the needle probe: A typical probe consists of a thin-walled metallic tube closed at one end. The length-to-diameter ratio should be in excess of 100 to ensure radial heat flow. A typical size for laboratory measurements is of length 200 mm and diameter 0.6–2 mm. Long (>150 mm) probes are necessary for nonhomogeneous materials. The tube contains a heater (a double-folded or spiraled wire) and a thermocouple insulated from each other and from the tube. The hot junction of the thermocouple is adjacent to the heater wire halfway along the tube. The other junction (cold or reference junction) is positioned near the top end of the tube. The heater wire is terminated a few centimeters away from the cold junction in order not to heat the cold junction.

- Sample container: A typical sample container is cylindrical with a height of 250 mm and sufficient internal diameter (100–150 mm) to allow time to measure a wide range of materials without the heat flux from the probe reaching the container walls. The container is closed with a screw cap (for liquids or powders) with the probe built into the cap to allow cleaning of the measurement cell. Solid samples would be cut to fit the cylindrical shape of the container, with a hole along the axis for the probe. The anticipated pressure in the container is 1–3 bars, and therefore, the endcap needs to be sealed against the body of the container. The electrical leads could be built into the probe head design. The container or sample would be inserted into the environmental chamber and held in a vertical position.
- Environmental chamber: The chamber housing the sample with the measurement probe should be able to cover the usual temperature range or measurement of foods, −40°C to 150°C. The temperature stability should be better than 0.1°C. The chamber houses the measurement container and is therefore preferably cylindrical with a diameter of at least 200 mm and depth of about 300–350 mm. Lead-throughs for power and temperature measurements have to be furnished. To effect temperature control, a cooling/heating tube is coiled around the chamber carrying circulating fluid from a thermostated bath covering the required temperature range. An alternative is to immerse the chamber in a continuously stirred liquid maintained at a constant temperature. The environmental chamber should be of copper or stainless steel and insulated from the surroundings. The internal walls should have thermocouples to monitor the temperature distribution within the chamber.
- Data-acquisition system: The data acquisition ideally needs a minimum of eight channels, with the minimum resolutions of the voltmeter stated in parentheses: temperature rise (10 nV), reference temperature (0.1 μV), heater voltage (0.1 V), heater current (10 mV across a series resistor) to measure the power stability to 0.01%, and two or three channels for measurement of temperature uniformity within the environmental chamber (0.1 μV).

For the usual PC data-acquisition cards, a direct current (dc) amplifier with gain of 10^5 would be required for the temperature rise signal to provide the 10 nV sensitivity. The number of readings required per second depends on the viscosity of the sample (the time taken to initiate convection currents in liquid samples) and for low viscosity samples may be of the order of 100 readings per second.

7.6.2 Reference Materials

Reference materials are useful for calibration of measuring apparatus and for comparison of thermal conductivity values. The projects described in Refs. [2,8], and [70] concerned reference materials and interlaboratory comparisons. Alumina grit, agar gel, Ottawa sand,

Table 7.5 Values of Thermal Conductivity of Reference Materials

Reference Material	Temperature (°C)	Thermal Conductivity (W m^{-1} K^{-1})
Alumina grinding grit[a]		0.251 ± 0.022
Glass ballotini,[b] 150 μm	23	0.168 ± 0.025
Carrageenan gel, 2% in water[c]	25	0.609 ± 0.050
Agar gel, 0.4% in water[a]	23	0.652
Pure ice[a]	−15	2.63
Olive oil[d]	23	0.164
Silicone oil[a]	−15	0.135
Silicone oil[a]	25	0.133

[a] Ref. [70] – Report for the UK DTI Project 3.4.
[b] The mean of conductivity values from four laboratories in the COST90 project.
[c] COST90 results, average of four laboratories.
[d] Olive oil from COST93 project, the mean of values of five laboratories.

and paraffin wax were identified as appropriate reference materials. Table 7.5 gives values of thermal conductivity for some of the reference materials.

7.7 OTHER PROPERTIES RELEVANT TO THERMAL PROCESSING OF FOODS

7.7.1 Compressibility and Thermal Expansion

High-pressure (600–800 MPa) processing of foods has emerged in the last decade as one of the nonthermal ways of inactivating microorganisms and has other interesting applications, such as pressure shift thawing and freezing. The adiabatic temperature rise during high-pressure processing is an unwanted effect, present to a degree determined by the coefficient of cubical expansion, β, of the food, given by the following equation:

$$\left(\frac{\partial T}{\partial P}\right)_S = \frac{T\beta}{\rho c_p}$$

Morley [27] has measured the quantities in this equation in beef at the initial freezing point. These quantities can also be associated with compressibility using thermodynamic relationships [71].

7.7.2 Glass Transitions

Glass transitions have no direct effect on macroscopic heat transfer. However, the concept of glass transitions originating from polymer science [72,73] has great significance for thermal processing of foods because it explains the behavior of foods in many food processes and the stability of food products in storage. The glass transition, occurring in the region of the glass transition temperature, T_g, is a second-order phase transition and as such, does not involve any latent heat [74,75], but the transition is detected by observing changes in various physical properties associated with changes in molecular mobility and viscosity. These effects are seen in dielectric, mechanical, and thermodynamic (enthalpy, free volume, heat capacity, and thermal expansion coefficient) properties [74–76]. DSC is the most common method used to determine T_g. DSC detects the change in heat capacity, c_p, occurring over the transition temperature range [74,77,78].

In foods containing water, the glass transition occurs at low temperatures, below freezing, as water is removed from the food matrix in the form of ice (refer to Chapter 8), a process equivalent to drying. In intermediate moisture and dry food, the glass transition occurs above room temperature.

7.7.3 Sorption and Hydration Properties

The heat of hydration evolves in mixing powdered ingredients (starches, flours, additives, etc.) with water (e.g., when preparing batters). This raises the temperature of the mix and has to be considered for temperature control. Conversely, during drying, energy has to be supplied to overcome the binding of water molecules, equivalent to the heat of sorption, ranging from 100 to 1000 kJ/kg [79].

7.8 CONCLUSIONS

Data on thermal properties of foods are essential for designing and controlling thermal processing of foods and thereby ensuring the quality and safety of foods. The correct use of measurement methods and the application of thermal knowledge in industrial applications are often nontrivial and difficult tasks. Data and standards are needed [80], both for certified reference materials (such as the Karlsruhe test substance but capable of withstanding temperatures above 100°C and high pressures) and for methods of measurement (standard protocols).

Measurements of thermal properties are expensive and cannot keep up with the new food products being continually developed. Therefore, efforts to develop predictive equations will continue. The primary problem is to incorporate not only the chemical composition but also the descriptors of the structure of the foods into predictive models [81]. Elements of artificial intelligence (case-based reasoning) and neural networks could eventually assist with this task [13].

The growing body of data and knowledge relating to thermal properties of foods, coupled with increasing understanding of related processes, such as mass transfer, and the increasing modeling capability of software packages are contributing to optimization of thermal processing of foods and improvements in their quality and safety. It is also important to enable the users of thermal knowledge in the food and other industries to access the knowledge, for example, through the EVITHERM project [82].

SYMBOLS, NAMES, AND DIMENSIONS

a Thermal diffusivity ($m^2 s^{-1}$)
c_p Specific heat capacity (kg kJ^{-1}°C^{-1})
H Enthalpy (kJ kg^{-1})
k Thermal conductivity (W m^{-1} K^{-1})
M_{air} Molecular weight of air, 0.02895 kg mol^{-1}
β Coefficient of cubic expansion (°C^{-1})
ε Porosity (–)
ε_i Volume fraction of the ith component
ρ Density (kg m^{-3})
p Pressure (Pa)
R The universal gas constant (8.3 J mol^{-1} K^{-1})
t Temperature (°C)
T Temperature (K)
T_g Glass transition temperature (K)
τ Heating time of the needle probe (s)
$\Delta\theta$ Temperature rise of the needle probe (°C)
x Mass fraction (–)
x_f Fat content (mass wet basis) (%)
x_p Protein content (mass wet basis) (%)

REFERENCES

1. Hallström, B., Skjöldebrand, C., and Trägårdh, C. 1988. *Heat Transfer and Food Products.* Elsevier Applied Science, New York.
2. Miles, C.A., van Beek, G., and Veerkamp, C.H. 1983. Calculation of thermophysical properties of foods. In *Physical Properties of Foods.* Jowitt, R., Escher, F., Hallström, B., Meffert, H.F.T., Spiess, W.E.L., and Vos, G., eds. Applied Science Publishers, New York. pp. 269–312.
3. Miles, C.A. and Morley, M.J. 1997. Estimation of the thermal properties of foods: A revision of some of the equations used in COSTHERM. In *Proceedings of a Conference–Workshop— Practical Instruction Course on Modelling of Thermal Properties of Foods during Production, Storage and Distribution.* Final workshop of the EU project CIPA-CT93-0243, Faculty of Mechanical Engineering, Czech Technical University, Prague. 23–25 June, 1997. pp. 135–143.
4. Nesvadba, P. 1982. Methods for the measurement of thermal conductivity and diffusivity of foodstuffs. *Journal of Food Engineering,* 1, 93–113.
5. Ohlsson, T. 1983. The measurement of thermal properties. In *Physical Properties of Foods* (Eds. Jowitt, R., Escher, F., Hallström, B., Meffert, H.F.T., Spiess, W.E.L., and Vos, G.). Applied Science Publishers, New York. pp. 313–328.
6. Jowitt, R., Escher, F., Hallström, B., Meffert, H.F.T., Spiess, W.E.L., and Vos, G., eds. 1983. *Physical Properties of Foods.* Applied Science Publishers, New York.
7. Kent, M., Christiansen, K., van Haneghem, I.A., Holtz, E., Morley, M.J., Nesvadba, P., and Poulsen, K.P. 1984. COST90 collaborative measurements of thermal properties of foods. *Journal of Food Engineering,* 3, 117–150.
8. Spiess, W.E.L., Walz, E., Nesvadba, P., Morley, M., van Haneghem, I.A., and Salmon, D. 2001. Thermal conductivity of food materials at elevated temperatures. *High Temperatures High Pressures,* 33, 693–697.

9. Sanz, P.D., Alonso, M.D., and Mascheroni, R.H. 1987. Thermal properties of meat products: General bibliographies and experimental measurement. *Transactions of the ASAE*, 30, 283.
10. Adam, M. 1969. *Bibliography of Physical Properties of Foodstuffs*. Czech Academy of Agriculture, Prague.
11. Houška, M., Adam, A., Celba, J., Havlíček, Z., Jeschke, J., Kubešová, A., Neumannová, J., Pokorný, D., Šesták, J., and Šrámek, P. 1994. *Thermophysical and Rheological Properties of Foods— Milk, Milk Products, and Semiproducts*. Food Research Institute, Prague.
12. Houška, M., Adam, A., Celba, J., Havlíček, Z., Jeschke, J., Kubešová, A., Neumannová, J., Nesvadba, P., Pokorný, D., Šesták, J., and Šrámek, P. 1997. *Thermophysical and Rheological Properties of Foods—Meat, Meat Products, and Semiproducts*. Food Research Institute, Prague.
13. Nesvadba, P., Houška, M., Wolf, W., Gekas, V., Jarvis, D., Sadd, P.A., and Johns, A.I. 2004. Database of physical properties of agro-food materials. *Journal of Food Engineering*, 61(4), 497–503.
14. Singh, R.P. and Heldman, D.R. 1993. Heat transfer in food processing. In *Introduction to Food Engineering*. 2nd ed. Singh, R.P. and Heldman, D.R., eds. Academic Press, London. pp. 129–224.
15. Rha, C. 1975. Thermal properties of food materials. Theory, determination and control of physical properties of food materials. *Series in Food Material Science*, 1, 311–355. UDC; 641.
16. Kessler, H.G. 2002. *Food and Bioprocess Engineering: Dairy Technology*. 5th edition. Publishing House A. Kessler, Munich, 2002. ISBN 3-9802378-5-0.
17. Kostaropoulos, A.E. 1971. Wärmeleitzahlen von Lebensmitteln und Methoden zu deren Bestimmung [Thermal conductivities of foods and methods of their determination]. *Bereitsheft 16 der Fachgemeinschaft Lufttechishe und Trocknung-Anlagen im VDMA*. Maschinenbau-Verlag GmbH, Frankfurt/Main.
18. Mohsenin, N.N. 1980. *Thermal Properties of Foods and Agricultural Materials*. Gordon and Breach, London.
19. Rahman, S.M. 1996. *Food Properties Handbook*. CRC Press, Boca Raton, FL.
20. Qashou, M.S., Vachou, R.I., and Touloukian, Y.S. 1972. Thermal conductivity of foods. *Transactions of ASHRAE*, 78, 165.
21. Lentz, C.P. 1961. Thermal conductivity of meats, fats, gelatin gels, and ice. *Food Technology*, 15, 243–247.
22. Hill, J.E., Leitman, J.D., and Sunderland, J.E. 1967. Thermal conductivity of various meats. *Food Technology*, 21, 1143.
23. Mellor, J.D. 1976. Thermophysical properties of foodstuffs: Introductory review. *Bulletin of the International Institute of Refrigeration*, 56, 551.
24. Mellor, J.D. 1979. Thermophysical properties of foodstuffs: Measurement. *Bulletin of the International Institute of Refrigeration*, 59, 51.
25. Morley, M.J. 1966. Thermal conductivity of muscle, fats and bones. *Journal of Food Technology*, 1, 303–311.
26. Morley, M.J. 1972. Thermal properties of meat—Tabulated data. Special Report No. 1. Agricultural Research Council, Meat Research Institute, Langford, Bristol, UK.
27. Morley, M.J. 1986. Derivation of physical properties of muscle tissue from adiabatic pressure-induced temperature measurements. *Journal of Food Technology*, 21, 269–277.
28. Morley, M.J. and Miles, C.A. 1997. Modelling the thermal conductivity of starch–water gels. *Journal of Food Engineering*, 33, 1–14.
29. Pham, Q.T. and Willix, J. 1989. Thermal conductivity of fresh lamb meat, offals, and fat in the range −40 to +30°C: Measurements and correlations. *Journal of Food Science*, 54, 508–515.
30. Sweat, V.E. 1974. Experimental values of thermal conductivity of selected fruits and vegetables. *Journal of Food Science*, 39, 1080–1083.
31. Sweat, V.E. 1975. Modelling of thermal conductivity of meats. *Transactions of the ASAE*, 18, 564.

32. Sweat, V.E. 1985. Thermal properties of low- and intermediate-moisture food. *Transactions of ASHRAE*, 91, 369–389.
33. Mayer, Z. and Houška, M. 1999. Bank of information on physical properties of foods—BIPPF Prague. In: *Modelling of Thermal Properties and Behaviour of Foods during Production, Storage, and Distribution: Proceedings of a Conference, Prague, June 1997*. Food Research Institute, Prague, pp. 93–97.
34. Mayer, Z. 2013. Database of Physical Properties of Foods. Department of Food Engineering, Food Research Institute Prague.
35. Krokida, M.K., Panagiotou, N.M., Maroulis, Z.B., and Saravacos, G.D. 2001. Thermal conductivity: Literature data compilation for foodstuffs. *International Journal of Food Properties*, 4, 111–137.
36. Krokida, M.K., Michailidis, P.A., Maroulis, Z.B., and Saravacos, G.D. 2002. Literature data of thermal conductivity of foodstuffs. *International Journal of Food Properties*, 5, 63–111.
37. Singh, R.P. 1995. *Food Properties Database. Version 2.0 for Windows*. CRC Press, Boca Raton, FL.
38. Nesvadba, P. 2003. Construction of a database of physical properties of foods, EU project FAIR CT96-1063. http://www.nelfood.com.
39. Choi, Y. and Okos, M.R. 1986. Effects of temperature and composition on the thermal properties of foods. In *Food Engineering and Process Applications, Vol. 1— Transport Phenomena*. Le Maguer, M. and Jelen, P., eds., Elsevier, New York. pp. 93–101.
40. Kaye, G.W.C. and Laby, T.H. 1966. *Tables of Physical and Chemical Constants*, 13th ed., Longman, London. p. 58.
41. Rusz, J. and Kopalová, M. 1976. Express determination of fat content in basic raw materials in meat manufacture [in Czech]. Report of the Research Institute of Meat Industry, Brno, Czech Republic, 1976.
42. Young, F.V.K. 1986. The chemical and physical properties of crude fish oils for refiners and hydrogenerators. *Fish Oil Bulletin No. 18 of the IAFMM (International Association of Fish Meal Manufacturers)*. 1–18.
43. Mohsenin, N.N. 1970. *Physical Properties of Plant and Animal Materials*. Gordon & Breach, New York.
44. Watson, P.D. and Tittsler, R.P. 1961. The density of milk at low temperatures. *Journal of Dairy Science*, 44, 416–424.
45. Sweat, V.E. and Parmelee, C.E. 1978. Measurement of thermal conductivity of dairy products and margarines. *Journal of Food Process Engineering*, 2(3), 187–197.
46. Fasina, O.O., Farkas, B.E., and Fleming, H.P. 2003. Thermal and dielectric properties of sweet potato puree. *International Journal of Food Properties*, 6, 461–472.
47. Svarovsky, L. 1987. *Powder Testing Guide: Methods of Measuring the Physical Properties of Bulk Powders*. Elsevier Science, London.
48. ASTM B527–06. *Standard Test Method for Determination of Tap Density of Metallic Powders*. ASTM International, Pennsylvania, USA.
49. Eunson, C. and Nesvadba, P. 1984. Moisture and temperature dependence of thermal diffusivity of cod minces. *Journal of Food Technology*, 19, 585–592.
50. Washburn, E. W. 1926–33. *International Critical Tables of Numerical Data, Physics, Chemistry, and Technology*. McGraw-Hill, New York. Vol. 5, p. 95.
51. Riedel, L. 1955. Kalorimetrische Untersuchungen über das Schmelzverhalten von Fetten und Ölen [Calorimetric studies of the melting behaviour of fats and oils]. *Fette-Seifen-Anstrichmittel*, 57, 771.
52. Riedel, L. 1957. Kalorimetrische Untersuchungen über das Gefrieren von Fleisch [Calorimetric investigations of the freezing of meat]. *Kältetechnik*, 9(2), 38–40.
53. Riedel, L. 1978. Eine Formel zur Berechnung der Entalpie fettarmer Lebensmitteln in Abhängigkeit von Wassergehalt und Temperatur [A formula for calculating the enthalpy

of low-fat foods as a function of water content and temperature]. *Chemie, Mikrobiologie und Technologie der Lebensmittel,* 5(5), 129–133.
54. Buhri, A.B. and Singh, R.P. 1994. Thermal property measurements of fried foods using differential scanning calorimeter. In *Developments in Food Engineering.* Proceedings of the 6th International Congress on Engineering and Food, Chiba, May 1993, Vol. 1. Yano, T., Matsuno, R., and Nakamura, K., eds. Blackie, Glasgow. pp. 283–285.
55. Ali, S.D., Ramaswamy, H.S., and Awuah, G.B. 2002. Thermophysical properties of selected vegetables as influenced by temperature and moisture content. *Journal of Food Process Engineering,* 25, 417–433.
55a. Bäckström, E.H.M. and Emblik, E. 1965. *Kältetechnik,* 3rd ed., Verlag G. Braun, Karlsruhe, Germany.
55b. Rao, M.A. and Rizvi, S.S.H., eds. 1986. *Engineering Properties of Foods.* Marcel Dekker, New York. p. 66.
56. Latyshev, V.P. and Ozerova, T.M. 1976. Specific heat and enthalpy of rendered beef and pork fat [in Russian]. *Kholodil'najaTekhnika,* 5, 37–39.
57. Ratcliffe, E.H. 1962. On thermal conductivity of ice. *Philosophical Magazine,* 7(79), 1197–1203.
58. Mazurenko, A.G., Fedorov, V.G., Chernaya, T.V., and Pavlenko, V.I. 1984. Complex determination of thermophysical characteristics of meat [in Russian]. *Myasnaya Indystriya [Meat Industry],* 1, 32–34.
59. Wang, N. and Brennan, J.G. 1992. Thermal conductivity of potato as a function of moisture content. *Journal of Food Engineering,* 17(2), 153–160.
60. Eucken, A. 1940. Allgemeine Gesetzmässigkeiten für das Wärmeleitvermögen verschiedener Stoffarten und Aggregatzustände [General laws for the thermal conductivity of various materials and aggregates]. *Forschung auf dem Gebiete des Ingenieurwesens,* 11(1), 6–20, ISSN 00157899.
61. Sakiyama, T., Matsushita, Y., Shiinoki, Y., and Yano, T. 1990. Finite-element analysis on the effective thermal conductivity of dispersed systems of spheres of various sizes. *Journal of Food Engineering,* 11(4), 317–331.
62. Sakiyama, T. and Yano, T. 1994. Finite element analysis on the effective thermal conductivity of dispersed systems. In *Developments in Food Engineering.* Proceedings of the 6th International Congress on Engineering and Food, Chiba, May 1993, Vol. 1. Yano, T., Matsuno, R., and Nakamura, K., eds. Blackie, Glasgow, pp. 146–148.
63. van Beek, G. 1974. Heat transfer through layers of agricultural products of near spherical shape. *Bulletin of the International Institute of Refrigeration,* Annex 3, 183.
64. Fikiin, A.G., Fikiin, K.A., and Triphonov, S.D. 1999. Equivalent thermophysical properties and surface heat transfer coefficient of fruit layers in trays during cooling. *Journal of Food Engineering,* 40, 7–13.
65. Salmon, D., Hammerschmidt, U., van Haneghem, I.A., Kubičár, L., Gustafsson, S., and Tye, R. 2003. Draft standard test protocol for measurement of thermophysical properties of materials using contact transient methods based on a common principle. National Physical Laboratory, Teddington, Middlesex, UK.
66. Gustafsson, S.E. 1991. Transient plane source technique for thermal conductivity and thermal diffusivity measurements of solid materials. *Review of Scientific Instruments,* 62, 797–804.
67. Murakami, E.G., Sweat, V.E., Sastry, S.K., Kolbe, E., Hayakawa, K., and Datta, A. 1996. Recommended design parameters for thermal conductivity probes for nonfrozen food materials. *Journal of Food Engineering,* 27, 109–123.
68. Tagawa, A., Murata, S., and Hinosawa, H. 1995. Measurements of effective thermal conductivity of azuki beans by transient heat flow method using a probe. *Nippon-Shokuhin-Kagaku-Kogaku-Kaishi,* 42(2), 93–99.
69. Voudouris, N. and Hayakawa, K.I. 1995. Probe length and filling material effects on thermal conductivity determined by a maximum slope data reduction method. *Journal of Food Science,* 60, 456–460.

70. Salmon, D., Boumaza, T., Lockmuller, N., and Nesvadba, P. 2001. Thermal conductivity standards for powders, sludges, and high moisture materials. National Physical Laboratory Report CBTLM S49 for the DTI National Measurement Project 3.4.
71. Pippard, A.B. 1961. *Elements of Classical Thermodynamics.* Cambridge University Press, Cambridge, UK.
72. Slade, L., Levine, H., Ievolella, J., and Wang, M. 1993. The glassy state phenomenon in applications for the food industry: Application of the food polymer science approach to structure–function relationships of sucrose in cookie and cracker systems. *Journal of the Science of Food and Agriculture,* 63, 133–176.
73. Slade, L. and Levine, H. 1991. Beyond water activity: Recent advances based on an alternative approach to the assessment of food quality and safety. *CRC Critical Reviews in Food Science and Nutrition,* 30(2–3), 115–360.
74. Wunderlich, B. 1981. The basis of thermal analysis. In *Thermal Characterization of Polymeric Materials,* Turi, E.A., ed. Academic Press, Inc., New York. pp. 91–234.
75. Sperling, L.H. 1986. *Introduction to Physical Polymer Science.* John Wiley and Sons, Inc., New York.
76. White, G.W. and Cakebread, S.H. 1966. The glassy state in certain sugar-containing food products. *Journal of Food Technology,* 1, 73–82.
77. Kalichevsky, M.T., Jaroszkiewicz, E.M., Ablett, S., Blanshard, J.M.V., and Lillford, P.J. 1992. The glass transition of amylopectin measured by DSC, DMTA, and NMR. *Carbohydrate Polymers,* 18, 77–88.
78. Roos, Y.H. 1992. Phase transitions and transformations in food systems. In Heldman, D.R. and Lund, D.B., eds. *Handbook of Food Engineering,* Marcel Dekker, New York. p. 145.
79. Iglesias, H.A. and Chirife, J. 1982. *Handbook of Food Isotherms.* Academic Press, New York.
80. Nesvadba, P. 1996. Thermal and other physical properties of foods: Needs for data and for standards. *International Journal of Food Science and Technology,* 31, 295–296.
81. Carson, J.K. 2011. Predictive modelling of thermal properties of foods. In Comeau, M.A., ed., *New Topics in Food Engineering.* Nova Science Publishers, Inc., New York, pp. 261–278.
82. Evitherm. 2003. *The European Virtual Institute for Thermal Metrology.* EU Project. http://www.evitherm.org.

8

Thermal Properties of Frozen Foods

Gail Bornhorst, Arnab Sarkar,[*] and R. Paul Singh

Contents

8.1	Introduction	248
8.2	Experimental Approaches to Measuring the Thermal Properties of Frozen Foods	248
	8.2.1 Initial Freezing Point and Unfrozen Water	249
	8.2.2 Density	249
	8.2.3 Thermal Conductivity	249
	8.2.4 Enthalpy	250
	8.2.5 Specific Heat	250
	8.2.6 Thermal Diffusivity	255
8.3	General Observations on the Reliability of Experimental Data	255
8.4	Modeling of the Thermal Properties of Frozen Foods	258
	8.4.1 Prediction of Unfrozen Water during Freezing of Foods	258
	8.4.1.1 Density	263
	8.4.1.2 Thermal Conductivity	265
	8.4.1.3 Enthalpy	268
	8.4.1.4 Apparent Specific Heat	272
	8.4.1.5 Thermal Diffusivity	274
	8.4.2 Limitations of Predictive Models	274
List of Symbols		275
Greek Symbols		275
Subscripts		275
References		276

---

[*] Disclaimer: Arnab Sarkar is an employee of PepsiCo Inc. The views expressed in this article are those of the author and do not necessarily reflect the position or policy of PepsiCo, Inc.

8.1 INTRODUCTION

In designing food freezing equipment and processes, knowledge of physical and thermal properties of foods is essential. The computation of refrigeration requirements and freezing times can be done only when quantitative information on food properties is available. Considerable research has been conducted to measure and model properties of foods undergoing various processing treatments. The key properties of interest in food freezing include density, thermal conductivity, specific heat, and thermal diffusivity.

Physical and thermal properties of foods determined at temperatures above freezing are of limited use at freezing conditions. Many food properties show a unique dependence on the state of the water in a food material. During the freezing process, water changes gradually from the liquid phase to solid ice. Because the properties of ice are different from those of liquid water, the properties of food determined at temperatures above freezing are often not valid for subfreezing conditions. In addition, density, porosity, and solutes present have a major effect on thermal properties. The most dramatic change in these properties is observed at temperatures close to the freezing point. Therefore, the determination and modeling of thermal properties of foods under frozen conditions requires explicit knowledge of the state of water in the foods.

In this chapter, we first review some techniques that are commonly employed to measure thermal properties of frozen foods. Here we consider factors that one should be aware of in measuring food properties at subfreezing temperatures. References to important sources of known thermal property data for frozen foods are presented along with observations on the acceptability of such information. Several models have been proposed in the literature to predict thermal properties of frozen foods; some of the key models are discussed in this chapter.

8.2 EXPERIMENTAL APPROACHES TO MEASURING THE THERMAL PROPERTIES OF FROZEN FOODS

Many experimental techniques employed to measure properties of frozen foods are similar to those used for unfrozen foods. The differences in these techniques are mainly in how phase changes of water are accounted for in analyzing the property data.

Experimental determination of food properties has been a subject of numerous publications. Major reviews of published data are provided by Dickerson (1969), Reidy (1968), Qashou (1970), Woodams and Nowrey (1968), Mohsenin (1980), Lind (1991), and Baik et al. (2001). For easier access to the published literature on food properties, a computerized database of food properties containing more than 2000 food and property combinations was developed by Singh (2003). This database contains experimental values of food properties along with the appropriate literature citations.

Selected experimental techniques used for measurement of food properties and important sources of property data of frozen foods are discussed in the following sections.

8.2.1 Initial Freezing Point and Unfrozen Water

Food products may be considered to be solutions of various components in water. The presence of the solutes suppresses the beginning of the freezing process. As a result, the freezing point of foods is lowered as compared to that of pure water (0°C). Differential scanning calorimetry (DSC) techniques have been used for this purpose (Roos and Karel, 1991a; Roos, 2010). More recently, researchers have noted that there may be glass transition of food components, such as sugars, starches, and salts, during freezing, affecting thermal properties. DSC may not be sensitive enough to determine the initial freezing point and the unfrozen water (Laaksonen and Roos, 2001), and other methods, such as dynamic-mechanical analysis (DMA) and modulated differential scanning calorimetry (MDSC) may be suitable alternatives.

However, DSC does provide a method to measure the glass transition temperature that occurs during the freezing process. Several researchers have used DSC or MDSC to determine the frozen glass transition temperatures of foods such as surimi–trehalose mixtures (Ohkuma et al., 2008), mushroom, green cauliflower, navy bean, pea pod (Haiying et al., 2007), and basmati rice (Sablani et al., 2009).

The freezing point temperature decreases during the freezing process as some of the water turns into ice. Thus, the water becomes progressively more concentrated and at the end, some of the water does not freeze at all. The unfrozen water fraction has been difficult to obtain, experimentally. Roos and Karel (1991b), Ablett et al. (1992), Jury et al. (2007), and Li et al. (2008) have used DSC to determine the unfrozen water fraction. Others have used spectroscopic techniques to achieve the same goal (Nagashima and Suzuki, 1985; Lee et al., 2002; Rasanen et al., 1998). Herrera et al. (2007) used a combination of DSC, DMA, and pulsed nuclear magnetic resonance (NMR) techniques to characterize the frozen and unfrozen water fractions, as well as the water mobility in aqueous sucrose and fructose solutions with the addition of polysaccharides.

8.2.2 Density

The density of a food product is measured by weighing a known volume of the product. Since food products are of different shapes and sizes, the accurate measurement of volume can be challenging. Mohsenin (1978) offers several techniques to determine the volume of foods. However, the published literature contains few applications of these methods in measuring the density of frozen foods. Brennvall (2007) presented a method that uses a position sensor to measure dimensional changes in food products during freezing as a means to determine food density. Bantle et al. (2010) used this method to determine the density *Calanus finmarchicus* (a zooplankton species) during freezing.

8.2.3 Thermal Conductivity

Measurement of thermal conductivity has involved the use of both steady-state and nonsteady-state methods. These techniques are discussed in Chapter 3. A steady-state procedure incorporating a guarded plate method was used by Lentz (1961) to measure the thermal conductivity of frozen foods. A nonsteady-state procedure using a probe has been used extensively to measure the thermal conductivity of food materials by Sweat et al. (1973) and Hough and Calvelo (1978). Thompson et al. (1983) used the probe method to measure the thermal conductivity of frozen corn, and Ramaswamy and Tung (1981) used it for measuring the thermal conductivity of frozen apples.

A comprehensive review of research studies on thermal conductivity was published by Murakami and Okos (1989). For additional data on thermal conductivity, the following references are recommended.

Fruits and vegetables: Drusas and Saravacos (1985), Hsieh et al. (1977), Kethley et al. (1950), Marin et al. (1985), Ramaswamy and Tung (1981), Smith et al. (1952), Sweat (1974), and Telis et al. (2007).

Meat products: Morley (1972), Fleming (1969), Levy (1982), Gogol et al. (1972), Zaritzky (1983), Lind (1991), Succar (1989), Farag et al. (2008), and Kumcuoglu et al. (2010).

Fish and seafoods: Smith et al. (1952), Matuszek et al. (1983), Annamma and Rao (1974), Levy (1982), and Bantle et al. (2010).

Poultry and egg products: Gogol et al. (1972) and Smith et al. (1952).

Miscellaneous foods: Baik et al. (2001), Meffert (1984), Woodams and Nowrey (1968), Heldman (1982), Polley et al. (1980), Succar and Hayakawa (1983), Mellor (1976, 1980), Qashou et al. (1972), Cuevas and Cheryan (1978), Van den Berg and Lentz (1975), Jowitt et al. (1983), ASHRAE (1985), Fikiin (1974), Rolfe (1968), Kumcuoglu et al. (2007), Singh et al. (1989), Sweat (1975), Tressler (1968), and Jury et al. (2007).

8.2.4 Enthalpy

The enthalpy of frozen foods has been mostly determined using calorimetric methods. Riedel (1951, 1956, 1957a,b) conducted pioneering studies in this area. Riedel (1951) found that the enthalpy of fruit juices was dependent on the dry matter content. He noted that the actual make-up of the dry matter did not appear to influence the enthalpy values. He prepared charts to express enthalpy values of fruit and vegetable juices as a function of temperature and the fraction of dry matter content. Using a similar approach, he analyzed data on the enthalpy of meats, fish, and egg products. Tabulated data from Riedel's charts are shown in Table 8.1 (Dickerson, 1969). Thompson et al. (1983) and Wang and Kolbe (1990) used calorimetric methods to determine enthalpy of corn on the cob and seafoods, respectively. Additional data on enthalpy values of foods can be obtained from the following references.

Fruits and vegetables: Hsieh et al. (1977) and Singh (1982).

Meat products: Lind (1991) and Succar (1989).

Miscellaneous foods: Rahman (1995), Baik et al. (2001), Chang and Tao (1981), Chen (1985), Schwartzberg (1976), Rolfe (1968), ASHRAE (1977), Mellor (1976, 1980), Succar and Hayakawa (1983), Heldman (1982), Li et al. (2008), and Bantle et al. (2010).

8.2.5 Specific Heat

Specific heat is most commonly determined by the use of calorimetric methods. However, these methods are more useful in determining specific heat when the phase change

Table 8.1 Enthalpy of Frozen Foods[a]

Product	Water Content (wt%)	Mean Specific Heat[b] at 4–32°C [kJ/(kg·°C)]		Temperature (°C)																		
				−40	−30	−20	−18	−16	−14	−12	−10	−9	−8	−7	−6	−5	−4	−3	−2	−1	0
Fruits and Vegetables																					
Applesauce	82.8	3.73	Enthalpy (kJ/kg)	0	23	51	58	65	73	84	95	102	110	120	132	152	175	210	286	339	343
			% water unfrozen		6	9	10	12	14	17	19	21	23	27	30	37	44	57	82	100	—
Asparagus, peeled	92.6	3.98	Enthalpy (kJ/kg)	0	19	40	45	50	55	61	69	73	77	83	90	99	108	123	155	243	381
			% water unfrozen						5	6	—	7	8	10	12	15	17	20	29	58	100
Bilberries	85.1	3.77	Enthalpy (kJ/kg)	0	21	45	50	57	64	73	82	87	94	101	110	125	140	167	218	348	352
			% water unfrozen				7	8	9	11	14	15	17	18	21	25	30	38	57	100	—
Carrots	87.5	3.90	Enthalpy (kJ/kg)	0	21	46	51	57	64	72	81	87	94	102	111	124	139	166	218	357	361
			% water unfrozen				7	8	9	11	14	15	17	18	20	24	29	37	53	100	—
Cucumbers	95.4	4.02	Enthalpy (kJ/kg)	0	18	39	43	47	51	57	64	67	70	74	79	85	93	104	125	184	390
			% water unfrozen									5					11	14	20	37	100
Onions	85.5	3.81	Enthalpy (kJ/kg)	0	23	50	55	62	71	81	91	97	105	115	125	141	163	196	263	349	353
			% water unfrozen		5	8	10	12	14	16	18	19	20	23	26	31	38	49	71	100	—
Peaches	85.1	3.77	Enthalpy (kJ/kg)	0	23	50	57	64	72	82	93	100	108	118	129	146	170	202	274	348	352

continued

Table 8.1 (continued) Enthalpy of Frozen Foods[a]

Product	Water Content (wt%)	Mean Specific Heat[b] at 4–32°C [kJ/(kg·°C)]		Temperature (°C)																	
				-40	-30	-20	-18	-16	-14	-12	-10	-9	-8	-7	-6	-5	-4	-3	-2	-1	0
Plums	80.3	3.65	% water unfrozen	—	6	9	10	12	14	17	19	21	23	26	29	35	43	54	80	100	—
			Enthalpy (kJ/kg)	0	25	57	65	74	84	97	111	119	129	142	159	182	214	262	326	329	333
Plums without stones	82.7	3.73	% water unfrozen	—	8	14	16	18	20	23	27	29	33	37	42	50	61	78	100	—	—
Raspberries			Enthalpy (kJ/kg)	0	20	47	53	59	65	75	85	90	97	105	115	129	148	174	231	340	344
			% water unfrozen	—	—	7	8	9	10	13	16	17	18	20	23	27	33	42	61	100	—
Spinach	90.2	3.90	Enthalpy (kJ/kg)	0	19	40	44	49	54	60	66	70	74	79	86	94	103	117	145	224	371
			% water unfrozen	—	—	—	—	—	—	6	7	—	—	9	11	13	16	19	28	53	100
Strawberries	89.3	3.94	Enthalpy (kJ/kg)	0	20	44	49	54	60	67	76	81	88	95	102	114	127	150	191	318	367
			% water unfrozen	—	—	5	—	6	7	9	11	12	14	16	18	20	24	30	43	86	100
Sweet cherries without stones	77.0	3.60	Enthalpy (kJ/kg)	0	26	58	66	76	87	100	114	123	133	149	166	190	225	276	317	320	324
			% water unfrozen	—	9	15	17	19	21	26	29	32	36	40	47	55	67	86	100	—	—
Tall peas	75.8	3.56	Enthalpy (kJ/kg)	0	23	51	56	64	73	84	95	102	111	121	133	152	176	212	289	319	323
			% water unfrozen	—	6	10	12	14	16	18	21	23	26	28	33	39	48	61	90	100	—

THERMAL PROPERTIES OF FROZEN FOODS

Food	Water %		Property																			
Tomato pulp	92.9	4.02	Enthalpy (kJ/kg)	0	20	42	47	52	57	63	71	75	81	87	93	103	114	131	166	266	382
			% water unfrozen	—	—	—	—	5	—	6	7	8	10	—	14	16	18	24	33	65	100
Eggs			% water unfrozen	—	—	10	—	—	—	—	13	—	—	—	18	20	23	28	40	82	100
Egg white	86.5	3.81	Enthalpy (kJ/kg)	0	18	39	43	48	53	58	65	68	72	75	81	87	96	109	132	210	352
Egg yolk	40.0	2.85	Enthalpy (kJ/kg)	0	19	40	45	50	56	62	68	72	76	80	85	92	99	109	128	182	191
			% water unfrozen	20	—	—	22	—	24	—	27	28	29	31	33	35	38	45	58	94	100
Whole egg with shell[d]	66.4	3.31	Enthalpy (kJ/kg)	0	17	36	40	45	50	55	61	64	67	71	75	81	88	98	117	175	281

Fish and Meat

Food	Water %		Property																			
Cod	80.3	3.69	Enthalpy (kJ/kg)	0	19	42	47	53	66	74	79	84	89	96	105	118	137	177	298	323	
			% water unfrozen	10	10	11	12	12	13	14	16	17	18	19	21	23	27	34	48	92	100
Haddock	83.6	3.73	Enthalpy (kJ/kg)	0	19	42	47	53	59	66	73	77	82	88	95	104	116	136	177	307	337
			% water unfrozen	8	8	9	10	11	11	12	13	14	15	16	18	20	24	31	44	90	100
Perch	79.1	3.60	Enthalpy (kJ/kg)	0	19	41	46	52	58	65	72	76	81	86	93	101	112	129	165	284	318
			% water unfrozen	10	10	11	12	12	13	14	15	16	17	18	20	22	26	32	44	87	100
Beef, lean fresh[e]	74.5	3.52	Enthalpy (kJ/kg)	0	19	42	47	52	58	65	72	76	81	88	95	105	113	138	180	285	304
			% water unfrozen	10	10	11	12	13	14	15	16	17	18	20	22	24	31	40	55	95	100
Beef, lean dried	26.1	2.47	Enthalpy (kJ/kg)	0	19	42	47	53	62	66	70	—	74	—	79	—	84	—	89	—	93

continued

Table 8.1 (continued) Enthalpy of Frozen Foods[a]

Product	Water Content (wt%)	Mean Specific Heat[b] at 4–32°C [kJ/(kg·°C)]		Temperature (°C)															
				−40	−30	−20	−18	−16	−14	−12	−10	−9	−8	−7	−6	−5	−4	−3	−2	−1	0
Bread White	37.3	2.60	Enthalpy (kJ/kg)	0	17	35	39	44	49	56	67	75	83	93	104	117	124	128	131	134	137
Whole wheat	42.4	2.68	Enthalpy (kJ/kg)	0	17	36	41	48	56	66	78	86	95	106	119	135	150	154	157	160	163

Source: Dickerson, R.W.J. 1981. Enthalpy of frozen foods. *Handbook and Product Directory Fundamentals*. American Society of Heating, Refrigeration, and Air Conditioning, New York.

[a] Above −40°C.
[b] Temperature range limited to 0–20°C for meats and 20–40°C for egg yolk.
[c] Total weight of unfrozen water = (total weight of food)(% water content/100)(water unfrozen/100).
[d] Calculated for a weight composition of 58% white (86.5% water) and 32% yolk (50% water).
[e] Data for chicken, veal, and venison very nearly matched the data for beef of the same water content.

occurs at a fixed temperature. During freezing of foods, the phase change occurs over a range of temperature, and as a result the calorimetric procedures are of limited application. An alternative approach involves experimentally determining the enthalpy values of foods for a range of temperatures and then calculating an apparent specific heat from the collected data. Duckworth (1971) applied differential thermal analysis to foods, and specific heat values using this approach have been reported by Ramaswamy and Tung (1981). Additional data on specific heat values can be obtained from the following references.

Fish and seafoods: Singh (1982), Rahman (1993), and Bantle et al. (2010).
Meat Products: Tavman et al. (2007) and Farag et al. (2008).
Miscellaneous food items: Rahman (1995), Baik et al. (2001), Meffert (1984), Polley et al. (1980), Heldman (1982), Mellor (1976, 1980), Staph (1949), Fikiin (1974), Schwartzberg (1976), and Chen (1985).

8.2.6 Thermal Diffusivity

Most published data on thermal diffusivity are based on calculating thermal diffusivity from known values of thermal conductivity, density, and specific heat. Direct measurement of thermal diffusivity is not very common. A review of measurement methods, previously reported values, and sources of measurement error were reported by Nesvadba (1982).

One method to calculate the thermal diffusivity is by conducting a nonlinear regression analysis of time–temperature histories. This method has been used to determine the thermal diffusivity in model foods, mashed potatoes, shrimp by Albin et al. (1979), ricotta cheese by Nunes et al. (2008), and frozen pizza and puff pastry by Kumcuoglu and Tavman (2007). Experimentally determined thermal diffusivities were reported by Annamma and Rao (1974) for fresh fish and by Farag et al. (2008) for beef meat blends.

8.3 GENERAL OBSERVATIONS ON THE RELIABILITY OF EXPERIMENTAL DATA

It is not uncommon to find conflicting results when comparing experimentally determined properties. In order to systematically compare the property values reported in the literature, Heldman and Singh (1986) suggested an alternative format to tabulate the available data. Their proposed format is presented in Table 8.2 for the published values for frozen fruits and vegetables. According to this format, the property values and the range of temperatures used during the measurements are included. This type of information is essential for the user to avoid mistakes in extrapolation of these values to other conditions. In addition, the format used for Table 8.2 allows for easy comparison between different reported values.

As shown in Table 8.2, the density values do not change significantly unless the product temperature is within 2–5°C of the initial freezing temperature of the product. In most published studies, either thermal conductivity is expressed as a single measurement or an equation is provided for a range of temperatures. Often, linear regression is used to develop an equation; therefore, caution must be exercised for interpolating within the indicated

Table 8.2 Thermal Properties of Fruits and Frozen Doughs

Fruit	Moisture Content (%)	Product Characteristics[a]	Temperature (°C)	Density (kg/m³)	Thermal Conductivity W/(m °C)	R^2	Specific Heat kJ/(kg °C)	R^2	Enthalpy kJ/kg	R^2	Ref.[b]
Apple	85.0	—	0	—	—		2.093		—		2
	84.1	—	—	—	—		1.884		—		1
	83.7	—	−40 to −20	—	—		2.763 + 0.042T	(0.88)	—		7
	84.6	—	−40 to −20	—	—		3.433 + 0.042T	(0.87)	—		8
	87.4	10.8 (s.s.)	−40 to −1	785–791	1.289−0.095T		2.84 + 0.0138T		—		6
	85.7	12.15 (s.s.)	−40 to −1	789–787	1.066−0.0111T		2.5 + 0.0118T		—		6
Apricot	85.4	—	0	—	—		1.926		—		2
Applesauce	82.8	—	−40 to −2	—	—				236 exp (0.08T)	(0.96)	3
Blackberries	85.0	—	0	—	—		1.926		—		2
Bilberries	85.1	—	−40 to −2	—	—				189 exp (0.08T)	(0.97)	3
Cherries	82.0	—	0	—	—		1.926		—		2
Currants	84.7	—	0	—	—		1.884		—		1
Figs	90.0	—	−40 to −20	—	—		3.475 + 0.042T	(0.97)	—		7
Grapes	79.3	—	−40 to −20	—	—		4.396 + 0.042T	(0.91)	—		7
	82.7	—	−40 to −20	—	—		3.852 + 0.042T	(0.75)	—		8
Grapefruit	89.0	—	0	—	—		2.01		—		2
Melon	92.6	—	−40 to −20	—	—		2.97 + 0.042T		—		10
Oranges	80.7	—	−40 to −20	—	—		3.6 + 0.042T		—		9
Orange juice	—	20°Brix	−30 to −15	1009	0.917−0.087T	(0.87)	5.945 + 0.987T	(1.0)	—		4
	—	40°Brix	−30 to −15	1121	−0.571−0.069T	(0.99)	10.424 + 0.209T	(0.87)	—		4
	—	60°Brix	−30 to −15	1266	—		8.373 + 0.126T	(0.36)	—		4
Peaches	89.6	—	−40 to −20	—	—		3.76 + 0.042T	(0.98)	—		7
	85.1	—	−40 to −2	—	—				227.9 exp (0.08T)	(0.96)	3
Pears	79.4	—	−40 to −20	—	—		3.6 + 0.042T	(0.88)	—		10
	83.8	—	−40 to −20	—	—		—		233.7 exp (0.08T)	(0.96)	3

Food	s.s.[a]	Note	Temperature (°C)	ρ	c	k		Ref.[b]
Plums	—	—	−13 to −17	577	0.294	—		9
	—	—	−14 to −19	609	0.242	—		9
	80.3	—	−40 to −2	—	—	287.3 exp (0.08T)	(0.97)	3
Raspberries	80.7	—	0	—	—	—		2
	82.7	—	−40 to −2	—	—	—	3	
Strawberries	—	Tightly packed	−12 to −19	801	1.123	—		9
	—	—	−8 to −20	801	1.097	—		9
	—	—	−6 to −17	801	1.111	—		9
	—	—	−3 to −10	801	1.089	—		9
	—	Sucrose syrup	−13 to −17	801	0.969	—		9
	—	Large	−14 to −19	641	0.537	—		9
	—	Mixed size	−14 to −19	641	0.537	—		9
	—	Small	−15 to −21	481	0.584	—		9
	—	—	−12	801	1.073	—		7
	90.9	—	−40 to −20	—	—	2.847 + 0.02T		7
	89.3	—	−40 to −2	—	—	170 exp (0.07T)	(0.97)	3
Cherries (sweet)	77.0	—	−40 to −2	—	—	296.6 exp (0.08T)	(0.97)	5
Bread dough	43.5	—	−43.5	1100	0.920	1.760	—	11
Bread dough	43.5	—	−28.5	1100	0.900	1.940	—	11
Bread dough	46.1	—	−38.0	1100	1.030	1.760	—	11

[a] s.s. = Soluble solids.
[b] 1, Anderson (1959); 2, ASHRAE (1977); 3, Dickerson (1981); 4, Keller (1956); 5, Larkin et al. (1983); 6, Ramaswamy and Tung (1981); 7, Short and Bartlett (1944); 8, Short and Staph (1951); 9, Smith et al. (1952); 10, Staph (1949); 11, Baik et al. (2001).

range of temperatures. The specific heat data are usually provided either as a single measurement or as an equation expressing the relationship between the specific heat and temperature. The range of temperatures for the specific heat values is –20 to –15°C, which is a rather narrow range. This small temperature range is used to avoid the nonlinear relationship between specific heat and temperature near the initial freezing temperature of the product. The enthalpy data are presented in Table 8.2 in the form of exponential relationships between enthalpy and temperature. The justification for the exponential relationships is based on the work of Heldman (1974, 1982). A reference temperature of –40°C is used for the experimental values in the regression equations shown in Table 8.2.

From the data presented in Table 8.2, it is evident that the available information on densities is meager; in fact, experimental values are presented only for frozen orange juice concentrate and strawberries. Similarly, the available information on thermal conductivity of frozen foods is limited. As noted by Heldman and Singh (1986), the data for strawberries and plums indicate a lower thermal conductivity than expected because these products have high water content. This result can be attributed to the measurement of thermal conductivity of a bulk of fruit that would contain air spaces between the fruit pieces.

Specific heat has been measured and reported for several food materials. As shown in Table 8.2, the mathematical relationships indicate high correlation coefficients. When the data are presented without identifying the corresponding temperature, the comparison between various reported data becomes difficult. It is important that investigators report product conditions in addition to the property data. As suggested by Heldman and Singh (1986), the format used in Table 8.2 can be extended to other foods for similar analysis.

8.4 MODELING OF THE THERMAL PROPERTIES OF FROZEN FOODS

As foods are frozen, there is no sharp phase change occurring at a fixed temperature; instead, the transition of water to ice takes place over a range of temperatures. Most food products begin to freeze between –1 and –3°C. The major change in phase occurs at temperatures of 4–10°C below the initial freezing temperature, and the phase change process may be considered to be complete only when the temperatures drop below –40°C. Since most foods contain large amounts of water, the phase change of water to ice has a dramatic influence on the thermal properties of the food.

8.4.1 Prediction of Unfrozen Water during Freezing of Foods

The thermal properties of frozen foods depend significantly on their water content in the frozen state (Heldman, 1982). Several investigators have developed models to predict thermal properties of frozen foods by first modeling the changes in the unfrozen water content of a food as it undergoes the freezing process (Heldman and Gorby, 1975; Hsieh et al., 1977; Heldman, 1982; Larkin et al., 1983). Their method requires knowledge of the more readily available initial freezing temperature (Table 8.3) and the thermal properties of the unfrozen product. The method is simple and can be easily programmed on a desktop computer. The results from a number of prediction models have generally found good agreement with experimentally obtained results.

Table 8.3 Initial Freezing Temperatures of Fruit, Vegetables, Juices, Meat, and Seafood

Product	Water Content (wt%)	Initial Freezing Temperature (°C)
Apple juice	87.2	−1.44
Apple juice concentrate	49.8	−11.33
Applesauce	82.8	−1.67
Asparagus	92.6	−0.67
Beef muscle	74.0	−1.75
Beef sirloin[a]	75.0	−1.20
Bilberries	85.1	−1.11
Bilberry juice	89.5	−1.11
Carrots	87.5	−1.11
Cherry juice	86.7	−1.44
Chicken[b]	76.0	−0.79
Cod[a]	78.0	−2.20
Grape juice	84.7	−1.78
Ham[a]	56.0	−1.70
Lamb muscle[a]	74.5	−1.74
Onions	85.5	−1.44
Orange juice	89.0	−1.17
Peaches	85.1	−1.56
Pears	83.8	−1.61
Plums	80.3	−2.28
Pork muscle[a]	74.5	−1.75
Raspberries	82.7	−1.22
Raspberry juice	88.5	−1.22
Shrimp[c]	70.8	−2.20
Spinach	90.2	−0.56
Strawberries	89.3	−0.89
Strawberry juice	91.7	−0.89
Sweet cherries	77.0	−2.61
Tall peas	75.8	−1.83
Tomato pulp	92.9	−0.72
Veal[a]	74.5	−1.75

Source: Heldman, D.R. and Singh, R.P. 1981. *Food Process Engineering.* 2nd ed. AVI Publishing Co., Westport, CT.

[a] Murakami and Okos (1989).
[b] Dickerson (1969).
[c] ASHRAE (1967).

In order to understand the development of this method, it is important to understand the concept of freezing point depression. Initially, a food consists of product solids and water. As sensible heat is removed, the temperature of the mix containing solid and water decreases. Just below the initial freezing point, the water begins to convert into ice. The food system then has three components: product solids, ice, and water. As more heat is removed, more of the water converts into ice, and the remaining solution becomes more concentrated in terms of the product solids. Because of the higher solids concentrations, the temperature at which freezing will occur is depressed. Thus, the removal of latent heat occurs over a range of temperature instead of at one fixed temperature as in the case of a pure water system.

For an ideal binary solution, Moore (1972) presented the following equation to describe the freezing point depression

$$\frac{\lambda}{R}\left[\frac{1}{T_P} - \frac{1}{T}\right] = \ln X_w \tag{8.1}$$

where λ is the heat of fusion per mole for pure water; R is the universal gas constant; T_P is the freezing point of pure water (273.15 K); T is the product temperature below T_P (K); and X_W is the mole fraction of water in the product.

The mole fraction of water in the product, X_W in Equation 8.1, is expressed as

$$X_W = \frac{M_W/W_W}{M_W/W_W + \sum M_j/W_j} \tag{8.2}$$

where M_W is the mass fraction of water; W_W is the molecular weight of water; M_j is the mass fraction of the jth component of the soluble solids; and W_j is the molecular weight of the jth component of the soluble solids.

The initial freezing point temperature, T_Z, can be obtained by combining Equations 8.1 and 8.2

$$\frac{1}{T_Z} = \frac{1}{T_P} - \frac{R}{\lambda}\ln\left[\frac{M_{WZ}/W_W}{M_{WZ}/W_W + \sum M_j/W_j}\right] \tag{8.3}$$

where M_{WZ} is the mass fraction of water at the initial freezing temperature. Equation 8.3 is used to calculate the initial freezing point when mass fractions of all soluble solids are known. Choi and Okos (1984) provide data on mass fractions in food systems.

If T_Z is close to T_P, then the following linear equation is obtained from Equation 8.3:

$$T_P - T_Z = \frac{RT_P^2}{\lambda}\sum \frac{M_j}{W_j} \tag{8.4}$$

The above equations are used in determining the unfrozen water in a food at any temperature. The procedure involves the following steps. Assuming that the initial freezing temperature is known, its value is substituted in Equation 8.3 or 8.4 and the term representing soluble solids is calculated. This term is then substituted in Equations 8.1 and 8.2 to determine the unfrozen water fraction in the food, M_W, at any desired temperature.

THERMAL PROPERTIES OF FROZEN FOODS

The soluble solid term can also be eliminated between Equations 8.3 and 8.4 to obtain the expression

$$M_W = M_{WZ} \frac{F_Z - F_P}{F - F_P} \qquad (8.5)$$

where

$$F = F\{T\} = \exp\left[\frac{\lambda}{RT}\right] \qquad (8.6)$$

During the freezing process, the ice and water fractions remain constant. The ice fraction can then be obtained from simple mass balance as

$$M_I = M_{WZ} - M_W \qquad (8.7)$$

This procedure was used by Heldman (1974) to predict the magnitude of freezing point depression for a food system. He found good agreement with several experimental results obtained by Dickerson (1969). The procedure can be easily programmed into a spreadsheet for calculation of unfrozen water content at any desired temperature. The appropriate equations are entered into a spreadsheet as shown in Figure 8.1. This spreadsheet calculation is useful for determining both freezing point depression and the unfrozen water content at a desired temperature.

	A	B	C	D	E
1					
2	**Freezing point depression**				
3	Given				
4	Solids content, fraction	0.15			
5	Water content, fraction	0.85			
6	Molecular weight of solids	180			
7					
8	Mole fraction of water, Xw	0.9827			
9	Temperature of ice crystal formation (K)	271.21			
10	Freezing point depression (K)	1.79	= 273−B9		
11	**Percent water unfrozen**				
12	Given				
13	Temperature (°C)	−10			
14	Temperature (K)	263	= 273+B13		
15	Initial freezing temperature (°C)	−1.22			
16	Initial freezing temperature (K)	271.78	= 273+B15		
17	Water content, fraction	0.827			
18					
19	Xw	0.9882	= EXP ((6003/8.314)*(1/273−1/B16))		
20	Effective molecular weight	315.3	= (1−B17)/B17/18/B19−		
21	Apparent mole fraction, Xu	0.904	= EXP ((6003/8.314)*(1/273−1/B14))		
22	Unfrozen water fraction	0.093	= (1−B17)*B21/(B20*(1/18−		
23	Percent of original water fraction	11.29	= B22/B17*10		
24					

Figure 8.1 A spreadsheet calculation of freezing point depression and percent unfrozen water content.

The above formulation does not account for unfreezable water content in a food system. For foods that contain high amounts of unfreezable water, this omission can result in a larger disagreement between predicted and experimental results. The following correction was made in Equation 8.5 to account for the unfreezable water:

$$M_W = (M_{WZ} - M_A)\frac{F_Z - F_P}{F - F_P} + M_A \qquad (8.8)$$

Lescano (1973) and Heldman (1974) found this procedure to provide satisfactory agreement with experimentally obtained values. Figure 8.2 shows predicted results for unfrozen water fraction of raspberries and experimental values.

Since the latent heat of ice decreases by 27% for a temperature decrease from 0°C to −40°C, the following expressions are suggested by Mannapperuma and Singh (1989) to account for this variation. They assumed that the latent heat variation in the temperature range 0°C to −40°C is linear; therefore,

$$\lambda = \lambda_0 + \lambda_1 T \qquad (8.9)$$

$$M_W = (M_{WZ} - M_A)\frac{F'_Z - F'_P}{F' - F'_P} + M_A \qquad (8.10)$$

$$F' = F'\{T\} = T^{-(\lambda_1/R)} \exp\left[\frac{\lambda_0}{RT}\right] \qquad (8.11)$$

Expressions 8.9 through 8.11 are recommended (instead of Equations 8.6 and 8.8) when T_Z is much lower than T_P.

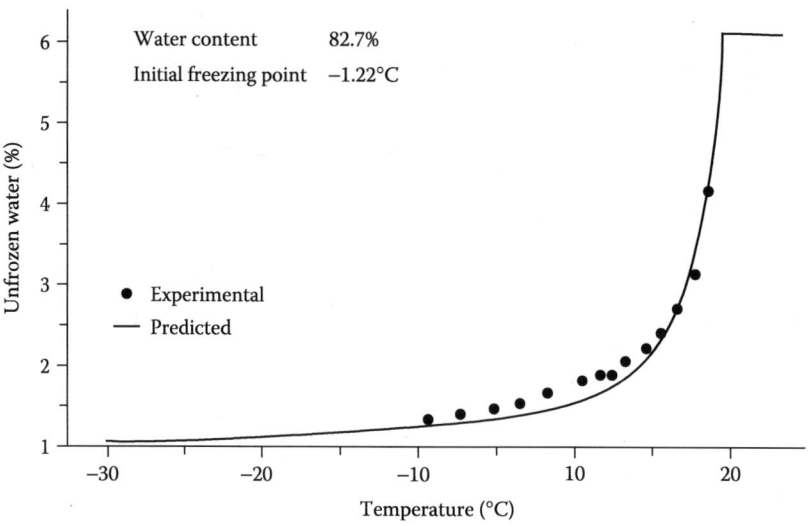

Figure 8.2 Percent unfrozen water content in raspberries. (From Heldman, D.R. 1974. *Transactions of ASAE.* 17(1):63–66.)

The relationships described here come from theoretical considerations based on the concept of ideal solutions. But food materials do not always behave as ideal solutions because of molecular and chemical interactions. Hence, the freezing point depression predicted by these relationships may be different, especially if there is strong solute–water interaction. As an alternative, some researchers have tried to empirically predict freezing point depression by curve-fitting large amounts of experimental data (Chen, 1986, 1987a; Chen and Nagy, 1987; Rahman, 1994; Rahman and Driscoll, 1994; Sanz et al., 1989). Chen (1987b) and Chen and Nagy (1987) proposed modifications to Equation 8.3 for liquid products to incorporate the nonideal nature of the solution. Succar and Hayakawa (1990) have suggested an iterative method based on enthalpy correlations above the freezing point for determining the initial freezing point. Boonsupthip and Heldman (2007) proposed a model to determine the frozen water fraction in food products based on the food composition, specifically, the concentration and molecular weight of food components. Their model has shown good agreement to experimental results.

8.4.1.1 Density

Hsieh et al. (1977) presented the following expression to predict the density of a frozen food:

$$\frac{1}{\rho} = M_U\left(\frac{1}{\rho_U}\right) + M_S\left(\frac{1}{\rho_S}\right) + M_r\left(\frac{1}{\rho_I}\right) \tag{8.12}$$

where ρ is density; ρ_U is the density of unfrozen water; ρ_S is the density of product solids; and ρ_I is the density of ice.

This model requires knowledge of the mass fractions of unfrozen water, product solids, and ice as inputs (Heldman, 1982; Heldman and Singh, 1981). The density values of each of the fractions are also provided as inputs to the model.

An example of the results obtained for predicted values of density of strawberries are shown in Figure 8.3. It is evident that between the temperatures of initial freezing and −10°C there is a major dependence of density on temperature. It is evident that the density of strawberries decreases from about 1050 to 960 kg/m³ as the product freezes and the temperature is lowered to −40°C.

When the predicted values of density are compared with the experimental data (as shown in Table 8.2), the predicted values are greater than those reported by investigators. The difference may be attributed to the value of initial or unfrozen product density used in the calculations and the actual density of the product in experiments. The prediction model for density can be used for situations that may be complex such as when one is predicting the density of different components of corn on the cob (Figure 8.4). The experimental values in Figure 8.4 were obtained by Thompson et al. (1983).

Since the porosity of a food material can strongly influence its density, Equation 8.12 may be modified to incorporate porosity (Mannapperuma and Singh, 1990)

$$\frac{1}{\rho} = \frac{1}{1-\varepsilon}\sum_i \frac{M_i}{\rho_i} \tag{8.13}$$

where ε is porosity and i denotes the ith component in the food system.

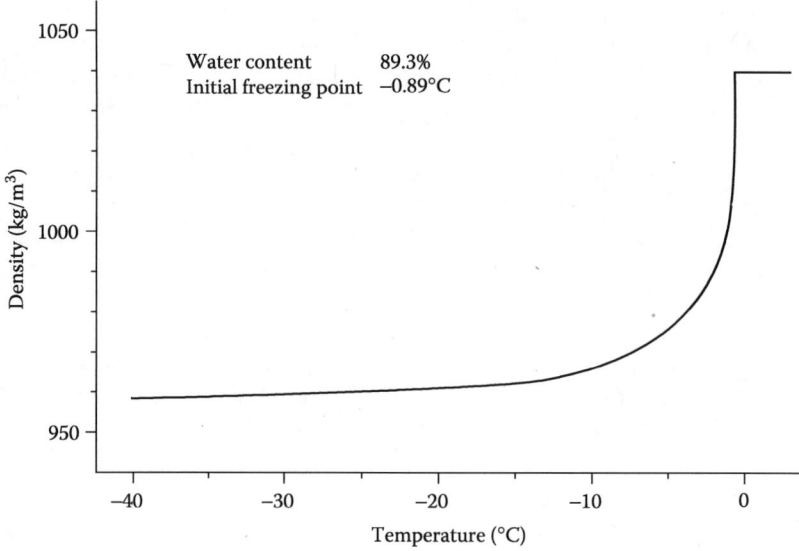

Figure 8.3 Density of strawberries as a function of temperature. (From Heldman, D.R. 1982. *Food Technology*. 36(2):92–96.)

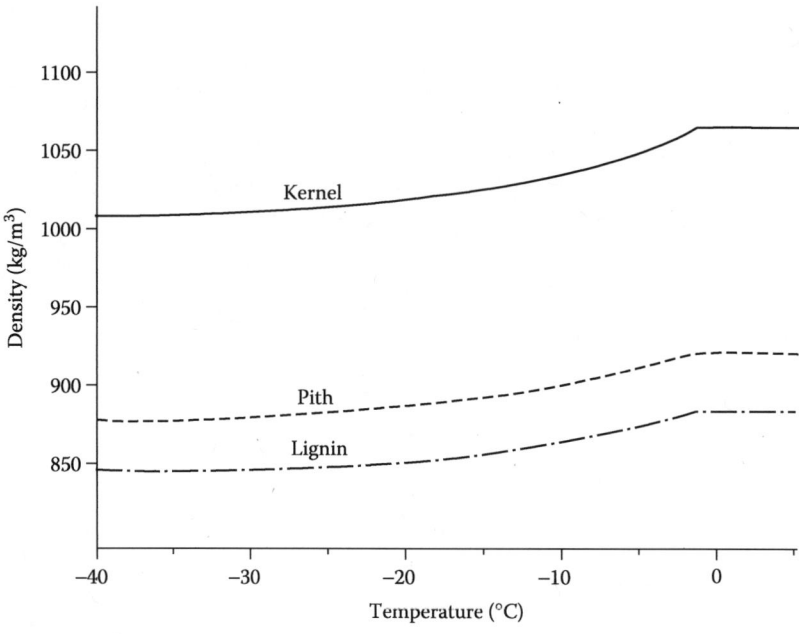

Figure 8.4 Density of components of corn on the cob as a function of temperature. (From Heldman, D.R. and Singh, R.P. 1986. Thermal properties of frozen foods. *Physical and Chemical Properties of Food*. Ed. Okos, M.R. American Society of Agricultural Engineers, St. Joseph, MI.)

Bantle et al. (2010) used Equation 8.12 to estimate the density of *Calanus finmarchicus* and found a good agreement with their measured values (<1% difference). In addition, an empirically derived polynomial equation may be used to estimate product density during freezing. Moraga et al. (2011) used a fifth-order polynomial to estimate the density of meat cylinders during freezing. To account for changes in density due to temperature, they varied the polynomial model coefficients over certain temperature ranges.

8.4.1.2 Thermal Conductivity

The structure of a food product has a major influence on the food's thermal conductivity. Foods that contain fibers exhibit different thermal conductivities parallel to the fibers compared to conductivities perpendicular to the fibers. Like density, porosity has a major influence on thermal conductivity of a food material. The freezing process may significantly alter the porosity of a food material; thus, the prediction of changes in thermal conductivity during freezing becomes more complicated.

Jason and Long (1955) and Lentz (1961) have used the Maxwell–Euken models for predicting changes in thermal conductivity during the freezing process. Considering that a food system constitutes a continuous phase and a dispersed phase, the following equation may be written to predict thermal conductivity:

$$k = k_c \left[\frac{(3 - \zeta) - 2\zeta V_d}{(3 - \zeta) + \zeta V_d} \right] \quad (8.14)$$

where

$$\zeta = 1 - \frac{k_d}{k_c} \quad (8.15)$$

k_d is the thermal conductivity of the dispersed phase; k_c is the thermal conductivity of the continuous phase; and V_d is the volume of the dispersed phase.

For unfrozen foods, the food solids are considered the dispersed phase and water the continuous phase. Jason and Long (1955) used a two-stage model to estimate thermal conductivity. In the first stage they considered ice dispersed in water, and in the second phase they considered food solids as dispersed in an ice–water mixture. Their prediction of thermal conductivity of codfish is shown with experimental values in Figure 8.5.

A more comprehensive treatment of modeling the thermal conductivity of foods was presented by Kopelman (1966). He considered three different models representing the foods: homogeneous, fibrous, and layered foods. He developed models for each of these systems and presented them as follows:

Homogeneous system:

$$k = k_c \left[\frac{1 - \zeta V_d^{2/3}}{1 - \zeta V_d^{2/3}(1 - V_d^{1/3})} \right] \quad (8.16)$$

Fibrous system:

$$k_p = k_c (1 - \zeta V_d) \quad (8.17)$$

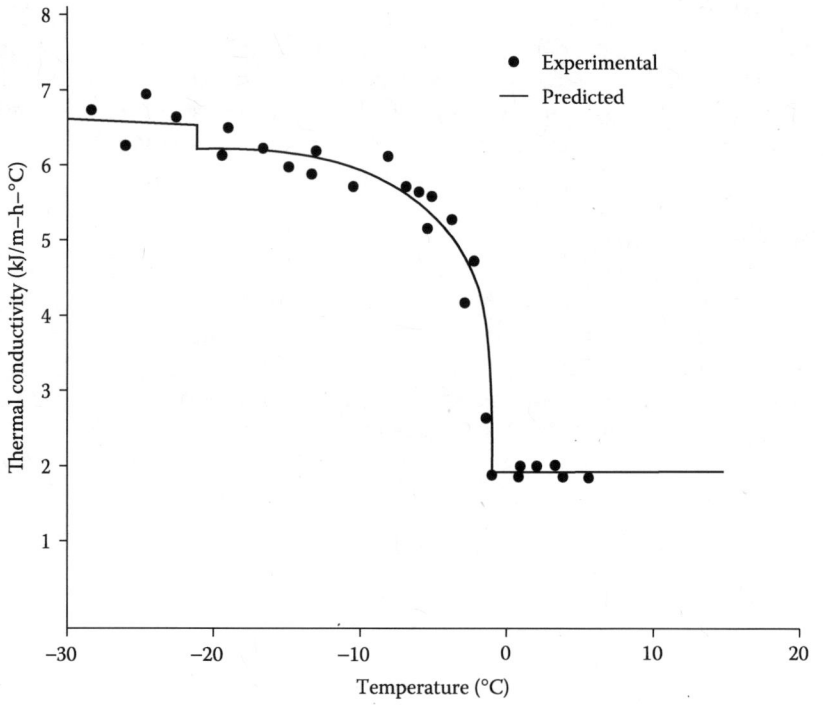

Figure 8.5 Thermal conductivity of codfish as a function of temperature. (From Jason, A.C. and Long, R.A.K. 1955. *IX International Congress of Refrigeration.* 2(1):160.)

$$k_\perp = k_c \left[\frac{\zeta - V_d^{1/2}}{\zeta - V_d^{1/2}(1 - V_d^{1/2})} \right] \tag{8.18}$$

Layered service:

$$k_p = k_c(1 - \zeta V_d) \tag{8.19}$$

$$k_\perp = k_c \left[\frac{\zeta - 1}{\zeta - 1 + \zeta V_d} \right] \tag{8.20}$$

Heldman and Gorby (1975) used the Kopelman models to predict the thermal conductivity of frozen foods. An illustration of the Kopelman predictive model is shown in Figure 8.6. They showed the variation in the thermal conductivity of beef, parallel and perpendicular to the direction of the fibers. The agreement with experimental data is quite satisfactory.

The experimental data of Thompson et al. (1983) for corn on the cob were used by Heldman and Singh (1986) to predict the thermal conductivities for various product

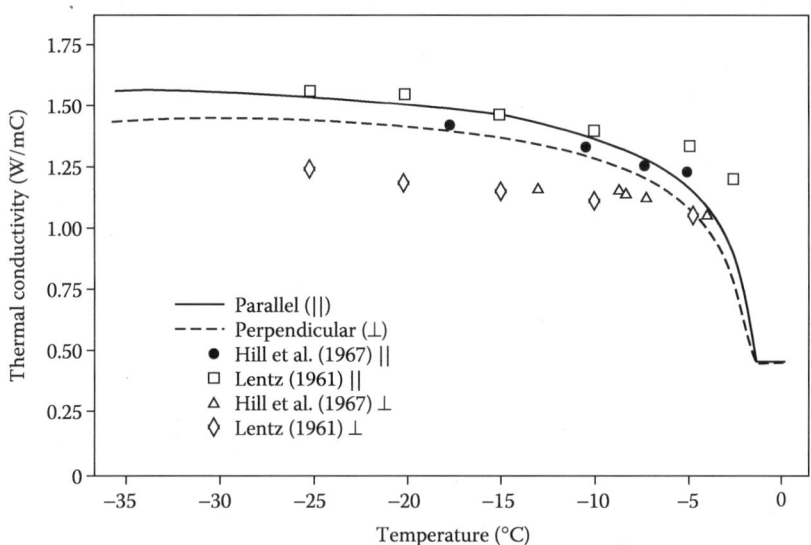

Figure 8.6 Thermal conductivity of frozen lean beef. (From Heldman, D.R. and Gorby, D.P. 1975. *Transactions of ASAE.* 18(4):740–744.)

components (Figure 8.7). As shown in the preceding figures, in a narrow temperature range close to the initial freezing point the thermal conductivity changes considerably.

Another approach to predicting thermal conductivity involves using thermal conductivities of food components and combining them according to the volume fraction of each component (Choi and Okos, 1984). This simple model has been used by Miles et al. (1983) and Mannapperuma and Singh (1989) to predict the thermal conductivity of a number of foods. This model is expressed as

$$k = \rho \sum k_i \frac{M_i}{\rho_i} \qquad (8.21)$$

where i denotes the ith component in the food system.

The influence of changing porosity during the freezing process can be incorporated into this model by first predicting the density value using Equation 8.13 and then substituting the calculated value of density in Equation 8.21. Figure 8.8 shows predicted thermal conductivities and experimental values for milk, juice, and sausage. As seen in this figure, good agreement between experimental and predicted values is obtained.

Additional approaches to model thermal conductivity of frozen foods can be found in the following references:

Meat products: Mascheroni et al. (1977), Moraga et al. (2011), van der Sman (2008), Kumcuoglu et al. (2010), Amos et al. (2008), and Reddy and P (2010).
Fish and seafood products: Bantle et al. (2010), van der Sman (2008), Amos et al. (2008), and Reddy and P (2010).

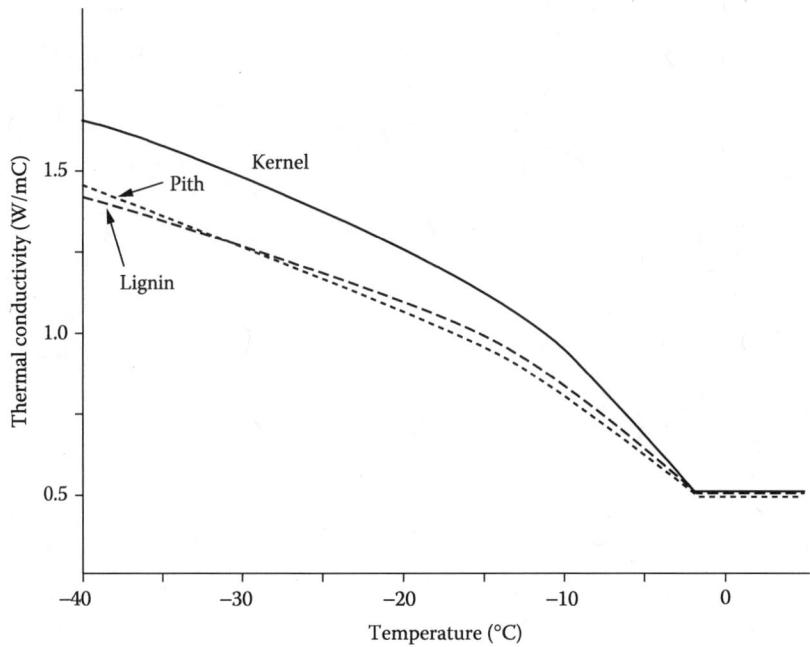

Figure 8.7 Thermal conductivity of components of corn on the cob as a function of temperature. (From Heldman, D.R. and Singh, R.P. 1986. Thermal properties of frozen foods. *Physical and Chemical Properties of Food*. Ed. Okos, M.R. American Society of Agricultural Engineers, St. Joseph, MI.)

Fruit and vegetable products: Mariani et al. (2009), Perussello et al. (2011), Telis et al. (2007), Amos et al. (2008), and Reddy and P (2010).

Bread and dough products: Kumcuoglu et al. (2007) and Jury et al. (2007).

8.4.1.3 Enthalpy

Predictive models for enthalpy of frozen foods have been suggested by Heldman and Singh (1981), Levy (1982), and Larkin et al. (1983). An example of predictions of enthalpy of sweet cherries is shown in Figure 8.9. There is a good agreement between the predicted and experimental values, although the predicted values are higher than experimental values for enthalpy close to the initial freezing temperature.

Using the approach of freezing point depression and estimating the ice and water fractions in a food, several investigators have presented models for enthalpy (Schwartzberg, 1976; Miles et al., 1983; Chen, 1985; Mannapperuma and Singh, 1989). The following equation for apparent enthalpy was derived by Schwartzberg (1976):

$$H_F = (T - T_D)\left[c_U + (M_A - M_{WZ})(c_W - c_I) \right.$$

$$\left. + (1 - X_{WZ})\frac{W_W}{W_S}\left\{\frac{RT_P^2}{(T_P - T)(T_P - T_D)} - 0.8(c_W - c_i)\right\}\right] \quad (8.22)$$

THERMAL PROPERTIES OF FROZEN FOODS

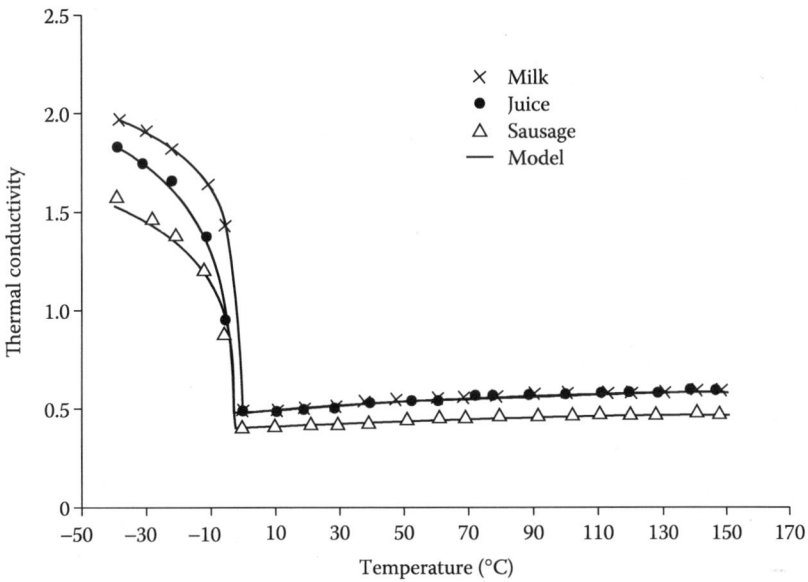

Figure 8.8 Prediction of thermal conductivity of milk, juice, and sausage as a function of temperature. (From Choi, Y. and Okos, M.R. 1984. Effect of temperature and composition on the thermal properties of foods. In *Food Engineering and Process Applications,* Vol. 1, pp. 93–101. M. Le Maguer and P. Jelen, Eds. Elsevier, New York.)

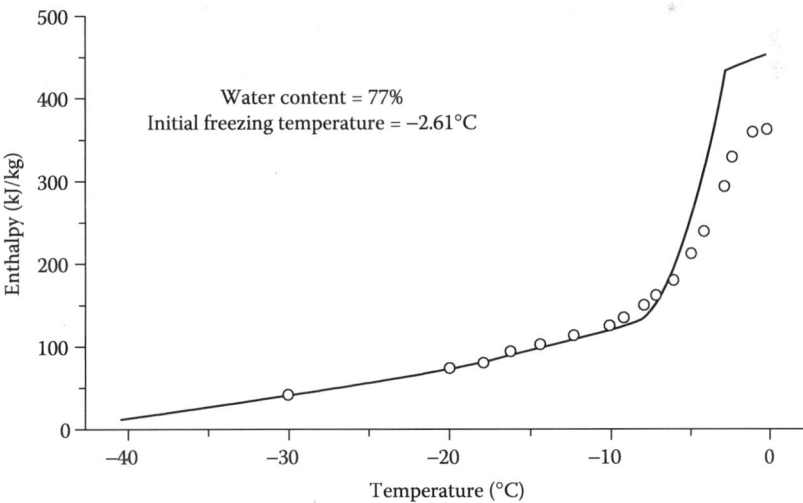

Figure 8.9 Enthalpy of sweet cherries as a function of temperature. (From Heldman, D.R. 1982. *Food Technology.* 36(2):92–96.)

Chen (1985) proposed the following two equations to estimate the enthalpy of foods below and above the initial freezing points:

$$\frac{H_F}{\psi} = (T - T_D)\left[0.37 + 0.3M_S + M_S \frac{RW_W T_P^2}{W_S(T - T_P)(T_D - T_P)}\right] \quad (8.23)$$

$$\frac{H_U}{\psi} = H_Z + (T - T_Z)(1 - 0.55M_S - 0.15M_S^3) \quad (8.24)$$

Good agreement was found when enthalpy values calculated from equations proposed by Chen (1985) and Schwartzberg (1976) were compared with the experimental values of Riedel (1956), as shown in Table 8.4. van der Sman (2008) and Bantle et al. (2010) also

Table 8.4 Predicted Values of Enthalpy and Apparent Specific Heat of Cod Fish

Temperature (°C)	Apparent Specific Heat [kJ/(kg · K)]			Enthalpy (kJ/kg)		
	Exptl[a]	Predicted[b]	Predicted[c]	Exptl[a]	Predicted[b]	Predicted[c]
−40	1.8	2.3	1.9	0	0	0
−30	2.0	2.4	2.0	19.2	23.3	19.6
−20	2.5	2.7	2.3	42.1	48.5	41.6
−18	2.7	2.8	2.5	47.5	54.1	45.9
−16	2.9	3.0	2.6	53.2	60.0	51.0
−14	3.2	3.2	2.9	59.3	66.3	56.6
−12	3.6	3.6	3.3	66.0	73.3	62.8
−10	4.1	4.3	4.0	73.6	81.3	69.7
−9	4.6	4.8	5.1	78.0	85.5	74.4
−8	5.3	5.5	5.3	82.9	91.3	79.3
−7	6.2	6.5	6.4	88.6	97.5	85.1
−6	7.7	8.1	8.0	95.5	105.1	92.3
−5	6.2	10.8	10.8	104.3	114.8	101.6
−4	15.3	15.6	15.9	116.7	129.1	114.6
−3	26.8	26.1	26.9	136.4	149.3	135.2
−2	67.4	55.9	58.2	176.4	189.3	174.6
−1	108.6	217.4	227.5	302.4	304.3	289.3
0	4.1	3.8	3.8	330.2		326.3
10	3.7	3.8	3.8	366.9		363.8
20	3.7	3.8	3.8	403.8		401.3

Source: Compiled by Mannapperuma, J.D. and Singh, R.P. 1990. Developments in food freezing. In Biotechnology and Food Process Engineering. Eds. Schwartzberg, H.G. and Rao, A. Marcel Dekker, New York.
[a] From Riedel (1956).
[b] From Schwartzberg (1976).
[c] From Chen (1985).

found a good agreement when enthalpy values were calculated based on equations from Schwartzberg (1976) and compositional data of meat products. Perussello et al. (2011) used a similar additive model considering the specific heat and mass fraction of each food component to predict the enthalpy of green beans. Their predictions showed a good fit with experimental data.

Mannapperuma and Singh (1989) used the unfrozen water fraction to model the enthalpy of frozen foods.

$$H_U = H_{FZ} + (c_W M_W + c_B M_B)(T - T_Z) \tag{8.25}$$

$$\begin{aligned} H_F &= (1 - M_{WZ})c_B(T - T_D) \\ &+ M_{WZ}\left[c_{IA}(T - T_D) + \frac{1}{2}c_{IB}(T^2 - T_D^2)\right] \\ &+ \left[(M_{WZ} - M_A)\frac{F'_Z - F'_P}{F' - F'_P} + M_A\right](\lambda_0 + \lambda_1 T) \\ &- \left[(M_{WZ} - M_A)\frac{F'_Z - F'_P}{F'_D - F'_P} + M_A\right](\lambda_0 + \lambda_1 T_D) \end{aligned} \tag{8.26}$$

A comparison of enthalpy values predicted for codfish with the experimental values of Riedel (1956) is shown in Figure 8.10. It is evident from this figure that a dramatic increase

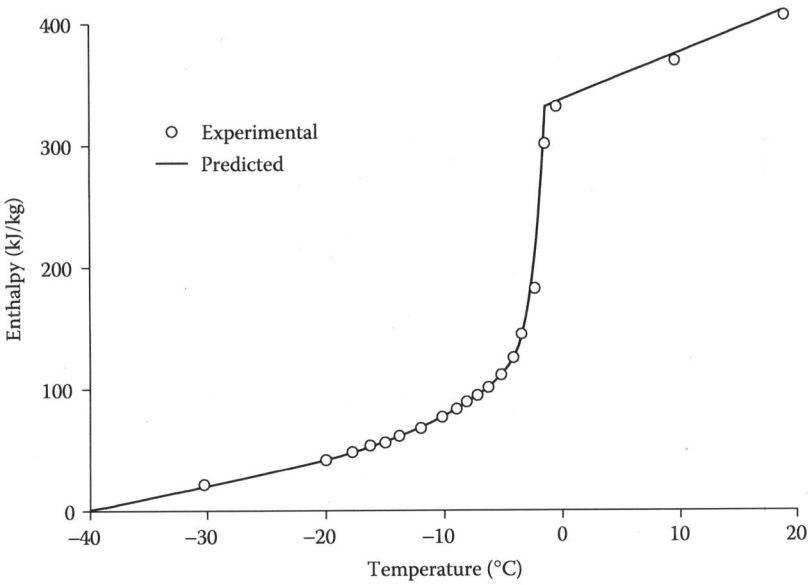

Figure 8.10 Enthalpy of codfish as a function of temperature. (From Mannapperuma, J.D. and Singh, R.P. 1990. Developments in food freezing. In *Biotechnology and Food Process Engineering*. Eds. Schwartzberg, H.G. and Rao, A. Marcel Dekker, New York.)

in enthalpy occurs close to the freezing point, when latent heat is removed along with the sensible heat.

8.4.1.4 Apparent Specific Heat

For temperatures above freezing, the prediction of the specific heat and enthalpy of a food is relatively easy. Knowing the composition of the food, the properties of each component are summed up in gravimetric proportions. The specific heat of a food may be determined using the expression

$$c_U = \sum c_i M_i \qquad (8.27)$$

This equation is valid in a temperature range where there is no phase change. If there is a phase change, such as in freezing, then the latent heat involved during the phase change must be incorporated. This is accomplished by using a new term called the *apparent specific heat*. The apparent specific heat is obtained by differentiating the enthalpy of the frozen food (which includes both latent and sensible heat) with respect to temperature. Heldman (1982) used this approach for frozen cherries; and his results are shown in Figure 8.11. As expected, the apparent specific heat increases dramatically near the initial freezing temperature.

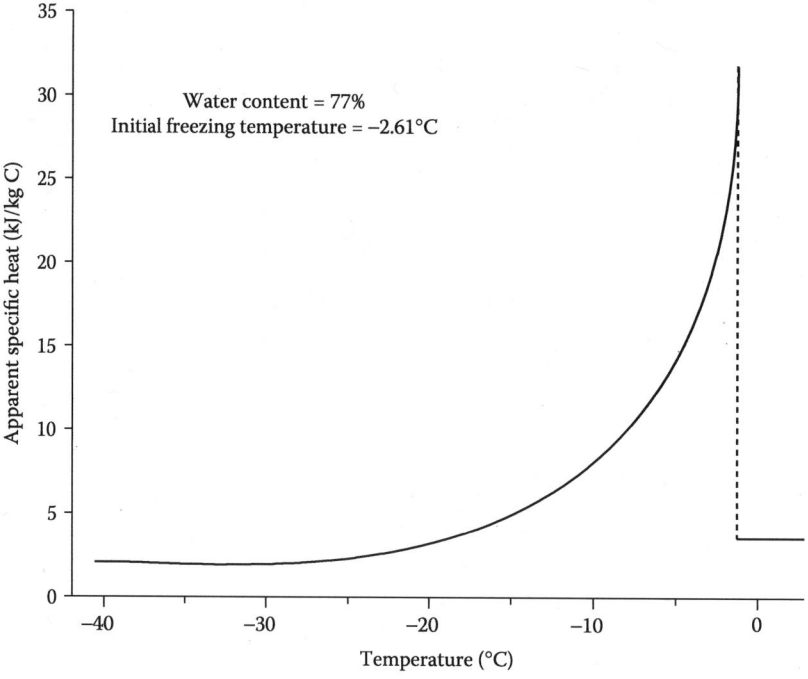

Figure 8.11 Apparent specific heat of sweet cherries as a function of temperature. (From Heldman, D.R. 1982. *Food Technology*. 36(2):92–96.)

Schwartzberg (1976) used the following expression for the prediction of apparent specific heat:

$$c_F = c_U + (M_A - M_{Wz})(c_W - c_I)$$
$$+ (1 - M_{WZ})\frac{W_W}{W_S}\left[\frac{RT_P^2}{(T_P - T)^2} - 0.8(c_W - c_I)\right] \quad (8.28)$$

Chen (1985) proposed the following equations:

$$\frac{c_F}{\psi} = 0.37 + 0.3M_S + M_S \frac{RW_W T_P^2}{W_S(T - T_P)^2} \quad (8.29)$$

$$\frac{c_U}{\psi} = 1 - 0.55M_S - 0.15M_S^3 \quad (8.30)$$

A comparison of predicted values of apparent specific heat of codfish obtained with the above models of Schwartzberg (1976) and Chen (1985) and experimental values obtained from Riedel (1956) are shown in Table 8.4 (Mannapperuma and Singh, 1990). The models provide good agreement with experimental values.

Using the unfrozen water fractions and specific heat of ice as a linear function of temperature, Mannapperuma and Singh (1990) proposed the following equations to predict the apparent specific heat:

$$c_I = c_{IA} + c_{IB}T \quad (8.31)$$

$$c_U = c_B M_B + c_W M_W \quad (8.32)$$

$$c_F = (1 - M_{WZ})c_B + M_{WZ}(c_{IA} + c_{IB}T)$$
$$+ (M_{WZ} - M_A)\left[\frac{F'[\lambda_0 + \lambda_1 T]^2}{RT^2[F' - F_F']} + \lambda_1\right]\frac{F_Z' - F_P'}{F' - F_P'} + M_A \lambda_1 \quad (8.33)$$

A computer program that encompasses these models was developed by Mannapperuma and Singh (1989) to predict the thermal properties of foods during freezing.

Another simple method can be used to determine specific heat. This method is based on the relationship between specific heat and enthalpy and uses enthalpy data for determining specific heat. Enthalpy can be determined using calorimetric techniques at various temperatures (Table 8.2). From this enthalpy data, the apparent specific heat at a given temperature can be determined using Equation 8.34

$$c = \frac{1}{\rho}\frac{H_T - H_{-40}}{(T + 40)} \quad (8.34)$$

Since for most practical food applications H_{-40} is assumed as zero, the above equation simplifies to

$$c = \frac{1}{\rho}\frac{H_T}{(T + 40)} \quad (8.35)$$

More recently, several researchers have attempted to correlate predicted values for apparent specific heat with experimentally determined values. Additional information on other models used to predict apparent specific heat can be found in the following references:

Meat products: Moraga et al. (2011), Tavman et al. (2007), and van der Sman (2008).
Fish and seafood products: Bantle et al. (2010) and van der Sman (2008).
Vegetables: Perussello et al. (2011).

8.4.1.5 Thermal Diffusivity

Thermal diffusivity values are commonly calculated based on predicted or measured values of thermal conductivity, density, and specific heat using Equation 8.36

$$\alpha = \frac{k}{\rho c_p} \qquad (8.36)$$

However, depending on the values used for the thermal conductivity, density, and specific heat, the thermal diffusivity will change during the freezing process as the water moves from an unfrozen to frozen state. Farag et al. (2008) and Kumcuoglu and Tavman (2007) reported predicted values that were greater than experimentally measured thermal diffusivity values for frozen meat and dough products, respectively.

8.4.2 Limitations of Predictive Models

When experimental data for the thermal properties of frozen foods are compared with the values obtained from predictive models, several disagreements are noted. Both experimental and predictive models have errors associated with them. The errors associated with experimental methods at temperatures above the freezing point are discussed by Reidy (1968). These errors become even more pronounced at or below the initial freezing temperature of a food product. The change of phase (either freezing or melting) caused by the experimental conditions brings about unwanted changes in the food product that introduce errors into the measured values. It is also difficult to obtain property data without causing a change in the unfrozen water fraction due to the experimental conditions.

The predictive methods generally require input values for initial freezing temperature, product moisture content, and the thermal properties of unfrozen product. Any errors associated with these values can lead to erroneous predictions. Fortunately, food properties under unfrozen conditions can be measured quite accurately using well-established procedures.

One of the main limitations of the mathematical models to predict thermal properties of foods discussed in this chapter is that they do not account for the changes in the characteristics of food components other than water during freezing. The latest work in this regard has shown that during the process of freezing, components other than water, such as sugars and starches, may also go through phase changes and glass transition, affecting the overall thermal properties of the food (Laaksonen and Roos, 2001; Roos and Karel, 1991a,b). Saad and Scott (1996) have attempted to model thermal properties of frozen foods using the Box–Kanemasu method. Their results agreed well for aqueous solutions with low concentration and simple solutes, such as sugar. The method failed for higher concentrations and solutes with complex molecular structures, such as gluten and methylcellulose. Further research in this area may provide interesting possibilities.

LIST OF SYMBOLS

c	specific heat, kJ/(kg·K)
F	defined in Equation 8.6
F'	function defined in Equation 8.11
H	specific enthalpy, kJ/kg
i	ith component in a food system
k	thermal conductivity, W/(m·°C)
M	mass fraction, dimensionless
R	the universal gas constant, J/(kg·K)
T	temperature, °C
V	volume fraction, dimensionless
W	molecular weight, dimensionless
X	mole fraction, dimensionless

Greek Symbols

ε	porosity, dimensionless
λ	heat of fusion per mole for pure water, (J/mole)
ρ	density (kg/m^3)
ψ	conversion factor (4184 J/cal)

Subscripts

A	unfreezable water
B	product solids
c	continuous phase
D	at the datum
d	dispersed phase
F	frozen food
I	ice
i	ith component of the food system
j	jth component of the soluble solids
P	at the initial freezing point of pure water
s	soluble solids
U	unfrozen food
W	water in the food
WZ	water at the initial freezing point of the food system
Z	at the initial freezing point of the food system

REFERENCES

Ablett, S. 1992. Overview of NMR applications in food science. *Trends in Food Science and Technology.* 3(8–9):246–250.
Albin, F.V., Badari Narayana, K., Srinivasa Murthy, S., and Krishna Murthy, M.V. 1979. Thermal diffusivities of some unfrozen and frozen food models. *Journal of Food Technology.* 14(4):361–367.
Amos, N.D., Willix, J., Chadderton, T., North, M.F., 2008. A compilation of correlation parameters for predicting the enthalpy and thermal conductivity of solid foods within the temperature range of –40°C to +40°C. *International Journal of Refrigeration.* 31(7):1293–1298.
Anderson, S.A. 1959. *Automatic Refrigeration.* Mac Lauren and Sons Ltd. for Danfoss, Norborg, Denmark.
Annamma, T.T. and Rao, C.V.N. 1974. Studies on thermal diffusivity and conductivity of fresh and dry fish. *Fishery Technology.* 11(1):28.
ASHRAE. 1977. *ASHRAE Handbook of Fundamentals.* American Society of Heating, Refrigerating, and Air-Conditioning Engineers, New York.
ASHRAE. 1985. *ASHRAE Handbook. Fundamentals.* Inch-pound ed. American Society of Heating, Refrigerating, and Air-Conditioning Engineers, Atlanta, GA.
Baik, O.D., Marcotte, M., Sablani, S.S., and Castaigne, F. 2001. Thermal and physical properties of bakery products. *Critical Reviews in Food Science and Nutrition.* 41(5):321–352.
Bantle, M., Eikevik, T.M., Brennvall, J.E., 2010. A novel method for simultaneous and continuous determination of thermal properties during phase transition applied to *Calanus finmarchicus. Journal of Food Science.* 75(6):E315–E322.
Boonsupthip, W., Heldman, D.R., 2007. Prediction of frozen food properties during freezing using product composition. *Journal of Food Science.* 72(5):E254–R263.
Brennvall, J., 2007. New techniques for measuring thermal properties and surface heat transfer applied to food freezing, *Dept. of Energy and Process Engineering.* Norwegian University of Science and Technology, Trondheim.
Chang, H.D. and Tao, L.C. 1981. Correlations of enthalpies of food systems. *Journal of Food Science.* 46(5):1493–1497.
Chen, C.S. 1985. Thermodynamic analysis of the freezing and thawing of foods: Ice content and Mollier diagram. *Journal of Food Science.* 50(4):1163–1166.
Chen, C.S. 1986. Effective molecular weight of aqueous solutions and liquid foods calculated from the freezing point depression. *Journal of Food Science.* 51(6):1537–1539.
Chen, C.S. 1987a. Sorption isotherm and freezing point depression equations of glycerol solutions. *Transactions of ASAE.* 30(1):278–282.
Chen, C.S. 1987b. Relationship between water activity and freezing point depression of food systems. *Journal of Food Science.* 52(2):433–435.
Chen, C.S. and Nagy, S. 1987. Prediction and correlations of freezing point depression of aqueous solutions. *Transactions of ASAE.* 30(4):1176–1180.
Choi, Y. and Okos, M.R. 1984. Effect of temperature and composition on the thermal properties of foods. In *Food Engineering and Process Applications,* Vol. 1, pp. 93–101. M. Le Maguer and P. Jelen, Eds. Elsevier, New York.
Cuevas, R. and Cheryan, M. 1978. Thermal conductivity of liquid foods—A review. *Journal of Food Process Engineering.* 2(4):283–306.
Dickerson, R.W.J. 1969. Thermal properties of food. *The Freezing Preservation of Foods,* 4th Ed. Eds. Tressler, D.K., Van Arsdel, W.B., and Copley, M.J. AVI Publishers, Westport, CT.
Dickerson, R.W.J. 1981. Enthalpy of frozen foods. *Handbook and Product Directory Fundamentals.* American Society of Heating, Refrigeration, and Air Conditioning, New York.
Drusas, A.E. and Saravacos, G.D. 1985. Thermal conductivity of tomato paste. *Journal of Food Engineering.* 4(3):157–168.

Duckworth, R.B. 1971. Differential thermal analysis of frozen food systems. I. The determination of unfreezable water. *Journal of Food Technology.* 6(3):317–327.

Farag, K.W., Lyng, J.G., Morgan, D.J., Cronin, D.A., 2008. Dielectric and thermophysical properties of different beef meat blends over a temperature range of −18 to +10°C. *Meat Science.* 79(4):740–747.

Fikiin, A.G. 1974. On the thermophysical parameters of frozen foodstuffs. *Bulletin of the International Institute of Refrigeration.* 2(1):173.

Fleming, A.K. 1969. Calorimetric properties of lamb and other meats. *Journal of Food Technology.* 4(1):199–215.

Gogol, E., Gogol, W., and Staniszewski, B. 1972. Bulletin of Institute of Refrigeration. *Annex.* 1(3):505.

Haiying, W., Shaozhi, Z., Guangming, C., 2007. Experimental study on the freezing characteristics of four kinds of vegetables. *LWT—Food Science and Technology.* 40(6):1112–1116.

Heldman, D.R. 1974. Predicting the relationship between unfrozen water fraction and temperature during food freezing using freezing point depression. *Transactions of ASAE.* 17(1):63–66.

Heldman, D.R. 1982. Food properties during freezing. *Food Technology.* 36(2):92–96.

Heldman, D.R. and Gorby, D.P. 1975. Prediction of thermal conductivity in frozen foods. *Transactions of ASAE.* 18(4):740–744.

Heldman, D.R. and Singh, R.P. 1981. *Food Process Engineering. 2nd ed.* AVI Publishing Co., Westport, CT.

Heldman, D.R. and Singh, R.P. 1986. Thermal properties of frozen foods. *Physical and Chemical Properties of Food.* Ed. Okos, M.R. American Society of Agricultural Engineers, St. Joseph, MI.

Herrera, M., M'Cann, J., Ferrero, C., Hagiwara, T., Zaritzky, N., Hartel, R., 2007. Thermal, mechanical, and molecular relaxation properties of frozen sucrose and fructose solutions containing hydrocolloids. *Food Biophysics.* 2(1):20–28.

Hough, G.E. and Calvelo, A. 1978. Thermal conductivity measurement parameters in frozen foods using the probe method. *Latin American Journal of Heat and Mass Transfer.* 2(1):71.

Hsieh, R.C., Lerew, L.E., and Heldman, D.R. 1977. Prediction of freezing times in foods as influenced by product properties. *Journal of Food Process Engineering.* 1(2):183–197.

Jason, A.C. and Long, R.A.K. 1955. The specific heat and thermal conductivity of fish muscle. *IX International Congress of Refrigeration.* 2(1):160.

Jowitt, R., Escher, F., Hallstrom, B., Meffert, H.F.T., and Voss, G.E. 1983. *Physical Properties of Foods.* Elsevier, London.

Jury, V., Monteau, J.-Y., Comiti, J., Le-Bail, A., 2007. Determination and prediction of thermal conductivity of frozen part baked bread during thawing and baking. *Food Research International.* 40(7):874–882.

Keller, G. 1956. Predicting temperature changes in frozen liquids. *Industrial Engineering Chemistry.* 48(2):188–196.

Kethley, T.W., Cown, W.B., and Bellinger, F. 1950. An estimate of thermal conductivity of fruits and vegetables. *Refrigeration Engineering.* 58:49.

Kopelman, I.J. 1966. Transient Heat Transfer and Thermal Properties in Food Systems. Ph.D. Thesis. Michigan State University, East Lansing, MI.

Kumcuoglu, S., Tavman, S., 2007. Thermal diffusivity determination of pizza and puff pastry doughs at freezing temperatures. *Journal of Food Processing and Preservation.* 31(1):41–51.

Kumcuoglu, S., Tavman, S., Nesvadba, P., Tavman, I.H., 2007. Thermal conductivity measurements of a traditional fermented dough in the frozen state. *Journal of Food Engineering.* 78(3):1079–1082.

Kumcuoglu, S., Turgut, A., Tavman, S., 2010. The effects of temperature and muscle composition on the thermal conductivity of frozen meats. *Journal of Food Processing and Preservation.* 34(3):425–438.

Laaksonen, T.J. and Roos, Y.H. 2001. Thermal and dynamic–mechanical properties of frozen wheat doughs with added sucrose, NaCl, ascorbic acid, and their mixtures. *International Journal of Food Properties.* 4(2):201–213.

Larkin, J.W., Heldman, D.R., and Steffe, J.F. 1983. An analytical approximation of frozen food enthalpy as a function of temperature. IFT Annual Meeting, New Orleans, LA.

Le Maguer, M. and Jelen, P. 1986. *Food Engineering and Process Applications*. Elsevier Applied Science Publishers, New York.

Lee, S., Cornillon, P., and Kim, Y.R. 2002. Spatial investigation of the nonfrozen water distribution in frozen foods using NMR SPRITE. *Journal of Food Science*. 67(6):2251–2255.

Lentz, C.P. 1961. Thermal conductivity of meats, fats, gelatin, gel, and ice. *Food Technology*. 15(5):243.

Lescano, C.E. 1973. Predicting Freezing Curves in Codfish Fillets Using the Ideal Binary Solution Assumptions. MS thesis, Michigan State University, East Lansing, MI.

Levy, F.L. 1982. Calculating the thermal conductivity of meat and fish in the freezing range. *International Journal of Refrigeration*. 5(3):149–154.

Li, J., Chinachoti, P., Wang, D., Hallberg, L.M., Sun, X.S., 2008. Thermal properties of ration components as affected by moisture content and water activity during freezing. *Journal of Food Science*. 73(9):E425–E430.

Lind, I. 1991. The measurement and prediction of thermal properties of food freezing and thawing— A review with particular reference to meat and dough. *Journal of Food Engineering*. 13(4):285–319.

Mannapperuma, J.D. and Singh, R.P. 1989. A computer-aided method for the prediction of properties and freezing/thawing of foods. *Journal of Food Engineering*. 9:275–304.

Mannapperuma, J.D. and Singh, R.P. 1990. Developments in food freezing. In *Biotechnology and Food Process Engineering*. Eds. Schwartzberg, H.G. and Rao, A. Marcel Dekker, New York.

Mariani, V.C., Do Amarante, Á.C.C., Dos Santos Coelho, L., 2009. Estimation of apparent thermal conductivity of carrot purée during freezing using inverse problem. *International Journal of Food Science & Technology*. 44(7):1292–1303.

Marin, M., Rios, G.M., and Gibert, H. 1985. Use of time–temperature data during fluidized bed freezing to determine frozen food. *Journal of Food Process Engineering*. 7(4):253–264.

Mascheroni, R.H., Ottino, J., and Calvelo, A. 1977. A model for the thermal conductivity of frozen meat. *Meat Science*. 1(1):235–243.

Matuszek, T., Niesteruk, R., and Ojanuga, A.G. 1983. Temperature conductivity of krill, shrimp, and squid over the temperature range 240–330 K. *Proceedings of the 6th International Congress of Food Science-and-Technology*, Dublin, Ireland, 1:221.

Meffert, H.F.T. 1984. Cost 90: Results of an international project on thermal properties. *International Journal of Refrigeration—Revue Internationale du Froid*. 7(1):21–26.

Mellor, J.D. 1976. Thermophysical properties of foodstuffs. I. Introductory review. *Bulletin of Institute of Refrigeration. Annex*. 56(3):551–563.

Mellor, J.D. 1980. Thermophysical properties of foodstuffs. 4. General bibliography. *Bulletin of the International Institute of Refrigeration*. 3:80.

Miles, C.A., Beck, G.V., and Veerkamp, C.H. 1983. Calculation of thermo-physical properties of foods. In *Physical Properties of Foods*. Eds. Jowitt, R., Escher, F., Hallstrom, B., Meffert, H.F.T., Spiess, W.E.L., and Vos, G. Applied Science Publishers, New York.

Mohsenin, N.N. 1978. *Physical Properties of Plant and Animal Materials: Structure, Physical Characteristics and Mechanical Properties*. 2nd ed., Gordon and Breach Science Publishers, London.

Mohsenin, N.N. 1980. *Thermal Properties of Foods and Agricultural Materials*. Gordon and Breach, New York.

Moore, W.J. 1972. *Physical Chemistry*. 4. Prentice-Hall, Englewood Cliffs, NJ.

Moraga, N., Vega-Gálvez, A., Lemus-Mondaca, R., 2011. Numerical simulation of experimental freezing process of ground meat cylinders. *International Journal of Food Engineering*. 7(6):1–18.

Morley, M.J. 1972. Thermal Properties of Meat—Tabulated Data. Special Report No. 1. Bristol, UK.

Murakami, E.G. and Okos, M.R. 1989. Measurement and prediction of thermal properties of foods. *Food Properties and Computer-Aided Engineering of Food Processing Systems*. Eds. Singh, R.P. and Medina, A.G. Kluwer Academic, Amsterdam.

Nagashima, N. and Suzuki, E. 1985. Computed instrumental analysis of the behavior of water in foods during freezing and thawing. *Properties of Water in Foods in Relation to Quality and Stability.* Eds. Simatos, D., Dordrecht, J.L.M., and Nijhoff, M. Kluwer Academic, Amsterdam.

Nesvadba, P., 1982. Methods for the measurement of thermal conductivity and diffusivity of foodstuffs. *Journal of Food Engineering.* 1(2):93–113.

Nunes, L.d.S., Duarte, M.E.M., Florêncio, I.M., Araújo, D.R.d., Campelo, I.K.M., 2008. Effective thermal diffusivity of ricotta in function of the freezing temperature. Presented at the *CIGR International Conference of Agricultural Engineering XXXVII*. Rio de Janeiro, Brazil.

Ohkuma, C., Kawai, K., Viriyarattanasak, C., Mahawanich, T., Tantratian, S., Takai, R., Suzuki, T., 2008. Glass transition properties of frozen and freeze-dried surimi products: Effects of sugar and moisture on the glass transition temperature. *Food Hydrocolloids.* 22(2):255–262.

Perussello, C.A., Mariani, V.C., do Amarante, Á.C., 2011. Combined modeling of thermal properties and freezing process by convection applied to green beans. *Applied Thermal Engineering.* 31(14–15):2894–2901.

Polley, S.L., Snyder, O.P., and Kotnour, P. 1980. A compilation of thermal properties of foods. *Food Technology.* 34(11):76.

Qashou, M.S., Vachon, R.I., and Touloukian, Y.S. 1972. Thermal conductivity of foods. *ASHRAE Transactions.* 78(1):165–183.

Qashou, S. 1970. Compilation of thermal conductivity of foods. MS thesis, Auburn University, Auburn, AL.

Rahman, M.S. 1993. Specific heat of selected fresh seafood. *Journal of Food Science.* 56(2):522–524.

Rahman, M.S. 1994. The accuracy of prediction of the freezing point of meat from general models. *Journal of Food Engineering.* 21(1):127–136.

Rahman, M.S. and Driscoll, R.H. 1994. Thermal conductivity of sea foods: Calamari, octopus, and prawn. *Food Australia.* 43(8):356.

Rahman, S. 1995. *Food Properties Handbook.* CRC Press, Boca Raton, FL.

Ramaswamy, H.S. and Tung, M.A. 1981. Thermophysical properties of apples in relations to freezing. *Journal of Food Science.* 46(3):724–728.

Rasanen, J., Blanshard, J.M.V., Mitchell, J.R., Derbyshire, W., and Autio, K. 1998. Properties of frozen food doughs at subzero temperatures. *Journal of Cereal Science.* 28(1):1–14.

Reddy, K.S., P, K., 2010. Combinatory models for predicting the effective thermal conductivity of frozen and unfrozen food materials. *Advances in Mechanical Engineering* 2010: 1–14, doi:10.1155/2010/901376.

Reidy, G.A. 1968. Thermal properties of foods and methods of their determination. MS thesis, Michigan State University, East Lansing, MI.

Riedel, L. 1951. The refrigeration effect required to freeze fruits and vegetables. *Refrigeration Engineering.* 59(2):670.

Riedel, L. 1956. Calorimetric investigations of the freezing of fish meat. *Kaltechnik.* 8(12):374–377.

Riedel, L. 1957a. Calorimetric investigations of the meat freezing process. *Kaltechnik.* 9(1):38–40.

Riedel, L. 1957b. Calorimetric investigations of the freezing of egg whites and yolks. *Kaltechnik.* 9(11):342–345.

Rolfe, E.J. 1968. The chilling and freezing of foodstuffs. *Biochemical and Biological Engineering Science.* Ed. Blakeborough, N. Academic Press, New York.

Roos, Y.H., 2010. Glass transition temperature and its relevance in food processing. *Annual Review of Food Science and Technology.* 1:469–496.

Roos, Y.H. and Karel, M. 1991a. Phase transitions of amorphous sucrose and frozen sucrose solutions. *Journal of Food Science.* 56(1):266–267.

Roos, Y.H. and Karel, M. 1991b. Nonequilibrium ice formation in carbohydrate solutions. *Cryo-Letter.* 12(1):367–376.

Saad, Z. and Scott, E.P. 1996. Estimation of temperature-dependent thermal properties of basic food solutions during freezing. *Journal of Food Engineering.* 28(1):1–19.

Sablani, S.S., Bruno, L., Kasapis, S., and Symaladevi, R.M., 2009. Thermal transitions of rice: Development of a state diagram. *Journal of Food Engineering.* 90(1):110–118.

Sanz, P.D., Dominguez, M., and Mascheroni, R.H. 1989. Equations for the prediction of thermo physical properties of meat products. *Latin American Applied Research.* 19(1):155–160.

Schwartzberg, H.G. 1976. Effective heat capacities for freezing and thawing of foods. *Journal of Food Science.* 41(1):152–156.

Short, B.E. and Bartlett, L.H. 1944. The specific heat of foodstuffs. The University of Texas Publ. No. 4432. Bur. Eng. Research, Eng. Res. Ser. No. 40.

Short, B.E. and Staph, L.H. 1951. The energy content of foods. *Ice Refrigeration.* 121(5):23.

Singh, R.P. 1982. Thermal diffusivity in food processing. *Food Technology.* 36(2):87–91.

Singh, R.P. 2003. *Food Properties Database.* Version 3.1. RAR Press, Davis, CA.

Singh, R.P. and Heldman, D.R. 2001. *Introduction to Food Engineering.* 3rd ed. Academic Press, San Diego.

Singh, R.P., Medina, A.G., and North Atlantic Treaty Organization. Scientific Affairs Division. 1989. *Food Properties and Computer-Aided Engineering of Food Processing Systems.* Kluwer Academic, Dordrecht, The Netherlands.

Smith, J.G., Ede, A.J., and Gane, R. 1952. Thermal conductivity of frozen foodstuffs. *Modern Refrigeration.* 55(1):254.

Staph, H.E. 1949. Specific heat of foodstuffs. *Refrigeration Engineering.* 57:767.

Succar, J. 1989. Heat transfer during freezing and thawing of foods. In *Developments in Food Preservation-5.* Ed. Thorne, S. Elsevier Applied Science, London.

Succar, J. and Hayakawa, K.I. 1983. A method for determining the apparent thermal diffusivity of spherical foods. *Lebensmittel Wissenschaft Technologie.* 16(6):373.

Succar, J. and Hayakawa, K. 1990. A method to determine initial freezing point of foods. *Journal of Food Science.* 55(6):1711–1713.

Sweat, V.E. 1974. Experimental value of thermal conductivity of selected fruits and vegetables. *Journal of Food Science.* 39(2):1080–1083.

Sweat, V.E. 1975. Modeling the thermal conductivity of meats. *Transactions of ASAE.* 18(3):564–568.

Sweat, V.E., Haugh, C.G., and Stadelman, W.J. 1973. Thermal conductivity of chicken meat at temperatures between –75 and 20° Centigrade. *Journal of Food Science.* 38(1):158–160.

Tavman, S., Kumcuoglu, S., and Gaukel, V., 2007. Apparent specific heat capacity of chilled and frozen meat products. *International Journal of Food Properties.* 10(1):103–112.

Telis, V.R.N., Telis-Romero, J., Sobral, P.J.A., and Gabas, A.L., 2007. Freezing point and thermal conductivity of tropical fruit pulps: Mango and papaya. *International Journal of Food Properties.* 10(1):73–84.

Thompson, D.R., Hung, Y.C., and Norwig, J.F. 1983. The influence of raw material properties on the freezing of sweet corn. *3rd International Congress on Engineering and Food*, Dublin, Ireland.

Tressler, D.K., Van Arsdel, W.B., and Copley, M.J. 1968. *The Freezing Preservation of Foods.* 4th ed. AVI Publishing Co., Westport, CT.

Van Den Berg, L. and Lentz, C.P. 1975. Effect of composition on thermal conductivity of fresh and frozen foods. *Journal of Canadian Institute of Food Technology.* 8(2):79–83.

van der Sman, R.G.M., 2008. Prediction of enthalpy and thermal conductivity of frozen meat and fish products from composition data. *Journal of Food Engineering.* 84(3):400–412.

Wang, D.Q. and Kolbe, E. 1990. Thermal conductivity of surimi—Measurement and modeling. *Journal of Food Science.* 55(5):1217–1221.

Woodams, E.E. and Nowrey, J.E. 1968. Literature values of thermal conductivity of foods. *Food Technology.* 22(4):150.

Zaritzky, N.E. 1983. Mathematical simulation of the thermal behaviour of frozen meat during its storage and distribution. *Journal of Food Process Engineering.* 6(1):15–36.

9

Properties Relevant to Infrared Heating of Food

Ashim K. Datta and Marialuci Almeida

Contents

9.1	Introduction	282
9.2	Fundamentals of Infrared Interactions with Materials	282
	9.2.1 Electromagnetic Spectrum and Near-, Mid-, and Far-Infrared Electromagnetic Waves	282
	9.2.2 Interaction between Infrared Radiation and Food Materials	283
	9.2.3 Sources of Infrared Radiation in Heating Applications	284
	9.2.4 Emission and Emissivity	284
	9.2.5 Reflection, Absorption, and Transmission	287
	9.2.6 Absorptivity and Emissivity	290
	9.2.7 Attenuation or Extinction	291
9.3	Measurement of Radiative Properties of Food	292
9.4	Radiative Property Data for Food Systems	294
	9.4.1 Radiative Property Data for Water, Ice, and Water Vapor	294
	9.4.2 Properties of Other Pure Food Components	295
	9.4.3 Spectral Variation of Radiative Property Data: Potato Tissue as an Example	295
	9.4.4 Moisture Dependence of Radiative Property Data	297
	9.4.5 Temperature Dependence of Radiative Property Data	298
	9.4.6 Dependence of Radiative Property Data on Food Structure	301
	9.4.7 How Processing Can Change Food Radiative Properties	301
9.5	Use of the Radiative Properties in Modeling of Heat Transfer and Sensitivity of Temperature to Radiative Properties	302
	9.5.1 Simple Models	303
	9.5.2 More Realistic Models	304
	9.5.2.1 Model Based on Monte Carlo Method	305

9.5.2.2 Model Based on Radiative Transport Equation 305
9.5.3 Sensitivity of Temperature to Radiative Property: Effect of Emissivity 306
9.6 Summary 307
Acknowledgments 308
References 308

9.1 INTRODUCTION

The infrared portion of the electromagnetic spectrum is extremely useful in food processing in various ways:

- Food processes involving heating
- Spectroscopic measurement of chemical composition (analytical applications) of food
- Noncontact temperature measurement of food

Although these three types of applications involve some of the same properties, this chapter is primarily intended for the applications of infrared in heating of food, such as drying [1], baking [2], roasting [3], blanching, and surface pasteurization [4]. For applications involving composition measurement, the reader is referred to reviews such as those by Williams and Norris [5] or Buning–Pfaue [6], and for noncontact temperature measurement, to books such as by Michalski et al. [7].

Infrared radiation in food surfaces involves relatively little penetration; however, for some food and wavelength combinations, the penetration can be significant, as will be discussed later. Thus, in general, these surfaces are semitransparent, and radiation is not just a surface property (as in opaque materials such as metal) since the entire volume of the food may interact with the material. Foods, like other materials, can exhibit behavior that varies with wavelength. Dependence of radiation properties on wavelength, food composition, and other factors will be the subject of this chapter. Although some definitions will be provided, readers are referred to undergraduate texts on heat transfer, such as Incropera and Dewitt [8] or specialized books on radiative heat transfer [e.g., 9–12] for further details on the properties as well as the use of the properties in modeling of radiative heat-transfer processes outside of specific food processing applications.

With the exceptions of two books by Russian authors that have a large amount of information [13, 14], data on radiative properties of foods is quite limited. Every effort has been made here to represent the entire published literature. Thus, as scant as data might seem, the chapter includes most of what is available outside the two books.

9.2 FUNDAMENTALS OF INFRARED INTERACTIONS WITH MATERIALS

9.2.1 Electromagnetic Spectrum and Near-, Mid-, and Far-Infrared Electromagnetic Waves

Infrared waves are part of the electromagnetic spectrum, as shown in Figure 9.1. The exact lower and the upper limits of wavelengths defining infrared are not consistently

PROPERTIES RELEVANT TO INFRARED HEATING OF FOOD

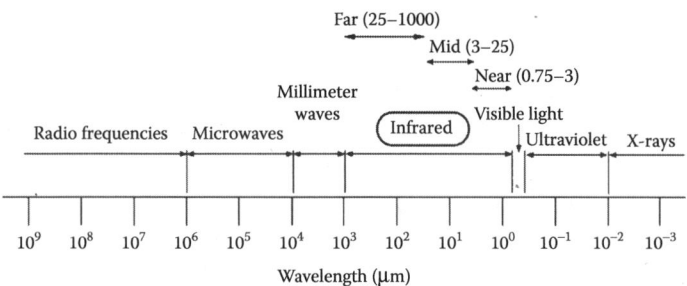

Figure 9.1 The electromagnetic spectrum, showing the region of infrared and its subregions near-, mid-, and far infrared.

mentioned in the literature. Although the wavelength range for infrared radiation is mentioned as 0.75–1000 µm [15], *thermal radiation* is generally considered to be in the range of 0.1–100 µm [8]. The entire infrared range is further divided into near-infrared (0.75–3 µm), mid-infrared (3–25 µm), and far-infrared (25–1000 µm) regions, also mentioned in Figure 9.1.

9.2.2 Interaction between Infrared Radiation and Food Materials

Interactions of food materials in the near- and mid-infrared range of electromagnetic waves primarily involve vibrational energy levels of molecules, whereas in the far-infrared range, their interaction involves rotational energy levels of molecules. During absorption, energy is transferred from the electromagnetic wave to a molecule (or atom), causing it to move to an excited state. If a molecule or atom is subjected to electromagnetic radiation of different wavelengths, it will only absorb photons at those wavelengths that correspond to exact differences between two different energy levels within the material. Foodstuffs are complex mixtures of different large biochemical molecules (simple sugars, amino acids, etc.), biochemical polymers (complex sugars, proteins, lipids, etc.), inorganic salts, and water. Infrared absorption bands relevant to food heating are shown in Table 9.1 [16]. These components have their individual signatures in the absorption of infrared, as illustrated in Figure 9.2 [17]. Amino acids, proteins, and nucleic acids reveal two strong absorption

Table 9.1 Infrared Absorption Bands Relevant to Food Heating

Chemical Group	Absorption Wavelength (µm)	Relevant Food Component
Hydroxyl group (O–H)	2.7–3.3	Water, carbohydrates
Aliphatic carbon–hydrogen bond	3.25–3.7	Fats, carbohydrates, and proteins
Carbonyl group (C=O) (ester)	5.71–5.76	Fats
Carbonyl group (C=O) (amide)	ca. 5.92	Proteins
Nitrogen–hydrogen group (–NH–)	2.83–3.33	Proteins
Carbon–carbon double bond (C=C)	4.44–4.76	Unsaturated fats

Source: I Rosenthal. *Electromagnetic Radiations in Food Science*, New York: Springer-Verlag, 1992.

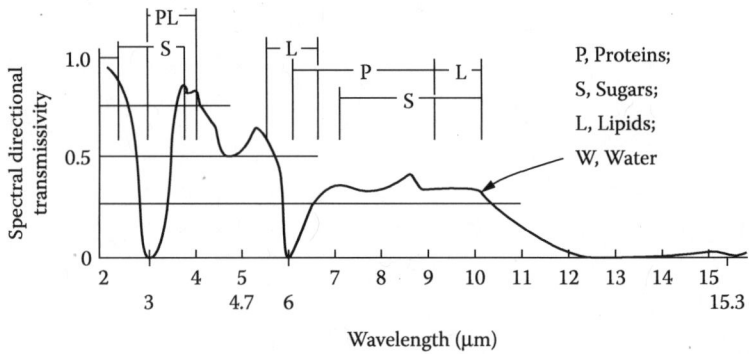

Figure 9.2 Spectral directional transmissivity of infrared in various food components, signifying the bands over which more interactions occur. (From C Sandu. *Biotechnology Progress* 2(3):109–119, 1986.)

bands localized at 3–4 and 6–9 μm. Lipids are strong absorbers over the entire infrared radiation spectrum, with three stronger absorption bands at 3–4, 6, and 9–10 μm. Sugars give two strong absorption bands centered at 3 and 7–10 μm. For more discussion on the interactions, see references on spectroscopy as applied to foods [e.g., 5,18].

9.2.3 Sources of Infrared Radiation in Heating Applications

Radiative properties of food depend on the wavelength of radiation incident on the food, which in turn depends on the emission characteristics of the source of the radiation. Thus, it is important to know the characteristics of common sources (emitters) used for thermal radiation. Infrared emitters can be made of various materials, such as quartz glass, ceramic, and metal. Generally speaking [19], shorter-wavelength emitters (e.g., tungsten filament) operate at temperatures above 2000°C, medium wavelength emitters (e.g., quartz tube) operate at around 700–1150°C, and long wavelength emitters (e.g., ceramic) operate below 800°C; this follows from Equation 9.3, discussed later, since the peak of the emission decreases with increasing temperatures). Figure 9.3 shows the typical spectral distribution of radiation from such emitters. Solar radiation is superimposed on this figure for comparison, since solar radiation is also used in processes such as drying. By using bandpass optical filters, specific spectral regions can be obtained that have been suggested for selective heating of foods [20,21].

9.2.4 Emission and Emissivity

The properties of interest in studying radiative heat transfer of foods are *emissivity, reflectivity, absorptivity,* and *transmissivity.* Emissivity concerns emitted radiation. A body above absolute zero emits radiation in all directions over a wide range of wavelengths. The quantity (amount) and quality (spectral distribution) of emitted energy depends on the temperature. Additionally, for real surfaces, it also depends on its emissivity. Total energy emitted by a perfect (black) body is given by

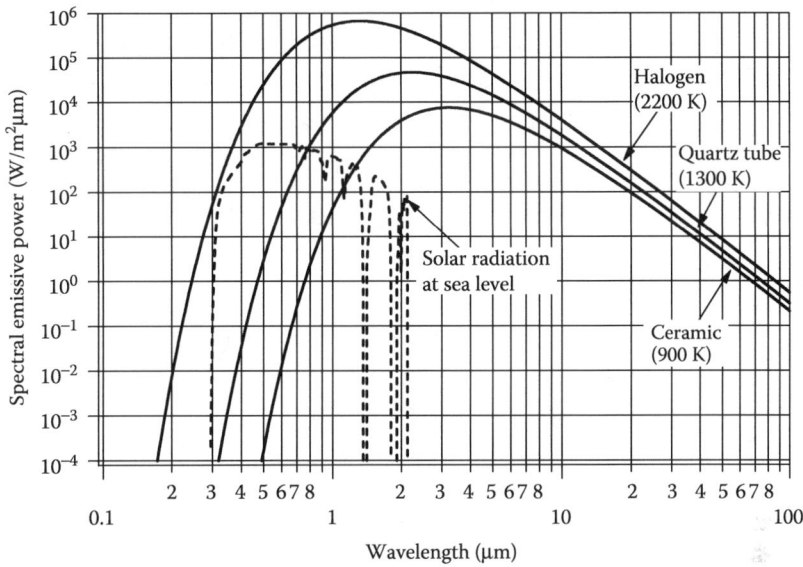

Figure 9.3 Spectral emissive powers of three classes of emitters at their typical temperatures (blackbody radiation at the noted temperatures, following Equation 9.2) in the range of thermal radiation (0.1–100 µm). Measured average solar radiation at sea level is superimposed (the missing bands in solar radiation are due to absorption in atmospheric gases including ozone, oxygen, water vapor, and carbon dioxide). (Data on solar radiation are from PR Gast. Solar electromagnetic radiation. In *Handbook of Geophysics and Space Environments*. New York: McGraw-Hill, Inc., 1965.)

$$E_b = \sigma T^4 \tag{9.1}$$

where T is the absolute temperature of the surface and $\sigma = 5.670 \times 10^{-8}$ W/m²K⁴ is the Stefan–Boltzmann constant. The spectral distribution of this energy (i.e., emitted energy as a function of wavelength) for a blackbody is given by the Planck's law of radiation (the reader is referred to a heat-transfer textbook for more details):

$$E_{b,\lambda}(\lambda, T) = \frac{2\pi h c_0^2}{\lambda^5 [\exp(hc_0/\lambda k T) - 1]} \tag{9.2}$$

where $h = 6.6256 \times 10^{-34}$ Js and $k = 1.3805 \times 10^{-23}$ J/K are the universal Planck and Boltzmann constants, respectively, $c_0 = 2.998 \times 10^8$ m/s is the speed of light in vacuum, and T is the absolute temperature of the blackbody, in K. Example of a plot of Planck's law of radiation (Equation 9.2) is shown in Figure 9.3. In this figure, the distribution of energy from an ideal surface at typical food temperatures is compared with that from a typical emitter and the solar radiation. Note that Equation 9.1 is obtained by integrating Equation 9.2 over all wavelengths.

From Figure 9.3, it can be seen that at any wavelength, the magnitude of the emitted radiation increases with increasing temperature. Also, as the temperature increases, the spectral region where most of the radiation is concentrated occurs at shorter wavelength. This relationship between the temperature and the peak of the curve is given by the Wien's displacement law

$$\lambda_{max} T = 2897.8 \; \mu m \cdot K \tag{9.3}$$

where λ_{max} is the peak of the curves in Figure 9.3.

The emissivity of a surface is defined as the ratio of the radiation emitted by a real surface, $E(T)$, to the radiation emitted by a perfectly radiating ideal surface (called a blackbody), $E_b(T)$, that is,

$$\varepsilon(T) = \frac{E(T)}{E_b(T)} = \frac{E(T)}{\sigma T^4} \tag{9.4}$$

Emissivity of a surface varies with temperature, wavelength, and direction of emitted radiation. Emissivity at a specified wavelength is called the spectral emissivity and is denoted as $\varepsilon \lambda$. An example of spectral emissivity is shown in Figure 9.4, for distilled water. For plant materials such as leaves, for example, the emissivity is typically above 0.97.

Likewise, the emissivity in a specified direction is called the directional emissivity, denoted by $\varepsilon \theta$, where θ is the angle between the direction of radiation and normal to the surface. The emissivity of a surface averaged over all wavelengths is called the hemispherical emissivity, and the emissivity averaged over all wavelengths is called the total emissivity. Thus, the total hemispherical emissivity $\varepsilon(T)$ of a surface is simply the average emissivity over all directions and wavelengths.

Such temperature-, wavelength-, and direction-dependent emissivity data are generally not available for food materials. Inclusion of such variations in radiative heat-transfer analysis can make it quite complex and almost intractable. Thus, radiative

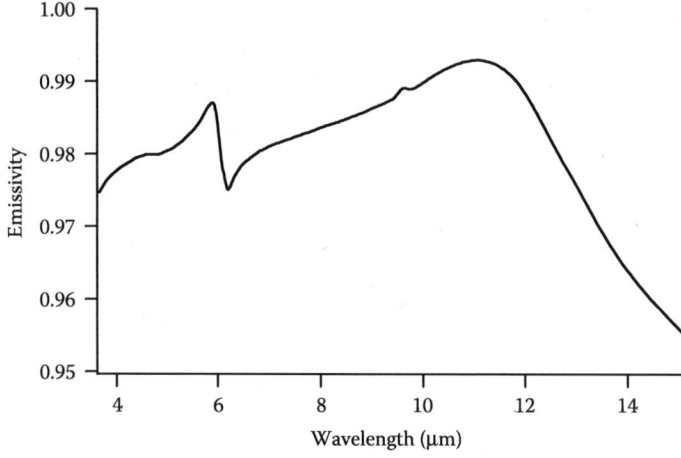

Figure 9.4 Spectral emissivity of distilled water. (Data from http://www.icess.ucsb.edu/modis/EMIS/html/water.html; accessed September 23, 2003.)

PROPERTIES RELEVANT TO INFRARED HEATING OF FOOD

heat-transfer calculations commonly use two approximations called *gray* and *diffuse*. A surface is called *diffuse* if its properties are independent of direction and *gray* if its properties are independent of wavelength. Thus, emissivity of a gray, diffuse surface is simply the total hemispherical emissivity of the surface, independent of wavelength and direction.

9.2.5 Reflection, Absorption, and Transmission

When electromagnetic radiation, such as infrared, strikes a surface, part of it is reflected, part of it is absorbed, and the remaining, if any, is transmitted. This is shown schematically in Figure 9.5. Here, G is the total radiation energy incident on the surface per unit area per unit time, also known as *irradiation*. The quantities G_{ref}, G_{abs}, and G_{trans} are the total reflected, absorbed, and transmitted energies, respectively. Since the total energy is conserved, these parts add up to the total amount of incident energy, that is,

$$G = G_{ref} + G_{abs} + G_{trans} \tag{9.5}$$

Dividing by G yields

$$\frac{G_{ref}}{G} + \frac{G_{abs}}{G} + \frac{G_{trans}}{G} = 1 \tag{9.6}$$

The quantities absorptivity, reflectivity, and transmissivity are defined as

$$\begin{aligned} \text{Reflectivity} &= \frac{\text{Reflected radiation}}{\text{Incident radiation}} = \frac{G_{ref}}{G} = \rho \\ \text{Absorptivity} &= \frac{\text{Absorbed radiation}}{\text{Incident radiation}} = \frac{G_{abs}}{G} = \alpha \\ \text{Transmissivity} &= \frac{\text{Transmitted radiation}}{\text{Incident radiation}} = \frac{G_{trans}}{G} = \tau \end{aligned} \tag{9.7}$$

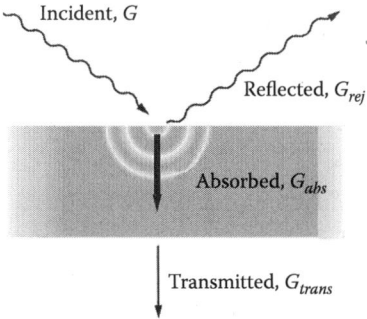

Figure 9.5 Incidence, reflection, and transmission of electromagnetic radiation from a surface. (From AK Datta. *Biological and Bioenvironmental Transport Processes*. New York: Marcel Dekker, 2002.)

ENGINEERING PROPERTIES OF FOODS

Thus, the properties reflectivity, absorptivity, and transmissivity are related as

$$\rho + \alpha + \tau = 1 \tag{9.8}$$

Note from the definition of these properties, their values lie between 0 and 1, that is,

$$\begin{aligned} 0 \leq \rho \leq 1 \\ 0 \leq \alpha \leq 1 \\ 0 \leq \tau \leq 1 \end{aligned} \tag{9.9}$$

These properties, however, are dependent on wavelength and direction, as is emissivity. An example of variation of specular reflectivity with incident angle can be seen in Figure 9.6 for a water surface. An example of reflectivity and transmissivity, averaged over all directions, can be seen in Figure 9.7, for water. As defined in Equation 9.9, the properties are total hemispherical properties, that is, they are the average properties for the material over all wavelengths and all directions. Although these properties can be defined for a specific wavelength or direction, only a small amount of data are available for the wavelength dependence and very little or no data are available for the directional dependence over food surfaces. Even if data on directional dependence were available, inclusion of them in radiative heat-transfer analysis would make an already complex problem much worse, and it is doubtful whether inclusion of such detailed properties would lead to significantly different results of practical consequence in food processing.

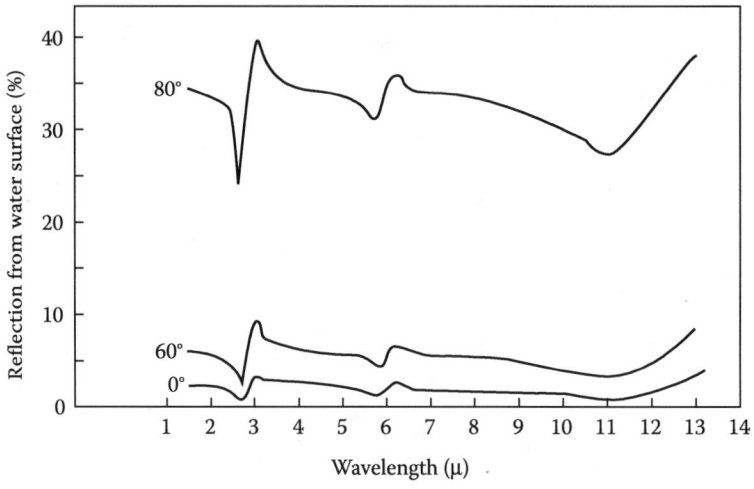

Figure 9.6 Spectral reflectance (in percent) of water as a function of incident angle and wavelength. (From WL Wolfe. *Handbook of Military Infrared Technology*. Washington, DC: Office of Naval Research, Department of the Navy, 1965.)

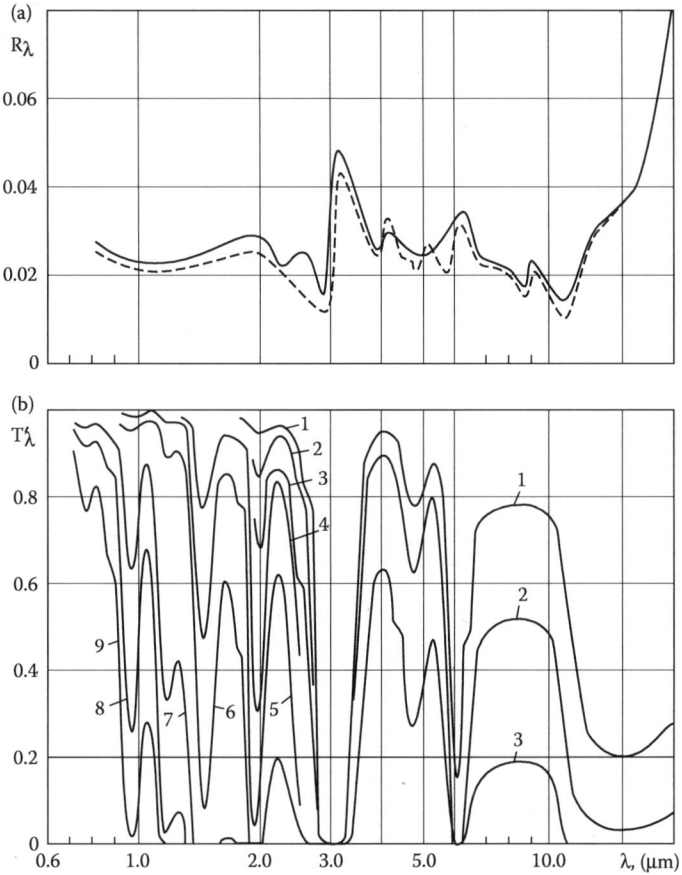

Figure 9.7 Spectral reflectance (a) and transmittance (b) of distilled water for various thicknesses: (1) 0.005; (2) 0.01; (3) 0.03; (4) 0.1; (5) 0.3; (6) 1.0; (7) 10.0; (8) 30.0; (9) 100.0. Dashed line in (a) is for NaCl solution and for water from the Black Sea. (From SG Il'yasov and VV Krasnikov. *Physical Principles of Infrared Irradiation of Foods*. 1st ed. New York: Hemisphere Publishing Corp, 1991.)

As mentioned, wavelength dependence of food properties data is available to a limited extent. Wavelength dependence of the properties is called spectral properties. Thus, spectral reflectivity, absorptivity, and transmissivity are defined as

$$\rho_\lambda = \frac{G_{\lambda,ref}}{G_\lambda}$$
$$\alpha_\lambda = \frac{G_{\lambda,abs}}{G_\lambda} \quad (9.10)$$
$$\tau_\lambda = \frac{G_{\lambda,trans}}{G_\lambda}$$

where G_λ is the radiation energy incident at wavelength λ, and $G_{\lambda,ref}$, $G_{\lambda,abs}$, and $G_{\lambda,trans}$ are the radiation energy reflected, absorbed, and transmitted, respectively. They are related to their average counterparts as

$$\rho = \frac{\int_0^\infty \rho_\lambda G_\lambda d\lambda}{\int_0^\infty G_\lambda d\lambda}$$

$$\alpha = \frac{\int_0^\infty \alpha_\lambda G_\lambda d\lambda}{\int_0^\infty G_\lambda d\lambda} \qquad (9.11)$$

$$\tau = \frac{\int_0^\infty \tau_\lambda G_\lambda d\lambda}{\int_0^\infty G_\lambda d\lambda}$$

For an example of the spectral transmissivity through a slice of potato, see Section 9.4.3.

Two limiting cases of reflection from a surface area are called *specular* and *diffuse*. In specular reflection, the angle of incidence is equal to the angle of reflection. In diffuse reflection, however, the intensity of the reflected radiation is the same at all angles of irradiation and reflection. Although real surfaces are neither totally specular nor totally diffuse, but somewhere in between, they are assumed to be one of the two limiting cases for simplicity. Polished and smooth surfaces exhibit near-specular properties. Relevant data on food surfaces are hard to find. Most food surfaces are likely to be rough, leading to a diffuse behavior, that is, the intensity of reflected radiation will be about the same at all angles.

9.2.6 Absorptivity and Emissivity

Absorptivity depends on the spectral distribution of incident radiation. It can be shown (Kirchoff's law; see, e.g., Incropera and Dewitt [8]) that when the material and the source of incident radiation are at the same temperature, emissivity is equal to absorptivity, that is,

$$\varepsilon_\lambda(T) = \alpha_\lambda(T) \qquad (9.12)$$

Note that directional dependence of emissivity or absorptivity is ignored here. In practice, average values are also considered equal, that is,

$$\varepsilon(T) = \alpha(T) \qquad (9.13)$$

Use of this law (Equation 9.12) when temperatures of the two surfaces exchanging radiation are equal is an approximation. The error due to such approximation depends on the problem and is generally considered small if the temperature differences are less than a few hundred degrees [25].

9.2.7 Attenuation or Extinction

As the electromagnetic waves move through a food material, part of its energy is absorbed or scattered. Electromagnetic energy is attenuated due to this combined effect of absorption and scattering. This attenuation is also called extinction and is typically represented as

$$q = q_0 e^{-x/\delta} \tag{9.14}$$

where q_0 is the incident energy flux and q is the energy flux at a distance x from the incident surface. This relationship is also known as Beer–Lambert's law in many contexts. Note that similar attenuation of energy also occurs in some restricted situations of microwave processing. The penetration depth, δ, describes the attenuation or extinction and is a complex function of [17]:

- The chemical composition of the food
- The physicochemical state of the irradiated medium, that is, solid, liquid, or powder; frozen or unfrozen; dispersion, emulsion, or solution; and so on
- Physical properties such as density, porosity, and water content

Table 9.2 shows typical penetration depths for some food materials from the work of Ginzberg [13]. See Section 9.4.3 for an example of the spectral variation of penetration depth in potato tissue.

Table 9.2 Penetration Depth in Some Typical Food Materials, to Be Used with Equation 9.14

Material	Penetration Depth (mm)	λ_{max} of Incident Radiation (μm)
Apple	1.8	1.16
	2.6	1.65
	3.2	2.35
Bread, rye	3.0	~0.88
Bread, wheat	4.8–5.2	~1
Bread, dried	1.7	~1
	5.2	~0.88
Carrots	0.65	Not available
Dough, macaroni	1–1.1	Not available
Dough, wheat (44% moisture)	1.7	~1
Potato, dry	6.5–7.8	~0.88
Potato, raw	2.6	~1
Tomato, paste (70 to 85% moisture)	0.4	~1
Wheat, grains	0.9	~1

Note: Recomputed from Ginzberg [13] to be consistent with Equation 9.14, where the penetration depth is defined as the distance over which the energy flux drops to $1/e$ of its incident value.

9.3 MEASUREMENT OF RADIATIVE PROPERTIES OF FOOD

Radiative properties are generally measured by having a source emitting the electromagnetic waves and a detector able to capture the waves that either passes through or reflects from the surface. There is, though, the directional nature of the waves to be considered. Thus, as most spectrometers measure only the transmitted spectral irradiance corresponding to a collimated beam of incident monochromatic radiation, they typically measure the spectral directional transmissivity [17].

Measurement of radiative properties of food started with Ginzberg [13]. The process consisted of a source irradiating at a known temperature and a monochromator. The detectors used were either pyrometers or pyroelectric detectors. Dagerskog [26,27] reproduced Ginzberg's experiments and added the rotating chopper to eliminate the influence of background irradiation. Il'yasov and Krasnikov [14] provided significant data on food properties and also discussed principles governing radiative energy transport in food systems, including details such as the two-dimensional effect of electromagnetic energy advancing through a material.

Some of the studies [e.g., 14,26–28] mention precautions to be taken in obtaining accurate radiative properties data for foodstuffs such as:

- The sample should be exposed to a monochromatic flux; this reduces to a minimum heat-up of the sample and radiation-induced changes in the physicochemical properties of foodstuffs.
- The incident radiation should be modulated so that the radiation coming from the sample will have no effect on the measurements.
- The radiation detector should register more than 80% of the total radiant energy reflected or transmitted by the layer.
- The effect of moisture content and temperature in determining the radiative properties in foodstuffs is critical. Thin (e.g., 0.2 mm) samples of vegetable tissue (e.g., potato) lose moisture very rapidly, and care should be taken to avoid this. At a minimum, sample moisture content before and after the experiment should be noted.

The most frequently employed methods for measuring radiative properties of food use a specular hemisphere (most spectroscopy instruments, in general), a specular rotational ellipsoid, or an integrating sphere.

In studying the radiative properties of paper, Ojala and Lampinen [29] were very successful in taking into account the moisture content of the samples measured. They made use of a Fourier transform spectrometer and an integrating sphere before the detector to make sure that the properties were no longer directional (Figure 9.8). They used three different integral sphere coatings for each of the wavelength intervals studied, from 0.4 to 20 µm.

Almeida [28] measured spectral hemispherical reflectance and transmittance using an infrared spectroradiometer system (Model 746 from Optronics Laboratories, Inc., Orlando, Florida) shown in Figure 9.9. The system created unidirectional, monochromatic, incident radiation (1.2 cm beam diameter) impinging on cylindrical, 2.5 cm-diameter test samples of varying thickness. Incident radiation was centered on the cylinder axis and directed approximately along the cylinder axis (10° off-normal angle per system design

PROPERTIES RELEVANT TO INFRARED HEATING OF FOOD

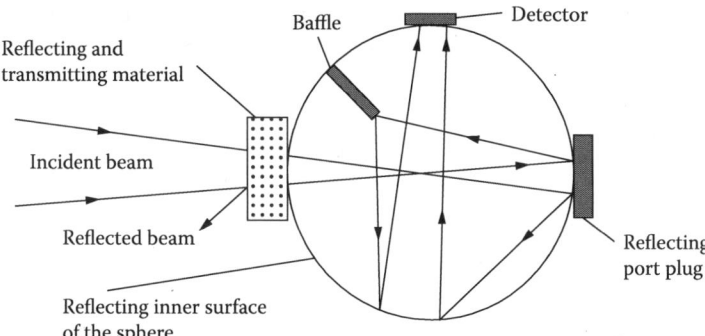

Figure 9.8 Schematic of transmission measurement using an integrating sphere. The inner surface of the integrating sphere is coated with a diffusing coating of high reflectance. (From KT Ojala, MJ Lampinen. *Modeling, Measurements, and Efficiencies of Infrared Dryers for Paper Drying*, volume 2 of Handbook of Industrial Drying, New York: Marcel Dekker, Inc., 1995, pp. 931–976.)

specifications). Test sample external boundaries were nonreflective. All measurements were taken at room temperature. Measurement uncertainties in reflectance and transmittances were ±10%. This was determined by measurement repeatability upon sample rotation, and using different spectroradiometer configurations in overlapping wavelength bands.

The system comprised a dual source attachment, a monochromator, an integrating sphere, and detectors with head and module for different ranges in the spectra. The source attachment was able to produce the stable irradiance (free of noise) required for detector spectral response in the reflectance and transmittance measurements. The dual source

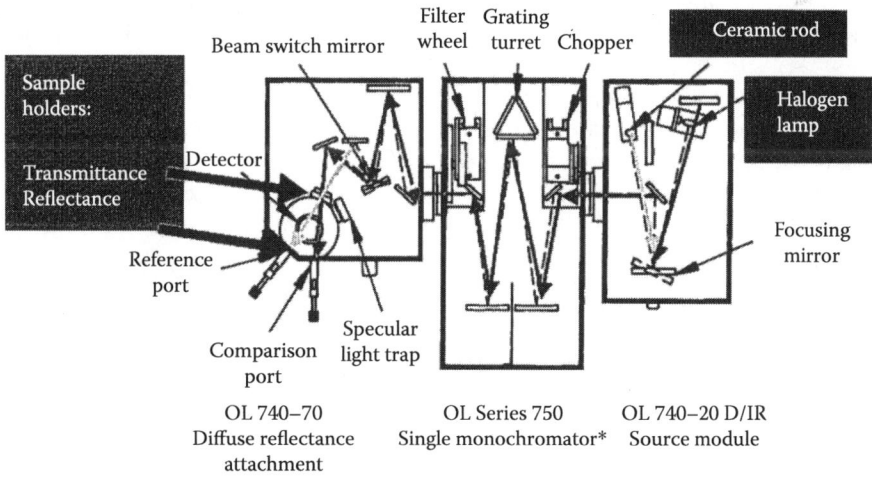

Figure 9.9 Schematic of a spectroradiometer. (Optronic Laboratories, Inc., Orlando, FL.)

unit incorporated both a 150 W quartz halogen lamp suitable for the 0.25–3.5 μm range and a ceramic rod glower, suitable for the 2.5–10 μm range.

Reflectance measurements were carried out through a comparison method using a calibrated sample of known reflectance in a different port; to complete one measurement the equipment first scanned the comparisons port and then scanned the sample port. The transmittance measurement required one calibration scan with the empty transmittance port (100% calibration), and then the samples would be scanned and given a percentage of transmittance compared to the calibration one.

The electromagnetic signals captured in the detectors were read in an oscilloscope connected to a PC for data logging and data retrieval. Samples were obtained by slicing a raw potato cylinder of 2.5 cm in layers as thin as 0.2 mm to 2 cm. The wavelength range of interest was 0.75–2.5 μm. Some results from this study are discussed in Sections 9.4.3 through 9.4.5.

9.4 RADIATIVE PROPERTY DATA FOR FOOD SYSTEMS

This section will discuss some of the data for food materials and their ingredients, with discussions on the parameters that influence the data. For additional data and processes in the food context, the reader is referred to the two excellent books mentioned earlier [13,14]. Infrared radiative properties of foodstuffs depend primarily on the physicochemical nature of the food, water content, and thickness.

9.4.1 Radiative Property Data for Water, Ice, and Water Vapor

Water is often the major constituent in food products. Fruits and vegetables that consist mainly of water have their radiative properties greatly influenced by the water content [14]. The state of water also affects the interaction in the infrared range. Therefore, it is instructive to study the infrared interaction with water, water vapor, and ice. Another reason to discuss water properties is that they have been studied in considerable detail, unlike food properties. Thus, it may be possible to develop a qualitative sense of a food radiative property by looking into the corresponding property of water.

Throughout the infrared spectral region, water exhibits a strong absorption and weak scattering of radiation. The absorption bands are due to the presence of a hydroxyl group held through hydrogen bonding. In the near- and mid-infrared region, absorption and scattering correspond to the vibrational modes of energetic transitions, whereas in the far infrared, they correspond to rotational transitions [30]. For the three states of water, the following is a quick overview of infrared absorption bands [17]:

- For liquid water, the absorption bands are centered around 1.19, 1.43, 1.94, 2.93, 4.72, 6.10, and 15.3 μm at 25°C; the last four are the principal ones. It is also noted that temperature has no significant effect on these absorption bands, while solutes and hydrates can shift these bands slightly.
- For ice, the absorption bands are located at about the same wavelengths as for liquid water.
- For water vapor, the absorption bands are centered at 1.14, 1.38, 1.87, 2.7, and 6.3 μm.

Specular reflectivity, ρλ, and transmissivity, τλ, of water were shown in Figure 9.7. Examples of additional data on the radiative properties of water and water vapor can be seen in Hale and Querry [31] and Edwards et al. [32].

9.4.2 Properties of Other Pure Food Components

An example of spectral variation of absorption in various food components is shown in Figures 9.10 (for protein and starch) and 9.11 (for protein and glucose). Such composition dependence of the spectral variation has been suggested for use in selective heating of food components [20,21]. Spectral data on other pure food components are available in the context of spectroscopic measurements [5].

9.4.3 Spectral Variation of Radiative Property Data: Potato Tissue as an Example

Spectral hemispherical reflectance, absorptance, and transmittance of potato tissue from the work of Almeida [28] are shown in Figures 9.12, 9.13, and 9.14, respectively. The variations with moisture content are explained later. Figure 9.14 shows how transmittance changes near zero at approximately 1.4 μm. Additional experimental data for change of spectral transmittance with moisture content can be found in Almeida (2004). Spectral variation of penetration depth, calculated from the transmittance data, is shown in Figure 9.15.

Figure 9.16 shows reflectance data for a number of food products in various physical states, from the work of Il'yasov and Krasnikov (14). Reflectance of different food products can differ by 20–60% in the wavelength ranges of 0.4–0.8 and 1.5–2.7 μm and by 5–20% in the range 3.5–5.8 μm.

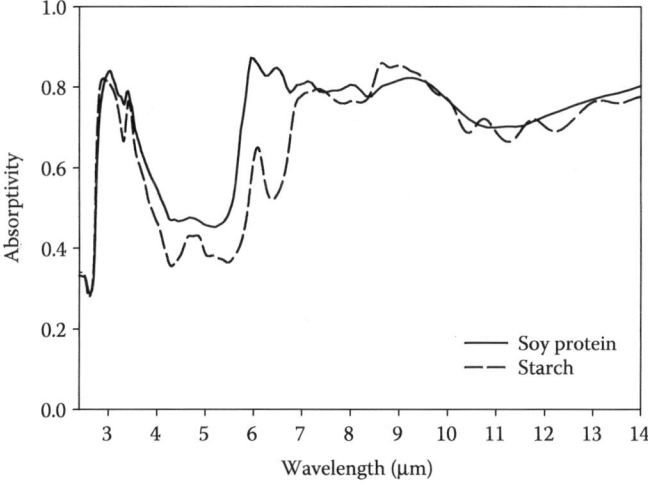

Figure 9.10 Spectral absorptivity of soy protein and starch. (From S Jun, J Irudayaraj. *Journal of Drying Technology* 21(1):69–82, 2003.)

ENGINEERING PROPERTIES OF FOODS

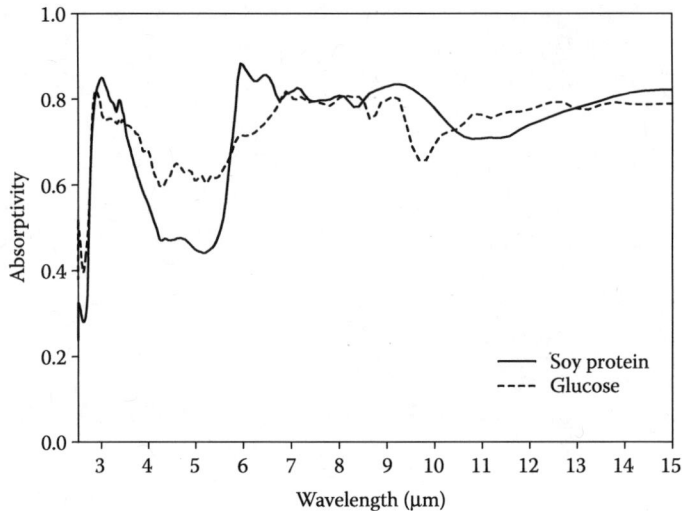

Figure 9.11 Spectral absorptivity of soy protein and glucose. (From S Jun, J Irudayaraj. *Journal of Drying Technology* 21(1):51–67, 2003.)

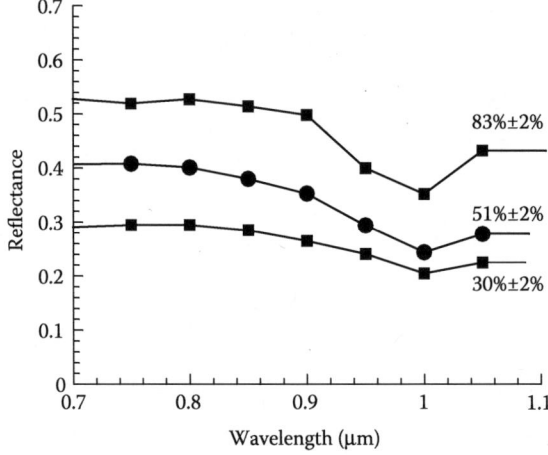

Figure 9.12 Spectral reflectance of potato tissue is shown as a function of moisture content. Measurements are at room temperature (~25°C), in samples of 1 cm thickness and 2.5 cm diameter. (From MF Almeida. Combination microwave and infrared heating of foods. PhD dissertation, Cornell University, Ithaca, NY, 2004.)

PROPERTIES RELEVANT TO INFRARED HEATING OF FOOD

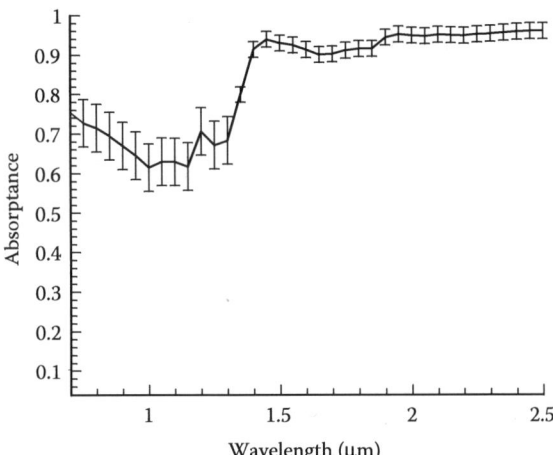

Figure 9.13 Spectral absorptance in potato tissue at 51% moisture content.

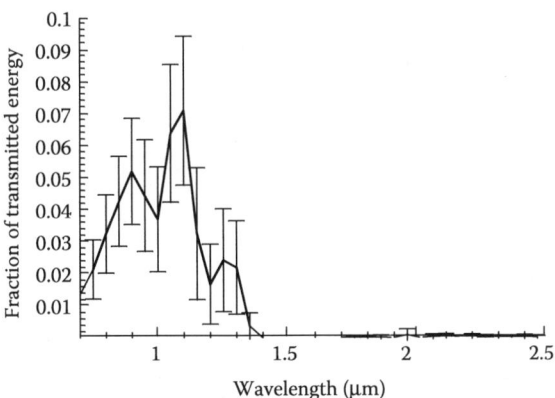

Figure 9.14 Fraction of transmitted energy in a sample of raw potato at 86% moisture content and 1 cm thickness. Sample diameter is 2.5 cm, and the temperature is approximately 26°C. (From MF Almeida. Combination microwave and infrared heating of foods. PhD dissertation, Cornell University, Ithaca, NY, 2004.)

9.4.4 Moisture Dependence of Radiative Property Data

Moisture content is a crucial variable when determining the radiative properties of food. Although the importance of moisture content was noted in the past [13], measurements on food products have been reported only recently [28]. Detailed measurements on paper [29], shown in Figure 9.17, are quite instructive in this respect. They show that the reflectance decreases with added moisture. This can be seen in a common observation of asphalt surfaces. When the asphalt surface is wetted, it becomes darker, that is, reflectance and scattering from the surface decrease.

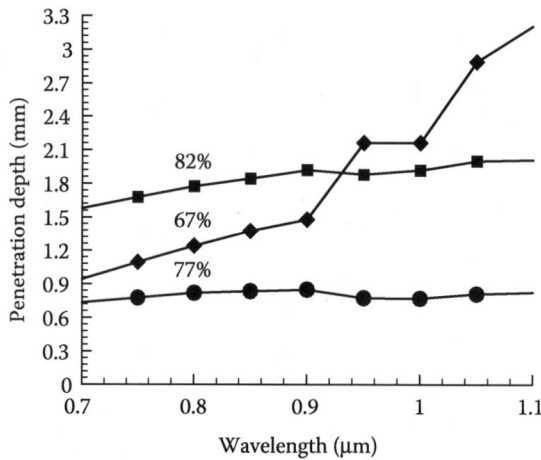

Figure 9.15 Spectral variation of penetration depth, δ, for potato samples at various moisture contents (shown in percentage). Penetration depth is calculated by fitting an exponential curve through the transmittance versus thickness data.

Change in structure as moisture is added or removed, however, can produce different effects. In the work of Almeida [28] shown in Figures 9.12 through 9.14, reflectance decreases with moisture (the samples were dried starting from higher moisture content). This decrease in reflectance is explained by considering the change of the porous matrix in potato tissue. The cellular pores that contain the starch are mainly sustained by a complex matrix of cellulose and water. As water is removed from the matrix, a collapse is expected, resulting in both smaller cellulose pore and starch granules, which results in less scattered energy at the surface (Figure 9.12). In the case of potato, the change in color intensity is a good indicator of this phenomenon—from light to deep and darker yellow as it is dried.

Dependence of spectral transmittance on moisture content, however, cannot be explained on the same basis. Although we would expect transmittance to increase with less water in the energy path, the collapse of pores right at the surface would change the matrix, allowing for absorption at the surface to change.

9.4.5 Temperature Dependence of Radiative Property Data

Temperature dependence of food radiative properties can be highly related to its dependence on moisture content (through the microstructure of the food). Unfortunately, during measurement, isothermal conditions are difficult to reproduce inside the spectroradiometer or spectral hemisphere cameras (or spots), making it difficult to isolate the temperature dependence. Thus, radiative properties of food as functions of temperature have not been reported, although the importance of knowing temperature dependence in practical applications has been emphasized [29]. Thus, further research is needed to study the temperature dependence of radiative properties of food.

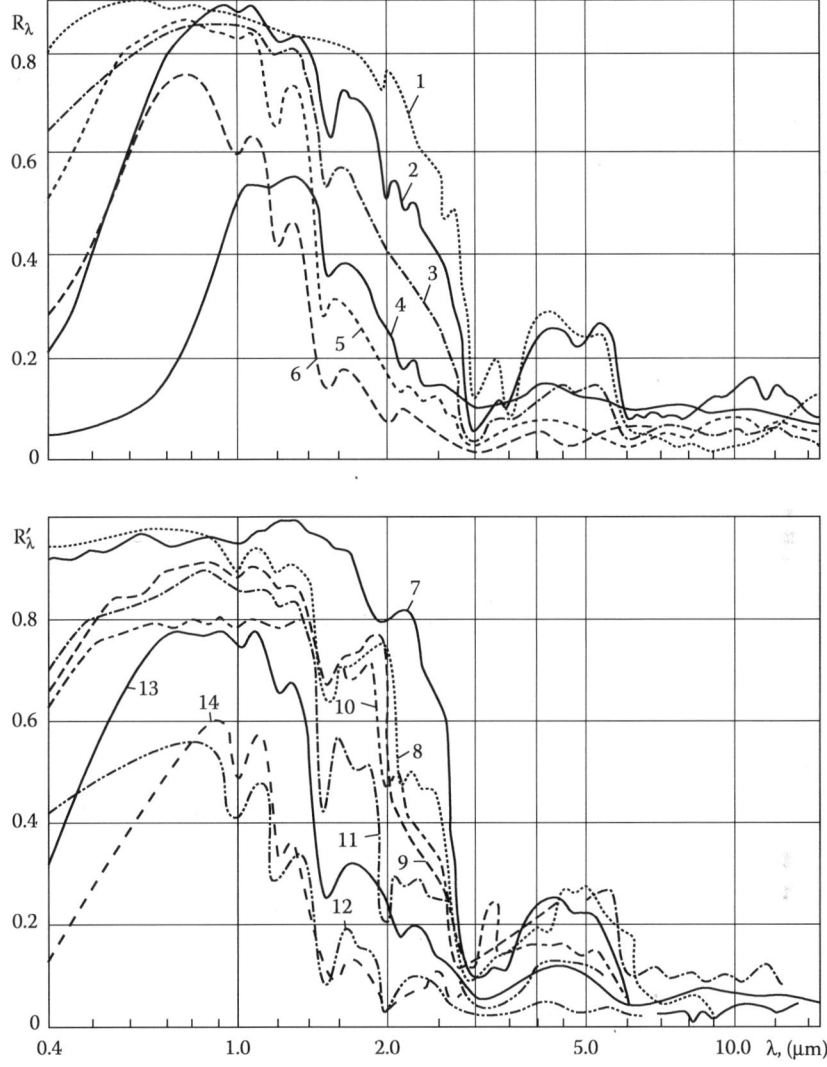

Figure 9.16 Reflection spectra of different substances: 1–enamel VL-55; 2–pinewood (MC = 6.2%); 3–flour (grade 1, MC = 8.1%); 4–baked brown bread crust together with soft part of bread (l = 40 mm); 5–dried potato (l_{raw} = 10 mm); 6–dough made of wheat flour (l = 40 mm); 7–MgO; 8–confectioner's sugar (MC = 0.06%); 9–fruit candy after it has gelled (MC = 30.0%); 10–outer layer of a silk cocoon; 11–potato starch (MC = 11.8%); 12–potato starch (MC = 76.5%); 13–dried pulp of pear (l_{raw} = 10 mm); 14–durum wheat dough (MC = 31.2%). MC represents moisture content. (Reproduced from SG Il'yasov, VV Krasnikov. *Physical Principles of Infrared Irradiation of Foods.* 1st ed. New York: Hemisphere Publishing Corp, 1991, with permission.)

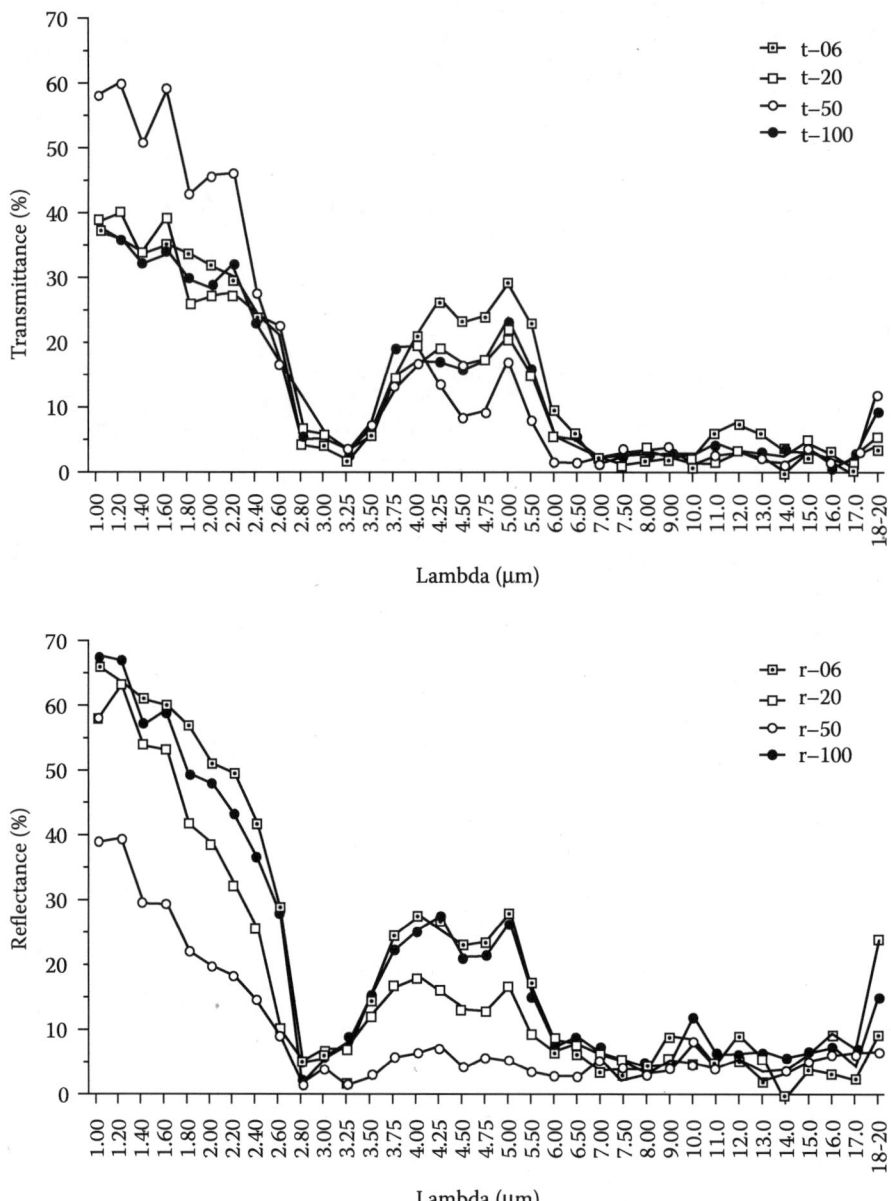

Figure 9.17 Spectral transmittance and reflectance of base paper at different moisture contents over the wavelength range of 1–20 µm. Dry weight is 41.1 g/m^2, and moisture contents are 6.0%, 20.8%, 52.5%, and 102.2%, as shown in the legend in rounded figures. (From KT Ojala, MJ Lampinen. *Modeling, Measurements, and Efficiencies of Infrared Dryers for Paper Drying,* volume 2 of Handbook of Industrial Drying, New York: Marcel Dekker, Inc., 1995, pp. 931–976.)

9.4.6 Dependence of Radiative Property Data on Food Structure

The porous structure of materials is known to affect their radiative properties [33]. This has not been studied for food materials. One of the materials where this has been well studied is reticulated porous ceramics (RPC) in Hendricks and Howell [34], where the same spectroradiometer as described above was used. In this concept, a new radiative property is proposed called the direct transmitted fraction, f_{dt}, which is obtained experimentally. The direct transmitted fraction is essentially the fraction of the incident radiative intensity that penetrates a depth, without any interaction with the internal structure. The quantity $(1-f_{dt})$ is then the fraction of radiative intensity that interacts with the structure of the material through normal absorption and scattering processes.

9.4.7 How Processing Can Change Food Radiative Properties

Processing is expected to change the radiative food properties. Details of such information are generally unavailable. As an example, consider Figure 9.18, which shows how the radiative properties of potato changed as it was heated in boiling water (under microwaves) for increasing duration. In these cases, the starch gelatinization stages were likely responsible for the decreasing reflectance at the potato surface. In general, changes in radiative properties would relate to changes in structure and chemical changes during processing.

Another example can be seen in data for crumb and crust portions of bread (Figure 9.19), which show significant differences. Transmittance values for bread crust are less than for crumbs, due to structure and surface property differences in the two states of gluten and starch matrix.

In another example, emissivity of chicken meat was measured during cooking (Figure 9.20) and it showed a 30% reduction in value during the particular cooking process. This

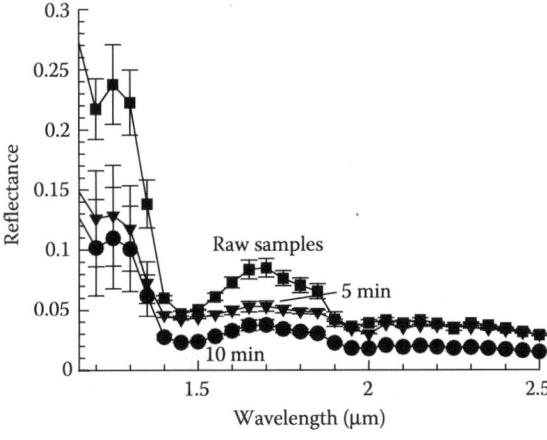

Figure 9.18 Change in spectral reflectance of potato samples due to various heat treatments (microwaved in boiling water for the duration specified). Samples have a thickness of 1 cm, diameter of 2.5 cm, and moisture content of 87%, and are at a temperature of approximately 26°C.

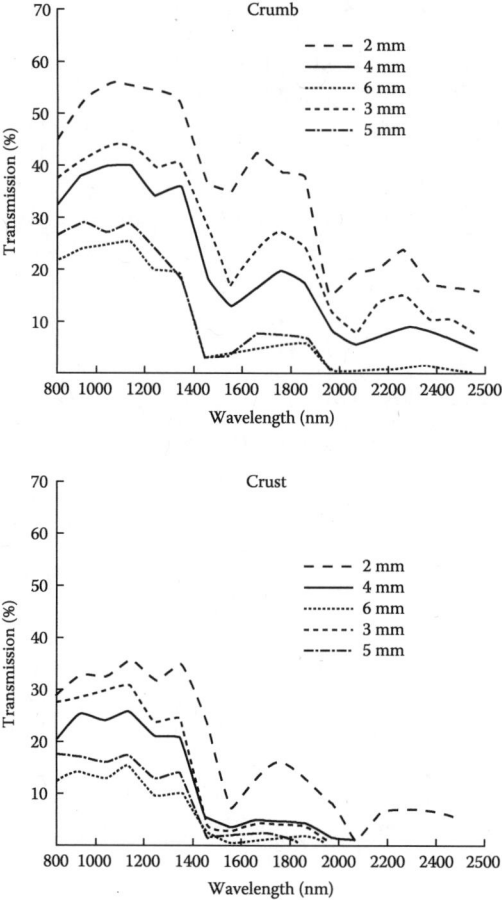

Figure 9.19 Spectral transmittance for bread crust and crumb of various thickness values. (From C Skjoldebrand, C Ellbjär, C Anderson. *Journal of Food Engineering* 8:129–139, 1988.)

was attributed to physical and chemical changes of the chicken meat during cooking. The cooling process after the end of cooking resulted in very little change in the emissivity.

9.5 USE OF THE RADIATIVE PROPERTIES IN MODELING OF HEAT TRANSFER AND SENSITIVITY OF TEMPERATURE TO RADIATIVE PROPERTIES

Exchange of radiative energy between two or more bodies is often a fairly complex problem. Modeling of radiative heating can range from highly simplified, as in undergraduate texts [8,25] to very realistic ones, as in specialized texts [11], which are briefly noted below:

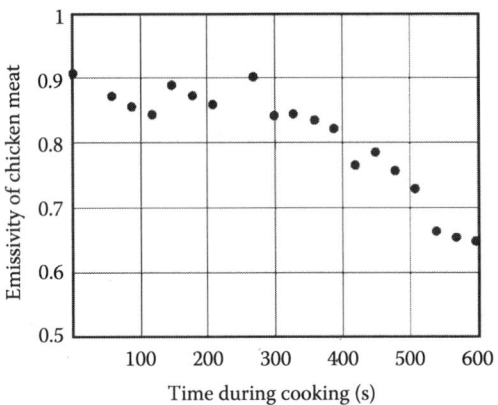

Figure 9.20 Variation in the emissivity of chicken meat during a cooking process. Time zero corresponds to raw meat, and the end of cooking was at 600 s. The spectral band used was 3.4–5 μm. (From JG Ibarra, et al. *Applied Engineering in Agriculture* 16(2):143–148, 2000.)

9.5.1 Simple Models

In perhaps the simplest situation of radiative heat transfer between an oven surface and a small food object placed inside the oven such that the food is completely enclosed by the oven surface, the net radiative exchange is given by

$$q_{1-2} = \varepsilon_1 \sigma A_1 (T_1^4 - T_2^4) \tag{9.15}$$

where q_{1-2} is the net radiative energy transfer between bodies 1 and 2 in W, ε_1 is the emissivity of the food surface, A_1 is the surface area of the food, and T_1 and T_2 are the temperatures of the food and the oven surface, respectively. When two bodies exchange radiation, the radiative exchange also depends on their size and shape and the relative orientation of their respective surfaces. The size, shape, and orientation factors are lumped in a parameter called the *configuration factor* or the *view factor*. In terms of the view factors, net radiative exchange between two blackbodies is given by

$$q_{1-2} = \sigma A_1 F_{1-2} (T_1^4 - T_2^4) \tag{9.16}$$

where F_{1-2} is the view factor that stands for the fraction of radiation leaving surface 1 that is intercepted by surface 2. For a large number of surface configurations, F_{1-2} can be found from either textbooks [e.g., 8] or specialized sources [9,10].

In an even simpler approach, Equation 9.16 is further simplified. For the special case in which temperatures T_1 and T_2 are close, we can write Equation 9.16 as

$$\begin{aligned} q_{1-2} &= \sigma A_1 F_{1-2} 4 T_1^3 (T_1 - T_2) \\ &= A_1 F_{1-2} h_r (T_1 - T_2) \end{aligned} \tag{9.17}$$

where h_r, given by

$$h_r = 4\sigma T_1^3 \qquad (9.18)$$

is termed the radiative heat-transfer coefficient, analogous to the convective heat-transfer coefficient. Note that the units for the radiative heat-transfer coefficient are W/m²K, the same as those for the convective heat-transfer coefficient.

To solve for temperatures inside a food material (or any other solid), the heat flux q, obtained from Equations 9.15, 9.16, or 9.17, provides the boundary condition (specified surface heat flux) for the energy equation. Solving the energy equation provides the temperature within the food.

If the depth of penetration, as defined earlier, is significant, a slightly different formulation is required. Instead of specifying the radiant heat flux, q, radiant heating is included as a volumetric heat source term. Equation 9.14 can be used to derive an expression for volumetric heat generation, Q, as

$$\begin{aligned} Q &= -\frac{dq}{dx} \\ &= \frac{q_0}{\delta} e^{-x/\delta} \end{aligned} \qquad (9.19)$$

which is known from the knowledge of surface radiant heat flux, q_0, and the penetration depth, δ. Such a formulation using Equation 9.19 is not the most fundamental approach when the medium (food) is absorbing radiation but is reasonable and avoids a much more complex problem formulation.

9.5.2 More Realistic Models

In a realistic food heating situation involving emitter, food, and the enclosing surfaces, modeling radiative heating is considerably more complex. Two approaches to radiative energy transport modeling are generally pursued. These are either to solve the general radiative exchange Equation 9.10 numerically or using Monte Carlo ray-tracing method where the energy from a physical surface in terms of groups of photons is traced [10]. The problem becomes progressively more complex as the properties are considered spectral and directional, and eventually if the medium through which radiation is transporting is participating, that is, it absorbs and scatters radiation. Also, temperature increases and moisture is lost during heating of a food, which changes its radiative properties and will contribute to additional complexities. However, spectral properties of food are generally not available (an exception is [37]) and directional properties are also not available. Also, in food materials that generally contain significant amounts of water, penetration depths are small compared to the size. Thus, the food does not need to be considered a participating media and instead the radiative flux, calculated from the radiative energy transport modeling, is used as boundary condition for heat conduction (or convection) inside the food, as already discussed. In an oven, for example, the air is generally considered not participating but a liquid or solid food will absorb most of the energy at its surface [38].

9.5.2.1 Model Based on Monte Carlo Method

One of the very few examples of modeling food heating that used the Monte Carlo technique for radiative transport is surface heating of strawberries for decontamination [39]. As mentioned above, here the Monte Carlo technique provides the boundary condition for heat conduction on a food surface and is combined with heat conduction equation inside the food. It used constant emissivity of 0.95 for the food (strawberry) surface with zero penetration, that is, directional and spectral variations of properties are ignored, presumably due to nonavailability of such data. Also, for short-term heating, temperature and moisture effects on the properties were ignored.

9.5.2.2 Model Based on Radiative Transport Equation

Likewise, perhaps the only example of solving the radiative energy transport equation (the other choice in modeling) in heating of food is the work of Dhall et al. [38], for infrared only heating in a combination microwave-infrared oven where only infrared was used (Figure 9.22). The solution to the radiative transport equation provides the heat flux boundary condition for heat conduction in the food assuming zero penetration of infrared in it. Spectral dependence of the emissivity of food from measured data is used in this study, as shown in Figure 9.21. Directional variations in properties are ignored. For short-term heating used in this study, temperature and moisture effects are ignored as well. As shown in Figure 9.22, the oven has a halogen source from which the infrared energy first goes through a glass (the glass is participating) and enters the oven cavity.

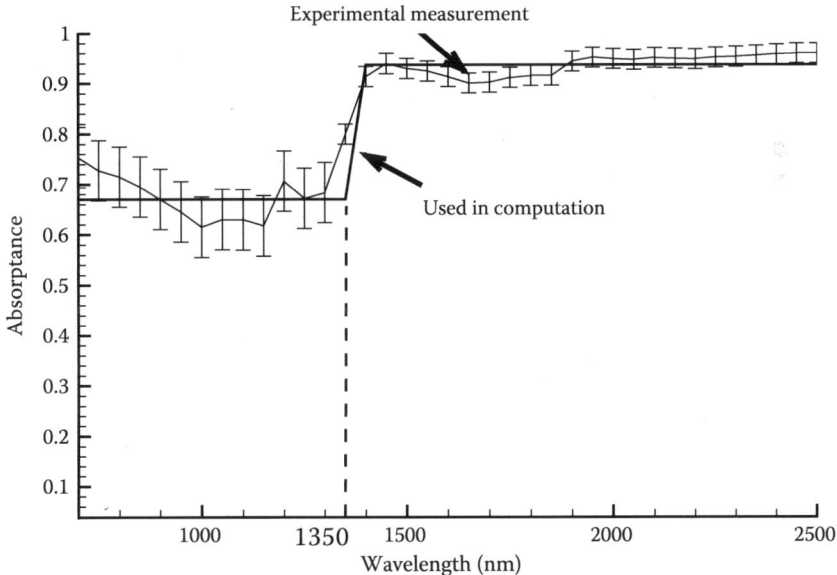

Figure 9.21 Illustration of simplification of the spectral dependence in radiative property data for use in radiative exchange calculations. (From MF Almeida. Combination microwave and infrared heating of foods. PhD dissertation, Cornell University, Ithaca, NY, 2004.)

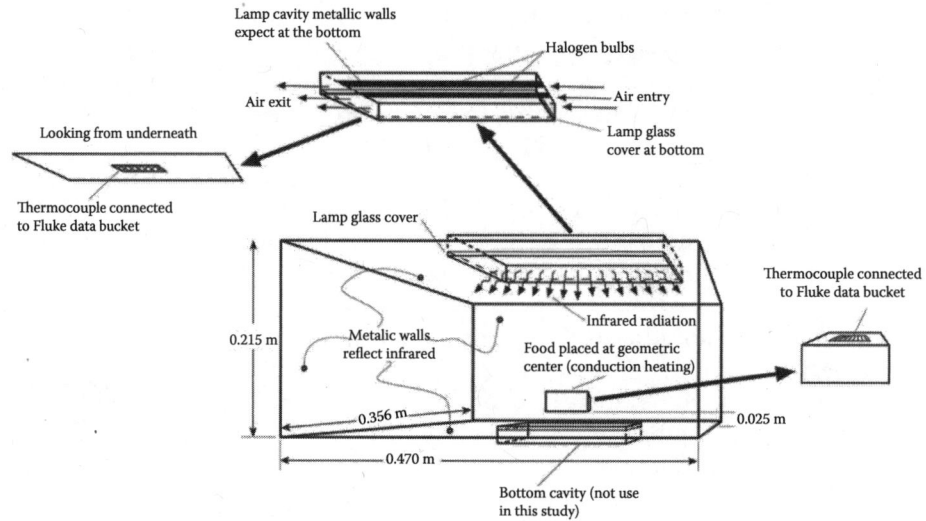

Figure 9.22 Schematic of a combination microwave-infrared oven with only radiative heating present. The geometry of the oven is rectangular, of size shown. The food inside the oven is a potato slab of geometry 0.0470 m × 0.0356 m × 0.0215 m. Food is placed at 2.5 cm above the geometric center of the oven's bottom surface and parallel, resting on a quartz glass tray. (From A Dhall, K E Torrance, A K Datta. *American Institute of Chemical Engineers Journal*, 55(9):2448–2460, 2009.)

Direct and indirect amounts of radiative energy reflected from the walls of the oven combine to produce the infrared fluxes on the food surface and its temperatures, as shown in Figure 9.23.

9.5.3 Sensitivity of Temperature to Radiative Property: Effect of Emissivity

Effect of food emissivity value on heat flux and temperature is shown in Figure 9.23 for three different food surface emittances. Measured data (Figure 9.21) for the potato surface exhibit nongray behavior. An average emissivity value over the wavelength can also be defined. Such averaging is only accurate when all the radiation sources have identical temperatures. When surfaces at different temperatures are present (e.g., source lamps, glass, and oven walls in this study), the use of such an average may not be appropriate. Simulations were carried out for three different food surface emittance combinations: 0.67 for all wavelengths; nongray surface (0.67 for wavelengths <1350 nm and 0.96 for wavelengths >1350 nm); and 0.96 for all wavelengths. Calculated heat flux and temperature profiles at the center of the top surface of the food sample are shown in Figure 9.23. Early in the transients, the flux and temperature profiles for $\varepsilon_{food} = 0.67$, which is gray, and the nongray case nearly coincide. This is because almost all the heat coming to the food is from the source lamp, that is, with wavelengths <1350 nm, for which ε_{food} in both the cases is 0.67. However, as the oven walls get heated and start contributing significantly to the heat delivered to the food, the profiles with $\varepsilon_{food} = 0.67$ start to fall below the nongray case.

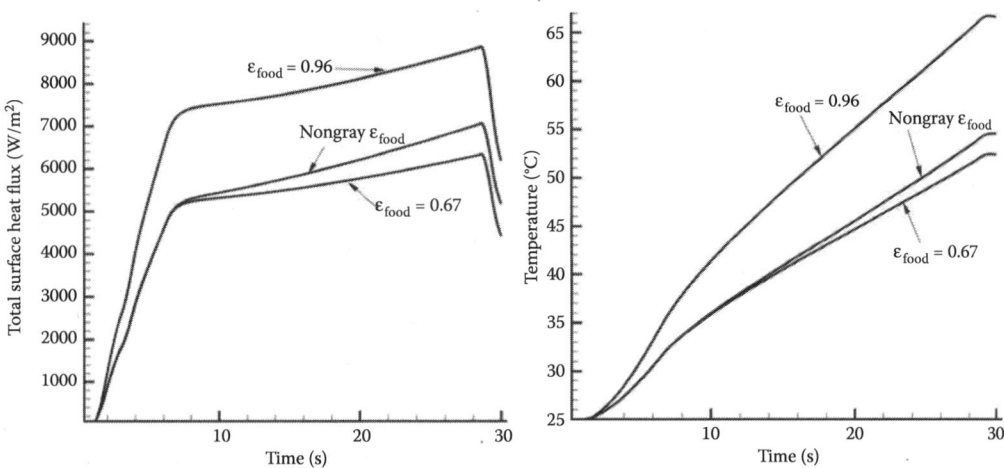

Figure 9.23 Effect of food surface emittance values on total heat flux and temperature at the center position of the top food surface, for the oven shown in Figure 9.22, computed from a model described in Section 9.5.2. Heating is continuous for 30 s. The emissivity values are discussed in Section 9.4.3. (From A Dhall, K E Torrance, A K Datta. *American Institute of Chemical Engineers Journal*, 55(9):2448–2460, 2009.)

This is because the oven walls emit radiation with longer wavelengths, for which the food emittance should be 0.96 and not 0.67. This effect will be much more pronounced at later times when emission from the oven walls becomes larger and also when the lamps are completely shut off (during cyclic heating) and all the heating is provided by the oven walls. When used for the whole duration of heating, the $\varepsilon_{food} = 0.96$ value gives much larger food surface absorbed fluxes and temperatures for obvious reasons.

Further discussion of radiative heat-transfer models [38–44] is beyond the scope of this chapter and the reader is referred to these studies.

9.6 SUMMARY

Comprehensive radiative property data on various food products are generally unavailable. For practical use, we often have to work with hemispherical total properties, that is, properties averaged over all directions and relevant wavelengths. The penetration depth, in particular, can be quite sensitive to the range of wavelength of interest, and thus whenever average penetration depths are used; it should be for the relevant wavelength range. For an even simpler formulation, when penetration depths are expected to be small over the wavelength range of interest, it may be possible to skip penetration altogether and use a surface heat flux boundary condition, as was discussed. Even if spectral dependence of data becomes available, modeling of radiative heat exchange may require simplification of such data into as few bands as possible (see, e.g., Figure 9.21) and averaging of the properties over all directions (the latter called the diffuse approximation).

While technological developments continue in infrared heating in terms of newer equipment design and applications, research in infrared heating of foods is somewhat stagnant. In particular, measurement of properties is not an area showing much activity. Lack of activity is also somewhat true in the area of modeling of infrared heating of foods, where the number of new articles average less than one per year. Several recent reviews of infrared heating of foods have appeared in the literature [45–47], suggesting continued interest in infrared radiation by the food community but as mentioned, these do not report new data on properties or any significant amount of new modeling.

ACKNOWLEDGMENTS

The authors gratefully acknowledge the valuable contributions made by reviewers Dr. Soojin Jun of The Pennsylvania State University, Dr. Kevin Keener of North Carolina State University, and Dr. Constantine Sandu of ConAgra Foods to the original version in the third edition of the book.

REFERENCES

1. V Macaluso. Infrared drying technology applications. *Cereal Foods World* 46(8):355–356, 2001.
2. F Martinez-Bustos, SE Morales, YK Chang, A Herrera-Gomez, MJL Martinez, L Banos, ME Rodriguez, MHE Flores. Effect of infrared baking on wheat flour tortilla characteristics. *Cereal Chemistry* 76(4):491–495, 1999.
3. S Cenkowski, J-T Hong, MG Scanlon, SD Arntfield. Development of a mathematical model for high intensity infrared processing (micronization) of peas. *Transactions of the ASAE*, 46(3):705–713, 2003.
4. J Sawai, K Sagara, A Hashimoto, H Igarashi, M Shimizu. Inactivation characteristics shown by enzymes and bacteria treated with far-infrared radiative heating. *International Journal of Food Science and Technology* 38:661–667, 2003.
5. P Williams, KH Norris. *Near-Infrared Technology in the Agricultural and Food Industries*. St. Paul, Minnesota: American Association of Cereal Chemists, 2001.
6. H Buning-Pfaue. Analysis of water in food by near infrared spectroscopy. *Food Chemistry* 82(1):107–115, 2003.
7. L Michalski, K Eckersdorf, J Kucharski, J McGhee. *Temperature Measurements*. Chichester: John Wiley & Sons, 2001.
8. FP Incropera, DP Dewitt. *Introduction to Heat Transfer*. New York: John Wiley & Sons, 1996.
9. MF Modest. *Radiative Heat Transfer*. New York: McGraw-Hill, 1993.
10. JR Howell. *A Catalog of Radiation Configuration Factors*. NewYork: McGraw-Hill, 1982.
11. R Siegel, JR Howell. *Thermal Radiation Heat Transfer*. New York: McGraw-Hill, 1981.
12. EM Sparrow, RD Cess. *Radiation Heat Transfer*. Washington, D.C.: Hemisphere Publishing Corp., 1978.
13. AS Ginzberg. *Application of Infra-Red Radiation in Food Processing*. London: Leonard Hill Books, 1969.
14. SG Il'yasov and VV Krasnikov. *Physical Principles of Infrared Irradiation of Foods*. 1st ed. New York: Hemisphere Publishing Corp, 1991.
15. SP Parker. *McGraw-Hill Dictionary of Scientific and Technical Terms*. New York: McGraw-Hill, 1984.
16. I Rosenthal. *Electromagnetic Radiations in Food Science*. New York: Springer-Verlag, 1992.
17. C Sandu. Infrared radiative drying in food engineering. *Biotechnology Progress* 2(3):109–119, 1986.

18. SS Nielsen. *Food Analysis*. Gaithersburg, MD: Aspen Publishers, 2003.
19. AC Metaxas. *Foundations of Electroheat: A Unified Approach*. Chichester, UK: John Wiley & Sons, 1996.
20. S Jun, J Irudayaraj. Selective far infrared heating system—Design and analysis (Part I). *Journal of Drying Technology* 21(1):51–67, 2003.
21. S Jun, J Irudayaraj. Selective far infrared heating system—Spectral manipulation (Part II). *Journal of Drying Technology* 21(1):69–82, 2003.
22. PR Gast. Solar electromagnetic radiation. In *Handbook of Geophysics and Space Environments*. New York: McGraw-Hill, Inc., 1965.
23. AK Datta. *Biological and Bioenvironmental Transport Processes*. New York: Marcel Dekker, 2002.
24. WL Wolfe. *Handbook of Military Infrared Technology*. Washington, DC: Office of Naval Research, Department of the Navy, 1965.
25. YA Cengel. *Heat Transfer: A Practical Approach*. New York: McGraw-Hill, 1998.
26. M Dagerskog. Infrared radiation for food processing II. Calculation of heat penetration during infrared frying of meat products. *Lebensmittel-Wissenschaft u. Technologie* 12:252–257, 1979.
27. M Dagerskog, L Österström. Infra-red radiation for food processing I: A study of the fundamental properties of infra-red radiation. *Lebensmittel-Wissenschaft u. Technologie* 12:237–242, 1979.
28. MF Almeida. Combination microwave and infrared heating of foods. PhD dissertation, Cornell University, Ithaca, NY, 2004.
29. KT Ojala, MJ Lampinen. *Modeling, Measurements, and Efficiencies of Infrared Dryers for Paper Drying*, volume 2 of *Handbook of Industrial Drying*, New York: Marcel Dekker, Inc., 1995, pp. 931–976.
30. DJ McClements. Spectroscopy: Instrumental techniques for food analysis. Document on the Web at http://www.unix.oit.umass.edu/~mcclemen/581Toppage.html, accessed February 4, 2004.
31. M Hale, MR Querry. Optical constants of water in the 200-nm and 200 μm wavelength region. *Applied Optics* 12(3):555–563, 1973.
32. DK Edwards, BJ Flornes, LK Glassen, W Sun. Correlation of absorption by water vapor at temperatures from 300 K to 1100 K. *Applied Optics* 4(6):715–721, 1965.
33. M Petterson, S Stenström. Absorption of infrared radiation and the radiation transfer mechanism in paper. Part I: Theoretical model. *Journal of Pulp and Paper Science* 24(11):349–355, 1998.
34. TJ Hendricks, JR Howell. New radiative analysis approach for reticulated porous ceramics using discrete ordinates method. *Journal of Heat Transfer* 118:911–917, 1996.
35. C Skjoldebrand, C Ellbjär, C Anderson. Optical properties of bread in the near-infrared range. *Journal of Food Engineering* 8:129–139, 1988.
36. JG Ibarra, Y Tao, AJ Cardarelli, J Shultz. Cooked and raw chicken meat: Emissivity in the mid-infrared region. *Applied Engineering in Agriculture* 16(2):143–148, 2000.
37. M Almeida, KE Torrance, AK Datta. Measurement of optical properties of foods in near- and mid-infrared radiation. *International Journal of Food Properties*, 9(4):651–664, 2006.
38. A Dhall, KE Torrance, AK Datta. Radiative heat exchange modeling inside an oven. *American Institute of Chemical Engineers Journal*, 55(9):2448–2460, 2009.
39. F Tanaka, K Morita, K Iwasaki, P Verboven, N Scheerlinck, B Nicolai. Monte Carlo simulation of far infrared radiation heat transfer: Theoretical approach. *Journal of Food Process Engineering* 29(4):349–361, 2006.
40. WJ Mao, Y Oshima, Y Yamanaka, M Fukuoka, N Sakai. Mathematical simulation of liquid food pasteurization using far infrared radiation heating equipment. *Journal of Food Engineering* 107(1):127–133, 2011.
41. N Sakai, A Fujii, T Hanzawa. Heat transfer analysis in a food heated by far-infrared radiation. In Japanese. *Nippon Shokuhin Kogyo Gakkaishi* 40(7):469–477, 1993.

42. N Sakai, T Hanzawa. Applications and advances in far-infrared heating in Japan. *Trends in Food Science and Technology* 5:357–362, 1994.
43. N Sakai, N Morita, P Qiu, T Hanzawa. Two-dimensional heat transfer analysis of the thawing process of tuna by far-infrared radiation. In Japanese. *Nippon Shokuhin Kogyo Gakkaishi* 42(7):524–530, 1995.
44. P Verboven, AK Datta, NT Anh, N Scheerlinck, BM Nicolai. Computation of airflow effects on heat and mass transfer in a microwave oven. *Journal of Food Engineering* 59:181–190, 2003.
45. S Noboru, M Weijie. *Infrared Heating. Thermal Food Processing.* Boca Raton, FL, CRC Press: 529–554, 2012.
46. NK Rastogi. Chapter 13 - Infrared heating of fluid foods. In: *Novel Thermal and Non-Thermal Technologies for Fluid Foods.* PJ Cullen, KT Brijesh, BKT Vasilis Valdramidis A2 - PJ Cullen, V Vasilis eds. San Diego, Academic Press: 411–432, 2012.
47. NK Rastogi. Recent trends and developments in infrared heating in food processing. *Critical Reviews in Food Science and Nutrition* 52(9):737–760, 2012.

10

Mass Transfer Properties of Foods

George D. Saravacos and Magda Krokida

Contents

10.1 Introduction	312
10.2 Phase Equilibria	313
10.2.1 Vapor–Liquid Equilibria	314
10.2.2 Gas–Liquid Equilibria	318
10.2.3 Liquid–Liquid and Liquid–Solid Equilibria	319
10.2.4 Gas–Solid and Vapor–Solid Equilibria	320
10.2.4.1 Water Activity	320
10.3 Diffusion	322
10.3.1 Diffusion in Gases	323
10.3.2 Diffusion in Liquids	324
10.3.3 Diffusion in Solids	325
10.3.3.1 Introduction	325
10.3.3.2 Diffusion in Polymers	327
10.3.3.3 Molecular Simulations	328
10.3.4 Estimation of Diffusivity in Solids	328
10.3.4.1 Sorption Kinetics	328
10.3.4.2 Permeation Measurements	329
10.3.4.3 Distribution of Penetrant	330
10.3.4.4 Drying Rate	330
10.4 Interphase Mass Transfer	331
10.4.1 Mass Transfer Coefficients	331
10.4.2 Penetration Theory	334
10.4.3 Analogies of Heat and Mass Transfer	335
10.4.4 Effect of Surfactants	335
10.5 Mass Transfer in Foods	336
10.5.1 Moisture Transport	336
10.5.1.1 Moisture Diffusion	336

10.5.1.2 Diffusion in Porous Foods	338
10.5.1.3 Interphase Moisture Transfer	342
10.5.2 Diffusion of Solutes	342
10.5.3 Diffusion of Aroma Compounds	344
10.6 Other Mass Transfer Processes	346
10.6.1 Extraction	346
10.6.2 Distillation and Gas Absorption	347
10.6.3 Crystallization	349
10.6.4 Food Packaging	350
Acknowledgments	352
List of Symbols	352
Greek Symbols	353
References	353

10.1 INTRODUCTION

Mass transfer plays a very important role in basic unit operations of food processing, such as drying, extraction, distillation, and absorption. In these physical operations, the resistance to mass transfer is usually the rate-limiting factor, although heat transfer and fluid flow may also be involved. Evaporation of water in concentration processes of liquid foods is normally controlled by heat transfer and it is not classified as a mass transfer operation. However, the stripping of volatile components from liquid foods may depend mainly on mass transfer within the liquid phase or at the liquid/vapor interphase.

Mass transfer is also involved in several physical, chemical, and biological food processes, such as salting, sugaring, oxygen absorption, deaeration, carbonation, crystallization, and cleaning of process equipment. It is important in food packaging and storage, where transfer of moisture, vapors or gases, and flavor components may adversely influence food quality.

Many food processes involving mass transfer are still based on empirical design and operation, because mass transfer theory in foods is not well advanced and mass transfer properties are not readily available. In contrast, the basic mass transfer operations of the chemical industry are well developed, and accurate methods of process and equipment design are available for physical operations such as distillation and gas absorption (Perry and Green, 1997; Treybal, 1980).

The difficulties of applying mass transfer theory to food processes arise from the complex physical structure and chemical composition of foods, which may vary even within the same food and may change during processing or storage. The difficulties are more pronounced in solid foods because, as a rule, transport processes are more complex in solids than in liquids (Saravacos and Maroulis, 2001).

Mass transfer processes involve the transfer of various components within a phase and between phases by molecular diffusion and natural or forced convection. Mass is transferred by concentration or partial pressure gradients, in contrast to the bulk transport of mass by mechanical (flow) energy.

Table 10.1 Mass Transfer Processes Applied to Foods

Process	Phases	Transfer Principle	Typical Applications
Distillation	Vapor/liquid	Difference in volatility	Ethanol separation, Aroma recovery
Absorption	Gas/liquid	Difference in solubility	Aeration, carbonation
Extraction	Liquid/liquid	Difference in solubility	Oil refining
Leaching	Liquid/solid	Difference in solubility	Sugar/oil extraction
Drying	Vapor/solid	Difference in volatility	Dehydration
Adsorption	Vapor/solid	Adsorption potential	Sorption isotherms

Most mass transfer operations applied to chemical engineering are, in effect, physical separation processes in which one or more components are separated and transferred from one phase to another (King, 1982). Typical mass transfer processes of importance in food engineering are shown in Table 10.1. The engineering analysis of mass transfer operations and equipment requires data on phase equilibria, diffusional mass transfer within a phase, and interphase mass transfer.

Mass transfer operations involving gases and liquids are well developed, in both theory and engineering design. Processing of solids is based partially on empirical knowledge. Most foods contain solid components in various forms, and analysis of mass transfer in these instances becomes more complicated than in simple homogeneous systems.

The efficient design and operation of mass transfer processes requires reliable data on the mass transfer properties of foods. These properties must be determined experimentally for each food system because mathematical prediction is still not feasible. The difficulties of prediction and determination of mass transfer properties have been recognized in the literature (Karel, 1975). Efforts have been made to obtain more reliable data on the thermophysical properties of foods, such as water activity and rheological, thermal, and diffusion properties. International cooperation in this area is very helpful, as shown by the results of the projects COST 90 and COST 90bis of the European Economic Community (Jowitt et al., 1983, 1987).

In Sections 10.2 through 10.4, the theory of mass transfer processes is discussed briefly, with an emphasis on those properties that are important to foods. Phase equilibria, diffusion within a phase, and interphase mass transfer are essential elements for the quantitative analysis of a mass transfer process. In Section 10.5, some of the more important applications of mass transfer to food systems are described. Selected literature values of mass transfer properties of various foods are presented in a number of tables.

10.2 PHASE EQUILIBRIA

Phase equilibria are important in calculations involving mass transfer between phases. Thus, vapor–liquid equilibria are needed in distillation, liquid–liquid equilibria are needed in extraction, and so on. Because of the industrial importance of distillation, vapor–liquid equilibria have received the most attention, in both theory and experimental measurement.

Two phases, I and II, are in thermodynamic equilibrium if there is no net transfer of mass at the interface and if the following conditions exist (Prausnitz et al., 1999):

$$G_i(I) = G_i(II) \quad \text{and} \quad T(I) = T(II) \tag{10.1}$$

where G_i is the partial free energy of component i in a mixture and T is the temperature of the phase. Chemical potential can be used instead of free energy in Equation 10.1.

10.2.1 Vapor–Liquid Equilibria

Vapor–liquid equilibria are useful in the analysis of several mass transfer operations of food processing, such as recovery of aroma compounds, distillation of ethanol, and dehydration (Le Maguer, 1992).

The required conditions for equilibrium of a liquid (L) and a vapor (V) phase are stated in Equation 10.1, which can be written as follows (Thijssen, 1974):

$$f_i(L) = f_i(V), \quad T(L) = T(V) \tag{10.2}$$

where f_i is the fugacity of component i, which is defined by the equation

$$G_i = G_{io} + RT \ln \frac{f_i}{f_{io}} \tag{10.3}$$

G_{io} and f_{io} are the free energy and the fugacity of pure component i at a standard state, respectively.

Equation 10.3 applies to both the liquid and vapor (or gas) phases. In ideal mixtures, the fugacity (f_i) is equal to the partial pressure of the component (p_i).

In most food processes, the pressure of the system is near or below 1 atm, and the vapor phase can be considered an ideal gas. However, the liquid phase in most cases is a nonideal solution because of the strong interaction of the food components with the water that is present in the food or added during processing. Under these conditions, the equilibrium is expressed by the equation

$$y_i P = \gamma_i x_i p_{io} \tag{10.4}$$

where x_i and y_i are the mole fractions of component i in the liquid and vapor phase, respectively; γ_i is the activity coefficient of i in the liquid phase; p_{io} is the vapor pressure of pure i at the system temperature; and P is the total pressure of the system.

An equilibrium can be expressed also by the partition coefficient or K factor:

$$K_i = \frac{y_i}{x_i} = \frac{\gamma_i p_{io}}{P} \tag{10.5}$$

The relative volatility of component i with respect to component j is defined by the equation

$$\alpha_{ij} = \frac{K_i}{K_j} = \frac{\gamma_i p_{io}}{\gamma_j p_{jo}} \qquad (10.6)$$

In food systems the solvent is water (W), and the relative volatility of an aroma compound, A, in dilute solution ($\gamma_j \to 1$) becomes

$$\alpha_{AW} = \frac{\gamma_A p_{Ao}}{p_{Wo}} \qquad (10.7)$$

Most food aroma compounds have high activity coefficients in the liquid (aqueous) phase, resulting in high relative volatilities.

The relative volatility is essential in distillation calculations. The equilibrium (y, x) diagram at a given total pressure is expressed by the equation

$$y = \frac{\alpha x}{1 + (\alpha - 1)x} \qquad (10.8)$$

Figure 10.1 shows three different y versus x diagrams of binary systems: a mixture of two soluble components with constant relative volatility; an azeotropic mixture (e.g., ethanol–water); and a mixture of two partially soluble components (e.g., ethyl acetate–water).

Although some experimental data for vapor–liquid equilibria are available in the literature (e.g., Hala et al., 1958; van Winkle, 1967), it often becomes necessary to extrapolate or predict such data by thermodynamic or empirical equations. Two very useful relations for

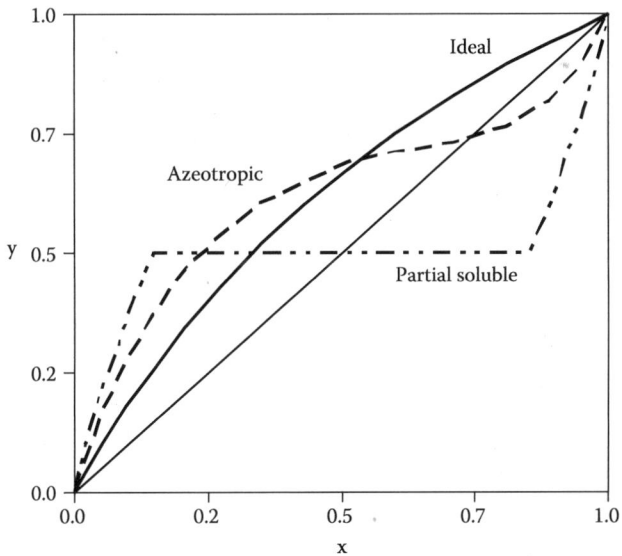

Figure 10.1 Vapor–liquid equilibria of three different binary systems: ideal, azeotropic, and partially soluble.

predicting activity coefficients of the liquid state (γ) are the van Laar and Wilson equations (van Winkle, 1967).

The van Laar equation can be applied to binary nonideal solutions and to partially soluble systems

$$T \ln \gamma_1 = \frac{B}{[1 + A(x_1/x_2)]^2}$$
$$T \ln \gamma_2 = \frac{AB}{[A + (x_2/x_1)]^2}$$
(10.9)

where T is the absolute temperature and A and B are the van Laar constants, which are determined experimentally.

The Wilson equation gives the activity coefficients as functions of the temperature, the molecular volumes of the liquid components, and the energies of interaction among the molecules. It can be applied to multicomponent solutions, and a modification known as the NRTL equation is applicable to partially soluble mixtures (Bruin, 1969). Recently, the computer-based models UNIQUAC and UNIFAC have been applied successfully to food systems (Le Maguer, 1992; Reid et al., 1987; Sancho and Rao, 1997; Saravacos et al., 1990b).

The volatile (aroma) components of foods are generally present at low concentrations, and the system can be considered a binary mixture of the particular compound and water. At higher concentrations, interactions between the aroma compounds become important, and the system should be treated as a multicomponent mixture.

The relative volatility of a partially water-soluble compound, A, can be estimated with the equation (Robinson and Gilliland, 1950)

$$\alpha_{AW} = \frac{x_A(A) p_{Ao}}{x_A(W) p_{Wo}}$$
(10.10)

where $x_A(W)$ is the solubility (mole fraction) of A in the aqueous phase; $x_A(A)$ is the solubility of A in the A phase; and p_{Ao} and p_{Wo} are the vapor pressures of A and water, respectively, at the system temperature. Thus, α_{AW} becomes very high for components having a high vapor pressure and a low solubility in water.

The vapor–liquid equilibria of aroma compounds are affected by the presence of dissolved components of foods. Prediction of equilibrium is not possible in such systems, and very few experimental data are available in the literature. Dissolved sucrose increased considerably the activity coefficients of typical aroma compounds, as shown in Figure 10.2 (Marinos-Kouris and Saravacos, 1975; Saravacos et al., 1990b). The equilibria (y, x) were measured in a static equilibrium still used (developed) by Bruin (1969). The estimated values of relative volatility of these compounds are shown in Table 10.2.

As a general rule, the relative volatility of a compound dissolved in water increases as the concentration of sugar or other soluble compounds is increased.

Electrolytes have a significant effect on the vapor–liquid equilibria of aqueous solutions. Inorganic salts (e.g., sodium chloride) increase the volatility of the more volatile components. This effect may be used in the disruption of the azeotrope of ethanol–water in an extractive distillation process (Meranda and Furter, 1974).

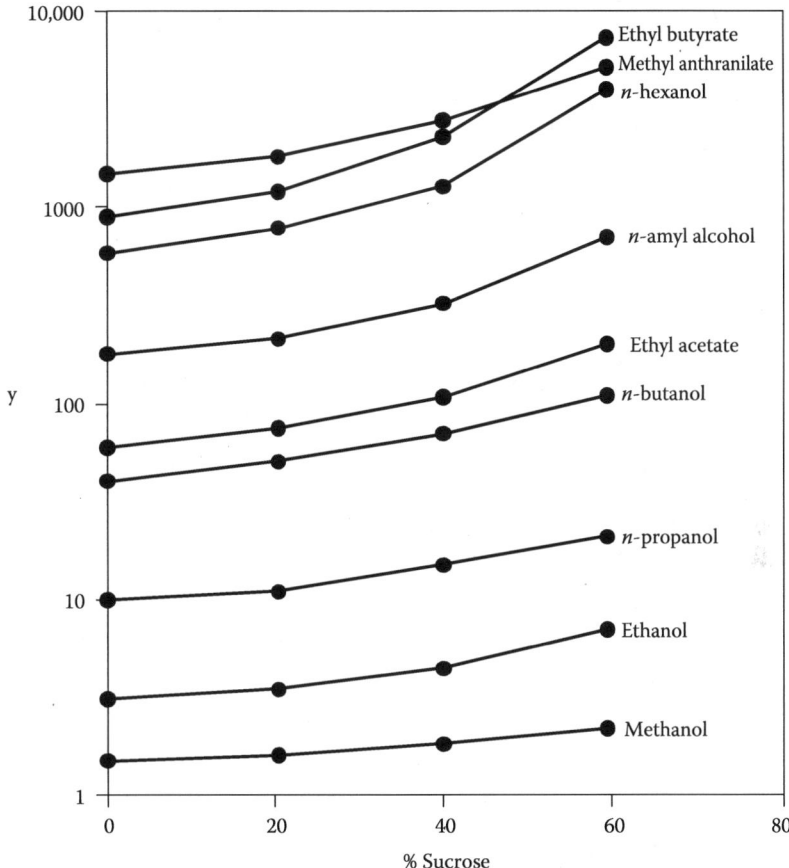

Figure 10.2 Activity coefficients (γ) of organic compounds in aqueous sucrose solutions at 25°C. (Data from Saravacos, G.D., Karathanos, V.T., and Marinos-Kouris, D. 1990. In *Flavors and Off-Flavors*. G. Charalambous (Ed.) pp. 729–733. Elsevier, Amsterdam.)

The relative volatility of organic aroma compounds in aqueous ethanol solutions decreases as the ethanol concentration is increased. Higher alcohols of aqueous fermentation liquors (fusel oils) are more volatile than water at low ethanol concentrations, but their relative volatility drops below unity near 70% ethanol. This change in volatility is used in the separation of fusel oils from ethanol in distillation columns (Robinson and Gilliland, 1950).

The typical relative volatilities of aroma compounds, pertinent to foods, are given in Table 10.3.

Temperature has a significant effect on the activity coefficients of aroma compounds in aqueous systems. However, the relative volatility of these compounds (with respect to water) does not change significantly with temperature, and constant values of α_{AW} can be assumed in most practical applications (Thijssen, 1974).

Table 10.2 Relative Volatilities (a_{AW}) of Aroma Compounds in Aqueous Sucrose Solutions (25°C)

Compound	Sucrose Concentration (%)			
	0	15	35	60
Methyl anthranilate	3.9	4.8	8.1	18.7
Methanol	8.3	8.9	9.8	14.5
Ethanol	8.6	9.0	10.4	16.7
1-Propanol	9.5	10.0	12.0	18.5
1-Butanol	14.1	15.0	21.0	43.0
n-Amyl alcohol	23.0	24.7	41.8	105.0
Hexanol	31.0	36.0	62.0	195.0
2-Butanone	76.0	96.0	112.0	181.0
Diethyl ketone	77.0	85.0	121.0	272.0
Ethyl acetate	205	265	368	986
Ethyl butyrate	643	855	1620	6500

Sources: Data from Marinos-Kouris, D. and Saravacos, G.D. 1975. Volatility of organic compounds in aqueous sucrose solutions. *5th Intl. Cong. Chem. Eng. CHISA 75*, Prague, CZ, paper No. f4.27 and Saravacos, G.D., Karathanos, V.T. and Marinos-Kouris, D. 1990b. Volatility of fruit aroma compounds in sugar solutions. In *Flavors and Off-Flavors*. G. Charalambous (Ed.) pp. 729–733. Elsevier, Amsterdam.

10.2.2 Gas–Liquid Equilibria

Gas–liquid equilibria are important in oxygen absorption during aerobic fermentation, in deaeration of food liquids, in absorption or stripping of carbon dioxide, and so on.

In dilute solutions, the gas is assumed not to react with the solvent, and Henry's law can be applied (King, 1982):

$$p_i = Hx_i \tag{10.11}$$

where p_i is the partial pressure of component i, x_i is the mole fraction of i in the liquid phase, and H is Henry's constant.

Because the gas phase of the food system can be considered ideal, Henry's law is written as

$$y_i = mx_i \tag{10.12}$$

where y_i is the mole fraction of i in the gas phase, and $m = H/P$ (P = total pressure of the system, in the same units as H, e.g., atmospheres or bars).

Concentrated or electrolyte solutions do not obey Henry's law, and experimental equilibrium data or empirical correlations may be found in the literature (Perry and Green, 1997; Reid et al., 1987). Table 10.4 shows some solubility data of gases in water that are relevant to food processing.

Table 10.3 Relative Volatilities (a_{AW}) of Aroma Compounds in Dilute Aqueous Solutions

Compound	Temperature (°C)	a_{AW}
Methyl anthranilate	81	3.4
Ethyl acetate	25	195
Ethyl butyrate	25	620
Isobutyl alcohol	25	22.5
Acetal	25	173
n-Hexanal	25	146
2-Propanol	—	8.5
Propionaldehyde	—	149
Allyl sulfide	—	2453
Methyl propyl sulfide	—	4040
Propanethiol	—	19,413

Sources: Data from Saravacos, G.D. and Moyer, J.M. 1968a. *Food Technol.* 22(5): 89–95; Thijssen, H.A.C. 1974. Fundamentals of concentration processes. In *Advances in Preconcentration and Dehydration of Foods.* A. Spicer (Ed.), pp. 13–43, Applied Science, London; and Mazza, G. and Le Maguer, M. 1980. Flavor retention during dehydration of onion. In *Food Process Engineering.* P. Linko, Y. Malkki, J. Olkku, and J. Larinkari (Eds.) pp. 399–406. Applied Science, London.

Table 10.4 Solubilities of Some Gases in Dilute Aqueous Solutions at 25°C

Gas	Solubility $(1/H)$[a]
Oxygen	2.28×10^{-5}
Nitrogen	1.16×10^{-5}
Carbon dioxide	6.1×10^{-4}
Sulfur dioxide	2.8×10^{-2}
Ammonia	0.32

[a] Solubilities are given in mole fractions per atmosphere $(1/H)$, where H is Henry's constant.

10.2.3 Liquid–Liquid and Liquid–Solid Equilibria

Liquid–liquid and liquid–solid equilibria are needed in the analysis of extraction processes that are used to remove or recover various food components during food processing.

The equilibrium of component i between two liquid phases can be expressed by the equation

$$y_i = Kx_i \qquad (10.13)$$

where K is the partition coefficient and y_i, x_i are the concentrations of i in the solvent (extract) and residue (raffinate), respectively. The concentrations are expressed as mass fractions (kg of solute per kg of solution). Liquid–liquid equilibria of two partially miscible phases (the usual system in extraction) can be represented in triangular diagrams (equilateral or right-angled). Each of the three corners of the triangle represents a pure component, and the two-phase region is enveloped by a curved equilibrium line (King, 1982). The tie-lines of these diagrams connect the composition of the raffinate (residue) and the extract layers. The two phases merge into one at the plait point.

In food systems, the concentration of the solute (component i) is usually low compared to that of the solvent and the aqueous solution. Under these conditions, a y versus x diagram can be used, on which the equilibrium is represented by a straight line (Equation 10.13). The prediction of liquid–liquid equilibria is not as well developed as with vapor–liquid systems; the van Laar, Wilson, and UNIQUAC equations have been suggested for this purpose. A large amount of experimental data is available in the literature (Perry and Green, 1997).

Liquid–solid equilibrium is defined between the supernatant solvent and the liquid adhering to the solid matrix. This condition is approached when the solvent is mixed thoroughly with the solid particles, and sufficient time has allowed the solute to reach equilibrium between the phases. Thus, equilibrium is expressed by the equation

$$y_i = x_i \tag{10.14}$$

and the equilibrium line coincides with the diagonal of the y versus x diagram. Liquid–solid equilibria are usually determined experimentally, and some data are available in the literature. The equilibrium data may be correlated by some equations describing the sorption isotherms, for example, the Langmuir equation.

10.2.4 Gas–Solid and Vapor–Solid Equilibria

Gas–solid and vapor–solid equilibria are very important in food systems. The sorption of water vapor is treated separately (see Section 10.2.4.1). The sorption of oxygen on solid foods is related to the oxidation of lipids and other labile food components. The interaction of organic vapors and solid food components has an important effect on the aroma retention during food dehydration.

The sorption of gases and vapors on solids is characterized by the sorption isotherms, which are determined experimentally. The equilibrium data can be fitted to semiempirical relations, such as the Freundlich, the Langmuir, and the Brunauer–Emmett–Tetter (BET) equations (Perry and Green, 1997). Little information is available in the literature on the sorption of gases and organic vapors on food components.

10.2.4.1 Water Activity

The sorption of water vapor by foods has received much attention because of its importance in dehydration processes, in packaging, and in quality changes during storage. The water activity $á_w$ of the vapor–solid system is defined as the ratio of the partial pressure, p_w, to the vapor pressure of water, p_{wo}, at equilibrium

$$\alpha_w = \frac{p_w}{p_{wo}} = \frac{\%RH}{100} \tag{10.15}$$

where %RH is the percent relative humidity in the surrounding atmosphere.

The equilibrium moisture content of the food material (X_e) corresponding to a given water activity and temperature is determined experimentally and is usually presented as the water or moisture sorption isotherm of the material. A collection of sorption isotherms of foods is given by Iglesias and Chirife (1982). A detailed study of the water activity of foods was undertaken by the European Economic Community (COST 90), and the conclusions were reported by Spiess and Wolf (1983). The gravimetric static method was used, in which the food samples were equilibrated in closed jars of constant relative humidity maintained by saturated salt solutions. Microcrystalline cellulose was used as the reference material of the cooperative study.

Several equations have been suggested to describe the water sorption isotherms of foods. Among them, the BET equation has been used extensively

$$\frac{\alpha_w}{(1-\alpha_w)X_e} = \frac{1}{X_m C} + \frac{\alpha_w(C-1)}{X_m C} \tag{10.16}$$

where X_m is the moisture content when a monomolecular layer is adsorbed and C is an empirical constant.

The COST 90 study has shown that most sorption isotherms of model and real foods can be expressed analytically by the Guggenheim–Anderson–de Boer (GAB) equation (Bizot, 1983):

$$\frac{\alpha_w}{X_e} = a\alpha_w^2 + b\alpha_w + c \tag{10.17}$$

where

$$a = \frac{k}{X_m}\left(\frac{1}{C} - 1\right), \quad b = \frac{1}{X_m}\left(1 - \frac{2}{C}\right), \quad c = \frac{1}{X_m C k}$$

The constants C and k are related to the heat of sorption, the heat of condensation of water, and the temperature of the system. X_m is the monomolecular moisture content. Equation 10.17 is convenient for computer calculations and fitting of experimental data.

The water activity and sorption isotherms are affected by the composition of the food and the temperature of the system. In general, polymers sorb more water than sugars and other soluble components at low water activities. However, the soluble components sorb more water above a certain water activity, for example, above 0.8 for sugars. Temperature has a negative effect on the equilibrium moisture content at low water activities. At high values of α_w, the soluble components, such as the sugars, sorb more water, and temperature has a positive effect because of the dissolving effect of water (Saravacos and Stinchfield, 1965).

Figure 10.3 shows two typical sorption isotherms of a food polymer (starch) and a dried fruit, which can be considered a mixture of polymeric and soluble materials (mainly

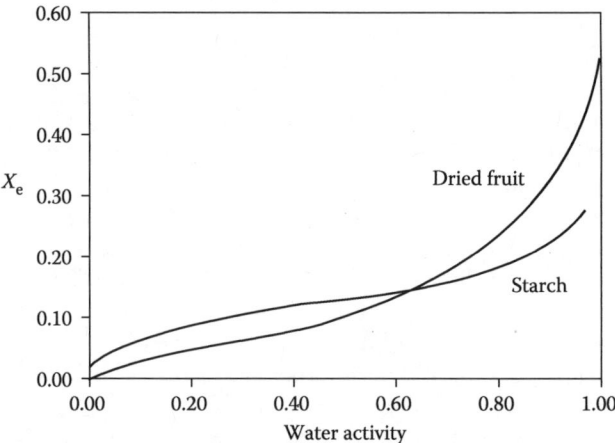

Figure 10.3 Moisture sorption isotherms of starch and a dried fruit at 30°C. X_e, equilibrium moisture content.

sugars). At high water activities, the dried fruit sorbs more water than the starch does, because of the higher sugar content. The GAB model has been applied successfully to several dried fruits (Maroulis et al., 1988).

10.3 DIFFUSION

Molecular diffusion is the transfer of mass caused by random movement of molecules. In gases and liquids, mass transport may also result from bulk motion of the fluid (convection).

In a binary system (A, B) the mass transport rate of component A per unit surface area and time (flux, J_A) is given by Fick's first law

$$J_A = -D_{AB} \frac{dC_A}{dz} \tag{10.18}$$

where C_A is the concentration of component A, z is the diffusion path, and D_{AB} is the diffusivity or diffusion coefficient of A relative to component B. The concentration is expressed in either kmol/m³ or kg/m³ and the corresponding diffusion fluxes in kmol/(m² · s) or kg/(m² s). The diffusivity has units of m²/s.

The diffusivity (D) is a physical property of the system (diffusing substance and medium). Diffusion in gases can be treated in terms of kinetic theory, but an empirical treatment is necessary for diffusion in solids.

The transient-state diffusion of component A in a binary system (A, B) is expressed by the continuity equation for component A (also called Fick's second law):

$$\frac{\partial C_A}{\partial t} = \frac{\partial}{\partial z}\left(D_{AB} \frac{\partial C_A}{\partial z}\right) \tag{10.19}$$

In most food systems, binary diffusion is assumed (diffusing component–food matrix), which considerably simplifies the calculations. When the diffusion of component A is affected by the presence of other components, multicomponent diffusion should be considered (Cussler, 1976).

10.3.1 Diffusion in Gases

Two basic diffusion processes are usually considered in gases. They are expressed by simple equations that can be applied to several practical problems (Geankoplis, 1993): (1) The diffusion of component A through a stagnant film of B. The flux of A is given by the equation

$$J_A = \frac{D_{AB} P}{R T p_{BM}} \frac{\Delta p_A}{\Delta z} \tag{10.20}$$

where D_{AB} is the diffusivity, Δp_A is the difference of partial pressure of A, p_{BM} is the log mean partial pressure of B across the film, P is the total pressure, Δz is the film thickness, T is the absolute temperature, and R is the gas constant in appropriate units.

(2) The other diffusion equation often considered in practice is the counterdiffusion of components A and B. This is given by the equation

$$J_A = -J_B = \frac{D_{AB} \Delta p_A}{R T \Delta z} \tag{10.21}$$

The diffusivities of the two components are equal. In mixtures of low concentration A, Equations 10.20 and 10.21 coincide because p_{BM} is approximately equal to the total pressure P.

The diffusivity of A in a binary gas mixture at low pressures can be predicted by the Chapman–Enskog equation, which is based on the kinetic theory of gases. For practical applications, semiempirical relations such as the Fuller equation (Geankoplis, 1993; Reid et al., 1987) are more convenient

$$D_{AB} = \frac{1.0 \times 10^{-7} T^{1.75} (1/M_A + 1/M_B)^{1/2}}{P\left[\left(\sum V_A\right)^{1/3} + \left(\sum V_B\right)^{1/3}\right]^2} \tag{10.22}$$

where T is the absolute temperature, P is the pressure (atm), M_A and M_B are the molecular weights of A and B, respectively, and ΣV_A and ΣV_B are the sums of structural volume increments for each of the components (Cussler, 1997). The diffusivity calculated from Equation 10.22 is obtained in m²/s. It should be noticed that in a given gas mixture the product $D_{AB} P$ is proportional to $T^{1.75}$. This relationship can be used in converting diffusivities to other pressures and temperatures. It means that for constant temperature the diffusivity is inversely proportional to the pressure (high diffusivities at low pressures).

The literature contains a large quantity of data on the diffusivities of gases. Some typical values of interest to food systems are given in Table 10.5. The diffusivities of gases at atmospheric pressure are relatively high (0.1–1 cm²/s, or 0.1×10^{-4}–1×10^{-4} m²/s).

Table 10.5 Diffusivity (D) of Gases and Vapors in Air at Atmospheric Pressure

Gas or Vapor	Temperature (°C)	$D \times 10^5$ m²/s
Hydrogen	0	6.11
Oxygen	0	1.78
Carbon dioxide	0	1.38
Water	25	2.60
Ethanol	25	1.06
Acetic acid	0	1.33
Ethyl acetate	0	0.71
n-Hexane	21	0.80
n-Butyl alcohol	0	0.70

Sources: Data from Geankoplis, C.J. 1993. *Transport Processes and Unit Operations.* 3rd ed., Prentice-Hall, New York and Perry, R.H. and Green, D.W. 1997. *Chemical Engineers' Handbook* 7th ed., McGraw-Hill, New York.

10.3.2 Diffusion in Liquids

The diffusivities in liquids are much smaller than in gases, because the density and the resistance to diffusion are higher. The flux of component A through a stagnant film of B is given by the equation

$$J_A = \frac{D_{AB} C_{av}}{x_{BM}} \frac{\Delta x_A}{\Delta z} \tag{10.23}$$

where Δx_A is the concentration difference of A (expressed in mole fraction), C_{av} is the average concentration of the solution, and x_{BM} is the log mean mole fraction B across the film.

In dilute solutions, x_{BM} is near unity, and the flux is similar to the counterdiffusion of A and B

$$J_A = -J_B = \frac{D_{AB} C_{av} \Delta X_A}{\Delta z} \tag{10.24}$$

The diffusivity of solute A in a solvent B can be predicted by empirical relations such as the Wilke–Chang equation (Reid et al., 1987)

$$D_{AB} = \frac{7.4 \times 10^{-12} (\varphi M_B)^{1/2} T}{\eta V_A^{0.6}} \quad (m^2/s) \tag{10.25}$$

where M_B is the molecular weight of B, V_A is the solute molar volume at the boiling point (given in literature tables), T is the absolute temperature, η is the viscosity of B in centipoises (cP), and φ is an association parameter of the solvent (2.6 for water, 1.5 for ethanol, and 1.0 for unassociated solvents). For large molecules of solute and small particles at low

Table 10.6 Diffusivity (D) in Dilute Aqueous Solutions

Solute	Temperature (°C)	$D \times 10^9$ (m²/s)
Oxygen	25	2.41
Carbon dioxide	25	2.00
Sulfur dioxide	25	1.70
Ethanol	25	1.24
Acetic acid	25	1.26
Urea	25	1.37
Glucose	25	0.69
Sucrose	25	0.56
Lactose	25	0.49
Sodium chloride	25	1.61
Caffeine	25	0.63
Myoglobin 25	25	0.113
Soybean protein	25	0.03
Catalase	25	0.041
Peroxidase	25	0.012

Sources: Data from Geankoplis, C.J. 1993. *Transport Processes and Unit Operations.* 3rd ed., Prentice-Hall, New York and Cussler, E.L. 1997. *Diffusion and Mass Transfer in Fluid Systems.* 2nd ed. Cambridge University Press, Cambridge.

concentrations, the Einstein–Stokes equation can be applied (assuming spherical molecules that move through a continuum of the solvent) (Cussler, 1997)

$$D_{AB} = \frac{k_B T}{6\pi \eta r_B} \quad (\text{m}^2/\text{s}) \tag{10.26}$$

where η is the viscosity of the solution (Pa · s), T is the absolute temperature (K), r_B is the particle radius (m), and k_B is the Boltzmann constant (1.38×10^{-23} J/molecule · K).

The diffusivity of electrolytes (ions) such as sodium chloride in dilute aqueous solutions can be estimated by the Nernst–Haskell equation (Sherwood et al., 1975). The diffusivity is expressed as a function of the temperature, the valences and ionic conductances of the ions, and the Faraday constant (Cussler, 1997).

Some useful diffusivities of solutes in solutions pertinent to foods are given in Table 10.6.

10.3.3 Diffusion in Solids

10.3.3.1 Introduction

Diffusion of gases, vapors, and liquids in solid media is a more complex process than diffusion in fluids. The solids usually have a heterogeneous structure, and they may interact with the diffusing compounds. As a result, the diffusivity of small molecules in solids is much lower than in liquids, and this may affect the rates of the various physical and chemical processes involving mass transfer.

Diffusion in solids can be treated mathematically in a manner similar to the conduction of heat. The nonsteady-state diffusion equation (Equation 10.19) has been solved for various shapes of the solid and various boundary conditions (Crank, 1975). Diffusion in polymers has been studied extensively because of its importance in the processing and applications of plastic materials (Crank and Parker, 1968; Frisch and Stern, 1983; Vieth, 1991). A similar treatment can be applied to foods that contain various types of biopolymers.

Solutions of the nonsteady-state (transient) diffusion equation for constant diffusivity are available in graphical form for the basic shapes of slab, infinite cylinder, and sphere (Crank, 1975; Sherwood et al., 1975). They are similar to the solutions of the Fourier equation for nonsteady-state heat conduction. For each geometric shape, the concentration ratio is given as a function of the Fourier number, Dt/L^2, where D is the diffusivity, t is the time, and L is the slab half-thickness or the radius of the cylinder or sphere.

If the resistance to mass transfer at the surface of the solid is significant, compared to the resistance inside the solid, it must be taken into consideration. For this purpose the Biot number (Bi) is used that, for mass transfer, is defined by the equation

$$\text{Bi} = \frac{k_c L}{D} \tag{10.27}$$

where k_c is the surface mass transfer coefficient (see Section 10.4) in m/s. The Biot number is incorporated in graphs (e.g., King, 1982) of transient diffusion in various geometric shapes. High Biot numbers can be obtained by increasing the mass transfer coefficient. Mass transport in porous solids may take place via mechanisms other than molecular diffusion, depending on the size, shape, and connection of pores. Knudsen diffusion occurs when the mean free path of the diffusing molecules is large in comparison with the diameter of the capillary. In porous solids, diffusion can be expressed in terms of an effective diffusivity D_e, which is smaller than the molecular diffusivity (D) and is defined as (Geankoplis, 1993)

$$D_e = \frac{\varepsilon D}{\tau} \tag{10.28}$$

where ε is the porosity (void fraction) of the solid and τ is the tortuosity, a factor that corrects for the long tortuous path through the pores (varying from 1.5 to 5). Since the estimation of tortuosity in solids is difficult, the effective diffusivity of solutes is usually determined experimentally.

Small solute molecules can be transported through solids by molecular diffusion. However, depending on the physical structure of the solid, other transport mechanisms may be important, such as surface diffusion, Knudsen diffusion, molecular effusion, and thermal diffusion. For engineering (practical) purposes, the overall transport may be considered as molecular diffusion, caused by a concentration gradient, and overall (effective) diffusivity is usually estimated.

The diffusivity of the various compounds in the solid depends on the temperature, and the following form of the Arrhenius equation has been applied:

Table 10.7 Diffusivity (D) of Various Molecules in Solids

Diffusant	Solid	Temperature (°C)	$D \times 10^{10}$ (m²/s)
Oxygen	Rubber	25	2.1
Carbon dioxide	Rubber	25	1.1
Nitrogen	Rubber	25	1.5
Water	Cellulose acetate		
	5% moisture	25	0.020
	12% moisture	25	0.032
Sodium chloride	Ion-exchange resin (Dowex 50)	25	0.95
Cyclohexanol	Potatoes (high solids content)	20	2.0
Sucrose	Agar gel	5	2.5

Source: Saravacos, G.D. and Maroulis, Z.B. 2001. *Transport Properties of Foods*. Marcel Dekker, New York.

$$\frac{d \ln D}{dT} = -\frac{E}{RT^2} \quad (10.29)$$

where E is the energy of activation for diffusion, which may vary with the concentration of the diffusant and the structure of the solids. Table 10.7 gives literature diffusivities in various solids.

10.3.3.2 Diffusion in Polymers

The transport of small molecules (solutes) in polymer materials is important in food processes, such as drying, film packaging, and food storage. The diffusivity of solutes depends on the molecular size, and the microstructure and morphology of the polymer, which may change considerably during mechanical and thermal processing. Experimental measurements and phenomenological correlations have been made for various polymer systems (Petropoulos, 1994).

Solid polymers are amorphous materials, which exist in the glassy or the rubbery state, with a characteristic glass transition temperature (Vieth, 1991). The solute diffusivity in glassy polymers is very small (e.g., 1×10^{-18} m²/s), while higher diffusivities characterize the rubbery state (e.g., 1×10^{-10} m²/s).

The transport properties of polymers much above and below the glass transition temperature (T_g) are affected by temperature according to the Arrhenius equation (10.29). In the temperature range of T_g to ($T_g + 100°C$) the Williams–Landel–Ferry (WLF) equation is used (Levine and Slade, 1992). The WLF equation predicts a sharp change of the transport properties (viscosity, diffusivity) above T_g.

$$\log\left(\frac{D}{D_g}\right) = \frac{C_1(T - T_g)}{C_2 + (T - T_g)} \quad (10.30)$$

where $C_1 = 17.44$ and $C_2 = 51.6$ (empirical constants) and D, D_g are the diffusivities at T and T_g, respectively.

Various theories have been proposed to predict the solute diffusivity in polymeric materials, for example, the dual-sorption and the free volume methods (Vieth, 1991).

10.3.3.3 Molecular Simulations

Molecular dynamics is used in materials science to design materials of tailored separation and barrier properties, such as polymer membranes, used in reverse osmosis and ultrafiltration.

Molecular simulations of solute diffusion in amorphous and glassy polymers have been developed using computer techniques (Theodorou, 1996). Molecular dynamics assumes that in polymers the solute (penetrant) moves into channels of the sorption sites, created by small fluctuations of the polymer configuration. Computer calculations, using Monte–Carlo algorithms, can estimate thermodynamic and transport constants, related to solute diffusivity.

10.3.4 Estimation of Diffusivity in Solids

The diffusivity of various substances in solid media is of special interest to food engineering because most mass transfer operations in the food industry involve solid or semisolid foods. Literature data are very limited, and they vary considerably because of the complex structure of foods and the lack of a standard method for determination of diffusivity. On the other hand, diffusivities in gases and liquids are more readily available in the literature, and equations are available for the prediction of diffusivity at specified conditions.

Diffusivity in solids can be determined using the following methods, which have been developed primarily for polymeric materials (Zogzas et al., 1994; Saravacos and Maroulis, 2001).

10.3.4.1 Sorption Kinetics

This method is based on the assumption that the adsorption and desorption rates follow the nonsteady-state diffusion equation through the solid sample (Crank, 1975); the surface resistance to mass transfer is assumed to be negligible. The sorption rate is measured with a sorption balance (spring or electrical). The fractional uptake or loss of diffusant in the sample (m/m_e) is plotted versus $(t/L^2)^{1/2}$, as shown in Figure 10.4. Here, m is the mass adsorbed or desorbed after time t, m_e is the equilibrium value at infinite time, and L is the half-thickness of the slab or film. Diffusion is assumed to take place from both flat surfaces of the sample. Solution of the diffusion equation (Equation 10.19) for constant D and appropriate initial and boundary conditions gives the following approximate relation (Crank, 1975; Vieth, 1991):

$$D = \frac{0.196}{(t/L^2)_{1/2}} \qquad (10.31)$$

where $(t/L^2)_{1/2}$ corresponds to half-equilibrium: $m/m_e = 0.5$ (Figure 10.4).

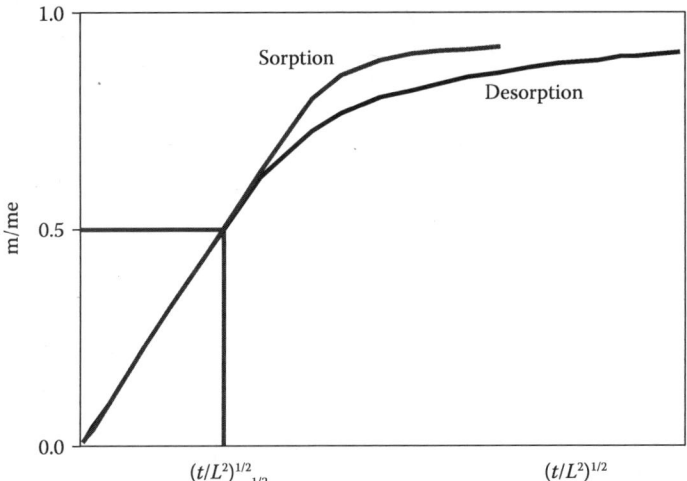

Figure 10.4 Sorption kinetics curves of a slab or membrane: L, half-thickness; m, penetrant adsorption or loss after time t; m_e, equilibrium sorption after infinite time.

The sorption kinetics method can be used to determine the diffusivity at various concentrations of the diffusant, $D(c)$, by carrying out experiments at different ranges of concentration.

A simple test of the diffusion equation is to plot the sorption ratio (m/m_e) versus the square root of time $(\sqrt{t})$. If a straight line is obtained, the process is Fickian diffusion. If the accumulated sorption is plotted versus time, a curve is obtained initially, which becomes linear when steady state is reached. Extrapolation of the linear portion to the time axis yields the time lag t_L, which is related to the diffusivity D by the simple equation

$$t_L = \frac{L^2}{1.5D} \tag{10.32}$$

10.3.4.2 Permeation Measurements

This is a steady-state method applied to a film of material through which the penetrant fluid diffuses under a partial pressure or concentration potential. The measurement is made in a diffusion cell, and the flow rate is estimated by measuring the penetrant concentration in the low pressure side of the cell. The flux of the penetrant at steady state is given by the equation

$$J = \frac{PM\Delta p}{\Delta z} = \frac{DS\Delta p}{\Delta z} \tag{10.33}$$

where PM is the permeability, Δp is the pressure difference across the film of thickness Δz, D is the diffusivity, and S is the solubility of the penetrant in the solid. S is the inverse of Henry's constant (H) and is defined as $S = C/p$, where C is the concentration in the solid and p is the partial pressure.

This method has been applied successfully to polymer films, but it cannot be applied to most foods, which have a heterogeneous structure, and it is difficult to prepare a uniformly thin sample with no pinholes.

10.3.4.3 Distribution of Penetrant

This method is based on nonsteady-state diffusion in a semi-infinite solid (e.g., a long cylinder) through a surface maintained at a constant concentration of the diffusant. The solution of the diffusion equation for this system at constant diffusivity is (Crank, 1975):

$$\frac{C - C_o}{C_e - C_o} = \text{erfc}\left[\frac{z}{(4Dt)^{1/2}}\right] \quad (10.34)$$

where C_o is the initial concentration of the diffusant in the sample, C is the concentration after time t, C_e is the equilibrium concentration, and z is the distance of penetration. The equilibrium concentration (C_e) can be calculated from the medium concentration (C_M) using the relation $C_e = C_M/K$, where K is the equilibrium distribution coefficient. The error functions erf and erfc (=1–erf) are given in the literature. The distribution of the diffusant in the solid at various time intervals is determined by chemical analysis of thin slices of the sample. Diffusion can be either from the medium to the sample or vice versa.

The principle of penetrant distribution is also applied to the distance–concentration method in two contacted long (cylindrical) samples (Equation 10.34). This method has been applied to the diffusion of sodium chloride in model gels and cheese (Gros and Ruegg, 1987) and to the diffusion of water in starch materials (Karathanos et al., 1991).

10.3.4.4 Drying Rate

The diffusivity of moisture in solids and foods of definite shape can be estimated from drying rate data under specified conditions (Zogzas and Maroulis, 1996a). It is assumed that during the falling rate period of drying, moisture is transferred mainly by molecular diffusion. The nonsteady-state diffusion Equation 10.19 for an infinite slab of half-thickness L, drying from both flat surfaces, yields the simplified solution

$$\frac{X - X_e}{X_c - X_e} = \frac{8}{\pi^2}\exp\left(-\frac{\pi^2 D_e t}{4L^2}\right) \quad (10.35)$$

where X is the mean moisture content after time t, X_c is the critical moisture at the beginning of the falling rate period, and X_e is the equilibrium moisture content for the air conditions existing in the drying chamber. Similar simplified solutions of the diffusion equation for spherical and cylindrical samples are available in the literature (Crank, 1975).

Thus, the effective diffusivity (D_e) can be estimated from the slope of a semilog plot of the moisture ratio versus time (Figure 10.5). D_e can also be calculated by comparing the slope of the experimental plot (Figure 10.5) to the slope of a theoretical plot (series solution) of the same moisture ratio versus the Fourier number Dt/L^2 (Perry and Green, 1997).

The drying method can be applied to the determination of variable effective diffusivities. The simplified method of slopes is based on the application of the diffusion equation at various moisture contents. Similar results are obtained with a computer simulation

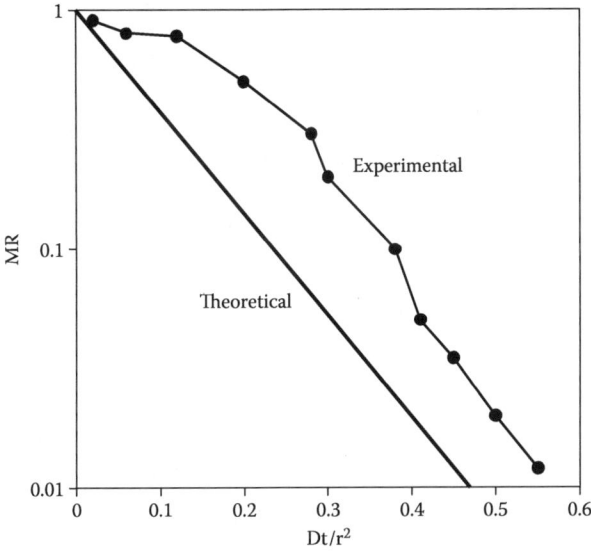

Figure 10.5 Experimental drying curve and theoretical diffusion plot. MR = moisture ratio = $(X-X_e)/(X_o-X_e)$; X_o, X, X_e, moisture contents, initial, after time t, and equilibrium, respectively; D, diffusivity; r, radius or half-thickness of sample.

method that estimates the variable moisture diffusivity by comparing the experimental data to assumed values of diffusivity (Karathanos et al., 1990).

Mathematical models have been proposed that relate the moisture diffusivity to the moisture content and temperature by various functions (exponential, gamma, etc.). A model-fitting procedure (nonlinear regression) can be applied to all experimental data for the drying of a food product obtained under various drying conditions. This method yields smoothed diffusivity values as exponential functions of the moisture and temperature (Kiranoudis et al., 1993).

The drying curve of food materials can be analyzed by the regular regime method to obtain the effective diffusivity at various moisture contents (Bruin and Luyben, 1980; Tong and Lund, 1990; Gekas, 1992). This method assumes that a stable moisture profile, which moves toward the center of the product, is established during the last (regular regime) period of drying. The diffusivity is assumed to be a power function of the moisture content, and the two fitting parameters of the model are determined by linear regression of the experimental data.

10.4 INTERPHASE MASS TRANSFER

10.4.1 Mass Transfer Coefficients

Mass transfer between phases may be visualized by the two-film theory, as applied to the absorption of a gas by a liquid (Figure 10.6). A gas component, A, of concentration

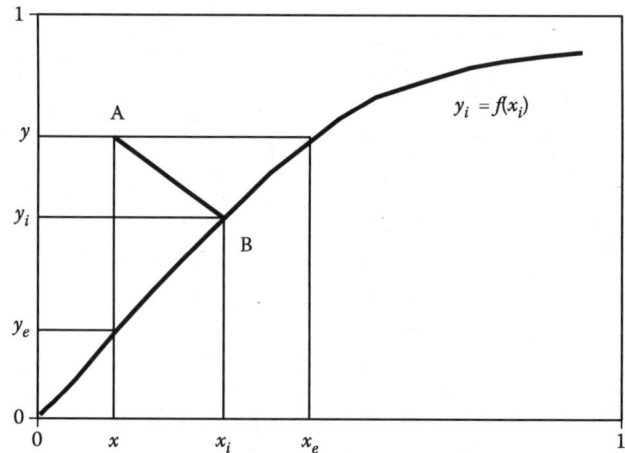

Figure 10.6 Absorption of a gas A (y, x) into a liquid.

y is absorbed by a liquid that is maintained at a concentration x. The concentrations are expressed as mole fractions y and x, and the equilibrium line $y_i = f(x_i)$ is plotted in the y versus x diagram. The equilibrium line is assumed to be known, either from theoretical prediction or from experimental measurements.

The two-film theory assumes that equilibrium exists at the interface (y_i, x_i) and that no mass accumulates at the interface. The rate of mass transfer of the gas component (A) from the gas phase to the interface (B) and to the bulk of the liquid will be

$$J_A = k_y(y - y_i) = k_x(x_i - x) \tag{10.36}$$

where J_A is the mass transfer rate of A, and k_y and k_x are the mass transfer coefficients of the gas phase and liquid phase, respectively.

If J_A is expressed in kmol/(m² · s), the mass transfer coefficients have units of kmol/(m² · s · mole fraction). If the concentration in the gas phase is expressed as partial pressure (Pa), and in the liquid phase as kmol/m³, the corresponding mass transfer coefficients k_G and k_c have units of kmol/(m² · s · Pa) and m/s, respectively. The two types of coefficients are interrelated by the equations

$$k_G = \frac{k_y}{P}, \quad k_c = \frac{k_x}{C} \tag{10.37}$$

where P is the total pressure and C is the concentration of the liquid phase.

Because the concentrations at the interface (y_i, x_i) are not easily determined, the mass transfer rate is usually expressed in terms of the overall mass transfer coefficients K_y, K_x (or K_G, K_c)

$$J_A = K_y(y - y_e) = K_x(x_e - x) \tag{10.38}$$

where y_e and x_e are the equilibrium concentrations corresponding to the bulk concentrations x and y of the other phase (Figure 10.6).

The overall mass transfer coefficients can be calculated from the partial mass transfer coefficients (k_y and k_x) and the equilibrium constant m (Equation 10.12) of a gas–liquid or vapor–liquid system as follows:

$$\frac{1}{K_y} = \frac{1}{k_y} + \frac{m}{k_x}, \quad \frac{1}{K_x} = \frac{1}{k_x} + \frac{1}{mk_y} \tag{10.39}$$

In mass transfer operations of interest to food processing, such as evaporation and drying, the driving force is either partial pressure difference ($p-p_i$) in (Pa) or concentration difference (C_i-C) in kg/m³. Thus, the transfer rate is expressed in kg/m² · s) and Equation 10.36 becomes

$$J_A = k_G(p - p_i) = k_c(C_i - C) \tag{10.40}$$

where k_G and k_c have the units of kg/(m² · s · Pa) and m/s, respectively.

The two mass transfer coefficients (k_G and k_c) are related by the equation

$$k_G = k_c/RT \tag{10.41}$$

where T is the absolute temperature (K) and R is the gas constant $R = 8.31$ Pa · m³/(MW · K), where MW is the gram molecular weight of the diffusant.

The mass transfer coefficients of various systems can be estimated by empirical correlations of the literature (Perry and Green, 1997; Sherwood et al., 1975). For some systems experimental values of the coefficients are available. The wetted-wall column is a convenient experimental setup for determining mass transfer coefficients in gas–liquid and vapor–liquid systems.

Prediction of mass transfer coefficients is analogous to the prediction of heat-transfer coefficients (h). For example, inside pipes dimensionless equations similar to the Sieder–Tate (turbulent flow) and Graetz (laminar flow) equations can be applied (Geankoplis, 1993). In these equations, the Nusselt and Prandtl numbers have been substituted by the Sherwood (Sh) and Schmidt (Sc) numbers, respectively, which are defined as

$$\mathrm{Sh} = \frac{k_c d}{D}, \quad \mathrm{Sc} = \frac{\eta}{\rho D} \tag{10.42}$$

where k_c is the mass transfer coefficient (m/s), d is the pipe diameter (m), D is the diffusivity (m²/s), η is the viscosity [kg/(m · s)], and ρ is the density of the fluid (kg/m³).

The mass transfer coefficients depend on the geometry of the system, the fluid velocity, and the thermophysical properties of the fluid phases. For mass transfer between a fluid and a solid particle, the following equation has been applied (Loncin and Merson, 1979; Sherwood et al., 1975):

$$\mathrm{Sh} = 2 + 0.6\,\mathrm{Re}^{0.50}\,\mathrm{Sc}^{0.33} \tag{10.43}$$

where the Sherwood (Sh) and Reynolds (Re) numbers are based on the particle diameter. Similar empirical equations have been suggested for mass transfer between gases and liquid drops and between liquids and gas bubbles.

In fluid–fluid mass transfer operations, the interface area cannot be determined accurately, and volumetric mass transfer coefficients are more convenient to use. Thus, the mass transfer rate in bubble columns is given by the equation

$$J_A = K_y \alpha V(y - y_e) \tag{10.44}$$

where $K_y\alpha$ is the volumetric mass transfer coefficient in kmol/(m³ · s · mole fraction), α is the interfacial area for mass transfer (m²/m³), and V is the active volume of the column (m³).

10.4.2 Penetration Theory

The two-film theory has been applied successfully to the design of various mass transfer processes and equipment. It predicts that the mass transfer coefficient (k_c) is directly proportional to the diffusivity (D) of the particular component in the film. However, in several systems, k_c is proportional to D^n, where $0.5 \leq n \leq 1$.

The penetration theory assumes that mass is transferred between phases by a short time contact of an element of one phase (e.g., liquid) with the other phase (e.g., gas). After the short time contact, the element (eddy) is mixed thoroughly in the main (liquid) phase, while a new element of uniform composition comes into contact with the other (gas) phase (Loncin and Merson, 1979; Sherwood et al., 1975).

During the short time contact at the interphase, it is assumed that nonsteady-state diffusion in a semi-infinite body takes place (Equation 10.34). Differentiating Equation 10.34 and taking into consideration Fick's first law (Equation 10.18) results in the equation

$$J_A = \left(\frac{D}{\pi t}\right)^{1/2} (C_i - C) \tag{10.45}$$

where t is the short contact time. By comparing Equation 10.45 with the equations defining the mass transfer coefficients (Equations 10.36 and 10.37), it follows that

$$k_c = \left(\frac{D}{\pi t}\right)^{1/2} \tag{10.46}$$

Thus, the penetration theory predicts that the mass transfer coefficient is proportional to $D^{0.5}$. Calculation of k_c from Equation 10.46 is not possible because the contact time (t, in seconds) cannot be determined experimentally. The contact time is known to decrease considerably with increasing turbulence.

The surface-renewal theory is an improvement over the penetration theory. It assumes that the replacement of the elements (eddies) at the interphase is random. If the fractional rate of replacement, s (s⁻¹) is constant, the mass transfer coefficient is given by the equation

$$k_c = (Ds)^{1/2} \tag{10.47}$$

10.4.3 Analogies of Heat and Mass Transfer

In some important cases, mass transfer coefficients can be estimated from heat-transfer data that are already available, or they can be obtained experimentally. For this purpose, the Chilton–Colburn analogies can be used (Geankoplis, 1993):

$$j_M = j_H = \frac{f}{2} \tag{10.48}$$

where f is the well-known friction factor of the Fanning equation for fluid flow, and j_M and j_H are the mass and heat-transfer factors, defined as

$$j_M = \frac{k_y}{u\rho} Sc^{0.67}, \quad j_H = \frac{h}{u\rho C_p} Pr^{0.67} \tag{10.49}$$

where u is the fluid velocity (m/s), h is the heat-transfer coefficient [W/(m² · K)], C_p is the specific heat of the fluid [J/(kg · K)], and ρ is the density (kg/m³).

The Chilton–Colburn analogies apply to flow in pipes and past a flat plate. For flow in packed beds, only the relation $j_M = j_H$ holds. It should be noted that the analogies predict that the mass transfer coefficient is proportional to $D^{0.67}$.

A useful empirical relation for flow of gases or vapors parallel to flat plate is

$$j_M = 0.036\, Re^{-0.2} \tag{10.50}$$

In air conditioning and the evaporation of water, the Lewis relation can be applied:

$$\frac{h}{k_c \rho} = C_p \tag{10.51}$$

For mass transfer at the interface in drying, Loncin and Merson (1979) suggest the empirical relation

$$\frac{h}{k_G \Delta H} = 64.7\, kg/(m \cdot s^2 \cdot K) \tag{10.52}$$

where k_G is the mass transfer coefficient expressed in kg/(m² · s · Pa), h is the heat-transfer coefficient in W/(m² · K), and ΔH is the heat of evaporation of water in J/kg.

10.4.4 Effect of Surfactants

Surface-active agents (surfactants) may affect interphase mass transfer in various ways, depending on the system. In aqueous systems, these compounds tend to concentrate at the liquid surface, forming a film, which influences mass transfer.

Evaporation from a water surface can be reduced significantly if a suitable surfactant, such as hexadecanol, is added. Surfactants may increase the drying rate of some foods in the first period of drying. This may be caused by increasing the evaporation surface of water by wetting in porous foods such as apples (Saravacos and Charm, 1962a,b) or by reducing the resistance to water transfer of the surface skin of foods such as corn (Suarez et al., 1984) and grapes (Saravacos et al., 1988).

Treatment of agar gels with surfactants reduced the drying rate significantly (Roth, 1992). This effect was attributed to the reduction of the water activity at the surface of the material, and it has been suggested that the treatment be used to reduce moisture losses of foods during storage. The surfactants may have different effects on gas–liquid and vapor–liquid operations. Thus, the presence of surfactants decreases the rate of absorption of oxygen in aqueous systems. However, addition of suitable surfactants may significantly increase the evaporation rate of water because of foaming. The presence of surfactants may improve the operation of distillation of aqueous solutions because of the improved wetting of the sieve plates.

10.5 MASS TRANSFER IN FOODS

The various mass transfer operations of food processing and storage can be analyzed on the basis of the physical and engineering principles outlined in the previous sections of this chapter. Exact application of these principles to food systems is usually difficult because of the complex and heterogeneous structure of foods and the physical, chemical, and biological changes that may take place during processing (Karel, 1975).

In this section, some of the more important applications of mass transfer to food systems are outlined. They include moisture transfer, diffusion in porous foods, diffusion of solutes, and diffusion of aroma compounds.

10.5.1 Moisture Transport

10.5.1.1 Moisture Diffusion

Moisture transport involves diffusion of moisture in solid foods and interphase moisture transfer in food processing and storage.

Diffusion in solid foods during drying, rehydration, or storage is a complex process that may involve molecular diffusion, capillary flow, Knudsen flow, hydrodynamic flow, or surface diffusion (Bruin and Luyben, 1980). The experimental data are used to estimate an effective or apparent diffusivity of moisture (D_e) at a specified temperature. Effective moisture diffusivities reported in the literature have been estimated usually from drying or sorption rate data. Compilations of moisture diffusivity in various foods have been published by Gekas (1992), Okos et al. (1992), Zogzas et al., 1996b, Mittal (1999), and Sablami et al. (2000).

In general, comparison between diffusivities reported in the literature is difficult because of the different methods of estimation and the variation of food composition and physical structure. The need for more reliable data on diffusivity is obvious. With this in mind, the European Economic Community has investigated the diffusion properties of foods, within Project COST 90bis (Jowitt et al., 1987).

The physical structure of food plays a very important role in the diffusion of water and other small molecules. A porous structure, produced, for example, by freeze-drying, significantly increases the diffusivity of moisture, as illustrated in Figure 10.7 (Saravacos,

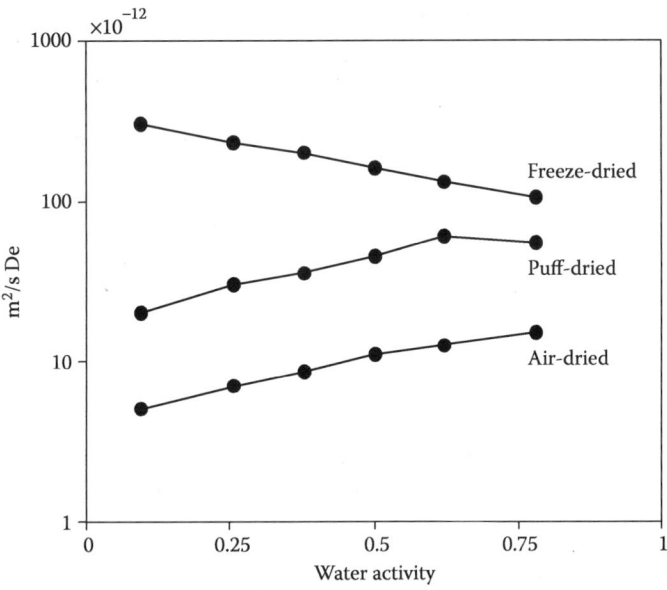

Figure 10.7 Effective moisture diffusivity (D_e) from sorption data in dried potato at 30°C. (Data from Saravacos, G.D. 1967. *J. Food Sci.* 32: 81–84.)

1967). The diffusivity (D_e) in air-dried potato increased with the moisture content (water activity) because of the swelling of the starch structure. The (D_e) of the freeze-dried product was very high at low moisture content, evidently due to the highly porous structure, but decreased gradually at higher moisture, evidently because of a collapse of the porous network. The (D_e) of the puff-dried product attained intermediate values. King (1968) explained this behavior by considering differences in heat and mass transfer in porous foods. Most experimental values of (D_e) in food dehydration have been obtained by assuming negligible surface resistance to mass transfer, which may be approached during the falling rate period of drying individual pieces or thin layers of foods at relatively high air velocities and temperatures. In industrial operations, interphase mass transfer may be very important and should be taken into consideration by including the Biot number ($k_c L/D$) in the calculations of nonsteady-state diffusion.

The presence of fats in food significantly decreases the diffusivity of moisture. This is illustrated with measurement of (D_e) in whole and defatted soybeans, which varied from 1×10^{-12} to 3×10^{-12} m²/s in the whole bean, whereas it remained at a constant higher value of 5.4×10^{-12} m²/s in defatted samples under the same environmental conditions of 30°C and water activity of 0.2–0.8 (Saravacos, 1969).

The variation of (D_e) with the moisture content of foods is a complex function, depending mainly on the physical structure of the food. Reported diffusivities are usually mean values of (D_e) in a certain range of moisture. In some cases, two diffusivities of moisture (bimodal diffusion) may characterize the drying of a food material, for example, fish muscle (Jason and Peters, 1973). In general, (D_e) varies considerably with the moisture content and the temperature.

The diffusivity decreases significantly at low moisture contents, as shown by sorption studies on starch gels (Fish, 1958). Air-drying measurements on starch gels have shown a similar behavior at low moistures, but a maximum of (D_e) was noticed in the range of 60–70% moisture content (Saravacos and Raouzeos, 1984).

Pressure has a negative effect on water diffusivity, since the diffusivity of water vapor is inversely proportional to the pressure, as shown in Equation 10.22 (Karathanos et al., 1991). Mechanical compression reduces the porosity and the effective moisture diffusivity.

The moisture diffusivities in wheat and corn pasta were found to be 0.35×10^{-10} and 2.6×10^{-10} m²/s, respectively (Andrieu et al., 1988). Expansion (puffing) of regular wheat pasta increased the moisture diffusivity from 0.3×10^{-10} to 1.2×10^{-10} m²/s (Xiong et al., 1991).

The moisture diffusivities in white and brown rice were determined to be 0.52×10^{-10} and 0.2×10^{-10} m²/s, respectively (Engels et al., 1986). Higher values of (D) were obtained in parboiled rice ($2–10 \times 10^{-10}$ m²/s), by Elbert et al. (2001). The diffusivity of moisture in wheat kernels was found to be a complex function of moisture and temperature, ranging from 0.8×10^{-10} to 2.5×10^{-10} m²/s (Jaros et al., 1992). Regression analysis of drying data for potato, carrot, onion, and green pepper yielded smooth exponential curves of effective moisture diffusivity versus moisture content in the range of 0.2×10^{-10}–10×10^{-10} m²/s (Kiranoudis et al., 1993). Low moisture diffusivities ($0.2–0.4 \times 10^{-10}$ m²/s), due to chemical composition and physical structure, were obtained in pine nut seeds (Karatas and Pinarli, 2001).

The moisture diffusivity of minced meat at 60 °C was determined as 1.0×10^{-10} m²/s for the raw and 1.8×10^{-10} m²/s for the heated product (Motarjemi and Hallstrom, 1987).

Protein-based coatings have low moisture diffusivity, and they are used to reduce moisture loss from wet foods (Gennadios, 2002).

A statistical analysis of the published moisture diffusivity of foods was presented by Saravacos and Maroulis (2001). Empirical correlations are proposed for the diffusivity (D) as a function of the temperature and moisture content. A few of these correlations can be very useful. Typical values of (D) are shown in Table 10.8a and some additional data on the moisture diffusivity of foods are shown in Table 10.8b.

10.5.1.2 Diffusion in Porous Foods

Most dehydrated foods have a porous structure, which is developed during the process of drying, especially when water is removed as a vapor. The porous structure can be characterized by the bulk porosity or void fraction ε of the material, which is estimated from the equation (Marousis and Saravacos, 1990):

$$\varepsilon = 1 - \frac{\rho_b}{\rho_p} \qquad (10.53)$$

where ρ_b is the bulk density and ρ_p the particle (solid) density.

Porosity plays a dominant role in determining the effective moisture diffusivity in starch-based systems (Leslie et al., 1991). Figure 10.8 shows the moisture diffusivity versus moisture content of granular, gelatinized, and extruded starch materials (Marousis et al., 1991). Higher diffusivities are observed in granular and extruded (puffed) starches than

Table 10.8a Typical Effective Moisture Diffusivities (D) in Food Materials at 30°C

Food Material	Moisture, kg/kg dm	$D \times 10^{10}$ m²/s	E, kJ/mol
Apple	0.50	2.0	60
Raisins	0.40	1.5	60
Potato	0.30	5.0	45
Carrot	0.30	2.0	45
Freeze-dried fruit	0.08	50.0	10
Corn kernel	0.20	0.40	40
Wheat kernel	0.20	0.30	40
Rice	0.20	0.40	40
Dough	0.40	5.0	40
Bread	0.30	2.0	40
Pasta	0.15	0.3	40
Minced meat	0.60	1.0	35
Pork sausage	0.20	0.5	35
Codfish	0.50	2.0	35
Mackerel fish	0.40	0.5	35

Source: Data from Saravacos, G.D. and Maroulis, Z.B. 2001. *Transport Properties of Foods.* Marcel Dekker, New York.
Note: E = Energy of activation for diffusion.

Table 10.8b Effective Moisture Diffusivity (D) of Various Foods (30°C)

Material	Moisture Diffusivity (m²/s)	Reference
Bananas	6.7×10^{-10}	da Silva et al. 2012
Coconut	4.8×10^{-10}	da Silva et al. 2013
Quince	1.0×10^{-10}	Noshad et al. 2012
Tomato	1.1×10^{-10}	Telis et al. 2004
Salted tomato	1.3×10^{-10}	Xanthopoulos et al. 2012

in gelatinized or pressed materials, evidently due to the differences in bulk porosity of the samples. The variation of the effective moisture diffusivity at 60°C and moisture $X = 0.25$ was from 0.75×10^{-10} (gelatinized starch) to 43×10^{-10} m²/s (extruded starch).

The development of bulk porosity during the drying of granular and gelatinized starch (high amylopectin) is shown in Figure 10.9 (Saravacos et al., 1990a). Wet starch materials ($X > 1$) have a very small porosity, which increases considerably during drying, particularly in granular starches. The porosity of the dried granular starch reaches nearly 0.45, while the dried gelatinized samples have a more compact structure with a porosity of less than 0.1.

Prediction of the effective moisture diffusivity is difficult because of the complex structure of most foods. Rotstein (1987) proposed a complex procedure for estimating the effective moisture diffusivity in cellular foods, based on the composition and physical

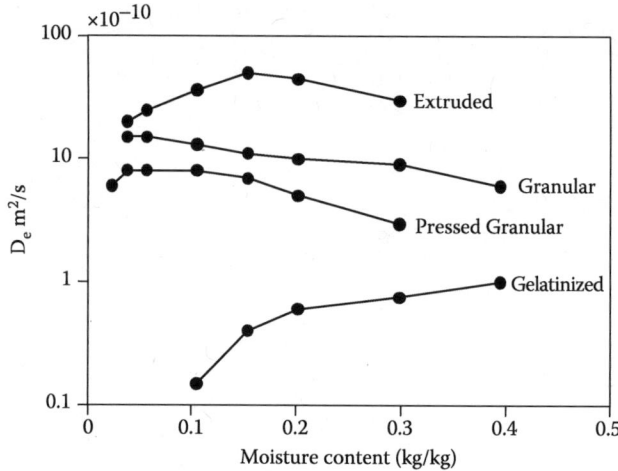

Figure 10.8 Effective moisture diffusivity (D_e) in various starch materials from air-drying data at 60°C. (Data from Marousis, S.N., Karathanos, V.T. and Saravacos, G.D. 1991. *J. Food Proc. Preserv.* 15: 183–195.)

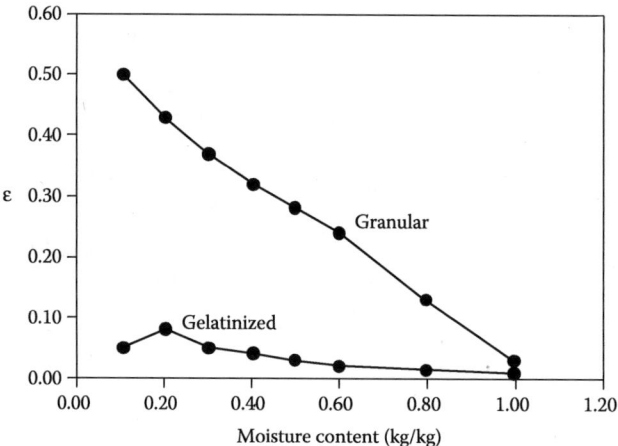

Figure 10.9 Development of bulk porosity (ε) during air drying of starch materials. (Data from Saravacos, G.D., et al. 1990a. Effect of gelatinization on the heat and mass transport properties of starch materials. In *Engineering and Food*. Vol. 1. W.E.L. Spiess and H. Schubert (Eds.), pp. 390–398, Elsevier, London.)

properties of the food materials. The model involves estimation of the porosity, tortuosity, mass conductivity, and water chemical potential. The procedure has been applied for the prediction of the rate of apple drying.

In porous (granular, extruded, puffed, etc.) foods, the effective moisture diffusivity increases gradually as the moisture content is reduced to about $X = 0.1$, evidently due to the development of porosity. The mechanism of water transport may change from liquid diffusion at high moisture to vapor diffusion, which is much faster. At low moisture ($X < 0.1$), the diffusivity drops sharply because of the difficulty of removing the strongly sorbed water molecules from the solid polymer matrix (Leslie et al., 1991; Saravacos and Maroulis, 2001).

Figure 10.10 shows a plot of experimental moisture diffusivities versus porosity for various starch materials and mixtures at 60°C and $X = 0.2$. Regression analysis of the effective moisture diffusivities (D_e) versus bulk porosity (ε), moisture content (X, dry basis) and absolute temperature (T, K) resulted in the following semiempirical equation (Marousis et al., 1991):

$$D_e = 10^{-10}\left\{4.842 + 0.5735X^{-4.34} + 34.2\left[\frac{\varepsilon^3}{(1-\varepsilon)^2}\right]\right\}\exp\left(\frac{-4.5}{RT}\right) \quad (10.54)$$

The effective moisture diffusivity depends mainly on the porosity and temperature. Moisture content has a minor effect. However, moisture content has a significant indirect effect on the diffusivity, through the porosity, which is a strong function of moisture (Figure 10.9). The mean activation energy for diffusion in porous starch materials is relatively low (18.8 kJ/mol). The activation energy for diffusion of water in gelatinized and sugar-containing starches is much higher than in granular/porous materials. Sugars significantly reduce the porosity and effective moisture diffusivity in granular starches due to precipitation in the interparticle pores during drying. The reduction in diffusivity is a function of the molecular weight of the water-soluble sugars, with dextrins being more effective than glucose. The sugars have a smaller structural effect in gelatinized than in granular starches, as evidenced by microscopic observations (Marousis and Saravacos, 1990; Marousis et al., 1989).

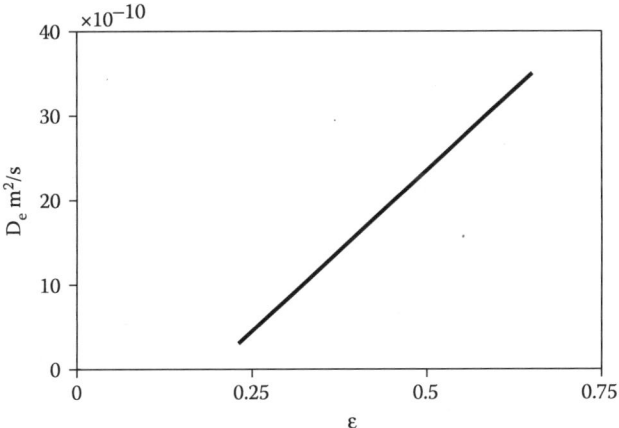

Figure 10.10 Effective moisture diffusivity (D_e) as a function of bulk porosity (ε) of various starch materials. (Data from Marousis, S.N., Karathanos, V.T. and Saravacos, G.D. 1991. *J. Food Proc. Preserv.* 15: 183–195.)

10.5.1.3 Interphase Moisture Transfer

Interphase moisture transfer is important in evaporation, drying, freezing, and storage of foods. The literature contains very few experimental data on mass transfer coefficients pertinent to food systems. The Chilton–Colburn analogy of heat and mass transfer and Equations 10.51 and 10.52 can be used if heat-transfer coefficients are available.

Regression analysis of mass transfer coefficient data in food systems resulted in the following empirical equation for the mass transfer factor (j_M):

$$j_M = 1.11 \text{Re}^{-0.54} \tag{10.55}$$

A similar equation was derived for the heat-transfer factor (j_H) in food systems (Saravacos and Maroulis, 2001)

$$j_H = 0.344 \text{Re}^{-0.423} \tag{10.56}$$

The two transfer factors (j_M and j_H) are close but not similar in food systems. The difference may be due to the different mechanisms of heat and mass transfer in solid food materials.

The interphase mass transfer coefficient of Equation 10.40 for air-drying of spherical starch samples was found to be $k_c = 0.035$ m/s (Saravacos et al., 1988), which is in good agreement with the predicted value of Equation 10.43. The desiccation of unprotected foods during frozen storage can be predicted by Equation 10.40. Pham and Willix (1984) corrected this equation for two additional resistances to mass transfer—the desiccated surface layer and the radiation cooling effect on the frozen food surface.

Interphase mass transfer rates can be increased by increasing the air velocity and/or temperature. Centrifugal force may increase the mass transfer rate, and a centrifugal fluidized bed has been proposed to accelerate air drying of fruits and vegetables (Lazar and Farkas, 1971).

10.5.2 Diffusion of Solutes

The diffusion of solutes and other food components within foods is very important in food processing and storage. Typical examples are the diffusion of salt in meat and pickles, the diffusion of sugars and fats, and the diffusion of flavor compounds in various foods. Table 10.9 gives typical diffusivities of various compounds in foods (Saravacos and Maroulis, 2001).

The diffusivities of sodium chloride in cheese, meat, fish, and pickles have been measured using the nonsteady-state diffusion of chloride ions in cylindrical samples (Equation 10.34). Due to the heterogeneous structure of the foods, the reported values should be considered as effective or apparent diffusivities.

The diffusivity of sodium chloride in gels, cheese, meat, and pickles is relatively high, close to the salt diffusivity in water (16.1×10^{-10} m²/s, Table 10.6). The lower salt diffusivity in green olives may be caused by the higher resistance to mass transfer of the skin and the oil-containing flesh of the olives (Drusas et al., 1988). Typical diffusivities of sodium chloride in foods include pickles, 11.1×10^{-10} m²/s (Pflug et al., 1975); cheese,

Table 10.9 Diffusivity (D) of Solutes in Gels and Foods

Solute	Substrate	Temperature (°C)	$D \times 10^{10}$ (m²/s)
Glucose	Agar gel, 0.79% solids	5	3.3
Sucrose	Agar gel, 0.79% solids	5	2.5
Sorbic acid	Agar gel		
	1.5% agar	25	7.35
	1.5% agar + 8% NaCl	25	4.92
Sodium chloride	Agar gel, 3% solids	25	13
Sodium chloride	Cheese	20	1.9
Sodium chloride	Pickles	18.9	11.1
	Green olives	25	3.2
Sodium chloride	Meat muscle		
	Fresh	2	2.2
	Frozen-thawed	2	3.9
Sodium chloride	Fish (herring)	20	2.3
Acetic acid	Same	20	4.5
Nitrite	Beef	5	1.8
Tripalmitin	MCC-gum arabic		
	3.6% moisture	50	0.0045
	12.4% moisture	50	0.35

Source: Data from Saravacos, G.D. and Maroulis, Z.B. 2001. *Transport Properties of Foods.* Marcel Dekker, New York.

MCC = microcrystalline cellulose.

1.9×10^{-10} m²/s (Gros and Ruegg, 1987); herring, 2.3×10^{-10} m²/s (Rodger et al., 1984); and salmon, 2.3×10^{-10} m²/s (Wang et al., 2000).

Diffusion of sugars and fats in solid foods is important in extraction operations, as discussed in Section 10.6.1. The diffusion process has been investigated by use of ^{14}C-labeled sugars or fats. The very low diffusivity of tripalmitin (a fatty compound) in a model system of dry microcrystalline cellulose and gum arabic was determined via the nonsteady-state diffusion method, Equation 10.34, by Naessens et al. (1981). It increased sharply to 0.35×10^{-10} m²/s at 12.4% moisture.

The apparent diffusivity of sorbic acid in agar gels was measured by Guilbert et al. (1983), using the distribution of the penetrant in cylinders (Equation 10.34) and in cubes (an equation similar to Equation 10.34). The diffusivity (about 7×10^{-10} m²/s) was found to be affected by the water activity (α_w) rather than by the moisture content, and it increased considerably at high values of α_w.

Mass transfer rates and transport properties are important in osmotic dehydration of foods. During the osmosis process, water diffuses from the food piece into the concentrating medium (usually sugar solution), while sugar diffuses into the food. Mathematical modeling of the process was reported by Marcotte et al. (1991) and Marcotte and Le Maguer (1992).

10.5.3 Diffusion of Aroma Compounds

The diffusion of volatile aroma compounds is very important in food processing and storage, and an understanding of the mechanism of mass transfer will help in maintaining the food quality. Retention of the characteristic aroma components is essential in drying food products such as coffee and tea extracts, and in fruit juices.

Most of the aroma compounds of foods are more volatile than water, because of a combination of high vapor pressure and low solubility in aqueous solutions. The volatility in food systems is expressed as the relative volatility (α_{AW}) of the aroma compound (A) compared to water (W) at equilibrium at the vapor–liquid interphase. Typical values of α_{AW} are given in Table 10.3.

On the basis of high relative volatility, high losses of volatiles would be expected during evaporation and drying of food products. However, these components may be retained at a relatively high percentage in the dried products because of the presence of soluble and insoluble solids in the food. The retention of aroma compounds during the drying operation has been investigated extensively.

The retention of some typical volatile aroma compounds during the vacuum drying of pectin solutions is illustrated in Figure 10.11 (Saravacos and Moyer, 1968a). Similar retention curves were obtained in the vacuum drying of grape juice and in the freeze-drying of pectin solutions and apple slices (Saravacos and Moyer, 1968b). At the beginning of the drying, the loss of volatiles was very rapid and depended mainly on the relative volatility of the compound, which was estimated to be 3.4 for methyl anthranilate, 195 for ethyl acetate, and 620 for ethyl butyrate. As drying progressed, the loss of volatiles decreased, and there was very little loss above 40% water evaporation. Final retention in the dried samples was about 70% of methyl anthranilate, 18% of ethyl butyrate, and 10% of ethyl acetate.

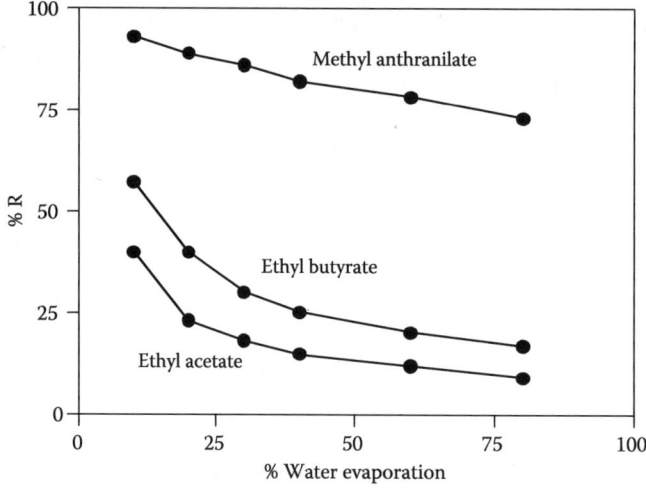

Figure 10.11 Retention (%R) of volatile aroma compounds on vacuum drying of pectin/glucose (5%/50%) solutions at 13 mbar. (Data from Saravacos, G.D. and Moyer, J.M. 1968a. *Food Technol.* 22(5): 89–95.)

Ethyl butyrate, with a relative volatility higher than that of ethyl acetate, was retained at a higher percentage, presumably because of its higher molecular weight and therefore its lower diffusivity in the dried product. Similar results were reported by Menting and Hoagstad (1967).

Two theories have been proposed to explain the high retention of volatile components in food drying processes:

- The selective diffusion theory of Thijssen (King, 1988; Rulkens, 1973)
- The microregion entrapment theory (Flink and Karel, 1970)

The mechanism of retention of volatiles in foods was reviewed by Voilley and Simatos (1980). The diffusivity of volatiles in carbohydrate solutions is reduced from about 10×10^{-10} m^2/s to 0.1×10^{-10} m^2/s as the sugar concentration is increased from 0% to 60% (Voilley and Roques, 1987).

The selective diffusion theory explains the retention by the lower diffusivity of the aroma compounds (D_A) compared to the diffusivity of moisture (D_W) in the food, during drying. The ratio D_A/D_W decreases rapidly as drying progresses to lower moisture contents, resulting in high retention of volatiles. In spray drying, retention of the aroma components may be relatively high owing to the formation of a semipermeable surface skin, which allows the diffusion of moisture but retains the volatiles in the food particle (Kerkhof and Schoeber, 1974).

The microregion retention theory assumes that volatiles are immobilized in the food matrix by a trapping mechanism. Carbohydrates and sugars are known to "lock in" volatile flavors. It should be noted that physical adsorption of organic volatiles on solid food components does not play an important role in aroma retention unless some chemical interaction takes place between the volatile component and the food substrate.

Retention of aroma compounds depends on the concentration and nature of the food solids (e.g., carbohydrates). Nonevaporative methods of food concentration—for example, reverse osmosis or freeze-concentration—may improve aroma retention in the dried product when they are used before the drying process.

The importance of relative volatility in the retention of volatile compounds at the beginning of drying was pointed out by Mazza and Le Maguer (1980) and King (1982). The selective diffusion theory explains aroma retention in the advanced stages of drying. Increasing the drying temperature may increase the retention of volatiles via a reduction in relative volatility and the formation of a semipermeable surface film.

During the fast evaporation of food liquids or wet solid foods, thermodynamic equilibrium is not reached with the very volatile components. As a result, the relative volatility will be less than the thermodynamic equilibrium value, and the losses of volatiles may be reduced (Bomben et al., 1973).

Encapsulation and controlled release technology has recently received much attention in the pharmaceutical and chemical industry (Vieth, 1991). Special polymeric compounds can be used to encapsulate various drugs, fertilizers, pesticides, and special chemicals, which can be released gradually by controlled diffusion through the polymer matrix. This technique can be applied to some important food systems, as in the controlled release of flavors, nutrients, and food additives. The mechanism of controlled release may be related to the retention of volatiles in drying (Karel, 1990).

The free-volume theory can be used to predict the retention and release of flavor compounds in food polymers (Yildiz and Kokini, 1999). It is assumed that elements of free volume exist within the material, through which solute molecules can be transported. Temperature and water activities affect the retention and release of flavor compounds in food biopolymers. The diffusivity of hexanol, hexanal, and octanoic acid in soy flour was predicted to decrease sharply as the temperature is reduced until reaching the glass transition temperature (T_g). High water activities and cooking sharply increase the flavor diffusivity.

10.6 OTHER MASS TRANSFER PROCESSES

10.6.1 Extraction

Liquid/liquid and liquid/solid extraction are important mass transfer operations applied to several food processes. Complete engineering analysis and design of an extraction process requires phase equilibrium data, mass transfer rates, and calculation of the contact stages and equipment needed to achieve a specified separation. The equilibrium relations were treated in Section 10.2.3. Mass transfer rates are considered here. Computation methods of extraction processes and equipment can be found in chemical engineering reference books and in Saravacos and Kostaropoulos (2002).

Most of the mass transfer literature in extraction concerns solid foods (leaching), where diffusion in the solid is usually the rate-controlling factor. Mass transfer in liquid/liquid extraction is facilitated by thorough mixing of the two phases in a series of contact operations. Examples of liquid/solid extractions are the leaching of sucrose from sugar beets and the extraction of oils from seeds and beans. Liquid/liquid extraction is applied to the refining of vegetable oils, the extraction of caffeine from aqueous solutions, and so on. Typical solute diffusivities in various solvents are given in Table 10.10. Interphase mass transfer may play an important role in some leaching operations, and the Biot number (Bi) should be used in mass transfer calculations (Besson, 1983). When Bi ≤ 50, interphase mass transfer may be very slow, due to a liquid film or skin effect at the surface of the solid. At higher values of Bi, the surface resistance to mass transfer can be neglected, and leaching can be considered a diffusion-controlled process (Schwartzberg and Chao, 1982).

The mass transfer coefficients in extraction operations are used in calculations of rate processes and separation stages. The overall mass transfer coefficient (K_c) in the form of the Sherwood number ($K_c\,x/D$) has been correlated with the stripping factor mV/L in graphs presented by Spaninks (1983). Here, D is the diffusivity of the solute, x is the characteristic dimension of mass transfer, m is the equilibrium partition coefficient, V is the amount of solvent, and L is the amount of liquid raffinate.

In the leaching of sugar beets with water and of soybeans or seeds with an organic solvent, the major resistance to mass transfer is found in the plant cells. Thermal denaturation of the cell membranes at 65–75 °C significantly increases the permeability of the cells, facilitating the leaching of sucrose from the beet pieces (Aguilera and Stanley, 1999).

The diffusion of sodium hydroxide through the skins of vegetables is important in the alkali peeling of tomato, potato, and pepper. The diffusivity of sodium hydroxide in tomato and pepper skins is 2×10^{-12} and 5.5×10^{-12} m^2/s, respectively (Floros and Chinnan, 1989).

Table 10.10 Solute Diffusivities (D) in Solvent Extraction of Solids

Solute	Solid/Solvent	Temperature (°C)	$D \times 10^{10}$ (m²/s)
Sucrose	Sugar beets (slices, 2 mm thick)/water	75	5.30
Coffee solubles	Coffee beans (particles, 8 mm)/water	97–100	3.05
Caffeine	Green coffee beans/CH_2Cl_2	30	0.47
Sodium chloride	Pickles (28.6 mm diameter)/water	21	8.40
Sodium hydroxide, 2M solution	Tomato skin/alkali solution	72	0.02
Lactose	Cottage cheese (particles, 3.78 mm)/water	25	3.91
Oil	Soybean flakes (0.43 mm)/hexane	69	1.08
Zein	Corn endosperm (dry-milled)/ethanol	25	0.0037
Hexane	Rapeseed meal (defatted)/hexane vapor	60	0.0017

Source: Data from Schwartzberg, H.G. and Chao, R.Y. 1982. *Food Technol.* 36(2): 73–86 and Saravacos, G.D. and Maroulis, Z.B. 2001. *Transport Properties of Foods*. Marcel Dekker, New York.

In oil extraction from vegetable materials, the flaking process is used, which mechanically reduces the thickness of the diffusion path of the solvent. Mechanical treatment of solid foods—for example, by milling, crushing, extrusion, or checking a surface skin—can increase the mass transfer rate.

Extraction of sucrose from sugar beets has been performed in continuous countercurrent extractors, which were proposed for the extraction of apple juice, replacing mechanical presses (Binkley and Wiley, 1978). Higher recoveries of soluble solids from apples were obtained than with mechanical pressing.

Supercritical fluid extraction (SCF) with carbon dioxide at high pressures is one method for recovering flavor and other components of foods. The main advantage of the method is that carbon dioxide, when used as a solvent, is removed easily from the extract and is not toxic. Equilibrium and mass transfer data for food systems are essential for the design of the process (Saravacos and Kostaropoulos, 2002).

Food applications of the SCF extraction process include decaffeination of coffee, extraction of flavors from spices and hops, and removal of cholesterol from lipids (McHugh, 1990).

10.6.2 Distillation and Gas Absorption

Distillation is a mass transfer operation used in the removal and recovery of volatile components from liquid foods (Bomben et al., 1973), in the production of ethanol and alcoholic beverages from fermentation liquors, in the purification of solvents, and so on. The analysis of the distillation process involves vapor–liquid equilibria (Section 10.2.1), calculation of the number of equilibrium stages, and hydraulics of the distillation column (van Winkle, 1967).

Mass transfer is important in the stripping of various volatiles from liquid foods and in the efficient operation of distillation trays. The tray efficiency is affected by the diffusion and mixing of the components, the vapor and liquid flow rates, and the geometry of the system. The efficiency of distillation is, in general, high because of good vapor–liquid contact. However, in dilute aqueous solutions, tray efficiencies are relatively low, mainly because of the high surface tension of the solution and the poor mixing.

Batch or differential distillation is applied for the removal or recovery of volatile food components. The percentage of evaporation required for the removal of a portion of the volatile matter from a liquid food is given by the Rayleigh equation (King, 1982)

$$\ln \frac{L_2}{L_1} = \int_{x_1}^{x_2} \frac{dx}{y - x} \qquad (10.57)$$

where L_1 and L_2 are the initial and final amounts of the liquid in the still (kmol), and x_1 and x_2 are the corresponding mole fractions of the volatile component; y is the mole fraction of the vapor phase, which is assumed to be in instantaneous equilibrium with the liquid. If the relative volatility is constant, the equilibrium equation (Equation 10.8) can be used, and the integral of Equation 10.57 can be evaluated (van Winkle, 1967). Otherwise, a graphical integration may be used.

According to the Rayleigh equation, very volatile components can be removed from a liquid by evaporation of a small portion of the liquid. This effect is used in the essence recovery process, where flash evaporation of a small percentage of a juice will remove the largest portion of the volatiles.

The distillation of volatile aroma compounds from aqueous solution in an agitated film evaporator was investigated by Marinos-Kouris and Saravacos (1975). Removal of the volatiles (ethanol, 1-butanol, and methyl anthranilate) was a function of the percent evaporation, the relative volatility, and the mixing of the liquid in the evaporator. The mixing was expressed as the Peclet number (uL/D), where u is the vertical film velocity, L is the length of the evaporating film, and D is the effective diffusivity of the component in the liquid film. High Peclet numbers resulted in the removal of higher percentages of the volatile component.

Steam distillation is applied to the removal of undesirable flavors from liquid foods such as milk and edible oils. It can be performed by vacuum treatment of the hot liquid food, by steam injection, or by a combination of the two. The degree of flavor removal depends on the relative volatility and the percentage evaporation of the liquid.

Absorption of gases in liquid foods is important in some food processing operations such as aerobic fermentation (oxygen transfer), deaeration of liquid foods, and carbonation of beverages (transfer of carbon dioxide). Analysis of the aeration processes and equipment follows well-established chemical engineering methods (Perry and Green, 1997; Sherwood et al., 1975).

In gas absorption, mass transfer potentials are usually expressed as partial pressures of the component, and the volumetric mass transfer coefficient $K_y\alpha$ should be used in Equation 10.44. Correlations of the mass transfer coefficients with agitation and the fluid properties in absorption systems are given in the literature. In general, the efficiency of absorption equipment (agitated tanks, columns) is very low (near 10%), because of poor gas/liquid mixing and consequently low mass transfer rates. Distillation and gas absorption equipment used in food processing is discussed by Saravacos and Kostaropoulos (2002).

10.6.3 Crystallization

Crystallization is a separation process used primarily for the production of pure crystalline compounds, such as sucrose and sodium chloride, from solutions. The crystallization of ice from fruit juices can be used for the concentration of juices at low temperature without loss of volatile aroma components. Crystallization plays an important role in food freezing, in food and dairy processing, and in food storage. Depending on the food product, the formation of crystals of some component may be desirable or undesirable, and storage conditions should preserve the desired structure (Hartel, 1992, 2002).

Industrial crystallization for the production of crystalline materials from solutions or melts is treated in chemical engineering texts (e.g., Mullin, 1993). It involves phase equilibria, crystallization kinetics, and equipment design and operation. Phase equilibria are found in the literature in the form of tables and phase diagrams of temperature versus saturation concentration. Crystallization kinetics includes data and empirical correlations of nucleation and crystal growth rates.

Crystallization takes place either from solution, where the driving force is a concentration difference (supersaturation), or from melt, where the driving force is a temperature difference (supercooling). Ice formation can be considered as crystallization from melt (water), where heat and mass transfer play equally important roles.

Nucleation is the formation of small crystal nuclei, which are essential for the growth of crystals. It can be either homogeneous or heterogeneous, depending on the crystallizing system. Thermodynamic equations can predict the nucleation rate as a function of temperature, supersaturation, intersurface tension, energy of formation of a nucleus, and the Boltzmann constant (Mullin, 1993).

In practice, the nucleation rate (N) is given by empirical equations of the form

$$N = k' \Delta C^i \tag{10.58}$$

where k' and i are the nucleation constants, and $\Delta C = C - C_e$ is the supersaturation.

In most practical applications, heterogeneous nucleation predominates and is caused by particles of the same or other materials. In some industrial applications, nucleation is induced by seeding, that is, by injection of small crystals in the supersaturated solution (sugar manufacture).

Crystal growth involves mass transfer from the solution or the melt to the crystal surfaces and is expressed by the rate equation

$$\frac{m}{A} = K \Delta C \tag{10.59}$$

where m is the crystal growth rate (kg/s), A is the interface area (m²), and K is the overall mass transfer coefficient (m/s), which is defined by the relation $1/K = 1/k_D + 1/k_s$. Here, k_d is the mass transfer coefficient for diffusion of the crystallizing component in the solution, and k_s is the mass transfer coefficient for diffusion on the crystal surface.

If the growth rate of crystals is independent of crystal size (McCabe or constant ΔL law), the growth rate equation (Equation 10.59) is written as

$$G = \frac{dL}{dt} = K \frac{\Delta C}{\rho} \tag{10.60}$$

where G is the crystal growth rate (m/s), L is a characteristic dimension of the crystal (m), and ρ is the crystal density (kg/m^3).

Crystallization kinetics (nucleation and crystal growth rates) can be estimated experimentally in mixed suspension–mixed product removal (MSMPR) crystallizers. In these continuous-type reactors, population (total number of particles) balances are applied in addition to the normal material and energy balances (Randolph and Larson, 1971).

The crystallization of ice from water and sugar solutions has been studied by Huige (1972) in connection with freeze-concentration. Conditions favoring low nucleation rates and large crystals are desirable for efficient separation of ice from the concentrated solution.

Crystallization kinetics of ice, involving nucleation and mass transfer, is important in food freezing processes because it affects food quality (Hartel, 2002). With supercooling, the rate of crystal growth increases slowly, whereas the nucleation rate increases sharply. Thus, a fast decrease of temperature results in a very large number of nuclei that cannot grow fast and therefore the frozen food can maintain its physical structure, which might otherwise be damaged by the formation of large crystals.

The state of solid foods (crystalline, glassy, or rubbery) affects various mass transfer processes and the storage stability of dehydrated foods. Transition from the glassy (amorphous) to the rubbery (fluid-like) state increases the mobility of small molecules (e.g., water, flavor compounds), affecting product quality. The glass transition temperature is related to the "sticky point" and the collapse temperature of dehydrated foods (Roos and Karel, 1991).

10.6.4 Food Packaging

Packaging is a special field of food science and technology and encompasses various aspects ranging from the properties of packaging materials to food quality. Mass transfer is involved in the transmission of water vapor and other vapors and gases through packaging materials and in the transport of moisture and other food components within the package. Transmission properties are pertinent to polymeric materials (plastic films and paper), whereas mass transfer within the package relates to metal containers as well.

Two transport properties of the packaging materials, permeability and migration, are of importance. The mechanism of gas permeation through polymers was reviewed by Vieth (1991).

The transport of gases through rubbery polymers is better understood than that through glassy polymers. At temperatures above the glass transition temperature T_g (rubbery polymers), gases quickly reach equilibrium with the polymer surface, and diffusion through the polymer controls the transport process. At temperatures below T_g (glassy state), the polymer does not reach true equilibrium with the gas because of the slower motions of the polymer chains. Because permeation of gases depends on both solution and diffusion, the state of the polymer is of considerable importance.

The permeability (PM) of the packaging films to gases and moisture is measured using the basic permeation equation (Equation 10.33). The driving force for gases is the partial pressure difference of the penetrant; for water vapor, it is the humidity difference. Measurements of the permeability reflect both the solubility (S) and the diffusivity (D) of the penetrant into the film, because $PM = SD$.

Standard methods are used to measure water vapor transmission (WVT) and gas transmission (GT). The characterization of polymeric materials has been reviewed by Miltz (1992). The SI units of permeability PM are kg/m · s · Pa or g/m · s · Pa. However, various units are used in packaging, reflecting the established methods of measurement. Table 10.11 shows the conversion factors for the various units, used in packaging, into SI units.

Table 10.12 shows some typical permeabilities of various packaging and food films and the corresponding diffusivities of water vapor (Saravacos and Maroulis, 2001).

Table 10.11 Conversion Factors to SI Permeability Units (g/m · s · Pa)

Conversion from/to (g/m · s · Pa)	Multiplying Factor
cm³ (STP)mil/100 in² · day · atm	6.42×10^{-17}
cm³ (STP)mil/100 in² · day · atm	4.14×10^{-18}
cm³ (STP)^m/m² · day · kPa	1.65×10^{-17}
g · p,m/m² · day · kPa	1.16×10^{-14}
g · mm/m² · day · kPa	1.16×10^{-11}
g · mil/m² · day · atm	2.90×10^{-15}
g · mil/m² · day · mmHg	2.20×10^{-12}
g · mil/m² · day · (90% RH, 100°F)	4.50×10^{-14}
g · mil/100 in² · day · (90% RH, 100°F)	7.00×10^{-13}
perm (ASTM)	1.45×10^{-9}

Note: 1 mil = 0.001 in. = 2.54×10^{-3} m, 1 mmHg = 133.3 Pa   pressure drop across a film at 90% RH, 100°F AP = 6500 Pa.

Table 10.12 Typical Permeabilities (PM) and Diffusivities (D) of Water Vapor

Film or Coating	$PM \times 10^{10}$, g/m · s · Pa	$D \times 10^{10}$, m²/s
HDPE	0.002	0.005
LDPE	0.014	0.010
PP	0.010	0.010
PVC	0.041	0.050
Cellophane	3.70	1.00
Protein films	0.10–10.0	0.100
Polysaccharide films	0.10–1.0	0.100
Lipid films	0.003–0.100	0.010
Chocolate	0.11	0.001
Gluten	5.00	1.00
Corn pericarp	1.60	0.10

Source: Data from Saravacos, G.D. and Maroulis, Z.B. 2001. *Transport Properties of Foods*. Marcel Dekker, New York.

Note: HDPE = high-density polyethylene, LDPE = low-density polyethylene, PP = polypropylene, PVC = polyvinyl chloride.

Differential permeability of packaging materials is important in maintaining a controlled atmosphere within the package during storage. Thus, polymer films may permit the partial removal of carbon dioxide while maintaining the water vapor in the package during the storage of packaged fresh fruits and vegetables.

Transport processes of food components within the package are pertinent to the quality of the stored food product. Transfer of moisture from high- to low-moisture regions may cause agglomeration (caking) of hygroscopic food powders. Transfer of oxygen may induce oxidation reactions and loss of nutrients and organoleptic quality. The transfer of moisture during in-packaging desiccation of dehydrated foods can be predicted by a modified form of the rate equation (Equation 10.40) that includes the mass transfer resistances of the product, the package atmosphere, and the desiccant (Hendel et al., 1958).

Simulation of food quality losses during storage and processing can be performed via computer techniques. Karel (1983) cited as examples predictions of moisture increase in a packaged product, quality changes due to moisture changes, and browning of a dehydrated product.

Migration of polymers and trace components of packaging materials into the packaged food product is critical because it can affect quality and have possible toxicological implications. Migration is a complex process, depending in part on the diffusivity of the migrating component. Few quantitative data on phase equilibria and diffusion are available in the literature (Miltz, 1992).

ACKNOWLEDGMENTS

We acknowledge the contributions of our associates and graduate students at the School of Chemical Engineering of National Technical University of Athens, and the Department of Food Science of Rutgers University.

LIST OF SYMBOLS

A   Surface area, m^2
αw Water activity
Bi   Biot number ($k_c L/D$)
C   Concentration, kg/m^3 or $kmol/m^3$
C_p Specific heat, J/kg K
D   Diameter, m
D   Diffusivity, m^2/s
E   Energy of activation, kJ/mol
F   Fugacity, Pa
Fo   Fourier number (Dt/L^2)
G   Free energy, kJ
H   Henry's constant, Pa/mole fraction or $m^3 Pa/kg$
H   Heat-transfer coefficient, W/m^2 K
ΔH Heat of evaporation, kJ/kg
J   Mass transfer or heat flux, $kg/m^2 s$ or W/m^2

k_c	Mass transfer coefficient, m/s
j_H	Heat-transfer factor
j_M	Mass transfer factor
k_x, k_y	Mass transfer coefficients, kmol/m² s (mole fraction)
K	Partition coefficient, equilibrium constant
L	Half-thickness, m
M	Equilibrium constant
M	Mass of diffusant, kg
M	Molecular weight, kg/kmol
p, P	Pressure, Pa
PM	Permeability, kg/m s Pa
Pr	Prandtl number ($C_p\eta/\lambda$)
R	Radius, m
R	Gas constant, 8.31 kJ/kmol K
Re	Reynolds number ($u\rho L/\eta$)
RH	Relative humidity ($100(p/p_o)$)
S	Solubility (C/p), kg/m³ Pa
Sc	Schmidt number ($\eta/\rho D$)
Sh	Sherwood number ($k_c d/D$)
t	Time, s
T	Temperature, K
u	Velocity, m/s
X	Moisture content, dry basis, kg/kg (dry matter)
X_e	Equilibrium moisture content, dry basis
x	Mole fraction, liquid phase
y	Mole fraction, vapor phase
z	Distance, m

Greek Symbols

α_{AW}	Relative volatility
γ	Activity coefficient
ε	Bulk porosity (void fraction)
η	Viscosity, Pa s or kg/m s
λ	Thermal conductivity, W/m K
ρ	Density, kg/m³
τ	Tortuosity

REFERENCES

Andrieu, J., Jallut, C., Stamatopoulos, A. and Zafiropoulos, M. 1988. Identification of water apparent diffusivity for drying of corn-based pasta. *Proceedings IDS'88*, Versailles, France, 71–75.

Aguilera, J.M. and Stanley, D.W. 1999. *Microstructural Principles in Food Processing and Engineering.* 2nd ed., Aspen Publishers, Gaithersburg, MD.

Besson, A. 1983. Mathematical modeling of leaching. In *Progress in Food Engineering.* C. Cantanelli and C. Peri, (Eds.) pp. 147–156. Forster-Verlag AG, Kusnacht, Switzerland.

Binkley, C.R. and Wiley, R.C. 1978. Continuous diffusion–extraction method to produce apple juice. *J. Food Sci.* 43: 1019–1023.

Bizot, H. 1983. Using the G.A.B. model to construct sorption isotherms. In *Physical Properties of Foods*. R. Jowitt, F. Escher, B. Hallstrom, H.F.T. Meffert, W.E.L. Spiess, and G. Vos (Eds.), pp. 43–54, Applied Science, London.

Bomben, J.L., Bruin, S., Thijssen, H.A.C. and Merson, R.L. 1973. Aroma recovery and retention in concentration and drying. *Adv. Food Res.* 20: 1–111.

Bruin, S. 1969. *Activity Coefficients and Plate Efficiencies in Distillation of Multicomponent Aqueous Solutions.* H. Veenmam and Zonen, Wageningen, The Netherlands.

Bruin, S. and Luyben, K.Ch.A.M. 1980. Drying of food materials. In *Advances in Drying*. Vol. 1. A.S. Mujumdar (Ed.) pp. 155–215, Hemisphere, New York.

Crank, J. 1975. *The Mathematics of Diffusion*. Oxford University Press, Oxford, UK.

Crank, J. and Parker, G.S. (Eds.). 1968. *Diffusion in Polymers*. Academic Press, New York.

Cussler, E.L. 1976. *Multicomponent Diffusion*. Elsevier, Amsterdam.

Cussler, E.L. 1997. *Diffusion and Mass Transfer in Fluid Systems*. 2nd ed. Cambridge University Press, Cambridge.

da Silva Wilton Pereira, Cleide, M. D. P. S. e Silva, Vera, S. O. Farias, and Josivanda P. Gomes. 2012 Diffusion models to describe the drying process of peeled bananas: Optimization and simulation. *Drying Technology*, 30: 164–174.

da Silva Wilton Pereira, Denise Silva do Amaralb, Maria Elita M. Duarteb, Mário E.R.M.C. Matab, Cleide M.D.P.S. e Silvaa, Rubens M.M. Pinheiroc, Taciano Pessoac. 2013, Description of the osmotic dehydration and convective drying of coconut (*Cocos nucifera* L.) pieces: A three-dimensional approach. *J. Food Eng.* 115(1), 121–131.

Drusas, A., Vagenas, G.K. and Saravacos, G.D. 1988, Diffusion of sodium chloride in green olives. *J. Food Eng.* 7: 211–222.

Elbert, G., Tolaba, M.P., Aguerre, R.J. and Suarez, C. 2001. A diffusion model with a moisture dependent diffusion coefficient for parboiled rice. *Drying Technol.* 19(1): 155–166.

Engels, C., Hendrickx, M., De Samblanx, S., De Gryze, I. and Tobback, P. 1986. Modeling water diffusion during long-grain rice soaking. *J. Food Eng.* 5: 55–73.

Fish, B.P. 1958. Diffusion and thermodynamics of water in potato starch gels. In *Fundamental Aspects of the Dehydration of Foodstuffs*. pp. 143–157. Soc. Chem. Ind., London.

Flink, J.M. and Karel, M. 1970. Retention of organic volatiles in freeze-dried solution of carbohydrates. *J. Agric. Food Chem.* 18: 259–297.

Floros, J.D. and Chinnan, M.S. 1989. Determining the diffusivity of sodium hydroxide through tomato and capsicum skins. *J. Food Eng.* 9: 128–141.

Frisch, H.L. and Stern, S.A. 1983. Diffusion of small molecules in polymers. *CRC Crit. Rev. Solid State Mater. Sci.* 2: 123–187.

Geankoplis, C.J. 1993. *Transport Processes and Unit Operations*. 3rd ed., Prentice-Hall, New York.

Gekas, V. 1992. *Transport Phenomena of Foods and Biological Materials*. CRC Press, Boca Raton, Florida.

Gennadios, A. (Ed.), 2002. *Protein-Based Films and Coatings*. CRC Press, Boca Raton, FL.

Gros, J.B. and Ruegg, M. 1987. Determination of the apparent diffusion coefficient of sodium chloride in model foods and cheese. In *Physical Properties of Foods-2*. R. Jowitt, F. Escher, M. Kent, B. McKenna, and M. Roques (Eds.) pp. 71–108, Elsevier, London.

Guilbert, S., Giannakopoulos, A. and Gurevitz, A. 1983. Diffusivity of sorbic acid in food gels at high and intermediate water activities. *3rd Intl. Symp. on Properties of Water*, Beaune, France, Sept. 11–16.

Hala, E., Pick, J., Fried, V. and Vilim, O, 1958. *Vapor–Liquid Equilibrium*. Pergamon Press, New York.

Hartel, R.W. 1992. Solid–liquid equilibrium: Crystallization of foods. In *Physical Chemistry of Foods*. H.G. Schwartzberg and R.W. Hartel (Eds.), pp. 47–81, Marcel Dekker, New York.

Hartel, R.W. 2002. *Crystallization in Foods*. Kluwer Academic, New York.

Hendel, C.E., Legault, R.R., Talburt, W.F., Burr, H.K. and Wilke, C.R. 1958. Water-vapor transfer in the in-package desiccation of dehydrated foods. In *Fundamental Aspects of the Dehydration of Foodstuffs*. pp. 89–99. Soc. Chem. Industry, London.

Huige, N.J.J. 1972. Nucleation and growth of ice crystals from water and sugar solutions in continuous stirred tank crystallizers. Doctoral thesis, Technical University of Eindhoven, The Netherlands.

Iglesias, H.A. and Chirife, J. 1982. *Handbook of Food Isotherms*. Academic Press, New York.

Jaros, M., Cenkonski, S., Jayas, D.S. and Pabis, S. 1992. A method for determination of the diffusion coefficient based on kernel moisture content and temperature. *Drying Technol.* 10: 213–225.

Jason, A.C. and Peters, G.R. 1973. Analysis of bimodal diffusion of water in fish muscle. *J. Phys. (D) Appl. Phys.* 6: 512–521.

Jowitt, R., Escher, F., Hallstrom, B., Meffert, H.F.T., Spiess, W.E.L. and Vos, G. (Eds.). 1983. *Physical Properties of Foods*. Applied Science, London.

Jowitt, R., Escher, F., Kent, M., B., McKenna, B. and Roques, M. (Eds.). 1987. *Physical Properties of Foods-2*. Elsevier, London.

Karatas, S. and Pinarli, I. 2001. Determination of moisture diffusivity of pine nut seeds. *Drying Technol.* 19(3&4): 791–808.

Karathanos, V.T., Villalobos, G. and Saravacos, G.D. 1990. Comparison of two methods of estimation of the effective moisture diffusivity from drying data. *J. Food Sci.* 55: 218–233.

Karathanos, V.T., Vagenas, G.K. and Saravacos, G.D. 1991. Water diffusivity of starch at high temperatures and pressures. *Biotechnol. Progr.* 7: 178–184.

Karel, M. 1975. Properties controlling mass transfer in foods and related systems. In *Theory, Control and Determination of if Physical Properties of Food Materials*. C. Rha (Ed.) pp. 221–250. Reidel, Dordrecht, The Netherlands.

Karel, M. 1983. Quantitative analysis and simulation of food quality losses during processing and storage. In *Computer-Aided Techniques in Food Technology*. I. Saguy (Ed.). pp. 117–135, Marcel Dekker, New York.

Karel, M. 1990. Encapsulation and controlled release of food components. In *Biotechnology and Food Process Engineering*. H.G. Schwartzberg, and M.A. Rao (Eds.), pp. 277–293. Marcel Dekker, New York.

Kerkhof, P.J.A.M. and Schoeber, W.J.A. 1974. Theoretical modeling of the drying behavior of droplets in spray dryers. In *Advances in Preconcentration and Dehydration of Foods*. A. Spicer (Ed.), pp. 349–397. Applied Science, London.

King, C.J. 1968. Rates of moisture sorption and desorption in porous dried foodstuffs. *Food Technol.* 22: 165–171.

King, C.J. 1982. *Separation Processes*, 2nd ed. McGraw-Hill, New York.

King, C.J. 1988. Spray drying of food liquids, and volatile retention. In *Preconcentration and Drying of Food Materials*. S. Bruin (Ed.) pp. 147–162, Elsevier, Amsterdam.

Kiranoudis, C.T., Maroulis, Z.B., Marinos-Kouris, D. and Saravacos, G.D. 1993. Estimation of the effective moisture diffusivity from drying data. In *Developments in Food Engineering ICEF6*. Part 1. T. Yano, R. Matsuno, and K. Nakamura (Eds.). pp. 340–342. Blackie Academic and Professional, London.

Lazar, M.E. and Farkas, D.F. 1971. The centrifugal fluidized bed. 2. Drying studies on piece-form foods. *J. Food Sci.* 36: 315–319.

Le Maguer, M. 1992. Thermodynamics and vapor–liquid equilibria. In *Physical Chemistry of Foods*. H.G. Schwartzberg, and R.W. Hartel (Eds.) pp. 1–45, Marcel Dekker, New York.

Leslie, R.B., Carillo, P.J., Chung, T.Y., Gilbert, S.G., Hayakawa, K., Marousis, S., Saravacos, G.D. and Solberg, M. 1991. Water diffusivity in starch-based systems. In *Water Relationships in Foods*. H. Levine, and L. Slade (Eds.) pp. 365–390. Plenum, New York.

Levine, H. and Slade, L. 1992. Glass transitions in foods. *In Physical Chemistry of Foods.* H.G. Schwartzberg, and R.W. Hartel (Eds.) pp. 83–221, Marcel Dekker, New York.

Loncin, M. and Merson, R.L. 1979. *Food Engineering.* Academic Press, New York.

Marcotte, M. and Le Maguer, M. 1992. Mass transfer in cellular tissues. Part II.: Computer simulations vs. experimental data. *J. Food Eng.* 17: 177–199.

Marcotte, M., Tooupin, C.J. and Le Maguer, M. 1991. Mass transfer in cellular tissues. Part I: The mathematical model. *J. Food Eng.* 13: 199–220.

Marinos-Kouris, D. and Saravacos, G.D. 1975. Volatility of organic compounds in aqueous sucrose solutions. *5th Intl. Cong. Chem. Eng. CHISA 75,* Prague, CZ, paper No. f4.27.

Maroulis, Z.B., Tsami, E., Marinos-Kouris, D. and Saravacos, G.D. 1988. Application of the G.A.B. model to the moisture sorption isotherms of dried fruits. *J. Food Eng.* 7: 3–78.

Marousis, S.N. and Saravacos, G.D. 1990. Density and porosity in drying starch materials. *J. Food Sci.* 55: 1367–1372.

Marousis, S.N., Karathanos, V.T. and Saravacos, G.D. 1989. Effect of sugars on the water diffusivity of hydrated granular starches. *J. Food Sci.* 54: 1496–1500.

Marousis, S.N., Karathanos, V.T. and Saravacos, G.D. 1991. Effect of physical structure of starch materials on water diffusivity. *J. Food Proc. Preserv.* 15: 183–195.

Mazza, G. and Le Maguer, M. 1980. Flavor retention during dehydration of onion. In *Food Process Engineering.* P. Linko, Y. Malkki, J. Olkku, and J. Larinkari (Eds.) pp. 399–406. Applied Science, London.

McHugh, M.A. 1990. Supercritical fluid extraction. In *Biotechnology and Food Process Engineering.* H.G. Schwartzberg and M.A. Rao (Eds.), pp. 203–212, Marcel Dekker, New York.

Menting, L.C. and Hoagstad, B. 1967. Volatiles retention during the drying of carbohydrate solutions. *J. Food Sci.* 32: 87–90.

Meranda, D. and Furter, W.F. 1974. Salt effects on vapor–liquid equilibrium. *AIChE J.* 20: 103–108.

Miltz, J. 1992. Food packaging. In *Handbook of Food Engineering.* D.R. Heldman and D.B. Lund (Eds.), pp. 667–718, Marcel Dekker, New York.

Mittal, G.S. 1999. Mass diffusivity of food products. *Food Rev. Intl.* 15(1): 19–66.

Motarjemi, Y. and Hallstrom, B. 1987. A study of moisture transport in minced beef. In *Physical Properties of Foods-2.* R. Jowitt, F. Escher, M. Kent, B. McKenna, and M. Roques, (Eds.). pp. 61–64, Elsevier, London.

Mullin, J. 1993. *Crystallization,* 3rd ed. Butterworths, London.

Naessens, W., Bresseleers, G. and Tobback, P. 1981. A method for the determination of diffusion coefficients of food components in low and intermediate moisture systems. *J. Food Sci.* 46: 1446–1449.

Noshad M., Mohebbat M., Fakhri S., and Seyed A. M.. 2012. Kinetic modeling of rehydration in sir-dried quinces pretreated with osmotic dehydration and ultrasonic. *Journal of Food Processing and Preservation* 36: 383–392.

Okos, M.R., Narsimham, G., Singh, R.K. and Weitnauer, A.C. 1992. Food dehydration. In *Handbook of Food Engineering.* D.R. Heldman and D.B. Lund (Eds.), pp. 437–562, Marcel Dekker, New York.

Perry, R.H. and Green, D.W. 1997. *Chemical Engineers' Handbook.* 7th ed., McGraw-Hill, New York.

Petropoulos, J.H. 1994. Diffusion in polymers. In *Polymeric Gas Separation Membranes.* D.R. Paul and Y.P. Yampolski (Eds). CRC Press, Boca Raton, Florida.

Pflug, J.J., Fellers, P.J. and Gurevitz, D. 1975. Diffusion of salt in the desalting of pickles. *Food Technol.* 21: 1634–1638.

Pham, G.T. and Willix, J. 1984. A model for food desiccation in frozen storage. *J. Food Sci.* 49: 1275–1294.

Prausnitz, J.M., Lichtenthaler, R.N. and de Azevedo, E.G.,1999. *Molecular Thermodynamics of Fluid-Phase Equilibria.* 3rd ed. Prentice-Hall, Englewood Cliffs, NJ.

Randolph, A. and Larson, M. 1971. *Theory of Particulate Processes.* Academic Press, New York.

Reid, R.C., Prausnitz, J.M. and Poling, B.E. 1987. *The Properties of Gases and Solids.* 4th ed., McGraw-Hill, New York.

Robinson, C.S. and Gilliland, E.R. 1950. *Elements of Fractional Distillation*. McGraw-Hill, New York.

Rodger, G., Hastings, R., Cryne, C. and Bailey, J. 1984. Diffusion properties of salt and acetic acid into herring. *J. Food Sci.* 49: 714–720.

Roos, Y. and Karel, M. 1991. Applying state diagrams to food processing and development. *Food Technol.* 45(2): 66–71.

Roth, T. 1992. Reduction of the rate of drying of foods by the use of surface active agents. *Zeit. Lebensm. Tech. Verfahrenstechnik* 33: 497–507.

Rotstein, E. 1987. The prediction of diffusivities and diffusion-related transport properties in the drying of cellular foods. In *Physical Properties of Foods-2*. R. Jowitt, F. Escher, M. Kent, B. McKenna, and M. Roques, (Eds.). pp. 131–145, Elsevier, London.

Rulkens, W.H. 1973. Retention of volatile trace components in drying aqueous carbohydrate solutions. Doctoral thesis, Technical University of Eindhoven, NL.

Sablami, S., Rahman, S. and Al-Habsi, H. 2000. Moisture diffusivity of foods. A review. In *Drying Technology in Agriculture and Food Sciences*. A.S. Mujumdar (Ed.). pp. 35–59, Science Publishers, Enfield, NH.

Sancho, M.F. and Rao, M.A. 1997. Infinite dilution activity coefficients of apple juice aroma compounds. *J. Food Eng.*, 34: 145–158.

Saravacos, G.D. 1967. Effect of the drying method on the water sorption of dehydrated apple and potato. *J. Food Sci.* 32: 81–84.

Saravacos, G.D. 1969. Sorption and diffusion of water in dry soybeans. *Food Technol.* 23: 145–147.

Saravacos, G.D. and Charm, S.E. 1962a. A study of the mechanism of fruit and vegetable dehydration. *Food Technol.* 16: 78–81.

Saravacos, G.D. and Charm, S.E. 1962b. Effect of surface active agents on the dehydration of fruits and vegetables. *Food Technol.* 16: 91–93.

Saravacos, G.D. and Moyer, J.M. 1968a. Volatility of some aroma compounds during vacuum-drying of fruit juices. *Food Technol.* 22(5): 89–95.

Saravacos, G.D. and Moyer, J.M. 1968b. Volatility of some flavor compounds during freeze-drying of foods. *Chem. Eng. Progr. Symp. Ser.* 64(86): 37–42.

Saravacos, G.D. and Raouzeos, G.S. 1984. Diffusivity of moisture in air-drying of starch gels. In *Engineering and Food*. Vol. 1. B. M. McKenna (Ed.) pp. 499–507. Elsevier, London.

Saravacos, G.D. and Stinchfield, R.M. 1965. Effect of temperature and pressure on sorption of water vapor by freeze-dried food materials. *J. Food Sci.* 30: 779–786.

Saravacos, G.D., Marousis, S.N. and Raouzeos, G.S. 1988. Effect of ethyl oleate on the rate of air-drying of foods. *J. Food Eng.* 7: 263–270.

Saravacos, G.D., Karathanos, V.T., Marousis, S.N., Drouzes, A.E. and Maroulis, Z.B. (1990a). Effect of gelatinization on the heat and mass transport properties of starch materials. In *Engineering and Food*. Vol. 1. W.E.L. Spiess and H. Schubert (Eds.), pp. 390–398, Elsevier, London.

Saravacos, G.D., Karathanos, V.T. and Marinos-Kouris, D. (1990b). Volatility of fruit aroma compounds in sugar solutions. In *Flavors and Off-Flavors*. G. Charalambous (Ed.) pp. 729–733. Elsevier, Amsterdam.

Saravacos, G.D. and Maroulis, Z.B. 2001. *Transport Properties of Foods*. Marcel Dekker, New York.

Saravacos, G.D. and Kostaropoulos, A.E. 2002. *Handbook of Food Processing Equipment*. Kluwer Academic Publishers, New York.

Schwartzberg, H.G. and Chao, R.Y. 1982. Solute diffusivities in the leaching processes. *Food Technol.* 36(2): 73–86.

Sherwood, T.K., Pigford, R.L. and Wilke, C.R. 1975. *Mass Transfer*. McGraw-Hill, New York.

Spaninks, J.A.M. 1983. Calculation methods for solid–liquid extractors. In *Progress in Food Engineering*. C. Cantanelli, and C. Peri (Eds.), pp. 109–124, Forster-Verlag AG, Kusnacht, Switzerland.

Spiess, W.E.L. and Wolf, W.R. 1983. The results of the COST 90 project on water activity. In *Physical Properties of Foods*. R. Jowitt, F. Escher, B. Hallstrom, H.F.T. Meffert, W.E.L. Spiess, and G. Vos. (Eds.), pp. 65–88, Applied Science, London.

Suarez, C., Loncin, M. and Chirife, J. 1984. Preliminary study on the effect of ethyl oleate dipping treatment on drying rate of grain in corn. *J. Food Sci.* 49: 236–238.

Telis, V.R.N., Murari, R.C.B.D.L. and Yamashita, F. 2004. Diffusion coefficients during osmotic dehydration of tomatoes in ternary solutions. *J. Food Eng.* 61: 253–259.

Theodorou, D.N. 1996. Molecular simulation of sorption and diffusion in amorphous polymers. In *Diffusion in Polymers*. P. Neogi (Ed.), pp. 67–142, Marcel Dekker, New York.

Thijssen, H.A.C. 1974. Fundamentals of concentration processes. In *Advances in Preconcentration and Dehydration of Foods*. A. Spicer (Ed.), pp. 13–43, Applied Science, London.

Tong, C.H. and Lund, D.B. 1990. Effective moisture diffusivity in porous materials as a function of temperature and moisture content. *Biotechnol. Progr.* 6: 67–75.

Treybal, R. 1980. *Mass Transfer Operations*. McGraw-Hill, New York.

Van Winkle, M. 1967. *Distillation*. McGraw-Hill, New York.

Vieth, W.R. 1991. *Diffusion in and Through Polymers*. Hanser Publ., Munich.

Voilley, A. and Roques, M.A. 1987. Diffusivity of volatiles in water in the presence of a third substance. In *Physical Properties of Foods-2*. R. Jowitt, F. Escher, M. Kent, B. McKenna, and M. Roques (Eds.). pp. 109–121, Elsevier, London.

Voilley, A. and Simatos, D. 1980. Retention of aroma during freeze and air-drying. In *Food Process Engineering*. P. Linko, Y. Malkki, J. Olkku, and J. Larinkari (Eds.) pp. 371–384. Applied Science, London.

Wang, D., Tang, J. and Correia, L.R. 2000. Salt diffusivities and salt diffusion in farmed Atlantic salmon muscle as influenced by rigor mortis. *J. Food Eng.* 43(2): 115–123.

Xanthopoulos, G., Yanniotis, S. and Talaiporou, E. 2012. Influence of salting on drying kinetics and water diffusivity of tomato halves. *Int. J. Food Properties*, 15: 847–863.

Xiong, X., Narsimham, G. and Okos, M.R. 1991. Effect of composition and pore structure on binding energy and effective diffusivity of moisture in porous foods. *J. Food Eng.* 15: 187–208.

Yildiz, M.E. and Kokini, J.L. 1999. Development of a prediction methodology to determine the diffusion of small molecules in food polymers. In *Proceedings of 6th Conference of Food Engineering CoFE 99*. G.V. Barbosa-Canovas, and S.P. Lombardo (Eds.), pp. 99–105, AIChE, New York.

Zogzas, N.P., Maroulis, Z.B. and Marinos-Kouris, D. 1994. Moisture diffusivity methods of experimental determination. A review. *Drying Technol.* 12(3): 435–515.

Zogzas, N.P. and Maroulis, Z.B. 1996a. Effective moisture diffusivity estimation from drying data. A comparison between various methods of analysis. *Drying Technol.* 14(7&8): 1543–1573.

Zogzas, N.P., Maroulis, Z.B. and Marinos-Kouris, D. 1996b. Moisture diffusivity data compilation in foodstuffs. *Drying Technol.* 14(10): 2225–2253.

11

Thermodynamic Properties of Foods in Dehydration

Syed S. H. Rizvi

Contents

11.1 Introduction	360
11.2 Thermodynamics of Food–Water Systems	361
11.2.1 Chemical Potential and Phase Equilibria	362
11.2.2 Fugacity and Activity	363
11.2.3 Water Activity in Foods	366
11.2.4 Measurement of Water Activity	370
11.2.4.1 Measurements Based on Colligative Properties	370
11.2.4.2 Measurements Based on Psychrometry	374
11.2.4.3 Measurements Based on Isopiestic Transfer	375
11.2.4.4 Measurements Based on Suction (Matric) Potential	375
11.2.5 Adjustment of Water Activity	375
11.2.6 Moisture Sorption Isotherms	379
11.2.6.1 Theoretical Description of MSIs	380
11.2.6.2 Effect of Temperature	386
11.3 Sorption Energetics	389
11.3.1 Differential Quantities	390
11.3.2 Integral Quantities	393
11.3.4 Hysteresis and Irreversibility	398
11.3.5 Kinetic Aspects	402
11.4 Dehydration Principles and Processes	405
11.4.1 Drying Behavior	406
11.4.2 Constant-Rate Period	409
11.4.3 Falling-Rate Period	411

11.4.4 Equilibrium Moisture Content	416
11.4.5 Energy Requirements	417
11.5 Conclusion	422
List of Symbols	423
Greek Symbols	424
Subscripts	425
Superscripts	425
References	425

11.1 INTRODUCTION

As the most abundant and the only naturally occurring inorganic liquid substance on Earth, water is known to exhibit unique and anomalous behavior. Held together by a random and fluctuating three-dimensional network of hydrogen bonds, no single theory to date has been able to explain the totality of its unusual molecular nature. Yet, throughout history, people have learned that either removing water or making it unavailable via binding to appropriate matrices can extend the period of usefulness of perishable products. It otherwise provides the critical environmental factor necessary for the ubiquitous biological, biochemical, and biophysical processes that degrade foods and ultimately render them unfit for human consumption. Any reduction in water content that retards or inhibits such processes will indeed preserve the food. Thus, dehydration as a means of preserving the safety and quality of foods has been at the forefront of technological advancements in the food industry. It has greatly extended the consumer-acceptable shelf life of appropriate commodities from a few days and weeks to months and years. The lower storage and transportation costs associated with the reduction of weight and volume due to water removal have provided additional economic incentives for widespread use of dehydration processes. The expanding variety of commercial dehydrated foods available today has stimulated unprecedented competition to maximize their quality attributes, to improve the mechanization, automation, packaging, and distribution techniques, and to conserve energy.

Knowledge of thermodynamic properties involved in sorption behavior of water in foods is important to dehydration in several respects. First, the thermodynamic properties of food relate the concentration of water in food to its partial pressure, which is crucial for the analysis of mass and heat-transport phenomena during dehydration. Second, they determine the end point to which foods must be dehydrated in order to achieve a stable product with optimal moisture content. Third, the enthalpy of sorption yields a figure for the theoretical minimum amount of energy required to remove a given amount of water from food. Finally, knowledge of thermodynamic properties can provide insights into the microstructure associated with a food, as well as theoretical interpretations for the physical phenomena occurring at food–water interfaces.

In this chapter, some of the fundamental thermodynamic functions are described, with emphasis on discussion of how they arise and what assumptions are made in their measurement. Practical aspects of these and other related properties in air dehydration of foods are examined and discussed.

11.2 THERMODYNAMICS OF FOOD–WATER SYSTEMS

On the basis of its thoroughly studied thermodynamic properties, it is generally agreed that there is no other chemical that could assume the central role of water in sustaining life. The remarkable properties of water as the best of all solvents, in combination with its unusually high specific heat, enthalpy of phase transformation, dielectric constant, and surface tension properties, make it uniquely fit to support biochemical processes, even under adverse conditions. There are two basic theories of bulk liquid water structure, both of which indicate that liquid water does not exist purely as monomers but rather consists of a transient, hydrogen-bonded network. The continuum model theory assumes that water is a continuously bonded structure similar to ice, with a very short relaxation time (10^{-12}–10^{-13} s) to account for the monomeric properties, or that it has certain defects in it that account for the perturbations. The other theory is the mixture model theory, in which it is suggested that water exists as a mixture of monomers and polymers of various sizes. The simplified cluster model presumes that there exists a strong cooperative hydrogen bonding among water molecules in bulk water with assemblies of ice-like, six-membered rings in rapid exchange with free water molecules, having a relaxation time of about 5×10^{-12} s. It is estimated that in liquid water at 0°C, the extension of the Hbonded network may be several hundred molecules, at 50°C about 100, and at 100°C about 40. In the cluster model, the H_2O–H interactions produce about 14.5–21.0 kJ/mol and hence strongly influence the liquid structure [1,2].

In biological systems such as foods, water is believed to exist with unhindered or hindered mobility and is colloquially referred to as free water (similar to liquid water) and as bound water. This has come about from a recognition of the fact that there exists a time-, number-averaged population of water molecules that interact with macromolecules and behave differently, both thermodynamically and kinetically, from bulk water; the use of the term *bound water* thus depends on the chosen time frame [3].

Bound water is generally defined as sorbent- or solute-associated water that differs thermodynamically from pure water [4]. It has been suggested that the water is bound to stronger hydrogen bond acceptors than liquid water (possibly with favored hydrogen bond angles) as well as to water-solvating nonpolar groups. According to Luck [2], bound water has a reduced solubility for other compounds, causes a reduction of the diffusion of water-soluble solutes in sorbents, and exhibits a decrease in its diffusion coefficient with decreasing moisture content. The decreased diffusion velocity impedes drying processes because of slower diffusion of water to the surface. Thus, the energetics of the sorption centers, topological and steric configurations, types of interactions between water and food matrix, pH, temperature, and other related factors exert a cumulative influence on foods in accordance with their changing values during the course of pretreatments and drying operations. They not only influence mechanisms of moisture transport, process kinetics, and energy requirements but also are decisive in defining such quality descriptors for dehydrated foods as organoleptic and nutritive values, bulk density, hygroscopicity, wettability, rehydratability, sinkability, and caking.

How water behaves when confined between macromolecular surfaces like in capillaries and narrow pores has been the subject of several recent articles [5–7]. Apart from shifting the phase diagram, hydrophobic surfaces have been reported to produce a thin layer of low-density fluid at the interface, often leading to the formation of a gas-like layer. These local changes in water density are very likely to affect such important parameters as pH, salt concentration, ionic strength, and so on, in the vicinity of the macromolecular interface and help explain the mysteries of the long-range hydrophobic attractions.

In view of the current uncertainties regarding the interactions between water molecules and their structure and behavior in bulk water, development of quantitative theoretical models for the behavior of water in food is beyond realization at present. However, the thermodynamic approach has been, within limits, successfully used to study water–solid equilibria, particularly sorption behavior of water in foods as it relates to dehydration and storage strategies for quality maximization. In the following section, therefore, thermodynamic considerations of fundamental importance to dehydrated foods are discussed.

11.2.1 Chemical Potential and Phase Equilibria

In searching for another spontaneity criterion to replace the awkward-to-use entropy maximization principle, Josiah Willard Gibbs noticed that in a number of cases the system changes spontaneously to achieve a low state of chemical energy or enthalpy (H) as well as a state of higher disorder or entropy (S). At times these properties work together, and sometimes are opposed, and the overall driving force for a process to occur or not is determined by some combination of enthalpy and entropy. Physically, enthalpy represents the total energy available to do useful work, whereas entropy at any temperature (T) provides lost work and gives a measure of energy not available to perform work. Thus, the energy that is available to do work is the difference between these two quantities. This idea is qualitatively expressed as *Gibbs free energy* (G) = total energy (enthalpy factor) − unavailable energy (entropy factor). Quantitatively, it can thus be written as

$$G = H - TS \tag{11.1}$$

In differential form, the expression becomes

$$dG = dH - T\,dS - S\,dT \tag{11.2}$$

When one recalls that $H = E + PV$ or $dH = dE + P\,dV + V\,dP$ and substitutes for dH, the expression for the differential dG becomes

$$dG = dE + P\,dV + V\,dP - T\,dS - S\,dT \tag{11.3}$$

For a reversible change where only pressure–volume work occurs, the first and second laws of thermodynamics give $dE = T\,dS - P\,dV$. Substituting the value of dE into Equation 11.3 and canceling like terms gives as the final equation for differential changes in Gibbs free energy:

$$dG = V\,dP - S\,dT \tag{11.4}$$

The above equation applies to any homogeneous system of constant composition at equilibrium where only work of expansion takes place. In real life, however, multicomponent systems with varying compositions are frequently encountered. The aforementioned method of calculating the total Gibbs free energy is therefore not adequate unless some consideration is given to the composition. The Gibbs free energy of a multicomponent system undergoing any change will depend not only on temperature and pressure as defined by Equation 11.4 but also on the amount of each component present in the system. The number of moles of each component must therefore be specified as field variables in addition to the natural variables of each thermodynamic state function of such systems. For a multicomponent system of varying composition, if the series $n_1, n_2, n_3, \ldots, n_i$ indicates the numbers of moles of components $1, 2, 3, \ldots, i$, then

$$G = G(P, T, n_i) \tag{11.5}$$

A complete differential for the above equation would then be

$$dG = \left(\frac{\partial G}{\partial P}\right)_{T,n_i} dP + \left(\frac{\partial G}{\partial T}\right)_{P,n_i} dT + \sum_i \left(\frac{\partial G}{\partial n_i}\right)_{T,P,n_{j\neq 1}} dn_i \tag{11.6}$$

The partial molar Gibbs free energy $(G/n_i)_{T,P,nj} \neq i$ is called the chemical potential of component i, μ_i, and represents the change in the total free energy per mole of component i added, when the temperature, total pressure, and numbers of moles of all components other than i are held constant. In other words,

$$\mu_i = \left(\frac{\partial G}{\partial n_i}\right)_{T,P,n_{j\neq 1}} \tag{11.7}$$

The expression for the differential dG for a reversible change is then given as

$$dG = V\,dP - S\,dT + \sum_i \mu_i\,dn_i \tag{11.8}$$

For systems with constant composition ($dn_i = 0$), as in pure substances or in systems with no chemical reaction occurring, Equation 11.8 reduces to the original Equation 11.4.

In his celebrated article, "On the Equilibrium of Heterogeneous Substances," Gibbs showed that for a simple system of one component and one phase or for a complex system of more than one component and existing in more than one phase, the necessary and sufficient condition for equilibrium is

$$\mu_i^I = \mu_i^{II} = \mu_i^{III} = L \tag{11.9}$$

where the superscripts refer to different phases. For coexisting phases in equilibrium, it must then also be true that

$$d\mu_i^I = d\mu_i^{II} = d\mu_i^{III} = L \tag{11.10}$$

11.2.2 Fugacity and Activity

As the term implies, the *chemical potential* is a driving force in the transfer of material from one phase to another and provides a basic criterion for phase equilibrium. Although extremely useful, the chemical potential cannot be conveniently measured. For any component in a system of fixed composition, the chemical potential is given by

$$d\mu_i = \bar{V}_i dP - \bar{S}_i dT \tag{11.11}$$

At a constant temperature, $dT = 0$, and hence

$$\left(\frac{\partial \mu_i}{\partial P}\right)_T = \bar{V}_i \tag{11.12}$$

In 1923, G. N. Lewis expressed the chemical potential of ideal gas in terms of easily measurable functions and then generalized the results for real systems. Combining Equation 11.12 with the ideal gas equation $P\bar{V} = RT$ gives

$$d\mu_i = RT \frac{dP}{P} \tag{11.13}$$

It is convenient indeed to express not the *change* in chemical potential but rather the chemical potential *itself*. However, because the absolute value of the chemical potential is not known, it must be given relative to the chemical potential of some standard or reference state—a state of defined temperature, pressure, and composition. If μ^0_i is the chemical potential at the chosen standard pressure (P^0) condition of 1 atm, then the free energy of component i in the gas phase at pressure P_i atm, when behaving ideally, is given by

$$\mu_i = \mu_i^0 + RT \ln P_i \tag{11.14}$$

This equation is also applicable to a mixture of ideal gas components. It must be recognized that the reference state must be at the same temperature as that of the system under consideration. Equation 11.14 is not obeyed by real gases except when $P \to 0$. In order to make it applicable to real systems whether mixed or pure, ideal or real, Lewis proposed a new function, f, called the fugacity, such that Equation 11.14 is correct for all values of pressure. The general equation then is

$$\mu_i = \mu_i^0 + RT \ln f_i \tag{11.15}$$

The only condition imposed on the fugacity is that when the gas is very dilute ($P \to 0$) and the ideal gas law is obeyed, it is identical to the partial pressure. This is achieved by defining

$$\lim_{P \to 0}\left(\frac{f_i}{P_i}\right) = 1 \tag{11.16}$$

By substituting the real measure of gas concentration, the partial pressure, with the fugacity, deviations from ideality have been corrected. The fugacity is thus nothing but corrected or fake pressure, which has the virtue of giving the true chemical potential in Equation 11.15. For convenience, the relation between the fugacity and pressure of the respective component is defined by a parameter called the *fugacity coefficient*, γ_f, which is simply the ratio of the fugacity to the pressure and is dimensionless. In general terms,

$$\gamma_{fi} = \frac{f_i}{P_i} \quad \text{or} \quad f_i = \gamma_{fi} P_i \tag{11.17}$$

Ideal solutions follow Raoult's law and exhibit behavior analogous to that of ideal gases in terms of physical models and the resulting thermodynamic equations. For an ideal solution of any volatile solvent, i, in equilibrium with its vapors, the chemical potential of i must be identical in both phases, as indicated by Equation 11.9. Thus,

$$\mu_i(\text{soln}) = \mu_i(\text{vapor}) = \mu_i^0(\text{vapor}) + RT \ln P_i \quad (11.18)$$

According to Raoult's law for an ideal solution, $P_i = x_i P^\Sigma_i$. Substituting for P_i in Equation 11.18 yields*

$$\mu_i(\text{soln}) = \mu_i^0(\text{vapor}) + RT \ln(x_i P_i^\bullet) \quad (11.19)$$
$$= \mu_i^0(\text{vapor}) + RT \ln x_i + RT \ln P_i^\bullet$$

From the above relation, it is apparent that when pure i is present ($x_i = 1$), the chemical potential becomes that of pure $i(\mu^\Sigma_i)$. If the standard-state chemical potential for the solution is defined as that of pure component in solution, then

$$\mu_i^*(\text{soln}) = \mu_i^*(\text{vapor}) + RT \ln P_i^\bullet \quad (11.20)$$

Substituting the above definition of standard-state chemical potential for solutions in Equation 11.19 gives the expected chemical potential of an ideal solution:

$$\mu_i(\text{soln}) = \mu_i^*(\text{soln}) + RT \ln x_i \quad (11.21)$$

Like the chemical potential of an ideal gas, the validity of Equation 11.21 depends on strict adherence to Raoult's law. Consequently, the chemical potential calculation will not give correct results for real solutions if used as such except when $x_i \to 1$. To correct for the nonideal behavior of real solutions, use is made of a quantity similar to the fugacity, termed the *activity* (*a*). Replacing mole fraction in the ideal Equation 11.21 with the activity gives

$$\mu_i(\text{soln}) = \mu_i^*(\text{soln}) + RT \ln a_i \quad (11.22)$$

The activity, a_i, in the above expression is a quantity whose value is such that Equation 11.22 is correct for all concentrations of component i. Because Raoult's law applies to pure solvents and solvents of very dilute solutions, the limiting value of activity is established by defining

$$\lim_{x_i \to 1}\left(\frac{a_i}{x_i}\right) = 1 \quad (11.23)$$

Like the fugacity coefficient, the activity coefficient γ_a may then be defined as the dimensionless ratio of the activity to the mole fraction:

$$\gamma_{ai} = \frac{a_i}{x_i}, \text{ or } a_i = \gamma_{ai} x_i \quad (11.24)$$

---

* The superior bullet (•) designates "pure."

11.2.3 Water Activity in Foods

To grasp the role of fugacity and activity in food–water–vapor interaction, consider the general setup simulating a food system shown in Figure 11.1. The solvent in the system is water, and food constituents such as salts, sugars, proteins, carbohydrates, and others are the solutes. At a constant temperature, all components and water in the food are in thermodynamic equilibrium with each other in both the adsorbed and vapor phases. Considering only the water component in the two phases, their chemical potentials can be equated, given by Equation 11.9:

$$\mu_w(\text{vapor}) = \mu_w(\text{food}) \tag{11.25}$$

From Equations 11.15 and 11.22, it then follows that

$$\mu_w^0 + RT \ln f_w = \mu_w^\bullet + RT \ln a_w \tag{11.26}$$

The chemical potential of pure water in liquid phase is obtained by substituting $a_w = 1$ as provided for by Equation 11.23. Denoting the fugacity of the water vapor in equilibrium with pure water by f^Σ_w, the standard-state chemical potential of pure water becomes

$$\mu_w^\bullet = \mu_w^0 + RT \ln f_w^\bullet \tag{11.27}$$

Substituting the above for μ^Σ_w in Equation 11.26 yields

$$RT \ln a_w = RT \ln f_w - RT \ln f_w^\bullet$$

which on solving for a_w gives

$$a_w = \left(\frac{f_w}{f_w^\bullet}\right)_T \tag{11.28}$$

Thus, the activity of water or any other component in foods or in any real gas–liquid–solid system is given by the ratio of the fugacity of the respective component in the mixture to its fugacity at the reference state, both taken at the same temperature. Because the fugacity of water vapor in equilibrium with pure water, defined as the *standard-state*

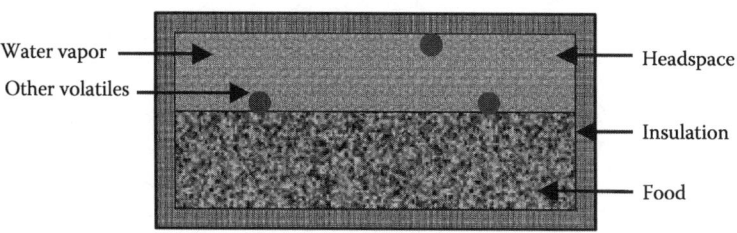

Figure 11.1 Schematic representation of a closed food–water-vapor system at constant temperature.

fugacity, equals the vapor pressure exerted by pure water, water activity becomes

$$a_w = \left(\gamma_f \frac{P_w}{P_w^\bullet}\right)_T \tag{11.29}$$

The fugacity coefficient of water vapor (γ_f) as a function of pressure for several temperatures is listed in Table 11.1. It is observed that in the temperature and pressure ranges considered, the fugacity coefficient approximates unity, indicating negligible deviation from ideality. Thus, the activity of water in foods is closely approximated by

$$a_w = \left(\frac{P_w}{P_w^\bullet}\right)_T \tag{11.30}$$

At high pressures, however, deviations from ideality occur, as is shown in Table 11.2 for several temperatures and pressures. In high-pressure and high-temperature food processing operations such as extrusion cooking, departures from ideality need to be taken into account.

According to Equation 11.24, water activity is also given by

$$a_w = \gamma_a x_w \tag{11.31}$$

The activity coefficient (γ_a) of water is a function of the temperature and composition of the mixture in liquid phase. Several equations have been proposed for binary mixtures of nonelectrolytes. The Van Laar and Margules equations are the oldest ones, and the better newer models include the Wilson and UNIQUAC equations. It is indeed difficult to compute the water activity coefficient in heterogeneous mixtures such as foods. For very dilute solutions, which follow Raoult's law, the activity coefficient approaches unity. Only in such cases is activity approximate by the mole fraction.

For prediction of water activity of multicomponent solutions, the following equation has been proposed [10]:

$$\log a_w = \log x_w - \left(k_2^{1/2} x_2 + k_3^{1/2} x_3 + k_4^{1/2} x_4 + \ldots\right)^2 \tag{11.32}$$

where k_2, k_3, and so on, are the constants for the binary mixtures. Also, for high-moisture mixtures ($a_w > 0.75$) the following equation has been suggested for use with both liquids and solids or mixtures [11].

$$a_w = a_{w1} a_{w2} a_{w3} \ldots \tag{11.33}$$

Based on the simplified Gibbs–Duhem equation, the above equation is essentially a solution equation and has been shown to be inaccurate for systems containing substantial amounts of solids [12]. The following modification, based on the molality of the components and the mixture, has been reported to predict a_w more accurately [13]:

$$a_w = \left[a_{w1}(m)\right]^{m1/m} \left[a_{w2}(m)\right]^{m2/m} \left[a_{w3}(m)\right]^{m3/m} \ldots \tag{11.34}$$

where $a_{w1}, a_{w2}, \ldots$ and $m1, m2, \ldots$ are the component water activities and molalities, respectively; m is the molality of the solution. The equation can also be applied to dilute

Table 11.1 Fugacity and Fugacity Coefficients of Water Vapor in Equilibrium with Liquid at Saturation and at 1 Atm Pressure

Temperature (°C)	Pressure (bars)[a] (±0.1%)	At Saturation Fugacity (bars) (±0.1%)	Fugacity Coefficient (±0.1%)	Fugacity (bars) at 1 atm (1.01325 bars) (±0.1%)
0.01[b]	0.00611	0.00611	0.9995	
10.00	0.01227	0.01226	0.9992	
20.00	0.02337	0.02334	0.9988	
30.00	0.04242	0.04235	0.9982	
40.00	0.07376	0.07357	0.9974	
50.00	0.12336	0.12291	0.9954	
60.00	0.19920	0.19821	0.9950	
70.00	0.31163	0.30955	0.9933	
80.00	0.47362	0.46945	0.9912	
90.00	0.70114	0.69315	0.9886	
100.00	1.0132	0.99856	0.9855	0.9986
110.00	1.4326	1.4065	0.9818	1.0004
120.00	1.9853	1.9407	0.9775	1.0019
130.00	2.7011	2.6271	0.9726	1.0031
140.00	3.6135	3.4943	0.9670	1.0042
150.00	4.7596	4.5726	0.9607	1.0051
160.00	6.1804	5.8940	0.9537	1.0059
170.00	7.9202	7.4917	0.9459	1.0066
180.00	10.027	9.3993	0.9374	1.0072
190.00	12.552	11.650	0.9282	1.0077
200.00	15.550	14.278	0.9182	1.0081
210.00	19.079	17.316	0.9076	1.0085
220.00	23.201	20.793	0.8962	1.0089
230.00	27.978	24.793	0.8842	1.0092
240.00	33.480	29.181	0.8716	1.0095
250.00	33.775	34.141	0.8584	1.0098
260.00	46.940	39.063	0.8445	1.0100
270.00	55.051	45.702	0.8302	1.0103
280.00	64.191	52.335	0.8153	1.0105
290.00	74.448	59.554	0.7999	1.0106
300.00	85.916	67.367	0.7841	1.0108

Source: Hass, J. L. 1970. Fugacity of H_2O from 0° to 350° at liquid–vapor equilibrium and at 1 atmosphere. *Geochim. Cosmochim. Acta* 34: 929–934.

[a] 1 bar = 0.9869 atm.
[b] Ice–liquid–vapor triple point.

Table 11.2 Fugacity Coefficients of Water Vapor at High Pressures and Temperatures

Pressure (bars)[a]	Fugacity Coefficient								
	200°C	300°C	400°C	500°C	600°C	700°C	800°C	900°C	1000°C
1	0.995	0.998	0.999	0.999	1.000	1.000	1.000	1.000	1.000
100	.134[b]	.701[b]	.874	.923	.952	.969	.980	.987	.993
200	.0708	.372[b]	.737	.857	.909	.938	.960	.975	.985
300	.0497	.259	.602	.764	.867	.913	.941	.962	.979
400	.0392	.204	.483	.735	.830	.886	.923	.950	.963
500	.0331	.172	.412	.655	.789	.861	.906	.943	.956
600	.0290	.150	.361	.599	.751	.838	.890	.931	.951
700	.0261	.135	.325	.555	.747	.818	.881	.927	.948
800	.0253	.124	.299	.517	.693	.800	.868	.919	.947
900	.0225	.115	.281	.472	.666	.783	.861	.916	.944
1000	.0213	.109	.264	.455	.642	.765	.851	.910	.944
1100	.0203	.103	.251	.436	.622	.750	.842	.907	.944
1200	.0196	.0991	.241	.420	.604	.737	.834	.904	.945
1300	.0190	.0956	.233	.406	.589	.725	.827	.901	.946
1400	.0185	.0928	.226	.395	.577	.715	.821	.899	.948
1500	.0181	.0906	.220	.385	.566	.706	.816	.898	.950
1600	.0178	.0887	.216	.377	.556	.698	.811	.897	.952
1700	.0176	.0872	.211	.370	.548	.691	.807	.897	.955
1800	.0174	.0860	.208	.365	.542	.685	.803	.896	.957
1900	.0173	.0850	.205	.360	.535	.680	.800	.896	.959
2000	.0172	.0843	.203	.356	.530	.676	.798	.896	.962

Source: Holser, W. T. 1954. Fugacity of water at high temperatures and pressure. *J. Phys. Chem.* 58: 316–317.

[a] 1 bar = 0.9864 atm.
[b] Two-phase region; do not interpolate.

electrolyte solutions by replacing molality with ionic strength. Recently, Lilley and Sutton [248] described an improvement on the Ross method that allows the prediction of a_w of multicomponent systems from the properties of solutions containing one and two solutes. For ternary solute systems their equation becomes

$$a_w(1,2,3) = \frac{a_w(1,2) \cdot a_w(1,3) \cdot a_w(2,3)}{a_w(1) \cdot a_w(2) \cdot a_w(3)} \qquad (11.35)$$

The utility of this approach was established by the authors by showing significantly better predictions at higher solute concentrations than those given by the Ross method.

The control of water activity for the preservation of food safety and quality is a method of widespread importance in the food industry. Although the water activity criterion for biological viability is of questionable value [14], limiting water activities for the growth of microorganisms have been reported in the literature (Table 11.3). The lower limits of water activity for microbial growth also depend on factors other than water. Such environmental factors as temperature, pH, oxidation–reduction potential, nutrient availability, presence of growth inhibitors in the environment, and the type of solute used to lower water activity influence the minimal water activity for growth [15]. The viability of microorganisms during storage in dry foods ($a_w < 0.6$) is reported to be greatly affected by temperature [16] and by the pH of the food (Christian and Stewart, 1973). It is also worth noting that inhibitory effects of many organic acids are much greater than would be predicted based on the pH values only.

In addition to possible microbial spoilage, dehydrated food represents a concentrated biochemical system prone to deterioration by several mechanisms. The limiting factors for biochemically useful storage life of most dehydrated foods are oxygen and moisture, acting either independently or in concert. Nonenzymatic browning in products such as dry milk and some vegetables is not a problem in the absence of water. Initially satisfactory moisture levels are no guarantee against browning. The increase in moisture content increases the rate of browning, and thus water plays an indirect role in such instances. Oxygen-sensitive foods develop rancidity during storage, whereas enzymatic oxidation of polyphenols and of other susceptible compounds causes enzymatic browning if the enzymes are not inactivated. The effect of water activity on the rate of oxygen uptake of dehydrated foods indicates that the rate of oxidation is high at very low water activity and reaches a minimum in the range of $a_w = 0.2$–0.4. At water activities higher than 0.4, the reacting chemical species become soluble and mobile in the solvent water, and the oxidation rate increases. This makes control of moisture content very critical during dehydration and subsequent packaging and storage of dehydrated foods. Ideally, this problem requires general formulas for temperature and moisture gain as functions of time and location within the food material and a set of thermophysical properties to describe the food. One of the major obstacles to this approach has been the lack of numerical values for such thermophysical properties as sorption enthalpy, thermal conductivity, moisture diffusivity, and heat capacity.

11.2.4 Measurement of Water Activity

The measurement of water activity in foods has been the subject of many studies, and a variety of methods have been used and reported in the literature [17–24]. The choice of one technique over another depends on the range, accuracy, precision, and speed of measurement required. The two major collaborative studies on the accuracy and precision of various water activity measuring devices showed considerable variations among different techniques. The accuracy of most methods lies in the range of 0.01–0.02 a_w units. On the basis of the underlying principles, the methods for water activity measurement have been classified into four major categories:

11.2.4.1 Measurements Based on Colligative Properties
11.2.4.1.1 Vapor Pressure Measurement
Assuming water vapor fugacity to be approximately equal to its pressure, direct measurement of pressure has been extensively used to measure the water activity of foods. The

THERMODYNAMIC PROPERTIES OF FOODS IN DEHYDRATION

Table 11.3 Water Activity and Growth of Microorganisms in Food

Range of a_W	Microorganisms Generally Inhibited by Lowest a_W in This Range	Foods Generally within This Range
0.20	No microbial proliferation	Whole-milk powder containing 2–3% moisture; dried vegetables containing approx. 5% moisture; corn flakes containing approx. 5% moisture; fruit cake; country-style cookies, crackers
0.30	No microbial proliferation	Cookies, crackers, bread crusts, etc., containing 3–5% moisture
0.40	No microbial proliferation	Whole-egg powder containing approx. 5% moisture
0.50	No microbial proliferation	Pasta containing approx. 12% moisture; spices containing approx. 10% moisture
0.60–0.65	Osmophilic yeasts (*Saccharomyces rouxii*), few molds (*Aspergillus echinulatus, Monoascus bisporus*)	Dried fruits containing 15–20% moisture; some toffees and caramels; honey
0.65–0.75	Xerophilic molds (*Aspergillus chevalieri, A. candidus, Wallemia sebi*), *Saccharomyces bisporus*	Rolled oats containing approx. 10% moisture, grained nougats, fudge, marshmallows, jelly, molasses, raw cane sugar, some dried fruits, nuts
0.75–0.80	Most halophilic bacteria, mycotoxigenic aspergilli	Jam, marmalade, marzipan, glacé fruits, some marshmallows
0.80–0.87	Most molds (mycotoxigenic penicillia), *Staphyloccus aureus*, most *Saccharomyces* (*bailii*) spp., *Debaryomyces*	Most fruit juice concentrates, sweetened condensed milk, chocolate syrup, maple and fruit syrups; flour, rice, pulses containing 15–17% moisture; fruit cake; country-style ham, fondants, high-ratio cakes
0.87–0.91	Many yeasts (*Candida, Torulopsis, Hansenula*), *Micrococcus*	Fermented sausage (salami), sponge cakes, dry cheeses, margarine; foods containing 65% (w/v) sucrose (saturated) or 15% sodium chloride
0.91–0.95	*Salmonella, Vibrio parahaemolyticus, Clostridium botulinum, Serratia, Lactobacillus, Pediococcus*, some molds, yeasts, (*Rhodotorula, Pichia*)	Some cheeses (cheddar, swiss, muenster, provolone), cured meat (ham), some fruit juice concentrates; foods containing 55% (w/w) sucrose or 12% sodium chloride
0.95–1.00	*Pseudomonas, Escherichia, Proteus, Shigella, Klebsiella, Bacillus, Clostridium perfringens*, some yeasts	High perishable (fresh) foods and canned fruits, vegetables, meat, fish, and milk; cooked sausages and breads; foods containing up to approx. 40% (w/w) sucrose or 7% sodium chloride

Source: Adapted from Beuchat, L. R. 1981. Microbial stability as affected by water activity. *Cereal Foods World* 26: 345–349.

measurement using a vapor pressure manometer (VPM) was first suggested by Makower and Meyers [25]. Taylor [26], Sood and Heldman [27], Labuza et al. [28], and Lewicki et al. [29] studied design features and provided details for a precision and accuracy of the VPM. A schematic diagram of the system is shown in Figure 11.2. The method for measurement of water vapor pressure of food consists of placing 10–50 g of sample in the sample flask

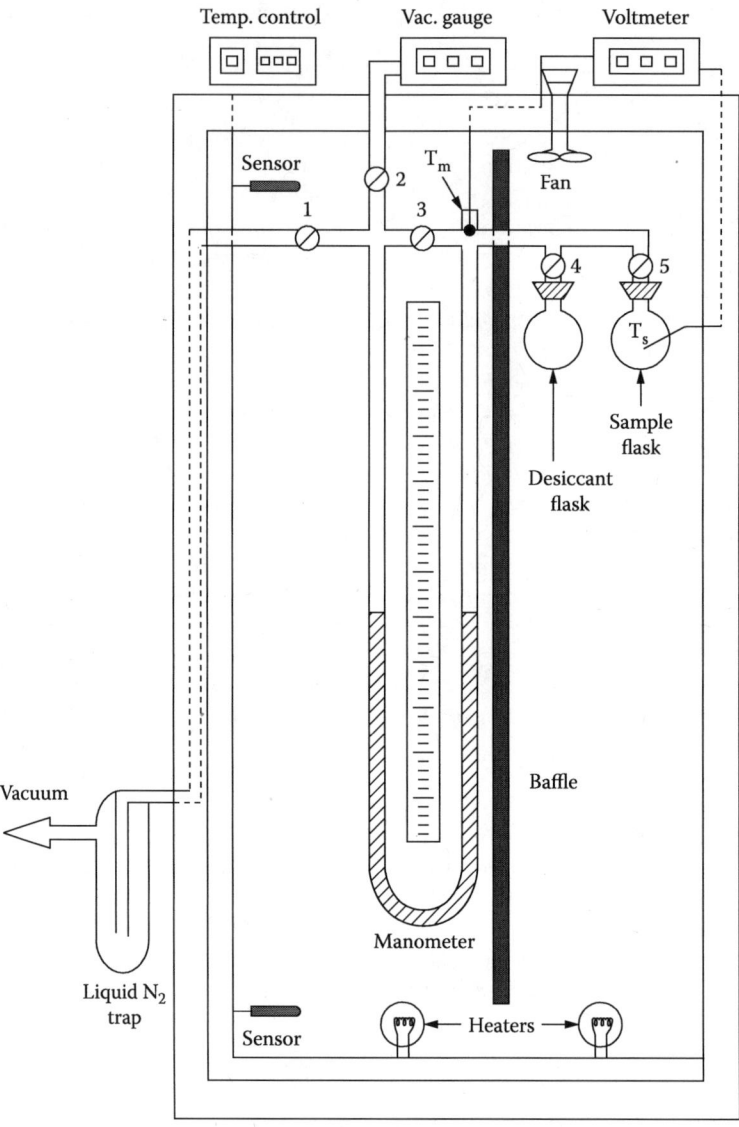

Figure 11.2 Schematic diagram of a thermostated vapor pressure manometer apparatus.

while the desiccant flask is filled with a desiccant material, usually $CaSO_4$. Keeping the sample flask isolated, the system is evacuated to less than 200 μmHg; this is followed by evacuation of the sample for 1–2 min. The vacuum source is then isolated by closing the valve between the manometer legs. Upon equilibration for 30–50 min the pressure exerted by the sample is indicated by the difference in oil manometer height (ΔL_1). The sample flask is subsequently excluded from the system, and the desiccant flask is opened. Water vapor is removed by sorption onto the desiccant, and the pressures exerted by volatiles and gases are indicated by the difference in manometer legs (ΔL_2) when they attain constant levels. For precise results it is necessary that the whole system be maintained at constant temperature, that the ratio of sample volume to vapor space be large enough to minimize changes in a_w due to loss of water by vaporization, and that a low-density, low-vapor-pressure oil be used as the manometric fluid. Apiezon B manometric oil (density 0.866 g/cm³) is generally used as the manometric fluid, and temperatures of the sample (T_s) and the vapor space in the manometer (T_m) are recorded. In the absence of any temperature gradient within the system ($T_m = T_s$), water activity of the sample, after thermal and pressure equilibration are attained at each measurement, is calculated from formula [29]

$$a_w = \frac{\Delta L_1 - \Delta L_2}{P_w^\bullet} \rho g \tag{11.36}$$

If T_s and T_m are different, water activity is corrected as

$$a_w = \left(\frac{\Delta L_1 - \Delta L_2}{P_w^\bullet}\right)\left(\frac{T_s}{T_m}\right)\rho g \tag{11.37}$$

Troller [22] modified the VPM apparatus by replacing the oil manometer with a capacitance manometer. This modification made the manometric device more compact and thus made the temperature control less problematic by eliminating temperature gradients that might otherwise exist in a larger setup. Nunes et al. [23] showed that both accuracy and precision of the VPM could be significantly improved by taking into account the change in volume that occurs when water vapor is eliminated from the air–water vapor mixture. This is achieved either by internal calibration, with P_2O_5 replacing the sample, or by plotting readings at different ΔL_2 and extrapolating the water activity values to $\Delta L_2 = 0$, the intersect. The authors also suggested a design criterion for building a VPM apparatus for a given error in a_w measurements that eliminates the need for correction. If the acceptable error level between corrected and uncorrected a_w values is Δa_w such that $\Delta a_w = a_{wu} - a_w$, then the criterion for improving the measurements becomes

$$\frac{\Delta a_w P_w^\bullet}{\Delta L_2 \rho g} \geq \frac{V_d}{V_1} \tag{11.38}$$

where V_1 and V_d are void volumes corresponding to the sample flask and the desiccant flask, respectively. Although used as the standard method, the VPM is not suitable for materials either containing large amounts of volatiles or undergoing respiration processes.

11.2.4.1.2 Freezing Point Depression Measurements

For $a_w > 0.85$, freezing point depression techniques have been reported to provide accurate results [12,30–33]. For real solutions, the relationship between the freezing point of an aqueous solution and its a_w is [34]

$$-\ln a_w = \frac{1}{R}\left[\left(\Delta \bar{H}_{\text{fus}} - JT_0\right)\left(\frac{1}{T_f} - \frac{1}{T_0}\right)\right] + \frac{J}{R}\ln\frac{T_0}{T_f} \quad (11.39)$$

The above expression can be numerically approximated to [12]

$$-\ln a_w = 9.6934 \times 10^{-3}\left(T_0 - T_f\right) + 4.761 \times 10^{-6}\left(T_0 - T_f\right)^2 \quad (11.40)$$

This method is applicable only to liquid foods and provides the a_w at freezing temperature instead of at room temperature, although the error is reported to be relatively small (0.01 at 25°C). This method, however, has the advantage of providing accurate water activity in the high range (>0.98) and can be effectively applied to systems containing large quantities of volatile substances. Other colligative properties such as osmotic pressure and boiling point elevation have not been used for food systems.

11.2.4.2 Measurements Based on Psychrometry

Measurements of dew point [35] and wet-bulb depression [36] of thermocouple psychrometers along with hair and electric hygrometers [37–39] have been used in the measurements of a_w. In dew point temperature measuring instruments, an airstream in equilibrium with the sample to be measured is allowed to condense at the surface of a cooled mirror. From the measured dew point, equilibrium relative humidity (ERH) of the sample is computed using psychrometric parameters. Modern instruments, based on Peltier-effect cooling of mirrors and the photoelectric determination of condensation on the reflecting surface via a null-point type of circuit, give very precise values of dew point temperatures. When interfaced with microprocessor-based systems for psychrometric analysis, direct water activity or ERH values may be obtained [21]. Dew point measuring devices are reported to have an accuracy of $0.03 a_w$ in the range of 0.75–0.99 [40]. At lower water activity levels, there is not sufficient vapor in the headspace to cover the reflecting surface, and the accuracy of these instruments is diminished.

The wet-bulb temperature indicates the temperature at which equilibrium exists between an air–vapor mixture and water and is dependent on the amount of moisture present in the air–vapor mixture. From a knowledge of the wet and dry-bulb temperatures of air in equilibrium with food, the relative humidity can be determined. This method is not very conducive to a_w measurement of small samples and has been used primarily to determine the relative humidity of large storage atmospheres and commercial dehydrators [19,41]. Small psychrometers, especially designed for use in foods, are commercially available [40,42,43]. Major limitations of psychrometric techniques are condensation of volatile materials, heat transfer by conduction and radiation, and the minimum wind velocity requirement of at least 3 m/s.

Several hygrometers are commercially available for indirect determination of water activity. They are based either on the change in length of a hair (hair hygrometer) or on the conductance of hygrosensors coated with a hygroscopic salt such as LiCl or sulfonated polystyrene as a function of water activity. The hair hygrometers, although inexpensive, are not very sensitive. They are best suited for range finding and rough estimates of higher a_w (0.7–0.95) levels [39]. However, electric hygrometers are reported to be precise, with a coefficient of variation ranging from 0 to 0.53% and a standard deviation ranging from 0 to 0.004 [38]. These instruments are portable, are convenient to use, and need a relatively short equilibration time. Their major drawbacks include sensor fatigue and sensor poisoning by volatiles such as glycol, ammonia, acetone, and other organic substances. Troller [21] stated that some types of sensor fatigue and contamination can be reversed by holding the device at a constant temperature and in vacuum under some circumstances until the error corrects itself.

11.2.4.3 Measurements Based on Isopiestic Transfer

The isopiestic method relies on the equilibration of the water activities in two materials in a closed system. Transfer of moisture may take place either through direct contact of the materials, thus allowing for movement of bulk, microcapillary, and gaseous water [44,45], or by maintaining the two materials separately, thus permitting transfer to occur only through the vapor phase. Analysis of the concentration of water in some reference material such as microcrystalline cellulose or a protein at equilibrium permits determination of water activity from the calibration curve [37,46–50]. Because only discrete a_w values of the reference material are used, sensitivity of the method is highly dependent on the accuracy of the standard calibration curve. A new standard curve must be established each time a different lot of reference material is used. This technique is not accurate at a_w levels of less than 0.50 or over 0.90 [21].

11.2.4.4 Measurements Based on Suction (Matric) Potential

The water potential of soil [26], the capillary suctional potential of gel [51], and the matric potentials of food gels have been determined using the principle of a tensiometer. Accurate for high a_w range, the technique is useful for materials that bind large quantities of water.

11.2.5 Adjustment of Water Activity

It is often necessary to adjust the a_w of food samples to a range of values in order to obtain the sorption data. The two principal techniques used for the adjustment of a_w are the integral and differential methods [52]. The integral method involves placing several samples each under a controlled environment simultaneously and measuring the moisture content on the attainment of constant weight. The differential method employs a single sample, which is placed under successively increasing or decreasing relative humidity environments; moisture content is measured after each equilibration. The differential method has the advantage of using only a single sample; hence, all sample parameters are kept constant, and the only effect becomes that of the environment. Because equilibration can take on the order of several days to occur, the sample may undergo various degenerative changes. This is particularly true for samples that rapidly undergo changes in their vulnerable quality factors. The integral method avoids this problem because each sample is discarded after the appropriate measurement is made; thus, the time for major deteriorative changes to occur is limited. One does not, however, have the convenience of dealing with only one sample for which environmental conditions affect the thermodynamic results.

The common sources used for generating environments of defined conditions for adjustment of a_w of foods as well as for calibration of a_w-measuring devices consist of solutions of saturated salts, sulfuric acid, and glycerol. Although saturated salt solutions are most popular, they are limited in that they provide only discrete a_w values at any given temperature.

Although the ERH values of various saturated salt solutions at different temperatures have been tabulated and reviewed in the literature [28,53–57], the reported values do not always agree. The a_w of most salt solutions decreases with an increase in temperature because of the increased solubility of salts and their negative heats of solution. The values of selected binary saturated aqueous solutions at several temperatures compiled by Greenspan [56] are generally used as standards. These were obtained by fitting a polynomial equation to literature data reported between 1912 and 1968, and the values for some of the selected salt solutions at different temperatures are given in Table 11.4. Recognizing the uncertainties and instrumental errors involved in obtaining the reported values, Labuza et al. [58] measured the a_w values of eight saturated salt solutions by the VPM method and found them to be significantly different from the Greenspan data. The temperature effect on the a_w of each of the eight solutions studied was obtained by regression analysis, using the least-squares method on ln a_w versus $1/T$, with r^2 ranging from 0.96 to 0.99 (Table 11.5). In view of the previous uncertainties about the exact effect of temperature on a_w salt solutions, the use of these regression equations should provide more uniform values.

Sulfuric acid solutions of varying concentrations in water are also used to obtain different controlled humidity environments. Changes in the concentration of the solution and the corrosive nature of the acid require caution in their use. Standardization of the

Table 11.4 Water Activities of Selected Salt Slurries at Various Temperatures

Salt	Water Activity (a_w)						
	5°C	10°C	20°C	25°C	30°C	40°C	50°C
Lithium chloride	0.113	0.113	0.113	0.113	0.113	0.1120	0.111
Potassium acetate	—	0.234	0.231	0.225	0.216	—	—
Magnesium chloride	0.336	0.335	0.331	0.328	0.324	0.316	0.305
Potassium carbonate	0.431	0.431	0.432	0.432	0.432	—	—
Magnesium nitrate	0.589	0.574	0.544	0.529	0.514	0.484	0.454
Potassium iodide	0.733	0.721	0.699	0.689	0.679	0.661	0.645
Sodium chloride	0.757	0.757	0.755	0.753	0.751	0.747	0.744
Ammonium sulfate	0.824	0.821	0.813	0.810	0.806	0.799	0.792
Potassium chloride	0.877	0.868	0.851	0.843	0.836	0.823	0.812
Potassium nitrate	0.963	0.960	0.946	0.936	0.923	0.891	0.848
Potassium sulfate	0.985	0.982	0.976	0.973	0.970	0.964	0.958

Source: Adapted from Greenspan, L. 1977. *J. Res. Natl. Bur Std. AL Phys. Chem.* 81A(1): 89–96.

Table 11.5 Regression Equations for Water Activity of Selected Salt Solutions at Selected Temperatures[a]

Salt	Regression Equation	r^2
LiCl	$\ln a_w = (500.95 \times 1/T) - 3.85$	0.976
$KC_2H_3O_2$	$\ln a_w = (861.39 \times 1/T) - 4.33$	0.965
$MgCl_2$	$\ln a_w = (303.35 \times 1/T) - 2.13$	0.995
K_2CO_3	$\ln a_w = (145.0 \times 1/T) - 1.3$	0.967
$MgNO_3$	$\ln a_w = (356.6 \times 1/T) - 1.82$	0.987
$NaNO_2$	$\ln a_w = (435.96 \times 1/T) - 1.88$	0.974
NaCl	$\ln a_w = (228.92 \times 1/T) - 1.04$	0.961
KCl	$\ln a_w = (367.58 \times 1/T) - 1.39$	0.967

Source: From Labuza, T. P., Kaanane, A., and Chen, J. Y. 1985. *J. Food Sci.* 50: 385–391.

[a] Temperature (T) in kelvins.

solution is needed for accurate ERH values. Ruegg [59] calculated the water activity values of sulfuric acid solutions of different concentrations as a function of temperature, and his values are presented in Table 11.6. Tabulated values for glycerol solutions in water are also frequently employed for creating a defined humidity condition (Table 11.7). As with sulfuric acid solutions, glycerol solutions must also be analyzed at equilibrium to ensure correct concentration in solution and therefore accurate ERH in the environment. For calibration of hygrometers, Chirife and Resnik [60] proposed the use of unsaturated sodium chloride solutions as isopiestic standards. Their recommendation was based on the excellent agreement between various literature compilations and theoretical models for the exact a_w values in NaCl solutions. It has also been shown that the water activity of NaCl solutions is invariant with temperature in the range of 15–50°C, which makes their use attractive. Table 11.8 shows theoretically computed a_w values at 0.5% weight intervals. Other means of obtaining controlled humidity conditions include the use of mechanical humidifiers where high accuracy is not critical and the use of desiccants when a very low RH environment is needed.

In adjusting the water activity of food materials, test samples are allowed to equilibrate to the preselected RH environment in the headspace maintained at a constant temperature. Theoretically, at equilibrium the a_w of the sample is the same as that of the surrounding environment. In practice, however, a true equilibrium is never attained because that would require an infinitely long period of time, and the equilibration process is terminated when the difference between successive weights of the sample becomes less than the sensitivity of the balance being used; this is called the *gravimetric method* [20]. Several investigators have suggested techniques for accelerating this approach to equilibrium. Zsigmondy [62] and many investigators since have reported increased rates of moisture exchange under reduced total pressures. Rockland [63] and Bosin and Easthouse [64] suggested the use of headspace agitation. Resultant equilibration times were 3–15 times shorter than with a stagnant headspace. Lang et al. [65] increased the rate of moisture exchange by minimizing

Table 11.6 Water Activity of Sulfuric Acid Solutions at Selected Concentrations and Temperatures

Percent H_2SO_4	Density at 25°C (g/cm³)	Water Activity a_w							
		5°C	10°C	20°C	23°C	25°C	30°C	40°C	50°C
5.00	1.0300	0.9803	0.9804	0.9806	0.9807	0.9807	0.9808	0.9811	0.9
10.00	1.0640	0.9554	0.9555	0.9558	0.9559	0.9560	0.9562	0.9565	0.9
15.00	1.0994	0.9227	0.9230	0.9237	0.9239	0.9241	0.9245	0.9253	0.9
20.00	1.1365	0.8771	0.8779	0.8796	0.8802	0.8805	0.8814	0.8831	0.8
25.00	1.1750	0.8165	0.8183	0.8218	0.8229	0.8235	0.8252	0.8285	0.8
30.00	1.2150	0.7396	0.7429	0.7491	0.7509	0.7521	0.7549	0.7604	0.7
35.00	1.2563	0.6464	0.6514	0.6607	0.6633	0.6651	0.6693	0.6773	0.6
40.00	1.2991	0.5417	0.5480	0.5599	0.5633	0.5656	0.5711	0.5816	0.5
45.00	0.3437	0.4319	0.4389	0.4524	0.4564	0.4589	0.4653	0.4775	0.4
50.00	1.3911	0.3238	0.3307	0.3442	0.3482	0.3509	0.3574	0.3702	0.3
55.00	1.4412	0.2255	0.2317	0.2440	0.2477	0.2502	0.2563	0.2685	0.2
60.00	1.4940	0.1420	0.1471	0.1573	0.1604	0.1625	0.1677	0.1781	0.1
65.00	1.5490	0.0785	0.0821	0.0895	0.0918	0.0933	0.0972	0.1052	0.1
70.00	1.6059	0.0355	0.0377	0.0422	0.0436	0.0445	0.0470	0.0521	0.0
75.00	1.6644	0.0131	0.0142	0.0165	0.0172	0.0177	0.0190	0.0218	0.0
80.00	1.7221	0.0035	0.0039	0.0048	0.0051	0.0053	0.0059	0.0071	0.0

Source: From Ruegg, M. 1980. *Lebensm.-Wiss. Technol.* 13: 22–24.

Table 11.7 Water Activity of Glycerol Solutions at 20°C

Concentration (kg/L)	Refractive Index	Water Activity
—	1.3463	0.98
—	0.3560	0.96
0.2315	1.3602	0.95
0.3789	1.3773	0.90
0.4973	1.3905	0.85
0.5923	1.4015	0.80
0.6751	1.4109	0.75
0.7474	1.4191	0.70
0.8139	1.4264	0.65
0.8739	1.4329	0.60
0.9285	1.4387	0.55
0.9760	1.4440	0.50
—	1.4529	0.40

Source: From Grover, D. W., and Nicol, J. M. 1940. *J. Soc. Chem. Ind. (Lond.)* 59: 175–177.

Table 11.8 Water Activity of NaCl Solutions in the Range of 15–50°C

Conc. (% w/w)	a_w	Conc. (% w/w)	a_w	Conc. (% w/w)	a_w	Conc. (% w/w)	a_w
0.5	0.997	7.0	0.957	13.5	0.906	20.0	0.839
1.0	0.994	7.5	0.954	14.0	0.902	20.5	0.833
1.5	0.991	8.0	0.950	14.5	0.897	21.0	0.827
2.0	0.989	8.5	0.946	15.0	0.892	21.5	0.821
2.5	0.986	9.0	0.943	15.5	0.888	21.5	0.821
3.0	0.983	9.5	0.939	16.0	0.883	22.5	0.808
3.5	0.980	10.0	0.935	16.5	0.878	23.0	0.802
4.0	0.977	10.5	0.931	17.0	0.873	23.5	0.795
4.5	0.973	11.0	0.927	17.5	0.867	24.0	0.788
5.0	0.970	11.5	0.923	18.0	0.862	24.5	0.781
5.5	0.967	12.0	0.919	18.5	0.857	25.0	0.774
6.0	0.964	12.5	0.915	19.0	0.851	25.5	0.766
6.5	0.960	13.0	0.911	19.5	0.845	26.0	0.759

Source: From Chirife, J., and Resnik, S. L. 1984. *J. Food Sci.* 49: 1486–1488.

the physical distance between the sample and the equilibrant salt solution by reducing the ratio between the headspace volume and solution surface area. They reported a greater than threefold decrease in equilibration times compared to the conventional desiccator technique.

The common problem with the above methods is the long waiting time needed for attainment of apparent equilibrium, which may alter the physicochemical and microbiological nature of the food under study. In recognition of these problems, Rizvi et al. [66] developed an accelerated method that does not depend on equilibration with a preselected water vapor pressure. Instead, pure water or a desiccant is used to maintain saturated or zero water vapor pressure conditions, thereby increasing the driving force for moisture transfer throughout the sorption processes. Sample exposure time to these conditions is the mechanism by which the amount of moisture uptake is controlled. Inverse gas chromatography has also been used effectively for the determination of water sorption properties of starches, sugars, and dry bakery products [67,68].

11.2.6 Moisture Sorption Isotherms

The relationship between total moisture content and the corresponding a_w of a food over a range of values at a constant temperature yields a moisture sorption isotherm (MSI) when graphically expressed. On the basis of the van der Waals adsorption of gases on various solid substrates reported in the literature, Brunauer et al. [69] classified adsorption isotherms into five general types (Figure 11.3). Type I is the Langmuir, and type II is the sigmoid or S-shaped adsorption isotherm. No special names have been attached to the three other types. Types II and III are closely related to types IV and V, except that the maximum

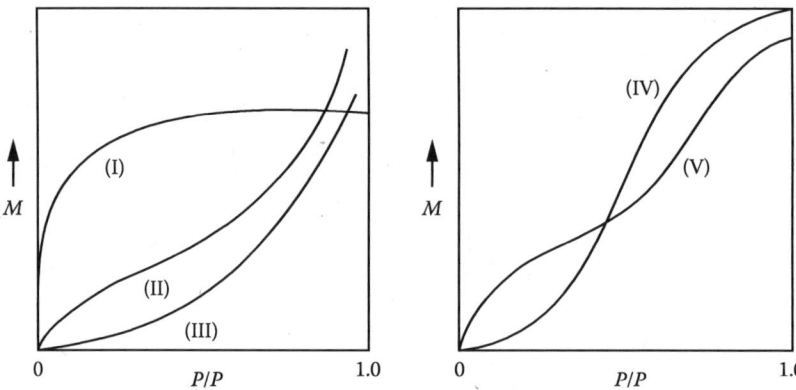

Figure 11.3 The five types of van der Waals adsorption isotherms. (From Brunauer, S., Deming, L. S., Deming, W. E., and Teller, E. 1940. *Am. Chem. Soc. J.* 62: 1723–1732.)

adsorption occurs at some pressure lower than the vapor pressure of the gas. MSIs of most foods are nonlinear, generally sigmoidal in shape, and have been classified as type II isotherms. Foods rich in soluble components, such as sugars, have been found to show type III behavior. Another behavior commonly observed is that different paths are followed during adsorption and desorption processes, resulting in a hysteresis. The desorption isotherm lies above the adsorption isotherm, and therefore more moisture is retained in the desorption process compared to adsorption at a given ERH.

11.2.6.1 Theoretical Description of MSIs

The search for a small number of mathematical equations of state to completely describe sorption phenomena has met with only limited success despite a relatively strong effort. For the purposes of modeling the sorption of water on capillary porous materials, about 77 different equations with varying degrees of fundamental validity are available [70]. Labuza [71,72] noted that the fact that no sorption isotherm model seems to fit the data over the entire a_w range is hardly surprising, because water is associated with a food matrix by different mechanisms in different activity regions. As the a_w of a food approaches multilayer regions, a mechanical equilibrium is set up. In porous adsorbents, adsorptive filling begins to occur, and Kelvin's law effects become significant [73]. Labuza [71,72] further noted that the use of the Kelvin equation alone is inappropriate, as capillary filling begins only after the surface has adsorbed a considerable quantity of water, though there is a significant capillary effect in a_w lowering, especially in the smaller radii. Manifestations of hysteresis effects make it difficult to speak of the universal sorption isotherm of a particular product. Some of the MSIs are therefore understandably and necessarily described by semiempirical equations with two or three fitting parameters. The goodness of fit of a sorption model to experimental data shows only a mathematical quality and not the nature of the sorption process. Many of these seemingly different models turn out to be the same after some rearrangement [74]. In their evaluation of eight two-parameter equations in describing MSIs for 39 food materials, Boquet et al. [75] found the Halsey [76] and Oswin [77] equations to be the most versatile. Of the larger number of models available in the literature [78], a few that are commonly used in describing water sorption on foods are discussed below.

11.2.6.1.1 The Langmuir Equation
On the basis of unimolecular layers with identical, independent sorption sites, Langmuir [79] proposed the following physical adsorption model:

$$a_w\left(\frac{1}{M} - \frac{1}{M_a}\right) = \frac{1}{CM_o} \tag{11.41}$$

where M_o is the monolayer sorbate content, and C is a constant. The relation described by the above equation gives the type I isotherm described by Brunauer et al. [69].

11.2.6.1.2 The Brunauer–Emmett–Teller Equation
The BET isotherm Equation 11.82 is the most widely used model and gives a good fit to data for a range of physicochemical systems over the region $0.05 < a_w < 0.35 - 0.5$ [81]. It provides an estimate of the monolayer value of moisture adsorbed on the surface and has been approved by the commission on Colloid and Surface Chemistry of the IUPAC (International Union of Pure and Applied Chemistry) [82] for standard evaluation of monolayer and specific areas of sorbates. The BET equation is generally expressed in the form

$$\frac{a_w}{(1 - a_w)M} = \frac{1}{M_oC} + \frac{C-1}{M_oC}a_w \tag{11.42}$$

The above equation is often abbreviated as

$$\frac{a_w}{(1 - a_w)M} = ba_w + c \tag{11.43}$$

where the constants are defined as $b = (C-1)/(M_oC)$ and $c = 1/(M_oC)$ are obtained from the slope and intercept of the straight line generated by plotting $a_w/(1-a_w)M$ against a_w. The value of monolayer can be obtained from $M_0 = 1/(b+c)$ and $C = (b+c)/c$.

Although not well defined, the monolayer (M_0) is often stated to represent the moisture content at which water attached to each polar and ionic group starts to behave as a liquid-like phase and corresponds with the optimal moisture content for stability of low-moisture foods [24,83].

The monolayer moisture content of many foods has been reported to correspond with the physical and chemical stability of dehydrated foods [84–86]. The theory behind the BET equation has been faulted on many grounds such as its assumptions that (1) the rate of condensation on the first layer is equal to the rate of evaporation from the second layer, (2) the binding energy of all of the adsorbate on the first layer is the same, and (3) the binding energy of the other layers is equal to those of pure adsorbate. Further assumptions of uniform adsorbent surface and absence of lateral interactions between adsorbed molecules are known to be incorrect in view of the heterogeneous food surface interactions. However, the equation has been found useful in defining an optimum moisture content for drying and storage stability of foods and in the estimation of surface area of a food. Iglesias and Chirife [87] reported 300 monolayer values corresponding to about 100 different foods and food components.

11.2.6.1.3 The Halsey Equation

The following equation was developed by Halsey [76] to account for condensation of multilayers, assuming that the potential energy of a molecule varies inversely as the rth power of its distance from the surface.

$$a_w = \exp\left(-\frac{A}{RT\theta^r}\right) \qquad (11.44)$$

where A and r are constant and $\theta = M/M_0$.

Halsey [76] also stated that the value of the parameter r indicates the type of adsorbate–adsorbent interaction. When r is small, van der Waals-type forces, which are capable of acting at greater distances, are predominantly involved. Large values of r indicate that attraction of the solid for the vapor is presumably very specific and is limited to the surface.

Recognizing that the use of the RT term does not eliminate the temperature dependence of A and r, Iglesias et al. [88] and Iglesias and Chirife [135] simplified the original Halsey equation to the form

$$a_w = \exp\left(-\frac{A'}{M^r}\right) \qquad (11.45)$$

where A' is a new constant. Iglesias et al. [88] and Iglesias and Chirife [89] reported that the Halsey equation could be used to describe 220 experimental sorption isotherms of 69 different foods in the range of $0.1 < a_w < 0.8$. Starch-containing foods [90] and dried milk products [91] have also been shown to be well described in their sorption behavior by this equation.

Ferro-Fontan et al. [92] have shown that when the isosteric heat of sorption (Q_{st}) varies with moisture content as $Q_{st} \sim M^{-r}$, integration of the Clausius–Clapeyron equation leads to the three-parameter isotherm equation

$$\ln\left(\frac{\gamma}{a_w}\right) = A(M)^{-r} \qquad (11.46)$$

where γ is a parameter that accounts for the "structure" of sorbed water. Chirife et al. [93] investigated the applicability of Equation 11.46 in the food area. They reported that the equation is able to describe the sorption behavior of 18 different foods (oilseeds, starch foods, proteins, and others) in the "practical" range of a_w (0.10–0.95), with only 2–4% average error in the predicted moisture content.

11.2.6.1.4 The Henderson Equation

The original equation, empirically developed by Henderson [94], is written as

$$\ln(1 - a_w) = -C_1 T M^n \qquad (11.47)$$

where C_1 and n are constants. At a constant temperature, the equation is simplified to

$$\ln(1 - a_w) = C'_1 M^n \tag{11.48}$$

where $C'_1 = C_1 T$, a new constant.

A linearized plot of $\ln[-\ln(1 - a_w)]$ versus moisture content has been reported to give rise to three "localized isotherms" [95,96], which do not necessarily provide any precise information on the physical state of water as was originally thought. Henderson's equation has been widely applied to many foods [97–100], but when compared to the Halsey equation, its applicability was found to be less versatile [81].

11.2.6.1.5 The Chung and Pfost Equation

Assuming a direct relationship between moisture content and the free energy change for sorption, Chung and Pfost [101–103] proposed the equation

$$\ln a_w = -\frac{C_1}{RT} \exp(-C_2 M) \tag{11.49}$$

where C_1 and C_2 are constants. The inclusion of temperature in the above equation precludes evaluation of the temperature dependence of parameters C_1 and C_2. When the RT term is removed, the modified version becomes equivalent to the Bradley Equation 11.106 and is given as [104]

$$\ln a_w = -C'_1 \exp(-C_2 M) \tag{11.50}$$

Young [100] evaluated the applicability of this equation in describing the sorption isotherms of peanuts and found reasonable fit to experimental data.

11.2.6.1.6 The Oswin Equation

This model is based on Pearson's series expansion for sigmoidal curves applied to type II isotherms by Oswin [77]. The equation is given as

$$M = C_1 \left(\frac{a_w}{1 - a_w} \right)^n \tag{11.51}$$

where C_1 and n are constants. Tables of parameter values for various foods and biomaterials were compiled by Luikov [105] for the Oswin equation. This equation was also used by Labuza et al. [106] to relate moisture contents of nonfat dry milk and freeze-dried tea up to $a_w = 0.5$.

11.2.6.1.7 The Chen Equation

The original model developed by Chen [107] is a three-parameter equation based on the steady-state drying theory for systems where diffusion is the principal mode of mass transport. This equation is written as

$$a_w = \exp\left[C_1 - C_2 \exp(-C_3 M) \right] \tag{11.52}$$

where C_1, C_2, and C_3 are temperature-dependent constants. Chen and Clayton [98] applied this equation to experimental water sorption data of a number of cereal grains and other field crops with a reasonable goodness of fit. They found that the values of the constant C_1 were close to zero and thus reduced this equation to the two-parameter model:

$$a_w = \exp\left[-C_2 \exp(-C_3 M)\right] \tag{11.53}$$

This simplified equation has been shown [81] to be mathematically equivalent to the Bradley Equation 11.106.

11.2.6.1.8 The Iglesias–Chirife Equation
While studying the water sorption behavior of high-sugar foods, such as fruits, where the monolayer is completed at a very low moisture content and dissolution of sugars takes place, Iglesias and Chirife [108] noticed the resemblance of the curve of a_w versus moisture content to the sinh curve [type III in the Brunauer et al. [69] classification] and proposed the empirical equation

$$\ln\left[M + \left(M^2 + M_{0.5}\right)^{1/2}\right] = C_1 a_w + C_2 \tag{11.54}$$

where $M_{0.5}$ is the moisture content at $a_w = 0.5$ and C_1 and C_2 are constants. Equation 11.54 was found to adequately describe the behavior of 17 high-sugar foods (banana, grapefruit, pear, strawberry, etc.).

11.2.6.1.9 The Guggenheim–Anderson–de Boer Equation
The three-parameter GAB equation, derived independently by Guggenheim [109], Anderson [110], and de Boer [111], has been suggested to be the most versatile sorption model available in the literature and has been adopted by the European Project Cost 90 on physical properties of food [78,112]. Fundamentally, it represents a refined extension of the Langmuir and BET theories, with three parameters having physical meanings. For sorption of water vapors, it is mathematically expressed as

$$\frac{M}{M_o} = \frac{CKa_w}{(1 - Ka_w)(1 - Ka_w + CKa_w)} \tag{11.55}$$

where C and K are constants related to the energies of interaction between the first and distant sorbed molecules at the individual sorption sites. Theoretically, they are related to sorption enthalpies as follows:

$$C = c_o \exp\left[\left(\bar{H}_o - \bar{H}_n\right)/RT\right] = c_o \exp\left(\Delta \bar{H}_c\right)/RT \tag{11.56}$$

$$K = k_o \exp\left[\left(\bar{H}_n - \bar{H}_l\right)/RT\right] = k_o \exp\left(\Delta \bar{H}_k\right)/RT \tag{11.57}$$

where c_o and k_o are entropic accommodation factors; $\bar{H}_o, \bar{H}_n$, and $\bar{H}_l$ are the molar sorption enthalpies of the monolayer, the multiplayer and the bulk liquid, respectively. When K is unity, the GAB equation reduces to the BET equation.

The GAB model is a refined version of the BET equation. While sharing the two original constants, the monolayer capacity (M_o) and the energy constant I, with the BET model, the GAB model derives its uniqueness from the introduction of a third constant, k. However, use of the two models with experimental data has been shown to yield different values of M_o and C, invariably giving inequality [113]: $M_{o(BET)} < M_{o(GAB)}$ and $C_{(BET)} > C_{(GAB)}$. The difference has been attributed to the mathematical nature of the two equations and not the physicochemical nature of the sorption systems to which they are applied [114].

The GAB Equation 11.55 can be rearranged as follows:

$$a_w/M = (k/M_o)\left[1/C - 1\right]a_w^2 + (1/M_o)\left[1 - 2/C\right]a_w + 1/(M_o kC)$$

If $\alpha = (k/M_o)[(1/C) - 1]$, $\beta = (1/M_o)[1 - (2/C)]$ and $\gamma = (1/M_o Ck)$ the equation transforms to a second-order polynomial form:

$$a_w/M = \alpha a_w^2 + \beta a_w + \gamma \qquad (11.58)$$

The constants are obtained by conducting a linear regression analysis on experimental values. The quality of the fit is judged from the value of the relative percentage square (%RMS):

$$\%\mathrm{RMS} = \left[\sum_1^N \left\{(M_i - M_i^*)/M_i\right\}^2 \Big/ N\right]^{1/2} \times 100$$

where

M_i = Experimental moisture content
M_i^* = Calculated moisture content
N = Number of experimental points

A good fit of an isotherm is assumed when the RMS value is below 10% [115]. The numerical values of the equation parameters are then obtained as follows:

$$k = \frac{\sqrt{\beta^2 - 4\alpha\gamma} - \beta}{2\gamma}, \quad C = \frac{\beta}{\gamma k} + 2, \quad \text{and} \quad m_o = \frac{1}{\gamma kC}$$

A summary on the use of the GAB equation to a variety of food materials studied by a number of researchers has been compiled by AlMuhtaseb et al. [116]. Some examples include cured beef [117], casein [112], fish [58], onions [118], pasta [119], potatoes [115,120], peppers [121], rice and turkey [113], and yogurt powder [122].

The major advantages of the GAB model are that [123] (1) it has a viable theoretical background; (2) it describes sorption behavior of nearly all foods from zero to 0.9 a_w; (3) it has a simple mathematical form with only three parameters, which makes it very amenable to engineering calculations; (4) its parameters have a physical meaning in terms of the sorption processes; and (5) it is able to describe some temperature effects on isotherms by means of Arrhenius-type equations.

Labuza et al. [58] applied the GAB equation to MSIs for fish flour and cornmeal in the range of $0.1 < a_w < 0.9$ at temperatures from 25°C to 65°C and reported an excellent fit. The monolayer moisture contents obtained with the GAB equation, however, did not differ from those obtained with the BET equation at the 95% level of significance for each temperature. In food engineering applications related to food–water interactions, the GAB model should prove to be a reliable and accurate equation for modeling and design work.

Iglesias and Chirife [124] tabulated the mathematical parameters of one or two of nine commonly used isotherm equations for over 800 MSIs. A brief description of the technique and the statistical significance associated with the equation employed to fit the data provide valuable information on analysis of sorption data.

11.2.6.2 Effect of Temperature

A knowledge of the temperature dependence of sorption phenomena provides valuable information about the changes related to the energetics of the system. The variation of water activity with temperature can be determined by using either thermodynamic principles or the temperature terms incorporated into sorption equations. From the well-known thermodynamic relation, $\Delta G = \Delta H - T\Delta S$, if sorption is to occur spontaneously, ΔG must have a negative value. During adsorption, ΔS will be negative because the adsorbate becomes ordered upon adsorption and loses degrees of freedom. For ΔG to be less than zero, ΔH will have to be negative, and the adsorption is thus exothermic. Similarly, desorption can be shown to be endothermic. Also, because ΔH decreases very slightly with temperature, higher temperatures will cause a corresponding decrease in ΔS, resulting in a reduction in adsorbed molecules. In principle, therefore, one can say that generally adsorption decreases with increasing temperature. Although greater adsorption is found at lower temperature, the differences are usually small. At times, however, larger differences are observed. There is no particular trend in these differences, and caution is required when one tries to interpret them because temperature changes can affect several factors at the same time. For instance, an increase in temperature may increase the rates of adsorption, hydrolysis, and recrystallization reactions. A change in temperature may also change the dissociation of water and alter the potential of reference electrodes. Foods rich in soluble solids exhibit antithetical temperature effects at high a_w values (>0.8) because of their increased solubility in water.

The constant in MSI equations, which represents either temperature or a function of temperature, is used to calculate the temperature dependence of water activity. The Clausius–Clapeyron equation is often used to predict a_w at any temperature if the isosteric heat and a_w values at one temperature are known. The equation for water vapor, in terms of isosteric heat (Q_{st}), is

$$d(\ln P) = -\frac{Q_{st}}{R} d\frac{1}{T} \tag{11.59}$$

Subtracting the corresponding relation for vapors in equilibrium with pure water at the same temperature gives

$$d(\ln P) - d(\ln P_w^{\bullet}) = -\frac{Q_{st} - \Delta \bar{H}_{vap}}{R} d\left(\frac{1}{T}\right) \tag{11.60}$$

In terms of water activities a_{w1} and a_{w2} at temperatures T_1 and T_2, the above relationship gives

$$\ln \frac{a_{w2}}{a_{w1}} = \frac{q_{st}}{R}\left(\frac{1}{T_1} - \frac{1}{T_2}\right) \quad (11.61)$$

where q_{st} = net isosteric heat of sorption (also called excess heat of sorption) = $Q_{st} - \bar{H}_{vap}$. Accurate estimates of q_{st} require measurements of water activities at several temperatures in the range of interest, although a minimum of only two temperatures are needed. The use of the above equation implies that, as discussed later, the moisture content of the system under consideration remains constant and that the enthalpy of vaporization of pure water (as well as the isosteric heat of sorption) does not change with temperature. For biological systems, extrapolation to high temperatures would invalidate these assumptions because irreversible changes and phase transformations invariably occur at elevated temperatures. The temperature dependence of MSIs in the higher temperature range (40–80°C) was evaluated from experimentally determined MSIs of casein, wheat starch, potato starch, pectin, and microcrystalline cellulose [125] and of cornmeal and fish flour [58]. The shift in a_w with temperature predicted by the Clausius–Clapeyron equation was found to fit the experimental data well. However, according to a large survey on the subject [126] and based on Equation 11.60, isosteres correlation coefficients in the range of 0.97–0.99 indicated only slight temperature dependence but a strong temperature dependence below 0.93.

The BET equation contains a temperature dependence built into the constant term C such that

$$C = \frac{a_1 b_2}{b_1 a_2} \exp \frac{\bar{H}_o - \bar{H}_L}{RT} \quad (11.62)$$

where $\bar{H}_o$ is the molar sorption enthalpy of the first layer, $\bar{H}_l$ is the molar sorption enthalpy of the adsorbate, and a_1, a_2, b_1, and b_2 are constants related to the formation and evaporation of the first and higher layers of adsorbed molecules.

In the absence of any evidence, the authors of the BET theory assumed that $b_1/a_1 = b_2/a_2$, making the preexponential of Equation 11.62 equal to unity. This simplifying assumption has been a source of considerable discussion, and notably different values of the preexponential factor are reported in the literature, ranging from 1/50 [127] to greater than unity [128]. Kemball and Schreiner [129] demonstrated that the value of preexponential may vary from 10^{-5} to 10, depending on the entropy changes accompanying the sorption process and thus on the sorbant–sorbate system. Aguerre et al. [130] calculated the preexponential factors for eight different dry foods and found them to range from 10^{-3} for sugar beet to 10^{-6} for marjoram. However, with the assumption that the preexponential is unity, the simplified expression for C has been used in the literature to evaluate the net BET heat of sorption $(\bar{H}_o - \bar{H}_l)$ of a wide variety of foods [108,131–133]. This value, when added to the heat of condensation, provides an estimate of the heat required to desorb one monolayer. Iglesias and Chirife [87,108] also noted that as C approaches unity, the monolayer moisture content and C become increasingly dependent on one another, and a small error in its calculation may result in a very significantly different value for the enthalpy of sorption. In a related work, Iglesias and Chirife [134] summarized values of $\bar{H}_o - \bar{H}_l$ reported by several authors

ENGINEERING PROPERTIES OF FOODS

for various foods and compared them to corresponding values of the net isosteric heat (q_{st}) taken at the monolayer moisture coverage obtained from vapor pressure data. They concluded that $\bar{H}_o - \bar{\bar{H}}_l$ computed from the BET correlation to date is almost invariably lower than the net isosteric heat, often by more than a factor of 10, and recommended that the BET equation not be used to estimate the heat of water sorption in foods. This energy difference has been a source of some confusion. It is not meant to be equal to the isosteric heat, and efforts to compare the two numbers directly are misdirected. The proper way to compare the two is by computing $d \ln a_w / d(1/T)$ at a constant moisture from the BET equation directly. The result of this differentiation is [73]

$$q_{st} = \frac{C(1-a_w)^2 (\bar{H}_o - \bar{\bar{H}}_l)}{\left[1+(C-1)a_w\right]^2 + (C-1)(1-a_w)} \tag{11.63}$$

The behavior of the isosteric heat with moisture coverage is shown in Figure 11.4 for desorption of water from pears, calculated from Equation 11.63, with BET constants

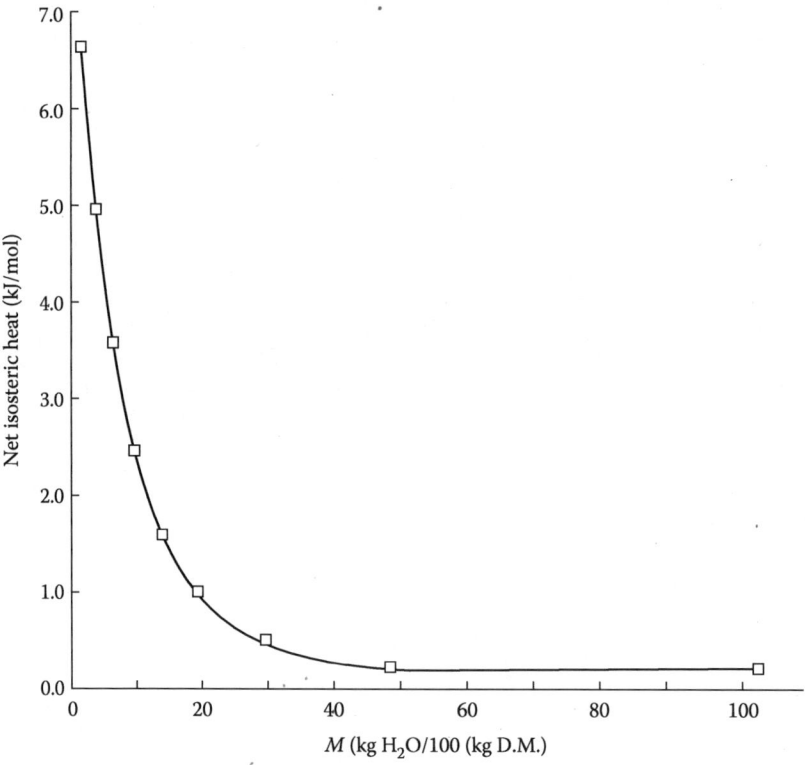

Figure 11.4 Net isosteric heat of sorption (q_{st}) computed from BET theory for pears at 20°C. (BET constants from Iglesias, H. A., and Chirife, J. 1976. Isosteric heats of water vapor sorption on dehydrated foods. I. Analysis of the differential heat curves. *Lebensm.-Wiss. Technol.* 9: 116–122.)

from Iglesias and Chirife [135]. The plot shows the characteristic sharp decline in q_{st} with increasing moisture and is, at least in a qualitative manner, of the same type found by applying the Clausius–Clapeyron equation directly to the data.

A survey of literature data on the effect of temperature on different dehydrated food products shows that monolayer moisture content decreases with increasing temperature [88,98,136,137]. This behavior is generally ascribed to a reduction in the number of active sites due to chemical and physical changes induced by temperature; the extent of decrease, therefore, depends on the nature of the food. In a note Iglesias and Chirife [138] proposed the following simple empirical equation to correlate the BET monolayer values with temperature:

$$\ln(M_o) = C_1 T + C_2 \qquad (11.64)$$

where C_1 and C_2 are constants, M_o is monolayer moisture content, and T is temperature in degree Celsius. They fitted the equation to 37 different foods in the approximate temperature range of 5–60°C, with mostly less than 5% average error. For foods exhibiting type III sorption behavior, the equation failed to reproduce the behavior of BET values with temperature. The relative effect of temperature on monolayer moisture content of dehydrated foods is important in dehydration and in shelf-life simulation and storage studies of foods. In view of the absence of any definite quantitative model, BET values corresponding to storage temperatures are directly computed from sorption data and are used in shelf-life correlation studies [139].

11.3 SORPTION ENERGETICS

Thermodynamic parameters such as enthalpy and entropy of sorption are needed both for design work and for an understanding of food–water interactions. However, there is a good degree of controversy regarding the validity of the thermodynamic quantities computed for the food systems. As will be discussed shortly, thermodynamic calculations assume that a true equilibrium is established and that all changes occurring are reversible. Therefore, any thermodynamic quantity computed from hysteresis data is suspect because hysteresis is a manifestation of irreversibility [128,130,140,141]. Nevertheless, as Hill [128] pointed out, hysteresis data yield upper and lower limits for the "true" isotherms and hence place bounds on the thermodynamic quantities as well.

The classical thermodynamics of sorption was established in its present form in the late 1940s and early 1950s, culminating in the famous series of articles by Hill [128,142–144], Hill et al. [145], and Everett and Whitton [146]. The relationship developed by these authors have been used to compute thermodynamic functions of sorbed species on various inorganic materials [73,145,147,148]. Thermodynamic properties of water on materials of biological origin have been computed for sugar beet root [135], onion bulbs [149], horseradish root [150], and dehydrated peanut flakes [86]. The calculation of the total energy required to drive off water from a food has been discussed by Keey [151] and in some detail by Almasi [85]. The effect of bound water on drying time has been investigated and is discussed by Gentzler and Schmidt [152], Roman et al. [133], Ma and Arsem [153], and Albin et al. [154].

The energetics of sorption can be expressed in two ways, each useful in its own content. The integral heat of sorption (Q_{int}) is simply the total amount of heat per unit weight of adsorbent for n_s moles of adsorbate on the system. The differential heat of sorption (q_{diff}), as the name implies, represents the limiting value of the quantity dQ_{int}/dn_s [155].

Several terms and definitions, none of which is universally accepted, are presently in use to describe sorption energetics. Enthalpy of water binding [156], isosteric heat of sorption [135], force of water binding [157], partial enthalpy [158], integral heat of sorption [159], and heat of adsorption [80] are among the many terms found in the literature. The reason for this is that a number of variables are involved in the sorption process, and several energy terms can be defined depending on which variables are kept constant. For exact definition of energy terms, the variables kept constant during the sorption process must therefore be specified.

In this section, we present the basic derivation of thermodynamic functions. No attempt will be made to present complete derivations, as these are presented elsewhere for the general case [143] and applied to biological materials [160,161]. We also discuss general principles associated with hysteresis and how these affect thermodynamic calculations.

11.3.1 Differential Quantities

Using solution thermodynamics and following the notations of Hill [142], if subscript l refers to adsorbate (water) and A indicates the adsorbent (nonvolatile food matrix), then the internal energy change of the system is given by the equation

$$dE = T\,dS - P\,dV + \mu_l dn_l + \mu_A dn_A \tag{11.65}$$

from which, in terms of Gibbs free energy,

$$dG = V\,dP - S\,dT + \mu_l dn_l + \mu_A dn_A \tag{11.66}$$

In terms of adsorbate chemical potential, it becomes

$$d\mu_l = \hat{V}_l dP - \hat{S}_l dT + \left(\frac{d\mu_l}{d\Gamma}\right)_{T,P} d\Gamma \tag{11.67}$$

where $\Gamma = n_l/n_A$ and the carets indicate differential quantities, for example, $\hat{S}_l = (S/n_l)_{n_A, P, T}$ and $\hat{V}_l = (V/n_l)_{n_A, P, T}$. It should be noted that Γ is proportional to the moisture content in food and that the number of moles of adsorbent, n_A (if such a thing exists), is constant and proportional to the surface area of the food. When the equation is used consistently, the resulting thermodynamic functions are equal, regardless of whether n_A or the area is used.

For the gas phase, it is easily shown that [162]

$$d\mu_g = \bar{V}_g dP - \bar{S}_g dT \tag{11.68}$$

where the bars indicate molar quantities. At equilibrium, $d\mu_g = d\mu_l$ and from Equations 11.67 and 11.68,

$$\hat{V}_l dP - \hat{S}_l dT + \left(\frac{d\mu_l}{d\Gamma}\right)_{T,P} d\Gamma = \bar{V}_g dP - \hat{S}_g dT$$

If moisture content (hence, Γ) is held constant, we have

$$\left(\frac{\partial P}{\partial P}\right)_\Gamma = \frac{\bar{S}_g - \bar{S}_l}{\bar{V}_g - \hat{V}_l} \tag{11.69}$$

This equation assumes the total hydrostatic pressure to be the vapor pressure of water over the food. Now assuming that the ideal gas law holds and that $V_g @ \hat{V}_l$ and recognizing that at equilibrium, $\bar{H}_g - \hat{H}_l = T(\bar{S}_g - \hat{S}_l)$, Equation 11.69 yields

$$\left(\frac{dP}{dT}\right)_\Gamma = \frac{(\bar{H}_g - \hat{H}_l)P}{RT^2} \tag{11.70}$$

From the above expression, the differential or *isosteric heat of sorption* (Q_{st}) is given as

$$\left[\frac{d\ln P}{d(1/T)}\right]_\Gamma = -\frac{\bar{H}_g - \hat{H}_l}{R} = -\frac{\Delta \hat{H}}{R} = \frac{Q_{st}}{R} \tag{11.71}$$

As shown earlier, the excess or "net" isosteric heat of sorption is obtained by subtracting from Equation 11.71 the corresponding relation for water:

$$\left(\frac{\partial \ln a_w}{\partial(1/T)}\right)_\Gamma = \frac{\bar{H}_o - \hat{H}_l}{R} = -\frac{q_{st}}{R} \tag{11.72}$$

In Table 11.9, there are some values of net isosteric (q_{st}) heat of sorption of selected foods reported in the literature.

The change in the Gibbs free energy of absorbed water is given by the equation

$$\Delta \hat{G} = RT \ln a_w \tag{11.73}$$

The energy of adsorption can therefore be computed as

$$\Delta \hat{S} = \frac{\Delta \hat{H} - \Delta \hat{G}}{R} \tag{11.74}$$

Assuming a standard state pressure of 1 atm, Morsi et al. [167] determined the standard differential values of free energy ($\Delta \hat{G}^0$), enthalpy at constant adsorption ($\Delta \hat{H}^0$), and entropy of adsorption ($\Delta \hat{S}^0$) for cotton cellulose, potato starch, corn amylose, amylose, amylomaize, and retrograded potato starch. The behavior of the computed differential thermodynamic functions with moisture content for amylomaize at 25°C is shown in Figure 11.5. The authors observed that the moisture content at which maximum $-\Delta \hat{H}^0$ and $-\Delta \hat{S}^0$ occur coincides with the monolayer moisture content of amylomaize. Other workers in this area reported observing similar trends with other food systems [86,149,157]. Bettleheim et al.

ENGINEERING PROPERTIES OF FOODS

Table 11.9 Net Isosteric Heat of Sorption of Selected Foods

Food	q_{st} (kJkG^{-1})	Reference
Apple	83–1112	Roman et al. [133]
Beef	356–1374	Iglesias and Chirife [124]
Carrot	567–1594	Kiranoudis et al. [163]
Chicken	249–2662	Iglesias and Chirife [124]
Corn	111–426	Cenkowski et al. [164], Iglesias and Chirife [124]
Eggs	95–491	Iglesias and Chirife [124]
Milk	34–395	Iglesias and Chirife [124]
Peppers	722–1961	Kiranoudis et al. [163] and Kaymak-Ertekin and Sultanoglu [121]
Potato	461–1933	Wang and Brennan [165], Kiranoudis et al. [163], McLaughlin and Magee [115], McMinn and Magee [120]
Rice	142–445	Iglesias and Chirife [124], Cenkowski et al. [164]
Tapioca	83–889	Soekarto and Steinberg [166]
Tomato	411–2383	Kiranoudis et al. [163]

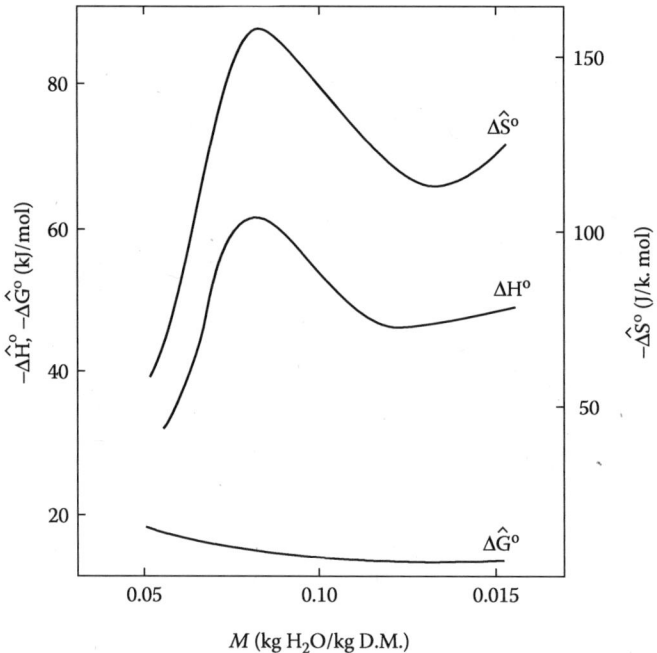

Figure 11.5 Standard differential thermodynamic functions for adsorption of water vapor on amylomaize as a function of moisture. (Adapted from Morsi, M. K. S., Sterling, C., and Volman, D. H. 1967. Sorption of water vapor B pattern starch. *J. Appl. Polym. Sci.* 11: 1217–1225.)

[168] rationalized this experimental observation by suggesting that water molecules are initially sorbed on the most accessible primary sites (i.e., on the energetically favorable polar groups), which have a relatively low polymer segment density; then they are sorbed on the primary sites, which have larger segment density in their neighborhood. This causes the heat of sorption to increase with moisture content. Finally, the water vapor is sorbed on primary sites in the least accessible region with the highest polymer segment density. Thus, a maximum heat of sorption is encountered just before the completion of a monolayer.

11.3.2 Integral Quantities

The integral quantities permit qualitative interpretation of the sorption phenomena and are more descriptive of the energy relationship involved in any adsorption or desorption process. The procedure for calculation of integral functions is described by the first law of thermodynamics, as given by Equation 11.64. When Equation 11.64 is applied to pure adsorbent only (here subscripted as "0A"), it gives

$$dE_{0A} = T\,dS_{0A} - P\,dV_{0A} + \mu_{0A}\,dn_A \tag{11.75}$$

Subtracting Equation 11.75 from Equation 11.65 gives

$$d(E - E_{0A}) = T\,d(S - S_{0A}) - P\,d(V - V_{0A}) \\ -(\mu_{0A} - \mu_A)dn_A + \mu_l\,dn_l \tag{11.76}$$

Now defining the energy, entropy, and volume of the sorbed species as $E_s = E - E_{0A}$; $S_s = S - S_{0A}$; and $V_s - V_{0A}$, and with a surface potential $\phi = \mu_{0A} - \mu_A$ playing the role of a second pressure, Equation 11.76 gives

$$dE_s = T\,dS_s - P\,dV_s - \phi\,dn_A + \mu_l\,dn_l \tag{11.77}$$

By defining the Gibbs free energy of the sorbed species as $G_s = n_{l\mu l} = H_s - TS_s$ and $H_s = E_s + PV_s + \phi n_A$, after some manipulation one obtains

$$d\mu_l = \bar{V}_s dP - \bar{S}_s dT + \frac{1}{\Gamma}d\phi \tag{11.78}$$

where the bars symbolize molar (not differential) quantities. Equating Equations 11.68 and 11.78 for equilibrium conditions gives

$$\bar{V}_s dP - \bar{S}_s dT + \frac{1}{\Gamma}d\phi = \bar{V}_g dP - \bar{S}_g dT \tag{11.79}$$

As in the case of isosteric heat, assuming ideality in the gas phase and $\bar{V}_g @ \bar{V}_S$ and applying $\bar{H}_g - \bar{H}_S = T(\bar{S}_g - \bar{S}_S)$ at constant ϕ, the equilibrium enthalpy of sorption [128] is given as

$$\left[\frac{d\ln P}{d(1/T)}\right]_\phi = \frac{\bar{H}_g - \bar{H}_s}{R} \tag{11.80}$$

The surface potential may be calculated by considering Equation 11.79 at constant temperature:

$$d\phi = \Gamma(\bar{V}_g - \bar{V}_s)dP \tag{11.81}$$

which, assuming $V_g \gg V_s$ and ideality in the gas phase, yields upon integration

$$\phi = RT \int_{P \to 0}^{P} \Gamma \, d\ln P \tag{11.82}$$

The technique for calculating the equilibrium enthalpy of sorption thus involves first graphically determining, through use of Equation 11.82, the surface potential as a function of pressure for various constant temperatures and then, at constant ϕ, determining values of P and T to use in Equation 11.80. The same procedure applies when the surface potential is expressed in terms of a_w as

$$\phi = RT \int_{a_w \to 0}^{a_w} \Gamma \, d\ln a_w \tag{11.83}$$

Because the enthalpy is computed at constant surface potential, the numerical value of ϕ is immaterial as long as the same value of ϕ is used throughout the calculations [143]. Also, the accurate determination of the number of moles of adsorbent for use in Equation 11.82 is immaterial, as the resulting thermodynamic functions are not influenced by its numerical value.

The equilibrium enthalpy value thus computed is analogous to the molar enthalpy of vaporization of pure water. The net equilibrium enthalpy of sorption is obtained on subtracting the corresponding equation for pure water as follows:

$$\left[\frac{\partial(\ln a_w)}{\partial(1/T)}\right]_\phi = -\frac{\bar{H}_l - \bar{H}_s}{R} \tag{11.84}$$

Rizvi and Benado [169] evaluated the equilibrium enthalpy of sorption for three food products (grain sorghum, horseradish root, and yellow globe onion) at normalized surface potentials (Table 11.10). The trends seem to generally be in the same direction, and the magnitudes of the enthalpies of adsorption seem to be similar. The equilibrium enthalpy of sorption of horseradish roots seems to flatten out at low moisture values, whereas the other two undergo a fairly rapid and constant decline.

Another integral quantity, the entropy of the sorbed species, can be calculated with reference to liquid water at the same temperature:

$$\bar{S}_s - \bar{S}_L = \frac{\Delta \bar{H}_{vap} - (\bar{H}_g - \bar{H}_s)}{\bar{T}} - R \ln \bar{a}_w \tag{11.85}$$

Hill et al. [143] suggested, on a more or less arbitrary basis, that the average temperature for use in Equation 11.85 be computed at $\bar{T} = [1/2(1/T_1 + 1/T_2] - 1$, and the average water activity as $\bar{a}_w = \exp[1/2(\ln a_{w1} + \ln a_{w2})]$. The calculation procedure is as follows.

Table 11.10 Variation of Integral Molar Enthalpy of Sorption ($\bar{H}_L - \bar{H}_S$) (kJ/kg) of Water with Normalized Surface Potential

Normalized Potential (ϕ)	Grain Sorghum[a]	Horseradish Root[b]	Yellow Globe Onion[c]
1	2174	1742	1948
2	1272	1442	1848
3	590	1292	1542
4	427	700	1142
5	332	592	842

[a] From Rizvi, S. S. H., and Benado, A. L. 1983–84. *Drying Technol.* 2(4): 471–502.
[b] From Mazza, G. 1980. *Lebensm.-Wiss. Technol.* 13: 13–17.
[c] From Mazza, G., and La Maguer, M. 1978. *Can. Inst. Food Sci. Technol. J.* 11: 189–193.

From a plot of Γ/P (or Γ/a_w) versus P (or a_w), one graphically integrates Equation 11.82 or (11.83). The existence of a numerical instability at $P = 0$ requires that a linear isotherm $M = \sigma a_w$ be taken as $a_w \to 0$ so that the integration may be carried out. The fact that the "molecular weight of the food" (if such a thing exists) is unknown poses no problem in that as long as it is constant, the relative position of the ϕ versus a_w plot will not change. Because the results do not depend on the absolute magnitude of ϕ, we define $K' = M_f / M_w$, and Equation 11.83 becomes

$$\phi = K' RT \int_{a_w \to 0}^{a_w} M \, d\ln a_w \tag{11.86}$$

where M is the moisture content of the food, M_f and M_w are the molecular weights of food and water, respectively, and K' is an unknown constant. The surface potential, ϕ, is thus computed in arbitrary units. At a constant ϕ, a_w is read off at two temperatures and inserted into Equations 11.84 and 11.85 to obtain the enthalpy and entropy values.

The entropy value calculated in this manner can be interpreted in terms of randomness, mobility, and other surface phenomena [170]. The entropy of adsorption can be interpreted in these terms, if one recognizes that $S_s = k \ln \Omega_s$, Ω_s being the total number of possible configurations [143]. In foods, there tend to be two opposing entropic contributions upon adsorption—a loss of entropy from localization of water and an increase in entropy due to incipient solution formation, that is, structural transformations in the food arising from solubilization and swelling [171].

Table 11.11 compares the molar entropy of water sorbed on grain sorghum [169] with that of water sorbed on soil [160] and full-fat peanut flakes [86], again taken at a series of normalized surface potentials. On soil at 25°C, the entropy of sorbed water behaves in much the same manner as does water on grain sorghum at 28°C, although the increase in entropy with moisture is more gradual.

Table 11.11 Variation in Molar Entropy of Sorption ($\bar{S}_s - \bar{S}_L$) [kJ/kg·°C)] of Water with Normalized Surface Potential

Normalized Potential (ϕ)	Grain Sorghum[a]	Soil[b]	Peanut Flakes[c]
1	−49.5	−8.6	−4.4
2	−28.1	−6.7	−4.7
3	−12.6	−4.9	−5.4
4	−9.3	−4.4	−6.0
5	−7.3	−3.0	−5.6

[a] From Rizvi, S. S. H., and Benado, A. L. 1984. *Food Technol.* 38(3): 83–92.
[b] From Taylor, S. A., and Kijne, J. W. 1963. Evaluating thermodynamic properties of soil water. In *Humidity and Moisture*, pp. 335–340. *Papers International Symp.*, Washington, DC.
[c] From Hill, P. E., and Rizvi, S. S. H. 1982. *Lebensm.-Wiss. Technol.* 15: 185–190.

The entropy of water sorbed on full-fat peanut flakes exhibits considerably different behavior. It initially decreases with increasing moisture, then reaches a minimum value, and then increases, finally approaching the entropy of free water at high moisture contents. This is indicative of some mobility and vibrational/rotational freedom at low moisture coverage, followed by a decrease in entropy due to localization as the first layer builds up. Finally, the entropy of sorbed water increases with the higher mobility associated with the buildup of layers further removed from the food surface. This may be due to the fact that full-fat peanut flakes contain a considerable amount of liquids that are not strong water binders, whereas soil and grain contain more polar groups that bind water more strongly. Iglesias et al. (1975) calculated the entropy of sorbed water on sugar beet root shown in Figure 11.6. With liquid water as the reference, the entropy of the adsorbed water is plotted as a function of the number of monolayers from BET analysis. The entropy of sorption goes through a minimum around the monomolecular layer coverage, because sorbed water is increasingly localized (and the degree of randomness is thus lowered) as the first layer is covered. At higher moisture contents, the entropy of the sorbed species begins to approach the entropy of pure water at the same temperature; hence, $S_s - S_L$ approaches zero.

Numerous investigators have expressed enthalpy of sorption by a generic term, *binding energy*. It is therefore hardly surprising to find several binding energies in the literature. Almasi [85] reviewed several of these binding energies. According to him, binding energy is the heat required to remove water during freeze-drying beyond the heat of vaporization of pure water. He compared binding energies calculated from calorimetric, isotherm, and BET heats for beef at the monolayer moisture content (Table 11.12). Temperature dependence of the calorimetric data was calculated using the Reidel [173] equation for the heat of binding, which gives the binding energy at any temperature T_2 as a function of the binding energy at any temperature T_1:

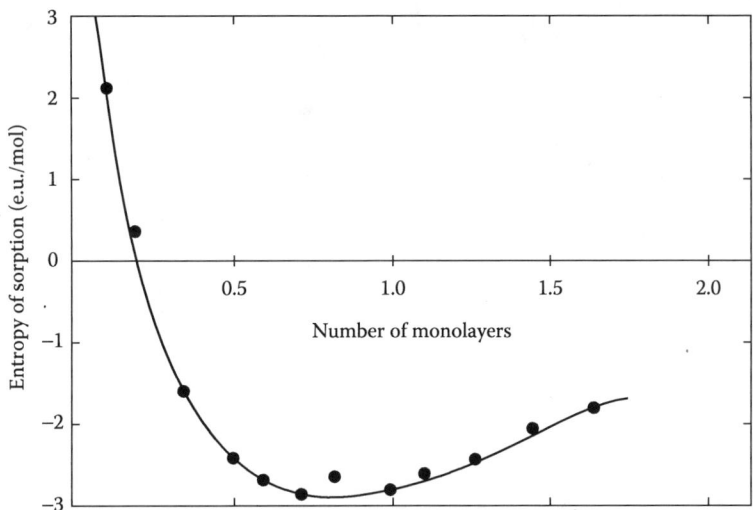

Figure 11.6 Entropy of sorbed water (with reference to the liquid at same temperature) on sugar beet root [e.u. = entropy units = cal/(mol ·°C)]. (From Iglesias, H. A., Chirife, J., and Viollaz, P. 1976. *J. Food Technol.* 11: 91–101.)

Table 11.12 Heat of Binding of Water at the Monolayer Moisture Content of Beef

Temperature (°C)	Heat of Binding of Water (kJ/kg)		
	Calorimetric	BET	Isosteric
0	502.4	178.2	477.2
−20	478.6	186.8	440.0
−40	457.6	182.2	420.7
−60	439.2	195.6	405.9
−80	423.7	215.4	375.5
−100	420.8	231.4	341.5
−140	333.3	—	—
−180	386.5	—	—

Source: From Almasi, E. 1978. *Acta Aliment.* 7: 213–255.

$$E(T_2) = E(T_1) + \int_{T_1}^{T_2} \left(C_{pw} - C_{pbw}\right) dT \qquad (11.87)$$

where C_{pw} is the heat capacity of pure water and C_{pbw} is the heat capacity of water associated with the matrix. Table 11.12 shows clear correspondence between the isosteric and calorimetric heats, whereas the BET terms are much lower.

Soekarto and Steinberg [166] postulated that water is associated with a food in three different ways (i.e., primary, secondary, and tertiary) and calculated characteristic binding energies for each fraction, using an energy-balance approach. They applied the Langmuir isotherm to the lower part of their sorption data, which yielded values of the Langmuir constant C at different temperatures. The primary fraction binding energy was then determined by application of Equation 11.61, and the secondary and tertiary binding energies were found by applying an energy balance. By plotting the product of the moisture content and the isosteric heat of desorption against the moisture content, Soekarto and Steinberg [166] obtained a line with a sharp break in it, corresponding to the transition between secondary and tertiary binding. The slope of each segment gave the binding energy for that fraction of water. It is noteworthy that their secondary fraction gave a negative value for the heat of binding, implying that although the secondary adsorption process is exothermic, it is slightly less so than the condensation of pure water. The authors noted that this is predicted in the theories of Brunauer [174]. Bettelheim et al. [168] reported qualitatively similar results and explained them by noting that this type of behavior is possible if water sorbed on hydrophilic sites is bound by only one hydrogen bond rather than several hydrogen bonds, as is the case in pure water. Soekarto and Steinberg [166] found tertiary binding to be very slightly stronger than the heat of condensation of pure water. Tertiary binding is characterized by dissolution of solutes, binding by macromolecules, and capillary phenomena. It may be concluded that the term *binding energy* can be considered a measure of the affinity of a food to water.

11.3.4 Hysteresis and Irreversibility

The applications of reversible thermodynamic principles to gain a fundamental understanding of food–water interactions for dehydration and storage of foods have met only limited success. The reason for this is the fact that sorption hysteresis is present to some degree in almost every food studied to date, which has prevented the exact calculation of thermodynamic functions such as enthalpy and entropy of water associated with foods because hysteresis is a manifestation of irreversibility. As La Mer [141] pointed out, reproducibility is a necessary but not a sufficient condition for defining an equilibrium, hence reversible, state, no matter how good the reproducibility of the data. Many theories concerning the origin of hysteresis can be found in the literature, several of which have been shown to have an effect on isotherms in a qualitative way.

Hysteresis is related to the nature and state of the components in a food. It may reflect their structural and conformational rearrangements, which alter the accessibility of energetically favorable polar sites and thus may hinder the movement of moisture. To date, however, no model has been found to quantitatively describe the hysteresis loop of foods, and there will probably be no final solution to this problem for a long time to come.

In his discussion of hysteresis, Hill [129] stated that for first-order phase changes, the adsorption branch represents the true equilibrium up to a certain point in the isotherm, and that the desorption branch never represents the true equilibrium. He also noted that for ordinary porous materials, such as foods, the region on the adsorption branch that represents equilibrium is very limited or nonexistent. The reason is that with a wide distribution of pore size it is impossible to determine with any certainty where capillary effects begin to exert a significant influence in vapor pressure lowering, although for the smallest pores it probably occurs quite early in the adsorption process. Among the factors that play a role in this type of phenomenon are the nature of the pore-size distribution [24,71] and the driving force involved in changing the water activity [175].

Gregg and Sing [148] argued that the desorption branch, having the lower pressure and hence the lower chemical potential, is closer to equilibrium than the adsorption branch. On the basis of this argument, Iglesias and Chirife [135] and others did their calculations. Rao [176], however, established that the hysteresis loop is crossed when a sample is moved from adsorption to desorption but not when moved the opposite way, indicating that the ERH during desorption depends much more on the starting moisture content. This behavior implies that the adsorption branch of MSIs is probably closer to the true equilibrium than the desorption branch. La Mer [141], however, maintained that "no amount of hypothetical speculations can make a computation thermodynamically acceptable if the basic data do not involve initial and final states which are independent of the direction of approach." From the arguments of Hill [128], if the equilibrium lies between the adsorption and desorption branches, the true thermodynamic function will lie somewhere between the adsorption and desorption "thermodynamic" functions.

The existence of irreversible steps implies that entropy is produced or lost during the sorption process. This production of entropy will typically be characterized by the conversion of work into heat. The *isosteric heat of desorption* will, therefore, usually be greater in magnitude than its adsorption counterpart. This was found to be the case by Iglesias and Chirife [135] in their review of isosteric heats reported in the literature. A typical example of their findings is shown in Figure 11.7, where the *isosteric heats* of adsorption and desorption for cooked chicken are plotted against moisture content. As expected, the heat of desorption is higher than the heat of adsorption, considerably so at low moisture contents. Several workers have also attempted to determine which branch of an isotherm that exhibits hysteresis represents the true equilibrium condition. Mazza and La Maguer [149] and Mazza [150] used the desorption branch as the true equilibrium in their calculations of the entropy of water on various foods. By taking dry protein as the reference state, Bettelheim et al. [168] arrived at the opposite conclusion for a swelling polymer network; their conclusion would appear to be supported by the data of Pixton and Warburton [175], who found no effect of method of adjustment of a_w on the adsorption branch of the sorption, isotherm of grain. Hill [128] and La Mer [141] showed that for complex phase changes, neither branch of the isotherm necessarily represents a true equilibrium state, and that therefore thermodynamic properties extracted from hysteresis sorption data alone cannot be calculated with any certainty.

La Mer [141] noted, however, that this does not necessarily make thermodynamics useless. One way to determine whether any one of the branches of a sorption isotherm corresponds to the equilibrium state is by thermodynamic testing. From Equation 11.87, it becomes clear that the isosteric heat is related to its calorimetric counterpart. Close correspondence between the two implies that the branch of the isotherm used to calculate the isosteric heat corresponds to the equilibrium value.

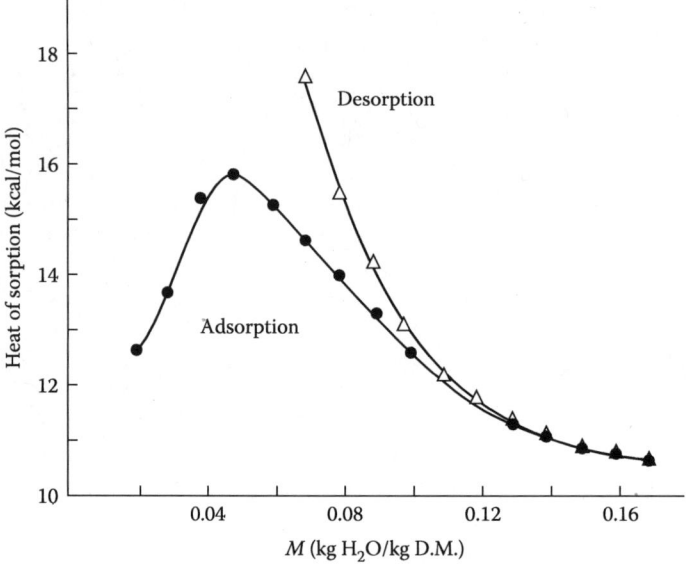

Figure 11.7 Net isosteric heats of adsorption and desorption on cooked chicken computed for each branch of isotherm. (From Iglesias, H. A., and Chirife, J. 1976. *Lebensm.-Wiss. Technol.* 9: 116–122.)

Insight may also be gained into the nature of the irreversibilities involved in the sorption hysteresis process by considering the arguments of Hill [128]. Subscripting adsorption and desorption with a and d, respectively, we have for Gibbs free energy,

$$G_a = H_a - TS_a \quad \text{and} \quad G_d = H_d - TS_d \tag{11.88}$$

Because desorption occurs at a lower water activity, and considering that $G = RT \ln a_w$, we have

$$G_d - G_a = RT \ln \frac{a_{wd}}{a_{wa}} = H_d - TS_d - H_a + TS_a \tag{11.89}$$

We know that $a_{wd} < a_{wa}$, $G_d < G_a$, and $H_d - TS_d < H_a - TS_a$; therefore,

$$H_d - H_a < T(S_d - S_a) \tag{11.90}$$

If $S_d < S_a$—that is, if entropy is irreversibly lost after adsorption—then $H_d < H_a$ as is so often found experimentally. Phenomena that lead to a lower entropy of water while desorption occurs include the entrapment of water in microcapillaries and matrix collapse. Equation 11.90 also points out, however, that the heat of sorption of water need not always be greater on desorption than on adsorption. For example, if $S_a < S_d$, then the heat of sorption may be either positive or negative and still satisfy the condition of Equation 11.90. For this case, chain rupture or irreversible swelling may be the cause for a heat of desorption being smaller than a heat of adsorption.

A useful measure of the relative amount of irreversibility is the so-called uncompensated heat, Q', which measures the work lost in a sorption process. The total uncompensated heat is given by

$$Q' = -RT \int M\, d(\ln a_w) \qquad (11.91)$$

The irreversible entropy change ΔS_{irr} may be found by dividing the uncompensated heat by the absolute temperature. Therefore, the total uncompensated heat and irreversible entropy are directly proportional to the area enclosed by the hysteresis loop on a plot of moisture content versus $\ln a_w$ of the material.

Benado and Rizvi [177] reported the thermodynamic functions of water on dehydrated rice as computed when rice exhibited hysteresis as well as when hysteresis was eliminated through two successive adsorption/desorption cycles (Figure 11.8). The isosteric heat calculated from the hysteresis-free cycle 2 isotherm (Figure 11.9) was found to lie between the isosteric heats calculated from the cycle 1 isotherms, which exhibited hysteresis. It is quite similar in shape to the isosteric heat of adsorption calculated from the cycle 1 isotherms, the maxima in the two curves both appearing at about 12% moisture. The equilibrium heat of sorption (Figure 11.10) and, similarly, the entropy of sorbed water calculated from reversible isotherm data also lie between the corresponding adsorption and desorption

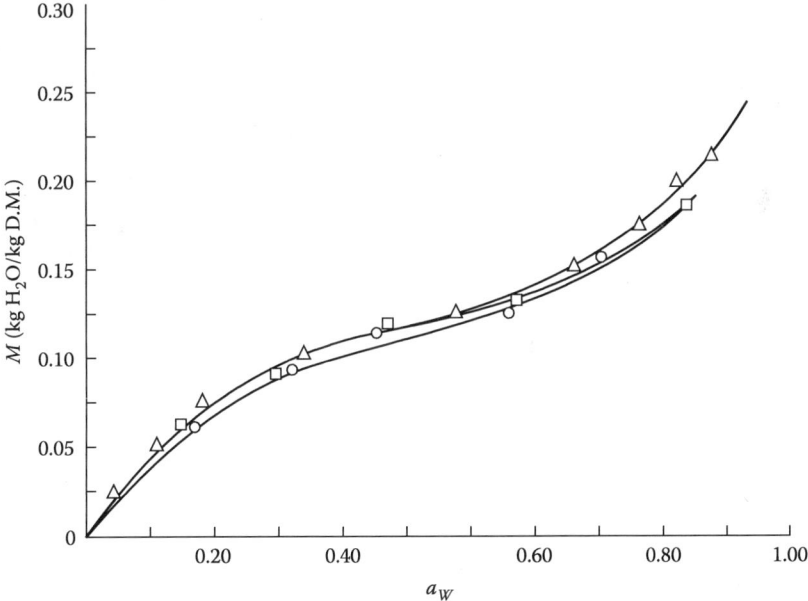

Figure 11.8 MSI of dehydrated rice at 30°C, with hysteresis (cycle 1) and with hysteresis removed (cycle 2). (From Benado, A. L., and Rizvi, S. S. H. 1985. *J. Food Sci.* 50: 101–105.) Cycle 1 (V) adsorption and (M) desorption; (n) cycle 2.

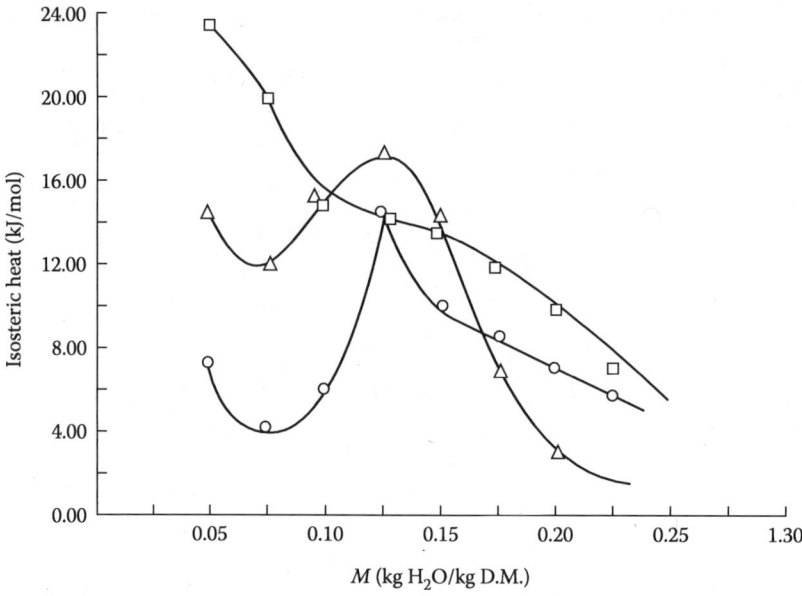

Figure 11.9 Isosteric heat of adsorption and desorption of dehydrated rice for cycle 1 (showing hysteresis) compared with the isosteric heat of sorption for cycle 2 (hysteresis free) at 20°C. Cycle 1 (V) adsorption and (M) desorption; (n) cycle 2. (From Benado, A. L., and Rizvi, S. S. H. 1985. *J. Food Sci.* 50: 101–105.)

quantities calculated from hysteresis data. The general shapes of the curves for adsorption functions, however, correspond more closely in a qualitative way to those calculated from reversible isotherms.

As indicated above, the precise magnitude of the heats of sorption cannot be determined with certainty; the knowledge of an upper bound for the isosteric heat of sorption permits the evaluation of drying processes in a "worst-case analysis." By using the isosteric heat of desorption in designing and evaluating drying equipment, the highest possible energy requirement and drying time will be designed for. Any error introduced by hysteresis then is in favor of higher throughput and lower energy expenditures.

11.3.5 Kinetic Aspects

From the preceding discussion, it is clear that the concept of a_w should apply only to systems in true thermodynamic equilibrium. Low-moisture foods exist in a state of pseudo-equilibrium as evidenced by the existence of hysteresis. In such situations, a_w is used to indicate the relative vapor pressure or relative humidity for developing processing and packaging protocols of immediate practical utility. For long-term storage stability, kinetic factors become more important, and clearly some other means are needed for dealing with mechanical relaxation and chemical reaction rates of an aqueous food matrix. In

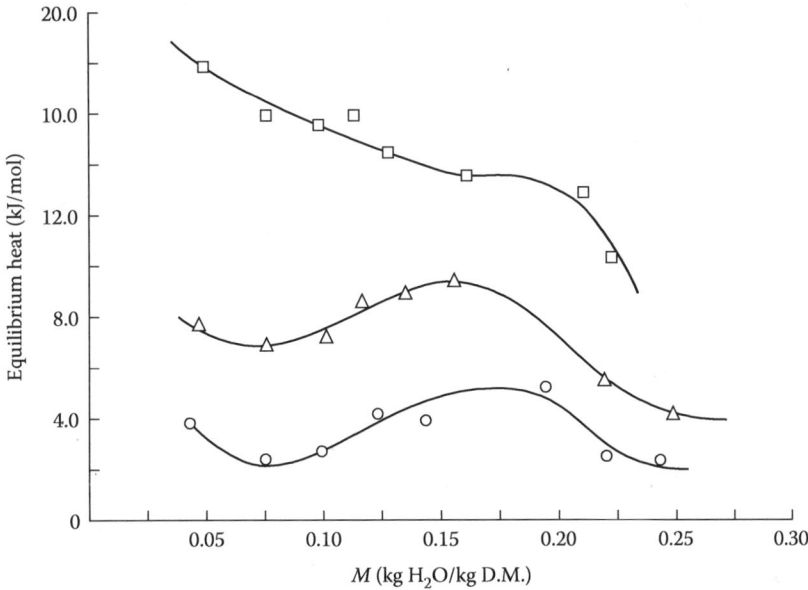

Figure 11.10 Equilibrium heat of adsorption and desorption of dehydrated rice for cycle 1 (showing hysteresis) compared with the equilibrium heat of sorption for cycle 2 (hysteresis free) at 20°C. Cycle 1 (V) adsorption and (M) desorption; (n) cycle 2. (From Benado, A. L., and Rizvi, S. S. H. 1985. *J. Food Sci.* 50: 101–105.)

1958, Rey proposed the glass transition temperature (T_g), at which glass–rubber transition occurs, as a threshold of instability for storage of biological tissues at low temperatures [178]. The properties of polymers near their glass transition temperatures have interested scientists for many years, and in the 1980s research activity on food materials using similar concepts was initiated by several groups [179]. This is logical in view of the dynamic behavior of nonequilibrium systems and the role of T_g as a physicochemical parameter for control of ultimate quality attributes of foods. A glass–rubber transition corresponds to a border between the solid and liquid states and is characterized by large changes in system properties such as viscosity, specific gravity, and diffusion. For multicomponent systems such as foods, T_g may well be a range of temperatures. The role of water as a plasticizer in lowering the T_g and as a determinant of the physical structure also becomes pivotal in the dynamic aspects of food behavior. T_g is operationally defined as the temperature at which the liquid viscosity reaches 10^{13}–10^{14} Pa s [180]. In the glassy state, below T_g, water is not available to support deteriorative reactions and foods are stable for extended time periods.

Figure 11.11 shows a schematic supplemental state diagram of an ideal system. T_m is the liquidus (liquid and solid coexisting in equilibrium) and is the only true equilibrium transition in the diagram. To distinguish it from an equilibrium phase diagram this is called a "state diagram." T'_g is the T_g at the maximal concentration of solute and its transformation into a glassy (amorphous) state. The time and temperature dependence

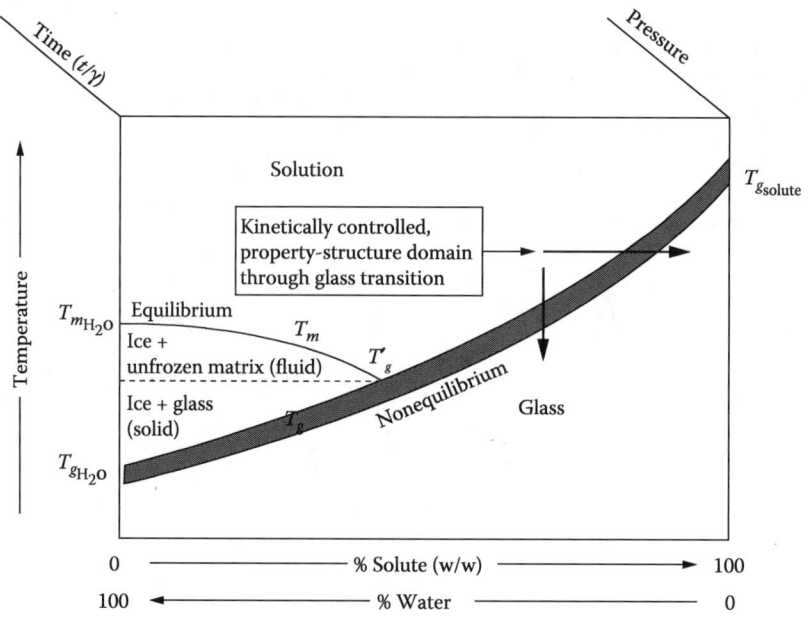

Figure 11.11 Idealized four-dimensional state diagram of a glass-forming system illustrating energetics of equilibrium and dynamics of nonequilibrium rubbery and glassy states. (From diagrams in Slade, L., and Levine, H. 1988. *Pure Appl. Chem.* 60: 1841–1846 and Gould, G. W., and Christian, J. H. B. 1988. Characterization of the state of water in foods—biological aspects. In *Food Preservation by Moisture Control*, C. C. Seow (ed.), pp. 43–56. Elsevier, London.)

of interrelationships among composition, structure, thermomechanical properties, and functional behavior have been conceptualized by Slade and Levine [179] by adding time (t/τ, where τ is relaxation time) and pressure scales to the state diagram to generate a four-dimensional "dynamic map." This permits accounting for the relationship between the experimental time scale and time frame of the relaxation experienced by the system. Pressure has been included as another possible critical variable of potential technological importance for dealing with processes occurring at various pressures. It is critically important to realize that the glass transition is very complicated, and significantly different values of T_g have been reported depending on the microstructure and the drying conditions of the material [181].

Studies have shown dramatic differences in viscosities of model solutions of different molecular weights and composition at equal a_w [184]. The water activity alone is not a useful parameter for following dynamic changes that may be influenced by diffusion and mobility in such systems. The glass transition parameters, however, in multicomponent, multiphase, and complex food are not easy to measure and are often undetectable. An integrated approach, including both a_w and T_g may be needed to understand and quantify the role of water in nonequilibrium, reduced moisture food systems. Over the water activity range of 0.1–0.8 a linear relationship between a_w and T_g has been reported for several

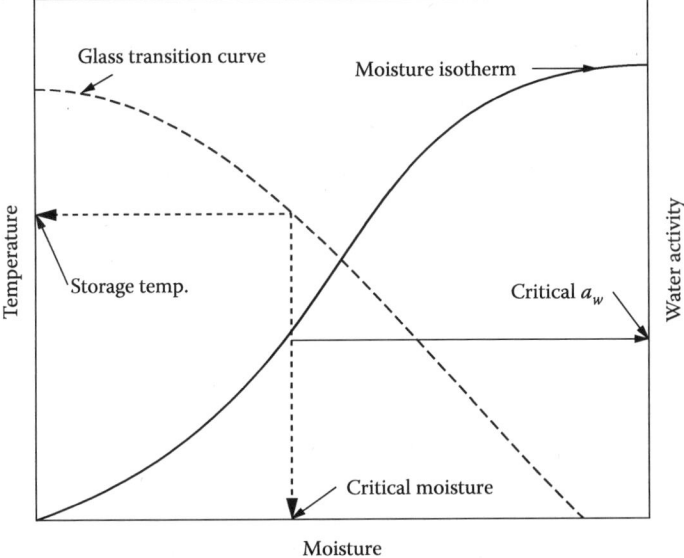

Figure 11.12 A schematic of integrated water activity and glass transition state diagram for storage of intermediate- and low-moisture foods.

amorphous food solids and food components and a schematic state diagram (Figure 11.12) showing depression of the glass transition temperature with increasing water activity for selection of storage conditions for low- and intermediate-food has also been proposed [185,186].

11.4 DEHYDRATION PRINCIPLES AND PROCESSES

Food dehydration is an energy-intensive unit operation, requiring 1000–2000 kJ per kilogram of water while the operating costs range from 4 to 15 U.S. cents per kg of water evaporated [187]. Dehydration involves simultaneous transfer of heat, mass, and momentum in which heat penetrates into the product and moisture is removed by evaporation into an unsaturated gas phase. Owing to the complexity of the process, no generalized theory yet exists to explain the mechanism of internal moisture movement. The desorption or adsorption of water is invariably accompanied by adsorption or evolution of heat that is conducted through the solid, resulting in temperature changes.

The major rate-limiting step is generally accepted to be internal mass transfer [188]. On the basis of the drying behavior, Bruin and Luyben [70] divided foods into three categories: (1) liquid solutions and gels such as milk, fruit juices, gelatinized products, and solutions containing dissolved materials; (2) capillary-porous rigid foods such as wheat and corn; and (3) capillary-porous colloidal foods as meats, vegetables, and tissues. In liquid solutions and gels, moisture moves by molecular diffusion from the interior to the surface, where it is removed by evaporation. In the case of capillary-porous materials, interstitial spaces, capillaries, and gas-filled cavities exist within the food matrix, and water transport can take place by several possible mechanisms, acting in various combinations. The possible mechanisms proposed by many workers include liquid diffusion due to concentration gradients, liquid transport due to capillary forces, vapor diffusion due to partial vapor pressure gradients, liquid or vapor transport due to the difference in total pressure caused by external pressure and temperature, evaporation and condensation effects, surface diffusion, and liquid transport due to gravity.

Additionally, Luikov [189] showed that moisture may also be transported inside a material if a suitable temperature gradient exists (thermogradient effect), because of thermodynamic coupling of heat and mass transport processes. Useful details on the mechanisms and theories involved are given in several reviews [70,190–194]. The relative contribution of the above mechanisms to any given food changes as drying progresses and the given set of conditions under which some of the mechanisms dominate are not well established. Furthermore, data on the thermal diffusivity and water diffusion coefficient for such complex materials are very sparse. In practice, therefore, the experimental approach to studying the problem related to drying remains the preferred alternative. Shortly before his death, Luikov [195] predicted that by 1985, advances in the understanding of the thermodynamics of moist materials and better solutions to the equations of change would obviate the need for empiricism in selecting optimum drying conditions. Poersch [196] succinctly concluded: "It is possible, though, to dry a product from experience and without having any theoretical base knowledge. But one cannot, without experience, design a dryer on the basis of the available theoretical knowledge." This is still true today.

Dehydration of foods generally involves a series of interdependent unit operations. In the last few years, research in this area has accelerated, and emphasis is now on combining of heat, mass, and momentum models of drying with product quality models to control the drying process more effectively. In most drying process optimization studies, overall moisture contents of foods are used as mass transfer potential. However, it has been recognized that different food constituents exhibit different affinities to moisture, and thus new simulation models accounting for such differences have now become available [197,198]. In this section, only the case of air-drying of foods is summarized and important properties are illustrated.

11.4.1 Drying Behavior

In air-drying processes, two drying periods are usually observed: an initial constant-rate period in which drying occurs as if pure water were being evaporated, and a falling-rate period where moisture movement is controlled by internal resistances. Figure 11.13 illustrates this by showing the moisture content as a function of time, where segment AB represents the initial unsteady state, warming-up period, and BC the constant-rate period. Figure 11.13 illustrates the derivative of the curve given in Figure 11.14 and shows the drying rate as a function of time. The same points are marked in Figure 11.15, where the drying rate is plotted against the moisture content. During the constant-rate period regime, the drying surface is saturated with water, and drying occurs at the wet-bulb temperature of the environment. The mechanism of internal liquid movement and consequently the structure of the food being dried determine the extent of the constant-rate period. In food

THERMODYNAMIC PROPERTIES OF FOODS IN DEHYDRATION

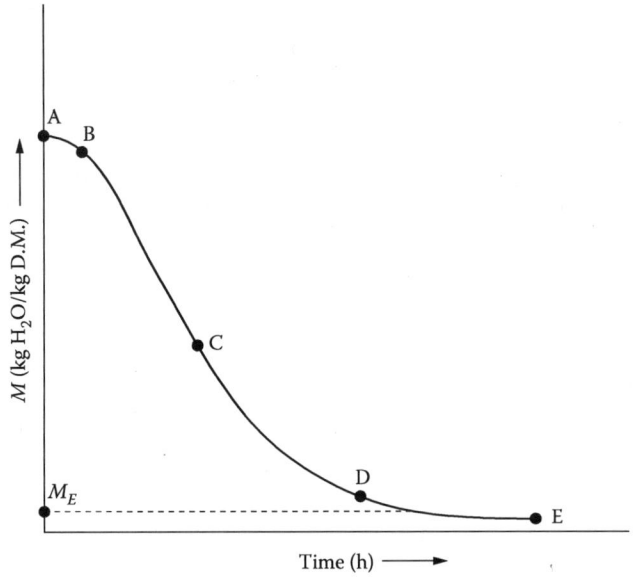

Figure 11.13 Drying curve, showing moisture content as a function of drying time.

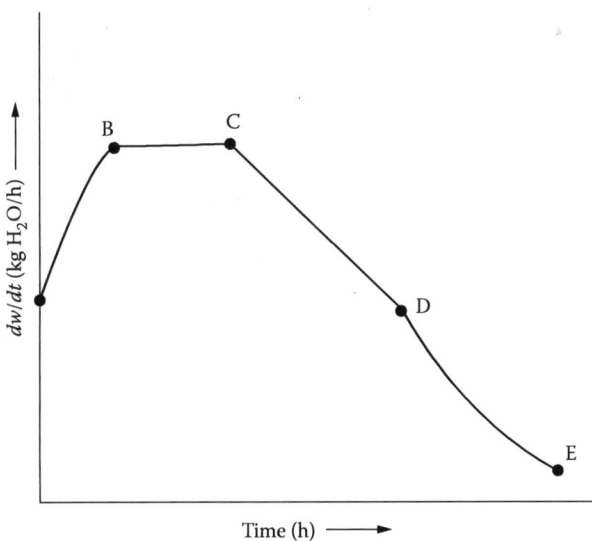

Figure 11.14 Drying rate curve showing rate of moisture removal as a function of drying time.

systems, where liquid movement is likely to be controlled by capillary and gravity forces, a measurable constant-rate period is found to exist. With structured foods, liquid movement is by diffusion, and therefore the water that is evaporated from the surface is not immediately replenished by movement of liquid from the interior of the food. Such foods are likely to dry without exhibiting any constant-rate period. Constant-rate drying periods have been reported for sweet potato, carrot, agar gel [199], fish [200], and several fruits and vegetables [201]. Under typical drying conditions, the absence of constant-rate periods has also been reported for air-drying of apples [202], tapioca [203], sugar beet root [204], and avocado [205].

The transition moisture content at which the departure from constant-rate drying is first noticed is termed the *critical moisture content* (M_c), indicated by point C in Figure 11.15. At this point, the moisture content of the food is not sufficient to saturate the entire surface. The critical moisture content generally increases with the thickness of the material and with the drying rate.

The drying period represented between points C and D in Figure 11.15 is termed the *first falling-rate period*. During this period, the rate of liquid movement to the surface is less than the rate of evaporation from the surface, and the surface becomes continually depleted in liquid water. Parts of the surface dry up by convective transfer of heat from the drying air, the surface temperature begins to rise, and vapor from inside the material starts diffusing into the gas stream until point D, where all evaporation occurs from the interior of the food. Beyond point D, the path for transport of both the heat and mass becomes longer and more tortuous as the moisture content continues to decrease. This period is called the *second falling-rate period*. Finally, the vapor pressure of the food becomes

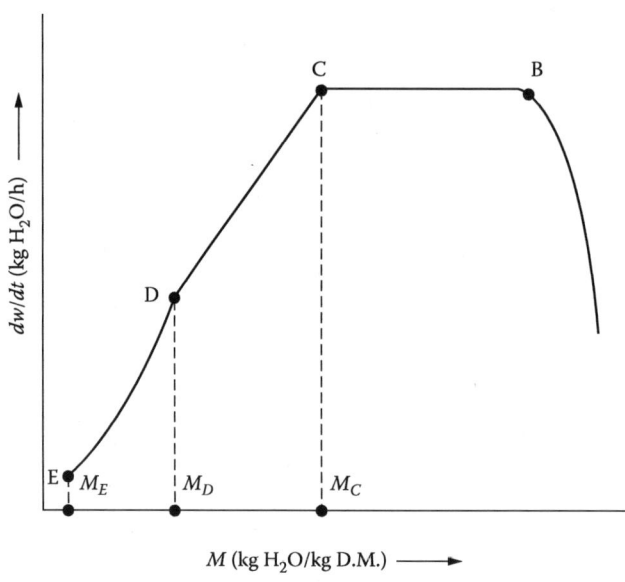

Figure 11.15 Drying rate as a function of moisture content.

equal to the partial vapor pressure of the drying air, and no further drying takes place. The limiting moisture content at this stage to which a material can be dried under a given drying condition is referred to as the *equilibrium moisture content* (M_e).

The fraction of the total drying time during which any of these four periods is operative depends on the ease of moisture transport through the food (i.e., the nature of the material) and the drying conditions. For example, if the rate of heat input is high, the constant-rate period may be too short to be noticed with some products. However, when the rate of input is low, the initial adjustment period may extend to the critical moisture content and no constant-rate period will be evident.

11.4.2 Constant-Rate Period

The rate of drying during the constant-rate period may be computed using either the mass transfer or heat-transfer equation. Because the surface of the material is maintained in a saturated condition and the temperature of the material is the wet-bulb temperature of the drying air, if we neglect heat transfer by conduction and radiation, the rate of drying is given as

$$-\frac{dw}{dt} = \frac{hA(T_d - T_w)}{\Delta \bar{H}_{vap}} = k_g A \left(P_w^{\bullet} - P_{wa} \right) \tag{11.92}$$

In terms of moisture content, on dry basis, it becomes

$$-m_s \frac{dM}{dt} = \frac{hA(T_d - T_w)}{\Delta \bar{H}_{vap}} = k_g A \left(P_w^{\bullet} - P_{wa} \right) \tag{11.93}$$

In practice, the heat-transfer equation gives a more reliable estimate of the drying rate than the mass transfer equation. Although correlations for calculation of the mass and heat-transfer coefficients have been proposed in the literature, few data are available to allow the constants in these correlations to be fixed with certainty. These coefficients are functions of the thickness of the air film around the products being dried, and are thus affected by the velocity of the air and its angle of impingement on the dry surfaces. Expressed in terms of the mass flow of air (G), the heat-transfer coefficient for the airflow parallel to the product surface is given as

$$h = C_1 G^{0.8} \tag{11.94}$$

and for the airflow at right angles to the drying surface or in through-flow drying the heat-transfer coefficient is given by

$$h = C_2 G^{0.4} \tag{11.95}$$

where C_1 and C_2 are constants.

The above equations indicate that increasing the mass flow of air will accelerate the drying rate. Additionally, a higher velocity minimizes the humidity differences between the inlet and outlet to the drying system, and thus a product of more uniform moisture content is generally obtained. Suzuki et al. [199] reported that the heat-transfer coefficient during the constant-rate period of drying of sweet potato, carrot, and agar gel is given by the following Nusselt-type equation:

$$\text{Nu} = b\,\text{Re}^{1/2}\,\text{Pr}^{1/3} + 2.0 \tag{11.96}$$

The mass transfer coefficient, k_m, is difficult to measure during the constant-rate drying period and is generally estimated by assuming that flow conditions influence heat and mass transfer in a similar fashion, and hence the Lewis number (Le = $h/k_m C_p$) equals unity. When the specific heat (C_p) of drying air is 1.21 kJ/kg °C), the relationship gives $k_m > 0.8$ h.

In food dehydration, it is often difficult to determine exactly the critical moisture content because of the shrinkage that occurs during drying. However, critical moisture contents at the end of the constant-rate period have been found to vary from 3.5 to 5 kg water/kg dry matter in vegetables and from 5.5 and 7.7 kg in fruits when a single layer of material in aluminum trays suspended in a tunnel dryer was dried [201]. With the reported critical moisture contents being close to the initial moisture contents, the relative importance of the constant-rate period becomes academic.

In spray drying, below a Reynold's number of 20 for spherical particles, the Nusselt number is generally given as

$$\text{Nu} = \frac{hD_p}{k_f} - 2 \tag{11.97}$$

The rate of evaporation in terms of change in moisture content is then expressed as

$$m_s \frac{dM}{dt} = \frac{2\pi k_f D_p (T_d - T_w)}{\Delta \bar{H}_{vap}} \tag{11.98}$$

The drying time in the constant-rate period is given by integrating the above equation:

$$t_c = \frac{m_s \Delta \bar{H}_{vap}}{2\Phi k_f (T_d - T_w)} \int_{M_2}^{M_1} \frac{dM}{D_p} \tag{11.99}$$

In many spray-drying systems, $T_d - T_w$ may not be constant, and a log mean temperature difference is substituted in its place. Solution of Equation 11.99 is easy when the liquid drops form a rigid structure by case hardening during the drying operation. When the volume of the drop is variable and decreases significantly with evaporation of water, the relationship between solids and moisture content is used to estimate the volume:

$$\left(\frac{M}{\rho_w} + \frac{1}{\rho_s}\right) m_s = \frac{\pi}{6} D_p^3 \tag{11.100}$$

Inserting the value of D_p in Equation 11.99 and integrating gives

$$t_c = \frac{3 m_s^{2/3} \Delta \bar{H}_{vap}}{4(6)^{1/3} \pi^{2/3} k_f (T_d - T_w)} \left[\left(\frac{M_1}{\rho_w} + \frac{1}{\rho_s}\right)^{2/3} - \left(\frac{M_2}{\rho_w} + \frac{1}{\rho_s}\right)^{2/3} \right] \tag{11.101}$$

The changes in surface areas to shrinkage during drying of foods have been investigated on a limited basis. A few quantitative investigations are reported in the literature [199,204,206]. Suzuki et al. [207] postulated three drying models for the formulation of the relation between the changes of the surface areas and moisture contents for air-drying of root vegetables (carrots, potatoes, sweet potatoes, and radishes). The core drying model, which assumes the formation of a dried layer at the outer side of the sample and the existence of the undried core at the center, was found to be in better agreement with the experimental data.

11.4.3 Falling-Rate Period

The first falling-rate period is the period of unsaturated surface dehydration. During this period, increasingly larger proportions of dry spots appear on the surface as drying progresses. To estimate the average drying time during the first falling-rate period, Fick's second law of diffusion has been used by several workers [158,208–211]. Assuming a constant diffusion coefficient, the partial differential equation for one-dimensional diffusion is given as

$$\frac{\partial M}{\partial t} = D_{\text{eff}} \left(\frac{\partial^2 M}{\partial r^2} + \frac{C}{r} \frac{\partial M}{\partial r} \right) \quad (11.102)$$

where C is a constant equal to 0 for planari, 1 for cylindrical, and 2 for spherical geometries. The initial and boundary conditions generally used are

$M(r,0) = M_i$ at $t = 0$
$M(0,t) = M_e$ at $r = r_0$ (at the surface)
$M(0,t) = $ finite   at $r = 0$ (at the center)

Assuming a uniform initial moisture distribution and in the absence of any external resistances, the analytical solutions of Fick's law for the most simple geometries are given in the form of infinite series [212,221]:

1. For an infinite slab:

$$\frac{\bar{M} - M_e}{M_i - M_e} = \frac{8}{\pi^2} \sum_{n=0}^{\infty} \frac{1}{(2n+1)^2} \exp\left[-\frac{(2n+1)^2}{4} \pi^2 X^2 \right] \quad (11.103)$$

2. For an infinite cylinder:

$$\frac{\bar{M} - M_e}{M_i - M_e} = \sum_{n=1}^{\infty} \frac{4}{\gamma n^2} \exp\left[-\frac{\gamma n^2}{4} X^2 \right] \quad (11.104)$$

3. For a sphere:

$$\frac{\bar{M} - M_e}{M_i - M_e} = \frac{6}{\pi^2} \sum_{n=1}^{\infty} \frac{1}{n^2} \exp\left[-\frac{n^2 \pi^2}{9} X^2 \right] \quad (11.105)$$

where $X = (A/V)(D_{\text{eff}} \, t)^{1/2}$.

For long drying times and unaccomplished moisture ratios $[(\bar{M} - M_e)/(M_i - M_e)]$, less than 0.6, generally only the first term of Equation 11.103, 11.104, or 11.105 is used to estimate the drying rate. The expressions thus reduce to the straight-line equation

$$\frac{\bar{M} - M_e}{M_i - M_e} = Ce^{-Kt} \qquad (11.106)$$

where C is a constant and K is called the dehydration constant (h^{-1}).

Thus, the drying time for the first falling rate versus the unaccomplished moisture ratio shows a linear relationship on semilogarithmic coordinates. This permits calculation of an effective diffusion coefficient (D_{eff}). Values of D_{eff} for several foods calculated during the first falling-rate period are shown in Table 11.13 and generally range from 10^{-11} to 10^{-9} m²/s. The reported values also illustrate the influence of variety and composition of foods on D_{eff}. The negative correlation of fat content with D_{eff} has been attributed to the hydrophobicity of fat [205,214]. The influence of temperature on D_{eff} has been attempted via an Arrhenius type of equation for different dry-bulb temperatures, and the activation energy values obtained are also given in Table 11.13. In view of the complexity and diversity of the drying mechanisms involved, it is rather difficult to provide a theoretical explanation for the behavior of activation energy. Despite the complexity of moisture transport mechanisms, Vaccarezza et al. [204] found that Fick's law can be used to predict with reasonable accuracy the average drying time, internal moisture distribution, and sample temperature during dehydration of sugar beet root. Figure 11.16 shows experimental and theoretical moisture distribution curves for various drying times of sugar beet root.

For situations in which the diffusion coefficient is dependent on concentration, Schoeber and Thijssen [215] reported the development of a regular regime method as an alternative technique for calculation of drying processes: The *regular regime period* is defined as the time during a transient diffusion process in which the concentration changes with time are taken into account but the effect of the initial condition on the process is neglected. A requirement for the application of this technique is knowledge of the regular regime curve at constant surface concentration and at a desired temperature. The curve characterizes the internal diffusion process at sufficiently long drying times that the concentration profiles inside the material are no longer dependent on the initial moisture content. This shortcut method is practical for calculating drying curves for many process conditions in which diffusion is concentration-dependent. Using this approach, Bruin and Luyben [70] successfully measured drying curves for a number of liquid and solid foods (glucose–water, skim milk, coffee extract, apple pieces, and potato pieces) with reasonable accuracy.

The drying analysis presented above is based on the assumption that the heat-transfer effects can be neglected and drying can be treated as a purely diffusion-controlled mass transport phenomenon with a constant effective diffusion coefficient. This approach is based on several experimental studies, which indicate the existence of very small internal temperature gradients within foods during drying [200,201,204,210,216–220]. During the second falling-rate period, drying occurs at a moisture content where the ERH is below saturation. In some instances, case hardening has been reported to occur as the drying rate changes from the first falling rate period to the second falling rate period [221]. For porous dried foods at less than saturation water vapor pressure, vapor-phase diffusion has been

Table 11.13 Effective Diffusion Coefficients and Activation Energy for Moisture Diffusion in Various Foods during the Initial Phase of the Falling-Rate Period

Food	Temperature (°C)	D_{eff} (m^2/s)[a]	E_a (kJ/mol)	Ref.
Aloe Vera	30	5.6E – 10	24.4	224
Apple				
Granny Smith	30	2.6E – 10	—	225
	76	3.6E – 09	—	226
McIntosh	66	1.1E – 09	—	205
Avocado 14.7% oil	31	1.1E – 10	39.8	227
Cheese	20	3.2–10	—	228
Fish				
Dogfish, 4.0% fat	30	2.2E – 10	—	218
Dogfish, 14.6% fat	30	1.3E – 10	—	218
Herring, 2% fat	30	1.9E – 10	—	218
Herring, 12.5% fat	30	3.9E – 11	—	218
Swordfish, 2–3% fat	40	3.0E – 10	15.1	229
Mullet roe	55	3.9E – 10	—	229
	20	4.2E – 10	37.2	230
	40	1.2E – 9	—	230
Fish Muscle				
Cod, 0.05% fat	30	3.4E – 10	29.7	218
Whiting, 0.036% fat	25	8.2E – 11	—	218
Starch Gel	25	2.4E – 11	18.8	160
Sugar beet root	47	3.8E – 10	28.9	207
	60	7.0E – 10	—	207
Tapoica root	55	3.5E – 10	22.6	220
	84	6.7E – 10	—	220

[a] E – 10 = 10^{-10}

proposed as a likely mechanism of moisture transport [222]. A few studies [158,223] have been made to calculate a moisture-dependent diffusion coefficient based on the solution of Fick's second law of diffusion of the form

$$\frac{\partial M}{\partial t} = \frac{\partial}{\partial x}\left(D_{eff}\frac{\partial M}{\partial x}\right) \quad (11.107)$$

The variable diffusivity given in Equation 11.107 can be solved by analytical and numerical techniques. Crank [213] presented methods for determining the functional dependence of D_{eff} on moisture content. Chirife [224] compiled values of moisture diffusivity for

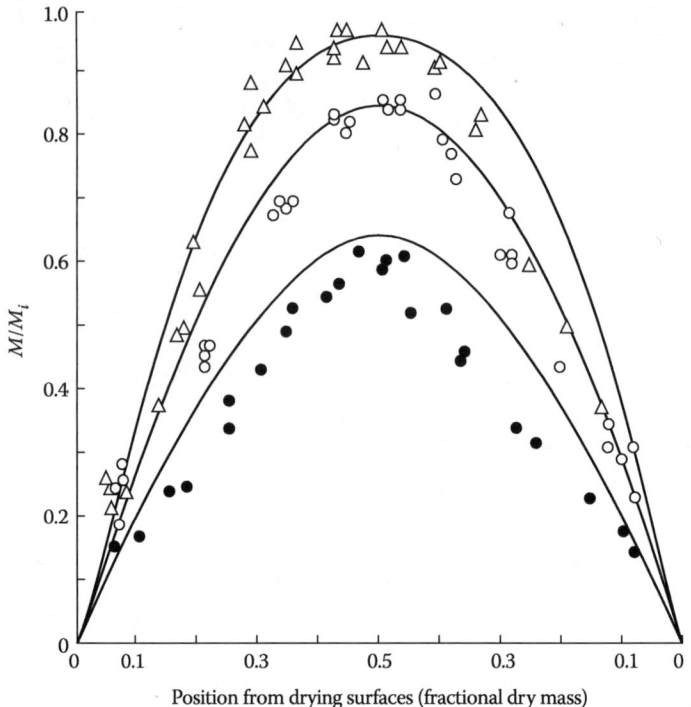

Figure 11.16 Comparison of theoretical and experimental moisture distribution curves for various times of sugar beet root drying at 81°C dry-bulb temperature. Experimental: (n) 45 min; (V) 65 min; (v) 120 min. Theoretical: solid line curves. (From Vaccarezza, L. M., Lambardi, J. L., and Chirife, J. 1974. *Can. J. Chem. Eng.* 52: 576–579.)

various foods in the last phase of the falling-rate period (Table 11.14) and noted that D_{eff} values are generally about 4–8 times lower than those found in the first falling-rate period.

King [222] suggested that in drying analysis, when the ERH is less than 100%, heat transfer should be considered along with mass transfer because desorption of moisture requires consumption of substantial amounts of heat. He showed that the effective diffusion coefficient is related to such physical properties of the food as thermal conductivity, bulk density, and enthalpy of desorption, as well as to environmental conditions. The equation developed by King [222], which relates D_{eff} to various properties, can be transformed into Equation [225]

$$D_{eff} = \frac{M_w}{\rho_{ps}} B \left(\frac{\partial a_w}{\partial M}\right)_T P_w^\bullet \frac{\alpha}{1+\alpha} \tag{11.108}$$

where

$$\alpha = \frac{RT^2 k}{Ba_w P_w^\bullet Q_{st}^2} \tag{11.109}$$

Table 11.14 Effective Diffusion Coefficients of Moisture during the Last Phases of Falling-Rate Period

Food		Temperature (°C)	D_{eff} (m²/s)[a]	Ref.
Apple				
<0.13–0.15% moisture		30	4.9E – 11	226
Fish				
Cod	0.05% fat	30	8.1E – 11	218
Catfish	0.10% fat		8.0E – 11	218
Haddock	0.105% fat	30	6 E – 11	218
Halibut	0.208% fat	30	5.8E – 11	218
Whiting	0.036% fat	30	4.8E – 11	218
Herring	2.9% fat	30	7.9E – 11	218
	12.5% fat	30	1.6E – 11	218
	16.2% fat	30	1.3E – 11	218
Potato				
<15% moisture		65	2 E – 10	234
Pepperoni				
13.3% fat		12	5.7E – 11	232
25.1% fat		12	4.7E – 11	232
Starch gel				
0.8% moisture		25	1 E – 14	160
6.3% moisture		25	1.5E – 13	160
14.1% moisture		25	3.6E – 12	160
Turkey, solid phase				
4% moisture		22	0.8E – 14	235
Wheat				
12–30% moisture		20.8	6.9E – 12	236
		50.0	5.7E – 11	236
		79.5	2.8E – 10	236

Source: Adapted from Chirife, J. 1983. Fundamentals of the drying mechanism during air dehydration of foods. In *Advances in Drying*, Vol. 1, A. K. Majumdar (ed.), pp. 73–102. Hemisphere, Washington, DC.

[a] E – 11 = 10^{-11}.

The effective diffusion coefficient given by the above expression has been rationalized by King [222] through a simple physical concept. The term $\alpha/1(1+\alpha)$ in Equation 11.108 determines the degree of heat or mass transfer control. For values of α ! 1, the process is totally controlled by heat transfer, and if α @ 1, the process is then entirely controlled by mass transfer. A more accurate and precise analysis of drying processes requires solution of coupled differential equations for moisture content and temperature history of the subject food with a variable diffusion coefficient, and the model becomes rather cumbersome to use because an analytical solution in the falling-rate period cannot be given. Harmathy [226] and Husain et al. [227] presented numerical solutions of a set of differential equations to predict the moisture and temperature histories during drying. Subsequent studies [228–232] have demonstrated the usefulness of the finite-element method of analysis to model the water diffusion coefficient as a function of moisture concentration, size, shape, and diffusional potential.

Whitakar and Chou [194] attempted to show that diffusion-like theories of drying cannot describe the complete spectrum of moisture transport mechanisms that occur during the drying of a granular porous medium and that gas-phase momentum should be included in any comprehensive theory of drying granular material. Their theory incorporates the liquid- and vapor-phase continuity equations, combines the liquid-, solid-, and vapor-phase thermal energy equations into a single temperature equation, and makes use of Darcy's law for the liquid phase to account for moisture transport due to capillary action. It will be interesting to see results from the application of such a refined mathematical approach to the drying of food and other related biomaterials.

A modified Crank method has been used by Gekas and Lamberg [233] to determine the diffusion coefficient in systems where volume change occurs during drying. Preliminary results indicate that volume changes are neither one-dimensional nor isotropically three-dimensional but follow a fractal–dimension relationship to the change of thickness. Despite numerous attempts by researchers around the world, it is still not possible to provide a complete description of the behavior of food materials during drying, partly because of the complexity and heterogeneity of such materials and partly due to changes in volume and structures of the product [234]. Availability of rapid methods of monitoring the state of water during drying has aided modeling of transport and degradation phenomena. One- and two-dimensional transient moisture profiles have been measured nondestructively by two- and three-dimensional Fourier transform nuclear magnetic resonance (NMR) imaging [235,236]. A methodology for the use of proton NMR in the rapid detection of various states of water during drying of carrots has been presented by Marques et al. The moisture profiles of foods in drying processes have also been obtained using scanning neutron radiography [237] and computer simulation with a mathematical model [238,239] with limited success.

11.4.4 Equilibrium Moisture Content

The moisture content remaining in a dry material when the drying rate drops to zero at specified conditions of the drying medium is called the *equilibrium moisture content*. It is in equilibrium with the vapor contained in the drying gas, and its magnitude is a function of the structure and type of the subject food and of the prevailing drying conditions. The equilibrium moisture values predicted by the *static* and *dynamic* moisture sorption [terms introduced by Becker and Sallans [136,240] to differentiate between the surface moisture content obtained from equilibrium isotherms and drying experiments, respectively] do not always agree over the whole range of relative humidity of the drying air. In a study [130] on the analysis of the interface conditions during drying of rough rice, it was shown that the equilibrium moisture given by the static desorption isotherm is valid only when drying is done with air of high relative humidity (43–59% for a drying temperature ranging from 40°C to 70°C). The reported experimental results also indicated that the surface moisture approaches the monolayer moisture content given by the BET isotherm when drying is done with air of low relative humidity (7–14% for the same temperature range). The high value of the heat of desorption related to the monolayer moisture content was offered as a possible explanation of the effects observed. The authors concluded that when drying takes place with air of relative humidity less than that corresponding to the equilibrium value with the monolayer moisture content, the equilibrium conditions no longer hold and the static and dynamic moisture sorption values show differences. Roth and Loncin [241] reported three methods for estimating the superficial water activity of foods during drying by measuring the steady-state temperatures and/or the drying rates of samples. These techniques are good for estimating the surface water activity in the constant drying rate period and for computing the effects of surface treatment on drying rates.

Gal [242] pointed out that the so-called marking points on MSIs of foods, as suggested by Heiss [243], serve very useful purposes in determining the end point of drying that provides the optimum moisture content in the final dried product. As illustrated in Figure 11.17, drying has two target water activities, one corresponding to the prevailing ambient relative humidity while drying, and the other, to the intended storage temperature. These points enable the correct end point of the drying process to be set so that food can be dried to its optimum moisture content for maximum storage stability. Knowledge of the drying end point also permits estimation of the maximum allowable humidity of the drying air. The moisture contents corresponding to the water activity at drying and storage temperatures are also indicated on the desorption isotherm in Figure 11.17. A sample of recommended specification (minimum) moisture contents for dehydrated foods representing a variety of compositions was compiled by Salwin [244] and is shown in Table 11.15, along with the corresponding monolayer moisture values.

During storage of dry foods, the adsorption branch of the isotherm is generally important, and Figure 11.17 shows the marking points corresponding to the point at the end of the first curved part of the isotherm (M_b), the maximum allowable moisture content (M_x) at which the food becomes unacceptable, and the ambient point (M_a) corresponding to equilibrium with the ambient environmental atmosphere. Because these marking points are temperature dependent, knowledge of the storage condition becomes important for their precise determination. On the basis of his work, Heiss [245] has listed maximum permissible ERHs for many foods corresponding to point M_x (Table 11.16). These values are important from the storage and packaging point of view and are useful in establishing the permeability requirements of polymeric films.

11.4.5 Energy Requirements

Among the most important parameters in the evaluation, design, and specification of drying systems are the energy requirements and drying times. A large number of analytical expressions for drying time and energy requirements have been developed, as have several numerical solutions to the more complex head and mass transfer equations. With few exceptions, the models developed have taken the heat of vaporization (or sublimation) to

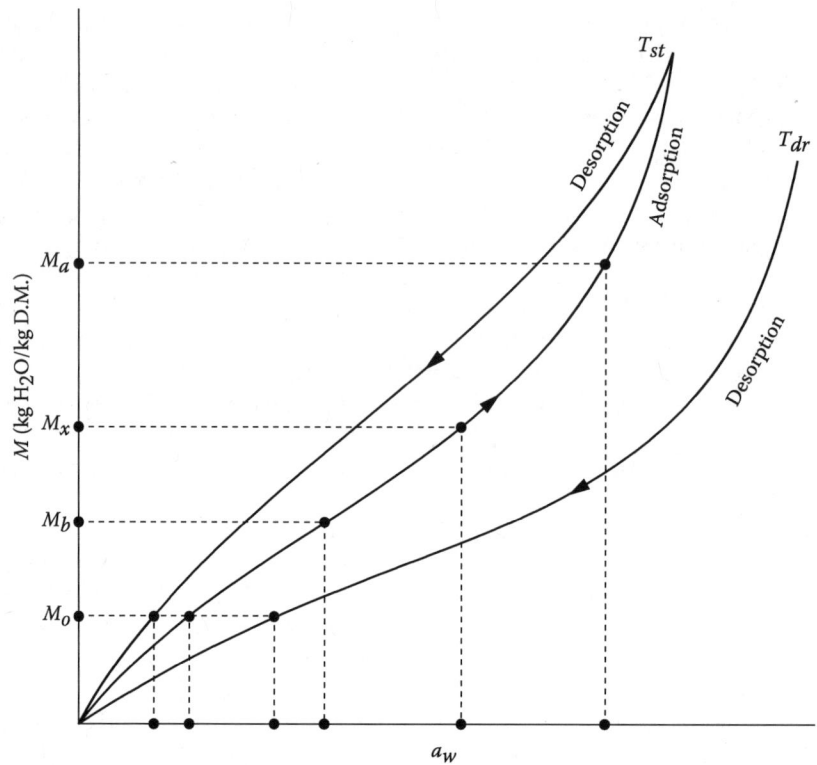

Figure 11.17 Typical marking points on a schematic sorption isotherm for use in drying and storage calculations.

be constant and invariant with moisture content. This is in fact true in most drying situations. As Keey [151] pointed out, even with markedly hygroscopic materials the enthalpy of evaporation does not differ significantly from the latent heat of vaporization of pure water until moisture contents of 0.1 kg/kg or less are reached. Most commercial operations do not dry materials to that degree. There are, however, some situations where drying may go below 0.1 kg/kg in moisture content.

The isosteric heat of sorption (Q_{st}) has been used in estimating the drying times. Assuming constant air density, the energy equation is given by (Roman et al., 133)

$$\rho_{eff}(C_p)_{eff}\frac{\partial T}{\partial t} + C_{pv}N_v\frac{\partial T}{\partial z} + \left(Q_{st} + M\frac{\partial Q_{st}}{\partial M}\right)_{T,P}\rho_{ps}\frac{\partial M}{\partial t} = \frac{\partial}{\partial z}\left(k\frac{\partial T}{\partial z}\right) \qquad (11.110)$$

Roman et al. [133] performed simulations based on this equation and found that the moisture loss rate during the drying of apples was not much affected by a 20% variation in Q_{st}. This corresponded to a moisture content of 12% on a dry basis, or an a_w of approximately

Table 11.15 Comparison of BET Mononuclear Layer of Adsorbed Water with Analytical and Specification Moisture Contents of Dehydrated Precooked Foods

	Percent Water, On an As-Is Basis		
Food item	BET Monolayer	Analytical Value	Specification Limit
Potato dice	5.46	5.84	6
Small red beans	4.50	4.73	4
Lima beans	5.37	3.93	4
Navy beans	5.21	3.19	4
Onion powder	3.58	4.10	4
Crackers	4.23	5.04	5
Instant macaroni	5.87	6.99	—
Instant starch	5.68	6.28	—
Dry whole milk	1.97	1.87	2.25
Nonfat dry milk	2.98	3.42	3.50
Instant nonfat dry milk	3.52	4.19	3.50
Spray dried cheese	2.22	1.82	2.50
Cocoa beverage powder	2.37	2.92	3
Beef soup and gravy base	2.38	2.78	4
Chicken soup and gravy base	1.68	2.43	4
Shrimp	5.56	3.09	2.5
Chicken	5.48	1.53	1.5
Ground beef[a]	6.19	0.78	2.25

Source: From Salwin, H. 1959. *Food Technol.* 13: 594–595.
[a] Fat-free basis.

0.6. In order to dry apples below this a_w, the increase in isosteric heat must be accounted for if the model is to predict drying times and energy requirements accurately.

The general case of drying of food materials involves energy inputs to meet the following energy requirements:

1. Removal of free water through sublimation or evaporation
2. Removal of water associated with the food matrix
3. Superheating of water vapor sublimed or evaporated as it passes through the food
4. Internal energy changes, that is, the supply of sensible heat to the foodstuff as it changes temperature

The energy for superheating the vapor and changing the internal energy of the food can usually be neglected inasmuch as the supply of sensible heat is usually minimal, on the order of the magnitude of the heat of vaporization/sublimation. The energy required to remove water from the food matrix will thus be given by the sum of the first two items. For drying at temperatures above the freezing point of water, the first two terms are combined

Table 11.16 Maximum Permissible Water Activity Values for Some Unpackaged Dry Foods at 20°C

Food	a_w	Food	a_w
Baking soda	0.45	Sugars	
Crackers	0.43	Pure fructose	0.63
Dried eggs	0.30	Pure dextrose	0.89
Gelatin	0.43–0.45	Pure sucrose	0.85
Hard candies	(0.25–)0.30	Maltose	0.92
Chocolate, plain	0.73	Sorbitol	0.55–0.65
Chocolate, milk	0.68	Dehydrated meat	0.72
Potato flakes	0.11	Dehydrated vegetables	
Flour	0.65	Peas	0.25–0.45
Oatmeal	(0.12–)0.25	Beans	0.8–0.12
Dried skim milk	0.30	Dried fruits	
Dry milk	(0.20–)0.30	Apples	0.70
Beef-tea granules	0.35	Apricots	0.65
Dried soups	0.60	Dates	0.60
Roast coffee	(0.10–)0.30	Peaches	0.70
Soluble coffee	0.45	Pears	0.65
Starch	0.60	Plums	0.60
Wheat preparations (macaroni, noodles, spaghetti, vermicelli)	0.60	Orange powder	0.10

Source: Adapted from Heiss, R. 1958. *Mod. Packag.* 31(8): 119–124, 172, 176.

to give the isosteric heat of sorption. To dry food from any moisture M_1 to another moisture M_2 thus requires the integration of the isosteric heat between the limits of the two moisture contents. On a per-kilogram basis, then [151]

$$Q_{req} = \frac{1}{M_2 - M_1} \int_{M_1}^{M_2} Q_{st} dM \tag{11.111}$$

Rizvi and Benado [169] compared the cumulative energy required to remove water from grain sorghum with the corresponding isosteric heat and the heat of vaporization of water in its pure form. Their results are shown in Figure 11.18, along with cumulative heats for chamomile and nutmeg calculated from the isosteric heat curves reported by Iglesias and Chirife [135]. The cumulative energy curves show the amount of heat required to remove water from a food starting from a moisture content of 0.25 kg/kg. At 0.25 kg/kg moisture, the isosteric heat of sorption of all these materials is very close to the heat of vaporization of water. As drying proceeds, the isosteric heat begins to rise at a rapid rate. The cumulative energy requirement rises at a slower rate because the less tightly bound water is removed first, making the overall energy expended at a given point in the drying

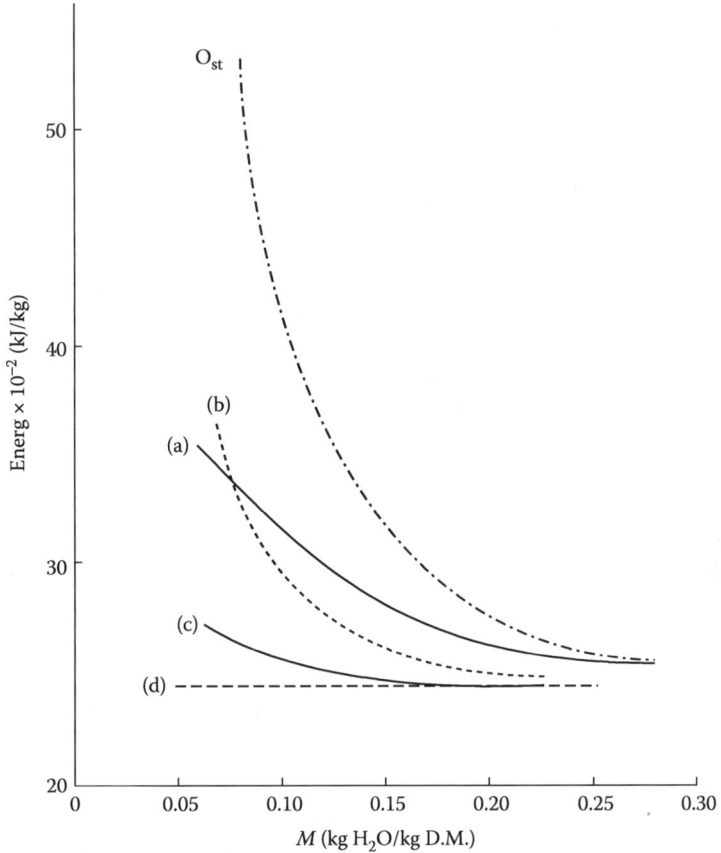

Figure 11.18 Cumulative energy requirement for water removal from selected foods: (a) grain sorghum; (b) nutmeg; (c) chamomile; (d) pure water. (From Rizvi, S. S. H., and Benado, A. L. 1983–84. *Drying Technol.* 2(4): 471–502.)

process lower than the isosteric heat, which is the instantaneous energy of binding. If the calculations had been started at a higher moisture content, the effect would have been to lower the cumulative energy curve more. Thus, although the isosteric heat rises at a very fast rate, the cumulative energy requirement lags behind considerably. Figure 11.18 shows, however, that the increase in energy required to remove water can be significant.

The choice of drying system depends on the type of food, the quantity to be dried, and the energy-saving potentials. Various types of dryers are available commercially, each with its own specialty. Nonhebel and Moss [246] give a classification of dryers (Table 11.17) that may serve as a guide for selection. In addition to the heat transfer by forced convection during air drying, conduction and radiation contribute significantly in some types of drying systems. The overall heat-transfer coefficients for some of the dryers are given in Table 11.18.

Table 11.17 Classification of Dryers by Scale of Production

Small Scale (to 20–50 kg/h)	Medium Scale, 50–1000 kg/h		Large Scale (>1000 kg/h) Continuous
Batch	Batch	Continuous	Continuous
Vacuum tray	Agitated	Fluidized bed	Indirect rotary
Agitated	Through-circulation	Vacuum bed	Spray
Convection tray	Fluidized bed	Indirect rotary	Pneumatic
Through-circulation		Spray	Direct rotary
Fluidized bed		Pneumatic	Fluidized bed
		Band conveyer	
		Tray	
		Through-circulation	

Source: Adapted from Nonhebel, G., and Moss, A. A. H. 1971. *Drying of Solids in the Chemical Industry.* Butterworths, London.

Table 11.18 Overall Heat-Transfer Coefficients (U) for Various Types of Drying Equipment

Type of Dryer	$U[W/(m^2 \cdot K)]$
Vacuum shelf	5–6
Indirect rotary	11–57
Jacketted trough	11–85
Rotary vacuum	28–284
Agitated tray	28–340
Drum	1135–1700

Source: Adapted from Williams-Gardner, A. 1971. *Industrial Drying.* Leonard Hill, London.

11.5 CONCLUSION

As a measure of chemical potential, the activity of water or any other quantity represents a thermodynamic function of state and thus should not depend on the path taken to achieve a given condition. Existence of hysteresis loops in the MSIs of foods is indicative of a non-equilibrium state, no matter how reproducible the data. Thermodynamic functions computed from hysteresis data are therefore not rigorously accurate. However, they provide upper and lower bounds on thermodynamic quantities and serve useful practical purposes. For instance, the variation of drying enthalpy indicates the level at which the interaction between water and food molecules is larger than the interaction between water molecules. Similarly, the variation of entropy with moisture content is related to the order and disorder concepts manifested during such changes as crystallization, swelling, dissolution, and similar phenomena. The monolayer moisture content, despite its oversimplified interpretation, provides a practical guide for drying food to minimize vulnerable quality factors like fat oxidation, nonenzymatic "browning," aroma loss, and enzymatic degradation, among others.

From a theoretical point of view, dehydration of porous materials, such as foods, is a rather complex process. It involves interactions not only between heat and mass transport processes occurring within the food itself, but also between the food and the drying medium under circulation around the solid matter. Successful solution of these competing phenomena requires simultaneous solution of the separate differential equations for heat, mass, and momentum transport within the food system being dried and in the external drying medium. Coupling of the two processes at the surface of the solid for general theoretical solutions to the overall drying problem, along with the dependence of the transport coefficients on the values of driving forces, presents serious computational problems. With the advent of sophisticated computing systems, limited progress has been made in predicting the drying profile and characteristics of a few simple systems. Further complications arise as a result of the lack of thermophysical property data for real foods. The sophisticated theories of drying and calculation procedures being developed will require some time to replace the empiricism currently applied in dryer design. Clearly, much work is needed toward obtaining numerical or analytical solutions of the basic differential equations for drying of real foods under practical drying conditions.

LIST OF SYMBOLS

A	Transfer area, m²
a	Activity
a_w	Water activity
a_{w1}	Water activity of component 1
a_{wu}	Uncorrected water activity
B	Vapor–space permeability, kg · mol/(m · h · atm)
b	Constant
C	Constant
C_p	Specific heat at constant pressure, kJ/(kg · °C)
D_p	Diameter of sphere, m
D_{eff}	Effective diffusion coefficient, m²/h
E	Internal energy, kJ
E_a	Activation energy, kJ/mol
g	Gravitational constant, m/s²
G	Gibbs free energy, kJ
H	Enthalpy, kJ
h	Convective heat-transfer coefficient, W/(m² · °C)
J	Difference in molal heat capacities of liquid water and ice, kJ/(mol · °C)
K	Drying constant, h⁻¹
k	Thermal conductivity, W/(m · °C)
k_f	Thermal conductivity of the gas (for dry air, approx. $9.1 \times 10^{-7} T^{1.8}$, T in kelvins), W/(m · °C)
k_g	Mass transfer coefficient, kg/(m² · h · Pa)
L_1	Manometer leg height with the sample flask connected, m
L_2	Manometer leg height with the desiccator flask connected, m

Le	Lewis number, dimensionless
M	Moisture content, kg water/kg dry matter
$\bar{M}$	Average moisture content, kg water/kg dry matter
M_o	Monolayer moisture content, kg water/kg dry matter
M_f	Molecular weight of food, kg/kg mol
M_w	Molecular weight of water, 18 kg/kg mol
M_s	Mass of dry solids, kg
m	Solution molality
N	Number of moles
Nu	Nusselt number, dimensionless
n	Integer
P	Pressure, torr, Pa, atm
P_{wa}	Partial pressure of water vapor in air, Pa
Pr	Prandtl number, dimensionless
Q_{st}	Isosteric enthalpy of sorption, kJ/mol
q_{st}	Net isosteric enthalpy of sorption, kJ/mol
R	Gas constant, m$^3 \cdot$ Pa/(kg $\cdot$ K)
Re	Reynolds number, dimensionless
r	Product coordinate, m
S	Entropy, kJ/K
T	Temperature, K, °C
T_d	Dry-bulb temperature, °C
T_{dr}	Drying temperature, °C
T_f	Freezing point of solution, K
T_m	Temperature of manometer leg, K
T_0	Freezing point of water, K
T_s	Temperature of sample, K
T_{st}	Temperature of storage, °C
T_w	Wet-bulb temperature, °C
t	Time, h
t_c	Time in the constant rate period, h
V	Volume, m^3
x	Mole fraction
Z	Distance coordinate, m

Greek Symbols

ϕ	Surface potential, kJ/mol
μ	Chemical potential, kJ/mol
Γ	Number of moles of water per "mole" of food
σ	Sloper of the linear part of the MSI, kg water/(kg $\cdot a_w$)
γ_a	Activity coefficient
γ_f	Fugacity coefficient
γ_n	Roots of the zero-order Bessel function
ρ	Density, kg/m^3

Subscripts

A	Adsorbent
b	End of the first curved part of the MSI
a	Adsorption
d	Desorption
e	Equilibrium
eff	Effective
fus	Fusion
G	Gas phase
i	Chemical species, initial condition
L	Liquid phase
l	Adsorbate
m	Minimum
n	Number of layers in the BET modification
o	Monolayer
$0A$	Bare surface
p	Pressure
ps	Dry, solid particles
req	Required
s	Sorbed species
v	Vapor
vap	Vaporization
w	Water
x	Maximum

Superscripts

0	Standard
□	Pure
—	Molar
^	Differential

REFERENCES

1. Frank, H. S. 1970. The structure of ordinary water. *Science* 169: 535–641.
2. Luck, W. A. P. 1981. Structure of water in aqueous systems. In *Water Activity: Influences on Food Quality*, L. B. Rockland and G. F. Stewart (eds.), pp. 407–434. Academic, New York.
3. Given, P. S. 1991. Molecular behavior of water in a flour-water baked model system. In *Relationships in Foods*, H. Levine and L. Slade (eds.), pp. 465–483. Plenum, New York.
4. Berlin, E. 1981. Hydration of milk proteins. In *Water Activity: Influence on Food Quality*, L. B. Rockland, and G. F. Stewart (eds.), pp. 467–488. Academic Press, New York.
5. Steitz, R., Gutberlet, T., Hauss, T., Klosgen, B., Krastev, R., Schemmel, S., Simonson, A. C., and Findenegg, G. H. 2003. Nanobubbles and their precursor layer at the interface of water against a hydrophobic substrate. *Langmuir* 19: 2409–2418.

6. Jenson, T. R., Jensen, M. O., Reitzel, N., Balashev, K., Peters, G. H., Kjaer, K., and Bjornholm, T. 2003. Water in contact with extended hydrophobic surfaces: Direct evidence of weak dewetting. *Phys Rev Lett* 0: 086–101.
7. Yamanisky, V. and Ohnishi, S. 2003. Physics of hydrophobic cavities. *Langmuir* 19: 1970–1976.
8. Hass, J. L. 1970. Fugacity of H_2O from 0° to 350° at liquid–vapor equilibrium and at 1 atmosphere. *Geochim. Cosmochim. Acta* 34: 929–934.
9. Holser, W. T. 1954. Fugacity of water at high temperatures and pressure. *J. Phys. Chem.* 58: 316–317.
10. Norrish, R. S. 1966. An equation for the activity coefficients and equilibrium relative humidities of water in confectionery syrups. *J. Food Technol.* 1: 25–31.
11. Ross, K. D. 1975. Estimation of water activity in intermediate foods. *Food Technol.* 29: 26–34.
12. Ferro-Fontan, C. F., and Chirife, J. 1981a. The evaluation of water activity in aqueous solutions from freezing point depression. *J. Food Technol.* 16: 21–30.
13. Ferro-Fontan, C. F., and Chirife, J. 1981b. A refinement of Ross' equation for predicting the water activity of non-electrolyte mixtures. *J. Food Technol.* 16: 219–221.
14. Franks, F. 1982. Water activity as a measure of biological viability and quality control. *Cereal Foods World* 27: 403–407.
15. Troller, J. A. 1980. Influence of water activity on microorganisms in foods. *Food Technol.* 34(5): 76–80, 82.
16. Beuchat, L. R. 1981. Microbial stability as affected by water activity. *Cereal Foods World* 26: 345–349.
17. Weiderhold, P. R. 1975a. Humidity measurements, Part II: Hygrometry. *Instrum. Tech.* 22(8): 45–50.
18. Weiderhold, P. R. 1975b. Humidity measurements, Part I: Psychrometers and recent R. H. sensors. *Instrum. Tech.* 22(6): 31–37.
19. Troller, J. A., and Christian, J. H. B. 1978. *Water Activity and Food*. Academic, New York.
20. Gal, S. 1981. Techniques for obtaining complete sorption isotherms. In *Water Activity: Influences on Food Quality*, L. B. Rockland and G. F. Stewart (eds.). Academic, New York.
21. Troller, J. A. 1983a. Methods to measure water activity. *J. Food Protect.* 46: 129–134.
22. Troller, J. A. 1983b. Water activity measurement with a capacitance manometer. *J. Food Sci.* 48: 739–741.
23. Nunes, R. V., Urbicain, M. J., and Rotstein, E. 1985. Improving accuracy and precision of water activity measurements with a vapor pressure monometer. *J. Food Sci.* 50: 148–149.
24. Bell, L. N. and Labuza, T. P. 2000. *Moisture Sorption: Practical Aspects of Isotherm Measurement and Use*. American Association of Cereal Chemists, St. Paul, MN.
25. Makower, B., and Meyers, S. 1943. A new method for the determination of moisture in dehydrated vegetables. *Proc. Inst. Food Technol.*, Fourth Annual Meeting, p. 156.
26. Taylor, A. A. 1961. Determination of moisture equilibria in dehydrated foods. *Food Technol.* 15: 536–540.
27. Sood, V. C., and Heldman, D. R. 1974. Analysis of a vapor pressure manometer for measurement of water activity in nonfat dry milk. *J. Food Sci.* 39: 1011–1013.
28. Labuza, T. P., Acott, K., Tatini, S. R., Lee, R. Y., Flink, J., and McCall, W. 1976). Water activity determination: A collaborative study of different methods. *J. Food Sci.* 41: 910–917.
29. Lewicki, P. P., Busk, G. C., Peterson, P. L., and Labuza, T. P. 1978. Determination of factors controlling accurate measurement of a_w by the vapor pressure manometric technique. *J. Food Sci.* 43: 244–246.
30. Wodzinski, R. J., and Frazier, W. C. 1960. Moisture requirements of bacteria. 1. Influence of temperature and pH on requirements of *Pseudomonas fluorescens*. *J. Bacteriol.* 79: 572–578.
31. Strong, D. H., Foster, E. M., and Duncan, C. L. 1970. Influence of water activity on the growth of *Clostridium perfringens*. *Appl. Microbiol.* 19: 980–987.

32. Rey, D. K., and Labuza, T. P. 1981. Characterization of the effect of solute on the water-binding and gel strength properties of carrageenan. *J. Food Sci.* 46: 786.
33. Lerici, C. R., Piva, M., and Dalla Rosa, M. 1983. Water activity and freezing-point depression of aqueous solutions and liquid foods. *J. Food Sci.* 48: 1667–1669.
34. Robinson, R. A., and Stokes, R. M. 1965. *Electrolyte Solutions*, 2nd ed. Butterworth, London.
35. Leistner, L., and Rodel, W. 1975. The significance of water activity for microorganisms in meats. In *Water Relations of Foods*, R. B. Duckworth (Ed.). Academic, London.
36. Riggle, F. R., and Slack, D. C. 1980. Rapid determination of soil water characteristics by thermocouple psychrometry. *Trans. ASAE* 23: 99–103.
37. Vos, P. T., and Labuza, T. P. 1974. Technique for measurement of water activity in the high a_w range. *J. Agric. Food Chem.* 22: 326–327.
38. Troller, J. A. 1977. Statistical analysis of a_w measurements obtained with the sinascope. *J. Food Sci.* 42: 86–90.
39. Labuza, T. P., Kreisman, L. N., Heinz, C. A., and Lewicki, P. P. 1977. Evaluation of the Abbeon cup analyzer compared to the VPM and FettVos methods for water activity measurement. *J. Food Proc. Preserv.* 1: 31–41.
40. Prior, B. A., Casaleggio, C., and Van Vuuren, H. J. J. 1977. Psychrometric determination of water activity in the high a_w range. *J. Food Protect.* 40: 537–539.
41. Van Arsdel, W. B. 1963. *Food Dehydration*. AVI, Westport, Conn.
42. Weibe, H. H., Kidambi, N., Richardson, G. H., and Ernstrom, C. A. 1981. A rapid psychrometric procedure for water activity measurement of foods in the intermediate moisture range. *J. Food Protect.* 44: 892–895.
43. Stamp, J. A., Linscott, S., Lomauro, C., and Labuza, T. P. 1984. Measurement of water activity of salt solutions and foods by several electronic methods as compared to direct vapor pressure measurement. *J. Food Sci.* 49: 1139–1142.
44. Dushchenko, V. P., Panchenko, M. S., and D'yachenko, S. F. 1969. Sorption of water vapor by capillary porous substances in relation to temperature. *J. Eng. Phys. (USSR)* 16(1): 67–71.
45. Fulford, G. D. 1969. A survey of recent Soviet research on the drying of solids. *Can. J. Chem. Eng.* 47: 378–382.
46. Landrock, A. H., and Proctor, B. E. 1951. Measuring humidity equilibria. *Mod. Packag.* 24: 123–130, 186.
47. Fett, H. M. 1973. Water activity determination in foods in the range of 0.80 to 0.99. *J. Food Sci.* 38: 1097–1098.
48. Multon, J. L., Savet, B., and Bizot, H. 1980. A fast method for measuring the activity of water in foods. *Lebensm.-Wiss. Technol.* 13: 271–273.
49. McCune, T. D., Lang, K. W., and Steinberg, M. P. 1981. Water activity determination with the proximity equilibration cell. *J. Food Sci.* 46: 1978–1979.
50. Northolt, M. D., and Heuvelman, C. J. 1982. The salt crystal liquification test—a simple method for testing the water activity of foods. *J. Food Sci.* 45: 537–540.
51. Labuza, T. P., and Lewicki, P. P. 1978. Measurement of gel water-binding capacity by capillary suction potential. *J. Food Sci.* 43: 1264–1269.
52. Neuber, E. E. 1981. Evaluation of critical parameters for developing moisture sorption isotherms of cereal grains. In *Water Activity: Influences on Food Quality*, L. B. Rockland and G. F. Stewart (eds.), pp. 199–222. Academic, New York.
53. Carr, D. S., and Harris, B. L. 1949. Solutions for maintaining constant relative humidity. *Ind. Eng. Chem.* 41: 2014–2015.
54. Stokes, R. M., and Robinson, R. A. 1949. Standard solutions for humidity control at 25°C. *Ind. Eng. Chem.* 41: 2013.
55. Rockland, L. B. 1960. Saturated salt solution for static control of relative humidity between 5 and 40°C. *Anal. Chem.* 32: 1375–1376.

56. Greenspan, L. 1977. Humidity fixed points of binary saturated aqueous solutions. *J. Res. Natl. Bur Std. AL Phys. Chem.* 81A(1): 89–96.
57. Chirife, J., Guillemo, F., Constantino, F. F., and Silvia, L. R. 1983b. The water activity of standard saturated salt solutions in the range of intermediate moisture foods. *Lebensm.-Wiss. Technol.* 16: 36–38.
58. Labuza, T. P., Kaanane, A., and Chen. J. Y. 1985. Effect of temperature on the moisture sorption isotherms and water activity of two dehydrated foods. *J. Food Sci.* 50: 385–391.
59. Ruegg, M. 1980. Calculation of the activity of water in sulfuric acid solutions at various temperatures. *Lebensm.-Wiss. Technol.* 13: 22–24.
60. Chirife, J., and Resnik, S. L. 1984. Saturated solutions of sodium chloride as reference sources of water activity at various temperatures. *J. Food Sci.* 49: 1486–1488.
61. Grover, D. W., and Nicol, J. M. 1940. The vapor pressure of glycerine solutions at 20°C. *J. Soc. Chem. Ind. (Lond.)* 59: 175–177.
62. Zsigmondy, R. 1911. Structure of gelatinous silicic acid. Theory of dehydration. *J. Anorg. Chem.* 71: 356–360.
63. Rockland, L. B. 1957. A new treatment of hygroscopic equilibria: Application to walnut and other foods. *Food Res.* 22: 604–628.
64. Bosin, W. A., and Easthouse, H. D. 1970. Rapid method of obtaining humidity equilibrium data. *Food Technol.* 24: 1155–1178.
65. Lang, K. W., McCune, T. D., and Steinberg, M. P. 1981. A proximity equilibration cell for rapid determination of sorption isotherms. *J. Food Sci.* 46: 936–938.
66. Rizvi, S. S. H., Santos, J., and Nigogosyan, N. 1984. An accelerated method for adjustment of equilibrium moisture content of foods. *J. Food Eng.* 3: 3–11.
67. Gilbert, S. G. 1984. Inverse gas chromatography. *Adv. Chromatogr.* 23: 199–228.
68. Helen, H. J., and Gilbert, S. G. 1985. Moisture sorption of dry bakery products by inverse gas chromatography. *J. Food Sci.* 50: 454–458.
69. Brunauer, S., Deming, L. S., Deming, W. E., and Teller, E. 1940. On a theory of the van der Waals adsorption of gases. *Am. Chem. Soc. J.* 62: 1723–1732.
70. Bruin, S., and Luyben, K. Ch. A. M. 1980. Recent developments in dehydration of food materials. In *Food Process Engineering*, Vol. 1, *Food Processing Systems*, P. Linko et al. (eds.), pp. 466–482. Applied Science, London.
71. Labuza, T. P. 1975. Interpretation of sorption data in relation to the state of constituent water. In *Water Relations in Foods*, R. Duckworth (ed.), pp. 155–172. Academic, New York.
72. Labuza, T. P. 1975. Sorption phenomena in foods: Theoretical and practical aspects. In *Theory, Determination and Control of Physical Properties of Food Materials*, C. K. Rha (ed.), pp. 197–219. Reidel, Boston.
73. Steele, W. A. 1974. *The Interaction of Gases and Solid Surfaces*. Pergamon, New York.
74. Boquet, R., Chirife, J., and Iglesias, H. A. 1980. On the equivalence of isotherm equations. *J. Food Technol.* 15: 345–349.
75. Boquet, R., Chirife, J., and Iglesias, H. A. 1978. equations for fitting water sorption isotherms of foods: II. Evaluation of various two-parameter methods. *J. Food Technol.* 13: 319–327.
76. Halsey, G. 1948. Physical adsorption on non-uniform surfaces. *J. Chem. Phys.* 16: 931–937.
77. Oswin, C. R. 1946. The kinetics of package life. III. The isotherm. *J. Chem. Ind. (Lond.)* 65: 419–423.
78. van den Berg, C., and Bruin, S. 1981. Water activity and its estimation in food systems: theoretical aspects. In *Water Activity: Influence on Food Quality*, L. B. Rockland and G. F. Stewart (eds.). Academic, New York.
79. Langmuir, I. 1918. The adsorption of gases on plane surfaces of glass, mica and platinum. *J. Am. Chem. Soc.* 40: 1361–1402.
80. Brunauer, S., Emmett, P. H., and Teller, E. 1938. Adsorption of gases in multimolecular layers. *Am. Chem. Soc. J.* 60: 309 319.

81. Chirife, J., and Iglesias, H. A. 1978. Equations for fitting water sorption isotherms of foods: Part 1. A review. *J. Food Technol.* 13: 159–174.
82. IUPAC. Reporting physisorption data for gas/solid systems. Commision on colloidal and surface chemistry of the International Union of Pure and Applied Chemistry. *Pure Appl Chem* 57: 603–619, 1985.
83. Labuza, T. P. 1980. The effect of water activity on reaction kinetics of food deterioration. *Food Technol* 34: 36–41, 59.
84. Karel, M. 1973. Recent research and development in the field of low-moisture and intermediate-moisture foods. *CRC Crit. Rev. Food Sci. Technol.* 3: 329–373.
85. Almasi, E. 1978. Binding energy of bound water in foodstuffs. *Acta Aliment.* 7: 213–255.
86. Hill, P. E., and Rizvi, S. S. H. 1982. Thermodynamic parameters and storage stability of drum dried peanut flakes. *Lebensm.-Wiss. Technol.* 15: 185–190.
87. Iglesias, H. A., and Chirife, J. 1976. BET monolayer values in dehydrated foods and food components. *Lebensm.-Wiss. Technol.* 9: 107–113.
88. Iglesias, H. A., Chirife, J., and Lombardi, J. L. 1975. Comparison of water vapour sorption by sugar beet root components. *J. Food Technol.* 10: 385–391.
89. Iglesias, H. A., and Chirife, J. 1976. A model for describing the water sorption behavior of foods. *J. Food Sci.* 41: 984–992.
90. Crapiste, G. H., and Rotstein, E. 1982. Prediction of sorptional equilibrium data for starch-containing foodstuff. *J. Food Sci.* 47: 1501–1507.
91. Linko, P., Pollari, T., Harju, M., and Heikonen, M. 1981. Water sorption properties and the effect of moisture on structure of dried milk products. *Lebensm.-Wiss. Technol.* 15: 26–30.
92. Ferro-Fontaon, C., Chirife, J., Sancho, E., and Iglesias, H. A. 1982. Analysis of a model for water sorption phenomena in foods. *J. Food Sci.* 47: 1590–1594.
93. Chirife, J., Bouquet, R., Ferro-Fontan, C., and Iglesias, H. 1983. A new model for describing the water sorption isotherm of foods. *J. Food Sci.* 48: 1382–1383.
94. Henderson, S. M. 1952. A basic concept of equilibrium moisture. *Agric. Eng.* 33: 29–32.
95. Rockland, L. B. 1969. Water activity and storage stability. *Food Technol.* 23: 1241–1251.
96. Iglesias, H. A., and Chirife, J. 1976. On the local isotherm concept and modes of moisture binding in food products. *J. Agric. Food Chem.* 24(1): 77–79.
97. Agrawal, K. K., Clary, B. L., and Nelson, G. L. 1971. Investigation into the theories of desorption isotherms for rough rice peanuts. *J. Food Sci.* 36: 919–924.
98. Chen, C. S., and Clayton, J. T. 1971. The effect of temperature on sorption isotherms of biological materials. *Trans. ASAE* 14: 927–929.
99. Singh, R. S., and Ojha, T. P. 1974. Equilibrium moisture content of groundnut and chillies. *J. Sci. Food Agric.* 25: 451–459.
100. Young, J. H. 1976. Evaluation of models to describe sorption and desorption equilibrium moisture content isotherms of Virginia-type peanuts. *Trans. ASAE* 19: 146–150.
101. Chung, D. S., and Pfost, H. B. 1967. Adsorption and desorption of water vapor by cereal grains and their products. Part I. Heat and free energy changes of adsorption and desorption. *Trans ASAE* 10: 549–551.
102. Chung, D. S., and Pfost, H. B. 1967. Adsorption and desorption of water vapor by cereal grains and their products. Part II. Development of the general isotherm equation. *Trans. ASAE* 10: 552–555.
103. Chung, D. S., and Pfost, H. B. 1967. Adsorption and desorption of water vapor by cereal grains and their product. Part III. A hypothesis for explaining the hysteresis effect. *Trans. ASAE* 10: 556–557.
104. Bradley, R. S. 1936. Polymer adsorbed films. Part I. The adsorption of argon on salt crystals at low temperatures and the determination of surface fields. *J. Chem. Soc.* 1936: 1467–1474.
105. Luikov, A. V. 1955. *Experimentelle und Theoretische Grundlagen der Trocknung*, V. E. B. Verlag, Berlin.

106. Labuza, T. P., Mizrahi, S., and Karel, M. 1972. Mathematical models for optimization of flexible film packaging of foods for storage. *Trans. ASAE* 15: 150–155.
107. Chen, C. S. 1971. Equilibrium moisture curves for biological materials. *Trans. ASAE* 14: 924–926.
108. Iglesias, H. A., and Chirife, J. 1978. An empirical equation for fitting water sorption isotherms of fruits and related products. *Can. Inst. Food Sci. Technol. J.* 11: 12–15.
109. Guggenheim, E. A. 1966. *Applications of Statistical Mechanics.* Clarendon Press, Oxford.
110. Anderson, R. B. 1946. Modifications of the B.E.T. equation. *J. Am. Chem. Soc.* 68: 686–691.
111. de Boer, J. H. 1953. *The Dynamical Character of Adsorption.* Clarendon Press, Oxford.
112. Bizot, H. 1983. Using the "G.A.B." model to construct sorption isotherms. In *Physical Properties of Foods,* Jowitt et al. (eds.), pp. 43–54. Applied Science, New York.
113. Timmerman, O., Chirife, J., and Iglesias, H. A. 2001. Water sorption isotherms of foods and foodstuffs: BET or GAB parameters? *J Food Eng* 48: 19–31.
114. Timmermann, E. O. 2003. Multilayer sorption parameters: BET or GAB values? *Colloids and Surfaces A* 220: 235–260.
115. McLaughlin, C. P., and Magee, T. R. A. 1998. The determination of sorption istotherm and the isosteric heats of sorption for potatoes. *J Food Eng* 35: 267–280.
116. Al-Muhtaseb, A. H., McMinn, W. A. M., and Magee, T. R. A. 2002. Moisture sorption isotherm characteristics of food products: A review. *Trans IChemE Part C* 80: 118–128.
117. Degado, A. E., and Da-Wen, Sun. 2002. Desorption isotherms of cooked and cured beef and pork. *J Food Eng* 51: 163–170.
118. Adam, E., Muhlbauer, W., Esper, A., Wolf, W., and Spie, W. 2000. Effect of temperature on water sorption equilibrium of onion. *Drying Technol* 18: 2117–2129.
119. Lagoudaki, M., Demertzis, P. G., and Kontominas, M. G. 1993. Moisture adsorption behavior of pasta products. *Lebensm-Wiss-Technol* 26: 512–516.
120. McMinn, W. A. M., and Magee, T. R. A. 1999. Studies on the effect of temperature on the moisture sorption characteristics of potatoes. *J Food Proc Eng* 22: 113–128.
121. Kaymak-Ertekin, F., and Sultanoglu, M. 2001. Moisture sorption isotherm characteristics of peppers. *J Food Eng* 47: 225–231.
122. Kim, S.S., and Bhomilk, S. R. 1994. Moisture sorption isotherms of concentrated yoghurt and microwave dried yoghurt powder. *J Food Eng* 21: 157–176.
123. van den Berg, C. 1984. Description of water activity of foods for engineering purposes by means of the G.A.B. model of sorption. In *Engineering and Food,* Vol. 1, *Engineering Sciences in the Food Industry,* B. M. McKenna (ed.). Elsevier, New York.
124. Iglesias, H. A., and Chirife, J. 1982. *Handbook of Isotherms.* Academic, New York.
125. Bandyopadhyay, S., Weisser, H., and Loncon, M. 1980. Water adsorption isotherm of foods at high temperatures. *Lebensm.-Wiss. Technol.* 13: 182–185.
126. Iglesias, H. A., Chirife, J., and Fontan, C. E. 1989. On the temperature dependence of isosteric heats of water sorption in dehydrated foods. *J Food Sci* 54: 1620–1623, 1631.
127. Cassie, A. B. D. 1945. Multimolecular absorption. *Trans. Faraday Soc.* 41: 450–464.
128. Hill, T. L. 1949. Statistical mechanics of adsorption. V. Thermodynamics and heat of adsorption. *J. Chem. Phys.* 17: 520–535.
129. Kemball, C., and Schreiner, G. D. 1950. The determination of heats of adsorption by the Brunauer–Emmett–Teller single isotherm method. *J. Am. Chem. Soc.* 72: 5605–5607.
130. Aguerre, R. J., Suárez, C., and Viollaz, P. E. 1984. Calculation of the variation of the heat of desorption with moisture content on the basis of the BET theory. *J. Food Technol.* 19: 325–331.
131. Bettelheim, F. A., and Volman, D. H. 1957. Pectic substances—water. II. Thermodynamics of water vapor sorption. *J. Polym. Sci.* 24: 445–454.
132. Volman, D. H., Simons, J. W., Seed, J. R., and Sterling, C. 1960. Sorption of water vapor by starch: Thermodynamics and structural changes for dextrin, amylose and amylopectin. *J. Polym. Sci.* 46: 355–364.

133. Roman, G. N., Urbician, M. W., and Rotstein, E. 1982. Moisture equilibrium in apples at several temperatures: Experimental data and theoretical considerations. *J. Food Sci.* 47: 1484–1488, 1507.
134. Iglesias, H. A., and Chirife, J. 1976. Isosteric heats of water vapor sorption on dehydrated foods. Part II. Hysteresis and heat of sorption: Comparison with BET theory. *Lebensm.-Wiss. Technol.* 9: 123–127.
135. Iglesias, H. A., and Chirife, J. 1976. Isosteric heats of water vapor sorption on dehydrated foods. I. Analysis of the differential heat curves. *Lebensm.-Wiss. Technol.* 9: 116–122.
136. Becker, H. A., and Sallans, H. R. 1956. A study of the desorption isotherms of wheat at 25°C and 50°C. *Cereal Chem.* 33: 79–91.
137. Day, D. L., and Nelson, G. L. 1965. Desorption isotherms for wheat. *Trans. ASAE* 8: 293–297.
138. Iglesias, H. A., and Chirife, J. 1984. Technical note: Correlation of BET monolayer moisture content in foods with temperature. *J. Food Technol.* 19: 503–506.
139. Villota, R., Saguy, I., and Karel, M. 1980. An equation correlating shelf life of dehydrated vegetable products with storage conditions. *J. Food Sci.* 45: 398–401.
140. Kapsalis, J. G. 1981. Moisture sorption hysteresis. In *Water Activity: Influences on Food Quality*. L. B. Rockland, and G. F. Stewart (eds.), pp. 143–177. Academic, New York.
141. La Mer, V. K. 1967. The calculation of thermodynamic quantities from hysteresis data. *J. Colloid. Interface Sci.* 23: 297–301.
142. Hill, T. L. 1950. Statistical mechanics of adsorption. IX. Adsorption thermodynamics and solution thermodynamics. *J. Chem. Phys.* 18: 246–256.
143. Hill, T. L. 1951. Thermodynamics of adsorption. *Trans. Faraday Soc.* 47: 376–380.
144. Hill, T. L. 1952. Theory of physical adsorption. *Adv. Catal.* 4: 211–269.
145. Hill, T. L., Emmett, P. H., and Joyner, L. G. 1951. Calculation of thermodynamic functions of adsorbed molecules from adsorption isotherm measurements: Nitrogen on graphon. *J. Am. Chem. Soc.* 75: 5102–5107.
146. Everett, D. H., and Whitton, W. I. 1952. A general approach to hysteresis. *Trans. Farraday Soc.* 48: 749–757.
147. Flood, E. A. 1967. *The Gas Solid Interface*, Vol. 2. Marcel Dekker, New York.
148. Gregg, S. J., and Sing, K. S. W. 1967. *Adsorption Surface Area and Porosity*. Academic, New York.
149. Mazza, G., and La Maguer, M. 1978. Water sorption properties of yellow globe onion (*Allium cepa* L.). *Can. Inst. Food Sci. Technol. J.* 11: 189–193.
150. Mazza, G. 1980. Thermodynamic considerations of water vapor sorption by horseradish roots. *Lebensm.-Wiss. Technol.* 13: 13–17.
151. Keey, R. B. 1972. *Drying Principles and Practice*. Pergamon, New York.
152. Gentzler, G. L., and Schmidt, F. W. 1973. Effect of bound water in the freeze drying process. *Trans. ASAE* 16: 183–188.
153. Ma, Y. M., and Arsem, H. 1982. Low pressure sublimation in combined radiant and microwave freeze drying. In *Drying '82*, S. Majumdar (ed.), pp. 196. Hemisphere, Washington, DC.
154. Albin, F. V., Murthy, S. S., and Murthy, M. V. K. 1982. Analysis of the food freeze drying process with predetermined surface temperature variation. In *Drying 1982*, S. Majumdar (ed.), p. 151. Hemisphere, Washington, DC.
155. Ponec, V., Knor, Z., and Cerny, S. 1974. *Adsorption on Solids*. CRC Press, Cleveland.
156. Lang, K. W., Whitney, R., McCune, T. D., and Steinberg, M. P. 1982. A mass balance model for enthalpy of water binding by a mixture. *J. Food Sci.* 47: 110–113.
157. Leung, H. K., and Steinberg, M. P. 1979. Water binding of food constituents as studied by NMR, sorption, freezing and dehydration. *J. Food Sci.* 44: 1212–1216, 1220.
158. Fish, B. P. 1958. Diffusion and thermodynamics of water in potato starch gel. *Fundamental Aspects of Dehydration of Foodstuffs*, pp. 143–147. Soc. Chem. Ind., London: Macmillan Co., New York.
159. Othmer, D. F., and Sawyer, F. G. 1943. Correlating adsorption data. *Ind. Eng. Chem.* 35: 1269–1276.

160. Taylor, S. A., and Kijne, J. W. 1963. Evaluating thermodynamic properties of soil water. In *Humidity and Moisture Papers*, pp. 335–340. International Symposium, Washington, DC.
161. Rizvi, S. S. H., and Benado, A. L. 1984. Thermodynamic properties of dehydrated foods. *Food Technol.* 38(3): 83–92.
162. Young, D. M., and Crowell, A. D. 1962. *Physical Adsorption of Gases*. Butterworth, Washington, DC.
163. Kiranoudis, C. T., Tsami, E., Maroulis, Z. B., and Morunos-Kouris, D. 1993. Equilibrium moisture content and heat of desorption of some vegetables. *J Food Eng* 20: 55–74.
164. Cenkowski, S., Jaya, D. S., and Dao, D. 1992. Latent heat of vaporization for selected foods and crops. *Can Agric Eng* 34: 281.
165. Wang, N., and Brennan, J. G. 1991. Moisture sorption isotherm characteristics of potatoes at four temperatures. *J Food Eng* 14: 269–282.
166. Soekarto, S. T., and Steinberg, M. P. 1981. Determination of binding energy for the three fractions of bound water. In *Water Activity: Influences on Food Quality*, L. B. Rockland and G. F. Stewart (eds.), pp. 265–279. Academic, New York.
167. Morsi, M. K. S., Sterling, C., and Volman, D. H. 1967. Sorption of water vapor B pattern starch. *J. Appl. Polym. Sci.* 11: 1217–1225.
168. Bettleheim, F. A., Block, A., and Kaufman, L. J. 1970. Heats of water vapor sorption in swelling biopolymers. *Biopolymer* 9: 1531–1538.
169. Rizvi, S. S. H., and Benado, A. L. 1983–84. Thermodynamic analysis of drying foods. *Drying Technol.* 2(4): 471–502.
170. Ross, S., and Oliver, J. P. 1964. *On Physical Adsorption*. Interscience, New York.
171. Berlin, E., Kliman, P. G., and Pallansch, M. J. 1970. Changes in state of water in proteinaceous systems. *J. Colloid Interface Sci.* 34: 488–494.
172. Iglesias, H. A., Chirife, J., and Viollaz, P. 1976. Thermodynamics of water vapor sorption by sugar beet root. *J. Food Technol.* 11: 91–101.
173. Reidel, L. 1961. Zum problem des gebundenen wassers in fleisch [For a problem of the bound water into meat]. *Kaltetechnik* 9: 107–110.
174. Brunauer, S. 1945. *The Adsorption of Gases and Vapors*. Princeton University Press, Princeton, NJ.
175. Pixton, S. W., and Warburton, S. 1973. The influence of the method used for moisture adjustment on the equilibrium relative humidity of stored products. *J. Stored Prod. Res.* 9: 189–197.
176. Rao, K. S. 1941. Hysteresis in sorption. II. Scanning of the hysteresis loop. Titania–gel–water systems. *J. Phys. Chem.* 45: 506–512.
177. Benado, A. L., and Rizvi, S. S. H. 1985. Thermodynamic properties of water on rice as calculated from reversible and irreversible isotherms. *J. Food Sci.* 50: 101–105.
178. Simatos, D., and Blond, G. 1991. DSC studies and stability of frozen foods. In *Water Relationships in Foods*, H. Levine and L. Slade (eds.), pp. 29–101. Plenum, New York.
179. Slade, L., and Levine, H. 1991. A food polymer science approach to structure–property relationships in aqueous food systems: Non-equilibrium behavior of carbohydrate–water systems. In *Water Relationships in Foods*, H. Levine and L. Slade (eds.), pp. 29–101. Plenum, New York.
180. Van den Berg, C. 1991. Food–water relations: Progress and integration, comments and thoughts. In *Water Relationships in Foods*, H. Levine and L. Slade (eds.), pp. 21–28. Plenum, New York.
181. Park, J., Kim, D., Kim, C., Maeng, K., and Hwang, T. 1991. Effect of drying conditions on the glass transition of poly (acrylic acid). *Polymer Eng.* 31(12): 867–872.
182. Slade, L., and Levine, H. 1988. Non-equilibrium behavior of small carbohydrate–water systems. *Pure Appl. Chem.* 60: 1841–1846.
183. Gould, G. W., and Christian, J. H. B. 1988. Characterization of the state of water in foods—biological aspects. In *Food Preservation by Moisture Control*, C. C. Seow (ed.), pp. 43–56. Elsevier, London.
184. Anese, M., Shtylla, I., Torreggiani, D., and Maltini, E. 1996. Water activity and viscosity relations with glass transition temperatures in model food systems. *Termochem Acta* 275: 131–137.

185. Roos, Y. H. 1995. *Phase Transitions in Foods*. Academic Press: St. Louis, MO, pp. 167–170.
186. Maltini, E., Torreggiani, D., Venir, E., and Bertolo, G. 2003. Water activity and the preservation of plant foods. *Food Chem* 82: 79–86.
187. Bakker-Arkema, F. W. 1985. Heat and mass transfer aspects and modeling of dryers. A critical evaluation. In *Concentration and Drying of Foods*, D. MacCarthy (ed.), pp. 165–202. Elsevier, London.
188. King, C. J. 1977. Heat and mass transfer fundamentals applied to food engineering. *J. Food Proc. Eng.* 1: 3–14.
189. Luikov, A. V. 1966. *Heat and Mass Transfer in Capillary-Porous Bodies*. Pergamon, Oxford.
190. Fortes, M., and Okos, M. R. 1980. Drying theories: Their bases and limitations as applied to foods and grains. In *Advance in Drying*, Vol. 1, A. S. Majumdar (ed.), pp. 119–154. Hemisphere, Washington, DC.
191. Whitaker, S. 1980. Heat and mass transfer in granular porous media. In *Advances in Drying*, Vol. 1, A. S. Majumdar (ed.), pp. 23–61. Hemisphere, Washington, DC.
192. Bruin, S., and Luyben, K. Ch. A. M. 1979. Drying of food materials: a review of recent developments. In *Advances in Drying*, Vol. 1, A. K. Majumdar (ed.), pp. 155–215. Hemisphere, Washington, DC.
193. Toei, R. 1983. Drying mechanisms of capillary porous bodies. In *Advances in Drying*, Vol. 2, A. S. Majumdar (ed.), pp. 269–297. Hemisphere, Washington, D.C.
194. Whitaker, S., and Chou, W. T. H. 1983–84. Drying granular porous media—Theory and experiment. *Drying Technol.* 1(1): 3–33.
195. Luikov, A. V. 1970. A prognosis of the development of science of drying capillary-porous colloidal materials. *Int. Chem. Eng.* 10: 599–604.
196. Poersch, W. 1977. Present state of drying technology. IchE Solid Drying Course. Birmingham. Quoted by R. B. Keey 1980: Theoretical foundation of drying technology. *Adv. Drying* 2: 17.
197. Sakai, N., and Hayakawa, K. 1992. Two dimensional simultaneous heat and moisture transfer in composite food. *J. Food Sci.* 57: 475–478.
198. Sakai, N., and Hayakawa, K. 1993. Heat and moisture transfer in composite food—theoretical analysis of influence of surface conductance and component arrangement. *J. Food Sci.* 58: 1335–1339.
199. Suzuki, K., Ihara, K., Kubota, K., and Hosaka, H. 1977. Heat transfer of the constant rate period in drying of agar gel, carrot and sweet potato. *Nippon Shokuhin Kogyo Gokkaishi* 24: 387–393.
200. Jason, A. C. 1958. A study of evaporation and diffusion processes in the drying of fish muscle. In *Fundamental Aspects of the Dehydration of Foodstuffs*, Papers read at the conference held in Aberdeen, 25–27 March 1958, pp. 103–135. Soc. Chem. Ind., London and Macmillan, New York.
201. Saravocos, G. D., and Charm, S. E. 1962. A study of the mechanism of fruit and vegetable dehydration. *Food Technol.* 16: 78–81.
202. Labuza, T. P., and Simon, I. B. 1970. Surface tension effect during dehydration. 1. Air drying of apple slices. *Food Technol.* 24: 712–715.
203. Chirife, J., and Cachero, R. A. 1970. Through-circulation drying of tapioca root. *J. Food Sci.* 35: 364–368.
204. Vaccerezza, L. M., Lombardi, J. L., and Chirife, J. 1974. Kinetics of moisture movement during air drying of sugar beet roots. *J. Food Technol.* 9: 317–327.
205. Alzamora, S. M., and Chirife, J. 1980. Some factors controlling the kinetics of moisture movement during avocado dehydration. *J. Food Sci.* 45: 1649–1651.
206. Zogzas, N. P., Marousis, Z. B., and Marinos-Kouris, D. 1996. Moisture diffusivity data compilation in foodstuffs. *Drying Technol* 14: 2225–2253.
207. Suzuki, K., Kubota, K., Hasegawa, T., and Hosaka, H. 1976). Shrinkage in dehydration of root vegetables. *J. Food Sci.* 41: 1189–1193.
208. Chen, C. S., and Johnson, W. H. 1969. Kinetics of moisture movement in hygroscopic materials. I. Theoretical considerations of drying phenomena. *Trans ASAE* 12: 109–113.

209. Vaccarezza, L. M., Lambardi, J. L., and Chirife, J. 1974. Heat transfer effects on drying rate of food dehydration. *Can. J. Chem. Eng.* 52: 576–579.
210. Ramaswamy, H. S., Lo, K. V., and Satlely, L. M. 1982. Air drying of shrimp. *Can. Agric. Eng.* 24(2): 123–128.
211. Mowlah, G., Takano, K., Kamoi, I., and Obara, T. 1983. Water transport mechanism with some aspects of quality changes during air dehydration of bananas. *Lebensm.-Wiss. Technol.* 16:103–107.
212. Brooker, D. B., and Bakker-Arkema, F. W., and Hall, C. W. 1974. *Drying Cereal Grains*. AVI, Westport, CT.
213. Crank, J. 1956. *The Mathematics of Diffusion*. Oxford University Press, Oxford.
214. Jason, A. C. 1965. Effects of fat content on diffusion of water in fish muscle. *J. Sci. Food Agric.* 16: 281–288.
215. Schoeber, W. J. A. H., and Thijssen, H. A. 1977. A short-cut method for the calculation of drying rates for slabs with concentration-dependent diffusion coefficient. *AIChE Symp. Ser.* 73(163): 12–24.
216. Chirife, J. 1971. Diffusional process in the drying of tapioca root. *J. Food Sci.* 36: 327–330.
217. McMinn, W. A. M., and Magee, T. R. A. 1996. Air drying kinetics of potato cylinders. *Drying Technol* 14: 2025–2040.
218. Wang, Z. H., and Chen, G. 1999. Heat and mass transfer during low intensity convective drying. *Chem Eng Sci* 54: 3899–3908.
219. Coumans, W. J. 2000. Models for drying kinetics based on drying curve slabs. *Chem Eng Proc* 39: 53–68.
220. Simal, S., Femenia, A., Llull, P., and Rosello, C. 2000. Dehydration of aloe vera: simulation of drying curves and evaluation of functional properties. *J Food Eng* 43: 109–114.
221. Fan, C. C., Liaw, S. P., Fu, W. R., and Pan, B. S. 2003. Mathematical model for prediction of intermittent drying and pressing of mullet roe. *J Food Sci* 68: 886–891.
222. King, C. J. 1968. Rate of moisture and desorption in porous, dried foodstuffs. *Food Technol.* 22: 509–514.
223. Palumbo, S. A., Komanowsky, M., Metzer, V., and Smith, J. L. 1977. Kinetics of pepperoni drying. *J. Food Sci.* 42: 1029–1033.
224. Chirife, J. 1983. Fundamentals of the drying mechanism during air dehydration of foods. In *Advances in Drying*, Vol. 1, A. K. Majumdar (ed.), pp. 73–102. Hemisphere, Washington, DC.
225. Bluestein, P. M., and Labuza, T. P. 1972. Kinetics of water vapor sorption in a model freeze-dried food. *Am. Inst. Chem. Eng. J.* 18(4): 706–712.
226. Harmathy, T. 1969. Simultaneous moisture and heat transfer in porous systems with particular reference to drying. *Ind. Eng. Chem. Fundam.* 8: 92–103.
227. Hussain, A., Chen, C. S., and Clayton, J. 1973. Simultaneous heat and mass diffusion in biological materials. *J. Agric. Eng. Res.* 18: 343–354.
228. Zhang, T., Bakshi, A. S., Gustafson, R. J., and Lund, D. B. 1984. Finite element analysis of non-linear water diffusion during rice soaking. *J. Food Sci.* 49: 246–250, 277.
229. Lamauro, G. L., and Bakshi, A. S. 1985. Finite element analysis of moisture diffusion in stored foods. *J. Food Sci.* 50: 392–396.
230. Sereno, A. M., and Medeiros, G. L. 1990. A simplified model for the prediction of drying rates of foods. *J. Food Eng.* 12: 1–11.
231. Guillard, V., Broyart, B., Bonazzi, C., Guilbert, S., and Gontard, N. 2003. Evolution of moisture distribution during storage in a composite food: Modeling and simulation. *J. Food Sci.* 68: 958–966.
232. Guillard, V., Broyart, B., Bonazzi, C., Guilbert, S., and Gontard, N. 2003. Moisture diffusivity in sponge-cake as related to porous structure evaluation and moisture content. *J. Food Sci.* 68: 555–562.
233. Gekas, V., and Lamberg, I. 1991. Determination of diffusion coefficients in volume-changing systems—Application in the case of potato drying. *J. Food Eng.* 14: 317–326.

234. Marques, J. M., Rutledge, D. N., and Ducauze, C. J. 1991. Low resolution of NMR detection of the mobilization point of solutes during the drying of carrots. *Lebensm.-Wiss. Technol.* 24: 93–98.
235. Song, H., and Litchfield, J. B. 1998. Nondestructive measurement of transient moisture profiles in corn during drying using NMR imaging. ASAE Paper No. 88-6532. *American Society of Agricultural Engineers*, St. Joseph, MI.
236. Song, H., and Litchfield, J. B. 1990. Nuclear magnetic resonance imaging of three-dimensional moisture distribution in an ear of corn during drying. *Cereal Chem.* 67(6): 580–584.
237. Pel, L., Ketelaars, A. A. J., Adan, O. C. G., and Well, A. A. 1993. Determination of moisture diffusivity in porous media using scanning neutron radiography. *Int J Heat Mass Transfer* 36: 1261–1267.
238. Bray, Y. L., and Prat, M. 1999. Three-dimensional pore network simulation of drying in capillary porous media. *Int J Heat Mass Transfer* 42: 4207–4224.
239. Mulet, A. 1994. Drying modeling and water diffusivity in carrots and potatoes. *J Food Eng* 22: 329–348.
240. Becker, H. A., and Sallans, H. R. 1955. A study of internal moisture movement in drying of the wheat kernel. *Cereal Chem.* 32: 212–225.
241. Roth, T., and Loncin, M. 1984. Superficial activity of water. In *Engineering and Food*, Vol. 1, B. M. McKenna (ed.), pp. 433–443. Elsevier, Essex, England.
242. Gal, S. 1983. The need for, and practical applications of sorption data. In *Physical Properties of Foods*, Jowitt et al. (eds.), pp. 13–25. Applied Science, New York.
243. Heiss, R. 1968. *Haltbarkeit und Sorptionsverhalten Wasserarmer Lebensmittel*. Springer-Verlag, Berlin.
244. Salwin, H. 1959. Defining minimum moisture contents for foods. *Food Technol.* 13: 594–595.
245. Heiss, R. 1958. Shelf-life determinations. *Mod. Packag.* 31(8): 119–124, 172, 176.
246. Nonhebel, G., and Moss, A. A. H. 1971. *Drying of Solids in the Chemical Industry*. Butterworths, London.
247. Williams-Gardner, A. 1971. *Industrial Drying*. Leonard Hill, London.
248. Lilley, T. H., and Sutton, R. L. 1991. The prediction of water activities in multicomponent systems. In *Water Relationships in Foods*, H. Levine and L. Slade (eds.), pp. 291–304. Plenum, New York.

12

Physicochemical and Engineering Properties of Food in Membrane Separation Processes

Dipak Rana, Takeshi Matsuura, and Srinivasa Sourirajan

Contents

12.1 Introduction	439
12.2 Transport Theories	439
12.2.1 Case 1: Preferential Sorption of Water at the Membrane–Solution Interface	439
12.2.1.1 Basic Transport Equations	440
12.2.1.2 Relationship between $(D_{AM}/K\delta)_{NaCl}$ and $D_{AM}/K\delta$ for Other Solutes	441
12.2.1.3 RO Process Design	444
12.2.2 Case II: Surface Force–Pore Flow Model, Generation of Interfacial Surface Force Parameters, and Their Application	445
12.2.2.1 Analysis Fundamentals	445
12.2.2.2 Quantities R_a, R_b, and $\underline{d}$	446
12.2.2.3 Definition of Dimensionless Quantities	447
12.2.2.4 Basic Transport Equations	448
12.2.2.5 Liquid Chromatography for the Determination of Interfacial Interaction Force Parameters	452
12.2.2.6 Data on Interfacial Surface Force Parameters	453
12.2.2.7 Data on Pore Size and Pore Size Distribution	453
12.3 Enumeration of Problems Involved in Membrane Separation and Concentration of Liquid Foods	456
12.3.1 Application of Water Preferential Sorption Model	463
12.3.1.1 Separation of Undissociated Organic Solutes Such as Sugars Present in High Concentration	469

12.3.1.2 Separations of Undissociated Polar Organic Solutes Present in Low Concentrations	471
12.3.1.3 Separation of Partially Dissociated Organic Solutes Present in Low Concentration	473
12.3.1.4 Problem of Separations of Low Concentrations of Undissociated Organic Solutes in Concentrated Sugar Solutions	475
12.3.1.5 Separation of Solutions of Partially Dissociated Acids Present in Concentrated Sugar Solutions	476
12.3.2 Application of Transport Equations to Real Fruit Juice Concentration	478
12.3.3 Application of Transport Equations for the Concentration of Green Tea Juice	482
12.3.4 Some Illustrative Examples of the Surface Force–Pore Flow Model	486
12.3.4.1 Parametric Studies on Solute Separation and PR	486
12.3.4.2 Another Parametric Study on Solute Concentration Profile and Solution Velocity Profile	490
12.3.5 Some Data on the UF of Proteins	492
12.3.5.1 UF of BSA and α-Casein	492
12.3.5.2 Effects of Fouling on Membrane Performance and Pore Size and Pore Size Distribution	495
12.3.5.3 Fractionation of the Protein–Sugar System and the Protein–Protein System in the Aqueous Solutions	497
12.3.6 Application of PV in the Recovery and Concentration of Food Flavors	499
12.4 Recent Literature on Membrane Applications	501
12.4.1 Dairy Product Industry	502
12.4.1.1 Reverse Osmosis	502
12.4.1.2 Nanofiltration	502
12.4.1.3 Ultrafiltration	503
12.4.1.4 Microfiltration	504
12.4.1.5 Electro-Dialysis, Forward Osmosis, Membrane Distillation, Osmotic Distillation, and PV	504
12.4.2 Beverage Industry	504
12.4.2.1 Reverse Osmosis	505
12.4.2.2 Ultrafiltration	506
12.4.2.3 Microfiltration	506
12.4.2.4 Electro-Dialysis, Forward Osmosis, Membrane Distillation, Osmotic Distillation, and PV	507
12.4.3 Edible Oil Industry	510
12.4.3.1 Reverse Osmosis	510
12.4.3.2 Ultrafiltration	511
12.4.3.3 Microfiltration	512
12.4.4 Miscellaneous Food Products	513
12.4.4.1 Nanofiltration	513
12.4.4.2 Ultrafiltration	513
12.4.4.3 Electro-Dialysis, Forward Osmosis, Membrane Distillation, Osmotic Distillation, and PV	513

12.5 Conclusion	514
List of Symbols	515
Greek Symbols	517
References	518

12.1 INTRODUCTION

Since the development of the first cellulose acetate membrane for reverse osmosis (RO) desalination of seawater in 1960, the application of membrane separation processes, such as RO and ultrafiltration (UF), has been steadily growing. Currently, the application of membrane separation processes covers a wide range: seawater desalination, wastewater treatment, waste recovery, food processing, medical application, application to biotechnology, gas separation, and waste recovery from nonsolvents. Among these, the application in food processing is one of the most important and the most promising. Particularly, in the process of concentrating liquid food, many volatile food flavors escape by the conventional method of evaporation, whereas in the membrane process, these flavor components are preserved in the food because no heat needs to be supplied in the latter process. Moreover, because it involves no phase change, the membrane process is intrinsically an energy–saving process compared with the evaporation process. For these two reasons, serious consideration has been given to various aspects of membrane food processing from the very beginning of membrane development and application of membranes, and some of them have had considerable industrial success. However, there are also problems in the membrane application in food processing industries. The liquid food usually contains dispersed particles, colloidals, and aqueous macromolecular solutes, which precipitate on the top of the membrane surface or plug the pores on the membrane surface, resulting in a drastic decrease in the membrane flux. This phenomenon is commonly called fouling and is one of the most serious problems inherent in membrane food processing. Engineering skill in the design of the module is often required to prevent fouling problems. Furthermore, the choice of membrane materials, the design of the membrane pore size, and the pore size distribution can reduce the fouling caused by the blocking of the pore. This latter aspect relates to the rational design of the membrane and is described later in this chapter.

Although the experimental testing of membrane food processing, such as in fruit juice concentration, cheese whey concentration, egg white concentration, and treatment of alcoholic beverages, is well documented in the literature, very little work has been conducted to elucidate the fundamental principles involved therein. In this chapter, the fundamental aspects of membrane food processing are particularly emphasized.

12.2 TRANSPORT THEORIES

12.2.1 Case 1: Preferential Sorption of Water at the Membrane–Solution Interface

This case is concerned with aqueous RO membrane systems where there is no significant solute accumulation within the membrane pore during RO. More specifically, this case is applicable for systems where water is preferentially sorbed or the solute is only weakly sorbed to the membrane polymer material. A broad area in RO/UF separations, including liquid food processing, primarily involves such systems. Hence, the RO transport equations applicable to such systems are of major practical importance.

12.2.1.1 Basic Transport Equations

A complete RO experiment involves obtaining data on the pure water permeation rate (PWP), the membrane–permeated product rate (PR) with respect to a given area of membrane surface, and fraction solute separation (f) at any point in the RO system under the specified operating condition of temperature, pressure, solute concentration in the feed solution, and feed flow rate.

At any given operating temperature and pressure, each set of RO data can be analyzed on the basis that

1. PWP is directly proportional to the operating gauge pressure P.
2. The solvent flux N_B through the membrane is proportional to the effective driving pressure for fluid flow through the membrane (assumed to be practically the same as $P-\Delta\pi$).
3. The solute flux N_A through the membrane is due to the pore diffusion and is hence proportional to the concentration difference across the membrane.
4. The mass transfer coefficient k on the high–pressure side of the membrane is given by "film theory" [1]. This analysis, which is applicable to all membrane materials and membranes at all levels of solute separation, gives rise to the following basic transport equations for RO, where the viscosity of the membrane permeated produce solution is assumed to be practically the same as that of water [2]:

$$A = \frac{\text{PWP}}{M_B \times S \times 3600 \times P} \tag{12.1}$$

$$N_B = A[P - \pi(X_{A2}) + \pi(X_{A3})] \tag{12.2}$$

$$N_B = \left(\frac{D_{AM}}{K\delta}\right)\left(\frac{1 - X_{A3}}{X_{A3}}\right)(c_{A2}X_{A2} - c_{A3}X_{A3}) \tag{12.3}$$

$$N_B = c_1 k(1 - X_{A3})\ln\left(\frac{X_{A2} - X_{A3}}{X_{A1} - X_{A3}}\right) \tag{12.4}$$

All the symbols are defined in the list of symbols at the end of the chapter.

Equation 12.1 defines the PWP constant A for the membrane, which is a measure of its overall porosity; Equation 12.3 defines the solute transport parameter $D_{AM}/K\delta$ of the solute for the membrane, which is also a measure of the average pore size on the membrane surface on a relative scale. Under steady–state operating conditions, a single set of experimental PWP, PR, and f data enable one to calculate the quantities A, X_{A2}, $D_{AM}/K\delta$, and k at any point (position or time) in the RO system via Equations 12.1 through 12.4. Conversely, PWP, PR, and f can be calculated from a given set of A, $D_{AM}/K\delta$, and k data under a given operating condition of feed solute concentration and operating pressure [2]. Further, for very dilute feed solution, where the osmotic pressure of the feed solution is negligible, $D_{AM}/K\delta$ is given by [3]

$$\frac{D_{AM}}{K\delta} = \frac{PR}{3600\underset{\sim}{\rho}}\left(\frac{1-f}{f}\right)\left[\exp\frac{PR}{3600\,Sk\underset{\sim}{\rho}}\right] \quad (12.5)$$

Equation 12.5 shows that $D_{AM}/K\delta$ determines the solute transport through the membrane and consequently governs the solute separation f. Therefore, an attempt is made in the following section to relate $D_{AM}/K\delta$ of a given solute to that of a reference NaCl solute.

12.2.1.2 Relationship between $(D_{AM}/K\delta)_{NaCl}$ and $D_{AM}/K\delta$ for Other Solutes

For completely ionized inorganic and simple (i.e., where electrostatic effects are dominant compared with steric and nonpolar effects) organic solutes,

$$\left(\frac{D_{AM}}{K\delta}\right)_{solute} \propto \exp\left[n_c\left(-\frac{\Delta\Delta G}{\underset{\sim}{R}T}\right)_{cation} + n_a\left(-\frac{\Delta\Delta G}{\underset{\sim}{R}T}\right)_{anion}\right] \quad (12.6)$$

where n_c and n_a represent the number of moles of cations and anions, respectively, in 1 mol of ionized solute. Applying Equation 12.6 to $(D_{AM}/K\delta)_{NaCl}$,

$$\ln\left(\frac{D_{AM}}{K\delta}\right)_{NaCl} = \ln C^*_{NaCl} + \left[\left(-\frac{\Delta\Delta G}{\underset{\sim}{R}T}\right)_{Na^+} + \left(-\frac{\Delta\Delta G}{\underset{\sim}{R}T}\right)_{Cl^-}\right] \quad (12.7)$$

where $\ln G_{NaCl}$ is a constant representing the porous structure of the membrane surface in terms of $(D_{AM}/K\delta)_{NaCl}$. By using the data on $-\Delta\Delta G/RT$ for Na^+ and Cl^- ions for the membrane material–solution system involved, the value of $\ln G_{NaCl}$ for the particular membrane employed can be calculated from the specified value of $(D_{AM}/K\delta)_{NaCl}$. Using the value of $\ln G_{NaCl}$ so obtained, the corresponding value of $D_{AM}/K\delta$ for any completely ionized inorganic or simple organic solute can be calculated from the relation

$$\ln\left(\frac{D_{AM}}{K\delta}\right)_{solute} = \ln C^*_{NaCl} + \left[n_c\left(-\frac{\Delta\Delta G}{\underset{\sim}{R}T}\right)_{cation} + n_a\left(-\frac{\Delta\Delta G}{\underset{\sim}{R}T}\right)_{anion}\right] \quad (12.8)$$

Thus, for any specified value of $(D_{AM}/K\delta)_{NaCl}$, the corresponding values of $D_{AM}/K\delta$ for a large number of completely ionized solutes can be obtained from Equation 12.8 by using data on $-\Delta\Delta G/RT$ for the ions involved. Available data on $-\Delta\Delta G/RT$ for different ions applicable for cellulose acetate (acetyl content, 39.8%) membrane–aqueous solution systems are listed in Table 12.1 [4].

For a completely nonionized polar organic solute,

$$\ln\left(\frac{D_{AM}}{K\delta}\right)_{solute} = \ln C^*_{NaCl} + \ln \Delta^* + \left(-\frac{\Delta\Delta G}{\underset{\sim}{R}T}\right) \\ + \delta^* \Sigma E_s + \omega^* \Sigma_{s^*} \quad (12.9)$$

Table 12.1 Data on Free Energy Parameter $(-\Delta\Delta G/RT)_i$, for Some Inorganic Ions at 25°C, Applicable for Interfaces Involving Aqueous Solutions and Cellulose Acetate (CA-398) Membranes in RO/UF Transport

Inorganic Cations		Inorganic Anions	
Species	$(-\Delta\Delta G/RT)_i$	Species	$(-\Delta\Delta G/RT)_i$
H^+	6.34	OH^-	−6.18
Li^+	5.77	F^-	−4.91
Na^+	5.79	Cl^-	−4.42
K^+	5.91	Br^-	−4.25
Rb^+	5.86	I^-	−3.98
Cs^+	5.72	IO_3^-	−5.69
NH_4^+	5.97	$H_2PO_4^-$	−6.16
Mg^{2+}	8.72	BrO_3^-	−4.89
Ca^{2+}	8.88	NO_2^-	−3.85
Sr^{2+}	8.76	NO_3^-	−3.66
Ba^{2+}	8.50	ClO_3^-	−4.10
Mn^{2+}	8.58	ClO_4^-	−3.60
Co^{2+}	8.76	HCO_3^-	−5.32
Ni^{2+}	8.47	HSO_4^-	−6.21
Cu^{2+}	8.41	SO_4^{2-}	−13.20
Zn^{2+}	8.76	$S_2O_3^{2-}$	−14.03
Cd^{2+}	8.71	SO_3^{2-}	−13.12
Pb^{2+}	8.40	CrO_4^{2-}	−13.69
Fe^{2+}	9.33	$Cr_2O_7^{2-}$	−11.16
Fe^{3+}	9.82	CO_3^{2-}	−13.22
Al^{3+}	10.41	$Fe(CN)_6^{3-}$	−20.87
Ce^{3+}	10.62	$Fe(CN)_6^{4-}$	−26.83
Cr^{3+}	11.28		
La^{3+}	12.89		
Th^{4+}	12.42		

Referring to the quantities on the right-hand side of Equation 12.9, the quantity $\ln G_{NaCl}$ is obtained from data on $(D_{AM}/K\delta)_{NaCl}$ by the use of Equation 12.7, the quantity $\ln \Delta^*$ sets a scale for $\ln (D_{AM}/K\delta)_{solute}$ in terms of $\ln G_{NaCl}$, and the quantities δ^* and ω^* are the coefficients associated with the Taft steric parameter and the modified Small's number (nonpolar parameter) applicable for the membrane material–solvent–solute system involved.

Furthermore, ln Δ^*, δ^*, and ω^* all depend on the membrane material as well as on membrane porosity; the first two variables are functions of ln G_{NaCl} [or ln (G_{NaCl}/A)], whereas ω^* is independent of the porous structure of the membrane surface. The correlations of ln Δ^* versus ln G_{NaCl} and δ^* versus ln G_{NaCl} are given in the literature (Figures 1 through 4 of Ref. [5]) for cellulose acetate and aromatic polyamide membranes.

The quantity $\Delta\Delta G$ is defined as

$$\Delta\Delta G = \Delta G_I - \Delta G_B \tag{12.10}$$

where ΔG represents the free energy of hydration for the solute species, and the subscripts I and B represent the membrane–solution interface and the bulk solution phase, respectively. With Equation 12.10, the values of $-\Delta\Delta G/RT$ to be applied in Equation 12.9 can be calculated on the basis of the molecular structure of the solute together with the following relations for ΔG_B and ΔG_I:

$$\Delta G_B = \Sigma \gamma_B \text{ (structural group)} + \gamma_{B,0} \tag{12.11}$$

$$\Delta G_I = \Sigma \gamma_I \text{ (structural group)} + \gamma_{I,0} \tag{12.12}$$

Data on γ_B (structural group), $\gamma_{B,0}$, γ_I (structural group), and $\gamma_{I,0}$ are available in the literature with respect to both cellulose acetate and aromatic polyamide materials (Table 4 in Ref. [5]).

The value for Taft's steric parameter ΣE_S for a substituted group in a polar organic molecule involving a monofunctional group is simply the summation of the E_S values for the hydrocarbon substituent–group components involved. The available data on E_S are given in the literature [6,7]. For solute molecules having a polyfunctional group (glucose, sucrose, etc.), there is no simple way to compute the values of ΣE_S from the data given for the substituent groups. For such solutes, an empirical method has been established [5] for estimating the value of the parameter $\delta^* \Sigma E_S$ on the basis that the latter reaches a limiting value, designated as $(\delta^* \Sigma E_S)_{lim}$, when the average pore size on the membrane surface becomes sufficiently small. For a given membrane material, this quantity $(\delta^* \Sigma E_S)_{lim}$ is simply an additive function of the contribution of each of the structural units involved in the solute molecule, so that

$$(\delta^* \Sigma E_s)_{lim} = \Sigma \phi \text{(structural unit)} + \phi_0 \tag{12.13}$$

Available data are given in the literature [5,8] on ϕ and ϕ_0 that are used for aqueous solutions involving polyfunctional solutes and cellulose acetate membranes. The value of $\delta^* \Sigma E_S$ to be applied for the particular membrane used can then be computed from the data on $(\delta^* \Sigma E_S)_{lim}$ and the correlation given by the literature (Figure 5 of Ref. [5]) relating the ratio $(\delta^* \Sigma E_S)/(\delta^* \Sigma E_S)_{lim}$ and the average pore size on the membrane surface as represented by the quantity in G_{NaCl}.

The modified Small's number Σs^* (nonpolar parameter) for a hydrocarbon molecule on the hydrocarbon backbone of a polar organic molecule is obtained from its chemical structure via the additive property of s^*; values for various structural groups are given in the literature [9]. For interfaces involving aqueous solutions of C_1–C_7 monohydric alcohol solutes and cellulose acetate molecules, appropriate values of ω^* are also listed in the literature [10].

12.2.1.3 RO Process Design

Any RO system can be specified in terms of three dimensionless parameters, γ, θ, and λ, which are defined as follows:

$$\gamma = \frac{\pi(X_{A1}^0)}{P} = \frac{\text{osmotic pressure of initial feed solution}}{\text{operating pressure}} \quad (12.14)$$

$$\theta = \frac{D_{AM}/K\delta}{v_\omega^*} = \frac{\text{solute transport parameter}}{\text{pure water permeation velocity}} \quad (12.15)$$

$$\lambda = \frac{k}{D_{AM}/K\delta} = \frac{\text{mass transfer coefficient on the high-pressure side of the membrane}}{\text{solute transport parameter}} \quad (12.16)$$

where

$$v = AP/c \quad (12.17)$$

and the quantity $\pi(X_{A1})$ refers to the osmotic pressure of the feed solution at the membrane entrance in a flow process or start of the operation in a batch process. The quantities γ, θ, and $\lambda\theta$ (=κ/v) can be described as the osmotic pressure characteristic, membrane characteristic, and mass transfer characteristic, respectively, of the system under consideration. Then, under the following assumption and definitions, the RO system illustrated in Figure 12.1 can be completely specified by the quantities γ, θ, and λ, and any one of the six quantities (performance parameters) C_1, C_2, C_3, X or X', or τ, and Δ uniquely fixes all the other quantities [11].

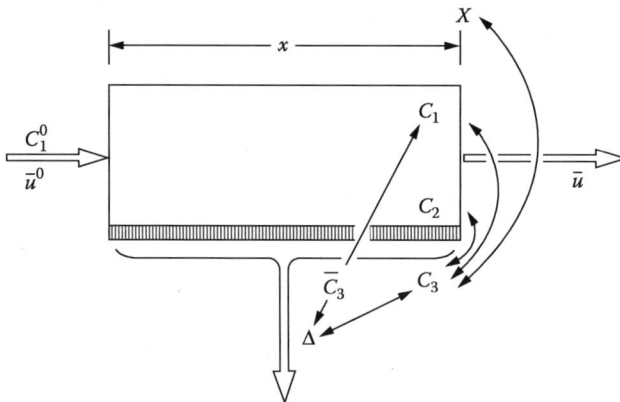

Figure 12.1 Performance parameters for a reverse osmosis system. (From H Ohya, S Sourirajan. *Reverse Osmosis System Specification and Performance Data for Water Treatment Applications*. Hanover, NH: The Thayer School of Engineering, Dartmouth College, 1971.)

PROPERTIES OF FOOD IN MEMBRANE SEPARATION PROCESSES

Assumptions

$$c_1 = c_2 = c_3 = c$$

$$\pi(X_A) \propto X_A$$

$$X_{A1} \ll 1$$

$D_{AM}/K\delta$ is independent of X_{A2}, and the longitudinal diffusive effect is negligible.

These assumptions are valid in practice for many RO systems in food processing applications.

The definition of dimensionless quantities is in order:

$$C = X_{A1}/X_{A1}$$

so that $C_1 = X_{A1}/X_{A1}$, $F_1 = 1$, $C_2 = X_{A2}/X_{A1}$, $F_2 = X_{A2}/X_{A1}$, $C_3 = X_{A3}/X_{A1}$, $F_3 = X_{A3}/X_{A1}$, and $C_3 = X_{A3}/X_{A1}$. Furthermore, for an RO system involving a longitudinal feed flow pattern in the module (such as spiral wound or tubular modules), let

$$X = \frac{v_\omega^*}{\bar{u}^0 h} \qquad (12.18)$$

and for an RO system involving a radial feed flow pattern in the module (such as in the DuPont hollow fiber module), let

$$X' = \frac{\alpha}{2}\left[\left(\frac{r_o}{r_i}\right)^2 - 1\right] \qquad (12.19)$$

where

$$\alpha = \frac{r_i v_w^*}{\bar{u}^0 h} \qquad (12.20)$$

and for an RO system involving a batch process, let

$$\tau = \frac{S v_w^*}{V_1^0} \qquad (12.21)$$

12.2.2 Case II: Surface Force–Pore Flow Model, Generation of Interfacial Surface Force Parameters, and Their Application

This case is general in scope, applicable to aqueous solution–RO membrane systems involving preferential sorption of either water or solute at the membrane–solution interface. This case is analyzed on the basis of the surface force–pore flow model, which is a quantitative expression of the preferential sorption–capillary flow mechanism for RO separations.

12.2.2.1 Analysis Fundamentals

In this model, the pores on the membrane surface are assumed equivalent to circular cylindrical pores with or without size distributions running across the top skin layer perpendicular

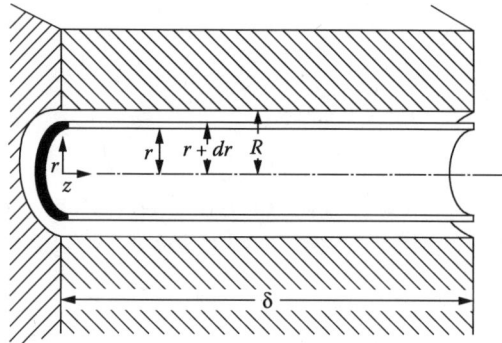

Figure 12.2 Cylindrical coordinates in a membrane pore. (From T Matsuura, S Sourirajan. *Ind Eng Chem Process Des Dev* 20:273–282, 1981.)

to the membrane surface, and the solute–membrane material interactions relative to water are expressed in terms of electrostatic or Lennard–Jones-type surface potential functions. The solute/solvent transport through the membrane pore is governed by such surface forces together with frictional force, which hinders the movement of the solute molecule. On the basis of these considerations, appropriate transport equations have been derived [12] for an individual pore of radius R_b to calculate the solute separation along with the ratio PR/PWP. The derivations are then extended to the multipore system of the actual membrane involving one or more equivalent average pore radii $\bar{R}_b$ and pore size distribution σ. The foregoing analysis results in a general expression for solute separation and fluid flux that is valid whether the solute is negatively or positively adsorbed at the membrane solution interface. Using the cylindrical coordinate system shown in Figure 12.2, the solute concentration and the velocity profile of the solution inside the membrane pore are analyzed in differential segments as a function of r and z, covering the entire pore region, under the steady-state operating conditions of the process. The derivations involved are given in detail elsewhere [4,12,13], and the resulting expressions are summarized below.

12.2.2.2 Quantities R_a, R_b, and $\underline{d}$

Since the size of the molecule involved is generally comparable to that of the membrane pore, a distinction is made between the radius of the membrane pore and that of the membrane pore available for fluid flow. For purposes of analysis, the location of a molecule is defined as the location of its center (assuming a spherical shape for the molecule); this means that in the region of the pore where the center of the molecule cannot exist, the molecule as an entity does not exist. Thus, denoting D_w as the radius of the water molecule (assumed equal to 0.87 Å in this work) and R_b as the radius of the membrane pore, the effective radius of membrane pore available for fluid flow is given by R_a, where

$$R_a = R_b - D_w \tag{12.22}$$

Similarly, if D represents the distance of steric repulsion for the solute from the pore wall of the surface of the membrane material, the center of the solute molecule cannot exist at distances less than D from the pore wall or surface of the membrane material. Consequently, solute concentration is effectively zero in the pore region up to a distance D from the pore wall or the membrane surface. Noting that r represents the radial distance of the molecule from the center of the pore and that $\underline{d}$ represents the distance of the molecule from the pore wall,

$$\underline{d} = R_b - r \tag{12.23}$$

12.2.2.3 Definition of Dimensionless Quantities

For ease of expression and analysis, the following dimensionless quantities are defined:

Dimensionless radial distance:

$$\rho = r/R_a \tag{12.24}$$

Dimensionless solute concentration at the pore outlet:

$$C_A = C_{A3}/C_{A2} \tag{12.25}$$

Dimensionless solution velocity in the pore:

$$\alpha(\rho) = u_B(r)\delta \frac{\chi_{AB}}{\underline{R}T} \tag{12.26}$$

Dimensionless solution viscosity:

$$\beta_1 = \frac{\eta}{\chi_{AB} R_a^2 c_{A2}} \tag{12.27}$$

Dimensionless operating pressure:

$$\beta_2 = \frac{P_i - P_o}{\underline{R} c_{A2}} \tag{12.28}$$

Dimensionless potential function:

$$\Phi(\rho) = \frac{\phi(r)}{\underline{R}T} \tag{12.29}$$

Dimensionless friction function:

$$b(\rho) = \frac{\chi_{AB} + \chi_{AM}(r)}{\chi_{AB}} \tag{12.30}$$

Note that $\phi(r)$ represents the local value of the potential function expressing the force exerted on the solute molecule by the pore wall or the membrane surface; when $\phi(r)$ is positive, the force is repulsive, and when $\phi(r)$ is negative, the force is attractive. The solute–solvent friction coefficient χ_{AB} is obtained from the relation

$$\chi_{AB} = \frac{\underline{R}T}{D_{AB}} \tag{12.31}$$

Further, solute separations are expressed as f' or f, defined as

$$f' = \frac{c_{A2} - c_{A3}}{c_{A2}} \tag{12.32}$$

$$f = \frac{c_{A1} - c_{A3}}{c_{A1}} \tag{12.33}$$

On the basis of the above definition and the mass transfer situation on the high-pressure side of the membrane, the quantities f and f' are related by the expression

$$f = \frac{f'}{f' + [(1-f')\exp(v_s/k)]} \tag{12.34}$$

12.2.2.4 Basic Transport Equations

On the basis of the detailed analysis [4,12] and the dimensionless quantities defined above, the following expressions have been derived.

Effective driving pressure for fluid flow through the membrane pore:

$$P(r,0) - P(r,\delta) = (P_i - P_o) - \underset{\sim}{R}T\{c_{A2} - c_{A3}(r)\} \\ \times \{1 - \exp[-\phi(r)/\underset{\sim}{R}T]\} \tag{12.35}$$

$$= (P_i - P_o) - \{\pi(c_{A2}) - \pi[c_{A3}(r)]\} \\ \times \{1 - \exp[-\phi(r)/\underset{\sim}{R}]\} \tag{12.36}$$

Solute separation:

$$f' = 1 - \frac{\int_0^1 C_A(\rho)\alpha(\rho)\rho d\rho}{\int_0^1 \alpha(\rho)\rho d\rho} \tag{12.37}$$

Radial velocity profile for the solution in the membrane pore:

$$\frac{d^2\alpha(\rho)}{d\rho^2} + \frac{1}{\rho}\frac{d\alpha(\rho)}{d\rho} + \frac{1}{\beta_1} + \frac{\beta_2}{\beta_1}\{1 - \exp[-\Phi(\rho)]\}\{C_A(\rho) - 1\} \\ - \frac{[b(\rho) - 1]\alpha(\rho)C_A(\rho)}{\beta_1} = 0 \tag{12.38}$$

where

$$C_A(\rho) = \frac{\exp[\alpha(\rho)]}{1 + \frac{b(\rho)}{\exp[-\Phi(\rho)]}\{\exp[\alpha(\rho)] - 1\}} \tag{12.39}$$

For solving Equation 12.38, the boundary conditions are

$$\text{At } \rho = 0, \quad \frac{d\alpha(\rho)}{d\rho} = 0 \tag{12.40}$$

At $\rho = 1$, $\alpha(\rho) = 0$ (12.41)

Solution flux through the pore as expressed by the PR/PWP ratio:

$$\frac{PR}{PWP} = \frac{2\int_0^1 \alpha(\rho)\rho \, d\rho}{\beta_2/\beta_1} \quad (12.42)$$

Equations 12.35 through 12.42 are applicable to both RO and UF transport; they predict f' and the PR/PWP ratio for any single–solute aqueous solution RO/UF system at any given set of process operating conditions provided the pore radius, applicable data on osmotic pressure, and surface force and friction force functions are known. Since there are a great many pores on the membrane surface, it is natural to introduce a pore size distribution. For practical purposes, one can represent the pore size distribution by one (or more) normal pore size distribution with an equivalent average pore radius $\bar{R}_b$ with a standard deviation σ. When there is more than one normal distribution, the pore size distribution function of the ith component is written as

$$Y_i(R_{b,i}) = \frac{1}{\sigma_i \sqrt{2\pi}} \exp\left[-\frac{(R_{b,i} - \bar{R}_{b,i})^2}{2\sigma_i^2}\right] \quad (12.43)$$

Also, we need a quantity

$$h_i = \frac{\text{number of pores in the } i\text{th normal distribution}}{\text{number of pores in the first normal distribution}} = \frac{n_i}{n_i} \quad (12.44)$$

to describe such a distribution completely. Further, we define

$$\bar{R}_{b,i+1} > \bar{R}_{b,i} \quad (12.45)$$

Then, Equations 12.37 and 12.42 become

$$f' = 1 - \frac{\sum_i h_i \int_{\bar{R}_{b,i}-3\sigma_i}^{\bar{R}_{b,i}+3\sigma_i} Y_i(R_{b,i}) \left[\int_0^1 \left\{\frac{\exp[\alpha(\rho)]}{1+[b(\rho)/e^{-\Phi(\rho)}](\exp[\alpha(\rho)]-1)}\right\}\alpha(\rho)\rho \, d\rho\right]_{R_{b,i}=R_{b,i}} dR_{b,i}}{\sum_i h_i \int_{\bar{R}_{b,i}-3\sigma_i}^{\bar{R}_{b,i}+3\sigma_i} Y_i(R_{b,i}) \left[\int_0^1 \alpha(\rho)\rho \, d\rho\right]_{R_{b,i}=R_{b,i}} dR_{b,i}}$$

(12.46)

$$\frac{PR}{PWP} = \frac{\sum_i h_i \int_{\bar{R}_{b,i}-3\sigma_i}^{\bar{R}_{b,i}+3\sigma_i} Y_i(R_{b,i}) \left[2\int_0^1 \alpha(\rho)\rho \, d\rho\right]_{R_{b,i}=R_{b,i}} dR_{b,i}}{\sum_i h_i \int_{\bar{R}_{b,i}-3\sigma_i}^{\bar{R}_{b,i}+3\sigma_i} Y_i(R_{b,i}) \left(\frac{\beta_2}{8\beta_1}\right)_{R_{b,i}=R_{b,i}} dR_{b,i}} \quad (12.47)$$

ENGINEERING PROPERTIES OF FOODS

Once again, with respect to the radial velocity of the solution in the pore, Equations 12.38 through 12.41 are applicable.

In a solute–solvent (water)–polymer membrane material system, the relative solute–membrane material interactions at the membrane–solution interface can be expressed as described below.

Ions in aqueous solution are repelled in the vicinity of membrane materials of low dielectric constant [14]. The potential function representing the electrostatic repulsion of ions at the membrane–solution interface due to relatively long-range Coulombic forces may be expressed as

$$\Phi(\underline{d}) = \frac{\underline{A}}{\underline{d}} \tag{12.48}$$

where A is the electrostatic repulsive force constant characteristic of the ionic solute.

The relative interaction force working between a nonionized organic solute and a membrane polymer material can be expressed as the sum of the short-range van der Waals forces (attractive or repulsive) and still shorter-range repulsive force (steric hindrance) arising from the overlap of electron clouds of interacting atoms and molecules. When the organic molecule is assumed to be spherical, such interaction between a point (molecule) and a flat surface (membrane material) can be given by a Lennard–Jones type of equation such as

$$\Phi(\underline{d}) = \begin{cases} \infty & \text{when } \underline{d} < \underline{D} \\ -\dfrac{\underline{B}}{\underline{d}^3} & \text{when } \underline{d} > \underline{D} \end{cases} \tag{12.49}$$

where B and D are constants expressing the magnitude of the van der Waals force and the steric hindrance, respectively. Combining Equations 12.22 and 12.23 with Equations 12.48 and 12.49, we obtain

$$\Phi(\rho) = \begin{cases} \text{very large,} & \text{when } \left(\dfrac{R_b}{R_a} - \rho\right) \leq \dfrac{\underline{D}}{R_a} \\ \dfrac{\underline{A}/R_a}{R_b/R_a - \rho}, & \text{when } \left(\dfrac{R_b}{R_a} - \rho\right) > \dfrac{\underline{D}}{R_a} \end{cases} \tag{12.50}$$

For nonionized organic solutes,

$$\Phi(\rho) = \begin{cases} \text{very large,} & \text{when } \left(\dfrac{R_b}{R_a} - \rho\right) \leq \dfrac{\underline{D}}{R_a} \\ \dfrac{-(\underline{B}/R_a^3)}{(R_b/R_a - \rho)^3}, & \text{when } \left(\dfrac{R_b}{R_a} - \rho\right) > \dfrac{\underline{D}}{R_a} \end{cases} \tag{12.51}$$

For the case of an organic solute, such as a dye molecule, which contains an ionic part causing an electrostatic force and one or more aromatic rings or other substituent groups causing a van der Waals attractive force, the potential function should involve both A and B, and then

$$\Phi(\rho) = \begin{cases} \text{very large,} & \text{when } \left(\dfrac{R_b}{R_a} - \rho\right) \leq \dfrac{D}{R_a} \\ \dfrac{(A/R_a)}{(R_b/R_a - \rho)} - \dfrac{-B/R_a^3}{(R_b/R_a - \rho)^3}, & \text{when } \left(\dfrac{R_b}{R_a} - \rho\right) > \dfrac{D}{R_a} \end{cases} \qquad (12.52)$$

Figure 12.3 is a schematic representation of the repulsive or attractive surface forces as a function of distance $\underline{d}$ from the pore wall or the membrane surface.

At this point, the physicochemical significance of the parameters A, B, and D must be understood. All three are interfacial parameters that depend on the chemical nature of the solute, solvent, and the membrane material and their mutual interactions. A is a measure of the resultant electrostatic repulsive force; it always has a positive sign. B is a measure of the resultant short-range van der Waals force; its sign can be positive or negative. D is a measure of the steric hindrance for the solute at the interface; its sign is always positive. Steric hindrance may be considered to reflect the effective size of the solute at the membrane material–solution interface. For example, the distance D may be regarded as the radius of the hydration sphere formed around an ion at the interface, the distance from the center of a molecule to the point where the molecule touches, or practically touches, the surface of the

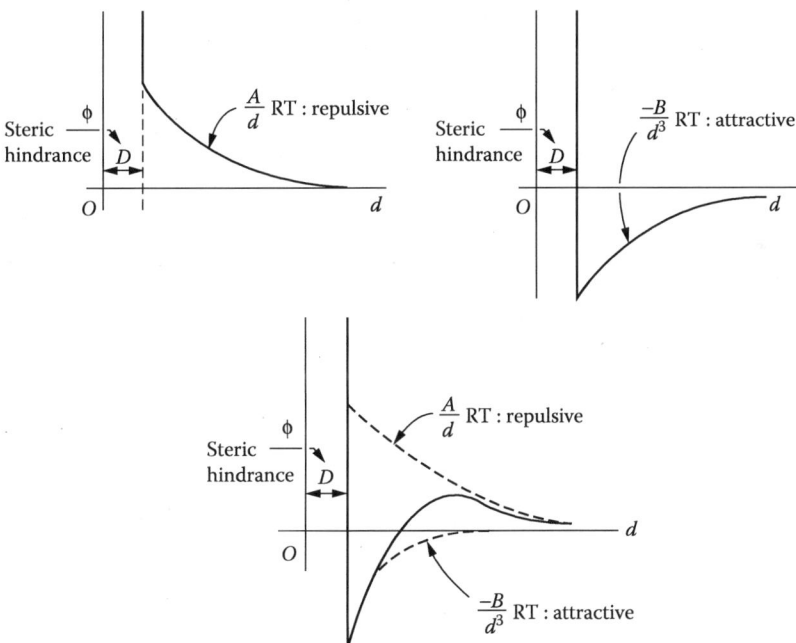

Figure 12.3 Potential curves expressing the interfacial forces working on the solute molecule from the polymer surface. (From S Sourirajan. *Lectures on Reverse Osmosis*. Ottawa: National Research Council Canada, 1983.)

membrane material, or the radius of an agglomerate of several molecules at the interface. It is often necessary to equate the numerical value of D as a first approximation to Stokes' radius, at least for some suitably chosen reference solutes for purposes of analysis; however, it must be recognized that, in principle, D is not identical with Stokes' radius.

From the foregoing discussion, it should be clear that the numerical values of A, B, and D govern the nature of the potential curves represented by the function $\Phi(\rho) = \varphi(r)/RT$, and the membrane performance represented by f' and the PR/PWP ratio can be predicted from the numerical values of R_b, A and/or B, and D if only the appropriate function $b(\rho)$ is known.

On the basis of the work of Faxen [15], Satterfield et al. [16], and Lane and Riggle [17], who have considered the problem of the friction experienced by a molecule as it moves through a narrow pore, the following empirical relationship has been developed [13]:

$$b(\rho) = \begin{cases} (1 - 2.104\lambda_f + 2.09\lambda_f^3 - 0.95\lambda_f^5)^{-1} & \text{when } \lambda_f \leq 0.22 \\ 44.57 - 416.2\lambda_f + 943.9\lambda_f^2 + 302.4\lambda_f^3 & \text{when } 1 > \lambda_f > 0.22 \end{cases} \quad (12.53)$$

where

$$\lambda_f = \frac{\tilde{D}}{R_b} \quad (12.54)$$

Equation 12.53 (which gives only an average value for $b(\rho)$ for the entire cross section of the pore) has proved to be adequate for the analysis of RO data involving an aqueous solution of nonionized polar organic solutes. Hence, the above equation has been used in this work with respect to such solutes.

12.2.2.5 Liquid Chromatography for the Determination of Interfacial Interaction Force Parameters

Liquid chromatography (LC), in which solutes are injected into a water stream (as the carrier solvent) flowing through a column packed with the polymer membrane material in powder form, is a useful tool for evaluating the nature and magnitude of the interaction force [18,19]. Defining the surface excess Γ_A as the positive or negative excess solute concentration at the polymer material–solution interface, a quantity $\Gamma_A/C_{A,b}$ can be obtained from the LC retention volume data by [20]

$$\frac{\Gamma_A}{c_{A,B}} = \frac{[V_R']_A - [V_R']_{\text{water}}}{A_t} \quad (12.55)$$

Further, via the Maxwell–Boltzmann equation, the interfacial concentration $C_{A,i}$ is related to the bulk concentration $C_{A,b}$ by

$$c_{A,i} = c_{A,b} \exp[-\Phi(\underline{d})] \quad (12.56)$$

Then, from the definition of Γ_A,

$$\Gamma_A = \int_{\underline{D}_\infty}^{\infty} (c_{A,i} - c_{A,b}) d(\underline{d}) \quad (12.57)$$

Combining Equations 12.56 and 12.57,

$$\frac{\Gamma_A}{c_{A,b}} = \int_{D_w}^{\infty} \{\exp[-\Phi(\underline{d})] - 1\} d(\underline{d}) \tag{12.58}$$

12.2.2.6 Data on Interfacial Surface Force Parameters

LC data on $\Gamma_A/C_{A,b}$ given in the literature (Tables II and III in Ref. [21]) and the experimental RO/UF data together offer a means of determining, on a relative scale, numerical values for the interfacial surface force parameters A, B, and D for different solute–solvent–polymer membrane materials. The details on the method of determination are given in the literature [4,13]. Briefly, the experimentally derived quantities are f' and $\Gamma_A/C_{A,b}$; the quantities to be determined are generally R_b, A, B, and D; and the equations available for such determinations are Equations 12.37 and 12.58, together with the associated expressions. Since there are only two equations to solve three unknowns, one has to fix one of the above three quantities at least for one solute, which then becomes a reference solute. This is done in this work by choosing glycerol as the reference solute and fixing its D value to be the same as its Stokes' radius r_A, obtained from the expression

$$r_A = \frac{kT}{6\pi\eta D_{AB}} \tag{12.59}$$

Since the D ($=r_A$) value for glycerol is thus known, its B value can be determined from Equation 12.58. Using these B and D values for glycerol, the R_b value for the RO membrane used can be obtained from Equation 12.37. Once the R_b value for the membrane is known, the B and D values for the other nonionized organic solutes, or the A and D values for completely ionized inorganic solutes, can be calculated from the corresponding RO and LC data applicable for the particular membrane. The data on A, B, and D so obtained for different solutes, membrane materials, and membranes are listed in Tables 12.2 through 12.4.

The results show that with respect to nonionized organic solutes, the values of B and D are independent of R_b for all the membrane materials reported. With respect to the ionized inorganic solutes, the values of A and D depend on R_b because the surface potential function is different for different ions, and the values reported are the mean values for the solute as a whole and are applicable only for the range of R_b values indicated.

12.2.2.7 Data on Pore Size and Pore Size Distribution

It was just pointed out how the value of R_b for an RO membrane can be obtained from the RO and LC data for glycerol, chosen as the reference solute. The values of R_b so obtained for a set of 24 different cellulose acetate membranes, along with their RO data for NaCl, are given in Table 12.5. These results show that solute separations varied from 94.4% to 62% with respect to NaCl and from 83.2% to 40% with respect to glycerol in the membranes used, and their average pore radii varied in the range 8.1–11.0 Å.

An extension of the method indicated above for determining R_b values also offers a method for determining both the average pore size $\P_b$ and its standard deviation σ on the surface of RO/UF membranes, corresponding to one or more equivalent normal pore size distributions. The details of the latter method (which also needs both LC and RO/UF

Table 12.2 Data on Surface Force Parameters and Stokes' Law Radii for Some Ionized Inorganic Solutes in Aqueous Solutions at the Polymer–Solution Interface

Solute	Stokes' Law Radius × 10^{10} (m)	CA-398 Polymer[a] $D \times 10^{10}$ (m)	CA-398 Polymer[a] $\underset{\sim}{A} \times 10^{10}$ (m)	CE Polymer[b] $\underset{\sim}{D} \times 10^{10}$ (m)	CE Polymer[b] $\underset{\sim}{A} \times 10^{10}$ (m)	PAH Polymer $\underset{\sim}{D} \times 10^{10}$ (m)	PAH Polymer $\underset{\sim}{A} \times 10^{10}$ (m)
LiF	2.01	4.67	0.949	—	—	2.91	0.465
LiCl	1.79	3.65	1.12	—	—	2.69	0.418
LiBr	1.77	3.52	1.25	4.00	2.686	2.57	0.868
NaF	1.74	3.91	1.15	4.04	1.707	2.93	0.230
NaCl	1.52	3.74	1.10	4.09	2.115	2.81	0.367
NaBr	1.50	3.80	0.895	3.95	2.879	2.80	0.238
NaI	1.51	3.53	0.686	4.04	1.951	2.80	0.039
KF	1.45	—	—	—	—	2.79	0.806
KCl	1.22	3.88	1.07	4.00	2.871	2.89	0.037
KBr	1.21	3.82	0.888	2.50	3.987	2.90	0.068
RbCl	1.19	3.97	0.757	4.12	1.578	—	—
CsCl	1.19	3.99	0.844	3.98	3.646	—	—
CsBr	1.18	3.91	1.058	3.85	3.474	—	—
NH_4Cl	1.23	3.99	0.845	—	—	—	—
$NaCH_3COO$	2.04	3.49	1.128	—	—	3.23	0.376
Na_2SO_4	1.99	—	—	4.87	2.497	—	—
$MgCl_2$	1.96	—	—	3.51	2.591	—	—
$CaCl_2$	1.83	4.05	1.021	3.50	2.326	–	–

[a] Data corresponding to pore radius $R_{\bar{b}}$ of 8.07×10^{-10} m. For other pore radii, use $\underset{\sim}{D} = -0.171 R_{\bar{b}} + (\underset{\sim}{D})_{R_{\bar{b}}=807 \times 10^{-10}} + 1.380 \times 10^{-10}$. $\underset{\sim}{A} = 0.041 R_{\bar{b}} + (\underset{\sim}{A})_{R_{\bar{b}}=8.07 \times 10^{-10}} - 0.33 \times 10^{-10}$.

[b] Values for $\underset{\sim}{D}$ and $\underset{\sim}{A}$ correspond to the boundary mole fraction of 8×10^{-5}.

data) are available in the literature [22]. Briefly, the method involves choosing a set of non-ionized polar organic compounds (of similar chemical structure and different molecular weights) as reference solutes, setting their D values to be the same as their corresponding Stokes' radii, and then determining their B values from LC data. Using the B and D values so obtained, and assuming values of $\P_b$ and σ for the particular membrane under study, data on RO/UF separation f' are calculated for all the reference solutes by means of Equations 12.43 and 12.46 and the associated expressions. The data on f' so calculated are then compared with the corresponding experimental data; if the calculated and experimental f' data for all the solutes are in reasonable agreement, the assumed R_b and σ values are deemed valid; otherwise, the calculations are repeated until reasonable agreement of the calculated and experimental f' values is obtained. A single normal distribution of pore sizes (involving one set of values of $\P_b$ and σ) is first assumed for the calculation of solute separations. If the calculated data on f' are too different from the experimental values,

Table 12.3 Data on Surface Force Parameters and Stokes' Law Radii for Some Nonionized Polar Organic Solutes in Aqueous Solutions at Polymer–Solution Interface

Solute	Stokes' Law Radius × 10^{10} (m)	CA-398 Polymer[a]		CE Polymer[b]		PAH Polymer	
		$D \times 10^{10}$ (m)	$B \times 10^{30}$ (m)	$D \times 10^{10}$ (m)	$B \times 10^{30}$ (m)	$D \times 10^{10}$ (m)	$B \times 10^{30}$ (m)
Methanol	1.45	1.85	8.31	—	—	1.41	5.71
Ethanol	2.05	2.03	21.54	2.05	−82.94	1.94	21.99
1-Propanol	2.12	2.15	37.77	—	—	2.13	36.40
2-Propanol	2.26	2.45	38.40	3.60	−164.8	3.06	89.74
1-Butanol	2.33	2.10	47.27	1.71	−277.3	2.57	95.50
2-Butanol	2.33	2.45	62.15	2.18	−328.9	3.14	134.6
2-Methyl-l-propanol	3.05	2.75	95.57	3.72	−136.8	3.78	243.6
2-Methyl-2-propanol	3.35	3.67	134.3	2.54	−371.3	3.79	132.2
1-Pentanol	2.63	—	—	3.51	−201.8	3.75	361.3
1-Hexanol	1.97	1.97	56.76	3.87	−102.7	—	—
Acetone	1.91	1.91	26.90	—	—	2.22	47.16
Methyl ethyl ketone	1.92	1.92	35.87	—	—	2.39	71.38
Methyl isopropyl ketone	2.59	2.45	83.67	3.86	−77.9	4.02	360.9
Methyl isobutyl ketone	2.86	2.45	97.88	—	—	—	—
Cyclohexanone	2.77	2.72	120.8	—	—	—	—
Diisopropyl ketone	3.11	2.62	131.7	—	—	—	—
Methyl acetate	2.05	1.73	27.88	—	—	2.58	85.88
Ethyl acetate	2.39	1.83	36.41	—	—	2.90	132.7
Propyl acetate	2.68	—	—	—	—	3.46	265.3
Ethyl propionate	2.68	1.88	45.63	—	—	—	—
Ethyl butyl ether	2.93	2.53	86.45	5.02	−20.0	—	—
Ethyl/-butyl ether	2.93	5.30	901.2	—	—	—	—
Isopropyl/-butyl ether	3.18	5.60	918.2	—	—	—	—
Propionamide	2.24	1.98	19.13	—	—	—	—
Acetonitrile	1.47	1.78	28.49	—	—	—	—
Propionitrile	1.85	1.78	34.20	—	—	—	—
Nitromethane	1.65	1.80	36.00	—	—	—	—
1-Nitropropane	2.00	2.05	65.76	—	—	—	—
Phenol	2.10	1.71	45.39	—	—	—	—
Resorcinol	2.74	1.73	45.16	—	—	—	—
Aniline	2.42	1.80	49.98	1.02	7.0	—	—
Dimethyl aniline	2.99	2.50	125.8	—	—	—	—
1,2-Ethanediol	2.11	2.20	−16.82	—	—	—	—
Glycerol	2.30	2.30	−52.30	3.64	−165.2	—	—
2,3-Butanediol	2.46	2.75	−11.97	—	—	—	—

continued

Table 12.3 (continued) Data on Surface Force Parameters and Stokes' Law Radii for Some Nonionized Polar Organic Solutes in Aqueous Solutions at Polymer–Solution Interface

Solute	Stokes' Law Radius × 10^{10} (m)	CA-398 Polymer[a]		CE Polymer[b]		PAH Polymer	
		$D \times 10^{10}$ (m)	$B \times 10^{30}$ (m)	$D \times 10^{10}$ (m)	$B \times 10^{30}$ (m)	$D \times 10^{10}$ (m)	$B \times 10^{30}$ (m)
1,2,6-Hexane-triol	3.07	2.82	−16.95	—	—	—	—
Dulcitol	3.31	—	—	—	—	2.16	−104.4
D-Sorbitol	3.30	4.64	−180.2	4.00	−52.4	—	—
D-Glucose	3.66	3.36	−203.1	—	—	2.29	−67.94
D-Fructose	3.22	4.51	−181.8	—	—	—	—
Sucrose	4.67	5.11	−343.2	3.99	−289.4	—	—
Maltose	4.98	4.98	−346.0	—	—	—	—

even with the best set of $\P_b$ and σ values, then a two–normal distribution of pores involving two sets of values for $\P_b$ and σ is assumed, the calculations for f' are repeated, and the results are compared with the experimental data. Available results show that the assumption of a two–normal pore size distribution is sufficient to obtain reasonable agreement between the calculated and experimental data.

Using polyethylene glycols (PEGs) of the molecular weight range 600–6000 as the reference solutes in the method indicated above, the average pore sizes and their distributions with respect to two polysulfone (Victrex) (PS-V), six cellulose acetate (CA-400), and three aromatic polyamide hydrazide (PAH) UF membranes were determined and the results obtained are given in Table 12.6 [22].

12.3 ENUMERATION OF PROBLEMS INVOLVED IN MEMBRANE SEPARATION AND CONCENTRATION OF LIQUID FOODS

In this section, we deal with the membrane separation of major components in liquid foods. Three typical liquid foods—fruit juices, green tea juice, and dairy foods—are considered. The primary components of fruit juices are carbohydrates such as food sugars, food acids, and low molecular weight flavor compounds, whereas those of dairy foods are fats, proteins, and carbohydrates. Tables 12.7 and 12.8 illustrate typical constituents of fruit juices and milk.

As is evident in Table 12.7, the major water-soluble components in fruit juices are sugars (up to 20%), with much smaller quantities (0.1 to 2%) of organic acids and inorganic salts and still smaller quantities (in the parts per million range) of many organic volatile flavor components including alcohols, aldehydes, ketones, and esters [23]. Therefore, one is concerned with the recovery of all of the above components in the concentration processes involving fruit juices.

Although the available cellulose acetate membranes are eminently suitable for the practical recovery of food sugars, these membranes are not equally efficient for the recovery

Table 12.4 Data on Surface Force Parameters and Stokes' Law Radii for Some Dyes, Macromolecules, and Proteins in Aqueous Solutions at Polymer–Solution Interfaces

	Stokes' Law	CA-398 Polymer			PAH Polymer		
Solute	Radius $\times 10^{10}$ (m)	$D \times 10^{10}$ (m)	$A \times 10^{10}$ (m)	$B \times 10^{30}$ (m^3)	$D \times 10^{10}$ (m)	$A \times 10^{10}$ (m)	$B \times 10^{30}$ (m^3)
Dyes							
Acridine orange	4.40	3.56	1.8	42.72	5.12	0.9	97.66
Methylene blue	4.43	3.06	1.8	39.06	4.42	1.0	133.5
Orange II	4.49	4.84	1.6	104.3	6.56	3.4	987.5
Acrid blue	4.85	4.44	2.4	117.8	4.62	~0	202.8
Indigo carmine	4.23	5.48	2.0	119.2	4.60	1.0	155.2
Amaranth	5.58	5.80	2.4	232.8	4.52	1.0	131.9
Coomassie blue	5.92	5.04	1.8	130.4	4.00	0.6	42.2
Naphthol Green B	7.19	5.46	3.2	278.5	4.06	~0	102.2
Alizarin red S	4.00	4.06	1.8	70.9	6.52	10.0	1499
Eriochrome Black T	5.06	7.96	0.8	997.6	5.44	1.2	116.2
Alizarol cyanine RC	5.80	4.90	1.6	83.45	4.40	0.4	50.0
Congo red	6.44	6.22	1.8	203.7	6.40	~0	644.7
Chlorazol	7.48	5.62	3.1	278.5	4.06	1.8	90.8
Black E Macromolecules[a]							
PEG −600	6.27	6.27[b]	0	69.6	6.27[b]	0	589.8
−1000	7.89	7.89[b]	0	254.6	—	—	—
−1500	9.95	—	—	—	9.95[b]	0	2194
−2000	11.43	11.43[b]	0	1119	11.43[b]	0	3313
−3000	14.06	14.06[b]	0	2408	14.06[b]	0	6012
−4000	15.34	15.34[b]	0	3053	15.34[b]	0	7536
−6000	25.00	25.00[b]	0	16283	25.00[b]	0	29575
Lignin[c]	30.90	0	37010	17.25	0	5247	—
Proteins	22.10	22.10[b]	0	1.03×10^5	—	—	—

(continued)

Table 12.4 Data on Surface Force Parameters and Stokes' Law Radii for Some Dyes, Macromolecules, and Proteins in Aqueous Solutions at Polymer–Solution Interfaces

Solute	Stokes' Law Radius $\times 10^{10}$ (m)	CA-398 Polymer $D \times 10^{10}$ (m)	CA-398 Polymer $A \times 10^{10}$ (m)	CA-398 Polymer $B \times 10^{30}$ (m³)	PAH Polymer $D \times 10^{10}$ (m)	PAH Polymer $A \times 10^{10}$ (m)	PAH Polymer $B \times 10^{30}$ (m³)
Bacitracin							
Pepsin	28.06	28.06[b]	0	2.19×10^5	—	—	—
α-Casein	36.71	36.71[b]	0	4.56×10^5	—	—	—
Bovine serum albumin	38.70	38.70[b]	0	5.28×10^5	—	—	—
γ-Globulin	56.28	50.00	0	10.71×10^5	—	—	—

[a] PEG = polyethylene glycol.
[b] D was equated to Stokes' law radius.
[c] Kraft lignin supplied by Lignosol Chemicals; pH of lignin solution was adjusted to 5.52.

Table 12.5 Average Pore Radius ($\bar{R}_b$) of Some Cellulose Acetate (CA-398) RO Membranes

Film No.	$A \times 10^7$ [kg · mol/(m^2 s kPa)]	$(D_{AM}/K\delta)_{NaCl}$ $\times 10^7$ (m/s)	$K_{NaCl} \times 10^6$ (m/s)	Solute Separation[a] (%) NaCl	Glycerol	$\bar{R}_b \times 10^{10}$ (m)
1	1.726	2.515	22.6	93.5	82.9	8.23
2	1.544	2.772	21.0	92.2	82.8	8.35
3	2.955	10.15	35.0	85.8	75.0	9.48
4	3.485	25.75	40.0	74.4	60.2	10.26
5	2.044	3.302	26.0	93.4	82.5	8.49
6	2.513	6.759	30.6	89.4	79.2	9.11
7	1.622	2.230	22.0	94.4	83.2	8.07
8	3.559	18.36	40.6	81.3	66.2	10.01
9	3.430	25.13	39.5	75.5	60.6	10.24
10	2.810	9.833	33.5	85.6	75.6	9.44
11	1.728	2.195	23.0	94.3	83.3	8.07
12	1.535	2.877	21.1	91.9	82.7	8.41
13	3.438	26.81	39.5	73.0	59.5	10.29
14	5.189	62.81	45.0	62.0	40.0	10.96
15	4.263	58.64	45.0	61.8	43.3	10.85
16	1.564	2.319	21.4	94.0	82.9	8.23
17	2.291	7.610	28.5	87.3	78.3	9.22
18	3.198	13.09	37.1	84.6	74.3	9.53
19	3.917	29.63	44.1	74.8	58.0	10.34
20	2.593	17.50	31.4	77.2	63.9	10.11
21	4.338	61.43	45.0	61.0	41.8	10.91
22	1.546	2.371	21.0	93.8	83.1	8.17
23	2.271	7.199	28.2	87.8	78.6	9.18
24	3.219	13.06	37.3	84.7	71.3	9.73

[a] Operating pressure, 1724 kPag (250 psig), NaCl concentration in feed, 0.026–0.06 molal.

of organic flavor compounds. RO separations of the latter components are generally better with aromatic polyamide and PAH membranes [24]. Further, these flavor components are better separated in RO from feed solutions that are essentially free of sugars [3]. For these reasons, it is preferable to carry out the fruit juice concentration process in two operations. In the first operation, which uses a cellulose acetate membrane, the primary object could be the recovery of most (>99%) of the sugars present in the fruit juice; a part of the acids and flavor components is also recovered along with sugars in this operation. In the second operation, which employs an aromatic polyamide (or PAH) membrane, the object could be a major recovery of flavor components by RO treatment of membrane-permeated fruit juice waters (obtained from the first-stage operation) under suitable experimental conditions. In the concentration of food sugars in which cellulose acetate membranes are used, one is also concerned with the fractionation of organics present in the solution. In terms of

Table 12.6 Pore Size Distributions in Some Membranes Studied

Membrane	$\bar{R}_{b,1}$ (nm)	$\sigma_1/\bar{R}_{b,1}$	$\bar{R}_{b,2}$ (nm)	$\Sigma_2/\bar{R}_{b,2}$	H_2
PS-V-1	2.60	0.002			
PS-V-2	2.81	0.005			
CA-400-1	2.20	0.200	10.10	0.250	0.010
CA-400-2	2.17	0.200	10.55	0.250	0.012
CA-400-3	2.29	0.200	10.40	0.250	0.019
CA-400-4	2.54	0.200	10.58	0.250	0.038
CA-400-5	2.82	0.200	10.22	0.250	0.040
CA-400-6	3.30	0.200	10.80	0.250	0.072
PAH-1	2.51	0.440	10.50	0.220	0.009
PAH-2	2.57	0.460	10.50	0.250	0.010
PAH-3	3.44	0.300	10.40	0.305	0.056

the fractionation and concentration of components in fruit juices, the operations described above involve one or more of the following fundamental separation problems in RO.

1. Separations of undissociated polar organic solutes such as sugars in high concentrations;
2. Separations of undissociated polar organic solutes (flavor components) present in low concentrations;
3. Separations of partially dissociated organic acids present in low concentrations;
4. Separations of low concentrations of undissociated organic solutes (flavor components) in concentrated sugar solutions;
5. Separations of partially dissociated organic acids present in concentrated sugar solutions.

Table 12.8 shows the composition of chemical components in typical cow's milk [25]. According to the table, 1 L of milk contains 30–50 g of milk fat (lipids). Although milk lipids have high nutritional value, membrane processing of milk and other dairy products necessitates the treatment of fat–free systems. Therefore, no further consideration will be given to milk fat. Milk normally contains 3.5% of total proteins. Broadly, these proteins can be classified into casein and whey proteins. Whereas the former precipitates from the solution when the pH value is adjusted to 4.6, the latter proteins do not. Casein is the major component of processed cheese, whereas whey proteins have so far been discarded into the effluent without being recovered. Whey proteins consist of three major constituent proteins, such as β-lactoglobulin, α-lactalbumin, and bovine serum albumin [25]. Another major component of milk is lactose. One liter of milk normally contains 45–50 g of lactose, and thus lactose is the principal carbohydrate of milk. In view of the milk constituents, therefore, the processing of milk and its products (such as whole milk concentration, skim milk concentration, whey concentration, whey fractionation, and concentration of permeate from whey UF) involves one or more of the following fundamental membrane separation problems:

1. Concentration of casein
2. Concentration of whey proteins

Table 12.7 Major Components in Solution in Fruit Juice

Component	Apple Juice	Pineapple Juice	Orange Juice	Grapefruit Juice	Grape Juice	Tomato Juice
Sugars (wt%)						
D-Glucose	1.3–2.0	2.1–2.4	2.6–5.8	3.5–5.0	11.5–19.3	~4.3[a]
D-Fructose	4.4–8.2	2.1–2.4			—	
Sucrose	1.7–4.2	8.4–9.5	3.1–5.1	1.3–3.0	0.2–2.3	
Acids (wt%)						
L-Lactic acid	Present				Present	
D-Malic acid	0.3–1.0	0–0.2			Present	
Citric acid	0–0.03	0.7–0.9	0.4–1.5	0.9–1.4	0.7–1.7	0.2–0.6
Tartaric acid	Present		Present			
Volatile flavor compounds (ppm)			Present[a]	NR	Present[a]	NR
Alcohols	46	Present				
Aldehydes	3	Present				
Ketones		Present				
Esters	1	22–414				
Acids		18–118				
Hydrocarbons		Present				
Fats (wt%)			0.2–0.5	~0.1		~0.2
Proteins (wt%)	0.4–0.5		0.6–0.8	0.3–0.6	0.2–0.9	~1.0
Vitamins (ppm)	110–116		300–800			~160
Inorganics (wt%)	0.2–0.5		0.5–0.9	0.2–0.4	0.3–0.4	~1.0
Water (wt%)			80–95			

[a] Reported as a group only.
NR = not reported.

3. Fractionation of whey proteins and lactose
4. Concentration of lactose (This is the same as problem 1)

In this section, the fundamental principles involved in the aforementioned eight problems (excluding problem 9) of membrane separation processes are presented on the basis of the water preferential sorption model and also of the surface force–pore flow model, which were described in the foregoing sections. For the first five problems, the former model was used as the transport theory, whereas in the last three cases the surface force–pore flow model was used. The object of this section is therefore to illustrate quantitatively how one can predict data on the membrane performance—that is, solute separation and membrane permeated PR—with reference to any specific membrane for solution systems considered in the problems enumerated above.

Table 12.8 Approximate Concentrations of the Major Constituents in Normal Cow's Milk

Constituent or Group of Constituents	Approximate conc. (wt[a]/L of milk)
Water	860–880
Lipids in emulsion	
Milk fat (a mixture of mixed triglycerides)	30–50
Phospholipids (lecithins, cephalins, etc.)	0.3
Sterols	0.1
Proteins in colloidal dispersion	
Casein (αs_1, β, y, and κ fractions)	25
β-Lactoglobulin(s)	3
α-Lactalbumin	0.7
Albumin (probably identical to blood serum albumin)	0.3
Euglobulin	0.3
Pseudoglobulin	0.3
Other albumins and globulin	1.3
Dissolved materials	
Carbohydrates	
Lactose (α and β)	45–50
Inorganic and organic ions and salts	
Calcium	1.25
Magnesium	0.10
Sodium	0.50
Potassium	1.50
Phosphates (as PO_4^{3-})	2.10
Citrates (as citric acid)	2.00
Chloride	1.00
Nitrogenous material, not protein or vitamins (as N)	250 mg
Ammonia (as N)	2–12 mg
Amino acids (as N)	35 mg
Urea (as N)	100 mg
Uracil-4-carboxylic acid	50–100 mg
Hippuric acid	30–60 mg

[a] In grams, except where indicated in milligrams.

12.3.1 Application of Water Preferential Sorption Model

As mentioned earlier, this analysis is applicable to systems in which either water is preferentially sorbed or the solute is only weakly sorbed to the membrane polymer material. For the purpose of illustration, a specific cellulose acetate membrane and a specific aromatic polyamide membrane were chosen to predict membrane performance data at a fixed operating pressure of 6895 kPag (= 1000 psig) and an operating temperature of 25°C. These membranes are specified by A and $(D_{AM}/K\delta)_{NaCl}$ given in Table 12.9 [3]. These data were obtained from a single set of RO data using 3500 ppm NaCl–H$_2$O feed solutions. The latter experimental data are also shown in the table. As pointed out already, in addition to A and $D_{AM}/K\delta$, appropriate values of k are needed to predict membrane performance. For the purpose of this work, four values of k_{NaCl} were chosen, namely $k_{NaCl} \times 10^4$ (in cm/s) = 17.4, 22.2, 126, and ∞. The first three of these values are all experimentally obtained by real membrane cell systems, whereas the last value represents the limiting condition at which solute separation and PR reach maximum values. The latter values are particularly useful for comparison with actual values at any finite value of k_{NaCl}. Thus, all four chosen k_{NaCl} values represent practically meaningful conditions as far as this work is concerned. In this section, the predicted data represent membrane performance at the start of the RO operation corresponding to infinitesimal volume change in feed solution; the PR data are given for an effective membrane area of 13.2 cm^2, which was the area actually used in the RO experiments with the NaCl–H$_2$O reference solution.

For the purpose of these illustrative calculations, 76 organic solutes that are of major interest in the RO concentration of fruit juices and food sugars were considered; they include 15 alcohol, 6 aldehyde, 3 ketone, 38 ester, 9 acid, and 5 sugar solutes listed in Table 12.10. For each one of these solutes, the values of ln $(D_{AM}/K\delta)$ were calculated for one or both of the films considered on the basis of their polar ($-\Delta\Delta G/RT$), steric ($\delta^*\Sigma E_s$),

Table 12.9 Specifications of Films Used for the Calculation[a]

Film Type	Cellulose Acetate[b]	Aromatic Polyamide
PWP constant $A\left(\dfrac{g \cdot mol\,H_2O}{cm^2 \cdot s \cdot atm}\right) \times 10^6$	1.029	0.206
Solute transport parameter		
$(D_{AM}/K\delta)_{NaCl}$ (cm/s) $\times 10^5$	1.467	0.180
ln C^*_{NaCl}	−12.5	−12.5
Solute separation (%)[c]	97.92	99.18
PR (g/h)[c]	55.92	11.47

[a] Operating pressure = 6895 kPag(=1000 psig).
[b] Cellulose acetate, batch 316 (10/30).
[c] Film area, 13.2 cm^2; mass transfer coefficient, $k = 22 \times 10^{-4}$ cm/s; NaCl concentration in feed, 3500 ppm.

Table 12.10 Physiochemical and Transport Parameter Data for Some Organic Solutes

Solute				$-\Delta\Delta G/RT$		ΣE_s or $(\delta^*\Sigma E_s)_{lim}$	$\omega^*\Sigma s^{*a}$	$\ln (D_{AM}/K\delta)^b$	
Name	Formula	$\Sigma\sigma^*$		Cellulose Acetate	Aromatic Polyamide			Cellulose Acetate	Aromatic Polyamide
Alcohols	R in R-OH								
2-Pentanol	2-C$_5$H$_{11}$	−0.230		4.54	2.12	−1.20	0	−7.96	−11.23
s-Butyl alcohol	S-C$_4$H$_9$	−0.210		4.66	2.42	−1.13	0	−7.84	−10.88
1-Propyl alcohol	i-C$_3$H$_7$	−0.190		4.78	2.69	−0.70	0	−7.72	−10.31
n-Hexyl alcohol	n-C$_6$H$_{13}$	−0.134		4.41	2.81	−0.40	1.83	−6.26	−9.97
n-Octyl alcohol	n-C$_8$H$_{17}$	−0.134		4.17	3.45	−0.33	2.41	−5.93	−9.28
n-Nonyl alcohol	n-C$_9$H$_{19}$	−0.134		4.05	3.77	−0.40		c	−9.01
n-Decyl alcohol	n-C$_{10}$H$_{21}$	−0.134		3.93	4.09	−0.40		c	−8.69
n-Pentanol	n-C$_5$H$_{11}$	−0.133		4.52	2.49	−0.40	1.57	−6.41	−10.29
n-Butyl alcohol	n-C$_4$H$_9$	−0.130		4.64	2.17	−0.39	1.29	−6.57	−10.61
i-Butyl alcohol	i-C$_4$H$_9$	−0.125		4.66	2.42	−0.93	0	−7.84	−10.74
n-Propyl alcohol	n-C$_3$H$_7$	−0.115		5.56	2.47	−0.36	0	−6.94	−10.29
Ethyl alcohol	C$_2$H$_5$	−0.100		6.26	2.76	−0.07	0	−6.24	−9.79

2-Methylbutan-1-ol	CH$_3$CH$_2$−CHCH$_2$−
	               |
	              CH$_3$
		−0.075		4.54	2.12	−1.25	0	−7.96	−11.27

3-Methylbutan-1-ol	CH$_3$−CHCH$_2$−CH$_2$−
	      |
	      CH$_3$
		−0.045		4.54	2.12	−0.35	0	−7.96	−10.63

Methyl alcohol	CH$_3$	0		7.10	3.06	0	0	−5.40	−9.44
Aldehydes	R in R-CHO								
1-Octanal	CH$_3$(CH$_2$)$_6$	−0.134		4.31	2.83	−0.40	2.13	−6.07	−10.65
1-Nonanal	CH$_3$(CH$_2$)$_7$	−0.134		4.19	3.15	−0.33	2.41	−5.91	−10.16
1-Decanal	CH$_3$(CH$_2$)$_8$	−0.134		4.07	3.47	−0.40			−10.01
1-Undecanal	CH$_3$(CH$_2$)$_9$	−0.134		3.95	3.79	−0.40			−9.69
1-Hexanal	CH$_3$(CH$_2$)$_4$	−0.133		4.54	2.19	−0.40	1.57	−6.39	−11.29
Acetaldehyde	CH$_3$	0		5.00	2.75	0	0	−7.50	−9.75

PROPERTIES OF FOOD IN MEMBRANE SEPARATION PROCESSES

Ketones

	R_1, R_2 in $R_1-\overset{\overset{O}{\|}}{C}-R_2$					
3-Pentanone	C_2H_5, C_2H_5	−0.200	5.45	2.18	−0.14	−7.12	−10.61
2-Pentanone	CH_3, C_3H_7	−0.115	5.45	2.18	−0.36	−7.23	−11.06
Acetone	CH_3, CH_3	0	5.67	2.47	0	−6.83	−10.03

Esters

	R_1, R_2 in $R_1-\overset{\overset{O}{\|}}{C}-O-R_2$					
2-Propyl 2-methyl propionate	$i\text{-}C_3H_7, i\text{-}C_3H_7$	−0.380		1.69	−1.40		−13.68
2-Propyl caproate	$n\text{-}C_5H_{11}, i\text{-}C_3H_7$	−0.323		1.50	−1.10		−13.26
Ethyl 2-methyl butyrate	$s\text{-}C_4H_9, C_2H_5$	−0.310		1.47	−1.20		−13.49
2-Propyl butyrate	$n\text{-}C_3H_7, i\text{-}C_3H_7$	−0.305		1.47	−1.06		−13.31
2-Propyl propionate	$C_2H_5, i\text{-}C_3H_7$	−0.290		1.75	−0.77		−12.33
Ethyl 2-methyl propionate	$i\text{-}C_3H_7, C_2H_5$	−0.290		1.75	−0.77		−12.33
Pentyl hexanoate	$n\text{-}C_5H_{11}, n\text{-}C_5H_{11}$	−0.266		1.31	−0.80		−12.83
n-Butyl caproate	$n\text{-}C_5H_{11}, n\text{-}C_4H_9$	−0.263		0.99	−0.79		−13.13
n-Butyl butyrate	$n\text{-}C_3H_7, n\text{-}C_4H_9$	−0.245		0.96	−0.75		−13.08
Ethyl heptanoate	$n\text{-}C_6H_{13}, C_2H_5$	−0.234		1.89	−0.47		−11.58
Ethyl octanoate	$n\text{-}C_7H_{15}, C_2H_5$	−0.234		2.21	−0.47		−11.26
Ethyl decanoate	$n\text{-}C_9H_{19}, C_2H_5$	−0.234		2.84	−0.47		−10.62
Ethyl caproate	$n\text{-}C_5H_{11}, C_2H_5$	−0.233		1.57	−0.47		−11.90
n-Butyl propionate	$C_2H_5, n\text{-}C_4H_9$	−0.230		1.25	−0.46		−12.20
Ethyl pentanoate	$n\text{-}C_4H_9, C_2H_5$	−0.230		1.25	−0.46		−12.20
Ethyl 3-methyl butyrate	$i\text{-}C_4H_9, C_2H_5$	−0.225		1.47	−1.00		−13.08
Ethyl butyrate	$n\text{-}C_3H_7, C_2H_5$	−0.215		1.54	−0.43		−11.85

continued

Table 12.10 *(Continued)* Physiochemical and Transport Parameter Data for Some Organic Solutes

Name	Formula	$\Sigma\sigma^*$	$-\Delta\Delta G/RT$ Cellulose Acetate	$-\Delta\Delta G/RT$ Aromatic Polyamide	ΣE_s or $(\delta^*\Sigma E_\theta)_{lim}$	$\omega^*\Sigma s^{*a}$	$\ln(D_{AM}/K\delta)^b$ Cellulose Acetate	$\ln(D_{AM}/K\delta)^b$ Aromatic Polyamide
Ethyl propionate	C_2H_5, C_2H_5	−0.200	1.82		−0.14			−10.97
2-Propyl acetate	$CH_3, i-C_3H_7$	−0.190	2.04		−0.70			−11.89
Methyl 2-methyl propionate	$i-C_3H_7, CH_3$	−0.190	2.04		−0.70			−11.89
Methyl heptanoate	$n-C_6H_{13}, CH_3$	−0.134	2.17		−0.40			−11.15
Methyl octanoate	$n-C_7H_{15}, CH_3$	−0.134	2.49		−0.40			−10.83
Methyl decanoate	$n-C_9H_{19}, CH_3$	−0.134	3.13		−0.40			−10.19
n-Hexyl acetate	$CH_3, n-C_6H_{13}$	−0.134	2.17		−0.40			−11.15
Methyl caproate	$n-C_5H_{11}, CH_3$	−0.133	1.85		−0.40			−11.47
n-Butyl acetate	$CH_3, n-C_4H_9$	−0.130	1.54		−0.39			−11.77
Methyl pentanoate	$n-C_4H_9, CH_3$	−0.130	1.54		−0.39			−11.77
2-Methyl-1-propyl acetate	$CH_3, i-C_4H_9$	−0.125	1.75		−0.93			−12.65
Methyl 3-methyl butyrate	$i-C_4H_9, CH_3$	−0.125	1.75		−0.93			−12.65
1-Propyl acetate	$CH_3, n-C_3H_7$	−0.115	1.82		−0.36			−11.42
Methyl butyrate	$n-C_3H_7, CH_3$	−0.115	1.82		−0.36			−11.42
Ethyl acetate	CH_3, C_2H_5	−0.100	2.11		−0.07			−10.54
Methyl propionate	C_2H_5, CH_3	−0.100	2.11		−0.07			−10.54
2-Methyl 1-butyl acetate	$CH_3, s-C_4H_9(CH_2)$	−0.075	1.47		−1.25			−13.60
3-Methyl 1-butyl acetate	$CH_3, i-C_4H_9(CH_2)$	−0.045	1.47		−0.35			−11.75
Methyl 4-methyl pentanoate	$i-C_4H_9(CH_2), CH_3$	−0.045	2.40		−0.35			−11.75
Methyl acetate	CH_3, CH_3	0			0			−10.10

PROPERTIES OF FOOD IN MEMBRANE SEPARATION PROCESSES

					$(\delta^{*}\Sigma E_{s})_{lim}°$	
Acids						
Benzoic acid	C_6H_5COOH	0.600	5.92			-11.82^d / -6.58^e
Acetic acid	CH_3COOH	0	5.69			-12.11^d / -6.81^e
Propionic acid	CH_3CH_2COOH	-0.100	5.57			-12.30^d / -6.93^e
Butyric acid	$CH_3CH_2CH_2COOH$	-0.115	5.45			-12.22^d / -7.05^e
Valeric acid	$CH_3(CH_2)_3COOH$	-0.130	5.34			-12.27^d / -7.16^e
Lactic acid	$CH_3CH(OH)COOH$ $\sigma^{*}_{COOH} = -0.100$ $\sigma^{*}_{OH} = -0.100$		5.65		-1.88	-12.46^e / -8.73^e
Malic acid	$HOOCCH(OH)CH_2\text{-}COOH$ $\sigma^{*}_{COOH} = -0.200$ $\sigma^{*}_{OH} = -0.100$	6.41	-3.57	-12.13^e / -9.66^e		
Tartaric acid	$HOOC(CHOH)_2COOH$ $\sigma^{*}_{COOH} = -0.200$ $\sigma^{*}_{OH} = -0.200$	6.50		-4.73	-12.56^d	-10.73^e
Citric acid	$HOOCCH_2C(OH)\text{-}(COOH)CH_2COOH$ $\sigma^{*}_{COOH} = -0.420$ $\sigma^{*}_{OH} = -0.190$	6.67	-6.05	-12.40^d / -11.88^e		
Sugars						
D-Glucose	$C_2H_{12}O_6$ $\sigma^{*}_{CHO} = -0.133$ $\sigma^{*}_{OH} = -0.951$	4.95	-5.42	-12.97		

continued

Table 12.10 (Continued) Physiochemical and Transport Parameter Data for Some Organic Solutes

Solute			−ΔΔG/RT				In $(D_{AM}/K\delta)$[b]	
Name	Formula	$\Sigma\sigma^*$	Cellulose Acetate	Aromatic Polyamide	ΣE_s or $(\delta^*\Sigma E_s)_{lim}$	$\omega^*\Sigma s^{*a}$	Cellulose Acetate	Aromatic Polyamide
	$\sigma^*_{CHO} = -0.133$							
	$\sigma^*_{OH} = -0.951$							
Sucrose	$C_{12}H_{22}O_{11}$	—	5.65	−7.86	−14.71			
Maltose	$C_{12}H_{22}O_{11}\ H_2O$	—	5.65	−7.86	−14.71			
Lactose	$C_{12}H_{22}O_{11}\ H_2O$	—	5.65	−7.86	−14.71			

[a] Data for cellulose acetate.
[b] Based on ln $C^*_{NaCl} = -12.5$.
[c] Preferential sorption of solute.
[d] For solutes in ionized form.
[e] Tor solutes in nonionized form.

and nonpolar ($\alpha * \Sigma_S^*$) parameters by Equations 12.7 through 12.9 and Equations 12.11 through 12.13 from data on $(D_{AM}/K\delta)_{NaCl}$ only, which are given in Table 12.9 for each film. All the calculated values of In $(D_{AM}/K\delta)$, along with the values of the parameters used, are included in Table 12.10. The values of $(D_{AM}/K\delta)$ listed in Table 12.10 for different solutes are specifically applicable to dilute solutions. It is known [5,26] that $(D_{AM}/K\delta)$ is independent of solute concentration with respect to both nonionized and ionized solutes for aromatic polyamide membranes; these facts are used in this work. Furthermore, for each value of k_{NaCl} the corresponding value for dilute solution was calculated from

$$k_{solute} = k_{NaCl} \left[\frac{(D_{AB})_{solute}}{(D_{AB})_{NaCl}} \right]^{2/3} \tag{12.60}$$

When necessary, k values for concentrated solutions were calculated by means of the relation [8]

$$k \propto \frac{D_{AB}^{0.67}}{v^{0.62}} \tag{12.61}$$

When a very small quantity of solute was mixed in a concentrated solution of sugar, the k value for the solute was obtained [27] from

$$k = k_{sugar} \left[\frac{D_{AB}}{(D_{AB})_{sugar}} \right]^{2/3} \tag{12.62}$$

where k_{sugar} and $(D_{AB})_{sugar}$ refer to k values for sugar solute and diffusivity of sugars, respectively, in the concentrated sugar solution, and D_{AB} refers to the diffusivity of the solute in the concentrated sugar solution.

The values of $D_{AM}/K\delta$ and k thus obtained for each solute, along with the values of A given in Table 12.9, were then used in the basic transport Equations 12.1 through 12.4 to predict solute separation and PR; in this prediction procedure, osmotic pressure data summarized in Tables 12.11 and 12.12 were used. When the solution was dilute and the osmotic pressure effect was negligible, Equation 12.5 was used.

12.3.1.1 Separation of Undissociated Organic Solutes Such as Sugars Present in High Concentration

This problem is represented by the RO separation of food sugars such as D-glucose, D-fructose, sucrose, maltose, and lactose in aqueous solutions. The cellulose acetate membrane can be used for these separations. Calculations showed (Table 12.10) that the value of In $(D_{AM}/K\delta)$ for D-glucose and D-fructose was −12.97, and the value for sucrose, maltose, and lactose was −14.71 for the cellulose acetate membrane specified in Table 12.9. Detailed prediction calculations for this membrane were then carried out for two sugar solutions, D-fructose and sucrose, representing each of the above In $(D_{AM}/K\delta)$ values; the results for other sugar solutes should be similar.

Data on solute separations and PRs in the concentration range 1–10 wt% for D-fructose and 1–40 wt% for sucrose were obtained for the four k values mentioned above by means

Table 12.11 Osmotic Pressure (kPa) Data for Aqueous Solutions of Some Solutes at 25°C

Molality	NaCl	Glucose	Fructose	Sucrose	Maltose	Lactose
0	0	0	0	0	0	0
0.1	462	259	253	248	214	214
0.2	917	517	496	503	455	455
0.3	1372	776	790	758	724	724
0.4	1820	1034	1013	1020	933	
0.5	2282	1293	1307	1282		
0.6	2744	1517	1611	1551		
0.7	3213	1744	1824	1827		
0.8	3682		2067	2103		
0.9	4158		2310	2379		
1.0	4640		2564	2668		
1.2	5612		3101	3241		
1.4	6612		3587	3840		
1.6	7646			4447		
1.8	8701			5061		
2.0	9784			5695		
2.2	10894					
2.4	12031					
2.6	13203					
2.8	14403					
3.0	15651			9128		
3.2	16913					
3.4	18278					
3.6	19540					
3.8	20919					
4.0	22326			12866		
4.2	23759					
4.4	25242					
4.6	26745					
4.8	28296					
5.0	29875					
5.2	31495					
5.4	33143					
5.6	34846					
5.8	36570					
6.0	38335					

Table 12.12 Osmotic Pressure (kPa) Data of Some Proteins at 25°C

Molality × 10^3	Bovine Serum Albumin	α-Casein
0	0	0
0.2	0.5	0.5
0.4	1.1	1.0
0.6	1.7	1.6
0.8	2.6	2.2
1.0	3.4	2.9
1.5	6.0	4.0
2.0	—	6.0
6.0	48	—

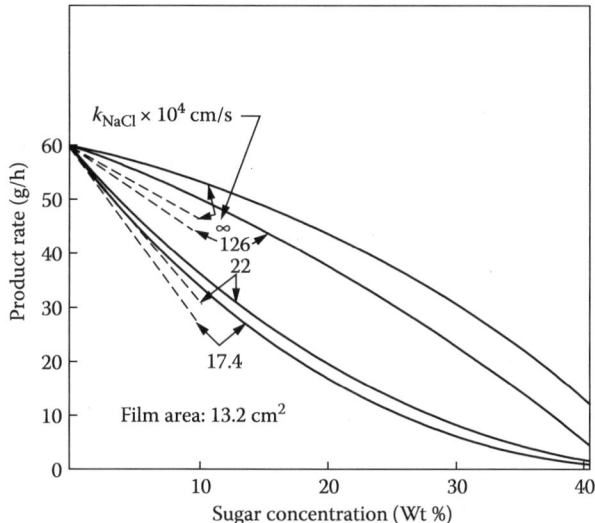

Figure 12.4 Effect of feed concentration on PR in the separations of food sugars. Film type, cellulose acetate [batch 316 (10/30)]; operating pressure, 6895 kPag (=1000 psig); solute separation, >99%. Solid line, sucrose; broken line, fructose. (From T Matsuura, S Sourirajan. *AIChE Symp Ser* 74:196–208, 1978.)

of the basic transport equations, Equations 12.1 through 12.4. Solute separations were 99.3% for D-fructose and >99.0% for sucrose at all the concentrations and k values investigated. The data obtained on PRs are given in Figure 12.4. The separation data illustrate the usefulness of the cellulose acetate membrane chosen for the concentration of D-fructose and sucrose, and the PR data given in Figure 12.4 illustrate the need for a high k value to increase product water flux in practical operations.

12.3.1.2 Separations of Undissociated Polar Organic Solutes Present in Low Concentrations

The problem is presented in RO separations of alcohol, aldehyde, ketone, and ester solutes, which are present in extremely small quantities (in the parts per million range) in fruit juices as flavor compounds. A list of such solutes is included in Table 12.10. This problem is relevant to the RO concentration of fruit juice waters, for which one may consider the use of either the cellulose acetate or the aromatic polyamide membrane specified in Table 12.9. Calculations showed (Table 12.10) that the values of ln $(D_{AM}/K\delta)$ for the alcohol, aldehyde, and ketone solutes listed in Table 12.10 were in the range −5.4 to −7.96 for the cellulose acetate membrane, and those for the above solutes and the ester solutes were in the range −8.69 to −13.68 for the polyamide membrane.

By means of Equation 12.5, data on solute separations were then calculated for both membranes in their respective ranges of ln $(D_{AM}/K\delta)$ values at all four chosen k values. The results obtained are given in Figures 12.5 and 12.6. With the above cellulose acetate membrane (Figure 12.5), the obtainable solute separations for the alcohol, aldehyde, and ketone solutes were in the range <4–78%; for the same solutes, the obtainable solute separations were in the range 54–95% with respect to the polyamide membrane (Figure 12.6); further, with the latter membrane, the obtainable solute separations for ester solutes were in the range 83–99.5%. These results show that from a practical point of view, the use of a polyamide membrane is preferable for the recovery of flavor components from fruit juice waters. Figures 12.5 and 12.6 also show that whereas the effect of k on solute separation is very significant for the cellulose acetate membrane, it is far less significant or practically insignificant for the polyamide membrane; this is because of the relatively low permeation velocity with respect to the latter membrane.

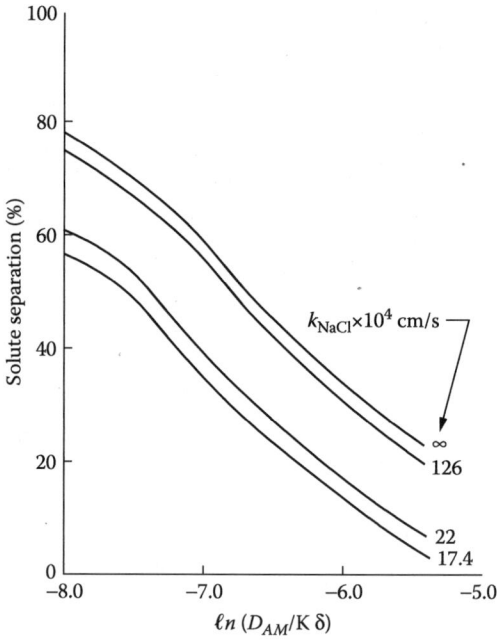

Figure 12.5 Effect of ln $(D_{AM}/K\delta)$ on the separations of alcohols, aldehydes, and ketones. Film type, cellulose acetate [batch 316 (10/30)1; operating pressure, 6895 kPag (=1000 psig). (From T Matsura, S Sourirajan. *AIChE Symp Ser* 74:196–208, 1978.)

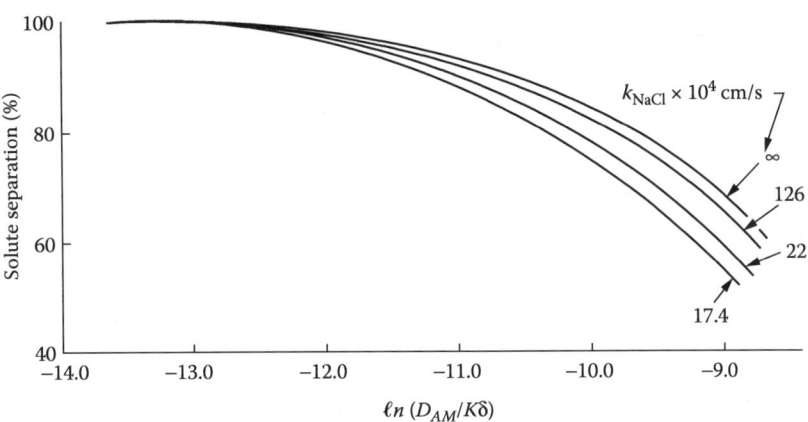

Figure 12.6 Effect of ln $(D_{AM}/K\delta)$ on the separations of alcohols, aldehydes, ketones, and esters. Film type, aromatic polyamide; operating pressure, 6895 kPag (=1000 psig). (From T Matsuura, S Sourirajan. *AIChE Symp Ser* 74:196–208, 1978.)

Because of the very low concentration of the flavor components, no significant osmotic pressure effects are involved in the RO treatment of fruit juice waters. Therefore, the obtainable PRs are essentially the same as the PWP rates, which are given by the respective values of A for the two membranes (Table 12.9) obtained by means of Equation 12.1.

12.3.1.3 Separation of Partially Dissociated Organic Solutes Present in Low Concentration

This problem is represented by the RO separation of monocarboxylic acids such as acetic, propionic, butyric, valeric, and benzoic acids, and hydroxycarboxylic acids such as lactic, malic, tartaric, and citric acids. For the purpose of illustration, the RO separations of all the above acids, each in the feed concentration range 1–1000 ppm, were calculated for the cellulose acetate membrane specified in Table 12.9. The derivation of the relevant transport equations (based on the water preferential sorption models) is given in the literature in detail [28]. In this derivation, the transport of the ionized species and that of the nonionized species (represented by the subscripts i and u, respectively) are treated separately, and the following relations applicable for dilute feed solutions are used.

$$N_t = c \exp\left(\frac{N_B}{kc}\right)\left[\left(\frac{D_{AM}}{k\delta}\right)_i (X_{i1} - X_{i3}) + \left(\frac{D_{AM}}{k\delta}\right)_u (X_{u1} - X_{u3})\right] \tag{12.63}$$

$$\frac{N_B}{N_t} = X_{A3} \tag{12.64}$$

$$\frac{N_B}{N_t} = X_{A3} \tag{12.65}$$

$$X_{A3} = X_{i3} + X_{u3} \tag{12.66}$$

$$X_{i1} = \frac{[K_a^2/4 + 1000c\, X_{A1}\, K_a]^{1/2} - K_a/2}{1000c} \quad (12.67)$$

$$X_{i3} = \frac{[K_a^2/4 + 1000c\, X_{A3}\, K_a]^{1/2} - K_a/2}{1000c} \quad (12.68)$$

In these equations, the quantities K_a (dissociation constant characteristic of acid), X_{A1} (from given feed concentration), N_B (=AP), $(D_{AM}/K\delta)_i$ (from Equation 12.8), $(D_{AM}/K\delta)$ (from Equation 12.9), c (the same as that for pure water = 0.05535 mol/cm), and k (appropriate for the chosen k_{NaCl} value) are known. The unknown quantities are N_t, X_{i1}, X_{u1}, X_{i3}, X_{u3}, and X_{A3}, which can be obtained by the simultaneous solution of the six equations, Equations 12.63 through 12.68. In this work, solute separations were calculated from the values of X_{A3} obtained by the computer solution of Equations 12.63 through Equation 12.68 by the Newton–Raphson iterative process. In the case of polycarboxylic acids, their lowest pK_a values were used in the above calculation.

Figure 12.7 gives the data on solute separation as a function of feed concentration for all nine acids mentioned above at k values corresponding to $k_{NaCl} = 22 \times 10^{-4}$ cm/s. Figure 12.8 gives similar data for lactic, benzoic, and acetic acids at k values corresponding to three other chosen k_{NaCl} values. These results illustrate that:

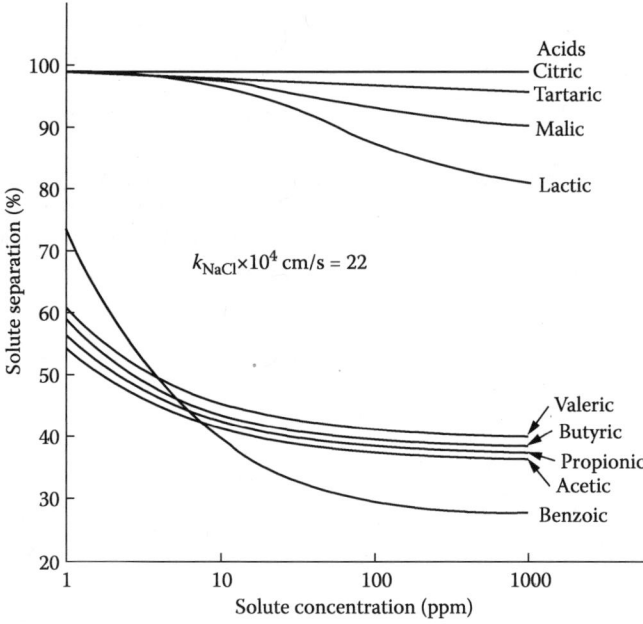

Figure 12.7 Effect of feed concentration on the separations of monocarboxylic acids and hydroxycarboxylic acids at $k_{NaCl} = 22 \times 10^{-4}$ cm/s. Film type, cellulose acetate [batch 316 (10/30)]; operating pressure, 6895 kPag (=1000 psig). (From T Matsuura, S Sourirajan. *AIChE Symp Ser* 74:196–208, 1978.)

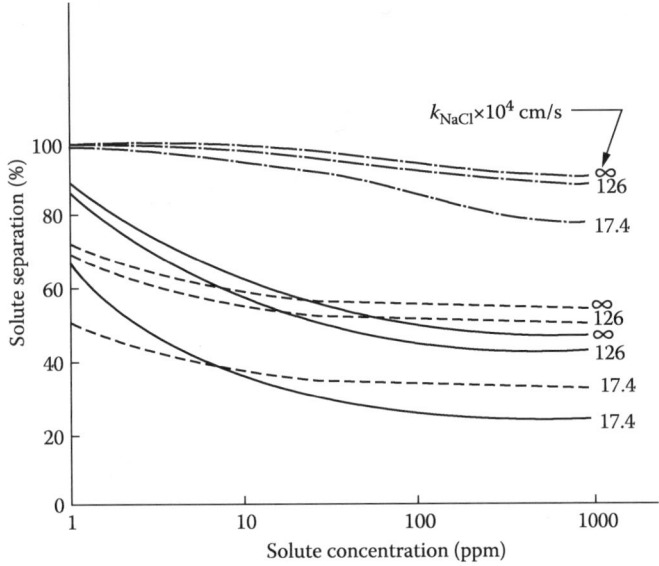

Figure 12.8 Effect of feed concentration on the separation of (----) lactic acid, (—) benzoic acid, and (---) acetic acid at different values of k. Film type, cellulose acetate [batch (10/30)]; operating pressure, 6895 kPag (=1000 psig). (From T Matsuura, S Sourirajan. *AIChE Symp Ser* 74:196–208, 1978.)

1. Solute separations for the hydroxycarboxylic acids are always higher than those for the monocarboxylic acids.
2. Higher solute separations are obtainable at lower feed concentration and higher k values.
3. The chosen cellulose acetate membrane is suitable for the separation of the above acids at practically useful levels.

In this problem also, no significant osmotic pressure effects are involved because of low feed concentrations; therefore, the obtainable PR is essentially the same as the pure water permeation rate.

12.3.1.4 Problem of Separations of Low Concentrations of Undissociated Organic Solutes in Concentrated Sugar Solutions

This problem is relevant to the recovery of flavor components from fruit juices in their primary concentration process. The presence of high concentrations of sugars in the feed solution tends to decrease RO separations of flavor components because of two factors: the decrease in water transport due to the high osmotic pressure of the sugar solution, and the decrease in k value for the flavor component because of the high viscosity of the sugar solution. Because these two factors are amenable to exact analysis, one can predict the RO separation of a flavor component present in low concentrations in concentrated sugar solutions. This prediction technique is illustrated in detail in the literature [27].

ENGINEERING PROPERTIES OF FOODS

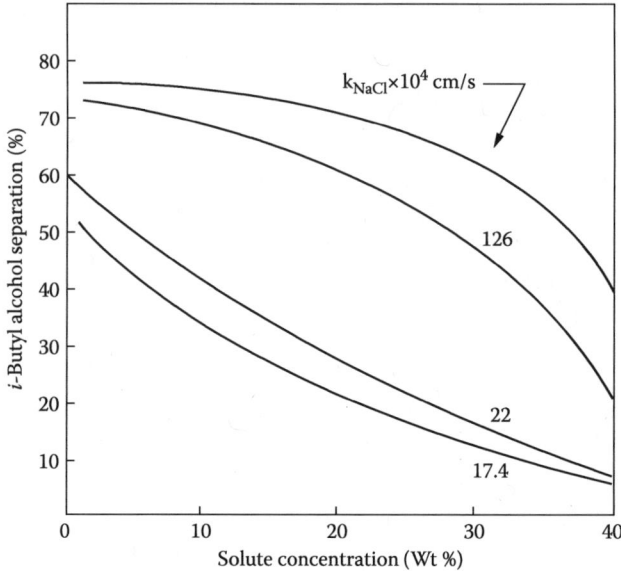

Figure 12.9 Effect of sucrose concentration on the separation of *i*-butyl alcohol. Film type, cellulose acetate [batch 316 (10/30)]; operating pressure, 6895 kPag (=1000 psig). (From T Matsuura, S Sourirajan. *AIChE Symp Ser* 74:196–208, 1978.)

Choosing, for illustration, isobutyl alcohol as the undissociated organic solute present in low concentrations (100 ppm), its RO separations from concentrated sucrose–water feed solutions were calculated for the cellulose acetate membrane specified in Table 12.9 for the different k_{NaCl} values, using the corresponding PR data given in Figure 12.4. The data obtained on alcohol separations are given in Figure 12.9, which shows precisely how alcohol separations decrease significantly with an increase in sucrose concentration in the feed and a decrease in the k_{NaCl} value. These results explain the common experience of low recovery of the flavor components in the primary concentration of fruit juices and the need for the RO treatment of fruit juice waters for higher recovery of the flavor components.

12.3.1.5 Separation of Solutions of Partially Dissociated Acids Present in Concentrated Sugar Solutions

The mixed-solute system under consideration is typical of that found in apple juices. There are three components involved here: sugar, ionized malic acid, and nonionized malic acid. Again, in the presence of highly concentrated sugar solute, the following problems are noted [29]:

1. A decrease in flux occurs because of the high osmotic pressure of the sugar solution.
2. A decrease in the *k* value occurs because of the high viscosity of the feed solution.

3. The dissociation of malic acid is suppressed in the solution of high sugar concentration because of a decrease in the dielectric constant.

Because all of these effects are amenable to exact analysis, a prediction of the RO performance was attempted for a system involving water, glucose, and malic acid [30]. The results are shown in Figure 12.10, which demonstrates good agreement between experimental and predicted values.

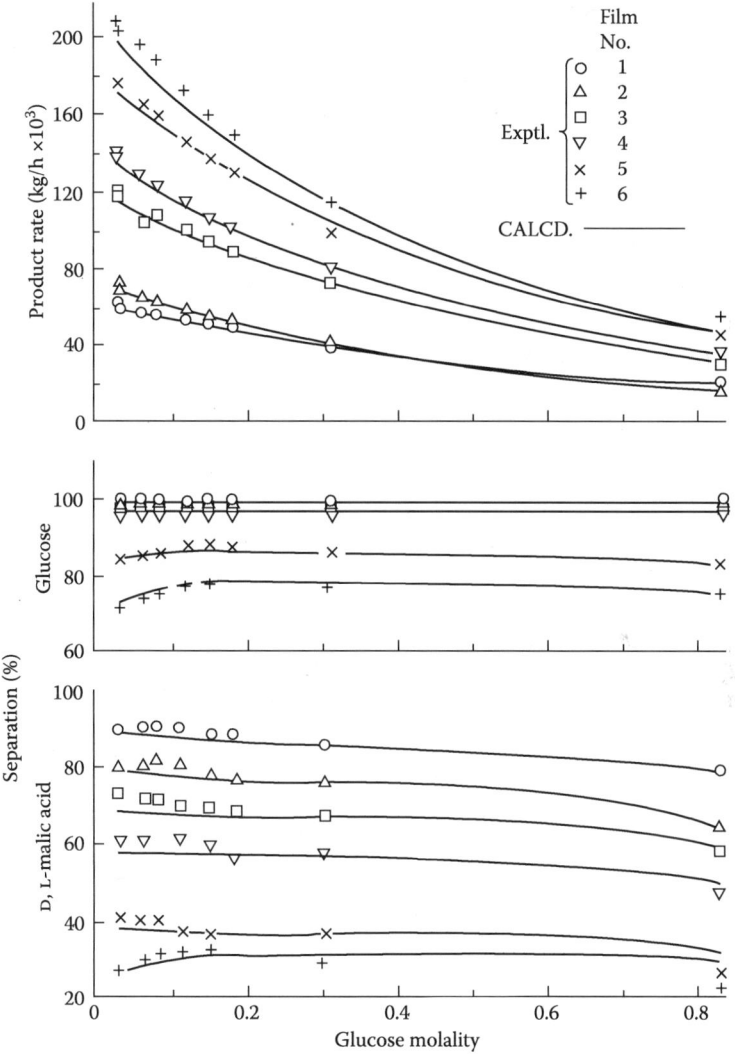

Figure 12.10 Glucose molality in feed versus solute separations of glucose and D, L-malic acid and PR. Membrane, cellulose acetate [batch 316 (10/30)]; operating pressure, 6895 kPag (=1000 psig); feed flow rate, 400 cm³/min; membrane area, 13.2 cm²; molality of glucose to malic acid, 4:1. (From P Malaiyandi, T Matsuura, S Sourirajan. *Ind Eng Chem Process Des Dev* 21:277–282, 1982.)

12.3.2 Application of Transport Equations to Real Fruit Juice Concentration

It may be recalled that Equations 12.1 through 12.4 involve quantities for which knowledge of the molecular weight of solute is needed. Because no definite value can be assigned to the molecular weight of solute in fruit juices, the above equations cannot be used directly to calculate $D_{AM}/K\delta$ and k in the RO processing of fruit juices. For the latter purpose, the set of Equations 12.69 through 12.72 can be written in analogous form as follows, in terms of quantities expressed in readily measurable weight units and carbon weight fraction [31]:

$$A_{(wt)} = \frac{PWP}{S \times 3600 \times P} \qquad (12.69)$$

$$N_{B(wt)} = A_{(wt)}[P - \pi(X_{c_2}) + \pi(X_{c_3})] \qquad (12.70)$$

$$= \left(\frac{D_{AM}}{K\delta}\right)\left(\frac{1 - X_{c_3}}{X_{c_3}}\right)[c_{2(wt)}X_{c_2} - c_{3(wt)}X_{c_3}] \qquad (12.71)$$

$$= kc_{1(wt)}(1 - X_{c_3})\ln\left(\frac{X_{c_2} - X_{c_3}}{X_{c_1} - X_{c_3}}\right) \qquad (12.72)$$

where $A_{(wt)}$, $N_{B(wt)}$, and $c_{(wt)}$ are in units of g/(cm²·s·atm), g/(cm²·s), and g/cm³, respectively, and X_c represents the carbon weight fraction in solution. It was confirmed experimentally that the $D_{AM}/K\delta$ values in Equations 12.3 and 12.71 and the k values in Equations 12.4 and 12.72 are numerically identical. In order to apply the above equations, osmotic pressure data of fruit juices as a function of carbon weight fractions are needed. It was found that the relation can be expressed by

$$\frac{\pi}{X_c} = a\pi + b \qquad (12.73)$$

The constants a and b are given in Table 12.13 [31,32]. Equations 12.69 through 12.72 further allow the calculation of $A_{(wt)}$, $D_{AM}/K\delta$, and k values from experimental data of PWP,

Table 12.13 Data on Constants a and b for Osmotic Pressure Calculation $\pi = X_C(a\pi + b)$ (psi)[a]

Juice	a	b
Lime	3.31	3997
Lemon	2.59	4442
Prune	3.31	4217
Carrot	4.93	3088
Tomato	8.95	4187
Other[b]	3.94	3560

[a] 1 psi = 6.895 kPa.
[b] Apple, pineapple, orange, grapefruit, and grape juices.

PR, and solute separation. Such data were obtained with respect to different fruit juices at different carbon weight fractions, and the results are given in Table 12.14 and Figure 12.11.

For the purpose of process design, a set of experiments on the RO concentration of fruit juices was carried out at 6895 kPag (=1000 psig) in the nonflow-type cell [31]. The quantity of product water removed in each case was 50% or more of the initial feed solution on volume basis. The data relating to these experiments are given in Tables 12.15 and 12.16.

Table 12.15 gives the specification of a cellulose acetate membrane used for apple juice concentration in terms of $A_{(wt)}$, $D_{AM}/K\delta$, and k. Using the above data and other necessary data listed in Table 12.15, the system specification for the concentration process could be given as $\gamma = 0.188$, $\theta = 0.001$, and $\lambda\theta = 0.45$. The Δ versus τ or X correlation obtained from the Ohya–Sourirajan analysis corresponding to the above data on system specification is given in Figure 12.12. The good agreement between the experimental data and analytical results indicates the validity of the system analysis discussed. The analytical results obtained on the effect of $\lambda\theta$ on the Δ versus τ or X correlation are also plotted in Figure 12.12. The latter give a quantitative illustration of the reductions obtainable in the value of τ or X for a given value of Δ by increasing the mass transfer coefficient. The system specification and performance data included in the Ohya–Sourirajan tables [11] may be used for similar parametric studies on fruit juice concentration systems.

Table 12.14 Effect of Feed Concentration on $D_{AM}/K\delta$ for Fruit Juice Solutes at 4137 kPag (=600 psig)[a]

Film Number	Feed Solution	Carbon Content in Feed Solution (ppm)	$(D_{AM}/K\delta) \times 10^5$ (cm/s)
J7	Apple juice	29,900	0.81
		43,800	0.84
		61,900	0.66
		84,800	0.36
J8	Pineapple juice	29,800	0.64
		47,300	0.43
		62,200	0.24
		80,400	0.35
J9	Orange juice	30,800	1.32
		45,000	0.97
		80,200	1.18
J10	Grapefruit juice	31,700	0.66
		45,900	0.35
		58,500	0.77
		86,900	0.43
J11	Grape juice	33,300	1.12
		48,100	0.63
		62,700	0.39
		81,500	0.69

[a] Experiments carried out in nonflow-type cell.

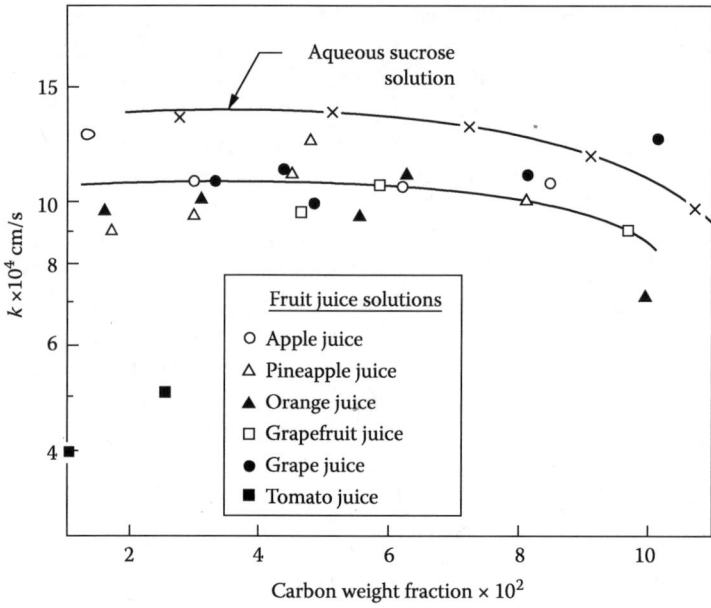

Figure 12.11 Mass transfer coefficients obtained during RO treatment of fruit juice solutions in the non-flow-type cell used. Film type, cellulose acetate [batch 316 (10/30)]; operating pressure, 6895 kPag (=1000 psig). (From T Matsuura, AG Baxter, S Sourirajan. *Acta Aliment* 2:109–150, 1973.)

Table 12.15 Concentration of Apple Juice: Film Number J13

Operating Conditions	
Operating pressure (atm)	68.0
Mass transfer coefficient, k (cm/s)	10.0×10^{-4}
Properties of feed solution	
Carbon content in feed (ppm)	43,800
Osmotic pressure of feed (atm)	12.82
Average density (g/cm³)	1.051
Film specification	
PWP constant, $A_{(Wt)}$ [g H_2O/(cm² · s · atm)]	34.24×10^{-6}
Solute transport parameter, $D_{AM}/K\delta$ (cm/s)	0.2×10^{-5}
System specification	
γ	0.188
θ	0.001
$\lambda\theta$	0.45

Table 12.16 Concentration of Fruit Juice

Experimental Details	Apple Juice	Pineapple Juice	Orange Juice	Grapefruit Juice	Grape Juice	Tomato Juice
Film number	J1	J2	J3	J4	J5	J6
Operating pressure (psig)	1000	1000	1000	1000	1000	1000
(kPag)	6895	6895	6895	6895	6895	6895
Feed						
Carbon weight fraction $\times 10^2$	4.38	5.30	5.24	5.38	5.98	2.91
Solid content (wt%)	10.75	11.45	10.35	13.90	15.40	5.85
Density (g/cm³)	1.038	1.046	1.042	1.052	1.063	1.011
pH	3.9	3.7	4.2	4.2	3.2	4.7
Product						
Carbon weight fraction $\times 10^2$	0.0720	0.0385	0.0552	0.0750	0.0740	0.0440
Concentrate						
Carbon weight fraction $\times 10^2$	8.51	9.40	10.41	9.50	11.03	5.92
Solid content (wt%)	19.35	21.65	22.45	25.25	25.70	12.85
Density (g/cm³)	1.074	1.089	1.081	1.105	1.107	1.100
Total carbon retained (%)	99.2	99.6	99.5	99.3	99.5	99.2
Feed volume/concentrate volume	2	2	2	2	2	2
Average PR (g/h)[a]	25.6	5.6	4.8	7.7	13.1	10.3
Processing capacity of film [gal/(day · ft²)]	31.4	6.9	5.9	9.5	15.8	13.8

[a] Film area = 9.6 cm².

For a batch concentration process, a quantity called the *processing capacity of the membrane for solute concentration* is a useful design parameter. This quantity is defined [2] as the volume of charge (feed solution) that 1 ft² of film surface can handle per day, $(V)_i/S_t$, in a batch concentration process under the specified experimental conditions. This quantity can be calculated from the following relation derived in the literature:

$$\frac{(V)_i}{St} = q_{av} \frac{\alpha_v}{\alpha_v(\rho_1)_i - (\rho_1)_f} \qquad (12.74)$$

where q_{av} is the average water flux (in weight units per unit area per unit time) during the concentration process, α_v is the volume ratio of the feed with respect to concentrate, $(V)_i/(V)_f$ and $(\rho_1)_i$ and $(\rho_1)_f$ are the densities of the feed and concentrate, respectively. Table 12.16 gives the processing capacities (in gallons per day per square foot, gal/(day.ft²)) of the membranes used in this work for the concentration of different fruit juices studied to give an α_v value of 2. These data show that for the type of apparatus and membranes used in this work, the processing capacity was highest [31.4 gal/(ft² · day)] for apple juice concentration and lowest [5.9 gal/(ft² · day)] for orange juice concentration at the operating pressure of 6895 kPag

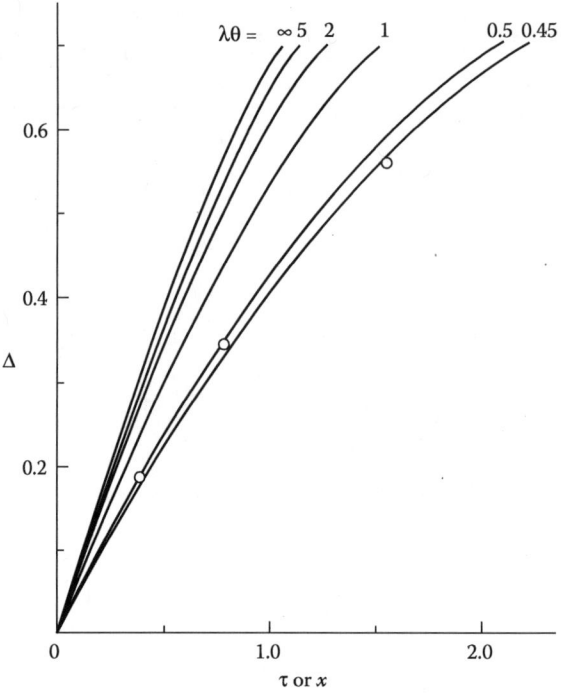

Figure 12.12 Some results of system analysis for apple juice concentration by RO. System specifications, $\gamma = 0.188$, $\theta = 0.001$, and $\lambda\theta = 0.45, 0.5, 1, 2, 5$, and ∞. (From T Matsuura, AG Baxter, S Sourirajan. *Acta Aliment* 2:109–150, 1973.)

(= 1000 psig). In view of the high economic value of the concentrated fruit juices, the above processing capacities are probably sufficiently high to be of practical interest in all cases.

12.3.3 Application of Transport Equations for the Concentration of Green Tea Juice

RO concentration of green tea juice was attempted at the feed tea juice concentration of 1000 ppm by cellulose acetate membranes of different pore sizes [33–35]. Figure 12.13 summarizes the data for the effect of pore size on the separation of various juice components and the PR. The decrease in the membrane pore size is represented by an increase in the sodium chloride separation and a decrease in the PR. The separation of typical juice components, such as amino acids, polyphenols, and caffeine, increases with an increase in sodium chloride separation. The separation based on the total organic carbon (TOC) content also increases with an increase in sodium chloride separation. The order in the separation of various tea juice components is as follows:

Amino acids > TOC ≥ polyphenols > caffeine

One of the objectives of the study was to find a membrane that enables the removal of caffeine into permeate while retaining polyphenols and other tea juice components in the

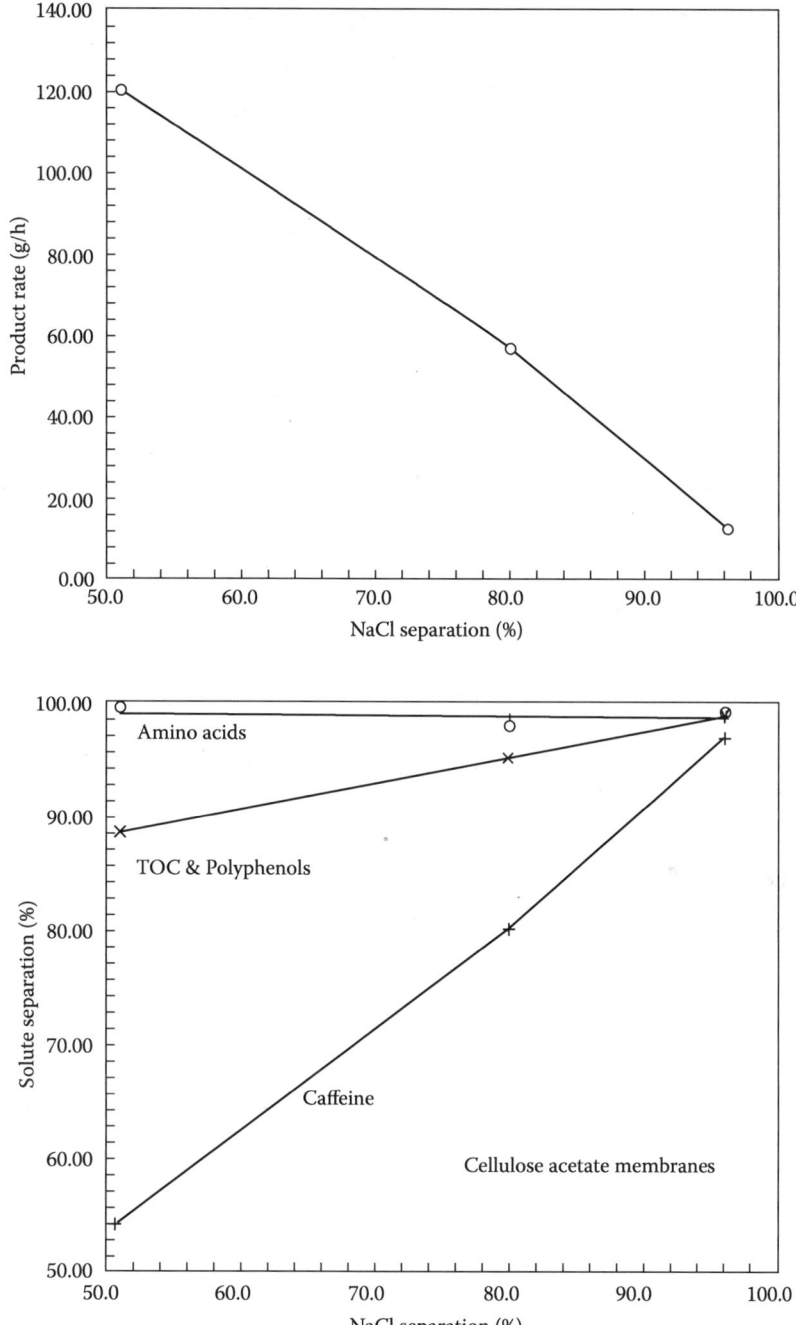

Figure 12.13 Separation of various tea juice components and PR versus sodium chloride separation. (From SQ Zhang, T Matsuura, K Chan. J Food Process Eng 14:85–105, 1991.)

retentate. The unshrunk cellulose acetate membrane, which showed 52% sodium chloride separation, seemed to be the best for this purpose among all cellulose acetate membranes studied, and therefore extensive investigations were performed on this particular membrane.

RO concentration of tea juice was performed with the static cell at feed concentrations of 1.09, 2.83, and 4.71 × 10 kg/m³ (about 1, 3, and 5 wt%); results are shown in Figure 12.14. The solute separations were consistently higher than those given in Figure 12.13. For example, both TOC and polyphenol separations were more than 98%. The separation was more than 90% in most cases for caffeine in contrast to the 52% value obtained when the feed concentration was 1000 ppm (Figure 12.13). Caffeine separation increased with an increase in the operating pressure. This membrane was therefore not effective for the fractionation of polyphenols and caffeine when the feed concentration was higher than 1%. Figure 12.14 also shows that the PR leveled off with an increase in the operating pressure and the asymptotic value decreased with an increase in the feed concentration.

Osmotic pressure data obtained experimentally are plotted versus carbon weight fraction in Figure 12.15. Surprisingly, osmotic pressure data of tea juice are exactly on the line that represents various fruit juices.

The data given in Figure 12.14 were further analyzed by transport equations, Equations 12.69 through 12.72, together with the osmotic pressure data given in Figure 12.15. In using transport equations, the weight fraction in terms of the total weight of the green tea juice components was employed during the calculation instead of the carbon weight fraction. Therefore, all the transport parameters reported hereafter are concerned with the total weight of the tea juice components unless otherwise specified. The correlation between log $(D_{AM}/K\delta)$ $(D_{AM}/K\delta$ in m/s$)$ and the tea juice concentration $c_{A1(wt)}$ (kg/m³) is given by the equations

$$\log\,(D_{AM}/K\delta) = \begin{cases} -0.016 c_{A1(wt)} - 7.540 & \text{at 1724 kPag} \\ -0.016 c_{A1(wt)} - 7.810 & \text{at 3448 kPag} \\ -0.016 c_{A1(wt)} - 8.350 & \text{at 6895 kPag} \end{cases}$$

A similar relationship was found between k (m/s) and the tea juice concentration $c_{A1(wt)}$, although the effect of the operating pressure was less. The relationship can be given by

$$\log\,k = \begin{cases} -0.026 c_{A1(wt)} - 5.020 & \text{at 1724 kPag} \\ -0.026 c_{A1(wt)} - 5.106 & \text{at 3448 kPag} \\ -0.026 c_{A1(wt)} - 5.179 & \text{at 6895 kPag} \end{cases}$$

As Figure 12.14 shows the separation of caffeine is lower than those of other tea juice components. The degree of caffeine permeation through the membrane seems to be one of the important factors in the RO design. Therefore, an attempt was made to obtain the solute transport parameter for caffeine, designated as $(D_{AM}/K\delta)_{cafferine}$. For this purpose, the mass transfer coefficient for caffeine, designated as $k_{cafferine}$, was calculated by an equation similar to Equation 12.62:

$$k_{caffeine} = k\left(\frac{0.656}{0.417}\right)^{2/3} \tag{12.62a}$$

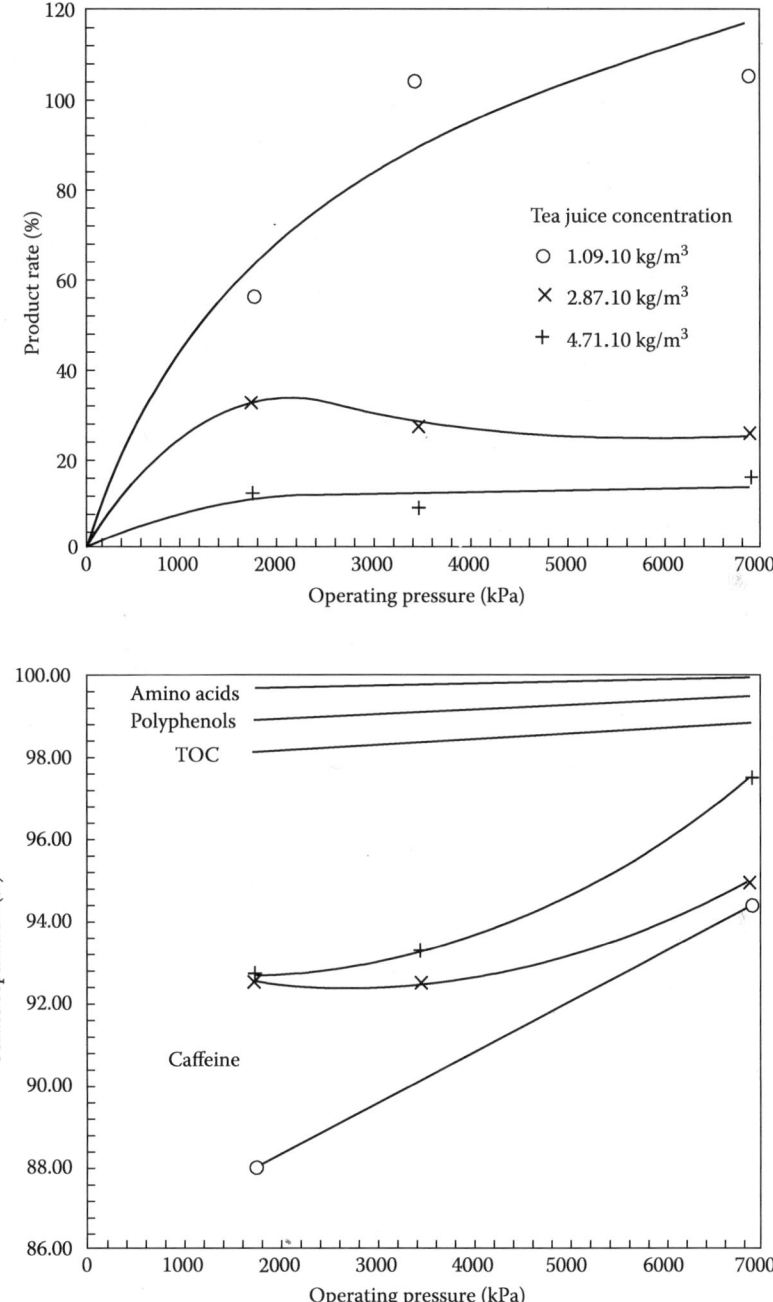

Figure 12.14 Separation of various tea juice components and PR for selected concentrations and operating pressures. (From SQ Zhang, T Matsuura, K Chan. *J Food Process Eng* 14:85–105, 1991.)

ENGINEERING PROPERTIES OF FOODS

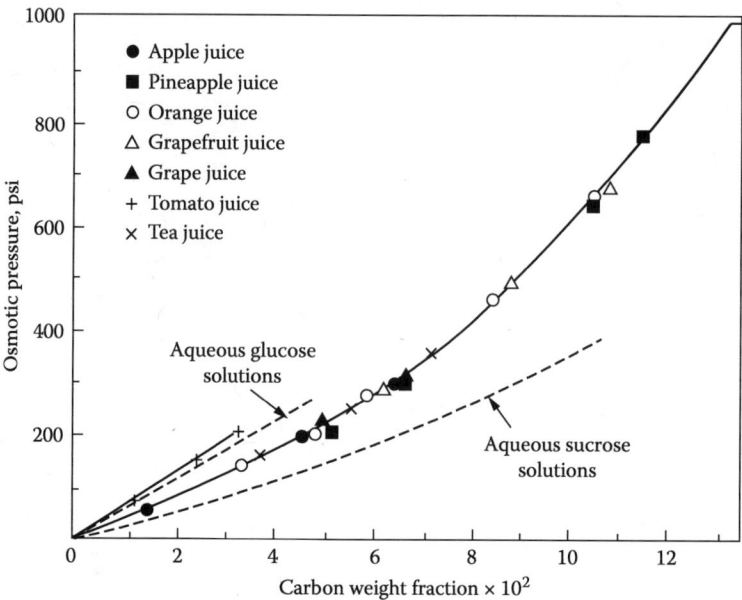

Figure 12.15 Comparison of tea juice osmotic pressure with those of various fruit juices. (From SQ Zhang, AE Fouda, T Matsuura, K Chan. *J Food Process Eng* 16:1–20, 1992.)

where k is the mass transfer coefficient for total juice components given by the foregoing equations. The fraction 0.656/0.416 represents the ratio of the diffusivities of caffeine and polyphenols, which constitute the major part of tea components. Using $k_{caffeine}$ so calculated, $(D_{AM}/K\delta)_{caffeine}$ can be obtained by applying Equation 12.5. The results are summarized by the equations

$$\log(D_{AM}/K\delta)_{caffeine} = \begin{cases} -0.026 c_{A1(wt)} - 6.288 & \text{at 1724 kPag} \\ -0.026 c_{A1(wt)} - 6.559 & \text{at 3448 kPag} \\ -0.026 c_{A1(wt)} - 7.062 & \text{at 6895 kPag} \end{cases}$$

Note that $D_{AM}/K\delta$ values are consistently higher for caffeine than for tea juice, indicating the higher permeation rate of caffeine solute. The validity of the parameters obtained above was examined by back calculating the PR data using the above parameters in the transport equations and comparing the results with the experimental PR values. Table 12.17 shows such a comparison. The agreement between the calculated and experimental values is reasonable, testifying to the validity of the numerical values for the parameter.

12.3.4 Some Illustrative Examples of the Surface Force–Pore Flow Model

12.3.4.1 Parametric Studies on Solute Separation and PR

The approach to RO/UF membrane transport represented by the surface force–pore flow model enables the calculation of solute separation and product permeation rate under a given set of experimental conditions when interfacial interaction force parameters such as, A, B, and

Table 12.17 Comparison of the Backcalculated and Experimental PR Data

$C_{A1(wt)}$ (kg/m³)	Pressure (kPag)	PR_{calcd} (g/h)[a]	PR_{exptl} (g/h)[a]
10.94	1724	25.69	20.13
	3448	33.16	37.44
	6895	34.09	37.78
28.73	1724	9.04	11.90
	3448	9.94	8.98
	6895	9.90	8.98
47.11	1724	2.59	4.22
	3448	2.92	3.06
	6895	2.99	5.23

[a] Effective film area, 9.6 cm².

D are known. The latter parameters are collected and listed in Tables 12.2 through 12.4 for several inorganic solutes, nonionized polar organic solutes, and some dye, macromolecular, and protein solutes in aqueous solutions at the polymer–solution interface that seem relevant to food processing [36]. The polymer membrane materials involved are cellulose acetate (CA398) and aromatic PAH; a limited amount of data relating to cellulose membrane material are also included in the table. Because it is difficult to discuss the RO/UF separation of each individual solute, an illustrative discussion is developed below, using some parametric studies.

From the foregoing discussion, it should be clear that when data on A or B, and D together with $\P_b$ (with or without associated σ value) are given, data on solute separation f and the PR/PWP ratio under specified RO/UF operating conditions can be predicted by the transport equations already described. Using a set of eight arbitrarily chosen combinations of B and D values (given in Table 12.18), the results of a few parametric studies are presented below [21]. In these studies, it is assumed that the feed solutions are dilute (i.e., the osmotic pressures of the feed and product solutions are negligible), and the mass transfer coefficient $k = \infty$ (i.e., $c_{A2} = c_{A1}$).

The arbitrarily chosen four values of D ($\times 10^{10}$ = 2, 4, 20, and 30 m) are close to Stokes' radii for ethyl alcohol (2.05 × 10⁻¹⁰ m), D-glucose and D-fructose (3.66 × 10⁻¹⁰ m), bacitracin

Table 12.18 Arbitrary Combinations of $\underset{\sim}{B}$ and $\underset{\sim}{D}$ Values Chosen for Parametric Studies

Case No.	$\underset{\sim}{B} \times 10^{30}$ (m³)	$\underset{\sim}{D} \times 10^{30}$ (m³)
I	100	2
II	100	4
III	105	20
IV	2 × 10⁵	30
V	– 500	2
VI	– 500	4
VII	– 500	20
VIII	– 500	30

(22×10^{-10} m), and pepsin (28×10^{-10} m), respectively; thus, the values have some relevance to the solutes involved in food and biological fluids. The first two solutes are separated by RO membranes, whereas the third and the fourth solutes are separated by UF membranes. For the first four cases in Table 12.18, the B values are positive, and for the last four cases, B values are negative; thus, the parameter analysis involves both attractive and repulsive interfacial forces.

The results of calculation of solute separation and the PR/PWP ratio as a function of pore radii on the membrane surface up to ~ 50×10^{-10} m are illustrated in Figure 12.16. When the size of the solute molecule is small (i.e., D is small) and the membrane material–solute attraction at the membrane solution interface is strong (case I), RO separations are positive in a relatively narrow range of small pore radii, and they turn negative, pass through a minimum, and gradually approach zero, with an increase in pore radius. This trend is generally applicable to phenolic solutes or benzoic acid solute (when the latter acid is nonionized) with respect to CA-398 membranes [37,38]. When the size of the solute molecule is increased (case II), both the magnitude and the range of pore size for positive solute

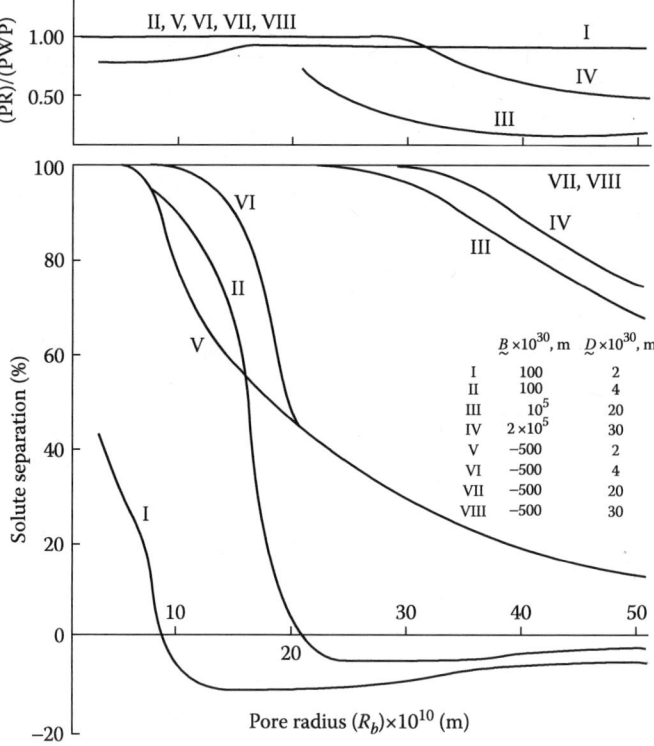

Figure 12.16 Effect of average pore size on solute separation and PR/PWP ratio under selected interaction forces. Operating pressure, 1724 kPag (=250 psig); solute concentration in feed, dilute solutions; mass transfer coefficient, k. (From T Matsuura, TA Tweddle, S Sourirajan. *Ind Eng Chem Process Des Dev* 23:674–684, 1984.)

separations also increase; however, with further increases in pore size on the membrane surface, solute separation ultimately becomes negative. This corresponds to the RO separation of bulky ether, ester, and ketone solutes that are often found as flavor components in fruit juices. However, when the solute is strongly repelled at the membrane–solution interface (cases V and VI), the magnitude of solute separation increases significantly, and positive separations prevail in the entire range of pore sizes studied. The general trend of cases V and VI is observed for sugars and polyhydric alcohol solutes with CA-398 membranes [39].

As for the PR/PWP ratio, that of case I is significantly lower than unity, whereas those of cases II, V, and VI are nearly equal to unity. The former result is similar to that observed for phenolic solute–water–CA-398 membrane systems [37,40]. The latter results reflect the higher D and/or the negative B values.

Cases III, IV, VII, and VIII represent solutes whose sizes correspond to those of macromolecules. With respect to such solutes, when B values are positive (cases III and IV), solute separations as well as PR/PWP ratios decrease sharply with an increase in pore radius beyond the corresponding D values; the latter decrease is commonly observed in the UF treatment of protein solutes. However, when B values are negative (cases VII and VIII), solute separations continue to be close to 100%, even up to a pore radius of 50×10^{-10} m, and PR/PWP ratios remain close to unity throughout, reflecting the effects of repulsive interfacial force.

The effect of operating pressure is illustrated in Figure 12.17 with respect to cases I, II, V, and VI and two values for pore radius (R_b) on the membrane surface, namely 3.87×10^{-10} m ($R_a = 3 \times 10^{-10}$ m) and 8.87×10^{-10} m ($R_a = 8 \times 10^{-10}$ m). With regard to cases II, V, and VI, solute separations are high, and they tend to increase with an increase in operating pressure; PR/PWP values are close to unity for both pore radii studied. These results are similar to those experimentally observed with respect to RO separations of sugars [41] or t-butyl alcohol [27] with CA-398 membranes. With respect to case I, the results reflect the attractive interfacial force; for the membrane with a smaller pore radius, solute separation is positive and tends to increase slightly with an increase in operating pressure; for the membrane with a larger pore radius, solute separation is less, and it decreases from a positive to a negative value with an increase in operating pressure. PR/PWP ratios are less than unity for both pore sizes throughout the operating pressure range studied; these results again are usually observable with phenolic or undissociated benzoic acid solute [37,38,40].

The effect of operating pressure on the separation of macromolecular solutes (cases III, IV, VII, and VIII) and on their PR/PWP ratio was studied using a membrane with a pore radius of 40.87×10^{-10} m ($R_a = 40 \times 10^{-10}$ m), and the results obtained are shown in Figure 12.18. When the macromolecular solutes are subject to strong attractive forces at the membrane–solution interfaces (cases III and IV), solute separations are significantly lower than 100%, and they tend to decrease with an increase in operating pressure; the corresponding PR/PWP ratios are also considerably less than unity, and they are practically unaffected by change in pressure. However, when the macromolecular solutes are subject to repulsive forces at the membrane–solution interfaces, solute separations are practically 100%, and the corresponding PR/PWP ratios are close to unity, throughout the range of operating pressures studied.

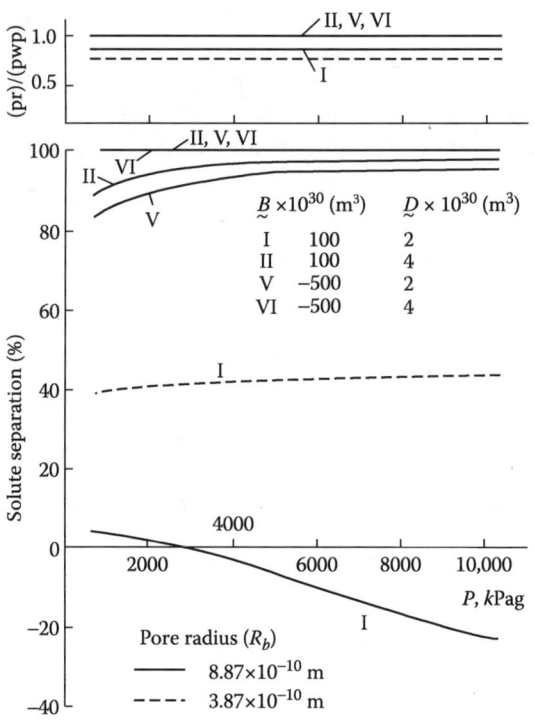

Figure 12.17 Effect of operating pressure on solute separation and PR/PWP ratio for small molecules. Feed concentration, dilute solutions; mass transfer coefficient, $k = \infty$. (From T Matsuura, TA Tweddle, S Sourirajan. *Ind Eng Chem Process Des Dev* 23:674–684, 1984.)

12.3.4.2 Another Parametric Study on Solute Concentration Profile and Solution Velocity Profile

Data on solute separation and the PR/PWP ratio illustrated in the previous section are the results of the superimposition of the solute concentration profile and the velocity profile, both in the radial direction within the membrane pore (Equations 12.25 and 12.26). Therefore, the disposition of the two profiles was studied, and the results are shown in Figures 12.19 and 12.20. It may be noted that the D value considered in this study (=36.7 Å) corresponds to the Stokes' radius of α-casein.

Figure 12.19 gives the correlation of $C_A(\rho) = c_{A3}(r)/c_{A2}$ as a function of $\underline{d}$ (=distance to the center of the solute molecule from the pore wall). Because the calculated $C_A(\rho)$ did not change in the operating pressure range 138–345 kPag (20–50 psig) and in the feed concentration range 50–18,600 ppm, the concentration profile given in Figure 12.19 represents all the calculated values in the above range of operating conditions. Apparently, the shape of the concentration profiles reflects that of the potential function, with positive B and D values shown in Figure 12.3.

Figure 12.20 gives the correlation of $\alpha(\rho)$ as a function of the distance from the center of the pore. Two pore radii, 27.1×10^{-10} m and 83.6×10^{-10} m, were chosen for study. Note

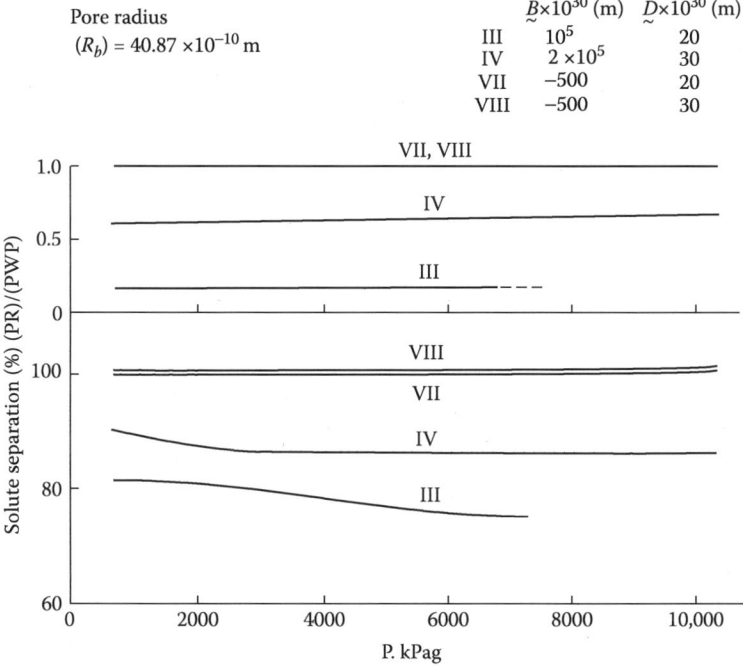

Figure 12.18 Effect of operating pressure on solute separation and PR/PWP ratio for macromolecules. Feed concentration, dilute solutions; mass transfer coefficient, $k = \infty$. (From T Matsuura, TA Tweddle, S Sourirajan. *Ind Eng Chem Process Des Dev* 23:674–684, 1984.)

that the former pore is too small, whereas the latter pore is sufficiently large to accommodate the solute under consideration. The results show that the solution velocity profile is not disturbed at all in the smaller pore, and it is a typical parabolic profile; this is natural because the solute does not exist in the pore, and hence the solution flow strictly obeys the Poiseuille law. In the case of the large pore, however, the flow pattern is disturbed by the presence of the solute in the pore. In the latter case, when the B value is 1×10^{-25} m³, the velocity profile is almost parabolic because the solute–membrane material interaction is so weak that there is practically no concentration buildup of solute inside the pore. As B increases, the solution velocity is gradually suppressed, and when B reaches 4×10^{-21} m³, an inflection point appears exactly at the position where the strongest adsorption takes place. When the B value increases further, the solution velocity is further depressed, and the inflection becomes deeper. At B values greater than 4.25×10^{-25} m³, the solution flow is practically divided into two regions on both sides of the solute adsorption. (It may be noted that the B value for α-casein with respect to CA-398 polymer material is 4.56×10^{-25} m³.) The distortion of the solution velocity profile and the suppression of solution velocity are expressions of the formation of less mobile macromolecular colloidal aggregates in the pore region. The tendency for the formation of such aggregates is enhanced with an increase in solute–polymer membrane material interaction, which may also result in a

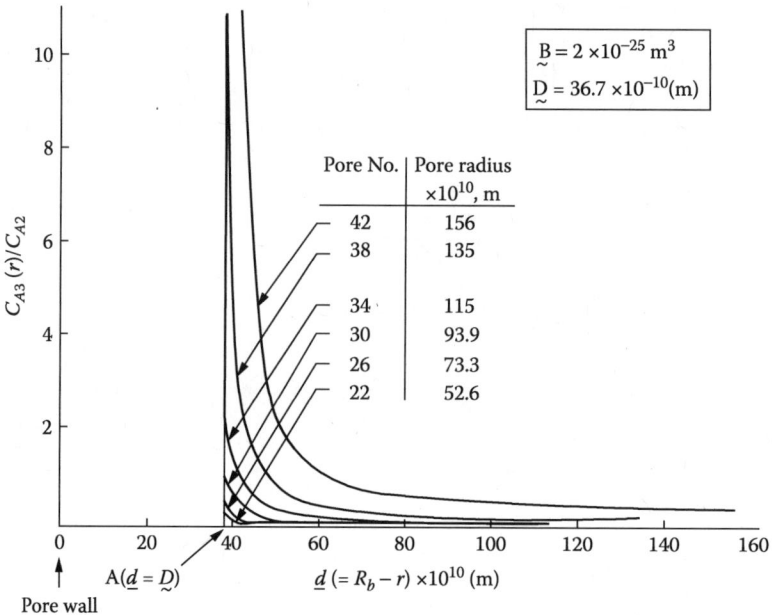

Figure 12.19 Dimensionless concentration at the pore outlet as a function of the distance between the pore wall and the center of the protein molecule for selected pore radii. Solute concentration, 50–18,600 ppm; operating pressure, 137–345 kPag (=20–50 psig); $B = 2 \times 10^{-25}$ m^3, $D = 36.7 \times 10^{-10}$ m; at point A, $\underline{d} = 36.7 \times 10^{-10}$ m.

significant decrease in the membrane permeated PR during UF. The foregoing observations are particularly relevant to the UF treatment of protein solutes in aqueous solutions.

12.3.5 Some Data on the UF of Proteins

12.3.5.1 UF of BSA and α-Casein

Figure 12.21 gives the experimental solute separation and PR data obtained with three different membranes (made from three different materials) in the UF treatment of bovine serum albumin (BSA)–water solutions involving different solute concentrations in the feed solutions and different operating pressures. The above experimental data can also be predicted with reasonable accuracy, as illustrated in the same figure, from the basic transport equations arising from the surface force-pore flow model and B and D values given for the systems under consideration, provided the necessary constraints are incorporated in them. The details of the prediction procedure are given in the literature [42]. Briefly, those constraints are

1. The use of realistic osmotic pressure and viscosity data in solving the velocity profile by Equation 12.38.
2. The representation of possible formation of colloidal aggregates by an effective change in the pore size distribution. Particularly, when the pores of the second

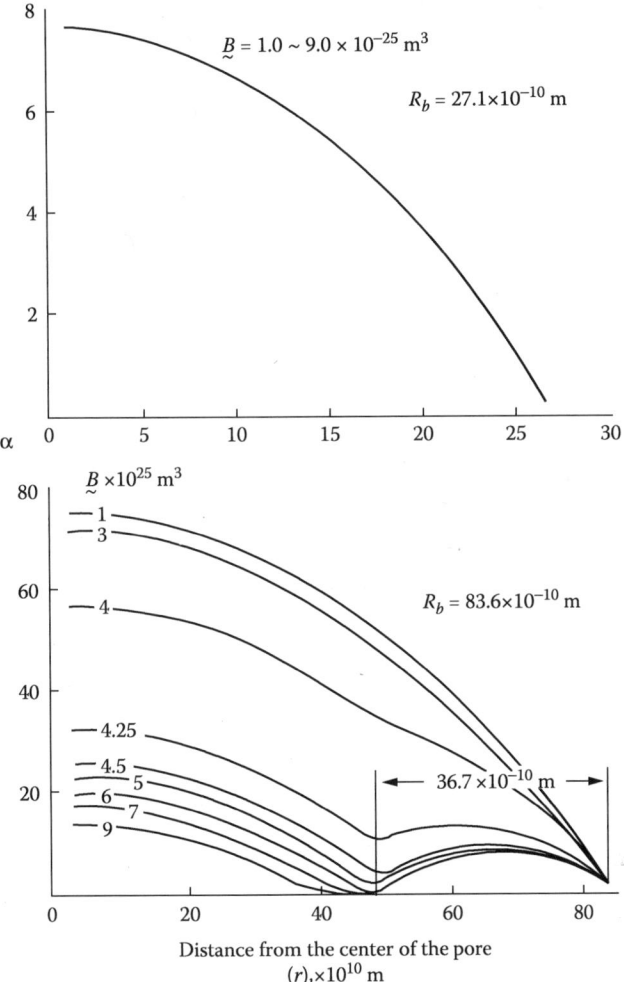

Figure 12.20 Dimensionless velocity versus r for various pore radii at a feed concentration of 50 ppm and operating pressure of 137 kPag (=20 psig). $D = 36.7 \times 10^{-10}$ m.

distribution can accommodate solute molecules whereas those of the first distribution cannot, the reduction of the average pore radius of the second distribution by r_A (Stokes' radius of bovine serum albumin) is most effective to predict the UF performance data.

The pore size distributions for the membranes used in the above work are given in Table 12.19. The B values for BSA with respect to the PS-V, CA-400, and PPPH 8273 (same as PAH) membrane materials are 60,300, 527,500, and 584,000, respectively. On the basis of the data and the constraints indicated above, the data on membrane performance given in Figure 12.21 were calculated. The results show good agreement between

ENGINEERING PROPERTIES OF FOODS

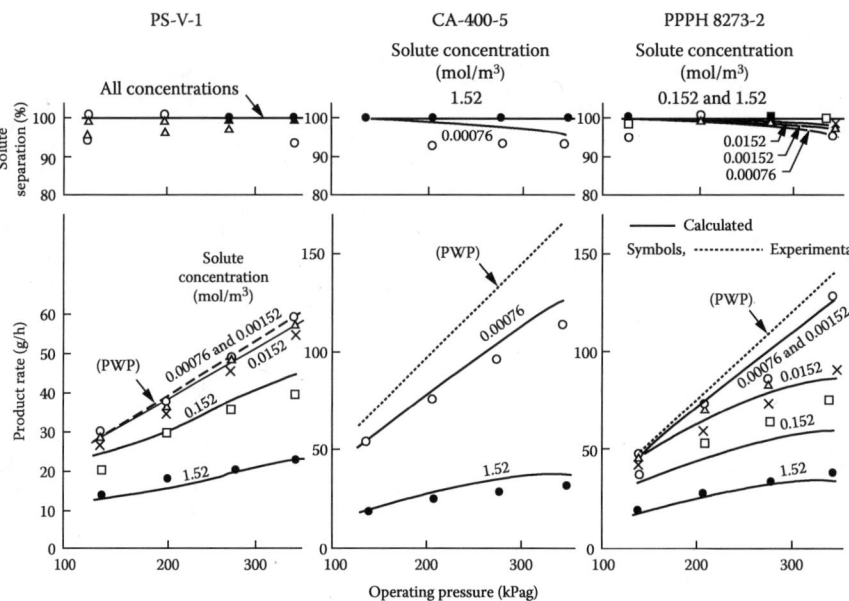

Figure 12.21 Experimental and calculated data on solute separation and PR for the separation of bovine serum albumin from aqueous solution. Effective membrane area, 14.1 cm²; feed flow rate, 2200 cm³/min. (From T Liu, K Chan, T Matsuura, S Sourirajan. *Ind Eng Chem Process Res Dev* 23:116–124, 1984.)

Table 12.19 Pore Size Distributions of Membranes

Membrane	$\bar{R}_{b,1} \times 10^{10}$ (m)	$\sigma_1/R_{b/1}$	$\bar{R}_{b,2} \times 10^{10}$ (m)	$\sigma_2/R_{b/2}$	H_2
PS-V-1[a]	26.0	0.002			
CA-400-5[a]	28.2	0.200	102.2	0.250	0.040
PPPH 8273-2[a]	25.7	0.460	105.0	0.250	0.010
CA-400-A[b]	29.8	0.190	106.0	0.180	0.025
CA-400-C[b]	45.0	0.350	110.0	0.420	0.070
CA-400-D[b]	30.0	0.470	137.0	0.140	0.025
CA-400-E[b]	30.0	0.440	135.0	0.150	0.022
CA-400-F[b]	29.8	0.400	135.0	0.100	0.013
CA-400-G[b]	29.9	0.400	135.0	0.100	0.015
CA-400-H[b]	30.0	0.440	136.2	0.140	0.022

[a] Specified by reference PG solutes.
[b] Specified by reference protein solutes.

the experimental and calculated values. The following features of BSA protein UF are revealed in Figure 12.21.

1. The PWP rate increases with the increase in the operating pressure almost linearly.
2. As the concentration of the feed BSA solution increases, the slope of the PR versus operating pressure decreases.
3. At the highest BSA concentration, the PR approaches an asymptotic value with the increase in the operating pressure. The latter value depends both on the membrane material and on the average pore size and the pore size distribution of the membrane.
4. The PR decrease from the pure water permeation rate is significantly less for PS-V membranes than for CA-400 and PPPH-8273 membranes.
5. The solute separation is generally above 90% in all membranes at the lowest BSA concentrations studied and reaches nearly 100% with an increase in the feed BSA concentration.

All of the above features, except the first, are predictable by transport equations. In particular, the higher PR/PWP ratio obtained for the PS-V membrane compared to the rest of the membranes at a given set of operating conditions is attributable to the single normal distribution structure of the PS-V membrane used.

Figure 12.22 shows a similar result with respect to the UF of an α-casein–water system.

12.3.5.2 Effects of Fouling on Membrane Performance and Pore Size and Pore Size Distribution

In this discussion, the term *fouling* refers to the permanent reduction in fluid flux through the membrane as a consequence of a stable reduction in pore size brought about by strong adsorption of BSA on the pore wall. Thus, the average pore radius of a membrane after fouling is smaller than that of the same membrane before fouling. The fouling of the membrane is brought about by UF treatment of a 50-ppm BSA–water solution continuously for 12 h at an operating pressure of 345 kPag (50 psig); the latter treatment is called here "BSA treatment." Membrane performance in the UF treatment of several 50-ppm protein–water solutions and average pore size and pore size distributions before and after BSA treatment were studied with a set of cellulose acetate UF membranes (CA-400-A to CA-400-H, listed in Table 12.19) using five reference protein solutes: bacitracin, pepsin, α-casein, BSA, and γ-globulin. The results obtained are given in Figures 12.23 and 12.24. It may be noted that all the membranes were characterized by two normal pore size distributions (Table 12.19), the average pore size in the second distribution being very much larger than that in the first distribution. The different pore sizes and pore size distributions in the original membranes (Table 12.19) were produced by using different concentrations of ethanol in the EtOH–water gelation in the process of making those membranes [43].

Figure 12.23 shows the performance of these membranes in the UF treatment of the reference protein solutions. Two observations are significant. Referring to the PWP data, the results show the occurrence of fouling by BSA treatment; referring to data on solute separations, there is reasonable agreement between the calculated and experimental values both before and after BSA treatment. Further, whereas the UF separations for bacitracin are lower, those for the other protein solutes studied are higher after BSA treatment. These

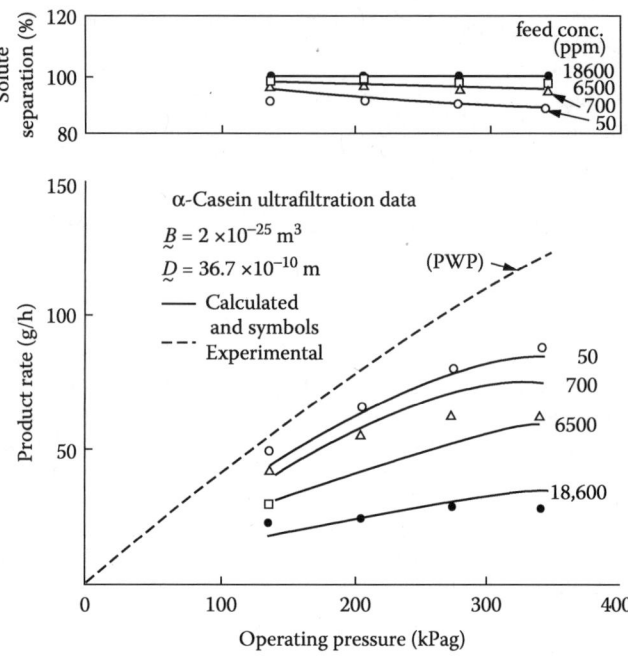

Figure 12.22 Experimental and calculated data on solute separation and PR for the separation of α-casein from aqueous solution. Effective membrane area, 14.1 cm²; feed flow rate, 2200 cm³/min. membrane compared to the rest of the membranes at a given set of operating conditions is attributable to the single normal distribution structure of the PS-V membrane used.

results show that membrane fouling can increase or decrease solute separations, depending on experimental conditions, and such changes in solute separations are predictable from the basic transport equations arising from the surface force–pore flow model.

Figure 12.24 shows the changes in pore size and pore size distribution brought about by membrane fouling with respect to three typical membranes. With respect to the CA-400-A membrane, the changes were very small, reflecting the least change in PWP and solute separation brought about by BSA treatment (Figure 12.23). With respect to the CA-400-G membrane, the major change occurred in the second distribution, in which the average pore radius was decreased by 28×10^{-10} m by BSA treatment. At the same time, the first distribution became less broad (presumably due to the inclusion of some pores in the second distribution in the procedure of mathematical optimization); however, the average pore radius in the first distribution remained practically unchanged. As a result of these changes, the separation of proteins (except bacitracin) increased significantly, and the PWP decreased to about two-thirds the initial value (Figure 12.23). With respect to CA-400-C membranes, the changes brought about by fouling were particularly significant. The average pore size decreased in both distributions. Because the average pore size in the first distribution was 45×10^{-10} m before BSA treatment, a large number of pores in the first distribution could accommodate BSA molecules and hence be susceptible to fouling (i.e., pore

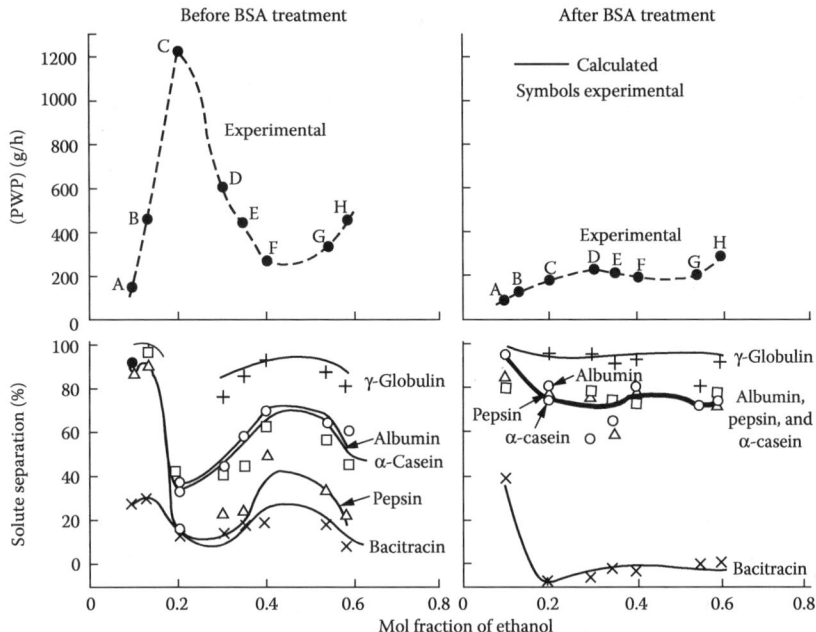

Figure 12.23 Experimental and calculated data on separations of various protein solutes and experimental data on pure water permeation rate for cellulose acetate UF membranes gelled in various mole fractions of ethanol in gelation bath. Membrane material, cellulose acetate Eastman E-400; operating pressure, 345 kPag (=50 psig); protein concentration in feed, 50 ppm; effective membrane area, 14.1 cm^2; feed flow rate, 2200 cm^3/min. (From T Liu, K Chan, T Matsuura, S Sourirajan. *Ind Eng Chem Process Res Dev* 23:116–124, 1984.)

size reduction). Because of pore size reduction in both distributions, solute separations increased for all the proteins studied except bacitracin, and PWP decreased to one-seventh of the initial value (Figure 12.23). The foregoing results also serve to illustrate the dominant part played by the second pore size distribution in membrane fouling and its effects.

12.3.5.3 Fractionation of the Protein–Sugar System and the Protein–Protein System in the Aqueous Solutions

Fractionation of sugars and volatile flavor compounds and that of sugars and food acids were discussed in relation to fruit juice concentrations. In a similar manner, the fractionations of protein–protein and protein–sugar are of interest in the membrane separation process of dairy products. Unfortunately, the latter problem has not yet been studied in a framework of transport theories; therefore, prediction of the fractionation data is at present not possible. Hence, some characteristic features of the above fractionation problems revealed experimentally are discussed below.

Table 12.20 shows the data of lactose separation by six cellulose acetate UF membranes of different porosities (experiment I) together with those of UF of a BSA–lactose mixture (experiment II). The PWP data from both experiments I and II indicate that there

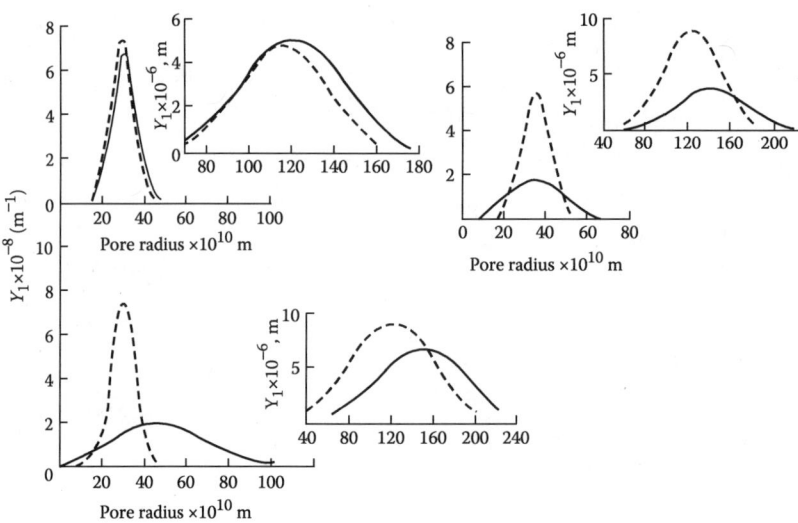

Figure 12.24 Pore size distribution of some UF membranes (———) before and (- - - -) after BSA treatment. (From T Liu, K Chan, T Matsuura, S Sourirajan. *Ind Eng Chem Process Res Dev* 23:116–124, 1984.)

was essentially no change in the porosity of the membranes between these two experiments. The significant decrease in PR data in experiment II reflects the presence of BSA in the feed. The BSA separation by the cellulose acetate membranes used for the study was almost complete, whereas lactose solutes permeated through the membrane significantly in both experiments I and II, and a slight increase in the lactose separation was observed in experiment II. The above results indicate that the membranes are useful for the fractionation of BSA and lactose solutes.

Table 12.20 UF Results of Lactose–Water and Lactose–Bovine Serum Albumin (BSA)–Water Systems[a]

		Experiment I: Lactose Only[b]		Experiment II: Lactose and BSA[c]		
					Solute Separation (%)	
Film No.	PWP (g/h)	PR (g/h)	Separation (%)	PR (g/h)	Lactose	BSA
1	63.21	50.1	2.59	15.55	2.92	99.9
2	26.90	16.5	21.47	7.97	24.2	99.9
3	70.68	56.52	2.59	17.55	5.0	99.5
4	21.27	12.90	24.06	7.22	29.2	~100
5	66.89	53.22	2.59	18.69	5.0	99.9
6	56.56	45.78	2.19	17.70	4.6	99.9

[a] Effective film area = 14.1 cm²; operating pressure = 138 kPag (20 psig).
[b] Feed lactose concentration, 5%.
[c] Feed lactose and BSA concentrations, 5% for each.

Table 12.21 shows the results of UF of porcine albumin (case I), γ-globulin (case II), and their mixture (case III). In every experiment, the concentration of albumin in the feed was about 2% and that of γ-globulin was about 1%. Additionally, less than 1% NaCl was added to each feed solution. The operating pressure of 100 psig was applied in each case. It is obvious from the data of the case I experiment that NaCl separation was zero and albumin separation was also very low, indicating that both solutes permeate through these membranes almost completely. The ratio of the PWP data before and after the albumin run indicates that there was no pore blocking during the run. The case II experiment, on the other hand, exhibits the complete retention of γ-globulin with severe pore blocking. The data from case I and case II studies therefore suggest that the fractionation of albumin and γ-globulin can be achieved by using these cellulose acetate membranes. However, the results from case III experiments show that albumin separation was increased to the 36–46% range with even more severe blocking of membrane pores. Apparently, the immobile colloidal aggregate of γ-globulin formed at the membrane pore narrowed the pore size, and thus the albumin separation was significantly increased in the presence of γ-globulin. The experimental result suggests that we are not able to judge whether a membrane is adequate for the fractionation of two protein solutes by the results of UF experiments performed with individual proteins.

12.3.6 Application of PV in the Recovery and Concentration of Food Flavors

Pervaporation (PV) is a membrane separation process in which the upstream side of the membrane is in contact with feed liquid while vacuum is applied on the downstream side of the membrane [44]. The permeant vaporizes somewhere between the upstream and downstream sides of the membrane; therefore, the permeate is obtained as vapor. The process consists of the following three steps according to the solution–diffusion model:

1. Sorption of the permeant from the feed liquid to the upstream side of the membrane
2. Diffusion of the permeant across the membrane
3. Desorption of the permeant from the downstream side of the membrane into the vapor phase

The pore flow mechanism was applied recently to the PV process, according to which the process consists of the following three steps:

1. Liquid transport from the pore inlet to the liquid/vapor phase boundary
2. Evaporation of liquid to vapor at the liquid/vapor phase boundary
3. Vapor transport from the liquid/vapor phase boundary to the pore outlet

Two types of membranes are being used commercially for PV, hydrophilic and hydrophobic. Hydrophilic membranes such as those made of polyvinyl alcohol and polyelectrolytes are water selective and are used for the dehydration of alcohols. Hydrophobic membranes such as those made of polydimethylsiloxane are selective to organic solutes from the aqueous solutions and are applied for the removal of organic pollutants from water. The potential of PV applications in food processing is found in the recovery and concentration of flavor compounds from liquid food by using the latter type of membrane.

Table 12.21 Experimental Results on Separation of Albumin, γ-Globulin, and Sodium Chloride

Case	Film No.	Fraction Product Recovery	Average PR × 10³ (kg/h)	PWP$_{before}$ × 10³ (kg/h)[b]	PWP$_{after}$/PWP$_{before}$	Feed Concentration (wt%)			Average Solute Separation (%)		
						Albumin	γ-Globulin	NaCl	Albumin	γ-Globulin	NaCl
I	4-a[a]	0.759	1066	1227.4	0.989	1.85	0.0	0.788	~0	—	~0
	5-a	0.759	565	751.8	0.898	1.85	0.0	0.788	3.0	—	~0
	6-a	0.759	480	629.0	0.948	1.85	0.0	0.788	3.9	—	~0
II	4-b	0.287	12.0	1214.1	0.025	0.0	0.936	0.796	—	~100	—
	5-b	0.431	18.0	675.1	0.104	0.0	0.936	0.796	—	~100	—
	6-b	0.201	8.4	596.6	0.037	0.0	0.936	0.796	—	~100	—
III	4-c	0.785	5.7	1360	0.011	2.37	0.790	0.672	42.0	~100	~0
	5-c	0.710	5.2	1080	0.013	2.37	0.790	0.672	46.0	~100	~0
	6-c	0.565	4.1	800	0.008	2.37	0.790	0.672	35.5	~100	~0

[a] Different film numbers indicate different membrane porosities. The letter following the film number indicates the membrane batch.
[b] Effective membrane area, 9.6 cm².

Table 12.22 GC-MS Analytical Results

Feed	Flavor Compound	Sample 1	Sample 2	Permeate[a]	Boiling Point (°C)
1	Acetic acid, ethyl ester	0.2		4.5	77.1
2	2-Methyl-1-propanol			6.8	108.0
3	1-Butanol	14.6	15.0	228.5	117.2
4	Propionic acid, ethyl ester	0.25	0.25	11.0	99.1
5	1,1-Diethoxyethane			53.3	
6	4-Methyl-1-hexene	1.4	0.7	34.0	87.5
7	Propionic acid	2.3	5.1	2.1	141
8	2-Methyl propionic acid, ethyl ester	0.2		15.3	109–111
9	Hexanal	2.3	1.5	33.1	128
10	2-Furfural	2.5	1.4	0.1	161.7
11	2-Methylpropanoic acid	5.0	7.2	11.2	153.2
12	2-Hexanal	9.8	9.2	224.7	146–147
13	3-Methylbutanoic acid, ethyl ester	0.5	0.5	4.0	
14	1-Hexanol	7.4	6.3	91.2	158
15	2-Ethyl-l,3-dioxolane			5.9	
16	2-Propyl-l,3-dioxolane			0.6	
17	3-Methyl-3-heptanol	0.6	0.5	5.3	163
18	1-Phenylethanone	100.0	100.0	100.0	
19	1,1-Diethoxyhexane			4.7	
20	l-(l-Ethoxyethoxy) hexane			1.2	
21	5-Hydroxymethyl-furfural	28.4	33.2		114–116

[a] Peak ratio (%) of flavor component to the internal standard (1-phenylethanone).

An attempt was made to concentrate the flavor components of apple essence (×500). The thickness of the polydimethylsiloxane membrane used was 25.4 μm. When vacuum was applied on the downstream side, a permeation flux of 3.76×10^{-5} kg/(m$^2 \cdot$ s) was obtained at room temperature. The concentrations of the flavor components in the feed and the permeate were determined by the GC-MS method. The results are summarized in Table 12.22. The ratio of the concentration in the permeate to that in the feed is strongly correlated to the boiling point of the flavor compounds. In particular, concentration ratios above 20 were achieved for flavor components with boiling points lower than 100°C.

12.4 RECENT LITERATURE ON MEMBRANE APPLICATIONS

The main advantage of the membrane technology is the reduction of costs associated with lowering the energy requirement of usual evaporation method as well as retaining the product quality without any damage. Four membrane separation technologies are generally applied in food processing industries depending on the conditions of applications: that is, RO, nanofiltration (NF), UF, and microfiltration (MF). However, other processes, such as electro-dialysis (ED), forward osmosis (FO), membrane distillation (MD), osmotic distillation (OD), and PV are also developing. RO membranes have pore sizes less than 1 nm and are used in concentration of food products to replace evaporation method. NF is, basically, combining RO and ultrafiltration separation techniques and is used for desalting and de-acidification of food products. UF membranes have pore sizes 1 to 100 nm and are used for separation and concentration of macromolecules and colloidal particles. In food processing industries, UF is used for concentration and purification of the food. MF is used for sterilization and fractionation of macromolecules in the food products. Some review articles regarding the applications of membrane technologies in food processing are also available in the literature [45–68] but current survey is still needed for the newest technological developments in the field. In this section, food products basically are divided into four categories: dairy product industry, beverage industry, edible oil industry, and miscellaneous food products.

12.4.1 Dairy Product Industry

Membrane technology is very popular in the dairy product industry due to low temperature processing, which minimizes the losses of volatile flavor as well as guards from adverse change by protein denaturation. Specific applications are concentration of milk, cheese, and whey, desalting of whey, manufacture of soft varieties of cheese, and standardization of milk for the manufacture of a range of cheese varieties. The texture of the dairy products is associated with coagulation properties of protein and crties of milk, which include different concentrations of protein and calcium. The byproducts of milk are buttermilk and milk powder.

12.4.1.1 Reverse Osmosis
Less volatile flavor components are lost during low temperature processing, and less heat sensitive proteins are denatured. For milk concentration processes, in general, temperature of 30°C for cellulose acetate and 45°C for polyacrylamide thin film composite membrane are used [45]. Pretreatment of milk by heating may reduce bacterial counts and also improve the flux. For example, heat treatment at 60°C for 5–15 min has increased the flux about twice compared with untreated skim milk. However, some physical properties of milk were affected under high pressures, which may have acted as a homogenizer during RO processing. The fat globules in raw milk were damaged [69].

12.4.1.2 Nanofiltration
NF is used for the fractionation of human milk oligosaccharides to produce biologically active oligosaccharides mixture with very little contaminating lactose. Milk oligosaccharides are known as effective anti-infective agents, which can prevent adhesion of microbial pathogens to host cells. The best separation of human milk oligosaccharides was achieved with cellulose acetate NF membrane [70]. Cross-flow NF appears to be a potential industrial scale method for purification and concentration of oligosaccharide mixtures. Two flat sheet asymmetric cellulose acetate membranes, a nominal 50% rejection of NaCl and

molecular weight cut-off (MWCO) of 1 kDa, were used. Purification of oligosaccharide gave a satisfactory yield value (98%). Both increasing pressure and increasing temperature caused decreased sugar rejection [71].

Combined nanofiltration, with membranes of MWCO 100–400 Da, and RO were used to separate and concentrate lactic acid from cheese whey fermentation broth [72]. Nanofiltration membrane could retain 97 ± 1% of lactose with mainly lactic acid and water in the permeate. The RO membrane could successfully separate nearly 100% of lactic acid from water. A concentrated lactose product containing reduced monovalent salts could be produced by nanofiltration process [73]. The dairy UF permeate containing both phosphate and citrate can form insoluble charged species with the calcium ions, reducing the net Ca^{2+} concentration and hence the overall effect of calcium on the negatively surface charged NF membrane. This ion interaction effect becomes more dominant at higher pH values, due to the greater tendency of calcium interacting with either phosphate or citrate at higher pH. An anionic-peptide-enriched fraction obtained by nanofiltration of whey protein trypsin-hydrolyzate was shown to inhibit the growth of two Gram-positive pathogenic bacteria, *Listeria monocytogenes* and *Staphylococcus aureus*, having significant impact on food safety [74].

12.4.1.3 Ultrafiltration

For the milk concentration process, polyethersulfone (PES) membrane is commonly accepted, compared to a cellulose acetate membrane, although the former shows more fouling behavior. This is because PES membranes can be used at extremely high or low pH, in the presence of chlorine, and also at higher temperature processing conditions [46]. Concentrated yoghurt (230 g/kg total solids) was produced by UF from normal yoghurt (160 g/kg total solids). The UF cartridge consisted of a bundle of tubular PES membranes (surface area 0.8 m^2 and MWCO 25 kDa) [75]. The influences of UF process on the treatment of milk and intensity of the heat treatment on the coagulation properties have been studied using a 10,000 M_r exclusion membrane. A slight decrease of rennet clotting time with increasing milk concentration was found if the milk was heated prior to UF, whereas the curd-firming rate of the retentates was increased [76]. Defatted milk was UF-diafiltered in a 20 kDa polysulfone (PS) flat membrane for obtaining concentrated milk with different protein/lactose ratios. The lactose content was found to be responsible for the coagulum formation. A lactose contents lower than 2% led to soft coagulum, and an increase in protein concentration (higher than 8%) led to a high viscosity product similar to cheese. Enriched-protein low-lactose fermented milk could be obtained using PS membrane of 20 kDa cutoff [77].

The galactooligosaccharides were produced in the membrane reactor with a ceramic UF membrane (150 kDa) in batch and continuous systems [78]. Continuous process offered better productivity than batch process due to the possibility of maintaining a permanent state, without loss of enzyme activity. Whey protein hydrolysates were produced using a bacterial protease in a continuous stirred tank membrane reactor consisting of a PES plate and frame UF module with a molecular weight cutoff of 3 kDa [79]. Hydrolysates with an average peptide chain length around 4 amino acids with low antigenicity were obtained. Protein–polymer and protein–protein interactions play a strong role in the initial fouling of skim milk filtration by UF and MF membranes [80]. A gel layer formed on the membrane surfaces is slightly compressible and densifies as the fouling progresses. The steady-state permeate flux is attained due to the permeability of the gel layer.

12.4.1.4 Microfiltration

The development of inorganic/ceramic membranes has opened a new era in milk processing. Attempts with polymeric membranes failed due to formation of a dynamic or secondary membrane of polarized solutes that could easily transfer to behave like UF membranes. The main applications of MF for milk processing are fat separation, bacterial removal, and caseinate concentration. The removal of micellar casein and soluble proteins from skim milk was possible using cross-flow MF (0.1 μm pore diameter). The MF of skim milk was extensively studied by using inorganic membranes [81]. Cross flow filtration could be used to separate casein-rich and whey protein-rich fractions from skim milk using ceramic membranes of pore size 0.05 μm. The optimal operating conditions for this separation were 5.4 ms^{-1} and 138 kPa at 50°C [82]. Separation of milk fat in small globules, diameter lower than 2 μm, and in large globules was possible using ceramic MF membranes of 2-μm average pore diameter, which gave a new possibility to adjust texture and maybe flavor of dairy products [83].

The separation of caseinomacropeptide (CMP) from native whey proteins was carried out with a ceramic tubular microfiltration membrane with an average pore diameter of 0.1 μm and an active layer of α-aluminium oxide [84]. CMP seems to play a significant role in the enzymatic crosslinking of native casein by transglutaminase, which is located on the surface of the casein micelle. Amaranth starch–milk was produced in a pilot scale by the Al-Hakkak process and treated by MF membrane made of 1000 kDa regenerated cellulose. All the starch granules (98%) and 67% protein could be retained. However, the retention of the nonstarch polysaccharides was low (14%) [85]. The high protein retention was due to the presence of some insoluble protein in the starch-milk.

12.4.1.5 Electro-Dialysis, Forward Osmosis, Membrane Distillation, Osmotic Distillation, and PV

High purity casein of 95% purity was produced from milk by ED with bipolar membrane (EDBM) separation technology instead of the 85% purity casein obtained by the conventional chemical acidification process [86]. The bipolar membranes decrease the pH homogeneously and the demineralization action is achieved with the homopolar membranes. The α-alumina (nanofiltration) and silica (molecular sieve) membranes were used side-by-side to compare performance potential and fundamental transport mechanisms of PV processes to dewater lactic acid [87]. The silica membrane showed a higher separation potential and a more stable flux, whereas alumina membrane initially displayed high fluxes but the flux dropped because of lactic acid adsorption on the alumina surface. The silica membrane did not exhibit this trend due to its surface neutrality and pore structure.

12.4.2 Beverage Industry

The main beverage industry applications are concentration and purification of fruit juice (orange, apple, grape, banana, pears, tomato, etc.), production of beers with low alcohol content, concentration of beers, and wine processing. Fruit juice is also processed with calcium and/or vitamins, which enhance the nutritional content of the beverage. The processing technique of tomato juice is different from that of conventional fruit juice due to the high pulp content (25% fiber) and high viscosity. Because of the fiber content and particle size, tubular modules will probably produce the best results with least fouling. Enzymatic hydrolysis combined with UF can produce beverage from vegetable proteins, for example, soya milk and vegetable soybean milk, which are substitute food for animal protein. In the beer industry, recovery of maturation and fermentation tank bottoms is applied on an industrial scale.

12.4.2.1 Reverse Osmosis
In direct osmosis of tomato juice, the osmotic flux is inversely proportional to the square root of the viscosity. Therefore, for the best performance low viscosity of the juice solution is required. According to the study, six different osmotic media [NaCl, $CaCl_2$, $Ca(NO_3)_2$, glucose, sucrose, PEG of average molecular weight 400 Da (PEG400)] were used and sodium chloride (NaCl) solution was found to be the best osmotic medium due to its very low viscosity [88]. Using tubular polyamide membrane (inner diameter 0.0125 m, length 1.2 m, total effective surface area 0.9 m^2, and 99% NaCl rejection) could concentrate apple juice with high retention of sugar (~100%) and malic acid (ranged between 96% and 98%). The sugar retention was considered as the main objective in fruit juice concentration, and the maintenance of a constant sugar/acid ratio in the apple juice was important to assure a final good flavor. A polyamide composite membrane could concentrate orange juice without a significant loss of aroma, sugar, or acids [89].

Four raw juices, peach, pear, apple, and mandarin, were clarified through tubular polysulfone UF membranes, MWCO 8 kDa, and finally were concentrated by RO membranes of tubular composite polyamide [90]. The use of a commercial pectinolytic enzyme provided increased permeate flux of approximately 40%, retaining the juice properties. The highest total soluble solids content increased from 12.2°Brix of clarified peach juice up to 30.5 and 21.52°Brix by using RO with transmembrane pressures of 4 and 2 MPa, respectively.

Two NF membranes were tested with the aim of increasing the sugar content of grape must for wine production [91]. The NF membranes provided high rejections, 77–97%, of sugars whereas malic acid was retained to a low extent, 2–14%. It is noted that the grape must enriched by NF resembles with minimal compositional modification of malic and tartaric acids, which is critical to preserve the quality of the wines, and also the total polyphenols, which affect the color of wine.

Commercially available prototype polymeric three NF and two UF membranes were used to separate yellow-orange oxygenated pigment of xanthophylls from solutions, such as lutein solutions and aqueous ethanol extracts of whole ground corn [92]. UF was used to separate ethanol-soluble protein and other large solutes from the extract, and xanthophylls-containing stream was concentrated by separating xanthophylls from the solvent using NF.

Autohydrolysis liquors were processed using UF and nanofiltration ceramic membranes, MWCO 1, 5, 15, and 50 kDa, to assess their ability for refining and/or concentrating oligosaccharides [93]. The dead-end MF and/or centrifugation were more favorable to remove suspended material than cross-flow filtration through a 50 kDa membrane.

12.4.2.2 Ultrafiltration

The use of polymeric membranes is widespread for the clarification of apple juice by UF. With tubular polymeric UF membranes, the flux for apple juice improved as the membrane MWCO increased from 9 to 200 kDa. Juice filtered through a 9-kDa membrane had lower soluble solids, flavanols, and yellow–brown pigments [94]. Apple juice of improved quality was produced by UF through a 10 kDa MWCO membrane due to removal of polyphenol oxidase or some other oxidase activity materials [95]. A banana extract containing no polyphenol oxidase was produced by UF using PS membranes with a cutoff of 20 kDa. The UF experiment at 600 kPa was better compared with 800 kPa due to a lower permeate flux decrease with time and concentration, indicating the continuation of the process for a longer time [96]. Hexanal, organoleptically known to possess green character, is of considerable interest in the food and beverage industry for flavors and aromas. The extraction was possible from tomato juice using hollow-fiber UF membrane (MWCO of 100 kDa). The hollow fiber cartridge consisted of 50 PS membrane fibers 1 mm in diameter and 36.2 cm long. The total area was 0.042 m^2 [97].

Raw depectinized apple juice was clarified in a lab-scale using ceramic tubular UF membranes, MWCO 30, 50, and 300 kDa, under various process parameters such as, transmembrane pressures (100–400 kPa), temperatures (20–55°C), and feed flow rates (0.1–0.9 L/min) [98]. The clarified juice has satisfactory clarity and color intensity value from all the studied membranes. Fresh kiwi fruit juice was clarified by polyvinylidene fluoride (PVDF) UF membrane, MWCO 15 kDa, on laboratory scale at operating pressure 90 kPa [99]. Cake layer and irreversible fouling resistances gave 2.23% and 2.75% contribution to the total resistance while the more significant 29.4% contribution from the reversible fouling was noticed due to combination of suspended particles and adsorbed macromolecular impurities. The UF process permitted 11.1% loss in total suspended solids of the fresh juice. A 7.8% reduction of the total antioxidant activity and 16% degradation of the initial content of ascorbic acid in the permeate stream with respect to the fresh juice was observed.

12.4.2.3 Microfiltration

In the wine industry, the cascade cross-flow MF (0.2-μm pore diameter) was used to allow limpidity and microbiological stability [100]. Prevention of physicochemical deterioration of fruit juices was a major problem. Cooxidation and polymerization reaction occurred, resulting in aroma changes and color change and clouding. Prior to MF, active filtration, using a pad of various mixture of diatomaceous earth with polyvinylpyrrolidone (PVPP) and activated charcoal, was carried out. The PVPP and activated charcoal partially decolored the oxidized juices and stabilized these products, removing a high percentage of phenolic compounds [101]. Membrane filtration was demonstrated to be an effective technique in terms of stabilization only if performed using low MWCO membranes. MF polymeric membranes were used for apple juice clarification. MF membranes of 0.2-μm pore size gave higher flux values for PVDF and PS compared to the PES and cellulose (CE). In addition, for PS and CE membrane, 10-kDa membranes had higher fouling layer resistances than 30- and 100-kDa membranes [94]. The results have been associated with membrane surface morphology rather than surface hydrophobicity. Smoother membranes produced a dense surface layer, whereas this same layer on rougher membranes was more open [102].

The preservation of the phytochemical property and flavor of the MF-clarified pineapple juice was comprehensively investigated using three storage temperatures, 4°C, 27°C, and 37°C and time points, 0–6 months [103]. Polysulfone hollow fiber membranes of pore size 0.2 μm were used. Storage at 4°C was proved to be most suitable, as it permitted the best retention in chemical, physical, and phytochemical properties of MF-clarified pineapple juice compared to the stored at 27°C and 37°C. The rejection of enzymes in honey by polysulfone UF membranes, MWCO 20, 25, 50, and 100 kDa, and by MF membranes, pore size 25, 100, and 450 nm, were attempted in the batch cell [104]. Clarified honey and enzyme-enriched honey were produced using a combination of MF and UF membranes process. Mosambi juice was treated using low-cost ceramic MF membrane, average pore diameter 0.285 μm [105]. Permeate fluxes of dead–end MF experiments of 5.65×10^{-6} and 21.45×10^{-6} m^3/m$^2 \cdot$ s were obtained after 45 min at 137.9 kPa transmembrane pressure drop for centrifuged mosambi juice and enzyme treated centrifuged mosambi juice, respectively. The physiochemical properties such as acidity, pH, and total soluble solids of mosambi juice did not change significantly for both enzymatic treatment and MF. The effects of membrane property on the permeate flux, membrane fouling and quality of clarified pineapple juice were studied using both MF, pore size of 0.1 and 0.2 μm, and UF, MWCO 30, and 100 kDa, membranes [106]. MF membrane with a pore size of 0.2 μm was considered to be the most suitable membrane for highest permeate flux, total phenolic content, and antioxidant capacity. Under the optimum conditions using batch concentration mode, an average flux of about 37 L m$^{-2} \cdot$ h^{-1} was obtained during the MF of pineapple juice.

12.4.2.4 Electro-Dialysis, Forward Osmosis, Membrane Distillation, Osmotic Distillation, and PV

Membrane-based integrated process for the concentration of fruit juice consists of initial clarification of freshly squeezed juice by UF, a preconcentration up to 25–30°Brix by RO, final concentration of about 60°Brix by osmotic distillation (OD) [107]. The process is very efficient in preserving the total antioxidant activity of the final product even at 60°Brix. A slight decrease in antioxidant components, ascorbic acid (~ −15%) and anthocyanins (~ −23%), was noticed in both UF–RO–OD and UF–OD configurations, which is lower than that observed with the traditional thermal treatment. The flavanones and hydroxycinnamic acids are very stable and concentrated juice retains its bright red color. The enzymatically treated apple juice was clarified by polymeric microfiltration membrane, an average pore diameter of 0.3 μm, and concentrated by TFC RO membrane, 98% rejection of NaCl, in plate and frame configuration and OD [108]. The coupling of RO and OD processes was proved to be efficient for the purpose of concentrating apple juice, resulting in a final solid concentration of approximately 530 g. kg^{-1}. It is noted that RO has been applied to many kinds of fruit juices. However, it limits their final concentration to 25–35°Brix, since the process stops when the juice's osmotic pressure becomes equal to the hydraulic pressure. OD can be considered a promising technique, since it allows juices' concentration up to 55–65°Brix, satisfactorily preserving their quality.

The performance of immobilized invertase for sucrose hydrolysis was evaluated in a continuous process using a membrane bioreactor (MB) coupled with an UF or a MF membrane [109]. Colorless high-fructose syrup with an average concentration of 71.5%, which is similar to the commercial high-fructose corn syrup, was attained.

The comparison of OD and membrane distillation (MD) processes was studied in terms of water flux and aroma retention using polypropylene (PP) hollow fiber, a nominal pore diameter of 0.2 µm, as a membrane contactor and sucrose solution as a model fruit juice [110]. Although a similar overall driving force was used, the flux of OD process is more than double that of MD process due to temperature polarization effects. A higher retention of two aroma compounds of orange juice, citral and ethyl butyrate, was observed with the OD process compared to MD. Transport of the aroma compounds was studied using flat PP membrane with a nominal pore diameter of 0.2 µm. The OD process has advantages over the MD process both in terms of water flux and retention of aroma compounds.

The effects of membrane filtration on the aromatic profile and phenolic quality of a Cabernet Sauvignon wine was studied using a pilot system with a 1.2 µm PP pre-filter and a 0.65 µm PVDF final filter system [111]. The concentration of tannins and anthocyanins was decreased by 4.8% and 2.4%, respectively. The total polyphenolic index was decreased by 10% due to membrane adsorptive phenomenon. A statistically significant variation was noticed in 12 of the aromatic compounds.

RO process for the concentration of black currant juice was carried out by flat sheet polyamide TFC membrane in cross-flow at 30–50 bar [112] and by tubular TFC membrane in pilot scale at 45 bar [113]. The utilization of enzyme treatment by *Panzym Super* E has been proven to increase the permeate flux rate and efficiency of the concentration in RO process.

The reduction of acidity of passion fruit juice was investigated by ED with bipolar membranes at the lab–cell and preindustrial stack using four states of juice: initial pulpy juice, juice clarified by tangential MF, twice-concentrated clarified juice, and centrifuged juice [114]. The combination of OD with centrifugation process exhibited better performance than direct deacidification of the pulpy juice by OD because of the high fouling observed during the OD.

The integrated membrane processes for the clarification and the concentration of citrus, orange and lemon, and carrot juices were studied using PVDF, nominal MWCO 15 kDa, tubular membrane module for UF; polyamide TFC, salt rejection minimum 99%, spiral-wound module for RO operating at 1–66 bar; PP hollow-fibers, mean pore diameter 30 nm, membrane contactor for OD [115]. The RO process was used to preconcentrate the permeate coming from the UF step up to 15–20 g total soluble solid (TSS)/100 g and OD yielded a concentration of the retentate coming from the RO up to 60–63 g TSS/100 g. Although in the OD and UF processes, total antioxidant activity was maintained with respect to the fresh blood orange juice, a little decrease of the total antioxidant activity (TAA) was observed in the RO treatment due to high pressure employed during the process. The potentiality of an integrated UF-OD membrane process for the clarification and the concentration of the cactus pear fruit juice and the juice quality in terms of TAA, ascorbic, citric, and glutamic acid, betalains were studied [116]. The polysulfone hollow-fibers, MWCO 10 kDa, for UF and MF PP hollow-fibers membrane contactor for OD, were used. The clarified fresh cactus pear juice with a TSS content of about 11°Brix was concentrated by UF and finally was concentrated up to 61°Brix by OD process. The integrated membrane process involving the clarification of the concentrated pomegranate juice by hollow-fiber UF membranes and the concentration of the clarified juice by OD were studied at ambient temperature [117]. Rejections of the UF membrane toward polyphenols and anthocyanins were 16.5% and 11.7%, respectively. A 3.2-fold concentration of total soluble solids was achieved by using OD without modifying the main physico-chemical parameters of the juice.

Aroma recovery of tropical pineapple juice by PV process was investigated using composite membranes, flat sheet, or hollow fiber, prepared in the laboratory. The composite flat sheet membranes were with ethylene–propylene–diene terpolymer (EPDM) or ethylene–vinyl acetate copolymer as selective layers. The composite hollow fiber membrane was with EPDM as selective layer. The dense silicone hollow-fiber commercial membrane was also used for comparison [118]. The composite EPDM hollow fiber showed the best performance having a very high enrichment of the most volatile components.

A model was developed based on heat and mass balances along a tubular membrane module concerning concentration of black currant juice by direct contact membrane distillation (DCMD) process [119]. The mass balance was based on the dusty gas model and the model included correction for the heat and mass transfer coefficients. If the membrane's geometrical data including the porosity and the tortuosity is known and the thermodynamic and transport properties of the feed and permeate liquids are known, then DCMD concentration of fruit juice can be accurately predicted as distillation in tray and packed-bed columns.

The in-depth studies of ED cell with ultrafiltration membranes (EDUF) technology were demonstrated regarding the effects of filtration area by stacking three UF membranes in an EDUF configuration, two distinct volume ratios, that is, volume of raw juice/volume of enriched juice, and polyphenols-UF membrane interactions [120–122]. PES UF membrane, MWCO 500-kDa, was used. EDUF technology selectively enriched anthocyanins in cranberry juice without affecting its sugar and vitamin C concentration.

A mass transfer-in-series resistance model was employed for osmotic membrane distillation (OMD) process, considering the resistance offered by the membrane as well the boundary layers, feed and brine sides, using a flat membrane module performance of pineapple/sweet-lime juice [123]. The model could predict the variation of transmembrane flux with respect to different process parameters, such as type, concentration, and flow rate of the osmotic agent; pore size of PP membranes in the range of 0.05–0.2 μm; temperature in the range of 30–38°C. The influence of the process parameters, such as osmotic agent concentration, flow rate of feed, and osmotic agent and membrane pore size on transmembrane flux on OMD was studied for phycocyanin colorant and sweet juice [124]. Phycocyanin solution was concentrated up to threefold and sweet-lime juice was concentrated from 5 to 55°Brix. The membrane mass transfer coefficient was estimated to be 0.079×10^{-5} m s^{-1} and it was in the transition region between Knudsen and molecular diffusion.

Apple juice treatment combining different regimes of enzymatic treatment and clarification methods resulted in a variation of biopolymer removal by MD [125]. Introduction of additional enzymatic deproteinization pretreatment and application of UF provided minimal biopolymer content in the MD feed. Pretreatment of juice potentially played detrimental role in MD process causing fouling of highly hydrophobic membranes. Increase in biopolymer removal enhanced transmembrane flux during concentration of clarified juice by MD. The recovery of seven characteristic black currant aroma compounds by vacuum membrane distillation (VMD) was evaluated at low temperatures (10–45°C) with varying feed flow rates (100–500 L/h) in a lab scale [126]. VMD at feed flow from 100 to 500 l/h at

30°C gave concentration factors from ~ 4 to 15. The concentration factors were increased from 21 to 31 when the juice temperature was decreased to 10°C for the highly volatile aroma esters. The recovered levels of the highly volatile aroma compounds ranged from 68 to 83 vol.% with a feed volume reduction of 5 vol.%

Direct osmotic performance was studied using flat sheet aromatic polyamide TFC RO membrane, in a pilot scale for glucose and sucrose solutions [127]. Membrane characteristics play a very significant role at low osmotic medium concentrations, however, they only play a minor role at higher concentrations. At high concentrations, the boundary layer resistance and poor contacting conditions could result in a dramatic drop of overall mass transfer coefficients. Based on the direct osmotic performance of a pilot scale and the fundamental equation for cross-flow mass transfer diffusion through membranes, a generalized mathematical model was developed in physical terms [128]. The quantitative model for a flat geometry direct osmotic concentrator was provided with an insight to the effect of several process parameters on the magnitude of osmotic flux, which revealed the important role of liquid viscosities in controlling direct osmotic flux values.

The OD process was applied to concentrate commercial noni juice using a hollow fiber membrane contactor at 30°C [129]. The transmembrane vapor water flux was experimentally determined from 0.090 up to 0.413 kg $\cdot$ h^{-1} $\cdot$ m^{-2}. Noni juice was concentrated from 8 to 32°Brix after 1 h treatment and the content of phenolic compounds of fruit was preserved after processing. Mass transfer through the membrane pore seems to be described by a transition mechanism between the Knudsen and molecular diffusion. The industrial pilot plant with a hydrophobic PP hollow-fiber membrane, an average pore diameter of 0.2 μm, was applied for concentrating vegetable extracts and fruit juices by OD [130]. The final TSS contents achieved were 660, 570, and 610 g $\cdot$ kg^{-1} for grape juice, apple juice, and roselle extract, respectively. Roselle extract could be concentrated, at temperatures between 35°C and 45°C, by OD to as much as 535–615 g TSS $\cdot$ kg^{-1}, with an average distillation flux of 1.1 to 1.5 kg $\cdot$ h^{-1} $\cdot$ m^{-2}. OD did not modify physico-chemical, biochemical, nutritional, and sensorial characteristics and contents of sugars, acids, polyphenols, or anthocyanins were unchanged for tropical fruit juices and plant extracts. RO membranes limit the final concentration of fruit juices to about 25–35°Brix, however, OD can concentrate juices to about 65°Brix.

12.4.3 Edible Oil Industry

Sunflower, corn, canola, peanut, soybean, cottonseed, rapeseed, mustard, and coconut oils are among the edible oils. Membrane technology is used to remove or reduce those contaminants that would adversely affect the quality of the oil. Most impurities are free fatty acids, partial glycerides, phosphatides, oxidation products, pigment, or compounds containing trace elements. These impurities are removed by conventional chemical refining, but this technique has many disadvantages. In contrast, the membrane process is very simple, and high purity of oil, which is important for food applications, can be obtained.

12.4.3.1 Reverse Osmosis
Pilot-scale treatment of edible vegetable oil industry effluent was performed using thin film composite polyamide membrane and was found to be a very suitable method [131].

Polymeric composite membranes having silicon as an active layer and polyimide and PS as the support layer were used for decolorization of vegetable oil. RO membranes made from cellulose acetate, polyamide, and polyvinyl alcohol were tested in the study [132].

The differential permeation of liquid oil constituents in nonporous (dense) polymeric composite membrane, silicone as active layer and polyimide as support layers, was studied for basic understanding of transport mechanism using triglycerides and oleic acid [133]. The behavior of these systems suggested that solution-diffusion is the predominant mechanism of transport of oil constituents through nonporous membranes. The effect of viscosity (temperature) on permeation suggested that convective flow exists.

12.4.3.2 Ultrafiltration

Phospholipids separation from crude vegetable oil using polyimide UF membrane was possible. A series of UF membranes were prepared by the phase inversion method and exhibited a very high flux when the MWCO obtained by PEG solutes was 20 kDa. Fouling of the membranes during UF treatment of soybean oil was due to multilayer adsorption in the membrane pores and at the membrane surface. As the flux increased, fouling also increased due to the faster diffusion rate of solutes toward the membrane surface [134]. UF technology was very useful in recovering pea whey protein using both spiral wound and hollow fiber membranes [135].

UF PES membranes were studied for their ability to selectively separate phospholipids miscella from triacylglycerols under various operating conditions and their resistance in the presence of hexane [136]. PES membrane exhibited 84–89% removal of phospholipids from crude soybean oil. A permeate flux of 61 $L.h^{-1}.m^{-2}$ was obtained even at 0.5 bar of pressure and for oils with higher phospholipids content when miscella was tested. The removal of phospholipids from sunflower oil miscella was studied by a membrane filtration setup, which is a cross-flow UF pilot-plant, using one of the two tubular PES membranes of two different MWCO (4 and 9 kDa) [137]. Although both membranes showed approximately the same rejection of phospholipids (95–97%), the 9 kDa membrane exhibited higher miscella permeate flux and rejection of lower oil and higher free fatty acids. The performance of asymmetric PVDF UF membranes, a MWCO of 6000 ± 800 Da and a mean pore radius of 27.2 ± 8.5 Å, was studied for degumming of vegetable oils, soybean, and sunflower [138]. The membrane yielded phospholipids retention higher than 95% and the red color reduction was significant (higher than 70%). The yellow color did not decrease in soybean miscella but in sunflower the decrease was significant. Degumming of sunflower oil produced a higher membrane fouling than degumming of soybean oil. The ceramic UF membrane with an average pore diameter 0.05 μm showed good performance for the removal of phospholipids using filtration of corn oil/hexane miscellas [139]. Permeate fluxes of crude corn oil were high, reaching values up to 120 $kg·h^{-1}·m^{-2}$ under the best operating conditions while the retention of phospholipids ranging between 64.7% and 93.5% was achieved. The optimization of soybean oil degumming process was studied using response surface methodology (RSM) considering the tangential velocity and transmembrane pressure as the independent variables [140]. The authors concluded that the transmembrane pressure influenced the reduction in phosphorous content and improved the permeate flux from their experimental RSM designed. The effects of fatty acids, palmitic, stearic, and oleic acid, on the flux reduction of UF membranes were investigated using a glycerol–water–fatty acid mixture [141]. Oleic acid with the longest carbon chain contributed to a severe flux decline for the PES membrane. The PVDF membrane exhibited higher fluxes and lower fouling propensity compared to the PES due to the higher hydrophobicity of the PVDF membrane. The pH of the mixture played a significant role in the adsorption of fatty acids onto the membrane surface. The most severe flux reduction occurred in acidic instead of alkaline conditions due to the increasing amount of undissociated acid resulting in an increase in the amount of fatty acid adsorption. Various commercial UF, MWCO from 15 to 300 kDa, and MF, pore size from 0.1 to 1 µm, membranes were used to verify the potential applications of membrane cross-flow filtration technologies for extra virgin olive oil [142]. Polarization gel tends to form on the membrane surface and its control is of the greatest importance on the mass transport. The fouling seems to be better controlled by initially removing large particles which foul the membrane surface.

12.4.3.3 Microfiltration

MF technique was used for soy protein isolation using two different pore size membranes, cellulose acetate with an average pore size 0.1 µm, and polypropylene (PP)/polyethylene (PE) with rectangular pores of 0.05×0.02 µm [143]. Membrane technology is used for removal of contaminants (i.e., trace amount of heavy metal including nickel, copper, manganese, and iron) and pigment (i.e., chlorophyll, carotenoids, xanthophylls, and their derivatives). Koseoglu and Vavra investigated the removal of the nickel catalysts from hydrogenated oils using three ceramic membranes with pore sizes ranging between 0.05 and 0.2 µm, two carbon-coated zirconia membranes with pore sizes between 0.08 and 0.14 µm, and a polyethylene imine (PI) membrane [144]. All membranes were capable of removing nickel catalysts, reducing at least by 100-fold the volume of catalyst-containing oil from hydrogenated soybean oil at reasonable flow rate. The high cost of bleaching earth and the associated oil losses led to an interest in the membrane technology to replace the traditional bleaching process. The membrane technology was applied for the removal of chlorophyll and β-carotene from sunflower oil and the decolorization of soybean oil using PE membrane of pore size 0.03 µm and PI composite membrane. The PE membrane gave very low rejections of chlorophyll (<4%), whereas the PI membrane gave over 95% rejection but permeate flux was very low (0.1–0.2 kg/m^2 · h). When food is fried in heated oil, many complex chemical reactions occur resulting in the production of degradation products. Membrane technology could be used to remove proteins, carbohydrates, and their decomposition products, as well as to improve the color of used oil [145]. Membrane processing could be used to enhance the life of used frying oils by removing the oil-soluble impurities [146–149]. The reduction of phospholipids in crude soyabean oil using PE membrane (pore size 30 nm) was in the range of 85.8–92.8% by MF [146]. A total polar material is used for the chemical index to determine the degree of cumulative degradation of the oil. Oil absorption by fried food increases with increasing oil deterioration. Experiments were carried out with an MF membrane and a polymer composite membrane. The viscosity of the used frying oil was reduced to 22%, and composite membranes were effective in reducing the soluble impurities as well as insoluble particulates [149].

12.4.4 Miscellaneous Food Products

12.4.4.1 Nanofiltration
Organic solvent resistant nanofiltration membrane, MWCO 200 Da, was used for concentration of fresh rosemary extracts using continuous or semi-batch cross-flow diafiltration process. The experimental results obtained were 25.7 L/m^2. h permeate flux at operating pressure 40 bar and 99.5% rejection of rosmarinic acid and other antioxidant constituents of the herb [150].

12.4.4.2 Ultrafiltration
Three UF membranes, MWCO 4, 8, and 20 kDa, and one NF membrane, MWCO 300 Da, were used to determine a process integrating an enzymatic reactor for the treatment of blue whiting peptide solutions [151]. High MWCO UF membrane was well suited for the separation of peptides and nonhydrolyzed proteins, whereas two other UF membranes seemed promising for the fractionation of peptides. NF membrane enables the concentration of low molecular weight marine peptides with reduced salt concentrations.

To compare protein yield, protein concentration and physicochemical characteristics of *Amaranth mantegazzianus* protein concentrates were studied in a pilot-scale by a conventional process (CP), alkaline extraction and isoelectric precipitation, and two alternative processes (AP): acid pretreatment process combined with isoelectric precipitation (IP) and acid pretreatment process combined with UF [152]. UF produced higher protein concentration, lower protein aggregation, and better nutritional quality (i.e., balanced amino acid composition and no limiting amino acid) than CP and IP. The authors concluded that UF process is a viable alternative to conventional processes and a promising method for producing amaranth protein concentrate. The potential of an integrated three-stage process consisting of MF, UF, and NF for the purification of sweeteners from *Stevia rebaudiana* Bertoni was evaluated [153]. The performance of different commercial and tailor-made PES UF membranes was studied. Starting from an extract purity of 11% sweeteners, a purity of 37% and a yield of 30% could be reached using the overall three-membranes process.

12.4.4.3 Electro-Dialysis, Forward Osmosis, Membrane Distillation, Osmotic Distillation, and PV
An integrated membrane process consisting of UF for clarification, RO for preconcentration, and OMD for final concentration was studied to concentrate anthocyanin, a natural red colorant from red radish [154]. Attempts were also made to compare the process performance of the three individual operations with the combined process. The integrated membrane process has the advantages of achieving higher concentration of anthocyanin compared to the individual membrane processes. The hybrid process achieved the concentration of anthocyanin from 0.4 to 9.8 g/L (25-fold increase in concentration from 1 to 26°Brix). Comparison of OMD and forward osmosis (FO) membrane processes for concentration of anthocyanin extract was studied [155]. The flux of FO was better than OMD.

Two-step integrated membrane process, combining desalination by ED and concentration by RO, was studied to recover natural marine aromas, specifically shrimp cooking juice [156]. The economic feasibility study and model development were attempted. Performance evaluation of concentration of sucrose solutions was carried out by FO process using RO membrane [157]. Sucrose solution concentration factors of 5.7 were possible with FO with a starting sucrose concentration of 0.29 M compared to reported values of 2.5 with RO. PV process was investigated for the recovery of aroma compounds from brown crab effluent using a hydrophobic membrane with a selective layer of polyoctylmethyl siloxane (POMS) [158]. Operating conditions such as feed flow rate, feed concentration, feed temperature, and permeate pressure were optimized to obtain PV permeates with a maximum organoleptic quality in its aroma profile. The authors also studied organophilic PV process using polydimethylsiloxane membrane, wherein total permeation flux was higher compared to the POMS membrane [159].

Ion-exchange membrane mediated ED treatment of scallop broth at various pH (2.5–6.4) conditions was evaluated in terms of changes in ions, free amino acids, and heavy metal profiles [160]. Reduction of the heavy metal and salt content was noticed after ED. Conventional ED and packed-bed ED (PBED) membrane stack configurations at a constant current were performed for desalination of fish meat extract [161]. The low membrane stack potential which led to the lower energy consumption for the same salt removal was achieved by the PBED system compared to the ED. PBED might be more suitable than ED for use in the fish meat extract desalination due to the effects of ion exchange membranes in enhancing ion electro-migration and mitigating membrane fouling. ED was used successfully in the pilot scale to reduce the salt content of fish sauce adequately maintaining its important characteristics such as content of total and amino nitrogen [162]. The enzymatic hydrolysis of β-lactoglobulin and the fractionation of bioactive peptides were simultaneously conducted using ED cell with ultrafiltration membranes (EDUF) stacked [163]. EDUF allowed a specific separation and concentration of anionic and cationic peptides in recovery compartments.

12.5 CONCLUSION

In this chapter, the fundamental principles underlying the liquid food concentration and the fractionation of liquid components are illustrated. Fundamental transport equations that are based on water preferential sorption models are applicable when solvent water is preferentially sorbed at the membrane–solution interface or when solutes are only weakly adsorbed on the membrane surface; they then enable one to predict membrane performance data quantitatively. They are also useful for the design of membrane module systems appropriate for a particular type of food processing. However, the transport equations based on the surface force–pore flow model are applicable for the more general case of preferential sorption of either solvent water or solute at the membrane–solution interface. These transport equations enable the prediction of membrane performance data on the basis of the average pore size and the pore size distribution that characterize a given membrane surface. Moreover, the latter approach enables one to design membranes appropriate for a particular food concentration and fractionation problem by producing membranes with desired pore structures. This approach also offers the means to choose membrane materials and to design the pore

PROPERTIES OF FOOD IN MEMBRANE SEPARATION PROCESSES

size distribution on the membrane surface in order to reduce the membrane fouling problems. Membrane separation processes other than RO and UF can also be applied for the processing of liquid food. As one such example, recovery and concentration of food flavors were described in this chapter.

LIST OF SYMBOLS

A	Pure water permeability constant, $g \cdot mol\ H_2O/(cm^2 \cdot s\ atm)$ or $kg \cdot mol\ H_2O/(m^2\ s \cdot kPa)$
A_t	Surface area of polymer powder in the chromatographic column, cm^2 or m^2
$\mathcal{A}$	Constant characterizing electrostatic repulsion force, cm or m
a	Constant involved in Equation 12.73
$\mathcal{B}$	Constant characterizing the van der Waals attraction force, cm^3 or m^3
b	Frictional function defined by Equation 12.30
b	Constant involved in Equation 12.73
C^0	C at the module entrance in the flow process and at the start of operation in the batch process
C	Average of C for the entire module
C_A	Dimensionless concentration defined by Equation 12.25
$\ln G_{NaCl}$	Constant defined by Equation 12.7
c	Molar density of solution, $g \cdot mol/cm^3$ or $g \cdot mol/m^3$
c_1, c_2, c_3	Molar density of feed solution, concentrated boundary solution, and product solution, respectively, $g \cdot mol/cm^3$ or $g \cdot mol/m^3$
c_A	Molar concentration of solute, $g \cdot mol/cm^3$ or $g \cdot mol/m^3$
c_{A1}, c_{A2}, c_{A3}	Molar concentration of solute, in the feed solution, concentrated boundary solution (or at pore inlet), and product solution (or at pore outlet), respectively, $g \cdot mol/cm^3$ or $g \cdot mol/m^3$
$c_{A,b}$	Bulk molar concentration of solute, $g \cdot mol/cm^3$ or $g \cdot mol/m^3$
$c_{A,i}$	Interfacial molar concentration of solute, $g \cdot mol/cm^3$ or $g \cdot mol/m^3$
D_{AB}	Diffusivity of solute in water, cm^2/s or m^2/s
D	Constant characterizing the steric repulsion at the interface, cm or m
D_w	Molecular radius of solvent water ($=0.87 \times 10^{-10}$ m in this work)
$D_{AM}/K\delta$	Solute transport parameter, cm/s or m/s
d	Distance between polymer material surface and the center of solute molecule, cm or m
ΣE_s	Taft's steric parameter for the substituent group in the organic molecule
f	Fraction solute separation based on the feed concentration
f'	Fraction solute separation based on the concentration in the boundary phase
$\Delta G_B, \Delta G_I$	Free energy of hydration of solute in the bulk solution phase, and that at the membrane–solution interface, respectively, $kcal/(g \cdot mol)$ or $kJ/(g \cdot mol)$
$-\Delta\Delta G/RT$	Free energy parameter
$(-\Delta\Delta G/RT)_i$	Free energy parameter for ionic species i
$1/h$	Membrane area per unit volume of channel space, cm^{-1} or m^{-1}

h_i	Quantity defined by Equation 12.44
K_a	Equilibrium dissociation constant of acid, $g \cdot mol/L$ or $g \cdot mol/m^3$
k	Mass transfer coefficient on the high-pressure side of the membrane, cm/s or m/s
k	Boltzmann constant
M_B	Molecular weight of solvent
N_B	Solvent water flux through membrane, $g \cdot mol/(cm^2 \cdot s)$ or $g \cdot mol/(m^2 \cdot s)$
N_t	Total flux of both ionized and nonionized solute through the membrane, $g \cdot mol/(cm^2 \cdot s)$ or $g \cdot mol/(m^2 \cdot s)$
P	Operating pressure, atm, kPa, or Pa
P_i	Operating pressure applied at pore inlet, atm, kPa, or Pa
P_o	Operating pressure prevailing at pore outlet, atm, kPa, or Pa
PWP	Pure water permeation rate through effective area of membrane surface, g/h or kg/h
PR	Product rate through effective area of membrane surface, g/h or kg/h
q_{av}	Average product rate, $g/(cm^2 s)$ or $kg/(m^2 s)$
R_a	$R_b - D_w$
R_b	Membrane pore radius defined as the distance from the wall of the pore, cm or m
$\P_b$	Average of R_b, cm or m
$R_{b,i}$	Pore radius belonging to ith distribution, cm or m
$\P_{b,i}$	Average of $R_{b,i}$, cm or m
R	Gas constant
r	Radial distance in cylindrical coordinate, cm or m
r_A	Stokes' law radius of molecule, cm or m
r_i	Outer radius of the distributor, cm or m
r_o	Outer radius of the hollow fiber bundle, cm or m
S	Effective membrane area, cm^2 or m^2
Σs^*	Modified Small's number, $cal^{1/2} \cdot cm^{3/2}/(g \cdot mol)$ or $J^{1/2} \cdot m^{3/2}/(g \cdot mol)$
T	Absolute temperature, K
t	Operational time, s
u	Average velocity of feed solution at the channel entrance, cm/s or m/s
u_B	Velocity of solvent in the pore, cm/s or m/s
V	Volume of bulk solution at the high-pressure side of the membrane, cm^3 or m^3
V_i, V_f	Initial and final values of V, respectively, cm^3 or m^3
$[V'_R]_A$	Retention volume of solute A, cm^3 or m^3
$[V'_R]_{water}$	Retention volume of water as represented by that of D_2O, cm^3 or m^3
v_s	Permeation velocity of product solution, cm/s or m/s
v	Pure water permeation velocity, cm/s or m/s
X_A	Mole fraction of solute
X_{A1}, X_{A2}, X_{A3}	Mole fraction of solute in feed solution, concentrated boundary solution, and product solution, respectively
X, X'	Dimensionless quantities defined by Equations 12.18 and 12.19, respectively
X_i, X_u	Mole fraction of ionized and nonionized solute, respectively

x	Longitudinal distance in the module, cm or m
$Y_i(R_b)$	Pore size distribution function of ith normal distribution, cm^{-1} or m^{-1}
z	Axial distance in cylindrical coordinate, cm or m

Greek Symbols

$\alpha(\rho)$	Dimensionless solution velocity profile in the pore defined by Equation 12.26
β_1	Dimensionless solution viscosity defined by Equation 12.27
β_2	Dimensionless operating pressure defined by Equation 12.28
Γ	Surface excess of solute, g . mol/cm^2 or g . mol/m^2
γ	Quantity defined by Equation 12.14
γ_B	(*structural* Incremental free energies of hydration for the structural group involved *group*) in the organic solute molecule, applicable for the bulk solution phase
γ_I	(*structural* and interfacial solution phase, respectively *group*)
$\gamma_{B,0}, \gamma_{I,0}$	Constants applicable for the bulk solution phase and the interfacial solution phase, respectively
Δ	Fraction product recovery
$\ln \Delta^*$	Scale factor
δ	Length of cylindrical pore, cm or m
δ^*	Coefficient associated with ΣE_S
η	Solution viscosity, poise or Pa.s
θ	Dimensionless quantity defined by Equation 12.15
λ	Dimensionless quantity defined by Equation 12.16
λ_f	Dimensionless quantity defined by Equation 12.54
ν	Kinematic viscosity, cm^2/s or m^2/s
π	Osmotic pressure of solution, atm or Pa
ρ	Dimensionless radial distance defined by Equation 12.24
$(\rho_1)_i, (\rho_1)_f$	Density of feed and concentrate solution, respectively, g/cm^3 or kg/m^3
ρ	Density of solution, g/cm^3 or kg/m^3
σ	Standard deviation of normal pore distributions, cm or m
σ_i	σ of pores belonging to the ith normal distribution, cm or m
τ	Quantity defined by Equation 12.21
$\Phi(\rho)$	Dimensionless potential defined by Equation 12.29 as a function of ρ
$\Phi(d)$	Dimensionless potential defined as a function of $\underline{d}$
φ	Potential function of the interaction force exerted on the solute from the pore wall, cal/(g . mol) or J/(g . mol)
φ	(*structural* Structural contribution to $(\delta^*\Sigma E_s)_{\lim}$ ***component***)
φ_0	Constant used in Equation 12.13
ω^*	Coefficient associated with Σ_s^*, g . mol/(cal$^{1/2}$. cm$^{3/2}$) or g . mol/(J$^{1/2}$. m$^{3/2}$)
χ_{AB}	Proportionality constant between friction on the solute and the difference in velocity between the solute and solvent, cal . s/(cm^2 . g . mol) or J . s/(cm^2 . g . mol)
χ_{AM}	Proportionality constant between friction on the solute and the difference in velocity between the solute and pore wall, cal . s/(cm^2 . g . mol) or J . s/(m^2 . g . mol)

REFERENCES

1. TK Sherwood. Mass transfer between phases. *33rd Annual Priestley Lectures*, Pennsylvania State Univ, 1959, p 38.
2. S Sourirajan. *Reverse Osmosis*. London: Logos, 1970.
3. T Matsuura, S Sourirajan. A fundamental approach to application of reverse osmosis for food processing. *AIChE Symp Ser* 74:196–208, 1978.
4. S Sourirajan. *Lectures on Reverse Osmosis*. Ottawa: National Research Council Canada, 1983.
5. T Matsuura, JM Dickson, S Sourirajan. Free energy parameters for reverse osmosis separations of undissociated polar organic solutes in dilute aqueous solutions. *Ind Eng Chem Process Des Dev* 15:149–161, 1976.
6. RW Taft Jr. Separation of polar, steric and resonance effects in reactivity. In: MS Newman, ed. *Steric Effects in Organic Chemistry*. New York: Wiley, 1956, pp 556–675.
7. T Matsuura, ME Bednas, JM Dickson, S Sourirajan. Polar and steric effects in reverse osmosis. *J Appl Polym Sci* 18:2829–2846, 1974.
8. EN Pereira, T Matsuura, S Sourirajan. Reverse osmosis separations and concentrations of food sugars. *J Food Sci* 41:672–680, 1976.
9. T Matsuura, S Sourirajan. Reverse osmosis separation of hydrocarbons in aqueous solutions using porous cellulose acetate membranes. *J Appl Polym Sci* 17:3683–3708, 1973.
10. T Matsuura, AG Baxter, S Sourirajan. Predictability of reverse osmosis separations of higher alcohols in dilute aqueous solutions using porous cellulose acetate membranes. *Ind Eng Chem Process Des Dev* 16:82–89, 1977.
11. H Ohya, S Sourirajan. *Reverse Osmosis System Specification and Performance Data for Water Treatment Applications*. Hanover, NH: The Thayer School of Engineering, Dartmouth College, 1971.
12. T Matsuura, S Sourirajan. Reverse osmosis transport through capillary pores under influence of surface forces. *Ind Eng Chem Process Des Dev* 20:273–282, 1981.
13. T Matsuura, Y Taketani, S Sourirajan. Estimation of interfacial forces governing reverse osmosis system: Nonionized polar organic solute–water–cellulose acetate membrane. In: AF Turbak, ed. *Synthetic Membranes: Volume II*. Washington, DC: American Chemical Society, ACS Symp Ser No 154, 1981, pp 315–338.
14. L Onsager, NNT Samaras. The surface tension of Debye–Hückel electrolytes. *J Chem Phys* 2:528–536, 1934.
15. H Faxen. About T Bohlin's paper: On the drag on rigid spheres, moving in a viscous liquid inside cylindrical tubes. *Kolloid Z* 167:146, 1959.
16. CN Satterfield, CK Colton, WH Pitcher Jr. Restricted diffusion in liquids within fine pores. *AIChE J* 19:628–635, 1973.
17. JA Lane, JW Riggle. Dialysis. *Chem Eng Prog Symp Ser* 55:127–143, 1959.
18. T Matsuura, S Sourirajan. Properties of polymer-solution interfacial fluid from liquid chromatography data. *J Colloid Interface Sci* 66:589–592, 1978.
19. Y Taketani, T Matsuura, S Sourirajan. Use of liquid chromatography for studying reverse osmosis and ultrafiltration. *Sep Sci Technol* 17:821–838, 1982.
20. NA Chudak, YA Eltekov, AV Kiselev. Study of adsorption from solutions on silica by liquid chromatography method. *J Colloid Interface Sci* 84:149–154, 1981.
21. T Matsuura, TA Tweddle, S Sourirajan. Predictability of performance of reverse osmosis membranes from data on surface force parameters. *Ind Eng Chem Process Des Dev* 23:674–684, 1984.
22. K Chan, T Matsuura, S Sourirajan. Interfacial forces, average pore size, and pore size distribution of ultrafiltration membranes. *Ind Eng Chem Process Res Dev* 21:605–612, 1982.
23. DK Tressler, MA Joslyn. *Fruit and Vegetable Juice Processing Technology*, 2nd ed. Westport, CT: AVI, 1971.
24. T Matsuura, AG Baxter, S Sourirajan. Reverse osmosis recovery of flavor compounds from apple juice water. *J Food Sci* 40:1039–1046, 1975.

25. RAM Delaney, JK Donnelly. Applications of reverse osmosis in the dairy industry. In: S Sourirajan, ed. *Reverse Osmosis and Synthetic Membranes: Theory, Technology, Engineering.* Ottawa: National Research Council Canada, 1977, pp 417–443.
26. JM Dickson, T Matsuura, P Blais, S Sourirajan. Some transport characteristics of aromatic polyamide membranes in reverse osmosis. *J Appl Polym Sci* 20:1491–1499, 1976.
27. T Matsuura, ME Bednas, S Sourirajan. Reverse osmosis separation of single and mixed alcohols in aqueous solutions using porous cellulose acetate membranes. *J Appl Polym Sci* 18:567–588, 1974.
28. T Matsuura, JM Dickson, S Sourirajan. Predictability of reverse osmosis separations of partially dissociated organic acids in dilute aqueous solutions. *Ind Eng Chem Process Des Dev* 15:350–357, 1976.
29. CB Monk. *Electrolytic Dissociation.* New York: Academic, 1961, pp 272–273.
30. P Malaiyandi, T Matsuura, S Sourirajan. Predictability of membrane performance for mixed solute reverse osmosis systems-system: Cellulose acetate membrane-D-glucose-D,L-malic acid-water. *Ind Eng Chem Process Des Dev* 21:277–282, 1982.
31. T Matsuura, AG Baxter, S Sourirajan. Concentration of fruit juices by reverse osmosis using porous cellulose acetate membranes. *Acta Aliment* 2:109–150, 1973.
32. T Matsuura, AG Baxter, S Sourirajan. Studies on reverse osmosis for concentration of fruit juices. *J Food Sci* 39:704–711, 1974.
33. SQ Zhang, AE Fouda, T Matsuura, K Chan. Some experimental results and design calculations for reverse osmosis concentration of green tea juice. *Desalination* 80:211–234, 1991.
34. SQ Zhang, T Matsuura, K Chan. Reverse osmosis concentration of green tea juice. *J Food Process Eng* 14:85–105, 1991.
35. SQ Zhang, AE Fouda, T Matsuura, K Chan. Reverse osmosis transport and model analysis for the green tea juice concentration. *J Food Process Eng* 16:1–20, 1992.
36. S Sourirajan, T Matsuura. *Reverse Osmosis/Ultrafiltration Process Principles.* Ottawa: National Research Council Canada, 1985, p 766.
37. T Matsuura, S Sourirajan. Reverse osmosis separation of phenols in aqueous solutions using porous cellulose acetate membranes. *J Appl Polym Sci* 16:2531–2554, 1972.
38. T Matsuura, S Sourirajan. Reverse osmosis separation of organic acids in aqueous solutions using porous cellulose acetate membranes. *J Appl Polym Sci* 17:3661–3682, 1973.
39. T Matsuura, S Sourirajan. Physicochemical criteria for reverse osmosis separation of monohydric and polyhydric alcohols and some related hydroxyl compounds in aqueous solutions using porous cellulose acetate membranes. *J Appl Polym Sci* 17:1043–1071, 1973.
40. JM Dickson, T Matsuura, S Sourirajan. Transport characteristics in the reverse osmosis system p-chlorophenol-water-cellulose acetate membrane. *Ind Eng Chem Process Des Dev* 18:641–647, 1979.
41. T Matsuura, S Sourirajan. Reverse osmosis separation of some organic solutes in aqueous solution using porous cellulose acetate membranes. *Ind Eng Chem Process Des Dev* 10:102–108, 1971.
42. T Liu, K Chan, T Matsuura, S Sourirajan. Determination of interaction forces and average pore size and pore size distribution and their effect on fouling of ultrafiltration membranes. *Ind Eng Chem Process Res Dev* 23:116–124, 1984.
43. O Kutowy, WL Thayer, S Sourirajan. High flux cellulose acetate ultrafiltration membranes. *Desalination* 26:195–210, 1978.
44. SQ Zhang, T Matsuura. Recovery and concentration of flavor compounds in apple essence by pervaporation, *J Food Process Eng* 14:291–296, 1991.
45. SS Koseoglu, KJ Guzman. Applications of reverse osmosis technology in the food industry. In: Z Amjad, ed. *Reverse Osmosis: Membrane Technology, Water Chemistry, and Industrial Applications.* New York: Van Nostrand Reinhold, 1993, pp 301–333.
46. M Cheryan, JR Alvarez. Food and beverage industry applications. In: RD Noble, SA Stern, ed. *Membrane Separations Technology: Principles and Applications.* Amsterdam: Elsevier, 1995, pp 415–465.

47. LJ Zeman, AL Zydney. *Microfiltration and Ultrafiltration: Principles and Applications.* New York: Marcel Dekker, 1996, pp 490–510, 524–543.
48. KSMS Raghavarao, N Nagaraj, G Patil, BR Babu, K Niranjan. Athermal membrane processes for the concentration of liquid food and natural colors. D-W Sun, ed. *Emerging Technologies for Food Processing.* San Diego: Elsevier, 2005, ch 10, pp 251–276.
49. S Nene, G Patil, KSMS Raghavarao. Membrane distillation in food processing. In: AK Pabby, SSH Rizvi, AM Sastre, eds. *Handbook of Membrane Separations: Chemical, Pharmaceutical, Food, and Biotechnological Applications.* Boca Raton: CRC Press, 2008, ch 19, pp 513–551.
50. Y Wan, J Luo, Z Cui. Membrane application in soy sauce processing. In: ZF Cui, HS Muralidhara, eds. *Membrane Technology: A Practical Guide to Membrane Technology and Applications in Food and Bioprocessing.* Oxford: Elsevier, 2010, ch 4, pp 45–62.
51. F Lipnizki. Cross-flow membrane applications in the food industry. In: KV Peinemann, SP Nunes, L Giorno, eds. *Membrane Technology: Membranes for Food Applications.* Weinheim: Wiley-VCH, 2010, vol 3, pp 1–20.
52. L Bazinet, A Doyen, C Roblet. Electrodialytic phenomena, associated electromembrane technologies and applications in the food, beverage and nutraceutical industries. In: SSH Rizvi, ed. *Separation, Extraction and Concentration Processes in the Food, Beverage and Nutraceutical Industries.* Cambridge: Woodhead Publishing Ltd, 2010, part 1, pp 202–218.
53. E Drioli, A Cassano. Advances in membrane-based concentration in the food and beverage industries: Direct osmosis and membrane contactors. SSH Rizvi, ed. In: *Separation, Extraction and Concentration Processes in the Food, Beverage and Nutraceutical Industries.* Cambridge: Woodhead Publishing Ltd, 2010, part 1, pp 244–283.
54. HH Himstedt, JA Hestekin. Membranes in the dairy industry. In: I Escober, B Van der Bruggen, eds. *Modern Applications in Membrane Science and Technology.* Washington: ACS Symp Ser, 2011, ch 11, pp 171–221.
55. LK Wang, NK Shammas, M Cheryan, Y-M Zheng, S-W Zou. Treatment of food industry foods and wastes by membrane filtration. In: LK Wang, JP Chen, Y-T Hung, NK Shammas, eds. *Handbook of Environmental Engineering, Vol 13: Membrane and Desalination Technologies.* New York: Springer, 2011, ch 6, pp 237–269.
56. B Jiao, A Cassano, E Drioli. Recent advances on membrane processes for the concentration of fruit juices: A review. *J Food Eng* 63:303–324, 2004.
57. J Shi, H Nawaz, J Pohorly, G Mittal, Y Kakuda, Y Jiang. Extraction of polyphenolics from plant material for functional foods—Engineering and technology. *Food Rev Int* 21:139–166, 2005.
58. B Krajewska. Membrane-based processes performed with use of chitin/chitosan materials. *Sep Purif Technol* 41:305–312, 2005.
59. GM Rios, M-P Belleville, D Paolucci-Jeanjean Membrane engineering in biotechnology: Quo vamus?. *Trends Biotechnol* 25:242–246, 2007.
60. TY Cath, AE Childress, M Elimelech. Forward osmosis: Principles, applications, and recent developments. *J Membrane Sci* 281:70–87, 2006.
61. R Xie, L-Y Chu, J-G Deng. Membranes and membrane processes for chiral resolution. *Chem Soc Rev* 37:1243–1263, 2008.
62. T Xu, C Huang. Electrodialysis-based separation technologies: A critical review. *AIChE J* 54:3147–3159, 2008.
63. C de Morais Coutinho, MC Chiu, RC Basso, APB Ribeiro, LAG Gonçalves, LA Viotto. State of art of the application of membrane technology to vegetable oils: A review. *Food Res Int* 42:536–550, 2009.
64. C Charcosset, Preparation of emulsions and particles by membrane emulsification for the food processing industry. *J Food Eng* 92:241–249, 2009.
65. E Ortega-Rivas, SB Perez-Vega. Solid-liquid separations in the food industry: Operating aspects and relevant applications. *J Food Nutrition Res* 50:86–105, 2011.

66. S-K Kim, M Senevirathne. Membrane bioreactor technology for the development of functional materials from sea-food processing wastes and their potential health benefits. *Membranes* 1:327–344, 2011.
67. V Sant'Anna, LDF Marczak, IC Tessaro. Membrane concentration of liquid foods by forward osmosis: Process and quality view. *J Food Eng* 111:483–489, 2012.
68. CM Galanakis. Recovery of high added-value components from food wastes: Conventional, emerging technologies and commercialized applications. *Trends Food Sci Technol* 26:68–87, 2012.
69. R De Bore, PFC Nooy. Concentration of raw whole milk by reverse osmosis and its influence of fat globules. *Desalination* 35:201–211, 1980.
70. DB Sarney, C Hale, G Frankel, EN Vulfson. A novel approach to the recovery of biologically active oligosaccharides from milk using a combination of enzymatic treatment and nanofiltration. *Biotechnol Bioeng* 69:461–467, 2000.
71. AK Goulas, PG Kapasakalidis, HR Sinclair, RA Rastall, AS Grandison. Purification of oligosaccharides by nanofiltration. *J Membrane Sci* 209:321–335, 2002.
72. Y Li, A Shahbazi, K Williams, C Wan. Separate and concentrate lactic acid using combination of nanofiltration and reverse osmosis membranes. *Appl Biochem Biotechnol* 147:1–9, 2008.
73. G Rice, AR Barber, AJ O'Connor, A Pihlajamaki, M Nystrom, GW Stevens, SE Kentish. The influence of dairy salts on nanofiltration membrane charge. *J Food Eng* 107:164–172, 2011.
74. V Demers-Mathieu, SF Gauthier, M Britten, I Fliss, G Robitaille, J Jean. Antibacterial activity of peptides extracted from tryptic hydrolyzate of whey protein by nanofiltration. *Int Dairy J* 28:94–101, 2013.
75. BH Ozer, RK Robinson. The behaviour of starter cultures in concentrated yoghurt (lebneh) produced by different techniques. *Food Sci Tech* 32:391–395, 1999.
76. NA Espinoza, MM Calvo. Effect of heat treatment and ultrafiltration process of cow's, ewe's, or goat's milk on its coagulation properties. *J Agric Food Chem* 46:1547–1551, 1998.
77. F Alvarez, M Arguello, FA Riera, R Alvarez, JR Iglesias, J Granda. Fermentation of concentrated skim-milk. Effects of different protein/lactose ratios obtained by ultrafiltration-diafiltration. *J Sci Food Agric* 76:10–16, 1998.
78. K Pocedičová, L Čurda, D Mišún, A Dryáková, L Diblíková. Preparation of galacto-oligosaccharides using membrane reactor. *J Food Eng* 99:479–484, 2010.
79. A Guadix, F Camacho, EM Guadix. Production of whey protein hydrolysates with reduced allergenicity in a stable membrane reactor. *J Food Eng* 72:398–405, 2006.
80. BJ James, Y Jing, XD Chen. Membrane fouling during filtration of milk—a microstructural study. *J Food Eng* 60:431–437, 2003.
81. DM Krstic, MN Tekic, MD Caric, SD Milanovic. The effect of turbulence promoter on cross-flow microfiltration of skim milk. *J Membrane Sci* 208:303–314, 2002.
82. P Punidadas, SSH Rizvi. Separation of milk proteins into fractions rich in casein or whey proteins by cross flow filtration. *Food Res International* 31:265–272, 1998.
83. H Goudedranche, J Fauquant, JL Maubois. Fractionation of globular milk fat by membrane microfiltration. *Lait* 80:93–98, 2000.
84. A Tolkach, U Kulozik. Fractionation of whey proteins and caseinomacropeptide by means of enzymatic crosslinking and membrane separation techniques. *J Food Eng* 67:13–20, 2005.
85. PG Middlewood, JK Carson. Extraction of amaranth starch from an aqueous medium using microfiltration: Membrane characterization. *J Membrane Sci* 405–406:284–290, 2012.
86. MP Mier, R Ibañez, I Ortiz. Influence of process variables on the production of bovine milk casein by electrodialysis with bipolar membranes. *Biochem Eng J* 40:304–311, 2008.
87. MC Duke, A Lim, SC da Luz, L Nielsen. Lactic acid enrichment with inorganic nanofiltration and molecular sieving membranes by pervaporation. *Food Bioproducts Processing* 86:290–295, 2008.

88. KB Petrotos, P Quantick, H Petropakis. A study of the direct concentration of tomato juice in tubular membrane—Module configuration. I. The effect of certain basic process parameters on the process performance. *J Membrane Sci* 150:99–110, 1998.
89. V Alvarez, S Alvarez, FA Riera, R Alvarez. Permeate flux prediction in apple juice concentration by reverse osmosis. *J Membrane Sci* 127:25–34, 1997.
90. AP Echavarría, V Falguera, C Torras, C Berdún, J Pagán, A Ibarz. Ultrafiltration and reverse osmosis for clarification and concentration of fruit juices at pilot plant scale. *LWT—Food Sci Technol* 46:189–195, 2012.
91. A Versari, R ferrarini, GP Perpinello, S Galassi. Concentration of grape must by nanofiltration membranes. *Trans IChemE Part C* 81:275–360, 2003.
92. EM Tsui, M Cheryan. Membrane processing of xanthophylls in ethanol extracts of corn. *J Food Eng* 83:590–595, 2007.
93. P Gullón, MJ González-Muñoz, Domínguez, JC Parajo. Membrane processing of liquors from *Eucalyptus globulus* autohydrolysis. *J Food Eng* 87:257–265, 2008.
94. B Girard, LR Fukumoto. Apple juice clarification using microfiltration and ultrafiltration polymeric membranes. *Food Sci Tech* 32:290–298, 1999.
95. L Gao, T Beveridge, CA Reid. Effects of processing and packaging conditions on haze formation in apple juices. *Food Sci Tech* 30:23–29, 1997.
96. P Tanada-Palmu, J Jardine, V Matta. Production of a banana (*Musa cavendishii*) extract containing no polyphenol oxidase by ultrafiltration. *J Sci Food Agric* 79:643–647, 1999.
97. BJ Cass, F Schade, CW Robinson, JE Thompson, RL Legge. Production of tomato flavor volatiles from a crude enzyme preparation using a hollow-fiber reactor. *Biotechnol Bioeng* 67:372–377, 2000.
98. GT Vladisavljević, P Vukosavljević, B. Bukvić. Permeate flux and fouling resistance in ultrafiltration of depectinized apple juice using ceramic membranes. *J Food Eng* 60:241–247, 2003.
99. A Cassano, C Conidi, R Timpone, M D'Avella, E Drioli. A membrane-based process for the clarification and the concentration of the cactus pear juice. *J Food Eng* 80:914–921, 2007.
100. G Daufin, JP Escudier, H Carrere, S Berot, L Fillaudeau, M Decloux. Recent and emerging applications of membrane processes in the food and dairy industry. *Food Bioproducts Processing* 79:89–102, 2001.
101. G Giovanelli, G Ravasini. Apple juice stabilization by combined enzyme-membrane filtration process. *Food Sci Tech* 26:1–7, 1993.
102. K Riedl, B Girard, RW Lencki. Influence of membrane structure on fouling layer morphology during apple juice clarification. *J Membrane Sci* 139:155–166, 1998.
103. A Laorko, Z Li, S Tongchitpakdee, S Chantachum, W Youravong. Effect of membrane property and operating conditions on phytochemical properties and permeate flux during clarification of pineapple juice. *J Food Eng* 100:514–521, 2010.
104. RS Barhate, R Subramanian, KE Nandini, HU Hebbar. Processing of honey using polymeric microfiltration and ultrafiltration membranes. *J Food Eng* 60:49–54, 2003.
105. BK Nandi, B Das, R Uppaluri, MK Purkait. Microfiltration of mosambi juice using low cost ceramic membrane. *J Food Eng* 95:597–605, 2009.
106. A Laorko, S Tongchitpakdee, W Youravong. Storage quality of pineapple juice non-thermally pasteurized and clarified by microfiltration. *J Food Eng* 116:554–561, 2013.
107. G Galaverna, G Di Silvestro, A Cassano, S Sforza, A Dossena, E Drioli, R Marchelli. A new integrated membrane process for the production of concentrated blood orange juice: Effect on bioactive compounds and antioxidant activity. *Food Chem* 106:1021–1030, 2008.
108. IB Aguiar, NGM Miranda, FS Gomes, MCS Santos, DGC Freitas, RV Tonon, LMC Cabral. Physicochemical and sensory properties of apple juice concentrated by reverse osmosis and osmotic evaporation. *Innovative Food Sci Emerging Technologies* 16:137–142, 2012.
109. EJ Tomotani, M Vitolo. Production of high-fructose syrup using immobilized invertase in a membrane reactor. *J Food Eng* 80:662–667, 2007.

110. VD Alves, IM Coelhoso. Orange juice concentration by osmotic evaporation and membrane distillation: A comparative study. *J Food Eng* 74:125–133, 2006.
111. JP Arriagada-Carrazana, C Sáez-Navarrete, E Bordeu. Membrane filtration effects on aromatic and phenolic quality of Cabernet Sauvignon wines. *J Food Eng* 68:363–368, 2005.
112. S Bánvolgyia, S Horvátha, É Stefanovits-Bányai, E Békássy-Molnára, G Vataia. Integrated membrane process for blackcurrant (*Ribes nigrum* L.) juice concentration. *Desalination* 241:281–287, 2009.
113. N Pap, E Pongrácz, M Jaakkola, T Tolonen, V Virtanen, A Turkki, Z Horváth-Hovorka, G Vatai, RL Keiski. The effect of pre-treatment on the anthocyanin and flavonol content of black currant juice (*Ribes nigrum* L.) in concentration by reverse osmosis. *J Food Eng* 98:429–436, 2010.
114. E Vera, J Sandeaux, F Persin, G Pourcelly, M Dornier, J Ruales. Deacidification of passion fruit juice by electrodialysis with bipolar membrane after different pretreatments. *J Food Eng* 90:67–73, 2009.
115. A Cassano, E Drioli, G Galaverna, R Marchelli, G Di Silvestro, P Cagnasso. Clarification and concentration of citrus and carrot juices by integrated membrane processes. *J Food Eng* 57:153–163, 2003.
116. A Cassano, L Donato, E Drioli. Ultrafiltration of kiwifruit juice: Operating parameters, juice quality and membrane fouling. *J Food Eng* 79:613–621, 2007.
117. A Cassano, C Conidi, E Drioli. Clarification and concentration of pomegranate juice (*Punica granatum* L.) using membrane processes. *J Food Eng* 107:366–373, 2011.
118. CC Pereira, JRM Rufino, AC Habert, R Nobrega, LMC Cabral, CP Borges. Aroma compounds recovery of tropical fruit juice by pervaporation: Membrane material selection and process evaluation. *J Food Eng* 66:77–87, 2005.
119. MB Jensen, KV Christensen, R Andrésen, LF Søtoft, B Norddahl. A model of direct contact membrane distillation for black currant juice. *J Food Eng* 107:405–414, 2011.
120. L Bazinet, C Cossec, H Gaudreau, Y Desjardins. Production of a phenolic antioxidant enriched cranberry juice by electrodialysis with filtration membrane. *J Agric Food Chem* 57:10245–10251, 2009.
121. L Bazinet, S Brianceau, P Dubé, Y Desjardins. Evolution of cranberry juice physico-chemical parameters during phenolic antioxidant enrichment by electrodialysis with filtration membrane. *Sep Purif Technol* 87:31–39, 2012.
122. E Husson, M Araya-Farias, Y Desjardins, L Bazinet. Selective anthocyanins enrichment of cranberry juice by electrodialysis with ultrafiltration membranes stacked. *Innovative Food Sci Emerging Technologies* 17:153–162, 2013.
123. N Nagaraj, G Patil, BR Babu, UH Hebbar, KSMS Raghavarao, S Nene. Mass transfer in osmotic membrane distillation. *J Membrane Sci* 268:48–56, 2006.
124. BR Babu, NK Rastogi, KSMS Raghavarao. Mass transfer in osmotic membrane distillation of phycocyanin colorant and sweet-lime juice. *J Membrane Sci* 272:58–69, 2006.
125. OS Lukanin, SM Gunko, MT Bryk, RR Nigmatullin. The effect of content of apple juice biopolymers on the concentration by membrane distillation. *J Food Eng* 60:275–280, 2003.
126. R Bagger-Jørgensen, AS Meyer, C Varming, G Jonsson. Recovery of volatile aroma compounds from black currant juice by vacuum membrane distillation. *J Food Eng* 64:23–31, 2004.
127. MI Dova, KB Petrotos, HN Lazarides. On the direct osmotic concentration of liquid foods. Part I: Impact of process parameters on process performance. *J Food Eng* 78:422–430, 2007.
128. MI Dova, KB Petrotos, HN Lazarides. On the direct osmotic concentration of liquid foods: Part II. Development of a generalized model. *J Food Eng* 78:431–437, 2007.
129. H Valdés, J Romero, A Saavedra, A Plaza, V Bubnovich. Concentration of noni juice by means of osmotic distillation. *J Membrane Sci* 330:205–213, 2009.
130. M Cissé, F Vaillant, S Bouquet, D Pallet, F Lutin, M Reynes, M Dornier. Athermal concentration by osmotic evaporation of roselle extract, apple and grape juices and impact on quality. *Innovative Food Sci Emerging Technologies* 12:352–360, 2011.

131. S Sridhar, A Kale, AA Khan. Reverse osmosis of edible vegetable oil industry effluent. *J Membrane Sci* 205:83–90, 2002.
132. KK Reddy, R Subramanian, T Kawakatsu, M Nakajima. Decolorization of vegetable oils by membrane processing. *Eur Food Res Technol* 213:212–218, 2001.
133. R Subramanian, KSMS Raghavarao, M Nakajima, H Nabetani, T Yamaguchi, T Kimura. Application of dense membrane theory for differential permeation of vegetable oil constituents. *J Food Eng* 60:249–256, 2003.
134. IC Kim, JH Kim, KH Lee, TM Tak. Phospholipids separation (degumming) from crude vegetable oil by polyimide ultrafiltration membrane. *J Membrane Sci* 205:113–123, 2002.
135. L Gao, KD Nguyen, AC Utioh. Pilot scale recovery of proteins from a pea whey discharge by ultrafiltration. *Food Sci Tech* 34:149–158, 2001.
136. JMLN de Moura, LAG Gonçalves, JCC Petrus, LA Viotto. Degumming of vegetable oil by microporous membrane. *J Food Eng* 70:473–478, 2005.
137. A García, S Álvarez, F Riera, R Álvarez, J Coca. Sunflower oil miscella degumming with polyethersulfone membranes: Effect of process conditions and MWCO on fluxes and rejections. *J Food Eng* 74:516–522, 2006.
138. C Pagliero, M Mattea, N Ochoa, J Marchese. Fouling of polymeric membranes during degumming of crude sunflower and soybean oil. *J Food Eng* 78:194–197, 2007.
139. MP de Souza, JCC Petrus, LAG Gonçalves, LA Viotto. Degumming of corn oil/hexane miscella using a ceramic membrane. *J Food Eng* 86:557–564, 2008.
140. APB Ribeiro, N Bei, LAG Gonçalves, JCC Petrus, LA Viotto. The optimisation of soybean oil degumming on a pilot plant scale using a ceramic membrane. *J Food Eng* 87:514–521, 2008.
141. INH Mohd Amin, AW Mohammad, M Markom, LC Peng. Effects of palm oil-based fatty acids on fouling of ultrafiltration membranes during the clarification of glycerin-rich solution. *J Food Eng* 101:264–272, 2008.
142. A Bottino, G Capannelli, A Comite, F Ferrari, F Marotta, A Mattei, A Turchini. Application of membrane processes for the filtration of extra virgin olive oil. *J Food Eng* 65:303–309, 2004.
143. BE Chove, AS Grandison, MJ Lewis. Emulsifying properties of soy protein isolates obtained by microfiltration. *J Sci Food Agric* 82:267–272, 2002.
144. SS Koseoglu, CJ Vavra. Catalyst removal from hydrogenated oil by using membrane processing. *INFORM* 3:536, 1992.
145. JB Snape, M Nakajima. Processing of agricultural fats and oils using membrane technology. *J Food Eng* 30:1–41, 1996.
146. R Subramanian, M Nakajima, T Kawakatsu. Processing of vegetable oils using polymeric composite membranes. *J Food Eng* 38:41–56, 1998.
147. R Subramanian, M Nakajima, A Yasui, H Nabetani, T Kimura, T Maekawa. Evaluation of surfactant-aided degumming of vegetable oils by membrane technology. *JAOCS* 76:1247–1253, 1999.
148. R Subramanian, KE Nandini, PM Sheila, AG Gopalakrishna, KSMS Raghavarao, M Nakajima, T Kimura, T Maekawa. Membrane processing of used frying oils. *JAOCS* 77:323–328, 2000.
149. R Subramanian, KSMS Raghavarao, H Nabetani, M Nakajima, T Kimura, T Maekawa. Differential permeation of oil constituents in nonporous denser polymeric membranes. *J Membrane Sci* 187:57–69, 2001.
150. D Peshev, LG Peeva, G Peeva, IIR Baptista, AT Boam. Application of organic solvent nanofiltration for concentration of antioxidant extracts of rosemary (*Rosmarinus officiallis* L.). *Chem Eng Res Des* 89:318–327, 2011.
151. L Vandanjon, R Johannsson, M Derouiniot, P Bourseau, P Jaouen. Concentration and purification of blue whiting peptide hydrolysates by membrane processes. *J Food Eng* 83:581–589, 2007.
152. V Castel, O Andrich, FM Netto, LG Santiago, CR Carrara. Comparison between isoelectric precipitation and ultrafiltration processes to obtain *Amaranth mantegazzianus* protein concentrates at pilot plant scale. *J Food Eng* 112:288–295, 2012.

153. J Vanneste, A Sotto, CM Courtin, V Van Craeyveld, K Bernaerts, J Van Impe, J Vandeur, S Taes, B Van der Bruggen. Application of tailor-made membranes in a multi-stage process for the purification of sweeteners from *Stevia rebaudiana*. *J Food Eng* 103:285–293, 2011.
154. G Patil, KSMS Raghavarao. Integrated membrane process for the concentration of anthocyanin. *J Food Eng* 78:1233–1239, 2007.
155. CA Nayak, NK Rastogi. Comparison of osmotic membrane distillation and forward osmosis membrane processes for concentration of anthocyanin. *Desal Water Treat* 16:134–145, 2010.
156. S Cros, B Lignot, P Jaouen, P Bourseau. Technical and economical evaluation of an integrated membrane process capable both to produce an aroma concentrate and to reject clean water from shrimp cooking juices. *J Food Eng* 77:697–707, 2006.
157. EM Garcia-Castello, JR McCutcheon, M Elimelech. Performance evaluation of sucrose concentration using forward osmosis. *J Membrane Sci* 338:61–66, 2009.
158. R Martínez, MT Sanz, S Beltrán. Concentration by pervaporation of representative brown crab volatile compounds from dilute model solutions. *J Food Eng* 105:98–104, 2011.
159. R Martínez, MT Sanz, S Beltrán. Concentration by pervaporation of brown crab volatile compounds from dilute model solutions: Evaluation of PDMS membrane. *J Membrane Sci* 428:371–379, 2013.
160. G Atungulu, S Koide, S Sasaki, W Cao. Ion-exchange membrane mediated electrodialysis of scallop broth: Ion, free amino acid and heavy metal profiles. *J Food Eng* 78:1285–1290, 2007.
161. S Shi, Y-H Lee, S-H Yun, PVX Hung, S-H Moon. Comparisons of fish meat extract desalination by electrodialysis using different configurations of membrane stack. *J Food Eng* 101:417–423, 2010.
162. J Jundee, S Devahastin, N Chiewchan. Development and testing of a pilot-scale electrodialyser for desalination of fish sauce. *Procedia Eng* 32:97–103, 2012.
163. A Doyen, E Husson, L Bazinet. Use of an electrodialytic reactor for the simultaneous β-lactoglobulinenzymatic hydrolysis and fractionation of generated bioactive peptides. *Food Chem* 136:1193–1202, 2013.

13

Electrical Conductivity of Foods

Sudhir K. Sastry and Pitiya Kamonpatana

Contents

13.1 Introduction	528
13.2 Basic Definitions	529
13.3 Liquid Foods	529
13.3.1 Theory of Electrolytic Conductivity	529
13.3.1.1 Strong Electrolytes	530
13.3.1.2 Weak Electrolytes	530
13.3.2 Relations between Electrical Conductivity and Other Transport Properties	531
13.3.3 Effect of Temperature	532
13.3.4 Effect of Electric Field Strength	534
13.3.5 Effect of Ingredients	534
13.3.5.1 Electrolytic Solutes	534
13.3.5.2 Inert Suspended Solids	534
13.3.5.3 Hydrocolloids	535
13.3.5.4 Phase Transitions of Suspended Solids	537
13.3.5.5 Effect of Nonelectrolytic Solutes	540
13.4 Solid Foods	540
13.4.1 Effect of Microstructure	540
13.4.2 Effect of Temperature and Electric Field Strength	542
13.4.2.1 Gels and Noncellular Solids	542
13.4.2.2 Solids with Undisrupted Cellular Structure	543
13.4.2.3 Modeling of Cell Membrane Breakdown	546
13.4.2.4 Reversibility and Repair of Pores	546
13.4.2.5 Extension to Eukaryotic Cells	547
13.4.3 Effect of Frequency	551
13.4.3.1 Relation to Dielectric Constant	553
13.4.4 Ingredient Effects	555

13.5 Solid–Liquid Mixtures	557
13.5.1 Models for Effective Electrical Conductivity	557
13.5.1.1 Maxwell Model	557
13.5.1.2 Meredith and Tobias (1960) Model	558
13.5.1.3 Series Model	558
13.5.1.4 Parallel Model	558
13.5.1.5 Kopelman Model	558
13.5.1.6 Probability Model (Palaniappan and Sastry, 1991c)	558
13.5.1.7 Comparison of Models	559
13.5.1.8 Effects of Solids in Tube Flow	559
13.6 Methods of Measurement of Electrical Conductivity	560
List of Symbols	565
Greek Letters and Other Symbols	566
Subscripts/Superscripts not Explained Elsewhere	566
Anions	566
References	567

13.1 INTRODUCTION

Interest in the electrical conductivity of foods, once primarily restricted to various testing applications, has increased in recent years, in response to the development of ohmic heating and pulsed electric field (PEF) processing technologies. Ohmic heating relies on the flow of alternating (or other waveform) current through a food material to heat it by internal generation. PEF processing applies high-intensity electric field pulses of short duration (~2–10 μs), to cause microbial inactivation via membrane rupture. Ohmic heating is an inevitable consequence of PEF processing, but is minimized by external cooling methods. In recent years, both these technologies have been explored for a variety of other applications; hence a class of processes known as moderate electric field (MEF) processes is emerging.

Equipment design and product safety assurance in both ohmic and PEF technologies depend on the electrical conductivity of the food in question. Indeed, it is safe to say that the development of equipment for these applications cannot be accomplished without basic knowledge of this property. Conversely, for a given equipment design, product formulation considerations must include electrical conductivity if the process is to work. Ideally, equipment designers and product developers should work together in producing an optimum solution.

The principal mode of conduction within electrolytes is via ionic conduction, in contrast to electronic conduction as occurs in common circuit materials. Thus, the presence of some ionic constituents is necessary for success. At higher frequencies (e.g., microwave and radio frequency heating), the contribution of dipole rotation of water molecules becomes significant. The combined effect of the ionic and dipole components is encapsulated within the effective dielectric loss factor of the material. This subject is discussed in greater detail in Chapter 14 "Dielectric Properties of Foods."

13.2 BASIC DEFINITIONS

The electrical conductivity (σ) of a material may be determined from the measurement of the current, voltage, and dimensions of a material. Referring to Figure 13.1, showing a conductor of constant cross-sectional area A, and length L; if a voltage of V is applied across the faces, a current I flows through the material. Then, from Ohm's law, the resistance is

$$R = \frac{V}{I} \tag{13.1}$$

The electrical conductivity may be determined from the resistance by the expression:

$$\sigma = \frac{L}{AR} \tag{13.2}$$

Methods for electrical conductivity determination are presented in a later section.

13.3 LIQUID FOODS

13.3.1 Theory of Electrolytic Conductivity

It is common in the physical chemistry literature to encounter the concept of molar conductivity Λ, which represents the electrical conductivity normalized for a system wherein one mole of an electrolyte is contained in between two parallel plates. This is visualized by considering a solution contained between two parallel plates of equal area, separated by unit distance (1 m), with 1 mole of electrolyte between the plates (see Figure 13.2). Thus, if the solution is of concentration C mol/m³, the volume of solution containing 1 mole would be $1/C$ m³/mol. Since the volume of the system is A m³, the area of the plates for a one-mole enclosure would be:

$$A \text{ (m}^2\text{)} \times 1 \text{ (m)}/1 \text{ (mole)} = \frac{1}{C}(\text{m}^3/\text{mole}) \tag{13.3}$$

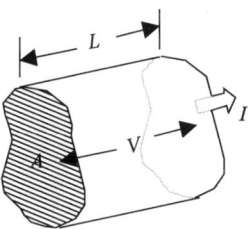

Figure 13.1 Sample conductor of length L, cross-sectional area A, with a voltage V applied across the faces. A current I flows perpendicular to the parallel faces.

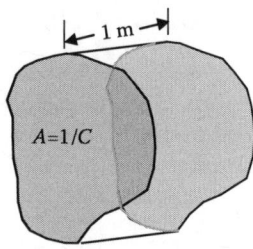

Figure 13.2 Illustration of molar conductivity—plates that are 1 m apart. A solution of concentration C contains 1 mole of solution between the plates if the plates are of area 1/C.

Thus,

$$A = 1/C \tag{13.4}$$

The molar conductivity Λ, is therefore the electrical conductivity of a system with a cross-sectional area of $1/C$, as compared to σ, which is normalized per unit area. Thus,

$$\frac{\Lambda \,(\text{S m}^2/\text{mol})}{1/C \,(\text{mol}/\text{m}^3)} = \sigma \,(\text{S/m}) \tag{13.5}$$

or

$$\Lambda = \frac{\sigma}{C} \tag{13.6}$$

The molar conductivity concept is useful in product formulations, and in determining the effect of individual ingredients on the overall electrical conductivity of a liquid phase. Some examples follow.

13.3.1.1 Strong Electrolytes

For strong electrolytes, the molar conductivity varies with the square root of electrolyte concentration, according to the empirical Kohlrausch relation (Crow, 1988):

$$\Lambda = \Lambda_\infty - k\sqrt{C} \tag{13.7}$$

Thus, applying Equation 13.6 the electrical conductivity at a concentration C is

$$\sigma = \Lambda_\infty C - kC^{1.5} \tag{13.8}$$

The above relation applies for dilute solutions up to 5 mol/m³. Notably, for solutions of low concentration, the electrical conductivity varies approximately linearly with concentration:

$$\sigma \cong \Lambda_\infty C \tag{13.9}$$

13.3.1.2 Weak Electrolytes

For weak electrolytes, the molar conductivity depends on the extent of dissociation. For example, consider a weakly dissociated electrolyte at a low concentration C, exhibiting an equilibrium wherein only a fraction α is dissociated, will have the ionized components of concentration $C\alpha$, while the undissociated part would have the concentration, $C(1 - \alpha)$, as follows:

$$BA \leftrightarrow B^+ + A^-$$
$$C(1-\alpha) \quad\quad C\alpha \quad C\alpha \tag{13.10}$$

This results in the equilibrium constant, given by the Ostwald dilution law:

$$K \approx \frac{[B^+][A^-]}{[BA]} = \frac{\alpha^2 C}{(1-\alpha)} \tag{13.11}$$

It was shown by Arrhenius (Crow, 1988) that

$$\alpha = \frac{\Lambda}{\Lambda_\infty} \tag{13.12}$$

Combining Equations 13.11 and 13.12 yields:

$$K = \frac{C\Lambda^2}{\Lambda_\infty(\Lambda_\infty - \Lambda)} = \frac{C\sigma^2}{\sigma_\infty(\sigma_\infty - \sigma)} \tag{13.13}$$

Thus, the electrical conductivity may be determined from the dissociation constant of the electrolyte in question.

The Kohlrausch law of independent migration of ions describes the molar conductivity of each electrolyte at infinite dilution as being the sum of the contributions of the individual ions, which behave independently of other ions. Thus, for a 1:1 electrolyte:

$$\Lambda_\infty = \lambda_+^\infty + \lambda_-^\infty \tag{13.14}$$

where it is necessary to define 1 mole of substance as that corresponding to a mole of unit charges.

13.3.2 Relations between Electrical Conductivity and Other Transport Properties

From an analysis of mass transport equations, it is possible (Crow, 1988) to establish the following relationship between electrical conductivity and mass diffusivity for a chemical species i.

$$\sigma_i = \frac{c_i D_i z_i^2 F^2}{RT} \tag{13.15}$$

Using the Kohlrausch law, the expression for a single electrolyte at infinite dilution becomes a form of the Nernst–Einstein equation:

$$\sigma_\infty = \frac{F^2}{RT}(v_+ D_+^\infty c_+ z_+^2 + v_- D_-^\infty c_- z_-^2) \tag{13.16}$$

Further, the mass diffusivity, D_i may be related to viscosity via the Stokes–Einstein equation, by considering ions as individual spherical particles:

$$D_i = \frac{RT}{6\pi r_i \eta N} \tag{13.17}$$

Substituting this relationship into Equation 13.15 yields a relation between electrical conductivity and viscosity:

$$\sigma_i = \frac{c_i z_i^2 F^2}{6\pi r_i \eta N} \tag{13.18}$$

It is possible to derive similar expressions for each ingredient of a solution, including colloidal particles (e.g., proteins). The balance between electrical and drag forces for a spherical particle (either ion or colloidal particle), yields the relation for electrophoretic mobility (μ_i) as

$$\mu_i = \frac{v_i}{E} = \frac{z_i F}{6\pi r_i \eta N} \tag{13.19}$$

Each such contribution could be included as the electrical conductivity contribution of each component.

Thus, the electrical conductivity may be related to the viscosity associated with addition of a given ingredient. The effective electrical conductivity of a pulp-free liquid food product could in principle, be modeled as a function of the various components, provided each of their dissociation constants and contribution to solution rheological properties were known. Addition of electrically neutral pulp solids would create shadow zones, which would require further characterization. Although the limitations of these theories in complex systems such as foods must be recognized, they may be useful to the product developer in determining the influence of the addition of various ingredients on the electrical conductivity. Experimental verification is always advisable.

13.3.3 Effect of Temperature

In general, the electrical conductivity of foods exhibits a linear increase with temperature. The only exceptions occur with components (e.g., starches) that may undergo phase transitions or significant structural changes during heating. This subject is dealt with in the section on the influences of individual ingredients.

Data on electrical conductivity–temperature relationships have been presented by a number of researchers, including Palaniappan and Sastry (1991a), Marcotte et al. (1998), and Ruhlman et al. (2001). Some of these data are presented in Figure 13.3 and Table 13.1. In the absence of phase transitions, the electrical conductivity of liquids is a linearly increasing function of temperature. This may be modeled by the relation:

$$\sigma = \sigma_0 [1 + mT] \tag{13.20}$$

The effect of temperature depends greatly on the nature of suspended solids. This is discussed in greater detail in the section on the effect of ingredients.

ELECTRICAL CONDUCTIVITY OF FOODS

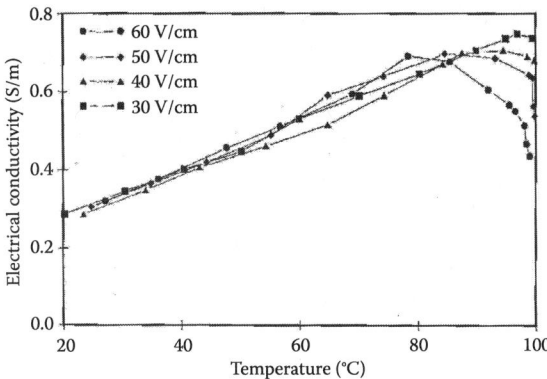

Figure 13.3 Electrical conductivity of orange juice under various electric field strengths. The decreases near 100°C are due to boiling effects. (From Palaniappan, S., and Sastry, S.K. 1991a. *J. Food Proc. Engr.* 14:247–260, reproduced with permission.)

Table 13.1 Average Values of Electrical Conductivity (S/m) of Liquid Foods at Various Temperatures

	Temperature					
	4°C	22°C	30°C	40°C	50°C	60°C
Product	Electrical Conductivity (S/m)					
Beer	0.08	0.143	0.16	0.188	0.227	0.257
Light beer	0.083	0.122	0.143	0.167	0.193	0.218
Black coffee	0.138	0.182	0.207	0.237	0.275	0.312
Coffee with milk	0.265	0.357	0.402	0.470	0.550	0.633
Coffee with sugar	0.133	0.185	0.210	0.250	0.287	0.323
Apple juice	0.196	0.239	0.279	0.333	0.383	0.439
Cranberry juice	0.063	0.090	0.105	0.123	0.148	0.171
Grape juice	0.056	0.083	0.092	0.104	0.122	0.144
Lemonade	0.084	0.123	0.143	0.172	0.199	0.227
Limeade	0.090	0.117	0.137	0.163	0.188	0.217
Orange juice	0.314	0.360	0.429	0.500	0.600	0.690
Carrot juice	0.788	1.147	1.282	1.484	1.741	1.980
Tomato juice	1.19	1.697	1.974	2.371	2.754	3.140
Veg. juice cocktail	1.087	1.556	1.812	2.141	2.520	2.828
Chocolate 3% fat milk	0.332	0.433	0.483	0.567	0.700	0.800
Chocolate 2% fat milk	0.420	0.508	0.617	0.700	0.833	1.000
Chocolate skim milk	0.532	0.558	0.663	0.746	0.948	1.089
Lactose-free milk	0.380	0.497	0.583	0.717	0.817	0.883
Skim milk	0.328	0.511	0.599	0.713	0.832	0.973
Whole milk	0.357	0.527	0.617	0.683	0.800	0.883

Source: Data from Ruhlman, K.T., Jin, Z.T., and Zhang, Q.H. 2001. Physical properties of liquid foods for pulsed electric field treatment. Chapter 3 in *Pulsed Electric Fields in Food Processing*. pp 45–56. Eds: G.V. Barbosa-Cánovas and Q.H. Zhang. Technomic Publishing Co., Lancaster, PA.

13.3.4 Effect of Electric Field Strength

Variations in electric field strength in the range from 0 to 100 V/cm have negligible effects on the electrical conductivity–temperature relationship of juices, as shown by Palaniappan and Sastry (1991a), and illustrated in Figure 13.3. This is to be expected when the solids are inert and unaffected by the electric field. Castro et al. (2003) observed no obvious field strength effects for 14.5° Brix strawberry pulp. However, some effects were observed for solid products, as will be discussed in a later section.

13.3.5 Effect of Ingredients

13.3.5.1 Electrolytic Solutes
Effects of electrolytes are as discussed in Section 13.3.1 on theory of electrolytic conductivity. The most notable electrolytes within foods are salts and acids; some gums and thickeners may also possess charged groups that would migrate toward electrodes and contribute to electrical conductivity.

13.3.5.2 Inert Suspended Solids
Suspended solids such as pulp and cellular solids are typically insulators, and will tend to reduce the electrical conductivity of the liquid media in which they are suspended. For tomato and orange juices, Palaniappan and Sastry (1991a) modeled solids content effects, using a 25°C reference temperature for the electrical conductivity of juice serum, as

$$\sigma_T = \sigma_{j25}[1 + K_1(T - 25)] - K_2 S \tag{13.21}$$

Values of the parameters are presented in Table 13.2.

The particle size distribution of suspended solids may have significant effects on the effective electrical conductivity of liquid containing suspended solids. A simplistic analysis of this effect may be made by considering the solids to be spheres of equal size, as illustrated in Figure 13.4.

In comparing two particle populations at equal volume fraction, the total volume of solids is given by

$$V_s = n\frac{4}{3}\pi r^3 \tag{13.22}$$

Table 13.2 Parameters for the Electrical Conductivity Model for Tomato and Orange Juices Given by Equation 13.21

Juice	σ_{j25} (S/m)	K_1 (°C^{-1})	K_2 (S/m% solids)
Tomato	0.863	0.174	0.101
Orange	0.567	0.242	0.036

Source: Data from Palaniappan, S., and Sastry, S.K. 1991a. *J. Food Proc. Engr.* 14:247–260.

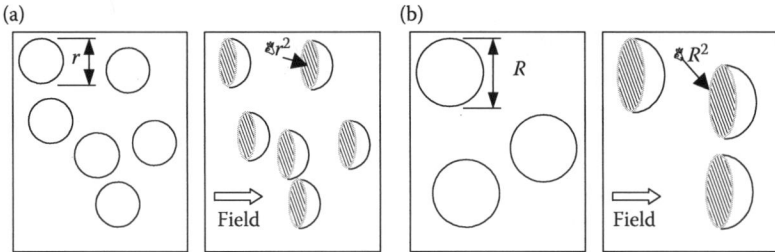

Figure 13.4 Illustration of particle size effects: (a) particles of radius r will result in a cross-section of Br^2 perpendicular to the field, and (b) particles of larger radius will similarly expose larger "faces" to the field.

The cross-sectional area exposed to an electric field is given (as in Figure 13.4) by

$$A_s = n\pi r^2 \tag{13.23}$$

Thus, for a constant volume fraction, the area is given by substituting Equation 13.23 into Equation 13.22, yielding

$$A_s = \frac{3V_s}{4r} \tag{13.24}$$

Thus, the total cross-sectional area of insulators that "block" the current increases with decreasing particle size. Based on such an analysis, fine-particle suspensions would be expected to exhibit lower electrical conductivity than coarser suspensions. This is borne out by the results of Palaniappan and Sastry (1991a), for polystyrene spheres within sodium phosphate solution. However, the opposite trend is noted for carrot solids within sodium phosphate solution (Palaniappan and Sastry, 1991a) as shown in Figure 13.5. This indicates that a simple model is unlikely to encapsulate the complexity of biological solids, where effects of particle shape and leaching of intracellular constituents may complicate the picture considerably.

Data on coarse solids (solid–liquid mixtures) will be treated in a later section.

13.3.5.3 Hydrocolloids

The influence of various hydrocolloids has been presented by Marcotte et al. (1998) who also studied the effects of concentration on electrical conductivities of hydrocolloids (starch, carrageenan, pectin, gelatin, and xanthan). They found that, as expected, a neutral polysaccharide such as starch showed the lowest electrical conductivity of the group. The more highly charged hydrocolloids such as carrageenan and xanthan exhibited the highest electrical conductivity, while pectin, which is less charged than these hydrocolloids, but more charged than starch, exhibited intermediate values of electrical conductivity. They fitted their data to the equation

$$\sigma = \sigma_{C,25} + K_T(T - 25) + K_{TC}(T - 25)C \tag{13.25}$$

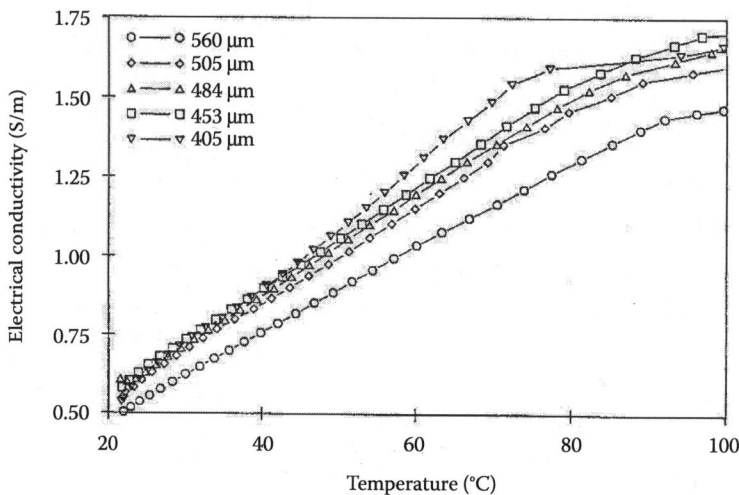

Figure 13.5 Electrical conductivity of sodium phosphate solution with suspended carrot solids of various mean particle sizes. (From Palaniappan, S., and Sastry, S.K. 1991a. *J. Food Proc. Engr.* 14:247–260, reproduced with permission.)

where

$$\sigma_{C,25} = K_{C,25} C \qquad (13.26)$$

Data on their parameters, $K_{C,25}$, K_T, and K_C are presented in Table 13.3.

In a later study, Marcotte et al. (2000) also investigated the effect of addition of salt and citric acid to hydrocolloid solutions which were adjusted to similar viscosities. Data were fitted to the model:

$$\sigma = \sigma_{25} + K_{\sigma T}(T - 25) \qquad (13.27)$$

Values of these parameters are presented in Table 13.4.

Table 13.3 Parameters for the Electrical Conductivity Model of Equation 13.25

Hydrocolloid Type	$K_{C,25}$ (S/m%)	K_T (S/m°C)	K_{TC} (S/m°C%)
Carrageenan	0.17	8.65×10^{-4}	3.90×10^{-3}
Xanthan	0.133	1.78×10^{-3}	2.46×10^{-3}
Gelatin	0.0299	5.29×10^{-5}	7.67×10^{-4}
Pectin	0.0305	5.76×10^{-4}	4.81×10^{-4}
Starch	0.013	1.95×10^{-4}	2.71×10^{-4}

Source: Data from Marcotte, M., Piette, J.P.G., and Ramaswamy, H.S. 1998. *J. Food Proc. Engr.* 21:503–520.

Table 13.4 Influence of Salt Concentration on the Electrical Conductivity Model Parameters (Equation 13.27) of Various Hydrocolloids

Type of Hydrocolloid	Salt Concentration (%)	σ_{25}	$K_{\sigma T}$
Carrageenan (1.7%)	0.25	0.848	0.0199
	0.50	1.371	0.0313
	0.75	1.914	0.0413
	1.0	2.173	0.0481
Xanthan (2%)	0.25	0.889	0.0181
	0.50	1.474	0.0305
	0.75	1.969	0.0396
	1.0	2.162	0.0419
Pectin (2.5%)	0.25	0.691	0.0153
	0.50	1.201	0.0261
	0.75	1.690	0.0349
	1.0	2.195	0.0455
Starch (4.3%)	0.25	0.582	0.0123
	0.50	1.066	0.0204
	0.75	1.544	0.0312
	1.0	2.109	0.0427

Source: Data from Marcotte, M., Trigui, M., and Ramaswamy, H.S. 2000. *J. Food Proc. Pres.* 24:389–406.

It should be noted that the above study appears to deal principally with ungelatinized starch. Gelatinization effects are treated in the following section.

13.3.5.4 Phase Transitions of Suspended Solids

Reports on the monitoring of starch gelatinization temperature by electrical conductivity measurement were made by Korobkov et al. (1978). Halden et al. (1990) noted a slight variation in the heating slope of a potato slice, and attributed it to starch gelatinization. Wang and Sastry (1997a), in studying the electrical conductivity of starch suspensions during ohmic heating, noted a negative peak in the electrical conductivity–temperature curve, which corresponded to the phase transition temperature in the differential scanning calorimeter (DSC) thermogram for the same material (Figure 13.6). This suggests that starch gelatinization may be detected using electrical conductivity measurements. Similar results have since been observed by Karapantsios et al. (2000).

The explanation is consistent with observed data on starch gelatinization, which occurs under the influence of water and heat, and involves the swelling of insoluble starch granules via incorporation of water, to many times their original size. This results in a great increase in volume fraction of the granular component during heating, and also serves to "block" the current path more effectively in doing so. Thus, there is a decrease in electrical conductivity during this phase. This process is followed by a collapse of

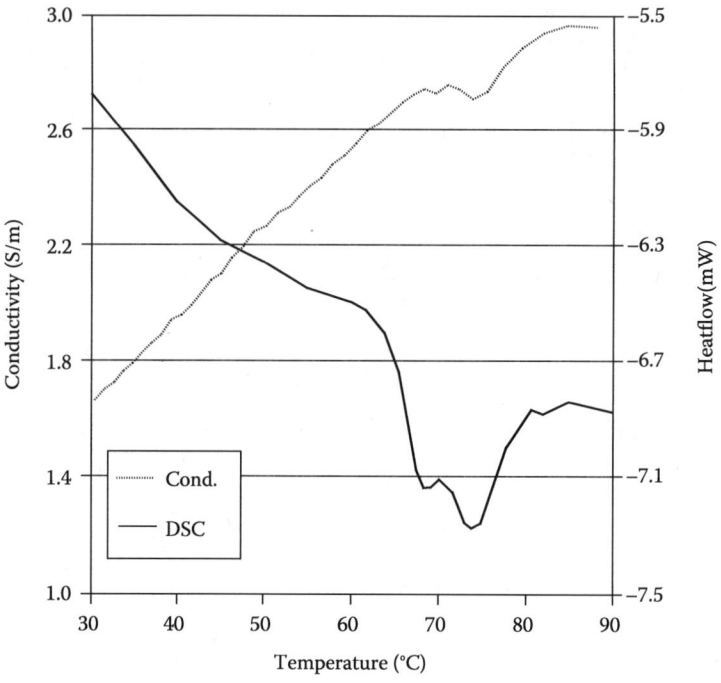

Figure 13.6 Electrical conductivity–temperature curve of potato starch in comparison with the DSC scan of the same material. (Adapted from Wang, W-C, and Sastry, S.K. 1997a. *J. Food Engr.* 34(3):225–242, with permission.)

the granules, whereupon the volume fraction of nonpolar solids once again decreases, causing the electrical conductivity to rise once more. This phenomenon is illustrated in Figure 13.7.

The above observation suggests that electrical methods could be developed as a supplement, or in some cases, even an alternative to DSC methods for monitoring starch gelatinization.

A slightly different approach to detection of starch gelatinization has been taken by Chaiwanichsiri et al. (2001), who determined the electrical conductivity of conventionally heated potato starch suspensions (without addition of salt) at 200 kHz. Their results, displayed in Figure 13.8, indicate that the electrical conductivity increases at the DSC onset temperature of starch gelatinization. They attributed this trend to ion release from starch granules during the gelatinization process. It was also explained that the difference between their study and that of Wang and Sastry (1997a) and Karapantsios et al. (2000) was due to their not having added any salt to their mixture, enabling them to detect the small difference in ionic concentration due to ion release. They also noted a relation between the final temperature of ion release and the temperature of rise in viscosity.

It appears from the various studies above that the monitoring of starch gelatinization via electrical conductivity measurement appears to have considerable potential.

ELECTRICAL CONDUCTIVITY OF FOODS

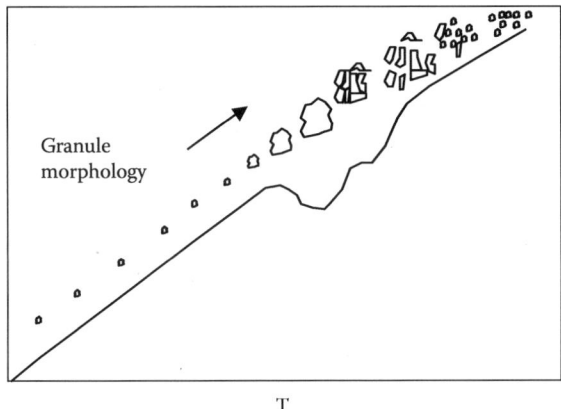

Figure 13.7 Illustration of change in starch granule morphology as electrical conductivity changes occur. The swollen granules more effectively insulate against current flow. After their collapse, the electrical conductivity continues its increase with temperature.

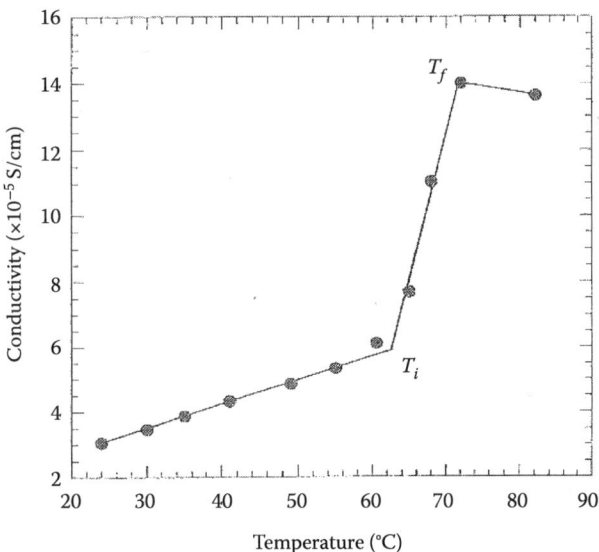

Figure 13.8 Electrical conductivity of potato starch suspension measured at 200 kHz during conventional heating. T_i and T_f refer to the onset and end of the zone of steep rise of electrical conductivity. (From Chaiwanichsiri, S. et al. 2001. *J. Sci. Food Agric.* 81:1586–1591, reproduced with permission.)

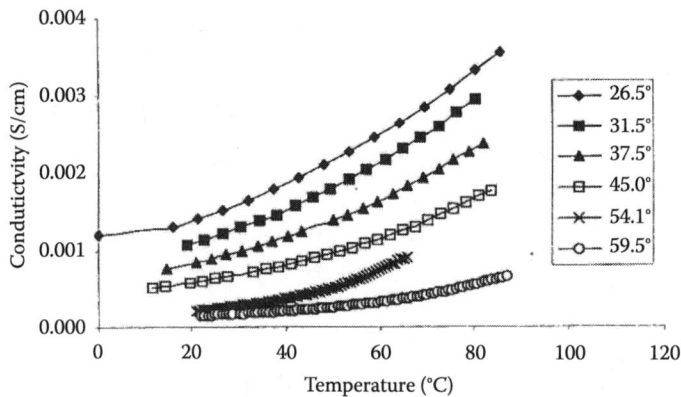

Figure 13.9 Influence of sugar content (°Brix) on the electrical conductivity of strawberry pulp. (From Castro, I., et al. 2003. *J. Food Proc. Engr.* 26(1):17–29, reproduced with permission.)

13.3.5.5 Effect of Nonelectrolytic Solutes

Some materials (e.g., sucrose) do not form electrically conducting ions in solution. Such substances result in decreased electrical conductivity of the solution. This effect has been clearly illustrated by Castro et al. (2003), who determined electrical conductivities of strawberry pulps that were adjusted to various degrees Brix levels by addition of sucrose. Their results, illustrated in Figure 13.9, show that increasing sugar content suppresses the electrical conductivity of the pulp.

13.4 SOLID FOODS

13.4.1 Effect of Microstructure

The electrical conductivity behavior of solid foods depends to a large extent on whether or not a cellular structure exists within the material. The behavior of gels and gel-like materials, or for foods in which a cell structure has been disrupted, are significantly different from materials with intact cells. It is necessary to treat these two categories separately.

The effect of tissue microstructure has been characterized by Wang et al. (2001), who determined average electrical conductivities (between 25°C and 95°C) of bamboo, sugarcane, lettuce stems, and mustard stems both along and across the stem. They found that conductivity along the stem was higher than that across the stem for bamboo shoots and sugarcane. However, the reverse was observed for lettuce and mustard stems. A microstructural examination revealed that two influences were important—the orientation of vascular bundles, and the shape of parenchyma cells. When both types of tissue were present, the vascular bundles dominated the trend in electrical conductivity, since these are the primary modes of water and nutrient transport within plants. However, in the absence of vascular tissue, the shape of the parenchyma cells were the dominant factor, explaining the different results between the different types of tissue. Fiber orientation of meat also affects electrical conductivity. Fibers oriented parallel show higher electrical conductivity than perpendicular to electric field (Saif et al., 2004; Zell et al., 2009). The reason might be that current flows along the muscle fibers easier than in the cross-fiber orientation. Shirsat et al. (2004) revealed that the electrical conductivity of leg was greater than that of shoulder lean. A possible explanation could be high density of muscle fiber structure and/or intramuscular fat in shoulder hindering the flow of current. The comparison of electrical conductivity between intact and minced meat was also investigated. Minced meat shows the greatest conductivity compared with intact meat (Shirsat et al., 2004; Zell et al., 2009). It is hypothesized that mincing meat destroys the myofibrillar cell tissue resulting in the release of water and inorganic content (Shirsat et al., 2004).

Meat composition also influences electrical conductivity. Fat is much less conductive than lean meat (Sarang et al., 2007; Shirsat et al., 2004). Adding fat to lean and then mincing accounted for the reduction of overall electrical conductivity (Shirsat et al., 2004). Fat by itself is nonconductive. When incorporated in lean meat, it might coat lean meat particles and block current flow, resulting in the decrease in the conductivity. Bozkurt and Icier (2010) observed that conductivity of minced beef–fat blends (2%, 9%, and 15% fat) increased with temperature until it arrived the critical cooking temperature and then considerably lowered (Figure 13.10). The decrease in the conductivity after the critical temperature might contribute to protein denaturation synchronized with or followed by an aggregation of protein, accounting for reduction of protein solubility.

Since biological cells consist of membranes that are largely lipid bilayers, they tend to act as capacitors enclosing electrolyte solutions. The length and orientation of such tissue can play a large role in determining electrical conductivity as well as dielectric properties. Solid materials that do not contain a cell structure do not possess similar capacitive properties, and their electrical conductivity behavior is simpler than for cellular tissue.

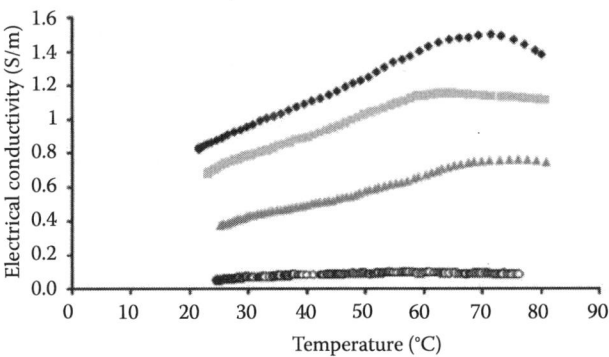

Figure 13.10 Electrical conductivity changes of minced beef-fat blends during ohmic cooking at 30 V/cm. (From Bozkurt, H., and Icier, F. 2010. *J. Food Engr.* 96:86–92, reproduced with permission.)

13.4.2 Effect of Temperature and Electric Field Strength

13.4.2.1 Gels and Noncellular Solids

The electrical conductivity of noncellular solids tends to increase with temperature. The trend is generally a linear one, as observed by Yongsawatdigul et al. (1995) for surimi pastes (Figure 13.11). Castro et al. (2003) have reported a slight nonlinearly increasing trend for strawberry jelly (Figure 13.12). Various reasons may be advanced for such effects, including the breakdown of the gel, resulting in lower drag on ions, and enhanced conductivity at higher temperatures. Electroosmotic effects are unlikely in such cases, since, as noted by Yongsawatdigul et al. (1995) products such as surimi have their cellular structures severely disrupted, thus no membranes or capillaries exist for such osmotic effects to take place. Yongsawatdigul et al. (1995) also noted that deviations from Ohm's law as well as significant electrode corrosion occurred (at 60 Hz frequency) when NaCl contents were

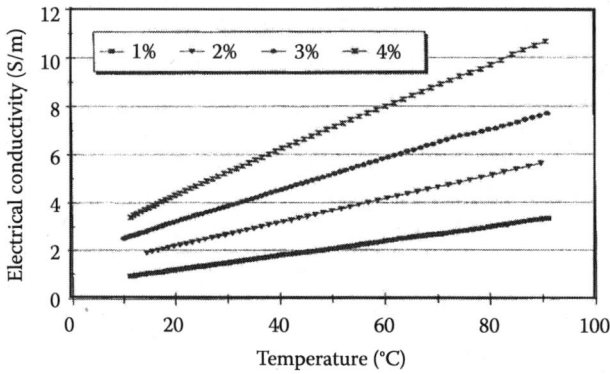

Figure 13.11 Electrical conductivity of surimi paste at various salt contents. (From Yongsawatdigul, J., Park, J.W., and Kolbe, E. 1995. *J. Food Sci.* 60(5):922–925,935, reproduced with permission.)

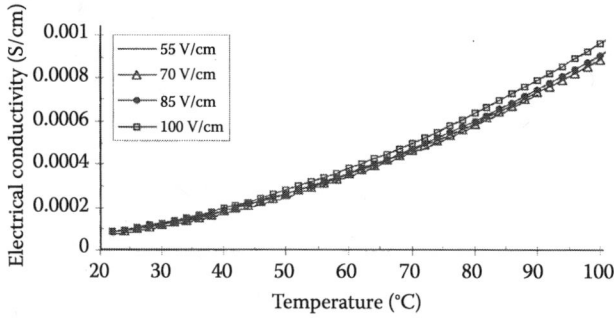

Figure 13.12 Electrical conductivity of strawberry jelly at various field strengths. (From Castro, I., et al. 2003. *J. Food Proc. Engr.* 26(1):17–29, reproduced with permission.)

3% or 4%. They suggested that it would be necessary to account for electrode polarization effects under these conditions.

Castro et al. (2003) have noted a slight influence of electric field strength on the electrical conductivity of strawberry jelly (Figure 13.12). Yongsawatdigul et al. (1995) suggested that any field strength effects observed in their study were not significant, and could be attributed to the influences of electrochemical reactions at electrode surfaces.

13.4.2.2 Solids with Undisrupted Cellular Structure

For solids with a cellular structure, such as fruits, vegetables, and intact muscle foods, the electrical conductivity depends on temperature as well as electric field strength. As illustrated by Palaniappan and Sastry (1991b), the electrical conductivity of a conventionally heated product undergoes little change with temperature until about 70°C, at which temperature, the cellular structure breaks down, and the electrical conductivity undergoes a significant increase (Figure 13.13). The electrical conductivity variation with temperature of fruits, vegetables, and meat is as shown in Table 13.5. As the electric field strength is increased, the change in electrical conductivity becomes more gradual, until at sufficiently high field strengths, the familiar linear electrical conductivity–temperature relation is seen (Figure 13.14). This suggests that under the influence of electricity, the cell structure is broken down at lower temperatures than for conventional heating. This phenomenon has been termed electroporation or electroplasmolysis.

The nature of the electric field effect is the subject of much investigation. Palaniappan and Sastry (1991b) attributed it to electroosmotic effects. However, more recently, it has been realized that there appears to be a cell-permeabilizing mechanism that results in localized electrical membrane breakdown and leakage of intracellular fluids to extracellular regions (Imai et al., 1995; Sastry and Barach, 2000; Kulshrestha and Sastry, 2003).

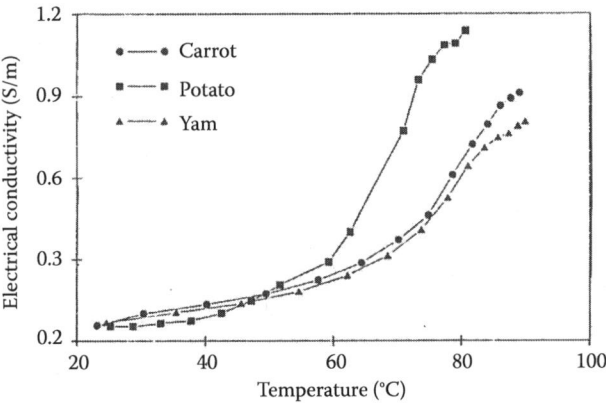

Figure 13.13 Electrical conductivity of vegetable tissue during conventional heating. (From Palaniappan, S., and Sastry, S.K. 1991b. *J. Food Proc. Engr.* 14:221–236, reproduced with permission.)

Table 13.5 Electrical Conductivity of Fruits, Meat, and Vegetables at Various Temperatures

Product	Temperature							Ref.
	25°C	40°C	60°C	80°C	100°C	120°C	140°C	
	Electrical Conductivity (S/m)							
Fruits								
Apple—golden	0.067	0.144	0.251	0.352	0.425	0.504	0.571	a
Apple—red	0.075	0.138	0.239	0.339	0.419	0.499	0.577	a
Pear	0.084	0.173	0.313	0.439	0.541	0.607	0.642	a
Pineapple	0.037	0.141	0.245	0.348	0.432	0.506	0.575	a
Strawberry	0.186	0.335	0.592	0.801	0.982	1.143	1.276	a
Meats								
Chicken—breast	0.665	0.873	1.142	1.386	1.678	1.948	2.212	a
Chicken—tender	0.549	0.766	0.979	1.207	1.436	1.696	1.960	a
Chicken—thigh	0.348	0.472	0.607	0.772	0.962	1.137	1.322	a
Chicken—drumstick	0.444	0.598	0.763	0.974	1.182	1.399	1.601	a
Chicken—separable fat	0.035	0.057	0.090	0.128	0.158	0.184	—	a
Pork—top loin	0.560	0.735	0.930	1.092	1.305	1.546	1.751	a
Pork—shoulder	0.532	0.696	0.886	1.085	1.316	1.544	1.717	a
Pork—tenderloin	0.584	0.750	0.957	1.155	1.407	1.695	1.961	a
Beef—bottom round	0.489	0.669	0.826	1.037	1.242	1.443	1.608	a
Beef—chuck shoulder	0.487	0.626	0.801	1.019	1.253	1.481	1.665	a
Beef—flank loin	0.371	0.502	0.710	0.960	1.240	1.464	1.696	a
Beef—top round	0.491	0.645	0.841	1.071	1.346	1.551	1.721	a
Vegetables								
Bean sprout	0.200	0.284	0.408	0.541	0.661	0.783	0.911	b
Mushroom	0.286	0.371	0.493	0.631	0.787	0.985	1.194	b
Celery	0.298	0.391	0.592	0.962	1.363	1.718	2.048	c
Canned water chestnut	0.286	0.355	0.447	0.539	0.631	0.723	0.815	d

[a] Sarang et al. (2007).
[b] Tulsiyan et al. (2008).
[c] Kamonpatana (2012).
[d] Kamonpatana et al. (2013).

The ability of electric fields to rupture cellular materials has been used to advantage in PEF technology, both from the standpoint of inactivation of bacteria (Sale and Hamilton, 1967a,b; Barbosa-Cánovas et al., 2000) and for rupturing cellular materials as a pretreatment for mass transfer processes (Bazhal et al., 2001; Taiwo et al., 2002). The fundamental principle is that biological cells contain electrolytic fluids, which are enclosed by membranes which are largely composed of lipid bilayers, which act as capacitors. As an electric field is applied (Figure 13.15), charges build up around the cell membranes. Opposing charges on opposite sides of the cell membrane attract one another, resulting in a compressive force. Further, the like charges on the same side of the membrane repel one another,

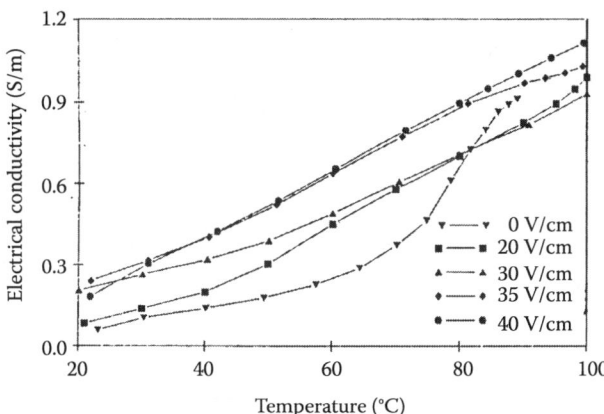

Figure 13.14 Electrical conductivity of carrot tissue at various electric field strengths. (From Palaniappan, S., and Sastry, S.K. 1991b. *J. Food Proc. Engr.* 14:221–236, reproduced with permission.)

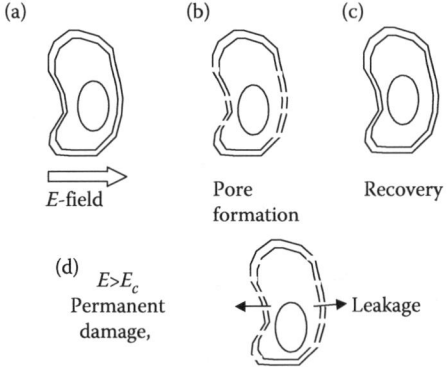

Figure 13.15 Mechanism of pore formation in a cell. (a) Electric field applied to intact cell; (b) reversible pore formation for low electric field strengths; (c) cell recovers if field strength is low; (d) for $E > E_c$ (critical membrane potential), the damage is permanent.

causing a tangential force on the membranes. Above a certain critical membrane potential (0.5 V per membrane in the path of the electric field, or approximately 1 V across a bacterial cell of approximately 1:m, corresponding typically to about 10 kV/m external field), pores form in the membrane. At relatively low field strengths, pore formation is reversible, and some recovery may occur. At higher field strengths, (around 25–30 kV/cm), pore formation is irreversible.

A detailed treatment of electroporation is beyond the scope of this chapter, and is covered in reviews such as that of Weaver and Chizmadzhev (1996). It is notable that direct evidence (e.g., imaging of membranes during pore formation) is remarkably elusive because of the extremely short time scale of the phenomena involved, and the inability to image the right location at the right time.

13.4.2.3 Modeling of Cell Membrane Breakdown

Considerable effort has been devoted to modeling the electroporation process. The cell membrane is typically regarded as a capacitor with a dielectric material of low dielectric constant compared to water; such that high charge concentrations cannot exist within membranes, although they could be accumulated at both surfaces (Zimmermann, 1986). In one of the earlier investigations, Crowley (1973) modeled the irreversible breakdown of a lipid bilayer membrane by considering it to be an isotropic elastic layer between two electrically conducting liquids. The calculation of the resulting electrical and elastic stresses showed the existence of a critical membrane potential at which rupture occurred. Subsequently, a number of models have been developed, including those of Zimmermann et al. (1974, 1977), which included mechanical external forces in the analysis. Further literature has included the consideration of surface tension effects, as well as the viscoelastic character of the cell membrane (Jain and Maldarelli, 1983; Dimitrov, 1984). The breakdown mechanism has been shown to consist of three parts.

1. The occurrence and growth of membrane shape fluctuations. This process has been shown to last on the order of microseconds, and can be described by the thin film model of Dimitrov (1984).
2. The increasing amplitude of the shape fluctuations results in decreasing membrane thickness at local points. This causes a rapid, nonlinear increase in the driving force due to the electric field. Then, molecular rearrangements, leading to the discontinuity of the membrane, can occur. This process is extremely short (on the order of nanoseconds).
3. The further growth of pores, result in irreversible mechanical breakdown of the membrane. This is a slow process, lasting on the order of milliseconds and more. This process has been studied by Chernomordik and Abidor (1980), among others. Sugar and Neumann (1984) have also modeled this phenomenon as a stochastic process.

13.4.2.4 Reversibility and Repair of Pores

Considerable effort has also been devoted to whether pores are actually formed in the cell membrane. The evidence for pore formation in lipid membranes has been presented by Benz et al. (1979), who noted a reversible change in lipid bilayers, with resealing times of about 2–20 s. However, cell membranes have been found to reseal over much longer times (up to 10 min). It is believed (Zimmermann, 1986) that conformational changes of proteins are responsible for the long-lived permeabilization in biological membranes.

13.4.2.4.1 Key Effects

Of particular interest from the food-processing standpoint, are the potential synergistic relationships between physical and physicochemical parameters, which may serve to markedly improve processes. In this context, we note the following effects that have been characterized in the literature and may help us to further optimize MEF processes.

13.4.2.4.2 Pressure Effects

Zimmermann (1986) and coworkers have shown that the critical membrane breakdown potential V_c depends on pressure, P, due to mechanical precompression of the membrane according to the relation:

$$V_c = V_{c(P=0)} \exp(-P/\gamma) \tag{13.28}$$

where γ is the effective elastic modulus. This equation indicates that the breakdown voltage decreases as pressure increases. Indeed, Zimmermann notes that at high pressures, the mechanical precompression is so great that membrane breakdown occurs at the natural resting potential. This may explain the permeabilization observed in high-pressure processing, as noted by Knorr and coworkers (Knorr, 1994), and also suggests a significant synergy between electric field and pressure processes.

13.4.2.4.3 Pulse Duration Effects

The work of Dimitrov (1984) shows that at longer pulse lengths, lower voltages are required to achieve breakdown. Deng et al. (2003) presented the result showing that JurKat cells could immediately penetrate propidium iodide (PI) when the longest pulses were applied. This is thought (Zimmermann, 1986) to be due to the increase of the compressive yield strength of the membrane with increasing electric field strength, because of the viscoelastic properties of the membrane. This would help explain our own observations of improved permeabilization with low frequencies (Lima and Sastry, 1999).

13.4.2.4.4 Frequency Effects

The effect of frequency on bacterial cell membranes was investigated via fluorescent nucleic acid stains (Loghavi and Sastry, 2009). The green, nonselective, cell membrane permeable SYTO 9 can be taken up by all cells which represents total cell population; while the penetration of the red, cell membrane impermeable PI indicates the permeabilization of cells. Loghavi and Sastry (2009) revealed that, in the presence of MEF, the red cell of *Lactobacillus acidophilus* showed the largest number at 45 Hz followed by treatment at 60 Hz. No PI uptake was observed at 1000 and 10,000 Hz. This finding is consistent with that of Deng et al. (2003).

13.4.2.4.5 Effect of Surfactants

The model of Dimitrov (1984) also predicts that breakdown voltage decreases with surface tension, even if the compressive modulus is very high. Experiments by Zimmermann's group have confirmed that this is indeed the case.

13.4.2.4.6 Effect of Proteolytic Enzymes

Ohno-Shosaku and Okada (1984) have shown that the presence of proteolytic enzymes can significantly decrease the breakdown potential of animal cell membranes. This is entirely predictable, since cell-wall breakdown would be expected to decrease its strength.

13.4.2.5 Extension to Eukaryotic Cells

While the original set of principles of electropermeabilization were established for individual bacterial cells, it has been realized more recently, that such pore formation can occur within eukaryotic cells as well. Since plant cells are much larger than bacteria, the electric field strength required for rupturing them is correspondingly lower. This has resulted in the use of MEFs (either in pulsed or alternating mode) for permeabilizing plant tissue (Bazhal et al., 2001; Imai et al., 1995; Lebovka et al., 2000, 2001; Taiwo et al., 2002).

The above phenomenon helps explain the trend in electrical conductivity observed by Palaniappan and Sastry (1991b) where the permeabilization effect is seen to occur even at relatively low temperatures (Figure 13.14). These data also suggest that at low electric field strengths, the extent of permeabilization is slight, translating to an electrical conductivity that is higher than that under conventional heating, but lower than that under high electric field strengths.

Recent data suggest that even electric fields of the order of a few volts/cm, applied for a few seconds are sufficient to cause permeation of vegetable cells. Kulshrestha (2002) studying beet tissue using light microscopy, noted that betanin pigments migrated out of beet cells with a short treatment. Kulshrestha and Sastry (2010) also revealed the permeabilization arising in the raw potato tissues treated by MEF at low frequency from 3°C to 25°C. The result showed that electrical conductivity of raw-MEF-treated potato was initially comparable to that of raw untreated tissues, and gradually increased to approach that of thawed and precooked potato over a 24 h measurement. In addition, the apparent dielectric constant of raw-MEF treated sample measured at both 100 Hz and 20 kHz over 24 h tended to approach the values for iso-conductive solution. Visualization of fresh mint leaves treated by MEF was conducted by Sensoy and Sastry (2004). Transmission electron microscopic images illustrate clean vacuoles, which may be due to molecular transport through tonoplast membranes in samples treated by MEF; while intracellular bodies were found dispersed in the vacuoles of samples that were conventionally heated (Figure 13.16). The results also reveal chloroplast compression shown in Figure 13.17. From the above findings, it is hypothesized that there are two critical electric field strengths. Low electric field strength causes plasma membrane breakdown while high electric field strength contributes to tonoplast membrane breakdowns (Angersbach et al., 1999). Asavasanti et al.

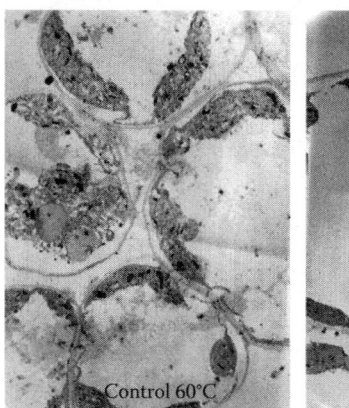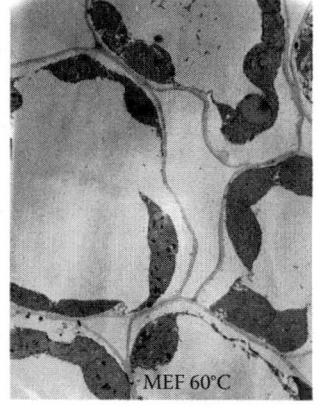

Figure 13.16 Transmission electron microscopic images of control- and MEF-treated fresh mint leaves, heated to 60°C. (From Sensoy, I., and Sastry, S. K. 2004. *J. Food Sci.* 69(1): 7–13, reproduced with permission.)

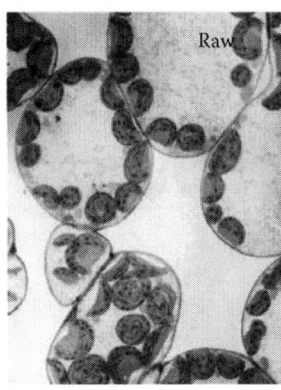 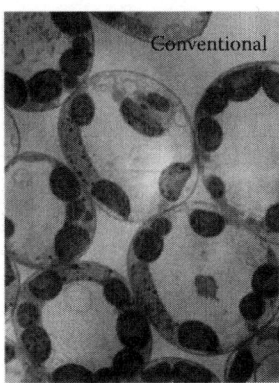 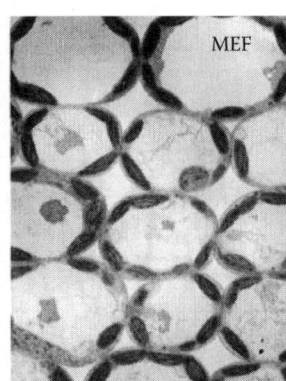

Figure 13.17 Transmission electron microscopic images of fresh mint leaves (magnification ×1800 zoom). C = chloroplast; N = nucleus. (From Sensoy, I., and Sastry, S. K. 2004. *J. Food Sci.* 69(1): 7–13, reproduced with permission.)

(2010) found the presence of plasma membrane rupture due to the application of an electric field of 67 V/cm for 10 pulses (100 μs/pulse). However, the tonoplast membrane breakdown occurs when the electric field of 200 V/cm for 10 pulses or 133 V/m for 100 pulses.

The implications of electropermeabilization is that, for previously unprocessed cellular tissue, the electrical conductivity is affected by the application of an electric field, or by heating to temperatures greater than 70°C for sufficient time. However, once the tissue has been permeabilized to its fullest extent (as discussed further in the section on frequency effects), no further permeabilization can occur, and the electrical conductivity does not change significantly thereafter. Kusnadi and Sastry (2012) investigated the effect of electrical and thermal treatment on celery and water chestnut under a light microscope and concluded that (1) the combination of electrical and thermal treatment impacts on cell permeability were greater than either treatment alone; (2) the combination of electric field (up to 1842 V/m) and low temperature (25°C) affects the cell structure roughly equal to heating at 80°C. Their study also showed that the effective diffusivity (D_{seff}) of water chestnut at 50°C treated from 1316 to 1842 V/m was much lower than that at 80°C treated from 1316 to 1842 V/m. It is likely that elevated temperature enhances the electroporation due to possible protein denaturation of the cell.

However, the amount of treatment for full permeabilization tends to vary with commodity. This has been observed by Wang and Sastry (1997b), who investigated the effects of multiple thermal treatments on the electrical conductivity of vegetable tissue. Their results are illustrated in Figures 13.18 and 13.19. Figure 13.18 shows the influence of thermal cycling on electrical conductivity of potato, where the electrical conductivity–temperature relationship tends to stabilize after two heating cycles. However, with products such as carrots, electrical conductivity continues to change beyond two cycles, as shown in Figure 13.19. Another study examining the effect of commodity was done by Kusnadi and Sastry (2012). The author studied the effect of MEF on effective diffusion coefficient (D_{seff}) of vegetables and found that D_{seff} of water chestnut was lower than that of celery and mushroom. This behavior might be attributable to cell structure of water chestnut, which is tightly packed and may hinder the movement of ions.

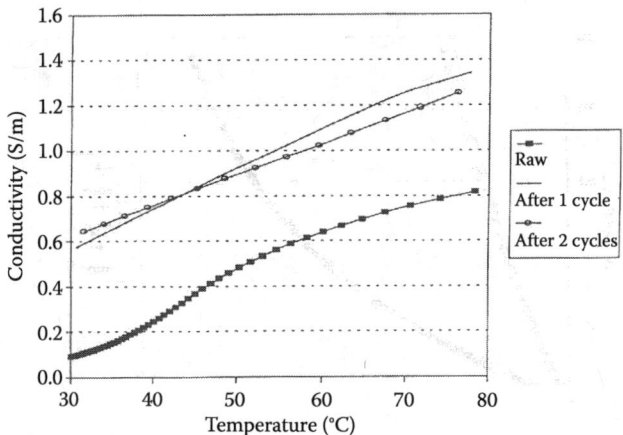

Figure 13.18 Electrical conductivity of potato tissue as affected by cyclic ohmic heating. (From Wang, W-C., and Sastry, S.K. 1997b. *J. Food Proc. Engr.* 20:499–516, reproduced with permission.)

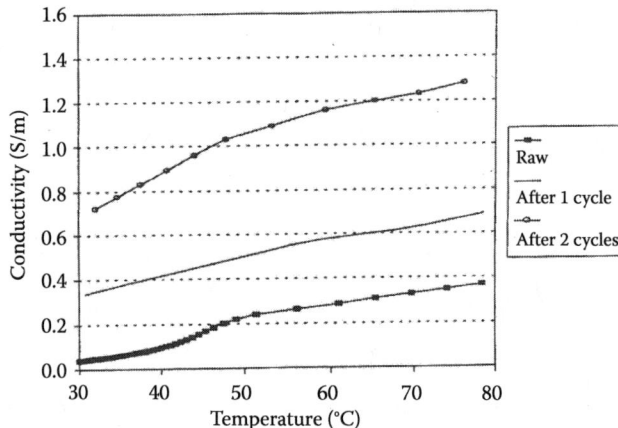

Figure 13.19 Electrical conductivity of carrot tissue as affected by cyclic ohmic heating. (From Wang, W-C., and Sastry, S.K. 1997b. *J. Food Proc. Engr.* 20:499–516, reproduced with permission.)

The nature of the changes in electropermeabilized tissue has been the subject of some investigation. Angersbach et al. (1999) have developed models for the electrical conductivity of intact and ruptured plant cells, using the low-frequency response of the tissue as an indicator of damage. Their approach has included the use of a conductivity ratio as an estimate of damage. Other models (Lebovka et al., 2001, 2002) have included considerations such as the probability of damage to a particular cell, and have also attempted to correlate the conductivity ratio to simulated damage. Most of these models consider a population of cells that are either intact or permeabilized, with or without significant extracellular spaces. While considerable progress has been made in the understanding of permeabilization phenomena when significant electric fields are applied, the understanding of the influences of low field strengths needs further work. In such cases, the challenges lie in defining the extent of "damage" of partly permeabilized tissue, or for cells that have recovered while having lost part of their intracellular constituents to the extracellular matrix.

13.4.3 Effect of Frequency

For materials with no cellular structure, or with a fully permeabilized cell structure, the electrical conductivity is generally considered to be independent of frequency. However, if intact cells are present, the electrical conductivity consists of three phases (Angersbach et al., 1999), as illustrated in Figure 13.20. At low frequencies, the electrical conductivity is nearly constant. At frequencies corresponding to the so-called β-dispersion (to be discussed later in this section), the electrical conductivity undergoes a remarkable increase with frequency, increasing to a constant value at high frequencies. At high frequencies, the electrical conductivities of ruptured or intact cells are not significantly different, since the impedance of the membrane at high frequencies is negligible (Angersbach et al., 1999). An investigation of the effect of frequency (50 Hz–1 MHz) and temperature (0.5–40°C) on electrical conductivity of damaged and intact tissues of peaches was performed by Shynkaryk et al. (2010). With this frequency range, the electrical conductivity falls into α- and β-dispersions. For both types of tissues, the electrical conductivity increased with increasing temperature (Figure 13.21). The conductivity changed very

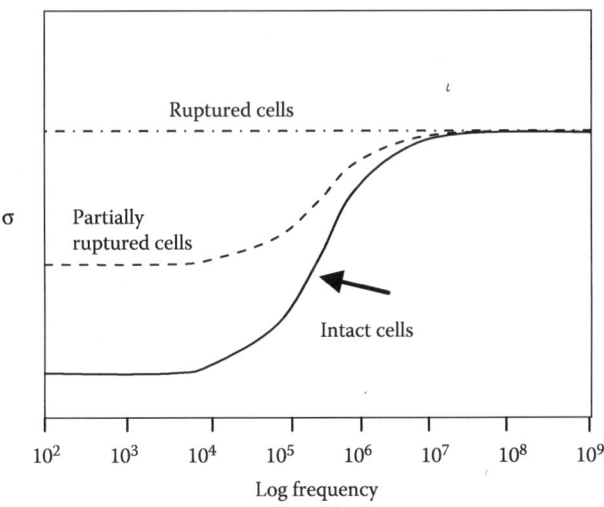

Figure 13.20 Illustration of electrical conductivity–frequency relationships for tissue samples with intact, partly ruptured, and fully ruptured cells.

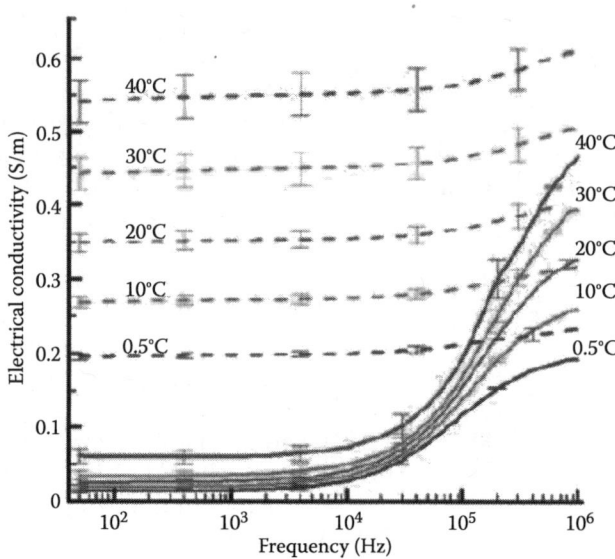

Figure 13.21 Typical electrical conductivity frequency dependence at different temperatures for the intact (solid line) and maximally damaged (dashed line) peach tissues. The error bars represent the standard data deviations of three observations. (From Shynkaryk, M.V., Ji, T., Alvarez, V.B., and Sastry, S.K. 2010. *J. Food Sci.* 75(7):E493–E500, reproduced with permission.)

little over the frequency range for damaged cell tissues while electrical conductivity was relatively low at low frequency and dramatically increased in the β-dispersion range for intact tissues.

These effects have been found to be useful as a means of assessing permeabilization of cellular tissue. Angersbach et al. (1999) have used the characteristic curves within the β-dispersion range (from 1 kHz to 100 MHz) to determine the extent of permeabilization of cellular tissue treated with PEFs and high pressure. They also developed an electrophysiological model, which permitted the assessment of damage due to processing. The criterion used was the ratio of conductivities:

$$Z = \frac{\sigma_l - \sigma_i}{\sigma_p - \sigma_i} \qquad (13.29)$$

The above approach was used with slight modifications by Angersbach et al. (1999) where $Z = 0$ for an intact cell and $Z = 1$ for a maximal damage of tissue. Further development of a damage index was done by Lebovka et al. (2002), to attempt to correlate the conductivity ratio with simulated damage using the empirical Archie equation:

$$Z = P^m \qquad (13.30)$$

where P is the damage index, defined as the ratio of the number of damaged cells to the total number of cells. A detailed discussion of physical damage is beyond the scope of this chapter, however, the effects of damage on electrical conductivity are of interest, and the succeeding discussion follows this spirit.

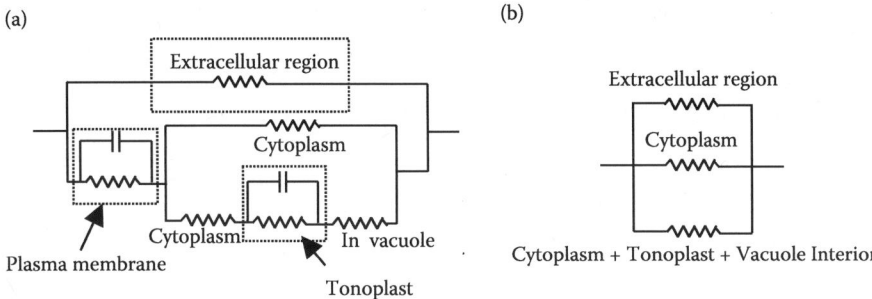

Figure 13.22 Model for cellular tissue. (a) Intact tissue; (b) ruptured tissue. (After Angersbach, A., Heinz, V., and Knorr, D. 1999. *Biotechnol. Prog.* 15:753–762.)

Angersbach et al. (1999) propose a model for multicellular biological tissue wherein the impedances of the extracellular spaces and cells are considered. The intact cell components include the resistance and capacitances of the cell membrane and the vacuole membrane (tonoplast), as illustrated in Figure 13.22. Ruptured cells are considered to be purely conductive elements, with no capacitance.

13.4.3.1 Relation to Dielectric Constant

The frequency relationship cannot be considered independently of that of the dielectric constant. Since the dielectric constant and loss factor* (which reduces to electrical conductivity when dipole effects are negligible) are related to each other via the Debye relation (Schwan, 1957),

$$\varepsilon^* = \varepsilon_{r\infty} + \frac{\varepsilon_{r0} - \varepsilon_{r\infty}}{1 + j\omega\tau} \tag{13.31}$$

In the absence of significant dipole effects, it can be shown that changes in the dielectric constant and electrical conductivity are related by (Schwan, 1957)

$$2\pi f(\varepsilon_{r0} - \varepsilon_{r\infty}) = (\sigma_\infty - \sigma_0) \tag{13.32}$$

Thus, an increase in electrical conductivity will be accompanied by a decrease in the dielectric constant and vice versa. This is not surprising, since they represent the extent of delay with which a system responds to an electrical stimulus; that is, they represent the in-phase and out-of-phase components of the response. The relation in Equation 13.32 indicates that as frequencies decrease, increases in conductance will be accompanied by more and more dramatic decreases in dielectric constant.

In biological systems, the dielectric constants undergo phases of significant decreases over various ranges of frequency. These are illustrated in Figure 13.23, as the α-, β-, and γ-dispersions.

---

* See definitions of dielectric constant and loss factor in Chapter 14, Dielectric Properties of Foods.

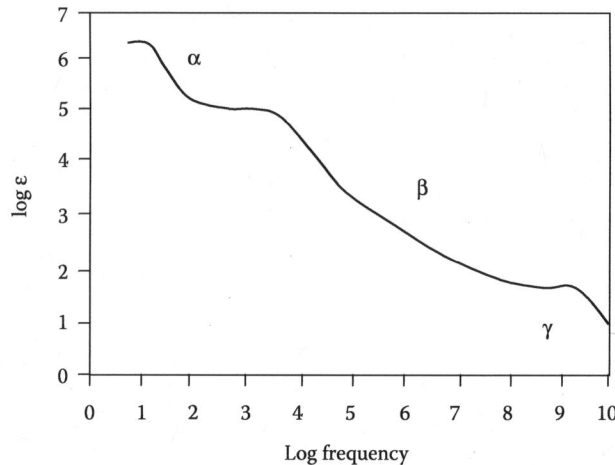

Figure 13.23 Dielectric constant as affected by frequency, showing the three major dispersion regions.

The reasons for the dispersions have been extensively researched but are still not very well understood. Attention has largely focused on the β-dispersion, and, to a lesser extent, the α-dispersion. At the γ- dispersions, dielectric constant is relatively low and the electrical conductivity of intact and disrupted cell is not a function of frequency. Detailed discussion is provided by Schwan (1957) and some additional information by Kuang and Nelson (1998).

The β-dispersion is attributed to membrane charging and discharging effects including ionic conductions. Various explanations have been sought for the α-dispersion. Among the explanations are the following.

1. The presence of a "gating" mechanism, whereby the ion channels within the cells widen and allow more conduction. The capacitance change is attributed to the relaxation phenomenon (i.e., change in response to changes in electrical conductivity).
2. The charging and discharging of the electric double layer of the cell membrane.
3. Ionic atmosphere around charged colloidal particles (possibly cells), which represent relatively large objects with a distributed charge. Such particles together with their ionic atmospheres will have sufficiently large time constants to exhibit α-dispersion. This is the explanation favored by Schwan (1957). This is the likely explanation for erythrocyte suspensions.

Kuang and Nelson further point out those measurements in the α-dispersion region are unreliable because of significant electrode polarization effects. Thus, understanding of the α-dispersion is limited.

Some further insight may be gained by recent work (Kulshrestha and Sastry, 2003), which suggests that vegetable tissue is maximally permeabilized by frequencies in the range of 10 Hz, which is at the low end of the α-dispersion mentioned by Schwan.

This suggests that the gating and membrane charging mechanisms discussed above may be more likely candidate explanations for this dispersion effect.

Some studies on minced products and gels have focused on sample impedances and heating rates rather than specific electrical properties. This approach is useful in tracking changes to samples over time. However, frequency dependencies in impedance or heating rates are not necessarily indicative of the actual behavior of the electrical properties themselves. Park et al. (1995) studied Alaska Pollock mince from 50 Hz to 10 kHz, and found the heating rate to increase with frequency. Wu et al. (1998) found decreasing impedance with increasing frequency for Pacific whiting surimi. In other studies, Imai et al. (1998) found that impedance of egg albumin solutions showed an increased heating rate accompanying gel formation, which was attributed to reduced heat loss because of the transition to conduction heat transfer. These researchers suggested that the heating rate results of Park et al. (1995) for Alaska Pollock might have been influenced by the same phenomena. Park et al. (1995), Wu et al. (1998), and Imai et al. (1998) found that sample impedance decreased with frequency to a nearly constant value. While neither Imai et al. nor Wu et al. reported the dielectric loss (or electrical conductivity) as a function of frequency, Park et al. (1995) did report a dielectric loss factor that showed a maximum around 10 kHz. Such frequency dependence in this frequency range is unexpected in a material of disrupted cellular structure unless disruption is incomplete.

It must also be noted that the observation of frequency-dependent sample impedance is, by itself not indicative of frequency-dependence either in electrical conductivity or dielectric constant of a material. Since most food materials function both as resistors (R) and capacitors (C), the impedance of an RC circuit is given by

$$Z = \sqrt{R^2 + \frac{1}{(2\pi f C)^2}} \qquad (13.33)$$

The capacitive reactance ($1/2\pi f C$) is dominant at low frequencies, but approaches zero at high frequencies. Under these circumstances, the sample impedance (Z) approaches the value of R at high frequencies. For a sample which shows no frequency-dependent dispersions, the impedance function should approach a constant value at high frequencies, as has been observed by the above researchers.

Indeed, there is a considerable body of evidence that suggests that the dielectric constant and electrical conductivity of noncellular materials, including liquids and gels show little or no frequency dependence below 1 GHz (e.g., Schwan, 1957; Chaiwanichsiri et al., 2001, shown in Figure 13.24, or the Chapter 14 on dielectric properties in this book), at which point the dielectric constant shows a decrease according to the Debye relation (Equation 13.31). Thus, sample impedance data alone is insufficient to show frequency dependence of properties of a material.

13.4.4 Ingredient Effects

The effects of various ingredients are similar to those for liquid foods. However, we note that generally, the preferred method of increasing the electrical conductivity of solid food particles is the addition of salt via an infusion process. This has been studied

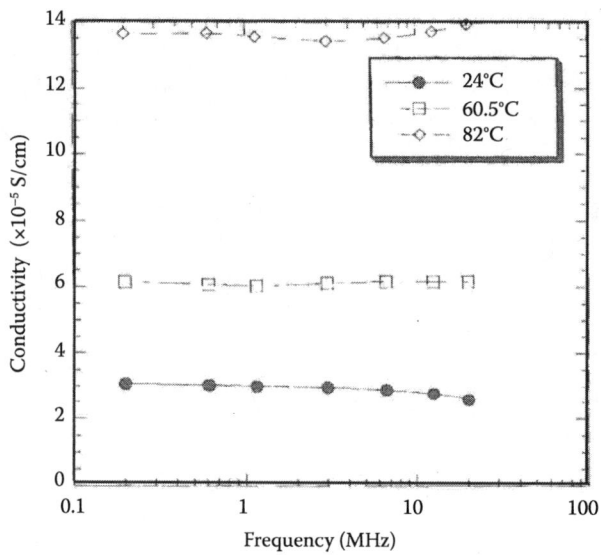

Figure 13.24 Electrical conductivity of potato starch solution as influenced by frequency and temperature. (From Chaiwanichsiri, S., Ohnishi, S., Suzuki, T., Takai, R., and Miyawaki, O. 2001. *J. Sci. Food Agric.* 81:1586–1591, reproduced with permission.)

by Wang and Sastry (1993a,b). Their results indicate the efficacy of salt infusion, but also note that vacuum infusion was effective only in penetrating outer layers of potato tissue which is appropriate for small particles. The effect of salt addition is shown in Figure 13.25.

A blanching method was developed by Sarang et al. (2007) in order to improve heating uniformity of chicken chow mein product (chicken, cut celery, cut mushroom, bean sprouts, and sliced water chestnut suspended in viscous sauce). The authors found that blanching vegetables in conductive sauce could raise the electrical conductivity; however, there was no significant increase in the conductivity of precooked chicken. The reason was determined to be its shrinkage and low permeability. Marinating raw meat shows potential in raising its electrical conductivity. Kamonpatana (2012) observed the limitation of increasing the conductivity of chicken by blanching alone, consistent with the study of Sarang et al. (2007) and suggested marination in conductive solution for the appropriate length of time. Celery could be made more conductive by blanching in sauce and salt solution. Cell wall breakdown, cell rupture, and tissue damage dominate the increase in conductivity via blanching. In case of mushroom, blanching treatment or vacuum infusion alone was ineffective in elevating electrical conductivity. Vacuum infusion followed by blanching in a conductive solution was proposed by Kamonpatana (2012). This may have been due to the expulsion of air from the mushroom via vacuum infusion and softer tissues through the blanching treatment.

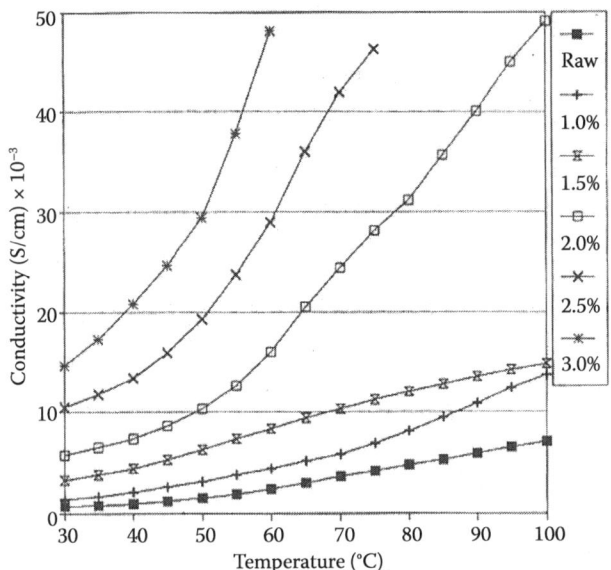

Figure 13.25 Effect of salt infusion at various concentrations on the electrical conductivity versus temperature relationship for potato tissue. (From Wang, W-C., and Sastry, S.K. 1993b. *J. Food Engr.* 20:299–309, reproduced with permission.)

13.5 SOLID–LIQUID MIXTURES

13.5.1 Models for Effective Electrical Conductivity

The original promise of ohmic heating technology was in the sterilization of solid–liquid mixtures. The sizing, and in some cases even the design, of ohmic heaters, depends on the effective electrical conductivity of a solid–liquid mixture. While experimental studies are doubtless important, the development of models is desirable, since once it is verified, it may be able to accommodate different product formulations without the need for repeated testing.

This subject has been studied in some detail by Palaniappan and Sastry (1991c), who compared several models for the effective electrical conductivity of solid–liquid mixtures. Six models were considered, five existing models, and one, the probability model which was developed by the authors.

13.5.1.1 Maxwell Model

This model (Maxwell, 1881) determines the effective electrical conductivity (σ_e) of a dispersion, which consists of a continuous phase of conductivity σ_c and a dispersed phase of conductivity σ_d, and volume fraction F:

$$\sigma_e = \sigma_c \left(\frac{1 - 2AF}{1 + 2AF} \right) \tag{13.34}$$

where

$$A = \frac{\sigma_c - \sigma_d}{2\sigma_c - \sigma_d} \tag{13.35}$$

13.5.1.2 Meredith and Tobias (1960) Model
This is a modification of the Maxwell model to account for interaction of fields around particles in an emulsion.

$$\sigma_e = \sigma_c \left(\frac{2(1+BF)}{2-BF}\right)\left(\frac{2+(2B-1)F}{2-(B+1)F}\right) \tag{13.36}$$

where

$$B = \frac{\sigma_d - \sigma_c}{2\sigma_c + \sigma_d} \tag{13.37}$$

13.5.1.3 Series Model
This has been used by Murakami and Okos (1988) in connection with thermal conductivity, and adapted to the electrical conductivity case.

$$\sigma_e = \frac{1}{((F/\sigma_d) + ((1-F)/\sigma_c))} \tag{13.38}$$

13.5.1.4 Parallel Model
Similarly extended from the thermal conductivity models mentioned by Murakami and Okos (1988).

$$\sigma_e = F\sigma_d + (1-F)\sigma_c \tag{13.39}$$

13.5.1.5 Kopelman Model
The Kopelman model (Sahin et al., 1999), developed from the thermal conductivity literature:

$$\sigma_e = \frac{\sigma_c(1-C)}{1-C(1-F^{1/3})} \tag{13.40}$$

where

$$C = F^{2/3}\left(1 - \frac{\sigma_d}{\sigma_c}\right) \tag{13.41}$$

13.5.1.6 Probability Model (Palaniappan and Sastry, 1991c)
This approach involves the determination of the probability that an incremental section of solid–liquid mixture consists of all liquid, all solid, or partially liquid and solid. A detailed

description of the model is provided by Palaniappan and Sastry (1991c) and will not be detailed here.

13.5.1.7 Comparison of Models
Palaniappan and Sastry (1991c) found that excepting the series model, all the others yielded satisfactory prediction of experimental data (<10% in most cases).

13.5.1.8 Effects of Solids in Tube Flow
All of the above models make assumptions regarding the arrangements of solids within a mixture. However, it is not clear whether these actually represent real arrangements in a flowing solid–liquid mixture. For this purpose, it is necessary to measure the solid area fraction of a flowing mixture. This has been done by Zitoun and Sastry (2003). A sampling of their results (Figure 13.26) indicates that the solid area fraction showed a normal distribution along the tube length.

Zitoun and Sastry also found that all cross-sections that they sampled contained some solids concentration (Figure 13.27). This explains the relative success of the simple parallel model, and suggests that it may be possible to use it in estimating the effective electrical conductivity of a system of sufficient solids concentration.

In particular, data on experimental solid area fraction distribution could be used to model the effective electrical conductivity of a mixture, increment by increment, by considering sections such as illustrated in Figure 13.28 using normally distributed area fraction data, and applying the parallel model to each section.

Relatively little experimental work exists to determine effective electrical conductivity of solid–liquid mixtures. One study is by Castro et al. (2003), who added solid pieces to strawberry pulp. Their results on the effect of solids concentration are shown in Figure 13.29.

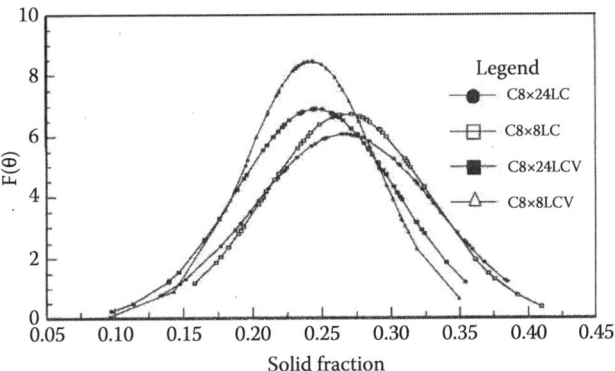

Figure 13.26 Distribution of solid area and volume fractions for cylindrical particles in a straight tube. (C: cylinders; 8 × 8 mm or 8 × 24 mm in size; LC: 30% solid concentration, and HC: 50% solid concentration.) (From Zitoun, K.B., and Sastry, S.K. 2003. *J. Food Engr.* 60:81–87, reproduced with permission.)

ENGINEERING PROPERTIES OF FOODS

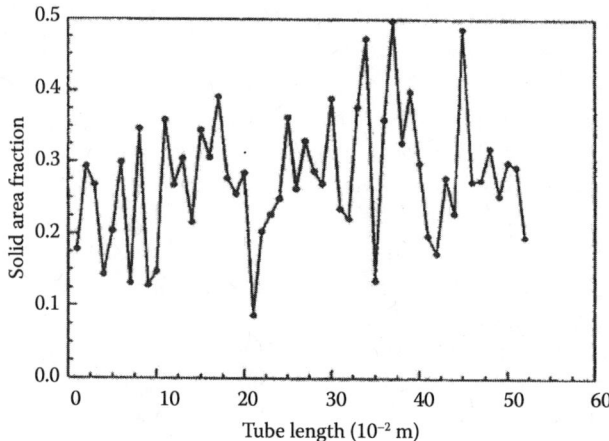

Figure 13.27 Experimental instantaneous solid area fraction distribution along the tube length for cubes of 8 mm in side and 30% solid concentration. (From Zitoun, K.B., and Sastry, S.K. 2003. *J. Food Engr.* 60:81–87, reproduced with permission.)

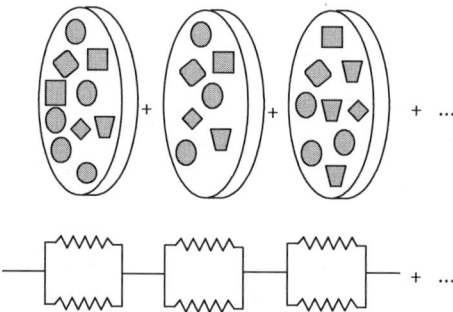

Figure 13.28 Illustration of application of parallel model together with area fraction data to calculation of effective electrical conductivity of a solid–liquid mixture.

For longer-term development of electrical and electrothermal processes, the characterization of models for effective electrical conductivity of solid–liquid mixtures would be useful.

13.6 METHODS OF MEASUREMENT OF ELECTRICAL CONDUCTIVITY

The general principle of measurement is simple, and is related to Equation 13.2. If a sample can be prepared of length L, and cross-sectional area A, then the measurement of the resistance across the length of the sample will provide sufficient information for determination of the electrical conductivity via Equation 13.2. If a sample cannot be shaped in this manner, the measurement becomes more complicated.

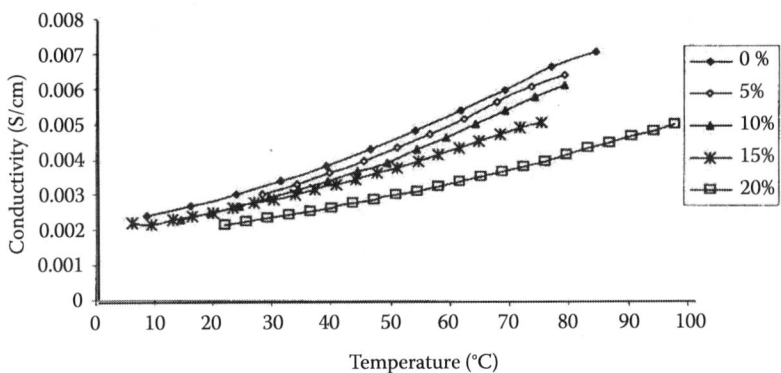

Figure 13.29 Electrical conductivity of strawberry pulp with added particles (average size 7.9 mm) during ohmic heating. (From Castro, I., et al. *J. Food Proc. Engr.* 26(1):17–29, reproduced with permission.)

The length-to-area ratio for the sample is designated as the cell constant, which is easy to determine for simple geometries as illustrated below in Figure 13.30.

One simple embodiment of a system for measuring the electrical conductivity of foods is illustrated in Figure 13.30. Here, the sample holder consists of a cylinder with two electrodes inserted from either end, with a port for temperature measurement. A slightly different (vertical) design for a sample holder device is described by Mitchell and de Alwis (1989).

Systems of the type illustrated in Figure 13.30 offer several advantages:

1. Samples are of precise shape and known dimensions, which simplifies the calculation of cell constants.
2. Thermocouple permits measurement of temperature. It is possible to include more than one thermocouple, although care must be taken to avoid shadows, as listed in

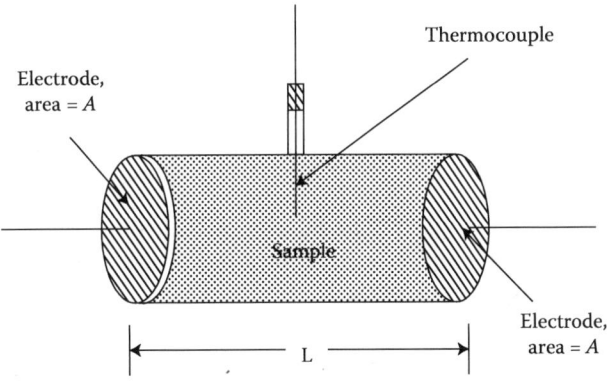

Figure 13.30 System for measurement of electrical conductivity of foods.

the precautions. Temperature measurement is critical, since electrical conductivity of foods is often a strong function of temperature.
3. Both liquid and solid samples may be used. Liquids will automatically conform to vessel dimensions, and most food samples may typically be cut to the appropriate dimensions.
4. Electrodes and thermocouples may be sealed by O-rings or packing glands, permitting operation of the system under pressures higher than atmospheric (typically temperatures greater than 100°C can be achieved). This is an advantage in ohmic heating applications.
5. Electrical conductivity may be measured continuously as a function of temperature by heating the sample *in situ*, and continuously logging voltage and current while the sample is heated.

Difficulties and precautions necessary for this system are:

1. Contact between electrode and sample must be assured. In practice, this requires a pressure on the sample, sufficient to result in adequate contact at the electrodes, while being not so great as to crush the sample. This may necessitate a spring-loaded assembly in some cases. In practice, for many high-moisture samples, a slight amount of moisture is exuded, which forms the necessary continuous contact.
2. Accurate temperature measurement is crucial. It is necessary to avoid electrical interference from the circuit; hence, the thermocouple may be insulated by coating with Teflon. This prevents electrical leakage at the cost of slow time response. This may be a problem in situations involving rapid ohmic heating. An alternative approach to speed up thermal response while avoiding electrical interference is to use uncoated thermocouples, but providing signal conditioning and isolation of thermocouples. Fiber-optic sensors (commonly used in microwave temperature measurement) may also be used.

The measurement of electrical conductivity is then accomplished by connecting the system to a power source, applying a brief and known electric field, and measuring the current and temperature. The electrical conductivity may then be calculated by combining Equations 13.1 and 13.2, yielding:

$$\sigma = \frac{LI}{AV} \tag{13.42}$$

If it is necessary to measure electrical conductivity as a function of temperature, this may be done in two ways.

For conventional heating applications, the sample is typically heated by hot fluid in a jacket, or by immersion of the sample cell in a controlled-temperature bath, until equilibration. The procedure above is then repeated.

For ohmic heating applications, the electric field is applied continuously, so that the sample heats ohmically. The voltage, current, and temperature are continuously logged over time and the electrical conductivity at each time determined by Equation 13.42. A schematic diagram of this experimental setup is provided in Figure 13.31.

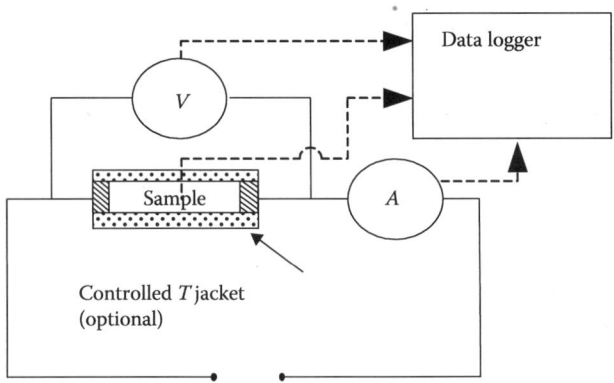

Figure 13.31 Schematic diagram of electrical conductivity measurement system.

Figure 13.32 Ohmic heating device for measurement of electrical conductivity of 10 samples.

To reduce setup time, use small particles while heating to sterilization temperatures, an ohmic heating device with 10 cylindrical cells (0.0078 m i.d. and 0.0079 m in length) equipped with platinized-titanium electrodes has been developed by Tulsiyan et al. (2008). Food samples may be placed in each cylindrical cell and pressurized. The electrical conductivity of 10 samples at a time could be continuously measured at temperatures up to 140°C (Figure 13.32).

If the sample is too delicate, and cannot withstand physical stresses, it may be necessary to use indirect methods of measurement. The sample itself could be placed in a similar sample holder, and the gap between the ends and the electrodes filled by a fluid of known electrical conductivity, as shown in Figure 13.33.

The system could then be modeled as a set of resistances in series, and the electrical conductivity of the sample calculated. Based on the approach of Palaniappan and Sastry (1991c), the effective electrical resistance (R_e) can be considered to have continuous (liquid) and discrete (solid) phases. The equivalent resistance of multicomponent mixtures is shown in Figure 13.34.

$$R_e = R_{lS} + R_P \tag{13.43}$$

$$R_p = \frac{1}{1/(R_{lP}) + \sum_{m=1}^{N}(1/R_{sPm})} \tag{13.44}$$

ENGINEERING PROPERTIES OF FOODS

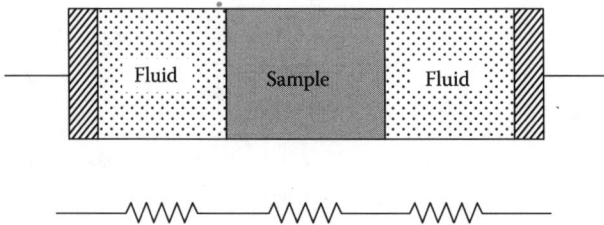

Figure 13.33 Diagram of measurement approach for fragile samples.

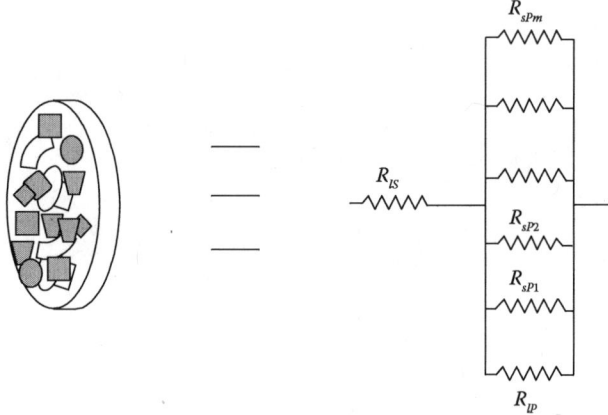

Figure 13.34 Equivalent circuit of multicomponent mixtures.

This procedure has the disadvantage that if the electrical conductivity of the fluid is different from the solid, diffusion could compromise the measurement. The experiment could be repeated for fluids of different electrical conductivity until the sample conductivity matched that of the fluid.

A number of alternative approaches exist for sample cell design. A number of these are described in the electrochemistry literature (see, e.g., Crow, 1988), and are designed for fluids within chemical glassware. These devices do not enclose the sample within well-defined dimensions. Thus, the cell constant is not easy to calculate, and must be determined by preliminary experimentation with fluids of known electrical conductivity.

Finally, it is also possible to measure sample resistance via the use of a bridge circuit, such as illustrated in Figure 13.35. These approaches take time for balancing the bridge circuit, and accordingly, require the use of small electric currents through the sample. Ohmic heating will alter the resistance of the sample, and destabilize the bridge circuit equilibrium—thus necessitating small currents. Such systems may not be suited to measure electrical conductivity of food samples undergoing significant *in situ* heating.

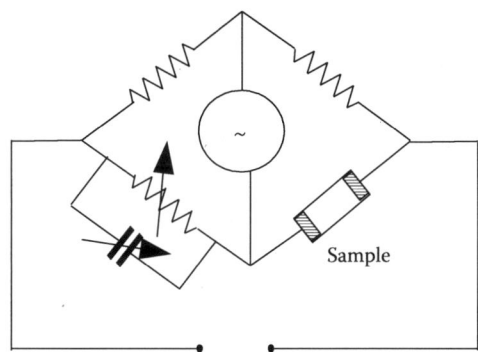

Figure 13.35 Bridge circuit for conductivity determination.

LIST OF SYMBOLS

A	Area (m²); name of chemical species
A_s	Cross-sectional area of solids exposed to electric current (m²)
B	Name of chemical species
C	Concentration (mol/m³), capacitance (Farads); parameter in Equation 13.41
c	Concentration (mol/m³)
D_i	Mass diffusivity of species i (m²/s)
E	Electric field strength (V/m)
f	Frequency (s⁻¹)
F	Faraday constant (9.6487 × 10⁴ Coulombs/mole); volume fraction of dispersed phase (in solid–liquid mixture models).
I	Electric current (A)
k	Constant of proportionality in Kohlrausch relation (Equation 13.7)
K_1	Temperature coefficient of electrical conductivity in Equation 13.21 (°C⁻¹)
K_2	Solids content coefficient in Equation 13.21 (S/m% solids)
K_T	Temperature coefficient in Equation 13.25 (S/m°C)
K_{TC}	Temperature and concentration coefficient in Equation 13.25 (S/m°C%)
$K_{\sigma T}$	Temperature coefficient in Equation 13.27 (S/m°C)
L	Length (m)
m	Temperature coefficient of electrical conductivity (°C⁻¹); Archie exponent (Equation 13.30)
n	Number of solid particles in suspension
N	Avogadro number (6.023 × 10²³ mol⁻¹); number of solid phases
P	Pressure (Pa); damage index (Equation 13.30)
r_i	"Radius" of ionic species I
r	Average radius of solids in suspension.
R	Electrical resistance (Ω), universal gas constant (8.314 J/°K mol)
S	Solids content (%)
T	Temperature (°C or K)

V	Voltage (V)
v	Velocity of ions (m/s)
V_C	Critical membrane breakdown potential
V_s	Total volume of solids
z	Charge of an electron
Z	Conductivity disintegration index, impedance (Equation 13.33)

Greek Letters and Other Symbols

α	Fraction of dissociated electrolyte
γ	Effective elastic modulus of membrane (Pa)
ε^*	Complex dielectric constant
ε_{r0}	Static dielectric constant
$\varepsilon_{r\infty}$	Dielectric constant at infinite frequency
η	Viscosity (Pa s)
λ	Molar conductivity of individual ionic species
Λ	Molar conductivity (S m²/mol)
Λ_∞	Molar conductivity at infinite dilution (S m²/mol)
μ_i	Electrophoretic mobility of species i
ν	Number of moles
σ	Electrical conductivity (S/m)
σ_0	Electrical conductivity at a reference temperature of 0°C (S/m)
σ_∞	Electrical conductivity at infinite dilution (S/m)
τ	Time constant (corresponding to relaxation frequency)
ω	Angular frequency (radians/s)

Subscripts/Superscripts not Explained Elsewhere

0	At zero frequency (DC)
25	At 25°C
∞	At infinite dilution, at infinite frequency
+	Cations

Anions

c	Continuous phase
C	Concentration
d	Dispersed phase
e	Effective
i	Species identifier index; intact cells (in Equation 13.29)
l	At low frequency
l	Liquid phase
m	Solid phase index
p	Ruptured cells (in Equation 13.29)
$P = 0$	At atmospheric pressure

P Parallel
s Solid
S Series

REFERENCES

Angersbach, A., Heinz, V., and Knorr, D. 1999. Electrophysiological model of intact and processed plant tissues: Cell disintegration criteria. *Biotechnol. Prog.* 15:753–762.

Asavasanti, S., Ersus, S., Ristenpart, W., Stroeve, P., and Barrett, D.M. 2010. Critical electric field strengths of onion tissues treated by pulsed electric fields. *J. Food Sci.* 75(7):E433–E443.

Barbosa-Cánovas, G.V., Pierson, M.D., Zhang, Q.H., and Schaffner, D.W. 2000. Pulsed electric fields. *J. Food Sci.* 65(8):65s–79s.

Bazhal, M.I., Lebovka, N.I., and Vorobiev, E. 2001. Pulsed electric field treatment of apple tissue during compression for juice extraction. *J. Food Engr.* 50:129–139.

Benz, R., Beckers, F., and Zimmermann, U. 1979. Reversible electrical breakdown of lipid bilayer membranes: A charge-pulse relaxation study. *J. Membrane Biol.* 48:191–204.

Bozkurt, H., and Icier, F. 2010. Electrical conductivity changes of minced beef-fat blends during ohmic cooking. *J. Food Engr.* 96:86–92.

Castro, I., Teixeira, J.A., Salengke, S., Sastry, S.K., and Vicente, A.A. 2003. The influence of field strength, sugar and solid content on electrical conductivity of strawberry products. *J. Food Proc. Engr.* 26(1):17–29.

Chaiwanichsiri, S., Ohnishi, S., Suzuki, T., Takai, R., and Miyawaki, O. 2001. Measurement of electrical conductivity, differential scanning calorimetry and viscosity of starch and flour suspensions during gelatinisation process. *J. Sci. Food Agric.* 81:1586–1591.

Chernomordik, L.V., and Abidor, I.G. 1980. The voltage-induced local defects in unmodified BLM. *Bioelectrochem. and Bioenerg.* 7:617–623

Crow, D.R. 1988. *Principles and Applications of Electrochemistry*. Chapman & Hall, London, New York.

Crowley, J.M. 1973. Electrical breakdown of bimolecular lipid membranes as an electromechanical instability. *Biophys. J.*13:711–724.

Deng, J., Schoenbach, K.H., Buescher, E.S., Hari, P.S., Fox, P.M., and Beebe, S.J. 2003. The effects of intense submicrosecond electrical pulses on cells. *Biophys. J.* 84:2709–2714.

Dimitrov, D.S. 1984. Electric field-induced breakdown of lipid bilayers and cell membranes: A thin viscoelastic film model. *J. Membrane Biol.* 78:53–60.

Halden, K., de Alwis, A.A.P., and Fryer, P.J.1990. Changes in the electrical conductivity of foods during ohmic heating. *Int. J. Food Sci. Technol.* 25:9–25.

Imai, T., Uemura, K., Ishida, N., Yoshizaki, S., and Noguchi, A. 1995. Ohmic heating of Japanese white radish *Rhaphanus sativus* L. *Int. J. Food Sci. Technol.* 30:461–472.

Imai, T., Uemura, K., and Noguchi, A. 1998. Heating rate of egg albumin solution and its change during ohmic heating. Chapter 10 in *Process-Induced Changes in Food*; F. Shahidi, C-T. Ho, and N. van Chuyen. Plenum Press, New York.

Jain, R.K., and Maldarelli, C. 1983. Stability of thin viscoelastic films with application to biological membrane deformation. *Ann. NY Acad. Sci.* 404:89–102.

Kamonpatana, P. 2012. Mathematical modeling and microbiological verification of ohmic heating of solid-liquid mixtures in continuous flow ohmic heater systems. PhD dissertation, the Ohio State University, Columbus, OH.

Kamonpatana, P., Mohamed, H.M.H., Shynkaryk, M., Heskitt, B., Yousef, A.E., and Sastry, S.K. 2013. Mathematical modeling and microbiological verification of ohmic heating of a solid liquid mixture in a continuous flow ohmic heater system with electric field perpendicular to flow. *J. Food Engr.* 118(3):312–325.

Karapantsios, T.D., Sakonidou, E.P., and Raphaelides, S.N. 2000. Electric conductance study of fluid motion and heat transport during starch gelatinization. *J. Food Sci.* 65:144–150.

Korobkov,, V.N., Zhushman, A.I., and Kostenko, V.G. 1978. Determination of starch gelation temperature on the basis of changes in electrical conductivity. *Sakharnaya Promyshlennost*, 7:67–70.

Knorr, D. 1994. Plant cell and tissue cultures as model systems for monitoring the impact of unit operations on plant food. *Trends Food Sci. Technol.* 5:328–331.

Kuang, W., and Nelson, S.O. 1998. Low-frequency dielectric properties of biological tissues: A review with some new insights. *Trans. ASAE.* 41(1):173–184.

Kulshrestha, S. 2002. Membrane permeability changes during ohmic heating of vegetable tissue. PhD dissertation, The Ohio State University, Columbus, OH.

Kulshrestha, S.A., and Sastry, S.K. 2010. Changes in permeability of moderate electric field (MEF) treated vegetable tissueover time. *Innov. Food Sci. Emerg. Technol.* 11:78–83.

Kulshrestha, S.A., and Sastry, S.K. 2003. Frequency and voltage effects on enhanced diffusion during moderate electric field (MEF) treament. *Innov. Food Sci. Emerg. Technol.* 4(2):189–194.

Kusnadi, C., and Sastry, S.K. 2012. Effect of moderate electric fields on salt diffusion into vegetable tissue. *J. Food Engr.* 110:329–336.

Lebovka, N.I., Bazhal, M.I., and Vorobiev, E. 2000. Simulation and experimental investigation of food material breakage using pulsed electric field treatment. *J. Food Engr.* 44(4):213–223.

Lebovka, N.I., Bazhal, M.I., and Vorobiev, E. 2001. Pulsed electric field breakage of cellular tissues: Visualization of percolative properties. *Innov. Food Sci. Emerg. Technol.* 2:113–125.

Lebovka, N.I., Bazhal, M.I., and Vorobiev, E. 2002. Estimation of characteristic damage time of food materials in pulsed electric fields. *J. Food Engr.* 54(4):337–346.

Lima, M., and Sastry, S.K. 1999. The effects of ohmic heating frequency on hot-air drying rate, desorption isotherms and juice yield. *J. Food Engr.* 41:115–119.

Loghavi, L., and Sastry, S.K. 2009. Effect of moderate electric field frequency and growth stage on the cell membrane permeability of *Lactobacillus acidophilus*. *Biotechnol. Prog.* 25(1):85–94.

Marcotte, M., Piette, J.P.G., and Ramaswamy, H.S. 1998. Electrical conductivity of hydrocolloid solutions. *J. Food Proc. Engr.* 21:503–520.

Marcotte, M., Trigui, M., and Ramaswamy, H.S. 2000. Effect of salt and citric acid on electrical conductivity and ohmic heating of viscous liquids. *J. Food Proc. Pres.* 24:389–406.

Maxwell, J.C. 1881. *A Treatise on Electricity and Magnetism.* 2nd Ed., Vol. 1. Clarendon Press, Oxford.

Meredith, R.E., and Tobias, C.W. 1960. Resistance to potential flow through a cubical array of spheres. *J. Appl. Phys.*, 31:1270–1273.

Mitchell, F.R.G., and de Alwis, A.A.P. 1989. Electrical conductivity meter for food samples. *J. Physics E. Sci. Instrum.* 22:554–556.

Murakami, E.G., and Okos, M.R. 1988. Measurement and prediction of thermal properties of foods. In: *Food Properties and Computer-Aided Engineering of Food Processing Systems*, R.P. Singh and A.G. Medina, Eds., pp. 3–48 NATO ASI Series E, Applied Sciences, Vol. 168.

Ohno-Shosaku,T., and Okada, Y. 1984. Facilitation of electrofusion of mouse lymphoma cells by the proteolytication of proteases. *Biochem. Biophys. Res. Commun.* 120:138–143.

Palaniappan, S., and Sastry, S.K.1991a. Electrical conductivity of selected juices: influences of temperature, solids content, applied voltage and particle size. *J. Food Proc. Engr.* 14:247–260.

Palaniappan, S., and Sastry, S.K. 1991b. Electrical conductivity of selected solid foods during ohmic heating. *J. Food Proc. Engr.* 14:221–236.

Palaniappan, S., and Sastry, S.K. 1991c. Modelling of electrical conductivity of liquid-particle mixtures. *Food and Bioproducts Proc., Part C., Trans. Instn. Chem Engrs.* (UK) 69:167–174.

Park, S.J., Dong, K., Uemura, K., and Noguchi, A. 1995. Influence of frequency on ohmic heating of fish protein gel. *Noppon Shokuhin Kagaku Kogaku Kaishi*, 42(8):569–574.

Ruhlman, K.T., Jin, Z.T., and Zhang, Q.H. 2001. Physical properties of liquid foods for pulsed electric field treatment. Chapter 3 in *Pulsed Electric Fields in Food Processing*. pp 45–56. Eds: G.V. Barbosa-Cánovas and Q.H. Zhang. Technomic Publishing Co. Lancaster, PA.
Sahin, S., Sastry, S.K., and Bayindirli, L. 1999. Effective thermal conductivity of potatoes during frying: measurement and modeling. *Intl. J. Food Properties*. 2(2):151–161.
Saif, S.M.H., Lan, Y., Wan, S., and Garcia, S. 2004. Electrical resistivity of goat meat. *Intl. J. Food Properties*. 7(3):463–471.
Sale, A.J.H., and Hamilton, W.A. 1967a. Effect of high electric fields on microorganisms. I. Killing of bacteria and yeast. *Biochim. Biophys. Acta* 148:781–788.
Sale, A.J.H., and Hamilton, W.A. 1967b. Effect of high electric fields on microorganisms. II. Mechanism of action of the lethal effect. *Biochim. Biophys. Acta* 148:789–800.
Sarang, S., Sastry, S.K., Gaines, J., Yang, T.C.S., and Dunne, P. 2007. Product formulation for ohmic heating: Blanching as a pretreatment method to improve uniformity in heating of solid-liquid food mixtures. *J. Food Sci.* 72(5):E227–E234.
Sarang, S., Sastry, S.K., and Knipe, L. 2008. Electrical conductivity of fruits and meats during ohmic heating. *J. Food Engr.* 87:351–356.
Sastry, S.K., and Barach, J.T. 2000. Ohmic and inductive heating. *J. Food Sci.* 65(8):42s–46s.
Schwan, H.P. 1957. Electrical properties of tissue and cell suspensions. In: *Advances in Biological and Medical Physics*; Laurence, J.H., and Tobias, CA., Eds.; Academic Press, New York. 5:147–209.
Sensoy, I., and Sastry, S.K. 2004. Extraction using moderate electric fields. *J. Food Sci.* 69(1):7–13.
Shirsat, N., Lyng, J.G., Brunton, N.P., and McKenna, B. (2004). Ohmic heating: Electrical conductivities of pork cuts. *Meat Sci.* 67:507–514.
Shynkaryk, M.V., Ji, T., Alvarez, V.B., and Sastry, S.K. 2010. Ohmic heating of peaches in the wide range of frequencies (50 to 1 MHz). *J. Food Sci.* 75(7):E493–E500.
Sugar, I.P., and Neumann, E. 1984. Stochastic model for electric field-induced membrane pores electroporation. *Biophys. Chem.* 19:211–225.
Taiwo, K.A., Angersbach, A., and Knorr, D. 2002. Influence of high intensity electric field pulses and osmotic dehydration on the rehydration characteristics of apple slices at different temperatures. *J. Food Engr.* 52:185–192.
Tulsiyan, P., Sarang, S., and Sastry, S.K. 2008. Electrical conductivity of multicomponent systems during ohmic heating. *Int. J. Food Prop.* 11:233–241.
Wang, C.S., Kuo, S.Z., Kuo-Huang, L.L., and Wu, J.S.B. 2001. Effect of tissue infrastructure on electric conductance of vegetable stems. *J. Food Sci.* 66(2):284–288.
Wang, W-C., and Sastry, S.K. 1993a. Salt diffusion into vegetative tissue as a pretreatment for ohmic heating: Determination of parameters and mathematical model verification. *J. Food Engr.* 20:311–323.
Wang, W-C., and Sastry, S.K. 1993b. Salt diffusion into vegetative tissue as a pretreatment for ohmic heating: Electrical conductivity profiles and vacuum infusion studies. *J. Food Engr.* 20:299–309.
Wang, W-C, and Sastry, S.K. 1997a. Starch gelatinization in ohmic heating. *J. Food Engr.* 34(3):225–242.
Wang, W-C., and Sastry, S.K. 1997b. Changes in electrical conductivity of selected vegetables during multiple thermal treatments. *J. Food Proc. Engr.* 20:499–516.
Weaver, J.C., and Chizmadzhev, Y.A. 1996. Theory of electroporation: A review. *Bioelectrochem. Bioenerg*, 41:135–160.
Wu, H., Kolbe, E., Flugstad, B., Park, J.W., and Yonsawatdigul, J. 1998. Electrical properties of fish mince during multi-frequency ohmic heating. *J. Food Sci.* 63(6):1028–1032.
Yongsawatdigul, J., Park, J.W., and Kolbe, E. 1995. Electrical conductivity of Pacific Whiting surimi paste during ohmic heating. *J. Food Sci.* 60(5):922–925,935.
Zell, M., Lyng, J.G., Cronin, D.A., and Morgan, D.J. 2009. Ohmic heating of meats: Electrical conductivities of whole meats and processed meat ingredients. *Meat Sci.* 83(3):563–570.

Zimmermann, U. 1986. Electrical breakdown, electropermeabilization and electrofusion. *Rev. Physiol. Biochem. Pharmacol.* 105:176–256.

Zimmermann, U., Pilwat, G., and Riemann, M. 1974. Dielectric breakdown of cell membranes. *Biophys. J.* 14:881–899.

Zimmermann, U., Beckers, F., and Coster, H.G.L. 1977. The effect of pressure on the electrical breakdown of membranes of *Valoniautricularis. Biochim Biophys. Acta* 464:399–416.

Zitoun, K.B., and Sastry, S.K. 2003. Solid area fraction distribution of solid-liquid food mixtures during flow through a straight tube. *J. Food Engr.* 60:81–87.

14
Dielectric Properties of Foods

Ashim K. Datta, G. Sumnu, and G. S. V. Raghavan

Contents

14.1	Introduction	572
14.2	Basic Principles	572
	14.2.1 RF versus Microwave Heating	580
14.3	Measurement Principles	580
	14.3.1 Waveguide and Coaxial Transmission Line Methods	580
	14.3.2 Short-Circuited Line Technique	581
	14.3.3 Open-Ended Probe Technique	581
	14.3.4 Time-Domain Reflectometry Method	582
	14.3.5 Free-Space Transmission Technique	583
	14.3.6 Microstrip Transmission Line	583
	14.3.7 Six-Port Reflectometer Using an Open-Ended Coaxial Probe	583
	14.3.8 Colloid Dielectric Probe (Hewlett-Packard)	584
	14.3.9 Test Cell with Boonton RX-Meter	584
	14.3.10 Cavity Perturbation Technique	585
	14.3.10.1 Solid Sample Preparation	586
	14.3.10.2 Liquid Sample Preparation	586
	14.3.10.3 Semisolid Samples	587
	14.3.11 Summary of Dielectric Property Measurement Techniques	587
14.4	Frequency and Temperature Dependence	587
	14.4.1 Frequency Dependence	587
	14.4.2 Frequency Dependence in Food Materials	590
	14.4.3 Temperature Dependence in Water, Salt Solutions, and Foods	592
	14.4.3.1 Dielectric Properties below Freezing and above Boiling Temperatures	593
	14.4.3.2 Temperature Dependence of Loss Factor and Runaway Heating	596
14.5	Composition Dependence	597
	14.5.1 Moisture Dependence	599

		14.5.2	Carbohydrate Dependence	601
			14.5.2.1 Starch	601
			14.5.2.2 Sugar	604
			14.5.2.3 Gums	606
		14.5.3	Protein Dependence	607
		14.5.4	Fat Dependence	609
		14.5.5	Dielectric Properties of Meats	612
		14.5.6	Dielectric Properties of Fish and Seafood	613
		14.5.7	Dielectric Properties of Fruits and Vegetables	615
		14.5.8	Dielectric Properties of Dairy Products	617
	14.6	Structure Dependence and Pureed Foods		618
	14.7	Dielectric Properties of Insect Pests		620
	14.8	Dielectric Properties of Packaging Materials		620
	14.9	Effects of Processing and Storage on Dielectric Properties of Foods		620
		14.9.1	Baking	621
		14.9.2	Drying	625
		14.9.3	Cooking	625
		14.9.4	Mixing	626
		14.9.5	Storage	626
	14.10	Assessment of Food Quality by Using Dielectric Properties		626
	14.11	Further Sources of Data		628
	Acknowledgment			629
	References			629

14.1 INTRODUCTION

Electromagnetic heating, such as microwave and radiofrequency (RF) heating, finds applications in several food processes in industry and at home, including reheating, precooking, cooking, tempering, baking, drying, pasteurization, and sterilization. Electromagnetic heating processes are governed by the material properties called dielectric properties. As microwave heating gains increasing use in food-processing systems in industry and in the home, knowledge of dielectric properties becomes increasingly critical for consistent and predictable product, process, and equipment development. In this chapter, dielectric properties are defined, their roles in heating systems are summarized, measurement of the properties is described, and frequency, temperature, and composition dependence of the properties is discussed for groups of foods from available literature data. Short compilations of data are also provided.

14.2 BASIC PRINCIPLES

Like light waves, microwaves are part of the electromagnetic spectrum. Figure 14.1 shows the electromagnetic spectrum. The frequencies allocated for microwave and RF heating are shown in Table 14.1.

When microwave energy is incident on a food material, part of the energy is absorbed by the food, leading to its temperature rise. The amount and distribution of microwave

DIELECTRIC PROPERTIES OF FOODS

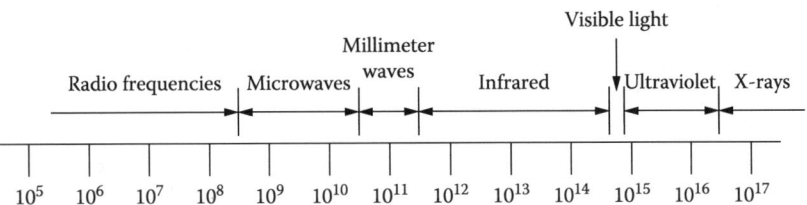

Figure 14.1 The electromagnetic spectrum showing the regions of microwave and radio frequencies. Only some of these frequencies are allocated for use in heating, as shown in Table 14.1. (From D Dibben. Electromagnetics: Fundamental aspects and numerical modelling In AK Datta, S Anantheswaran, Eds. *Handbook of Microwave Technology for Food Applications,* Marcel Dekker, New York, 2001.)

Table 14.1 Frequencies Assigned by the FCC for Industrial, Scientific, and Medical (ISM) Use[a]

	Frequency
RF	13.56 MHz ± 6.68 kHz
	27.12 MHz ± 160.00 kHz
	40.68 MHz ± 20.00 kHz
Microwaves	915 MHz ± 13 MHz
	2450 MHz ± 50 MHz
	5800 MHz ± 75 MHz
	24125 MHz ± 125 MHz

[a] Typically, microwave heating of foods is done at 2450 or 915 MHz and RF heating at 27.12 MHz.

energy absorption in a food material is described by Maxwell's equations of electromagnetics. Electromagnetic waves are composed of an electric field and a magnetic field. Maxwell's equations govern the propagation of electromagnetic waves in materials, and are written in time-harmonic form by

$$\nabla \times \mathbf{H} = j\omega\varepsilon\mathbf{E} \tag{14.1}$$

$$\nabla \times \mathbf{E} = -j\omega\mu\mathbf{H} \tag{14.2}$$

$$\nabla \cdot \mu\mathbf{H} = 0 \tag{14.3}$$

$$\nabla \cdot \varepsilon\mathbf{E} = \rho \tag{14.4}$$

Here $\mathbf{H}$ is the magnetic field vector and $\mathbf{E}$ is the electric field vector. The ε in the above equations is called the *permittivity* of the material (measured in farads per meter or F/m); it characterizes the interaction between the electric field of the microwaves and the material. The μ in the above equations is called the *permeability* of the material (measured in Henrys per meter or H/m); it characterizes the ability of the food to interact with the magnetic field of the microwaves. Both ε and μ can be complex quantities in general.

In practice, a relative complex permittivity of the medium, defined by $\varepsilon^* = \varepsilon/\varepsilon_0$, is used where $\varepsilon_0 = 8.8542 \times 10^{-12}$ F/m is the permittivity of the free space. This relative complex permittivity, ε^*, is composed of two different properties: ε' (called the dielectric constant) and ε'' (called the dielectric loss factor) given by

$$\varepsilon^* = \varepsilon' - j\varepsilon'' \tag{14.5}$$

Note that the dielectric constant and the dielectric loss factor are dimensionless, since they are defined relative to the permittivity of free space (Equation 14.5). In everyday use, the asterisk on ε^* is often not used, that is, Equation 14.5 is written as $\varepsilon = \varepsilon' - j\varepsilon''$, with the understanding that all quantities are relative to permittivity of free space, ε_0, and are therefore dimensionless.

Both the dielectric constant and the dielectric loss factor measure the ability of the material to interact with the electric field of the microwaves. The dielectric constant is a measure of the food material's ability to store electromagnetic energy, whereas the dielectric loss is the material's ability to dissipate electromagnetic energy (which results in heating). Since foods typically do not have components that would interact with a magnetic field, permeability of the food materials is generally assumed to be that of free space, given by $\mu = \mu_0 = 4\pi \times 10^{-7}$ H/m. It is worthwhile to note that the universal constants ε_0 and μ_0 are related to the velocity of electromagnetic wave, c (including light) in free space by the expression

$$c = \frac{1}{\sqrt{\varepsilon_0 \mu_0}} \quad \frac{m}{s} \tag{14.6}$$

Microwave absorption in a material occurs because of different dielectric mechanisms such as dipolar, electronic, atomic, and Maxwell–Wagner effects (Metaxas and Meredith, 1988). In addition, there are conductive or ohmic losses. Since most dielectric measuring techniques cannot differentiate between the different forms of dielectric losses or separate them from the conductive loss, all forms of losses are grouped together in practice, defining an effective loss factor ε''_{eff} as a function of frequency as

$$\varepsilon''_{eff}(\omega) = \varepsilon''_{combined}(\omega) + \frac{\sigma}{\varepsilon_0 \omega} \tag{14.7}$$

where ω is the angular frequency, $\varepsilon''_{combined}$ stands for the combined loss mechanisms due to dipolar and other sources mentioned earlier, and σ represents the conductive or ohmic loss mechanism. Thus, in Chapter 13 on Electrical Conductivity of Foods, the conductivity mentioned is primarily the component σ in Equation 14.7. In practice, the subscript on ε''_{eff} is dropped, and ε'' is used to denote the loss factor due to all loss mechanisms combined. This interpretation of ε'' will be used for the rest of this chapter.

Microwave energy absorption in foods involves primarily two mechanisms—dipolar relaxation and ionic conduction, as depicted in Figure 14.2 (see also publications such as Decareau, 1984 and Engelder and Buffler, 1991). These interactions are with the electric field of the RF and microwaves. Water in the food is often the primary component responsible

DIELECTRIC PROPERTIES OF FOODS

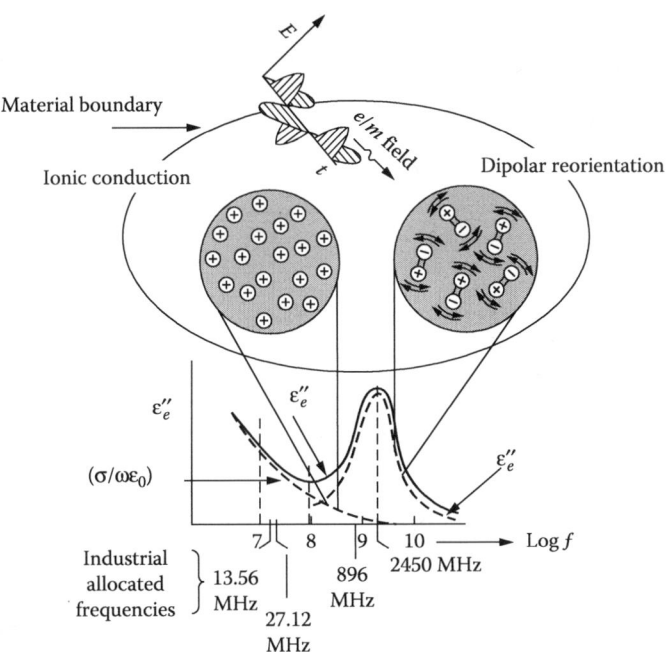

Figure 14.2 A schematic diagram depicting the dipolar and ionic loss mechanisms and their contributions to the dielectric properties as function of frequency. Some commonly used RF and microwave frequencies are noted (the 896-MHz frequency is used in the United Kingdom whereas 915-MHz is used in the United States). The dashed lines are contributions due to individual mechanisms, and the solid line stands for the combined effect. (From AC Metaxas. *Foundations of Electroheat.* John Wiley and Sons, Chichester, UK, 1996.)

for the dielectric heating. Because of their dipolar nature, water molecules try to follow the electric fields as they alternate at very high frequencies. Such rotations of the water molecules produce heat. Ions, such as those present in a salty food, migrate under the influence of the electric field, generating heat. This is the second major mechanism of heating in microwaves and RF energy. As mentioned, these and other mechanisms of energy losses are all considered to be included in the effective loss factor, ε''_{eff} (Equation 14.6) or simply ε''.

Some alternate quantities that are related to the dielectric constant and the loss factor just mentioned are often reported. Thus, the loss tangent, $\tan \delta$, is defined as

$$\tan \delta = \frac{\varepsilon''}{\varepsilon'} \tag{14.8}$$

The loss factor (i.e., the effective loss factor) is also related to an equivalent (effective) *conductivity*, σ, by

$$\sigma = \omega \varepsilon_0 \varepsilon'' \left[\frac{S}{m}\right] \tag{14.9}$$

The unit S/m stands for siemens per meter. Note that the equivalent conductivity σ now stands for any conductive or ohmic loss together with the conductive equivalent of dielectric loss. Figure 14.3 shows an overview of food dielectric property data with clustering for various groups of food products relative to each other. The characteristics of these food groups contributing to the dielectric properties in Figure 14.2 is discussed later in the chapter.

The rate of heat generation per unit volume, Q, at a location inside the food during microwave and RF heating is given by (this comes from the solution to Equations 14.1 through 14.4).

$$Q = 2\pi f \varepsilon_0 \varepsilon'' E^2 \left[\frac{W}{m^3}\right] \quad (14.10)$$

where E is the strength of electric field of the wave at that location and f is the frequency of the microwaves or the RF waves. Although only the dielectric loss appears in the above equation, the rate of energy absorption, Q, is also dependent on the dielectric constant, since that affects the electric field E in Equation 14.8.

Under the simplest of heating situations, where a plane wave is incident on a flat infinite surface, the solution to the Maxwell's equations can be written as

$$E = E_0 e^{-x/\delta}$$

where δ is the penetration depth (also called the skin depth for good conductors such as metals) in the context of electric field. Using Equation 14.8, the decay in energy deposition is written as

$$Q = Q_0 e^{-2x/\delta} = Q_0 e^{-x/\delta_p} \quad (14.11)$$

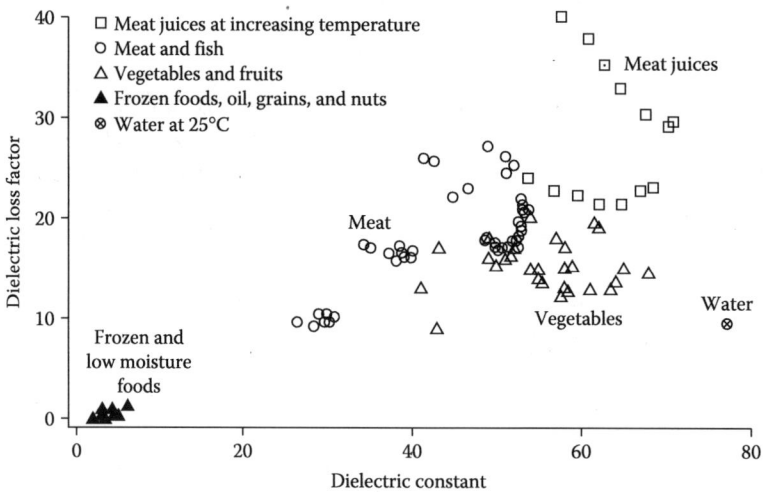

Figure 14.3 A scatter plot of some of the literature data for a variety of food materials showing some approximate grouping. Most of the data are from a frequency range of 2400–2500 MHz. Temperature varies between 5°C and 65°C for meats and meat juices, while the vegetable data is mostly at 23°C.

DIELECTRIC PROPERTIES OF FOODS

where δ_p is called the power penetration depth or often simply the penetration depth. In the food literature, penetration depth refers to the power penetration depth δ_p, as opposed to ε defined earlier, and this power penetration depth will be implied when referring to penetration depth in the remaining part of the chapter. It is the distance over which 63% of the power is dissipated and is related to the material properties by the equation

$$\delta_p = \frac{\lambda_0}{2\pi\sqrt{2\varepsilon'}} \left(\sqrt{1 + (\varepsilon''/\varepsilon')^2} - 1\right)^{-\frac{1}{2}} \quad (14.12)$$

where λ_0 is the wavelength of the microwaves in free space. Typical penetration depths for a range of food products corresponding to Figure 14.3 are shown in Figure 14.4.

The wavelength of microwaves in the food, λ, is given by

$$\lambda = \frac{\sqrt{2}}{f} \frac{c}{\sqrt{\varepsilon'}} \left[\sqrt{1 + \left(\frac{\varepsilon''}{\varepsilon'}\right)^2} + 1\right]^{-1/2} \quad (14.13)$$

The wavelength of the microwaves inside the food is important. For example, when a plane wave is incident on a surface at an angle, the angle of transmission into the material would depend on the wavelength. For a smaller wavelength, corresponding to a different set of properties in Equation 14.13, the angle of transmission is smaller, and this can lead to increased focusing effect in a curved food material such as a sphere (Zhang and Datta, 2003).

It is important to note that the exponential decay of microwave energy deposition, given by Equation 14.10, is only true for restricted microwave heating situations. For finite

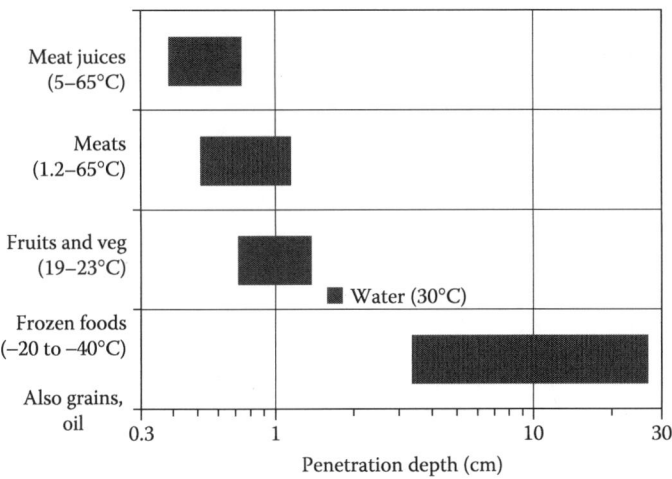

Figure 14.4 Typical ranges of penetration depths for various groups of food products (the dielectric properties correspond to those shown in Figure 14.3). Data are at frequencies near 2.45 GHz.

food sizes and oven configurations (which involve standing waves instead of traveling plane waves), such exponential decay cannot generally be assumed. In general, the amount of energy absorbed, as well as its spatial variation, is strongly dependent on the food shape, volume, surface area, dielectric properties, and equipment (oven) configuration. These dependencies are discussed in detail elsewhere (Zhang and Datta, 2001). For illustration, three of the important effects, those of volume, shape, and dielectric properties, are shown in Figures 14.5 through 14.7. Figure 14.5 shows how the amount of power absorbed in a food heated inside a microwave oven is dependent on the food volume. At smaller volumes, the amount of energy absorbed is smaller.

Figure 14.6 illustrates the combined effect of geometry and dielectric properties on spatial distribution of microwave energy absorption inside a food. For this spherical material, the penetration depth δ_p is large enough compared to its diameter, leading to focusing of energy near the center. The spatial distribution of energy will have higher values toward the center, opposite of what is given by Equation 14.10. Figure 14.7 shows how the relative amount of energy absorption varies in two different foods (of widely varying dielectric properties) heated simultaneously. It shows that at small volume, a food with higher dielectric loss is more efficient in absorbing energy than a lower loss material is. This reverses at higher volume, making the lower loss material absorb relatively more power.

Thus, dielectric properties alone (i.e., without solving Equations 14.1 through 14.4) cannot provide comprehensive information about the magnitude and distribution of

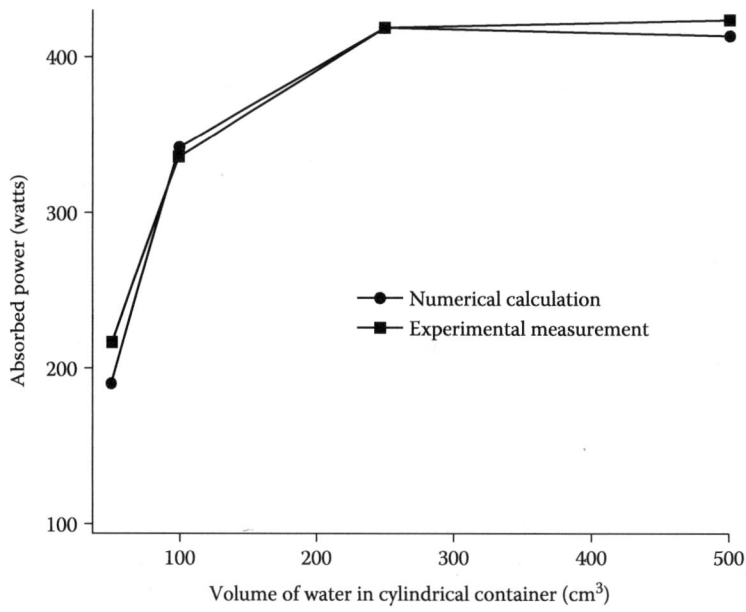

Figure 14.5 Absorbed power in a microwave oven changes with the volume of the food material being heated. (From H Zhang, AK Datta. Electromagnetics of microwave heating: Magnitude and uniformity of energy absorption in an over. In AK Datta, RC Anantheswaran, Eds. *Handbook of Microwave Technology for Food Applications*. Marcel Dekker, New York, 2001, pp. 33–68.)

DIELECTRIC PROPERTIES OF FOODS

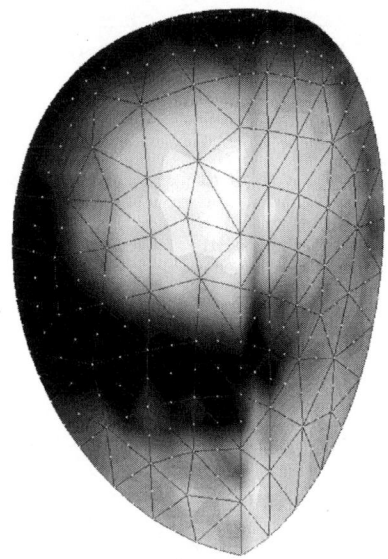

Figure 14.6 Combination of dielectric properties, size, and shape producing focusing effect with enhanced interior heating (lighter shade). From electromagnetic computations of food (egg white, $\varepsilon = 70-j15.8$; radius = 3 cm) heated in a microwave oven. (From H Zhang, AK Datta. Electromagnetics of microwave heating: Magnitude and uniformity of energy absorption in an over. In AK Datta and RC Anantheswaran, Eds. *Handbook of Microwave Technology for Food Applications*. Marcel Dekker, New York, 2001, pp. 33–68.)

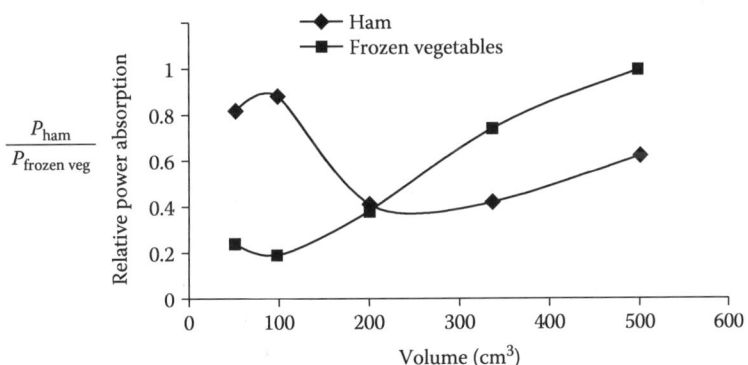

Figure 14.7 Effect of dielectric properties and volume on the relative heating rates during simultaneous heating of two materials. (From H Zhang, AK Datta. Electromagnetics of microwave heating: Magnitude and uniformity of energy absorption in an over. In AK Datta, RC Anantheswaran, Eds. *Handbook of Microwave Technology for Food Applications*. Marcel Dekker, New York, 2001, pp. 33–68.)

microwave energy absorption in foods. In some cases, however, dielectric properties may provide qualitative understanding of the relative rates of heating in different foods.

14.2.1 RF versus Microwave Heating

Over the range of temperature, composition, and frequencies of interest, dielectric properties for RF heating can be significantly different from those at microwave frequencies, although sometimes following a similar trend. In the following sections, RF properties are noted whenever available or appropriate. For example, for RF properties in baking, see Section 14.9.1 and for pasteurization and sterilization (higher temperatures), see Section 14.4.3.1.

14.3 MEASUREMENT PRINCIPLES

The methods for dielectric property measurement relevant for any desired application depend on the nature of the dielectric material to be measured, both physically and electrically, the frequency of interest, and the degree of accuracy required. Despite the fact that various techniques are in use, only instruments that can provide reliable measurements of the desired dielectric properties involving the unknown material in the frequency range of interest are to be considered (Nelson, 1999). The challenge in making accurate permittivity or dielectric property measurements lies in the design of sample holders for those measurements (RF and MW frequency ranges) and adequately modeling the circuit for reliable calculation of the permittivity from the electrical measurements. If one can estimate the RF circuit parameters, for example, the impedance or admittance, then the dielectric properties of the material at that particular frequency can be determined from equations relating material permittivity to the circuit parameters.

Microwave dielectric property measurement methods can be categorized as reflection or transmission types depending on resonant or nonresonant systems, with open or closed structures for sensing the properties of material samples (Kraszewski, 1980). Waveguide and coaxialline transmission methods represent closed structures, while the freespace transmission measurements and open-ended coaxialline systems represent open-structure techniques. Resonant structures can include either closed resonant cavities or open resonant structures operated as two-port devices for transmission measurements or as one-port devices for reflection measurements (Nelson, 1999).

14.3.1 Waveguide and Coaxial Transmission Line Methods

Early efforts to characterize the dielectric properties of materials were made at the Massachusetts Institute of Technology (von Hippel, 1954a, b). The values of ε' and ε'' were derived from transmission line theory, which indicated that these parameters could be determined by measuring the phase and amplitude of microwave signals reflected from or transmitted through a sample of material. For a waveguide structure, rectangular samples that fit into the dimensions of the waveguide at the frequency being measured are essential. For coaxial lines, an annular sample needs to be fabricated. Dielectric sample holder design for a particular material of interest is an important aspect of the measurement technique.

14.3.2 Short-Circuited Line Technique

The short-circuited line technique was originally reported by Roberts and von Hippel (1946). In a coaxial transmission line, by placing a terminating surface such as a metallic barrier, the transmitted electromagnetic wave is reflected back to the source. The field strength at any given point within the transmission line is simply a vector sum of the strengths of the incident and reflected waves. The standing wave ratio (SWR) is defined as the ratio of the vector sum of the strengths at maximum to that of minimum (Pace et al., 1968). The insertion of a dielectric material into the transmission line in contact with the short-circuit termination will cause changes in the position and the width of the standing wave nodes. The changes in the position of the node and the SWR are used to calculate the dielectric constant and the loss factor of the inserted dielectric materials. This technique could be used to measure the dielectric properties of liquid, powder, or solid samples. The limitation of this technique when dealing with a solid sample lies in the sample preparation. An annular sample has to be prepared, which may be time-consuming. This technique was used by Nelson (1972), and a general computer program was developed for low- or high-loss materials with measurements in short-circuited coaxial lines and cylindrical or rectangular waveguides (Nelson et al., 1974). At lower frequencies, coaxial lines are more practical because of the large size of waveguides required.

14.3.3 Open-Ended Probe Technique

A method that circumvents many disadvantages of the transmission line measurement technique was suggested by Stuchly and Stuchly (1980). The coaxial probe method is a modification of the transmission line method. It calculates the dielectric parameters from the phase and amplitude of the reflected signal at the end of an open-ended coaxial line inserted into a sample to be measured. Care must be exercised with this technique because errors are introduced at both very low frequencies and very high frequencies, as well as at low values of dielectric constant and loss factor. This technique is valid for 915 and 2450 MHz, especially for materials with loss factors greater than 1. Interpretation for lower loss materials, such as fats and oils, must be treated with caution. Typical open-ended probes utilize 3.5 mm (0.138") diameter coaxial line. For measurement of solid samples, probes with flat flanges may be utilized (Gabriel et al., 1986). A typical HP probe dielectric property measurement system includes an HP probe, a network analyzer, a coaxial cable, a PC, and the software. The Model HP 85070 probe can be used in the frequency range of 200 MHz to 20 GHz. In practice, the measurement band depends on the selection of the network analyzer. The accuracy of the HP probe was claimed to be ±5% for the dielectric constant and ±0.005 for the loss tangent (HP 85070B user's manual). This model can be used in the temperature range of −40°C to 200°C.

When using microwave or RF energy for pasteurization or sterilization processes for food products, the high-temperature measurement capacity becomes essential to study the dielectric properties of food material at temperatures above the boiling point of water.

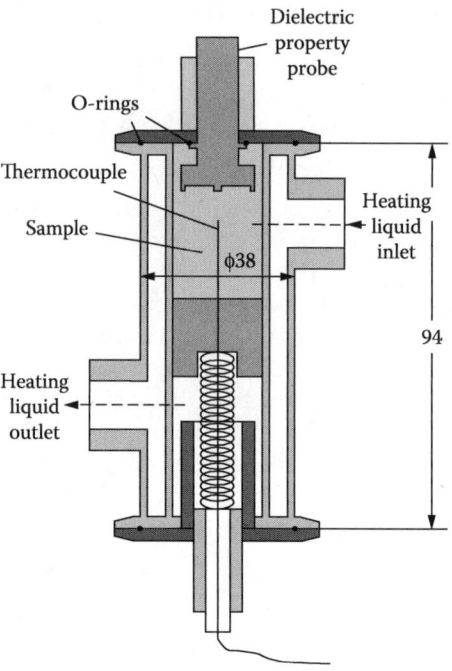

Figure 14.8 Diagram of pressure-proof dielectric test cell (stainless steel), dimensions in mm. (From Y Wang et al. *J Food Eng* 57: 257–268, 2003.)

Y. Wang et al. (2003) reported the measurement of dielectric properties of various food products with an HP probe (HP 85070B) using a custom-built temperature-controlled test cell (Figure 14.8). The test cell was constructed using two coaxial sections of 1- and 1.5-in. OD 304 stainless-steel sanitary tubing. Both parts were welded to a 1-in. sanitary ferrule at each end to serve as sample holder and water jacket for temperature control. The probe was installed, and a stainless-steel spring and a stainless-steel piston on the other end were used to make sure that the probe and sample had close contact. A thermocouple was inserted into the sample to monitor the temperature. With this system the dielectric properties of various food products were measured at temperatures from 20°C to 121.1°C under frequencies of 27, 40, 915, and 1800 MHz (Y. Wang et al., 2003).

14.3.4 Time-Domain Reflectometry Method

Time-domain reflectometry (TDR) (or spectroscopy) methods were developed in the 1980s and used for studies of the dielectric properties of food. Essentially, this method also utilizes the reflection characteristics of the material under test to compute the dielectric properties. They cover a frequency range from 10 MHz to 10 GHz. Measurement is very rapid and accuracy is high, within a few percent error. The sample size is very small and the substance measured must be homogeneous. Although these methods are expensive, they are excellent tools for advanced research on the interaction of the electromagnetic energy and materials over a wide frequency range (Mashimo et al., 1987).

14.3.5 Free-Space Transmission Technique

Freespace transmission technique is a nondestructive and noncontact measuring method. It does not require special sample preparation. Therefore, it is particularly suitable for materials at high temperatures and for inhomogeneous dielectrics. In addition, it may be easily implemented in industrial applications for continuous monitoring and control (Kraszewski et al., 1995). In a freespace transmission technique, a sample is placed between a transmitting antenna and a receiving antenna, and the attenuation and phase shift of the signal are measured. The results can be used to compute the material dielectric properties. Accurate measurement of the permittivity over a wide range of frequencies can be achieved by freespace techniques. In most systems, the accuracy of ε' and ε'' determined depends mainly on the performance of the measuring system and the validity of the equations used for the calculation. The usual assumption made for this technique is that a uniform plane wave is normally incident on the flat surface of a homogenous material, and that the planar sample has infinite extent laterally, so that diffraction effects at the edges of the sample can be neglected. Trabelsi et al. (1997) accounted for multiple reflections, mismatches, and diffraction effects at the edges of the sample, as they are generally considered the main sources of errors. To enhance the measurement accuracy, special attention must be paid to the choice of the radiating elements, the design of the sample holder, and the sample geometry and location between the two radiating elements.

14.3.6 Microstrip Transmission Line

The effective permittivity, represented by a combination of the substrate permittivity and the permittivity of the material above the line, of a microstrip transmission line (at least for thin width-to-height ratios) is strongly dependent on the permittivity of the region above the line. This effect has been utilized in implementing microwave circuits and to a lesser extent on the investigation of dielectric permittivity. Furthermore, the measurement of effective permittivity is relatively straightforward, and is well suited for implementation in industrial equipment. Such a system could be based on determining the effective permittivity of a microstrip line covered by an unknown dielectric substance (Keam and Holmes, 1995). Use of printed circuit boards, adding substrate materials to characterize materials, and measuring permittivity using algorithmic models have been reported. However, its applicability to food and agricultural material processing would still be an anticipatory issue at this stage.

14.3.7 Six-Port Reflectometer Using an Open-Ended Coaxial Probe

Ghannouchi and Bosisio (1989) have been working on nondestructive broadband permittivity measurements using open-ended coaxial lines as impedance sensors, which are of great interest in a wide variety of biomedical applications. An attempt is made to replace an expensive automatic network analyzer such as the HP8510B by combining the capabilities of personal computers with customized software to derive all the necessary information from less expensive components. The reported measuring system consists of a microwave junction designed to operate from 2 to 8 GHz and a number of standard microwave laboratory instruments (power meters, counters, sweepers, etc.) controlled by an IEEE 488 bus interface by a microcomputer (HP9816) to provide a precision low-cost automatic reflectometer suitable for permittivity measurements. The device is an open-ended coaxial test probe immersed in the test liquid kept at a constant temperature. Data acquisition and reduction are fully automatic. The complex reflection coefficient is calculated from the four power readings and the calibration parameters of the sixport reflectometer (SPR).

The SPR can provide nondestructive broadband permittivity measurements with an accuracy comparable to commercial Automated Network Analyzer accuracy but at a considerable reduction in equipment costs. This effective transmission line method used to represent the fringing fields in the test medium provides a good model to interpret microwave permittivity measurements in dielectric liquids. Using such a model, the precision expected on relatively high-loss dielectric liquid measurements is good. However, this method involves a more complex mathematical procedure in order to translate the signal characteristics into useful permittivity data.

14.3.8 Colloid Dielectric Probe (Hewlett-Packard)

Engineers at Hewlett-Packard (HP) have developed the first RF dielectric probe for evaluating colloidal liquids such as milk. The unit can quickly and accurately measure dielectric properties of these types of materials, offering the promise of improving a variety of food, chemical, pharmaceutical, and biochemical products.

The HP E5050A Colloid Dielectric Probe is designed for permittivity evaluation of colloidal liquid materials in the food, chemical, pharmaceutical, and biochemical industries. It operates from 200 kHz to 20 MHz with the HP4285A precision LCR meter and HP Vectra personal computer. The advanced sensing technique provides permittivity vs. frequency characteristics. Its electromagnetic technique eliminates the electrode polarization effect, which causes measurement error when ionic materials are measured with metal electrodes.

14.3.9 Test Cell with Boonton RX-Meter

Bengtsson et al. (1963) reported a test cell method for measuring dielectric properties of food materials in the range of 10–200 MHz. The equipment consists of a test cell and a Boonton RX-meter, which is in principle a modified Schering bridge circuit in combination with oscillator and null detector. The test cell is composed of two silver-coated copper electrodes covered and surrounded by Plexiglas to form a sample compartment. The dielectric properties of samples are determined using admittance of the fully loaded and the empty test cell. The RX-meter was also used by Jorgensen et al. (1970) with a coaxial sample holder for dielectric property measurements from 50 to 250 MHz.

DIELECTRIC PROPERTIES OF FOODS

14.3.10 Cavity Perturbation Technique

The cavity (TM or TE mode) perturbation technique is one of the most commonly used techniques for measuring dielectric properties of homogeneous food materials because of its simplicity, ease of data reduction, accuracy, and high-temperature capability (Bengtsson and Risman, 1971; de Loor and Meijboom, 1966; Metaxas and Meredith, 1988; Sucher and Fox, 1963). The technique is also well suited to low-loss materials (Hewlett-Packard, 1992; Kent and Kress-Rogers, 1987). It is based on the shift in resonant frequency and the change in absorption characteristics of a tuned resonant cavity due to insertion of a sample of target material. The measurement is made by placing a sample completely through the center of a waveguide (rectangular or circular) that has been made into a cavity. Changes in the center frequency and width due to insertion of the sample provide information to calculate the dielectric constant. Changes in the Qfactor (ratio of energy stored to energy dissipated) are used to estimate the dielectric loss. EM field orientation for two standard cavity modes (TE and TM) is shown in Figure 14.9.

The size of the cavity must be designed for the frequency of interest; the relationship is inverse (higher frequency, smaller cavity). Each cavity needs calibration, but once the calibration curves have been obtained, calculations are rapid. Sample preparation is relatively easy, and the permittivities of a large number of samples can be determined in a short time. This method is also easily adaptable to high (up to 140°C) or low (–35°C) temperatures (Risman and Bengtsson, 1971; Ohlsson and Bengtsson, 1975; Meda, 1996), and has been used to determine the dielectric properties of many agrifood products over a wide range of frequencies, temperatures, and compositions.

Meda (1996) used a cavity perturbation system for dielectric properties at 2450 and 915 MHz. As illustrated in Figure 14.9, the system consists of a dielectric analyzer, a

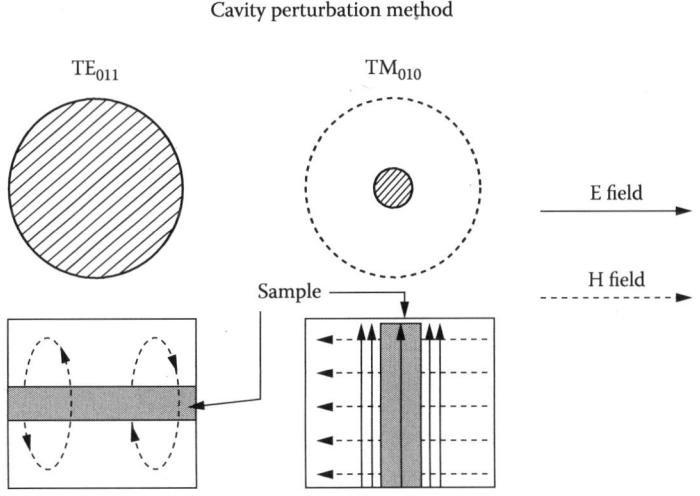

Figure 14.9 Schematic of a circular perturbation cavity in simple TE and TM modes.

measurement cavity, a heating/cooling unit, and a PC. The dielectric analyzer can handle both 2450 and 915 MHz; however, two cavities are needed for the two frequencies. Both cavities work in the TM_{010} mode. The dimensions for the two cavities are 249.7 mm I.D. and 40 mm in height for 915 MHz, and 96 mm I.D. and 40 mm in height for 2450 MHz. The materials used for constructing the cavities are copper and the combination of copper and brass for 915 and 2450 MHz, respectively. The heating and cooling unit is used to regulate the desired measurement temperature of the sample. The dielectric properties of the samples introduced to the cavities are determined by the following equations:

$$\varepsilon' = 1 + 0.539 \left(\frac{V_o}{V_s}\right)(\Delta f) \tag{14.14}$$

$$\varepsilon'' = 0.269 \left(\frac{V_o}{V_s}\right)\left(\frac{1}{Q_s} - \frac{1}{Q_o}\right) \tag{14.15}$$

$$\Delta f = \frac{f_o - f_s}{f_s} \tag{14.16}$$

where V_s and V_o are the volumes of the sample and the cavity, respectively; f_o and f_s are resonant frequencies of the empty and sample-loaded cavity, respectively; and Q_o and Q_s are the quality factors of the empty and sample-loaded cavity, respectively. The resonant frequencies and the Q factors are determined by the computer program based on the signal sent by the dielectric analyzer.

Before using the system, a calibration using a sample with known dielectric properties, for example, distilled water, must be performed. Sample preparation is relatively simple compared to the coaxial-line technique, but the volume of the sample is a critical factor. Here, sample preparation methods for different forms of samples are introduced.

14.3.10.1 Solid Sample Preparation
For solid materials, samples in the form of rods can be formed, molded, or machined directly from their material into microwave-transparent test tubes or tubing. While quartz is the best available material for this purpose, borosilicate glass is considered acceptable; ordinary glass should not be used. Wall thickness should be as thin as possible while having the required mechanical rigidity. Paper or plastic straws may be used if glass is not available. For a semisolid material such as Tylose™, the sample preparation is quite difficult; however, special micropipeting equipment for such gel-type materials has been successfully designed and built (Meda, 1996).

14.3.10.2 Liquid Sample Preparation
Liquids are filled into test tube sample holders with a pipette. Small-diameter pipettes themselves also make excellent sample holders. Two hundred microliter pipettes are suitable for low-loss materials and 10 μL pipettes for high-loss materials. Materials that can be melted are poured into sample holders and allowed to solidify. This technique is appropriate if the material does not change its properties following melting and resolidification.

14.3.10.3 Semisolid Samples

Sample preparation involves either filling the sample in its molten state and then solidifying or applying vacuum at one end while forcing the sample into thin, cylindrical-shaped holders. Since temperature measurements may be difficult due to the nature of the materials, such as cheese or butter, it is important to develop suitable fixtures to contain samples at different threshold conditions.

The system was tested for repeatability using 13 different samples with four replicates each at 22°C. The samples covered the dielectric constants from about 2 to about 80 and loss factors from 0.009 to 12 m with a relative standard error of 0.034–1.88% for the dielectric constant and 0.03–3.18% for the loss factor. The dielectric properties of many organic solvents were measured with this system, and they show good agreement with the values reported in the literature (Meda, 2002).

14.3.11 Summary of Dielectric Property Measurement Techniques

Although many techniques have been developed for dielectric property measurement, the selection of the proper technique depends on the nature of the dielectric material, the frequency of interest, the degree of accuracy required, and the availability of the measuring equipment. Automated network analyzers provide very accurate measurement for most food products, but the high cost of the equipment limits its application. For the measurement of dielectric properties close to the ISM frequencies, especially at the frequencies of 915 and 2450 MHz, the cavity perturbation technique, using a dielectric analyzer instead of a network analyzer, could be attractive because of its accuracy, ease of operation, simple sample preparation, convenient temperature control, and the affordability of owning this equipment.

14.4 FREQUENCY AND TEMPERATURE DEPENDENCE

As with many other food properties, first-principle-based prediction equations for dielectric properties as a function of temperature, composition, and frequency are not available. Empirical or semiempirical correlations are the only possibilities, but even these are scarce. In one attempt (Sun et al., 1995) to obtain empirical correlations of properties, when data from many different types of foods (meats, fruits, and vegetables) were considered together, very little correlation with composition was observed. This was attributed, among other factors, to variability in sample composition and measurement technique, and to the general unavailability of detailed composition data. Thus, correlations had to be restricted to a particular group of food products or even one particular food type (Calay et al., 1995; Sipahioglu and Barringer, 2003). Thus, when available, predictive equations will be discussed in this section under specific groups of food products.

14.4.1 Frequency Dependence

Dielectric properties can vary significantly with the frequency of the waves, which will now be examined in detail. Since heating applications use only a few specific frequencies (see Table 14.1), data at those frequencies suffice, and attention is focused more on composition and temperature dependence at these frequencies. In other applications, dependence of properties over a range of frequencies can provide significant information on the nature of the material. As mentioned earlier, dipolar and ionic conduction are the two primary mechanisms for microwave absorption in food materials. Thus, the frequency dependence of the dielectric properties arises from the frequency dependence of these two mechanisms. Of these two, most significant is the frequency dependence of the dipolar loss mechanism in polar materials such as water. The Debye model (Debye, 1929) describes the frequency dependence of dielectric properties of pure polar materials:

$$\varepsilon' = \frac{\varepsilon_s - \varepsilon_\infty}{1 + \omega^2 \tau^2} + \varepsilon_\infty \tag{14.17}$$

$$\varepsilon'' = \frac{(\varepsilon_s - \varepsilon_\infty)\omega\tau}{1 + \omega^2 \tau^2} \tag{14.18}$$

where ε_s is the dielectric constant at very low frequencies (d.c.), ε_∞ is the dielectric constant at high frequencies, τ is the relaxation time of the system that controls the build-up and decay of polarization, and ω is angular frequency.

The idealized relationships defined by Equations 14.17 and 14.18 are plotted in Figure 14.10. At low frequencies of the waves, the dipoles in the material (e.g., water) have time to follow the variations of the applied field, and the dielectric constant is at its maximum value ε_s. For example, the low-frequency dielectric constant of pure water (a dipole) at 20°C has a value of 80.2. The dielectric constant has a value of ε_∞ at high microwave frequencies, where dipoles are unable to follow rapid field reversals. For water at 20°C, this value is 5.6. The loss factor is zero at both high and low frequencies. At low frequency, dipoles do not oscillate at a high enough rate to produce significant heat generation, while at a very high frequency, the dipoles are not reacting to the electric field (they are unable to follow the

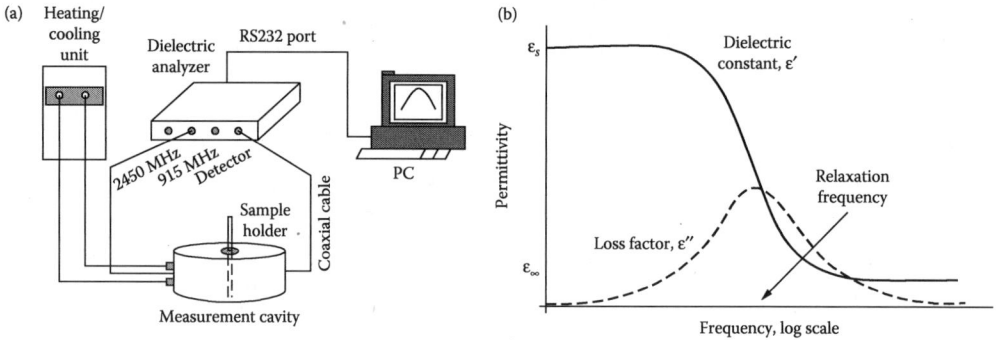

Figure 14.10 (a) Schematic of dielectric property measurement system at frequencies of 2450 and 915 MHz using cavity perturbation technique. (b) Idealized Debye dielectric relaxation (Equations 14.17 and 14.18) for polar molecules with a single relaxation time. (From P Debye. *Polar Molecules*. The Chemical Catalog Co., New York, 1929.)

DIELECTRIC PROPERTIES OF FOODS

rapid field reversal). At the relaxation frequency, loss factor is at its maximum. This Debye interpretation of the variation of dielectric constant and loss factor with frequency in terms of dipolar rotation against frictional forces in the medium has limitations, and large deviations can occur for solid dielectrics, for example.

Dielectric relaxation in liquid water has been studied extensively (Hasted, 1973). Water in the liquid state is a good example of a polar material, and the Debye equation describes the variation of its dielectric properties with frequency (called a dielectric spectrum) quite well in the frequency ranges of interest in food processing. For example, Figure 14.11 shows the measured data on frequency dependence of dielectric properties of water and is quite like the Debye behavior predicted in Figure 14.10. The relaxation parameters given in Table 14.2 can be used with Equations 14.17 and 14.18 to provide close estimates for dielectric properties of water for different temperatures and frequencies.

Often, industrial microwave heating involves regular tap water, as opposed to distilled and/or deionized water, shown in Figure 14.11. For example, water is often used as a terminal load to protect the magnetron (which produces the microwave energy). Properties of tap water can be significantly different from the data for distilled water discussed above, depending on the source of the water. Tap water properties can also be relevant in domestic microwave heating. Some data and an empirical model on the dielectric properties of tap water can be seen in Eves and Yakovlev (2002).

Frequency dependence of the ionic loss mechanisms has been reported for salt solutions (e.g., Mudgett, 1995). In the frequency range covering RF and microwave, the ionic component of the loss factor of salt solutions (at high enough concentration where ionic effects dominate) decreases with frequency (this is illustrated in the frequency response in Figure 14.2). As temperature increases, the curve shifts to the right, that is, the loss factor

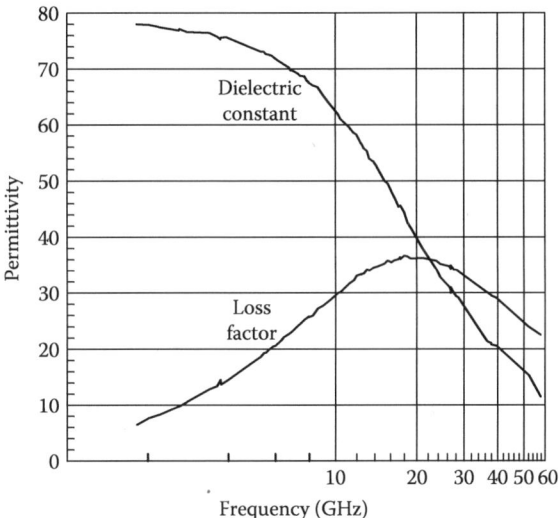

Figure 14.11 Measured dielectric properties of bidistilled and deionized water. (Data from U Kaatze. *J Chem Eng Data* 34: 371–384, 1989.)

ENGINEERING PROPERTIES OF FOODS

Table 14.2 Debye Dielectric Relaxation Parameters for Water

Temperature (°C)	ε_s	ε_∞	τ, ps	Relaxation Frequency (GHz)
0	87.9	5.7	17.67	9.007
10	83.9	5.5	12.68	12.552
20	80.2	5.6	9.36	17.004
30	76.6	5.2	7.28	21.862
40	73.2	3.9	5.82	27.346
50	69.9	4.0	4.75	33.506
60	66.7	4.2	4.01	39.690

Source: U Kaatze. *J Chem Eng Data* 34: 371–384, 1989.

due to ionic effects increases with temperature over the entire range of frequencies. Such plots can be seen in Mudgett (1995).

14.4.2 Frequency Dependence in Food Materials

Water in its pure liquid state is rarely found in food products. Instead, the water in foods contains dissolved constituents, physically absorbed in food capillaries or chemically bound to other molecules in the food. Therefore, it is difficult to understand and predict the dielectric behavior of food materials at different frequencies, temperatures, and moisture contents based on the behavior of water alone.

As an example, the frequency dependence of dielectric properties of a high-moisture material such as a potato is shown in Figure 14.12. The decrease in dielectric constant can

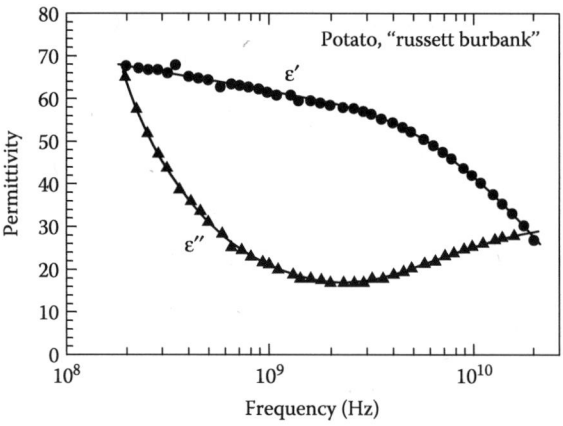

Figure 14.12 Frequency dependence of microwave dielectric properties of potato at 23°C. (From SO Nelson, WR Forbus, KC Lawrence. *J Microwave Power Electromagn Energy* 29: 81–93, 1994.)

be seen to follow that of water (Figure 14.11) in the frequency range involved. The loss factor, however, does not follow Figure 14.11 as closely. The decrease and increase in loss factor has been related to ionic conductivity at lower frequencies, to bound water relaxation, and to free water relaxation near the top of the frequency range (Nelson and Datta, 2001). The data illustrates that the dielectric properties of water alone are not sufficient to predict the properties of even high-moisture foods. Another illustration of property variation with frequency is shown in Figure 14.13 for a low-moisture material such as wheat grain over a frequency range of 250 Hz to 12 GHz. The dielectric constant decreases but the loss factor may increase or decrease with frequency. The loss factor of very dry wheat indicates probable bound water relaxation, but this is not very evident in the curves for dielectric constant (Nelson and Datta, 2001). Moisture content obviously has a very significant effect on these properties.

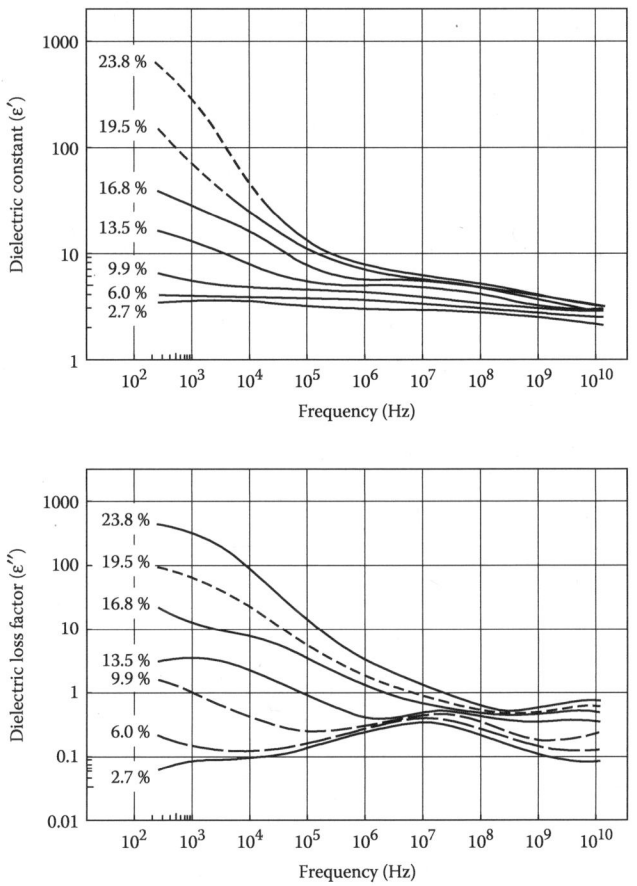

Figure 14.13 Frequency dependence of the dielectric properties of hard red winter wheat at 24°C and indicated moisture contents. (From SO Nelson, LE Stetson. *J Agric Eng Res* 21: 181–192, 1976.)

14.4.3 Temperature Dependence in Water, Salt Solutions, and Foods

Temperature generally has a significant influence on dielectric properties. In pure water, an increase in temperature increases the frequency at which the loss factor is at its maximum (relaxation frequency, see Table 14.2 presented earlier). Since the relaxation frequency of water at 0°C is about 9 GHz, and the curve shifts further to the right as temperature increases, the loss factor of water due to dipolar heating reduces as temperature increases for microwave frequencies of 2.45 GHz (or lower).

Salt solutions can be thought of as the next most complicated system beyond pure water and have been studied in detail (Hasted et al., 1948; Stogryn, 1971). Figure 14.14 illustrates the properties of salt solutions. The loss factor of a salt solution is the combined effect of two mechanisms, dipole loss and ionic loss. It was mentioned above that the dipolar loss decreases with temperature. The contribution to loss factor from ionic conduction, however,

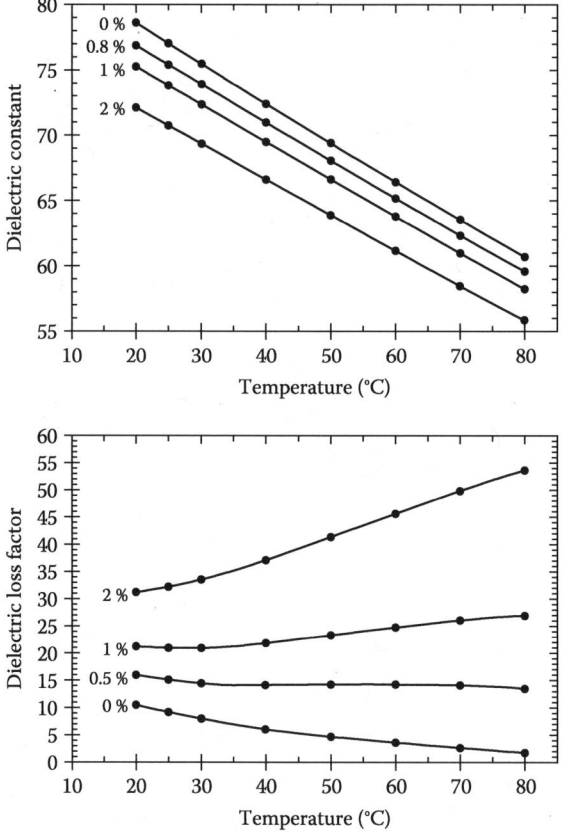

Figure 14.14 Permittivity of aqueous NaCl solutions of indicated salinity (salt content by weight) as a function of temperature at 2450 MHz. (From SO Nelson, AK Datta. Dielectric properties of food materials and electric field interactions. In AK Datta, RC Anantheswaran, Eds. *Handbook of Microwave Technology for Food Applications*. Marcel Dekker, New York, 2001, pp. 69–114.)

increases with temperature due to decreased viscosity of the liquid and increased mobility of the ions. The variation of loss factor in the salt solution depends on which mechanism dominates. At higher salt concentrations, ionic loss dominates, and the loss factor increases with temperature. At concentrations between 0.5% and 1%, the temperature coefficient changes from negative to positive for temperatures greater than 25°C (Figure 14.14).

The concentration of salt itself has significant effect on the dielectric properties of the salt solution, as expected. Increase in salt concentration decreases the dielectric constant but increases the loss factor, as shown in Figure 14.14.

Data on salt solutions are useful in studying food systems that are more complex. In a manner similar to a salt solution, the addition of salt to a food product reduces the dielectric constant due to the ability of salt to bind free water in the system. The binding force depends on the size and charge of the molecule. However, the addition of salt increases the loss factor above that of pure water since more ions are present and charge migration is increased. Also, since the dissociated or ionized forms of electrolytes interact with microwaves, pH and ionic strength can have significant effects on dielectric properties. For this reason, ionizable materials are used in browning formulations for microwaveable products (Shukla and Anantheswaran, 2001).

In foods, the variation of dielectric properties with temperature follows partly the trends discussed for salt solutions but can be complicated by the presence of bound water. Figure 14.15 shows the variation of dielectric properties of different foods with temperature at 2.8 GHz. The property variations below freezing are discussed separately below. The similarity of these food data (mostly at high moisture content) with salt solution data is obvious. In both cases, the dielectric constant decreases with temperature. The dielectric loss decreases with temperature except for ham, similar to the low salt concentration data for salt solutions. The data for ham have a trend comparable to the salt solution data for high salt concentration, which is consistent, since ham has a high salt concentration. Detailed information on the contribution from bound water (at low moisture contents) is less available. In data on wood, the loss factor increased with temperature (an effect opposite to that in Figure 14.15) at very low moisture contents for microwave frequencies; this was attributed to reduction in physical binding, whereby the dipoles are freer to reorient (Metaxas and Meredith, 1988). Generally, it is accepted that the contribution to dielectric loss from bound water increases with temperature for food materials at microwave frequencies.

14.4.3.1 Dielectric Properties below Freezing and above Boiling Temperatures

One of the sharpest possible changes in dielectric properties occurs during a freezing or thawing process. Accurate dielectric properties in frozen and partially frozen material are critical to determining the rates and uniformity of heating in operations involving frozen foods such as microwave thawing and tempering. As the ice in the food melts, absorption of microwaves increases tremendously. Thus, the portions of material that thaw first absorb significantly more microwaves and heat at increasing rates that can lead to undesirable runaway heating, described later. Dielectric properties of frozen food materials have been reported in the literature (Bengtsson et al., 1963; Bengtsson and Risman, 1971), but there are very few data points in the partially frozen range, where the properties can be a strong function of composition, particularly the total water and salt content. Salt affects

ENGINEERING PROPERTIES OF FOODS

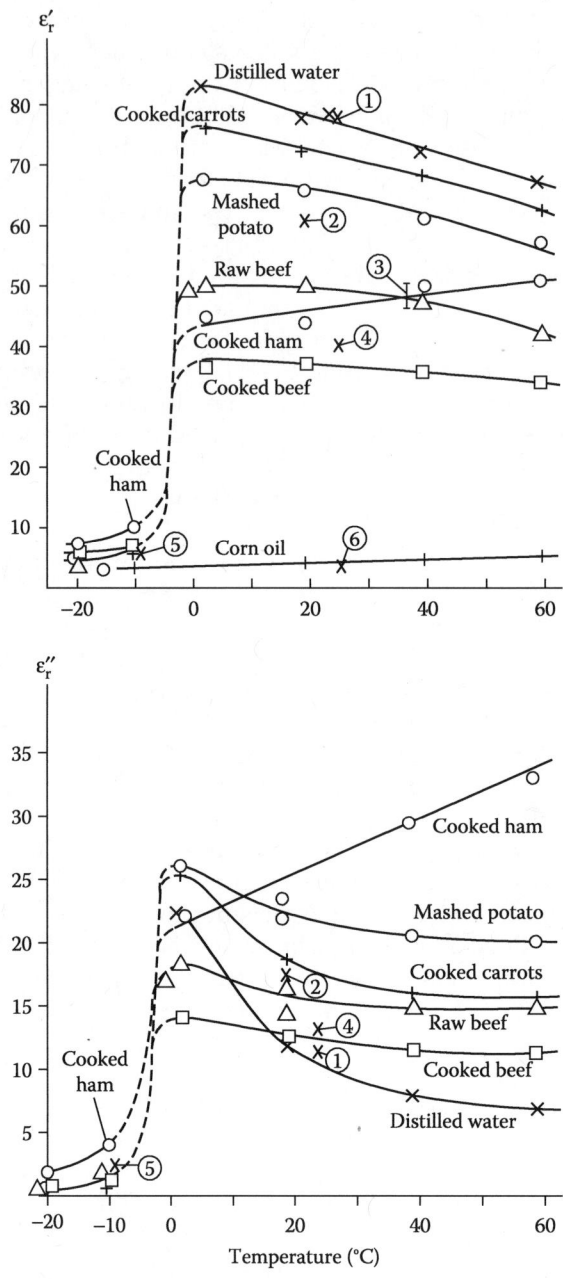

Figure 14.15 Temperature dependence of dielectric constant (top figure) and dielectric loss factor (bottom figure) of selected foods. (From NE Bengtsson, PO Risman. *J Microwave Power* 6: 107–123: 1971.)

dielectric behavior through the freezing point depression, leaving more water unfrozen at any temperature. Salt also increases the ionic content and therefore the interactions with microwaves.

Chamchong and Datta (1999) reported dielectric properties of tylose, a food analog, covering the frozen range, as shown in Figure 14.16. By measuring the apparent specific heat using differential scanning calorimetry (DSC), the fraction of water frozen at any temperature is measured directly as shown in Figure 14.17. The frozen fraction is distinctly different for the three salt concentrations in tylose, with higher salt concentration leading to less frozen water at any temperature. The dielectric constant and loss in the partially frozen region (Figure 14.16) is predicted as a linear function of the fraction of unfrozen water obtained experimentally (Figure 14.17) using the following equations.

$$\varepsilon'(T) = 56.98\, f(T) + 3.4402[1 - f(T)] \tag{14.19}$$

$$\varepsilon''(T) = 33.79\, f(T) + 0.7450[1 - f(T)] \tag{14.20}$$

where $f(T)$ is the fraction of unfrozen water at any temperature T. Both the dielectric constant and the loss factor decrease significantly as more water freezes. Since the fraction of unfrozen water is a nonlinear function of temperature, the increase in dielectric properties of the partially frozen material is also nonlinear with temperature. Above

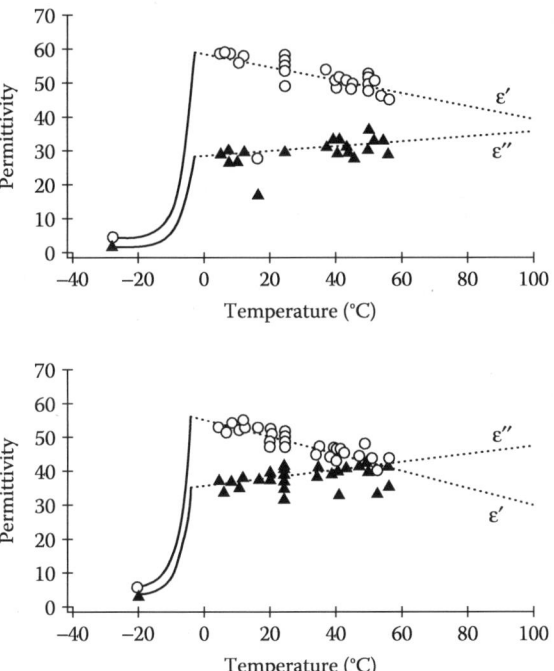

Figure 14.16 Dielectric properties of tylose at 2.45 GHz with 2 (top figure) and 4% salt (bottom figure). (From M Chamchong, AK Datta. *J Microwave Power Electromagn Energy* 34: 9–21, 1999.)

ENGINEERING PROPERTIES OF FOODS

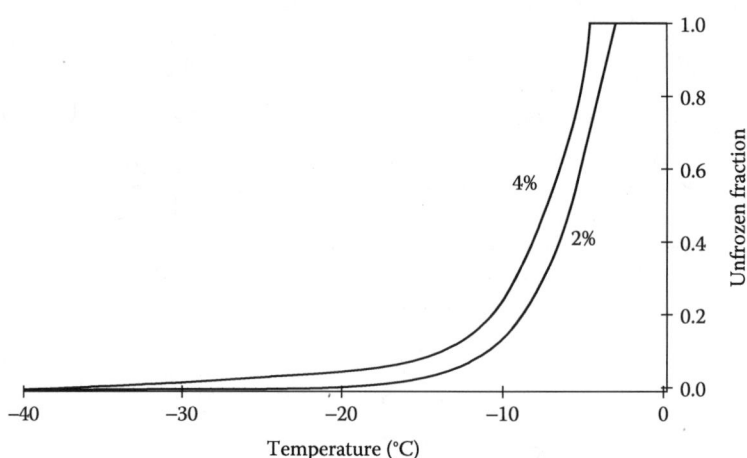

Figure 14.17 Unfrozen fraction of water in tylose (a food analog) as a function of temperature for two different salt contents. (From M Chamchong, AK Datta. *J Microwave Power Electromagn Energy* 34: 9–21, 1999.)

the freezing range, the dielectric constant of tylose decreased linearly with temperature, while the dielectric loss of tylose increased linearly with temperature. With the addition of salt, the dielectric constant decreased while the dielectric loss increased.

As in the freezing range, high temperatures can also lead to a significant change in dielectric properties, especially in foods containing salt. Such high temperatures occur, for example, in microwave pasteurization and sterilization. Data at high temperatures have been scarce (Ohlsson and Bengtsson, 1975; Y. Wang et al., 2003). Higher temperature data for a number of food products from the study of Ohlsson and Bengtsson (1975) are shown in Figures 14.18 and 14.19. Higher temperature data on macaroni and cheese (Y. Wang et al., 2003) are shown in Figure 14.20. Data at the commonly used microwave frequency of 2450 MHz were not measured in this study but were extrapolated from the data measured at other frequencies for a number of food materials. One of the observations from these data is that runaway heating is more likely to occur at radiofrequencies than at microwave frequencies.

In addition to temperature effects *per se*, physical and chemical changes such as gelatinization of starch (Miller et al., 1991) and denaturation of protein leading to release of water and shrinkage (Bircan and Barringer, 2002a) at higher temperature can significantly change dielectric properties. These data are discussed in Section 14.5.3.

14.4.3.2 Temperature Dependence of Loss Factor and Runaway Heating

An increase in the dielectric loss factor with temperature, as shown, for example, in Figures 14.14, 14.16, and 14.19, can lead to what is commonly referred to as runaway heating. Runaway heating is the material's ability to absorb increasing amounts of microwave energy as its temperature increases; thus, the rate of temperature rise progressively increases as heating progresses. In frozen foods, for example, the regions that will thaw

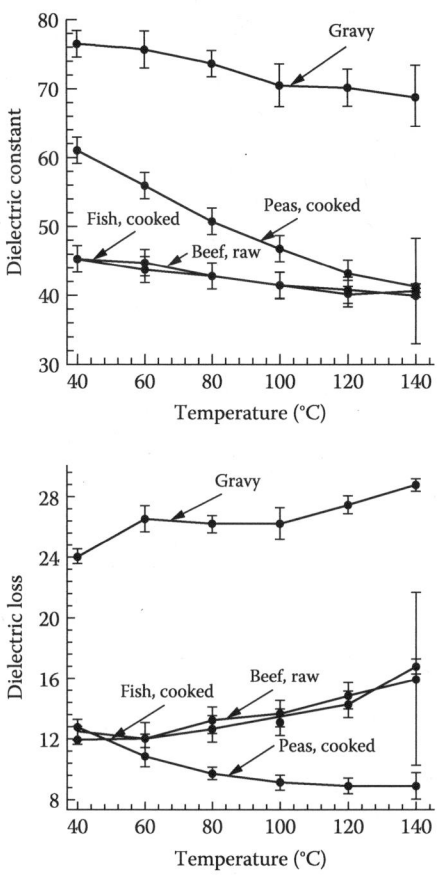

Figure 14.18 Dielectric constant and loss factors at 2800 MHz of several food products as a function of temperature covering sterilization temperatures. The compositions are raw beef (74.9% water, 1.3% fat), cooked fish (77.2% water, 0.2% fat), gravy (89.8% water, 4.4% fat), and cooked peas (78.6% water). (From T Ohlsson, NE Bengtsson. *J Microwave Power* 10(1): 93–108, 1975.)

sooner will absorb an increasing amount of microwave energy (following Figure 14.16) and can be boiling while other regions within the same food are still frozen. Another example is Figure 14.20, which shows that for these materials (macaroni and cheese), runaway heating is more likely in RF heating as compared to microwave heating (Tang et al., 2002).

14.5 COMPOSITION DEPENDENCE

Dielectric properties of food products obviously depend on composition. Moisture, salt content, carbohydrate, protein, and fat are some of the major relevant components. The dielectric constant and loss factor are affected by the presence of free and bound water,

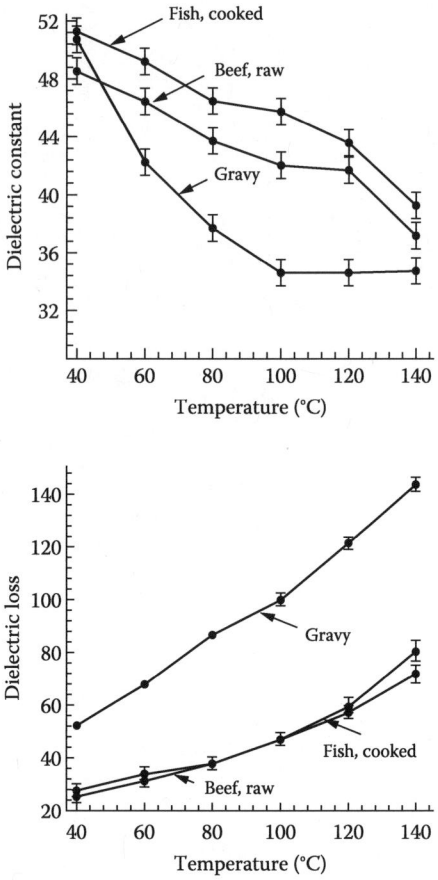

Figure 14.19 Dielectric constant and loss factors at 900 MHz of several food products as a function of temperature covering sterilization temperatures. The compositions are raw beef (74.9% water, 1.3% fat), cooked fish (77.2% water, 0.2% fat), gravy (89.8% water, 4.4% fat), and cooked peas (78.6% water). (From T Ohlsson, NE Bengtsson. *J Microwave Power* 10(1): 93–108, 1975.)

surface charges, electrolytes, nonelectrolytes, and hydrogen bonding in the food product. The physical changes that take place during processing, such as moisture loss and protein denaturation, also affect dielectric properties. Therefore, investigation of the dielectric behavior of major food components and effects of processing on dielectric properties are needed to design microwave food products, processes, and equipment.

Before going into detail, a qualitative picture of relative dielectric loss values of various components in food materials is useful (Table 14.3). As mentioned earlier, general correlations relating composition to dielectric properties have been unsuccessful (Calay et al., 1995; Sun et al., 1995), perhaps due to the complex interaction between the various components in contributing to dielectric properties of the overall system.

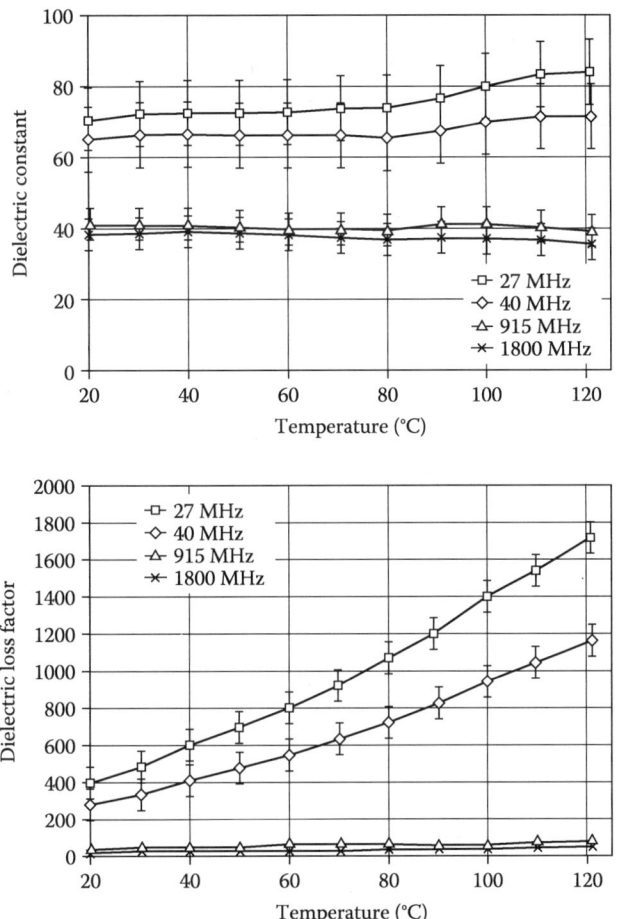

Figure 14.20 Dielectric constant and loss factors of macaroni and cheese as a function of temperature covering sterilization temperatures. (From Y Wang et al. *J Food Eng* 57: 257–268, 2003.)

14.5.1 Moisture Dependence

As has been discussed under dielectric properties of water, liquid water is very polar and can easily absorb microwave energy because of dipolar rotation. For this reason, moisture content is one of the major determinants of food dielectric properties. Water can exist in food systems either in free form or in a loosely defined state called the bound state. Free water is found in capillaries, and bound water is physically adsorbed to the surface of dry material. Although the loss factor is affected by both free and bound water, since relaxation of bound water takes place below the microwave frequencies, its effects are small in microwave heating. The stronger the binding forces between water and protein or carbohydrates, the smaller is the contribution of the bound water to the dielectric constant or the loss factor.

ENGINEERING PROPERTIES OF FOODS

Table 14.3 Qualitative Picture of Dielectric Loss of Major Food Components at Microwave Frequencies

Food Components	Relative Activity
Bound water	Low
Free water	High
Protein	Low
Triglycerides	Low
Phospholipids	Medium
Starch	Low
Monosaccharides	High
Associated electrolytes	Low
Ions	High

Source: Shukla and Anantheswaran. *Handbook of Microwave Technology for Food Applications.* Marcel Dekker, New York, 2001, pp. 355–395.

Increase in the dielectric constant and loss factor of food systems with moisture content were shown in various studies (Bengtsson and Risman, 1971; Guo et al., 2010a, b; Ndife et al., 1998; Nelson et al., 1991; Wang et al., 2011). The increase in water content increases the polarization, increasing both dielectric constant and loss factor. As an example, Figure 14.21 shows the variation of dielectric constant and loss factor of potato puree as moisture is added (Wang et al., 2011). At moisture contents less than 20%, dielectric properties were almost constant, showing that water and salt were in tightly bound form. Between moisture contents of 20% and 76%, there was a rapid increase in dielectric properties due to the increase in freely available water.

However, in frozen state, both dielectric constant and loss factor of samples were small and not much affected with increase in moisture content. This is attributed to the presence of moisture and salt in the sample in bound form.

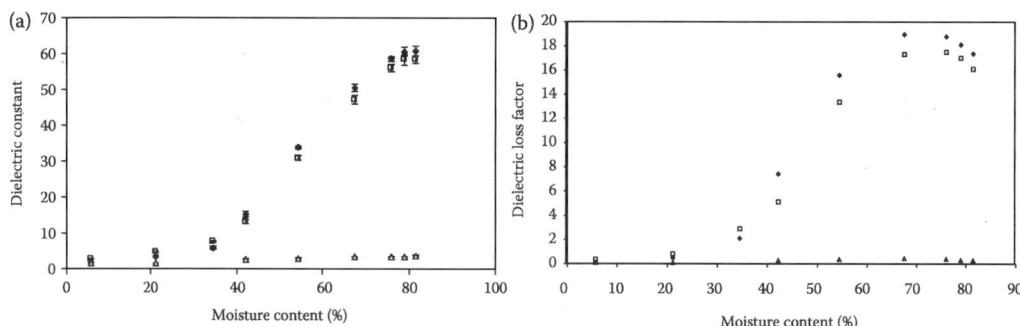

Figure 14.21 Variation of dielectric constant (a) and loss factor (b) of potato puree at different temperature and moisture content (◆): 20°C, (□): 65°C, (△): −20°C. (From R Wang et al. *J Food Eng* 106: 290–297, 2011.)

Table 14.4 Variation of Dielectric Properties of Cake Batter with Moisture Content and Temperature

Moisture Content (kg/kg⁻¹ db)	Temperature (°C)							
	20		40		60		80	
	ε'	ε''	ε'	ε''	ε'	ε''	ε'	ε''
0.429	16.5 ± 0.4	7.8 ± 0.3	16.3 ± 1.1	6.8 ± 0.9	16.3 ± 2.0	6.5 ± 1.8	15.6 ± 2.4	5.3 ± 1.8
0.500	18.6 ± 0.8	8.8 ± 0.5	19.7 ± 0.8	8.1 ± 1.0	20.2 ± 1.6	7.7 ± 1.2	19.2 ± 1.8	6.1 ± 1.9
0.600	22.8 ± 0.1	10.5 ± 0.1	22.8 ± 0.6	9.0 ± 0.7	21.6 ± 1.1	7.9 ± 1.0	22.8 ± 1.2	7.4 ± 1.2
0.700	25.3 ± 0.3	11.1 ± 0.3	25.3 ± 1.5	9.5 ± 1.1	25.7 ± 1.5	9.0 ± 1.7	24.3 ± 1.8	7.7 ± 1.3
0.800	27.5 ± 0.7	11.8 ± 0.8	28.6 ± 1.1	10.3 ± 0.9	26.8 ± 1.4	9.0 ± 1.0	26.3 ± 1.8	7.8 ± 1.3
0.900	30.8 ± 0.9	12.5 ± 0.7	31.9 ± 0.7	11.0 ± 0.5	28.0 ± 1.2	9.3 ± 0.8	28.5 ± 1.4	8.6 ± 0.9
1.000	32.6 ± 1.1	12.6 ± 0.6	3.8 ± 1.0	11.1 ± 0.6	28.8 ± 1.9	9.2 ± 1.3	31.5 ± 2.4	10.1 ± 1.8

As another example of increase in dielectric properties with moisture content, Table 14.4 shows data for cake batter at several temperatures (Al-Muhtaseb et al. 2010).

Since correlations with moisture are not possible when all kinds of foods are considered together, correlations with moisture (and temperature) are available only for specific groups of food products. See, for example, equations in Section 14.5.5 for meats and in Section 14.5.7 for fruits and vegetables.

14.5.2 Carbohydrate Dependence

The major carbohydrates that are present in food systems are starches, sugars, and gums. Carbohydrates do not show appreciable dipolar polarization at microwave frequencies (Ryynänen, 1995). Therefore, for carbohydrate solutions, the effect of free water on dielectric properties becomes significant. Hydrogen bonds and hydroxyl group water interactions also play a significant role in dielectric properties of high sugar, maltodextrin, starch hydrolysate, and lactose-like disaccharide-based foods (Roebuck and Goldblith, 1972).

14.5.2.1 Starch

Various researchers have studied the dielectric properties of starch in the solid state and/or in suspension form (Guardena et al., 2010; Moteleb, 1994; Ndife et al., 1998; Roebuck and Goldblith, 1972; Ryynänen et al., 1996). When the dielectric properties of different starches in powder form were measured at 2450 MHz, both the dielectric constant and the loss factor increased with temperature (Ndife et al., 1998). The difference between the loss factors of different starches in powder form can be explained by the differences in their bulk densities (Ndife et al., 1998). The lower the bulk density, the lower the loss factor observed. Loss factors of other granular materials were found to be dependent on bulk density (Calay et al., 1995; Nelson, 1983).

For starch suspensions, the effect of free water on dielectric properties becomes significant. The dielectric constant and the loss factor of different starch suspensions were shown to decrease with increasing temperatures and increasing starch concentrations

(Ndife et al., 1998; Ryynänen et al., 1996). The dielectric properties of aqueous solutions are negatively related to temperature in the absence of ions. In attempting to align with the electric field, the hydrogen bonds between the water molecules are disrupted, utilizing energy from the field. At high temperatures hydrogen bonds become rare. Therefore, less energy is required at high temperatures to overcome the intermolecular bond, which causes a negative relationship between temperature and dielectric loss factor (Prakash, 1991). The increase in starch concentration decreases both the dielectric constant and the loss factor because starch molecules bind water and reduce the amount of free water in the system. This trend with starch concentration has also been shown in white sauce model systems (Guardena et al., 2010).

The dielectric loss factor of different starch suspensions is also shown to be a function of starch type (Ndife et al., 1998). Wheat, rice, and corn starches had significantly higher loss factors than tapioca, waxymaize, and amylomaize starches did (Figure 14.22), which may be related to the moisture-binding properties of these starches. It is advisable to use starches having high dielectric properties in microwave-baked products, where poor starch gelatinization resulting from short baking time needs to be avoided. High dielectric properties of starch should be accompanied with low thermal properties such as gelatinization enthalpy and specific heat capacity to achieve sufficient gelatinization in the product during baking.

Gelatinization of starch is an important physical phenomenon that affects dielectric properties. Dielectric constant of gelatinized starch was lower than ungelatinized starch for different concentrations of rice flour slurry (Ahmed et al., 2007). In starch which was gelatinized and cooled down, mobility of the available free water became less which might

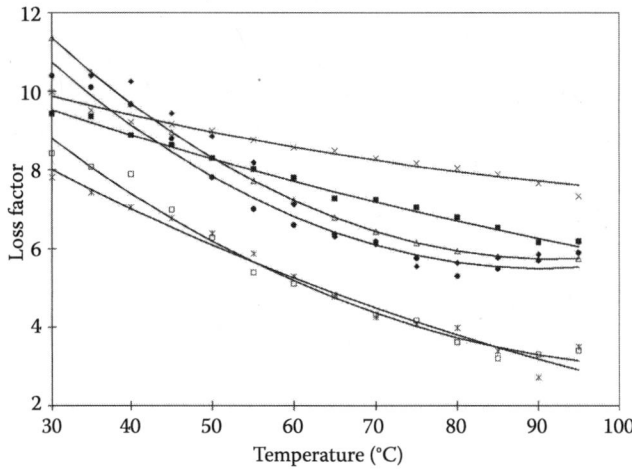

Figure 14.22 Variation of loss factor of starches with temperature for starch:water ratio of 1:2 (□): waxymaize[d], (∗): amylomaize[d], (■): corn[b], (Δ): wheat[b], (●): tapioca[c], (x): rice[a]. Starches with different letters have significantly different loss factor values. (From M Ndife, G Sumnu, L Bayindrli. *Food Res Int* 31: 43–52, 1998.)

explain the lower dielectric constant of gelatinized starch (Figure 14.23). However, the loss factor of gelatinized rice flour slurry was higher than that of ungelatinized one. When the concentration of rice flour slurry increased to 50%, gelatinization did not affect the loss factor significantly.

Dielectric properties of rice flour slurry were shown to be dependent on rice flour concentration (Ahmed et al., 2007). An increase in flour content, which means a decrease in moisture content, significantly reduced both dielectric constant and loss factor values. As explained before, the increase in moisture content increases the mobility of molecules which are reflected to the high dielectric constant and loss factor values. Dependence of

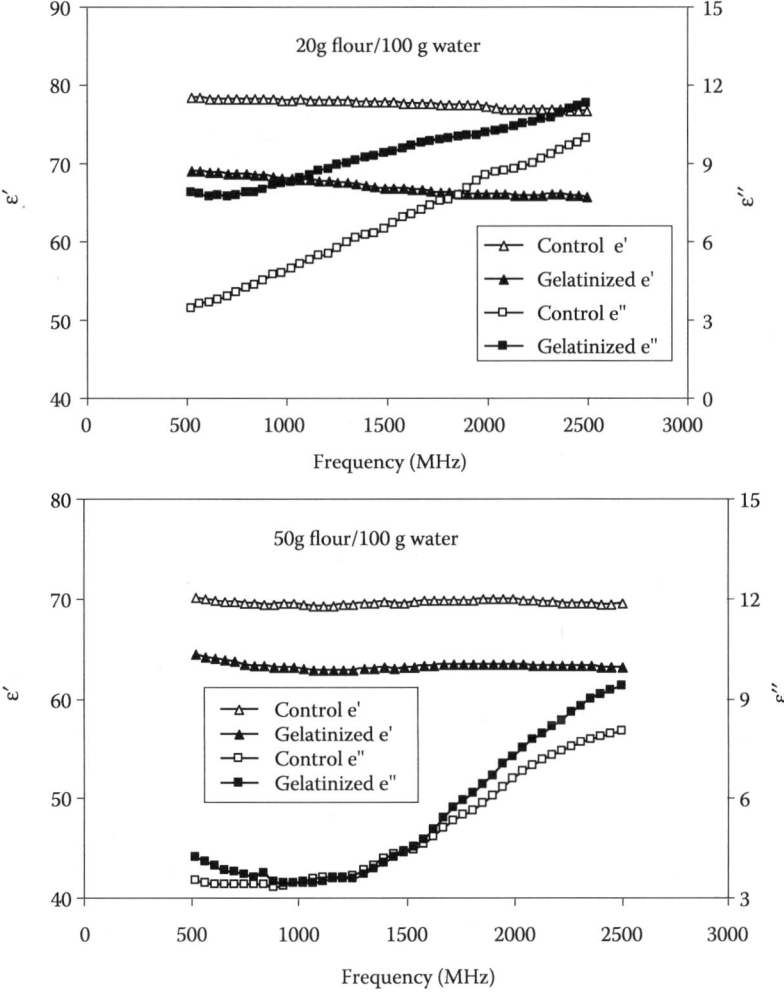

Figure 14.23 Effect of gelatinization on dielectric properties of Basmati rice slurry at selected concentration at 20°C. (From J Ahmed, HS Ramaswamy, GSV Raghavan. *J. Food Eng* 80: 1125–1133, 2007.)

dielectric parameters with temperature could also be influenced by amylose and amylopectin ratio of starch. While starch is heated in the presence of water, amylose could leach out of the granules, which could affect dielectric properties (Tsai et al., 1997).

14.5.2.2 Sugar

Sugar is an important microwave-absorbing food ingredient as compared to other hydrocolloids. Therefore, sugars can be used both for surface heating and for creating a high-loss shield that prevents the next layer of food from heating. Most browning compositions used in microwaveable foods consist of sugars (Shukla and Anantheswaran, 2001).

Sugars modify the dielectric behavior of water. The hydroxyl water interactions stabilize liquid water by hydrogen bonds and affect the dielectric properties of sugar solutions. The degree of microwave interaction depends on the extent of hydrogen bonding. Hydroxyl groups of glucose are more accessible for hydrogen bonding as compared to those of starches. In starches, fewer hydroxyl groups are exposed to water, and fewer stable hydrogen bonds are formed. Therefore, the loss factors of starch solutions were reported to be lower than those of sugar solutions (Roebuck and Goldblith, 1972).

Dielectric properties of sugar solutions have been studied by various researchers (Al-Muhtaseb et al., 2010; Liao et al., 2003; Roebuck and Goldblith, 1972). The dielectric properties of glucose and sucrose solutions having different concentrations were found to be a function of temperature and composition (Al-Muhtaseb et al., 2010; Liao et al., 2003). The dielectric constant of glucose solutions increased but loss factor decreased with temperature. The variation of dielectric properties of sucrose solutions with temperature followed the same trend with glucose solutions (Figures 14.24 and 14.25). The increase in dielectric constant with temperature was more significant at moisture content less than 0.9 kg kg^{-1}db. The most significant decrease in loss factor can be seen for sucrose samples with a moisture content greater than 0.6 kg kg^{-1}db.

According to the study of Liao et al. (2003), the increase in sugar concentration either increased or decreased the loss factor of sugar solutions depending on temperature. This might be due to the increase in solubility of sugar with increase in temperature. At higher temperatures (>40°C) in the unsaturated range, loss factor increased with increase of concentration, since more hydrogen bonds were stabilized by the presence of more hydroxyl groups of sugars. However, at lower temperatures, glucose solution became saturated at a lower concentration and the loss factor decreased with concentration. This shows that there is a critical sugar concentration that affects the dielectric behavior of sugar solution.

Guo et al. (2010b) added different concentrations of sugar to honey to study honey adulteration. It was shown that there was a linear correlation between dielectric constant and total soluble solids of honey at 25°C and for a wide range of frequencies (10–4500 MHz).

Dielectric properties of strawberries osmotically dehydrated using sucrose solution (30%, 40%, and 50% w/w) were measured by Changrue et al. (2008). Sucrose concentration, temperature of solution, and time of treatment had a significant effect on dielectric constant. The increase in treatment time, sucrose concentration, and temperature decreased dielectric constant. The decrease in dielectric constant can be explained by the moisture removal during osmotic dehydration. However, no effect of moisture removal and sugar gain was observed on dielectric loss factor.

DIELECTRIC PROPERTIES OF FOODS

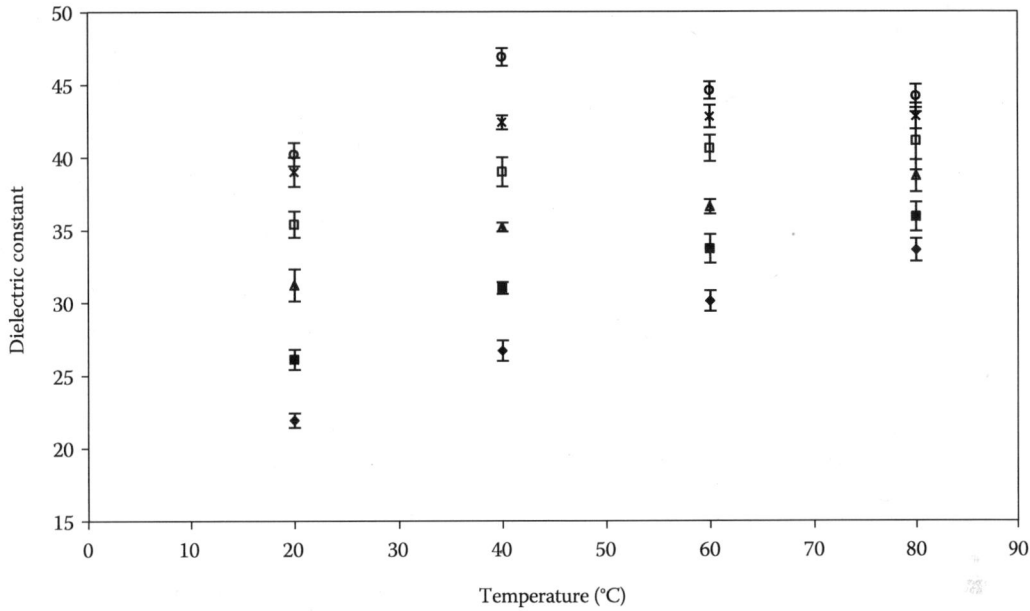

Figure 14.24 Variation of dielectric constant of sugar at 2450 MHz with different temperatures and moisture contents (kg. kg^{-1}db) (♦):0.5 (■): 0.6, (Δ): 0.7, (□): 0.8, (*): 0.9, (o): 1.0 (From AH Al-Muhtaseb et al. *J Food Eng* 98: 84–92, 2010.)

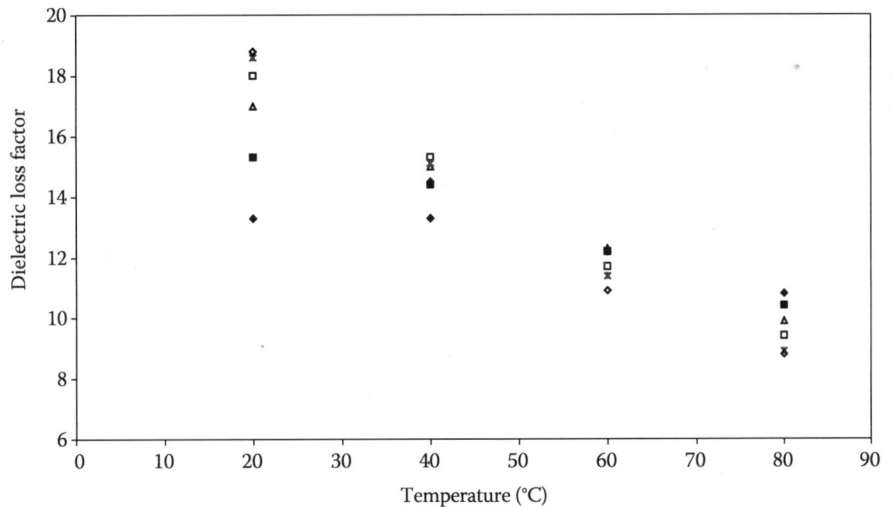

Figure 14.25 Variation of dielectric loss factor of sugar at 2450 MHz with different temperatures and moisture contents (kg. kg^{-1}db) (♦):0.5 (■): 0.6, (Δ): 0.7, (□): 0.8, (*): 0.9, (o): 1.0 (From AH Al-Muhtaseb et al. *J Food Eng* 98: 84–92, 2010.)

14.5.2.3 Gums

Gums are long-chain polymers that dissolve or disperse in water to give a thickening or viscosity-building effect. In addition to their texturizing capabilities, they can be used for stabilization of emulsions, control of crystallization, inhibition of syneresis, and formation of a film (Glicksman, 1982). Gums have the ability to bind a large amount of free water in the system. Therefore, depending on the amount of moisture bound to the gums, the dielectric constant and loss factor of the system change.

Predictive models were developed (Prakash et al., 1992) by expressing both the dielectric constant and the loss factor of gums in powdered form as a function of moisture, temperature, and stoichiometric charge of the molecule as

$$\varepsilon' = 2.1256 - 0.00125CT + 0.0010TM - 0.01565MC + 0.00220M^2 \tag{14.21}$$

$$\varepsilon'' = 0.1295 - 0.00370CM + 0.000436TM - 0.000993M^2 - 0.14469C^{1/3} \tag{14.22}$$

where C is stoichiometric charge (moles of charge/kg), T is temperature (°C), and M is moisture (% wet basis).

In water-limited systems, the effect of charge on dielectric properties may be due to the fact that water associated with highly hydrophilic charged groups may not be free to interact with microwaves. As the charge increases, the amount of moisture bound to charged groups increases, which lowers the dielectric constant and loss factor (Prakash et al., 1992). In the absence of water, the effect of charge disappears.

For microwaveable food formulations, it is important to have information on the water-binding capacity of the gums and viscosity of the solution to have an idea about the dielectric properties and microwave heatability of these formulations. When hydrocolloids are used in the range of 0.1–2.0% they can immobilize 25–60% water (Shukla and Anantheswaran, 2001). Since hydrocolloids can bind different amounts of water, food formulations containing different hydrocolloids are expected to have different amounts of free water in the system, which can affect polarization. Therefore, interaction of food with microwaves is expected to change in the presence of gums.

When dielectric properties of breads formulated with different gums were compared, addition of κ-carragenan gum resulted in the highest dielectric properties, as shown in Figure 14.26 (Keskin et al., 2007). This may be due to the ionic nature of κ-caragenan gum. Another reason may be due to the low porosity of κ-carragenan gum containing breads. Dielectric properties increase as porosity decreases since lower porosity means less air (having drastically lower dielectric constant and loss factor) is present in the sample.

Addition of different types of gums to cake batter was shown to affect dielectric properties significantly (Turabi et al., 2010). Gluten-free rice cake batters containing xanthan and guar gum gave the highest dielectric constant. HPMC and locust bean gum containing cake batters were lower than the other batters. This may be due to the low dielectric constant values of these gums in powder form (Table 14.5). Similar to dielectric constant values, xanthan gum, and xanthan-guar gum blend containing cake batters gave the highest values.

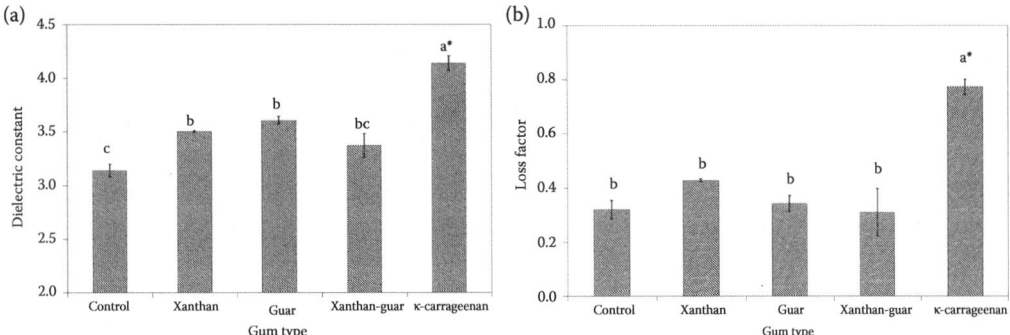

Figure 14.26 The effects of different gums on (a) dielectric constant and (b) dielectric loss factor of breads baked in infrared-microwave combination oven. (*Bars with different letters (a, b, c, d) are significantly different $p \leq 0.05$.) (From SO Keskin, G Sumnu, S Sahin. *Eur Food Res Technol* 224: 329–334, 2007.)

Table 14.5 Dielectric Constants and Dielectric Loss Factors of Gum Types in Powder Form and Rice Flour

Hydrocolloid	ε'	ε''
Xanthan gum	2.855	0.202
Guar gum	2.908	0.383
Locust bean gum	2.418	0.279
HPMC	1.728	0.078
Carrageenan	2.951	0.232
Rice flour	3.108	0.491

Source: E Turabi et al. 2010. *Int J Food Prop* 13: 1199–1206.

14.5.3 Protein Dependence

Proteins are relatively inert and do not interact significantly with microwaves (Table 14.3). Proteins are partly soluble and partly insoluble with ionizable surface regions that may bind water or salts to give rise to zeta potential and double-layer effects associated with free surface charge (Mudgett and Westphal, 1989). These have little effect on dielectric behavior at microwave frequencies. The solvated or hydrated form of protein, protein hydrolysates, and polypeptides are much more microwave reactive. The dielectric properties of proteins depend on their side chains, which can be nonpolar with decreasing order of alanine, glycine, leucine, isoleucine, methionine, phenylalanine, and valine, or polar with decreasing order of thyrosine, tryptophan, serine, threonine, proline, lysine, arginine, aspartic acid, aspergine, glutamic acid, glutamine, cysteine, and histidine (Shukla and Anantheswaran, 2001).

Free amino acids are dielectrically reactive (Pething, 1979). Free amino acids and polypeptides contribute to an increase in dielectric loss factor. Since protein dipole moments are a function of their amino acids and the pH of the medium, the dielectric properties and microwave reactivity of cereal, legume, milk, meat, and fish proteins are expected to be different. The water adsorbed on proteins also affects their dielectric properties (Shukla and Anantheswaran, 2001).

The dielectric activity of proteins can be categorized in four origins:

- High activity due to charge effects of ionization of carboxyl, sulfhydryls, and amines,
- Hydrogen and ion binding as affected by pH,
- Net charges on dissolved proteins,
- Relatively low activity due to relaxation and conductive effects.

Such activities are important for hydrolyzed proteins and free amino acids (Shukla and Anantheswaran, 2001).

Since most of the protein-containing foods are consumed in heated or cooked form, it is important to determine dielectric properties during denaturation of proteins to understand the microwave heating of these foods. Protein denaturation is defined as the physical change of the protein molecule due to heat, ultraviolet (UV), or agitation, which results in a reduction in protein solubility, a loss of crystallinity and an increase in solution viscosity (McWilliams, 1989). During denaturation of proteins, since the protein structure is disturbed, the asymmetry of the charge distribution will increase. This will result in a large dipole moment and polarization, which will affect the dielectric properties. Moreover, since the amount of free water in the system changes during denaturation, dielectric properties of foods are affected. Moisture is either bound by the protein molecule or released to the system during denaturation. Various studies show that the dielectric properties can be used to understand protein denaturation (Ahmed and Luciano, 2009; Brunton et al., 2006).

In the study of Brunton et al. (2006), protein denaturation of beef muscle was determined both by using dielectric properties and differential scanning calorimeter (DSC). DSC endotherms related to the denaturation of proteins, as shown in Figure 14.27. DSC

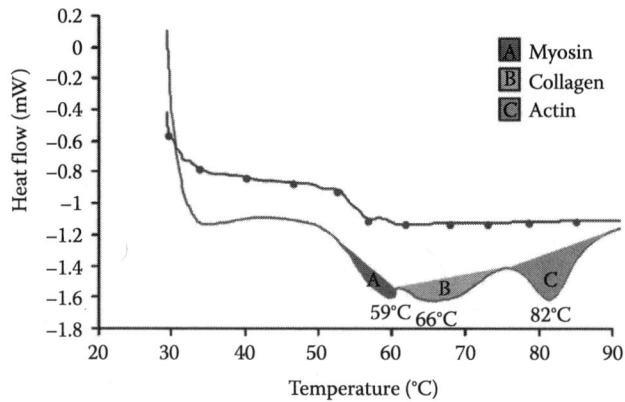

Figure 14.27 Differential scanning calorimetry thermogram of uncooked (–) and cooked (·) beef muscle. (From NP Brunton et al. *Meat Sci* 72: 236–244, 2006.)

endotherms between 56°C and 81°C corresponds to the denaturation of myosin, collagen, and actin. The changes in dielectric properties took place at denaturation temperatures (Figures 14.28 and 14.29). As can be seen in Figure 14.28, there was an increase in dielectric constant in temperatures above 65–66°C, which is related to the water released during denaturation. The temperature at which dielectric constant increases is close to the peak related to the collagen denaturation measured by DSC. The fluid released from collagen and myosin denaturation resulted in an increase in dielectric loss factor. Calcium and magnesium ions released from myosin during denaturation might have also increased the loss factor. Loss factor at RF is more dependent on ionic mobility than loss factor at microwave frequencies. However, meat structural protein denaturation is accompanied with physical shrinkage which can also influence dielectric properties.

Dielectric properties of gluten protein were also affected by heating (Umbach et al., 1992). The dielectric constant and loss factor of a heated gluten–starch mixture were found to be less than those of the unheated mixture. As the amount of gluten protein in the system increased, the dielectric constant decreased, but the loss factor remained constant. The interaction of gluten with microwaves has been known to have an adverse effect on the texture of microwave-baked breads (Yin and Walker, 1995). Microwave-baked breads containing a small amount of gluten were softer than those containing a larger amount of gluten (Ozmutlu et al., 2001).

Ahmed and Luciano (2009) studied the dielectric properties of β-lactoglobulin and found that temperature (particularly values causing protein denaturation) and concentration of β-lactoglobulin dispersions affected the dielectric properties significantly. Both dielectric constant and loss factor significantly increased at temperatures above 80°C. There was an inverse relationship between dielectric constant and pH. The change of pH had a significant effect on the electrostatic charge on the protein surface which finally affected the dielectric constant at higher temperatures. In the presence of low pH and high temperature, loss factor increased significantly. Since isoelectric point of β-lactoglobulin is at pH = 5.3, a decrease in pH to 4 can minimize intermolecular electrostatic repulsion. Thus, increase in loss factor corresponded to the denaturation temperature. Dielectric properties of soy protein isolate dispersion were also shown to be dependent on frequency, concentration, and pH (Ahmed et al., 2008).

14.5.4 Fat Dependence

Lipids are hydrophobic, except for ionizable carboxyl groups of fatty acids, and do not interact much with microwaves (Mudgett and Westphal, 1989). Therefore, the dielectric properties of fats and oils are very low (e.g., Lizhi et al., 2008). The effect of fat on dielectric properties of food systems is mainly due to their dilution effect in the system. The increase in fat content reduces the free water content in the system, which reduces the dielectric properties (Ryynänen, 1995). Increase in fat content from 12.4% to 29.7% decreased dielectric constant in comminuted meat blend but the reduction in loss factor was not significant (Zhang et al., 2007).

The dielectric constants and loss factors of different fats and oils at temperatures of 25–82°C and frequencies of 300, 1000, and 3000 MHz are given in Table 14.6. The loss

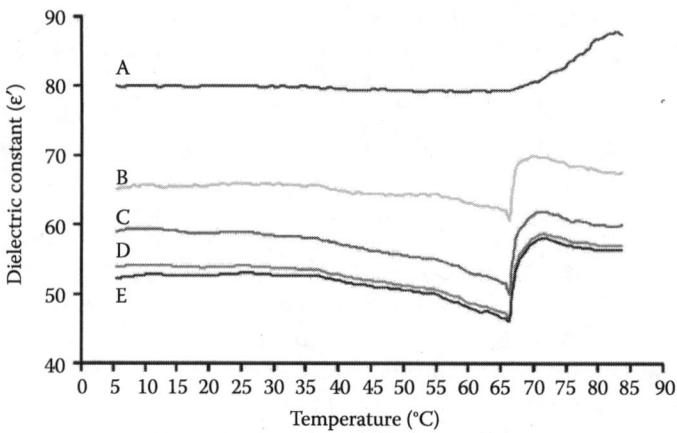

Figure 14.28 Dielectric constant of beef sample at different frequencies: (a) 27.12 MHz, (b) 300 MHz, (c) 915 MHz, (d) 2450 MHz, (e) 3000 MHz. (From NP Brunton et al. *Meat Sci* 72: 236–244, 2006.)

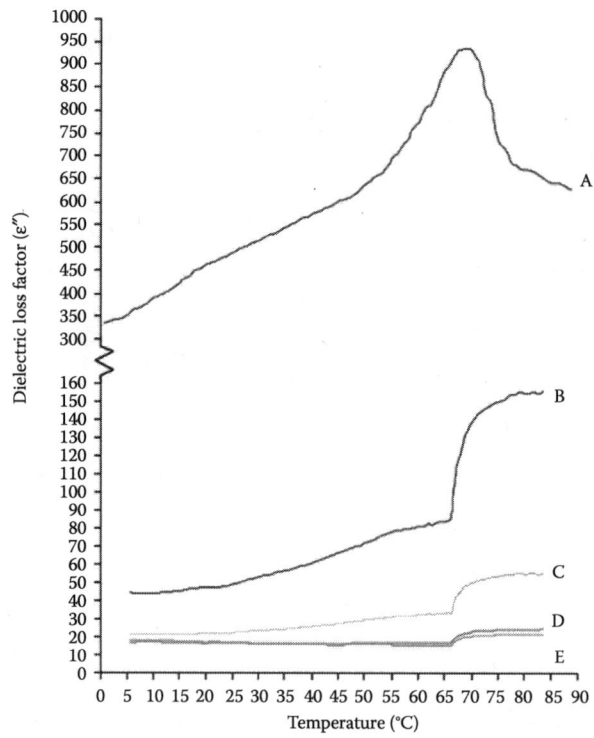

Figure 14.29 Dielectric loss factor of beef sample at different frequencies: (a) 27.12 MHz, (b) 300 MHz, (c) 915 MHz, (d) 2450 MHz, (e) 3000 MHz. (From NP Brunton et al. *Meat Sci* 72: 236–244, 2006.)

DIELECTRIC PROPERTIES OF FOODS

Table 14.6 Dielectric Data on Eight Commercial Fats and Oils

Sample		300 MHz			1000 MHz			3000 MHz		
		25°C	48°C	82°C	25°C	48°C	82°C	25°C	48°C	82°C
Soybean salad oil	ε'	2.853	2.879	2.862	2.612	2.705	2.715	2.506	2.590	2.594
	ε''	0.159	0.138	0.092	0.168	0.174	0.140	0.138	0.168	0.160
Corn oil	ε'	2.829	2.868	2.861	2.638	2.703	2.713	2.526	2.567	2.587
	ε''	0.174	0.134	0.103	0.175	0.174	0.146	0.143	0.166	0.163
Cotton seed cooking oil	ε'	2.825	2.859	2.834	2.629	2.669	2.673	2.515	2.536	2.554
	ε''	0.171	0.132	0.103	0.174	0.171	0.146	0.143	0.165	0.160
Lard	ε'	2.718	2.779	2.770	2.584	2.651	2.656	2.486	2.527	2.541
	ε''	0.153	0.137	0.109	0.158	0.159	0.137	0.127	0.154	0.148
Tallow	ε'	2.603	2.772	2.765	2.531	2.568	2.610	2.430	2.454	2.492
	ε''	0.126	0.141	0.105	0.147	0.146	0.134	0.118	0.143	0.144
Hydrogenated vegetable shortening	ε'	2.683	2.777	2.772	2.530	2.654	2.665	2.420	2.534	2.550
	ε''	0.141	0.140	0.103	0.147	0.153	0.137	0.117	0.146	0.146
Conventionally rendered bacon fat	ε'	2.753	2.799	2.767	2.615	2.655	2.637	2.498	2.539	2.526
	ε''	0.172	0.149	0.099	0.163	0.161	0.144	0.133	0.152	0.148
Microwave rendered bacon fat	ε'	2.742	2.796	2.772	2.601	2.655	2.660	2.487	2.536	2.546
	ε''	0.158	0.129	0.098	0.162	0.161	0.143	0.126	0.152	0.150

factors of oils and fats were different at 25°C due to the differences in the phase of samples, whether they were in solid or liquid form (Pace et al., 1968). Loss factors were greater in more liquid samples, such as corn oil and cottonseed oil, as compared to lard and tallow fats. At low temperatures, high internal viscosity produced little dipole rotation, so low values were obtained for the dielectric constant and loss factor. At 3000 MHz, as temperature increased from 25°C to 48°C, viscosity decreased and rotation increased, which increased the dielectric constant and loss factor. As the temperature reached 82°C, the relaxation time was too small, which increased the dielectric constant and decreased the loss factor.

In a baked product such as a cake, increase in fat content increases dielectric constant and loss factor, as shown in Figures 14.30a and 14.30b (Sakiyan et al., 2007). At higher fat content, porosities are significantly lower (Figure 14.31), leading to higher dielectric properties as explained earlier.

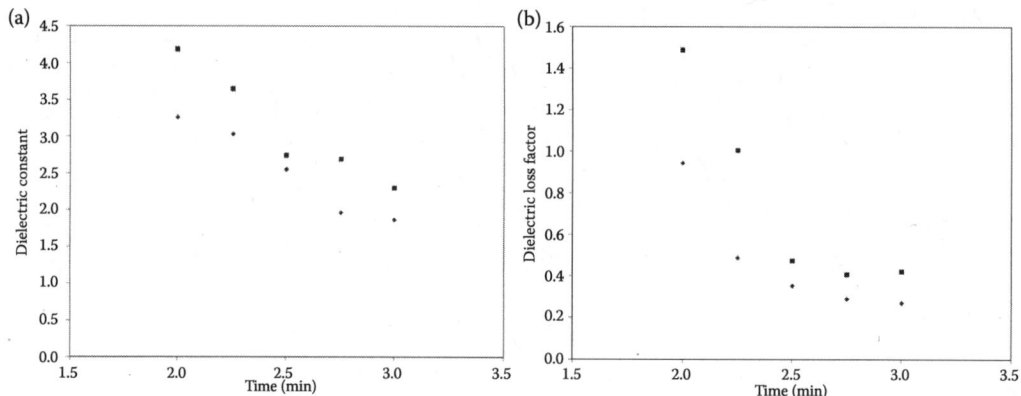

Figure 14.30 Variation of dielectric constant (a) and loss factor (b) with baking time for cakes with different formulation baked in microwave oven ■ 25% fat, ♦ non-fat. (From O Sakiyan et al. *J Food Sci* 72: E205–e213, 2007.)

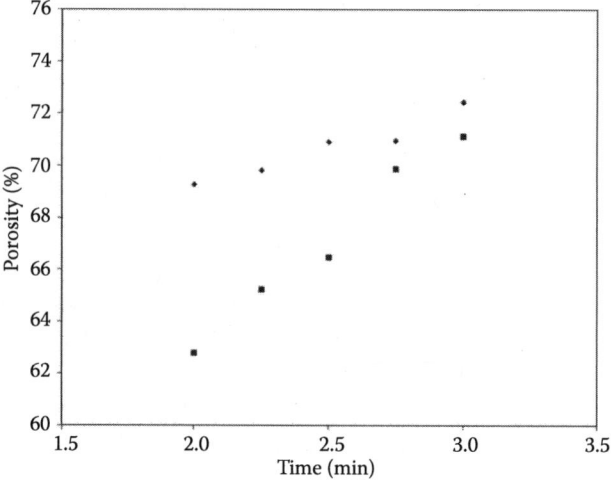

Figure 14.31 Variation of porosity with baking time for cakes with different formulation baked in microwave oven ■ 25% fat, ♦ nonfat. (From O Sakiyan et al. *J Food Sci* 72: E205–213, 2007.)

14.5.5 Dielectric Properties of Meats

The dielectric properties of raw and cooked meats were measured at 2.8 GHz, as shown in Figure 14.15, with a resonant cavity, as a function of temperature (Bengtsson and Risman, 1971). The dielectric properties of raw and cooked meat decreased as temperature increased for temperatures greater than freezing temperatures. However, the loss factor of brined ham increased with increasing temperature as a result of added salt.

An earlier study also investigated composition dependence in model meat emulsions (Ohlsson et al., 1974).

In one study (Sun et al., 1995), dielectric properties of raw beef, beef juice, raw turkey, and turkey juice (data from To et al., 1974) were correlated to moisture and ash (indicator of salts) as

$$\varepsilon'_{meats} = m_{water}(1.0707 - 0.0018485T) + m_{ash}(4.7947) + 8.5452 \quad (14.23)$$

$$\varepsilon''_{meats} = m_{water}(3.4472 - 0.01868T + 0.000025T^2) + m_{ash}(-57.093 + 0.23109T) - 3.5985 \quad (14.24)$$

Heating of meats leads to changes in dielectric properties beyond those due to temperature alone. For beef samples, dielectric properties at 915 and 2450 MHz increased abruptly between 70°C and 75°C at the denaturation temperature of collagen, as shown in Figures 14.28 and 14.29. When proteins denature, they shrink and juice is expelled. The sudden increase of dielectric properties occurred in the same temperature range as the sudden increase of drip loss, indicating that the water expelled during heating was responsible for the increase of dielectric properties. The decrease of frequency increased both the dielectric constant and the loss factor. For the loss factor, the extent of changes at denaturation temperatures was greater at lower frequencies than at higher frequencies because at lower frequencies, the loss factor was more sensitive to changes in the mobility of salts. When the sample from Figure 14.32 was reheated (Figure 14.33), no changes in the dielectric properties occurred at the temperature at which an increase in dielectric properties occurred with fresh samples (Figure 14.32). This finding is consistent with the irreversibility of protein denaturation.

14.5.6 Dielectric Properties of Fish and Seafood

Dielectric properties of fish and seafood are generally in the same range as that of meat. Available dielectric property data include those for codfish (10–200 MHz, –25 to 10°C; Bengtsson et al., 1963), tilapia (Zhang et al., 2012), fish meal (10 GHz, 10–90°C; Kent, 1972, 1977), three common Japanese fish varieties (Liu et al., 2012), and sea cucumbers (Cong et al., 2012).

Dielectric properties of raw nonmarinated and marinated catfish and shrimp measured at 915 and 2450 MHz and at temperatures from 10°C to 90°C showed that the dielectric constant decreased but the loss factor increased with temperature (Zheng et al., 1998). Marination increased both the dielectric constant and the loss factor.

The dielectric properties of fresh, frozen, and thawed surimi measured at 2450 MHz are shown in Table 14.7. The dielectric constant of fresh surimi was the same as or greater than that of frozen and thawed surimi except for fresh plain surimi. This was explained by the low ion concentration of plain surimi paste since no solutes were added to the plain surimi paste. In addition, because of the washing cycles and dewatering during surimi making, ions originally present in fish were removed with the wash water, resulting in a low ion concentration in surimi (Wu et al., 1988).

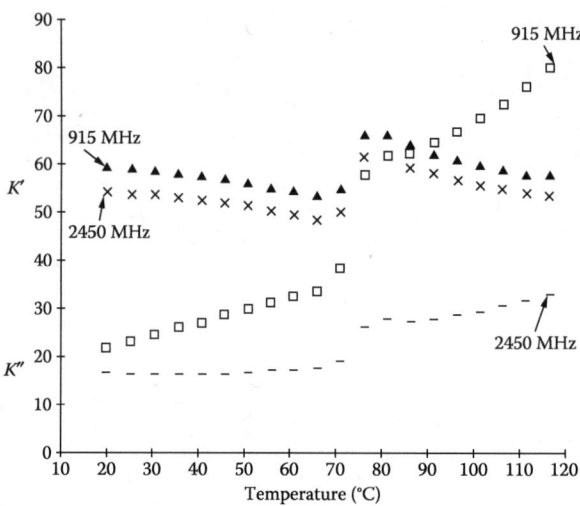

Figure 14.32 Temperature dependence of dielectric properties of beef measured at 915 and 2450 MHz. Symbols κ′ and κ″ in the figure are the same as ε′ and ε″, respectively. (From C Bircan, SA Barringer. *J Food Sci* 67: 202–205, 2002a.)

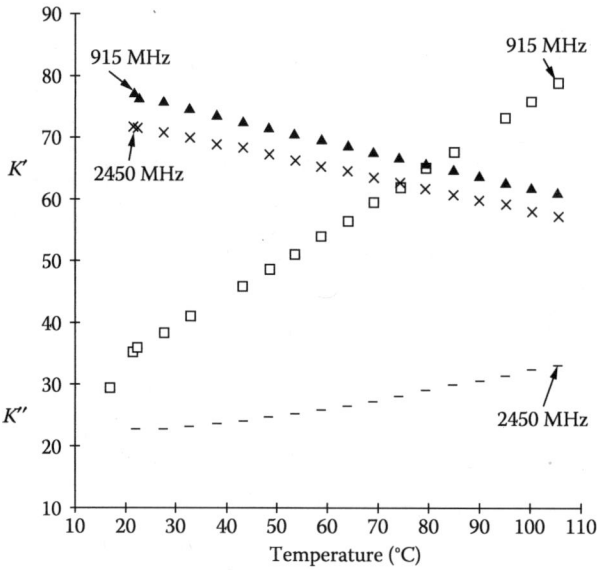

Figure 14.33 Dielectric properties of cooked beef measured at 915 and 2450 MHz during reheating. Symbols κ′ and κ″ in the figure are the same as ε′ and ε″, respectively. (From C Bircan, SA Barringer. *J Food Sci* 67: 202–205, 2002a.)

Table 14.7 Dielectric Constants and Loss Factors of Surimi

Surimi Sample	Fresh		Frozen		Thawed	
	ε'	ε''	ε'	ε''	ε'	ε''
Plain surimi paste	57.38	10.52	22.82	5.90	59.86	13.51
6% sucrose, 5% D-sorbitol	59.39	12.80	17.27	6.81	51.48	11.32
6% sucrose, 6% D-sorbitol	60.32	12.79	29.93	8.37	54.95	12.17
6% sucrose, 5% D-sorbitol, 2% NaCl	56.39	26.31	55.25	23.5	52.99	23.11
7% sucrose, 6% D-sorbitol, 1% NaCl	57.54	20.12	23.27	8.97	50.66	18.80
8% sucrose, 7% D-sorbitol, 1% NaCl	57.25	18.37	20.64	8.38	57.61	17.91
8% sucrose, 5% D-sorbitol, 2% NaCl	55.82	22.43	36.68	13.63	54.65	25.36
9% sucrose, 8% D-sorbitol, 3% NaCl	52.32	28.47	35.46	18.12	46.25	23.72

Source: P Yagmaee, TD Durance. Predictive equations for dielectric properties of NaCl, D-Sorbitol and sucrose solutions and surimi at 2450 MHz. *J Food Sci* 67: 2207–2211, 2002.

14.5.7 Dielectric Properties of Fruits and Vegetables

The dielectric properties of various fruits and vegetables have been reported in a number of studies (Alfaifi et al., 2013; Nelson, 1982; Nelson et al., 1994; Seaman and Seals, 1991; Tran et al., 1984). The dielectric properties of 23 different fruits and vegetables were reported over a frequency range of 3–20 GHz at room temperature (Kuang and Nelson, 1997). Sipahioglu and Barringer (2003) measured the dielectric properties of 15 fruits and vegetables at 2450 MHz over the temperature range 5°C to 130°C by using an open-ended coaxial probe. The dielectric constants of these fruits and vegetables decreased with temperature (Figure 14.34) since most of the water in fruits and vegetables exists as free water, and the dielectric constant of free water decreases with temperature. As expected, the dielectric constant was positively related with moisture content. The dielectric constant of vegetables decreased with ash content. Ash, which is composed of salts, is capable of binding water, and the decrease in the amount of available water decreases the dielectric constant. However, the ash content was not significant in affecting the dielectric constant of fruits due to the low concentration of ash in fruits. The dielectric behavior of garlic was found to be somewhat different from that of the other vegetables. The dielectric constant of garlic increased up to 55°C and then decreased as temperature increased. Garlic contains 30% oligofructosaccharides in the form of insulin, which binds water (Van Loo et al., 1995). For most of the fruits and vegetables, the loss factor decreased with increasing moisture content at temperatures less than 34°C and then the loss factor increased with moisture content above that temperature (Figure 14.35). Dielectric properties at 9.85 GHz for pulverized coriander showed increase in the properties with packing density (Karhale and Kalamse, 2012). Dielectric properties of ground almond shells, in the context of pasteurization of in-shell almonds was reported by Gao et al. (2012).

Overall predictive equations were developed by combining the data for different fruits and vegetables to express dielectric properties as a function of temperature at 2450 MHz (Equations 14.25 and 14.26).

ENGINEERING PROPERTIES OF FOODS

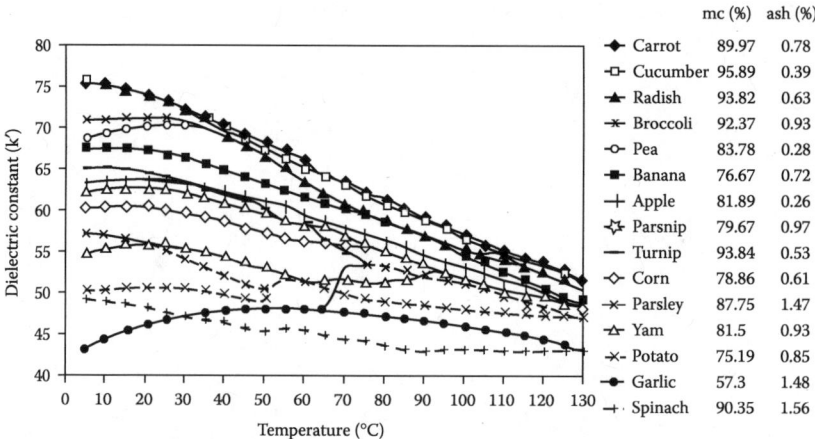

Figure 14.34 The effect of ash and moisture concentration on the dielectric constant of vegetables and fruits at 2450 MHz. (From O Sipahioglu, SA Barringer. *J Food Sci* 68: 234–239, 2003.)

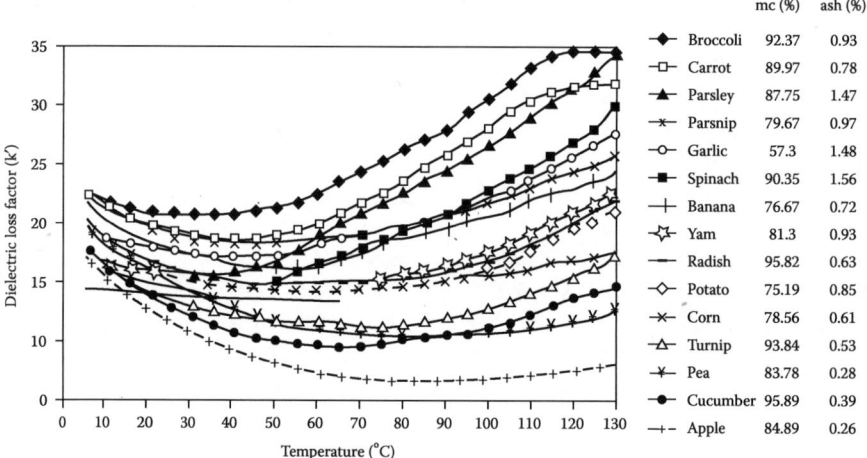

Figure 14.35 The effect of ash and moisture concentration on dielectric loss factor of vegetables and fruits at 2450 MHz. (From O Sipahioglu, SA Barringer. *J Food Sci* 68: 234–239, 2003.)

$$\varepsilon' = 38.57 + 0.1255T + 0.456M - 14.54A \\ -0.0037MT + 0.07327AT \quad (14.25)$$

$$\varepsilon'' = 17.72 - 0.4519T + 0.001382T^2 - 0.07448M \\ +22.93A - 13.44A^2 + 0.002206MT + 0.1505AT \quad (14.26)$$

14.5.8 Dielectric Properties of Dairy Products

The dielectric properties of dairy products have been studied for milk, dairy powders with added water, cheese, and butter (Ahmed et al., 2007; Datta, 1994; Green, 1997; Herve et al., 1998; Kudra et al., 1992; Mudgett et al., 1971, 1974; O'Connor and Synnott, 1982; Rzepecka and Pereira, 1974). The dielectric properties of milk and its constituents at 2450 MHz are shown in Table 14.8 (Kudra et al., 1992).

The dielectric properties of cheese were found to be dependent on the amount of moisture present (Figure 14.36) (Green, 1997). As the moisture content increased, both the dielectric constant and loss factor increased. For processed cheese (Everard et al., 2006), dielectric constants generally decreased with increasing temperature up to temperatures of 55–75°C. Dielectric loss factors generally increased with increasing temperature, except for low moisture/fat ratio process cheese below 55°C, where the loss factors tended to decrease with increasing temperature.

Similarly, the dielectric constants of cottage cheeses were dependent on composition (Figure 14.37). The cheese with the highest fat content had the lowest dielectric constant since the presence of higher fat content means lower moisture content and lower dielectric constant. The dielectric constant of cottage cheese decreased slightly when the temperature increased and was affected slightly by frequency. Loss factors of 2%- and 4%-fat cottage cheese were close but they were smaller than that of 0%-fat cottage cheese (Figure 14.37). The loss factor of cottage cheese at different fat contents decreased as temperature or frequency increased.

The dielectric constants and loss factors of processed cheese at different compositions for temperatures of 20°C and 70°C are shown in Table 14.9. At higher moisture and lower fat contents, the loss factor increases somewhat with temperature. However, the dielectric

Table 14.8 Dielectric Properties of Milk and Its Constituents

Description	Fat (%)	Protein (%)	Lactose (%)	Moisture (%)	ε'	ε''
1% Milk	0.94	3.31	4.93	90.11	70.6	17.6
3.25% Milk	3.17	3.25	4.79	88.13	68.0	17.6
Water + lactose I[a]	0	0	4.0	96.00	78.2	13.8
Water + lactose II	0	0	7.0	93.00	77.3	14.4
Water + lactose III	0	0	10.0	90.00	76.3	14.9
Water + sodium caseinate I	0	3.33	0	96.67	74.6	15.5
Water + sodium caseinate II	0	6.48	0	93.62	73.0	15.7
Water + sodium caseinate III	0	8.71	0	91.29	71.4	15.9
Lactose (solid)	0	0	100	0	1.9	0.0
Sodium caseinate (solid)	0	100	0	0	1.6	0.0
Milk fat (solid)	100	0	0	0	2.6	0.2
Water, distilled	0	0	0	100	78.0	13.4

Source: T Kudra. *J Microwave Power Electromagn Energy* 27: 199–204, 1992.
[a] Level of concentration.

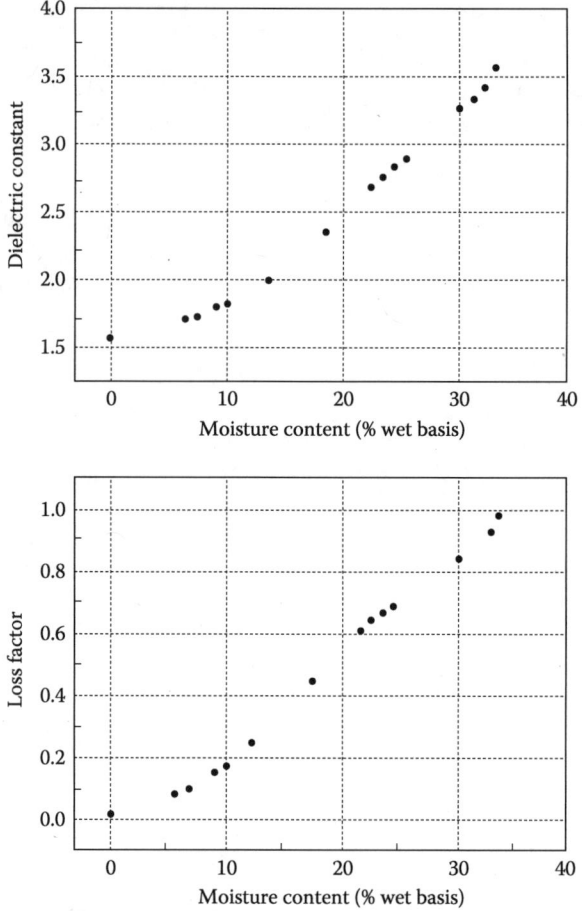

Figure 14.36 Permittivity of grated and chopped cheddar cheese with a bulk density of 0.39 g/cm^3 at 3950 MHz at 20°C as a function of moisture content. (From AD Green. *J Microwave Power Electromagn Energy* 32: 16–27, 1997.)

constant and loss factor of processed cheese are not generally temperature dependent (Table 14.9).

14.6 STRUCTURE DEPENDENCE AND PUREED FOODS

Dependence of dielectric properties on food structure is generally not the focus of property measurement studies, although discussion of some of the structure effects are embedded in the various categories of foods discussed above. Effect of fiber direction on the dielectric properties of beef semitendinosus muscle at radio and microwave frequencies

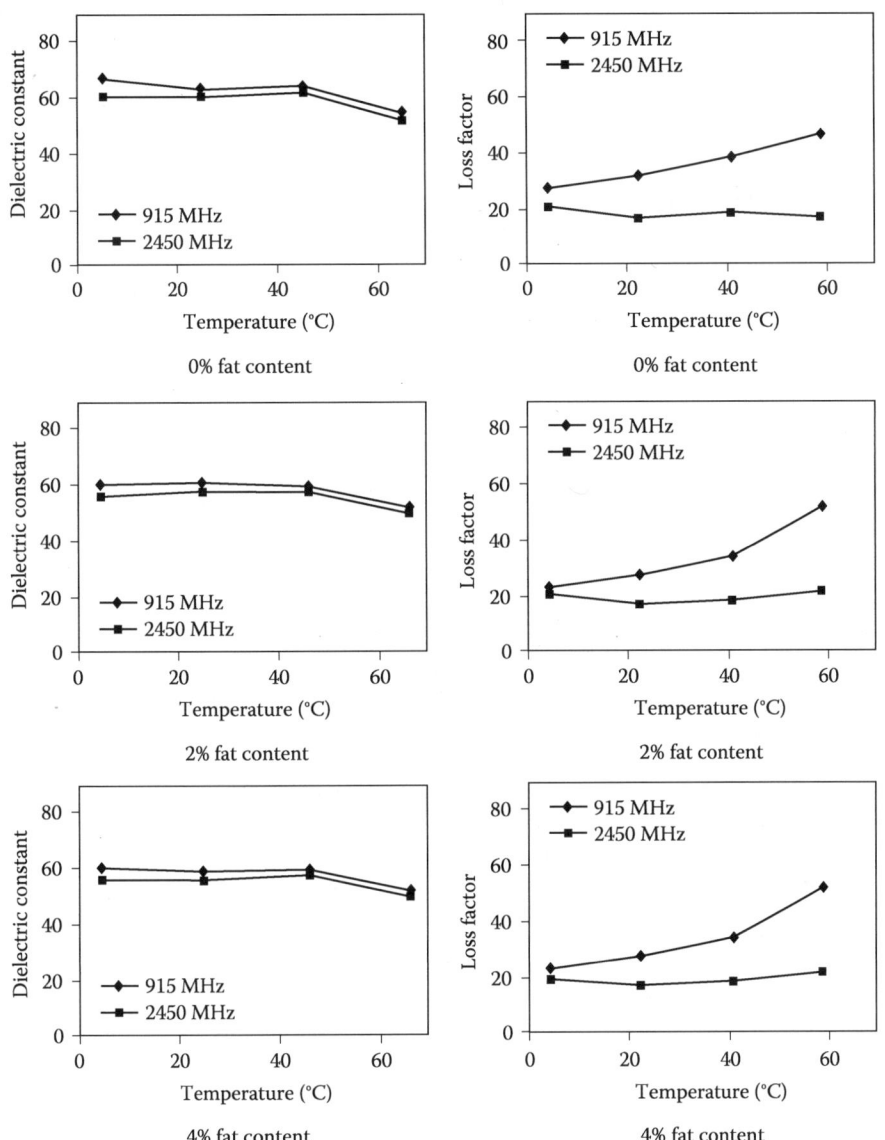

Figure 14.37 Dielectric constants and loss factors of three different cottage cheeses at 2450 and 915 MHz. (From AG Herve et al. *J Food Eng* 37: 389–410, 1998.)

was studied by Basaran-Akgul et al. (2008). They showed that the dielectric constant and dielectric loss factors were generally higher with the muscle fiber in a parallel orientation to the probe compared to a perpendicular orientation at the same frequency and temperature. Tissue orientation appeared to have a greater effect on dielectric loss values at lower frequencies.

Table 14.9 Dielectric Properties of Processed Cheese at 2450 MHz as Related to Composition

Composition		Temperature (°C)			
		20		70	
% Fat	% Moisture	ε'	ε''	ε'	ε''
0	67	43	29	43	37
12	55	30	21	32	23
24	43	20	14	22	17
36	31	14	8	13	9

Source: AK Datta. Effect of composition, temperature, and storage on dielectric properties of cheese. Unpublished data, 1994.

Dielectric properties of a number of liquids and purees were studied in the context of continuous flow microwave heating by Coronel et al. (2008). Properties varied with respect to temperature in a similar manner as the properties of solid foods. Data on pureed baby foods is available in Ahmed (2007).

14.7 DIELECTRIC PROPERTIES OF INSECT PESTS

Exposures of grain infested by stored-grain insects to RF energy can control the insect infestations by selective dielectric heating of insect species (Guo et al., 2011; Nelson, 1996; Nelson and Charity, 1972; Wang et al., 2003a,b). Representative data on insect pests is shown in Figures 14.38 and 14.39. Note that the loss factors are in a much higher range for the insect pests as compared to grain (Figure 14.38). At lower frequencies, the differences in dielectric loss between the grain and the insects are even greater.

14.8 DIELECTRIC PROPERTIES OF PACKAGING MATERIALS

Relevant packaging materials for microwave heating or dual (microwave + convection) heating of materials are paper, PET (polyethylene terephthalate), CPET (crystalline PET), and polypropylene. Electrical properties, particularly at given frequencies, are hard to locate. There are also variations in these materials from one manufacturer to another. The following representative data (Table 14.10) are provided for convenience. Data on glass, paperboard, and other materials are also included here from the literature. Note that many of the data are not at microwave frequencies. Information on susceptor material electrical properties is still harder to find.

14.9 EFFECTS OF PROCESSING AND STORAGE ON DIELECTRIC PROPERTIES OF FOODS

Dielectric properties of foods are expected to change because of physical changes such as moisture loss, starch gelatinization, and protein denaturation during processing and

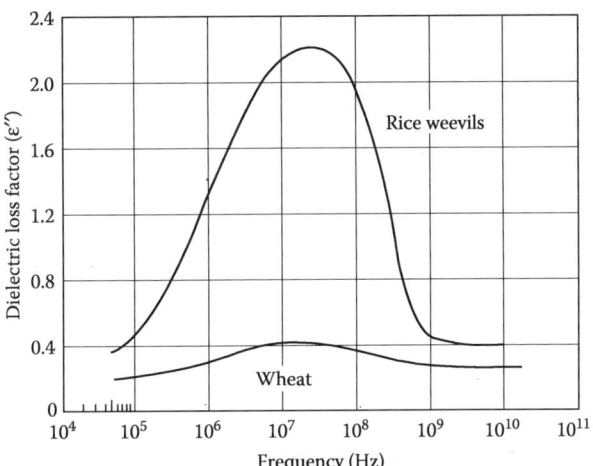

Figure 14.38 Frequency dependence of the dielectric loss factor of bulk samples of adult rice weevils and of hard red winter wheat at 10.6% moisture content at 24°C. (From SO Nelson, LF Charity. *Trans ASAE*, 15: 1099–1102, 1972; SO Nelson. *Trans ASAE*, 39: 1475–1484, 1996.)

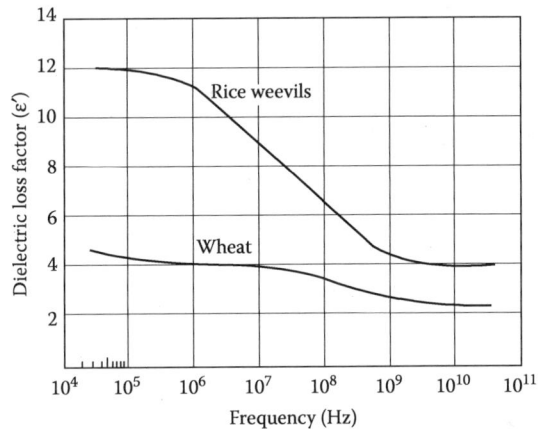

Figure 14.39 Frequency dependence of the dielectric constant of bulk samples of adult rice weevils and of hard red winter wheat at 10.6% moisture content at 24°C. (From SO Nelson, LF Charity. *Trans ASAE*, 15: 1099–1102, 1972; SO Nelson. *Trans ASAE*, 39: 1475–1484, 1996.)

storage. Measurement of dielectric properties will be useful for design and control of microwave food processes and for online monitoring of quality during processing.

14.9.1 Baking

Although baking is one of the useful applications of dielectric heating, microwave baking of bread has not been successful. The reasons for quality changes in a microwave-baked

Table 14.10 Dielectric Properties of Some Packaging and Related Materials

Material and Frequency	Dielectric Constant	Dielectric Loss Factor
PET (1 kHz)[a]	3.5	0.01
PET (1 MHz)[b]	2.9–3.2	0.010–0.020
CPET (1 kHz)[c]	3.3	0.002
CPET (1 MHz)[d]	3.2	0.021
Polypropylene (1 kHz)[e]	2.2–2.6	0.0005–0.0018
Polypropylene (1 MHz)[f]	2.2–2.6	0.0003–0.0005
Polyethylene (23°C; 3 GHz)[g]	2.25	0.0026
Polythene (24°C; 3 GHz)[g]	2.25	0.0007
Plexiglass, perspex (27°C; 3 GHz)[g]	2.6	0.015
Polytetrafluoroethylene, Teflon (22°C; 3 GHz)[g]	2.1	0.0003
Paper (royal gray, 82°C, 0% moisture, 1 GHz)[g]	3.0	0.216
Paper (royal gray, 82°C, 0% moisture, 3 GHz)[g]	2.94	0.235
Paperboard, box (230 g/m^2; E parallel to web; 22°C, 5% moisture, 0.1 GHz)[g]	2.8	0.3
Paperboard, box (230 g/m^2; E parallel to web; 22°C, 5% moisture, 3 GHz)[g]	2.7	0.3
Glass (fused silica; 25°C, 1 and 3 GHz)[g]	3.78	0.0002
Glass (96% SiO$_2$; 25°C, 3 GHz)[g]	3.84	0.0026

[a] http://www.azom.com/details.asp?ArticleID = 796
[b] http://www.loctite.com/pdf/pbg50-51.pdf
[c] http://www.azom.com/details.asp?ArticleID = 795
[d] http://www.kern-gmbh.de/cgi-bin/riweta.cgi?nr = 1301andlng = 2
[e] http://www.sdplastics.com/polypro.html
[f] http://www.goodfellow.com/csp/active/static/E/PP30.HTML
[g] AC Metaxas, RJ Meredith. *Industrial Microwave Heating*. Peter Peregrinus, London, 1988.

bread were stated to be insufficient starch gelatinization, microwave-induced gluten changes, and rapidly generated gas and steam caused by the heating mode (Yin and Walker, 1995). Other reasons are the differences between microwave and other heating mechanisms and specific interactions of each component in the formulation with microwave energy (Goebel et al., 1984). Dielectric properties during baking of foods can provide insights into possible improvements in baking. In biscuit baking, dielectric properties of biscuit dough (flour–water mixture) were measured at 27 MHz at different temperatures (24–125°C) while it was being baked in a parallel-plate capacitor at 200°C, as shown in Figure 14.40. Although the properties increase with moisture, as expected, there are some differences (Kim et al., 1998). The dielectric constant gradually increased with moisture content and temperature. The dielectric loss factor showed a sudden and exponential increase with moisture content beyond a certain moisture content value. Temperature affected the dielectric loss factor beyond this point. The ionic conductivity and the bound water relaxation are considered the dominant loss mechanisms in

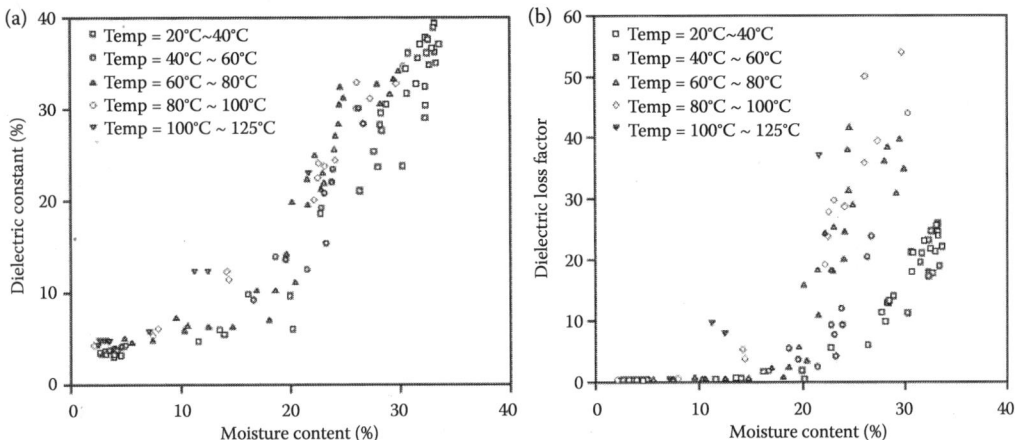

Figure 14.40 Dielectric properties of baked dough at various moisture contents and temperatures. (From YR Kim et al. *Journal of Microwave Power and Electromagnetic Energy* 33: 184–194, 1998.)

the baked dough. Temperature, in combination with water mobility, affects the ionic conductivity.

Models of the dielectric properties of baked dough (Kim et al., 1998) were also provided as a function of moisture content, bulk density, and temperature as

$$\varepsilon'(\rho_b, M, T) = \left[1 + \frac{\rho_b}{\rho_s(M)}\left((0.33T^{1/3}\varepsilon'_2(M))^{1/2} - 1\right)^2\right] \quad (14.27)$$

$$\varepsilon''(\rho_b, M, T) = \left(\frac{\rho_b}{\rho_s(M)}\right)^2 (1 + 0.00073M^3T^3)\varepsilon''_2(M) \quad (14.28)$$

where M is moisture content (g water/g total), T is temperature (°C), ρ_b is measured bulk density of the flour–water mixture (g/cm³), $\rho_s(M)$ is solid density as function of moisture content (g/cm³), $\varepsilon'_2(M)$ is dielectric constant of the flour–water mixture at solid density as a function of moisture content, and $\varepsilon''_2(M)$ is the loss factor of the flour–water mixture at solid density as a function of moisture content. Equations for $\rho_s(M)$, $\varepsilon'_2(M)$, and $\varepsilon''_2(M)$ were obtained through linear regression with the measured and estimated data.

Sumnu et al. (2007) showed that porosity increase during baking was responsible for the decrease in dielectric properties. As can be seen in Figure 14.41, both dielectric constant and loss factor decreased suddenly at 2–3 min of baking and then remained constant. This decrease can be related to the increase in specific volume (porosity) at the initial periods of baking (Figure 14.42).

Like bread dough, cake batter undergoes various physical, chemical, and structural changes during baking which affects dielectric properties. Artifical neural network has been recently used to estimate dielectric properties of cakes depending on porosity, moisture content, and formulations (Boyaci et al., 2009). Dielectric constant of cakes decreased

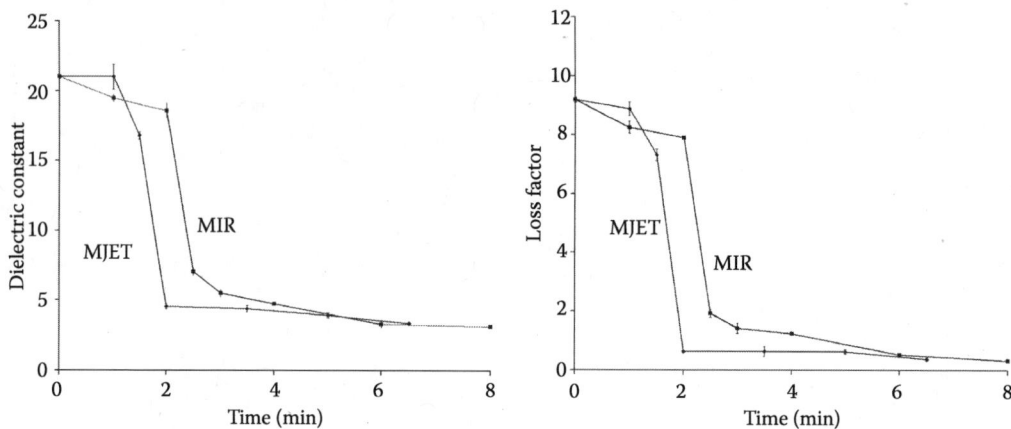

Figure 14.41 Dielectric constant (a) and loss factor (b) of breads measured at the center region of crumb during heating in different microwave combination ovens. MJET, microwave jet impingement oven; MIR, microwave-infrared combination oven. (From G Sumnu et al. *J Food Eng* 78: 1382–1387, 2007.)

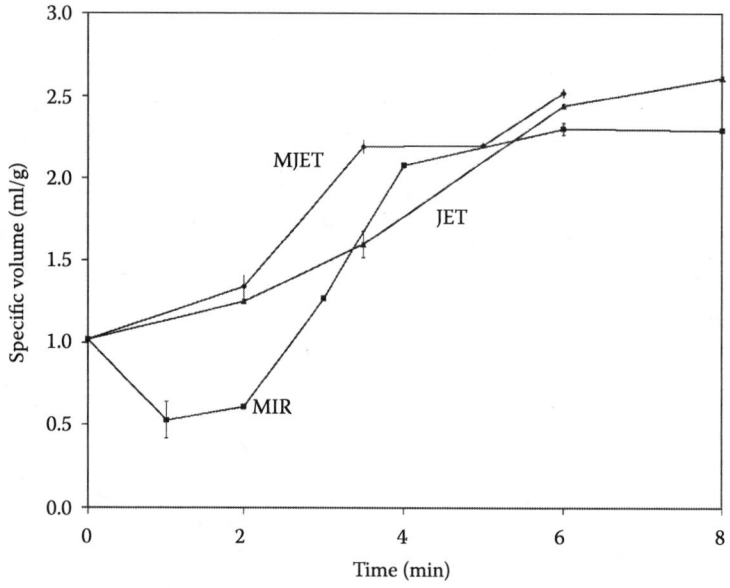

Figure 14.42 Variation of specific volume of breads during baking in different ovens. (From G Sumnu et al. *J Food Eng* 78: 1382–1387, 2007.)

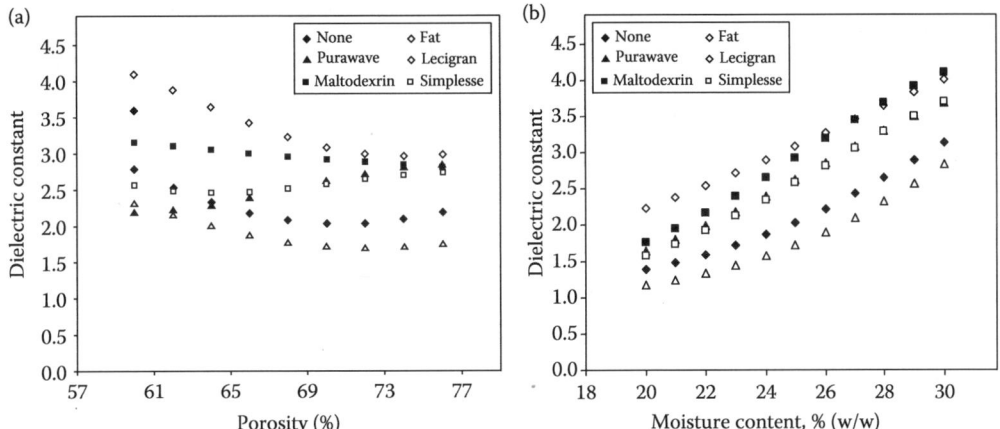

Figure 14.43 Variation of predicted dielectric constant data with (a) porosity and (b) moisture content. (From IH Boyaci, G Sumnu, O Sakiyan. *Food Bioprocess Technol* 2: 353–360, 2009.)

(Figure 14.43) as porosity increased and moisture content decreased during baking. Dielectric loss factor of cakes formulated with and without fat decreased as porosity increased.

14.9.2 Drying

Microwave drying offers an advantage over conventional hot-air heating by selectively heating more moist regions where microwave absorption is higher. This results in the popularly known moisture-leveling effect. Since the loss factor is mostly related to moisture content, the wet parts of a material will absorb more microwave energy, which leads to higher evaporation and will tend to level off the initial nonuniform moisture distribution (Metaxas and Meredith, 1988). The dry parts will not absorb as much of the microwave energy. For example, a strong moisture-leveling effect was noted in apples when they were dried from 50% to 4% at 60°C (Feng et al., 2002).

Thus, dielectric properties are needed as a function of moisture to describe quantitatively the spatial variation of microwave absorption in a microwave drying process. Variation of dielectric properties with moisture has already been discussed in Section 14.5.1.

14.9.3 Cooking

During cooking of muscle foods, dielectric properties were shown to change because of denaturation of proteins, which affected the availability of water or minerals for polarization or ionic conduction (Bircan and Barringer, 2002b). In muscle tissue, water is released during denaturation, so the dielectric properties of meat decrease. Detailed discussion of the protein denaturation effect on dielectric properties is provided in Section 14.5.3, dealing with the temperature effect on dielectric properties.

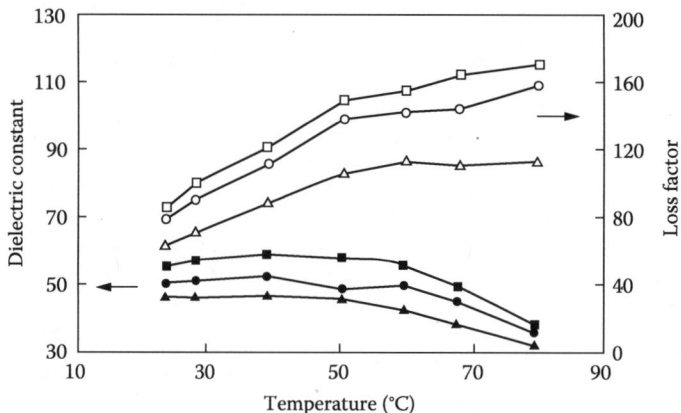

Figure 14.44 Dielectric constant and loss factor of dough samples at different mixing time and temperatures at 10 MHz. (From YR Kim, P Cornillon. *Lebensm Wiss u Technol* 34: 417–423, 2001.)

14.9.4 Mixing

The extent of mixing of various components in a food can affect dielectric properties. For example, dough mixing was shown to affect the dielectric properties (Kim and Cornillon, 2001). As mixing time increased, the dielectric constant of wheat dough decreased because of the small amount of mobile water in the sample after mixing (Figure 14.44). The loss factor also decreased during mixing since mixing decreased the amount and mobility of dissolved ions and water.

14.9.5 Storage

Studies showing the effects of storage on dielectric properties of foods are limited. The dielectric properties of cheese were shown to change significantly during storage (Datta, 1994). Both the dielectric constant and loss factor decreased during storage, which was likely due to composition changes as a result of strong proteolysis.

14.10 ASSESSMENT OF FOOD QUALITY BY USING DIELECTRIC PROPERTIES

Dielectric properties can be used for monitoring physiological processes (Narayan Jha et al., 2011). One area of application is assessment of fish and meat freshness. Muscle goes through a rigor mortis phase a few hours after death, which is defined as the temporary rigidity of muscles that develops after death of an animal (McWilliams, 1989). The level of glycogen stored in the animal at the time of slaughter is important in determining the onset of rigor mortis and the palatability factors of the meat when it is ready to be marketed. A postmortem decrease in pH brings proteins closer to their isoelectric point and

DIELECTRIC PROPERTIES OF FOODS

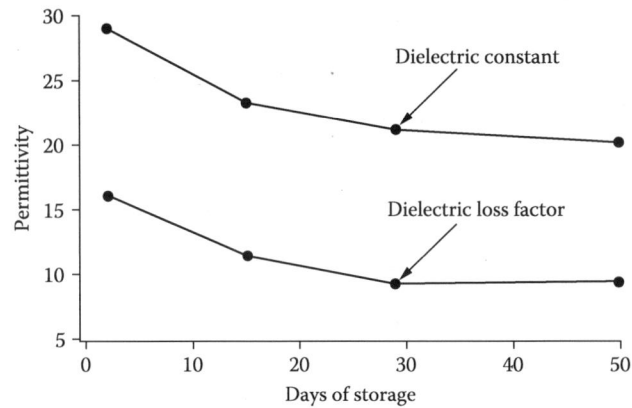

Figure 14.45 Effect of storage on dielectric properties of mozzarella cheese at 2.45 GHz and 24°C. (AK Datta, unpublished data.)

causes a decrease in water-holding capacity (Parisi et al., 2002). Therefore, the dielectric properties are expected to change during rigor mortis. The measurement of dielectric properties of fish was found to be a promising alternative to sensory analysis in evaluating the state of freshness of sea bass. Haddock exhibited significant changes in dielectric properties during rigor mortis (Martinsen et al., 2000). Firmness and ripeness of banana were also predicted using a 100 kHz signal (Soltani et al., 2011) and relationship between ripening and dielectric properties of mango has also been studied (Sosa-Morales et al., 2009). Freshness of leafy vegetables have also been shown to correlate with dielectric properties (Wen et al., 2007).

There is significant ongoing activity on using more detailed electromagnetic measurements for nondestructive evaluation of food quality. Two methods in particular seem to be popular—these are dielectric spectroscopy and dielectric TDR. Microwave dielectric spectroscopy (1 Hz–20 GHz) has been used to determine the sugar content in yogurt (Bohigas et al., 2008), maturity of apples (Castro-Giráldez et al., 2010), internal quality of peaches (Lleó et al., 2007) and watermelons (Guo et al., 2013), added water in processed prawns, cod, pork, and poultry meats (e.g., Kent et al., 2001, 2007; Samuel et al., 2012), pork meat quality (Castro-Giráldez et al., 2010), and quality of wine (Harley et al., 2011). TDR measures dielectric properties by the interaction of an electromagnetic step containing a wide range of frequencies at the same time with the sample (Miura et al., 2003). In this case, the obtained response is a time-domain curve which integrates the information of all the emitted frequencies. It has been used to measure loss of quality of stored fish (Kent et al., 2007) and dry-cured meat products (Fulladosa et al., 2013).

Dielectric properties can also be used for evaluation of frying oil quality that affects the quality of fried foods (Fritsch et al., 1979; Inoue et al., 2002; Shi et al., 1998). An example of change in dielectric properties due to the effect of degradation over time when heated at a higher temperature is shown in Figure 14.46. Such data have been suggested as the basis for real-time and continuous measurement of frying oil quality.

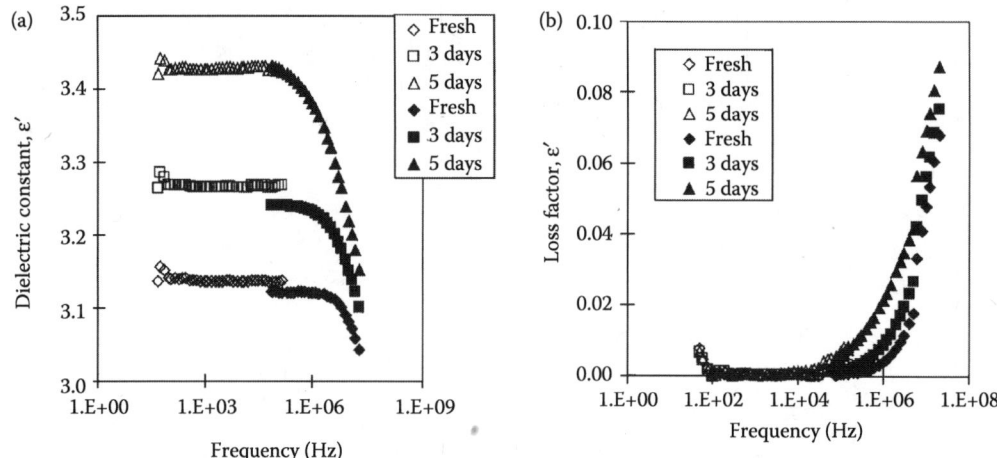

Figure 14.46 Dielectric constant and loss factor of soybean oil for various continuous heating times at 220°C, showing the effect of degradation over time. The data are measured at 16.5 ± 0.5°C. (From C Inoue et al. *J Food Sci* 67: 1126–1129, 2002.)

Because the dielectric properties of foods are dependent on moisture content, they have been studied extensively in that context (Chen and Sun, 1991; Nelson, 1984, 1985, 1986, 1987; McKeown et al., 2012). The dielectric properties provide a rapid means for nondestructive sensing of moisture content of agricultural products (Nelson, 1991). Various studies showed that it was possible to model the relationship of dielectric constant with moisture content, frequency, and bulk density so that moisture of agricultural products could be determined indirectly by measuring dielectric properties (Nelson, 1987). An example of such data can be seen in Figure 14.13.

Dielectric properties have been related to water activity and thus proposed as a measure for the same (Clerjon et al., 2003; Henry et al., 2003a). Prediction of dielectric properties for bound water and free water has been provided (Henry et al., 2003b). Likewise, dielectric properties have been correlated to color of poultry breast meat in an effort to measure its quality (Samuel et al., 2012).

Detection of honey adulteration using dielectric properties (Guo et al., 2010b) has been proposed based on linear correlation between dielectric constant and total soluble solids of honey.

14.11 FURTHER SOURCES OF DATA

Because of limited space in this publication, only representative data were presented that would provide the general trend for certain types of food products. Dielectric data are widely dispersed throughout the food literature; large compilations of data for food materials are available starting with von Hippel (1954b), for agricultural products and particularly grains (Nelson, 1973), several food materials (Datta et al., 1995; Kent, 1987), and a more recent article covering most of the food categories (Sosa-Morales et al., 2010). An Internet-based properties database has also been developed (Nesvadba et al., 2004), that will have an increasing amount of data available in the future. For now, large bibliographic databases are perhaps the best source of data for specific food products. Data in other biological literature (e.g., Gabriel et al., 1996) can also be quite relevant to food processing. Engineering analysis and design can often be performed with the representative data for similar systems, combined with sensitivity analysis (showing the effect of property variation on the process). When accurate data are needed for a specific composition and condition, carefully conducted experiments are suggested.

ACKNOWLEDGMENT

The authors respectfully acknowledge the important contributions made by the reviewers of the third edition version of this manuscript. They include Dr. Stuart Nelson, U.S. Department of Agriculture (USDA); Dr. Lília Ahrné, SIK; Professor Nils Bengtsson, SIK; Dr. Birgitta Raholt, SIK; and Professor Juming Tang, Washington State University.

REFERENCES

J Ahmed, HS Ramaswamy, GSV Raghavan. Dielectric properties of Indian Basmati rice flour slurry. *J. Food Eng* 80: 1125–1133, 2007.

J Ahmed, HS Ramaswamy, GSV Raghavan. Dielectric properties of soybean protein isolate dispersions as a function of concentration, temperature and pH. *LWT- Food Sci Technol* 41: 71–81, 2008.

J Ahmed, G Luciano. Dielectric properties of beta-lactoglobulin as influenced by pH, concentration and temperature. *J Food Eng* 95: 30–35, 2009.

J Ahmed. Rheological, thermal and dielectric properties of pureed baby foods. *Stewart Postharvest Review* 3: 1–12, 2007.

J Ahmed, HS Ramaswamy, VGS Raghavan. Dielectric properties of butter in the MW frequency range as affected by salt and temperature. *Journal of Food Engineering* 82: 351–358, 2007.

B Alfaifi, S Wang, J Tang, B Rasco, S Sablani, Y Jiao. Radio frequency disinfestation treatments for dried fruit: Dielectric properties. *LWT - Food Science and Technology* 50: 746–754, 2013.

AH Al-Muhtaseb, MA Hararah, EK Megahey, WAM Mcminn, TRA Magee. Dielectric properties of microwave-baked cake and its constituents over a frequency range of 0.915–2.450 GHz. *J Food Eng* 98: 84–92, 2010.

N Basaran-Akgul, P Basaran, BA Rasco. Effect of temperature (−5 to 130°C) and fiber direction on the dielectric properties of beef Semitendinosus at radio frequency and microwave frequencies. *Journal of Food Science* 73: E243–E249, 2008.

NE Bengtsson, J Melin, K Remi, S Soderlind. Measurements of the dielectric properties of frozen and defrosted meat and fish in the frequency range 10–200 MHz. *J Sci Food Agric* 14: 593–604, 1963.

NE Bengtsson, PO Risman. Dielectric properties of food at 3 GHz as determined by a cavity perturbation technique. II. Measurements on food materials. *J Microwave Power* 6: 107–123, 1971.

C Bircan, SA Barringer. Determination of protein denaturation of muscle foods using dielectric properties. *J Food Sci* 67: 202–205, 2002a.

C Bircan, SA Barringer. Use of dielectric properties to detect egg protein denaturation. *J Microwave Power Electromagn Energy* 37: 89–96, 2002b.

X Bohigas, R Amigó, J Tejada. Characterisation of sugar content in yoghurt by means of microwave spectroscopy. *Food Research International* 41: 104–109, 2008.

IH Boyaci, G Sumnu, O Sakiyan. Estimation of dielectric properties of cakes based on porosity, moisture content, and formulations using statistical methods and artificial neural networks. *Food Bioprocess Technol* 2: 353–360, 2009.

NP Brunton, JG Lyng, L Zhang, JC Jacquier. The use of dielectric properties and other physical analyses for assessing protein denaturation in beefbiceps femorismuscle during cooking from 5 to 85°C. *Meat Sci* 72: 236–244, 2006.

RK Calay, M Newborough, D Probert, PS Calay. Predictive equations for dielectric properties of foods. *Int J Food Sci Technol* 29: 699–713, 1995.

M Castro-Giráldez, MC Aristoy, F Toldrá, P Fito. Microwave dielectric spectroscopy for the determination of pork meat quality. *Food Res Int* 43: 2369–2377, 2010.

M Castro-Giráldez, PJ Fito, C Chenoll, P Fito. Development of a dielectric spectroscopy technique for the determination of apple (Granny Smith) maturity. *Innovative Food Sci Emerging Technol* 11: 749–754, 2010.

M Chamchong, AK Datta. Thawing of foods in a microwave oven: I. Effect of power levels and power cycling. *J Microwave Power Electromagn Energy* 34: 9–21, 1999.

V Changrue, V Orsat, GSV Raghavan, D Lyew. Effect of osmotic dehydration on the dielectric properties of carrots and strawberries. *J Food Eng* 88: 280–286, 2008.

P Chen, Z Sun. A review of nondestructive methods for quality evaluation and sorting of agricultural products. *J Agric Eng Res* 49: 85–98, 1991.

S Clerjon, J-D Daudin, J-L Damez. Water activity and dielectric properties of gels in the frequency range 200 MHz–6 GHz. *Food Chem*, 82: 87–97, 2003.

H Cong, F Liu, Z Tang, C Xue. Dielectric properties of sea cucumbers (*Stichopus japonicus*) and model foods at 915 MHz. *J Food Eng* 109: 635–639, 2012.

P Coronel, , J Simunovic, KP Sandeep, P Kumar. Dielectric properties of pumpable food materials at 915 MHz. *Int J Food Properties* 11: 508–518, 2008.

AK Datta. Effect of composition, temperature, and storage on dielectric properties of cheese. Unpublished data, 1994.

AK Datta, E Sun, A Solis. Food dielectric property data and their composition based prediction. In MA Rao, SSH Rizvi, Eds. *Engineering Properties of Foods*. Marcel Dekker, New York, 1995, pp. 457–494.

GP De Loor, FW Meijboom. The dielectric constant of foods and other materials with high water contents at microwave frequencies. *J Food Technol* 1: 313–322, 1966.

P Debye. *Polar Molecules*. The Chemical Catalog Co., New York, 1929.

RV Decareau. Microwaves in food processing. *Food Technol Aust* 36: 81–86, 1984.

D Dibben. Electromagnetics: Fundamental aspects and numerical modelling. In AK Datta, S Anantheswaran, Eds. *Handbook of Microwave Technology for Food Applications*, Marcel Dekker, New York, 2001.

SD Engelder, RC Buffler. Measuring dielectric properties of food products at microwave frequencies. *Microwave World* 12: 6–14, 1991.

CD Everard, CC Fagan, CP O'Donnell, DJ O'Callaghan, JG Lyng. Dielectric properties of process cheese from 0.3 to 3 GHz. *Journal of Food Engineering* 75 :415–422, 2006.

E Eves, V Yakovlev. Analysis of operational regimes of a high power water load. *J Microwave Power Electromagn Energy* 37(3): 127–144, 2002.

H Feng, J Tang, RP Cavalieri. Dielectric properties of dehydrated apples as affected by moisture and temperature. *Trans ASAE* 45: 129–135, 2002.

CW Fritsch, DC Egberg, JS Magnusun. Changes in dielectric constant as a measure of frying oil deterioration. *J Am Oil Chem Soc* 56: 746–750, 1979.

Fulladosa, E., P. Duran-Montgé, X. Serra, P. Picouet, O. Schimmer, and P. Gou. Estimation of dry-cured ham composition using dielectric time domain reflectometry. *Meat Sci* 93: 873–879, 2013.

C Gabriel, S Gabriel, E Corthout. The dielectric properties of biological tissues: I. Literature survey. *Phys Med Biol* 41: 2231–2249, 1996.

C Gabriel, EH Grant, IR Young. Use of time domain spectroscopy for measuring dielectric properties with a coaxial probe. *J Phys E Sci Instrum* 19: 843–846, 1986.

M Gao, J. Tang, JA Johnson, S Wang. Dielectric properties of ground almond shells in the development of radio frequency and microwave pasteurization. *J Food Eng* 112: 282–287, 2012.

FM Ghannouchi, RG Bosisio. Measurement of microwave permittivity using a sixport reflectometer with an openended coaxial line. *IEEE Trans Instrum Meas* 38: 505–508, 1989.

M Glicksman. Origins and classification of hydrocolloids. In M Glicksman, Ed. *Food Hydrocolloids*. CRC Press, Boca Raton, FL, 1982, p 3.

NK Goebel, J Grider, EA Davis, J Gordon. The effects of microwave energy and conventional heating on wheat starch granule transformations. *Food Microstruct* 3: 73–82, 1984.

AD Green. Measurement of the dielectric properties of cheddar cheese. *J Microwave Power Electromagn Energy* 32: 16–27, 1997.

W Guo, S Wang, G. Tiwari, JA Johnson, J Tang. Temperature and moisture dependent dielectric properties of legume flour associated with dielectric heating. *LWT- Food Sci Technol.* 43: 193–201, 2010a.

W Guo, X Zhu, Y Liu, H. Zhuang. Sugar and water contents of honey with dielectric property sensing. *J Food Eng* 97: 275–281, 2010b.

W Guo, X Zhu, SO Nelson. Permittivities of watermelon pulp and juice and correlation with quality indicators. *International Journal of Food Properties* 16: 475–484, 2013.

W Guo, X Wu, X Zhu, S Wang. Temperature-dependent dielectric properties of chestnut and chestnut weevil from 10 to 4500 MHz. *Biosystems Engineering* 110: 340–347, 2011.

S.J. Harley, V. Lim, M.P. Augustine. Using low frequency dielectric absorption to screen full
intact wine bottles. *Anal Chim Acta* 702(2); 188-194, (2011).

JB Hasted. *Aqueous Dielectrics*. Chapman & Hall, London, 1973.

JB Hasted, DM Ritson, CH Collie. Dielectric properties of aqueous ionic solutions. Parts I and II. *J Chem Phys* 16: 1–21, 1948.

F Henry, LC Costa, M Serpelloni. Dielectric method for determination of a_w. *Food Chem* 82: 73–77, 2003a.

F Henry, M Gaudillat, LC Costa, F Lakkis. Free and/or bound water by dielectric measurements. *Food Chem* 82: 29–34, 2003b.

AG Herve, J Tang, L Luedecke, H Feng. Dielectric properties of cottage cheese and surface treatment using microwaves. *J Food Eng* 37: 389–410, 1998.

Hewlett-Packard Co. HP 85070 A Dielectric Probe Kit, Data Sheet, # 5952–2381. HP 85071 A Material measurement software, data sheet, # 5952–2382, 1992.

C Inoue, Y Hagura, M Ishikawa, K Suzuki. The dielectric property of soybean oil in deep-fat frying and effect of frequency. *J Food Sci* 67: 1126–1129, 2002.

JL Jorgensen, AR Edison, SO Nelson, LE Stetson. A bridge method for dielectric measurements of grain and seed in the 50 to 250 MHz range. *Trans ASAE*, 13(1): 18–20, 24, 1970.

U Kaatze. Complex permittivity of water as a function of frequency and temperature. *J Chem Eng Data* 34: 371–384, 1989.

GA Karhale, GM Kalamse. Dielectric measurement of Pulverized coriander (*Coriandrum sativum* Linn.) at microwave frequency. *Curr Bot* 3: 4–8, 2012.

RB Keam, WS Holmes. Uncertainty analysis of measurement of complex permittivity using microstrip transmission line. *SBMO/IEEE MTTS IMOC '95 Proceedings*, 137–142, 1995.

M Kent. Microwave dielectric properties of fish meal. *J Microwave Power* 7: 109–116, 1972.

M Kent. Complex permittivity of fish meal: A general discussion of temperature, density, and moisture dependence. *J Microwave Power* 12: 341–345, 1977.

M Kent. *Electrical and Dielectrical Properties of Food Materials*. Science and Technology Publishers, Essex, England, 1987.

M Kent, R Knöchel, F. Daschner, U-K Berger. Composition of foods including added water using microwave dielectric spectra. *Food Control* 12: 467–482, 2001.

M Kent, R Knöchel, F Daschner, O Schimmer, J Oehlenschläger, S Mierke-Klemeyer, M Kroeger et al. Intangible but not intractable: The prediction of fish "quality" variables using dielectric spectroscopy. *Meas Sci Technol* 18(4): 1029–1037, 2007.

M Kent, K Rogers. Microwave moisture and density measurements in particulate solids. *Trans INST M C* July–Sept 8(3): 167–168, 1987.

SO Keskin, G Sumnu, S Sahin. A study on the effects of different gums on dielectric properties and quality of breads baked in infrared-microwave combination oven. *Eur Food Res Technol* 224: 329–334, 2007.

YR Kim, P Cornillon. Effects of temperature and mixing time on molecular mobility in wheat dough. *Lebensm Wiss u Technol* 34: 417–423, 2001.

YR Kim, MT Morgan, MR Okos, RL Stroshine. Measurement and prediction of dielectric properties of biscuit dough at 27 MHz. *J Microwave Power Electromagn Energy* 33: 184–194, 1998.

AW Kraszewski. Microwave aquametry—A review. *J Microwave Power* 15: 209–220, 1980.

AW Kraszewski, S Trabelsi, SO Nelson. Microwave dielectric properties of wheat. *Proceedings of 30th Microwave Power Symposium*, July 9–12, Denver, CO, pp 90–93, 1995.

W Kuang, SO Nelson. Dielectric relaxation characteristic of fresh fruits and vegetables from 3 to 20 GHz. *J Microwave Power Electromagn Energy* 32: 114–122, 1997.

T Kudra, V Raghavan, C Akyel, R Bosisio, F van de Voort. Electromagnetic properties of milk and its constituents at 2.45 GHz. *J Microwave Power Electromagn Energy* 27: 199–204, 1992.

XJ Liao, GSV Raghavan, J Dai, VA Yaylayan. Dielectric properties of α-D glucose aqueous solutions at 2450 MHz. *Food Res Int* 36: 485–490, 2003.

SX Liu, M Fukuoka, N Sakai.Dielectric properties of fish flesh at microwave frequency. *Food Sci Technol Res* 18: 157–166, 2012.

H Lizhi, K Toyoda, I Ihara. Dielectric properties of edible oils and fatty acids as a function of frequency, temperature, moisture and composition. *J Food Eng* 88: 151–158, 2008.

L Lleó, M Ruiz-Altisent, N Hernández, P Gutiérrez. Application of microwave return loss for sensing internal quality of peaches. *Biosystems Eng* 96: 525–539, 2007.

OG Martinsen, S Grimnes, P Mirtaheri. Noninvasive measurements of postmortem changes in dielectric properties of haddock muscle—A pilot study. *J Food Eng* 43: 189–192, 2000.

S Mashimo, S Kuwabara, S Yagihara, K Higasi. Dielectric relaxation time and structure of bound water in biological materials. *J Phys Chem* 91: 6337–6338, 1987.

MS McKeown, S Trabelsi, EW Tollner, SO Nelson. Dielectric spectroscopy measurements for moisture prediction in Vidalia onions. *J Food Eng* 111: 505–510, 2012.

M McWilliams. *Foods: Experimental Perspectives*. MacMillan Publishing Company, New York, 1989, pp 161–162, 277–279.

V Meda. Cavity perturbation technique for measurement of dielectric properties of some agrifood materials, MSc thesis, McGill University, Macdonald Campus, Canada, 1996.

V Meda, Integrated dual frequency permittivity analyzer using cavity perturbation concept, PhD thesis, McGill University, Macdonald Campus, Canada, 2002.

AC Metaxas, RJ Meredith. *Industrial Microwave Heating*. Peter Peregrinus, London, 1988.

AC Metaxas. *Foundations of Electroheat*. John Wiley and Sons, Chichester, UK, 1996.

LA Miller, J Gordon, EA Davis. Dielectric and thermal transition properties of chemically modified starches during heating. *Cereal Chem* 68: 441–448, 1991.

N Miura, S Yagihara, S Mashimo. Microwave dielectric properties of solid and liquid foods investigated by time-domain reflectometry. *J Food Sci* 68: 1396–1403, 2003.

MMA Moteleb. Some of the dielectric properties of starch. *Polymer Int* 35: 243–247, 1994.

RE Mudgett. Electrical properties of foods. In MA Rao, SSH Rizvi, Eds. *Engineering Properties of Foods*. Marcel Dekker, New York, 1995.

RE Mudgett, WB Westphal. Dielectric behavior of an aqueous cation exchanger. *J Microwave Power* 24: 33–37, 1989.

RE Mudgett, AC Smith, DIC Wang, SA Goldblith. Prediction of dielectric properties in nonfat milk at frequencies and temperatures of interest in microwave processing. *J Food Sci* 39: 52–54, 1974.

RE Mudgett, AC Smith, DIC Wang, SA Goldblith. Prediction of relative dielectric loss factor in aqueous solutions of non-fat dried milk through chemical simulation. *J Food Sci* 26: 915–918, 1971.

S Narayan Jha, K Narsaiah, AL Basediya, S Rajiv, J Pranita, K Ramesh, B Rishi. Measurement techniques and application of electrical properties for nondestructive quality evaluation of foods–A review. *J Food Sci Technol* 48: 387–411, 2011.

M Ndife, G Sumnu, L Bayindrli. Dielectric properties of six different species of starch at 2450 MHz. *Food Res Int* 31: 43–52, 1998.

SO Nelson. A system for measuring dielectric properties at frequencies from 8.2 to 12.4 GHz. *Trans ASAE* 15: 1094–1098, 1972.

SO Nelson. Electrical properties of agricultural products—A critical review. *Trans ASAE* 16: 384–400, 1973.

SO Nelson. Dielectric properties of some fresh fruits and vegetables at frequencies of 2.45 to 22 GHz. ASAE Paper 82–3053, 1982.

SO Nelson. Density dependence of dielectric properties of particulate materials. *Trans ASAE* 26: 1823–1825, 1829, 1983.

SO Nelson. Moisture, frequency, and density dependence of the dielectric constant of shelled, yellow-dent field corn. *Trans ASAE* 27: 1573–1578, 1585, 1984.

SO Nelson. A mathematical model for estimating the dielectric constant of hard red winter wheat. *Trans ASAE* 28: 234–238, 1985.

SO Nelson. Mathematical models for the dielectric constants of spring barley and oats. *Trans ASAE* 29: 607–610, 615, 1986.

SO Nelson. Models for the dielectric constants of cereal grains and soybeans. *J Microwave Power* 22: 35–39, 1987.

SO Nelson. Dielectric properties of agricultural products—Measurements and applications. *IEEE Trans Elect Insulation*, 26: 845–869, 1991.

SO Nelson. Review and assessment of radiofrequency and microwave energy for stored-grain insect control. *Trans ASAE*, 39: 1475–1484, 1996.

SO Nelson. Dielectric properties measuring techniques and applications. *Trans ASAE* 42: 523–529, 1999.

SO Nelson, LF Charity. Frequency dependence of energy absorption by insects and grain in electric fields. *Trans ASAE*, 15: 1099–1102, 1972.

SO Nelson, AK Datta. Dielectric properties of food materials and electric field interactions. In AK Datta, RC Anantheswaran, Eds. *Handbook of Microwave Technology for Food Applications*. Marcel Dekker, New York, 2001, pp 69–114.

SO Nelson, WR Forbus, KC Lawrence. Permittivity of fresh fruits and vegetables from 0.2 to 20 GHz. *J Microwave Power Electromagn Energy* 29: 81–93, 1994.

SO Nelson, A Prakash, K Lawrence. Moisture and temperature dependence of the permittivities of some hydrocolloids at 2.45 GHz. *J Microwave Power Electromagn Energy* 26: 178–185, 1991.

SO Nelson, LE Stetson. Frequency and moisture dependence of the dielectric properties of hard red wheat. *J Agric Eng Res* 21: 181–192, 1976.

SO Nelson, LE Stetson, CW Schlaphoff. A general computer program for precise calculation of dielectric properties from short-circuited waveguide measurements. *IEEE Trans Instrum Meas* 23: 455–460, 1974.

P Nesvadba, M Houska, W Wolf, V Gekas, D Jarvis, PA Sadd, A I Johns. Database of physical properties of agro-food materials, *J Food Eng* 61: 497–503. 2004.

J F O'Connor, EC Synnott. Seasonal variation in dielectric properties of butter at 15 MHz and 4°C. *J Food Sci Technol* 6: 49–59, 1982.

T Ohlsson, M Henriques, NE Bengtsson. Dielectric properties of model meat emulsions at 900 and 2800 MHz in relation to their composition. *J Food Sci* 39: 1153–1156, 1974.

T Ohlsson, NE Bengtsson. Dielectric food data for MW sterilization processing. *J Microwave Power* 10(1): 93–108, 1975.

O Ozmutlu, G Sumnu, S Sahin. Effects of different formulations on the quality of microwave-baked bread. *Eur Food Res Technol* 213: 38–42, 2001.

WE Pace, WB Westphal, SA Goldblith. Dielectric properties of commercial cooking oils. *J Food Sci* 33: 30–36, 1968.

G Parisi, O Franci, BM Poli. Application of multivariable analysis to sensorial and instrumental parameters of freshness in refrigerated sea bass (*Dicentrarchus labrax*) during shelf life. *Aquaculture* 214: 153–167, 2002.

R Pething. *Dielectric and Electronic Properties of Biological Materials*. Wiley, New York, 1979.

A Prakash. The effect of microwave energy on the structure and function of food hydrocolloids. M.S. thesis, The Ohio State University, Columbus, OH, 1991.

A Prakash, SO Nelson, ME Mangino, PMT Hansen. Variation of microwave dielectric properties of hydrocolloids with moisture content, temperature and stoichiometric charge. *Food Hydrocolloids* 6: 315–322, 1992.

PO Risman, NE Bengtsson. Dielectric properties of food at 3 GHz as determined by a cavity perturbation technique. I. measuring technique. *J Microwave Power* 6: 101–106, 1971.

S Roberts, A von Hippel. A new method for measuring dielectric constant and loss in the range of centimeter waves. *J Appl Phys* 17: 610–616, 1946.

BD Roebuck, SA Goldblith. Dielectric properties of carbohydrate–water mixtures at microwave frequencies. *J Food Sci* 37: 199–204, 1972.

S Ryynänen. The electromagnetic properties of food materials: A review of basic principles. *J Food Eng* 26: 409–429, 1995.

S Ryynänen, PO Risman, T Ohlsson. The dielectric properties of native starch solutions: A research note. *J Microwave Power Electromagn Energy* 31: 50–53, 1996.

MA Rzepecka, RR Pereira. Permittivity of some dairy products at 2450 MHz. *J Microwave Power* 9: 277–288, 1974.

O Sakiyan, G Sumnu, S Sahin, V Meda. Investigation of dielectric properties of different cake formulations during microwave and infrared-microwave combination baking. *J Food Sci* 72: E205–213, 2007.

D Samuel, S Trabelsi, AB Karnuah, NB Anthony, SE Aggrey. The use of dielectric spectroscopy as a tool for predicting meat quality in poultry. *Int J Poultry Sci* 11: 551–555, 2012.

D Samuel, S Trabelsi. Influence of color on dielectric properties of marinated poultry breast meat. *Poultry Sci* 91: 2011–2016, 2012.

R Seaman, J Seals. Fruit pulp and skin dielectric properties for 150 MHz to 6400 MHz. *J Microwave Power Electromagn Energy* 26: 72–81, 1991.

LS Shi, BH Lung, HL Sun. Effects of vacuum frying on the oxidative stability of oils. *J Am Oil Chem Soc* 75: 1393–1398, 1998.

TP Shukla, RC Anantheswaran. Ingredient interactions and product development. In AK Datta, RC Anantheswaran, Eds. *Handbook of Microwave Technology for Food Applications*. Marcel Dekker, New York, 2001, pp 355–395.

O Sipahioglu, SA Barringer. Dielectric properties of vegetables and fruits as a function of temperature, ash and moisture content. *J Food Sci* 68: 234–239, 2003.

M Soltani, R Alimardani, M Omid. Evaluating banana ripening status from measuring dielectric properties. *J Food Eng* 105: 625–631, 2011.

ME Sosa-Morales, G Tiwari, S Wang, J Tang, HS Garcia, A Lopez-Malo. Dielectric heating as a potential post-harvest treatment of disinfesting mangoes, Part I: Relation between dielectric properties and ripening. *Biosystems Eng* 103: 297–303, 2009.

ME Sosa-Morales, L Valerio-Junco, A López-Malo, HS García. Dielectric properties of foods: Reported data in the 21st Century and their potential applications. *LWT—Food Sci Technol* 43: 1169–1179, 2010.

A Stogryn. Equations for calculating the dielectric constant of saline water. *IEEE Trans Microwave Theory Techn* 19: 733–736, 1971.

MA Stuchly, SS Stuchly. Dielectric properties of biological substances—Tabulated. *J Microwave Power* 15: 19–26, 1980.

G Sumnu, AK Datta, S Sahin, SO Keskin, V Rakesh. Transport and related properties of breads baked using various heating modes. *J Food Eng* 78: 1382–1387, 2007.

M Sucher, J Fox. *Handbook of Microwave Measurements*. Polytechnic Press of the Polytechnic Institute of Brooklyn, Brooklyn, NY, 1963.

E Sun, AK Datta, S Lobo. Composition-based prediction of dielectric properties of foods. *J Microwave Power Electromagn Energy* 30(4): 205–212, 1995.

J Tang, JF Hao, M Lau. Microwave heating in food processing. In XH Yang, J Tang, Eds. *Advances in Bioprocessing Engineering*. World Scientific, NJ, 2002, pp. 1–44.

EC To, RE Mudgett, DIC Wang, SA Goldblith. Dielectric properties of food materials. *J Microwave Power* 9: 303–315, 1974.

S Trabelsi, AW Kraszewski, SO Nelson. A new density independent function for microwave moisture content determination in particulate materials. *IEEE Instrumentation and Measurement Technology Conference*, Ottawa, Canada, May 19–21, 1997.

VN Tran, SS Stuchly, A Kraszewski. Dielectric properties of selected vegetables and fruits 0.1–10.0 GHz. *J Microwave Power* 19: 251–258, 1984.

ML Tsai, CF Li, CY Lii. 1997. Effects of granular structures on pasting behaviours of starches. *Cereal Chem* 74: 750–757.

E Turabi, M Regier, G Sumnu, S Sahin, M Rother. 2010. Dielectric and thermal properties of rice cake formulations containing different gums types. *Int J Food Prop* 13: 1199–1206.

SL Umbach, EA Davis, J Gordon, PT Callaghan. Water self-diffusion coefficients and dielectric properties determined for starch–gluten–water mixtures heated by microwave and conventional methods. *Cereal Chem* 69: 637–642, 1992.

J Van Loo, P Coussement, L de Leenheer, H Hoebregs, G Smits. On the presence of inülin and oligofructose as natural ingredients in Western diet. *Crit Rev Food Sci Tehnol* 35: 525–552, 1995.

AR Von Hippel. *Dielectrics and Waves*, MIT, Cambridge, MA, 1954a.

AR Von Hippel. *Dielectric Materials and Applications*. The Technology Press of MIT and John Wiley, New York, 1954b.

R Wang, M Zhang, AS Mujumdar, H Jiang. Effect of salt and sucrose content on dielectric properties and microwave freeze drying behaviour of re-structured potato slices. *J Food Eng* 106: 290–297, 2011.

SJ Wang, JA Tang, E Johnson, JD Mitcham, G Hallman, SR Drake, Y Wang. Dielectric properties of fruits and insect pests as related to radio frequency and microwave treatments. *J Food Eng* 57: 257–268, 2003a.

S Wang, J Tang, RP Cavalieri and DC Davis. Differential heating of insects in dried nuts and fruits associated with radio frequency and microwave treatments. *Trans ASAE* 46(4): 1175–1182, 2003.

Y Wang, TD Wig, J Tang, LM Hallberg. Dielectric properties of foods relevant to RF and microwave pasteurization and sterilization. *J Food Eng* 57: 257–268, 2003.

Q Wen, JS Ye, WT Xue, AX Huang, Y Hugura, K Suzuki. Correlation between dielectric properties and freshness of leaf vegetables. *J Food Process Preserv* 31: 736–750, 2007.

H Wu, E Kolbe, B Flugstad, JW Park, J Yongsawatdigul. Electrical properties of fish mince during multifrequency ohmic heating. *J Food Sci* 63: 1028–1032, 1988.

P Yagmaee, TD Durance. Predictive equations for dielectric properties of NaCl, D-Sorbitol and sucrose solutions and surimi at 2450 MHz. *J Food Sci* 67: 2207–2211, 2002.

Y Yin, CE Walker. A quality comparison of breads baked by conventional versus nonconventional ovens: A review. *J Sci Food Agric* 67: 283–291, 1995.

H Zhang, AK Datta. Electromagnetics of microwave heating: Magnitude and uniformity of energy absorption in an oven. In AK Datta and RC Anantheswaran, Eds. *Handbook of Microwave Technology for Food Applications*. Marcel Dekker, New York, 2001, pp. 33–68.

H Zhang, AK Datta. Heating concentrations of microwaves in spherical and cylindrical foods. Part one: In plane waves. *Food Bioprod Process* 83(C1): 6–13, 2005.

L Zhang, JG Lyng, NP Brunton. The effect of fat, water and salt on the thermal and dielectric properties of meat batter and its temperature following microwave or radio frequency heating. *J Food Eng* 80: 142–151, 2007.

B Zhang, H Yu, Y Cheng, Y Jin. Effects of temperature, frequency, and moisture content on the dielectric properties of tilapia. *Journal of Fisheries of China* 36: 1785–1792, 2012.

M Zheng, YW Huang, SO Nelson, P Bartley, KW Gates. Dielectric properties and thermal conductivity of marinated shrimp and channel catfish. *J Food Sci* 63: 668–672, 1998.

15

Ultrasound Properties of Foods

Donghong Liu and Hao Feng

Contents

15.1 Introduction	638
15.2 Basic Ultrasound Physics	639
15.2.1 Properties of an Ultrasound Wave	639
15.2.1.1 Speed of Sound	639
15.2.1.2 Attenuation	642
15.2.1.3 Impedance, Reflection, and Refraction	642
15.2.2 Interactions of Ultrasound with Foods	643
15.3 Food Property Measurement with Ultrasound	644
15.3.1 Ultrasonic Measurement Systems	644
15.3.1.1 Pulse-Echo	644
15.3.1.2 Transmission	645
15.3.1.3 Ultrasonic Imaging	645
15.3.1.4 Air-Coupled	647
15.3.2 Ultrasonic Signal Processing Methods	647
15.3.2.1 Conventional Ultrasound Parameters	647
15.3.2.2 Spectrum Analysis	649
15.3.2.3 Pattern Recognition	650
15.4 Compilation of Acoustic Properties of Foods	651
15.4.1 Composition	651
15.4.1.1 Solutions	651
15.4.1.2 Emulsions and Suspensions	653
15.4.1.3 Muscle Foods	655
15.4.2 Phase Transition	657
15.4.2.1 Freezing	657
15.4.2.2 Crystallization	657
15.4.2.3 Gelation	658

15.4.3 Texture	659
15.4.3.1 Firmness of Fruits/Vegetables	659
15.4.3.2 Cheese	659
15.4.3.3 Starch Products	660
15.4.4 Viscosity	661
15.4.4.1 Ultrasound Doppler Velocity Profiling and the UVP–PD Method	661
15.4.4.2 Pulse-Echo or Transmission-Through Method	663
15.4.4.3 Measurement of the Viscosity–Density Product	663
15.5 Applications	664
15.5.1 Vegetable and Fruit Quality Monitoring	664
15.5.2 Quality Evaluation of Meat Products	666
15.5.3 Ultrasonic Measurement for Liquid Food	666
15.5.3.1 Particle Size Analysis	666
15.5.3.2 Bubbly Liquids	667
15.5.4 Detection of Leaks in Food Packages	668
List of Symbols	670
Greek Symbols	670
Subscripts/Superscripts	671
References	671

15.1 INTRODUCTION

The consumer's need for fresh, nutritious, and safe foods has driven the food industry to pursue fast online quality-monitoring strategies. Low-intensity, high-frequency ultrasound (<1 W/cm^2 and >100 kHz) has been tested as a promising tool to achieve the rapid, accurate, inexpensive, simple, and nondestructive online measurement of food quality indexes. Ultrasonic detection enables real-time measurement so that food composition can be adjusted, process parameters modified, and quality assured.

Research activity on the utilization of low-intensity ultrasound for food characterization can be traced back over 60 years. Various techniques have been developed over time, including pulse-echo (PE) transmission, process tomography, and air-coupled methods. These techniques utilize the relationships between ultrasonic properties (velocity, attenuation coefficient, and impedance) and food physicochemical properties (composition, geometry, microstructure, and state) to provide quantitative data that can be directly used in process control. Low-power ultrasound (LPU) has been tested for evaluating the composition of raw and fermented meat products, fish, and poultry. It has also been investigated for use in the quality control of fruits and vegetables, cheese, cooking oils, breads and cereals, beverages, bulk and emulsified fat-based food products, food gels, aerated foods, and frozen foods. Other applications include the detection of honey adulteration, defects in food packages, and assessment of the aggregation state, size, and type of proteins (Awad et al., 2012).

In ultrasonic characterization of foods, the velocity of the ultrasound in the food, the wave-amplitude attenuation, and the impedance the sound waves experience when they travel through the food are the pertinent characteristics. However, compared to the work on transport process-related food properties, the research activities on the ultrasonic measurement and prediction of food properties are scattered and unsystematic. In this chapter, after a brief introduction of the basics of ultrasound and related measurement methods, the ultrasonic properties of certain selected foods will be presented, followed by a few applications. It was not the intention here to present a complete collection of the data currently available in the literature or a comprehensive review of the technology. Interested readers can refer to the excellent article by McCarthy et al. (2005) published in the third edition of this book, and the reviews of Povey (1989), McClements (1997), and Coupland (2004) for more information.

15.2 BASIC ULTRASOUND PHYSICS

15.2.1 Properties of an Ultrasound Wave

15.2.1.1 Speed of Sound
Sound velocity refers to the distance traveled during a unit time by an acoustic wave propagating through an elastic medium. For sinusoidal sound waves of infinite length propagating in one dimension, a simple relation can be used to characterize the relation between sound speed, frequency, and wavelength

$$c = \lambda f \tag{15.1}$$

where c is speed of the sound wave, f is frequency, and λ is wavelength.

It should be noted that the concept of sound velocity is different from the concept of particle velocity. Ultrasound wave can be viewed as the transport of energy through a medium by the motion of a disturbance, while particle velocity is the physical speed of a parcel of fluid that vibrates around its equilibrium position with a relatively small velocity. The particle velocity is also not the velocity of individual molecules. This section will respectively introduce the formulas and factors influencing ultrasonic velocity in gas, liquid, and solids.

15.2.1.1.1 Speed of Sound in Gas
Since gas only has a bulk modulus without a shear modulus, sound propagates in it in the form of longitudinal waves, meaning that the material particle oscillate in the direction of the wave propagation. A longitudinal sound wave propagates in gas by virtue of compressions and ratifications (or pressure pulses) in the direction of travel.

The expression for sound velocity in gas is complicated. Many factors affect it, including molecular weight, specific heat, equation of state, and so on. Equations considering all these factors are usually used for the accurate calculation and measurement of sound speed, or for measuring the progress of sound propagation and its effect on the medium. However, in engineering applications, the sound velocity in an ideal gas is often used

$$c = \sqrt{\frac{1}{k_s \rho_0}} = \sqrt{\frac{\gamma P_0}{\rho_0}} = \sqrt{\frac{\gamma RT}{M}} \tag{15.2}$$

ENGINEERING PROPERTIES OF FOODS

where k_s is the gas adiabatic compression coefficient, P_0 is ambient pressure, R is the molar gas constant, T is absolute temperature, M is gas molecular weight, and γ is the ratio of the gas constant pressure specific heat C_p to gas constant volume specific heat C_v.

Under the conditions of $T = 273.16$ K and $P_0 = 105$, the velocity of sound in air, with 0.3 M carbon dioxide and no water, is $C_0 = (331.45 \pm 0.05)$ m/s.

When other conditions remain unchanged, the relationship between temperature and sound velocity is

$$C = C_0 \sqrt{\frac{T}{T_0}} \qquad (15.3)$$

15.2.1.1.2 Speed of Sound in Liquid

Like gas, liquid also does not have a shear modulus, and sound propagates in it likewise in the form of longitudinal waves. Under linear acoustic conditions, the expression of sound velocity in liquid is

$$C = \sqrt{\frac{1}{k'_s \rho_0}} \qquad (15.4)$$

where k'_s is the liquid adiabatic compression coefficient. For water at 20°C, $\rho_0 = 998$ kg/m³, $k'_s = 45.8 \times 10^{-11}$ m²/N, and $C_0 = 1480$ m/s.

Since the relationship between pressure and density of fluid is relatively complex, theoretical analysis of the relationship between sound speed and temperature is difficult; thus, we usually use the relations obtained experimentally. Figure 15.1 shows an experimental curve of sound velocity as a function of temperature in pure water. As can be seen from the figure, sound velocity reaches its maximum when the temperature is around 74°C. When temperatures are in the range of 0–73°C, sound velocity increases monotonously, and hence we can obtain the temperature by measuring the sound velocity.

The correlation between the speed of sound, temperature, and concentration of a solution is more involved. Less precise but simple empirical formulas are often used. The concentration of a sodium chloride solution as influenced by temperature and sound speed (Figure 15.2), for example, is given by

$$S_{NaCl} = -106.87 + 0.75537C + 1.148T + 5.8 \times 10^{-4} CT + 1.4 \times 10^{-3} T^2 \qquad (15.5)$$

where S_{NaCl} is the concentration of the solution. Through this formula, we may calculate the concentration of a sodium chloride solution when temperature and sound velocity are known.

15.2.1.1.3 Speed of Sound in Solids

In solids, in addition to a bulk modulus, there are shear, bending, and torsional elasticity. Therefore, sound propagates in solids in the form of both longitudinal and transverse waves (shear waves). These different waves may have different speeds at the same frequency. The speed of sound in solids is somewhat greater than in liquid and much greater than in gas.

ULTRASOUND PROPERTIES OF FOODS

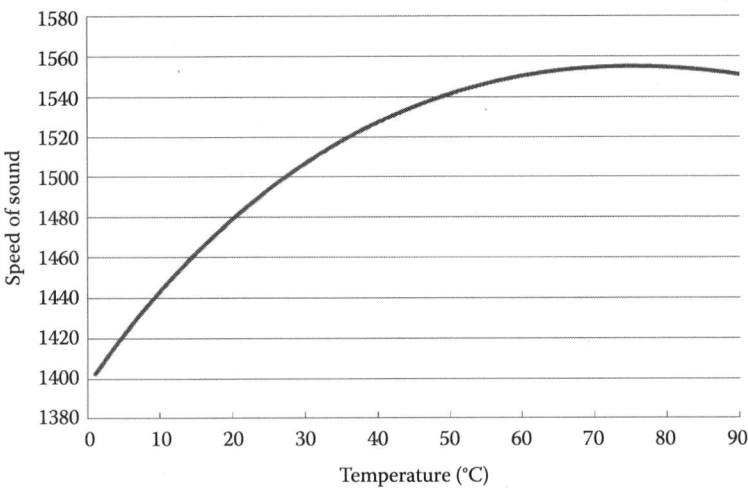

Figure 15.1 Experimental curve of sound velocity as a function of temperature in pure water. (From Bilaniuk, N., and Wong, G.S.K. 1993. *Journal of the Acoustical Society of America*. 93:1609–1612.)

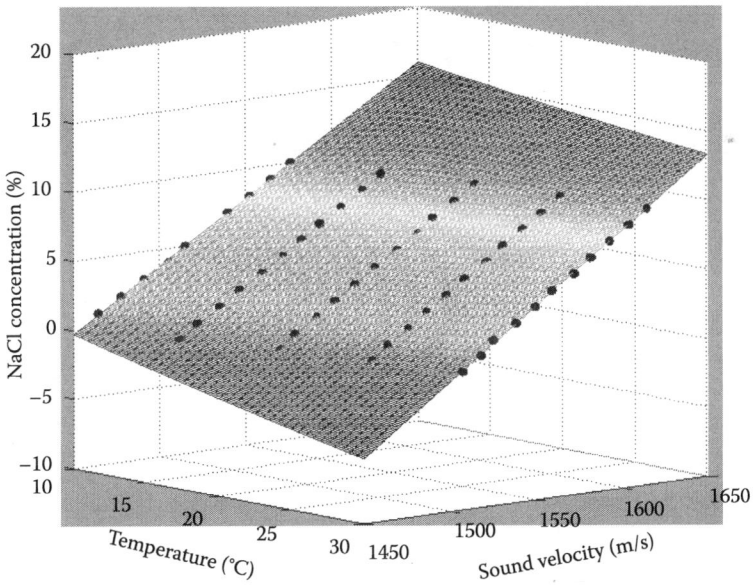

Figure 15.2 Sodium chloride concentration as influenced by temperature and sound speed. (From Meng, R.F. 2012. Concentration measurement for food aqueous solution based on ultrasonic signal transformation. Zhejiang University, 2012.)

15.2.1.2 Attenuation

When an ultrasonic wave propagates through a medium, a reduction of pressure amplitude and hence a loss of energy in the ultrasonic wave can be observed, which is referred to as attenuation. Factors affecting the amplitude and waveform include ultrasonic beam spreading, energy absorption, dispersion, nonlinearity, transmission at interfaces, scattering by inclusions and defects, the Doppler effect, and so on. To characterize ultrasonic attenuation, an attenuation coefficient α is defined as follows:

$$A = A_0 e^{-\alpha x} \tag{15.6}$$

where A is the peak amplitude of the wave at propagation distance x and A_0 is the initial peak amplitude. The attenuation coefficient α is experimentally determined from the variation of the peak amplitude, and it can be given in decibels per meter (dB/m) or in nepers per meter (Np/m). In general, the attenuation coefficient highly depends on frequency. Since this frequency dependence reflects the microstructure of materials, it can be used to characterize microscopic material properties relating to chemical reactions and mechanical processes.

The acoustic absorption in simple liquids (e.g., solutions) is predominantly influenced by bulk viscosity η_v and shear viscosity η_s.

$$\alpha = \frac{2\pi^2 f^2}{\rho c^3}\left(\frac{4}{3}\eta_s + \eta_v\right) \tag{15.7}$$

$$\alpha = \alpha_{viscous} + \alpha_{thermal} + \alpha_{relaxation} + \alpha_{scattering}$$

For complex liquid mixtures, further effects must be considered, including, for instance, thermal losses $\alpha_{thermal}$ and losses due to relaxation $\alpha_{relaxation}$. If there are particles, or bubbles in the liquid, the effect of sound scattering $\alpha_{scattering}$ usually predominates. For these reasons, an accurate determination of the acoustic absorption would be more difficult, but it can be determined by an amplitude measurement.

15.2.1.3 Impedance, Reflection, and Refraction

Acoustic impedance indicates how much sound pressure is generated by a vibration at a given frequency. For a wave in a weakly absorbing medium, the acoustic impedance Z can be written as

$$Z = \rho c \tag{15.8}$$

where ρ is density and c is wave velocity. The transmission of an ultrasonic wave from one medium to another is maximum when the acoustic impedances of the two media are equal. Acoustic impedance plays an important role in determining the acoustic transmission and reflection at the boundary of two media having different material properties.

When an ultrasonic wave obliquely impinges onto an interface between two materials as shown in Figure 15.3, reflection and refraction happen. The ratio of the amplitude of the reflected wave A_r to that of the incident wave A_i is called the reflection coefficient R.

Considering a balance of stresses and a continuity of velocities on both sides of the interface, the reflection coefficients R can be given as follows:

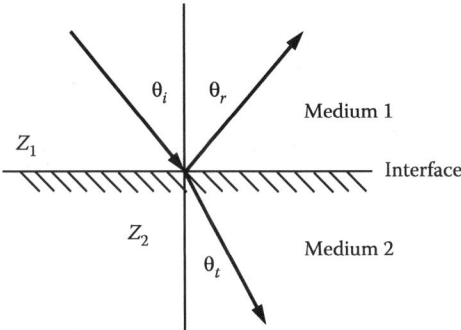

Figure 15.3 Ultrasonic wave obliquely impinges on an interface between two media.

$$R = \frac{A_r}{A_i} = \frac{Z_1 - Z_2}{Z_1 + Z_2} \qquad (15.9)$$

where subscripts 1 and 2 refer to media 1 and 2, respectively, and Z is the acoustic impedance.

The refraction is caused by the velocity difference on either side of the interface. When the sound velocities in the two media and the incidence angle are known, the refraction angle can be calculated from Snell's law:

$$\frac{\sin \theta_i}{\sin \theta_t} = \frac{c_1}{c_2} = \varepsilon \qquad (15.10)$$

It should be noted that mode conversion occurs in refraction that generates one type of wave from another. For example, a longitudinal wave incident on an interface between a liquid and a solid is transmitted partially as a refracted longitudinal wave and partially as a converted shear wave in the solid.

15.2.2 Interactions of Ultrasound with Foods

Food products are heterogeneous materials often composed of materials in different phases, making the propagation of sound waves in food materials more complicated. The velocity c at which an ultrasonic wave travels through a food product depends on its elastic modulus E and density ρ, that is, $1/c^2 = \rho/E$. The moduli and densities of materials depend on their structure, composition, and physical state. Ultrasonic velocity is constant for a material under given operational conditions and depends on its physicochemical properties. The ultrasonic velocity can thus be used to characterize the chemical composition of samples at the molecular level (Singh et al., 2004). Distortion of ultrasound waves propagating through food materials is induced by the specific chemical composition of the food. Changes in the concentration and viscoelastic properties of a solution produce measurable changes in the velocity of propagating waves (Dion and Burns, 2011; Resa et al., 2005). Food components could affect the ultrasonic velocity. In a multiphase product, ultrasonic velocity is related to the density, adiabatic compressibility, and volume fraction of the phases (McClements and Fairley, 1991). For a product rich in fat, the degree of fat crystallinity, fat mass fraction, and solid fat content could affect the ultrasonic velocity (Singh et al., 2004). The ultrasonic velocity could be mathematically related to the moisture content and textural parameters of food. For multicomponent heterogeneous system, the relationship between ultrasonic velocity and the textural properties and moisture content was shown in the following equation: $v = B\,(TP)^{1/2} + C\varphi_w + D$ (B, C, and D are constants, TP is a textural parameter, and Φ_w is the moisture content) (Benedito et al., 2001). Ultrasonic attenuation and velocity will change according to particle size in an emulsion. For example, sound waves in a sunflower oil-in-water emulsion with smaller droplets attenuate more strongly and have a lower velocity than ones in an emulsion with large droplets when the droplet size range is larger than ~0.6 μm (Wang and Povey, 1999).

Attenuation is the best variable for characterizing dispersed phase composition and particle sizes in a food product (Demetriades and McClements, 1999). The major causes of attenuation are adsorption and scattering, measurements of which provide valuable information about the physicochemical properties of food materials, including concentration, viscosity, molecular relaxation, and microstructure (McClements 1995). High signal attenuation can arise from a high solid content, as well as from air pockets created by solidification and contraction. It can be affected by micelles or the submicelles formed by free surfactant molecules. This might be due to the relaxation arising from exchanges between surfactants and solvents (Wang and Povey, 1999). Numerical calculations have shown that the ultrasonic attenuation spectra of emulsions is determined by the percentage of flocculated droplets, the size of the flocs, and the packing of the droplets within the flocs. At low frequencies, the attenuation coefficients of emulsions decrease as the percentage of droplets that are flocculated increases, whereas the opposite is true at high frequencies (Demetriades and McClements, 1999). How to convert ultrasonic attenuation spectra into the size, structure, and concentration of the flocs present in an emulsion should be further studied.

Acoustic impedance is a fundamental physical characteristic, which is influenced by the composition and microstructures of a food product. Measurements of acoustic impedance can therefore be used to provide valuable information about the properties of foods.

15.3 FOOD PROPERTY MEASUREMENT WITH ULTRASOUND

15.3.1 Ultrasonic Measurement Systems

15.3.1.1 Pulse-Echo

The PE (or *reflection*) technique is used for the detection and characterization of defects in a sample in which ultrasound pulses are transmitted and received on the same side of a specimen after being reflected from the opposite face. Defects cause a decrease in the reflection amplitude.

In the PE method (Figure 15.4), a piezoelectric transducer, installed on or near the top of a specimen and with its longitudinal axis placed perpendicular to the specimen can be used to transmit as well as receive ultrasonic power. The ultrasonic waves are reflected

ULTRASOUND PROPERTIES OF FOODS

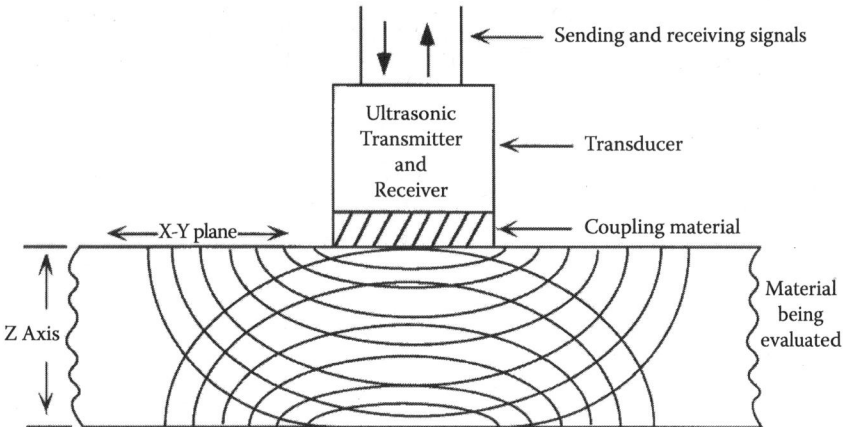

Figure 15.4 Schematic of pulse-echo method. (From http://www.nasa.gov/offices/oce/llis/0765.html.)

from the opposite face of the material or from discontinuities, layers, voids, or inclusions within the material, received by the same transducer in which the reflected energy is changed into an electrical signal. The electrical signal is computer-prepared for display on a video monitor or even TV screen.

15.3.1.2 Transmission

In the transmission method, an ultrasonic transmitter is used on one side of the material and a detector placed on the opposite side. Scanning of the material using this method will result in internal characteristics of the food product, in the X–Y plane (Figure 15.5). This is a nondestructive method for testing food materials.

In both the PE method and transmission methods, changes in the acoustic properties (velocity, attenuation, impedance) of a food are used to detect the structures and compositions. Generally, PE is used when the detected material is hard to penetrate with ultrasonic waves. Examples of the PE and transmission methods are shown in Table 15.1.

15.3.1.3 Ultrasonic Imaging

Ultrasonic imaging is a relatively new method that acquires internal images for use in food quality evaluation. There are two modes for the ultrasonic imaging of biological tissues. One is A-mode (amplitude modulation), and the other is B-mode (brightness modulation). While A-mode is one dimensional and limited to measuring the depth of tissue, B-mode allows for the characterization of tissue with different densities. The basic principle of B-mode imaging involves transmitting small pulses of ultrasound from a transducer into the material. As the waves penetrate a food material having different acoustic impedances, some are reflected back to the transducer as echo signals and some continue to penetrate deeper. Sequential coplanar echo signals returned from pulses are processed to generate an image. Thus, an ultrasound transducer functions both as a speaker (generating sound waves) and a microphone (receiving sound waves). The ultrasound pulses are quite short

ENGINEERING PROPERTIES OF FOODS

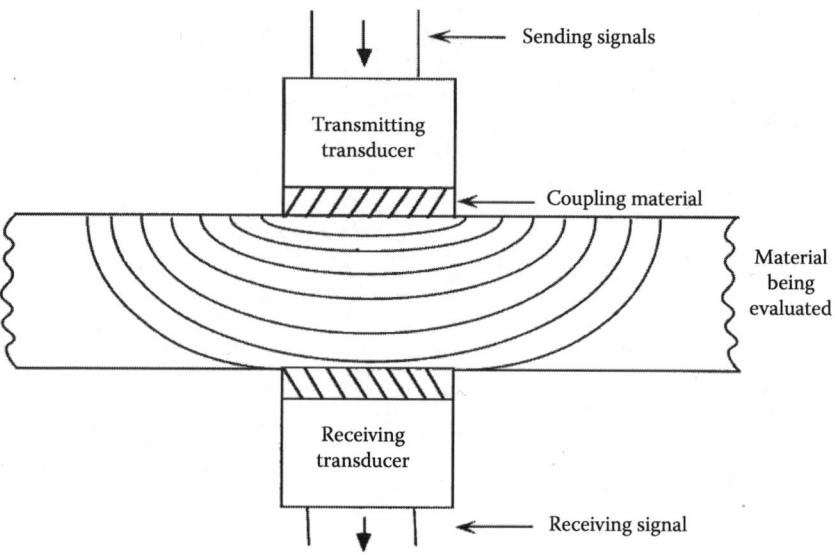

Figure 15.5 Schematic of transmission method. Each transmitted/received wave represents a pulse of energy. (From http://www.nasa.gov/offices/oce/llis/0765.html.)

Table 15.1 Selected Applications of Ultrasonic Detection with Pulse-Echo Method or Transmission Method

Application	Reference
Sugar and alcohol content of beverages	Krause et al., 2011
Deterioration of edible oil during frying	Benedito et al., 2002
Solid content of semicrystalline fats	Coupland, 2004
Mechanical properties of cheeses	Cho and Irudayaraj, 2003
Sugar content of melon	Mizrach et al., 1991
Solid fat index	McClements, 1997
Volume fraction of some components in foods	Coupland, 2004

and traverse in a straight path. They are often referred to as ultrasound beams. If the ultrasound propagates along the beam line, it is called the *axial direction*, while the perpendicular direction is called the lateral direction. Normally, only a small fraction of the ultrasound pulses return as echoes after reaching a material interface, while the remainder continues along the beam line deeper into the material.

Real-time ultrasound (RTU) is a specialized version of B-mode, which produces image of moving objects almost instantaneously. Ultrasonic imaging has been used as a cost-effective and reliable method of quality evaluation in the meat industry. Ultrasonic image analysis has been used to measure fat thickness, estimate yield, and assess the quality of meat. Table 15.2 summarizes ultrasound image applications for meat quality evaluation.

Table 15.2 Selected Applications of Ultrasound Image Applications for Meat Quality Evaluation

Application	Reference
Selection for lean weight in a meat line of sheep	Kvame and Vangen, 2007
Image texture analysis for characterizing intramuscular fat content	Kim et al., 1998
Predicting percentage of intramuscular fat	Hassen et al., 2001
Measurements of back fat and loin muscle area	Moeller and Christian, 1998
Classification in pork	Busk et al., 1999
Grading carcasses	Fortin et al., 2003
Carcass grading	Brødum et al., 1998

15.3.1.4 Air-Coupled

Air-coupled ultrasonic detection is a noncontact technique, which has become increasingly common for nondestructive testing, as food products must not be contaminated during manufacturing by couplants used in regular ultrasonic testing. The air-coupled ultrasonic technique is very efficient for the testing of large areas. Plate waves can cover large distances, and the absence of water columns allows for high scan velocities. The technique is simple and inexpensive compared to a laser technique.

The biggest challenge in the application of the air-coupled ultrasonic technique is the large acoustic impedance mismatch between air and solid materials. The solution is to maximize the acoustic output and receiver sensitivity, minimize path losses, and use signal processing to improve the signal-to-noise ratio.

Bhardwaj (2000) and Cho (2001, 2003) reported the use of piezoelectric transducers for the air-coupled inspection of food materials. In their reports, a quarter wavelength matching layer was used for the proper transduction of the ultrasound in air. Some success was also reported in the use of such transducers for detecting glass and metal fragments using amplitude data (Cho, 2003). More recently, capacitive transducers were used in air-coupled material inspection for evaluating properties of milk-based products with changing pH levels and those of palm oil at different temperatures (Gan et al., 2006; Meyer et al., 2006; Schindel et al., 1995). Capacitive transducers were also successfully employed for the detection of foreign bodies in polymer-based bottled soft drinks and the measurement of coagulation and agglomeration in milk-based beverages (Gan et al., 2002; Meyer et al., 2006). Pallav et al. (2008) developed a rapid measurement system for easy implementation in a commercial environment, which can be applied to canned products.

15.3.2 Ultrasonic Signal Processing Methods

15.3.2.1 Conventional Ultrasound Parameters

There are three basic parameters in ultrasonic measuring techniques: acoustic velocity c, attenuation α, and impedance Z. The time-of-flight (TOF) is often used to determine the ultrasonic velocity c and can also be applied to directly correlate it with some process parameters for the purpose of property characterization. The sound amplitude can be related to properties of the sample, including attenuation α and acoustic impedance Z, which are related to material characteristics. Some conventional ultrasound parameter measurement methods are introduced in this section.

Velocity measurements normally involve longitudinal waves propagating through gases, liquids, or solids. In solids, transverse (shear) waves are also present and sometimes need to be considered. The longitudinal velocity is independent of the sample geometry when the sample size in a plane perpendicular to the beam is large compared to the beam area and wavelength. Similarly, the physical dimensions of a sample have little effect on transverse velocity.

15.3.2.1.1 Pulse-Echo and Pulse-Echo-Overlap Methods

A simple estimation of ultrasonic velocity can be carried out by measuring the time it takes for a pulse of ultrasound to travel from one transducer to another (pitch-catch) or to return to the same transducer (PE). It can also be determined by comparing the phase of a detected sound wave with a reference signal. With this method, slight changes in the transducer separation are seen as slight phase changes, and the sound velocity can be calculated with the phase change information. These two methods are suitable for estimating acoustic velocity to a precision of 1%. Standard practice for measuring velocity in materials is detailed in ASTM E494-2010 (Standard Practice for Measuring Ultrasonic Velocity in Materials).

15.3.2.1.2 Precision Velocity Measurements (EMATs)

Electromagnetic-acoustic transducers (EMAT) generate ultrasound using electromagnetic mechanisms without the need to contact the sample. There are two components in an EMAT transducer, a magnet and an electric coil. The coil is driven by an alternating current (AC) electric signal at an ultrasonic frequency (20 kHz to 10 MHz) and placed near to the surface of an electrically conducting object. Eddy currents will then be induced in a near surface region. If a static magnetic field from a magnet is also present, these currents will experience Lorentz forces, which are given by

$$F = J \times B \tag{15.11}$$

where F is a body force per unit volume, J is the induced dynamic current density, and B is the static magnetic induction. A major advantage of ultrasonic nondestructive measurement with EMAT is that it does not require a couplant. Couplant-free transduction allows operation without contacting the food product and hence reduces the chance of contamination, and the measurement can be done from q remote location. EMATs can also be designed to excite complex wave patterns and polarizations difficult to achieve with fluid-coupled piezoelectric probes.

15.3.2.1.3 Attenuation Measurements

Ultrasonic attenuation, an indication of energy loss during propagating through a product, is mainly dependent on the damping capacity and scattering from boundaries between food components and particles. There are two types of attenuation measurements. Relative measurements examine the exponential decay of reflections from boundary surfaces.

However, significant variations in product microstructures and properties often produce only a relatively small change in wave velocity and attenuation. Absolute measurements of attenuation are difficult because the amplitude depends on various factors. The most common method for a quantitative measurement is to use an ultrasonic source and detector transducer separated by a known distance. By varying the distance, the attenuation can be obtained from the changes in the amplitude. To avoid the problems encountered in conventional ultrasonic attenuation measurements, ultrasonic spectral parameters are also used for frequency-dependent attenuation measurements.

15.3.2.2 Spectrum Analysis

In ultrasound spectrum analysis, a complex waveform with time is used as the independent variable. It contains a series of different frequencies and phases of the sine (or cosine) wave. Frequency domain signals are different from time domain signals, with the latter setting the frequency range as the coordinates of the horizontal axis and the desired frequency dependence variable as the vertical-axis value. The relationships among amplitude, phase, and power spectra are shown in Figure 15.6.

Two spectrum signal generation methods for ultrasound that have found practical applications in food processing are the sweep and pulse methods. The frequency sweep method produces a series of sinusoidal signals of different frequencies and requires professional ultrasonic spectrum equipment to perform cyclic scan of the material. The detection

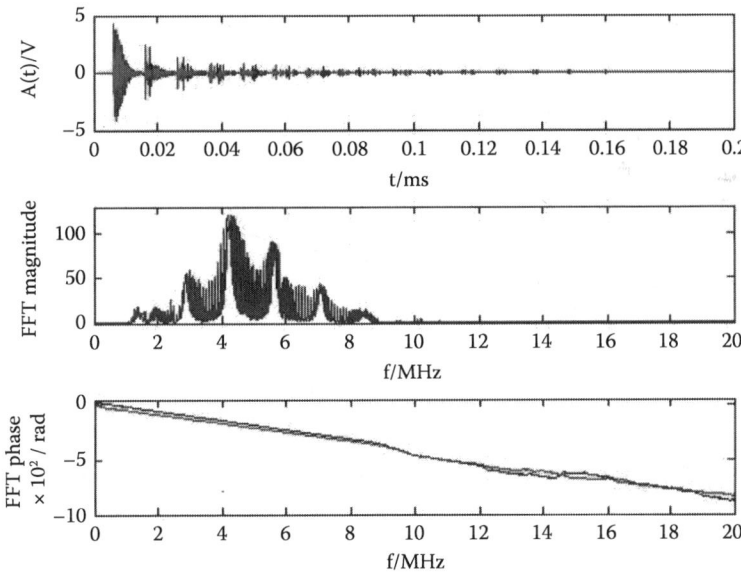

Figure 15.6 Time domain ultrasound signal and its amplitude spectrum and phase spectrum by fast Fourier transform. (From Meng, R.F. 2012. Concentration measurement for food aqueous solution based on ultrasonic signal transformation. Zhejiang University, 2012.)

Table 15.3 Selected Applications of Ultrasound Spectrum on Food Detecting

Application	Reference
Whey-protein-aggregation studies	Bryant et al., 1999
Particle-size-distribution analysis of casein in water	Povey et al., 1999
Monitoring the flocculation of droplets in protein-stabilized emulsions	Demetriades et al., 1999
Studies of the acid gelation of milk	Dalgleish et al. 2004
Evaluation of heat stability of calcium-fortified milk	O'Driscoll et al., 2003
Alcohol and carbohydrate content in commercial beverages	Dion et al., 2011
Characterization of bread crumb	Lagrain et al., 2006

method is easy to execute and has a large detection frequency range. The limitations are that the scan cycle is long and equipment more costly. However, the pulse method uses a broadband ultrasound pulse containing a certain frequency segment. The method is easy to realize and cost less, but the signal processing is relatively complex, and an ultrasonic probe can only detect a relatively small period of the ultrasound frequency segment. Table 15.3 summarizes some applications based on ultrasonic spectrum analysis for food property detection.

15.3.2.3 Pattern Recognition

Pattern recognition is a method used in machine learning, a branch of artificial intelligence, and is generally categorized according to the type of learning procedure used to generate the output values. An example of pattern recognition in ultrasonic detecting is classification, which attempts to assign each input value to one of a given set of classes, that is, determining whether a given beef sample is fresh or stale. Pattern recognition algorithms aim to seek a reasonable answer for all possible inputs, in contrast to pattern-matching algorithms aiming to find exact matches between the input and preexisting patterns. Common pattern-matching algorithms predict sucrose and ethanol concentration by examining the patterns of a given ultrasonic parameters and the relevant concentrations in the solutes (Resa et al., 2005; Schoeck et al., 2010). Pattern matching is not a machine-learning method, although pattern-matching algorithms can sometimes provide good output of the sort provided by pattern-recognition algorithms.

A stable pattern recognition method in ultrasound detection applications must involve two key factors. One is that acoustic parameters such as velocity, attenuation, impedance, spectrum, and so on, which are relevant to the output value, should be carefully chosen. The other is the selection of an appropriate pattern-recognition algorithm. The artificial neural network (ANN) and some classic classification algorithm such as principal component analysis (PCA) or partial least-squares (PLS) are mostly used in this field. Pattern recognition was used by Wallhäußer et al. (2011) in the detection of fouling in dairy processing.

15.4 COMPILATION OF ACOUSTIC PROPERTIES OF FOODS

15.4.1 Composition

15.4.1.1 Solutions

There is an increasing interest in the real-time measurement of the concentrations of carbohydrates (such as sucrose) and ethanol for many food processing operations, especially in fermentation and commercial beverage processes. Conventional methods for determination of the concentrations are mostly offline measurements and can be costly. For the food and drink industries, where consistency must be balanced with high throughput, rapid, *in situ* analysis techniques would be useful. In recent years, more attention has been paid to the use of ultrasound for the analysis of sample properties including the determination of solution concentrations (McClements, 1997; Resa et al., 2005; Van Sint Jan et al., 2008). For instance, ultrasound characterization of aqueous solutions with varying sugar and ethanol contents was conducted by Krause et al. (2011) with multivariate regression methods.

Ultrasound can measure the contents and concentrations of sugar and ethanol simultaneously. It can also detect their concentrations in mixtures of different carbohydrates. A typical relationship between ultrasound wave speed and sucrose concentration is shown in Figure 15.7. The outputs from ultrasound speed measurements in sucrose solutions showed a good fit to the linear relation given in Equation 15.12 ($R^2 = 0.9995$), where υ is the sound speed in (m/s) and s is the sucrose concentration in % w/w. The relationship between sound speed and ethanol concentration (<16%) also showed a linearity ($R^2 = 0.9972$) and could be represented by Equation 15.12 (Van Sint Jan et al., 2008).

$$\upsilon = 3.14s + 1485 \tag{15.12}$$

$$\upsilon = 7.48e + 1487$$

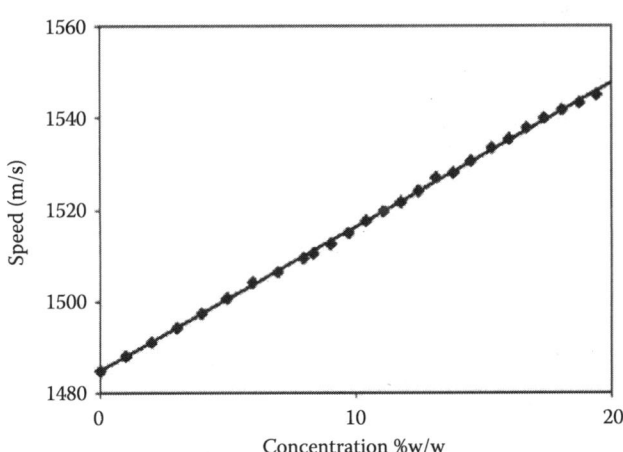

Figure 15.7 Speed of ultrasound in a sugar solution. (From Van Sint Jan, M. 2008. Ultrasound based measurements of sugar and ethanol concentrations in hydroalcoholic solutions. *Food Control*. 19:31–35.)

The velocity of sound c (m/s) in a pure liquid is given by

$$\frac{1}{c^2} = \left(\frac{\partial \rho}{\partial P}\right) = \rho \kappa \qquad (15.13)$$

where ρ (kg/m³) and κ (Pa⁻¹) are the density and adiabatic compressibility, respectively. In a liquid mixture of two components, A and B, this velocity is often predicted using the linear dependence of the density and compressibility with the volume fraction Φ (1):

$$\rho = \Phi_A \rho_A + \Phi_B \rho_B \qquad (15.14)$$

$$\kappa = \Phi_A \kappa_A + \Phi_B \kappa_B \qquad (15.15)$$

leading finally to the following expression as a function of the mass concentration (ω):

$$\frac{1}{c^2} = \frac{\Phi_A^2}{\omega_A c_A^2} + \frac{\Phi_B^2}{\omega_B c_B^2} \qquad (15.16)$$

which is known as the Uric equation.

In a study of Resa et al. (2009), in order to accurately compute the velocity of sound in a ternary mixture (water–sugar–alcohol), the excess volume of the corresponding binary mixtures, v_{AB}^E was introduced:

$$v = v_A + v_B + v_C + v_{AB}^E + v_{AC}^E \qquad (15.17)$$

where A is the major component. Substituting this expression into the molar fractions, the equation for three components results in

$$\frac{1}{c^2} = \frac{\Phi_A^2}{w_A c_A^2} + \frac{\Phi_B^2}{w_B c_B^2} + \frac{\Phi_C^2}{w_C c_C^2} + \rho \left(\Phi_{AB}^E \kappa_{AB}^E + \Phi_{AC}^E \kappa_{AC}^E\right) \qquad (15.18)$$

where the quantities in excess, represented by the superscript E, can be calculated from the measurements of the corresponding binary mixtures.

Cha and Hitzmann (2004) used a multilinear regression model to predict ultrasonic velocity from concentrations in a *Saccharomyces cerevisiae* culture. Resa et al. (2005, 2007) calculated the concentrations of the main components from the sound velocity during alcoholic and lactic acid fermentations. Lately, Resa et al. (2009) observed changes in ultrasound velocity as a consequence of monosaccharide catabolism, polysaccharide hydrolysis, gas production, and microorganism growth.

Attenuation is the reduction in intensity and amplitude of the pulsed ultrasound. The attenuation of sound in fluid systems is caused by different mechanisms: $\alpha = \alpha_0 + \alpha_{vis} + \alpha_{therm} + \alpha_{sc} + \alpha_{relax}$. Here, α_0 is the "classical" absorption by the emulsion or dispersion medium, α_{vis} describes the viscous losses, α_{therm} the thermal losses, α_{sc} the scattering losses, and α_{relax} losses due to relaxation (Hauptmann et al., 2002).

The relationship between sound velocity and chemical composition is not always linear or monotonic. Further, estimating concentrations in multicomponent mixtures is limited as the measured velocity may have a nonunique solution. Therefore, only measuring

ULTRASOUND PROPERTIES OF FOODS

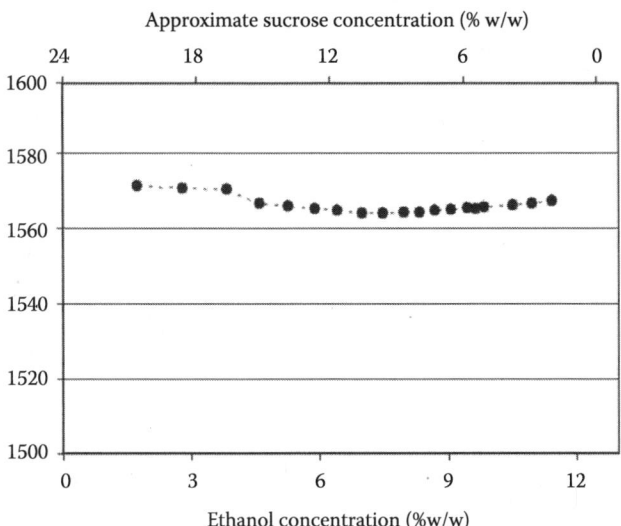

Figure 15.8 Ultrasound wave speed in a mimic fermentation. (From Van Sint Jan, M. et al. 2008. *Food Control*. 19:31–35.)

the speed of a sound alone is not enough to determine concentrations accurately even if the temperature is known (Figure 15.8). Dion and Burns (2011) developed a methodology to monitor the volume fraction of ethanol and the carbohydrate concentrations in liquid mixtures using ultrasound frequency analysis. Distortion created by the hydrogen bonding among water, ethanol, and sucrose can be monitored in the frequency domain using 5 MHz wideband ultrasonic transducers. Multilinear regression was used to quantify both ethanol and sucrose over a wide range of concentrations with correlation coefficients (r^2) greater than 0.98.

15.4.1.2 Emulsions and Suspensions

Food emulsions are suspensions of small oil droplets in aqueous dispersion media (milk, mayonnaise, ice cream mix, etc.) or vice versa (butter, margarine, and chocolate). The properties that are used to characterize them include particle size distribution (PSD) and oil concentration (Coupland and McClements, 2001). It is difficult to directly evaluate these parameters by conventional light microscopy because the droplets are too small. Laser light scattering is only suitable for very dilute systems ($\Phi < 0.05$ wt%), and the use of nuclear magnetic resonance is expensive and difficult to operate. Ultrasonic sensors, on the other hand, are reliable, rapid, and convenient for food emulsions under a range of process conditions. They also allow online, noninvasive measurements for process control.

The size of oil droplets in an emulsion can be determined from ultrasound velocity and attenuation spectra. Wang and Povey (1999) reported that smaller droplets in a sunflower oil-in-water emulsion attenuated more and had lower velocity than that in an emulsion with larger oil droplets. The latter had a lower attenuation and high sound speed when the droplet sizes were larger than ~0.6 μm.

Ultrasonic techniques can detect microorganism growth in a liquid medium. Elvira et al. (2005) developed an eight-channel ultrasonic device to detect microorganism growth in ultra high temperature (UHT) milk without opening the packages. Changes in the liquid medium caused by the microorganism growth produced changes in the ultrasonic propagation parameters, allowing for noninvasive detection of developing contamination. The changes in the velocity and amplitude of the ultrasonic wave showed the growth signature of these microorganisms. As a result, ultrasound detection as a noninvasive technique could substitute for conventional analysis like pH or acidity measurements (Elvira et al., 2006).

Ultrasound has been tested for determining other properties of emulsions. Telis-Romero et al. (2011) assessed fresh cheese composition by measuring ultrasonic velocity in cheese and cheese blends at different temperatures. Ultrasonic velocity was not only heavily dependent on the composition of the cheese but also on its structures. They developed a model that provided good results for the assessment of fat and water content at six temperatures ranging from 3°C to 29°C. The properties of concentrated emulsions can also be determined *in situ* by ultrasonic spectroscopy (Demetriades and McClements, 1999).

A collection of sound velocity data in selected liquid and semiliquid foods is given in Table 15.4.

Table 15.4 Sound Speed of Selected Liquid and Semiliquid Food Products

Product	Temperature (°C)	Frequency (MHz)	Average Velocity (m/s)	Reference
Apple juice	25	3.6	1543	Zacharias and Parnell, 1972a
Apricot juice	25	3.6	1550	Zacharias and Parnell, 1972a
Cranberry juice	25	3.6	1556	Zacharias and Parnell, 1972b
Grape juice	25	3.6	1563	Zacharias and Parnell, 1972b
Grapefruit juice	25	3.6	1531	Zacharias and Parnell, 1972b
Lemon juice	25	3.6	1524	Zacharias and Parnell, 1972b
Prune juice	25	3.6	1574	Zacharias and Parnell, 1972a
Tomato juice	25	3.6	1526	Zacharias and Parnell, 1972b
Apricot juice	25	3.6	1550	Zacharias and Parnell, 1972a
Mixed vegetable juice	25	3.6	1529	Zacharias v Parnell, 1972a
Apple sauce	25	3.6	1584	Zacharias and Parnell, 1972b
Corn syrup	25	3.6	1956	Zacharias and Parnell, 1972b
Maple syrup	25	3.6	1858	Zacharias and Parnell, 1972b
Chocolate syrup	25	3.6	1900	Zacharias and Parnell, 1972a
Sauterne wine	25	3.6	1575	Zacharias and Parnell, 1972a
Egg white	20	2.1	1530	Javanaud et al., 1984
Egg yolk	21.5	2.1	1500	Javanaud et al., 1984
Homogenized milk	25	1	1522	Hueter et al., 1953
Skimmed milk	28	1	1505	Konoplev et al., 1974

15.4.1.3 Muscle Foods

Ultrasound has been used in the livestock industry as a means of assessing aspects of body composition since the 1950s. In the past two decades, its use has been extended to estimation of intramuscular fat, or marbling, and thus encompasses meat quality as well as tissue proportion prediction (Fisher, 1997). The speed and attenuation of ultrasound are the major parameters in determining the properties of muscle foods. Attenuation is dependent on the tissue characteristics and the ultrasound frequency used.

Based on the relationship between the ultrasonic properties of fish tissue and its composition, ultrasound is widely used for determining the location and quantity of fish swimming under water, and has been used to detect the presence of parasites in fish tissues. Ultrasound can be used to determine fish composition (Ghaedian et al., 1998). Ultrasonic velocity in fish tissue is related to its physical properties by Equation 15.19.

Equation 15.19 assumes that a material can be treated as a liquid. In multicomponent materials, such as fish tissue, it is necessary to take into account the influence of composition on ultrasound velocity. Ultrasonic properties of a multicomponent material can be described by

$$\frac{1}{c^2} = \sum_j^n \Phi_j \rho_j \sum_{j=1}^n \Phi_j K_j \tag{15.19}$$

where ρ_j, K_j, and Φ_j are the density, adiabatic compressibility, and volume fraction of component j, respectively.

The ultrasonic velocity of a multicomponent material can be calculated by the following equation:

$$\frac{1}{c^2} = \sum_{j=1}^n \frac{\Phi_j}{c_j^2} \tag{15.20}$$

where Φ_j and c_j are the mass percentage and ultrasonic velocity in component j, respectively.

Shannon et al. (2004) used ultrasonic techniques to measure the composition of fish during processing. They indicated that the velocities should be grouped into "high fat," "medium fat," and "low fat" categories because a precise prediction of fat content is difficult due to inhomogeneity in the structure of the muscle.

Benedito et al. (2001) used the multicomponent equation at two different temperatures and took into account that the sum of the components was 100% of the product; consequently, the following three-equation system was obtained to describe the relation between the ultrasonic velocity and the composition of a pork mixture:

$$\frac{100}{c_{T_1}^2} = \frac{\Phi_f}{c_{fT_1}^2} + \frac{\Phi_w}{c_{WT_1}^2} + \frac{\Phi_{p+0}}{c_{p+0\,T_1}^2}$$

$$\frac{100}{c_{T_2}^2} = \frac{\Phi_f}{c_{fT_2}^2} + \frac{\Phi_w}{c_{WT_2}^2} + \frac{\Phi_{p+0}}{c_{p+0\,T_2}^2} \tag{15.21}$$

$$\Phi_f + \Phi_\omega + \Phi_{p+0} = 100$$

where c_{T1} is the ultrasonic velocity at temperature T_1 and c_{fT1}, c_{wT1}, and c_{p+0T1} are the ultrasonic velocities of fat, water, and protein + others, respectively, at temperature T_1, Φ_f, Φ_w, and Φ_{p+0} are the percentages of fat, water, and protein + others, respectively, being the variables to be obtained. Ultrasonic velocity increased with temperature for lean tissue and decreased for fat, thus demonstrating the possibility of assessing the composition of meat products. The sound velocity and attenuation of selected biological tissues are summarized in Table 15.5.

Llull et al. (2002a) examined the relationship between the textural properties of a meat-based product and ultrasonic velocity. The results showed that ultrasonic velocity could be mathematically related to the textural parameters such as hardness and compression work. In addition, velocity decreased when temperature increased. The moisture and fat contents of the meat-based product (%var = 91% and 93%, respectively) could be determined by the slopes of the temperature–velocity straight lines. Llull et al. (2002b) employed ultrasonic velocity measurement to evaluate the moisture content and textural properties of sobrassada de Mallorca (a raw cured-pork product). The ultrasonic velocity was related to the moisture content and textural parameters (hardness, compression work, maximum puncture force, and puncture work).

Ultrasound has also been used to determine the composition of meat-based products. Simal et al. (2003) estimated the composition of a fermented meat-based product (sobrassada)

Table 15.5 Acoustic Properties of Biological Tissue

Tissue or Organ	Temperature (°C)	Average Velocity (m/s)	Attenuation (dB/cm)	Reference
Muscle—porcine		1621	1.02	Koch et al., 2011a
Skin—porcine		1682	2.5	Koch et al., 2011b
Fat—porcine	20	1450	1.89	Koch et al., 2011b
Backfat—porcine		1502	1.59–2.27	Koch et al., 2011b
Liver—bovine		1586–1609	1.0	Akaslii et al., 1995
Fat—bovine		1441–1450	2.3	Akaslii et al., 1995
Heart—bovine		1565–1570	0.8	Akaslii et al., 1995
Fat—bovine	10	1680		Bamber and Hill, 1979
Fat—bovine	20	1575		Bamber and Hill, 1979
Fat—bovine	30	1490		Bamber and Hill, 1979
Fat—bovine	40	1390		Bamber and Hill, 1979
Round—bovine	21–24	1579–1600		Marcus and Carstensen, 1975
Chuck—bovine	21–24	1579–1603		Marcus and Carstensen, 1975
Steak—bovine	21–24	1401–1579		Marcus and Carstensen, 1975
Fat—sheep		1430		Wolcott and Allen, 2005
Muscle—sheep		1600		Wolcott and Allen, 2005
Chicken breast	26.4	1601	1.67	Shishitani et al., 2010

by ultrasound. The explained variance was 98.0% for protein + others, 97.6% for fat, and 95.6% for moisture. Mörlein et al. (2005) investigated the fat content of porcine *longissimus* muscle through acoustic parameters obtained by spectral analysis of ultrasound echo signals. Ultrasonic techniques have showed promise in the characterization and classification of backfat samples from Iberian pigs, and prediction of intramuscular fat content (IMF) in the loin muscle of pig carcasses (Lakshmanan et al., 2012; Niñoles et al., 2007).

15.4.2 Phase Transition

15.4.2.1 Freezing

Freezing is important for extending the shelf life and preserving the quality of many food products. Miles and Cutting (1974) reported that the speed of sound in a frozen meat is related to the amount of unfrozen water in it (Figure 15.9). Sigfusson et al. (2004) also tested ultrasound for monitoring food (gelatin, chicken, and beef) freezing. In their study, a reasonable prediction of the time for complete freezing was proposed. Aparicio et al. (2008) determined the temperature of food and its ice content by measuring the speed of sound. According to these authors, the ultrasonic method is quick and suitable for the online monitoring of frozen, freezing, and thawing food systems.

15.4.2.2 Crystallization

Ultrasonic techniques are very effective in monitoring emulsion crystallization, which is important for the quality of many food emulsion products such as butter, whipped cream,

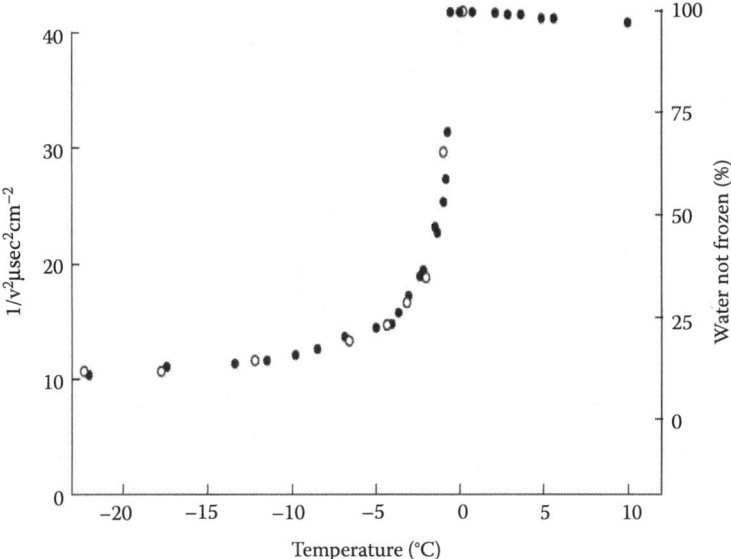

Figure 15.9 Comparison of the velocity of ultrasound (as $1/c^2$) (●) to unfrozen water (○) as a function of temperature. (From Miles, C.A., and Cutting, C.L. 1974. *Journal of Food Technology*. 9:119–122.)

and ice cream. Based on the difference in sound velocity propagating through solids versus liquids, velocity measurement has been successfully used to monitor the phase transition and crystallization of palm oil, palm mid-fraction, and palm kernel fats in O/W emulsion systems. Research results have provided insights into the mechanism of crystallization acceleration induced by the template films of high-melting emulsifier molecules (Awad, 2012).

Povey et al. (2007) assessed emulsion stability through ultrasound velocity measurements (UVMs), specifically, the relation between the velocity and the time used in cycles of cooling and heating (Povey et al., 2007). UVMs were also used for studying crystallization kinetics (Povey et al., 2009). The volume fraction of crystallized material, nucleation rate, and the crystallization kinetics of a "monodispersed" cocoa butter emulsion during crystallization were measured.

Awad et al. (2008) used temperature scanning UVM to monitor the complex thermal transitions that occur during the crystallization and melting of triglyceride solid lipid nanoparticles (SLNs). They concluded that temperature scanning UVM may prove to be a useful alternative to conventional differential scanning calorimetry (DSC) techniques for monitoring phase transitions in colloidal systems. UVM is cheaper and more convenient than DSC, and it can be useful for automated testing and quality control.

15.4.2.3 Gelation

The behavior of aqueous solutions of different types of carrageenan was investigated with ultrasonic velocity and attenuation measurements. Decrease in ultrasonic velocity was correlated with the aggregation of carrageenan molecules in ordered conformation and the increase in attenuation was related to the friction between the gel network and water molecules. The results indicated that ultrasound may be used as a suitable technique for the control of gel quality by monitoring carrageenan molecular properties, enabling differentiation of the gelation behavior of different carrageenan systems (Wang et al., 2005).

Ting et al. (2009) investigated the gelation process in heat-induced $CaSO_4 \cdot 2H_2O$ tofu curd by measuring the ultrasonic velocity and attenuation at MHz using noncontact and nondestructive LPU waves. They reported that the attenuation followed first-order kinetics in a way similar to the firmness changes determined by textural analysis. It is thus possible to use ultrasonic attenuation as an alternative to textural or rheological analysis in the characterization of tofu gel. They suggested that ultrasonic measurements that are noncontact and nondestructive in nature could be used as a real-time indicator of tofu maturity on a production line.

Knorr et al. (2011) studied rennet-induced pregelation and gelation processes in milk with ultrasonic analysis. They found that high-resolution ultrasonic spectroscopy was a powerful tool for the analysis of the microscopic processes at work in the formation of milk gel. The method allowed the characterization of the pregelation processes, such as hydrolysis and aggregation, and the initial stages in the formation of the gel network, as well as the monitoring of the microscopic evolution in the gel in the postgelation stage.

Parker and Povey (2012) mapped ultrasonic (8 MHz) speed and attenuation to study the phase diagram, hysteresis, and kinetics of edible-grade gelatin in water during the gelation. Their results demonstrate that ultrasound provides a versatile approach to studying gelation, offering rapid data acquisition, insights into viscoelastic behavior, and the ability to probe small sample quantities.

15.4.3 Texture

15.4.3.1 Firmness of Fruits/Vegetables

The firmness of fruits and vegetables is determined by measuring ultrasonic parameters such as velocity of sound and attenuation and correlating them to product firmness changes. The velocities of sound at 50 kHz and other frequencies in selected fruits and vegetables are listed in Table 15.6. Mizrach et al. (2000) employed ultrasound for firmness measurements in studying the effects of storage time and temperature on the softening of avocados. It was found that firmness decreased during storage, and wave attenuation increased (Figure 15.10). However, the velocity changes in the avocados showed a nonmonotonic and complex relationship and were found to be not sensitive enough to the physiological changes in the sample. A nondestructive ultrasonic method was also used to determine the maturity of plum fruits during storage (Mizrach, 2004) and monitor the firmness and sugar content of greenhouse tomatoes (cv. 870) during their shelf life (Mizrach, 2007). A linear relationship was observed between attenuation and firmness until the end of the softening process. Kim et al. (2009) evaluated the potential use of ultrasonic parameters for the determination of apple firmness and developed calibration equations for the purpose.

15.4.3.2 Cheese

Ultrasonic velocity was related to sensory and instrumental texture measurements of Mahon cheese by Benedito et al. (2000). The results indicated that it could be used nondestructively to assess the instrumental and sensory properties of Mahon cheese texture and was therefore suitable for grading purposes.

Noncontact ultrasound parameters, such as velocity and the attenuation coefficient, were correlated with Young's modulus, hardness, and toughness in different types of cheeses.

Table 15.6 Sound Wave Velocity in Selected Fresh Fruits and Vegetables

Crop	Frequency (kHz)	Velocity (m/s)	Reference
Apple	50	184	Mizrach et al., 1989
Avocado	50	383	Mizrach et al., 1989
Carrot	37	396	Nielsen et al. 1998
Carrot (longitudinal)	50	341	Mizrach et al., 1989
Cucumber (longitudinal)	50	371	Mizrach et al., 1989
Melon	50	242	Mizrach et al., 1989
Potato	50	380	Mizrach et al., 1989
Potato	250	824	Cheng and Haugh, 1994
Potato	500	700–850	McClements, 1997
Pumpkin	50	267	Mizrach et al., 1989

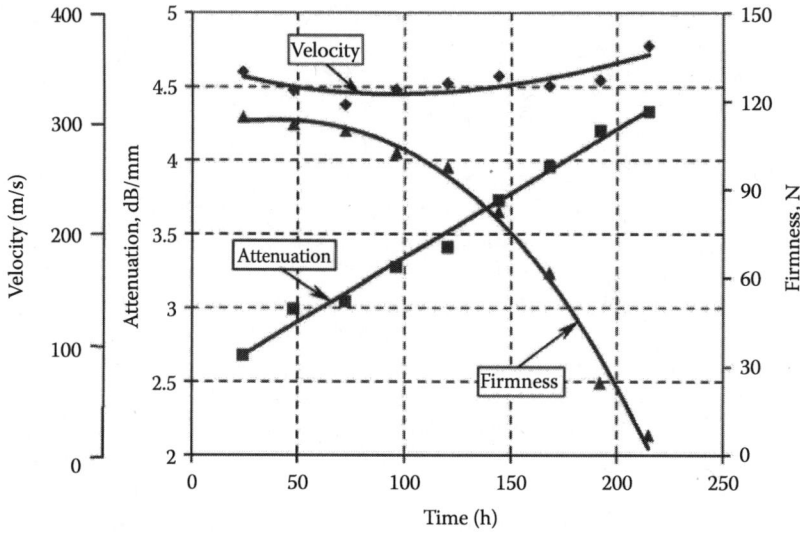

Figure 15.10 The mean values of firmness, wave attenuation, and velocity versus storage time for avocado, and the suggested model curves. (From Mizrach, A. 2000. *Ultrasonics*. 38:717–722.)

It was found that ultrasound velocity was highly correlated with the mechanical properties of the cheeses. The ultrasound technique achieved an accuracy of 99.98% (standard error = 0.089 mm) in the measurement of cheese thicknesses (Cho and Irudayaraj, 2003).

Benedito et al. (2006) reported that ultrasonic velocity related well to textural parameters of cheese such as compression work and hardness (R^2 = 0.843 and 0.826, respectively). The experiments carried out with surface probes attached to a texture analyzer showed a good relationship between the internal and external textural parameters.

Ultrasonic velocity measurements were applied for fresh cheese quality assessment (Telis-Romero et al., 2008). The ultrasonic velocity was found to relate well to textural parameters such as hardness (R^2 = 0.920), compression work (R^2 = 0.886), and Young's modulus (R^2 = 0.920).

15.4.3.3 Starch Products

In recent years, ultrasonic technique has been used to investigate the gelatinization (Lehmann et al., 2004) and retrogradation (Lionetto et al., 2006) of starch. Ross et al. (2004) used both ultrasound and conventional rheology techniques to examine the effects of mixing on different dough systems. They reported agreement between measurement methods. To explore the use of a low-intensity ultrasonic sensor in the control of a cake making process, Gómez et al. (2008) attempted to correlate physical properties of batters (density, viscosity, and rheology) and cakes (volume, symmetry, volume index, height, and density) with ultrasonic measurements. Significant correlations were obtained between the acoustic impedance and the batter consistency (R^2 = 0.53), G'' (R^2 = 0.66), and G^* (R^2 = 0.53).

15.4.4 Viscosity

A major limitation of commercial process rheometers is that they are offline measurement tools, often based on intrusive techniques that disturb the flow and result in hygiene problems. They only provide the viscosity at one or at most a few shear rates, and often are unable to monitor transient flows in real time (Wiklund and Stading, 2008). In contrast, LPU emitted from noncontact equipment is well suited for food property determination, as it does not introduce extraneous contamination or alter the properties of the substance (Cartwright, 1998; Gan et al., 2006). There are a number of methods based on ultrasound for measuring rheological properties, especially viscosity.

15.4.4.1 Ultrasound Doppler Velocity Profiling and the UVP–PD Method

Ultrasound Doppler velocity profiling (UVP) was originally a high-resolution technique capable of measuring velocity profiles in a liquid flowing along the pulsed ultrasonic beam axis (Takeda, 1986). UVP was later extended to measure flow profiles in fluids and has been accepted as an important tool for this purpose (Takeda, 1991).

The combination of the UVP technique with pressure difference (PD) measurements has recently been investigated for inline rheometry (Birkhofer et al., 2008; Dogan et al., 2005; Müller et al., 1997; Wunderlich and Brunn, 1999; Young et al., 2008). The UVP–PD method has been tested for the inline rheological characterization of particle suspensions, fibers, emulsions, and colloidal polymers dispersed in a continuous phase flowing through a pipe. For instance, UVP was used to monitor the displacement of yogurt by water, and the results were in good agreement with computational fluid dynamics (CFD) simulations (Regner et al., 2007).

Rheological properties were derived from the velocity profiles and corresponding pressure drops for samples in a pipe, using a UVP–PD system containing a flow-adaptor with an ultrasonic transducer and two pressure sensors. A schematic of the inline adaptor is given in Figure 15.11. The average velocity profile of a xanthan solution measured with the UVP–PD system, using a time-domain algorithm to calculate the data, is shown in Figure 15.12 (Wiklund et al., 2010).

A nonlinear regression of the velocity profiles was performed using a power-law model (Equation 15.22) and the Herschel–Bulkley model (Equation 15.23). The latter enabled the determination of the power-law exponent n. The parameter K was determined using the pressure drop in the pipe. The viscosity and shear stress distributions in the pipe were then calculated. The detailed theoretical background can be found in Wiklund et al. (2007).

$$V_x(r) = \left(\frac{\Delta P}{2LK}\right)^{\frac{1}{n}} \frac{R^{1+\frac{1}{n}}}{1+\frac{1}{n}} \left(1 - \left(\frac{r}{R}\right)^{1+\frac{1}{n}}\right) \tag{15.22}$$

$$V_x(r) = \left(\frac{\Delta P}{2LK}\right)^{\frac{1}{n}} \frac{1}{1+\frac{1}{n}} \left((R - R^*)^{1+\frac{1}{n}} - (r - R^*)^{1+\frac{1}{n}}\right) \tag{15.23}$$

ENGINEERING PROPERTIES OF FOODS

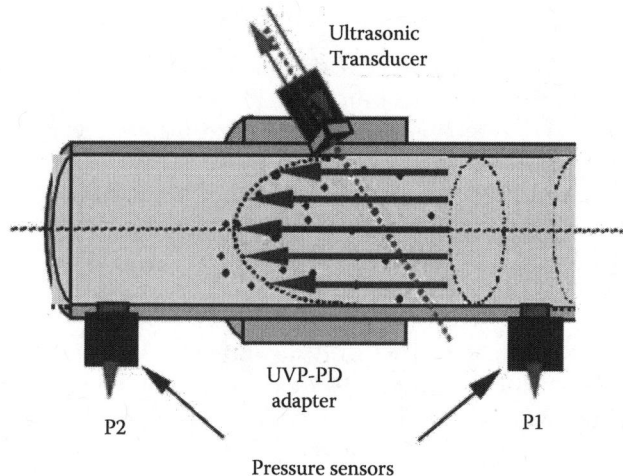

Figure 15.11 Schematic representation of the in-line adapter. (From Wiklund, J. et al. 2001. *Annual Transactions—Nordic Rheology Society.* 8:128–130.)

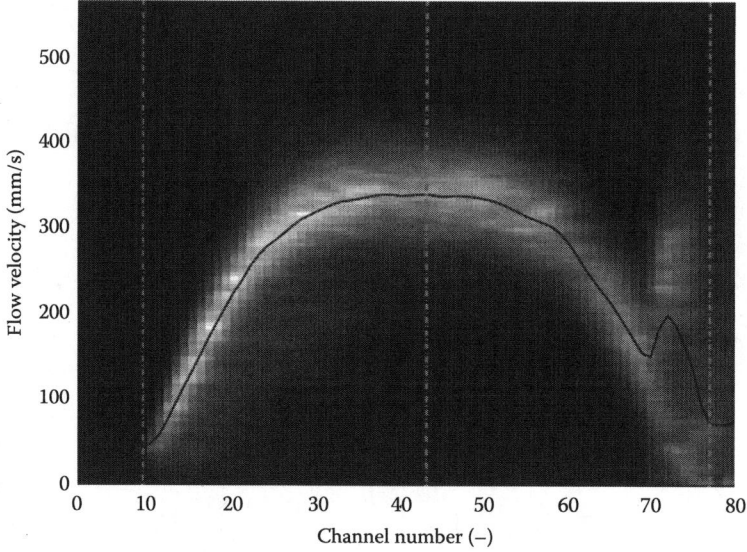

Figure 15.12 Arithmetic average velocity profiles (solid line) measured in steady-state flow of xanthan solution. (From Wiklund, J., Stading, M., and Trägårdh, C. 2010. *Journal of Food Engineering.* 99:330–337.)

Based on the UVP–PD technique, Wang et al. (2004) investigated the effect of temperature gradient on velocity measurement. Nonisothermal conditions have two important consequences: changed flow behavior and increased ultrasound Doppler velocimetry (UDV) measurement errors due to the dependence of sound speed on temperature. In the temperature range from ambient temperature to 80°C, variations in the speed of sound in most aqueous solutions are small, while those in most oils are significant (Contreras et al., 1992; McClements and Povey, 1988; Zacharias and Parnell, 1972). Therefore, the temperature effect for UDV measurements in aqueous solution flows may not be an issue. However, corrections are recommended for ultrasound sound when nonisothermal conditions exist. There are two correction methods used in experimental conditions, (1) use the average speed of sound for the entire pipe in the UDV data processing, or (2) incorporate the known temperature gradient into the velocity data processing algorithm.

15.4.4.2 Pulse-Echo or Transmission-Through Method

Velocity measurement by the PE method is simple and accurate. However, using the velocity of ultrasound to obtain viscosity information requires a governing law, which predicts the changes of viscosity as a function of sound velocity. Zhao et al. (2003) developed such a law for tomato and orange juices in an explicit form. They reported a quadratic relationship between apparent viscosity and velocity of ultrasound for juice (Equation 15.24)

$$\mu = a_2 c + a_1 c + a_0 \tag{15.24}$$

where μ is apparent viscosity, c is sound speed in the juice, and a_0, a_1, and a_2 are coefficients.

Kuo et al. (2008) applied PE and TT methods to determine sugar content and viscosity of reconstituted orange juice. The experimental results of both methods showed that power attenuation is ineffective for the determination of sugar content and viscosity, and that the velocity of ultrasound responded linearly to sugar content and exponentially to viscosity. The velocity exhibited a high linear correlation with Brix in orange juice ($R^2 = 0.988$ in the PE mode).

15.4.4.3 Measurement of the Viscosity–Density Product

The use of a quartz wedge to measure the viscosity of a liquid was first proposed and tested by Mason et al. (1949) and O'Neil (1949), and later by McSkimin (1960). Later on, scientists at Pacific Northwest National Laboratory developed an online computer-controlled sensor, based on ultrasound reflection measurements, to determine the product of the viscosity and density of a liquid or slurry for Newtonian fluids, and the shear impedance of a liquid for non-Newtonian fluids (Greenwood and Bamberger, 2002; Greenwood et al., 2005, 2006). The measurement was based on multiple reflections of an ultrasonic shear horizontal (SH) wave within a quartz wedge (Figure 15.13). It is known that liquids do not support shear waves easily. However, as the viscosity of a liquid increases, a larger fraction of incident ultrasound is transmitted into the liquid. Consequently, the reflection coefficient at the quartz liquid interface is a measure of the liquid viscosity. As the viscosity increases, the reflection coefficient decreases (Greenwood et al., 2005). The acoustic impedance for a shear wave traveling in a liquid was given by (Greenwood and Bamberger, 2002).

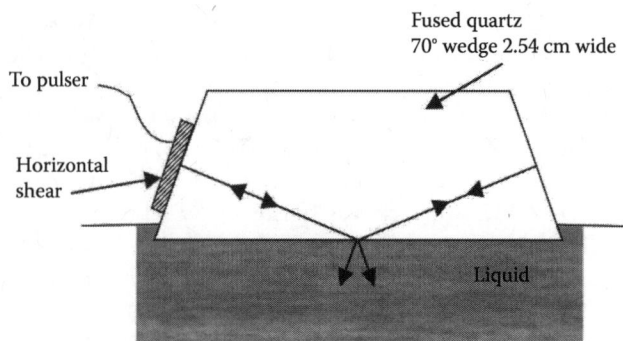

Figure 15.13 Diagram of fused quartz wedge and illustration of multiple reflections of a horizontal shear wave. (From Greenwood, M.S., Adamson, J.D., and Bond, L.J. 2005. *AIP Conference Proceedings.* 760(1):1690.)

$$Z = \left(\frac{\omega\rho\mu}{2}\right)^{0.5}(1+j) \tag{15.25}$$

where ρ and μ are the density and viscosity of the liquid, respectively. The complex acoustic impedance of a liquid at a solid–liquid interface was obtained by O'Neil (1949):

$$Z_{liq} = Z_s \frac{1-R}{1+R} + jZ_s \frac{2R\sin\theta}{(1+R)^2} \tag{15.26}$$

where Z_s is the acoustic impedance of the quartz and R is the reflection coefficient. Equating the real parts of Equations 15.25 and 15.26 results in

$$(\rho\mu)^{0.5} = \rho_s c_{TS}\left(\frac{2}{\omega}\right)^{0.5}\left(\frac{1-R}{1+R}\right) \tag{15.27}$$

where ρ_s is the density of the quartz and c_{TS} is shear-wave velocity in the quartz. When the densities of the liquid and quartz and c_{TS} are known, with measured R, Equation 15.27 can be used to obtain the viscosity of the liquid. The viscosity of a sugar solution obtained by this method was in good agreement with the values published in the *Handbook of Chemistry and Physics*, as shown in Figure 15.14 (Greenwood et al., 2005).

15.5 APPLICATIONS

15.5.1 Vegetable and Fruit Quality Monitoring

Physicochemical indexes such as dry weight content (DW), oil content, total soluble solids (TSS), and acidity are often used as indicators of fruit maturity and preharvest quality. Ultrasound measurement has been investigated as a rapid and nondestructive method to evaluate fruit maturity before harvesting (Mizrach et al., 1999, 2000). The measurements

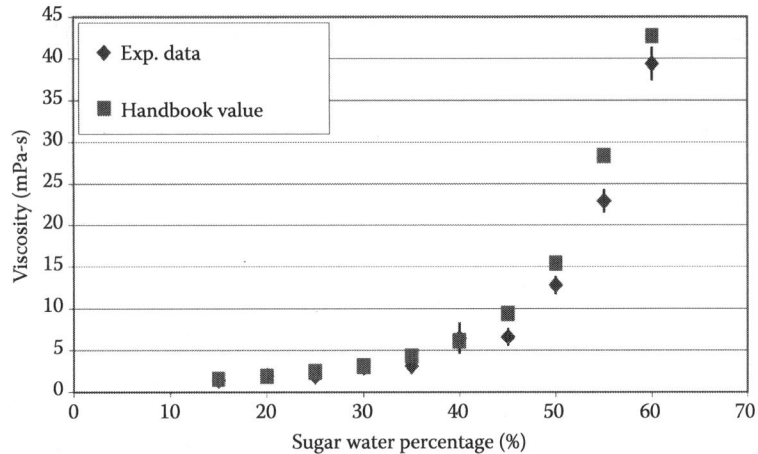

Figure 15.14 Comparison of the viscosity measurements by for sugar water solutions with handbook value. (From Greenwood, M.S., Adamson, J.D., and Bond, L.J. 2005. *AIP Conference Proceedings*. 760(1):1690.)

were based on correlations between sound attenuation in the fruit and selected quality indexes. A high-power, low-frequency transmitter–receiver system was developed by Mizrach et al. (1999) to measure ultrasound attenuation in fruits. Two transducers (50 kHz), one acting as the transmitter and the other as the receiver, were mounted with an angle of about 120°C between their axes. The attenuation of the ultrasonic signal was calculated from the measurements according to the exponential expression: $A = A_0\, e^{-\alpha l}$, where l was the distance between the two probes, A and A_0 were the ultrasonic amplitudes at the beginning and the end of a distance l along the propagation path of the ultrasonic wave, respectively, and α was the apparent attenuation coefficient of the signal. The DW was determined by a convection oven method and correlated to the sound attenuation. Similarly, the oil content of the samples was determined by a refractive index technique, the acidity by an extraction method, and both were correlated with sound attenuation in avocados and mangos (Mizrach et al., 2000).

The use of ultrasound to detect the internal defects of fruits and other products is another potential application. Ultrasonic measurements were done with a through-transmission system applied to whole potato tubers to investigate the existence of hollow hearts (Cheng and Haugh, 1994). The authors found that transmitted waveforms through a potato with hollow hearts and those with sound ones were different. Esehaghbeygi (2011) developed a nondestructive ultrasound method to find potato tubers with mechanical damage (pressure and impact stresses) for three potato varieties (Agria, Marfona, and Diamant) using four pairs of ceramic ultrasonic transducers (25, 32.8, 40, and 50 kHz). Ultrasound attenuation was used to determine the tuber damage conditions. The results of the 25-kHz tests showed accuracies of 83%, 94.5%, and 89% in differentiating damaged tubers from intact ones for Agria, Marfona, and Diamant potatoes, respectively.

15.5.2 Quality Evaluation of Meat Products

The moisture content of cod fillets (and possibly the protein content by deduction) can be rapidly and accurately evaluated using ultrasonic velocity measurements. Ghaedian et al. (1997) applied the PE technique and examined the relationships between the composition and ultrasonic properties of Atlantic cod fillets. The results showed that the fat and mineral contents of cod remained fairly constant and it was thus possible to estimate the protein content from the moisture content determined by ultrasound measurements. They envisioned the development of hand-held ultrasonic devices or automated online devices for grading fish according to their composition.

The IMF of the *longissimus dorsi* muscle is widely recognized as an important parameter influencing the sensory characteristics of meat. The meat industry has sought objective and reliable real-time procedure to nondestructively determine IMF in livestock and carcasses (Mörlein et al., 2005). Several methods have been tested to predict IMF in living animals and meat tissues. The TOF of transmitted ultrasound waves has been used to predict the fat content of raw meat mixtures and excised beef muscles (Benedito et al., 2001). However, the TOF method can only be used *ex vivo* (Koch et al., 2011a). Ultrasonic B-mode image analysis using different statistical parameters of gray value distribution has been successfully used in the prediction of IMF in living steers (Hassen et al., 2003). This method, however, reportedly had difficulties in handling samples with wider ranges of IMF, such as in pork loin (1.1–11.2%). A third method is ultrasound spectral analysis, which can provide more comprehensive information about tissue constitution. Lakshmanan et al. (2012) proposed a novel method for IMF estimation via spectral analysis of backscattered acoustic signals obtained from pig carcasses. The IMF was predicted from backscatter data collected with a customized commercial hand-held ultrasound device. The researchers developed several algorithms for ultrasound data preprocessing and applied prior to acoustic parameter estimation. A unique feature of this method was that no amplitude parameters and fewer acoustic parameters were needed to predict the IMF. As a result, the IMF prediction was much less affected by transmission or reflection losses along the propagation path from the transducer surface to the muscle region of interest.

15.5.3 Ultrasonic Measurement for Liquid Food

15.5.3.1 Particle Size Analysis

Ultrasound spectrometry is one of the very few methods that can measure particles sizes over the range of 10 nm to 1 mm (Povey, 2013). The particle sizing is based on scattering of ultrasound by particles by measuring the velocity and/or the attenuation of the ultrasound as a function of frequency or angular dependence of the intensity of the scattered ultrasound (McClements, 1997). From the attenuation data, the PSD can be obtained by finding the best fit of the measured and predicted properties with an appropriate model, which requires a number of physical properties for the scatterers and medium.

The PSD of milkfat droplets of raw and homogenized milk was determined by Meyer et al. (2006) with an ultrasound spectrometer. The ultrasonic attenuation caused by the intrinsic absorption and the scattering of an ensemble of droplets is given by a model based on the ECAH model. The simultaneous measurement of nano-sized casein micelles

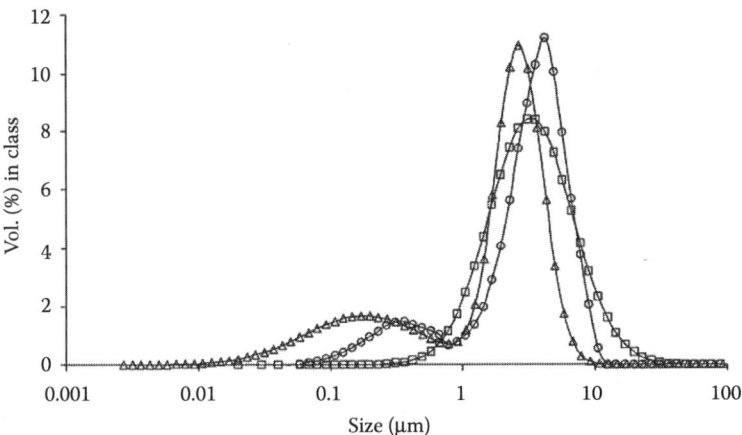

Figure 15.15 Simultaneous measurement of nano-sized casein micelles and micron sized oil particles in undiluted milk. Particle size distribution of raw milk measured by (O) laser light technique, (Δ) ultrasound technique, (□) PSD of milkfat by the ultrasound technique when the part due to the protein was removed from the calculations. (From Meyer, S. et al. 2006. *Measurement Science and Technology.* 17:1838–1846.)

and micro-sized oil particles in undiluted milk was achieved (Figure 15.15). Since the attenuation in whole milk is due to the milkfat, proteins, and milk serum, the PSDs plotted in Figure 15.15 did not provide information about the PSD of the individual components in the milk, that is, casein and oil. This is partially due to the fact that proteins both scatter sound and exhibit relaxation (Povey et al., 2011). However, Povey et al. (2011) reported that ultrasound attenuation spectroscopy was sensitive to protein–protein interactions such as dimerization, trimerization, and stages of denaturation, including gelation. Therefore, this method can provide useful information about protein–protein and protein–solvent interactions at the relatively high protein concentration found in foods.

15.5.3.2 Bubbly Liquids

Many liquid and solid food products, such as ice cream, bread dough, and frozen foods, contain air bubbles or cells. The size and concentration of bubbles in these foods have a significant effect on the shelf life, texture, and appearance of these products (Kulmyrzaev et al., 2000). It has been a challenge to characterize bubbly foods because they are optically opaque or have delicate structures that are easily damaged by contact measurement methods. Efforts have been made in recent years to use ultrasound technology to measure bubble size distribution and bubble fraction. Kulmyrzaev et al. (2000) applied an ultrasonic reflectance spectroscopy where the acoustic impedance spectra of aerated liquids were measured by reflecting ultrasonic waves from their surfaces to characterize a mixture of xanthan (1.4%) and whey protein (8%) with different bubble concentrations (up to 49%). They reported that the method was sensitive to changes in bubble size and concentration, but that there was a need to explore means to acquire quantitative data.

Cents et al. (2004) developed a setup that consisted of an arbitrary waveform generator (AWG) and eight pairs of broadband immersion transducers (100 kHz to 80 MHz).

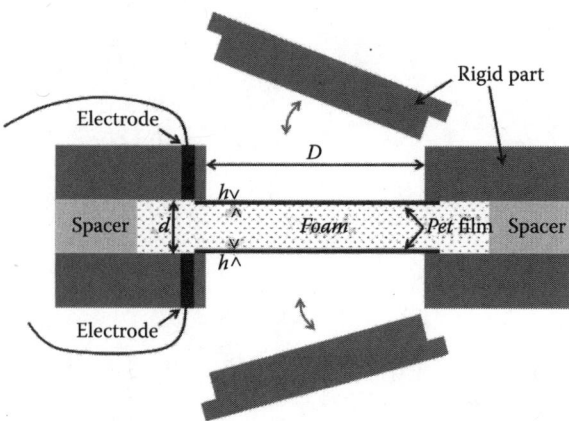

Figure 15.16 The liquid foam is contained in a thin-walled cell whose thickness d is well defined. Two rigid plates are used to rigidify the walls when the cell is being filled. A pair of electrodes is also included in the cell for electrical conductivity measurements, which give access to the liquid content of the foam. (From Pierre, J., Elias, F., and Leroy, V. 2013. *Ultrasonics*. 53:622–629.)

At each frequency, a narrow-band tone-burst was used to extract phase velocity and the attenuation by fast Fourier transform (FFT). An inverse method was solved with a preselected bubble size distribution to obtain the size distributions and the volume fractions of the dispersed phases (2D and 3D) from the experimental values of the velocity and the attenuation coefficient. The method was validated with a laser scattering technique for the size distribution of glass bead with and without the presence of 6.1% bubbles. Bubble size distribution in an air–water system with a gas holdup of 5.85% was determined.

An air-borne ultrasound set up was developed by Pierre et al. (2013) to measure ultrasound transmission through liquid foams produced by the "double-syringe" method. A special sample cell was designed to hold the liquid foam and to minimize impedance mismatch between the air and the walls (Figure 15.16). While the liquid fraction and bubble size distribution were determined by a conductivity measurement, the set up was able to deduce ultrasound attenuation and phase velocity from the amplitude and phase data for a liquid foam sample with a liquid fraction of 6%. Obviously, in future work, the problem of how to utilize attenuation and phase velocity to obtain liquid fraction and bubble size distribution needs to be resolved.

15.5.4 Detection of Leaks in Food Packages

Channel leaks in the seal region of flexible food packages provide potential pathways for microbial penetration that eventually results in spoilage of the product or proliferation of infiltrated pathogenic organisms with accompanied food safety risk. Detection of flaws in food packages is accomplished through visual inspection or destructive testing. Since the minimum leak size through which a microorganism might penetrate could be very small considering the size of a bacterium (i.e., 1–2 μm in diameter for *Escherichia coli*), plus the fact that human ocular resolution is only 50 μm, visual inspection cannot detect flaws behind naked eye detection limit and thus is less useful. However, destructive tests used in the food industry are costly and only provide a statistical assurance of integrity. To address this issue, several nondestructive inspection methods, including ultrasound, have been proposed and tested for sealing leak detection.

Two ultrasonic techniques used for flaw detection are transmission and PE (reflection) techniques. Safvi et al. (1997) reported a scanning laser acoustic microscopy (SLAM) method for detection of leaks in seal region of a polyethylene film and plastic microwaveable retort pouch. When operating at 100 MHz, the method enabled reliable detect and imaged channel defects as small as 10 μm in diameter. However, for transmission-mode imaging, the inspected sample has to be placed between the transmitting and receiving components of the system, making it difficult for real-world applications due to the variety of different sizes and shapes of food packages. For the high frequencies required to detect 10 μm diameter defects, the insertion loss may be too large; further limiting the method to very thin samples. An ultrasonic PE backscattered amplitude integral method (BAI) was developed by Raum et al. (1998). The BAI value for each scan point was calculated by integrating the envelope of the received echo signal. The BAI value was then proportional to the square root of the backscattered energy. When the seal region of the package was flawless and its surfaces were smooth and parallel, the BAI value would be virtually constant. However, any discontinuities could cause the value to vary, and the variation was visualized by a gray mode BAI image. The BAI method demonstrated reliable detection of subwavelength defects (38 μm diameter) when the wavelength was ≈86 μm (17.3 MHz). Since the BAI technique cannot reliably detect leaks smaller than 38 μm, an RF sampling (RFS) method and an RF correlation technique (RFC) that showed the ability to detect defects with size of 15 μm or smaller was developed by the same group (Faizer et al., 2000). Figure 15.17 shows an

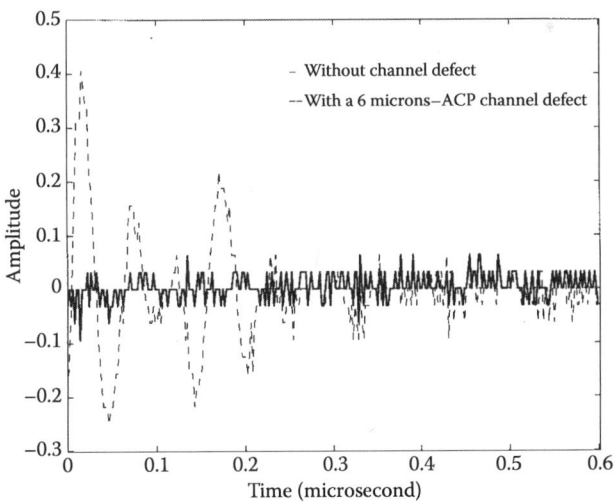

Figure 15.17 Difference of RF signals and reference signal in the correlation window. (From Fazier, C.H. et al.2000. *IEEE Transactions on Ultrasonics, Ferroelectrics and Frequency Control.* 47(3):530–539.)

example of signals from RFS. The dashed line is an RF signal passing through the channel defect minus the reference signal, and the solid line is an RF signal not passing through the channel defect minus the reference signal. In these plots, the channel defect is 6-μm ACP trilaminate film. It is clear that the difference is relatively large for the RF signal that passes through the channel defect compared with the difference for the signal that does not pass through the channel defect. Some recent studies utilized similar techniques with reported detection limit in the range of 50 μm (He et al., 2008; Liu, 2002).

LIST OF SYMBOLS

a	Coefficient
A	Peak amplitude of sound wave
B	Static magnetic induction or constant
c	Sound velocity (m/s)
C	Sound velocity (m/s) or constant
C_p	Gas constant pressure specific heat
C_v	Gas constant volume specific heat
D	Constant
e	Ethanol concentration
E	Elastic modulus
F	Frequency or body force
J	Induced dynamic current density
ks	Gas adiabatic compression coefficient in Equation 15.4
ks'	Liquid adiabatic compression coefficient in Equation 15.1
K	Consistency index in Equation 15.22
L	Length of pipe between pressure sensors in Equation 15.22
M	Gas molecular weight
n	Flow index
P_0	Ambient pressure
R	Gas constant or reflection coefficient
s	Sucrose concentration
S_{NaCl}	NaCl concentration of a solution
T	Temperature
TP	Texture parameter
v	Sound velocity
Z	Acoustic impedance

Greek Symbols

α	Attenuation coefficient
ε	Refraction angle
γ	Ratio of C_p and C_v
η_s	Shear viscosity
η_v	Bulk viscosity

φ	Volume fraction
φ_w	Moisture content
κ and K	Adiabatic compressibility
λ	Wavelength (m)
μ	Apparent viscosity
ρ	Density of air, liquid, or solid (subscript a, l, or s)
θ	Angle

Subscripts/Superscripts

A	Major component
B	Component
f	Fat
i	Incident
p	Protein
r	Reflected
s	Solid
t	Transmitted
W	Water

REFERENCES

Aparicio, C., Otero, L., Guignon, B., Molina-García, A.D., and Sanz, P.D. 2008. Ice content and temperature determination from ultrasonic measurements in partially frozen foods. *Journal of Food Engineering*. 88(2):272–279.

Awad, T.S., Helgason, T., Kristbergsson, K., Decker, E.A., Weiss, J., and McClements, D. 2008. Temperature scanning ultrasonic velocity study of complex thermal transformations in solid lipid nanoparticles. *Langmuir*. 24(22):12779–12784.

Awad, T.S., Moharram, H.A., Shaltout, O.E., Asker, D., and Youssef, M.M. 2012. Applications of ultrasound in analysis, processing and quality control of food: A review. *Food Research International*. 48:410–427.

Benedito, J., Carcel, J.A., Gonzalez, R., and SanJuan, N. 2000. Prediction of instrumental and sensory textural characteristics of Mahon cheese from ultrasonic measurements. *Journal of Texture Studies*. 31(6):631–643.

Benedito, J., Carcel, J., Rosselló, C., and Mulet, A. 2001. Composition assessment of raw meat mixtures using ultrasonics. *Meat Science*. 57(4):365–370.

Benedito, J., et al. 2006. Manchego cheese texture evaluation by ultrasonics and surface probes. *International Dairy Journal*. 16(5):431–438.

Bhardwaj, M.C. 2000. High transduction piezoelectric transducers and introduction of non-contact analysis, NDT.net 5.

Birkhofer, B.H., Jeelani, S.A.K., Windhab, E.J., Ouriev, B., Lisner, K.J., Braun, P., and Zeng, Y. 2008. Monitoring of fat crystallization process using UVP–PD technique. *Flow Measurement and Instrumentation*. 19(3–4):163–169.

Brødum, J., Egebo, M., Agerskov, C., and Busk, H. 1998. On-line pork carcass grading with the AutoFom ultrasound system. *Journal of Animal Science*. 76:1859–1868.

Cartwright, D. 1998. Off-the-shelf ultrasound instrumentation for the food industry. In *Ultrasound in Food Processing*. Chapter 2. Povey, M.J.W., and Mason, T.J., Ed. London: Blackie.

Cents, A.H.G., Brilman D.W.F., and Versteeg, G.F. 2004. Measuring bubble, drop and particle sizes in multiphase systems with ultrasound. *AIChE Journal*. 50:2750–2762.

Cha, Y.L., and Hitzmann, B. 2004. Ultrasonic measurements and its evaluation for the monitoring of *Saccharomyces cerevisiae* cultivation. *Bioautomation*. 1:16–29.

Cheng, Y., and Haugh, C.G. 1994. Detecting hollow heart in potatoes using ultrasound. *Transactions of the ASAE*. 37:217–222.

Cho, B., Irudayaraj, J., and Bhardwaj, M.C. 2001. Rapid measurement of physical properties of cheddar cheese using a non-contact ultrasound technique. *American Society of Agricultural Engineers*. 44(6):1759–1762.

Cho, B.K., and Irudayaraj, J.M.K. 2003. A noncontact ultrasound approach for mechanical property determination of cheeses. *Journal of Food Science*. 68(7):2243–2247.

Cho, B.-K., and Irudayaraj, J.M.K. 2003. Foreign object and internal disorder detection in food materials using noncontact ultrasound imaging. *Journal of Food Science*. 68:967–974.

Contreras, N.I., Fairley, P., McClements, D.J., and Povey, M.J.W. 1992. Analysis of the sugar content of fruit juice and drinks using ultrasonic velocity measurements. *International Journal of Food Science and Technology*. 27:515–529.

Coupland, J.N., and McClements, D.J. 2001. Droplet size determination in food emulsions: Comparison of ultrasonic and light scattering techniques. *Journal of Food Engineering*. 50:117–120.

Coupland, J.N. 2004. Low intensity ultrasound. *Food Research International*. 37:537–543.

Dalgleish, D., Alexander, M., and Corredig, M. 2004. *Food Hydrocolloids*. 18(5):747.

Demetriades, K., and McClements, D.J. 1999. Ultrasonic attenuation spectroscopy study of flocculation in protein stabilized emulsions. *Colloids and Surfaces A: Physicochemical and Engineering Aspects*. 150:45–54.

Dion, J.R., and Burns D.H. 2011. Simultaneous determination of alcohol and carbohydrate content in commercial beverages by ultrasound frequency analysis. *Talanta*. 86:384–392.

Dogan, N., McCarthy, M.J., and Powell, R.L. 2005. Application of an in-line rheological characterization method to chemically modified and native corn starch. *Journal of Texture Studies*. 36:237–254.

Elvira, L., Sampedro, L., Matesanz, J., Gómez-Ullate, Y., Resa, P., Iglesias, J.R., Echevarría, F.J., and Montero de Espinosa, F. 2005. Non-invasive and non-destructive ultrasonic technique for the detection of microbial contamination in packed UHT milk. *Food Research International*. 38:631–638.

Elvira, L., Sampedro, L., Montero de Espinosa, F., Matesanz, J., Gómez-Ullate, Y., Resa, P., Echevarría, F.J., and Iglesias, J.R. 2006. Eight-channel ultrasonic device for non-invasive quality evaluation in packed milk. *Ultrasonics*. 45:92–99.

Esehaghbeygi, A., Raghami, N., and Kargar, A. 2011. Detection of internal defects in potato based on ultrasound attenuation. *American Journal of Potato Research*. 88(2):160–166.

Fazier, C.H., Tian, Q., Ozguler, A., Morris, S.A., and O'Brien, W.R. 2000. High contrast ultrasound images of defects in food package seals. *IEEE Transactions on Ultrasonics, Ferroelectrics and Frequency Control*. 47(3):530–539.

Fisher, A.V. 1997. A review of the technique of estimating the composition of livestock using the velocity of ultrasound. *Computers and Electronics in Agriculture*. 17:217–231.

Gan, T.H., Hutchins, D.A., and Billson, D.R. 2002. Preliminary studies of a novel air-coupled ultrasonic inspection system for food containers. *Journal of Food Engineering*. 53:315–323.

Gan, T.H., Pallav, P., and Hutchins, D.A. 2006. Non-contact ultrasonic quality measurements of food products. *Journal of Food Engineering*. 77:239–247.

Ghaedian, R., Decker, E.A., and McClements, D.J. 1997. Use of ultrasound to determine cod fillet composition. *Journal of Food Science*. 62:500–504.

Ghaedian, R., Coupland, J.N., Decker, E.A., and McClements, D.J. 1998. Ultrasonic determination of fish composition. *Journal of Food Engineering*. 35:323–337.

Gómez, M., Oliete, B., García-Álvarez, J., Ronda, F., and Salazar, J. 2008. Characterization of cake batters by ultrasound measurements. *Journal of Food Engineering*. 89:408–413.

Greenwood, M.S., and Bamberger, J.A. 2002. Measurement of viscosity and shear wave velocity of a liquid or slurry for on-line process control. *Ultrasonics.* 39(9):623–630.
Greenwood, M.S., Adamson, J.D., and Bond, L.J. 2005. Measurement of the viscosity-density product using a quartz wedge. *AIP Conference Proceedings.* 760(1):1690.
Greenwood, M.S., Adamson, J.D., and Bond, L.J. 2006. Measurement of the viscosity–density product using multiple reflections of ultrasonic shear horizontal waves. *Ultrasonics.* 44:e1031–e1036.
Hassen, A., Wilson, D. E., and Rouse, G. H. 2003. Estimation of genetic parameters for ultrasound-predicted percentage of intramuscular fat in Angus cattle using random regression models. *Journal of Animal Science.* 81:35 – 45.
Hauptmann, P., Hoppe, N., and Püttmer, A. 2002. Application of ultrasonic sensors in the process industry. *Measurement Science and Technology.* 13:R73–R83.
He, C.F., Yuan, H.M., Wu, B., and Song G. 2008. An ultrasonic system for detecting channel defects in flexible packages. *IEEE Intelligent Computation Technology and Automation.* 1974–1977.
McClements, D.J. 1997. Ultrasonic characterization of foods and drinks: Principles, methods, and applications. *Critical Reviews in Food Science and Nutrition.* 37:1–46.
Kim, K.B., et al. 2009. Determination of apple firmness by nondestructive ultrasonic measurement. *Postharvest Biology and Technology.* 52(1):44–48.
Knorr, D., Froehling, A., Jaeger, H., Reineke, K., Schlueter, O., and Schoessler, K. 2011. Emerging technologies in food processing. *Annual Review of Food Science and Technology.* 2:203–235.
Koch, T., Brand, S., Sannachi, L., Raum, K., Wicke, M., and Mörlein, D. 2011a. Ultrasound velocity and attenuation of porcine soft tissues with respect to structure and composition: I. Muscle. *Meat Science.* 88:51–58.
Koch, T., Brand, S., Sannachi, L., Raum, K., Wicke, M., and Mörlein, D. 2011b. Ultrasound velocity and attenuation of porcine soft tissues with respect to structure and composition: II. Skin and backfat. *Meat Science.* 88:67–74.
Krause, D., Schock, T., Hussein, M.A., and Becker, T. 2011. Ultrasonic characterization of aqueous solutions with varying sugar and ethanol content using multivariate regression methods. *Journal of Chemometrics.* 25:216–223.
Kulmyrzaev A., Cancellier C.e, and McClements D.J. 2000. Characterization of aerated foods using ultrasonic reflectance spectroscopy. *Journal of Food Engineering.* 46:235–241.
Kuo, F.J., Sheng, C.T., and Ting, C.H. 2008. Evaluation of ultrasonic propagation to measure sugar content and viscosity of reconstituted orange juice. *Journal of Food Engineering.* 86(1):84–90.
Lakshmanan, S., Koch, T., Brand, S., Männicke, N., Wicke, M., Mörlein, D. et al. 2012. Prediction of the intramuscular fat content in loin muscle of pig carcasses by quantitative time-resolved ultrasound. *Meat Science.* 90:216–225.
Lehmann, L., Kudryashov, E., and Buckin, V. 2004. Ultrasonic monitoring of the gelatinization of starch. *Trends in Colloid and Interface Science XVI.* p. 136–140.
Lionetto, F., Maffezzoli, A., Ottenhof, M.A., Farhat, I.A., and Mitchell, J.R. 2006. Ultrasonic investigation of wheat starch retrogradation. *Journal of Food Engineering.* 75(2):258–266.
Liu, W.G. 2002. *MATLAB Programming and Application*, Beijing: Higher Education Press.
Llull, P., Simal, S., Benedito, J., and Rosselló, C. 2002a. Evaluation of textural properties of a meat-based product (sobrassada) using ultrasonic techniques. *Journal of Food Engineering.* 53:279–285.
Llull, P., Simal, S., Femenia, A., Benedito, J., and Rosselló, C. 2002b. The use of ultrasound velocity measurement to evaluate the textural properties of sobrassada from Mallorca. *Journal of Food Engineering.* 52:323–330.
McCarthy, M.J., Wang, L., and McCarthy, K.L. 2005. Ultrasound properties. In: *Engineering Properties of Foods.* 3rd ed., pp. 567–610. Rao, M.A., Rizvi, S.S., and Datta, A.K., Eds. New York: Taylor & Francis Group.
Mason, W.P., Baker, W.O., McSkimin, H.J., and Heiss, J.H. 1949. Measurement of shear elasticity and viscosity of liquids at ultrasonic frequencies. *Physical Review* 75:936–946.

McClements, D.J., and Povey, M.J.W. 1988. Ultrasonic velocity measurements in some liquid triglycerides and vegetable oils. *Journal of the American Oil Chemists' Society*. 65:1787–1790.

McClements, D.J., and Fairley, P. 1991. Ultrasonic pulse echo reflectometer. *Ultrasonics*. 29:58–62.

McClements, D.J. 1994. Ultrasonic determination of depletion flocculation in oil-in-water emulsions containing a non-ionic surfactant. *Colloids and Surfaces A: Physicochemical and Engineering Aspects*. 90(1):25–35.

McClements, D.J. 1995. Advances in the application of ultrasound in food analysis and processing. *Trends in Food Science and Technology*. 6:293–299.

McClements, D.J. 1997. Ultrasonic characterization of foods and drinks: Principles, methods, and applications. *Critical Reviews in Food Science & Nutrition*. 37(1):1–46.

McSkimin, H.J. 1960. Measurement of dynamic properties of materials. U.S. Patent 2966058.

Meyer, S., Hindle, S.A., Sandoz, J.P., Gan, T.H., and Hutchins, D.A. 2006. Non-contact evaluation of milk based products using air-coupled ultrasound. *Measurement Science and Technology*. 17:1838–1846.

Meyer, S., Berrut, S., Goodenough, T.I.J., Rajendram, V.S., Pinfield, V.J., and Povey, M.J.W. 2006. A comparative study of ultrasound and laser light diffraction techniques for particle size determination in dairy beverages. *Measurement Science and Technology*. 17:289–297.

Miles, C.A., and Cutting, C.L. 1974. Technical note: Changes in the velocity of ultrasound in meat during freezing. *Journal of Food Technology*. 9:119–122.

Mizrach, A., Flitsanov, U., El-Batsri, R., and Degani, C. 1999. Determination of avocado maturity by ultrasonic attenuation measurements. *Scientia Horticulturae*. 80:173–180.

Mizrach, A. 2000. Determination of avocado and mango fruit properties by ultrasonic technique. *Ultrasonics*. 38:717–722.

Mizrach, A., et al., 2000. Monitoring avocado softening in low-temperature storage using ultrasonic measurements. *Computers and Electronics in Agriculture*. 26(2):199–207.

Mizrach, A. 2004. Assessing plum fruit quality attributes with an ultrasonic method. *Food Research International*. 37(6):627–631.

Mizrach, A. 2007. Nondestructive ultrasonic monitoring of tomato quality during shelf-life storage. *Postharvest Biology and Technology*. 46:271–274.

Mörlein, D., Rosner, F., Brand, S., Jenderka, K.-V., and Wicke, M. 2005. Non-destructive estimation of the intramuscular fat content of the longissimus muscle of pigs by means of spectral analysis of ultrasound echo signals. *Meat Science*. 69:187–199.

Müller, M., Brunn, P., and Harder, C. 1997. New rheometric technique: The gradient-ultrasound pulse Doppler method. *Applied Rheology*. 7(5):204–210.

Niñoles, L., Clemente, G., Ventanas, S., and Benedito, J. 2007. Quality assessment of Iberian pigs through backfat ultrasound characterization and fatty acid composition. *Meat Science*. 76:102–111.

O'Neil, H.T. 1949. Reflection and refraction of plane shear waves inviscoelastic media. *Physical Review*. 75:928–935.

Pallav, P., Hutchins, D.A., and Gan, T.H. 2009. Air-coupled ultrasonic evaluation of food materials. *Original Research Article Ultrasonics*. 49:244–253.

Parker, N.G., and Povey, M. 2012. Ultrasonic study of the gelation of gelatin: Phase diagram, hysteresis and kinetics. *Food Hydrocolloids*. 26(1):99–107.

Pierre, J., Elias, F., and Leroy, V. 2013. A technique for measuring velocity and attenuation of ultrasound in liquid foams. *Ultrasonics*. 53:622–629.

Povey, M.J.W. 1989. Ultrasonics in food engineering. Part II: Applications. *Journal of Food Engineering*. 9:1–20.

Povey, M.J.W., Awad, T.S., Huo, R., and Ding, Y. 2007. Crystallization in monodisperse emulsions with particles in size ranges 20–200 nm. In *Food Colloids: Self-Assembly and Material Science*. Vol. 302. (pp. 399–412).

Povey, M.J.W., Awad, T.S., Huo, R., and Ding, Y. L. 2009. Quasi-isothermal crystallisation kinetics, non-classical nucleation and surfactant-dependent crystallisation of emulsions. *European Journal of Lipid Science and Technology*. 111(3):236–242.

Povey M.J.W., Moore J.D., Braybrook J., Simons H., Belchamber R., Raganathan M., and Pinfiel V. 2011. Investigation of bovine serum albumin denaturation using ultrasonic spectroscopy. *Food Hydrocolloids*. 25:1233–1241.

Povey, M.J.W. 2013. Ultrasound particle sizing: A review. *Particuology*. 11:135–147.

Raum, K., Ozguler, A., Morris, S.A., and O'Brien, W.D. 1998. Channel defect detection in food packages using integrated backscatter ultrasound imaging *IEEE Transactions on Ultrasonics, Ferroelectrics, and Frequency Control*. 45:30–40.

Regner, M., Henningsson, M., Wiklund, J., Ostergren, K., and Tragardh, C. 2007. Predicting the displacement of yoghurt by water in a pipe using CFD. *Chemical Engineering Technology*. 30(7):844–853.

Resa, P., Luis, E., Montero de Espinosa, F.R., and Gómez-Ulltate, Y. 2005. Ultrasonicvelocity in water–ethanol–sucrose mixtures during alcoholic fermentation. *Ultrasonics*. 43(4):247–252.

Resa, P., Bolumar, T., Elvira, L., Pérez, G., and Montero de Espinosa, F. 2007. Monitoring of lactic acid fermentation in culture broth using ultrasonic velocity. *Journal of Food Engineering*. 78(3):1083–1091.

Resa, P., Elvira, L., Montero de Espinosa, F., and JoséBarcenilla, R.G. 2009. On-line ultrasonic velocity monitoring of alcoholic fermentation kinetics. *Bioprocess and Biosystems Engineering*. 32:321–331.

Ross, K.A., Pyrak-Nolte, L.J., and Campanella, O.H. 2004. The use of ultrasound and shear oscillatory tests to characterize the effect of mixing time on the rheological properties of dough. *Food Research International*. 37(6):567–577.

Safvi, A.A., Meerbaum, J., Morris, S.A., Harper, C.L., and O'Brien, W.D. 1997. Acoustic imaging of defects in flexible food packages. *Journal of Food Protection*. 60(3): 309–314.

Schindel, D.W., Hutchins, D.A., Zou, L., and Sayer, M. 1995. The design and characterization of micromachined air-coupled capacitance transducer. *IEEE Transaction on Ultrasonics, Ferroelectrics and Frequency Control*. 42(1).

Schoeck, T., Hussein, M.A., and Becker, T. 2010. Konzentrationsbestimmung in Wasser-Zucker-Ethanol-GemischenmittelsadiabatischerKompressibilität und Dichte. *Tm-TechnischesMessen*. 77(1):30–37.

Shannon, R.A., Probert-Smith, P.J., Lines, J., and Mayia, F. 2004. Ultrasound velocity measurement to determine lipid content in salmon muscle; The effects of myosepta. *Food Research International*. 37:611–620.

Sigfusson, H., Ziegler, G.R., and Coupland, J.N. 2004. Ultrasonic monitoring of food freezing. *Journal of Food Engineering*. 62(3):263–269.

Simal, S., Benedito, J., Clemente, G., Femenia, A., and Rosselló, C. 2003. Ultrasonic determination of the composition of a meat-based product. *Journal of Food Engineering*. 58:253–257.

Singh, A.P., McClements, D.J., and Marangoni, A.G. 2004. Solid fat content determination by ultrasonic velocimetry. *Food Research International*. 37:545–555.

Takeda, Y. 1986. Velocity profile measurement by ultrasound Doppler shift method. *The International Journal of Heat and Fluid Flow*. 7(4):313–318.

Takeda, Y. 1991. Development of an ultrasound velocity profile monitor. *Nuclear Engineering and Design*. 126:277–284.

Telis-Romero, J., Benedito, J., Cerveró, R.P., Pérez-Muelas, M.N., and Mulet, A. 2008. *Textural Properties of Fresh Cheese Evaluated by Ultrasounds*. http://www.cabdirect.org/abstracts/20103078562.html;jsessionid=34D774598131C72AD9211C6A2818F72D.

Telis-Romero, J., Váquiro, H.A., Bon, J., and Benedito, J. 2011. Ultrasonic assessment of fresh cheese composition. *Journal of Food Engineering*. 103:137–146.

Ting, C.H., et al. 2009. Use of ultrasound for characterizing the gelation process in heat induced $CaSO_4 \cdot 2H_2O$ tofu curd. *Journal of Food Engineering*. 93(1):101–107.

Van Sint Jan, M., Guarini, M., Guesalaga, A., Pérez-Correa, J.R., and Vargas, Y. 2008. Ultrasound based measurements of sugar and ethanol concentrations in hydroalcoholic solutions. *Food Control*. 19:31–35.

Wallhäußer, E., Hussein, W.B., Hussein, M.A., JörgHinrichs, and Becker, T.M. 2011. On the usage of acoustic properties combined with an artificial neural network—A new approach of determining presence of dairy fouling. *Journal of Food Engineering*. 103:449–456.

Wang, L., McCarthy, K.L., and McCarthy, M.J. 2004. Effect of temperature gradient on ultrasonic Doppler velocimetry measurement during pipe flow. *Food Research International*. 37(6):633–642.

Wang, Q., et al. 2005. Gelation behaviour of aqueous solutions of different types of carrageenan investigated by low-intensity-ultrasound measurements and comparison to rheological measurements. *Innovative Food Science & Emerging Technologies*. 6(4):465–472.

Wang, Y., and Povey, M.J.W. 1999. A simple and rapid method for the determination of particle size in emulsions from ultrasound data. *Colloids and Surfaces B: Biointerfaces*. 12:417–427.

Wiklund, J., Shahram, I., and Stading, M. 2007. Methodology for in-line rheology by ultrasound Doppler velocity profiling and pressure difference techniques. *Chemical Engineering Science*. 62:4277–4293.

Wiklund, J., and Stading, M. 2008. Application of in-line ultrasound Doppler based UVP-PD rheometry method to concentrated model and industrial suspensions. *Flow Measurement and Instrumentation*. 19:171–179.

Wiklund, J., Stading, M., and Trägårdh, C. 2010. Monitoring liquid displacement of model and industrial fluids in pipes by in-line ultrasonic rheometry. *Journal of Food Engineering*. 99:330–337.

Wunderlich, T. and Brunn, P.O. 1999. Ultrasound pulse Doppler method as a viscometer for process monitoring. *Flow Measurement and Instrumentation*. 10(4):201–205.

Young, N., Wassell, P., Wiklund, J., and Stading, M. 2008. Monitoring struturants of fat blends with ultrasound based in-line rheometry (UVP–PD). *International Journal of Food Science and Technology*. 43(10):2083–2089.

Zacharias, E.M., Jr. and Parnell, R.A., Jr. 1972. Measuring the solids content of foods by sound velocimetry. *Food Technology*. 26:160–162,164,166.

Zhao, B., Basir, O.A., and Mittal, G.S. 2003. Correlation analysis between beverage apparent viscosity and ultrasound velocity. *International Journal of Food Properties*. 6(3):443–448.

16

Kinetic Data for Biochemical and Microbiological Processes during Thermal Processing

Ann Van Loey, Stefanie Christiaens, Ines Colle, Tara Grauwet, Lien Lemmens, Chantal Smout, Sandy Van Buggenhout, Liesbeth Vervoort, and Marc Hendrickx

Contents

16.1 Introduction	678
16.2 General Aspects of Kinetic Modeling	680
16.2.1 Identification of Primary Kinetic Models	681
16.2.1.1 Zero-Order Model	682
16.2.1.2 First-Order Model	682
16.2.1.3 Biphasic Model	683
16.2.1.4 Fractional Conversion Model	684
16.2.2 Identification of Secondary Kinetic Models	684
16.2.2.1 Influence of Temperature on the Reaction Rate Constant	684
16.2.2.2 Selection of a Temperature Coefficient Model	685
16.3 Measurement Techniques	686
16.4 Specific Data on Properties	688
16.4.1 Microbial Inactivation	688
16.4.2 Enzyme Inactivation	690
16.4.3 Texture Degradation	690
16.4.4 Color Degradation	693
16.4.5 Flavor Degradation	694
16.4.6 Nutrient Degradation	696
16.5 From Single-Response toward Multi-Response Modeling	697
References	700

16.1 INTRODUCTION

The safety and/or quality of a food is determined by the effect of all reactions occurring in the product, integrated over the full history of the product until the moment of consumption. In Figure 16.1, the idea of a "preservation reactor" is presented. This concept applies to the entire process of manipulating a food product—preparation and packaging, processing, distribution to storage—as well as to a single unit operation or processing step in the production chain. The rates at which desired (e.g., microbial inactivation) and undesired (e.g., nutrient destruction) reactions (related to food safety and quality) take place are function of intrinsic (i.e., food specific) properties and extrinsic (i.e., process specific) factors [1,2].

As thermal processing has been and is still one of the most widely used physical methods of food preservation, the objective of this chapter is to particularly focus on elevated temperature as a main extrinsic factor affecting chemical, biochemical, and microbiological changes in food products. The impact of a thermal process on a food safety or quality attribute is usually quantified using the concept of an "equivalent time at reference temperature," commonly referred to as process-value F representing the integrated microbial lethality, or as cook-value C representing the integrated effect on a quality characteristic. This concept translates the time–temperature variable product profile into an equivalent time at a chosen constant reference temperature that will affect the food safety or quality attribute in the same way as the actual time–temperature variable profile to which the food (i.e., the attribute of interest) was subjected.

The impact of a heat treatment on a product safety or quality attribute depends on the rates of the heat-induced reactions that affect this attribute, and on the time interval during which these reaction rates occur. As in practice isothermal heating profiles almost never

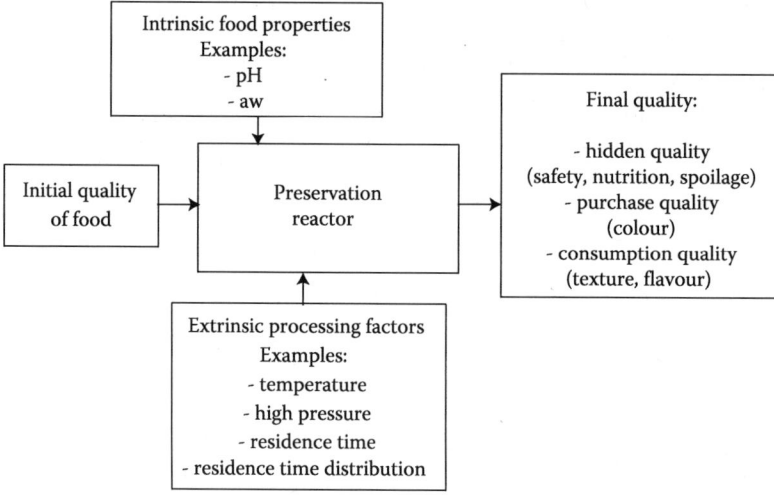

Figure 16.1 The preservation reactor. Examples of intrinsic food properties and extrinsic process factors that can influence the safety/quality of the product are given. (From A Van Loey et al. *Trends Food Sci Technol* 7(1):16–26, 1996.)

occur, reaction rates are varying as a function of processing time. The process impact is defined as the integral over time of the rate at each encountered temperature relative to the rate at a chosen reference temperature T_{ref} denoted as subscript (Equation 16.1).

$$^{E_a}F_{T_{ref}} = \int_0^t \frac{k}{k_{ref}} dt \tag{16.1}$$

where F is the process value, k the rate constant at T, and k_{ref} the rate constant at reference temperature T_{ref}.

An analogue expression for the process value can be obtained by combination of Equation (16-valid for first-order reactions (see later)) with Equation (16.1):

$$^{z}F_{T_{ref}} = \int_0^t \frac{D_{ref}}{D} dt \tag{16.2}$$

where F is the process value, D the decimal reduction time at T, and D_{ref} the decimal reduction time at reference temperature T_{ref}.

Both the rate constant k and the decimal reduction time D are varying with temperature, and the temperature dependence can either be described by the Arrhenius model (Equation 16.20) or the Thermal Death Time (TDT) concept (Equation 16.21) (see later). Combination of Equations 16.1 and 16.20—in case of the Arrhenius terminology—or Equations 16.2 and 16.21—in case of the TDT terminology—gives a first set of expressions (Equations 16.3 and 16.4) to determine the impact of a thermal process on a specific product aspect, characterized by its activation energy or z-value, denoted as superscript.

$$^{z_{safety}}F_{T_{ref}} = \int_0^t 10^{\left(\frac{T-T_{ref}}{z_{safety}}\right)} dt$$

$$^{z_{quality}}C_{T_{ref}} = \int_0^t 10^{\left(\frac{T-T_{ref}}{z_{quality}}\right)} dt \tag{16.3}$$

$$^{Ea_{safety}}F_{T_{ref}} = \int_0^t \exp\left(\frac{Ea_{safety}}{R_g}\left(\frac{1}{T_{ref}} - \frac{1}{T}\right)\right) dt$$

$$^{Ea_{quality}}C_{T_{ref}} = \int_0^t \exp\left(\frac{Ea_{quality}}{R_g}\left(\frac{1}{T_{ref}} - \frac{1}{T}\right)\right) dt \tag{16.4}$$

Hence, based on the temperature history of the product safety or quality aspect, either recorded using temperature sensors or simulated by mathematical modeling, combined with knowledge on the kinetics of the target safety or quality attribute, the thermal process impact on that aspect can be calculated, an approach that is commonly referred to as the physical–mathematical method.

ENGINEERING PROPERTIES OF FOODS

The impact of a thermal process on a specific safety or quality attribute can also be determined relying solely on the initial and final status of the attribute of interest (e.g., microbial count, nutrient concentration) and on its kinetics. The general expression of a first-order decay for nonisothermal process conditions can be written as Equation 16.5 and of an nth order reaction ($n \neq 1$) as Equation 16.6 (see later).

$$\log\left(\frac{X}{X_0}\right) = -\int_0^t \frac{dt}{D} = -\int_0^t \frac{k}{2.303} dt \tag{16.5}$$

$$\frac{X^{1-n} - X_0^{1-n}}{1-n} = -\int_0^t k\, dt \tag{16.6}$$

By the combination of Equation 16.2 with Equation 16.5 and of Equation 16.1 with Equation 16.6, Equations 16.7 and 16.8 can be obtained to determine the process value F or cook value C for a parameter that obeys first-order decay or nth order ($n \neq 1$) decay, respectively.

$$^{z(Ea)\text{safety}} F_{T_{\text{ref}}} = (D_{\text{ref}})_{\text{safety}} \log\left(\frac{N_0}{N}\right) = \frac{1}{(k_{\text{ref}})_{\text{safety}}} \ln\left(\frac{N_0}{N}\right)$$

or

$$^{z(Ea)\text{quality}} C_{T_{\text{ref}}} = (D_{\text{ref}})_{\text{quality}} \log\left(\frac{X_0}{X}\right) = \frac{1}{(k_{\text{ref}})_{\text{quality}}} \ln\left(\frac{X_0}{X}\right) \tag{16.7}$$

$$^{Ea\text{safety}} F_{T_{\text{ref}}} = \frac{1}{(k_{\text{ref}})_{\text{safety}}} \left(\frac{N^{1-n} - N_0^{1-n}}{n-1}\right)$$

$$^{Ea\text{quality}} C_{T_{\text{ref}}} = \frac{1}{(k_{\text{ref}})_{\text{quality}}} \left(\frac{X^{1-n} - X_0^{1-n}}{n-1}\right) \tag{16.8}$$

Hence, based on the response status of a safety (N) or quality (X) attribute before (N_0 or X_0) and after (N or X) thermal treatment, combined with its kinetics, the process impact can be calculated using Equation 16.7 or 16.8 depending on the order of the heat-induced reaction occurring to this attribute.

From the above, the importance of kinetic data for heat-induced chemical, biochemical, and microbiological changes in food products is obvious for adequate thermal process design, evaluation, and optimization.

16.2 GENERAL ASPECTS OF KINETIC MODELING

Kinetics of an inactivation/degradation process describe its progress in time. The way in which the inactivation or degradation progresses as a function of time is expressed by the mathematical form of the kinetic model. The rate of inactivation is reflected by the numerical values of the kinetic parameter estimates. Dealing with heat-inactivation/degradation kinetic studies, the data-analysis procedure involves (i) the identification of an adequate primary model or rate equation (e.g., first order, biphasic, nth order) and (ii) the identification of a secondary model (e.g., TDT model, Arrhenius model) describing the temperature dependence of the rate equation.*

16.2.1 Identification of Primary Kinetic Models

In general, an nth order reaction rate equation can be written as

$$\frac{dX}{dt} = -kX^n \tag{16.9}$$

where X is the response value at time t (e.g., microbial count, nutrient concentration), k the reaction rate constant, and n the order of the reaction.

The order of reaction can be determined by many different methods [3,4]. Under simple initial and boundary conditions, the integration of differential equation 16.9 results in analytical solutions describing response property (X) as a function of time. In case the rate constant k is not varying with time, Equation 16.10 or Equation 16.11 are obtained depending on the reaction order (with respect to time).

$$n = 1 \qquad X = X_0 \exp(-kt) \tag{16.10}$$

$$n \neq 1 \qquad X^{(1-n)} = X_0^{(1-n)} + (n-1)kt \tag{16.11}$$

By trial-and-error procedures an estimate of the reaction order n (with respect to time) can be obtained by analyzing graphically the trends and/or deviations from a linear behavior. However, a complication in determining the order of a reaction from the method of integration is that little or no distinction can be made between zero-, first-, and second-order reactions with respect to time when the response value changes only little [5]. Even when a linear plot is obtained, conclusions must be drawn cautiously from the data, especially if the data points correspond to no more than 10–20% conversion because many mathematical functions are roughly linear over a sufficiently small range of variables. Some authors suggest that one should perform at least experimental runs in which data are taken at 40%, 50%, or higher conversions [3], while others even suggest that reactions should be followed, where possible, through at least 90–99% conversion (or 1–2 log reductions) [6]. If the method of analysis is not sufficient to measure response values as low as this, the longest heating times possible should be used [7]. In the next section, several primary kinetic models that are often applied to model heat-induced changes in food products, will be described.

---

* The primary and secondary kinetic models that will be described are engineering tools for thermal process evaluation, design, and optimization; they do not necessary allow for a mechanistic interpretation of the reactions occurring during heat inactivation.

16.2.1.1 Zero-Order Model

A zero-order reaction is a reaction whose rate is independent of the concentration of reactants (Equation 16.12). Consequently, there is a linear relationship between the concentration and time in zero-order reactions.

$$X = X_0 - kt \quad (16.12)$$

where X is response property at time t; X_0 is initial response property; t is treatment time; and k is the zero-order inactivation rate constant.

Zero-order reactions are not frequently encountered to describe changes in food safety or quality aspects during thermal processing. They are sometimes applied to describe thermal degradation of color or to describe flavor formation in food products during thermal processing.

In a kinetic experiment, a plot of X versus treatment time yields a straight line in case of zero-order model and the rate constant at given process conditions can be estimated by linear regression analysis of X versus inactivation time. The validity of a zero-order reaction can be examined by plotting the residual response property X versus treatment time and evaluation of the goodness-of-fit by means of, for example, lack-of-fit test, coefficient of determination (R^2), and analysis of the distribution of residuals.[*] Residuals of an appropriate fit represent only the experimental error and should therefore be randomly distributed. The existence of trends in residuals (with respect to either the independent or dependent variable(s)) suggests that some systematic behavior is present in the data that is not accounted for by the model [8], in this case by a zero-order reaction.

The reaction rates are sometimes characterized in terms of the half-life time ($t_{1/2}$), which is the time required to reduce the response property X to half of its initial value. For zero-order reactions, the half-life time is given by Equation 16.13.

$$t_{1/2} = \frac{X_0}{2k} \quad (16.13)$$

16.2.1.2 First-Order Model

Thermal inactivation or degradation of many food constituents can often be described by a first-order reaction. Under constant processing conditions, a first-order reaction can be expressed as Equation 16.10, which can be linearized by a logarithmic transformation, yielding

$$\ln(X) = \ln(X_0) - kt \quad (16.14)$$

where X is the response property at time t; X_0 is the initial response property; t is treatment time; and k is the first-order inactivation rate constant.

In many cases, thermal inactivation of microorganisms and enzymes, as well as thermal degradation of nutrients, color, and texture are described by a first-order model. In case of a first-order inactivation model, a plot of $\ln(X)$ versus treatment time yields a

---

[*] Residual are the differences between experimentally observed dependent variable values and the ones predicted by the regression equation.

straight line and the rate constant at given process conditions can be estimated by linear regression analysis of ln(X) versus inactivation time (Equation 16.14), or alternatively by nonlinear regression analysis on the non-log-transformed inactivation data (Equation 16.10). The selection between a linear or a nonlinear kinetic model has previously been discussed [9].

For first-order reactions, the half-life time is given by Equation 16.15. Hence, the half-life of a first-order reaction may be calculated from the first-order rate constant and vice versa.

$$t_{1/2} = \frac{0.693}{k} \qquad (16.15)$$

In the area of food science and technology, it is common to characterize first-order reactions using the TDT concept. The decimal reduction time (D-value) is defined as the time at given constant process conditions, needed for a 90% reduction of the initial response value, or in other words, the time at constant temperature necessary to traverse one log cycle. For first-order reactions, D-values and rate constants are directly related (Equation 16.16).

$$D = \frac{\ln(10)}{k} \qquad (16.16)$$

Substitution of Equation 16.16 into Equation 16.14 yields

$$\log(X) = \log(X_0) - \frac{t}{D} \qquad (16.17)$$

The decimal reduction time at a given inactivation temperature can be calculated from the slope of a linear regression analysis of log(X) versus inactivation time at constant inactivation temperature (Equation 16.17), or alternatively by nonlinear regression analysis on the nonlog-transformed inactivation data.

The validity of a first-order inactivating behavior can be examined by plotting residual response property versus treatment time on a semilogarithmic scale and evaluation of the goodness-of-fit in a similar way as for a zero-order model.

16.2.1.3 Biphasic Model

Thermal inactivation or degradation curves can often be subdivided into two (or more) fractions with different processing stability, for example, one more thermal resistant than the other and both inactivating according to a first-order decay kinetic model. For constant extrinsic and intrinsic factors and assuming that the inactivation of both fractions is independent of each other, the inactivation can be modeled according to

$$X = X_l \exp(-k_l t) + X_s \exp(-k_s t) \qquad (16.18)$$

where X_s is the response property of the stable enzyme fraction, X_l the response property of the labile enzyme fraction, k the first-order inactivation rate constant where the subscripts s and l for k denote thermostable and thermolabile, respectively.

A biphasic model has most often been applied for thermal inactivation of enzymes and for texture degradation during thermal processing. As to enzyme inactivation, the biphasic inactivation behavior has been attributed to the occurrence of isozymes with different thermostabilities, whereas for texture degradation, the biphasic behavior has been attributed to two simultaneous first-order kinetic mechanisms acting on two substrates [10].

By plotting the residual response property after different time intervals versus time, the inactivation rate constant of the labile fraction (k_l-value), the inactivation rate constant of the stable fraction (k_s-value), and the response values of both fractions (X_l and X_s) can be estimated using nonlinear regression analysis (Equation 16.18).

16.2.1.4 Fractional Conversion Model

Fractional conversion refers to an (first-order) inactivation process that takes into account a nonzero response value upon prolonged heating. A fractional conversion model can be expressed mathematically as

$$X = X_\infty + (X_0 - X_\infty)\exp(-kt) \tag{16.19}$$

where X is the response property at time t; X_0 is the initial response property; X_∞ the response property upon prolonged processing; t is the treatment time; and k is the rate constant.

The fractional conversion model is often applied to describe thermal degradation of a physical property of food such as texture or color.

By plotting residual response values after different time intervals versus time, the inactivation rate constant (k) and the remaining response value after prolonged treatment (X_∞) can be estimated using nonlinear regression analysis (Equation 16.19). The reaction should be followed for prolonged heating times in order to accurately estimate the X_∞-value. The nonzero response value (X_∞) may or may not be a function of applied temperature.

16.2.2 Identification of Secondary Kinetic Models

16.2.2.1 Influence of Temperature on the Reaction Rate Constant

Because in thermal processing of foods (e.g., pasteurization, sterilization, blanching), temperature is the main extrinsic factor for guaranteeing safety and quality during the production (and storage) of food products, the discussion is limited to systems for which temperature is the rate determining extrinsic factor.

Two terminologies are commonly used to quantify the influence of temperature on the inactivation rate of safety as well as of quality aspects: the Arrhenius model [11] usually used in the chemical kinetics area and the TDT model [12], especially to describe first-order heat-inactivation kinetics and usually used in thermobacteriology and in the thermal processing area.

The most well-known and perhaps most frequently used theory in the area of biological engineering and chemical kinetics is that proposed by Arrhenius, which is applicable to reactions in solutions and heterogeneous processes. According to Arrhenius, the temperature dependence of the rate constant k can be expressed as Equation 16.20.

$$k = k_{\text{ref}} \exp\left(\frac{E_a}{R_g}\left(\frac{1}{T_{\text{ref}}} - \frac{1}{T}\right)\right) \tag{16.20}$$

where E_a is the activation energy, k_{ref} is the rate constant at reference temperature (T_{ref}), and R_g is the universal gas constant. Equation 16.20 can be linearized by a log transformation which allows the activation energy to be estimated based on linear regression analysis of the natural logarithm of k versus the reciprocal of the absolute temperature. Alternatively, the activation energy can be estimated based on the nonlinearized Arrhenius relationship (Equation 16.20) using nonlinear regression analysis.

In thermobacteriology (related to thermal processing of foods and pharmaceuticals), however, the TDT-concept of Bigelow [12] (D- and z-value) is commonly used to describe heat-inactivation kinetics of first-order reactions. Bigelow observed that if the logarithm of the decimal reduction times (D-values) are plotted versus temperature on an arithmetic scale, the result over the usual range of temperatures of interest can be represented by a straight line. The temperature dependence of the D-value is given by the z-value which is the temperature increase necessary to obtain a tenfold decrease of the D-value (Equation 16.21).

$$D = D_{\text{ref}} 10^{\left(\frac{T_{\text{ref}} - T}{z}\right)} \tag{16.21}$$

where D_{ref} is the decimal reduction time at reference temperature T_{ref}, T the actual temperature and z the z-value of the system. After a linearizing log transformation of Equation 16.21, the z-value can be estimated based on linear regression analysis of the 10-based logarithm of the decimal reduction time versus temperature. Alternatively, the z-value can be estimated directly using nonlinear regression analysis based on Equation 16.21.

Food technologists sometimes use the Q_{10}-factor to describe the temperature dependence of a reaction rate. The temperature dependence parameter Q_{10} is defined as the factor by which the reaction rate is increased if the temperature is raised by 10°C

$$Q_{10} = \frac{v_{(T+10)}}{v_T} \tag{16.22}$$

where v is the reaction rate and T is the temperature.

16.2.2.2 Selection of a Temperature Coefficient Model
In the Arrhenius equation, the logarithm of the reaction rate constant is related to the reciprocal of the absolute temperature with the activation energy (E_a) representing the slope index of the semilogarithmic curve. In the TDT method, decimal reduction times are described as a direct exponential function of temperature with the z-value being the negative reciprocal slope of the semilogarithmic curve. Hence, these two concepts rely on mathematically different models with the kinetic parameters being proportional to temperature in the TDT model and to its reciprocal in the Arrhenius model, both concepts being valid within a finite temperature range [13].

Comparison of both temperature coefficient models allows Equation 16.23, that relate E_a to z, to be derived.

$$E_a = \frac{2.303\ R_g T_1 T_2}{z} \qquad (16.23)$$

Although E_a and z-values are assumed to be temperature independent, the two parameters are related by temperature T_1 and T_2 by Equation 16.23. Equation 16.23 implies that if z is constant over a temperature range, then E_a cannot be constant in that temperature range and vice versa. Lund noted that the two concepts were reconcilable over small temperature ranges where T could be considered proportional to $1/T$. He suggested to use the selected reference temperature as T_1 and a temperature z degrees less than T_1 as T_2 [14]. Ramaswamy and coworkers demonstrated that the conversion of E_a to z or vice versa is strongly influenced by the associated reference temperature and the temperature range. They suggested the use of the upper and lower limits of the experimental temperature range used for kinetic data acquisition to convert E_a into z-values or vice versa [15]. An additional advantage of this suggested approach is that it automatically restricts the conversion to be carried out within the limits of the experimental temperature range. Anyway this relationship is to be used cautiously only to estimate the activation energy order of magnitude from a known z-value and vice versa when raw kinetic data are no longer available [13].

In an attempt to reach a decision regarding which temperature-coefficient model should be used, several studies have directly compared the two models and suggested that the two models fit experimental kinetic data reasonably well [16–20]. Both TDT and Arrhenius concepts have merit and have been proven to be adequate to study degradation kinetics. Although theoretical approaches based on thermodynamic arguments have confirmed the Arrhenius relationship for simple gaseous systems and solutions, the relationship is empirical in nature, just as the TDT concept, for more complex systems. Preference for either model can only be justified if its statistical accuracy for describing experimental data points is superior to the other model. The choice of which model to apply depends entirely on the raw kinetic data [13].

16.3 MEASUREMENT TECHNIQUES

Thermal inactivation or degradation kinetics can be determined using either steady-state or unsteady-state procedures, the steady-state procedure being the most straightforward approach to determine the kinetics of destruction of a heat-labile food safety or quality characteristic.

A classical steady-state experiment consists of subjecting a sample to a square-wave temperature profile for various treatment times. Both batch or flow methods can be used for sample heating and cooling. Whatever method is being used, care has to be taken to ensure that heating and cooling is quasiimmediate, or else appropriate compensation for thermal lags has to be taken into account [e.g., 6,21], especially when the heating or cooling lag is not sufficiently small relative to the half-life of the reaction. For isothermal batch treatments, samples are usually enclosed in small, preferably highly conductive, vials or tubes (e.g., glass capillaries, thermal death tubes, thermal death cans) to minimize heating and cooling lags. The samples are immersed in a temperature-controlled heating bath at constant inactivation temperature for predetermined time intervals. Immediately upon withdrawal from the heating bath, the samples are cooled in ice water to stop the thermal inactivation, and the residual response value is measured. The steady-state method, due to its simplicity, has been most frequently applied to study thermal inactivation kinetics. For data analysis, it is common practice to estimate kinetic inactivation parameters by an individual or two-step regression method: one estimates at first inactivation rate constants (and possibly other kinetic parameters such as enzyme fraction, reaction order) from inactivation data at constant temperature by linear or nonlinear regression analysis. In a second step, one estimates temperature coefficients (activation energy or z-value) from regression analysis of the obtained inactivation rate constants as a function of temperature. Kinetic inactivation parameters can also be estimated in a global or one-step regression approach. To model a global inactivation data set, the temperature coefficient model is being incorporated in the inactivation rate equation. Using nonlinear regression analysis, one gets estimates of the inactivation rate constant at reference conditions and a temperature coefficient. The selection between an individual or global approach has previously been discussed [9].

Several unsteady-state (i.e., nonisothermal) methods for kinetic parameter estimation have been described in literature [22–25]. A main advantage of unsteady-state procedures for kinetic parameter estimation is the avoidance of interference by thermal lag effects. The use of nonisothermal methods is limited, however, due to increased complexity of data analysis. Hayakawa and coworkers [22] described a procedure to determine by an iteration procedure inactivation kinetics based on a programmed temperature history curve and response values at different heating times. Lenz and Lund [23] applied an unsteady-state procedure to determine the kinetic parameters for heat destruction of thiamin in conductive pea puree, based on an assumption of first-order inactivation. Rhim and coworkers [24] developed and experimentally validated a differential kinetic model for determination of kinetic parameters based on a linearly increasing temperature profile. Unsteady-state methods have also been applied to quantify thermal inactivation of several quality aspects such as thiamin and surface color of canned tuna [26], texture of green asparagus [27], and lipoxygenase in green beans [28].

The choice of which procedure to use (steady state or unsteady state) depends to some degree on the available equipment and availability of methods of data analysis. Besides, the choice of the method may also be dependent on the half-life of the component under study, relative to the thermal lags experienced in the sample holder. If the half-life is relatively small, then significant destruction will occur during the lag period so that an unsteady-state procedure may be recommended.

In the high-temperature range above 125°C (relevant for ultra-high-temperature [UHT] processes), kinetic data are scarce. A computer-controlled thermoresistometer has been applied to study heat-inactivation kinetics under high-temperature short-time conditions for microbial inactivation [e.g., 29], enzyme inactivation [e.g., 30], or quality degradation [e.g., 31]. A microminiaturized tubular heating system with continuous heating and cooling has been especially designed for kinetic studies at high temperatures (110–160°C) [32]. Some kinetic studies in the high-temperature region have been performed using a pilot-scale UHT plant, especially for milk products [e.g., 33,34]. Due to this lack of kinetic data in the high-temperature range, one often relies on extrapolation of kinetic data obtained at lower temperatures. However, extreme caution must be employed in extrapolating data outside the experimental temperature domain in which original kinetic data were obtained, since this can lead to major discrepancies.

16.4 SPECIFIC DATA ON PROPERTIES

Kinetic parameter values for thermal inactivation/degradation of microorganisms, enzymes, chemical constituents, or physical properties are largely varying depending on the reaction environment. Factors such as homogeneity and purity of the material, origin, presence of stabilizing agents (e.g., sugars or polyols), pH value, moisture content, and so on are largely influencing the observed heat-inactivation kinetics. Knowledge of these environmental conditions are essential to allow critical evaluation and use of the kinetic data, but, unfortunately, they are not always reported consistently because we have only a limited feeling for them. Due to this wide spread in kinetic values, we will refer to detailed review articles on thermal degradation of different food constituents, and present for each food component an exemplifying table covering a range of kinetic parameter values for common model systems or real food products.

16.4.1 Microbial Inactivation

Kinetic parameters for heat inactivation of microorganisms, both in model systems and real food products have been reviewed [35–38]. Table 16.1 presents some examples of kinetic parameter values for thermal inactivation of microorganisms. In thermobacteriology, most kinetic data are reported applying the TDT concept. Z-values for microbial inactivation in general range from 4°C to 12°C. Thermostability of spore-forming microorganisms is much higher than of vegetative cells.

In the past, heat inactivation of microorganisms has been described by a first-order inactivation model. The assumption that microbial heat inactivation follows first-order kinetics is the basis of safety evaluation of thermally processed foods [39]. However, many isothermal inactivation curves do clearly not conform to a first-order reaction (Figure 16.2). The most commonly encountered nonlinear semilogarithmic survival curves are characterized by an upward or downward concavity, and sigmoid curves with alternating downward and upward concavity (shoulder and tailing effects) or vice versa. Cell clumping, heterogeneity of the microbial population, acquired heat resistance, multiple hit theories and simultaneous activation and inactivation theories have been suggested to account for this. The shortcomings of the first-order concept have been known for years, and several methods have been proposed to rectify them [40–48]. Modeling of microbial nonlinear thermal inactivation kinetics has been reviewed [49,50]. Proposed models for nonlinear microbial heat inactivation include a Weibull distribution model [e.g., 51,52], a modified Gompertz equation [e.g., 53–55], the Baranyi model [e.g., 55,56], a log logistic transformation vitalistic approach [e.g., 57], or the Casolari

KINETIC DATA FOR BIOCHEMICAL AND MICROBIOLOGICAL PROCESSES

Table 16.1 Kinetic Parameter Values for Thermal Inactivation of Microorganisms

Microorganism	Medium	D_T (min)	z (°C)	Ref.
Bacillus cereus	Phosphate buffer (pH 7)	$D_{85°C} = 33.8$	9.7	38
	Milk-based infant formula (pH 6.3)	$D_{95°C} = 2.7$	8.1	38
	milk	$D_{95°C} = 1.8$	9.4	38
Bacillus subtilis	Buffer (pH 6.8)	$D_{121°C} = 0.57$	9.8	37
Clostridium botulinum spores				
Type A	Phosphate buffer (pH 7)	$D_{110°C} = 4.3$	9.4	38
	Canned peas (pH 5.2)	$D_{110°C} = 0.61$	7.6	38
	Macaroni (pH 7)	$D_{110°C} = 2.48$	8.8	38
Type B (proteolytic)	Phosphate buffer (pH 7)	$D_{120°C} = 0.14$	11.0	38
	Canned peas (pH 5.6)	$D_{110°C} = 3.07$	10.1	38
	Canned corn	$D_{110°C} = 2.15$	9.6	38
Type B (non-prot.)	Phosphate buffer (pH 7)	$D_{82.2°C} = 32.3$	9.7	38
Yersinia enterocolitica	Milk	$D_{51.7°C} = 23.4–29.9$	5.1–5.8	38
Escherichia coli	Broth	$D_{56°C} = 4.5$	4.9	37
Staphylococcus aureus	Pea soup	$D_{60°C} = 10.4$	4.6	37
Salmonella senftenberg	Pea soup	$D_{60°C} = 10.6$	5.7	37
Listeria monocytogenes	Carrot	$D_{70°C} = 0.27$	6.7	37

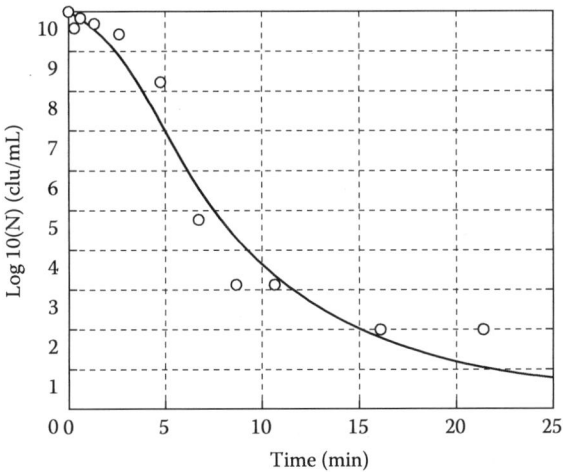

Figure 16.2 Thermal inactivation of *E. coli* O157:H7 at 60°C, modeled by the second model of Casolari. (From AH Geeraerd, CH Herremans, JF Van Impe. *Int J Food Microbiol* 59:185–209, 2000.)

model [40]. Sapru and coworkers [58,59] derived a model to describe the inactivation of microbial spores during sterilization process, which implies an initial increase of the activated spore population.

16.4.2 Enzyme Inactivation

Enzyme inactivation during heat processing has been reviewed [60,61]. Table 16.2 presents some examples of thermal inactivation kinetics of food-related enzymes.

Although enzyme inactivation is a complex process involving several events, such as formation and/or disruption of different interactions and/or bonds, decomposition of amino acids, aggregation and/or dissociation, enzyme inactivation by thermal processing often follows first-order kinetics (e.g., polyphenoloxidase [80,81,84], lipoxygenase [67,68], alkaline phosphatase [62], tomato pectinmethylesterase [75], and polygalacturonase [78,79]). Frequently encountered non-first-order mathematical models for thermal enzyme inactivation are (i) a biphasic inactivation behavior, where a fast inactivation period is followed by a decelerated decay, which has been attributed to the occurrence of isozymes with different thermostabilities (e.g., lipoxygenase [85], tomato polygalacturonase [86] (Figure 16.3), orange pectinmethylesterase [72,87], carrot pectinmethylesterase [88], peroxidase [71,76,77]), or (ii) a fractional conversion model which has been attributed to the presence of a very thermoresistant enzyme fraction which does not inactivate in the investigated temperature region (e.g., orange pectinmethylesterase [72], tomato polygalacturonase [86]). In case nonfirst-order inactivation behavior of enzymes is observed, purification of the different isozyme fractions and characterization of their thermal inactivation kinetics, might explain the observed inactivation behavior in the more complex systems (e.g., enzyme crude extracts, real food systems, enzyme mixtures).

16.4.3 Texture Degradation

Studies on kinetics of thermal softening of food products have been reviewed [89–91]. Firmness has been mostly used in quantifying kinetics of texture degradation because it best relates to the consumer perception [92]. Kinetic parameter values for thermal degradation of food texture are presented in Table 16.3.

Several inactivation models have been applied in literature to describe thermal texture degradation of food products. Many published studies have indicated that texture degradation follows first-order kinetics [e.g., 93–95,105,102]. However, in case texture degradation is evaluated at longer treatment times, a biphasic model indicating the occurrence of two simultaneous first-order reactions has been proposed [e.g., 10,96,97,106–108]. Rizvi and Tong [109] reexamined published data supporting the biphasic thermal texture degradation of vegetables. They concluded that a fractional conversion model provides an alternate inactivation model that is more accurate and reliable in determining the texture degradation kinetics of vegetables. Hereafter, the fractional conversion model has been applied in literature to model heat-induced texture degradation of several food products (Figure 16.4) [e.g., 99–101,104,110,111].

Table 16.2 Kinetic Parameter Values for Thermal Inactivation of Food-Related Enzymes

Enzyme	Source/Medium	D_T (min)	z (°C)	k_T (min^{-1})	E_a (kJ/mol)	Ref.
Alkaline phosphatase	Bovine milk	$D_{60°C} = 24.6$	5.4			62
	Raw milk	$D_{71.7°C} = 0.157$	5.4			63
	Glycine buffer	$D_{60°C} = 36.8$	23.1	$k_{60°C} = 0.0626$	97.2	64
	Pasteurized milk	$D_{60°C} = 162.2$	15.2	$k_{60°C} = 0.014$	149.9	64
	Raw milk (cow)	$D_{60°C} = 24.68$	11.8	$k_{60°C} = 0.093$	207.8	64
	Equine milk	$D_{60°C} = 7.77$	5.31		155	65
Lactoperoxidase	Bovine milk	$D_{70°C} = 57.2$	3.7			62
	Goat milk	$D_{73°C} = 8.9$	3.38	$k_{73°C} = 0.263$	679.0	66
	Cow milk	$D_{73°C} = 13.59$	3.58	$k_{73°C} = 0.170$	641.6	66
Lipoxygenase	Soybean/TrisHCl buffer pH 9.0			$k_{64°C} = 0.05$	319.8	67
	Green peas			$k_{70°C} = 0.642$	584.3	68
	Peas/phosphate buffer (pH 6)			$k_{70°C} = 0.162$	581.0	69
	Peas/dry flour			$k_{130°C} = 0.252$	126.2	70
	Broccoli florets (enzyme extract)[a]			$k_{80°C/1} = 0.773$ $k_{80°C/2} = 0.004$	61 55	71
	Asparagus tip (enzyme extract)[a]			$k_{90°C/1} = 7.806$ $k_{90°C/2} = 0.028$	76 65	71
	Asparagus stem (enzyme extract)[a]			$k_{90°C/1} = 4.187$ $k_{90°C/2} = 0.069$	78 56	71
Pectinesterase	Oranges/citric acid buffer (pH 3.7)			$k_{65°C} = 0.288$	389.3	72
	Strawberry/TrisHCl buffer (pH 7)			$k_{60°C} = 0.127$	206.7	73
	Banana/TrisHCl buffer (pH 7)	$D_{70°C} = 41.7$	5.9	$k_{70°C} = 0.055$	379.4	74
	Tomato juice	$D_{68°C} = 5.3$	6.2	$k_{68°C} = 0.436$	363.8	75
Peroxidase	Broccoli florets (enzyme extract)[a]			$k_{80°C/1} = 2.976$ $k_{80°C/2} = 0.214$	75 58	71
	Asparagus tip (enzyme extract)[a]			$k_{80°C/1} = 7.199$ $k_{80°C/2} = 0.032$	67 43	71
	Asparagus stem (enzyme extract)[a]			$k_{80°C/1} = 4.422$ $k_{80°C/2} = 0.064$	61 53	71
	Carrot cortex (enzyme extract)[a]			$k_{80°C/1} = 4.5$ $k_{80°C/2} = 0.008$	95 86	71
	Carrot core (enzyme extract)[a]			$k_{80°C/1} = 5.58$ $k_{80°C/2} = 0.004$	97 83	71

continued

Table 16.2 (continued) Kinetic Parameter Values for Thermal Inactivation of Food-Related Enzymes

Enzyme	Source/Medium	D_T (min)	z (°C)	k_T (min^{-1})	E_a (kJ/mol)	Ref.
Poly-galacturonase	Carrot (crude extract)[a]	$D_{55°C/1} = 79.7$		$k_{55°C/1} = 0.029$ $k_{55°C/2} = 0.002$	98.6 148	76
	Watercress (enzyme extract)[a]	$D_{55°C/2} = 1206$		$k_{84.6°C/1} = 18$ $k_{84.6°C/2} = 0.24$	421 352	77
	Carrot			$k_{80°C} = 0.522$	411	78
	Tomato/sodium acetate buffer	$D_{58°C} = 5.92$	7.7	$k_{58°C} = 0.386$	270.6	79
Polyphenol-oxidase	Avocado/phosphate buffer (pH 7)			$k_{70°C} = 0.133$	319.3	80
	Mushroom/phosphate buffer (pH 6.5)	$D_{60°C} = 4.7$	6.5		319.1	81
	Apple/phosphate buffer (pH 6.8)	$D_{78°C} = 8.1$	8.9		256.0	82
	Grape	$D_{60°C} = 41.2$	9.41	$k_{60°C} = 0.060$	225	83

[a] The data were modeled using a biphasic inactivation model.

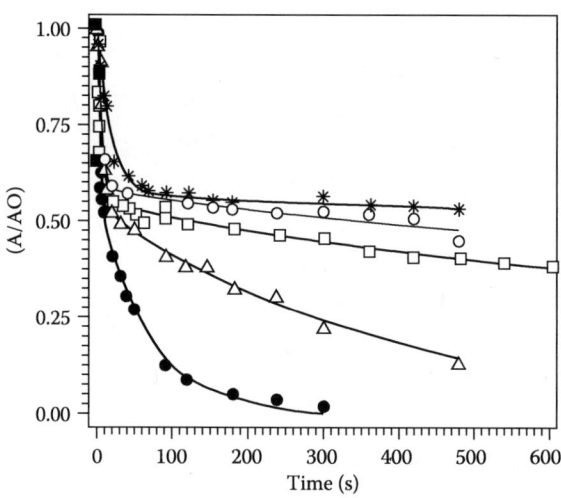

Figure 16.3 Thermal inactivation of crude tomato polygalacturonase extract in 40 mM Na-acetate buffer (pH 4.4) modeled using a biphasic model at different temperatures. 70°C (∗); 75°C (O); 80°C (□); 85°C (△); 90°C (●). (From D Fachin et al. *J Food Sci* 67(5):1610–1615, 2002.)

Table 16.3 Kinetic Parameter Values for Thermal Degradation of Food Texture

Product	Mathematical Model	T-range	D_T (min)	z (°C)	k_T (min^{-1})	E_a (kJ/mol)	Ref.
Apricots	First order	60–90°C				96.6	93
Asparagus	First order	70–98°C				100.3	94
	First order	100–130°C				79.4–96.1	95
	Biphasic	70–100°C			$k_{85°C/1} = 1.047$ $k_{85°C/2} = 0.057$	1:40.0 2:85.4	96
Beets	Biphasic	104–121.1°C			$k_{121.1°C/1} = 0.428$ $k_{121.1°C/2} = 0.003$	1:94.5 2:53.9	10
	Biphasic	70–100°C				1:82–92 2:82–92	97
Carrot (slices)	First order	90–120°C	$D_{121°C} = 5.5$	22.2	$k_{120°C} = 0.381$	116.6	98
	Biphasic	104–121°C			$k_{121.1°C/1} = 0.234$ $k_{121.1°C/2} = 0.001$	1:63.5 2:21.3	10
	Fractional conversion	80–110°C			$k_{100°C} = 0.222$	117.6	99
	Fractional conversion	95–105°C			$k_{100°C} = 0.296$	152.1	100
	Fractional conversion	90–110°C			$k_{100°C} = 0.123$	114.4	101
Garlic	First order	80–95°C	$D_{85°C} = 61.7$	17.4	$k_{85°C} = 0.037$	139.4	102
Peas		98.9–126.7°C	$D_{121°C} = 9.2$	36.7	$k_{121°C} = 0.250$	77.3	103
	Biphasic	100–110°C			$k_{110°C/1} = 0.211$ $k_{110°C/2} = 0.004$	1:113.3 2:102.0	10
Pumpkin	Fractional conversion	75–95°C			$k_{85°C} = 0.39$	72.21	104
Turnip	Biphasic	70–100°C				1:141–167 2:60–83	97

16.4.4 Color Degradation

Colored pigments that are naturally present in food include for plant-derived food products chlorophylls, anthocyanins, betalains, carotenoids, and for meat-derived products mainly myoglobin. Besides kinetic studies on destruction of these naturally occurring pigments during thermal processing, kinetics of brown pigment formation resulting from nonenzymatic browning reactions have been investigated. Some reviews on kinetics of color degradation during thermal processing have been published [90,91,112]. Published data on kinetics of thermal color degradation of some fruits and vegetables have been reviewed [113]. Table 16.4 presents some kinetic parameter values for thermal degradation of food color.

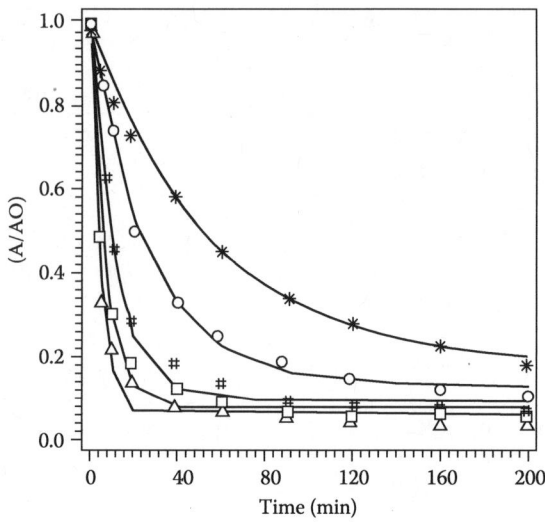

Figure 16.4 Thermal inactivation of texture of carrot dices modeled using a fractional conversion model at different temperatures. 80°C (∗), 85°C (O), 90°C (#), 95°C (□), and 100°C (△). (From TS Vu et al. Effect of preheating on thermal degradation kinetics of carrot texture. *Innovative Food Sci Technol*, 5:37–44, 2003.)

In many studies, color degradation is evaluated based on changes in tristimulus values (L—measure of lightness, a—change from green to red, b—change from blue to yellow) and/or total color difference values. Sometimes the degradation of the pigment under study is quantified using HPLC or spectroscopic techniques, especially the conversion of chlorophylls to pheophytins, anthocyanins, betanines, and the degradation and isomerization of carotenoids have been investigated on a chemical basis. Few studies rely on trained taste panels to quantify color degradation.

Color degradation kinetics during heating is usually described by a zero-order reaction for total color difference [130,135], a first-order reaction [94,113,116,118,121, 122,124,127,128,130,131,133–135,141] or a fractional conversion model [104,115,125,126,132, 142,143]. Figure 16.5 presents an example of the degradation of the green color in broccoli juice, modeled using a fractional conversion model. Kinetics of brown pigment formation can be described using a two-stage model: the first stage describes formation of color by Maillard reaction and the second stage describes color destruction [144,145].

16.4.5 Flavor Degradation

Kinetics of flavor changes have been reviewed [90,146]. Flavor measurement is usually based on gas chromatography (for volatile components), HPLC analysis, or sensory methods by untrained consumer panels or trained taste panels. Flavor destruction and flavor formation have been the least studied in terms of kinetics during thermal processing. Typically, the reaction order for flavor degradation in foods follow first-order kinetics

Table 16.4 Kinetic Parameter Values for Thermal Degradation of Food Color

Attribute	Medium	D_T (min)	z (°C)	k_T (min^{-1})	E_a (kJ/mol)	Ref.
Color pigments						
Chlorophyll a	Pureed spinach	$D_{121°C} = 13.2$	22.7	$k_{121°C} = 0.178$	105.3	114
	Pureed green peas			$k_{80°C} = 0.017$	81.5	115
	Green peas (pH 5.5)			$k_{90°C} = 0.118$	58.6	116
	Broccoli juice			$k_{120°C} = 0.236$	63.3	117
	Pureed coriander leaves (pH 5.5)			$k_{125°C} = 0.0187$	91.1	118
Chlorophyll b	Pureed spinach	$D_{121.1°C} = 28.1$	25.1	$k_{121°C} = 0.085$	94.1	114
	Pureed green peas			$k_{80°C} = 0.008$	71.5	115
	Green peas (pH 5.5)			$k_{90°C} = 0.003$	41.9	116
	Broccoli juice			$k_{120°C} = 0.091$	80.1	117
	Pureed coriander leaves (pH 5.5)			$k_{125°C} = 0.024$	96.0	118
Anthocyanin	Raspberry juice (pH 3.3)	$D_{121.1°C} = 45$	28	$k_{108°C} = 0.020$	91.9	119
	Cherry juice (pH 3.5) (cyanidin-3,5-diglucoside)	$D_{121.1°C} = 26$	24	$k_{108°C} = 0.029$	106.2	120
	Raspberry paste (cyanidin-3-sophoroside)			$k_{110°C} = 0.021$	75.3	121
	Strawberry paste (pelargonidin-3-glucoside)			$k_{110°C} = 0.029$	94.4	122
Betalain	Beet juice (pH 5.2)	$D_{121.1°C} = 6.1$	33.4	$k_{100°C} = 0.098$	77.7	123
β-Carotene	Citrus juice	$D_{90°C} = 434.3$	22.5	$k_{90°C} = 0.005$	110	124
	Carrot puree (isomerization)			$k_{110°C} = 0.035$	10.5	125
β-Cryptoxanthin	Citrus juice	$D_{90°C} = 532.3$	15.9	$k_{90°C} = 0.004$	156	124
Color parameters						
A	Peas				76.1	115
	Broccoli juice			$k_{100°C} = 0.055$	69.0	126
	Peach puree			$k_{122.5°C} = 0.03$	106	113
	Plum puree				27.8	127
	Spinach leaves				28.8	128
	Mustard leaves				41.6	128
	Pumpkin			$k_{85°C} = 0.12$	117.9	104
a/b	Asparagus		41.7		75.6	129

continued

Table 16.4 (continued) Kinetic Parameter Values for Thermal Degradation of Food Color

Attribute	Medium	D_T (min)	z (°C)	k_T (min^{-1})	E_a (kJ/mol)	Ref.
	Green beans		38.9		82.8	129
	Green peas		39.4		63.5	129
	Green peas	$D_{121°C} = 13.2$	38.3		73.2	103
	Concentrated tomato paste				28.7	130
	Broccoli			$k_{80°C} = 0.005$	28.8	131
L	Apple pulp				66.3	132
	Peach pulp				45.0	132
	Plum pulp				67.7	132
	Pumpkin			$k_{85°C} = 0.12$	120.3	104
La/b	Pea puree	$D_{121°C} = 31.1$	42.9	/	67.9	133
	Onion puree				27.6	134
Lab	Plum puree				30.7	127
B	Pumpkin			$k_{85°C} = 0.15$	111.0	104
C	Pumpkin			$k_{85°C} = 0.14$	112.4	104
TCD	Concentrated tomato paste				42.6	130
	Apple pulp				28.5	132
	Peach pulp				39.6	132
	Plum pulp				36.0	132
	Peach puree			$k_{122.5°C} = 0.009$	119	113
	Pumpkin			$k_{85°C} = 0.14$	98.8	104

TCD: total color difference.

(e.g., 147,148). Flavor formation, however, can often be described by "pseudo" zero-order kinetics [149–151].

16.4.6 Nutrient Degradation

Thermal processing of foods has been demonstrated to have, in general, a negative effect on the stability of vitamins. Vitamins investigated in terms of thermostability include ascorbic acid (vitamin C), thiamin (vitamin B1), riboflavin (vitamin B2), pyridoxine (vitamin B6), folic acid and its derivatives, pantothenic acid, and the fat-soluble vitamin A, although the majority of the work has been focusing on thiamin. Thermostability data of these vitamins have been reviewed [90,91]. Some representative data on thermal degradation of nutrients are reported in Table 16.5.

In most cases, thermal destruction of ascorbic acid in model and real food systems has been described as a first-order reaction, both under aerobic and anaerobic conditions [31,103,124,156,158,159,160,169–171]. Also the kinetics of thermal thiamin degradation

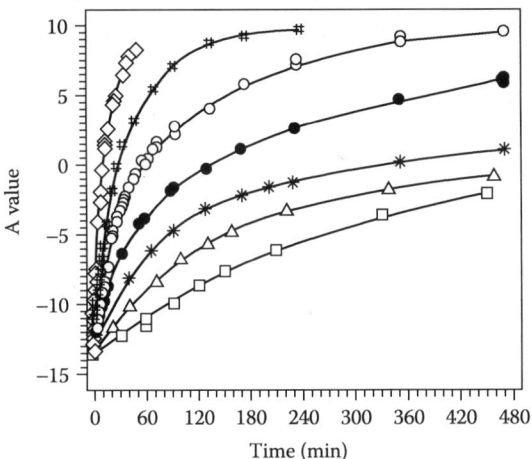

Figure 16.5 Degradation of the green color in broccoli juice, modeled using a fractional conversion model at 60°C (□), 70°C (Δ), 80°C (*), 90°C (·), 100°C (O), 110°C (#), and 120°C (◊). (From CA Weemaes et al. *J Agric Food Chem* 47(6):2404–2409, 1999.)

have been described as a first-order reaction by several authors [20,26,152,153,159,172–174]. Thermal degradation of pantothenic acid in model systems and in real food products followed first-order reaction kinetics [175,176]. More recently, research has focused on the process stability of folic acid and its derivatives due to its specific role in the prevention of neural tube defects. First-order kinetics have been suggested for the thermal destruction of folic acid, 5-methyltetrahydrofolic acid, and 5-formyltetrahydrofolic acid in the presence of oxygen [162–166, Figure 16.6], but second-order kinetics in case of limited oxygen supply [177,178].

16.5 FROM SINGLE-RESPONSE TOWARD MULTI-RESPONSE MODELING

Depending on the number of responses one takes into account for modeling a food reaction, a distinction can be made between (i) single-response modeling and (ii) multi-response modeling.

Single-response models require limited input information as they are based on determination of only one single response (e.g., nutrient concentration, enzyme activity, a single quality attribute). They are in principle straightforward in their formulation, and are mostly empirical in nature. Single-response models represent a useful pragmatic approach in modeling food processing unit operations, but care has to be taken with regard to the transferability of the model and its parameters to reaction conditions other than those for which the models were derived (due to the complex reaction networks and the influence of the reaction environment inherent in the empirical parameters or apparent reaction orders, rate constants, and activation energies). Based on properly designed experiments, they can be used under static as well as dynamic temperature conditions.

Table 16.5 Kinetic Parameter Values for Thermal Degradation of Nutrients

Attribute	Medium	D_T (min)	z (°C)	k_T (min^{-1})	E_a (kJ/mol)	Ref.
Thiamine	Phosphate buffer	$D_{121.1°C} = 156.8$	25		122.9	152
	Water	$D_{121.1°C} = 350$	26.4	$k_{121°C} = 0.006$	103–118	20
	Pea puree	$D_{121.1°C} = 278$	22	$k_{121°C} = 0.008$	116.2	152
	Beef puree	$D_{121.1°C} = 257$	23	$k_{121°C} = 0.009$	115.0	152
	Restructured beef			$k_{121°C} = 0.019$	110.8	153
	Soymilk	$D_{120°C} = 178$	30	$k_{120°C} = 0.013$	97	154
Ascorbic acid	Canned peas	$D_{121.1°C} = 1003$	16	$k_{121°C} = 0.002$	164.3	155
	Grapefruit juice (diff. Brix)	$D_{121.1°C} = 82–577$	54–124	$k_{121°C} = 0.004–0.028$	20.8–47.2	156
	Squeezed tomatoes	$D_{120°C} = 325.3$	27.7	$k_{120°C} = 0.007$	115.0	157
	Squeezed oranges	$D_{121°C} = 302.9$	27.2	$k_{120°C} = 0.008$	117.5	157
	Green asparagus - aerobic			$k_{125°C} = 2.82$	51.4	31
	Green asparagus - anaerobic			$k_{125°C} = 0.246$	25.5	
	Cupuacu nectar			$k_{80°C} = 0.032$	74.0	158
	Rose hip pulp	$D_{90°C} = 192$	53	$k_{90°C} = 0.012$	47.5	159
	Citrus juice	$D_{80°C} = 1222$	64	$k_{80°C} = 0.002$	36	124
	Orange juice	$D_{75°C} = 27.0$	24.4	$k_{75°C} = 0.085$	39.8	160
Vitamin A	Beef liver puree	$D_{121.1°C} = 41.7$	23	$k_{121°C} = 0.055$	112.4	161
5MTHF	Citric acid buffer (pH 4)			$k_{100°C} = 0.192$	71.1	162
	Citric acid buffer (pH 6)			$k_{100°C} = 0.104$	82.8	162
	Water (pH 7)			$k_{78°C} = 0.015$	39.7	163
	Phosphate buffer (pH 7)			$k_{70°C} = 0.013$	80.0	164
	Phosphate buffer (pH 7)			$k_{60°C} = 0.047$	89.9	165
	Apple juice			$k_{70°C} = 0.25$	32.8	162
	Tomato juice			$k_{100°C} = 0.37$	44.3	162
5FTHF	Acetate buffer (pH 5)			$k_{90°C} = 0.003$	62.8	166
PGA	Phosphate buffer (pH 7)			$k_{160°C} = 0.0047$	51.7	164
	Citric acid buffer (pH 3)			$k_{100°C} = 0.001$		167
	Apple juice (pH 3.4)	$D_{121.1°C} = 524$	31	$k_{121°C} = 0.005$	83.6	168
	Tomato juice (pH 4.3)	$D_{121.1°C} = 636$	31	$k_{121°C} = 0.003$	82.3	168

5MTHF: 5 methyltetrahydrofolic acid.
5FTHF: 5 formyltetrahydrofolic acid.
PGA: Pteroylglutamic acid.

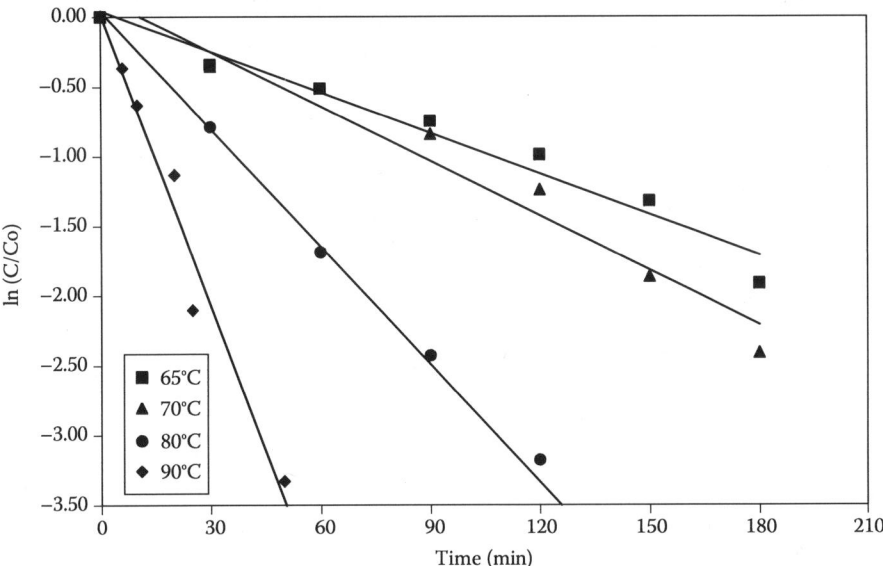

Figure 16.6 First-order thermal degradation of 5-methyltetrahydrofolic acid (10 μg/mL) in phosphate buffer (0.2 M, pH 7) at temperatures from 65°C to 90°C. (From MT Nguyen, Indrawati, M Hendrickx. *J Agric Food Chem* 51:3352–3357, 2003.)

However, reactions in foods often do not occur isolated but rather within a complex reaction network, in which several reactants, intermediates, and end products take part. Measuring multiple responses of a reaction system at the same time and using all the available information simultaneously when modeling experimental data, is much more powerful than solely considering a single response [179]. Such multiresponse approach requires the proposition of a reaction pathway. This reaction pathway is translated into a mathematical model by setting up several differential equations. Afterwards, the mathematical model which is proposed based on the postulated reaction mechanism, is fitted to the experimental data using an appropriate fit criterion. If the fit proves inadequate, the postulated model should be adjusted and the fitting procedure should be repeated until an acceptable model is found.

Multiresponse models rely on more elaborated mechanistic insight, are based on completer reaction networks and require responses of different precursors, reaction intermediates, and end products. Such models require more extensive experiments (as compared to single-response models) and are well suited to help understand the mechanism involved as they allow more hypothesis-driven model validation. As for the single-response models, they can be used both under static and dynamic reaction conditions (in terms of temperature and reaction environment), but elaborated studies are required to build such models. Advantages of multiresponse over single-response modeling are: (i) kinetic models can be tested much more rigorously because all the information contained in the data is linked and used, (ii) better results in terms of precision of parameters, (iii) useful to gain insight into the actual reaction network as hypothesis-driven reaction networks can be evaluated [179,180].

Several examples of multiresponse modeling of reactions in food during thermal processing are described in literature (e.g., acrylamide formation and elimination in model systems [180,181,182]; vitamin C degradation in red fruit [183]; carotene degradation and isomerization in carrots and tomatoes [184]).

REFERENCES

1. G Maesmans, M Hendrickx, Z Weng, A Keteleer, P Tobback. Endpoint definition, determination and evaluation of thermal processes in food preservation. *Belgian J Food Chem Biotechnol* 45(5):179–192, 1990.
2. A Van Loey, M Hendrickx, S De Cordt, T Haentjens, P Tobback. Quantitative evaluation of thermal processes using time temperature integrators. *Trends Food Sci Technol* 7(1):16–26, 1996.
3. CG Hill, RA Grieger-Block. Kinetic data: Generation, interpretation, and use. *Food Technol* 34(2): 56–65, 1980.
4. KJ Laidler. *Chemical Kinetics*. New York: Harper & Row, 1987.
5. MAJS van Boekel, P Walstra. Use of kinetics in studying heat-induced changes in foods. In: *Monograph on Heat-Induced Changes in Milk* (ed. PF Fox) *Bulletin of the International Dairy Federation*, Brussels, 2nd edition, IDF special issue 9501, 22–50, 1995.
6. A Arabshahi, DB Lund. Considerations in calculating kinetic parameters from experimental data. *J Food Process Eng* 7: 239–251, 1985.
7. MK Lenz, DB Lund. Experimental procedures for determining destruction kinetics of food components. *Food Technol* 34(2): 51–55, 1980.
8. M Straume, ML Johnson. Analysis of residuals: Criteria for determining goodness-of-fit. *Methods Enzymol* 210:87–105, 1992.
9. A Van Loey, Indrawati, C Smout, M Hendrickx. Inactivation of enzymes: From experimental design to kinetic modelling. In: *Handbook for Food Enzymology* (eds. J Whitaker, F Voragen, D Wong). Marcel Dekker Inc., New York, 49–58, 2003.
10. YT Huang, MC Bourne. Kinetics of thermal softening of vegetables. *J Texture Stud* 14(1):1–9, 1983.
11. S Arrhenius. Über die Reaktionsgeschwindigkeit bei der Inversion von Rohrzucker durch Säuren. *Zeitschrift für Physikalische Chemie* 4(2):226–248, 1889.
12. WD Bigelow. The logarithmic nature of thermal death time curves. *J Infect Dis* 29(5):528–536, 1921.
13. M Hendrickx, G Maesmans, S De Cordt, J Noronha, A Van Loey, P Tobback. Evaluation of the integrated time temperature effect in thermal processing of foods. *CRC Crit Rev Food Sci Nutr* 35(3):231–262, 1995.
14. DB Lund. Heat processing. In *Principles of Food Science, Part II, Physical Principles of Food Preservation* (eds. M Karel, OR Fennema, DB Lund). Marcel Dekker, New York, 1975.
15. HS Ramaswamy, FR Van de Voort, S Ghazala. An analysis of TDT and Arrhenius methods for handling process and kinetic data. *J Food Sci* 54(5):1322–1326, 1989.
16. U Jonsson, B Snygg, B Harnulv, T Zachrisson. Testing two models for the temperature dependence of the heat inactivation rate of *Bacillus stearothermophilus* spores. *J Food Sci* 42(5):1251–1252, 1263, 1977.
17. AC Cleland, GL Robertson. Determination of thermal processes to ensure commercial sterility of food cans. In *Developments in Food Preservation* (ed. Thorne, S.), Elsevier Applied Science, London, 1985.
18. B Manji, FR van de Voort. Comparison of two models for process holding time calculations: Convection systems. *J. Food Protect* 48(4):359–363, 1985.
19. JRD David, RL Merson. Kinetic parameters for inactivation of *Bacillus stearothermophilus* at high temperatures. *J Food Sci* 55(2):488–493, 1990.

20. HS Ramaswamy, S Ghazala, FR Van de Voort. Degradation kinetics of thiamine in aqueous systems at high temperatures. *Can Inst Food Sci Technol J* 23(2/3):125–130, 1990.
21. AG Perkin, H Burton, HM Underwood, FL Davies. Thermal death kinetics of *Bacillus stearothermophilus* spores at ultra high temperatures. II. Effect of heating period on experimental results. *J Food Technol* 12:131–148, 1977.
22. K Hayakawa, PG Schnell, DH Kleyn. Estimating thermal death time characteristics of thermally vulnerable factors by programmed heating of sample solution or suspension. *Food Technol* 23:104–108, 1969.
23. MK Lenz, DB Lund. The lethality-Fourier number method. Experimental verification of a model for calculating average quality factor retention in conduction-heating canned foods. *J Food Sci* 42:997–1001, 1977.
24. JW Rhim, RV Nunes, VA Jones, KR Swartzel. Determination of kinetic parameters using linearly increasing temperature. *J Food Sci* 54(2):446–450, 1989.
25. H Nasri, P Simpson, J Bouzas, JA Torres. An unsteady-state method to determine kinetic parameters for heat inactivation of quality factors: Conduction-heated foods. *J Food Eng* 19(3):291–301, 1993.
26. JR Banga, AA Alonso, JM Gallardo, RI Perez-Martin. Kinetics of thermal degradation of thiamine and surface colour in canned tuna. *Zeitschrift fuer Lebensmittel Untersuchung und Forschung* 197(2):127–131, 1993.
27. C Rodrigo, A Mateu, A Alvarruiz, F Chinesta, M Rodrigo. Kinetic parameters for thermal degradation of green asparagus texture by unsteady-state method. *J Food Sci* 63(1):126–129, 1998.
28. RL Garrote, ER Silva, RA Bertone. Kinetic parameters for thermal inactivation of cut green beans lipoxygenase calculated using unsteady-state methods. *Int J Food Sci Technol*, 36(4):377–385, 2001.
29. S Condon, MJ Arrizubieta, FJ Sala. Microbial heat resistance determinations by the multipoint system with the thermoresistometer TR-SC: Improvement of this methodology. *J Microbiol Methods* 18(4):357–366, 1993.
30. C Rodrigo, A Alvarruiz, A Martinez, A Frigola, M Rodrigo. High-temperature short-time inactivation of peroxidase by direct heating with a five-channel computer-controlled thermoresistometer. *J Food Protect* 60(8):967–972, 1997.
31. MJ Esteve, A Frigola, L Martorell, C Rodrigo. Kinetics of green asparagus ascorbic acid heated in a high-temperature resistometer. *Zeitschrift fuer Lebensmittel Untersuchung und Forschung* 208(2):144–147, 1999.
32. U Viberg, R Oeste. Development and evaluation of microscale apparatus for the generation of kinetic data at high temperatures, applied to the degradation of thiamine. *Food Chem* 52(1):29–33, 1995.
33. RRB Singh, GR Patil. Kinetics of whitening of milk during UHT processing. *Milchwissenschaft* 45(6):367–369, 1990.
34. DJ Oldfield, S Harjinder, MW Taylor, KN Pearce. Kinetics of denaturation and aggregation of whey proteins in skim milk heated in an ultra-high temperature (UHT) pilot plant. *Int Dairy J* 8(4):311–318, 1998.
35. JF Norwig, DR Thompson. Microbial population, enzyme and protein changes during processing. In: *Physical and Chemical Properties of Food*, American Society of Agricultural Engineers, Michigan, pp. 202–265, 1986.
36. GD Betts. The microbiological safety of sous-vide processing. Technical manual n° 39; Campden & Chorleywood Food Research Association, Chipping Campden, UK, 1992.
37. GD Betts, JE Gaze. Food pasteurization treatments. Technical manual n° 27; Campden & Chorleywood Food Research Association, Chipping Campden, UK, 1992.
38. TA Roberts, AC Baird-Parker, RB Tompkin. Micro-organisms in foods—Microbiological specifications of food pathogens. Blackie Academic & Professional, London, p. 513, 1996.

39. IJ Pflug. Using the straight-line semilogarithmic microbial destruction model as an engineering design model for determining the F-value for heat processes. *J Food Protect* 50(4):342, 1987.
40. A Casolari. Microbial death. In: *Physiological Models in Microbiology 2* (eds. MJ Bazin, JI Prosser). CRC Press, Boca Raton, FL, pp. 1–44, 1988.
41. G LeJean, G Abraham, E Debray, Y Candau, G Piar. Kinetics of thermal destruction of *Bac. stearothermophilus* spores using a two reactions model. *Food Microbiol* 11:229–237, 1994.
42. RC Whiting. Microbial modeling of foods. *Crit Rev Food Sci Nutr* 35:467–494, 1995.
43. WF Anderson, PJ McClure, AC Baird-Parker, MB Cole. The application of a log-logistic model to describe the thermal inactivation of *Clostridium botulinum* 213B at temperatures below 121.1°C. *J Appl Bacteriol* 80:283–290, 1996.
44. I Körmendy, L Körmendy. Considerations for calculating heat inactivation processes when semilogarithmic thermal inactivation models are non-linear. *J Food Eng* 34:33–40, 1997.
45. JC Augustine, V Carlier, J Rozjier. Mathematical modelling of the heat resistance of *L. monocytogenes*. *J Appl Microbiol* 84:185–191, 1998.
46. M Peleg, MB Cole. Reinterpretation of microbial survivor curves. *Critical Reviews in Food Science and Nutrition* 38(5):353–380, 1998.
47. M Peleg. Microbial survivor curves—the reality of flat shoulders and absolute thermal death times. *Food Res Int* 33:531–538, 2000.
48. OH Campanella, M Peleg. Theoretical comparison of a new and the traditional method to calculate *Clostridium botulinum* survival during thermal inactivation. *J Sci Food Agric* 81:1069–1076, 2001.
49. AH Geeraerd, CH Herremans, JF Van Impe. Structural model requirements to describe microbial inactivation during a mild heat treatment. *Int J Food Microbiol* 59:185–209, 2000.
50. MA Casadei, K Jewell. Modelling of microbial non-linear thermal inactivation kinetics: A review. *Campden & Chorleywood Food Research Association Review* n° 26, 2001.
51. CR Loss, JH Hotchkiss. Effects of dissolved carbon dioxide on thermal inactivation of microorganisms in milk. *J Food Protect* 65(12):1924–1929, 2002.
52. A Fernandez, C Salmeron, PS Fernandez, A Martinez. Application of a frequency distribution model to describe the thermal inactivation of two strains of *Bacillus cereus*. *Trends Food Sci Technol* 10(4/5):158–162, 1999.
53. RH Linton, WH Carter, MD Pierson, CR Hackney. Use of a modified Gompertz equation to model nonlinear survival curves for *Listeria monocytogenes* Scott A. *J Food Protect* 58(9):946–954, 1995.
54. AT Chhabra, WH Carter, RH Linton, MA Cousin. A predictive model to determine the effects of pH, milk fat, and temperature on thermal inactivation of *Listeria monocytogenes*. *J Food Protect* 62(10):1143–1149, 1999.
55. R Xiong, G Xie, AE Edmondson, RH Linton, MA Sheard. Comparison of the Baranyi model with the modified Gomperts equation for modelling thermal inactivation of *Listeria monocytogenes* Scott A. *Food Microbiol* 16(3):269–279, 1999.
56. J Baranyi, A Jones, C Walker, A Kaloti, TP Robinson, BM Mackey. A combined model for growth and subsequent thermal inactivation of *Brochothrix thermosphacta*. *Appl Environ Microbiol* 62:1029–1035, 1996.
57. MB Cole, KW Davies, G Munro, CD Holyoak, DC Kilsby. A vitalistic model to describe the thermal inactivation of *Listeria monocytogenes*. *J Indus Microbiol* 12:232–239, 1993.
58. V Sapru, AA Teixeira, GH Smerage, JA Lindsay. Predicting thermophilic spore population dynamics for UHT sterilization processes. *J Food Sci* 575:1248–1252, 1992.
59. V Sapru, GH Smerage, AA Teixeira, JA Lindsay. Comparison of predictive models for bacterial spore population resources to sterilization temperature. *J Food Sci* 58(1):223–228, 1993.
60. JB Adams. Review: Enzyme inactivation during heat processing of food-stuffs. *Int J Food Sci Technol* 26(1):1–20, 1991.
61. L Ludikhuyze, A Van Loey, Indrawati, C Smout, M Hendrickx. Effects of combined pressure and temperature on enzymes related to quality of fruits and vegetables: From kinetic information to process engineering aspects. *CRC Critical Reviews in Food Science and Nutrition* 43(5):527–586, 2003.

62. WL Claeys, LR Ludikhuyze, AM Van Loey, ME Hendrickx. Inactivation kinetics of alkaline phosphatase and lactoperoxidase, and denaturation kinetics of β-lactoglobulin in raw milk under isothermal and dynamic temperature conditions. *J Dairy Res* 68:95–107, 2001.
63. G Murthy, J Bradshaw, J Peeler. Thermal inactivation of phosphatase by the AOAC-V method. *J Food Prot* 53:969–971, 1990.
64. S Fadiloglu, O Erkmen, G Sekeroglu. Thermal inactivation kinetics of alkaline phosphatase in buffer and milk. *J Food Process Pres* 30:258–268, 2006.
65. S Marchand, M Merchiers, W Messens, K Coudijzer, J De Block. Thermal inactivation kinetics of alkaline phosphatase in equine milk. *Int Dairy J* 19:763–767, 2009.
66. L Dumitrascu, N Stanciuc, S Stanciu, G Râpeanu. Thermal inactivation of lactoperoxidase in goat, sheep and bovine milk—A comparative kinetic and thermodynamic study. *J Food Eng* 113:47–52, 2012.
67. L Ludikhuyze, Indrawati, I Van den Broeck, C Weemaes, M Hendrickx. Effect of combined pressure and temperature on soybean lipoxygenase. 1. Influence of extrinsic and intrinsic factors on isobaric-isothermal inactivation kinetics. *J Agric Food Chem* 46(10):4047–4080, 1998.
68. Indrawati, AM Van Loey, LR Ludikhuyze, ME Hendrickx. Pressure–temperature inactivation of lipoxygenase in green peas (*Pisum sativum*): A kinetic study. *J Food Sci* 66(5):686–693, 2001.
69. SG Svensson, CE Eriksson. Thermal inactivation of lipoxygenase from peas (*Pisum sativum* L.). I. Time–temperature relationships and pH-dependence. *Lebensm Wiss u Technol* 5:118–123, 1972.
70. HM Henderson, G Blank, H Sustackova. Thermal inactivation of pea flour lipoxygenase. *J Food Biochem* 15:107–115, 1991.
71. EF Morales-Blancas, VE Chandia, L Cisneros-Zevallos. Thermal inactivation kinetics of peroxidase and lipoxygenase from broccoli, green asparagus and carrots. *J Food Sci* 67(1):146–154, 2002.
72. I Van den Broeck, LR Ludikhuyze, CA Weemaes, AM Van Loey, ME Hendrickx. Thermal inactivation kinetics of pectinesterase extracted from oranges. *J Food Proc Preserv* 23:391–406, 1999.
73. B Ly Nguyen, A Van Loey, D Fachin, I Verlent, T Duvetter, TS Vu, C Smout, M Hendrickx. Strawberry pectin methylesterase: Purification, characterisation, thermal and high-pressure inactivation. *Biotechnol Progr* 18:1447–1450, 2002.
74. B Ly Nguyen, A Van Loey, D Fachin, I Verlent, M Hendrickx. Purification, characterization, thermal, and high-pressure inactivation of pectin methylesterase from bananas (cv Cavendish). *Biotechnol Bioeng* 78(6):683–691, 2002.
75. D Fachin, A Van Loey, B Ly-Nguyen, I Verlent, Indrawati, M Hendrickx. Comparative study of the inactivation kinetics of pectinmethylesterase in tomato juice and purified form. *Biotechnol Progress* 18(4):739–744, 2002.
76. Ç Soysal, Z Söylemez. Kinetics and inactivation of carrot peroxidase by heat treatment. *J Food Eng* 68:349–356, 2005.
77. RMS Cruz, MC Vieira, CLM Silva. Effect of heat and thermosonication treatments on peroxidase inactivation kinetics in watercress (*Nasturtium officinale*). *J Food Eng* 72:8–15, 2006.
78. GE Anthon, DM Barrett. Kinetic parameters for the thermal inactivation of quality-related enzymes in carrots and potatoes. *J Agric Food Chem* 50:4119–4125, 2002.
79. D Fachin, C Smout, I Verlent, B Ly Nguyen, AM Van Loey, ME Hendrickx. Inactivation kinetics of purified tomato polygalacturonase by thermal and high-pressure processing. *J Agric Food Chem* 52:2697–2703, 2004.
80. CA Weemaes, LR Ludikhuyze, I Van den Broeck, ME Hendrickx. Kinetics of combined pressure-temperature inactivation of avocado polyphenoloxidase. *Biotechnol Bioeng* 60(3):292–300, 1998.
81. C Weemaes, P Rubens, S De Cordt, L Ludikhuyze, I Van den Broeck, M Hendrickx, K Heremans, P Tobback. Temperature sensitivity and pressure resistance of mushroom polyphenoloxidase. *J Food Sci* 62:261–266, 1997.

82. A Yemenicioglu, M Özkan, B Cemeroglu. Heat inactivation kinetics of apple polyphenoloxidase and activation of its latent form. *J Food Sci* 62:508–510, 1997.
83. G Rapeanu, A Van Loey, C Smout, M Hendrickx. Biochemical characterization and process stability of polyphenoloxidase extracted from Victoria grape (*Vitis vinifera* spp *sativa*). *Food Chem* 94:253–261, 2006.
84. CA Weemaes, L Ludikhuyze, I Van den Broeck, M Hendrickx. Effect of pH on pressure and thermal inactivation of avocado polyphenol oxidase. *J Agric Food Chem* 46(7):2785–2792, 1998.
85. Indrawati, AM Van Loey, LR Ludikhuyze, ME Hendrickx. Single, combined, or sequential action of pressure and temperature on lipoxygenase in green beans (*Phaseolus vulgaris* L.): A kinetic inactivation study. *Biotechnol Progr* 15:273–277, 1999.
86. D Fachin, A Van Loey, Indrawati, L Ludikhuyze, M Hendrickx. Thermal and high-pressure inactivation of tomato polygalacturonase: A kinetic study. *J Food Sci* 67(5):1610–1615, 2002.
87. I Van den Broeck, LR Ludikhuyze, AM Van Loey, CA Weemaes, ME Hendrickx. Thermal and combined pressure–temperature inactivation of orange pectinesterase: Influence of pH and additives. *J Agric Food Chem* 47(7):2950–2958, 1999.
88. B Ly-Nguyen, A Van Loey, D Fachin, I Verlent, Indrawati, M Hendrickx. Partial purification, characterization, and thermal and high-pressure inactivation of pectin methylesterase from carrots (*Daucus carota* L.). *J Agric Food Chem* 50(19):5437–5444, 2002.
89. MA Rao, DB Lund. Kinetics of thermal softening of foods—A review. *J Food Proc Preserv* 10(4):311–329, 1986.
90. R Villota, JG Hawkes. Kinetics of nutrients and organoleptic changes in foods during processing. In: *Physical and Chemical Properties of Food*, American Society of Agricultural Engineers, Michigan, pp. 266–366, 1986.
91. SD Holdsworth. Kinetic data—What is available and what is necessary. In: Field, R.W., Howell, J.A. *Processing and Quality of Foods. 1 HTST Processing*, Elsevier, London, 74–78, 1990.
92. MC Bourne. *Food Texture and Viscosity*. Academic Press, New York, 1982.
93. P Varoquaux, M Souty, F Varoquaux. Water blanching of whole apricots. *Sciences des Aliments* 6(4):591–600, 1986.
94. MH Lau, J Tang, BG Swanson. Kinetics of textural and color changes in green asparagus during thermal treatments. *J Food Eng* 45(4):231–236, 2000.
95. C Rodrigo, M Rodrigo, SM Fiszman. The impact of high temperature short time thermal treatment on texture and weight loss of green asparagus. *Food Res Technol* 205(1):53–58, 1997.
96. C Rodrigo, M Rodrigo, S Fiszman, T Sanchez. Thermal degradation of green asparagus texture. *J Food Protect* 60(3):315–320, 1997.
97. AR Taherian, HS Ramaswamy. Kinetic considerations of texture softening in heat treated root vegetables. *Int J Food Prop* 12(1):114–128, 2009.
98. K Paulus, I Saguy. Effect of heat treatment on the quality of cooked carrots. *J Food Sci* 45(2):239–241, 245, 1980.
99. TS Vu, C Smout, DN Sila, B Ly Nguyen, AML Van Loey, MEG Hendrickx. Effect of preheating on thermal degradation kinetics of carrot texture. *Innovative Food Sci Technol* 5:37–44, 2003.
100. A De Roeck, J Mols, T Duvetter, A Van Loey, M Hendrickx. Carrot texture degradation kinetics and pectin changes during thermal versus high-pressure/high-temperature processing. A comparative study. *Food Chem* 120:1104–1112, 2010.
101. DN Sila, X Yue, S Van Buggenhout, C Smout, A Van Loey, M Hendrickx. The relation between (bio-)chemical, morphological, and mechanical properties of thermally processed carrots as influenced by high-pressure pretreatment condition. *Eur Food Res Technol* 226:127–135, 2007.
102. L Rejano, AH Sánchez, A Montano, FJ Casado, A de Castro. Kinetics of heat penetration and textural changes in garlic during blanching. *J Food Eng* 78:465–471, 2007.
103. MA Rao, CY Lee, J Katz, HJ Cooley. A kinetic study of the loss of vitamin C, color, and firmness during thermal processing of canned peas. *J Food Sci* 46(2):636–637, 1981.

104. EM Gonçalves, J Pinheiro, M Abreu, TRS Brandao, CLM Silva. Modelling the kinetics of peroxidase inactivation, colour and texture changes of pumpkin (*Cucurbita maxima* L.) during blanching. *J Food Eng* 81:693–701, 2007.
105. MA Anzaldua, A Quintero, R Balandran. Kinetics of thermal softening of six legumes during cooking. *J Food Sci* 61(1):167–170, 1996.
106. A Van Loey, A Fransis, M Hendrickx, G Maesmans, P Tobback. Kinetics of thermal softening of white beans evaluated by a sensory panel and the FMC tenderometer. *J Food Proc Preserv* 18:407–419, 1994.
107. ES Lazos, DC Servos, D Parliaros. Kinetics of texture degradation in apples during thermal processing. *Chimika Chronika* 25(1):11–27, 1996.
108. MD Alvarez, W Canet. Kinetics of thermal softening of potato tissue heated by different methods. *Eur Food Res Technol* 212(4):454–464, 2001.
109. AF Rizvi, CH Tong. Fractional conversion for determining texture degradation kinetics of vegetables. *J Food Sci* 62(1):1–7, 1997.
110. IMLB Avila, C Smout, CLM Silva, M Hendrickx. Development of a novel methodology to validate optimal sterilization conditions for maximizing the texture quality of white beans in glass jars. *Biotechnol Progr* 15(3):565–572, 1999.
111. TR Stoneham, DB Lund, CH Tong. The use of fractional conversion technique to investigate the effects of testing parameters on texture degradation kinetics. *J Food Sci* 65(6):968–973, 2000.
112. CLM Silva, P Ignatiadis. Modelling food colour degradation kinetics—A review. *Proceedings of the first main meeting of the concerted action CIPA* (CT94-0195, pp. 76–81), 1995.
113. IMLB Avila, CLM Silva. Modelling kinetics of thermal degradation of colour in peach puree. *J Food Eng* 39(2):161–166, 1999.
114. SJ Schwartz, JH von Elbe. Kinetics of chlorophyll degradation to pyropheophytin in vegetables. *J Food Sci* 48(4):1303–1306, 1983.
115. JA Steet, CH Tong. Degradation kinetics of green color and chlorophylls in peas by colorimetry and HPLC. *J Food Sci* 61(5):924–927, 1996.
116. N Koca, F Karadeniz, HS Burdurlu. Effect of pH on chlorophyll degradation and colour loss in blanched green peas. *Food Chem* 100:609–615, 2006.
117. A Van Loey, V Ooms, C Weemaes, I Van den Broeck, L Ludikhuyze, Indrawati, S Denys, M Hendrickx. Thermal and pressure-temperature degradation of chlorophyll in broccoli (*Brassica oleracea* L. *italica*) juice: A kinetic study. *J Agric Food Chem* 46(12):5289–5294, 1998.
118. SG Rudra, BC Sarkar, US Shivhare. Thermal degradation kinetics of chlorophyll in pureed coriander leaves. *Food Bioprocess Technol* 1:91–99, 2008.
119. SS Tanchev. Kinetics of the thermal degradation of anthocyanins of the raspberry. *Z Lebensm Unter Forsch* 150(1):28–30, 1972.
120. N Ioncheva, S Tanchev. Kinetics of thermal degradation of some anthocyanidin-3,5-diglucosides. *Z Lebensm Unter Forsch* 155(5):257–262, 1974.
121. L Verbeyst, K Van Crombruggen, I Van der Plancken, M Hendrickx, A Van Loey. Anthocyanin degradation kinetics during thermal and high pressure treatments of raspberries. *J Food Eng* 105:513–521, 2011.
122. L Verbeyst, I Oey, I Van der Plancken, M Hendrickx, A Van Loey. Kinetic study on the thermal and pressure degradation of anthocyanins in strawberries. *Food Chem* 123:269–274, 2010.
123. I Saguy. Thermostability of red beet pigments (betanine and vulgaxanthin-I): Influence of pH and temperature. *J Food Sci* 44(5):1554–1555, 1979.
124. C Dhuique-Mayer, M Tbatou, M Carail, C Caris-Veyrat, M Dornier, MJ Amiot. Thermal degradation of antioxidant micronutrients in *citrus* juice: Kinetics and newly formed compounds. *J Agric Food Chem* 55:4209–4216, 2007.
125. L Lemmens, K De Vleeschouwer, KRN Moelants, IJP Colle, AM Van Loey, ME Hendrickx. β-Carotene isomerization kinetics during thermal treatments of carrot puree. *J Agric Food Chem* 58:6816–6824, 2010.

126. CA Weemaes, V Ooms, AM Van Loey, M Hendrickx. Kinetics of chlorophyll degradation and color loss in heated broccoli juice. *J Agric Food Chem* 47(6):2404–2409, 1999.
127. J Ahmed, US Shivhare, GSV Raghavan. Thermal degradation kinetics of anthocyanin and visual colour of plum puree. *Eur Food Res Technol* 218:525–528, 2004.
128. J Ahmed, A Kaur, US Shivhare. Color degradation kinetics of spinach, mustard leaves and mixed puree. *J Food Sci* 67(3):1088–1091, 2002.
129. K Hayakawa, GE Timbers. Influence of heat treatment on the quality of vegetables: Changes in visual green colour. *J Food Sci* 42(3):778–781, 1977.
130. JA Barreiro, M Milano, AJ Sandoval. Kinetics of colour changes of double concentrated tomato paste during thermal treatment. *J Food Eng* 33(3/4):359–371, 1997.
131. EM Gonçalves, J Pinheiro, C Alegria, M Abreu, TRS Brandao, CLM Silva. Degradation kinetics of peroxidase enzyme, phenolic content, and physical and sensorial characteristics in broccoli (*Brassica oleracea* L. ssp. *Italica*) during blanching. *J Agric Food Chem* 57:5370–5375, 2009.
132. JE Lozano, A Ibarz. Colour changes in concentrated fruit pulp during heating at high temperatures. *J Food Eng* 31:365–373, 1997.
133. S Seonggyun, SR Bhowmik. Thermal kinetics of color changes in pea puree. *J Food Eng* 24(1):77–86, 1995.
134. J Ahmed, US Shivhare, GSV Raghavan. Color degradation kinetics and rheological characteristics of onion puree. *Trans ASAE* 44(1):95–98, 2001.
135. IMLB Avila, CLM Silva. Mathematical modelling of thermal degradation kinetics of peach puree total carotenoids and colour. IFT annual meeting, Book of abstracts (ISSN 1082–1236), p. 18, 1996.
136. FL Canjura, SJ Schwartz, RV Nunes. Degradation kinetics of chlorophylls and chlorophyllides. *J Food Sci* 56(6):1639–1643, 1991.
137. AH Sanchez, L Rejano, A Montano. Kinetics of the destruction by heat of colour and texture of pickled green olives. *J Sci Food Agric* 54(3):379–385, 1991.
138. A Van Loey, A Fransis, M Hendrickx, G Maesmans, P Tobback. Kinetics of quality changes of green peas and white beans during thermal processing. *J Food Eng* 24(3):361–377, 1995.
139. J Ahmed, US Shivhare, GSV Raghavan. Rheological characteristics and kinetics of colour degradation of green chilli puree. *J Food Eng* 44(4):239–244, 2000.
140. J Ahmed, US Shivhare. Thermal kinetics of color change, rheology and storage characteristics of garlic puree/paste. *J Food Sci* 66(5):754–757, 2001.
141. J Ahmed, US Shivhare, K Mandeep. Thermal colour degradation kinetics of mango puree. *Int J Food Properties* 5(2):359–366, 2002.
142. J Ahmed, US Shivhare, HS Ramaswamy. A fraction conversion kinetic model for thermal degradation of color in red chili puree and paste. *Lebensmittel Wissenschaft Technol* 35(6):497–503, 2002.
143. J Ahmed, US Shivhare, KS Sandhu. Thermal degradation kinetics of carotenoids and visual color of papaya puree. *J Food Sci* 67(7):2692–2695, 2002.
144. S Garza, A Ibarz, J Pagan, J Giner. Non-enzymatic browning in peach puree during heating. *Food Res Int* 32(5):335–343, 1999.
145. A Ibarz, J Pagan, S Garza. Kinetic models for colour changes in pear puree during heating at relatively high temperatures. *J Food Eng* 39(4):415–422, 1999.
146. LA Wilson. Kinetics of flavor changes in foods. In: *Physical and Chemical Properties of Food*, M. R. Okos (Ed.), American Society of Agricultural Engineers, Michigan, pp. 382–407, 1986.
147. T Matoba, M Kuchiba, M Kimura, K Hasegawa. Thermal degradation of flavor enhancers, inosine 5′-monophosphate and guanosine 5′-monophosphate in aqueous solution. *J Food Sci* 53(4):1156–1159, 1988.
148. FM Silva, C Sims, MO Balaban, CLM Silva, S O'Keefe. Kinetics of flavour and aroma changes in thermally processed cupuacu (*Theobroma grandiflorum*). *J Sci Food Agric* 80(6):783–787, 2000.
149. F Jousse, T Jongen, W Agterof, S Russell, P Braat. Simplified kinetic scheme of flavor formation by the Maillard reaction. *J Food Sci* 67(7):2534–2542, 2002.

150. W Chobpattana, IJ Jeon, J Scott-Smith. Kinetics of interaction of vanillin with amino acids and peptides in model systems. *J Agric Food Chem* 48(9):3885–3889, 2000.
151. DR Cremer, K Eichner. Formation of volatile compounds during heating spice paprika (*Capsicum annuum*) powder. *J Agric Food Chem* 48(6):2454–2460, 2000.
152. EA Mulley, CR Stumbo, WM Hunting. Kinetics of thiamin degradation by heat. Effect of pH and form of the vitamin on its rate of destruction. *J Food Sci* 40(5):989–992, 1975.
153. JA Steet, CH Tong. Thiamin degradation kinetics in pureed restructured beef. *J Food Proc Preserv* 18(3):253–262, 1994.
154. KC Kwok, YW Shiu, CH Yeung, K Niranjan. Effect of thermal processing on available lysine, thiamine and riboflavin content in soymilk. *J Sci Food Agric* 77:473–478, 1998.
155. PJ Lathrop, HK Leung. Rates of ascorbic acid degradation during thermal processing of canned peas. *J Food Sci* 45(1):152–153, 1980.
156. I Saguy, IJ Kopelman, S Mizrahi. Simulation of ascorbic acid stability during heat processing and concentration of grapefruit juices. *J Food Process Eng* 2(3):231–225, 1978.
157. I Van den Broeck, LR Ludikhuyze, CA Weemaes, AM Van Loey, ME Hendrickx. Kinetics of isobaric-isothermal degradation of L-ascorbic acid. *J Agric Food Chem* 46(5):2001–2006, 1998.
158. MC Vieira, AA Teixeira, CLM Silva. Mathematical modelling of the thermal degradation kinetics of vitamin C in cupuacu (*Theobroma grandiflorum*) nectar. *J Food Eng* 43(1):1–7, 2000.
159. M Karhan, M Aksu, N Tetik, I Turhan. Kinetic modeling of anaerobic thermal degradation of ascorbic acid in rose hip (*Rosa canina*) pulp. *J Food Quality* 27:311–319, 2004.
160. VB Vikram, MN Ramesh, SG Prapulla. Thermal degradation kinetics of nutrients in orange juice heated by electromagnetic and conventional methods. *J Food Eng* 69:31–40, 2005.
161. SA Wilkinson, MD Earle, AC Cleland. Kinetics of vitamin A degradation in beef liver puree on heat processing. *J Food Sci* 46(1):32–33, 1981.
162. AP Mnkeni, T Beveridge. Thermal destruction of 5-methyltetrahydrofolic acid in buffer and model food systems. *J Food Sci* 48(2):595–599, 1983.
163. TS Chen, RG Cooper. Thermal destruction of folacin: Effect of ascorbic acid, oxygen and temperature. *J Food Sci* 44(3):713–716, 1979.
164. MT Nguyen, Indrawati, M Hendrickx. Model studies on the stability of folic acid and 5-methyltetrahydrofolic acid degradation during thermal treatment in combination with high hydrostatic pressure. *J Agric Food Chem* 51:3352–3357, 2003.
165. I Oey, P Verlinde, M Hendrickx, A Van Loey. Temperature and pressure stability of L-ascorbic acid and/or [6S] 5-methyltetrahydrofolic acid: A kinetic study. *Eur Food Res Technol* 223:71–77, 2006.
166. MT Nguyen, I Oey, M Hendrickx, A Van Loey. Kinetics of (6R,S) 5-formyltetrahydrofolic acid isobaric-isothermal degradation in a model system. *Eur Food Res Technol* 223:325–331, 2006.
167. B Paine-Wilson, TS Chen. Thermal destruction of folacin: Effect of pH and buffer ions. *J Food Sci* 44(3): 717–722, 1979.
168. AP Mnkeni, T Beveridge. Thermal destruction of pteroylglutamic acid in buffer and model systems. *J Food Sci* 47(6):2038–2041, 1982.
169. NS Kincal, C Giray. Kinetics of ascorbic acid degradation in potato blanching. *Int J Food Sci Technol* 22(3):249–254, 1987.
170. CC Ariahu. Kinetics of heat/enzymic degradation of ascorbic acid in fluted pumpkin (*Telfairia occidentalis*) leaves. *J Food Proc Preserv* 21(1):21–32, 1997.
171. MJ Esteve, A Frigola, L Martorell, C Rodrigo. Kinetics of ascorbic acid degradation in green asparagus during heat processing. *J Food Protect* 61(11):1518–1521, 1998.
172. EA Mulley, CR Stumbo, WM Hunting. Thiamine: A chemical index of the sterilization efficacy of thermal processing. *J Food Sci* 40(5):993–996, 1975.
173. Suparno, AJ Rosenthal, SW Hanson. Kinetics of the thermal destruction of thiamin in the white flesh of rainbow trout (*Salmo gairdneri*). *J Sci Food Agric* 53(1):101–106, 1990.

174. CC Ariahu, AO Ogunsua. Thermal degradation kinetics of thiamine in periwinkle based formulated low acidity foods. *Int J Food Sci Technol* 35(3):315–321, 2000.
175. DJ Hamm, DB Lund. Kinetic parameters for thermal inactivation of pantothenic acid. *J Food Sci* 43(2):631–633, 1978.
176. YR Pyun, HJ Park, HY Cho, YY Lee. Kinetic studies on the thermal degradation of pantothenic acid. *Korean J Food Sci Technol* 13(3):188–193, 1981.
177. JE Ruddick, J Vanderstoep, JF Richards. Kinetics of thermal degradation of methyl tetrahydrofolic acid. *J Food Sci* 45(4):1019–1022, 1980.
178. BPF Day, JF Gregory. Thermal stability of folic acid and 5-methyltetrahydrofolic acid in liquid model food systems. *J Food Sci* 48(2):581–587, 1983.
179. MAJS van Boekel. Statistical aspects of kinetic modeling for food science problems. *J Food Sci*, 61(3):477–486, 1996.
180. K De Vleeschouwer, I Van der Plancken, A Van Loey, ME Hendrickx. Modelling acrylamide changes in foods: From single-response empirical to multiresponse mechanistic approaches. *Trends Food Sci Technol*, 20:155–167, 2009.
181. K De Vleeschouwer, I Van der Plancken, A Van Loey, ME Hendrickx. Investigation of the influence of different moisture levels on acrylamide formation/elimination reactions using multiresponse analysis. *J Agric Food Chem*, 56(15):6460–6470, 2008.
182. K De Vleeschouwer, I Van der Plancken, A Van Loey, ME Hendrickx. Role of precursors on the kinetics of acrylamide formation and elimination under low moisture conditions using a multiresponse approach—Part I: effect of the type of sugar. *Food Chem*, 114(1):116–126, 2009.
183. L Verbeyst, R Bogaerts, I Van der Plancken, ME Hendrickx, A Van Loey. Modelling of vitamin C degradation during thermal and high-pressure treatments of red fruit. *Food Bioprocess Technol*, 6:1015–1023, 2013.
184. IJP Colle, L Lemmens, G Knockaert, A Van Loey, ME Hendrickx. Carotene degradation and isomerization during thermal processing: A review on the kinetic aspects. *Crit Rev Food Sci Nutr*, Accepted.

17

Kinetics and Process Design for High-Pressure Processing

Tara Grauwet, Stijn Palmers, Liesbeth Vervoort, Ines Colle, Marc Hendrickx, and Ann Van Loey

Contents

17.1 Basic Principles of HPP	710
17.1.1 Le Châtelier Principle	710
17.1.2 Pascal Principle	711
17.1.3 Adiabatic Heat of Compression	711
17.2 High-Pressure Pasteurization versus High-Pressure Sterilization	711
17.3 Theoretical Considerations on the Analysis of Kinetics under HPP	713
17.3.1 Experimental Design for Kinetic Studies under Static Processing Conditions	713
17.3.2 Modeling Process-Induced Changes under Static Processing Conditions	715
17.3.2.1 Determination of the Reaction Rate Equation	716
17.3.2.2 Determination of a Temperature and/or Pressure Coefficient Model	717
17.3.2.3 Selection of a Regression Method and Fitting Criterion for Kinetic Parameter Estimation	719
17.3.2.4 Model Evaluation	720
17.4 Case Studies on Food Safety and Quality-Related Kinetics	720
17.4.1 Microbial Inactivation	720
17.4.2 Enzyme Inactivation	729
17.4.3 Quality Conversion	729
17.5 From Kinetics to Process Design	730
17.6 Conclusion and Future Perspectives	731
Acknowledgments	732
References	732

In general, the result of the cumulative effect of the product-treatment history is defined as the "impact of a process on a product." The impact of a process on a product can be determined at the end of the process (end-point measurements). However, this is a blackbox impact evaluation and no information is obtained in the origin of the impact. To obtain this information, the changes in the product characteristics need to be quantified as a function of the process parameters. In other words, there is a need for systematic kinetic studies on the evolution of important aspects during processing, such as microorganisms, enzymes, quality factors, and structural properties (Sizer et al., 2002; Heinz and Buckow, 2010).

In this chapter, the terminology "process design" refers to the characterization of the process-induced changes in the three-dimensional pressure, temperature, time frame, which are the three process variables to take into account in high pressure processing (HPP). To provide a background on HPP, the chapter begins with the description of the basic principles of HPP (see Section 17.1) and its major food applications in the context of shelf-life extension (see Section 17.2). Consequently, theoretical considerations on the analysis of kinetics under pressure, temperature conditions including aspects of experimental design and modeling are discussed. In a second part, literature data on kinetics of some target food safety and quality attributes are summarized. A lot has been done on studying high pressure (HP) process-induced changes in proteins (in particular enzymes), but also some examples of other process-induced changes are given.

Finally, process design is interpreted in the context of selecting, based on kinetic data, processing conditions that result in both the desired safety and quality level.

17.1 BASIC PRINCIPLES OF HPP

Three general principles underlie the effect of HPP on food attributes (e.g., safety and/or quality attributes) and are the basis for HPP as a diversely employable food-processing technique (see Section 17.2). As can be concluded from these three basic principles, pressure (p), time (t), and temperature (T) are the processing variables that can be important in HPP.

17.1.1 Le Châtelier Principle

Pressure effects are governed by Le Châtelier principle, which states that if a chemical system at equilibrium experiences a change in temperature, concentration, and/or volume, the equilibrium will shift to counteract the impact change and a new equilibrium will be established. Consequently, an increase in pressure directs the equilibrium toward reduction of the volume of a system and vice versa (Cheftel, 1995).

In case of a reaction with negative activation volume of the activated complex (i.e., volume of the activated complex is smaller than that of the initial reactants), a p-increase will increase the reaction rate constant and vice versa (Cheftel et al., 1992). Consequently, and in contrast to the effect of temperature on a reaction rate (at 0.1 MPa), pressure can, depending on the reaction under consideration, have an increasing or decreasing effect.

In other words, interactions among biomolecules (e.g., hydrogen bonds, electrostatic hydrophobic interactions) are modified in one way or another depending on the volume change related to these interactions. One would expect that temperature has an antagonistic effect on p-induced reactions because increasing temperature results in a volume increase due to disordering (microscopic principle). However, the reaction rate increases with increasing temperature according to Arrhenius' law, making this statement case dependent.

17.1.2 Pascal Principle

According to the Pascal principle, pressure applied during HPP is set quasiinstantaneously and uniformly throughout the HP vessel and product, whether in direct contact with the pressurizing medium or insulated in a flexible container (Delgado and Hartmann, 2003). Thus, the effect of pressure on a food is independent of product size and geometry (Knorr, 1993). If a food product contains sufficient moisture and not too much air, pressure will not damage the product at the macroscopic level as long as the pressure is applied uniformly in all directions (Balasubramaniam et al., 2008). Consequently, high moisture foods are the most obvious products to be studied.

17.1.3 Adiabatic Heat of Compression

The conversion of the work of compression into internal energy results in a temperature increase of compressible materials. Conversely, the temperature will drop during decompression. Under adiabatic conditions, this temperature change is termed "adiabatic heat of compression" and can be defined by (e.g., Zemansky, 1957):

$$\frac{dT}{dp} = \frac{T\alpha_{(T,p)}}{\rho_{(T,p)} c_{p(T,p)}} \tag{17.1}$$

In Equation 17.1, α represents the volumetric thermal expansion coefficient (K^{-1}), ρ the density (kg/m³), c_p the specific heat (J/kg.K) at a particular temperature T (K), and pressure p (Pa). In this equation, the thermophysical properties as well as their p,T-dependencies are component dependent. For particular processing conditions, due to compression heating, p,T-conditions can be reached in which temperature exerts an important effect on food attributes. Differences in compression heating between vessel wall, pressure medium, food components, food packaging material, HP baskets, and so on, and subsequent heat transfer are the basis for temperature nonuniformity issues during HPP (Grauwet et al., 2009, 2012).

17.2 HIGH-PRESSURE PASTEURIZATION VERSUS HIGH-PRESSURE STERILIZATION

By analogy with conventional thermal processing and from the viewpoint of microbial destruction, based on the intensity of the HP treatment, a distinction can be made between HP-pasteurization (HP-P) (400–600 MPa; initial temperature 5–25°C; 1–15 min) and HP-sterilization conditions (HP-S) (500–800 MPa; process temperature 90–120°C; 1–10 min) (Ramaswamy and Marcotte, 2006). Important factors in determining the required intensity of the treatment are the desired shelf-life under particular storage conditions following the process, the food pH, and the combined p,T-resistance of the target attribute (e.g., vegetative microorganism, spore, enzyme). Pasteurization is a mild treatment intended to destroy selected vegetative microbial species (particularly pathogens). In most cases, pasteurized foods have a rather limited shelf-life (several days to several months) and need to be stored under conditions minimizing microbial growth and preventing spore germination (e.g., refrigeration, low pH). In theory, under sterilization conditions, the destruction of both vegetative microorganisms and spores is targeted, resulting in shelf-stable foods (several years at room temperature). In practice, a commercial food sterilization process is designed to destroy all pathogenic and only most spoilage microorganisms while preventing the growth of more resistant bacteria. As the target attribute for safety in HP-S has still not been clearly defined (Ganzle et al., 2007; Patazca et al., 2006), in this work, the term "sterilization" will refer to processing conditions that inactivate spores to a considerable extent rather than to conditions leading to shelf-stable food products.

Typical HPP conditions needed for pasteurization (Figure 17.1a) and sterilization (Figure 17.1b) clearly illustrate the potential of HPP as an alternative for thermal processing of high moisture foods when keeping in mind the three suggestions for optimizing the balance between food safety and quality: (i) lower processing temperature; (ii) reduction of the treatment time; (iii) using other physical process parameters which do not/less affect the food quality in a negative way. For reasons of comparison, processing temperatures used in conventional thermal treatment (0.1–0.3 MPa) are also marked. As can be seen in Figure 17.1a, using HP, products can be pasteurized at much lower temperatures than at atmospheric pressure. HP-P is frequently termed "cold pasteurization." Since quality degradation is generally reduced at lower temperature, lower quality degradation is expected. Unlike temperature, pressure only influences the volume of a system according to Le Châtelier principle without affecting the internal energy. Consequently, pressure at room temperature does not affect the covalent bond. In conclusion, low-molecular-weight

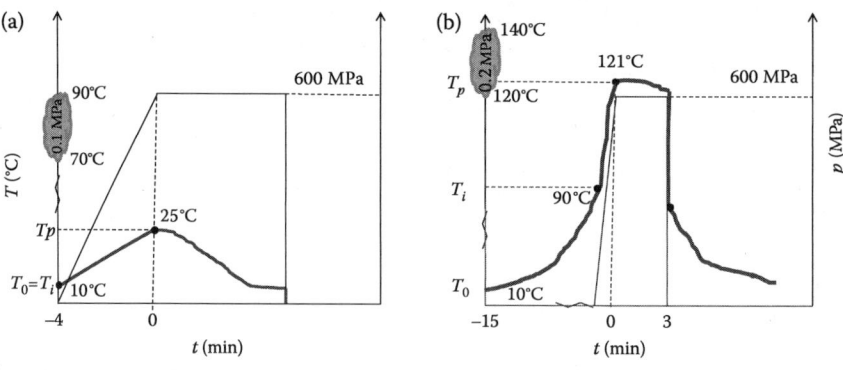

Figure 17.1 Schematic representation of pressure (p) (blue), temperature (T), (red) time (t) profiles resulting in (a) HP-P and (b) HP-S intensities. Processing temperatures used in conventional thermal treatment are marked in gray. (T_0: start temperature; T_i: initial temperature: i.e., temperature at start p-build-up; T_p: process temperature: i.e., maximal temperature reached at holding).

food components such as pigments, flavor compounds, and vitamins should be largely preserved after a HP-P treatment.

HP-S is a combined high pressure–high temperature (HPHT) treatment. As depicted in Figure 17.1b, sterilization conditions are reached in a temperature range comparable to conventional thermal processing. In HP-S, the phenomenon of compression heating (T_i to T_p) is exploited. Due to the rapid and volumetric compression heating and cooling, a HPHT process allows applying the high temperature short time (HTST) principle to products that heat slowly (e.g., by conductive heat transport) under conventional conditions (de Heij et al., 2003; de Heij et al., 2005; Heinz and Knorr, 2005; Hoogland et al., 2001; Sizer et al., 2002; Ting et al., 2002). This should lead to shorter processing times and, under the condition that pressure has little accelerating effect on the quality degradation reaction rate, to a product of which various quality aspects are superior to those of conventionally heated products of equivalent safety. However, starting from room temperature (T_0), using only compression heating, the food product temperature cannot be raised sufficiently. Therefore, a preheating step at atmospheric pressure to a well-defined T_i needs to precede the actual HPHT treatment.

17.3 THEORETICAL CONSIDERATIONS ON THE ANALYSIS OF KINETICS UNDER HPP

Kinetic data describe the evolution of a reaction (e.g., inactivation, denaturation) as a function of time. The time dependency of a reaction can be expressed by the reaction rate equation (see Section 17.3.2.1). How the reaction can be affected by pressure and/or temperature can be described by a temperature and/or pressure coefficient model (see Section 17.3.2.2).

Kinetic data can be obtained under static (i.e., isobaric–isothermal) or dynamic (i.e., temperature and/or pressure varies as a function of time) processing conditions, the static procedure giving more direct insight in the individual effect of the process parameters on the reaction rate and being more straightforward (Ludikhuyze et al., 2003; Van Loey et al., 2002). Indeed, the rate equation of an nth-order degradation reaction (e.g., inactivation, denaturation) (Equation 17.2) can be solved analytically under the assumption that all intrinsic (e.g., pH) and extrinsic (e.g., pressure) processing conditions are independent of time:

$$r = \frac{dA}{dt} = -kA^n \tag{17.2}$$

with r the rate of reaction, A the activity or concentration of the compound of concern at treatment time t, n the order of reaction, and k the reaction rate constant.

17.3.1 Experimental Design for Kinetic Studies under Static Processing Conditions

Many kinetic studies are performed in food model systems (e.g., compound of concern in buffer system) (see Section 17.4). Food model systems can be completely characterized with regard to composition (i.e., information about intrinsic properties) and the target can be studied in a smaller volume (e.g., microtube, bag). The latter can be necessary to obtain reliable information on the extrinsic processing conditions (e.g., exact process temperature) (Balasubramaniam et al., 2004).

It has to be kept in mind that acid–base equilibria, responsible for the pH and pOH of a solution, are affected both by temperature and pressure (Kitamura and Itoh, 1987). To circumvent this issue, pressure- (e.g., MES buffer) or temperature-stable (e.g., phosphate buffer) buffers can be selected. However, no buffer system is currently available whose pK_a is not shifted under both increased pressure and temperature (Bruins et al., 2007; Mathys et al., 2008). Today, no experimental method is available to directly monitor pH under HPP conditions. However, methods are reported to calculate the pH changes of buffer systems based on information of (i) the reaction volume (ΔV^0) and pK_a^0 at reference conditions (25°C, 0.1 MPa), and (ii) $\Delta pH^0/°C$ of the buffer selected (Bruins et al., 2007). In this way, the pH shift of buffers can be predicted and depending on the reaction under consideration and its pH-sensitivity, a well-considered decision on the buffer choice of the model system can be made (Bruins et al., 2007; De Roeck et al., 2008; De Vleeschouwer et al., 2010).

In the case of nonconstant intrinsic properties, integration of Equation 17.2 is still possible if the reaction under consideration is little affected by intrinsic-property changes under HPP. In practice, the effect of intrinsic-property changes due to processing on a reaction as well as the effect of particular constant intrinsic properties (e.g., study in buffer solution in comparison to real food system), will be incorporated in the kinetic model and parameters estimated. Ideally, models and parameters obtained should be validated for each type of food separately (Sizer et al., 2002).

Obtaining static and uniform temperature conditions in the HP reactor can be addressed by two approaches (Van der Plancken et al., 2008): (i) using a slow p-build-up for kinetic studies of relatively slow processes; (ii) using a fast p-build-up for kinetics studies of relatively fast processes. Using a slow p-build-up, during the p-build-up phase, compression heat will be readily transferred to the vessel wall. The starting point of the kinetic study is then taken after an equilibrium period in which isobaric–isothermal conditions are attained.

In the second approach, the compression heat during fast p-build-up is used to attain a specified process temperature, which is kept constant by additional heat. In literature, different methods for gradual temperature increase of the vessel wall during build-up are discussed, using an (i) internal heater (Koutchma et al., 2005) or (ii) external heater (Ahn et al., 2007; Margosch et al., 2006). In addition, the use of a protocol in which the temperature of the vessel wall can be controlled separately from the temperature of sample and the pressure medium has been described in literature (De Roeck et al., 2009; Grauwet et al., 2010; Verbeyst et al., 2010). In this protocol, at the start of a process, the temperature of the vessel wall is preset to the process temperature under consideration, while the temperature of an isolating sample container filled with pressure medium and sample is equilibrated to a particular lower, initial temperature. The temperature from which the process temperature can be reached due to compression heating is selected as the corresponding initial temperature.

These approaches are merely allowed when first-order kinetics (including the special cases; Table 17.1) can be assumed.

Table 17.1 Overview of Different Mathematical Models to Describe Biphasic Inactivation Behavior of Enzymes under Static Processing Conditions

Model Type	Model Equation	n°
Distinct—isozyme model	$A = A_l \exp(-k_l t) + A_s \exp(-k_s t)$	(6)
Fractional—conversion model	$A = A_{fin} + (A_0 - A_{fin}) \exp(-kt)$	(7)
Consecutive—step model	$A = \left(A_1 - A_2 \left(\dfrac{k_1}{k_1 - k_2}\right)\right) \exp(-k_1 t) + A_2 \left(\dfrac{k_1}{k_1 - k_2}\right) \exp(-k_2 t)$	(8)

Note: With A_s and A_l the activity of the more stable and labile enzyme fraction, respectively; A_1 and A_2 the activity of step 1 and step 2, respectively. A_{fin} and A_0 the activity of the stabile and initial fraction, respectively. With k the rate constants of the corresponding inactivating fractions.

Next to problems with nonuniformity in the HP vessel, problems with nonuniformity in the sample under consideration could occur. As explained above, kinetic studies of target attributes are frequently performed in flexible, small containers that allow fast pressure and heat transfer. In this way, the T-field of the pressure medium can be assumed to be equally distributed throughout the treated sample.

In conclusion, under all conditions, when performing kinetic studies under HPP, accurate documentation of pressure, temperature, and time is indispensible. Under static conditions, evaluation of the p,T-profile in one coordinate is sufficient, which is currently technically achievable at lab-scale.

To obtain quantitative data for different p,T,t-combinations, specialized HP equipments, consisting of several individual thermostated pressure vessels, are often used (a.o. Fachin et al., 2002; Guan et al., 2005; Panagou et al., 2007; Tassou et al., 2008; Van Opstal et al., 2005). Such equipment allows simultaneous submission of several samples to treatments at a particular pressure and temperature for different preset times. However, kinetic studies performed in single vessel system have also been reported (a.o. Ahn et al., 2007; Dogan and Erkman, 2004; Ramaswamy et al., 2003).

After treatment, samples should be immediately transferred to a cooling bath to stop further reactions which should not be attributed to the processing conditions under consideration (Van Loey et al., 2002).

17.3.2 Modeling Process-Induced Changes under Static Processing Conditions

Modeling p,T-induced changes, involves a four-step (iterative) approach: (i) establishment of an adequate reaction rate equation (i.e., effect of time); (ii) identification of the temperature and/or pressure coefficient model; (iii) selection of an appropriate regression method to estimate the model parameters; (iv) model evaluation (van Boekel, 1996; Van Loey et al., 2002).

17.3.2.1 Determination of the Reaction Rate Equation

As explained above, in case of isobaric–isothermal (static) experiments, in which the rate constant k is not varying with time, Equation 17.2 has an analytical solution describing A as a function of time. Depending on the reaction order, a function of the form of Equation 17.3 or Equation 17.4 can be obtained for first-order and nth-order reactions, respectively. The reaction order is an empirical, not necessarily integer, parameter (van Boekel, 1996).

$$n = 1 \quad A = A_0 \exp(-kt) \tag{17.3}$$

$$n \neq 1 \quad A^{(1-n)} = A_0^{(1-n)} + (n-1)kt \tag{17.4}$$

with A_0 the initial activity or concentration at time $t_{iso} = 0$ min, k the reaction rate order at constant pressure and temperature.

Kinetics of inactivation of enzymes and microorganisms are frequently described by a first-order equation (Equation 17.3) (e.g., Balogh et al., 2004; Dogan and Erkman, 2004; Indrawati et al., 2005; Koutchma et al., 2005; Ramaswamy et al., 2009; Van Eylen et al., 2007; Wang et al., 2008; Weemaes et al., 1998; Zhu et al., 2008). This is remarkable for reactions which are a complex process involving several events. Apparent first-order processes have been attributed to the fact that one of these events predominates over the others. In the area of food science and technology, it is common to characterize first-order reactions using the thermal-death-time (TDT) concept (Bigelow, 1921), the direct relation between the decimal-reduction time (D-value) (min) and k (min^{-1}) being expressed by

$$D = \frac{\ln(10)}{k} \tag{17.5}$$

The decimal-reduction time expresses the time needed, at a given temperature and pressure, for a 90% reduction of the initial activity or concentration.

Often, more complex reaction rate equations are found, because the reaction may occur by consecutive or parallel processes or it is unlikely that a whole population behaves the same. Applying first-order concepts in these situations would result in over- or underestimation of the processing times for the compound of concern (Peleg and Cole, 1998).

Next to the general nth-order rate equation (e.g., as used by Buckow et al., 2009), several other mathematical expressions have been suggested for modeling nonfirst-order inactivation behavior of enzymes (Table 17.1).

A "biphasic" inactivation behavior where a fast inactivation period is followed by a decelerated decay, has been explained by the occurrence of isozymes with different stabilities (distinct-isozyme model; Equation 17.6; a.o. selected by Fachin et al. 2002), to the presence of a resistant enzyme fraction (fractional-conversion model; Equation 17.7; a.o. selected by Van den Broeck et al. (2000)) or to intermediate steps in the overall inactivation process (consecutive-step model; Equation 17.8; a.o. selected by Ludikhuyze et al. 1999).

By analogy, nth-order rate equations have been used to model kinetic data of microorganisms (Ananta et al., 2001; Margosch et al., 2006). In addition, biphasic inactivation behavior has been described, in which two subpopulations can be found which independently decay according to first-order kinetics (Cerf, 1977; Geeraerd et al., 2000) (Table 17.2;

Table 17.2 Overview of Different Mathematical Models to Describe the Inactivation Behavior of Microorganisms under Static Processing Conditions

Model Type	Model Equation	$n°$
Biphasic model	$\log\left(\dfrac{N_0}{N}\right) = \log\left(\dfrac{N_0}{N}\right) - \log\left[f\exp(-k_1 t) + (1-f)\exp(-k_2 t)\right]$	(9)
Log-logistic model	$LogN = \alpha + \dfrac{\varpi - \alpha}{1 + \exp\left(\dfrac{4\sigma(\tau - \log t)}{\varpi - \alpha}\right)}$	(10)
Modified Gompertz model	$\log\left(\dfrac{N}{N_0}\right) = C\exp(-\exp(BM)) - C\exp(-\exp(-B(t-M)))$	(11)
Weibull model	$\log\left(\dfrac{N}{N_0}\right) = -bt^n$	(12)

Note: With N_0 and N (CFU/mL) the number of living cells after treatment time $t = 0$ min or t, f the fraction of the initial population in a major subpopulation, $(1-f)$ the fraction of the initial population in a minor subpopulation and k_1 and k_2 the specific inactivation rates of the two populations (min^{-1}), respectively, α and ϖ the upper lower asymptote (CFU/mL), respectively, σ the maximum rate of inactivation (log (CFU/mL)/log min), τ the log time to the maximum rate of inactivation (log min), C the difference in value of the upper and lower asymptotes, B the relative inactivation rate at M, M the time at which the absolute inactivation rate is maximum, b the scale factor, and n the scale factor.

Equation 17.9). Specific models have been proposed describing deviation from linearity of the microbial inactivation curve in the form of shoulders and/or tailing, for example: log-logistic model (Table 17.2; Equation 17.10) (Cole et al., 1993), modified Gompertz model (Table 17.2—Equation 17.11) (Bhaduri et al., 1991), and Weibull model (Table 17.2; Equation 17.12) (Peleg and Cole, 1998), the latter being challenged in several recent kinetics studies under HPP conditions (Guan et al., 2005; Panagou et al., 2007; Pilavtepe-Celik et al., 2009; Rajan et al., 2006). A tool for testing nine different types of microbial-survival models on user-specific experimental data relating the evolution of the microbial population with time was developed by Geeraerd et al. (2005).

It has been described that models most accurately describing the inactivation data may depend on the processing conditions under study (Reyns et al., 2000; Tassou et al., 2008; Van Opstal et al., 2005). According to open literature, these log-logistic, Weibull, and so on models are only used in microbiological studies.

17.3.2.2 Determination of a Temperature and/or Pressure Coefficient Model

The Arrhenius model is often used to express the T-dependency of the rate constant k at constant pressure in terms of activation energy, E_a (Table 17.3; Equation 17.13) (Arrhenius, 1889). In case of first-order reactions, the TDT-model expresses the T-dependency of the D-value at constant pressure in terms of the z_T-value which can be interpreted as the

Table 17.3 Overview of Temperature (T) and Pressure (p) Coefficient Models

Model Type	Applicability	Model Equation	n°
Arrhenius model	all n constant p	$k_T = k_{refT} \exp\left(\dfrac{-E_a}{R}\left(\dfrac{1}{T} - \dfrac{1}{T_{ref}}\right)\right)$	(13)
TDT-model	$n = 1$ constant p	$D_T = D_{refT} 10^{\frac{T_{ref}-T}{z_T}}$	(14)
Eyring model	all n constant T	$k_p = k_{refp} \exp\left(\dfrac{-V_a}{RT}(p - p_{ref})\right)$	(15)
PDT-model	$n = 1$ constant T	$D_p = D_{refp} 10^{\frac{p_{ref}-p}{z_p}}$	(16)

Source: Arrhenius, A. 1889. *Zeitschrift Für Physikalische Chemie*, 4, 227–248; Bigelow, W. D. 1921. *Journal of Infectious Diseases*, 29, 528–536; Eyring, H. 1946. *Journal of Physical Chemistry*, 50, 453–464; Basak, S. and Ramaswamy, H. S. 1996. *Food Research International*, 29(7), 601–607.

Note: With n order of reaction, k and k_{ref} reaction rate constant at observed and reference p,T conditions, respectively; D and D_{ref} (min) decimal reduction time at observed and reference p,T-conditions, respectively; E_a (J/mol) and V_a (cm³/mol) activation energy and volume, respectively; z_T (K or °C) and z_p (MPa) T- or p-increase, respectively, to obtain a 10-fold decrease of decimal reduction time. T and T_{ref} (K), observed and reference temperature, respectively; p and p_{ref} (MPa) observed and reference temperature, respectively; R, the universal gas constant (8.314 J.(K mol)$^{-1}$ = 8.314 cm³.MPa.(K mol)$^{-1}$).

temperature increase necessary to obtain a 10-fold decrease of the D-value (Table 17.3; Equation 17.14) (Bigelow, 1921).

By analogy with temperature, models to express the p-dependency of the rate constant k at constant temperature have been defined. The Eyring model and the PDT-model (pressure-death-time model) express the p-dependency in terms of activation volume, V_a, and z_p-value, respectively (Table 17.3; Equations 17.15 and 17.16) (Basak and Ramaswamy, 1996; Eyring, 1946).

When the reaction rate is both pressure and temperature dependent, it is suggested to describe the combined p,T-dependency using one single model. In literature, the thermodynamics-based kinetic model equation (Equation 17.17; selected a.o. by Van Eylen et al. (2007)), modified Arrhenius–Eyring model equations (e.g., Equation 17.18; a.o. selected by de Heij et al. (2003)), as well as pure empirical-based model equations (e.g., Equation 17.19; selected by a.o. Buckow et al. (2009)) have been challenged (Table 17.4) to model combined p,T-dependent changes. For an overview of more mathematical models available in literature, the reader is referred to Ludikhuyze et al. (2003).

Table 17.4 Kinetic Models to Describe the Combined p,T-Dependency of the Reaction Rate Constant

Model Type	Model Equation	n°
Thermodynamic-based kinetic model	$\ln(k_{T,p}) = \ln(k_{\text{ref}T,p}) + \dfrac{\Delta S_0}{RT}(T - T_{\text{ref}})$ $- \dfrac{\Delta V_0}{RT}(p - p_{\text{ref}}) - \dfrac{2\Delta\beta}{RT}(p - p_{\text{ref}})(T - T_{\text{ref}})$ $- \dfrac{\Delta c_p}{RT}\left[T\left(\ln\dfrac{T}{T_{\text{ref}}} - 1\right) + T_{\text{ref}}\right] - \dfrac{\Delta\alpha'}{2RT}(p - p_{\text{ref}})^2$	(17)
Arrhenius–Eyring model	$k_{T,p} = k_{\text{ref}T,p} \exp\left(\dfrac{-E_a}{R}\left(\left(\dfrac{1}{T} - \dfrac{1}{T_{\text{ref}}}\right)\right) \exp\left(\dfrac{-V_a}{RT}(p - p_{\text{ref}})\right)\right)$	(18)
Elliptical equation-based model	$\ln k_{T,p} = A + B(p - p_{\text{ref}}) + C(T - T_{\text{ref}}) + D(p - p_{\text{ref}})^2$ $+ E(T - T_{\text{ref}})^2 + F(p - p_{\text{ref}})(T - T_{\text{ref}}) + G(T - T_{\text{ref}})^3$ $+ H(p - p_{\text{ref}})^3$	(19)

Note: T_{ref} and p_{ref} correspond to an arbitrarily chosen reference point (associated with k_{ref}), Δ indicates difference in the corresponding parameter between end and initial state, β is compressibility factor (cm^6/J.mol), α' is thermal expansivity factor (cm^3/mol.K), c_p is heat capacity (J/mol.K), V is the volume (cm^3/mol), and S the entropy (J/mol.K), A-H are empirical parameters.

17.3.2.3 Selection of a Regression Method and Fitting Criterion for Kinetic Parameter Estimation

In general, models should be selected based on results obtained for an extensive set of p,T,t-combinations. Their validity should not be extrapolated outside the experimental domain.

A distinction can be made between linear and nonlinear models. A model is linear if the first partial derivatives of the model with respect to the parameters are independent of the parameters, whereas in nonlinear models at least one of these derivatives depends on at least one of the parameters. Linear models can be solved by linear regression methods, providing an explicit analytical solution. In food science, kinetic problems are frequently nonlinear, which will need to be solved iteratively by linear approximation using starting values provided by the user (nonlinear regression) (Motulsky and Ransnas, 1987; van Boekel, 1996).

It is common practice to estimate kinetic parameters through a two-step regression approach. First, for each constant p,T-condition, the time-dependent changes are modeled. Second, temperature and/or pressure coefficients are estimated from regression analysis of the reaction rate constants obtained from the first step. Using this approach, the validity of the model parameters can be evaluated graphically. Alternatively, for example, by inserting Equation 17.18 into Equation 17.3, one-step regression can be performed (Van Loey et al., 2002).

The purpose of a fitting criterion is to find the most likely values of the parameters. The most commonly used criterion for both linear as well as nonlinear regression is the method of least-squares minimization. However, least-squares fitting can only be used if several assumptions, on which this criterion is based, are valid. One of these assumptions is that the variability of the dependent variables (e.g., residual activity) follows a Gaussian bell-shaped distribution. Given this assumption, it can be proven that the most likely values of parameters can be found by minimizing the sum of squares of the vertical distances (i.e., residuals) between the experimental data and the model (Motulsky and Christopoulos, 2004).

17.3.2.4 Model Evaluation

After selecting a regression model, method, and fitting criterion, the model can fit adequately, without however, being a correct representation of reality. Several techniques have been described to evaluate the performance of a model: scrutiny of the residuals, examination of a graph of a curve superimposed on the data points, interpretation of the scientific plausibility of the best-fit parameter values, checking the standard error values, and so on (Motulsky et al., 2004).

17.4 CASE STUDIES ON FOOD SAFETY AND QUALITY-RELATED KINETICS

In what follows, a selection of some recent, characteristic, kinetic studies on the p,T,t-effect on some target attributes (microorganisms vs. enzymes vs. quality target attributes) will be presented in Tables 17.5, 17.6, 17.7, and 17.8, respectively). In each table, the results are listed in chronological order. A selection of kinetic parameter estimates is reported. In Tables 17.5 and 17.7, as far as available, estimates of industrially relevant processing conditions are shown. In Table 17.6, selecting processing conditions illustrating the broad spectrum of inactivation rates was aimed at. In addition, in some cases, conditions were selected for comparison purposes. It needs to be remarked that (approximate) standard errors of parameter estimates are not specified to reduce complexity of the table. Studies reported below have modeled their data under the assumption of static p,T-conditions. For all cases, this assumption was cross-checked examining the described experimental design and has been marked in the tables by "iso."

17.4.1 Microbial Inactivation

As can be seen from Table 17.5, kinetic studies have been performed on the inactivation of vegetative cells of different food microorganisms (HP-P applications; $T_{iso} \leq 45°C$). Matrix studied and p,T-area investigated are displayed, as are the evaluated (X) and selected (**X** in bold) reaction rate equation(s) (B: biphasic; W: Weibull; L: log-logistic; M: modified Gompertz; F: first-order) (see Section 17.3.2.1). If the inactivation rate at constant temperature increased with pressure, the kinetics is indicated to be pressure dependent (p-dep). By analogy, kinetics is indicated to be T-dependent (T-dep) at constant pressure. If data were gathered at 0.1 MPa, these are not taken into consideration for the latter two statements.

Table 17.5 Selection of Kinetic Studies on Microbial Target Attributes (Vegetative Cells) under HP-T Conditions ($T_{iso} \leq 50°C$)

Micro-organism	Matrix	p,T-Area	Model Inactivation $f(t)$ B	W	L	M	F	Inactivation p-Dep	T-Dep	Remark	Reference
Zygosaccharomyces bailii (CBS1097)	Tris-HCl buffer pH 6.5	0–45°C 120–320 MPa Iso					X	X	Slowest inactivation at 10–20°C	• Tailing observed for prolonged treatment times, but these conditions were omitted in data analysis • More inactivation in orange juice experimentally determined as predicted by model • OGYE agar (no oxytetracycline) • Elliptical equation-based kinetic model ($LogD = f(p,T)$)	(Reyns et al., 2000)
Listeria monocytogenes (ScottA)	UHT whole milk	22–50°C 400–500 MPa Iso	5-Log: 20°C-300 MPa-14' $z_{p,20°C} = 66.5$ MPa; $z_{T,260 MPa,T \leq 20°C} = -28.9°C$; $z_{T,260 MPa,T \geq 20°C} = 21.7°C$ X X X				X	X	X	• Tailing observed • 5-Log: 22°C–500 MPa-7.5' • TSAYE agar	(Chen and Hoover D. C., 2003)
E. coli (29055)	Apple juice	25°C 150–400 MPa Iso					X	X	N.D.	• No tailing observed • 5-Log: 25°C–300 MPa-14.8'/9.3' for BHI (nonselective)/VRB (selective) agar. • $z_{p,25°C} = 126/140$ MPa for BHI/VRB	(Ramaswamy et al., 2003)
L. monocytogenes (KUEN136)	• Orange juice • Peach juice • Milk	25°C 200–700 MPa Iso					X X X	X X X	N.D. N.D. N.D.	• No tailing observed • 5-Log: 25°C–600 MPa-4.4'/7.6'/12.2' for orange/peace juice/milk $z_{p,25°C} = 576/506/480$ MPa for orange/peace juice/milk. • Listeria selective agar	(Dogan and Erkman, 2004)
E. coli (MG1655)	• HEPES-KOH buffer pH 7.0 • Carrot juice	5–45°C 150–600 MPa Iso	X X	X X				X X	Slowest inactivation at 20–30°C X	• 5-Log: 20°C; 500 MPa; 43'/12' (juice/buffer) • Juice: $z_{p,20°C} = 207$ MPa; $z_{T,400 MPa} = 26.5°C$	(Van Opstal et al., 2005)

continued

Table 17.5 *(Continued)* Selection of Kinetic Studies on Microbial Target Attributes (Vegetative Cells) under HP-T Conditions ($T_{iso} \leq 50°C$)

Micro-organism	Matrix	p,T-Area	Model Inactivation $f(t)$ B	W	L	M	F	Inactivation p-Dep	T-Dep	Remark	Reference
Pediococcus damnosus	• Phosphate buffer pH 6.7 • Gilt-head seabream	23°C 500–650 MPa Iso	X X	X X			X X	X X	N.D. N.D.	• 5-Log: 23°C–600 MPa-2′/2.7′ for buffer/seabream • Tailing above 600 MPa • MRS agar	(Panagou et al., 2007)
S. aureus	Ham model system	25; 45°C 450–600 MPa Iso	Data modeled by log-linear model which accounts for tailing (Geeraerd et al., 2005)					X	X	• 5-Log: 25°C–600 MPa-6′/ >17′ for BHI (nonselective)/BP (selective?) agar • $z_{p,25°C}$ = 154/303 MPa for BHI/BP. • Tailing observed at more extreme processing conditions	(Tassou et al., 2008)
S. aureus (485)	• Carrot juice • 0.1% peptone H_2O	40°C 200–400 MPa Iso	X X	X X			X X	X X	N.D N.D	• 5-Log: 40°C–400 MPa-7′/18′ for *S. aureus* in carrot juice/peptone H_2O	(Pilavtepe-Celik et al., 2009)
E. coli (O157:H7)	• Carrot juice • 0.1% peptone H_2O		X X	X X			X X	X X	N.D N.D	• 5-Log: 40°C–400 MPa-9′/3.5′ for *E. coli* in carrot juice/peptone H_2O • Tailing observed for *E. coli* under experimental conditions • TSAYE agar	

N.D.: not determined; B: biphasic; W: Weibull; L: log-logistic; M: modified Gompertz; F: first-order model.

Table 17.6 Selection of Kinetic Studies on Microbial Target Attributes (Spores) under HPHT Conditions ($T_{iso} \geq 80°C$)

Micro-organism	Matrix	p,T-Area	Model Inactivation f(t) B	W	L	M	F	Inactivation p-dep	T-dep	Remark	Reference
C. sporogenes (PA3679)	Phosphate buffer pH 7.0	91–108°C 600–800 MPa iso					X	X	X	• $D_{600\,MPa,91°C} = 6.1$ min; • $D_{800\,MPa,108°C} = 0.81$ min • $z_T = 23.7°C$ at 600–800 MPa • $z_p = 1500$ MPa at 91–108°C • TSAYE agar	(Koutchma et al., 2005)
Bacillus amyloliquefaciens (TMW2.479)	Tris-his buffer pH 5.15	80–120°C 600–1400 MPa iso			N.D.			X	X	• B. amyloliquefaciens (ST1 agar) less stable than C. botulinum (RCA) in all processing conditions tested	(Margosch et al., 2006)
C. botulinum (TMW2.357)			• n^{th}-order model; $n = 1.35$ • Elliptical equation-based kinetic model ($\ln k = f(p,T)$)					X	X	• Inactivation under HPHT conditions sometimes < corresponding HT (0.1 MPa) conditions: p-mediated protection depends on conditions and species	
Geobacillus stearothermophilus (ATCC10149)	Sterile, deionized H_2O	92–111°C 500–700 MPa					X	X	X	• $D_{500\,MPa,92°C} = 1.8$ min; • $D_{700\,MPa,111°C} = 0.10$ min • $z_T = 58.5/27.4°C$ at 500/700 MPa • $z_p = 352/216$ MPa at 92/111°C • Trypticase soy agar	(Patazca et al., 2006)
B. amyloliquefaciens (FAD82)	Egg patty mince	95–121°C 500–700 MPa iso		X			X	X	X	• $D_{500\,MPa,95°C} = 11.58$ min; • $D_{700\,MPa,121°C} = 0.08$ min • $z_T = 16.7/26.8°C$ at 500/700 MPa • $z_p = 170/332$ MPa at 95/121°C • Inactivation at HPHT > HT • Tailing observed for prolonged holding times • Trypticase soy agar	(Rajan et al., 2006)

continued

Table 17.6 (Continued) Selection of Kinetic Studies on Microbial Target Attributes (Spores) under HPHT Conditions ($T_{iso} \geq 80°C$)

Micro-organism	Matrix	p,T-Area	Model Inactivation $f(t)$					Inactivation		Remark	Reference
			B	W	L	M	F	p-dep	T-dep		
C. sporogenes	Deionized H_2O	105–121°C 700 MPa iso						N.D	X	• $D_{700\,MPa,105°C} = 0.6/0.4$ min; $D_{700\,MPa,121°C} = 0.3/0.1$ min for T. thermosacc./B. amyloliquefaciens. (Fad82) • T. thermosacc. and B. amyloliquefaciens (Fad82;Fad11/2) most HPHT resistent • Inactivation at HPHT conditions > at corresponding HT conditions • Trypticase soy agar or RCA	(Ahn et al., 2007)
C. tyrobutylicum					X		X	N.D	X		
#B. amylolio-uefaciens strains					X		X	N.D	X		
B. sphaericus								N.D	X		
C. sporogenes (PA3679 = ATCC7955)	Ground beef	80–100°C 700–900 MPa iso	*Elliptical equation-based kinetic model ($\log D = f(p,T)$)					X	X	• $D_{700\,MPa,80°C} = 15.8$ min; • $D_{900\,MPa,100°C} = 0.63$ min • $z_T = 19.7/19.1°C$ at 700/900 MPa • $z_p = 563/520$ MPa at 80/100°C • MPA3679 agar	(Zhu et al., 2008)
C. sporogenes (ATCC11437)	milk	80–100°C 700–900 MPa iso					X	X	X	• $D_{700\,MPa,80°C} = 17.0$ min; • $D_{900\,MPa,100°C} = 0.73$ min • $z_T = 16.5/18.2°C$ at 700/900 MPa • $z_p = 714/1250$ MPa at 80/100°C • Inactivation at HPHT conditions > corresponding HT conditions • TSAYE agar	(Ramaswamy et al., 2009)

N.D.: not determined; B: biphasic; W: Weibull; L: log-logistic; M: modified Gompertz; F: first-order model.

Table 17.7 Selection of Kinetic Studies on Enzymatic Target Attributes under HP-T Conditions

Enzyme	Matrix	p,T-Area	Model Inactivation $f(t)$					Inactivation			Remark	Reference
			D	Fr	C	n^{th}	F	p-Dep	T-Dep			
Pectinm-ethyles-terase (carrot)	• 0.1 M citrate buffer pH6.0 • 3 U in carrot juice • *In situ* in carrot pieces	10–25°C (iso) 650–800 MPa 10°C (iso) 700–800 MPa 40°C (iso) 700–800 MPa					X X X	X X X	X N.D. N.D.		• $D_{700\,MPa, 10°C}$ = 188.0 min • $z_{p, 10°C}$ = 127 MPa • $D_{700\,MPa, 10°C}$ = 171.6 min • $z_{p, 10°C}$ = 148 MPa • $D_{700\,MPa, 40°C}$ = 391.54 min • $z_{p, 10°C}$ = 140 MPa	(Balogh et al., 2004)
Polyphenol oxidase (strawberry)	Phosphate buffer 0.1 M pH 7.0	10–65°C Iso					X	X	X	• p,T,t-effect studied on stable enzyme fraction • Antagonistic effect if $T \geq 50°C$ and $p \leq 200$ MPa • Thermodynamics-based kinetic model	• $k_{600\,MPa, 25°C}$ = 0.0071 min^{-1}; $k_{600\,MPa, 50°C}$ = 0.0592 min^{-1} • E_a = 188/12 kJ/mol for 100/700 MPa • V_a = −61.7/ −2.81 cm³/mol for 25/65°C (no $p \leq 200$ MPa)	(Dalmadi, Rapeanu, Van Loey, Smout, & Hendrickx, 2006)
Myrosinase	Endogenous in broccoli juice	10–60°C 100–600 MPa iso					X	X	X	• Antagonistic effect if $T \geq 50°C$ and $p \leq 200$ MPa • Thermodynamics-based kinetic model	• $k_{500\,MPa, 10°C}$ = 0.005 min^{-1}; $k_{500\,MPa, 40°C}$ = 0.250 min^{-1} • E_a = 84/39 kJ/mol 300/500 MPa • V_a = −27.5/ −26.8 cm³/mol for 10/50°C (no $p \leq 200$ MPa)	(Van Eylen et al., 2007)
Myrosinase	Endogenous in broccoli tissue	15–60°C 50–500 MPa iso					X	X	X	• Antagonistic effect if $T \geq 50°C$ and $p \leq 150$ MPa • Thermodynamics-based kinetic model	• $k_{500\,MPa, 15°C}$ = 0.006 min^{-1}; $k_{500\,MPa, 45°C}$ = 0.159 min^{-1} • E_a = 77/24 kJ/mol for 300/500 MPa • V_a = −30.3/ −10.6 cm³/mol for 15/55°C (no $p \leq 150$ MPa)	(Van Eylen et al., 2008)
Lipoxy-genase	Endogenous soy milk	5–60°C 200–650 MPa iso					X	X	Slowest inactivation at 20°C	• Thermodynamics-based kinetic model	• $k_{550\,MPa, 20°C}$ = 0.013 min^{-1}; $k_{550\,MPa, 45°C}$ = 0.724 min^{-1} • E_a = 200/96 kJ/mol for 475/600 MPa • V_a = −65.5/ −68.6 cm³/mol (5/50°C)	(Wang et al., 2008)

continued

Table 17.7 (*Continued*) Selection of Kinetic Studies on Enzymatic Target Attributes under HP-T Conditions

Enzyme	Matrix	p,T-Area	Model Inactivation f(t)					Inactivation		Remark	Reference
			D	Fr	C	nth	F	p-Dep	T-Dep		
Lipoxy-genase	Endogenous crude soybean extract	5–60°C 200–650 MPa iso					X	X	Slowest inactivation at 20°C	• Thermodynamics-based kinetic model • $k_{550\,MPa,20°C} = 0.015$ min^{-1}; $k_{550\,MPa,45°C} = 0.779$ min^{-1} • $E_a = 204/99$ kJ/mol for 475/600 MPa • $V_a = -65.1/-70.9$ cm^3/mol (5/50°C)	(Wang et al., 2008)
Ficin (EC 3.4.22.3)	0.05 g/L in 0.05 M phosphate Buffer pH 7.0	50–80°C 500–900 MPa iso					X	X	X	• Combined Arrhenius–Eyring-based kinetic model • $k_{500\,MPa,50°C} = 0.0006$ min^{-1}; $k_{900\,MPa,80°C} = 0.1022$ min^{-1} • $E_a = 139/43$ kJ/mol for 500/900 MPa • $V_a = -24.7/-6.17$ cm^3/mol (50/80°C)	(Katsaros, Katapodis, & Taoukis, 2009a)
Papain (EC 3.4.22.2)	0.05 g/L in 0.05 M phosphate Buffer pH 7.0	50–80°C 500–900 MPa iso					X	X	X	• Combined Arrhenius–Eyring-based kinetic model • $k_{500\,MPa,50°C} = 0.0073$ min^{-1}; $k_{900\,MPa,80°C} = 0.1357$ min^{-1} • $E_a = 50/87$ kJ/mol for 500/900 MPa • $V_a = -1.34/-7.9$ cm^3/mol (50/80°C)	(Katsaros et al., 2009a)
Actinidin	Endogenous in kiwi juice	25–50°C 200–800 MPa iso					X	X	X	• Combined Arrhenius–Eyring-based kinetic model • $k_{600\,MPa,25°C} = 0.049$ min^{-1}; $k_{600\,MPa,50°C} = 1.265$ min^{-1} • $E_a = 73/111$ kJ/mol for 200/800 MPa • $V_a = -10.2/-16.4$ cm^3/mol (25/50°C)	(Katsaros, Katapodis, & Taoukis, 2009b)
Pectinmet-hylesterase	Endogenous in valencia orange juice	20–40°C 100–500 MPa iso	X					X	X	• Combined Arrhenius–Eyring-based kinetic model • $k_{500\,MPa,20°C} = 0.280$ min^{-1}; $k_{500\,MPa,40°C} = 10.2$ min^{-1} • $E_a = 61/135$ kJ/mol for 100/500 MPa • $V_a = -28/-42$ cm^3/mol (20/40°C)	(Katsaros, Tsevdou, Panagiotou, & Taoukis, 2010)

N.D.: not determined; D: distinct isozyme; Fr: fractional conversion; C: consecutive step; nth: n-order; F: first-order inactivation model.

Table 17.8 Selection of Kinetic Studies on Quality Target Attributes Under HP-T Conditions

Quality Characteristic	Matrix	p,T-Area	Model Inactivation $f(t)$						Inactivation		Remark	Reference
			D	Fr	C	n^{th}	F		p-dep	T-dep		
Stability of 5-methyl-tetrahydro-folic acid	0.2 μg/mL in 0.1 M sodium phosphate Buffer pH 7.0, 0.8 ppm beta-Mercapto-ethanol	30–45°C 200–700 MPa iso					X	• Thermodynamics-based kinetic model	X	X	• $k_{600\,MPa,30°C} = 0.0159$ min^{-1} • $E_a = 111/105$ kJ/mol for 200/500 MPa • $V_a = -5.5/-4.5$ cm^3/mol for 30/40°C	(Indrawati et al., 2005)
Degree of gelatinization	In situ in thai glutinous Rice	20–70°C 100–600 MPa iso			• Combined Arrhenius–Eyring-based kinetic model		X		X	X	• $k_{600\,MPa,20°C} = 0.0006*10^{-1}$ min^{-1}; • $k_{600\,MPa,70°C} = 0.0006$ min^{-1}; • $E_a = 31/34$ kJ/mol for 300/600 MPa • $V_a = -7.06/-11.1$ cm^3/mol for 20/70°C	(Ahromit et al., 2007)
Stability of isothiocyanates (sulforaphane and phenyethyl isothiocyanate)	Endogenous in broccoli juice	30–60°C 600–800 MPa iso⁻					X	Sulforaphane		X	• $k_{600\,MPa,40°C} = 21.30$ min^{-1} • $E_a = 84/81$ kJ/mol 600/800 MPa phenyethyl • $k_{600\,MPa,40°C} = 32.20$ min^{-1} • $E_a = 40/90$ kJ/mol 600/800 MPa	(Van Eylen et al., 2007)
Stability and mechanism of degradation of 5-methyltetrahydro-folic acid	0.4 μM in water	40–65°C 200–800 MPa iso			• Combined Arrhenius–Eyring-based kinetic model		X		X	X	• $k_{600\,MPa,40°C} = 0.0221$ min^{-1}; • $E_a = 54/27$ kJ/mol for 200/700 MPa • $V_a = -6.4/-3.3$ cm^3/mol for 40/60°C	(Verlinde et al., 2009)
Texture degradation	Carrot	95–110°C 600 MPa Iso		X						X	• $k_{600\,MPa,110°C} = 0.1342$ min^{-1}; • $E_a = 177$ kJ/mol for 600 MPa	(De Roeck et al., 2010)
Stability of anthocyanins (pelargonidine-3-glucoside)	Endogenous in strawberries	80–110°C 200–700 MPa iso					X	• Elliptic equation-based kinetic model	X	X	• $k_{700\,MPa,110°C} = 0.0072$ min^{-1}; • $E_a = 63/58$ kJ/mol for 200/700 MPa • $V_a = -2.6/-4.9$ cm^3/mol (80/110°C)	(Verbeyst et al., 2010)

N.D.: not determined; D: distinct isozyme; Fr: fractional conversion; C: consecutive step; nth: n-order; F: first-order inactivation model.

Moreover, short remarks on tailing, agar used, processing conditions to obtain a degree of inactivation and kinetic parameters are reported in the table. As can be seen from Table 17.5, processing conditions to obtain a degree of inactivation (5-Log) differ greatly among species, matrix, or culture medium selected. In most cases, authors report that bacteria are more resilient in a more complex matrix compared to a buffer at the same pH (Panagou et al., 2007; Van Opstal et al., 2005). However, Pilavtepe et al. (2009) detected a protective effect of carrot juice on *Escherichia coli*, whereas it had a sensitizing effect on *Staphylococcus aureus*. In addition, low pH seems to enhance the inactivation of vegetative bacteria by HP (Dogan and Erkman, 2004; Patterson, 2005). In this context, authors should consider the potential pH-shift effect of buffers when analyzing microbial inactivation data (Mathys et al., 2008) (see Section 17.3.1). Different stabilities are reported when treated microorganisms are grown on different culture media (Ramaswamy et al., 2003; Tassou et al., 2008). Extreme care should be taken in the interpretation of results obtained from plating on selective agar, as injured cells may not grow and therefore the level of inactivation may be overestimated (Balasubramaniam et al., 2004; Ramaswamy et al., 2003).

By analogy with Table 17.5, Table 17.6 was assembled based on kinetic studies on spores under HPHT conditions (HP-S applications; $T_{iso} \geq 80°C$). Similar conclusions on the effect of matrix and culture medium have been drawn. In contrast to kinetic studies on vegetative cells, only a minority of studies report tailing of the inactivation curves under HPHT conditions (Rajan et al., 2006). Consequently, most studies could model the inactivation data selecting a first-order rate equation. Since data analysis on spore inactivation is performed in a more uniform way throughout the studies, values of estimated kinetic parameters could be used as a basis for comparison.

A broad range of decimal-reduction times (D-values) is observed using different HPHT conditions. Comparison of thermal D-values (~0.1 MPa) with those derived during the isobaric–isothermal phase of HPHT processes only, might be inadequate to indicate synergistic or antagonistic effect of pressure on thermal inactivation. Several studies have shown that substantial spore inactivation may occur during the dynamic come-up time (Ahn et al., 2007; Bull et al., 2009; Margosch et al., 2006; Rajan et al., 2006). In most cases, inactivation rates have been determined to be (if only a little) faster under HPHT conditions than the inactivation at the same temperature at atmospheric pressure (Ahn et al., 2007; Rajan et al., 2006; Ramaswamy et al., 2009). Only Margosch et al. (2006) observed the protective effect of pressure on thermal inactivation. If the pH shift of the buffer used under HPHT conditions affects the kinetics, data should be interpreted with caution (Bull et al., 2009; Mathys et al., 2008).

Decrease in temperature/pressure sensitivities (higher z_T/z_p) has been reported at increasing pressure/temperature, respectively (Rajan et al., 2006; Ramaswamy et al., 2009). Consequently, for example, the higher the process temperature applied, the lesser is the synergism between pressure and temperature. In the context of applying the HTST principle on conduction-heating products, processing conditions at higher temperature will show greater potential. However, deviations from these trends in z_T/z_p are also reported (Patazca et al., 2006; Zhu et al., 2008).

In general, T-sensitivities under HP conditions of both vegetative cells (Table 17.5) and spores (Table 17.6) are clearly lower (as expressed as higher z_T-values) than the T-sensitivity generally observed in thermal food processing (generally $z_T \leq 10°C$ for vegetative cells and spores under high temperature at atmospheric pressure; Holdsworth, 2009; Marques de Silva, 2009).

Only in a few studies, a model describing the combined p,T-dependency of the inactivation rate constant ($\ln k$) of D-value ($\log D$) was searched for (Margosch et al., 2006; Reyns et al., 2000; Van Opstal et al., 2005; Zhu et al., 2008). These models have great power in process design by enabling calculation of different processing conditions leading to a particular level of inactivation. It should be kept in mind that the inactivation behavior, as well as the kinetic parameter estimates can be different depending on the matrix, strain, and so on studied. Kinetic data should be validated using a cocktail of microorganisms in real food systems (Balasubramaniam et al., 2004).

17.4.2 Enzyme Inactivation

By analogy with Tables 17.5 and Table 17.6, Table 17.7 was compiled based on a selection of representative, kinetic enzyme-inactivation studies. Ludikhuyze et al. (2003) published an extensive review on the combined p,T-effect on enzymes related to quality of fruits and vegetables. In Table 17.7, it was attempted to summarize more recent studies on important food-related enzymes. Different kinetic models are generally accepted for describing inactivation of enzymes in comparison to inactivation of microorganisms (D: distinct isozyme; Fr: fractional conversion; C: consecutive step; nth: n-order; F: first-order inactivation model) (see Section 17.3.2.1). In addition, it is more common to model the pressure/temperature dependency at constant temperature/pressure using the Arrhenius–Eyring concept, rather than the TDT concept and to look for a combined p,T-dependent model. As can be seen from Table 17.7, the stability of enzymes is enzyme, enzyme source, and matrix dependent. In most cases, the matrix has a protective effect, and the more complex the matrix, the more pronounced the effect is (Balogh et al., 2004; Van Eylen et al., 2007, 2008). Studying kinetics in buffer systems, the same caution needs to be taken on the effect of a pH-shift as applies for the inactivation of microorganisms.

By analogy with microorganisms, kinetic models should be validated in real food systems in order to allow reliable prediction. For many enzymes, an antagonistic effect of pressure on T-inactivation at elevated pressure ($p \geq 50$ MPa) was observed in the high-temperature, low-pressure domain. At higher pressures, enzyme inactivation is predominantly characterized by synergism between the HPP parameters. However, the synergism can either be reduced (smaller E_a;$|V_a|$) or enhanced (larger E_a;$|V_a|$) at higher pressures/temperatures as reported by, for example, Dalmadi et al. (2006), Katsaros et al. (2009a), or by Katsaros et al. (2009b) and Katsaros et al. (2010).

17.4.3 Quality Conversion

Only few kinetic studies on quality target attributes besides enzymes (e.g., micronutrient stability, texture, flavor retention) are reported in the HP-T window (Table 17.8). However, insight in their chemical mechanisms and kinetics are greatly needed for HP process design and optimization (Verlinde et al., 2009). In all studies of Table 17.8, an accelerating effect of temperature on the quality conversion reaction rate under HP was observed. The pressure dependency of the reaction rates could not be generalized: some authors reported a retarding effect of pressure on the quality conversion reaction rate (De Roeck et al., 2010; Van Eylen et al., 2007) when others estimated negative reaction volumes (Ahromit et al., 2007; Indrawati et al., 2005; Van Eylen et al., 2007; Verbeyst et al., 2010; Verlinde et al., 2009). Consequently, notwithstanding the fact that pressure seems to have a limited effect on the covalent bond (Balny et al., 1997) and several qualitative studies on flavor, color, micronutrients, and so on described that HPP insignificantly or only slightly affected quality during processing (Oey et al., 2008). This trend cannot be generalized since several oxidation reactions seem to be enhanced by pressure, resulting in, for example, significant losses of water-soluble vitamins such as folates and anthocyanins (Verbeyst et al., 2010; Verlinde et al., 2009). However, as has been mentioned before, the impact of HPP is the result of the integrated effect of pressure, temperature, and time. Even if pressure is enhancing a degradation reaction compared to traditional thermal processing, when the process time can be considerably reduced, the quality after processing might still be larger than its equivalently thermal treated counterpart. To assess the impact of HPP and other processing techniques on important quality conversion reactions incorporating mechanistic insights, the use of multiresponse kinetics should be favored over single-response modeling. In this context, Verlinde et al. (2009) reported the first multiresponse kinetic study under HPP.

For an overview of qualitative data on quality target attributes under HPP, the reader is referred to recent reviews (Oey et al., 2008; Sila et al., 2008). It has been described that the effect of processing on the quality attribute is clearly matrix dependent. In addition, the importance of the stability of quality components after processing during storage should not be overlooked.

17.5 FROM KINETICS TO PROCESS DESIGN

For graphical representation of the combined p,T-dependency, an isorate contour plot can be constructed, which connects static p,T-combinations resulting in particular k/D-values. In Figure 17.2, an example of such an isorate contour diagram is depicted for recently obtained data on the inactivation of (i) *Clostridium botulinum* in buffer system (Margosch et al., 2006); (ii) *Clostridium sporogenes* in ground beef (Zhu et al., 2008); (iii) *E. coli* in carrot juice (Van Opstal et al., 2005); (iv) lipoxygenase in soy milk (Wang et al., 2008); (v) myrosinase in broccoli juice (Van Eylen et al., 2007); and (vi) pelargonidin-3-glucoside (an anthocyanin) in strawberry paste (Verbeyst et al., 2010) encompassing kinetic data of undesired safety (i–iii) and quality (iv–v) target attributes as well as desired quality (vi) attributes, respectively.

From this graph and by analogy with Tables 17.5 through 17.8, it can be concluded that the p,T-domain of the kinetics of change as well as the p,T-sensitivities of safety and quality target attributes strongly differ. In addition, focusing on a selected target attribute, for instance, soy milk lipoxygenase, different p,T-sensitivities can be observed depending on the p,T-domain. In this context, four major domains can be considered for soy milk lipoxygenase: (i) below 10°C, where a p-increase enhances the rate of inactivation, while a T-increase retards it; (ii) between 10°C and 23°C where temperature exerts little effect on the reaction rate, while pressure enhances it; (iii) between 23°C and 47°C where both

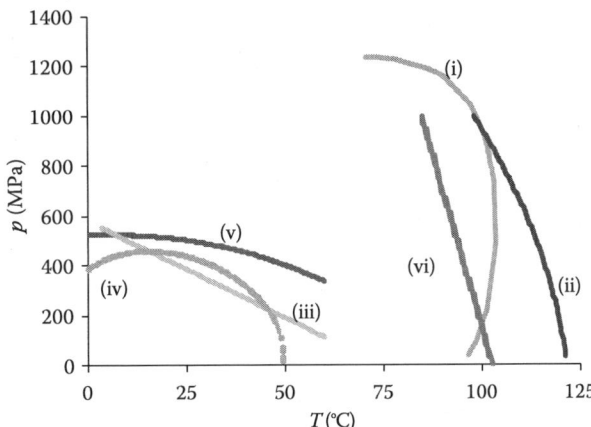

Figure 17.2 p,T-combinations leading to a particular reaction rate constant or decimal reduction time for the inactivation of (i) *C. botulinum* ($D = 0.5$ min); (ii) *C. sporogenes* ($D = 0.5$ min); (iii) *E. coli* ($D = 20$ min); (iv) *lipoxygenase* ($k = 0.1$ min^{-1}); (v) *myrosinase* ($k = 0.1$ min^{-1}); (vi) degradation of pelargonidin-3-glucoside ($k = 0.02$ min^{-1}).

pressure and temperature accelerate the kinetics; and (iv) above 47°C, where pressure exerts little effect on the reaction rate, while temperature enhances it.

In the end, process design implies the selection of processing conditions leading to, for example, both the desired inactivation level of the target microorganism (e.g., 5-Log *E. coli* inactivation) and target quality-degrading enzyme (e.g., 90% lipoxygenase inactivation) as well as the conservation of the desired intrinsic quality characteristic (e.g., x% retention of anthocyanin). In this context, Figure 17.2 can be transformed into a diagram presenting relevant processing conditions to reach these objectives. From the shelf-life extension point of view, successful p,T-combinations can be found for conditions which lead minimally to the targeted inactivation level. From these successful p,T-combinations, choosing the p,T-combination which results in maximal quality retention would be optimal. In addition, it can be necessary to select processing conditions in which processing-induced contaminants are not/minimally formed. However, especially for quality attributes and for sure in the HPHT domain, these kinetic data are mostly unexisting. Furthermore, selection and thus kinetic information of these processing conditions strongly depends on the food product or type under consideration. It was not in the scope of this work to give an overview of all these kinetic data.

17.6 CONCLUSION AND FUTURE PERSPECTIVES

Characterization of process-induced changes in HPP implies expressing the dependency on the three processing variables (p,T,t). Kinetic studies in a broad p,T-range enable prediction of processing conditions leading to the aimed level of the target attribute. In the end, process design implies selecting a p,T,t-combination resulting in the desired level of microbial or enzymatic inactivation while maximizing the food quality after processing. Kinetic models and parameters differ greatly between target attributes. Kinetic data on important targets under HPP in a broad p,T-range are far from complete. In addition, transferability of data from matrix to matrix can be difficult. Consequently, it is indispensable to validate kinetic models and parameters in real food systems under industrially relevant processing conditions. In future studies, the potential of performing kinetic studies under dynamic HPP conditions and obtaining multiresponse insight in complex food reactions should be fully exploited (De Vleeschouwer et al., 2010; Verlinde et al., 2009). Obtaining insight in food changes during processing is very relevant, but the importance of insight in changes during storage of food after processing should not be forgotten (Timmermans et al., 2011; Vervoort et al., 2011). As a last future perspective, the authors want to point out to fingerprinting approaches (e.g., using gas chromatography—mass spectrometry) which could be considered as a first step in studying food changes (during processing and/or storage) (Vervoort et al., 2012, 2013). Fingerprinting approaches can be defined as an untargeted analytical approach to perform comparative analysis among different samples to find differences. In this context, fingerprinting enables the selection of discriminant markers (i.e., compounds which are different among the sample classes to be compared) reducing the risk to overlook unintended reaction products. In a next step, specific kinetic studies on the selected markers can be set up to obtain full insight in the effect of process variables.

ACKNOWLEDGMENTS

Tara Grauwet is a postdoctoral researcher funded by the Research Foundation Flanders (FWO). Stijn Palmers is PhD-fellow funded by the Agency for Innovation by Science and Technology in Flanders (IWT571 Vlaanderen). This work was financially supported by KULeuven Research Fund.

REFERENCES

Ahn, J., Balasubramaniam, V. M., and Yousef, A. E. 2007. Inactivation kinetics of selected aerobic and anaerobic bacterial spores by pressure-assisted thermal processing. *International Journal of Food Microbiology*, 113, 321–329.

Ahromit, A., Ledward, D. A., and Niranjan, K. 2007. Kinetics of high pressure facilitated starch gelatinisation in Thai glutinous rice. *Journal of Food Engineering*, 79(1), 834–841.

Ananta, E., Heinz, V., Schlnter, O., and Knorr, D. 2001. Kinetic studies on high-pressure inactivation of *Bacillus stearothermophilus* spores suspended in food matrices. *Innovative Food Science & Emerging Technologies*, 2(4), 261–272.

Arrhenius, A. 1889. Studying the reaction rate of the inversion of sucrose by acids. *Zeitschrift Für Physikalische Chemie*, 4, 227–248.

Balasubramaniam, V. M., Farkas, D., and Turek, E. J. 2008. Preserving foods through high-pressure processing. *Food Technology*, 62(11), 32–38.

Balasubramaniam, V. M., Ting, E. Y., Stewart, C. M., and Robbins, J. A. 2004. Recommended laboratory practices for conducting high-pressure microbial inactivation experiments. *Innovative Food Science & Emerging Technologies*, 5(3), 299–306.

Balny, C., Mozhaev, V. V., and Lange, R. 1997. Hydrostatic pressure and proteins: Basic concepts and new data. *Comparative Biochemistry and Physiology*, *116A*(4), 299–304.

Balogh, T., Smout, C., Nguyen, B., Van Loey, A., and Hendrickx, M. E. 2004. Thermal and high pressure inactivation kinetics of carrot pectinmethylesterase: From model to real foods. *Innovative Food Science & Emerging Technologies*, *5*(1), 429–436.

Basak, S. and Ramaswamy, H. S. 1996. Ultra high pressure treatment of orange juice: A kinetic study on inactivation of pectin methyl esterase. *Food Research International*, *29*(7), 601–607.

Bhaduri, S., Smith, P. W., Palumbo, S. A., and Turner-Jones, C. O. 1991. Thermal destruction of *Listeria monocytogenes* in liver sausage slurry. *Food Microbiology*, *8*(1), 75–78.

Bigelow, W. D. 1921. The logarithmic nature of thermal death time curves. *Journal of Infectious Diseases*, *29*, 528–536.

Bruins, M., Matser, A., Janssen, A. E. M., and Boom, R. M. 2007. Buffer selection for HP treatment of biomaterials and its consequence for enzyme inactivation studies. *High Pressure Research*, *27*(1), 101–107.

Buckow, R., Weiss, U., and Knorr, D. 2009. Inactivation kinetics of apple polyphenol oxidase in different pressure-temperature domains. *Innovative Food Science & Emerging Technologies*, *10*, 441–448.

Bull, M., Olivier, A., van Diepenbeek, R., Kormelink, F., and Chapman, B. 2009. Synergistic inactivation of spores of proteolytic *Clostridium botulinum* strains by high pressure and heat is strain and product dependent. *Applied and Environmental Microbiology*, *75*(2), 434–445.

Cerf, O. 1977. A review. Tailing of survival curves of bacterial spores. *Journal of Applied Microbiology*, *42*(1), 19.

Cheftel, J. C. 1995. Review: High-pressure, microbial inactivation and food preservation. *Food Science and Technology International*, *1*(2–3), 75–90.

Cheftel, J. C. 1992. Effect of high hydrostatic pressure on food constituents: An overview. In C. Balny, R. Hayashi, K. Heremans, and P. Masson (Eds.), *High Pressure of Biotechnology* (pp. 195–209). Montrouge: John Libbey Eurotext.

Chen, H. and Hoover D. C. 2003. Modeling the combined effect of high hydrostatic pressure and mild heat on the inactivation kinetics of *Listeria monocytogenes* Scott A in whole milk. *Innovative Food Science & Emerging Technologies*, *4*(1), 25–34.

Cole, M. B., Davies, K. W., Munro, G., Holyoak, C. D., and Kilsby, D. C. 1993. A vitalistic model to describe thermal inactivation of *Listeria monocytogenes*. *Journal of Industrial Microbiology and Biotechnology*, *12*(3), 232–239.

Dalmadi, I., Rapeanu, G., Van Loey, A., Smout, C., and Hendrickx, M. 2006. Characterization and inactivation by thermal and pressure processing of strawberry (*Fragaria ananassa*) polyphenol oxidase: A kinetic study. *Journal of Food Biochemistry*, *30*(1), 56–76.

de Heij, W., Van den Berg, R., Van Schepdael, L., and Hoogland, H. 2005. Sterilisation—only better. *New Food*, (2), 56–61.

de Heij, W., van Schepdael, L., Moezelaar, R., Hoogland, H., Matser, A., and van den Berg R. 2003. High pressure sterilization: Maximizing the benefits of adiabatic heating. *Food Technology*, *57*(3), 37–41.

De Roeck, A., Duvetter, T., Fraeye, I., Van der Plancken, I., Sila, D. N., Van Loey, A., and Hendrickx, M. 2009. Effect of high pressure/high temperature processing on chemical pectin conversions in relation to fruit and vegetable texture. *Food Chemistry*, *115*, 207–213.

De Roeck, A., Mols, J., Duvetter, T., Van Loey, A., and Hendrickx, M. 2010. Carrot texture degradation kinetics and pectin changes during thermal versus high pressure high temperature processing: A comparative study. *Food Chemistry*, *120*, 1104–1112.

De Roeck, A., Sila, D. N., Duvetter, T., Van Loey, A., and Hendrickx, M. 2008. Effect of high pressure/high temperature processing on cell wall pectic substances in relation to firmness of carrot tissue. *Food Chemistry*, *107*(3), 1225–1235.

De Vleeschouwer, K., Van der Plancken, I., Van Loey, A., and Hendrickx, M. E. 2010. The effect of high pressure–high temperature processing conditions on acrylamide formation and other Maillard reaction compounds. *Journal of Agricultural and Food Chemistry*, *58*(1), 11740–11748.

Delgado, A. & Hartmann, C. 2003. Pressure treatment of food: Instantaneous but not homogeneous effect. Winter, R. (Ed.), *Advances in High Pressure Bioscience and Biotechnology II.* (pp. 459–464). Heidelberg: Springer.

Dogan, C. & Erkman, O. 2004. High pressure inactivation kinetics of *Listeria monocytogenes* inactivation in broth, milk and peach and orange juices. *Journal of Food Engineering*, 62(1), 47–52.

Eyring, H. 1946. Pressure and reactivity of proteins, with particular reference to invertase. *Journal of Physical Chemistry*, 50, 453–464.

Fachin, D., Van Loey, A., Indrawati, Ludikhuyze, L., and Hendrickx, M. 2002. Thermal and high pressure inactivation of tomato polygalacturonase: A kinetic study. *Journal of Food Science*, 67, 1610–1615.

Ganzle, M. G., Margosch, D., Buckow, R., Ehrmann, M. A., Heinz, V., and Vogel, R. F. 2007. Pressure and heat resistance of *Clostridium botulinum* and other endospores. In C. J. Doona and F. E. Feeherry, *High Pressure Processing of Foods* (pp. 95–114). Iowa: Blackwell Publishing.

Geeraerd, A. H., Heremans, C. H., and Van Impe, J. F. 2000. Structural model requirements to describe microbial inactivation during a mild heat treatment. *International Journal of Food Microbiology*, 59(1), 185–209.

Geeraerd, A. H., Heremans, C. H., and Van Impe, J. F. 2005. GinaFit, a freeware tool to assess non-log-linear microbial survivor curves. *International Journal of Food Microbiology*, 102(1), 95–105.

Grauwet, T., Rauh, C., Van der Plancken, I., Vervoort, L., Delgado, A., Hendrickx, M., and Van Loey, A. 2009. High hydrostatic pressure processing uniformity in the picture. *New Food Digital*, 1, 14–20.

Grauwet, T., Rauh, C., Van der Plancken, I., Vervoort, L., Hendrickx, M. E., Delgado, A., and Van Loey, A. 2012. Potential and limitations of methods for temperature uniformity mapping in high pressure thermal processing. *Trends in Food Science & Technology*, 23, 97–110.

Grauwet, T., Van der Plancken, I., Vervoort, L., Hendrickx, M., and Van Loey, A. 2010. Protein-based indicator system for detection of temperature differences in high pressure high temperature processing. *Food Research International*, 43(1), 862–871.

Guan, D., Chen, H., and Hoover D. C. 2005. Inactivation of *Salmonella typhimurium* DT 104 in UHT whole milk by high hydrostatic pressure. *International Journal of Food Microbiology*, 104(1), 145–153.

Heinz, V. and Buckow, R. 2010. Food Preservation by high pressure. *Journal of Consumer Protections and Food Safety*, 5(1), 73–81.

Heinz, V. and Knorr, D. 2005. High pressure assisted heating as a method for sterilizing foods. In G. V. Barbosa-Canovas, *Novel Food Processing* (pp. 207–232). Boca Raton, FL: CRC Press.

Holdsworth, S. D. 2009. Principles of thermal processing: Sterilization. In R. Simpson, *Engineering Aspects of Thermal Food Processing* (pp. 3–12). Boca Raton, FL: CRC Press.

Hoogland, H., de Heij, W., and Van Schepdael, L. 2001. High pressure sterilization: Novel technology, new products, new opportunities. *New Food*, 21–26.

Indrawati, Van Loey, A., and Hendrickx, M. 2005. Pressure and temperature stability of 5-methyltetrahydrofolic acid: A kinetic study. *Journal of Agricultural and Food Chemistry*, 53(8), 3081–3087.

Katsaros, G. I., Katapodis, P., and Taoukis, P. S. (2009b). Modeling the effect of temperature and high hydrostatic pressure on the proteolytic activity of kiwi fruit juice. *Journal of Food Engineering*, 94(1), 40–45.

Katsaros, G. I., Katapodis, P., and Taoukis, P. S. (2009a). High hydrostatic pressure inactivation kinetics of plant proteases ficin and papain. *Journal of Food Engineering*, 91(42), 48.

Katsaros, G. I., Tsevdou, M., Panagiotou, T., and Taoukis, P. S. 2010. Kinetic study of high pressure microbial and enzyme inactivation and selection of pasteurisation conditions for Valencia orange juice. *International Journal of Food Science and Technology*, 45(1), 1119–1129.

Kitamura, Y. and Itoh, T. 1987. Reaction volume of protonic ionization for buffering agents, prediction of pressure dependence of pH and pOH. *Journal of Solution Chemistry*, 16(9), 715–725.

Knorr, D. 1993. Effects of high-hydrostatic-pressure processes on food safety and quality. *Food Technology*, 156–163.

Koutchma, T., Guo, B., Patazca, E., and Parisi, B. 2005. High pressure-high temperature sterilization: From kinetic analysis to process verification. *Journal of Food Process Engineering*, 28(6), 610–629.

Ludikhuyze, L., Van Loey, A., Indrawati, Smout, C., and Hendrickx, M. 2003. Effects of combined pressure and temperature on enzymes related to quality and vegetables: From kinetic information to processing engineering aspects. *Critical Reviews in Food Science and Nutrition*, 43(5), 527–586.

Margosch, D., Ehrmann, M. A., Buckow, R., Heinz, V., Vogel, R. F., and Ganzle, M. G. 2006. High-pressure-mediated survival of *Clostridium botulinum* and *Bacillus amyloliquefaciens* endospores at high temperature. *Applied and Environmental Microbiology*, 72(5), 3476–3481.

Marques de Silva, F. V. 2009. Principles of thermal processing: Pasteurization. *Engineering Aspects of Thermal Processing*. New York: CRC Press, Taylor & Francis Group.

Mathys, A., Kallmeyer, R., Heinz, V., and Knorr, D. 2008. Impact of dissociation equilibrium shift on bacterial spore inactivation by heat and pressure. *Food Control*, 19(12), 1165–1173.

Motulsky, H. J. and Christopoulos, A. 2004. *Fitting Models to Biological Data Using Linear and Nonlinear Regression—A Practical Guide to Curve Fitting*. New York: Oxford University Press.

Motulsky, H. J. and Ransnas, L. A. 1987. Fitting curves to data using nonlinear regression: A practical and nonmathematical review. *FASEB Journal*, 1, 365–374.

Oey, I., Lille, M., Van Loey, A., and Hendrickx, M. 2008. Effect of high-pressure processing on colour, texture and flavour of fruit- and vegetable-based food products: A review. *Trends in Food Science & Technology*, 19(6), 320–328.

Oey, I., Van der Plancken, I., Van Loey, A., and Hendrickx, M. 2008. Does high pressure processing influence nutritional aspects of plant based food systems? *Trends in Food Science & Technology*, 19(6), 300–308.

Panagou, E. Z., Tassou, C. C., Manitsa, C., and Mallidis, C. 2007. Modelling the effect of high pressure on the inactivation kinetics of a pressure-resistant strain of *Pediococcus damnosus* in phosphate buffer and gilt-head seabram (*Sparus aurata*). *Journal of Applied Microbiology*, 102(1), 1499–1507.

Patazca, E., Koutchma, T., and Rwaswamy, H. S. 2006. Inactivation kinetics of *Geobacillus stearothermophilus* spores in water using high-pressure processing at elevated temperatures. *Journal of Food Science*, 71(3), M110–M116.

Patterson, M. F. 2005. Microbiology of pressure-treated foods. *Journal of Applied Microbiology*, 98, 1400–1409.

Peleg, M. and Cole, M. B. 1998. Reinterpretation of microbial survival curves. *Critical Reviews in Food Science and Nutrition*, 38(1), 353–380.

Pilavtepe-Celik, M., Buzrul, S., Alpas, H., and Bozoglu, F. 2009. Development of a new mathematical model for inactivation of *Escherichia coli* 0157:H7 and *Staphylococcus aureus* by high hydrostatic pressure in carrot juice and peptone water. *Journal of Food Engineering*, 90(388), 394.

Rajan, S., Ahn, J., Balasubramaniam, V. M., and Yousef, A. E. 2006. Combined pressure-thermal inactivation kinetics of *Bacillus amyloliquefaciens* spores in egg patty mince. *Journal of Food Protection*, 69(4), 853–860.

Ramaswamy, H. S. and Marcotte, M. 2006. Background basis. *Food Processing: Principles and Applications* (pp. 7–66). Boca Raton, FL: CRC Press.

Ramaswamy, H. S., Riahi, E., and Idziak, E. 2003. High pressure destruction kinetics of *E. coli* (29055) in apple juice. *Journal of Food Science*, 68(5), 1750–1756.

Ramaswamy, H. S., Shao, Y., and Zhu, S. 2009. High-pressure destruction kinetics of *Clostridium sporogenes* ATCC11437 spores in milk at elevated quasi-isothermal conditions. *Journal of Food Engineering*, doi:10.1016/j.foodeng.2009.07.019.

Reyns, K. M. F. A., Soontjens, C. C. F., Cornelis, K., Weemaes, C. A., Hendrickx, M. E., and Michiels, C. W. 2000. Kinetic analysis and modelling of combined high pressure temperature inactivation of the yeast *Zygosaccharomyces bailii*. *International Journal of Food Microbiology*, 56(1), 199–210.

Sila, D. N., Duvetter, T., De Roeck, A., Verlent, I., Smout, C., Moates, G. K., Hills, B. P., Waldron, K. K., Hendrickx, M., and Van Loey, A. 2008. Texture changes of processed fruits and vegetables: Potential use of high-pressure processing. *Trends in Food Science & Technology*, 19(6), 309–319.

Sizer, C. E., Balasubramaniam, V. M., and Ting, E. Y. 2002. Validating high pressure processes for low-acid foods. *Food Technology*, 56(2), 36–42.

Tassou, C. C., Panagou, E. Z., Samaras, F. J., Galiatsatou, P., and Mallidis, C. 2008. Temperature-assisted high hydrostatic pressure inactivation of *Staphylococcus aureus* in a ham model system: Evaluation in selective and nonselective medium. *Journal of Applied Microbiology*, 104(1), 1764–1773.

Timmermans, R. A. H., Mastwijk, H. C., Knol, J. J., Quataert, M. C. J., Vervoort, L., der Plancken, I. V., Hendrickx, M. E., and Matser, A. M. 2011. Comparing equivalent thermal, high pressure and pulsed electric field processes for mild pasteurization of orange juice. Part I: Impact on overall quality attributes. *Innovative Food Science & Emerging Technologies*, 12(3), 235–243.

Ting, E. Y., Balasubramaniam, V. M., and Raghubeer, E. 2002. Determining thermal effects in high pressure processing. *Food Technology*, 56(2), 31–35.

van Boekel, M. A. 1996. Statistical aspects of kinetic modeling for food science problems. *Journal of Food Science*, 61(3), 477–489.

Van den Broeck, I., Ludikhuyze, L., Van Loey, A. M., and Hendrickx, M. E. 2000. Inactivation of orange pectinesterase by combined high-pressure and temperature treatments: A kinetic study. *Journal of Agricultural and Food Chemistry*, 48, 1960–1970.

Van der Plancken, I., Grauwet, T., Oey, I., Van Loey, A., and Hendrickx, M. 2008. Impact evaluation of high pressure treatment on foods: Considerations on the development of pressure–temperature–time integrators (pTTIs). *Trends in Food Science & Technology*, 19(6), 337–348.

Van Eylen, D., Oey, I., Hendrickx, M., and Van Loey, A. 2007. Kinetics of stability of broccoli (*Brassica oleracea* Cv. *Italica*) myrosinase and isothiocyanates in broccoli juice during pressure/temperature treatments. *Journal of Agricultural and Food Chemistry*, 55(6), 2163–2170.

Van Eylen, D., Oey, I., Hendrickx, M., and Van Loey, A. 2008. Effects of pressure/temperature treatments on stability and activity of endogenous broccoli (*Brassica oleracea* L. cv. *Italica*) myrosinase and on cell permeability. *Journal of Food Engineering*, 89(2), 178–186.

Van Loey, A., Oey, I., Smout, C., and Hendrickx, M. 2002. Inactivation of enzymes. From experimental design to kinetic modeling. In J. R. Whitaker, A. G. J. Voragen, and D. W. S. Wong, *Handbook of Food Enzymology* (pp. 49–58). New York: Marcel Dekker Inc.

Van Opstal, I., Vanmuysen, S. C. M., Wuytack, E. Y., Masschalck, B., and Michiels, C. W. 2005. Inactivation of *Escherichia coli* by high hydrostatic pressure at different temperatures in buffer and carrot juice. *International Journal of Food Microbiology*, 98(2), 179–191.

Verbeyst, L., Oey, I., Van der Plancken, I., Hendrickx, M., and Van Loey, A. 2010. Kinetic study on the thermal and pressure degradation of anthocyanins in strawberries. *Food Chemistry*, 123(2), 269–274.

Verlinde, P. H. C. J., Oey, I., Deborggraeve, W. M., Hendrickx, M. E., and Van Loey, A. 2009. Mechanism and related kinetics of 5-methyltetrahydrofolic acid degradation during combined high hydrostatic pressure-thermal treatments. *Journal of Agricultural and Food Chemistry*, 576803, 6814.

Vervoort, L., Grauwet, T., Kebede, B. T., Van der Plancken, I., Timmermans, R., Hendrickx, M., and Van Loey, A. 2012. Headspace fingerprinting as an untargeted approach to compare novel and traditional processing technologies: A case-study on orange juice pasteurisation. *Food Chemistry*, 134(4), 2303–2312.

Vervoort, L., Grauwet, T., Njoroge, D. M., Van der Plancken, I., Matser, A., Hendrickx, M., and Van Loey, A. 2013. Comparing thermal and high pressure processing of carrots at different processing intensities by headspace fingerprinting. *Innovative Food Science & Emerging Technologies, 18*(0), 31–42.

Vervoort, L., Van der Plancken, I., Grauwet, T., Timmermans, R. A. H., Mastwijk, H. C., Matser, A. M., Hendrickx, M. E., and Van Loey, A. 2011. Comparing equivalent thermal, high pressure and pulsed electric field processes for mild pasteurization of orange juice: Part II: Impact on specific chemical and biochemical quality parameters. *Innovative Food Science & Emerging Technologies, 12*(4), 466–477.

Wang, R., Zhou, X., and Chen, Z. 2008. High pressure inactivation of lipoxygenase in soy milk and crude soybean extract. *Food Chemistry, 106*(1), 603–611.

Weemaes, C., Ludikhuyze, L., Van den Broeck, I., and Hendrickx, M. 1998. High pressure inactivation of polyphenoloxidases. *Journal of Food Science, 63*, 873-877.

Zemansky, M. W. 1957. *Heat and Thermodynamics* (pp. 248–253). New York: McGraw-Hill Book Company.

Zhu, S., Naim, F., Marcotte, M., Ramaswamy, H., and Shao, Y. 2008. High-pressure destruction kinetics of *Clostridium sporogenes* spores in ground beef at elevated temperatures. *International Journal of Food Microbiology, 126*(1–2), 86–92.

18

Gas Exchange Properties of Fruits and Vegetables

Bart M. Nicolaï, Jeroen Lammertyn, Wendy Schotsmans, and Bert E. Verlinden

Contents

18.1 Introduction	739
18.2 Fundamental Considerations	740
18.2.1 Respiration and Fermentation	740
18.2.1.1 Respiration Rate	742
18.2.1.2 Respiration and Fermentation Models	743
18.2.1.3 Gas Transport Properties	747
18.2.2 Measurement Techniques	748
18.2.2.1 Oxygen Consumption and Carbon Dioxide Production Rate	748
18.2.2.2 Measurement of O_2 and CO_2 Concentration	749
18.2.2.3 Measurement of Heat of Respiration	752
18.2.2.4 Skin Resistance and Gas Diffusion Properties	753
18.2.3 Gas Exchange Data for Selected Fruits and Vegetables	755
18.3 Applications	760
Acknowledgments	765
Symbols	765
References	766

18.1 INTRODUCTION

Horticultural products such as fresh fruits and vegetables are still metabolically active when harvested. Under certain conditions some anabolic processes such as photosynthesis and synthesis of flavor components may still take place. The major metabolic processes are of a catabolic nature and include respiration and fermentation and also other processes such as cell wall degradation. Respiration and fermentation are of particular interest to postharvest technologists because they are strongly associated with quality-associated processes taking place in fruits and vegetables. As a consequence, the reduction of the respiratory activity of horticultural products has been a major concern in postharvest technology. Appropriate refrigeration procedures, often in combination with the application of *controlled atmosphere* storage conditions, have therefore been developed to reduce the respiration rate and are applied nowadays on a large scale.

In this chapter, gas exchange properties of fresh fruits and vegetables will be considered. These properties are not only affected by the metabolic processes in the product after harvest, but also by the gas transport properties of the skin and flesh of the product. Both aspects will therefore be addressed. Several methods to measure respiration, fermentation, and gas diffusion will be outlined. Enzyme kinetics-based models to describe the respiratory activity will be reviewed briefly. Existing literature data on gas exchange properties of horticultural produce will be summarized. It will be shown how heat production rates, which must be known for cool room design purposes, can be calculated from respiration rates. As an illustration, controlled atmosphere storage of fresh fruits and vegetables will be described.

18.2 FUNDAMENTAL CONSIDERATIONS

18.2.1 Respiration and Fermentation

Respiration is the central metabolic process in living cells and is responsible for the energy production and synthesis of many biochemical precursors essential for growth and maintenance of the cellular organization and membrane integrity in living cells. Many biochemical and physiological processes are known to be related to respiration, including softening of fruit through degradation of the middle lamella (which glues cells together), production of aroma volatiles, hydrolysis of starch and other carbohydrate storage polymers to simple sugars, synthesis of acids, and pigmentation of fruit [1].

Respiration is basically the oxidation of a wide variety of compounds (e.g. starch, sugars, organic acids), which breaks down into water and carbon dioxide in presence of oxygen. It is a sequence of enzymatically controlled catabolic reactions. The most prominent substrate is glucose, and its oxidation can be described as

$$\text{Glucose} + 6O_2 \rightarrow 6CO_2 + 6H_2O + 2870 \text{ kJ} \tag{18.1}$$

A considerable amount of the energy that is produced in this reaction is incorporated into ATP (adenosine triphosphate), the universal energy carrier in the cell. Some of the energy is discharged as heat and must be removed.

As respiration is clearly an exothermic process, it is affected to a large extent by temperature. The respiration rate is usually maximal at moderate temperatures between 20°C and 30°C but decreases considerably to almost zero around 0°C, depending on the genus, species, and even cultivar. According to Le Châtelier's principle, increasing the partial pressure of CO_2 or decreasing the partial pressure of O_2 will also shift the reaction shown in Equation 18.1 to the left and reduce the respiration rate. However, if the O_2 concentration becomes too low, the respiration metabolism may be inhibited completely and fermentation will take place. This metabolic route is much less efficient from the energetic point of view and has ethanol as an end product:

$$\text{Glucose} \rightarrow 2CO_2 + 2 \text{ ethanol} + 234 \text{ kJ} \tag{18.2}$$

As ethanol causes off-flavors in fruits and vegetables, fermentation is to be avoided. Also, many physiological disorders such as core breakdown in pear (Figure 18.1) have been shown to be associated with fermentation. This is believed to be due to the very low energetic efficiency of the fermentation pathway (only 234 kJ instead of 2870 kJ for respiration) so that there is an insufficient amount of ATP for the cell to fulfill maintenance requirements.

The respiration rate is affected by the development stage of the fruit as well. For *climacteric* fruits such as pears and apples, the ripening process is associated with a rise in respiratory activity, often denoted as the climacteric rise (Figure 18.2). This rise is triggered by the plant hormone ethylene, which is autocatalytically produced. The respiratory climacteric can be seen as an indication of the natural end of a period of active synthesis and maintenance and the beginning of the actual senescence of the fruit. Climacteric fruit can usually be stored for a prolonged period if harvested before the climacteric rise. Further ripening is then initiated when the fruit temperature is raised (apples, pears), or when the ethylene concentration in the surrounding atmosphere is increased artificially (banana, kiwi).

In *nonclimacteric fruit* (e.g., citrus, pineapple), the respiration rate does not increase but progressively declines during senescence until microbial or fungal invasion (Figure 18.2). Also, the increase in respiration of nonclimacteric fruit is concentration dependent [1]. The natural ripening of nonclimacteric fruit therefore cannot be manipulated by modifying the ethylene concentration in the surrounding air.

Wounding of fruits and vegetables increases their stress and may provoke an increased respiration rate. Wounding can be caused by mechanical loading during harvesting,

Figure 18.1 Core breakdown in Conference pear. This disorder occurs during storage under excessively low oxygen concentrations or excessively high carbon dioxide concentrations.

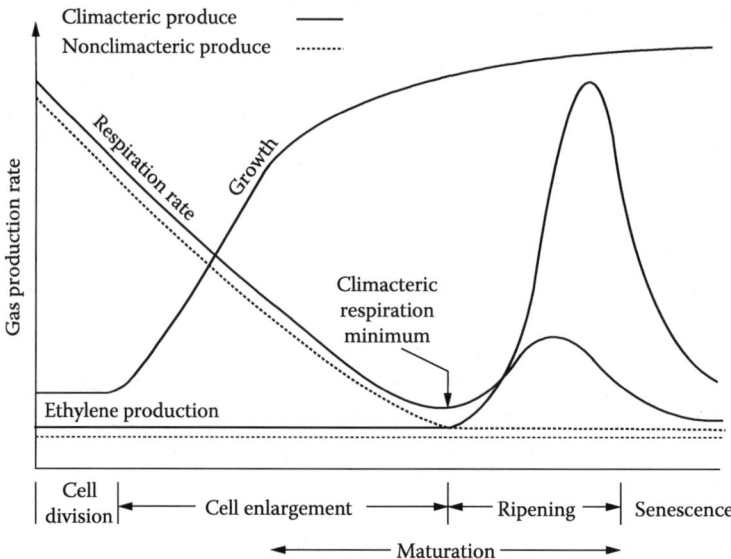

Figure 18.2 Respiration, fruit growth, and ethylene production in climacteric and nonclimacteric fruit.

grading, and transportation of fruits, but also through infections by microorganisms and insects.

18.2.1.1 Respiration Rate

Respirometry (measuring the respiration rate) is the most commonly applied method to determine the metabolic activity of horticultural commodities. It comprises the measurement of the change in carbon dioxide and oxygen concentrations from which the respiration rate is calculated. Kidd and West [2] proposed the CO_2 production rate as an index for respiration.

Since respiration, as shown in Equation 18.1, transforms one O_2 molecule in one CO_2 molecule, it was presumed that the CO_2 production rate and O_2 consumption rate were equal. However, in practice they are not since other reactions such as fermentation can take place or substrates other than glucose may be oxidized. Nowadays, both the CO_2 production rate and O_2 consumption rate are considered as respiration indices.

In a similar manner, the respiration heat production is expressed as $J\ kg^{-1}\ s^{-1}$ or $J\ s^{-1}$. The energy produced by the respiration reactions shown as Equations 18.1 and 18.2 is mainly stored as chemical energy in ATP. Depending on the efficiency of ATP-producing reactions, some of the energy is dissipated as heat. ATP is consumed to drive other metabolic reactions, and eventually all energy is dissipated as heat. At harvest, when fruits or vegetables need to be cooled, the main portion of heat to be removed is the field heat. Once the produce is cooled to the storage temperature, it is the heat generated by the respiration activity of the produce that needs to be evacuated.

In postharvest gas exchange studies, comparison of research results is often complicated because of the many different units used in presenting data. It is preferred to use the system of SI units presented by Banks et al. [3]. Rates of transfer of all gases of physiological interest in postharvest research should be expressed in absolute terms [3], either per unit mass (mol kg^{-1} s^{-1}) or for the entire system under consideration, such as fruits, package, or storage room (mol s^{-1}).

18.2.1.2 Respiration and Fermentation Models

Modified or extended Michaelis–Menten kinetics are widely used to describe the relationship between O_2 and CO_2 partial pressures or concentrations on the one hand, and O_2 consumption and CO_2 production rates on the other. The whole respiration pathway is assumed to be determined by one rate-limiting enzymatic reaction [4]. In this case the reaction rate can be described as Michaelis–Menten kinetics:

$$V_{O_2} = \frac{V_{m,O_2} P_{O_2}}{K_{m,O_2} + P_{O_2}} \tag{18.3}$$

with V_{m,O_2} the maximal O_2 consumption rate (mol kg^{-1} s^{-1}); K_{m,O_2} (kPa) the Michaelis–Menten constant for O_2 consumption; and P_{O_2} the O_2 partial pressure (kPa). This latter parameter can be interpreted as the O_2 partial pressure at which the oxygen consumption rate becomes half its maximal value.

Modern studies of respiration by plant mitochondria have shown that the electron transport system is branched, terminating in two different terminal oxidase systems: a cytochrome oxidase and an alternative cyanide-resistant oxidase, each with its own affinity for O_2 [5]. In such a case, the Michaelis–Menten model can still be used as a semiempirical model to describe the respiration characteristics, although its parameters should then be interpreted with caution.

Based on the interaction mechanism of the inhibitor with the enzyme, three types of CO_2 inhibition on the O_2 uptake rate are distinguished: competitive (Equation 18.4), uncompetitive (Equation 18.5), and noncompetitive (Equation 18.6) [6–8].

In *competitive inhibition*, the inhibitor competes with an enzyme's substrate for binding to the active site

$$V_{O_2} = \frac{V_{m,O_2} P_{O_2}}{K_{m,O_2} \left(1 + (P_{CO_2}/K_{mc,CO_2})\right) + P_{O_2}} \tag{18.4}$$

with P_{CO_2} the CO_2 partial pressure (kPa); and K_{mc,CO_2} the Michaelis–Menten constant for competitive CO_2 inhibition of O_2 consumption (kPa).

In *uncompetitive inhibition*, the inhibitor interacts with the enzyme–substrate complex at a site other than the active site:

$$V_{O_2} = \frac{V_{m,O_2} P_{O_2}}{K_{m,O_2} + P_{O_2} \left(1 + (P_{CO_2}/K_{mu,CO_2})\right)} \tag{18.5}$$

with K_{mu,CO_2} the Michaelis–Menten constant for uncompetitive CO_2 inhibition of O_2 consumption (kPa).

Noncompetitive inhibition is a special case of linear mixed inhibition in which the inhibitor interacts with both the free enzyme and the enzyme–substrate complex at a site other than the active site:

$$V_{O_2} = \frac{V_{m,O_2}\, P_{O_2}}{(K_{m,O_2} + P_{O_2})(1 + (P_{CO_2}/K_{mn,CO_2}))} \tag{18.6}$$

with K_{mn,CO_2} the Michaelis–Menten constant for noncompetitive CO_2 inhibition of O_2 consumption (kPa).

More recently, a mixed type of inhibition (Equation 18.7) was used by Peppelenbos and van't Leven [9] to model gas exchange of horticultural produce. It comprises both the uncompetitive and the competitive types of inhibition:

$$V_{O_2} = \frac{V_{m,O_2}\, P_{O_2}}{K_{m,O_2}(1 + (P_{CO_2}/K_{mc,CO_2})) + P_{O_2}(1 + (P_{CO_2}/K_{mu,CO_2}))} \tag{18.7}$$

The CO_2 production rate is composed of an oxidative and a fermentative part (Equation 18.8) [9]:

$$V_{CO_2} = RQ_{ox}\, V_{O_2} + \frac{V_{m,f,CO_2}}{(1 + (P_{O_2}/K_{m,f,O_2}))} \tag{18.8}$$

with V_{CO_2} the CO_2 production rate (mol kg^{-1} s^{-1}); RQ_{ox} the respiratory quotient; V_{m,f,CO_2} the maximal CO_2 fermentative production rate (mol kg^{-1} s^{-1}); and K_{m,f,O_2} (kPa) the Michaelis–Menten constant for O_2 inhibition on fermentative CO_2 production. The respiratory coefficient is defined as the ratio of CO_2 production rate to the O_2 production rate in the absence of fermentation.

All models have been used successfully to describe respiration of horticultural produce; however, due to the large variability of the measured respiration rates, it is often difficult to distinguish between the different types of inhibition and to find out which inhibition model better fits the data. In this case, two or more inhibition models will describe the respiration kinetics accurately. For pears, all models had a similar R^2 value, but the noncompetitive model is simplest [10].

The maximum O_2 consumption rate and fermentative CO_2 production rate are highly temperature dependent. This effect is described through Arrhenius' law (Equations 18.9 and 18.10). Michaelis–Menten constants are usually assumed to be temperature independent [7,10]:

$$V_{m,O_2} = V_{m,O_2,ref}\, \exp\left[\frac{E_{a,vm,O_2}}{R}\left(\frac{1}{T_{ref}} - \frac{1}{T}\right)\right] \tag{18.9}$$

$$V_{m,f,CO_2} = V_{m,f,CO_2,ref}\, \exp\left[\frac{E_{a,vm,f,CO_2}}{R}\left(\frac{1}{T_{ref}} - \frac{1}{T}\right)\right] \tag{18.10}$$

with $V_{m,O_2,ref}$ and $V_{m,f,CO_2,ref}$ the maximal O_2 consumption and fermentative CO_2 production rate at T_{ref} (mol kg^{-1} s^{-1}), respectively; E_{a,vm,O_2} and E_{a,vm,fO_2} the activation energies for O_2 consumption and fermentative CO_2 production (J mol^{-1}), respectively; T_{ref} a reference temperature; and R the universal gas constant (8.314 J mol^{-1} K^{-1}).

The respiration kinetics for Conference pears are illustrated in Figures 18.3 and 18.4. The O_2 consumption rate and CO_2 production rate, respectively, are shown as functions of O_2 partial pressure for different temperatures and CO_2 partial pressures. The effect of temperature is clearly the most important. The increase of V_{CO_2} at very low O_2 partial pressures is due to fermentation. The O_2 partial pressure at which V_{CO_2} reaches its minimum is called the anaerobic compensation point (ACP). This is often the target O_2 partial pressure that is applied in controlled atmosphere storage.

Respiration heat production can be considered to be proportional to the respiration rate.

$$Q = q_{O_2} V_{O_2} \tag{18.11}$$

in which Q is the respiration heat production expressed as J kg^{-1} s^{-1} and qO_2 is the proportionality constant. In aerobic conditions and when glucose as a substrate can be assumed, it follows from Equation 18.1 that it has the value 478.3 kJ mol^{-1} oxygen consumed.

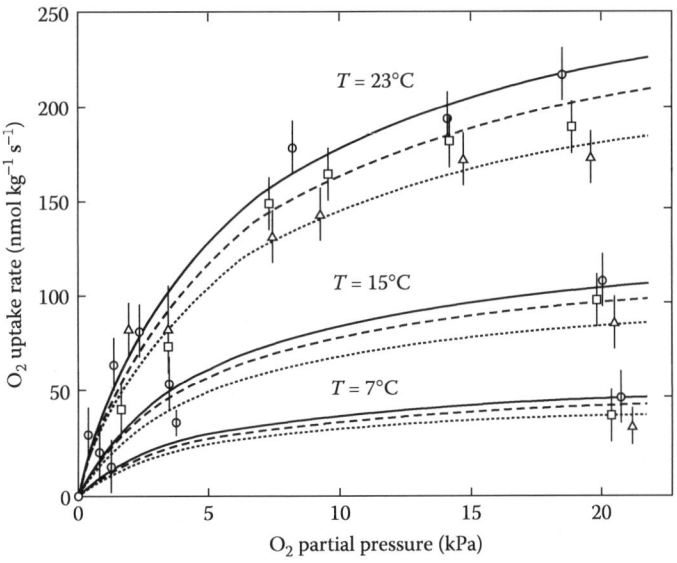

Figure 18.3 Oxygen uptake rate for Conference pears as a function of the temperature and O_2 and CO_2 partial pressures. Within one temperature, the upper, middle, and lower curves represent the modeled O_2 uptake rate at, respectively, 0, 5, and 15 kPa CO_2. Values represent means ($n = 8$) at, respectively, 0 (C), 5 (N), and 15 kPa (n) CO_2. Vertical bars indicate 95% confidence limits of the mean. (From J Lammertyn et al. *J Exp Bot* 52:1769–1777, 2001, with permission.)

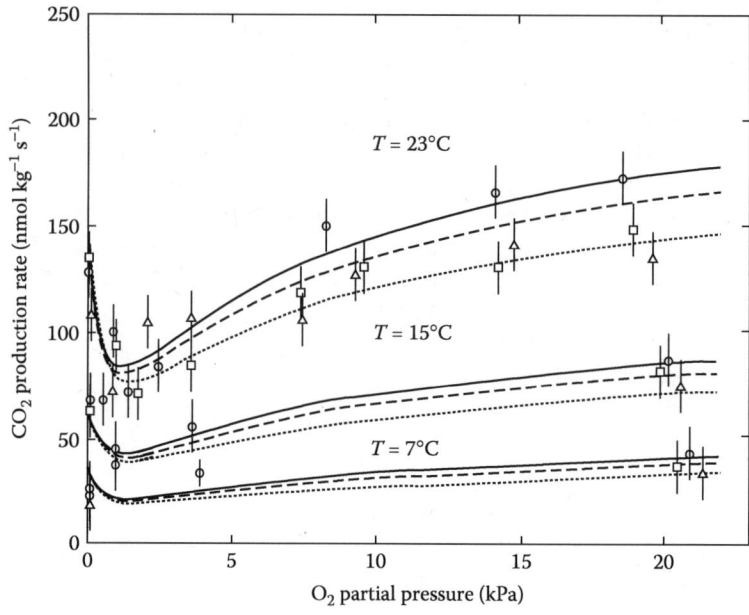

Figure 18.4 Carbon dioxide production rate for Conference pears as a function of the temperature and O_2 and CO_2 partial pressures. Within one temperature, the upper, middle, and lower curves represent the modeled CO_2 production rate at, respectively, 0, 5, and 15 kPa CO_2. Values represent means ($n = 8$) at, respectively, 0 (C), 5 (N), and 15 kPa (n) CO_2. Vertical bars indicate 95% confidence limits of the mean.

Respiration heat is often expressed in terms of carbon dioxide production (18.11). At high oxygen concentration conditions, when carbon dioxide production due to fermentation can be neglected, Equation 18.11 can be adapted as follows:

$$Q = q_{O_2} \frac{V_{CO_2}}{RQ_{ox}} \tag{18.12}$$

In low oxygen conditions, when the fermentative pathway becomes important, the following relation can be used:

$$Q = q_{O_2} V_{O_2} + q_{f,CO_2} \frac{V_{m,f,CO_2}}{1 + \dfrac{P_{O_2}}{K_{m,f,O_2}}} \tag{18.13}$$

In Equation 18.13, a second term is introduced to account for the heat produced by the fermentation metabolism. If it is assumed that mainly ethanol is produced according to Equation 18.2, q_{f,CO_2} has a value of 117 kJ mol^{-1} fermentative carbon dioxide produced.

18.2.1.3 Gas Transport Properties

Not only respiration but also the transport of the respiratory gases (O_2, CO_2) determine the gas exchange of the product. Burton [12] determined four main steps in gas exchange between an organ and its environment:

1. Transport in the gas phase through the outer integument or skin
2. Transport in the gas phase through the intercellular system
3. Exchange of gases between the intercellular atmosphere and the cellular solution
4. Transport in solution in the cell to or from centers of consumption or production

Several transport routes for gas exchange at the surface of the fruit have been suggested: through lenticels or stomata, the cuticle and cracks in the cuticle, and the pedicel opening or floral end [13,14]. Sealing the pedicel opening or floral end slowed down the CO_2 efflux considerably in the case of bell pepper [13,14] and tomato [13,15,16], while for other commodities (oranges, lemons, avocados, limes, pumpkins, bananas, plums, acorn squash, apples, pears), covering the pedicel opening or floral end had no effect [12]. Every cultivar has a specific set of skin properties (lenticels or stomata, the cuticle, and cracks in the cuticle), which may result in cultivar-dependent gas transport characteristics through the skin and contribute to a different reaction on surface coating [17]. Oxygen and carbon dioxide exchange happens mainly through the pores in the fruit skin [18,19]; nevertheless, the cuticular route can contribute substantially [20]. In young apples, the pores in the fruit skin are stomata with guard cells regulating their opening and closure; in mature apples stomata are no longer functional [21]. For mature apples, most stomata are closed [21], completely covered with wax [22], or transformed into lenticels. These lenticels do not function like stomata but close progressively due to the suberization of substomatal cells [21]. For most horticultural produce (oranges, lemons, avocados, limes, pumpkins, bananas, plums, acorn squash, apples, and pears) the most important barrier to gas exchange seems to be the skin of the fruit. Therefore, skin resistance of fruits and vegetables to gas diffusion has been determined by several authors in the past using steady or nonsteady state methods [13,23–27]. All these methods are based on Fick's first law of diffusion and assume that the skin represents the main barrier to gas exchange, and gas transport limitations in the fruit flesh are negligible. The product can be considered a hollow sphere, the internal atmosphere is uniform, and respiratory rates are at equilibrium.

Recently, it has been demonstrated that gas transport limitations in the fruit flesh are not always negligible [28–31]. In fleshy fruits, the transport in the fruit flesh takes place through an internal "ventilation" system consisting of continuous gaseous channels formed by intercellular spaces interconnected with narrow capillary tubes [32]. These intercellular spaces are formed during cell division and growth. Cell division in pome fruit is complete 4–6 weeks after bloom, and further growth of the fruit is due to parenchyma cell enlargement and increase in the size of the intercellular spaces [33]. Cells in plant tissue are surrounded by cell walls that are comparatively rigid and give mechanical support to the tissue [34]. Regulation of wall loosening (dissolving of the middle lamella) and pectin synthesis in growing plant cells is balanced so that newly secreted polymers form linkages, which maintain cell wall stability. In developing fruits and vegetables, degradation can exceed synthesis [35] so that structural changes occur in the middle lamella and primary cell wall, which lead to cell separation, formation of intercellular spaces, and softening of the tissue [36,37]. The most common intercellular spaces develop by the splitting apart of cells through the middle lamella; this starts in the corner, where more than two cells are adjoined [38]. As this process spreads to other wall parts and adjacent cells round off during enlargement, the individual spaces increase in size. This type of intercellular space is called schizogenous since it arises from splitting. Some intercellular spaces also result from breakdown of entire cells and are called lysigenous, or arising by dissolution [38]. The intercellular space volume is cultivar dependent [39,40], not evenly distributed [32], and is smaller in the core parenchyma [39,41].

Gases move from areas of high concentration to areas of low concentration by diffusion due to the random movement of the individual molecules caused by their kinetic energy. Gases may also move by bulk flow (all of the gases present move together rather than independently) caused by a pressure gradient, but in nature this type of transport is not of major importance. In postharvest gas exchange, O_2 in the gas phase diffuses through the skin to the intercellular system and is subsequently transported to the cytoplasm and the centers of O_2 consumption. Carbon dioxide is produced in the cell cytoplasm in dissolved form and follows the reverse path to oxygen. The rate of gas movement depends on the properties of the gas molecule and the physical properties of the intervening barriers [42].

Gas transport in fruit and other bulky storage organs can be macroscopically described with Fick's laws of diffusion ([13,15,26,43–47], see also Chapter 8). Fick's first law of diffusion states that the flux of a gas, J (mol m^{-2} s^{-1}), diffusing through a barrier, is dependent on the diffusivity of the gas, D (m^2 s^{-1}), and the concentration gradient over this barrier, $\partial C/\partial x$ (mol m^{-3} m^{-1}).

$$J = -D \frac{\partial C}{\partial x} \tag{18.14}$$

18.2.2 Measurement Techniques

18.2.2.1 Oxygen Consumption and Carbon Dioxide Production Rate

Whereas in principle every component of the respiration process (O_2, CO_2, water, the respiration heat, see Equation 18.1) can be used to measure the respiration rate; most often, the CO_2 production or O_2 consumption rate is measured. This is because the production of metabolic water is too small in comparison to the overall amount of water present in the product, and heat production measurements need to be carried out in adiabatic setups, which are not easy to realize.

In a *first technique*, the product is placed in a closed recipient at controlled temperature (Figure 18.5). The recipient is flushed with a well-specified gas mixture until the fruit is in

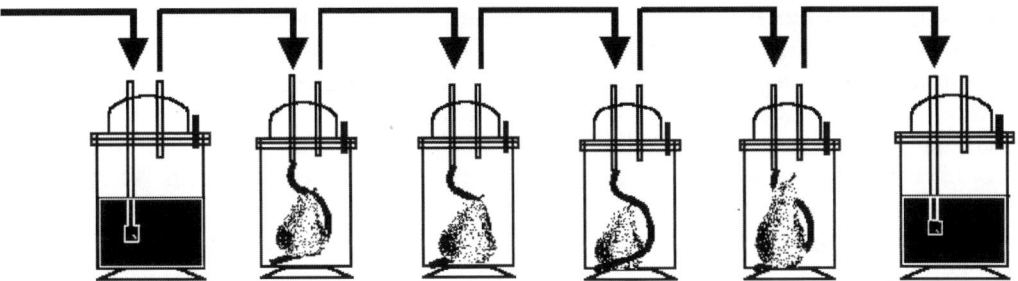

Figure 18.5 Schematic representation of a respiration measurement system. Four jars with fruit, placed in series, are preceded by an air humidifier and followed by a water lock at the end.

equilibrium with this atmosphere. The flushing is then stopped, and the decrease of O_2 or increase of CO_2 concentration inside the container is measured at time t_0 and $t_0 + \Delta t$. The time lag Δt between the two measurements should be chosen so that only a small change in gas concentration will have taken place. The oxygen consumption rate V_{O_2} or the carbon dioxide production rate V_{CO_2} can then be calculated from

$$V_{O_2} = -\frac{V}{m}\frac{d}{dt}C_{O_2}(t) \cong -\frac{V}{m\Delta t}(C_{O_2}(t_0 + \Delta t) - C_{O_2}(t_0)) \tag{18.15}$$

and

$$V_{CO_2} = \frac{V}{m}\frac{d}{dt}C_{CO_2} \cong \frac{V}{m\Delta t}(C_{CO_2}(t_0 + \Delta t) - C_{CO_2}(t_0)) \tag{18.16}$$

with m the mass of the product (kg); V the volume of the recipient minus the volume of the product (m³); and C_{O_2} and C_{CO_2} the concentrations of O_2 and CO_2, respectively (mol/m³).

The *second technique* is based on a flow-through system through which a continuous flow Φ (m³/s) of air with a well-defined temperature and composition is sent. From a simple mass balance it follows that

$$V_{O_2} = \frac{\Phi}{m}(C_{O_2,in} - C_{O_2,out}) \tag{18.17}$$

with $C_{O_2,in}$ and $C_{O_2,out}$ the oxygen concentration at the inlet and outlet, respectively.

18.2.2.2 Measurement of O_2 and CO_2 Concentration
18.2.2.2.1 Gas Chromatography
Gas chromatography (GC) is the most established technique to measure the headspace composition in a recipient. The headspace is sampled with a syringe and injected in a gas

chromatograph for analysis. For CO_2, a silica gel [48], an activated alumina column [47], or a Poraplot column (Varian Inc., Bergen-op-Zoom, Netherlands), combined with a thermal conductivity detector can be used. For O_2, a molecular sieve column combined with a thermal conductivity detector is suitable [47]. This technique requires the use of airtight needles and syringes, and no contamination may occur during transfer to the GC. Another drawback is the duration of the measurement, which can take up to 30 min.

An alternative technique is the micro-GC, which is generally equipped with a pump for automated sampling. The latter prevents contamination of the gas sample [49]. Only one sample is needed for the determination of both O_2 and CO_2 since the gas sample is simultaneously led through two to four modules, each with its own injector, column and column oven, and detector. This enables a much faster measurement, of typically 80 s. The drawback of this method is the larger sampling volume compared to the use of a syringe.

The output of a GC is expressed in percentages or mole fractions; to obtain the partial pressure, the total pressure in the system has to be monitored with a pressure sensor [3,49].

18.2.2.2.2 Amperometric and Potentiometric Methods

The O_2 concentration can be measured with a Clark-type amperometric oxygen-sensitive sensor (Figure 18.6) [28,50]. It is a two-electrode polarographic cell with the working

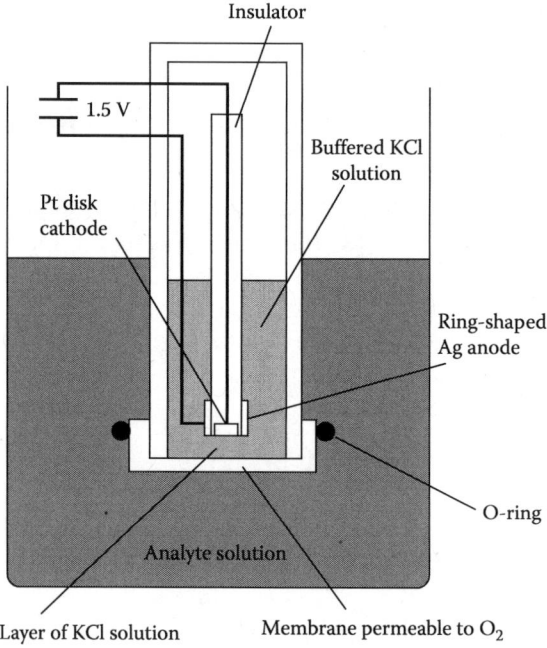

Figure 18.6 Working principle of a Clark-type amperometric oxygen-sensitive sensor.

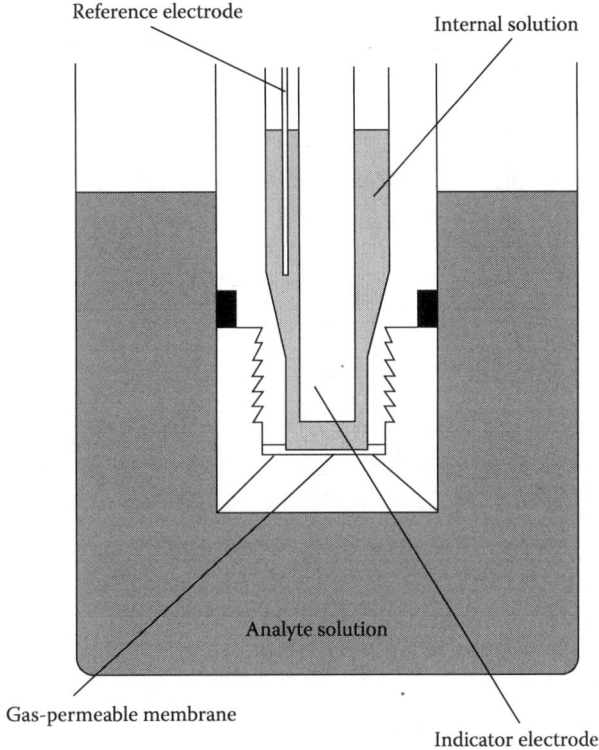

Figure 18.7 Working principle of a CO_2 electrode.

electrode (typically positioned at the end of a tubular structure) separated from the test solution by a thin membrane (PTFE), which is permeable to oxygen.

When a potential of about 0.7 V is applied between the anode (a reference electrode of silver–silver chloride) and the cathode (typically a tiny platinum or gold electrode beneath the membrane), dissolved gaseous oxygen is reduced at the cathode ($O_2 + 4H^+ + 4e^- \rightleftharpoons 2H_2O$). This produces a current and consumes the oxygen in the immediate vicinity of the exposed platinum cathode ($Ag + Cl^- \rightarrow AgCl(s) + e^-$). Oxygen in the sample volume diffuses through the membrane to the oxygen-poor electrolyte solution (typically an aqueous potassium chloride solution) between the membrane and the electrode. When a steady state is reached (which usually happens in 10–20 s), the electrode current is proportional to the rate of arrival of oxygen molecules at the cathode, which is in turn proportional to the concentration of oxygen outside the membrane. Calibration is required.

Carbon dioxide can be measured using a potentiometric sensor. It consists of a membrane module containing a CO_2-permeable membrane and filled with electrolyte solution and a glass pH electrode (Figure 18.7). The sensor has a gas-permeable membrane and an internal buffer solution. Carbon dioxide diffuses across the membrane and reacts with a buffer solution, changing its pH. This change is measured with a combination pH sensor within the housing. Due to the construction, no external reference electrode is needed.

18.2.2.2.3 Infrared Carbon Dioxide Sensor

Many gases absorb radiation at specific wavelengths and transform it into rotation and oscillation energy as described by the Lambert–Beer Law. The absorption at this specific wavelength makes the measurement principle selective. Dual wavelength technology, one to be a reference where no gases are absorbed, enables a compensation of drifting effects caused by the aging of the infrared (IR) source or variations of the measurement cell, resulting in a higher degree of selectivity and accuracy. The IR CO_2 sensor is mostly used to measure CO_2 in a continuous flow of gas [17–52] and is extremely accurate in the very low CO_2 concentration range (0.01 vol%). The beam passes through the atmosphere to be analyzed, and a detector converts the amount of absorbed radiation into an electric signal that is a measure of carbon dioxide concentration in the air stream.

18.2.2.2.4 Paramagnetic Oxygen Sensor

The paramagnetic oxygen sensor is also used for measurement in a continuous flow of gas [51,53]. Oxygen is strongly paramagnetic, and no other gases commonly present in air exert a magnetic influence; hence, this property can be used to measure the concentration of oxygen.

Two nitrogen-filled glass spheres are mounted onto a suspension, free to oscillate at a natural frequency within a nonuniform magnetic field. Any oxygen in the sample gas will be attracted into the strongest part of the magnetic field, thus altering the natural oscillation frequency. The difference between the oscillation period for an oxygen sample and that for nitrogen allows an accurate measurement to be made [54].

18.2.2.2.5 Fiber-Optic Probes

Fiber-optic sensors use a fluorescence technique to measure the absolute oxygen concentration. An optical fiber carries the light produced by a blue LED to the probe tip covered with a dye. The emission characteristic of this specific sensitive dye (e.g., a ruthenium complex) depends on the oxygen concentration of the surrounding medium. The fluorescence intensity of the dye decreases when it is quenched by molecular oxygen. The relation between the fluorescence intensity and lifetime in the absence and presence of oxygen concentration is given by the Stern–Volmer equation. The sensitivity of oxygen sensors based on luminescence quenching is high especially at low oxygen concentrations. The specific sensitive dye is usually immobilized in an oxygen-permeable membrane and typically fixed at a certain substrate layer [55].

18.2.2.3 Measurement of Heat of Respiration

Hayakawa et al. [56] developed a semiclosed calorimeter to determine the rates of respiration heat generation of fresh produce. However, measuring the heat of respiration in a direct way is cumbersome and not done very often. Mostly, the heat of respiration is determined indirectly from oxygen consumption and/or carbon dioxide production rates using Equations 18.11 through 18.13.

18.2.2.4 Skin Resistance and Gas Diffusion Properties

The diffusion properties of fruit skin and tissue can be measured using steady-state or nonsteady-state experiments. In steady-state experiments [13,16,23,24,57], biologically active gases (O_2, CO_2, C_2H_4) are used, and the concentration of the gas of interest inside the product is measured. The assumption is made that the internal atmosphere is uniform, respiratory rates are at equilibrium, and the product is a hollow sphere. In all methods, Fick's first law of diffusion is used to calculate diffusivity, resistance, permeability, or permeance from the difference between the external and internal gas composition.

The difference between these steady-state methods lies in the measurement of this internal gas composition. Destructive methods involve inserting a hypodermic needle [13,16,45,47] or a sensor [14] into the fruit to sample the internal atmosphere. Nondestructive methods include preloading the fruit with N_2 thereby reducing the internal concentration of the biologically active gases (O_2, CO_2, C_2H_4) to zero [24], attaching small vials to the outside of the fruit that equilibrate with the internal atmosphere, and taking samples from these vials [17,58]. The internal atmosphere can also be measured with the vacuum extraction method, in which the fruit is submerged in water and a vacuum is applied removing the internal gas from the fruit [57]. Measuring the internal atmosphere of a product might be easy for a product with an internal cavity (tomato, pepper); it is more difficult [48] for fruits that are more compact.

A simple nonsteady-state method was developed based on kinetic analysis of the efflux of a specific gas [15]. This so-called efflux method involves measuring changes in the concentration of an inert gas (e.g., neon) in a jar caused by diffusion of the gas out of a preloaded fruit. The change in concentration can be described using Fick's first law of diffusion. The speed of the method was enhanced by Banks [26] and used by several other authors to determine skin resistance to gas diffusion [45,46,49,59]. Schotsmans et al. [27] combined the accuracy of the original method with the speed of the enhanced method to measure the diffusion properties of apple. The fruit is placed in an airtight jar and after closing the jar, neon gas is injected and the fruit is stored overnight to allow the gas to diffuse into the fruit. Subsequently, at equilibrium the concentration in the jar is measured and the fruit is quickly transferred to a new jar of equal volume, filled with air without neon gas. The concentration change in the jar is then monitored over time. This technique is based on Fick's first law of diffusion (Equation 18.14). It is assumed that the change in concentration in the barrier (the skin of the fruit) is linear; hence, in Equation 18.14 the concentration gradient $\partial C/\partial x$ can be replaced by $\Delta C/\Delta x$. Since the skin resistance to gas diffusion, R_s (sec m^{-1}), is the parameter of interest, $\Delta x/D$ is replaced with this more commonly used resistance to gas diffusion:

$$J = -\frac{1}{R_s}\Delta C \qquad (18.18)$$

The concentration change in the free external volume with time, $dC_e(t)/dt$, is then given by

$$\frac{dC_e(t)}{dt} = -\frac{A}{R_s V_e}(C_e(t) - C_i(t)) \tag{18.19}$$

with V_e (m³), the free external volume, A (m²) the surface area, and $C_i(t)$, the gas concentration inside the fruit (on the inside of the barrier, which is the skin) [26].

A constant mass during the experiment can be assumed since the amount of gas taken out by sampling is very small and can therefore be neglected. Additionally, the second jar is initially free [$C_e(0) = 0$] of inert gas. The initial concentration of inert gas in the fruit, or more specifically in the internal free gas space volume in the fruit (V_i in m³), is $C_i(0)$ and after complete equilibration, the concentrations of the inert gas in the fruit and in the free space outside the fruit are equal [$C_i(\infty) = C_e(\infty)$]. The change in concentration of the inert gas in the jar, $dc_e(t)/dt$, is then given by Equation 18.20, which can be integrated to yield Equation 18.21.

$$\frac{dC_e(t)}{dt} = \frac{A(V_i + V_e)}{R_s V_i V_e}(C_e(\infty) - C_e(t)) \tag{18.20}$$

$$C_e(t) = C_e(\infty)(1 - e^{-A(V_i+V_e)t/R_s V_i V_e}) \tag{18.21}$$

The gas concentration typically changes exponentially in time, with an initial, virtually linear phase, and this has been confirmed experimentally by Cameron and Yang [15] and Banks [26]. In the original method, the measurements are log-transformed, resulting in a linear function (Equation 18.22) when plotted against time [15].

$$-\ln\left(1 - \frac{C_e(t)}{C_e(\infty)}\right) = \frac{A(V_i + V_e)}{R_s V_i V_e} t \tag{18.22}$$

From the slope (S) of the resulting linear function the resistance value (R) is calculated from Equation 18.23.

$$R_s = \frac{A(V_i + V_e)}{V_i V_e S} \tag{18.23}$$

In an attempt to find a faster method and because the log-transformation of the data did not always result in a linear plot, for instance, in the case of potato, Banks [26] suggested using the nontransformed efflux curve. When a sufficient quantity of samples is taken during the short initial virtually linear phase of the efflux process, the resistance value (R_s) is obtained from the slope (S_t) of the tangent of the nontransformed efflux curve at $t = 0$ as given by

$$R_s = \frac{C_i(0)A}{V_e S_t} \qquad (18.24)$$

However, when the sampling frequency is a restricting factor, it is impossible to limit sampling within the short initial linear part.

The short measurement time can be combined with enough accuracy if a nonlinear parameter estimation procedure is used, where the parameter (R_s) is directly estimated from Equation 18.21 using a nonlinear least-squares method [27].

Measurement of skin resistance to gas transport on whole fruit [26,27,45,46,59] is nondestructive but is based on the assumption that gas transport limitations in the fruit flesh are negligible. Although the skin represents the main barrier to gas exchange, and gas transport in the fruit flesh is 10–20 times faster, it does not exclude the fruit flesh as a possible barrier [52,60]. Therefore, measurement of gas diffusivity in the fruit flesh is needed. Burg and Burg [13] developed a method based on diffusion of gas through a tissue sample. A tissue sample is placed between two chambers, one chamber is flushed with air containing CO_2 and the other is flushed with CO_2-free air, both gas streams are then sampled, and the composition is determined. A similar method to determine oxygen diffusivity was used by Zhang and Bunn [61] and Lammertyn et al. [28]. Streif [62] used diffusion of an inert gas through a tissue sample, and the diffusivity of oxygen and carbon dioxide was then recalculated using Graham's law, which states a constant relation between the diffusivity of two gases; however, this law has since been shown not to apply to biological tissue [28,31]. In these methods respiration effects were not taken into account, although they do affect the measurements, as shown by Lammertyn et al. [28], who measured the respiration of the tissue and the diffusivity of the respiratory gases (O_2, CO_2) separately. Lammertyn et al. [28] used an adaptation of the method of Burg and Burg [13], in that one chamber was constantly flushed with a gas mixture with constant composition; the concentration change in the other chamber was then followed in time. In the method developed by Schotsmans et al. [31], neither chamber was constantly flushed, and the respiration of the tissue and the diffusivity of the respiratory gases (O_2, CO_2) were measured separately.

Gas diffusion measurements are discussed extensively in Chapter 8.

18.2.3 Gas Exchange Data for Selected Fruits and Vegetables

For many fruits and vegetables, gas exchange properties have been reported in the literature. In Table 18.1, they are tabulated for a number of horticultural products. Only for a few products are the constants of the more advanced Michaelis–Menten models with CO_2 inhibition known. Some values are given in Table 18.2.

ENGINEERING PROPERTIES OF FOODS

Table 18.1 Gas Exchange–Related Properties of Horticultural Produce

Botanical Species	Common Name	Cultivar	T_{ref}	V_{m,O_2}	V_{m,CO_2}	RQ_{ox}	K_{m,O_2}	E_{a,vm,O_2}
Agaricus bisporus	Mushroom		10	4.72×10^{-7}	3.61×10^{-7}			
Allium cepa L.	Onion	TG 1015Y	1		3.31×10^{-7}			
		TG 1015Y	15		1.08×10^{-6}			
		TG 1015Y	30		6.95×10^{-7}			
Asparagus officinalis L.	Asparagus	*altilis*	19	5.73×10^{-7}		0.84	2.80	
Brassica oleracea	Broccoli		16	6.47×10^{-3}		0.71		
		Italica	19	1.85×10^{-6}		0.68	4.78	
Brassica oleracea L.	Cabbage (shredded)			2.57×10^{-3}		0.80		
Capsicum annuum L.	Sweet pepper	Reflex	24	2.55×10^{-6}	2.22×10^{-6}		1.40	
	Green bell pepper	Tasty	15	1.45×10^{-7}		0.88	3.40	3.09×10^4
Chicorium intybus L.	Chicory (cut)	Foliosum	8	6.06×10^{-7}		0.99	3.73	
	Chicory		10	1.12×10^{-7}		0.90	2.70	6.71×10^4
Citrus sinensis L. Osbeck	Orange				1.37×10^{-10}			
Cucumis melo	Cantaloupe							
Fragaria × ananassa Duchesne	Strawberry	Elsanta	10	2.70×10^{-7}		0.91	2.63	7.48×10^4
Ipomea batatas (L.) Lam	Sweet potato	Coquinho			3.88×10^{-4}	4.13×10^{-4}		
Lycopersicon esculentum Mill.	Tomato	Kada			5.51×10^{-4}	5.44×10^{-4}		
		Platense	10	1.22×10^{-7}		0.91	23.20	6.73×10^4
			10	3.90×10^{-4}				
Malus sylvestris subsp. *mitis* (Wallr.) Mansf.	Apple	Blenheim Oragen	4	2.81×10^{-8}	4.41×10^{-8}			
		Braeburn	0	4.55×10^{-8}	4.11×10^{-8}	0.94	7.25	
		Braeburn	20		1.07×10^{-7}			
		Braeburn	30		2.10×10^{-7}			5.53×10^4
		Cox's Orange Pippin	3	5.09×10^{-8}	7.24×10^{-8}	1.42		
		Cox's Orange Pippin	18	1.76×10^{-7}	2.49×10^{-7}	1.01	3.00	

GAS EXCHANGE PROPERTIES OF FRUITS AND VEGETABLES

Table 18.1 Gas Exchange–Related Properties of Horticultural Produce

E_{a,vm,f,CO_2}	$C_{C_2H_4}$	P	P'_{o_2}	P'_{co_2}	$D^{skin}_{O_2}$	$D^{flesh}_{O_2}$	$D^{skin}_{CO_2}$	$D^{flesh}_{CO_2}$	Source
									(63)
									(64)
									(64)
									(64)
									(65)
									(66)
									(65)
									(66)
									(67)
			2.30×10^{-11}	2.37×10^{-10}					(52)
			6.50×10^{-10}	1.66×10^{-10}					(14)
3.99×10^4									(68)
									(65)
7.16×10^4									(69)
	3.44×10^{-13}		6.84×10^{-10}	7.20×10^{-10}					(25)
			1.17×10^{-9}						(70)
5.74×10^4									(69)
									(71)
							2.23×10^{-8}	2.50×10^{-7}	(45)
									(71)
				5.25×10^{-10}				2.40×10^{-8}	(16)
6.52×10^4									(69)
			3.80×10^{-10}						(44)
									(66)
									(72)
									(73)
	5.85×10^{-10}	900							(74)
									(20)
									(75)
									(76)

continued

Table 18.1 (continued) Gas Exchange–Related Properties of Horticultural Produce

Botanical Species	Common Name	Cultivar	T_{ref}	V_{m,O_2}	V_{m,CO_2}	RQ_{ox}	K_{m,O_2}	E_{a,vm,O_2}
		Cox's Orange Pippin	20	3.06×10^{-7}			2.20	
		Elstar	20	1.83×10^{-7}		0.98	4.67	
		Elstar						
		Empire						
		Fuji						
		Gala	20					
		Gala						
		Golden Delicious	10	1.06×10^{-7}		0.84	3.76	5.29×10^4
		Golden Delicious	19	2.80×10^{-7}		0.99	6.59	
		Golden Delicious		5.27×10^{-8}				
		Granny Smith	20	2.10×10^{-7}			4.20	
		Granny Smith						
		Granny Smith						
		Gravenstein						
		Jonagold						
		Jonathan	4	3.06×10^{-8}	4.77×10^{-8}			
		McIntosh						
		Newton Wonder	4	3.06×10^{-8}	4.65×10^{-8}			
		Red Delicious						
		Red Delicious						
		Rome Beauty						
		Spartan						
		Starking						
		Stayman	20					
		Sturmer Pippin	4	4.04×10^{-8}	5.75×10^{-8}			
		Tydeman's Late Orange	4	5.02×10^{-8}	6.61×10^{-8}			
		York Imperial	20					
Phaseolus aureus, syn mungo	Mung bean sprouts	Wilczek	8	2.01×10^{-7}	1.56×10^{-7}	0.88	1.21	
Prunus persica (L.) Batsch	Nectarine	Independence	18	3.27×10^{-7}		0.92	0.68	
Pyrus communis L.	Pear	Bartlett	20	4.31×10^{-7}	3.20×10^{-7}	1.00	2.16	
		Bartlett	25	4.04×10^{-7}				
		Bartlett						

Table 18.1 (continued) Gas Exchange–Related Properties of Horticultural Produce

E_{a,vm,f,CO_2}	$C_{C_2H_4}$	P	P'_{O_2}	P'_{CO_2}	$D^{skin}_{O_2}$	$D^{flesh}_{O_2}$	$D^{skin}_{CO_2}$	$D^{flesh}_{CO_2}$	Source
			4.56×10^{-10}						(77)
									(65)
		800							(78)
		815							(78)
		862							(78)
								1.41×10^{-7}	(45)
		841							(78)
5.24×10^4									(69)
									(65)
			1.19×10^{-10}	9.03×10^{-11}	7.40×10^{-12}	2.67×10^{-9}	5.50×10^{-12}	3.28×10^{-9}	(79)
			4.02×10^{-10}						(77)
									(59)
					3.74×10^{-9}	1.90×10^{-8}			(61)
		824							(80)
		816							(78)
									(72)
		788							(78)
									(72)
					3.71×10^{-9}	1.81×10^{-8}			(61)
		827							(78)
					3.10×10^{-9}	1.84×10^{-8}			(61)
		790							(78)
			8.14×10^{-10}						(81)
								1.01×10^{-7}	(45)
									(72)
									(72)
								1.44×10^{-7}	(45)
									(82)
									(65)
		958							(80)
									(83)
	4.04×10^{-10}								(84)
		1002							(80)

continued

Table 18.1 (continued) Gas Exchange–Related Properties of Horticultural Produce

Botanical Species	Common Name	Cultivar	T_{ref}	V_{m,O_2}	V_{m,CO_2}	RQ_{ox}	K_{m,O_2}	E_{a,vm,O_2}
		Beurre Bosc	20	2.94×10^{-7}	1.45×10^{-7}	1.00	8.55	
		Conference	12					
		Conference	18	2.09×10^{-7}	1.48×10^{-7}	0.71		
		Conference	20	8.71×10^{-8}			6.20	6.46×10^{4}
		Doyenne du Comice	15					
		Doyenne du Comice	20	1.69×10^{-7}	1.45×10^{-7}	1.00	1.13	
		Packham's Triumph	20	2.53×10^{-7}	1.45×10^{-7}	1.00	2.21	
Solanum tuberosum L.	Potato	Bintje		1.17×10^{-4}	1.32×10^{-4}			
		King Edward	10	6.03×10^{-7}				
		King Edward						
		King Edward Russet						
		Russet Burbank	10		2.89×10^{-8}			
		Russet Burbank	27		7.95×10^{-8}			

18.3 APPLICATIONS

The respiration rate of fruits can be used as an indicator to optimize storage conditions to increase the longevity of the fruit. Temperature has been shown to be by far the most significant environmental factor affecting the quality deterioration of fruits and vegetables during storage, since all biochemical reactions taking place during ripening are retarded at a lower temperature [1]. Moreover, low-temperature storage reduces the growth of fungi, bacteria, and insects [85]. A decreased O_2 concentration results in a slower metabolism in many vegetables and fruits, such as broccoli, carrots, peas, tomatoes [86], apples, peaches, and pears [9,87]. The role of CO_2 in the respiration metabolism of fruits is more ambiguous. It can act as both a suppressor and an inducer of respiration. Increased CO_2 levels have been shown to retard the respiration rate of apples, broccoli, Belgian endives, and sprouts [7,82], while no effect was noticed on the respiration rate of onions, lettuce, and spinach [88]. The mechanism by which elevated CO_2 concentrations affect the regulation of respiratory metabolism is still unclear; several hypotheses have been proposed for its mode of action [88].

The principle of raising the CO_2 and reducing the O_2 concentration in the storage atmosphere, in combination with a low temperature, finds its application in two techniques widely used to extend the storage life of fruits and vegetables: storage under controlled atmosphere (CA) and packaging under modified atmosphere (MA).

In *controlled atmosphere storage*, the O_2 and CO_2 concentrations are maintained at a well-defined set point level. This set point is often close to the ACP, the O_2 partial pressure

Table 18.1 (continued) Gas Exchange–Related Properties of Horticultural Produce

E_{a,vm,f,CO_2}	$C_{C_2H_4}$	P	P'_{O_2}	P'_{CO_2}	$D^{skin}_{O_2}$	$D^{flesh}_{O_2}$	$D^{skin}_{CO_2}$	$D^{flesh}_{CO_2}$	Source
									(83)
					3.29×10^{-10}	4.32×10^{-10}	4.34×10^{-10}	1.73×10^{-9}	(31)
									(49)
									(10)
									(31)
									(83)
									(83)
									(71)
									(24)
				3.73×10^{-8}					(58)
			1.16×10^{-10}						(12)
		1068							(80)
							6.24×10^{-9}	2.50×10^{-6}	(47)
							7.26×10^{-9}	2.65×10^{-6}	(47)

at which the CO_2 production rate is at its minimum. The optimal gas composition is a compromise between a respiration rate that is as low as possible, while avoiding quality disorders (core breakdown, production of off-flavors) that are related to elevated CO_2 or reduced O_2 concentrations and must be determined in large-scale storage trials. The optimal storage temperature depends on the produce and is, in general, the lowest temperature preventing the produce from chilling injury or freezing damage.

The oxygen can be reduced by respiration as such, and by injecting N_2 gas produced from ordinary air by removing ("scrubbing") the O_2 and CO_2. O_2 is typically separated from N_2 by means of a membrane separator, which is based on the selective permeability of membranes for different gases (Figure 18.8). The separator consists of a bundle of hollow membrane fibers in a cylindrical shell. The compressed air is fed to the inlet end of the separator and flows inside the hollow fibers toward the opposite end. On the way the air molecules start to permeate through the walls of the fibers according to their permeability. Oxygen, carbon dioxide, and water vapor permeate faster than nitrogen, and the result is an almost pure nitrogen stream at the outlet end. Carbon dioxide can be removed by leading the air over calcium hydroxide, which reacts irreversibly to form calcium carbonate. More recent scrubbers use active coal and molecular sieves to absorb CO_2.

Storage under controlled atmosphere is widely used to store all kinds of horticultural produce. In ultra low oxygen (ULO), the O_2 concentration is lower than 2%. Some optimal gas compositions for long-term storage of fruits and vegetables are summarized in Table 18.3.

Modified atmosphere packaging is a technique in which the produce is kept fresh for a longer period, using a gas-permeable foil allowing, to some extent, gas exchange between

Table 18.2 Constants of Michaelis–Menten Kinetics with Inhibition for Some Fruits and Vegetables

Botanical Species	Common Name	Cultivar	T_{ref}	V_{m,O_2}	V_{m,CO_2}	V_{mf,CO_2}	RQ_{ox}	K_{m,O_2}	K_{mf,O_2}	K_{mf,CO_2}	K_{mu,CO_2}	K_{mn,CO_2}	K_{mc,CO_2}	Ref.
Brassica oleracea L.	Broccoli	Italica	24	2.55×10^{-6}	2.22×10^{-6}			1.40		42.30	114.70			(67)
			19	1.85×10^{-6}		2.41×10^{-7}	0.68	4.78	89.17			11.65		(65)
Chicorium intybus L.	Chicory (cut)	Foliosum	8.1	6.06×10^{-7}		1.14×10^{-7}	0.99	3.73	0.31			13.68		(65)
Phaseolus aureus, syn *mungo*	Mung bean sprouts	Wilczek	18	3.27×10^{-7}		1.42×10^{-7}	0.92	0.68	0.37			14.39		(65)
			8	2.01×10^{-7}	1.56×10^{-7}		0.88	1.21	1.85	6.35		19.70		(82)
Asparagus officinalis L.	Asparagus	Altilis	19	5.73×10^{-7}			0.84	2.79	0.19			45.69		(82)
Malus sylvestris subsp. *mitis* (Wallr.) Mansf.	Apple	Braeburn	0	4.55×10^{-8}	4.11×10^{-8}		0.94	7.25	0.28		0.85			(73)
		Golden Delicious	19	2.80×10^{-7}		1.93×10^{-7}	0.99	6.59	0.24			64.95		(65)
		Golden Delicious	10	1.06×10^{-7}		1.78×10^{-7}	0.84	3.76	1.01	9.63			7.36	(69)
Pyrus communis L.	Pear	Conference	20	8.71×10^{-8}				6.20				70.70		(10)
Lycopersicon esculentum Mill.	Tomato		10	1.22×10^{-7}		8.17×10^{-8}	0.91	23.20	1.37	6.49	7.85		21.30	(69)

GAS EXCHANGE PROPERTIES OF FRUITS AND VEGETABLES

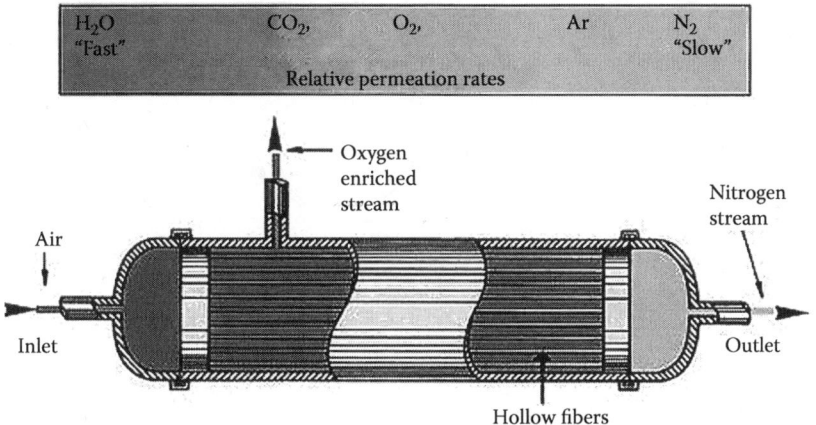

Figure 18.8 Membrane separator. (Courtesy of Air Products AS, Kristiansand, Norway.)

Table 18.3 Optimal Gas Compositions for Long-Term Storage of Fruits and Vegetables

Cultivar	Temperature (°C)	O_2 (%)	CO_2 (%)
Apple			
Braeburn	0.5	3	<1
Cox's orange pippin	4	1.3	0.7
Golden delicious	0.5	2.5	1
Red delicious	0 to 0.5	2.5	4.5
Jonagold	1	1	4.5
Pear			
Conference	−1	2.5	0.7
Packham's Triumph	−0.5	1.5	2.5
Williams Bon Chretien	0 to −0.5	1	0
Other Fruits			
Apricot	0 to 5	2 to 3	2 to 3
Avocado	5 to 13	2 to 5	3 to 10
Banana	12 to 16	2 to 5	2 to 5
Blueberry	0 to 5	2 to 5	12 to 20
Cherimoya	8 to 15	3 to 5	5 to 10
Sweet cherry	0 to 5	3 to 10	10 to 15
Grapefruit	10 to 15	3 to 10	5 to 10
Kiwi	0 to 5	1 to 2	3 to 5
Lemon, Lime	10 to 15	5 to 10	0 to 10
Lychee	5 to 12	3 to 5	3 to 5

continued

Table 18.3 (continued) Optimal Gas Compositions for Long-Term Storage of Fruits and Vegetables

Cultivar	Temperature (°C)	O_2 (%)	CO_2 (%)
Mango	10 to 15	3 to 7	5 to 8
Nectarine	0 to 5	1 to 2	3 to 5
Olive	5 to 10	2 to 3	0 to 1
Orange	5 to 10	5 to 10	0 to 5
Papaya	10 to 15	2 to 5	5 to 8
Pineapple	8 to 13	2 to 5	5 to 10
Plum	0 to 5	1 to 2	0 to 5
Raspberry, strawberry	0 to 5	5 to 10	15 to 20
Vegetables			
Asparagus	2	20	10 to 14
Broccoli	0	1 to 2	5 to 10
Brussels sprouts	0	1 to 2	5 to 7
Cabbage	0	2 to 3	3 to 6
Cantaloupes	3	3 to 5	10 to 20
Celeriac	0	2 to 4	2 to 3
Cucumber	12	1 to 4	0
Lettuce	0 to 5	1 to 3	0
Mushrooms	0	3 to 21	5 to 15
Onion	0	1 to 2	0 to 10
Pepper (bell)	8	2 to 5	2 to 5
Spinach	0	7 to 10	5 to 10
Tomato green	12	3 to 5	2 to 3
Tomato ripe	10	3 to 5	3 to 5
Witloof chicory	0	3 to 4	4 to 5
Fresh-cut vegetables			
Broccoli, florets	0 to 5	2 to 3	6 to 7
Cabbage, shredded	0 to 5	5	5
Lettuce, chopped	0 to 5	0.5 to 3	5 to 10
Onion, sliced or diced	0 to 5	2 to 5	10 to 15
Potato, sliced or peeled	0 to 5	1 to 3	6 to 9
Spinach, cleaned	0 to 5	0.8 to 3	8 to 10

Source: A Kader, D Seaver. *Optimal Controlled Atmospheres for Horticultural Perishables.* Davis, CA: Postharvest Technology Research and Information Center, 2001.

the internal micro atmosphere in the package and the external atmosphere [67]. The micro atmosphere is usually generated through respiration of the produce, but often a gas mixture is injected into the package to obtain a faster equilibrium.

ACKNOWLEDGMENTS

Jeroen Lammertyn and Wendy Schotsmans are postdoctoral fellows of the Fund for Scientific Research-Flanders (F.W.O.-Vlaanderen) (Belgium) and the IWT-Vlaanderen, respectively.

SYMBOLS

A	Surface area (m²)
ACP	Anaerobic compensation point
C	Gas concentration (mol m⁻³)
CA	Controlled atmosphere
D	Diffusivity (m² s⁻¹)
E_{a,vm,f,CO_2}	Activation energy of fermentative CO_2 production (J mol⁻¹)
E_{a,vm,O_2}	Activation energy of O_2 consumption (J mol⁻¹)
e	External
i	Internal
in	Inlet
J	Molar flux (mol m⁻² s⁻¹)
K_{mc,CO_2}	Michaelis–Menten constant for competitive CO_2 inhibition of O_2 consumption (kPa)
K_{mc,fO_2}	Michaelis–Menten constant for O_2 inhibition on fermentative CO_2 production (kPa)
K_{mn,CO_2}	Michaelis–Menten constant for noncompetitive CO_2 inhibition of O_2 consumption (kPa), (%)
K_{m,O_2}	Michaelis–Menten constant for O_2 consumption (kPa), (mg L⁻¹)
K_{mu,CO_2}	Michaelis–Menten constant for uncompetitive CO_2 inhibition of O_2 consumption (kPa)
m	Mass of the product (kg)
MA	Modified atmosphere
n_{cell}	Number of cells
out	Outlet
Φ	Continuous flow (m³ s⁻¹)
P_{CO_2}	CO_2 partial pressure (Pa)
P_{O_2}	O_2 partial pressure (Pa)
P_{O_2}	Permeability to O_2 (mol m⁻² s⁻¹ Pa⁻¹)
PTFE	Polytetrafluoroethylene
Q	Respiration heat production (J kg⁻¹ s⁻¹)
q_{O_2}	Proportionality constant for heat produced by respiration (kJ mol⁻¹)
q_{f,CO_2}	Proportionality constant for heat produced by fermentation (kJ mol⁻¹)
R	Universal gas constant: 8.314 (J mol⁻¹ K⁻¹)
ref	At T_{ref}
ρ	Density (kg m⁻³)

RQ_{ox} Respiratory quotient
R_s Skin resistance to gas diffusion, (s m^{-1})
S Slope of linear regression line (s^{-1})
S_t Slope of tangent of nontransformed efflux curve at $t = 0$ (mol m^{-3} s^{-1})
T_{ref} Reference temperature (°C), (K)
V Volume of the recipient (m^3)
V_{CO_2} CO_2 production rate (mol kg^{-1} s^{-1})
V_e Free external volume (m^3)
V_i Internal free gas space volume in the fruit (m^3)
V_{m,CO_2} Maximal CO_2 production rate (mol kg^{-1} s^{-1})
V_{m,f,CO_2} Maximal fermentative CO_2 production rate (mol kg^{-1} s^{-1})
V_{mO_2} Maximal O_2 consumption rate (mol kg^{-1} s^{-1})
V_{O_2} O_2 consumption rate (mol kg^{-1} s^{-1})

REFERENCES

1. JS Kays. *Postharvest Physiology of Perishable Plant Products*. New York: Van Nostrand Reinhold, 1991.
2. F Kidd, C West. The refrigerated gas storage of apples. *The Department of Scientific and Industrial Research* 8: 1935.
3. NH Banks, DJ Cleland, AC Cameron, RM Beaudry, AA Kader. Proposal for a rationalized system of units for postharvest research in gas exchange. *J Am Soc Hortic Sci* 30: 1129–1131, 1995.
4. P Chevillotte. Relation between the reaction cytochrome oxidase–oxygen and oxygen uptake in cells *in vivo*. *J Theor Biol* 39: 277–295, 1973.
5. AH Millar, FJ Bergersen, DA Day. Oxygen affinity of terminal oxidase in soybean mitochondria. *Plant Physiol Biochem* 32: 847–852, 1994.
6. R Chang. *Physical Chemistry with Applications to Biological Systems*. New York: MacMillan Publishers, 1981.
7. MLATM Hertog, HW Peppelenbos, RG Evelo, LMM Tijskens. A dynamic and generic model on the gas exchange of respiring produce: The effects of oxygen, carbon dioxide, and temperature. *Postharvest Biol Technol* 14: 335–349, 1998.
8. AG Marangoni. *Enzyme Kinetics: A Modern Approach*. 1st ed., Hoboken, NY: John Wiley & Sons, 2003.
9. HW Peppelenbos, J van't Leven. Evaluation of four types of inhibition for modelling the influence of carbon dioxide on oxygen consumption of fruits and vegetables. *Postharvest Biol Technol* 7: 27–40, 1996.
10. J Lammertyn, C Franck, BE Verlinden, BM Nicolaï. Comparative study of the O_2, CO_2 and temperature effect on respiration between "Conference" pear cells in suspension and intact pears. *J Exp Bot* 52: 1769–1777, 2001.
11. In: Parsons RA, Ed. *1993 Ashrae Handbook Fundamentals*. SI ed., Atlanta: American Society of Heating, Refrigerating and Air-Conditioning Engineers, Inc., 1993.
12. WG Burton. *Post-Harvest Physiology of Food Crops*. London: Longman Scientific and Technical, 1982.
13. SP Burg, EA Burg. Gas exchange in fruits. *Physiol Plant* 18: 870–874, 1965.
14. J Bower, BD Patterson, JJ Jobling. Permeance to oxygen of detached *Capsicum annuum* fruit. *Aust J Exp Agric* 40: 457–463, 2000.
15. AC Cameron, SF Yang. A simple method for the determination of resistance to gas diffusion in plant organs. *Plant Physiol* 70: 21–23, 1982.
16. N Bertola, A Chaves, NE Zaritzky. Diffusion of carbon-dioxide in tomato fruits during cold-storage in modified atmosphere. *Int J Food Sci Technol* 25: 318–327, 1990.

17. C Amarante, NH Banks, S Ganesh. Relationship between character of skin cover of coated pears and permeance to water vapour and gases. *Postharvest Biol Technol* 21: 291–301, 2001.
18. NH Banks, BK Dadzie, DJ Cleland. Reducing gas exchange of fruits with surface coatings. *Postharvest Biol Technol* 3: 269–284, 1993.
19. NH Banks, JGM Cutting, SE Nicholson. Approaches to optimising surface coatings for fruits. *NZ J Crop Hortic Sci* 25: 261–272, 1997.
20. Q Cheng, NH Banks, SE Nicholson, AM Kingsley, BR MacKay. Effects of temperature on gas exchange of "Braeburn" apples. *NZ J Crop Hortic Sci* 26: 299–306, 1998.
21. YM Park. Seasonal changes in resistance to gas diffusion of "McIntosh" apples in relation to development of lenticel structure. *J Kor Soc Hortic Sci* 32: 329–334, 1991.
22. EA Veraverbeke, P Verboven, P Van Oostveldt, BM Nicolaï. Prediction of moisture loss across the cuticle of apple [*Malus sylvestris* subsp. *mitis* (Wallr.)] during storage Part 1. Model development and determination of diffusion coefficients. *Postharvest Biol Technol* 30(1): 75–88, 2003.
23. F Kidd, C West. Respiratory activity and duration of life of apples gathered at different stages of development and subsequently maintained at a constant temperature. *Plant Physiol* 20: 467–504, 1945.
24. WG Burton. The permeability to oxygen of the periderm of the potato tuber. *J Exp Bot* 16: 16–23, 1965.
25. S Ben-Yehoshua, SP Burg, R Young. Resistance of citrus fruit to mass transport of water vapor and other gases. *Plant Physiol* 79: 1048–1053, 1985.
26. NH Banks. Estimating skin resistance to gas diffusion in apples and potatoes. *J Exp Bot* 36: 1842–1850, 1985.
27. W Schotsmans, BE Verlinden, J Lammertyn, A Peirs, P Jancsók, N Scheerlinck, BM Nicolaï. Factors affecting skin resistance measurements in pipfruit. *Postharvest Biol Technol* 25: 169–179, 2002.
28. J Lammertyn, N Scheerlinck, BE Verlinden, W Schotsmans, BM Nicolaï. Simultaneous determination of oxygen diffusivity and respiration in pear skin and tissue. *Postharvest Biol Technol* 23: 93–104, 2001.
29. J Lammertyn, N Scheerlinck, P Jancsók, BE Verlinden, BM Nicolaï. A respiration–diffusion model for "Conference" pears I: Model development and validation. *Postharvest Biol Technol* 20(1): 31–44, 2003.
30. J Lammertyn, N Scheerlinck, P Jancsók, BE Verlinden, BM Nicolaï. A respiration–diffusion model for "Conference" pears II: Simulations and relation to core breakdown. *Postharvest Biol Technol* 30(1): 45–57, 2003.
31. W Schotsmans, BE Verlinden, J Lammertyn, BM Nicolaï. Simultaneous measurement of oxygen and carbon dioxide diffusivity in pear tissue. *Postharvest Biol Technol* 29: 155–166, 2003.
32. F Ruess, R Stösser. Über die dreidimensionale Rekonstruktion des Interzellularsystems von Apfelfrüchten [Over the three-dimensional reconstruction of the intercellular system of apple fruits.]. *Angew Bot* 67: 113–119, 1993.
33. KG Lapsley, FE Escher, E Hoehn. The cellular structure of selected apple varieties. *Food Struct* 11: 339–349, 1992.
34. IM Bartley, M Knee, MA Casimir. Fruit softening. 1. Changes in cell-wall composition and endo-polygalacturonase in ripening pears. *J Exp Bot* 33: 1248–1255, 1982.
35. M Knee. Fruit softening. II. Precursor incorporation into pectin by pear tissue slices. *J Exp Bot* 33: 1256–1262, 1982.
36. IM Bartley, M Knee. The chemistry of textural changes in fruit during storage. *Food Chem* 9: 47–58, 1982.
37. R Ben-Arie, N Kislev. Ultrastructural changes in the cell walls of ripening apple and pear fruit. *Plant Physiol* 64: 197–202, 1979.
38. K Esau. *Anatomy of Seed Plants*. New York: John Wiley & Sons, 1977.
39. RM Reeve. Histological investigations of texture in apples II. Structure and intercellular spaces. *Food Res* 604–617, 1953.

40. MC Goffinet, TLLAN Robinson. A comparison of "Empire" apple fruit size and anatomy in unthinned and hand-thinned trees. *J Hortic Sci* 70: 375–387, 1995.
41. WG Burton. Some biophysical principles underlying controlled atmosphere storage of plant material. *Ann Appl Biol* 78: 149–168, 1974.
42. AA Kader. *Respiration and Gas Exchange of Vegetables. Postharvest Physiology of Vegetables.* New York: Marcel Dekker, Inc., 1988, pp. 25–43.
43. CJ Geankoplis. *Transport Processes and Unit Operations.* 3rd ed. Englewood Cliffs, NJ: Prentice-Hall, Inc., 1993.
44. AC Cameron, SM Reid. Diffusive resistance: Importance and measurement in controlled atmosphere storage. In: Richardson DG, Meheriuk, M, Ed. *Controlled Atmosphere for Storage and Transport of Perishable Agricultural Commodities.* Beaverton, OR: Timber Press, 1982.
45. T Solomos. Principles of gas exchange in bulky plant tissues. *HortScience* 22: 766–771, 1987.
46. JP Emond, F Castaigne, CJ Toupin, D Desilets. Mathematical modelling of gas exchange in modified atmosphere packaging. *Trans ASAE* 34: 239–245, 1991.
47. AA Abdul-Baki, T Solomos. Diffusivity of carbon dioxide through the skin and flesh of "Russet Burbank" potato tubers. *J Am Soc Hortic Sci* 119: 742–746, 1994.
48. NH Banks. Evaluation of methods for determining internal gases in banana fruit. *J Exp Bot* 34: 871–879, 1983.
49. HPJ de Wild, HW Peppelenbos. Improving the measurement of gas exchange in closed systems. *Postharvest Biol Technol* 22: 111–119, 2001.
50. DA Skoog, DM West, FJ Holler. *Fundamentals of Analytical Chemistry.* 7th ed. Orlando, FL: Saunders College Publishing, 1996.
51. JC Fidler, CJ North. The respiration of apples in C.A. storage conditions. *Commisions* 4,5, 1966, pp. 1–8.
52. NH Banks, SE Nicholson. Internal atmosphere composition and skin permeance to gases of pepper fruit. *Postharvest Biol Technol* 18: 33–41, 2000.
53. H Bohling, B Bauer. Determination of the respiration activity of fruit and vegetables in controlled atmospheres using an automatic measuring device. *Acta Hortic* 343: 161–162, 1993.
54. M Brown, P Hammond, T Johnson. *Dictionary of Medical Equipment.* London: Chapman & Hall, 1986.
55. M Krihak, MRA Shahriari. A highly sensitive, all solid state fiber optic oxygen sensor based on the sol-gel coating technique. *Electron Lett* 32: 240–242, 1996.
56. K Hayakawa, D Biran, E Vaccaro, SG Gilbert. Development of a new procedure for direct determination of respiration heat generation by fresh produce. *Lebensm Wiss u Technol* 12: 189–193, 1979.
57. L Rodriguez, D Zagory, AA Kader. Relation between gas diffusion resistance and ripening in fruits. *Proceedings of the Fifth International Controlled Atmosphere Research Conference*, Washington, D.C., 1989, 2, pp. 1–7.
58. NH Banks, SJ Kays. Measuring internal gases and lenticel resistance to gas diffusion in potato tubers. *J Am Soc Hortic Sci* 113: 577–580, 1988.
59. M Knee. Rapid measurement of diffusion of gas through the skin of apple fruits. *HortScience* 26: 885–887, 1991.
60. T Solomos. A simple method for determining the diffusivity of ethylene in McIntosh apples. *Sci Hortic* 39: 311–318, 1989.
61. J Zhang, JM Bunn. Oxygen diffusivities of apple flesh and skin. *Trans ASAE* 43: 359–363, 2000.
62. J Streif. Gasdiffusionsmessungen an Früchten [Gas diffusion measurements of fruits.]. *Annual Meeting of the DGQ*, Freising-Weihenstephan, Germany, 1999, XXXIV.
63. P Varoquaux, B Gouble, C Barron, F Yildiz. Respiratory parameters and sugar catabolism of mushroom (*Agaricus bisporus* Lange). *Postharvest Biol Technol* 16: 51–61, 1999.
64. KS Yoo, CR Andersen, LM Pike. Internal CO_2 concentrations in onion bulbs at different storage temperatures and in response to sealing of the neck and base. *Postharvest Biol Technol* 12: 157–163, 1997.

65. HW Peppelenbos. The use of gas exchange characteristics to optimize CA storage and MA packaging of fruits and vegetables. PhD dissertation, Landbouwuniversiteit Wageningen, Wageningen, The Netherlands, 1996.
66. Y Makino, K Iwasaki, T Hirata. A theoretical model for oxygen consumption in fresh produce under an atmosphere with carbon dioxide. *J Agric Eng Res* 65: 193–203, 1996.
67. DS Lee, PE Haggar, KL Yam. Model for fresh produce respiration in modified atmospheres based on principles of enzyme kinetics. *J Food Sci* 56: 1580–1585, 1991.
68. X Chen, MLATM Hertog, NH Banks. The effect of temperature on gas relations in MA packages for capsicums (*Capsicum annuum* L., cv. Tasty): an integrated approach. *Postharvest Biol Technol* 20: 71–80, 2000.
69. MLATM Hertog, HAM Boerrigster, GJPM van den Boogaard, LMM Tijskens, ACR van Schaik. Predicting keeping quality of strawberries (cv "Elsanta") packed under modified atmospheres: an integrated model approach. *Postharvest Biol Technol* 15: 1–12, 1999.
70. JM Lyons, WB McGlasson, HK Pratt. Ethylene production, respiration and internal gas concentrations in cantaloupe fruits at various stages of maturity. *Plant Physiol* 37: 31–36, 1962.
71. AA Nery, AG Calbo. Adapting constant-volume manometry for studying gas exchange by bulky plant organs. *J Am Soc Hortic Sci* 119: 1222–1229, 1994.
72. JC Fidler. Controlled atmosphere storage of apples. *The Institute of Refrigeration*, 1965, pp. 1–7.
73. MLATM Hertog, SE Nicholson, NH Banks. The effect of modified atmospheres on the rate of firmness change in 'Braeburn' apples. *Postharvest Biol Technol* 23: 175–184, 2001.
74. CW Yearsley, NH Banks, S Ganesh. Temperature effects on the internal lower oxygen limits of apple fruit. *Postharvest Biol Technol* 11: 73–83, 1997.
75. M Knee. Physiological responses of apple fruits to oxygen concentrations. *Ann Appl Biol* 96: 243–253, 1980.
76. HW Peppelenbos, R Rabbinge. Respiratory characteristics and calculated ATP production of apple fruit in relation to tolerance of low O_2 concentrations. *J Hortic Sci* 71: 985–993, 1996.
77. BK Dadzie, NH Banks, DJ Cleland, EW Hewett. Changes in respiration and ethylene production of apples in response to internal and external oxygen partial pressures. *Postharvest Biol Technol* 9: 297–309, 1996.
78. OL Lau. Effect of growing season, harvest maturity, waxing, low O_2 and elevated CO_2 on flesh browning disorders in "Breaburn" apples. *Postharvest Biol Technol* 14: 131–141, 1998.
79. JD Mannapperuma, RP Singh, ME Montero. Simultaneous gas diffusion and chemical reaction in foods stored in modified atmospheres. *J Food Eng* 14: 167–183, 1991.
80. AG Calbo, NF Sommer. Intercellular volume and resistance to air flow of fruits and vegetables. *J Am Soc Hortic Sci* 112: 131–134, 1987.
81. G Andrich, R Fiorentini, A Tuci, A Zinnai. Skin permeability to oxygen of refrigerated apples. *Ital Food Bev Technol* 2: 23–27, 1993.
82. HW Peppelenbos, L Brien, LGM Gorris. The influence of carbon dioxide on gas exchange of mungbean sprouts at aerobic and anaerobic conditions. *J Sci Food Agric* 76: 443, 1998.
83. C Amarante, NH Banks, S Ganesh. Characterising ripening behaviour and the lower oxygen limit in relation to the internal atmosphere of coated pears. *Postharvest Biol Technol* 23: 51–59, 2001.
84. AA Kader. Mode of action of oxygen and carbon dioxide on postharvest physiology of "Bartlett" pears. *Acta Hortic* 258: 161–167, 1989.
85. AK Thompson. *Controlled Atmosphere Storage of Fruits and Vegetables*. Oxon, UK: CAB International, 1998.
86. J Weichmann. *Postharvest Physiology of Vegetables*. New York: Marcel Dekker, Inc., 1987.
87. DK Salunkhe, SS Kadam. *Handbook of Fruit Science and Technology. Production, Composition, Storage and Processing.* New York: Marcel Dekker, Inc., 1995.
88. FM Mathooko. Regulation of respiratory metabolism in fruits and vegetables by carbon dioxide. *Postharvest Biol Technol* 9: 247–264, 1996.

INDEX

A

Absorptivity, 287; *see also* Emissivity; Infrared heating
  and emissivity, 290
  radiation energy incident at wavelength, 290
  spectral, 289, 295, 296, 297
AC, *see* Alternating current (AC)
Acids, 46
Acoustic absorption in liquids, 642
Acoustic impedance, 642, 644
Acoustic properties of biological tissue, 656
Activation energy, 685
Activity, 365; *see also* Water activity
Adams consistometer, 149; *see also* Flow property measurement
Adiabatic heat of compression, 711
Adsorption energy, 391
AFM, *see* Atomic force microscopy (AFM)
Air
  -borne ultrasound set up, 668
  comparison pycnometer, 10
  -coupled ultrasonic detection, 647
  -drying processes, 406; *see also* Food dehydration
  -flow planimeter, 14
Alkaline phosphatase thermal inactivation kinetics, 691
α-Relaxations, 96; *see also* Glass transition temperature
Alternating current (AC), 648
Alternative processes (AP), 513
A-mode, *see* Amplitude modulation (A-mode)
Amorphous materials, 94; *see also* Glass transition
  molecular mobility, 96
  plasticizers, 104
  specific volume vs. temperature plot, 94
Amplitude modulation (A-mode), 645
ANN, *see* Artificial neural network (ANN)
Anthocyanin thermal degradation kinetics, 695
Antifade reagents, 70
AP, *see* Alternative processes (AP)

Apparent density, 4; *see also* Buoyant force method; Density; Gas pycnometer method; Liquid displacement method
  determination, 6
  geometric dimension method, 6–7
  solid displacement method, 11
  volume displacement method, 9
Apparent porosity, 5; *see also* Porosity
Apparent specific heat, 272; *see also* Specific heat; Thermal properties of frozen foods
  prediction of, 273
  simple methods, 273–274
  of sweet cherries, 272
Arbitrary waveform generator (AWG), 667
Aroma compounds
  diffusion of, 344
  distillation of, 348
  free-volume theory, 346
  microregion retention theory, 345
  retention of volatile, 344
  selective diffusion theory, 345
Arrhenius–Eyring model, 719
Arrhenius model, 684–685, 718
Artificial neural network (ANN), 650
Ascorbic acid thermal degradation kinetics, 698
Ash, 615
Atomic force microscopy (AFM), 85–86, 87
Attenuation, 642, 644, 652
  coefficient, 642
  measurements, 648–649
Autofluorescence, 69
AWG, *see* Arbitrary waveform generator (AWG)
Axial direction, 646

B

Bacillus thermal inactivation kinetics, 689
Back scattered electrons (BSE), 70
Backscattered amplitude integral method (BAI method), 669
BAI method, *see* Backscattered amplitude integral method (BAI method)
Baker compressimeter, 213; *see also* Food texture

INDEX

Baking effect on dielectric properties, 621
   dielectric constant, 622, 624
   dielectric property models for baked dough, 623
   loss factor, 624
   porosity and dielectric properties, 623
   at various moisture and temperatures, 623
   volume variation during baking, 624
Bases, 46
Beam damage, 72; see also Scanning electron microscopy (SEM)
Beer–Lambert's law, 291
Bend testing, 211; see also Food texture
   key analyzes, 212
   testing method, 212–213
Bending test, 184; see also Solid food quasistatic tests
BET, see Brunauer, Emmett and Teller (BET)
β-Carotene thermal degradation kinetics, 695
β-Cryptoxanthin thermal degradation kinetics, 695
Betalain thermal degradation kinetics, 695
β-Relaxations, 100
Beverage industry, 504
   mass transfer-in-series resistance model, 509
   membrane distillation, 507–508, 509–510
   microfiltration, 506–507
   osmotic distillation, 507–508, 510
   PV, 509
   reverse osmosis, 505
   ultrafiltration, 506
Binding energy, 396–398
Bingham relationship, 130; see also Rheological models for viscous food
Biot number, 326; see also Diffusion
Biphasic model, 683–684
   for inactivation behavior of enzymes, 715, 716
   for inactivation behavior of microorganisms, 717
Blackbody emitted energy, 286; see also Emissivity
B-mode, see Brightness modulation (B-mode)
Bostwick consistometer, 149–150; see also Flow property measurement
Boundary volume, 3; see also Volume
Bound water, 361, 599
Bovine serum albumin (BSA), 492; see also Ultrafiltration (UF)
   treatment, 495

Bread; see also Baking effect on dielectric properties
   dielectric constant and loss factor, 624
   dough dielectric properties, 623
   porosity and dielectric properties, 623
   volume variation during baking, 624
Brightness modulation (B-mode), 645
Browning effect, 604
Brunauer, Emmett and Teller (BET), 106; see also Gas–solid and vapor–solid equilibria; Moisture sorption isotherm (MSI)
   equation, 320, 321, 381–382
   monolayer values and temperature, 389
   net isosteric heat of sorption computed from, 388
   temperature dependence, 387
BSA, see Bovine serum albumin (BSA)
BSE, see Back scattered electrons (BSE)
Bubbly liquid food analysis, 667
Buckingham equation, 159, 160
Bulk density, 4; see also Density
Bulk porosity, 6; see also Porosity
Buoyant force method, 7; see also Apparent density
   analytical balance for, 8
   top-loading balance for, 7
Burgers model, 188; see also Solid food quasistatic tests
   creep curves for, 187

C

CA, see Cellulose acetate (CA); Controlled atmosphere (CA)
Capacitive transducers, 647
Capillary rise method, 51
Capillary viscometers, 140; see also Flow property measurement
   shear rate, 141
   sheer stress, 140
Carbohydrate dependence of dielectric properties, 601; see also Fat dependence of dielectric properties; Temperature dependence of dielectric properties
   dielectric constant, 602, 603, 605
   effects of gum on, 606, 607
   loss factor, 602, 604, 605
   of rice flour slurry, 603
Carboxy methylcellulose (CMC), 139
Carman-Kozeny equation, 15

Carrageenan, 607
Case hardening phenomenon, 28
Caseinomacropeptide (CMP), 504
Cassie state, 57
Cassie-Wenzel state transition, 57
Casson model, 129–130; *see also* Rheological models for viscous food
Cavity perturbation technique, 585
    liquid sample preparation, 586
    semisolid samples, 587
    in simple TE and TM modes, 585
    solid sample preparation, 586
CE, *see* Cellulose (CE)
Cell membrane breakdown, 546
    frequency effect, 547
    permeabilization effect in low temperature, 548
    pore formation, 545, 547–548
    pores repair, 546
    pressure effect, 547
    proteolytic enzyme effect, 547
    pulse duration effect, 547
    surfactants effect, 547
    tissue changes, 549, 550–551
Cellulose (CE), 506
Cellulose acetate (CA), 456
    specifications of, 463
CFD, *see* Computational fluid dynamics (CFD)
Channel rheometers, *see* Slit rheometers
Chapman–Enskog equation, 323; *see also* Diffusion
Characteristic temperature, 106; *see also* State diagram
Cheddar cheese permittivity, 618
Cheese
    dielectric properties, 619, 620
    permittivity of, 618
    quality assessment, 659
Chemical potential, 363; *see also* Thermodynamic properties in dehydration; Water activity
    of ideal gas, 364
    ideal solution, 365
    reference state, 364
    water, 366
Chen equation, 383–384; *see also* Moisture sorption isotherm (MSI)
Cherry–Burrell meter, 213; *see also* Food texture
Chilton–Colburn analogies, 335
Chlorophyll thermal degradation kinetics, 695
Chung and Pfost equation, 383; *see also* Moisture sorption isotherm (MSI)
CK equation, *see* Couchman and Karasz equation (CK equation)
Clark-type amperometric O_2-sensitive sensor, 750
Clausius–Clapeyron equation, 109, 382; *see also* Moisture sorption isotherm (MSI); State diagram
    for water vapor, 386–387
Climacteric fruits, 741, 742
Closed pore porosity, 5; *see also* Porosity
Clostridium botulinum thermal inactivation kinetics, 689
CLSM, *see* Confocal laser scanning microscopy (CLSM)
Cluster model, 362
CMC, *see* Carboxy methylcellulose (CMC)
cmc, *see* Critical micelle concentration (*cmc*)
CMP, *see* Caseinomacropeptide (CMP)
CO_2 production rate, 744–745, 746
Coefficient of internal friction, 195
Coefficient of viscosity, 181
Cold pasteurization, *see* High-pressure pasteurization (HP-P)
Collapse-related phenomena, 95
Colloid dielectric probe, 584
Color degradation kinetics, 693–694, 695–696, 697
Combination microwave-infrared oven, 306
Competitive inhibition, 743
Complex modulus, 125; *see also* Rheology of fluid food
Complex viscosity, 125; *see also* Rheology of fluid food; Rheological models for viscous food
Component separation, 482, 483
Computational fluid dynamics (CFD), 661
Computational tomography (CT), 213
Computer-aided tomography, 82; *see also* Micro-CT
Confocal laser scanning microscopy (CLSM), 66; *see also* Food microstructure analysis
    cheese images, 68
    contrast stains, 68
    dark chocolate image, 69
    in situ studies using, 70
    resolution, 67–68
    sample preparation, 68–69
Consecutive—step model, 715
Consistometer, 213; *see also* Food texture

Contact angle, 43; see also Surface properties
  analysis, 43
  critical surface tension, 44
  Dupre equation, 43, 44
  equation of state relationship, 47–48
  evaluation of, 51
  interfacial area reduction, 43
  liquid surface tension, 44
  polar and dispersive contributions, 44–47
  for solid materials, 59
  Zisman plot, 44, 45
Continuum model theory, 362
Controlled atmosphere (CA), 760
  oxygen removal, 761
  storage, 760, 761
Conventional process (CP), 513
Conventionally rendered bacon fat dielectric
    data, 611
Cooking effect on dielectric properties, 625
Cook-value C, 678
  determination, 680
Core breakdown, 741
Core drying model, 411; see also
    Food dehydration
Corn oil dielectric data, 611
Corona discharge methods, 58
COSTHERM, 226
Cottage cheeses dielectric properties, 619
Cotton seed cooking oil dielectric data, 611
Couchman and Karasz equation (CK equation),
    100–101; see also Glass transition
    temperature
Couette flow viscometers, 141; see also Flow
    property measurement
  commercial viscometers, 141
  concentric cylinder viscometer, 142
  non-Newtonian shear rates, 143
  shear rate, 143
Cox–Merz rule, 126; see also Rheology of
    fluid food
CP, see Conventional process (CP)
CPET, see Crystalline PET (CPET)
Creep curves, 187; see also Solid food
    quasistatic tests
Critical micelle concentration (cmc), 50
Critical moisture content, 408; see also
    Food dehydration
Cryo-SEM, 72; see also Scanning electron
    microscopy (SEM)
  advantages and disadvantages, 74
  etching, 74

freezing step, 73
images, 73, 74
Cryo-TEM, 81; see also Transmission electron
    microscopy (TEM)
Crystal growth, 349–350; see also Mass transfer
    processes
Crystalline PET (CPET), 620
Crystallization, 349–350; see also Mass transfer
    processes
  kinetics, 350
  monitoring, 657–658
CT, see Computational tomography (CT)
Current flow in conductor, 529

D

DABCO, see 1,4-diazobicyclo(2.2.2)
    octane (DABCO)
Dairy product industry, 502
  cheese dielectric properties, 618, 619, 620
  electro-dialysis, 504
  ion interaction effect, 503
  microfiltration, 504
  milk dielectric properties, 617
  nanofiltration, 502–503
  PV, 504
  reverse osmosis, 502
  ultrafiltration, 503–504
Darcy's equation, 15
Dashpot, 181; see also Newtonian liquid
dB/m, see Decibels per meter (dB/m)
dc, see Direct current (dc)
DCMD, see Direct contact membrane
    distillation (DCMD)
DE, see Dextrose equivalent (DE)
Debye dielectric relaxation parameters, 590
Decibels per meter (dB/m), 642
Decimal reduction time (D-value), 683, 716
  heat-inactivation kinetics of first-order
    reactions, 685
Deformation tensor, 123; see also Rheology of
    fluid food
Degradation kinetics, see Thermal inactivation
Dehydration, 360
Density, 3, 226, 249; see also Apparent
    density; Mass–volume–area-related
    properties; Material density; Thermal
    properties of frozen foods; Thermal
    properties of unfrozen foods
  apparent, 6
  bulk, 4, 14, 228

INDEX

of composite mixture, 26
of foodstuffs, 227, 263–265
as function of temperature, 264
of gases and vapors, 16
law of addition, 228
of liquid foods, 17–18
material, 3
measurement, 6
of multiphase system, 25
parameters of quadratic equations, 20
particle density, 4, 13
perfect gas law, 228
vs. porosity, 228
predictions, 16
of solid foods, 18–21
vs. temperature, 227
true, 3
Van der Waal's equation of state, 16
Denture tenderometer, 208; *see also* Food texture
DES, *see* Dielectric spectroscopy (DES)
Desired reactions in food, 678
DETA, *see* Dielectric thermal analysis (DETA)
Dew point temperature measuring instruments, 374; *see also* Water activity
Dextrose equivalent (DE), 102; *see also* Glass transition temperature
1,4-diazobicyclo(2.2.2) octane (DABCO), 70
Dielectric constant, 574
    of bread, 624
    dairy product, 611, 612
    frequency and, 553–555
    of gelatinized starch, 602, 603
    in partially frozen region, 595
    at sterilization temperatures, 598
    of sugar, 605
    temperature dependence of, 594
Dielectric loss factor, 574
    of bread, 624
    of starch solutions, 604
    of sugar, 605
    temperature and, 602, 604
Dielectric properties; *see also* Food dielectric properties
    baking effect on, 621–625
    carbohydrate dependence of, 601–607
    cooking effect on, 625
    of dairy product, 617–618
    drying effect on, 625
    fat dependence of, 609–612
    frequency dependence of, 587–591

    mixing effect on, 626
    protein dependence of, 607–609
    storage effect on, 626
    temperature dependence of, 592–599
    water effect on, 608
Dielectric property measurement, 580, 587; *see also* Cavity perturbation technique; Food dielectric properties
    colloid dielectric probe, 584
    free-space transmission technique, 583
    microstrip transmission line, 583
    open-ended probe technique, 581–582
    pressure-proof dielectric test cell, 582
    short-circuited line technique, 581
    six-port reflectometer, 583–584
    test cell with Boonton RX-meter, 584
    time-domain reflectometry method, 582–583
    waveguide and coaxial transmission line methods, 580–581
Dielectric relaxation, 115; *see also* Glass transition temperature
Dielectric spectroscopy (DES), 99; *see also* Glass transition temperature
Dielectric thermal analysis (DETA), 115, 116; *see also* Glass transition temperature
Differential interference contrast, 66
Differential method, 375; *see also* Water activity
Differential scanning calorimeter (DSC), 96, 537; *see also* Glass transition in solid foods; Glass transition temperature
    frozen glass transition temperature determination, 249
    frozen water fraction measurement, 595
    heat capacity, 205
    heating thermogram, 108, 110
    for monitoring phase transitions, 658
    plot, 205
    protein denaturation determination, 608
    thermal transitions of polymers, 204
    types, 112
Diffusion, 322; *see also* Aroma compounds; Diffusivity; Mass transfer processes
    Biot number, 326
    Chapman–Enskog equation, 323
    Einstein–Stokes equation, 325
    Fick's law, 322
    Fuller equation, 323
    in gases, 323–324
    in liquids, 324
    mass transport rate, 322
    molecular simulations, 328

Diffusion (*Continued*)
   in polymers, 327–328
   in porous foods, 338
   in solids, 325–327
   of solutes, 342–343
   of sugars and fats, 343
   transient-state diffusion, 322
   Wilke–Chang equation, 324
Diffusivity, 322; *see also* Diffusion; Diffusivity estimation in solids
   in air-dried potato, 337
   in dilute aqueous solutions, 325
   effective, 326
   effective moisture, 337
   of gases and vapors, 324
   in glassy polymers, 327
   pressure and, 338
   of sodium chloride, 342
   of sodium hydroxide, 346
   in solids, 327
   of solutes in gels and foods, 343
   in solvent extraction of solids, 347
   of water vapor, 351
Diffusivity estimation in solids, 328; *see also* Diffusion; Diffusivity
   drying curve and diffusion plot, 331
   drying rate, 330–331
   Fickian diffusion, 320
   model-fitting procedure, 331
   penetrant distribution, 330
   penetrant flux, 320
   permeation measurements, 329–330
   sorption kinetics, 320, 328
Dimensionless quantities, 447
Dimensions, 240–241
Dipole rotation of water molecules, 528
Direct contact membrane distillation (DCMD), 509
Direct current (dc), 238
Direct stress-strain tests, 193, 216–217; *see also* Solid food testing
   application of, 199
   coefficient of internal friction, 195
   dynamic modulus, 194
   dynamic test apparatus, 196
   empirical equation, 196
   food properties by, 200–202
   loss modulus, 195
   phase angle, 193
   storage modulus, 194
Direct transmitted fraction, 301

Dispersions, 122; *see also* Rheology of fluid food
Directional emissivity, 286; *see also* Emissivity
Distance–concentration method, 330; *see also* Diffusivity estimation in solids
Distillation, 347; *see also* Mass transfer processes
   Rayleigh equation, 348
   steam, 348
   tray efficiency, 347–348
   of volatile aroma compounds, 348
Distinct—isozyme model, 715
DLS, *see* Dynamic light scattering (DLS)
DMA, *see* Dynamic-mechanical analysis (DMA)
Dry weight content (DW content), 664
Dryers, 421, 422; *see also* Food dehydration
Drying, 406; *see also* Diffusivity estimation in solids; Food dehydration
   curve, 407
   dryers, 421, 422
   effect on dielectric properties, 625
   energy for water removal, 421
   grains, 215–216
   method, 330–331
   rate, 408, 409
DSC, *see* Differential scanning calorimeter (DSC)
Du Nouy ring tensiometry, 52–53
Dupre equation, 43, 44
D-value, *see* Decimal reduction time (D-value)
DW content, *see* Dry weight content (DW content)
Dynamic light scattering (DLS), 80
   size distributions and TEM images, 82
Dynamic-mechanical analysis (DMA), 113, 249; *see also* Glass transition temperature
Dynamic-mechanical thermal analysis (DMTA), 99, 113–114; *see also* Glass transition temperature
Dynamic properties, 191; *see also* Dynamic viscosity
Dynamic rheological tests, 144; *see also* Small-amplitude oscillatory shear (SAOS)
   methods, 216–217
   types of, 166–168
Dynamic viscosity, 125; *see also* Rheology of fluid food; Rheological models for viscous food

E

ED, *see* Electro-dialysis (ED)
EDBM, *see* Electro-dialysis with bipolar membrane (EDBM)

INDEX

Edible oil industry, 510
   microfiltration, 512
   reverse osmosis, 510–511
   RSM, 511
   ultrafiltration, 511–512
EDUF, see Electro-dialysis with ultrafiltration membranes (EDUF)
Effective conductivity, 575
Effective diffusivity, 326, 330; see also Diffusivity
Effective elelctric conductivity models
   comparison of, 559
   effects of solids in tube flow, 559–560
   Kopelman model, 558
   Maxwell model, 557–558
   Meredith and Tobias model, 558
   parallel model, 558, 560
   probability model, 558–559
   series model, 558
Effective loss factor, 574
Efflux tube viscometer, 150; see also Flow property measurement
Einstein–Stokes equation, 325; see also Diffusion
Elastic modulus, 181; see also Solid food quasistatic tests
Electrical conductivity, 530; see also Effective elelctric conductivity models; Electrolytic conductivity
   anions, 566
   in β-dispersion range, 552
   during conventional heating, 543
   determination, 529, 531
   under electric field strengths, 533, 542
   of foods, 532
   –frequency relationship, 551, 552
   to increase, 553
   and mass diffusivity, 531
   during no phase transitions, 532
   during ohmic cooking, 541, 550
   parameters for EC model, 534
   of potato starch solution, 539, 556
   related to viscosity, 532
   salt effect on, 542, 557
   of sodium phosphate solution, 536
   on starch granule morphology, 539
   of strawberry pulp, 561
   sugar influence on, 540
   symbols, 565–566
   –temperature curve, 538
   at various electric field strengths, 545
   at various temperatures, 533, 544, 562

Electrical conductivity measurement, 560
   Bridge circuit for, 565
   as function of temperature, 562
   modeled as set of resistances, 563
   ohmic heating device for, 563
   schematic diagram, 563
   system for, 561
Electro-dialysis (ED), 502, 514; see also Beverage industry; Dairy product industry; Edible oil industry
Electro-dialysis with bipolar membrane (EDBM), 503
Electro-dialysis with ultrafiltration membranes (EDUF), 509, 514
Electrolytes, strong, 530
Electrolytic conductivity, 529; see also Dielectric properties; Liquid food electrical conductivity; Molar conductivity; Solid food electrical conductivity; Solid–liquid mixture electrical conductivity
   dipole rotation of water molecules, 528
   frequency dependence at different temperatures, 552
   Kohlrausch law of independent migration of ions, 531
   mode of conduction in, 528
   strong electrolytes, 530
   weak electrolytes, 530–531
Electromagnetic-acoustic transducers (EMAT), 648
Electromagnetic heating, 572
Electromagnetic radiation; see also Absorptivity; Emissivity; Food radiative property; Infrared heating; Reflectivity; Transmissivity
   attenuation, 291
   blackbody emitted energy, 285
   directional emissivity, 284, 286
   emissivity of surface, 286
   irradiation, 287
   penetration depth, 291
   spectral emissive powers, 285
   spectral reflectance of water, 288
   spectrum, 282–283
Electromagnetic spectrum, 282–283
   assigned by FCC, 573
Electromagnetic waves, 573
Electrophoretic mobility, 532
Electroscan Environmental SEM (ESEM), 74; see also Variable pressure SEM (VP-SEM)

Elliptical equation-based model, 719
EM, *see* Young's modulus (EM)
EMAT, *see* Electromagnetic-acoustic transducers (EMAT)
Emissivity, 284; *see also* Absorptivity; Infrared heating; Spectral emissivity
   absorptivity and, 290
   black body energy emission, 284–285
   diffuse, 287
   directional, 286
   gray, 287
   hemispherical, 286
   Planck's law of radiation, 285
   powers, 285
   spectral, 286
   of surface, 286
   total, 286
   Wien's displacement law, 286
Emulsion, 122; *see also* Rheology of fluid food
   stability, 49
Enthalpy; *see also* Thermal properties of frozen foods
   apparent enthalpy, 268
   of cod fish, 270, 271
   determination, 250
   equations to estimate, 270
   of frozen foods, 251–254, 268
   of sweet cherries, 269
   unfrozen water, 271
Enzyme inactivation, 729
   kinetics, 690, 691–692
EPDM, *see* Ethylene–propylene–diene terpolymer (EPDM)
Equilibrium
   moisture content, 409, 416–417; *see also* Food dehydration
   spreading pressure, 41; *see also* Gibbs adsorption equation
Equilibrium relative humidity (ERH), 374
Equivalent time at reference temperature, 678
ERH, *see* Equilibrium relative humidity (ERH)
Escherichia coli thermal inactivation
   kinetics, 689
ESEM, *see* Electroscan Environmental SEM (ESEM)
Etching, 74; *see also* Cryo-SEM
Ethylene–propylene–diene terpolymer (EPDM), 509
EU, *see* European Union (EU)
European Union (EU), 225
Excess interfacial free energy, 38

Extensional flows, 146; *see also* Flow property measurement; Squeezing flow
   normal stress difference, 147
   strain rate, 147
   test setup, 148
Extinction, 291
Eyring model, 718

F

Failure in solid foods, 202; *see also* Rheology of solid food
   cheeses, 204
   fracture mechanisms, 203
   meat products, 203–204
   modulus and bulk density, 203
   vegetables and fruits, 203
Fanning friction factor, 160, 161
Fat dependence of dielectric properties, 609; *see also* Dielectric properties
   dielectric constant and loss factor, 611, 612
   dielectric data on fats and oils, 611
Fermentation, 740–741, 746; *see also* Gas exchange in fruits and vegetables; Respiration
Fiber-optic sensors, 752
Fickian diffusion, 320; *see also* Diffusivity estimation in solids
Fick's laws of diffusion, 322, 748; *see also* Diffusion
Firmness of fruits/vegetables, 659, 660
First-order reaction, 682
   model, 682–683
Five-element model relaxation curves, 189; *see also* Solid food quasistatic tests
Flaking process, 347; *see also* Mass transfer processes
Flavor degradation kinetics, 694
Flex testing, *see* Bend testing
Flexural testing, *see* Bend testing
Flory–Huggins model, 111; *see also* State diagram
Flow model of Cross, 130, 131; *see also* Rheological models for viscous food
Flow property determining mixers, 151; *see also* Flow property measurement
   dynamic yield stress, 152
   effective shear rate, 154
   energy dissipation rate, 153
   failure stress, 153
   food dispersion, 153

Metzner and Otto method, 151
mixing effect on dielectric properties, 626
mixture model theory, 362
power law parameters, 151–152
Rieger and Novak method, 151
static yield stress, 152
vane method, 152
viscous stress component, 153
yield stress, 152
Flow property measurement, 140; see also
　　Capillary viscometers; Couette flow
　　viscometers; Extensional flows;
　　Rheology measurement; Flow
　　property determining mixers; Parallel
　　plate geometry; Rheology of fluid
　　food; Slit rheometers
　Adams consistometer, 149
　Bostwick consistometer, 149–150
　categories, 140
　efflux tube viscometer, 150
　empirical methods, 149
　fundamental methods, 140
　imitative methods, 150
　plate-and-cone viscometers, 143–144
　Visco-Amylo-Graph, 150
Fluid food flow in tubes, 158; see also
　　Rheology of fluid food
　apparent viscosity, 160
　Bingham plastic model, 159
　Buckingham equation, 159, 160
　chart of friction and Reynolds number, 161
　Fanning friction factor, 160, 161
　friction loss, 160–162, 163–164
　Herschel-Bulkley model, 159
　isothermal flow, 158
　kinetic energy calculation, 164
　power law model, 159
　pressure drop, 162–163
　rheology and pressure drop, 160
　velocity profiles, 158
　volumetric flow rate, 158
　Weissenberg-Rabinowitsch-Mooney
　　equation, 161
Fluorescence recovery, 70
FMC pea tenderometer, 211; see also Food
　　texture
FO, see Forward osmosis (FO)
Food
　acoustic properties, 651; see also Ultrasonic
　　analysis in food
　categories, 405

dispersion, 153
preservation reactor, 678
products, 501
quality determinant, 678
Food dehydration, 405; see also
　　Membrane separation processes;
　　Thermodynamic properties in
　　dehydration
　activation energy, 413
　adsorption branch of isotherm, 417, 418
　air-drying processes, 406
　constant-rate period, 409–411
　core drying model, 411
　critical moisture content, 408
　dryers, 421, 422
　drying process, 406
　drying rate, 407, 408, 409
　effective diffusion coefficient, 412, 413,
　　415–416
　energy equation, 418
　energy for water removal, 421
　energy requirements, 417–422
　equilibrium moisture content, 409, 416–417
　evaporation rate, 410
　falling-rate period, 408, 411–416
　food categories, 405
　heat-transfer coefficient, 409
　maximum permissible water activity
　　values, 420
　modified Crank method, 416
　moisture distribution curves, 414
　Nusselt-type equation, 410
　rate-limiting step, 405
　regular regime method, 412
　specification moisture contents and BET
　　mononuclear layer, 417, 419
　thermogradient effect, 406
　variable diffusivity, 413
Food dielectric properties, 572; see also
　　Carbohydrate dependence of
　　dielectric properties; Dielectric
　　property measurement; Frequency
　　dependence of dielectric properties;
　　Microwave energy; Temperature
　　dependence of dielectric properties
　cooking effect on, 625
　data, 576, 628–629
　dielectric constant, 574
　dipolar and ionic loss mechanisms, 575
　drying effect on, 625
　effective conductivity, 575

Food dielectric properties (*Continued*)
   electromagnetic heating, 572
   electromagnetic waves, 573
   of fish and seafood, 613, 615
   of fruits and vegetables, 615–616
   of insect pests, 620, 621
   loss factor, 574, 600
   loss tangent, 575
   of meats, 612–613, 614
   mixing effect on, 626
   moisture dependence, 599–601
   packaging material effect on, 620, 622
   penetration depth, 576–577
   protein dependence, 607–609, 610
   and quality assessment, 626–628
   rate of heat generation, 576
   solution to Maxwell's equations, 576
   storage effect on, 626, 627
   structure dependence of, 618
Food microstructure analysis, 64, 88; *see also* Confocal laser scanning microscopy (CLSM); Light microscopy (LM); Micro-CT; Scanning electron microscopy (SEM); Transmission electron microscopy (TEM)
   atomic force microscopy, 85–86, 87
   autofluorescence, 69
   challenges of, 64
   image analysis, 84
   length scales in structural hierarchy of food materials, 64
   magnetic resonance techniques, 86–87
   microstructure quantification, 83–84
   optical coherence tomography, 87
   photobleaching, 69–70
   scattering techniques, 85
   structural analysis methods, 84
   structural features in food materials, 65
   x-ray microscopy, 83
Food process, 312; *see also* Membrane separation processes
   basic unit operations of, 312
   high retention of volatile components, 345
   operations, 224
Food radiative property, 284; *see also* Food radiative property data; Infrared heating; Radiative heating model
   measurement of, 292
   precautions, 292
   reflectance measurements, 294
   spectroradiometer, 293
   transmission measurement, 293
Food radiative property data, 294; *see also* Absorptivity; Emissivity; Food radiative property; Infrared heating; Radiative heating model
   absorptance, 297
   absorptivity, 295, 296
   dependence of, 301
   direct transmitted fraction, 301
   emissivity variation, 303
   moisture dependence of, 297–298
   penetration depth variation, 298
   processing and food radiative properties, 301–302
   pure food components, 295
   reflectance, 296, 300, 301
   reflection spectra of different substances, 299
   spectral dependence in, 305
   spectral variation, 295–297
   temperature dependence of, 298
   transmittance, 300, 302
   transmitted energy fraction, 297
   water, 294–295
Food texture, 207; *see also* Bend testing; Flow property measurement; Rheology of solid food
   Baker compressimeter, 213
   Cherry–Burrell meter, 213
   consistometer, 213
   CT x-ray scanning analysis, 210, 214
   denture tenderometer, 208
   empirical methods, 213
   FMC pea tenderometer, 211
   instron, 209
   instruments to study, 207
   Magness–Taylor pressure tester, 213
   penetrometer, 211
   pore size distribution, 215
   porosity, 215
   profile analysis, 208–209
   ridgelimeter, 213
   shear press, 210
   squeeze tester, 213
   structure, 213–215
   textural attributes, 207
   texture profile curve, 208–209
   texturometer, 208
   Warner-Bratzler shear, 210–211
5 Formyltetrahydrofolic acid (5FTHF), 698

Forward osmosis (FO), 502, 513; *see also* Beverage industry; Dairy product industry; Edible oil industry
Fouling, 439
 effect on membrane performance and pore size, 495–497
Four-element model, 190; *see also* Solid food quasistatic tests
 relaxation curves for, 189
Fourier law of heat conduction, 231; *see also* Thermal conductivity
Fox and Flory equation, 102; *see also* Glass transition temperature
Fractional conversion, 684
 model, 684, 715
Free
 -space transmission technique, 583
 volume, 96–97, 346; *see also* Aroma compounds
 water, 599
Freeze-drying method, 27–28
Freezing, 657; *see also* Water activity
 point depression techniques, 374
 process, 248
Frequency dependence of dielectric properties, 587; *see also* Food dielectric properties; Temperature dependence of dielectric properties
 bidistilled and deionized water, 589
 Debye dielectric relaxation parameters, 590
 in food materials, 590–591
 of hard red winter wheat, 591
 of ionic loss mechanisms, 589
 loss factor, 591
 of potato, 590
Friction function, dimensionless, 447
Frozen product analysis, 657
Fruit products
 components in juice, 456, 461
 rheological properties of, 128–129
FTC shear press, 210; *see also* Food texture
5FTHF, *see* 5 Formyltetrahydrofolic acid (5FTHF)
Fugacity, 364; *see also* Thermodynamic properties in dehydration; Water activity
 coefficient, 364
 coefficients of water vapor, 369
 of water vapor, 366–367, 368
Fully permeabilized tissue, 549

G

GAB equation, *see* Guggenheim–Anderson–de Boer equation (GAB equation)
Gas
 absorption, 348; *see also* Mass transfer processes
 adsorption method, 13; *see also* Material density
 composition for long-term storage, 763–764
 diffusion, 748, 753–754
Gas chromatography (GC), 749
Gas exchange in fruits and vegetables, 739; *see also* Fermentation; Respiration
 applications, 760
 barrier to, 747
 Clark-type amperometric O_2-sensitive sensor, 750
 data, 755, 756–760, 761
 fiber-optic sensors, 752
 Fick's laws of diffusion, 748
 gas concentration, 754
 gas diffusion, 748, 753–754
 Graham's law, 755
 infrared CO_2 sensor, 752
 lenticels, 747
 membrane separator, 763
 metabolically active state, 739
 O_2 and CO_2 concentration through GC, 749–750
 O_2 consumption and CO_2 production rate, 748–749
 paramagnetic oxygen sensor, 752
 potentiometric sensor, 751
 respiration heat measurement, 752
 skin resistance to, 755
 steps in, 747
 symbols, 765
 tissue softening, 747–748
Gas pycnometer method, 10; *see also* Apparent density
 air comparison pycnometer, 10
 Horiba helium pycnometer VM-100, 11
Gas–solid and vapor–solid equilibria, 320; *see also* Phase equilibria
 BET equation, 320, 321
 GAB equation, 321
 moisture sorption isotherms, 322
 water activity, 320
 water sorption isotherms, 321

INDEX

Gas transmission (GT), 351
Gates–Gaudin–Schuhmann function (GGS function), 29–30
GC, *see* Gas chromatography (GC)
Gel cure experiment, 145
Gelation monitoring, 658
General Foods texturometer, 208; *see also* Food texture
Generalized Maxwell model, 189; *see also* Solid food quasistatic tests
Geometric dimension method, 6–7; *see also* Apparent density
Geometric mean model, 233; *see also* Thermal conductivity
GGS function, *see* Gates–Gaudin–Schuhmann function (GGS function)
Gibbs adsorption equation, 40; *see also* Surface properties
   equilibrium spreading pressure, 41
   limitation, 42
   solid-liquid interface, 42
Gibbs free energy, 362–363; *see also* Thermodynamic properties in dehydration
   absorbed water, 391
Glass transition, 94; *see also* Glass transition temperature; State diagram
   α-relaxations, 96
   fundamental considerations, 95
   heat capacity, 95–96
   heat flow, 95
   and molecular mobility, 96
   phenomenon, 96
   relaxation below, 99
   specific volume vs. temperature plot, 94
   technological importance, 94–95
   theory, 27; *see also* Mass–volume–area-related properties
   thermal analysis methods, 112
   thermal processing of foods, 239–240
   of water, 105–106
Glass transition in solid foods, 204; *see also* Differential scanning calorimeter (DSC); Glass transition temperature; Rheology of solid food
   factors affecting, 204
   idealized-state diagram, 206
   importance of, 206
   measurement of, 204–205

Glass transition temperature, 94; *see also* Differential scanning calorimeter (DSC); Glass transition; State diagram
   β-relaxations, 100
   for biopolymer, 113
   CK equation, 100–101
   and DE, 102
   dielectric thermal analysis, 115, 116
   dynamic mechanical thermal analysis, 113–114
   factors affecting, 102
   Fox and Flory equation, 102
   free volume, 96–97
   glass transition of water, 105–106
   GT equation, 101
   k value, 101
   Kauzmann temperature, 98
   log viscosity, 97
   material relaxation, 99
   maximum enthalpy recovery, 99
   measurement techniques, 111
   of miscible components, 100–102
   molecular mobility, 96–100
   molecular weight, 102
   plasticizers, 104–105
   pressure effect on, 105
   for selected sugars, 99
   structure and architecture, 104
   thermomechanical analysis, 114, 115
   Willams–Watts equation, 100
   WLF equation, 97–98
Gordon–Taylor (GT equation), 101; *see also* Glass transition temperature
Graham's law, 755
Gravimetric method, 377; *see also* Water activity
Greek symbols
GT, *see* Gas transmission (GT)
GT equation, *see* Gordon–Taylor (GT equation)
Guar gum, 607
Guggenheim–Anderson–de Boer equation (GAB equation), 384; *see also* Gas–solid and vapor–solid equilibria; Moisture sorption isotherm (MSI)
   advantages of, 385
   to MSIs, 386
   RMS value, 385
   and sorption enthalpies, 384
   for sorption of water vapors, 384
Gums, 606

H

Hair hygrometers, 374–375; *see also* Water activity
Half-life time, 682
Halsey equation, 382; *see also* Moisture sorption isotherm (MSI)
Heat
    capacity, 95–96
    flow, 95
    -inactivation kinetics, 685
    probe method, 235; *see also* Thermal conductivity
    rate, 95
    treatment impact, 678
Hemispherical emissivity, 286; *see also* Emissivity
Henderson equation, 382–383; *see also* Moisture sorption isotherm (MSI)
Henry's law, 318; *see also* Gas–solid and vapor–solid equilibria
Herschel–Bulkley model, 127, 661; *see also* Rheological models for viscous food
    friction losses for, 163–164
    kinetic energy correction factor, 164
Hewlett-Packard (HP), 584
High pressure–high temperature treatment (HPHT treatment), 713
High-pressure pasteurization (HP-P), 711
High-Pressure Process (HPP), 710; *see also* Kinetic studies under static processing conditions; Safety and quality kinetics
    adiabatic heat of compression, 711
    HP-P, 711, 712, 713
    HP-S, 711, 712, 713
    kinetics, 730–731
    Le Châtelier principle, 710–711
    Pascal principle, 711
    pressure, temperature, time profiles, 712
    product-treatment history, 710
High-pressure sterilization (HP-S), 711
High-resolution ultrasonic spectroscopy, 658
High temperature short time (HTST), 713
Hookean solid, 181; *see also* Solid food quasistatic tests
Horiba helium pycnometer VM-100, 11
Hot air drying method, 27–28
HP, *see* Hewlett-Packard (HP)

HPHT treatment, *see* High pressure–high temperature treatment (HPHT treatment)
HPMC, 606, 607
HP-P, *see* High-pressure pasteurization (HP-P)
HPP, *see* High-Pressure Process (HPP)
HP-S, *see* High-pressure sterilization (HP-S)
HTST, *see* High temperature short time (HTST)
Hydrogenated vegetable shortening dielectric data, 611
Hygrometers, 374–375; *see also* Water activity
Hysteresis, 398

I

Ideal gas equation, 16, 364
Iglesias–Chirife equation, 384; *see also* Moisture sorption isotherm (MSI)
IMF, *see* Intramuscular fat content (IMF)
Industrial, Scientific, and Medical (ISM), 573
Infrared; *see also* Electromagnetic radiation
    absorption bands, 283
    CO_2 sensor, 752
    radiation, 282
    radiation sources, 284
    spectral directional transmissivity of, 284
    waves, 282
Infrared heating, 282, 307–308; *see also* Electromagnetic radiation; Radiative heating model
    absorption bands, 283
    in food materials, 283–284
    infrared interactions with materials, 282
Instron universal testing machine, 209; *see also* Food texture
Integral method, 375; *see also* Water activity
Interfaces, 38; *see also* Surface properties
Interfacial free energy, 38; *see also* Surface properties
Internal friction, coefficient of, 195
Interphase mass transfer, 331, 346; *see also* Mass transfer coefficients; Mass transfer processes
    Chilton–Colburn analogies, 335
    heat and mass transfer analogies, 335
    penetration theory, 334
    surface-renewal theory, 334
    effect of surfactants, 335–336
Intramuscular fat content (IMF), 657
Intrinsic viscosity, 139

IP, see Isoelectric precipitation (IP)
Irradiation, 287; see also Electromagnetic radiation
ISM, see Industrial, Scientific, and Medical (ISM)
Isoelectric precipitation (IP), 513
Isopiestic method, 375; see also Water activity
Isosteric heat of sorption, 391
  of selected foods, 392
Isothermal heating profiles, 678
Isotropic pressure, 124; see also Rheology of fluid food

J

JG β-relaxations, see Johari–Goldstein β-relaxations (JG β-relaxations)
Johari–Goldstein β-relaxations (JG β-relaxations), 100; see also Glass transition temperature

K

k value, 101; see also Glass transition temperature
Kauzmann temperature, 98; see also Glass transition temperature
  determination of, 99
KE, see Kinetic energy (KE)
Kelvin model, 186, 188; see also Solid food quasistatic tests
Kinetic energy (KE), 164
Kinetic parameter estimation, 719–720
Kinetic studies under static processing conditions, 713, 717, 731–732; see also Safety and quality kinetics
  biphasic inactivation, 715, 716, 717
  design for, 713–715
  elliptical equation-based model, 719
  kinetic models, 719
  kinetic parameter estimation, 719–720
  log-logistic model, 717
  model evaluation, 720
  modified Gompertz model, 717
  process-induced changes, 715
  reaction rate equation determination, 716–717, 719
  temperature and pressure coefficient models, 718
  Weibull model, 717
Kinetics of thermal inactivation of microorganisms, 689

Kohlrausch law of independent migration of ions, 531
Kopelman model, 558
  predictive model, 266
Kramer shear press, 210; see also Food texture

L

Lactoperoxidase thermal inactivation kinetics, 691
Langmuir equation, 381; see also Moisture sorption isotherm (MSI)
Lard dielectric data, 611
LC, see Liquid chromatography (LC)
Le Châtelier Principle, 710–711, 740
Leaks detection in food packages, 668–670
Lee Kramer shear press, 210; see also Food texture
Leidenfrost effect, 57
Lennard–Jones type of equation, 450
Lenticels, 747
Light microscopy (LM), 64; see also Food microstructure analysis
  comparison of micrographs from SEM and, 73
  disadvantages of, 67
  polarizing filters, 66
  resolution of, 65
  sample preparation, 66
  stains for contrast in, 66, 67
Light speed in vacuum, 285
Linear viscoelastic region (LVR), 113; see also Glass transition temperature
Lipid, 609
Lipoxygenase thermal inactivation kinetics, 691
Liquid chromatography (LC), 452
Liquid displacement method, 9; see also Apparent density
  mercury intrusion porosimetry, 10
  procedures, 9
  toluene, 9
Liquid food electrical conductivity; see also Electrolytic conductivity; Solid food electrical conductivity; Solid–liquid mixture electrical conductivity
  current flow in conductor, 529
  determination, 529
  electric field strength effect on, 534
  electrical conductivity, 530
  electrolytes effects on, 534
  electrophoretic mobility, 532

hydrocolloid effects on, 535–537
mass diffusivity and viscosity, 532
molar conductivity, 530
Nernst–Einstein equation, 531
nonelectrolytic solute effect on, 540
phase transitions effect on, 537–540
suspended solids effects on, 534–535, 536
temperature effect on, 532–533
Liquid foods, 456; *see also* Flow property measurement
Listeria monocytogenes thermal inactivation kinetics, 689
LM, *see* Light microscopy (LM)
LN function, *see* Log-Normal function (LN function)
Locust bean gum, 607
Log-logistic model, 717
Log-Normal function (LN function), 30
Lorentz forces, 648
Loss tangent, 125, 165; *see also* Rheology of fluid food
Lotus effect, 57
Low-power ultrasound (LPU), 638
Low vacuum SEM, *see* Variable pressure SEM (VP-SEM)
LPU, *see* Low-power ultrasound (LPU)
LVR, *see* Linear viscoelastic region (LVR)

M

MA, *see* Modified atmosphere (MA)
Magness–Taylor pressure tester, 213; *see also* Food texture
Magnetic resonance imaging (MRI), 86–87, 155; *see also* Rheology measurement
Magnetron, 589
Mark-Houwink relation, 139
Mass diffusivity and viscosity, 532
Mass transfer coefficients, 331, 333, 346; *see also* Interphase mass transfer; Mass transfer processes
  absorption of gas into liquid, 332
  for caffeine, 484
  transfer rate, 332, 333, 334
  two-film theory, 331
  wetted-wall column, 333
Mass transfer-in-series resistance model, 509
Mass transfer processes, 312, 346; *see also* Diffusion; Interphase mass transfer; Packaging; Phase equilibria

applied to foods, 313
crystallization, 349–350
distillation, 347–348
extraction, 346–347
flaking process, 347
in foods, 336
gas absorption, 347
interphase mass transfer, 346
mass transfer coefficients, 346
SCF, 347
of solids, 313
Mass–volume–area-related properties, 2, 31; *see also* Density; Porosity; Volume
  fundamental considerations, 3
  GGS function, 29–30
  glass transition concept, 27–28
  LN function, 30
  measurement techniques, 6
  mechanisms of collapse, 27
  modified beta function, 30–31
  modified Gaudin–Meloy, 30
  pore size distribution, 6
  Rahman's hypothesis, 28, 29
  Rosin-Rammler distribution, 30
  size distribution, 28
  specific data, 16
  surface area, 6
  test for skewness, 28–29
  theoretical prediction, 25
Massachusetts Institute of Technology (MIT), 208
Material density, 3; *see also* Density; Mercury porosimetry
  gas adsorption method, 13
  pycnometer method, 11
Maximum relative deflection (MRD), 212
Maximum tensile stress (MTS), 212
Maxwell–Eucken model, 234; *see also* Thermal conductivity
Maxwell model, 186, 188, 557–558; *see also* Solid food quasistatic tests
  creep curves for, 187
  generalized, 189
  relaxation curves for, 189
MB, *see* Membrane bioreactor (MB)
MD, *see* Membrane distillation (MD)
Meat products quality evaluation, 666
MEF processes, *see* Moderate electric field processes (MEF processes)
Membrane bioreactor (MB), 508

INDEX

Membrane distillation (MD), 502; *see also* Beverage industry; Dairy product industry; Edible oil industry
Membrane separation processes, 439, 514–515; *see also* Beverage industry; Dairy product industry; Edible oil industry; Pervaporation (PV); Reverse osmosis (RO); Surface force–pore flow model; Transport equations; Transport theories; Ultrafiltration (UF); Water preferential sorption model
   advantage of, 501
   bioreactor, 508
   capacity for solute concentration, 481
   distillation, 502
   fruit juice components, 456, 461
   membrane–solution interface, 450
   milk components, 460, 462
   problems involved in, 456
   separator, 763
   technology, 512
Mercury porosimetry, 12; *see also* Material density
   intrusion porosimetry, 10
   relation of intrusion pressure to pore diameter, 13
   Washburn equation, 12
Meredith and Tobias model, 558
Metabolically active fruits, 739
5 Methyltetrahydrofolic acid (5MTHF), 698, 700
Metzner and Otto method, 151; *see also* Flow property determining mixers
MF, *see* Microfiltration (MF)
Michaelis–Menten kinetics, 743, 755, 762
Microbial spore inactivation, 690
Micro-CT, 81; *see also* Food microstructure analysis
   application in food, 82
   synchrotron-based, 83
Microfiltration (MF), 502; *see also* Beverage industry; Dairy product industry; Edible oil industry
Microminiaturized tubular heating system, 687
Microregion retention theory, 345; *see also* Aroma compounds
Microstrip transmission line, 583
Microwave energy; *see also* Dielectric property measurement; Food dielectric properties
   absorption, 574, 578, 579
   angle of transmission, 577
   decay, 577
   dielectric property measurement, 580, 588
   dielectric relaxation in liquid water, 589
   on food, 572
   heating, 575
   magnetron, 589
   penetration depths, 577
   rendered bacon fat dielectric data, 611
   spatial distribution, 578, 579
   water as terminal load, 589
   wavelength inside food, 577
Milk
   components, 460, 462
   dielectric properties, 617
   processing of, 460–461
MIT, *see* Massachusetts Institute of Technology (MIT)
Mixed suspension–mixed product removal (MSMPR), 350
Mixture model theory, 362
Model-fitting procedure, 331; *see also* Diffusivity estimation in solids
Moderate electric field processes (MEF processes), 528
Modified atmosphere (MA), 760
   for long-term storage, 763–764
   packaging, 760
Modified beta function, 30–31
Modified Gaudin–Meloy, 30
Modified Gompertz model for inactivation behavior of microorganisms, 717
Modified Small's number, 443
Modulated differential scanning calorimetry (MDSC), 108, 249
Modulated temperature DSC (MTDSC), 112
Modulus of elasticity, 181; *see also* Solid food quasistatic tests
   by bending and Hertz's equations, 184
   of foods in rectangular beam form, 192
Modulus of rigidity, *see* Shear—modulus
Moisture diffusion, 336; *see also* Diffusion; Diffusivity; Moisture transport
   diffusivity in air-dried potato, 337
   effective moisture diffusivity, 337, 339
   pressure and water diffusivity, 338
Moisture sorption isotherm (MSI), 379; *see also* Brunauer, Emmett and Teller (BET); Water activity
   Chen equation, 383–384
   Chung and Pfost equation, 383
   Clausius–Clapeyron equation, 382

of dehydrated rice, 401
GAB equation, 384
Halsey equation, 382
Henderson equation, 382–383
Iglesias–Chirife equation, 384
Langmuir equation, 381
Oswin equation, 383
temperature dependence of, 387
theoretical description of, 380
van der Waals adsorption isotherms, 379, 380
Moisture transport, 336; *see also* Diffusion; Diffusivity; Mass transfer processes; Moisture diffusion; Porosity
development of bulk porosity, 339, 340
diffusion in porous foods, 338
effective moisture diffusivity vs. bulk porosity, 341
interphase moisture transfer, 342
moisture diffusivity vs. moisture content, 340
void fraction, 338
water transport mechanism, 341
Molar conductivity, 530; *see also* Electrolytic conductivity
equilibrium constant, 531
for strong electrolytes, 530
for weak electrolytes, 530–531
Molecular diffusion, 322; *see also* Diffusion
Molecular weight cut-off (MWCO), 503
Monte Carlo ray-tracing method, 304
MRD, *see* Maximum relative deflection (MRD)
MSI, *see* Moisture sorption isotherm (MSI)
MSMPR, *see* Mixed suspension–mixed product removal (MSMPR)
MTDSC, *see* Modulated temperature DSC (MTDSC)
5MTHF, *see* 5 Methyltetrahydrofolic acid (5MTHF)
MTS, *see* Maximum tensile stress (MTS)
Multiresponse model, 699–700
MWCO, *see* Molecular weight cut-off (MWCO)

N

NA, *see* Numerical aperture (NA)
Nanofiltration (NF), 502; *see also* Beverage industry; Dairy product industry; Edible oil industry
rosemary extraction, 513

Needle probe method, 235; *see also* Thermal conductivity
design specification, 237–238
heated needle probe, 235
operation criteria, 236
temperature rise, 236, 237
Nepers per meter (Np/m), 642
Nernst–Einstein equation, 531
Newtonian foods, 125; *see also* Rheology of fluid food
flow curves of, 127
Newtonian liquid, 181; *see also* Rheology of fluid food
NF, *see* Nanofiltration (NF)
NMR, *see* Nuclear magnetic resonance (NMR)
Nomarski contrast, *see* Differential interference contrast
Nonclimacteric fruit, 741, 742
Noncompetitive inhibition, 744
Non-Newtonian fluids, 125; *see also* Rheology of fluid food
flow curves of, 127
Non-Newtonian foods, 126; *see also* Rheology of fluid food
shear-thinning foods, 126
with time-dependent flow properties, 127
NPG, *see* n-propyl gallate (NPG)
Np/m, *see* Nepers per meter (Np/m)
n-propyl gallate (NPG), 70
NRTL equation, 316; *see also* Vapor–liquid equilibrium
nth order reaction rate equation, 681
Nuclear magnetic resonance (NMR), 108, 249, 416
Nucleation, 349; *see also* Mass transfer processes
Numerical aperture (NA), 67
Nusselt-type equation, 410
Nutrient degradation kinetics, 696–697
Nylon, 66, 225

O

OCT, *see* Optical coherence tomography (OCT)
OD, *see* Osmotic distillation (OD)
Ohmic heating, 528
Ohm's law, 529
OMD, *see* Osmotic membrane distillation (OMD)
Open-ended probe technique, 581–582
Open pore porosity, 5; *see also* Porosity
Operating pressure, dimensionless, 447

INDEX

Optical coherence tomography (OCT), 87
Order of reaction determination, 681
Osmotic distillation (OD), 502; *see also* Beverage industry; Dairy product industry; Edible oil industry
Osmotic membrane distillation (OMD), 509, 513
Osmotic pressure
    data for aqueous solutions, 470
    data of proteins, 471
    data on constants, 478
Oswin equation, 383; *see also* Moisture sorption isotherm (MSI)
Oxygen-sensitive foods, 370

P

Packaging, 350; *see also* Mass transfer processes
    conversion factors to SI permeability units, 351
    differential permeability, 352
    migration of packaging materials into packaged food, 352
    permeabilities and water vapor diffusivities, 351
    permeation equation, 329
    rate equation, 333
Packed-bed ED (PBED), 514
PAH, *see* Polyamide hydrazide (PAH)
Parallel model, 558, 560
Parallel plate geometry, 144; *see also* Flow property measurement
    tests to examine phase transitions, 145
Paramagnetic oxygen sensor, 752
Partial least-squares (PLS), 650
Particle density, 4; *see also* Density
Particle size distribution (PSD), 653
    analysis in liquid food, 666–667
Pascal Principle, 711
Pasteurization, 712
Pattern recognition, 650
PBED, *see* Packed-bed ED (PBED)
PCA, *see* Principal component analysis (PCA)
PD measurements, *see* Pressure difference measurements (PD measurements)
PDMS, *see* Polydimethylsiloxane (PDMS)
PDT-model, *see* Pressure-death-time model (PDT-model)
PE, *see* Polyethylene (PE)
PE technique, *see* Pulse-echo technique (PE technique)

PE transmission, *see* Pulse-echo transmission (PE transmission)
Peclet number, 348
Pectinesterase thermal inactivation kinetics, 691
PEF processing, *see* Pulsed electric field processing (PEF processing)
PEGs, *see* Polyethylene glycols (PEGs)
Pendant drop method, 51–52
Penetrant distribution, 330; *see also* Diffusivity estimation in solids
Penetration depth, 291, 576
    spectral variation of, 298
Penetration theory, 334; *see also* Interphase mass transfer
Penetrometer, 211; *see also* Food texture
Peppelenbos and van't Leven gas exchange model, 744
Permeability
    in low temperature, 548
    of material, 573
Permeation
    equation, 329
    measurements, 329–330; *see also* Diffusivity estimation in solids
Permittivity of material, 573, 592
Peroxidase thermal inactivation kinetics, 691
Pervaporation (PV), 499; *see also* Beverage industry; Dairy product industry; Edible oil industry; Membrane separation processes
    GC-MS analytical results, 501
    membranes used, 499
    pore flow mechanism, 499
    process, 499
PES, *see* Polyethersulfone (PES)
PET, *see* Polyethylene terephthalate (PET)
PGA, *see* Pteroylglutamic acid (PGA)
Phase angle, 193
Phase equilibria, 313; *see also* Gas–solid and vapor–solid equilibria; Mass transfer processes; Vapor–liquid equilibrium
    of component, 319–320
    gas–liquid equilibria, 318
    gas solubility data, 319
    Henry's law, 318
    liquid–liquid and liquid–solid equilibria, 319–320
Photobleaching, 69
    fluorescence recovery, 70
Physical-mathematical method, 679

PI, *see* Polyethylene imine (PI)
Planck's law of radiation, 285
PLAs, *see* Polylactides (PLAs)
Plasma polymerization processes, 58–59
Plasma treatments, 58
Plasticizers, 104–105; *see also* Amorphous materials; Glass transition temperature
Plate-and-cone viscometers, 143–144; *see also* Flow property measurement
PLS, *see* Partial least-squares (PLS)
Polarizing filters, 66
Polyamide hydrazide (PAH), 456
  specifications of, 463
Polydimethylsiloxane (PDMS), 54
Polyethersulfone (PES), 503
Polyethylene (PE), 512
Polyethylene glycols (PEGs), 456
Polyethylene imine (PI), 512
Polyethylene terephthalate (PET), 620
Polygalacturonase thermal inactivation kinetics, 692
Polylactides (PLAs), 104
Polymer surfaces, 54
Polyoctylmethyl siloxane (POMS), 514
Polyphenoloxidase thermal inactivation kinetics, 692
Polypropylene (PP), 508, 512
Polysulfone [Victrex] (PS-V), 456
Polyvinyl pyrrolidone (PVP), 106
Polyvinylidene fluoride (PVDF), 506
Polyvinylpyrrolidone (PVPP), 506
POMS, *see* Polyoctylmethyl siloxane (POMS)
Pore
  formation, 545, 547–548
  repair, 546
  volume, 3; *see also* Volume
Pore surface area measurement, 15; *see also* Surface area
  based on adsorption, 15
  based on fluid flow, 15–16
  Carman-Kozeny equation, 15
  Darcy's equation, 15
  mercury intrusion, 16
Porosity, 5, 338; *see also* Mass–volume–area-related properties; Moisture transport
  apparent, 5
  bulk, 6
  bulk-particle, 6
  closed pore, 5
  direct measurement method, 14
  at full turgor, 22
  as function of water content, 22
  measurement by density method, 14
  open pore porosity, 5
  optical microscopic measurement method, 14
  predictions of, 21–23
  total, 6
Potential function, dimensionless, 447
Potentiometric sensor, 751
Powell-Eyring model, 130, 131; *see also* Rheological models for viscous food
Power law model, 127, 130; *see also* Rheological models for viscous food
PP, *see* Polypropylene (PP)
PPD, *see* p-Phenylenediamine (PPD)
p-Phenylenediamine (PPD), 70
PR, *see* Product rate (PR)
Preservation reactor, 678
Pressure-death-time model (PDT-model), 718
Pressure difference measurements (PD measurements), 661
Pressure-proof dielectric test cell, 582
Principal component analysis (PCA), 650
Probability model, 558–559
Process
  design, 710
  -induced changes, 715
Process impact, 679; *see also* Thermal process kinetics
  physical–mathematical method, 679
  on safety or quality, 680
  thermal death time, 679
Process-value F, 678
  determination, 680
Product rate (PR), 440
  effect of feed concentration on, 471
  vs. operating pressures, 485
  vs. sodium chloride separation, 483
Protein adsorption, 48; *see also* Surface properties
  conformational changes, 49
  exchange rates, 48
  exchange reactions, 50
  factors affecting, 49
  interfacial behavior, 48, 50
  for liquid foods, 50
PSD, *see* Particle size distribution (PSD)
PS-V, *see* Polysulfone [Victrex] (PS-V)
Pteroylglutamic acid (PGA), 698
Pulsed electric field processing (PEF processing), 528

INDEX

Pulse-echo technique (PE technique), 644
Pulse-echo transmission (PE transmission), 638
Pure elastic behavior, 181
Pure viscous flow of liquid, 181
Pure water permeation rate (PWP), 440
PV, see Pervaporation (PV)
PVDF, see Polyvinylidene fluoride (PVDF)
PVP, see Polyvinyl pyrrolidone (PVP)
PVPP, see Polyvinylpyrrolidone (PVPP)
PWP, see Pure water permeation rate (PWP)
Pycnometer method, 11; see also Material density

Q

Q_{10}-factor, 685
Quality monitoring of vegetable and fruit, 664–665

R

Radial distance, dimensionless, 447
Radiative heating model, 302; see also Infrared heating; Food radiative property
   based on Monte Carlo method, 305
   based on radiative transport equation, 305–306
   food emissivity effect, 306–307
   microwave-infrared oven, 306
   Monte Carlo ray-tracing method, 304
   net radiative exchange, 303
   realistic models, 304
   simple models, 303–304
   spectral dependence in radiative property data, 305
   temperature sensitivity to radiative property, 306–307
   view factors, 303
   volumetric heat generation, 304
Radiofrequency correlation technique (RFC technique), 669
Radiofrequency heating (RF heating), 572, 580
   runaway heating, 596
Radiofrequency sampling (RFS), 669
Rahman's hypothesis, 28; see also Mass–volume–area-related properties
   on mechanism of pore formation, 29
Raoult's law for ideal solution, 365
Rapid transient methods, 234; see also Thermal conductivity

Rapid Visco Analyzer (RVA), 150
Rate equation, 333; see also Mass transfer processes
Reaction rate equation determination, 716–717, 719
Real-time ultrasound (RTU), 646
Reflection coefficient, 642–643
Reflection technique, see Pulse-echo technique (PE technique)
Reflectivity, 287; see also Absorptivity; Emissivity; Infrared heating
   of base paper, 300
   of distilled water, 289
   limiting cases of, 290
   of potato tissue, 296, 301
   radiation energy incident, 290
   spectral, 289
   of water, 288
Refraction angle, 643
Regular regime method, 412; see also Food dehydration
Relative complex permittivity of the medium, 574
Relaxation curve, 189; see also Solid food quasistatic tests
Residuals, 682
Resolution, 65
   lateral and axial, 67–68
Resonance methods, 191, 216; see also Solid food testing
   application of, 197
   dynamic tester, 194
   food property determination, 198
   modulus of elasticity, 192
   natural frequency, 191
   resonance parameters, 197–199
Respiration, 740; see also Fermentation; Gas exchange in fruits and vegetables
   climacteric and nonclimacteric fruit, 742
   CO_2 production rate, 744–745, 746
   core breakdown, 741
   heat measurement, 752
   heat production, 745, 746
   inhibitin of, 740, 746
   Le Chatelier's principle, 740
   Michaelis–Menten kinetics, 743, 755, 762
   O_2 consumption rate, 744–745
   Peppelenbos and van't Leven gas exchange model, 744
   rate, 741, 742–743

Response surface methodology (RSM), 511
Reticulated porous ceramics (RPC), 301
Reverse osmosis (RO), 439; *see also* Beverage industry; Dairy product industry; Edible oil industry; Membrane separation processes; Transport equations; Water preferential sorption model
   dimensionless parameters, 444, 445
   juice concentration, 482
   membranes, 502
   monocarboxylic acid separation, 473
   organic solute separations, 471–473
   performance parameters for, 444
   separation problems in, 460
   sugar separation, 469
Revolutions per minute (rpm), 141
RF heating, *see* Radiofrequency heating (RF heating)
RFC technique, *see* Radiofrequency correlation technique (RFC technique)
RFS, *see* Radiofrequency sampling (RFS)
Rheological models for viscoelastic foods, 135; *see also* Dynamic rheological tests; Rheological models for viscous food; Rheology of fluid food; Small-amplitude oscillatory shear (SAOS)
   creep compliance, 137
   mechanical models, 138
   rheological measurement, 164
   shear rate effect, 136
   stress data, 135–137
Rheological models for viscous food, 127; *see also* Dynamic rheological tests; Rheological models for viscoelastic foods; Rheology of fluid food
   apparent viscosity, 134
   Bingham relationship, 130
   Casson model, 129–130
   concentration effect, 134–135
   flow behavior index, 128
   flow model of Cross, 130, 131
   Herschel–Bulkley model, 127
   models for time-independent behavior, 127–131
   Powell-Eyring model, 130, 131
   power law model, 127, 130
   effect of temperature and shear rate, 132–134
   thermorheological models, 132–133
   thixotropic foods, 131–132

Rheology measurement, 154; *see also* Flow property measurement
   curve fit method, 156–157
   requirements, 154–155
   spatial location, 156
   tomographic techniques, 155–157
   UDV and pressure drop measurements, 157
   using ultrasonic Doppler velocity meter, 156
   vibrational viscometers, 157
Rheology of fluid food, 122, 169; *see also* Dynamic rheological tests; Fluid food flow in tubes; Flow property measurement; Rheological models for viscoelastic foods; Rheological models for viscous food
   activation energy for flow, 133
   complex modulus, 125
   complex viscosity, 125
   Cox–Merz rule, 126
   deformation tensor, 123
   dynamic rheological tests, 125
   dynamic viscosity, 125
   flow curves of, 127
   fruit product rheology, 128–129
   intrinsic viscosity, 139
   isotropic pressure, 124
   loss tangent, 125
   Mark-Houwink relation, 139
   physicochemical approach, 139
   reduced viscosity, 137
   rheological classification of, 123
   solution viscosity, 137–139
   stress and shear rate, 124
   structure of fluid food, 137
   viscosity function, 124, 125
Rheology of solid food, 180, 215–217; *see also* Failure in solid foods; Food texture; Glass transition in solid foods; Solid food testing; Solid food quasistatic tests
Rheometer, 140
Rice flour, 607
Ridgelimeter, 213; *see also* Food texture
Rieger and Novak method, 151; *see also* Flow property determining mixers
Rigor mortis, 626
RO, *see* Reverse osmosis (RO)
Rosin-Rammler distribution, 30
RPC, *see* Reticulated porous ceramics (RPC)
rpm, *see* Revolutions per minute (rpm)
RSM, *see* Response surface methodology (RSM)

RTU, *see* Real-time ultrasound (RTU)
Runaway heating, 596–597
RVA, *see* Rapid Visco Analyzer (RVA)

S

Safety and quality kinetics, 720
    enzymatic target under HP-T, 725–726
    enzyme inactivation, 729
    inactivation degree, 718
    microbial inactivation, 720
    on microbial target, 721–724, 728
    quality conversion, 729–730
    on quality target attributes onditions, 727
    T-sensitivities under HP of both vegetative cells and spores, 721–724, 728
Salmonella senftenberg thermal inactivation kinetics, 689
SAOS, *see* Small-amplitude oscillatory shear (SAOS)
Sc numbers, *see* Schmidt numbers (Sc numbers)
Scanning electron microscopy (SEM), 70; *see also* Cryo-SEM; Food microstructure analysis; Variable pressure SEM (VP-SEM)
    beam damage, 72
    components of, 70, 71
    image contrast mechanisms, 70–71
    micrographs, 71, 72, 73
    shrinkage, 72
Scanning laser acoustic microscopy (SLAM), 669
Scanning near-field optical microscopy (SNOM), 85
Scanning probe microscopy (SPM), 85
Scanning tunnelling microscopy (STM), 85
SCF, *see* Supercritical fluid extraction (SCF)
Schmidt numbers (Sc numbers), 333
SE, *see* Secondary electrons (SE)
Secondary electrons (SE), 70
Selective diffusion theory, 345; *see also* Aroma compounds
SEM, *see* Scanning electron microscopy (SEM)
Series model, 558
Sh numbers, *see* Sherwood numbers (Sh numbers)
SH wave, *see* Shear horizontal wave (SH wave)
Shear
    modulus, 181; *see also* Modulus of elasticity; Solid food quasistatic tests
    press, 210; *see also* Food texture
    rate, 144
    stress, 144
    -thinning foods, 126; *see also* Rheology of fluid food
Shear horizontal wave (SH wave), 663
Sherwood numbers (Sh numbers), 333
Short-circuited line technique, 581
Single-response models, 697
Sixport reflectometer (SPR), 583–584
Skin depth, *see* Penetration depth
SLAM, *see* Scanning laser acoustic microscopy (SLAM)
Slit rheometers, 145; *see also* Flow property measurement
    geometry, 146
    shear rate, 146
    shear stress, 145, 146
    velocity profile for Newtonian fluid, 145
SLNs, *see* Solid lipid nanoparticles (SLNs)
Small-amplitude oscillatory shear (SAOS), 164; *see also* Dynamic rheological tests
    complex modulus, 165
    loss modulus determination, 166
    loss tangent, 165
    phase difference, 164
    stress, 165
    viscosity, 165
SNOM, *see* Scanning near-field optical microscopy (SNOM)
Solid displacement method, 11; *see also* Apparent density
Solid food; *see also* Failure in solid foods
    empirical and imitative tests, *see* Food texture
    mass transfer processes, 313
    state of, 350
Solid food electrical conductivity; *see also* Cell membrane breakdown; Liquid food electrical conductivity; Solid–liquid mixture electrical conductivity
    β-dispersion, 554
    cellular tissue model, 553
    frequency and dielectric constant, 553–555
    frequency effect on, 551–553
    of gel, 542–543
    impedance of RC circuit, 555
    ingredient effects on, 555–557
    microstructure effect on, 540–541
    of undisrupted cellular structure, 543–546

Solid food quasistatic tests, 181; *see also* Rheology of solid food
  bending test, 184
  Burgers model, 188
  creep, 186–188
  curvature of axis of beam, 184
  elastic modulus, 181
  four-element model, 190
  generalized Maxwell model, 189
  Hookean solid, 181
  Kelvin model, 186
  Maxwell model, 186
  modulus of elasticity, 181, 185
  pure elastic behavior, 181
  relaxation, 188–190
  rheological modeling, 185–186
  shear modulus, 181
  simple tests, 182–185
  stress and strain, 182
  stress calculation, 183–184
  uniaxial compression/tension tests, 182, 183
  volumetric strain, 182
Solid food testing, 190; *see also* Direct stress-strain tests; Resonance methods; Rheology of solid food; Solid food quasistatic tests
  advantage of, 190
  dynamic properties, 191
  theoretical considerations, 191
  types, 191
Solid lipid nanoparticles (SLNs), 658
Solid–liquid interface, 42
Solid–liquid mixture electrical conductivity, 557; *see also* Effective elelctric conductivity models; Flow property measurement
Solute concentration at pore outlet, dimensionless, 447
Solution viscosity, dimensionless, 447
Sorption energetics, 389; *see also* Thermodynamic properties in dehydration
  binding energy, 396–398
  change in Gibbs free energy, 391
  differential quantities, 390–393
  energy of adsorption, 391
  enthalpy, 395
  entropy, 394, 396, 397, 401
  equilibrium heat, 403
  four-dimensional state diagram, 403–404
  glass–rubber transition, 403
  heat of binding of water, 397
  hysteresis, 398
  integral quantities, 393–398
  internal energy change of system, 390
  irreversibility, 399
  isosteric heat, 391, 392, 399, 400, 402
  kinetic aspects, 402
  MSI of dehydrated rice, 401
  sorption enthalpy, 394
  surface potential calculation, 394
  thermodynamic functions, 392
  uncompensated heat, 401
  water activity and glass transition, 404–405
Sorption kinetics, 328–329; *see also* Diffusivity estimation in solids
Sound wave; *see also* Ultrasound wave
  acoustic absorption in liquids, 642
  acoustic impedance, 642
  acoustic properties of biological tissue, 656
  in food, 654
  in fruits and vegetables, 659
  as longitudinal waves, 639
  refraction angle, 643
  velocity and chemical composition, 652
  velocity and temperature, 640, 641
  velocity in ideal gas, 639–640
  velocity in pure liquid, 652
Soybean salad oil dielectric data, 611
Specific heat, 250; *see also* Apparent specific heat; Thermal properties of frozen foods
  of food, 272
Specific heat capacity, 228; *see also* Thermal properties of unfrozen foods
  empirical equations, 229
  formula, 230
  latent heat of melting, 230
  of pork and beef fat, 230
  representative values of, 229
  specific and latent heat of fats, 230–231
Spectral emissivity, 286; *see also* Emissivity
Spectroradiometer, 193
Spectrum analysis, 649–650
SPM, *see* Scanning probe microscopy (SPM)
SPR, *see* Sixport reflectometer (SPR)
Spreading pressure, 41; *see also* Gibbs adsorption equation
Squeeze tester, 213; *see also* Food texture
Squeezing flow, 147; *see also* Extensional flows
  elongational viscosity, 148
  lubricated, 149

INDEX

Squeezing flow (*Continued*)
   strain rate, 148
   velocity field, 147
Standing wave ratio (SWR), 581
Staphylococcus aureus thermal inactivation
      kinetics, 689
Starch product analysis, 660
State diagram, 106; *see also* Glass transition
   advantages of, 106
   characteristic temperature, 106
   Clausius–Clapeyron equation, 109
   DSC heating thermogram, 108, 110
   Flory–Huggins model, 111
   of food, 106
   freezing curve, 109
   glass transition, 108–109, 110–111
   lines and points, 108
   macroregions, 106, 107
   maximal-freeze-concentration
      condition, 109–110
   parameters of, 103
   for solid foods, 206
   solids-melting characteristics, 111
Static uniaxial normal creep, 186; *see also*
      Solid food quasistatic tests
Steady-state method, 234, 686–687; *see also*
      Thermal conductivity
Sterilization, 712
STM, *see* Scanning tunnelling microscopy
      (STM)
Stokes' radius, 453
Storage effect on dielectric properties, 626
   of mozzarella cheese, 627
Stress relaxation, 188; *see also* Solid food
      quasistatic tests
Stress tensor, 124; *see also* Rheology of
      fluid food
Sugar, 604
   glass transition temperature, 99
   Kauzmann temperature, 99
   low-molecular weight, 102
Supercritical fluid extraction (SCF), 347
Superhydrophobic microtextured surface, 57
Surface-active agents, *see* Surfactants
Surface and colloid science, 37; *see also*
      Surface properties
Surface area, 6; *see also* Pore surface area
      measurement
   boundary surface area, 14–15
   cross-sectional area, 16
   euclidian geometry, 24

   measurement techniques, 14
   non-euclidian or irregular geometry, 24–25
   prediction of, 24
   relation for fractal-shaped objects, 25
   of some common shapes, 4
Surface energy, 38–39; *see also* Surface
      properties
   for pure liquids, 50
Surface force–pore flow model, 445, 486; *see
      also* Membrane separation processes;
      Transport theories
   analysis fundamentals, 445–446
   cylindrical coordinate system, 446
   dimensionless quantities, 447–448
   effective driving pressure, 448
   effective pore radius, 446
   friction on molecule, 452
   interfacial and bulk concentration, 452
   interfacial forces on solute, 451
   interfacial surface force, 453
   Lennard–Jones type of equation, 450
   liquid chromatography, 452–453
   location of molecule, 446
   membrane–solution interface, 450
   molecule distance from pore wall, 447
   parametric studies, 486–492
   pore radius, 459
   pore size and distribution, 453, 454, 456, 460
   PR/PWP ratio, 449
   radial velocity profile, 448
   solute separation, 447, 448
   Stokes' radius, 453
   surface force and Stokes' law radii, 454,
      455–456, 457–458
   transport equations, 448–452
Surface properties, 37, 59–60; *see also* Contact
      angle; Gibbs adsorption equation;
      Protein adsorption; Surface tension
   Cassie state, 57
   Cassie-Wenzel state transition, 57
   chemical and physical modifications, 58–59
   considerations, 38
   corona discharge methods, 58
   effects of adsorbed layer composition and
      structure on interfacial energy, 48
   emulsion stability, 49
   excess interfacial free energy, 38
   interfaces, 38
   interfacial free energy, 38
   Lotus effect, 57
   measurement techniques, 51

modification of, 56
plasma polymerization processes, 58–59
plasma treatments, 58
polymer surfaces, 54
superhydrophobic surfaces, 57
surface energy, 38–39
surface property data, 54–56
surface tension evaluation, 53
UV radiation, 59
wet-chemical etching methods, 58
work of adhesion and cohesion, 39
Surface-renewal theory, 334; see also Interphase mass transfer
Surface tension, 39–40; see also Surface properties; Surfactants
capillary rise method, 51
Du Nouy ring tensiometry, 52–53
ethanol solution, 56
evaluation of liquid, 51
of liquid water, 50
pendant drop method, 51–52
recorded for nonpolar diagnostic liquids, 55
Wilhelmy plate method, 52–53
Surfactants, 335
-rich media, 57
SWR, see Standing wave ratio (SWR)
Symbols, 31–33, 170–171, 240–241, 275, 352–353, 423–425, 515–517, 765

T

TAA, see Total antioxidant activity (TAA)
Tallow dielectric data, 611
TDR, see Time-domain reflectometry (TDR)
TDT, see Thermal Death Time (TDT)
TDT-model, 718
TEM, see Transmission electron microscopy (TEM)
Temperature and pressure coefficient determination, 717–719
Temperature coefficient model, 685–686
Temperature dependence of dielectric properties, 592; see also Carbohydrate dependence of dielectric properties; Dielectric properties; Frequency dependence of dielectric properties
concentration of salt, 593
of dielectric constant, 594, 598
dielectric loss factor, 593s, 594, 597, 598
in partially frozen region, 595
permittivity of NaCl solutions, 592
runaway heating, 596–597
in salt containing foods, 596
unfrozen water in tylose, 596
Test cell with Boonton RX-meter, 584
Texture profile, 208; see also Food texture
curve, 208–209
degradation kinetics, 690, 693, 694
Thermal conductivity, 231; see also Needle probe method; Thermal properties of frozen foods; Thermal properties of unfrozen foods
of codfish, 266
of corn, 268
Fourier law of heat conduction, 231
of frozen foods, 265, 267
geometric mean model, 233
heated probe method, 235
influence of structure of food on, 234
Kopelman predictive model, 266
Maxwell–Eucken model, 234
measurement methods, 234
of milk, juice, and sausage, 269
models for, 232
of multicomponent foods, 233
parallel model, 233
predictive equations, 231–234
rapid transient methods, 234
reference materials, 238–239
representative values of, 232
series model, 233
steady-state methods, 234
3D-finite-element model, 234
transient methods, 234–235
for two-component food, 233
Thermal Death Time (TDT), 679, 716
Thermal degradation kinetics, 695, 698
Thermal diffusivity, 274
Thermal inactivation, 682, 683; see also Thermal process kinetics
determination, 686–688
Thermal processing
equipment, 224
of foods, 684
Thermal process kinetics, 678; see also Thermal process primary kinetic model
color degradation, 693–694, 695–696, 697
enzyme inactivation, 690, 691–692
flavor degradation, 694
food preservation temperature, 678
heat treatment impact, 678
isothermal heating profiles, 678

Thermal process kinetics (*Continued*)
  microbial inactivation, 688–690
  of microorganisms, 688–690
  multiresponse model, 699–700
  nutrient degradation, 696–697
  preservation reactor, 678
  process impact, 679
  reactions in food, 678
  single-response models, 697
  texture degradation, 690, 693, 694
  thermal death time, 679
Thermal process kinetics measurement, 686
  procedure, 687
  steady-state experiment, 686–687
  unsteady-state methods, 687
Thermal process primary kinetic model, 680; *see also* Thermal process secondary kinetic model
  biphasic model, 683–684
  first-order model, 682–683
  fractional conversion model, 684
  half-life time, 682
  nth order reaction rate equation, 681
  order of reaction, 681, 682
  residuals, 682
  zero-order model, 682
Thermal process secondary kinetic model, 684; *see also* Decimal reduction time (D-value); Thermal process primary kinetic model
  activation energy, 685
  Arrhenius model, 684–685
  heat-inactivation kinetics, 685
  Q_{10}-factor, 685
  temperature coefficient model, 685–686
Thermal properties of frozen foods, 248; *see also* Apparent specific heat; Enthalpy; Unfrozen water content prediction
  data reliability, 255, 258
  density, 249, 263–265
  freezing point and unfrozen water, 249
  freezing temperatures of frozen foods, 259
  fruits and frozen doughs, 256–257
  Kopelman predictive model, 266
  limitations of predictive models, 274–275
  to measure, 248
  modeling of, 258
  specific heat, 250, 255
  thermal conductivity, 249–250, 265–268
  thermal diffusivity, 255, 274

Thermal properties of unfrozen foods, 224, 240; *see also* Density; Specific heat capacity; Thermal conductivity
  compressibility and expansion, 239
  COSTHERM, 226
  data, 225, 226
  glass transitions, 239–240
  literature on, 225–226
  measurement, 225
  prediction software, 226
  processes modeling and optimization, 224–225
  properties relevant to, 239
  quality and safety, 224
  sorption and hydration properties, 240
  thermal processes, 224
Thermal radiation, 283
Thermodynamic-based kinetic model, 719
Thermodynamic properties in dehydration, 360, 422–423; *see also* Chemical potential; Food dehydration; Sorption energetics; Water activity
  activity, 365
  closed food–water-vapor system, 366
  cluster model, 362
  continuum model theory, 362
  food–water system thermodynamics, 361
  fugacity, 364
  Gibbs free energy, 362–363
  mixture model theory, 362
Thermodynamic relation, 386
Thermogradient effect, 406
Thermomechanical analysis (TMA), 114, 115; *see also* Glass transition temperature
Thiamine thermal degradation kinetics, 698
Thixotropic foods, 131–132; *see also* Rheological models for viscous food
Three-dimensional finite-element model, 234; *see also* Thermal conductivity
Three-element model; *see also* Solid food quasistatic tests
  creep curves for, 187
  relaxation curves for, 189
Time-domain reflectometry (TDR), 582–583
Time-of-flight (TOF), 647; *see also* Rheology measurement
  meters, 155
Tissue softening, 747–748
TMA, *see* Thermomechanical analysis (TMA)
TOC, *see* Total organic carbon (TOC)
TOF, *see* Time-of-flight (TOF)

Toluene, 9
Total antioxidant activity (TAA), 508
Total emissivity, 286; *see also* Emissivity
Total organic carbon (TOC), 482
Total porosity, 6; *see also* Porosity
Total soluble solid (TSS), 508, 664
Transient methods, 234–235; *see also* Thermal conductivity
Transmission electron microscopy (TEM), 70; *see also* Food microstructure analysis
   cryo-TEM, 81
   freeze-fracture replication steps, 81
   images and DLS size distributions, 82
   sample preparation, 80
   staining techniques, 80
Transmission method, 645
Transmissivity, 287; *see also* Absorptivity; Emissivity; Infrared heating
   of base paper, 300
   for bread crust and crumb, 302
   of distilled water, 289
   radiation energy incident, 290
   spectral, 289
Transport equations, 440; *see also* Membrane separation processes; Reverse osmosis (RO)
   constants for osmotic pressure calculation, 478
   feed concentration, 479
   for fruit juice concentration, 478–482
   for green tea concentration, 482–486
   mass transfer coefficient, 480, 484
   membrane processing capacities, 481
   PR data, 483, 485, 487
   system analysis for apple juice concentration, 482
Transport theories, 439; *see also* Membrane separation processes; Reverse osmosis (RO); Surface force–pore flow model; Water preferential sorption model
Tripalmitin, 343
True density, 3; *see also* Density
TSS, *see* Total soluble solid (TSS)
Two-film theory, 331; *see also* Mass transfer coefficients

U

UDV meters, *see* Ultrasonic doppler velocity meters (UDV meters)
UF, *see* Ultrafiltration (UF)
UHT, *see* Ultra high temperature (UHT)
Ultrafiltration (UF), 439; *see also* Beverage industry; Dairy product industry; Edible oil industry; Membrane separation processes
   amaranth protein concentrate, 513
   of BSA and α-casein, 492–495
   BSA protein, 494, 495
   data, 494
   fouling, 495–497
   fractionation, 497–499, 500
   membranes, 502
   pore size distribution, 494, 498
   proteins, 492
Ultra high temperature (UHT), 654, 687
Ultrasonic analysis in food, 651; *see also* Ultrasonic measurement systems; Ultrasonic signal processing methods; Ultrasonic viscosity measurement
   air-borne ultrasound set up, 668
   bubbly liquid food analysis, 667
   cheese quality assessment, 659
   to detect microorganism, 654
   droplet size and ultra sound speed, 653
   emulsions and suspensions, 653
   firmness of fruits/vegetables, 659, 660
   frozen product analysis, 657
   gelation monitoring, 658
   leaks detection in packages, 668–670
   meat quality evaluation, 666
   monitoring emulsion crystallization, 657–658
   muscle foods, 655–657
   particle size analysis in liquid food, 666–667
   quality monitoring of vegetable and fruit, 664–665
   solutions, 651–653
   speed in mimic fermentation, 653–654
   starch product analysis, 660
   sugar and ethanol measurement, 651
   texture and ultrasonic velocity, 656
   Uric equation, 652
Ultrasonic doppler velocity meters (UDV meters), 155, 156, 663; *see also* Rheology measurement
Ultrasonic imaging, 645–647
Ultrasonic measurement systems, 644; *see also* Ultrasonic velocity measurements; Ultrasonic viscosity measurement
   air-coupled ultrasonic detection, 647
   capacitive transducers, 647

Ultrasonic measurement systems (*Continued*)
    PE technique, 644–645
    transmission method, 645
Ultrasonic signal processing methods, 647
    attenuation measurements, 648–649
    EMAT, 648
    Lorentz forces, 648
    pattern recognition, 650
    PE method, 648
    spectrum analysis, 649–650
Ultrasonic velocity measurements, 647; *see also* Ultrasonic analysis in food
    attenuation measurements, 648–649
    EMAT, 648
    Lorentz forces, 648
    PE method, 648
Ultrasonic viscosity measurement, 661; *see also* Ultrasonic measurement systems
    comparison, 665
    PE method, 663
    velocity profiling, 661–663
    viscosity–density measurement, 663–664
Ultrasound; *see also* Rheology measurement
    beams, 646
    to detect microorganism, 654
    spectrum analysis, 649
    velocity and chemical composition, 652
Ultrasound Doppler velocity profiling (UVP), 661–663
Ultrasound Doppler velocity profiling–pressure difference system (UVP–PD system), 661, 662
Ultrasound speed; *see also* Ultrasonic velocity measurements
    droplet size and, 653
    in mimic fermentation, 653–654
    in sugar solution, 651
Ultrasound velocity measurements (UVMs), 658
Ultrasound wave, 155, 639; *see also* Sound wave; Ultrasonic measurement systems; Ultrasonic signal processing methods; Ultrasonic velocity measurements
    acoustic impedance, 644
    applications, 664
    attenuation, 642, 644
    axial direction, 646
    emulsion particle size on, 644
    interaction with foods, 643–644
    reflection coefficient, 642–643
    spectrum analysis, 649–650
    speed in sugar solution, 651
    symbols, 670–671
    ultrasonic detection, 646
    ultrasound beams, 646
    velocity and textural properties and moisture content, 644
Ultraviolet (UV), 608
    radiation, 59
Uncompensated heat, 401
Uncompetitive inhibition, 743
Undesired reactions in food, 678
Unfrozen water content prediction, 258; *see also* Thermal properties of frozen foods
    in food, 260–263
    freezing point depression, 260, 261
    ice fraction, 261
    initial freezing point temperature, 260
    mole fraction of water, 260
Uniaxial compression/tension test, 182; *see also* Solid food quasistatic tests
Unsteady-state methods, 687
Uric equation, 652
UV, *see* Ultraviolet (UV)
UVMs, *see* Ultrasound velocity measurements (UVMs)
UVP, *see* Ultrasound Doppler velocity profiling (UVP)

V

Vacuum membrane distillation (VMD), 509
Van der Waal's
    constants for different gases, 17
    equation of state, 16
Van Laar equation, 316; *see also* Vapor–liquid equilibrium
Vane method, 152; *see also* Flow property determining mixers
Vapor–liquid equilibrium, 314; *see also* Phase equilibria
    activity coefficients, 317
    conditions for, 314
    electrolytes effect on, 316
    NRTL equation, 316
    relative volatility, 315, 316, 318, 319
    temperature effect on activity coefficients, 317
    van Laar equation, 316
    Wilson equation, 316

INDEX

Vapor pressure manometer (VPM), 372
Vapor pressure measurement, 370; *see also* Water activity
  criterion for improvement, 373
  thermostated vapor pressure manometer, 372
Variable pressure SEM (VP-SEM), 74; *see also* Scanning electron microscopy (SEM)
  differential pumping and pressure-limiting apertures, 75
  dynamic and *in situ* studies using, 76–78
  environmental SE detector, 75
  gas in sample chamber of, 76
  image of mayonnaise, 78
  sample preparation for, 76
  sequence of images, 77, 79
Vibrational viscometers, 157; *see also* Rheology measurement
Visco-Amylo-Graph, 150; *see also* Flow property measurement
Viscoelastic materials, 181
  experimental curves, 182
Viscoelastic parameters, 125; *see also* Rheology of fluid food; Rheological models for viscous food
Viscometer, 140
Viscosity; *see also* Rheology of fluid food; Rheological models for viscous food
  coefficient, 181
  complex, 125
  function, 125
Vitamin A thermal degradation kinetics, 698
VMD, *see* Vacuum membrane distillation (VMD)
Volatility retention, 344; *see also* Aroma compounds
  aroma compounds, 344
  relative volatility, 345
Volume, 3; *see also* Apparent density; Mass–volume–area-related properties
  boundary, 3
  of common shapes, 4
  displacement method, 9
  pore, 3
Volumetric strain, 182; *see also* Solid food quasistatic tests
VPM, *see* Vapor pressure manometer (VPM)
VP-SEM, *see* Variable pressure SEM (VP-SEM)

W

Warner-Bratzler shear, 210–211; *see also* Food texture
Washburn equation, 12; *see also* Mercury porosimetry
Water, 360
  in biological systems, 361
  bound, 361
  dielectric properties of bidistilled and deionized, 589
  dielectric relaxation in, 589
  dipole rotation of molecules, 528
  effect on dielectric properties, 608
  in food system, 599
  on loss factor, 599
  as terminal load in magnetron, 589
Water activity, 320–321, 366; *see also* Brunauer, Emmett and Teller (BET); Gas–solid and vapor–solid equilibria; Moisture sorption isotherm (MSI); Thermodynamic properties in dehydration; Vapor pressure measurement
  adjustment of, 375–379
  approximation, 367
  calculation, 373
  chemical potentials, 366
  dew point temperature measuring instruments, 374
  differential method, 375
  freezing point depression techniques, 374
  fugacity, 366–367, 368
  fugacity coefficients, 369
  of glycerol solutions, 378
  gravimetric method, 377
  hygrometers, 374–375
  integral method, 375
  isopiestic method, 375
  measurement of, 370
  measurements, 375
  and microbial growth, 369–370, 371
  of NaCl solutions, 379
  nonenzymatic browning, 370
  prediction of, 367–370
  and rate of oxygen uptake, 370
  regression equations, 377
  of salt slurries, 376
  standard-state chemical potential, 366
  of sulfuric acid solutions, 378

Water activity (*Continued*)
  effect of temperature on, 386
  wet-bulb temperature, 374
Water preferential sorption model, 439, 463; *see also* Membrane separation processes; Reverse osmosis (RO); Transport theories
  basic transport equations, 440–441
  data on free energy parameter, 442
  data on organic solutes, 464–468
  feed concentration on acid separation, 474, 475
  feed concentration on PR, 471
  film specifications used for calculation, 463
  glucose molality in feed vs. solute separations, 477
  k values, 469, 472, 473
  modified Small's number, 443
  osmotic pressure data, 470, 471
  partially dissociated acid solution, 476–477
  partially dissociated organic solute, 473–474
  RO data analysis, 440–441
  RO process design, 444
  separation problem, 475–476
  solute transport parameters, 441–444
  sucrose effect on *i*-butyl alcohol separation, 476
  Taft's steric parameter, 443
  undissociated organic solute, 469–473
Water vapor transmission (WVT), 351
Waveguide and coaxial transmission line methods, 580–581
Weak electrolytes, 530–531
Weibull model for inactivation behavior of microorganisms, 717
Weissenberg-Rabinowitsch-Mooney equation, 161
Wet-bulb temperature, 374; *see also* Water activity

Wet-chemical etching methods, 58
Whey proteins, 460
  CMP from, 504
Wien's displacement law, 286
Wilhelmy plate method, 52–53
Wilke–Chang equation, 324; *see also* Diffusion
Willams, Landel, and Ferry (WLF equation), 97, 327; *see also* Glass transition temperature; Diffusion
  Fermi's model, 98
Willams–Watts equation, 100; *see also* Glass transition temperature
Wilson equation, 316; *see also* Vapor–liquid equilibrium
WLF equation, *see* Willams, Landel, and Ferry (WLF equation)
Work of adhesion, 39; *see also* Surface properties
Work of cohesion, 39; *see also* Surface properties
WVT, *see* Water vapor transmission (WVT)

X

Xanthan gum, 606, 607
X-ray microscope (XRM), 83
XRM, *see* X-ray microscope (XRM)

Y

Yersinia enterocolitica thermal inactivation kinetics, 689
Young's modulus (EM), 212, *see* Modulus of elasticity

Z

Zero-order model, 682
Zero-order reaction, 682
Zisman plot, 44, 45
Z-value, 685